BRAIN & Behavior

3
EDITION

To Duejean

because you are the joy of my life.

BRAIN & Behavior

An Introduction to Biological Psychology

3 EDITION

BOB GARRETT

Visiting Scholar, California Polytechnic State University, San Luis Obispo

Contributions by

Gerald Hough
Rowan University

Beth Powell
Smith College

Los Angeles | London | New Delhi
Singapore | Washington DC

For information:

 SAGE Publications, Inc.
2455 Teller Road
Thousand Oaks, California 91320
E-mail: order@sagepub.com

SAGE Publications Ltd.
1 Oliver's Yard
55 City Road, London, EC1Y 1SP
United Kingdom

SAGE Publications India Pvt. Ltd.
B 1/I 1 Mohan Cooperative Industrial Area
Mathura Road, New Delhi 110 044
India

SAGE Publications Asia-Pacific Pte. Ltd.
33 Pekin Street #02-01
Far East Square
Singapore 048763

Printed in Canada

Library of Congress Cataloging-in-Publication Data

Garrett, Bob.
Brain & behavior: An introduction to biological psychology / Bob Garrett. — 3rd ed.
    p. cm.
Includes bibliographical references and index.
ISBN 978-1-4129-8168-2 (pbk.)
    1. Psychobiology—Textbooks. I. Title.

QP360.G375 2011
612.8—dc22   2010039735

Printed on acid-free paper

11   12   13   14   10   9   8   7   6   5   4   3   2

Acquiring Editor:	Vicki Knight
Associate Editor:	Lauren Habib
Production Editor:	Eric Garner
Copy Editor:	Paula L. Fleming
Typesetter:	C&M Digitals (P) Ltd.
Proofreader:	Wendy Jo Dymond
Indexer:	Molly Hall
Cover Designer:	Candice Harman
Permissions Editor:	Adele Hutchinson

Brief Contents

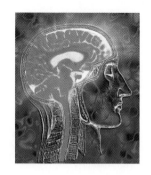

Preface   xiii
About the Author   xvi

CHAPTER 1. What Is Biopsychology?   1

PART I. Neural Foundations of Behavior: The Basic Equipment

CHAPTER 2. Communication Within the Nervous System   23
CHAPTER 3. The Organization and Functions of the Nervous System   53
CHAPTER 4. The Methods and Ethics of Research   91

PART II. Motivation and Emotion: What Makes Us Go

CHAPTER 5. Drugs, Addiction, and Reward   125
CHAPTER 6. Motivation and the Regulation of Internal States   155
CHAPTER 7. The Biology of Sex and Gender   189
CHAPTER 8. Emotion and Health   223

PART III. Interacting With the World

CHAPTER 9. Hearing and Language   253
CHAPTER 10. Vision and Visual Perception   293
CHAPTER 11. The Body Senses and Movement   329

PART IV. Complex Behavior

CHAPTER 12. Learning and Memory   363
CHAPTER 13. Intelligence and Cognitive Functioning   393
CHAPTER 14. Psychological Disorders   427
CHAPTER 15. Sleep and Consciousness   467

Glossary   504

References   513

Chapter-Opening Photo Credits   565

Author Index   566

Subject Index   595

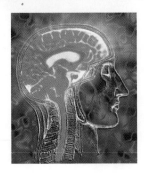

Detailed Contents

Preface   xiii

About the Author   xvi

CHAPTER 1. What Is Biopsychology?   1

The Origins of Biopsychology   3
 Prescientific Psychology and the Mind-Brain Problem   4
 Descartes and the Physical Model of Behavior   4
 Helmholtz and the Electrical Brain   6
 The Localization Issue   6

 In the News: The Mind-Brain Debate Isn't Over   8
Nature and Nurture   9
 The Genetic Code   9

 Application: A Computer Made of DNA   10
 Genes and Behavior   12
 The Human Genome Project   12
 Heredity: Destiny or Predisposition?   13

PART I. Neural Foundations of Behavior: The Basic Equipment

CHAPTER 2. Communication Within the Nervous System   23

The Cells That Make Us Who We Are   24
 Neurons   24

 Application: Targeting Ion Channels   32
 Glial Cells   33
How Neurons Communicate With Each Other   36
 Chemical Transmission at the Synapse   36
 Regulating Synaptic Activity   41
 Neurotransmitters   42
 Of Neuronal Codes, Neural Networks, and Computers   44

 Application: Agonists and Antagonists in the Real World   45

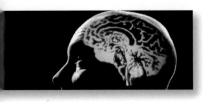

CHAPTER 3. The Organization and Functions of the Nervous System   53

The Central Nervous System   54
 The Forebrain   55

 Application: The Case of Phineas Gage   62
 The Midbrain and Hindbrain   66
 The Spinal Cord   67
 Protecting the Central Nervous System   70

The Peripheral Nervous System   71
    The Cranial Nerves   71
    The Autonomic Nervous System   71
Development and Change in the Nervous System   74
    The Stages of Development   74
    How Experience Modifies the Nervous System   78
    Damage and Recovery in the Central Nervous System   79

    In the News: Is The Brain Too Fragile for Sports?   81

    Application: Mending the Brain With Computer Chips   84

CHAPTER 4. The Methods and Ethics of Research   91

Science, Research, and Theory   92
    Theory and Tentativeness in Science   92
    Experimental Versus Correlational Studies   93
Research Techniques   95
    Staining and Imaging Neurons   95
    Light and Electron Microscopy   98
    Measuring and Manipulating Brain Activity   98
    Brain Imaging Techniques   102

    Application: Brain Implants That Move   103

    Application: Scanning King Tut   107

    Investigating Heredity   108
Research Ethics   112
    Plagiarism and Fabrication   112
    Protecting the Welfare of Research Participants   112
    Gene Therapy   115
    Stem Cell Therapy   116

    In the News: A Brave New World?   116

PART II. Motivation and Emotion: What Makes Us Go

CHAPTER 5. Drugs, Addiction, and Reward   125

Psychoactive Drugs   126
    Opiates   127
    Depressants   128
    Stimulants   130
    Psychedelics   134
    Marijuana   135

    In the News: Controversy Over Medical Marijuana Takes a New Turn   137

Addiction   137
    The Neural Basis of Addiction and Reward   138
    Dopamine and Reward   139

    Application: Is Compulsive Gambling an Addiction?   140

    Dopamine, Learning, and Brain Plasticity   141
    Treating Drug Addiction   142

The Role of Genes in Addiction   146
    Separating Genetic and Environmental Influences   146
    What Is Inherited?   147
    Implications of Addiction Research   149

CHAPTER 6. Motivation and the Regulation of Internal States   155

Motivation and Homeostasis   156
    Theoretical Approaches to Motivation   156
    Simple Homeostatic Drives   158
Hunger: A Complex Drive   160
    The Role of Taste   161
    Application: Predator Control Through Learned Taste Aversion   164
    Digestion and the Two Phases of Metabolism   164
    Signals That Start a Meal   168
    Signals That End a Meal   169
    Long-Term Controls   170
Obesity   172
    The Myths of Obesity   174
    The Contribution of Heredity   174
    Obesity and Reduced Metabolism   175
    Application: The Epigenetics of Weight Regulation   177
    Treating Obesity   178
Anorexia and Bulimia   181
    Environmental and Genetic Contributions   182
    The Role of Serotonin and Dopamine   182

CHAPTER 7. The Biology of Sex and Gender   189

Sex as a Form of Motivation   190
    Arousal and Satiation   191
    The Role of Testosterone   192
    Brain Structures and Neurotransmitters   193
    Odors, Pheromones, and Sexual Attraction   196
    Application: Of Love and Bonding   198
The Biological Determination of Sex   198
    Chromosomes and Hormones   199
    Prenatal Hormones and the Brain   201
Gender-Related Behavioral and Cognitive Differences   202
    Some Demonstrated Male-Female Differences   202
    Origins of Male–Female Differences   202
Sexual Anomalies   204
    Male Pseudohermaphrodites   204
    Female Pseudohermaphrodites   205
    Sex Anomalies and the Brain   206
    In the News: Sex, Gender, and Sports   207
    Ablatio Penis: A Natural Experiment   207

Sexual Orientation 209
    The Social Influence Hypothesis 210
    Genes and Sexual Orientation 211
    Hormonal Influence 212
    Brain Structures 213
    The Challenge of Female Homosexuality 215
    Social Implications of the Biological Model 215

CHAPTER 8. Emotion and Health   223

Emotion and the Nervous System 224
    Autonomic and Muscular Involvement in Emotion 224
    The Emotional Brain 227
    The Prefrontal Cortex 229
    Application: Why I Don't Jump Out of Airplanes 231
    The Amygdala 231
    Hemispheric Specialization in Emotion 233
Stress, Immunity, and Health 234
    Stress as an Adaptive Response 234
    Negative Effects of Stress 235
    Social and Personality Variables 237
    Application: One Aftermath of 9/11 Is Stress-Related Brain Damage 238
    Pain as an Adaptive Emotion 239
Biological Origins of Aggression 240
    Hormones and Aggression 241
    The Brain's Role in Aggression 242
    Serotonin and Aggression 243
    Heredity and Environment 245
    In the News: Aggression, Genes, and the Law 246

PART III. Interacting With the World

CHAPTER 9. Hearing and Language   253

Hearing 255
    The Stimulus for Hearing 255
    The Auditory Mechanism 256
    Frequency Analysis 260
    Locating Sounds With Binaural Cues 267
    Application: Cochlear and Brainstem Implants to Restore Hearing 267
Language 270
    Broca's Area 271
    Wernicke's Area 272
    The Wernicke-Geschwind Model 272
    Reading, Writing, and Their Impairment 275
    Mechanisms of Recovery From Aphasia 278
    A Language-Generating Mechanism? 279
    Language in Nonhumans 282
    Neural and Genetic Antecedents 283

CHAPTER 10. Vision and Visual Perception   293

Light and the Visual Apparatus   294
The Visible Spectrum   294
The Eye and Its Receptors   295
Pathways to the Brain   298

Application: Restoring Lost Vision   300

Color Vision   300
Trichromatic Theory   301
Opponent Process Theory   301
A Combined Theory   303
Color Blindness   306
Form Vision   307
Contrast Enhancement and Edge Detection   307
Hubel and Wiesel's Theory   309
Spatial Frequency Theory   312
The Perception of Objects, Color, and Movement   313
The Two Pathways of Visual Analysis   314
Disorders of Visual Perception   316
The Problem of Final Integration   321

Application: When Binding Goes Too Far   322

CHAPTER 11. The Body Senses and Movement   329

The Body Senses   330
Proprioception   330
The Skin Senses   330
The Vestibular Sense   332
The Somatosensory Cortex and the Posterior Parietal Cortex   333
Pain and Its Disorders   336

In the News: Getting an Anesthetic Shouldn't Hurt   338

Application: Treating Pain in Limbs That Aren't There   343

Movement   344
The Muscles   344
The Spinal Cord   345
The Brain and Movement   346
Disorders of Movement   351

Part IV. Complex Behavior

CHAPTER 12. Learning and Memory   363

Learning as the Storage of Memories   364
Amnesia: The Failure of Storage and Retrieval   365

In the News: The Legacy of HM   366

Mechanisms of Consolidation and Retrieval   366
Where Memories Are Stored   368
Two Kinds of Learning   370
Working Memory   371

Brain Changes in Learning 372

Long-Term Potentiation 372

How LTP Happens 374

Neural Growth in Learning 374

Consolidation Revisited 376

Changing Our Memories 377

Application: Total Recall 378

Learning Deficiencies and Disorders 379

Effects of Aging on Memory 379

Alzheimer's Disease 380

Application: Genetic Interventions for Alzheimer's 386

Korsakoff's Syndrome 386

CHAPTER 13. Intelligence and Cognitive Functioning 393

The Nature of Intelligence 394

What Does "Intelligence" Mean? 394

The Structure of Intelligence 395

The Biological Origins of Intelligence 396

The Brain and Intelligence 396

Specific Abilities and the Brain 400

Application: We Aren't the Only Tool Users 402

Heredity and Environment 403

Application: Enhancing Intelligence and Performance 406

Deficiencies and Disorders of Intelligence 407

Effects of Aging on Intelligence 407

Intellectual Disability 409

Autism 412

In the News: Childhood Vaccines and Autism 418

Attention Deficit Hyperactivity Disorder 419

CHAPTER 14. Psychological Disorders 427

Schizophrenia 429

Characteristics of the Disorder 429

Heredity 431

Two Kinds of Schizophrenia 433

The Dopamine Hypothesis 434

Beyond the Dopamine Hypothesis 434

Brain Anomalies in Schizophrenia 436

Affective Disorders 443

Heredity 444

The Monoamine Hypothesis of Depression 445

Electroconvulsive Therapy 447

Application: Electrical Stimulation for Depression 448

Antidepressants, ECT, and Neural Plasticity 448

Rhythms and Affective Disorders 449

Bipolar Disorder   450
Brain Anomalies in Affective Disorder   451
Suicide   452

Anxiety Disorders   454
Generalized Anxiety, Panic Disorder, and Phobia   454
Posttraumatic Stress Disorder   455

In the News: Virtual Reality Isn't Just for Video Games   456

Obsessive-Compulsive Disorder   457

Application: Of Hermits and Hoarders   459

CHAPTER 15. Sleep and Consciousness   467
Sleep and Dreaming   468
Circadian Rhythms   470
Rhythms During Waking and Sleeping   473
The Functions of REM and Non-REM Sleep   475
Sleep and Memory   476
Brain Structures of Sleep and Waking   478
Sleep Disorders   481

In the News: In the Still of the Night   483

Sleep as a Form of Consciousness   485

The Neural Bases of Consciousness   485
Awareness   486
Attention   488
The Sense of Self   490
Theoretical Explanations of Consciousness   496

Application: Determining Consciousness When It Counts   499

Glossary   504

References   513

Chapter-Opening Photo Credits   565

Author Index   566

Subject Index   595

Preface

A Message From the Author

Flip through this book and you'll see that its pages are chock-full of facts—just a sampling gleaned from a vast supply that grows too fast for any of us to keep up. But sifting through those facts and reporting them is neither the most difficult nor the most important function of a good textbook. A greater challenge is that most students fail to share their instructors' infatuation with learning; perhaps they lack the genes or the parental role models, or just the revelation that education is life enriching. At any rate, they can find a text like this intimidating, and it is the textbook's role to change their minds.

The colorful illustrations and intriguing case studies may capture students' interest, but interest alone is not enough. That's why I've adopted a "big-picture" approach in writing the text, one that marshals facts into explanations and discards the ones left standing around with nothing to do. When you put facts to work that way, you begin to see students look up and say, "That makes sense," or "I've always wondered about that, but I never thought of it that way," or "Now I understand what was going on with Uncle Edgar."

I believe education has the capacity to make a person healthy, happy, and productive, and it makes a culture strong. Education realizes that promise when it leads people to inquire and to question, when they *learn how to learn*. When 51% of the public believes in ghosts, and politics has become a game played by shouting the loudest or telling the most convincing lie, education more than ever needs to teach young people to ask, "Where is the evidence?" and "Is that the only possible interpretation?"

To those who would teach and those who would learn, this book is for you.

To the Instructor

When I first wrote *Brain and Behavior*, I had one goal, to entice students into the adventure of biological psychology. There were other good texts out there, but they read as though they were written for students preparing for their next biopsych course in graduate school. Those students will find this book adequately challenging, but I wrote it so anyone who is interested in behavior, including the newly declared sophomore major or the curious student who has wandered over from the history department, could have the deeper understanding that comes from a biological perspective as they take other courses in psychology.

It is not enough to draw students in with lively writing or by piquing their interest with case studies and telling an occasional story along the way; unless they feel they are learning something significant, they won't stay—they'll look for excitement in more traditional places. As I wrote, I remembered the text I struggled with in my first biopsychology class; it wasn't very interesting because we knew much less about the biological underpinnings of behavior than we do now. Since that time, we have learned how the brain changes during learning, we have discovered some of the genes and brain deficiencies that cause schizophrenia, and we are beginning to understand how intricate networks of brain cells produce language, make us intelligent, and help us play the piano or find a mate. In other words, biopsychology has become a lot more interesting. So the material is there; now it is my job to communicate the excitement I have felt in discovering the secrets of the brain and to make a convincing case that biopsychology has the power to answer the questions *students* have about behavior.

A good textbook is all about teaching, but there is no teaching if there is no learning. Over the years, my students taught me a great deal about what they needed to help them learn. For one thing, I realized how important it is for students to build on their knowledge throughout the course, so I made several changes from the organization I saw in other texts. First, the chapter on neuronal physiology precedes the chapter on the nervous system, because I believe that you can't begin to understand the brain until you know how its neurons work. And I reversed the usual order of the vision and audition chapters, because I came to understand that audition provides a friendlier context for introducing the basic principles of sensation and perception. The chapters on addiction, motivation, emotion, and sex follow the introduction to neurophysiology; this was done to build student motivation before tackling sensation and perception. Perhaps more significantly, some topics have been moved around among chapters so they can be developed in a more behaviorally meaningful context. So language is discussed along with audition, the body senses with the mechanisms of movement, the sense of taste in the context of feeding behavior, and olfaction in conjunction with sexual behavior. Most unique, though, is the inclusion of a chapter on the biology of intelligence and another on consciousness. The latter is a full treatment of recent developments in the field, rather than limited to the usual topics of sleep and split-brain behavior. These two chapters strongly reinforce the theme that biopsychology is personally relevant and capable of addressing important questions.

Brain and Behavior has several features that will motivate students to learn and encourage them to take an active role in their learning. It engages the student with interest-grabbing opening vignettes, illustrative case studies, and In the News items and Application boxes that take an intriguing step beyond the chapter content. Throughout each chapter, questions in the margins keep the student focused on key points, a Concept Check at the end of each section serves as a reminder of the important ideas, and On the Web icons point the way to related information on the Internet. At the end of the chapter, In Perspective emphasizes the importance and implications of what the student has just read, a summary helps organize that information, and Testing Your Understanding assesses the student's conceptual understanding as well as factual knowledge. Then, For Further Reading is a guide for students who want to explore the chapter's topics more fully. I have found over the years that students who use the study aids in a class are also the best performers in the course.

New in the Third Edition

As you would expect, the third edition of *Brain and Behavior* includes a number of changes. Reflecting the rapid advances in biological psychology and neuroscience, this edition contains 500 new references. More than 60 illustrations are also new, and another 50 have been revised to increase their informational and educational value. New topics have been added, entire sections have been rewritten, and more terms are now defined in the glossary. In addition, the Application and In the News features, which were well received in the first two editions, have been updated or replaced.

The new edition continues the theme of showcasing our rapidly increasing understanding of genetic influences on behavior; this time, however, there is a greater emphasis on the roles of rare mutations and epigenetic influences. There is also a shift away from considering particular brain areas as origins of behavior and instead looking at networks, for example, in intelligence, psychological disorders, and consciousness. And a previous theme that has been strengthened is the broader societal relevance of biopsychology, from the ethical implications of stem cell research to the cost of addictions and disorders to new strategies for treating brain and spinal cord damage.

To the Student

Brain and Behavior is my attempt to reach out to students, to beckon them into the fascinating world of biological psychology. These are exceptionally exciting times, comparable in many ways to the renaissance that thrust Europe from the Middle Ages into the modern world. In Chapter 1, I quote Kay Jamison's comparison of neuroscience, which includes biopsychology, to a "romantic, moon-walk sense of exploration." I know of no scientific discipline with greater potential to answer the burning questions about ourselves than neuroscience in general and biopsychology in particular. I hope this textbook will convey that kind of excitement as you read about discoveries that will revolutionize our understanding of what it means to be human.

I want you to succeed in this course, but, more than that, I want you to learn more than you ever imagined you could and to go away with a new appreciation for the promise of biological psychology. So, I have a few tips I want to pass along. First, try to sit near the front of the class, because those students usually get the best grades. That is probably because they stay more engaged and ask more questions; but to ask good questions you should *always* read the text assignment before you go to class. And so you'll know where you're going before you begin to read, take a look at "In this chapter you will learn," then skim the chapter subheadings, and read the summary. Use the questions in the margins as you go through, answer the Concept Check questions, and be sure to test yourself at the end. Computer icons like the one you see here will tell you which figures have been animated on the text's website to help sharpen your understanding, and numbered WWW icons in the margins will direct you to a wealth of additional information on the web. Then don't forget to look up some of the books and articles in For Further Reading. If you do all of these things, you won't just do better in this course; you will leave saying, "I really got something out of that class!"

I wrote *Brain and Behavior* with you in mind, so I hope you will let me know where I have done things right and, especially, where I have not (bgarrett@calpoly.edu) I wish you the satisfaction of discovery and knowledge as you read what I have written *for you.*

Supplemental Material

Student Study Guide

This affordable student study guide and workbook to accompany Bob Garrett's *Brain and Behavior, Third Edition* will help students get the added review and practice they need to improve their skills and master their course. Each part of the study guide corresponds to the appropriate chapter in the text and includes the following: chapter outline, chapter summary, study quiz, and a chapter posttest.

Student Study Site

This free student study site provides additional support to students using *Brain and Behavior, Third Edition.* The website includes animations of key figures in the text, links to On the websites and other Internet resources, e-flashcards, study quizzes (students can receive their score immediately), and relevant SAGE journal articles with critical thinking questions Visit the study site at www.sagepub.com/garrett3e.

Instructor's Resources on CD-ROM

This set of instructor's resources provides a number of helpful teaching aids for professors new to teaching biological psychology and to using *Brain and Behavior, Third Edition.* Included on the CD-ROM are PowerPoint slides, a computerized test bank to allow for easy creation of exams, lecture outlines, suggested class activities and critical thinking questions, and video and Internet resources for each chapter of the text.

Acknowledgments

Vicki Knight has been the editor for all three editions of Brain and Behavior. Without her support and vision, the book wouldn't have reached the second edition, much less the third. Vicki and I have been nobly aided by Eric Garner, production editor; Lauren Habib, associate editor; Paula Fleming, copy editor; and Marcy Lunetta, photo research and permissions. And I want to extend special recognition to Beth Powell of Smith College and Gerald Hough at Rowan University. Beth prepared the student study guide and PowerPoint slides, Gerald wrote the instructor's manual, and they both constructed the online test bank and contributed significantly to the revision of all the chapters. Kudos to Beth and Jerry for their keenness, talent, and inspiring commitment.

I have had a number of mentors along the way, to whom I am forever grateful. A few of those special people are Wayne Kilgore, who taught the joys of science along with high school chemistry and physics; Garvin McCain, who introduced me to the satisfactions of research; Roger Kirk, who taught me that anything worth doing is worth doing

over and over until it's right; and Ellen Roye and Ouilda Piner, who shared their love of language. These dedicated teachers showed me that learning was my responsibility, and they shaped my life with their unique gifts and quiet enthusiasm.

But of all my supporters, the most important has been my wife, Duejean; love and thanks to her for her patient understanding and her appreciation of how important this project is to me.

In addition, the following reviewers gave generously of their time and expertise throughout the development of this text; they contributed immensely to the quality of *Brain and Behavior*:

First Edition: Susan Anderson, University of South Alabama; Patrizia Curran, University of Massachusetts–Dartmouth; Lloyd Dawe, Cameron University; Tami Eggleston, McKendree College; James Hunsicker, Southwestern Oklahoma State University; Eric Laws, Longwood College; Margaret Letterman, Eastern Connecticut State University; Doug Matthews, University of Memphis; Grant McLaren, Edinboro University of Pennsylvania; Rob Mowrer, Angelo State University; Anna Napoli, University of Redlands; Robert Patterson, Washington State University; Joseph Porter, Virginia Commonwealth University; Jeffrey Stern, University of Michigan–Dearborn; Aurora Torres, University of Alabama in Huntsville; Michael Woodruff, East Tennessee State University; and Phil Zeigler, Hunter College.

Second Edition: M. Todd Allen, University of Northern Colorado; Patricia A. Bach, Illinois Institute of Technology; Wayne Brake, University of California–Santa Barbara; Steven I. Dworkin, University of North Carolina; Sean Laraway, San Jose State University; Mindy J. Miserendino, Sacred Heart University; Brady Phelps, South Dakota State University; Susan A. Todd, Bridgewater State College; and Elizabeth Walter, University of Oregon.

Third Edition: John A. Agnew, University of Colorado at Boulder; Michael A. Bock, American International College; Rachel E. Bowman, Sacred Heart University; Jessica Cail, Pepperdine University; Mary Jo Carnot, Chadron State College; Cheryl A. Frye, The University at Albany–State University of New York; Rebecca L. M. Fuller, Catholic University of America; Cindy Gibson, Washington College; Bennet Givens, Department of Psychology, Ohio State University; Robert B. Glassman, Lake Forest College; Gerald E. Hough, Rowan University; Joseph Nuñez, Michigan State University; and Kimberly L. Thomas, University of Central Oklahoma.

—Bob Garrett

About the Author

Bob Garrett is a Visiting Scholar at California Polytechnic State University, San Luis Obispo. He was Professor of Psychology at DePauw University in Greencastle, Indiana, and held several positions there, including Chairperson of the Department of Psychology, Faculty Development Coordinator, and Interim Dean of Academic Affairs. He received his BA from the University of Texas at Arlington and his MA and PhD from Baylor University. He received further training in the Department of Physiology at Baylor University College of Medicine and at the Aeromedical Research Primate Laboratory, Holloman Air Force Base. Bob and his wife, Duejean, live on a 3,200-acre ranch they share with 47 other families in the hills outside San Luis Obispo. Their two sons, daughter-in-law, and three beautiful grandchildren all live nearby.

About the Contributors

Beth Powell is currently a Lecturer at Smith College where she teaches in the Psychology Department and Neuroscience Program. Past positions include teaching in the PsyD program at Antioch/New England College and in the Psychology Department at the University of Massachusetts. At Smith College, Beth teaches courses in psychology and neuroscience research methods, physiology of behavior, and psychopharmacology, and a seminar in neuroscience, ethics, and policy. She recently received the Sherrerd Award at Smith College for distinguished teaching. She received her BA at Smith College and her MA and PhD in the Neuroscience and Behavior Department at the University of Massachusetts Amherst. In her spare time she enjoys gardening, volunteering at an animal shelter, and spending time with her two daughters and two dogs.

Gerald Hough is an Associate Professor of Biological Sciences and Psychology at Rowan University in Glassboro, New Jersey. He has taught undergraduate courses in both Departments on anatomy, animal behavior, research methods, and learning. He has served as the undergraduate advising coordinator for Psychology, the Chair of the IACUC, and curriculum development for the new Cooper Medical School at Rowan University. His research focus is in the neural bases of spatial and communication behaviors in birds. He received his BS from Purdue University and his MS and PhD from The Ohio State University.

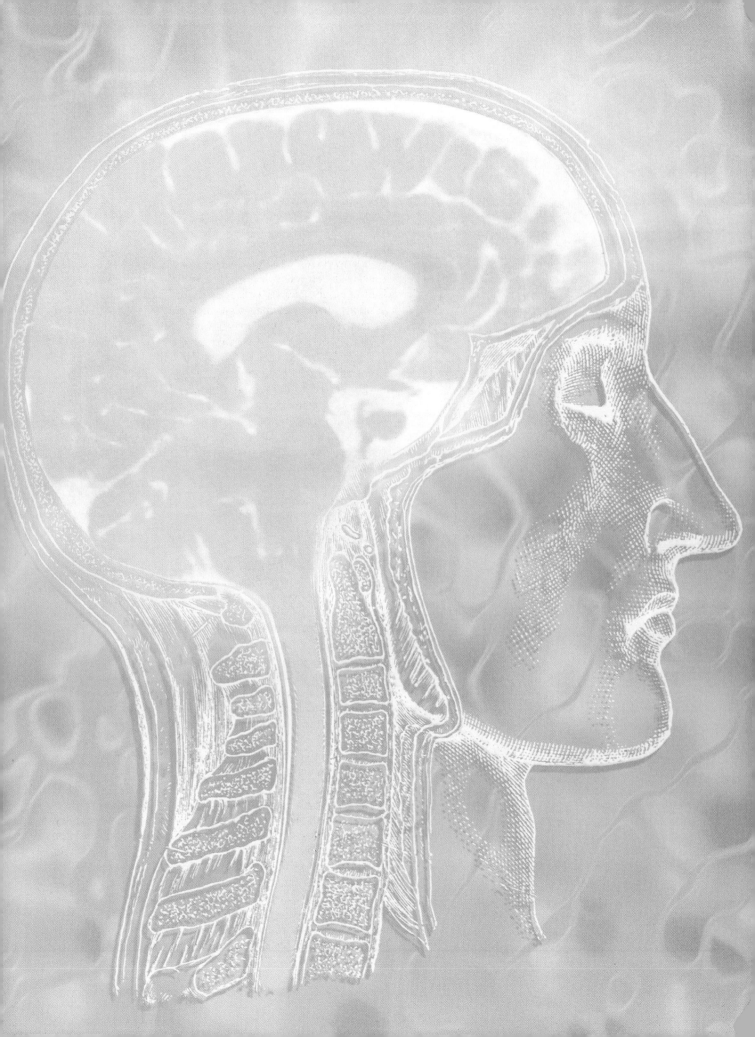

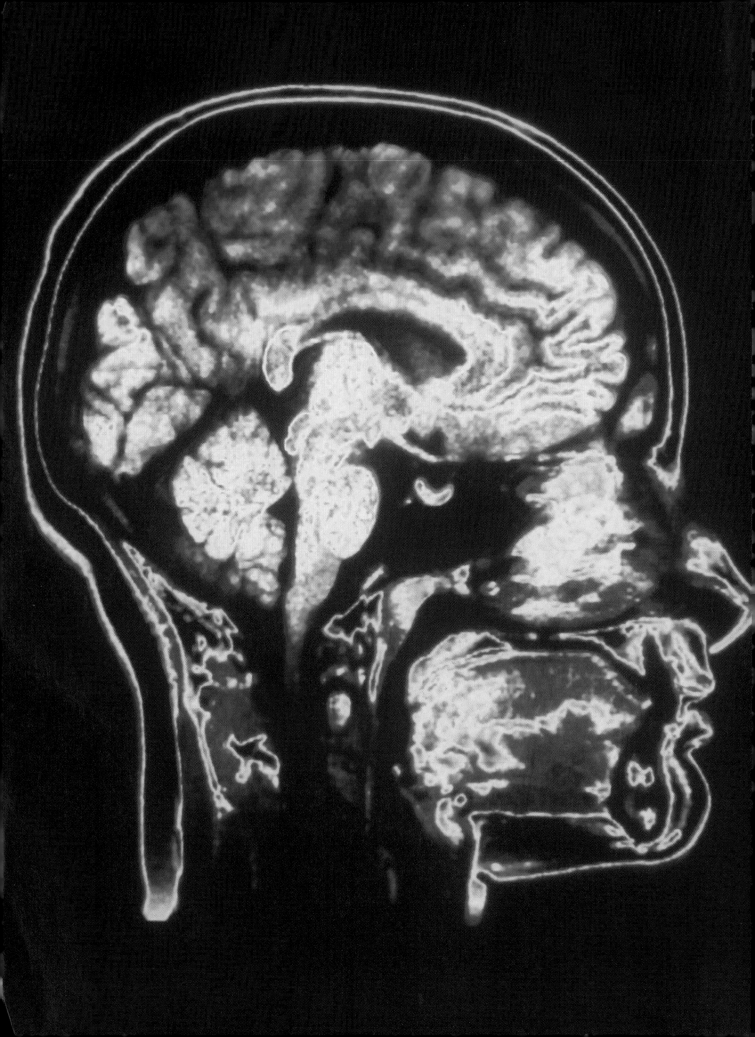

What Is Biopsychology?

1

The Origins of Biopsychology

Prescientific Psychology and the Mind-Brain Problem

Descartes and the Physical Model of Behavior

Helmholtz and the Electrical Brain

The Localization Issue

    IN THE NEWS: THE MIND-BRAIN DEBATE ISN'T OVER

    CONCEPT CHECK

Nature and Nurture

The Genetic Code

    APPLICATION: A COMPUTER MADE OF DNA

Genes and Behavior

The Human Genome Project

Heredity: Destiny or Predisposition?

    CONCEPT CHECK

In Perspective

Summary

Study Resources

In this chapter you will learn

- How biological psychology grew out of philosophy and physiology

- How brain scientists think about the mind-brain problem

- How behavior is inherited and the relationship between heredity and environment

---

There is a wonderful kind of excitement in modern neuroscience, a romantic, moonwalk sense of exploration and setting out for new frontiers. The science is elegant . . . and the pace of discovery absolutely staggering.

    —Kay R. Jamison, An Unquiet Mind

Neuroscience is the multidisciplinary study of the nervous system and its role in behavior. An interesting topic, surely, but *neuroscience is a romantic moonwalk?* To understand why Kay Jamison chose this analogy, you would need to have watched in astonishment from your backyard on an October night in 1957 as the faint glint of reflected light from Sputnik crossed the North American sky. The American people were stunned and fearful as the Russian space program left them far behind. But as the implications of this technological coup sank in, the United States set about constructing its own space program and revamping education in science and technology. Less than 4 years later, President Kennedy made his startling commitment to put an American astronaut on the moon by the end of the decade. But the real excitement would come on the evening of July 20, 1969, as you sat glued to your television set watching the *Eagle* lander settle effortlessly on the moon and the first human step onto the surface of another world (Figure 1.1). For Kay Jamison and the rest of us involved in solving the mysteries of the brain, there is a very meaningful parallel between the excitement of Neil Armstrong's "giant leap for mankind" and the thrill of exploring the inner space of human thought and emotion.

There is also an inescapable parallel between Kennedy's commitment of the 1960s to space exploration and Congress's declaration 30 years later that the 1990s would be known as the Decade of the Brain: Understanding the brain demands the same incredible level of effort, ingenuity, and technological innovation as landing a human on the moon. There were important differences between those two decades, though. President Kennedy acknowledged that no one knew what benefits would arise from space exploration. But as the Decade of the Brain began, we understood that we would not only expand the horizons of human knowledge but also advance the treatment of neurological diseases, emotional disorders, and addictions that cost the United States an estimated trillion dollars a year for care, lost productivity, and crime (Uhl & Grow, 2004).

FIGURE 1.1 The Original Romantic Moonwalk.

Space exploration and solving the mysteries of the brain offer similar challenges and excitement. Which do you think will have the greater impact on your life?

SOURCE: Courtesy of NASA.

Another difference was that the moon-landing project was born out of desperation and a sense of failure, while the Decade of the Brain was a celebration of achievements, both past and current. In the past few years, we have developed new treatments for depression, identified key genes responsible for the devastation of Alzheimer's disease, discovered agents that block addiction to some drugs, learned ways to hold off the memory impairment associated with old age, and produced a map of the human genes.

The United States could not have constructed a space program from scratch in the 1960s; the achievement was built on a long history of scientific research and technological experience. In the same way, the accomplishments of the Decade of the Brain had their roots in a 300-year scientific past, and in 22 centuries of thought and inquiry before that. For that reason, we will spend a brief time examining those links to our past.

The Origins of Biopsychology

The term *neuroscience* identifies the subject matter of the investigation rather than the scientist's training. A neuroscientist may be a biologist, physiologist, anatomist, neurologist, chemist, psychologist, or psychiatrist—even a computer scientist or a philosopher. Psychologists who work in the area of neuroscience specialize in biological psychology, or *biopsychology*, the branch of psychology that studies the relationships between behavior and the body, particularly the brain. (Sometimes the term *psychobiology* or *physiological psychology* is used.) For psychologists, *behavior* has a very broad meaning, which includes internal events such as learning, thinking, and emotion, as well as the overt acts everyone would identify as behavior. Biological psychologists attempt to answer questions like "What changes in the brain when a person learns?" "Why does one person develop depression, another becomes anxious, and another is normal?" "What is the physiological explanation for emotions?" "How do we recognize the face of a friend?" "How does the brain's activity result in consciousness?" Biological psychologists use a variety of research techniques to answer these questions, as you will see in Chapter 4. Whatever their area of study or their strategy for doing research, biological psychologists try to go beyond the mechanics of how the brain works to focus on the brain's role in behavior.

To really appreciate the impressive accomplishments of today's brain researchers, it is useful, perhaps even necessary, to understand the thinking and the work of their predecessors. Contemporary scientists stand on the shoulders of their intellectual ancestors, who made heroic advances with far less information at their disposal than is available to today's undergraduate student.

Writers have pointed out that psychology has a brief history, but a long past. What they mean is that thinkers have struggled with the questions of behavior and experience for more than two millennia, but psychology arose as a separate discipline fairly recently; the date most accept is 1879, when Wilhelm Wundt (Figure 1.2) established the first psychology laboratory in Leipzig, Germany. But biological psychology would not emerge as a separate subdiscipline until psychologists offered convincing evidence that the biological approach could answer significant questions about behavior. To do so, they would have to resolve an old philosophical question about the nature of the mind. Because the question forms a thread that helps us trace the development of biological psychology, we will orient our discussion around this issue.

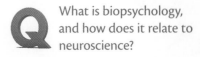

Q What is biopsychology, and how does it relate to neuroscience?

"
In the sciences, we are now uniquely privileged to sit side by side with the giants on whose shoulders we stand.

—*Gerald Holton*
"

FIGURE 1.2 Wilhelm Wundt (1832–1920).

SOURCE: Copyright © Archives of the History of American Psychology, University of Akron.

"
*The nature of the mind and soul is
bodily.*

—Lucretius, c. 50 BCE

*What we call our minds is simply a
way of talking about the functions of
our brains.*

—Francis Crick, 1966
"

How do monists and dualists
disagree on the mind-brain
question?

Prescientific Psychology and the Mind-Brain Problem

This issue is usually called "the mind-body problem," but it is phrased differently here to place the emphasis squarely where it belongs—on the brain. The *mind-brain problem* deals with what the mind is and what its relationship is to the brain. There can be no doubt that the brain is essential to our behavior, but does the mind control the brain, or is it the other way around? Alternatively, are mind and brain the same thing? How these questions are resolved affects how we ask all the other questions of neuroscience.

At the risk of sounding provocative, I will say that there is no such thing as mind. It exists only in the sense that, say, weather exists; weather is a concept we use to include rain, wind, humidity, and related phenomena. We talk as if there is *a weather* when we say things like, "The weather is interfering with my travel plans." But we don't really think that there is *a weather.* Most, though not all, neuroscientists believe that we should think of the mind in the same way; it is simply the collection of things that the brain does, such as thinking, sensing, planning, and feeling. But when we think, sense, plan, and feel, we get the compelling impression that there is *a mind* behind it all, guiding what we do. Most neuroscientists say this is just an illusion, that the sense of mind is nothing more than the awareness of what our brain is doing. Mind, like weather, is also just a concept; it is not a *something*; it does not *do* anything.

This position is known as monism from the Greek *monos*, meaning "alone" or "single." *Monism* is the idea that the mind and the body consist of the same substance. Idealistic monists believe that everything is nonphysical mind, but most monists take the position that the body and mind and everything else are physical; this view is called *materialistic monism.* The idea that the mind and the brain are separate is known as *dualism.* For most dualists, the body is material and the mind is nonmaterial. Most dualists also believe that the mind influences behavior by interacting with the brain.

This question did not originate with modern psychology. The Greek philosophers were debating it in the fifth century BCE (G. Murphy, 1949), when Democritus proposed that everything in the world was made up of atoms (*atomos*, meaning "indivisible"), his term for the smallest particle possible. Even the soul, which included the mind, was made up of atoms so it, too, was material. Plato and Aristotle, considered the two greatest intellectuals among the ancient Greeks, continued the argument into the 4th century BCE. Plato was a dualist, while his student Aristotle joined the body and soul in his attempt to explain memory, emotions, and reasoning.

Defending either position was not easy. The dualists had to explain how a nonphysical mind could influence a physical body, and monists had the task of explaining how the physical brain could account for mental processes such as perception and conscious experience. But the mind was not observable, and even the vaguest understanding of nerve functioning was not achieved until the 1800s, so neither side had much ammunition for the fight.

Descartes and the Physical Model of Behavior

Scientists often resort to the use of models to understand whatever they are studying. A *model* is a proposed mechanism for how something works. Sometimes, a model is in the form of a theory, such as Darwin's explanation that a species developed new capabilities because the capability enhanced the individual's survival. Other times, the model is a simpler organism or system that researchers study in an attempt to understand a more complex one. For example, researchers have used the rat to model everything from learning to Alzheimer's

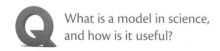

What is a model in science,
and how is it useful?

disease in humans, and the computer has often served as a model of cognitive processes.

In the 17th century, the French philosopher and physiologist René Descartes (Figure 1.3a) used a hydraulic model to explain the brain's activity (Descartes, 1662/1984). Descartes's choice of a hydraulic model was influenced by his observation of the statues in the royal gardens. When a visitor stepped on certain tiles, the pressure forced water through tubes to the statues and made them move. Using this model, Descartes then reasoned that the nerves were also hollow tubes. The fluid they carried was not water, but what he called "animal spirits"; these flowed from the brain and inflated the muscles to produce movement. Sensations, memories, and other mental functions were produced as animal spirits flowed through "pores" in the brain. The animal spirits were pumped through the brain by the pineal gland (Figure 1.3b). Descartes's choice of the pineal gland was based on his belief that it was at a perfect location to serve this function; attached just below the two cerebral hemispheres by its flexible stalk, it appeared capable of bending at different angles to direct the flow of animal spirits into critical areas of the brain. Thus, for Descartes, the pineal gland became the "seat of the soul," the place where the mind interacted with the body. Although Descartes assigned

FIGURE 1.3 Descartes (1596–1650) and the Hydraulic Model.

Descartes believed that behavior was controlled by animal spirits flowing through the nerves.

(a)   (b)

SOURCES: (a) Courtesy of the National Library of Medicine. (b) Corbis.

1

Mind and Body: Descartes to James

control to the mind, his unusual emphasis on the physical explanation of behavior foreshadowed the physiological approach that would soon follow.

Descartes lacked an understanding of how the brain and body worked, so he relied on a small amount of anatomical knowledge and a great deal of speculation. His hydraulic model not only represented an important shift in thinking, but it also illustrates the fact that a model or a theory can lead us astray, at least temporarily. Fortunately, this was the age of the Renaissance, a time not only of artistic expansion and world exploration, but of scientific curiosity. Thinkers began to test their ideas through direct observation and experimental manipulation as the Renaissance gave birth to science. In other words, they adopted the method of *empiricism*, which means that they gathered their information through observation rather than logic, intuition, or other means. Progress was slow, but two critically important principles would emerge as the early scientists ushered in the future. (The WWW icon in the margin indicates that you can find the address to an interesting Internet site on this topic at the end of the chapter.)

Q What two discoveries furthered the early understanding of the brain?

Helmholtz and the Electrical Brain

In the late 1700s, the Italian physiologist Luigi Galvani showed that he could make a frog's leg muscle twitch by stimulating the attached nerve with electricity, even after the nerve and muscle had been removed from the frog's body. A century later in Germany, Gustav Fritsch and Eduard Hitzig (1870) produced movement in dogs by electrically stimulating their exposed brains. What these scientists showed was that animal spirits were not responsible for movement; instead, *nerves operated by electricity!* But the German physicist and physiologist Hermann von Helmholtz (Figure 1.4) demonstrated that nerves do not behave like wires conducting electricity. He was able to measure the speed of conduction in nerves, and his calculation of about 90 feet/second (27.4 meters/second) fell far short of the speed of electricity, which travels through wires at the speed of light (186,000 miles/second or 299,000 kilometers/second). It was obvious that researchers were dealing with a biological phenomenon and that the functioning of nerves and of the brain was open to scientific study. Starting from this understanding, Helmholtz's studies of vision and hearing gave "psychologists their first clear idea of what a fully mechanistic 'mind' might look like" (Fancher, 1979, p. 41). As you will see in later chapters, his ideas were so insightful that even today we must refer to his theories of vision and hearing before describing the current ones.

FIGURE 1.4 Hermann von Helmholtz (1821–1895).

SOURCE: Getty Images.

The Localization Issue

The second important principle to come out of this period—localization—emerged over the first half of the 19th century. *Localization* is the idea that specific areas of the brain carry out specific functions. Fritsch and Hitzig's studies with dogs gave objective confirmation to physicians' more casual observations dating as far back as 17th-century BCE Egypt (Breasted, 1930), but it was two medical case studies that really grabbed the attention of the scientific community. In 1848, Phineas Gage, a railroad construction foreman, was injured when a dynamite blast drove an iron rod through his skull and the frontal lobes of his brain. Amazingly, he survived with no impairment of his intelligence, memory, speech, or movement. But he became irresponsible and profane and was unable to abide by social conventions (H. Damasio, Grabowski, Frank, Galaburda, & Damasio, 1994). Then, in 1861, the French physician Paul Broca (Figure 1.5) performed an autopsy on the brain of a man who had lost the ability to speak after a stroke. The autopsy showed that damage was limited to an area on the left side of his brain now known as Broca's area (Broca, 1861).

By the mid-1880s, additional observations like these had convinced researchers about localization (along with some humorists, as the quote shows!). But a few brain theorists took the principle of localization too far, and we should be on guard lest we make the same mistake. Near the end of the 18th century, when interest in the brain's role in behavior was really heating up, the German anatomist Franz Gall came up with an extreme and controversial theory of brain localization. According to *phrenology*, each of 35 different "faculties" of emotion and intellect—such as combativeness, inhabitiveness (love of home), calculation, and order—was located in a precise area of the brain (Spurzheim, 1908). Gall and his student Spurzheim determined this by feeling bumps on people's skulls and relating any protuberances to the individual's characteristics (Figure 1.6). Others, such as Karl Lashley (1929), took an equally extreme position at the other end of the spectrum; *equipotentiality* is the idea that that the brain functions as an undifferentiated whole. According to this view, the extent of damage, not the location, is what determines how much function is lost.

We now know that bumps on the skull have nothing to do with the size of the brain structures beneath and that most of the characteristics Gall and Spurzheim identified have no particular meaning at the physiological level. But we also know that the brain is not equipotential. The truth, as is often the case, lies somewhere between these two extremes.

Today's research tells us that functions are as much *distributed* as they are localized; behavior results from the interaction of many widespread areas of the brain. In later chapters, you will see examples of cooperative relationships among brain areas in language, visual perception, emotional behavior, motor control, and learning. In fact, you will learn that neuroscientists these days less frequently ask where a function is located than ask how the brain integrates activity from several areas into a single experience or behavior. Nevertheless, the locationists strengthened the monist position by showing that language, emotion, motor control, and so on are controlled by *relatively* specific locations in the brain (Figure 1.7). This meant that the mind ceased being *the explanation* and became *the phenomenon to be explained*.

Understand that the nature and role of the mind are still

FIGURE 1.5 Paul Broca (1825–1880).

SOURCE: Getty Images.

> "I never could keep a promise. . . . It is likely that such a liberal amount of space was given to the organ which enables me to make promises that the organ which should enable me to keep them was crowded out.
>
> —Mark Twain, in Innocents Abroad

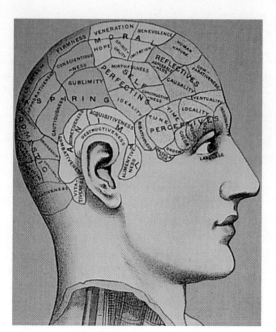

FIGURE 1.6 A Phrenologist's Map of the Brain.

Phrenologists believed that the psychological characteristics shown here were controlled by the respective brain areas.

SOURCE: Bettmann/Corbis.

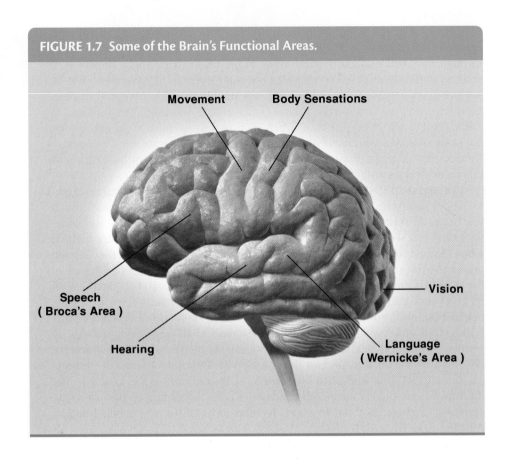

FIGURE 1.7 Some of the Brain's Functional Areas.

Movement   Body Sensations

Vision

Speech
(Broca's Area)

Hearing

Language
(Wernicke's Area)

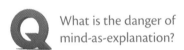

Q What is the danger of mind-as-explanation?

debated in some quarters (see "In the News: The Mind-Brain Debate Isn't Over"). But as you explore the rest of this text, you will see why most brain scientists are material monists: Brain research has been able to explain a great deal of behavior without any reference to a nonmaterial mind.

The Mind-Brain Debate Isn't Over

As neuroscience becomes increasingly successful at explaining what some would call "mind functions," a few self-proclaimed "maverick" neuroscientists are renewing the mind-brain debate. Some have struck up an alliance with the Discovery Institute, the U.S. headquarters for intelligent design. Intelligent design argues that biological life is too complex to have arisen by evolution and that an "intelligent cause" is responsible. The Discovery Institute also funds research into "nonmaterial neuroscience," which counts neuroscientist Jeffrey Schwartz among its adherents. Nonmaterial neuroscientists point to research like the study by Schwartz (see Chapter 14) in which psychological therapy resulted in altered brain functioning in obsessive-compulsive patients. Schwartz and others interpret this as evidence of the mind changing the brain; materialist neuroscientists see it as the brain changing the brain. Schwartz has also teamed up with physicist Henry Stapp to develop nonstandard interpretations of quantum mechanics to explain how the "nonmaterial mind" affects the physical brain.

Now nonmaterial neuroscientists are turning their focus toward consciousness. They believe that material neuroscience will be unable to explain how a material brain can generate conscious experience and that this will spell the final doom of materialism. (Consciousness is so important that the entire final chapter of this text is devoted to the topic.)

SOURCE: Gefter (2008).

Nature and Nurture

A second extremely important issue in understanding the biological bases of behavior is the *nature versus nurture* question, or how important heredity is relative to environmental influences in shaping behavior; like the mind-brain issue, this is one of the more controversial topics in psychology, at least as far as public opinion is concerned. The arguments are based on emotion and values almost as often as they appeal to evidence and reason. For example, some critics complain that attributing behavior to heredity is just a form of excusing actions for which the person or society should be held accountable. A surprising number of behaviors are turning out to have some degree of hereditary influence, so you will be running into this issue throughout the following chapters. Because there is so much confusion about heredity, we need to be sure you understand what it means to say that a behavior is hereditary be fore we go any further.

The Genetic Code

Q How are characteristics inherited?

The *gene* is the biological unit that directs cellular processes and transmits inherited characteristics. Most genes are found on the chromosomes, which are located in the nucleus of each cell, but there are also a few genes in structures outside the nucleus, called the mitochondria. Every body cell in a human has 46 chromosomes, arranged in 23 pairs (see Figure 1.8). Each pair is identifiably distinct from every other pair. This is important, because genes for different functions are found on specific chromosomes. The chromosomes are referred to by number, except for the sex chromosomes, which are designated X or Y. A female has two X chromosomes, while a male has an X and a Y chromosome. Notice that the members of a pair of chromosomes are similar, again with the exception that the Y chromosome is much shorter than the X chromosome.

Unlike the body cells, the male's sperm cells and the female's ova (egg cells) each have 23 chromosomes. When these sex cells are formed by the division of their parent cells, the pairs of chromosomes separate so each daughter cell receives only one chromosome from each pair. When the sperm enters the

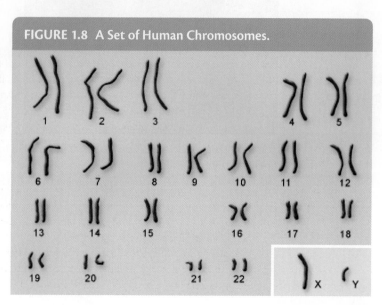

FIGURE 1.8 A Set of Human Chromosomes.

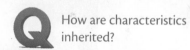

sex chromosomes

SOURCE: U.S. National Library of Medicine.

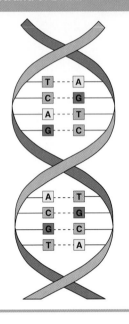

ovum during fertilization, the chromosomes of the two cells merge to restore the number to 46. The fertilized egg or *zygote* then undergoes rapid cell division and development on its way to becoming a functioning organism. For the first 8 weeks (in humans), the new organism is referred to as an *embryo* and from then until birth as a *fetus.*

The mystery of how genes carry their genetic instructions began to yield to researchers in 1953 when James Watson and Francis Crick published a proposed structure for the deoxyribonucleic acid that genes are made of. *Deoxyribonucleic acid (DNA)* is a double-stranded chain of chemical molecules that looks like a ladder that has been twisted around itself; this is why DNA is often referred to as the *double helix* (see Figure 1.9). Each rung of the ladder is composed of two of the four nucleotides—adenine, thymine, guanine, and cytosine (A, T, G, C). The order in which these appear on the ladder forms the code that carries all our genetic information. The four-letter alphabet these nucleotides provide is adequate to spell out the instructions for every structure and function in your body; the feature "Application: A Computer Made of DNA" will give you some appreciation of DNA's complexity and power.

We only partially understand how genes control the development of the body and its activities, as well as how they influence many aspects of behavior. However, we do know that genes exert their influence in a deceptively simple manner: They provide the directions for making proteins. Some of these proteins are used in the construction of the body, and others are enzymes; enzymes act as catalysts,

Application   A Computer Made of DNA

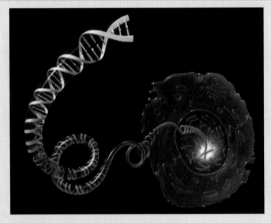

The complexity of DNA (left) permits Shapiro to build microscopic computers. Shapiro is holding a large-scale model.

SOURCE: (left) National Human Genome Research Institute. (right) Used with permission of Professor Ehud Shapiro of the Computer Science Department at the Weizmann Institute of Science.

When some people look at DNA, they are struck by its similarities with a computer. Ehud Shapiro's lab at the Weizmann Institute in Israel is taking that similarity to its logical conclusion by building computers out of DNA. They are so small that a single drop of water can hold a trillion of them. Basically, a strand of DNA serves as an "input" molecule, which is operated on by two enzymes and a preprogrammed "software" molecule to produce an "output" molecule. All this happens at the binary level, which means that the computer works by manipulating 1s and 0s.

These computers may be simple and tiny, but what they can do is surprising. One version can identify the RNA molecules associated with prostate cancer and release short DNA strands that kill the cancer cells; another can detect a form of lung cancer (Benenson, Gil, Ben-Dor, Adar, & Shapiro, 2004). So far, this has been done only in test tubes, but Shapiro and his colleagues think they can eventually produce DNA "medical kits" that operate inside the body, detecting a variety of diseases and treating them even before symptoms appear.

modifying chemical reactions in the body. It is estimated that the sequences of nucleotides that make up our DNA differ among human individuals by only about 0.5% (S. Levy et al., 2007); however, you will see throughout this text that this variation leads to enormous differences in development and behavior.

Because chromosomes are paired, most genes are as well; a gene on one chromosome is paired with a gene for the same function on the other chromosome—except for a few genes on the X chromosome that are not paired with genes on the shorter Y chromosome. Although paired genes have the same function, their effects often differ; these different versions of a gene are called *alleles*. In some cases the effects of the two alleles blend to produce a result; for example, a person with the allele for type A blood on one chromosome and the allele for type B blood on the other will have type AB blood.

In other cases, one gene may be dominant over the other. A *dominant* gene will produce its effect regardless of which gene it is paired with; a *recessive* gene will have an influence only when it is paired with the same recessive gene on the other chromosome. A simple behavioral example of dominance and recessiveness is hand clasping. Clasp your hands together, with the fingers interlaced, and notice whether your left thumb is over the right thumb or the other way around. Change positions and see which is more comfortable. The allele for the left-thumb-over-right position is dominant, and the allele for right-over-left is recessive; handedness has little or no effect.

Inheritance of hand clasping preference is illustrated in Figure 1.10. In the example on the left, note that one of the parents (shown outside the rectangle) is *homozygous* for the hand-clasping gene, which means that the two alleles of the gene are identical; as a result, all the offspring inherit at least one dominant allele. Although the children have different *genotypes* (combinations of genes), they all

---

FIGURE 1.10 Offspring of Parents Homozygous and Heterozygous for Hand Clasping.

On the left, a homozygous individual has mated with a heterozygous individual (shown on the outside of the rectangle; L = dominant, r = recessive). On the right, two heterozygous individuals have mated. (Note that all four parents have a left-thumb-over-right preference because they have at least one L allele.) In the first family, all children will be genetically predisposed to prefer the left-thumb-over-right position, while in the second case, approximately one fourth will be predisposed to prefer right-over-left.

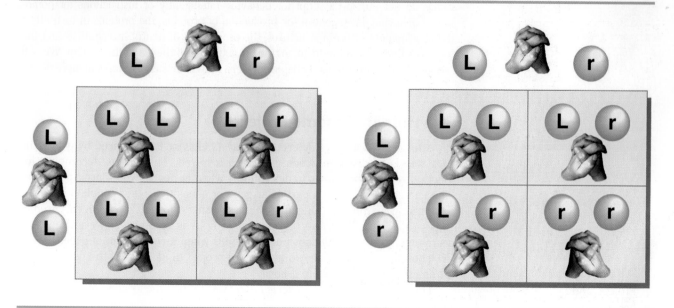

have the same *phenotype* (characteristic); that is, they all show left preference. Right preference occurs when a person receives two recessive alleles; in the second example, where both parents are *heterozygous* (each has one L and one r allele), about one out of four children will show right preference.

In the case of unpaired genes on the X chromosome, a recessive gene alone is adequate to produce an effect because it is not opposed by a dominant gene. A characteristic produced by an unpaired gene on the X chromosome is referred to as *X-linked*. X linkage explains why, for example, males are red-green color-blind eight times as frequently as females.

Some characteristics—such as hand clasping, blood type, and the degenerative brain disorder Huntington's disease—result from a single pair of genes, but many characteristics are determined by several genes; they are *polygenic*. Height is polygenic, and most behavioral characteristics such as intelligence and psychological disorders are also controlled by multiple genes.

Q Why do males more often show characteristics that are caused by recessive genes?

Genes and Behavior

We have known from ancient times that animals could be bred for desirable behavioral characteristics such as hunting ability or a mild temperament that made them suitable as pets. Charles Darwin helped establish the idea that behavioral traits can be inherited in humans as well, but the idea fell into disfavor as an emphasis on learning as the major influence on behavior became increasingly fashionable. But in the 1960s and 1970s, the tide of strict environmentalism began to ebb, and the perspective shifted toward a balanced view of the roles of nature and nurture (Plomin, Owen, & McGuffin, 1994). By 1992, the American Psychological Association was able to identify genetics as one of the themes that best represent the present and the future of psychology (Plomin & McClearn, 1993).

Of the behavioral traits that fall under genetic influence, intelligence is the most investigated. Most of the behavioral disorders, including alcoholism and drug addiction, schizophrenia, major mood disorders, and anxiety, are partially hereditary as well (McGue & Bouchard, 1998). The same can be said for some personality characteristics (T. J. Bouchard, 1994) and sexual orientation (J. M. Bailey & Pillard, 1991; J. M. Bailey, Pillard, Neale, & Agyei, 1993; Kirk, Bailey, Dunne, & Martin, 2000).

Q What are some of the inheritable behaviors?

However, you should exercise caution in thinking about these genetic effects. Genes do not provide a script for behaving intelligently or instructions for homosexual behavior. They control the production of proteins; the proteins in turn affect the development of brain structures, the production of neural transmitters and the receptors that respond to them, and the functioning of the glandular system. We will see specific examples in later chapters, where we will discuss this topic in more depth.

The Human Genome Project

After geneticists have determined that a behavior is influenced by genes, the next step is to discover which genes are involved. The various techniques for identifying genes boil down to determining whether people who share a particular characteristic also share a particular gene or genes that other people don't have. This task is extremely difficult if the researchers do not know which chromosomes to examine, because the amount of DNA is so great. However, the gene search received a tremendous boost in 1990 when a consortium of geneticists at 20 laboratories around the world began a project to identify all the genes in our chromosomes, or the human *genome*.

The goal of the *Human Genome Project* was to map the location of all the genes on the human chromosomes and to determine the genes' codes—that is, the order of

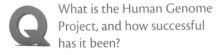

Q What is the Human Genome Project, and how successful has it been?

bases within each gene. In 2000—just 10 years after the project began—the project group and a private organization simultaneously announced "rough drafts" of the human genome (International Human Genome Sequencing Consortium, 2001; Venter et al., 2001). Three years later, the group had brought the map to 99% completion and reduced the number of gaps from 150,000 to 341 (International Human Genome Sequencing Consortium, 2004).

There are still questions, of course. For one, the map has revealed that we have only 20,000 to 25,000 functioning genes, just a few more than the roundworm; 97% of our DNA does not encode proteins and is frequently referred to as "junk" DNA. The number of genes is not correlated with an organism's complexity, but the amount of junk DNA is, so it must have an important function (Andolfatto, 2005; Siepel et al., 2005). Some noncoding DNA controls gene expression—whether the gene functions or doesn't (Pennacchio et al., 2006). For example, when a stretch of noncoding DNA known as *HACNS1*—which is unique to humans—is inserted into a mouse embryo, genes are activated in the "forearm" and "thumb" (Figure 1.11; Prabhakar et al., 2008). DNA from the same area in chimpanzees and Rhesus monkeys does not have that effect. It is speculated that *HACNS1* turns on genes that led to the evolutionarily important dexterity of the human thumb.

A second question is what the genes do. The gene map doesn't answer that question, but it does make it easier to find the genes responsible for a particular disorder or behavior. For example, when geneticists were searching for the gene that causes Huntington's disease in the early 1980s, they found that most of the affected individuals in a large extended family shared a couple of previously identified genes with known locations on chromosome 4 while the disease-free family members didn't. This meant that the Huntington's gene was on chromosome 4 and near these two *marker* genes (Gusella et al., 1983). Actually finding the Huntington gene still took another 10 years; even though the map is incomplete, it is already dramatically reducing the time required to identify genes.

Identifying the genes and their functions will improve our understanding of human behavior and psychological as well as medical disorders. We will be able to treat disorders genetically, counsel vulnerable individuals about preventive measures, and determine whether a patient will benefit from a drug or have an adverse reaction, thus eliminating delays from trying one treatment after another.

2
Gene Research Site

FIGURE 1.11 Human Junk DNA Turns on Genes in a Mouse Embryo's Paw.

To determine where the DNA was having an effect, it was paired with a gene that produces a blue protein when activated. The blue area indicates that *HACNS1* is targeting genes in the area analogous to the human thumb.

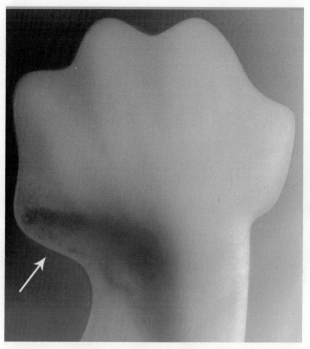

SOURCE: From "Human-Specific Gain of Function in a Developmental Enhancer," by S. Prabhakar et al., 2008, *Science, 321*, p. 1348. Reprinted with permission from AAAS.

Heredity: Destiny or Predisposition?

To many people, the idea that several, if not most, of their behavioral characteristics are hereditary implies that they are clones of their parents and their future is engraved in stone by their genes. This is neither a popular nor a comfortable view and creates considerable resistance to the concept of behavioral genetics. The view is also misleading; a hallmark of genetic influence is actually *diversity*.

 Do genes lock a person into a particular outcome in life?

Genes and Individuality

Although family members do tend to be similar to each other, children share only half of their genes with each of their parents or with each other. A sex cell receives a random half of the parent's chromosomes; as a result, a parent can produce 2^{23}, or 8 million, different combinations of chromosomes. Add to this the uncertainty of which sperm will unite with which egg, and the number of genetic combinations that can be passed on to offspring rises to 60 or 70 trillion! So sexual reproduction increases individuality in spite of the inheritability of traits. This variability powers what Darwin (Figure 1.12) called *natural selection*, which means that those whose genes endow them with more adaptive capabilities are more likely to survive and transmit their genes to more offspring (Darwin, 1859).

The effects of the genes themselves are not rigid; they can be variable over time and circumstances. Genes are turned on and turned off, or their activity is upregulated and downregulated, so they produce more or less of their proteins or different proteins at different times. If the activity of genes were constant, there would be no smoothly flowing sequence of developmental changes from conception to adulthood. A large number of genes change their functioning late in life, apparently accounting for many of the changes common to aging (Ly, Lockhart, Lerner, & Schultz, 2000), as well as the onset of diseases such as Alzheimer's (Breitner, Folstein, & Murphy, 1986). The functioning of some genes is even controlled by experience, which explains some of the changes in the brain that constitute learning (C. H. Bailey, Bartsch, & Kandel, 1996). For the past quarter century, researchers have puzzled over why humans are so different from chimpanzees, our closest relatives, considering that 95% to 98% of our DNA sequences are identical (R. J. Britten, 2002; M.-C. King & Wilson, 1975). Now, it appears that part of the answer is that we differ more dramatically in which genes are *expressed*—actually producing proteins—in the brain (Enard, Khaitovich, et al., 2002).

Genes also have varying degrees of effects; some determine the person's characteristics, and others only influence them. A person with the mutant form of the *huntingtin* gene *will* develop Huntington's disease, but most behavioral traits depend on many genes; a single gene will account for only a slight increase in intelligence or in the risk for schizophrenia. The idea of risk raises the issue of vulnerability and returns us to our original question, the relative importance of heredity and environment.

FIGURE 1.12 Charles Darwin (1809–1882).

SOURCE: Courtesy of Library of Congress.

Heredity, Environment, and Vulnerability

To assess the relative contributions of heredity and environment, there is a need to be able to quantify the two influences. *Heritability* is the percentage of the variation in a characteristic that can be attributed to genetic factors. The calculation of heritability is based on a comparison of how often identical twins share the characteristic with how often fraternal twins share the characteristic. The reason for this comparison is that identical twins develop from a single egg and therefore have the same genes, while fraternal twins develop from separate eggs and share just 50% of their genes, like nontwin siblings. Heritability estimates are around 50% for intelligence (Plomin, 1990), which means that about half of the population's differences in intelligence are due to heredity. Heritability has been estimated at 60% to 90% for schizophrenia (Tsuang,

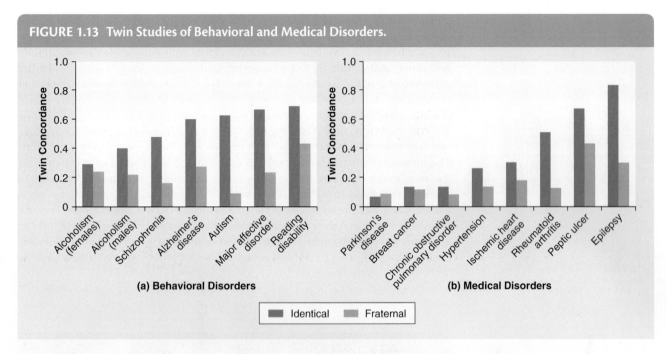

FIGURE 1.13 Twin Studies of Behavioral and Medical Disorders.

(a) Behavioral Disorders

(b) Medical Disorders

Identical   Fraternal

The concordance of (a) behavioral disorders and (b) medical disorders in identical and fraternal twins. Concordance is the proportion of twin pairs in which both twins have the disorder. Note the greater concordance in identical twins and the (generally) higher concordance for behavioral disorders than for medical disorders.

SOURCE: From "The Genetic Basis of Complex Human Behavior," by R. Plomin, M. J. Owen, and P. McGuffin, *Science*, 264, p. 1734. © 1994 American Association for the Advancement of Science. Reprinted with permission from AAAS.

Gilbertson, & Faraone, 1991) and 40% to 50% for personality characteristics and occupational interests (Plomin et al., 1994). The heritability for height is approximately 90% (Plomin, 1990), which makes the values for behavioral characteristics seem modest. On the other hand, the genetic influence on behavioral characteristics is typically stronger than it is for common medical disorders, as Figure 1.13 shows (Plomin et al., 1994).

Since about half of the differences in behavioral characteristics among people is attributable to heredity, then approximately half is due to environmental influences. Keep in mind that heritability is not an absolute measure but tells us the proportion of genetic influence relative to the amount of environmental influence. For example, adoption studies tend to overestimate the heritability of intelligence because adopting parents are disproportionately from the middle class. Because the children's environments are unusually similar, environmental influence will appear to be lower and heritability higher than typical (McGue & Bouchard, 1998). Similarly, heritability will appear to be lower if we look only at a group of closely related individuals.

Researchers caution us that "we inherit dispositions, not destinies" (R. J. Rose, 1995, p. 648). This is because the influence of genes is only partial. This idea is formalized in the vulnerability model, which has been applied to disorders such as schizophrenia (Zubin & Spring, 1977). *Vulnerability* means that genes contribute a predisposition for the disorder that may or may not exceed the threshold required to produce the disorder; environmental challenges such as neglect or emotional trauma may combine with a person's hereditary susceptibility to exceed that threshold. The general concept applies to behavior and abilities as well. For example, the combination of genes a person receives determine a broad range for the person's

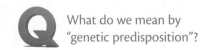

What do we mean by "genetic predisposition"?

potential intelligence; environmental influences then will determine where in that range the person's capability will fall. Psychologists no longer talk about heredity versus environment, as if the two are competing with each other for importance. Both are required, and they work together to make us what we are. As an earlier psychologist put it, "To ask whether heredity or environment is more important to life is like asking whether fuel or oxygen is more necessary for making a fire" (Woodworth, 1941, p. 1).

With increasing understanding of genetics, we are now in the position to change our very being. This kind of capability carries with it a tremendous responsibility. The knowledge of our genetic makeup raises the question whether it is better for a person to know about a risk that may never materialize, such as susceptibility to Alzheimer's disease. In addition, many worry that the ability to do genetic testing on our unborn children means that some parents will choose to abort a fetus because it has genes for a trait they consider undesirable. Our ability to plumb the depths of the brain and of the genome is increasing faster than our grasp of either its implications or how to resolve the ethical questions that will arise. We will consider some of the ethical issues of genetic research in Chapter 4.

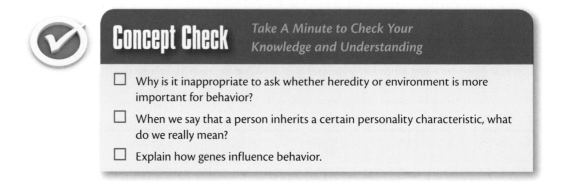

Concept Check   *Take A Minute to Check Your Knowledge and Understanding*

☐ Why is it inappropriate to ask whether heredity or environment is more important for behavior?

☐ When we say that a person inherits a certain personality characteristic, what do we really mean?

☐ Explain how genes influence behavior.

In Perspective

In the first issue of the journal *Nature Neuroscience,* the editors observed that brain science still has a "frontier" feel to it ("From Neurons to Thoughts," 1998). The excitement Kay Jamison talked about is real and tangible, and the accomplishments are remarkable for such a young discipline. The successes come from many sources: the genius of our intellectual ancestors, the development of new technologies, the adoption of empiricism, and, I believe, a coming to terms with the concept of the mind. Evidence of all these influences will be apparent in the following chapters.

Neuroscience and biopsychology still have a long way to go. For all our successes, we do not fully understand what causes schizophrenia, exactly how the brain is changed by learning, or why some people are more intelligent than others. Near the end of the Decade of the Brain, Torsten Wiesel (whose landmark research in vision you will read about later) scoffed at the idea of dedicating a decade to the brain as "foolish. . . . We need at least a century, maybe even a millennium" (quoted in Horgan, 1999, p. 18). As you read the rest of this book, keep in mind that you are on the threshold of that century's journey, that millennium of discovery.

4
More Useful Websites

Summary

The Origins of Biopsychology

- Biopsychology developed out of physiology and philosophy as early psychologists adopted empiricism.
- Most psychologists and neuroscientists treat mind as a product of the brain, believing that mental activity can be explained in terms of the brain's functions.
- Localization describes brain functioning better than equipotentiality, but a brain process is more likely to be carried out by a network of structures than by a single structure.

Nature and Nurture

- We are learning that a number of behaviors are genetically influenced. One does not inherit a behavior itself, but genes influence structure and function in the brain and body in a way that influences behavior.
- Behavior is a product of both genes and environment. In many cases, genes produce a predisposition, and environment further determines the outcome.
- With the knowledge of the genome map, we stand on the threshold of unbelievable opportunity for identifying causes of behavior and diseases, but we face daunting ethical challenges as well. ■

Study Resources

F For Further Thought

- Why, in the view of most neuroscientists, is materialistic monism the more productive approach for understanding the functions of the mind? What will be the best test of the correctness of this approach?
- Scientists were working just as hard on the problems of the brain a half century ago as they are now; why were the dramatic discoveries of recent years not made then?
- What are the implications of knowing what all the genes do and of being able to do a scan that will reveal which genes an individual has?
- If you were told that you had a gene that made it 50% likely that you would develop a certain disease later in life, would there be anything you could do?

T Testing Your Understanding

1. How would a monist and a dualist pursue the study of biopsychology differently?
2. What was the impact of the early electrical stimulation studies and the evidence that specific parts of the brain were responsible for specific behaviors?
3. Explain how two parents who have the same characteristic produce children who are different from them in that characteristic? Use appropriate terminology.
4. A person has a gene that is linked with a disease but does not have the disease. We have mentioned three reasons why this could occur; describe two of them.
5. Discuss the interaction between heredity and environment in influencing behavior, including the concept of vulnerability.

Select the best answer:

1. The idea that mind and brain are both physical is
   known as
   a. idealistic monism.
   b. materialistic monism.
   c. idealistic dualism.
   d. materialistic dualism.

2. A model is
   a. an organism or a system used to understand a
      more complex one.
   b. a hypothesis about the outcome of a study.
   c. an analogy, not intended to be entirely realistic.
   d. a plan for investigating a phenomenon.

3. Descartes's most important contribution
   was in
   a. increasing knowledge of brain anatomy.
   b. suggesting the physical control of behavior.
   c. emphasizing the importance of nerves.
   d. explaining how movement is produced.

4. Helmholtz showed that
   a. nerves are not like electrical wires because
      they conduct too slowly.
   b. nerves operate electrically.
   c. nerves do not conduct animal spirits.
   d. language, emotion, movement, and so on
      depend on the activity of nerves.

5. In the mid-1800s, studies of brain-damaged
   patients convinced researchers that
   a. the brain's activity was electrical.
   b. the mind was not located in the brain.
   c. behaviors originated in specific parts of the
      brain.
   d. the pineal gland could not serve the role
      Descartes described.

6. Localization means that
   a. specific functions are found in specific parts of
      the brain.

b. the most sophisticated functions are located in
   the highest parts of the brain.
c. any part of the brain can take over other
   functions after damage.
d. brain functions are located in widespread
   networks.

7. X-linked characteristics affect males more than
   females because
   a. the X chromosome is shorter than the
      Y chromosome.
   b. unlike males, females have only one
      X chromosome.
   c. the responsible gene is not paired with
      another gene on the Y chromosome.
   d. the male internal environment exaggerates
      effects of the genes.

8. Two parents are heterozygous for a dominant
   characteristic. They can produce a child with the
   recessive characteristic:
   a. if the child receives a dominant gene and a
      recessive gene.
   b. if the child receives 2 recessive genes.
   c. if the child receives 2 dominant genes.
   d. under no circumstance.

9. The Human Genome Project has
   a. counted the number of human genes.
   b. made a map of the human genes.
   c. determined the functions of most genes
   d. cloned most of the human genes.

10. Heritability is greatest for
    a. intelligence.   b. schizophrenia.
    c. personality.   d. height.

11. If we all had identical genes, the estimated
    heritability for a characteristic would be
    a. 0%.   b. 50%.
    c. 100%.   d. impossible to determine.

1.b, 2.a, 3.b, 4.a, 5.c, 6.a, 7.c, 8.b, 9.b, 10.d, 11.a.

Answers:

Online Resources

The following resources are available at
www.sagepub.com/garrett3e.

Chapter Resources

- Flash Cards
- Chapter Quiz
- Internet Resources and Exercises

On the Web

The following websites are coordinated with
the chapter's content. Their numbers correspond
with the numbers of the icons you saw throughout
the chapter. You can find links to these websites by
going to the web address above and selecting this
chapter's **Chapter Resources**, then choosing **Internet
Resources and Exercises**.

1. **Mind and Body** covers the history of the idea from René Descartes to William James. Most pertinent sections are I: 1–5 and II: 1–2.

2. You can search **Online Mendelian Inheritance in Man** by characteristic/disorder (e.g., schizophrenia), chromosomal location (e.g., 1q21–q22), or gene symbol (e.g., *SCZD9*) to get useful genetic information and summaries of research articles.

3. You can get updates on the Human Genome Project and news of genetic research breakthroughs from **Functional Genomics** and **Omics Gateway**.

4. The following journals are major sources of neuroscience articles (some require subscriptions to obtain full-text articles):
   Brain, Behavior, and Evolution
   Journal of Neuroscience
   Nature
   Nature Neuroscience
   New Scientist (nonprofessional; for the general reader)
   Science
   Scientific American Mind (nonprofessional; for the general reader)
   Trends in Neurosciences

5. General information sites:
   Brain Briefings (various topics in neuroscience), www.sfn.org/briefings
   Brain in the News (neuroscience news from media sources), www.dana.org/news/braininthenews
   Neuroguide (a small but growing offering of resources), www.neuroguide.com
   Science Daily (latest developments in science; see "Mind & Brain" and "Health & Medicine"), www.sciencedaily.com

Chapter Updates and Biopsychology News

R For Further Reading

1. "The Emergence of Modern Neuroscience: Some Implications for Neurology and Psychiatry" by W. Maxwell Cowan, Donald H. Harter, and Eric R. Kandel (*Annual Review of Neuroscience*, 2000, *23*, 343–391) describes the emergence of neuroscience as a separate discipline in the 1950s and 1960s, then details its most important accomplishments in understanding disorders.

2. "Neuroscience: Breaking Down Scientific Barriers to the Study of Brain and Mind" by E. R. Kandel and Larry Squire (*Science,* 2000, *290,* 1113–1120) is a briefer treatment of the recent history of neuroscience, with an emphasis on psychological issues; a timeline of events over more than three centuries is included.

3. *Behavior Genetics Principles: Perspectives in Development, Personality, and Psychopathology*, edited by Lisabeth F. Dilalla and Irving I. Gottesman (American Psychological Association, 2004), is a compilation of articles by the foremost researchers in the genetics of behavior.

4. "Tweaking the Genetics of Behavior" by Dean Hamer (available at http://apbio.savithasastry.com/Units/Unit%208/articles/cle_review_genesandbehavior.pdf) is a fanciful but thought-provoking story about a female couple in 2050 who have decided to have a child cloned and the decisions available to them for determining their baby's sex and her physical and psychological characteristics through genetic manipulation.

K Key Terms

alleles 11
biopsychology 3
deoxyribonucleic acid (DNA) 10
dominant 11
dualism 4
embryo 10
empiricism 6
equipotentiality 7
fetus 10
gene 9
genome 12
genotype 11
heritability 14
heterozygous 12
homozygous 11
Human Genome Project 12

localization 6

materialistic monism 4

mind-brain problem 4

model 4

monism 4

natural selection 14

nature versus nurture 9

neuroscience 1

phenotype 12

phrenology 7

polygenic 12

recessive 11

vulnerability 15

X-linked 12

zygote 10

Part 1

Neural Foundations of Behavior: The Basic Equipment

Chapter 2. Communication Within the Nervous System

Chapter 3. The Organization and Functions of the Nervous System

Chapter 4. The Methods and Ethics of Research

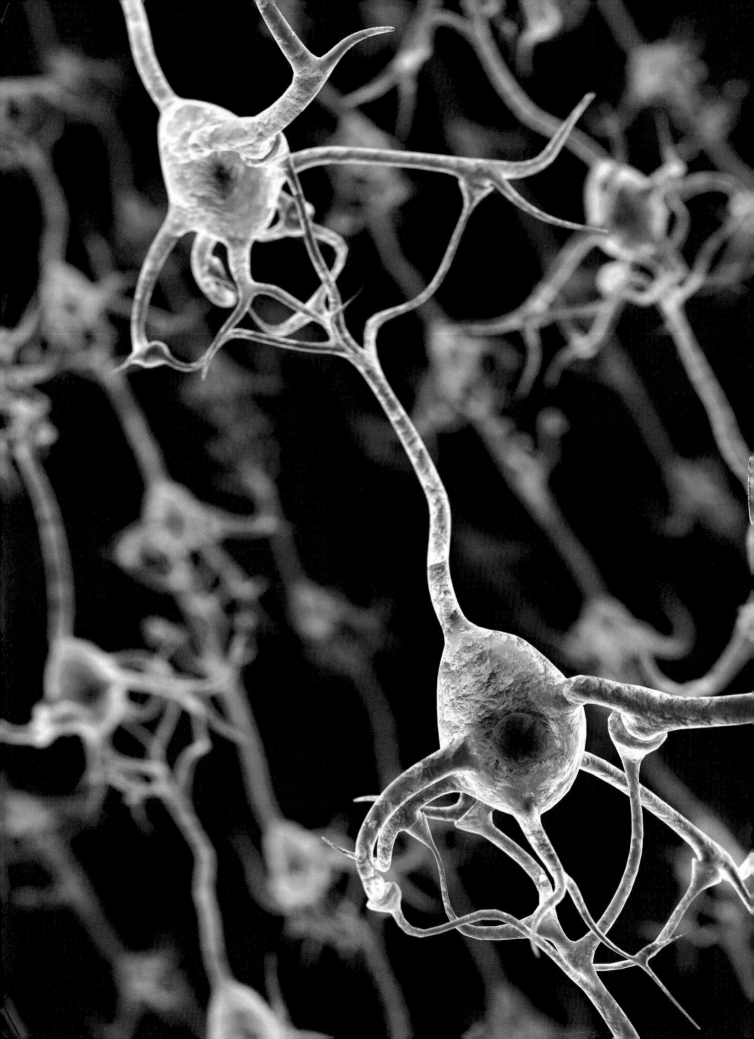

Communication Within the Nervous System

2

The Cells That Make Us Who We Are

Neurons
   APPLICATION: TARGETING ION CHANNELS
Glial Cells
   CONCEPT CHECK

How Neurons Communicate With Each Other

Chemical Transmission at the Synapse
Regulating Synaptic Activity
Neurotransmitters
   APPLICATION: AGONISTS AND
      ANTAGONISTS IN THE REAL WORLD
Of Neuronal Codes, Neural Networks, and Computers
   CONCEPT CHECK

In Perspective

Summary

Study Resources

In this chapter you will learn

- How neurons are specialized to conduct information

- How glial cells support the activity of neurons

- How neurons communicate with each other

- Strategies neurons use to increase their information capacity

- The functions of some of the major chemical transmitters

- How computer simulations of neural networks are duplicating many brain functions

Things were looking good for Jim and his wife. She was pregnant with their first child, and they had just purchased and moved into a new home. After the exterminating company treated the house for termites by injecting the pesticide chlordane under the concrete slab, Jim noticed that the carpet was wet and there was a chemical smell in the air. He dried the carpet with towels and thought no more about it, not realizing that chlordane can be absorbed through the skin. A few days later, he developed headaches, fatigue, and numbness. Worse, he had problems with memory, attention, and reasoning. His physician referred him to the toxicology research center of a large university medical school. His intelligence test score was normal, but the deficiencies he was reporting showed up on more specific tests of cognitive ability. Jim and his wife had to move out of their home. At work, he had to accept reduced responsibilities because of his difficulties in concentration and adapting to novel situations. The chlordane had not damaged the structure of his brain as you might suspect, but it had interfered with the functioning of the brain cells by impairing a mechanism called the sodium-potassium pump (Zillmer & Spiers, 2001). Jim's unfortunate case reminds us that the nervous system is as delicate as it is intricate. Only by understanding how it works will you be able to appreciate human behavior, to enhance human performance, and to treat behavioral problems such as drug addiction and psychosis.

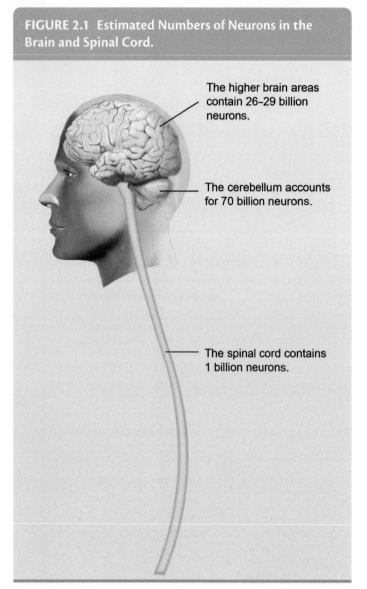

FIGURE 2.1 Estimated Numbers of Neurons in the Brain and Spinal Cord.

The higher brain areas contain 26–29 billion neurons.

The cerebellum accounts for 70 billion neurons.

The spinal cord contains 1 billion neurons.

The Cells That Make Us Who We Are

To understand human behavior and the disorders that affect it, you must understand how the brain works. And to understand how the brain works, you must first have at least a basic understanding of the cells that carry messages back and forth in the brain and throughout the rest of the body. *Neurons* are specialized cells that convey sensory information into the brain; carry out the operations involved in thought, feeling, and action; and transmit commands out into the body to control muscles and organs. It is estimated that there are about 100 billion neurons in the human brain (Figure 2.1; R. W. Williams & Herrup, 2001). This means that there are more neurons in your brain than stars in our galaxy. But as numerous and as important as they are, neurons make up only 10% of the brain's cells and about half its volume. The other 90% are glial cells, which we will discuss later in the chapter.

Neurons

Neurons have the responsibility for all the things we do—our movements, our thoughts, our memories, and our emotions. It is difficult to believe that anything so simple as a cell can measure up to this task, and the burden is on the neuroscientist to demonstrate that this is true. As you will see, the neuron is deceptively simple in its action but impressively complex in its function.

Basic Structure: The Motor Neuron

First let's look inside a neuron, because I want to show you that the neuron is a cell, very much like other cells in the body. Figure 2.2 is an illustration of the most prominent part of the neuron, the *cell body* or soma. The cell body is filled with a watery liquid called cytoplasm and contains a number of *organelles*. The largest of these organelles is the *nucleus*, which contains the cell's chromosomes. Other organelles are responsible for converting

What are the parts of the neuron?

nutrients into fuel for the cell, constructing proteins, and removing waste materials. So far, this could be the description of any cell; now let's look at the neuron's specializations that enable it to carry out its unique role. Figure 2.3 illustrates a typical neuron. "Typical" is used guardedly here, because there are three major kinds of neurons and variations within those types. This particular type is a *motor neuron*, which carries commands to the muscles and organs. It is particularly useful for demonstrating the structure and functions that neurons have in common.

Dendrites are extensions that branch out from the cell body to receive information from other neurons. Their branching structure allows them to collect information from many neurons. The *axon* extends like a tail from the cell body and carries information to other locations, sometimes across great distances. The myelin sheath that is shown wrapped around the axon supports the axon and provides other benefits that we will consider later. Branches at the end of the axon culminate in swellings called end bulbs or *terminals*. The terminals contain chemical *neurotransmitters*, which the neuron releases to communicate with a muscle or an organ or the next neuron in a chain. In our examples, we will talk as if neurons form a simple chain, with one cell sending messages to a single other neuron, and so on; in actuality, a neuron receives input from many neurons and sends its output to many others.

Neurons are usually so small that they can be seen only with the aid of a microscope. The cell body is the largest part of the neuron, ranging from 0.005 to 0.1 millimeter (mm) in diameter in mammals. (In case you are unfamiliar with metric measurements, a millimeter is about the thickness of a dime.) Even the giant neurons of the squid, favored by researchers for their conveniently large size, have cell bodies that are only 1 mm in diameter. Axons, of course, are smaller; in mammals, they range from 0.002 to 0.02 mm in diameter. Axons can be anywhere from 0.1 mm to more than a meter in length.

FIGURE 2.2 Cell Body of a Neuron.

Part of the membrane has been removed to show interior features.

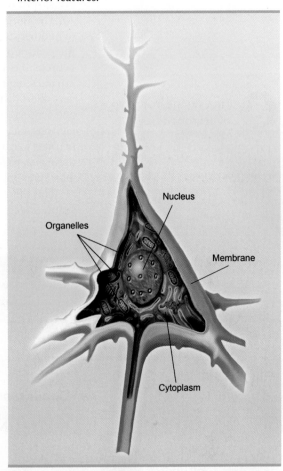

FIGURE 2.3 Components of a Neuron.

The illustration is of a motor neuron.

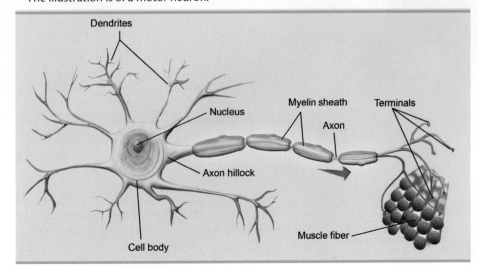

Other Types of Neurons

The second type of neuron is the sensory neuron. *Sensory neurons* carry information from the body and from the outside world into the brain and spinal cord. Motor and sensory neurons have the same components, but they are configured differently. A motor neuron's axon and dendrites extend in several directions from the cell body, which is why it is called a *multipolar* neuron. Sensory neurons can be either *unipolar* or *bipolar*. The sensory neuron in Figure 2.4a is called a unipolar neuron because of the single short stalk from the cell body that divides into two branches. Bipolar neurons have an axon on one side of the cell body and a dendritic process on the other (Figure 2.4b). Motor and sensory neurons are specialized for transmission over long distances; their lengths are not shown here in the same scale as the rest of the cell.

The third type is neither motor nor sensory. *Interneurons* connect one neuron to another in the same part of the brain or spinal cord. Notice in Figure 2.4c that this neuron is also multipolar, but its axon appears to be missing; for some interneurons this is so, and when they do have axons, they are often so short that they are indistinguishable from dendrites. Because interneurons make connections over very short distances, they do not need the long axons that characterize their motor and sensory counterparts. In the spinal cord, interneurons bridge between sensory neurons and motor neurons to produce a reflex. In the brain, they connect adjacent neurons to carry out the complex processing that the brain is noted for. Considering the major role they play, it should come as no surprise that interneurons are the most numerous neurons.

The different kinds of neurons operate similarly; they differ mostly in their shape, which fits them for their specialized tasks. We will examine how neurons

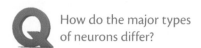

Q How do the major types of neurons differ?

FIGURE 2.4 Sensory Neurons and an Interneuron.

Compare the location of the soma in relation to the dendrites and axon in these and in the motor neuron.

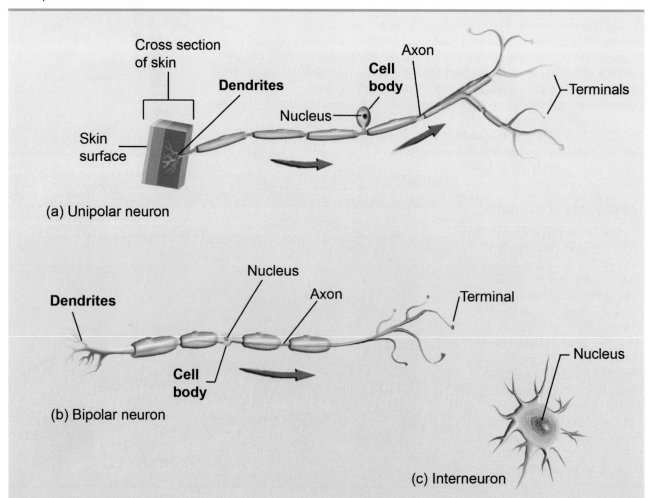

TABLE 2.1 The Major Types of Neurons.

Type	Function	Form and Location	Description
Motor	Conducts messages from brain and spinal cord to muscles and organs.	Multipolar; throughout nervous system.	Axon, dendrites extend in several directions from cell body.
Sensory	Carry information from body and world to brain and spinal cord.	Unipolar; outside brain. Bipolar; outside brain and spinal cord.	Single short stalk from cell body divides into two branches. Axon and dendritic processes are on opposite sides of cell body.
Interneuron	Conducts information between neurons in same area.	Multipolar; brain and spinal cord.	Has short axon or no axon.

work in the next few sections. The types of neurons and their characteristics are summarized in Table 2.1.

The Neural Membrane and Its Potentials

The most critical factor in the neuron's ability to communicate is the membrane that encloses the cell. The membrane is exceptionally thin—only about 8 micrometers (millionths of a meter) thick—and is made up of lipid (fat) and protein (see Figure 2.5). Each lipid molecule has a "head" end and a "tail" end. The heads of the molecules are water soluble, so they are attracted to the seawater-like fluid around and inside cells. The tails are water insoluble, so they are repelled by the fluid. Therefore, as the heads orient toward the fluid and the tails orient away from the fluid, the molecules turn their tails toward each other and form a double-layer membrane.

The membrane not only holds a cell together but also controls the environment within and around the cell. Some molecules, such as water, oxygen, and carbon dioxide, can pass through the membrane freely. Many other substances are barred from entry. Still others are allowed limited passage through protein channels (shown here in green) that open and close under different circumstances. This selective permeability contributes to the most fundamental characteristic of neurons, *polarization*, which means that there is a

FIGURE 2.5 Cross Section of the Cell Membrane of a Neuron.

Notice how the lipid molecules form the membrane by orienting their heads toward the extracellular and intracellular fluids.

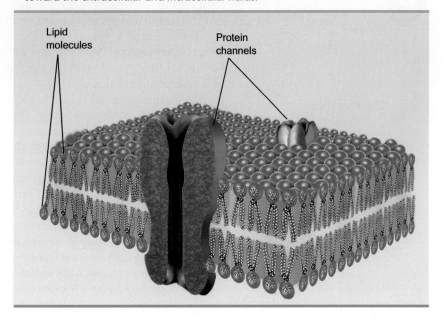

Lipid molecules

Protein channels

FIGURE 2.6 Recording Potentials in a Neuron.

Potentials are being recorded in the axon of a neuron, with an electrode inside the cell and one in the fluid outside. Due to the size of neurons, the electrodes have microscopically small tips. On the right is a highly magnified view of a microelectrode alongside neurons for size comparison. Electrodes for recording inside neurons are even smaller.

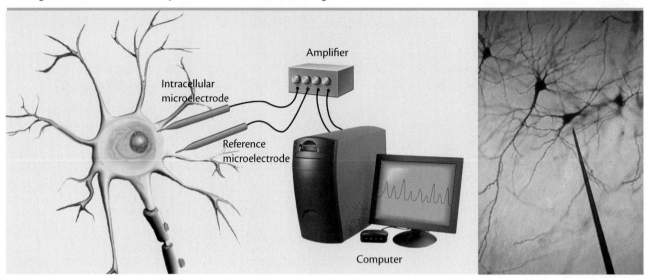

SOURCE: © Bob Jacobs, Colorado College.

 What accounts for the resting potential?

difference in electrical charge between the inside and the outside of the cell. A difference in electrical charge between two points, such as the poles of a battery or the inside and outside of a cell, is also called a *voltage.*

The Resting Potential. Just as you would measure the voltage of a battery, you can measure a neuron's voltage (see Figure 2.6). By arbitrary convention, the voltage is expressed as a comparison of the inside of the neuron with the outside. The difference in charge between the inside and outside of the membrane of a neuron at rest is called the *resting potential.* This voltage is negative and varies anywhere from −40 to −80 millivolts (mV) in different neurons but is typically around −70 mV. You should understand that neither the inside of the neuron nor the outside has a voltage, because a voltage is a *difference* and is meaningful only in comparison with another location. Note that this voltage is quite small—the voltage of a 1.5-V flashlight battery is 25 times greater. No matter; we're moving information, and very little power is required.

The resting potential is due to the unequal distribution of electrical charges on the two sides of the membrane. The charges come from *ions,* atoms that are charged because they have lost or gained one or more electrons. Sodium ions (Na^+) and potassium ions (K^+) are positively charged. Chloride ions (Cl^-) are negative, and so are certain proteins and amino acids that make up the organic anions (A^-). The fluid outside the neuron contains mostly Na^+ and Cl^- ions, and the ions inside the neuron are mostly K^+ and A^- (see Figure 2.7). The inside of the neuron has more negative ions than positive ions, while the ions on the outside are mostly positive, and this makes the resting potential negative.

If you remember from grade-school science that molecules tend to diffuse from an area of high concentration to one of low concentration, then you are probably wondering how this imbalance in ion distribution can continue to exist. In fact, two forces do work to balance the location of the ions. Because of the *force of diffusion*, ions move through the membrane to the side where they are less concentrated. And, as a result of *electrostatic pressure*, ions are repelled from the side that is similarly charged and attracted to the side that is oppositely charged.

In spite of these two forces, a variety of other influences keep the membrane polarized. Both forces would move the ions out, but they are too large to pass through the membrane. At the same time, the ions' negative charge repels the chloride ions, canceling out the force of diffusion, which would otherwise push them inside. The "real players" then turn out to be potassium and sodium ions. There is a slightly greater tendency for potassium to move outward (because its force of diffusion is stronger than its electrostatic pressure), while the force of both gradients attracts sodium inside. However, ions may cross the membrane only through channels like those in Figure 2.7 that are

FIGURE 2.7 Distribution of Ions Inside and Outside the Resting Neuron.

Ions on the outside are mostly Na⁺ and Cl⁻ ions; inside, the ions are mostly K⁺ ions and organic anions. The arrows represent the sodium-potassium pump, returning sodium ions to the outside and potassium ions to the inside.

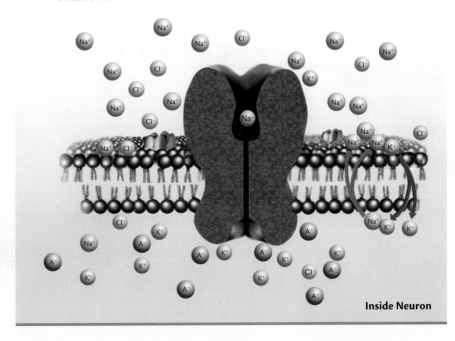

selective for particular ions. In the neuron's resting state, both the sodium channel and the potassium channel are closed, and only a few ions trickle through.

The few ions that do make it through are returned by the *sodium-potassium pump*, which consists of large protein molecules that move sodium ions through the cell membrane to the outside and potassium ions back inside. Its exchange rate of three sodium ions for every two potassium ions helps keep the inside of the membrane more negative than the outside. The pump is a metabolic process, which means that it uses energy; in fact, it accounts for an estimated 40% of the neuron's energy expenditure. But you will soon see that this energy is well spent, because the resting potential stores the energy to power the action potential.

The Action Potential. A neuron is usually excited by input that arrives on the neuron's dendrites and cell body from another neuron or from a sensory receptor. An excitatory signal causes a partial depolarization, which means that the polarity in a small area of the membrane is shifted toward zero. This partial depolarization disturbs the ion balance in the adjacent membrane, so the disturbance flows down the dendrites and across the cell membrane. This looks at first like the way the neuron might communicate its messages through the nervous system; however, because a partial depolarization is decremental—it dies out over distance—it is effective only over very short distances. For this reason, the partial depolarization is often called the local potential. Fortunately, the membrane of the axon has unique physical properties. If the local potential exceeds the threshold for activating that neuron, typically about 10 mV more positive than the resting potential, it will cause the normally closed sodium channels in that area to open, which triggers an action potential.

The *action potential* is an abrupt depolarization of the membrane that allows the neuron to communicate over long distances. (The following discussion is illustrated in Figure 2.8.) The voltage across the *resting* neuron membrane is stored energy,

1

Resting and Action Potentials

FIGURE 2.8 Ion Movement and Voltages During the Neural Impulse.

At point ❶ of (a), the membrane voltage is at the neuron's resting potential, –70 mV; as indicated in (b), there is an abundance of sodium ions (red) on the outside of the membrane and an abundance of potassium ions (blue) on the inside. (Chloride ions and organic anions are not shown here.) At ❷ a local potential has arrived at the axon hillock, partially depolarizing a small area of the axon membrane. At ❸ this partial depolarization has reached the neuron's threshold, which triggers the opening of sodium channels (c, left); sodium ions rush in, completely depolarizing the membrane. At ❹ the potassium channels begin to open, and potassium ions start flowing out (d, right); shortly after, at ❺ sodium channels close (d, left). With no more sodium ions entering the membrane and with potassium ions leaving, the membrane begins recovering its resting potential at ❻. At ❼ the membrane voltage continues to fluctuate around the resting potential for several milliseconds (unless there is another action potential).

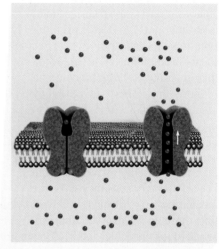

(d) Recovery

 (c) Depolarization

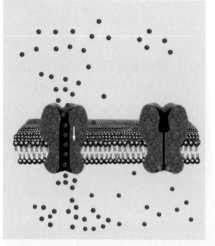

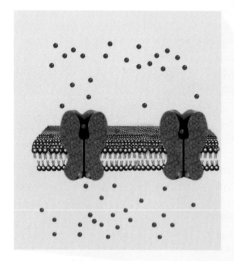

(b) Resting

This symbol indicates this figure is animated at www.sagepub.com/garrett3e.

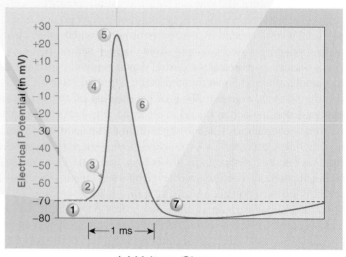

(a) Voltage Changes

just as the term resting *potential* implies. Imagine countless sodium ions being held outside the neuron against the combined forces of diffusion and electrostatic pressure. Opening the sodium channels allows the sodium ions in that area to rush into the axon at a rate 500 times greater than normal; they are propelled into the cell's interior so rapidly that the movement is often described as explosive. A small area inside the membrane becomes fully depolarized to zero; the potential even overshoots to around +30 or +40 mV, making the interior at that location temporarily positive. This depolarization is the action potential.

Just as abruptly as the neuron "fired," it begins to recover its resting potential. At the peak of the action potential, the sodium channels close, so there is no further depolarization. By that time the potassium channels have opened; the positive charge and the concentration of potassium ions inside the membrane combine to move potassium ions out. This outward flow of potassium ions returns the axon to its resting potential. The action potential requires about 1 millisecond (ms; one thousandth of a second) or so to complete; the actual duration varies among individual neurons.

The action potential causes nearby sodium channels to open as well. Thus, a new action potential is triggered right next to the first one. That action potential in turn triggers another farther along, creating a chain reaction of action potentials that move through the axon; thus, a signal flows from one end of the neuron to the other. Nothing physically moves down the axon. Instead, a series of events occurs in succession along the axon's length, much as a line of dominoes standing on end knock each other over when you tip the first one. When the action potential reaches the terminals, they pass the signal on to the next neuron in the chain (or to an organ or muscle). The transmission from neuron to neuron is covered later; for now the action potential needs to be examined a bit further.

Although the neuron has returned to its resting potential, a number of extra sodium ions remain inside, and there is an excess of potassium ions on the outside. Actually, only the ions in a very thin layer on either side of the membrane have participated in the action potential, so the dislocated ions are able to diffuse into the surrounding fluid. Eventually, though, the ions must be replaced or the neuron cannot continue firing. The sodium-potassium pump takes care of this chore. (Perhaps you can see now why Jim was in such a bad way after his bout with chlordane.)

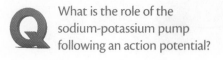

What is the role of the sodium-potassium pump following an action potential?

The action potential differs in two important ways from the local potential that initiates it. First, the local potential is a *graded potential*, which means that it varies in magnitude with the strength of the stimulus that produced it. The action potential, on the other hand, is *ungraded*; it operates according to the *all-or-none law*, which means that it occurs at full strength or it does not occur at all. A larger graded potential does not produce a larger action potential; like the fuse of a firecracker, the action potential depends on the energy stored in the neuron. A second difference is that the action potential is *nondecremental*; it travels down the axon without any decrease in size, propagated anew and at full strength at each successive point along the way. The action potential thus makes it possible for the neuron to conduct information over long distances.

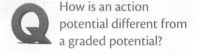

How is an action potential different from a graded potential?

However, because the action potential is all-or-none, its size cannot carry information about the intensity of the initiating stimulus. One way stimulus intensity is represented is in the number of neurons firing, because a more intense stimulus will recruit firing in neurons with higher thresholds. There is, though, a way in which the individual neuron can encode stimulus strength, as you will see in the discussion of refractory periods.

Refractory Periods

Right after the action potential occurs, the neuron goes through the *absolute refractory period*, a brief time during which it cannot fire again; this occurs

Application   Targeting Ion Channels

The Japanese delicacy *fugu*, or puffer fish, produces an exciting tingling sensation in the diner's mouth; improperly prepared, it causes numbness and weakness and, in some cases, a paralysis of the respiratory muscles that has claimed the lives of a few thousand culinary risk takers. The fish's natural poison, tetrodotoxin (TDT), blocks sodium channels and prevents neurons from firing (Kandel & Siegelbaum, 2000a). Other *neurotoxins* (neuron poisons) are found in snake venoms, which block either sodium or potassium channels (Benoit & Dubois, 1986; Fertuck & Salpeter, 1974), and scorpion venom, which keeps sodium channels open, prolonging the action potential (Catterall, 1984; Chuang, Jaffe, Cribbs, Perez-Reyes, & Swartz, 1998; Pappone & Cahalan, 1987).

Interfering with neuron functioning can be useful, though; for example, most local anesthetics prevent neuron firing by blocking sodium channels (Ragsdale, McPhee, Scheuer, & Catterall, 1994), and some general anesthetics hyperpolarize the neuron by opening potassium channels and allowing the potassium ions to leak out (Nicoll & Madison, 1982; A. J. Patel et al., 1999). The cone snail of the South Seas can penetrate a wet suit with its proboscis and inject toxins that will kill a human in half an hour, but its hundred or so toxins that target sodium, potassium, or calcium channels or block

neurotransmitter receptors are in demand by researchers developing pain relievers and drugs for epilepsy (L. Nelson, 2004).

An exciting new research strategy involves modifying channels so the neurons can be controlled by light. In one approach, light-sensitive proteins from other organisms are inserted into the membrane. A protein from green algae provides a channel that allows positive ions to enter the cell, producing action potentials; another from bacteria produces a pump that moves chloride ions inside and hyperpolarizes the neuron (see figure). These proteins are triggered by different colors of light, so the researcher can increase or decrease the firing rate in the neuron (Han & Boyden, 2007). In another technique, glutamate receptors are modified so that one color of light triggers action potentials and another terminates them (Szobota et al., 2007). Because these modifications can be targeted to specific types of neurons, this procedure promises more precise control of brain functions than electrical stimulation, which activates all neurons in the area. Light control of neurons has numerous possibilities, such as allowing researchers to determine just which neural pathways are involved in depression or to carry out precisely localized therapeutic brain stimulation through an implanted optic fiber.

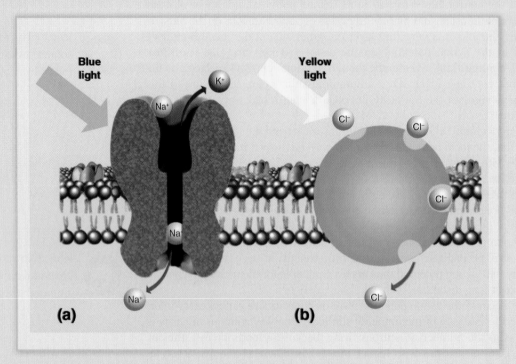

Modified Membrane Enables Light Control of Neuron Activity.

(a) Blue light activates a channel from green algae; the channel allows positive ions to pass, triggering neural impulses.

(b) Yellow light activates a chloride pump from bacteria; chloride ions hyperpolarize the neuron.

SOURCE: Adapted from "Controlling Neural Circuits With Light," by M. Häusser and S. L. Smith, 2007, *Nature, 446,* 617–619 (Figure 1a, p. 617).

because the sodium channels cannot reopen. This delay in responsiveness has two important effects. First, the absolute refractory period limits how frequently the neuron can fire. If a neuron takes a full millisecond to recover to the point where it can fire again, then the neuron can fire, at most, a thousand times a second; many neurons have much lower firing rate limits. A second effect of this recovery period is that the action potential will set off new action potentials only in front of it (the side toward the terminals), not on the side it has just passed. This is critical, because backward-moving potentials would block responses to newly arriving messages.

A second refractory period plays a role in intensity coding in the axon. The potassium channels remain open for a few milliseconds following the absolute refractory period, and the continued exit of potassium makes the inside of the neuron slightly more negative than usual (the "dip" in Figure 2.8). During the *relative refractory period*, the neuron can be fired again, but only by a stronger-than-threshold stimulus. A stimulus that is greater than threshold will cause the neuron to fire again earlier and thus more frequently. The axon encodes stimulus intensity not in the size of its action potential but in its firing rate, an effect called the *rate law*.

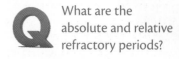

What are the absolute and relative refractory periods?

Glial Cells

Glial cells are nonneural cells that provide a number of supporting functions to neurons. The name *glia* is derived from the Greek word for glue, which gives you some idea how the role of glial cells has been viewed in the past. However, glial cells do much more than hold neurons together. One of their most important functions is to increase the speed of conduction in neurons.

Myelination and Conduction Speed

Survival depends in part on how rapidly messages can move through the nervous system, enabling the organism to pounce on its prey, outrun a predator, or process spoken language quickly. The speed with which neurons conduct their impulses varies from 1 to 120 meters (m) per second (s), or about 270 miles per hour. This is much slower than the flow of electricity through a wire, the analogy sometimes mistakenly used to describe neural conduction. Because conduction speed is so critical to survival, strategies have evolved for increasing it. One way is to develop larger axons, which provide less resistance to the flow of electrical potentials. By evolving motor neurons with a diameter of 0.5 mm, the squid has achieved conduction speeds of 30 m/s with its 0.5-mm-thick axons, compared with 1 m/s in the smallest neurons.

However, conduction speed does not increase in direct proportion to axon size. To reach our four-times-greater maximum conduction speed of 120 m/s, our axons would have to be $4^2 = 16$ times larger than the squid axon, or 8 mm in diameter! Obviously, your brain would be larger than you could carry around. In other words, if axon size were the only way to achieve fast conduction speed, *you* would not exist.

Another way to improve conduction speed would be to rely on graded local potentials in the axon, because graded potentials travel down the axon faster than action potentials. However, you will remember that graded potentials die out over distance. Vertebrates (animals with backbones) have developed a best-of-both-worlds solution, called myelination. Glial cells produce *myelin,* a fatty tissue that wraps around the axon to insulate it from the surrounding fluid and from other neurons. Only the axon is covered, not the cell body. Myelin is produced in the

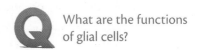

What are the functions of glial cells?

brain and spinal cord by a type of glial cell called *oligodendrocytes* and in the rest of the nervous system by *Schwann cells* (see Figure 2.9).

Because there are very few sodium channels under the myelin sheath, action potentials cannot occur there; conduction in myelinated areas is by graded potential (Waxman & Ritchie, 1985). However, myelin appears in segments about 1 mm long, with a gap of one or two thousandths of a millimeter between segments. The gaps in the myelin sheath are called *nodes of Ranvier* (see Figure 2.9 again). At each node of Ranvier, where the membrane is exposed and there are plenty of sodium channels, the graded potential triggers an action potential; action potentials thus jump from node to node in a form of transmission called *saltatory conduction.* So myelination and the resulting saltatory conduction increase conduction speed through graded potentials while retaining the benefits of nondecremental action potentials.

Myelination provides an additional boost to conduction speed because the insulating effect of myelin reduces an electrical effect called *capacitance*, which resists the movement of ions during a graded potential. The overall effect of myelination is the equivalent of increasing the axon diameter 100 times (Koester & Siegelbaum, 2000). And speed is not the only benefit of myelination; myelinated neurons use much less energy because there is less work for the sodium-potassium pump to do.

Some diseases, such as multiple sclerosis, destroy myelin. As myelin is lost, the capacitance rises, reducing the distance that graded potentials can travel before dying out. The individual is worse off than if the neurons had never been

2

Myelination and Conduction Speed.

FIGURE 2.9 Glial Cells Produce Myelin for Axons.

A single oligodendrocyte provides myelin for multiple segments of the axon and for multiple neurons. A Schwann cell covers only one segment of an axon.

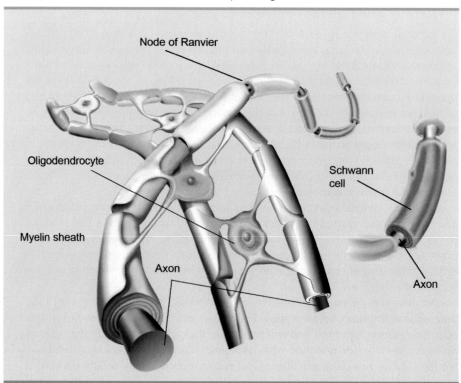

Node of Ranvier

Oligodendrocyte

Schwann cell

Myelin sheath

Axon

Axon

FIGURE 2.10 Glial Cells Increase the Number of Connections Between Neurons.

Neurons were cultured for 5 days in the absence of glial cells (a) and in the presence of glia (b). The number of neurons was similar in both cultures; the greater density on the right is due to increased connections among the neurons.

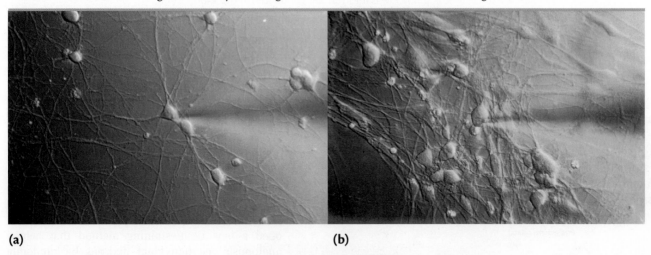

(a)   **(b)**

SOURCE: From "Synaptic Efficacy Enhanced by Glial Cells in vitro," by F. W. Pfrieger and B. A. Barres, *Science, 277,* p. 1684. © 1997. Used by permission of the author.

myelinated; due to the reduced number of sodium channels, action potentials may not be generated in the previously myelinated area. Conduction slows or stops in affected neurons.

Other Glial Functions

During fetal development, one kind of glial cells forms a scaffold that guides new neurons to their destination. Later on, glial cells provide energy to neurons and respond to injury and disease by removing cellular debris. Others contribute to the development and maintenance of connections between neurons. Neurons form seven times as many connections in the presence of glial cells, and if glial cells are removed from a laboratory dish, the neurons start to lose their synapses (Pfrieger & Barres, 1997; Ullian, Sapperstein, Christopherson, & Barres, 2001; see Figure 2.10). You will see later that glia play an important role in neural activity as well. An indication of the importance of glial cells is that as brain complexity increases across species, there is also a progressive increase in the ratio of astrocytes to neurons; astrocytes are the glial cells most intimately involved with neural activity (Figure 2.11).

FIGURE 2.11 Number of Astrocytes per Neuron in Various Species.

The ratio of astrocytes per neuron increases as behavioural complexity increases. Notice that the leech, frog, mouse, and rat all have less than one astrocyte per neuron, while the cat and humans have more astrocytes than neurons.

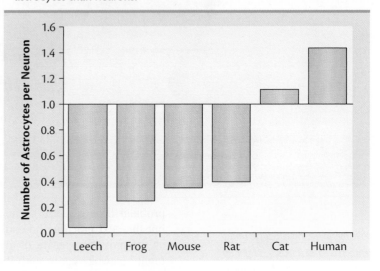

SOURCE: Based on data from "New Roles for Astrocytes: Redefining the Functional Architecture of the Brain," by M. Nedergaard, B. Ransom, and S. A. Goldman, *Trends in Neurosciences, 26,* pp. 523–530. © 2003.

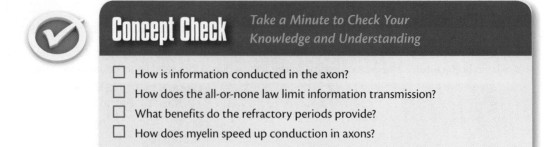

Concept Check
Take a Minute to Check Your Knowledge and Understanding

☐ How is information conducted in the axon?

☐ How does the all-or-none law limit information transmission?

☐ What benefits do the refractory periods provide?

☐ How does myelin speed up conduction in axons?

FIGURE 2.12 The Synapse Between a Presynaptic Neuron and a Postsynaptic Neuron.

Notice the separation between the presynaptic axon terminal and the postsynaptic neuron.

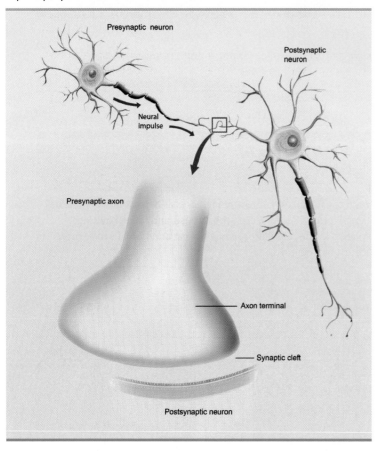

> *I awoke again, at three o'clock, and I remembered what it was. . . . I got up immediately, went to the laboratory, made the experiment . . . and at five o'clock the chemical transmission of the nervous impulse was conclusively proved.*
>
> —Otto Loewi

How Neurons Communicate With Each Other

Before the late 1800s, microscopic examination suggested that the brain consisted of a continuous web. At that point, however, Camillo Golgi developed a new tissue-staining method that helped anatomists see individual neurons by randomly staining some entire cells without staining others (see the discussion of staining methods in Chapter 4). With this technique, the Spanish anatomist Santiago Ramón y Cajal (1937/1989) was able to see that each neuron is a separate cell. The connection between two neurons is called a *synapse,* a term derived from the Latin word that means "to grasp." The neurons are not in direct physical contact at the synapse but are separated by a small gap called the *synaptic cleft.* Two terms will be useful to us in the following discussion: The neuron that is transmitting to another is called the *presynaptic* neuron; the receiving neuron is the *postsynaptic* neuron (see Figure 2.12).

Chemical Transmission at the Synapse

Until the 1920s, physiologists assumed that neurons communicated by an electrical current that bridged the gap to the next neuron. The German physiologist Otto Loewi believed that synaptic transmission was chemical, but he did not know how to test his hypothesis. One night Loewi awoke from sleep with the solution to his problem (Loewi, 1953). He wrote his idea down so he would not forget it, but the next morning he could not read his own writing. He recalled that day as the most "desperate of my whole scientific life" (p. 33). But the following night he awoke again with the same idea; taking no chances, he rushed to his laboratory. There he isolated the hearts of two frogs. He applied electrical stimulation to the vagus nerve attached to one of the hearts, which made the heart beat slower. Then he extracted a salt solution that he had placed in the heart beforehand and placed it in the second heart. If neurons used a chemical messenger, the chemical might have leaked into the salt solution. The second heart slowed, too, just as Loewi expected. Then he stimulated the accelerator nerve of the first heart, which caused the heart to beat faster. When he transferred salt solution from the first heart to the second, this time it speeded up (see Figure 2.13). So Loewi demonstrated that transmission at the synapse is chemical and that there are at least two different chemicals that carry out different functions.

It turned out later that some neurons do communicate electrically by passing ions through channels that connect one neuron to the next; their main function appears to be synchronizing activity in nearby neurons (Bennett & Zukin, 2004). In addition, some neurons release a gas transmitter. Still, Loewi was essentially correct because most synapses are chemical. By the way, if this example suggests to you that the best way to solve a problem is to "sleep on it," keep in mind that such insight occurs only when people have paid their dues in hard work beforehand!

At chemical synapses, the neurotransmitter is stored in the terminals in membrane-enclosed containers called *vesicles;* the term means, appropriately, "little bladder." When the action potential arrives at the terminals, it opens channels that allow calcium ions to enter the terminals from the extracellular fluid. The calcium ions cause the vesicles clustered nearest the membrane to fuse with the membrane. The membrane opens there, and the transmitter spills out and diffuses across the cleft (see Figure 2.14).

On the postsynaptic neuron, the molecules of neurotransmitter dock with specialized chemical receptors that match the molecular shape of the transmitter molecules (Figure 2.14). Activation of these receptors causes ion channels in the membrane to open. *Ionotropic receptors* open the channels directly to produce the immediate reactions required for muscle activity and sensory processing; *metabotropic receptors* open channels indirectly and slowly to produce longer-lasting effects. Opening the channels is what sets off the graded potential that initiates the action potential. You will see in the next section that the effect this has on the postsynaptic neuron depends on which receptors are activated.

The chemical jump across the synapse takes a couple of milliseconds; that is a significant slowing compared with transmission in the axon. In a system that places a premium on speed, inserting these gaps in the neural pathway has some compensating benefit. As you will see in the following sections, synapses add important complexity to the simple all-or-none response in the axon.

Q How does synaptic transmission differ from transmission in the axon?

FIGURE 2.13 Loewi's Experiment Demonstrating Chemical Transmission in Neurons.

Loewi stimulated the first frog heart. When he transferred fluid from it to the second heart, it produced the same effect there as the stimulation did in the first heart.

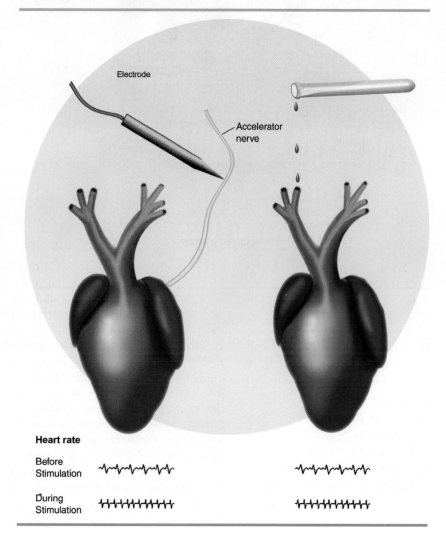

Electrode

Accelerator nerve

Heart rate

Before Stimulation

During Stimulation

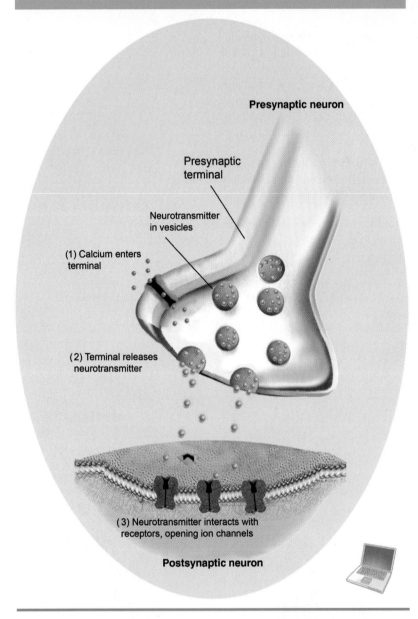

FIGURE 2.14 A Presynaptic Terminal Releases Neurotransmitter at the Synapse.

Presynaptic neuron

Presynaptic terminal

Neurotransmitter in vesicles

(1) Calcium enters terminal

(2) Terminal releases neurotransmitter

(3) Neurotransmitter interacts with receptors, opening ion channels

Postsynaptic neuron

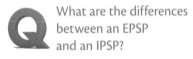

What are the differences between an EPSP and an IPSP?

Excitation and Inhibition

Opening ion channels on the dendrites and cell body has one of two effects: It can cause the local membrane potential to shift in a positive direction toward zero, partially depolarizing the membrane, or it can shift the potential farther in the negative direction. Partial depolarization, or *hypopolarization*, is excitatory and facilitates the occurrence of an action potential; increased polarization, or *hyperpolarization*, is inhibitory and makes an action potential less likely to occur. The value of excitation is obvious, but inhibition can communicate just as much information as excitation does. Also, the message becomes more complex because input from one source can partially or completely negate input from another. In addition, inhibition helps prevent runaway excitation; one cause of the uncontrolled neural storms that sweep across the brain during an epileptic seizure is a deficiency in an inhibitory transmitter system (Baulac et al., 2001).

What determines whether the effect on the postsynaptic neuron is facilitating or inhibiting? It depends on which transmitter is released and the type of receptors on the postsynaptic neuron. A particular transmitter can have an excitatory effect at one location in the nervous system and an inhibitory effect at another; however, some transmitters typically produce excitation, and others most often produce inhibition. If the receptors open sodium channels, this produces hypopolarization of the dendrites and cell body, which is an *excitatory postsynaptic potential (EPSP)*. Other receptors open potassium channels, chloride channels, or both; as potassium moves out of the cell or chloride moves in, it produces a hyperpolarization of the dendrites and cell body, or an *inhibitory postsynaptic potential (IPSP)*.

At this point, there is only a graded local potential. This potential spreads down the dendrites and across the cell body to the *axon hillock* (where the axon joins the cell body). At the axon, a positive graded potential that reaches threshold will produce an action potential; a negative graded potential makes it harder for the axon to fire. Most neurons fire spontaneously all the time, so EPSPs will increase the rate of firing and IPSPs will decrease the rate of firing (Figure 2.15). So now another form of complexity has been added at the synapse: The message to the postsynaptic neuron can be *bidirectional*, not just off-on.

You should not assume that excitation of neurons always corresponds to activation of behavior or that inhibition necessarily suppresses behavior. An EPSP may activate a neuron that has an inhibitory effect on other neurons, and an IPSP may reduce activity in an inhibitory neuron. An example of this paradox at the behavioral level is the effect of Ritalin. Ritalin and many other medications used to treat hyperactivity in children are in a class of drugs called stimulants, which increase activity in the nervous system. Yet, they calm hyperactive individuals and improve their ability to concentrate and focus attention (D. J. Cox, Merkel, Kovatchev, & Seward, 2000; Mattay et al., 1996). They probably have this effect by stimulating frontal areas of the brain where activity has been found to be abnormally low (Faigel, Szuajderman, Tishby, Turel, & Pinus, 1995).

Next you will see that the ability to combine the inputs of large numbers of neurons expands the synapse's contribution to complexity even further.

Postsynaptic Integration

The output of a single neuron is not enough by itself to cause a postsynaptic neuron to fire or to prevent it from firing. In fact, an excitatory neuron may depolarize the membrane of the postsynaptic neuron by as little as 0.2 to 0.4 mV (Kandel & Siegelbaum, 2000b); remember that it takes an approximately 10-mV depolarization to trigger an action potential. However, a typical neuron receives input from around a thousand other neurons (Figure 2.16); because of the branching of the terminals, this amounts to as many as 10,000 synaptic connections in most parts of the brain and up to 100,000 in the cerebellum (Kandel & Siegelbaum, 2000a).

Because a single neuron has such a small effect, the postsynaptic neuron must combine potentials from many neurons to fire. This requirement is actually advantageous: It ensures that a neuron will not be fired by the spontaneous activity of a single presynaptic neuron, and it allows the neuron to combine multiple inputs into a more complex message. These potentials are combined at the axon hillock in two ways. *Spatial summation* combines potentials occurring simultaneously at different locations on the dendrites and cell body. *Temporal summation* combines potentials arriving a short time apart. Temporal summation is possible because it takes a few milliseconds for a potential to die

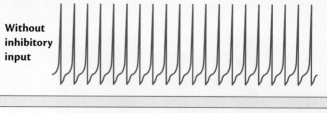

FIGURE 2.15 Effect of Inhibition on Spontaneous Firing Rate.

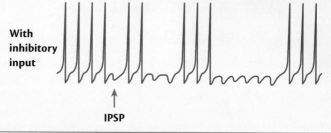

SOURCE: From *Principles of Neural Science*, 4th ed., by E. R. Kandel et al., pp. 207–208. © 2002, McGraw-Hill Companies, Inc. Used with permission.

FIGURE 2.16 A Cell Body Virtually Covered With Axon Terminals.

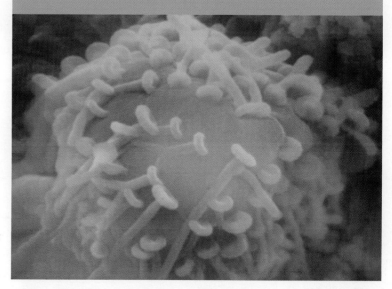

SOURCE: © Science VU/Lewis-Everhart-Zeevi/Visuals Unlimited.

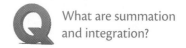

Q What are summation and integration?

FIGURE 2.17 Spatial and Temporal Summation.

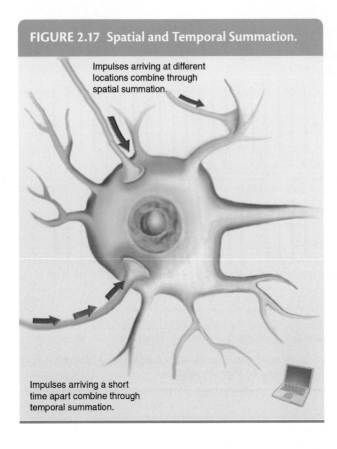

Impulses arriving at different locations combine through spatial summation.

Impulses arriving a short time apart combine through temporal summation.

out. Spatial summation and temporal summation occur differently, but they have the same result. Summation is illustrated in Figure 2.17.

As you can see in Figure 2.18, summation combines EPSPs so that an action potential is more likely to occur. Alternatively, summation of IPSPs drives the membrane's interior even more negative and makes it more difficult for incoming EPSPs to trigger an action potential. When both excitatory and inhibitory impulses arrive on a neuron, they will also summate, but algebraically. The combined effect will equal the difference between the sum of the hypopolarizations and the sum of the hyperpolarizations. Spatial summation of two excitatory inputs and one inhibitory input is illustrated in Figure 2.19. The effect from temporal summation would be similar.

Because the neuron can summate inputs from multiple sources, it rises above the role of a simple message conductor—it is an *information integrator*. And, using that information, it serves as a *decision maker*, determining whether to fire or not. Thus, the nervous system becomes less like a bunch of telephone lines and more like a computer. In subsequent chapters, you will come to appreciate how important the synapse is in understanding how we see, how we learn, and how we succumb to mental illness.

FIGURE 2.18 Temporal and Spatial Summation.

(1) An EPSP; (2) temporal summation of 2 EPSPs; (3) temporal summation of 3 EPSPs reaches threshold; (4) spatial summation of EPSPs reaches threshold; (5) an IPSP; (6) temporal summation of 2 IPSPs.

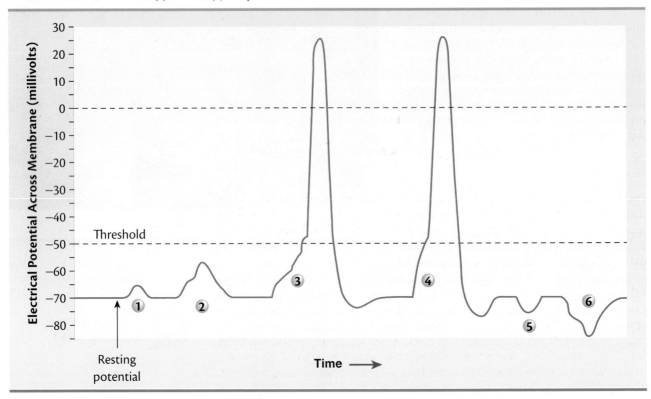

Terminating Synaptic Activity

The neurotransmitter's story does not end when it has activated the receptors. Usually, the transmitter must be inactivated to prevent it from "locking up" a circuit that must respond frequently or from leaking over to other synapses and interfering with their function. Typically, the transmitter is taken back into the terminals by membrane proteins called transporters in a process called *reuptake;* it is repackaged in vesicles and used again. At some synapses, the transmitter in the cleft is absorbed by glial cells. The neurotransmitter acetylcholine (ACh), on the other hand, is deactivated by acetylcholinesterase, an enzyme that splits the molecule into its components of choline and acetate. Choline is then taken back into the terminals and used to make more acetylcholine.

Controlling how much neurotransmitter remains in the synapse is one way to vary behavior, and many drugs capitalize on this mechanism. Cocaine blocks the uptake of dopamine; some antidepressant medications block the reuptake of serotonin, norepinephrine, or both, while others (*MAO inhibitors*) prevent monoamine oxidase from degrading those transmitters as well as dopamine and epinephrine; and drugs for treating the muscular disorder myasthenia gravis increase ACh availability by inhibiting the action of acetylcholinesterase.

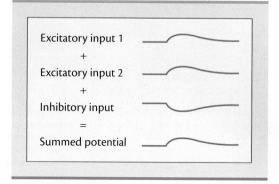

FIGURE 2.19 Spatial Summation of Excitatory and Inhibitory Potentials.

Note that inhibitory potentials cancel out excitatory potentials of equal strength (and vice versa).

Regulating Synaptic Activity

The previous description has been of a system that amounts to *neuron A stimulates neuron B, neuron B stimulates neuron C,* and so on. However, such a simple system cannot transmit the complex information required to solve a math equation, write a symphony, or care for a newborn. Not only that, but as messages flow from neuron to neuron, activity would soon drift out of control; some activity would fade out, while other activity would escalate until it engulfed an entire area of the brain. A nervous system that controls complex behavior must have several ways to regulate its activity.

One of the ways is through axoaxonic synapses. The synapses described so far are referred to as *axodendritic* and *axosomatic* synapses, because their targets are dendrites and cell bodies. At axoaxonic synapses, a third neuron releases transmitter onto the terminals of the presynaptic neuron (see #1 in Figure 2.20). The result is *presynaptic excitation* or *presynaptic inhibition,* which increases or decreases, respectively, the presynaptic neuron's release of neurotransmitter onto the postsynaptic neuron. One way an axoaxonic synapse adjusts a presynaptic terminal's activity is by regulating the amount of calcium entering the terminal, which, you will remember, triggers neurotransmitter release.

Neurons also regulate their own synaptic activity in two ways. *Autoreceptors* on the presynaptic terminals sense the amount of transmitter in the cleft; if the amount is excessive, the presynaptic neuron reduces its output (Figure 2.20, #2). Postsynaptic neurons participate in regulation of synaptic activity as well. When there are unusual increases or decreases in neurotransmitter release, postsynaptic receptors change their sensitivity or even their numbers to compensate (Figure 2.20, #3). You will see in a later chapter that receptor changes figure prominently in psychological disorders such as schizophrenia.

We are now learning that glial cells also contribute to the regulation of synaptic activity. They surround the synapse and prevent neurotransmitter from spreading to other synapses. More important, they sometimes absorb neurotransmitter in the synaptic cleft and recycle it for the neuron's reuse (Figure 2.21 on page 43); they influence synaptic activity by varying the amount of transmitter absorbed (Oliet, Piet, & Poulain, 2001). They even release the neurotransmitter glutamate in response to transmitter levels in the synapse; this stimulates the presynaptic terminal to enhance or depress further transmitter release (Newman, 2003).

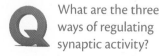

Q What are the three ways of regulating synaptic activity?

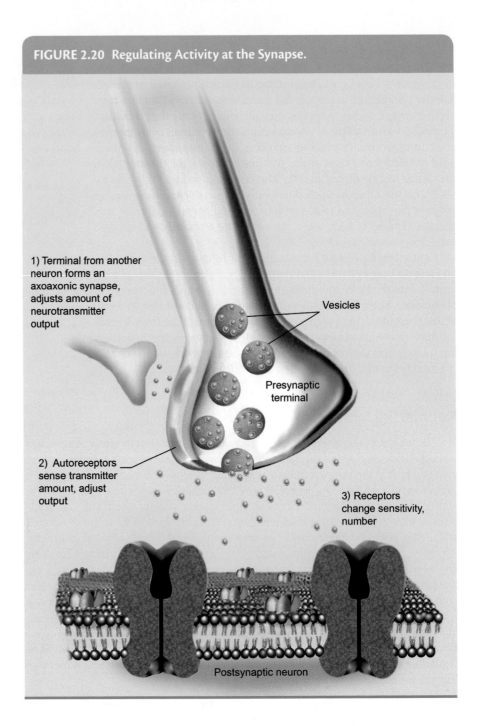

FIGURE 2.20 Regulating Activity at the Synapse.

1) Terminal from another neuron forms an axoaxonic synapse, adjusts amount of neurotransmitter output

Vesicles

Presynaptic terminal

2) Autoreceptors sense transmitter amount, adjust output

3) Receptors change sensitivity, number

Postsynaptic neuron

Neurotransmitters

Table 2.2 on page 44 lists several transmitters, grouped according to their chemical structure. This is an abbreviated list; there are other known or suspected transmitters, and there are doubtless additional transmitters yet to be discovered. This summary is intended to illustrate the variety in neurotransmitters and to give you some familiarity with the functions of a few of the major ones. You will encounter most of them again as various behaviors are discussed in later chapters.

Having a variety of neurotransmitters multiplies the effects that can be produced at synapses; the fact that there are different subtypes of the receptors adds even more. For example, two types of receptors detect acetylcholine: the nicotinic receptor, so called because it is also activated by nicotine, and the muscarinic receptor, named for the mushroom derivative that can stimulate it. Nicotinic receptors

FIGURE 2.21 Glial Cell Interacting With Neurons at the Synapse.

An astrocyte, a type of glial cell, encloses the synapse, where it absorbs the neurotransmitter glutamate (Glu) from the synaptic cleft. It recycles the transmitter into its precursor glutamine (Gln), which it returns to the presynaptic terminal for reuse. The glial cell can influence synaptic activity by granting or withholding transmitter absorption and by releasing in response to the neurotransmitter level in the synapse.

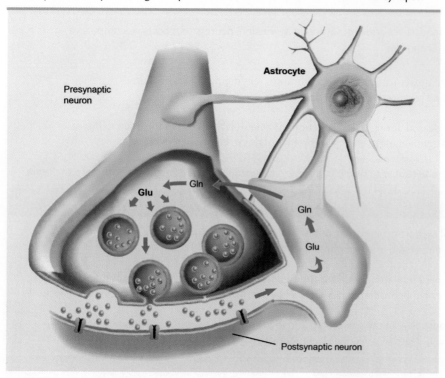

SOURCE: Adapted with permission from "Energy on Demand," by P. J. Magistretti et al., 1999, *Science, 283,* p. 497. Copyright © 1999. Reprinted with permission from AAAS.

are excitatory; they are found on muscles and, in lesser numbers, in the brain. Muscarinic receptors are more frequent in the brain, where they have an excitatory effect at some locations and an inhibitory one at others. Other transmitters have many more receptor subtypes than acetylcholine does.

For decades, neurophysiologists labored under the erroneous belief, known as *Dale's principle,* that a neuron was capable of releasing only one neurotransmitter. We learned only fairly recently that many neurons ply their postsynaptic partners with two to four and perhaps even more neurotransmitters. Since then, most researchers have thought that the combination invariably consisted of a single fast-acting "traditional" neurotransmitter and one or more slower-acting neuropeptides that prolong and enhance the effect of the main transmitter (Hökfelt, Johansson, & Goldstein, 1984). Peptides are chains of amino acids (longer chains are called proteins); neuropeptides, of course, are peptides that act as neurotransmitters.

Recent studies have found that some neurons release two fast transmitters (Rekling, Funk, Bayliss, Dong, & Feldman, 2000). Even more surprising, we have learned that the same neuron can release both an excitatory transmitter and an inhibitory transmitter (Duarte, Santos, & Carvalho, 1999; Jo & Schlichter, 1999). It appears that the two types of transmitters are released at *different* terminals (Duarte et al., 1999; Sulzer & Rayport, 2000). This corelease suggests that a neuron can act as a two-way switch (Jo & Schlichter, 1999); one example is in cells in the eye

 What are two additional ways synapses add information complexity?

TABLE 2.2 Some Representative Neurotransmitters.

Neurotransmitter	Function
Acetylcholine	Transmitter at muscles; in brain, involved in learning, etc.
Monamines	
Serotonin	Involved in mood, sleep, and arousal and in aggression, depression, obsessive-compulsive disorder, and alcoholism.
Dopamine	Contributes to movement control and promotes reinforcing effects of abused drugs, food, and sex; involved in schizophrenia and Parkinson's disease.
Norepinephrine	A hormone released during stress. Functions as a neurotransmitter in the brain to increase arousal and attentiveness to events in the environment; involved in depression.
Epinephrine	A stress hormone related to norepinephrine; plays a minor role as a neurotransmitter in the brain.
Amino Acids	
Glutamate	The principal excitatory neurotransmitter in the brain and spinal cord. Vitally involved in learning and implicated in schizophrenia.
Gamma-aminobutyric acid (GABA)	The predominant inhibitory neurotransmitter. Its receptors respond to alcohol and the class of tranquilizers called benzodiazepines. Deficiency in GABA or receptors is one cause of epilepsy.
Glycine	Inhibitory transmitter in the spinal cord and lower brain. The poison strychnine causes convulsions and death by affecting glycine activity.
Peptides	
Endorphins	Neuromodulators that reduce pain and enhance reinforcement.
Substance P	Transmitter in neurons sensitive to pain.
Neuropeptide Y	Initiates eating and produces metabolic shifts.
Gas	
Nitric Oxide	One of two known gaseous transmitters, along with carbon monoxide. Can serve as a retrograde transmitter, influencing the presynaptic neuron's release of neurotransmitter. Viagra enhances male erections by increasing nitric oxide's ability to relax blood vessels and produce penile engorgement.

that produce excitation when a viewed object moves in one direction and inhibition when movement is in the opposite direction (Duarte et al., 1999).

Of Neuronal Codes, Neural Networks, and Computers

Underlying this discussion has been the assumption that we can explain behavior by understanding what neurons do. But we cannot make good on that promise as long as we talk as if neural communication is limited to single chains of neurons that either fire or don't fire. In fact, neurons are capable of generating complex messages, which they send across intricate networks.

Application   Agonists and Antagonists in the Real World

Neurotransmitters are not the only substances that affect transmitters. Many drugs, as well as other compounds, mimic or increase the effect of a neurotransmitter and are called *agonists*. Any substance that reduces the effect of a neurotransmitter is called an *antagonist*. Practically all drugs that have a psychological effect interact with a neurotransmitter system in the brain, and many of them do so by mimicking or blocking the effect of neurotransmitters (S. H. Snyder, 1984).

You have already seen that the effect of acetylcholine is duplicated by nicotine and by muscarine at the two kinds of receptors. Opiate drugs such as heroin and morphine also act as agonists, stimulating receptors for opiate-like transmitters in the body. The drug naloxone acts as

Amazonian Indians Tip Their Darts With the Plant Neurotoxin Curare.

SOURCE: © Jack Fields/Corbis.

an antagonist to opiates, occupying the receptor sites without activating them; consequently, naloxone can be used to counteract an overdose.

The plant toxin curare blocks acetylcholine receptors at the muscle, causing paralysis (Trautmann, 1983). South American Indians in the Amazon River basin tip their darts with curare to disable their game (see the photo). A synthetic version of curare was used as a muscle relaxant during surgery before safer and more effective drugs were found (M. Goldberg & Rosenberg, 1987). Ironically, it has even been used in the treatment of tetanus (lockjaw), which is caused by another neurotoxin; a patient receiving this treatment has to be artificially respirated for weeks until recovery occurs to prevent suffocation.

Coding of Neural Messages

Neurons don't just produce a train of equally spaced impulses: They vary the intervals between spikes, they produce bursts of varying lengths, and the bursts can be separated by different intervals (see review in Cariani, 2004). But do these temporal (time-related) variations in firing pattern form a code that the brain can use, or are they just "noise" in the system? The best way to answer this question is to look at sensory processes, because the researcher can correlate firing patterns with sensory input and with behavior. A good example is an early study done by Patricia Di Lorenzo and her colleague Gerald Hecht (1993). First, they recorded the firing patterns in individual taste neurons of rats during stimulation with sucrose (sugar) solution and quinine; as you can see in Figure 2.22a, these flavors produce different neural activity. Then they duplicated these patterns in the form of electrical pulses (Figure 2.22b), which they used to stimulate the taste pathways of other rats. The assumption was that if the brain *uses* this information, the unanesthetized rats would behave as if they were actually *tasting* sweet sucrose or bitter quinine. As Figure 2.22c shows, that is exactly what happened: the rats licked a water tube at a high rate when they were receiving stimulation patterned after sucrose, but almost stopped licking—even though they were water deprived—when the stimulation was patterned after quinine.

However, this coding apparently is not sufficient by itself to carry the complex information involved in brain communication. An additional opportunity for coding is provided by the fact that neural information often travels over specialized pathways. For example, taste information is carried by at least five types of specialized fibers; Di Lorenzo and Hecht (1993) recorded the sucrose firing pattern from a

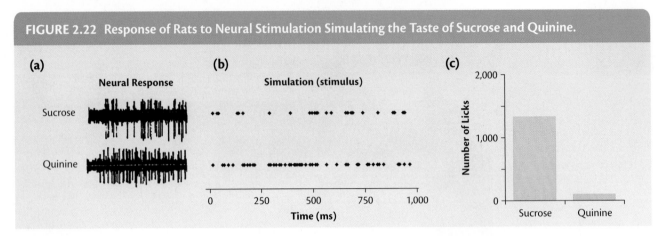

FIGURE 2.22 Response of Rats to Neural Stimulation Simulating the Taste of Sucrose and Quinine.

(a) Recordings from individual neurons during stimulation with sucrose and quinine. (b) Electrical stimulation mimicking the recorded neuronal activity; each diamond represents a single neural impulse. (c) The average number of times the rats licked a drinking tube for water during delivery of the quinine simulation and the sucrose simulation.

SOURCES: (a) and (b) adapted from Figure 7 of "Temporal Coding in the Gustatory System," by R. M. Hallock and P. M. Di Lorenzo, 2006, *Neuroscience and Biobehavioral Reviews, 30*, p. 1156. (c) adapted from Figure 4 of "Perceptual Consequences of Electrical Stimulation in the Gustatory System," by P. M. Di Lorenzo and G. S. Hecht, 1993, *Behavioral Neuroscience, 107*, p. 135.

"labeled line" specialized for sweet stimuli and the quinine pattern from another specialized for bitter stimuli. In later chapters you will see that not only taste but information about color and about the higher sound frequencies is also transmitted over a limited number of labeled lines. However, even with temporal coding and labeled lines, a significant burden remains for the brain if it is to make sense of this information; this leads us to the topic of neural networks.

Neural Networks

Individual neurons cannot carry enough information to determine the taste of a bite of food or the color of an object. Color processing, for example, depends on just four labeled lines carrying information about red, green, blue, and yellow light. However, we can distinguish millions of colors by comparing the relative activity in these four pathways. In other cases, the brain combines the activity from a number of neurons. A good example is the discrimination of sounds in the lower frequency range. Neurons from the ear "follow" frequencies up to about 5,000 cycles per second, but you know that a single neuron cannot fire more often than about 1,000 times per second. However, some neurons will be firing on each wave of the sound, and the brain must pool their activity to determine the sound's frequency.

This processing requires complex networks of neurons; *neural networks* are groups of neurons that function together to carry out a process. Neural networks are where the most complex neural processing—the "computing" work of the brain—is carried out. Sometimes these networks involve a relatively small number of neurons in a single area, such as groups of neurons in a part of the rat's brain called the hippocampus. During an experimenter-imposed delay in maze running, these store the rat's preceding choices and calculate its next choice. They perform so reliably that the researcher can use their activity to predict which way the rat will turn after the delay (Pastalkova, Itskov, Amarasingham, & Buzsáki, 2008). And as you will see in later chapters, other networks combine the activity of widespread brain areas to perform language functions (Chapter 9), to identify an object visually and locate it in space (Chapter 10), and, some researchers believe, to produce conscious awareness (Chapter 15).

Because researchers have found these networks to be discouragingly complex and rather inaccessible, some are creating *artificial neural networks* on computers.

An artificial neural network consists of multiple layers of simulated neurons that are highly interconnected. Rather than being programmed to perform in a specific way, these networks *learn* how to carry out their task, much as we do. The network is trained by giving it feedback about correct responses, and its neurons respond by adjusting the stimulating and inhibiting effects they have on their neighbors. As with the brain, the network's performance at first is random, but it improves with practice. For example, NETtalk (Sejnowski & Rosenberg, 1987), designed to read and speak English text, initially produced random sounds. These were replaced by babbling and then pseudowords, and after just 10 training trials, the speech was intelligible and sounded like a small child's.

Artificial neural networks have practical uses, such as detecting cancer cells in a biopsy, but it is their ability to mimic human and animal behavior that intrigues neuroscientists. One program simulated rats learning to find an escape platform hidden just below the surface of murky water (M. A. Brown &

FIGURE 2.23 A Lifelike Android.

Hiroshi Ishiguro of Osaka University with one of his lifelike androids.

SOURCE: © Everett Kennedy Brown/epa/Corbis.

Sharp, 1995). When the platform was deleted from the program, the virtual rats would "swim" past the platform's original location, then turn and swim back in the other direction. More visually remarkable are Japanese androids that learn to produce incredibly human gestures and facial expressions just by "watching" human models (Figure 2.23; Matsui, Minato, MacDorman, & Ishiguro, 2005). Artificial neural networks allow researchers to test hypotheses experimentally, and often they can learn how the network solved its problem by examining the new configuration of neurons in the hidden layer (Fleischer, Gally, Edelman, & Krichmar, 2007).

While we're waiting for neuroscientists to explain how the brain works, the idea of neural networks provides a useful way of thinking about mental processes. The next time you are trying to remember a person's name that is "on the tip of your tongue," imagine your brain activating individual components of a neural network until one produces the name you're looking for. If you visualize the person's face as a reminder, imagine that the name and the image of the face are stored in related networks so that activating one memory activates the other. This is not just speculation: Electrode recordings from patients preparing for brain surgery show that the information triggered by the photo of a familiar person and by the person's written name converge on the same neurons in a critical memory area of the brain (Quiroga, Kraskov, Koch, & Fried, 2009)—thanks, of course, to neural networks.

3
Neural Networks

In Perspective

It is impossible to understand the brain and impossible to understand behavior without first knowing the capabilities and the limitations of the neuron. Although more complexity is added at the synapse, a relatively simple off-or-on device is the basis for our most sophisticated capabilities and behaviors. However, what happens at the individual neuron is not enough to account for human behavior. Some researchers are using artificial neural networks to understand how neurons work together to produce thought, memory, emotion, and consciousness. In the next chapter, you will learn about some of the functional structures in the brain that are formed by the interconnection of neurons.

Summary

The Cells That Make Us Who We Are

- There are three major kinds of neurons: motor neurons, sensory neurons, and interneurons. Although they play different roles, they have the same basic components and operate the same way.
- The neural membrane is electrically polarized. This polarity is the resting potential, which is maintained by forces of diffusion and electrostatic pressure in the short term and by the sodium-potassium pump in the long term.
- Polarization is the basis for the neuron's responsiveness to stimulation, in the form of the graded potential and the action potential.
- The neuron is limited in firing rate by the absolute refractory period and in its ability to respond to differing strengths of stimuli by the all-or-none law. More intense stimuli cause the neuron to fire earlier during the relative refractory period, providing a way to encode stimulus intensity (the rate law).
- Glial cells provide the myelination that enables neurons to conduct rapidly while remaining small. They also help regulate activity in the neurons and provide several supporting functions for neurons.

How Neurons Communicate With Each Other

- Transmission from neuron to neuron is usually chemical in vertebrates, involving neurotransmitters released onto receptors on the postsynaptic dendrites and cell body.
- The neurotransmitter can create an excitatory postsynaptic potential, which increases the chance that the postsynaptic neuron will fire, or it can create an inhibitory postsynaptic potential, which decreases the likelihood of firing.
- Through temporal and spatial summation, the postsynaptic neuron integrates its many excitatory and inhibitory inputs.
- Regulation of synaptic activity is produced by axoaxonic synapses from other neurons, adjustment of transmitter output by autoreceptors, and change in the number or sensitivity of postsynaptic receptors.
- Leftover neurotransmitter may be broken down, taken back into the presynaptic terminals, or absorbed by glial cells.
- The human nervous system contains a large number of neurotransmitters, detected by an even greater variety of receptors. A neuron can release combinations of two or more neurotransmitters.
- The computing work of the brain is done in complex neural networks, and artificial neural networks are helping us understand these networks. ∎

Study Resources

F   For Further Thought

- What would be the effect if there were no constraints on the free flow of ions across the neuron membrane?

- What effect would it have on neural conduction if the action potential were decremental?

- Sports drinks replenish electrolytes that are lost during exercise. Electrolytes are compounds that separate into ions; for example, sodium chloride (table salt) dissociates into sodium and chloride ions. What implication do you think electrolyte loss might have for the nervous system? Why?

- Imagine what the effect would be if the nervous system used only one neurotransmitter.

- How similar to humans do you think computers are capable of becoming? How much is your answer based on how you think human behavior is controlled versus how capable you think computers are?

T   Testing Your Understanding

1. Describe the ion movements and voltage changes that make up the neural impulse, from graded potential (at the axon hillock) to recovery.

2. Discuss the ways in which the synapse increases the neuron's capacity for transmitting information.

3. Describe how artificial neural networks function like the brain and what humanlike behaviors they have produced.

Select the best answer:

1. The inside of the neuron is relatively poor in _____ ions and rich in _____ ions.
   a. chloride, phosphate   b. sodium, potassium
   c. potassium, sodium   d. calcium, sodium

2. The rate law
   a. explains how the intensity of stimuli is represented.
   b. does not apply to neurons outside the brain.
   c. describes transmission in myelinated axons.
   d. describes the process of postsynaptic integration.

3. Without the sodium-potassium pump, the neuron would become
   a. more sensitive because of accumulation of sodium ions.
   b. more sensitive because of accumulation of potassium ions.
   c. overfilled with sodium ions and unable to fire.
   d. overfilled with potassium ions and unable to fire.

4. There is a limit to how rapidly a neuron can produce action potentials. This is due to
   a. inhibition.
   b. facilitation.
   c. the absolute refractory period.
   d. the relative refractory period.

5. Saltatory conduction results in
   a. less speed and the use of more energy.
   b. greater speed with the use of less energy.
   c. less speed but with the use of less energy.
   d. greater speed but with the use of more energy.

6. General anesthetics open potassium channels, allowing potassium ions to leak out of the neuron. This
   a. increases firing in pain-inhibiting centers in the brain.
   b. increases firing in the neuron until it is fatigued.
   c. hypopolarizes the neuron, preventing firing.
   d. hyperpolarizes the neuron, preventing firing.

7. When the action potential arrives at the terminal button, entry of _____ ions stimulates release of transmitter.
   a. potassium   b. sodium
   c. calcium   d. chloride

8. All the following neurotransmitters are deactivated by reuptake except
   a. acetylcholine.   b. norepinephrine.
   c. serotonin.   d. dopamine.

9. An inhibitory neurotransmitter causes the inside of the postsynaptic neuron to become
   a. more positive.   b. more negative.
   c. more depolarized. d. neutral in charge.

10. Excitatory postsynaptic potentials are typically
    produced by movement of _____ ions, whereas
    inhibitory postsynaptic potentials are typically
    produced by movement of _____ ions.
    a. potassium; sodium or chloride
    b. potassium; sodium or calcium
    c. sodium; calcium or chloride
    d. sodium; potassium or chloride

11. Which of the following is not an example of
    regulation of synaptic activity?
    a. A neuron has its synapse on the terminals of
       another and affects its transmitter release.
    b. Autoreceptors reduce the amount of
       transmitter released.
    c. A presynaptic neuron inhibits a postsynaptic
       neuron.
    d. Postsynaptic receptors change in numbers or
       sensitivity.

12. The graph below shows three graded potentials
    occurring at the same time.

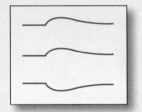

Assume that the resting potential is −70 mV and
that each graded potential individually produces
a 5-mV change. What is the membrane's voltage
after the graded potentials arrive?
a. −65 mV   b. −70 mV
c. −75 mV   d. +75 mV

13. The presence of synapses in a neuron chain
    provides the opportunity for
    a. increases in conduction speed.
    b. modification of neural activity.
    c. two-way communication in a pathway.
    d. regeneration of damaged neurons.

14. Artificial neural networks
    a. rely on prewired "neural" connections.
    b. solve problems in a couple of trials by
       insight.
    c. are preprogrammed.
    d. learn how to carry out the task themselves.

Answers:
1.b, 2.a, 3.c, 4.c, 5.b, 6.d, 7.c, 8.a, 9.b, 10.d, 11.c, 12.a, 13.a, 14.d.

Online Resources

The following resources are available at
www.sagepub.com/garrett3e.

Chapter Resources

- Flash Cards
- Chapter Quiz
- Internet Resources and Exercises

On the Web

The following websites are coordinated with
the chapter's content. Their numbers correspond
with the numbers of the icons you saw throughout
the chapter. You can find links to these websites by
going to the web address above and selecting this
chapter's **Chapter Resources**, then choosing **Internet
Resources and Exercises**.

1. **Neuroscience for Kids** (don't be put off by the
   name!) has a review of the resting and action
   potentials and an animation of their electrical
   recording.

2. **The Schwann Cell and Action Potential** is a
   visually beautiful animation of myelination and
   how it speeds conduction.

3. The video **The Origin of the Brain** describes
   how (and why) neurons and synapses evolved and
   ends with a very nice demonstration of how simple
   circuits can "remember" and make "decisions."
   Osaka University's **Intelligent Robotics
   Laboratory** site offers videos of several of its
   lifelike androids.
   A site with informative articles and links to
   research sites is the **American Association for
   Artificial Intelligence** page on neural networks.

Animations

- The Neural Impulse (Figure 2.8)
- Transmission at the Synapse (Figure 2.14)
- Spatial and Temporal Summation (Figure 2.17)

Chapter Updates and Biopsychology News

R For Further Reading

1. *Synaptic Self* by Joseph LeDoux (Penguin Books, 2003) takes the position that "your 'self,' the essence of who you are, reflects patterns of interconnectivity between neurons in your brain." A good read by a noted neuroscientist.

2. "Understanding Synapses: Past, Present, and Future" by Thomas Südhoff and Robert Malenka (*Neuron,* 2008, *60,* 469–476). This review will convince you of the importance of synapses and will provide a useful reference throughout the course.

3. "Tripartite Synapses: Astrocytes Process and Control Synaptic Information" by Gertrudis Perea, Marta Navarrete, and Alfonso Araque *(Trends in Neurosciences,* 2006, *32,* 421–431) reviews what we know about glial influence on synaptic activity.

4. "All My Circuits: Using Multiple Electrodes to Understand Functioning Neural Networks" by Earl Miller and Matthew Wilson (*Neuron,* 2008, *60,* 483–488) gives a good description of brainwide neural networks that play a variety of roles.

5. *Neurons and Networks: An Introduction to Behavioral Neuroscience* by John E. Dowling (Harvard University Press, 2001). Written by a well-known Harvard neuroscientist (you will see some of his work in Chapter 10), the first half of this book elaborates on the topics in this chapter. According to one student, the book "goes into depth without becoming murky."

6. "How Neural Networks Learn From Experience" by Geoffrey Hinton (*Scientific American*, March 1993, 145–151). This brief article explains neural networks and gives examples of their use.

7. "Debunking the Digital Brain" (*Scientific American*, February 1997). This brief feature article describes the view of Christof Koch at the California Institute of Technology that neurons are much more complicated than we give them credit for, which has implications for computer simulation.

8. *The Computational Brain* by Patricia Churchland and Terrence Sejnowski (Bradford Books, 1994). This treatment of the attempt to model brain functioning with computers is by two of the foremost experts in the field; it is a bit difficult, but lively and well written.

K Key Terms

absolute refractory period ... 31
action potential ... 29
agonist ... 45
all-or-none law ... 31
antagonist ... 45
autoreceptor ... 41
axon ... 25
cell body ... 24
Dale's principle ... 43
dendrite ... 25
electrostatic pressure ... 29
excitatory postsynaptic potential (EPSP) ... 38
force of diffusion ... 29
glial cell ... 33
graded potential ... 31
hyperpolarization ... 38
hypopolarization ... 38
inhibitory postsynaptic potential (IPSP) ... 38
interneuron ... 26
ion ... 28
ionotropic receptor ... 37
metabotropic receptor ... 37
motor neuron ... 25
myelin ... 33
neural network ... 46
neuron ... 24
neurotoxin ... 32
neurotransmitter ... 25
node of Ranvier ... 34
nondecremental ... 31
oligodendrocyte ... 34
polarization ... 27
postsynaptic ... 36
presynaptic ... 36
presynaptic excitation ... 41
presynaptic inhibition ... 41
rate law ... 33
relative refractory period ... 33
resting potential ... 28
reuptake ... 41
saltatory conduction ... 34
Schwann cell ... 34
sensory neuron ... 25
sodium-potassium pump ... 29
spatial summation ... 39
synapse ... 36
synaptic cleft ... 36
temporal summation ... 39
terminal ... 25
vesicle ... 37
voltage ... 28

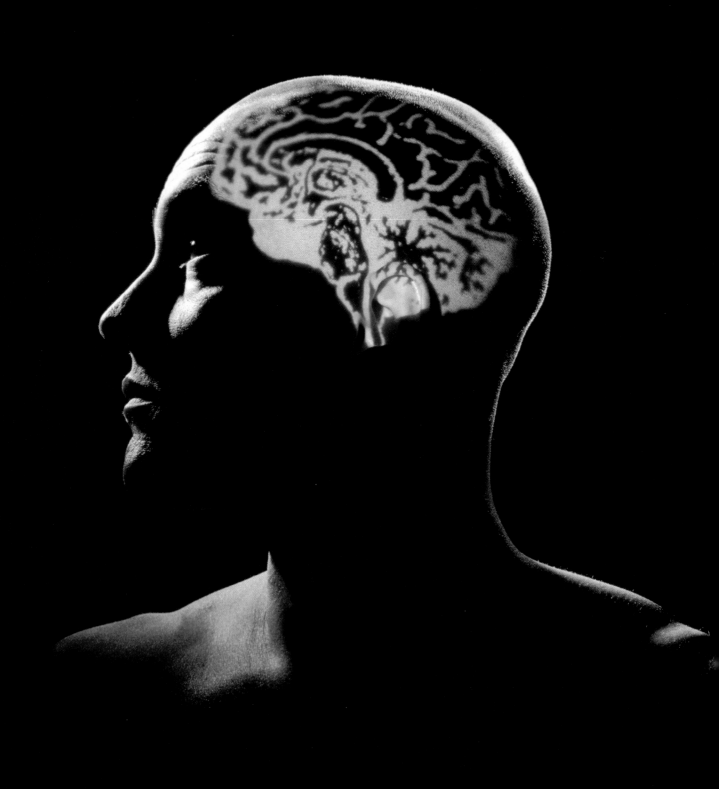

The Organization and Functions of the Nervous System

3

The Central Nervous System

The Forebrain

    APPLICATION: THE CASE OF PHINEAS GAGE

The Midbrain and Hindbrain

The Spinal Cord

Protecting the Central Nervous System

    CONCEPT CHECK

The Peripheral Nervous System

The Cranial Nerves

The Autonomic Nervous System

    CONCEPT CHECK

Development and Change in the Nervous System

The Stages of Development

How Experience Modifies the Nervous System

Damage and Recovery in the Central Nervous System

    IN THE NEWS: IS THE BRAIN TOO FRAGILE FOR SPORTS?

    APPLICATION: MENDING THE BRAIN WITH COMPUTER CHIPS

    CONCEPT CHECK

In Perspective

Summary

Study Resources

In this chapter you will learn

- The major structures of the nervous system and some of their functions

- How the nervous system develops and how it changes with experience

- Strategies for repairing damaged brains and spinal cords, and the obstacles

FIGURE 3.1 A Normal Brain and a Lissencephalic Brain.

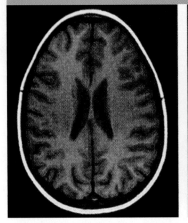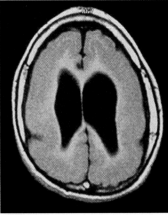

SOURCE: Photo Science. Yuanyi Feng and Christopher Walsh, "Protein-Protein interactions, cytoskeletal regulation and neuronal migration," *Nature Reviews Neurscience*, 2: 408-416, June 2001. Copyright © 2001.

Karen is a college graduate and holds a job with considerable responsibility. She is married and leads a normal life except for occasional epileptic seizures. When her doctors ordered a brain scan to find the cause of her seizures, they were astounded. The normal person's brain has many folds on its surface, so it is wrinkled like a walnut; Karen's is perfectly smooth. You can see how different her brain is by comparing it with a normal brain in Figure 3.1. Notice, too, that the dark areas in the middle of her brain are enlarged, which indicates that she has a deficiency in the amount of brain tissue. People with her disorder are usually not only *lissencephalic* (literally, *smooth-brained*) and epileptic like Karen, but severely impaired intellectually as well (Barinaga, 1996; Eksioglu et al., 1996). So what really amazed Karen's doctors was not how abnormal her brain is, but that she functions not only normally but well above average. How do we explain why some people are able to escape the consequences of what is usually a devastating developmental error? The answer is that we do not know why; it is one of the mysteries that neuroscientists are attempting to solve in order to understand the brain's remarkable resilience.

You are now well versed in the functioning of neurons and how they interact with each other. What you need to understand next is how neurons are grouped into the functional components that make up the nervous system. In the next few pages, we will review the physical structure of the nervous system so that you will have a road map for more detailed study in later chapters. And we will include an overview of major functions to prepare you for the more detailed treatments to come in later chapters. We will look first at the central nervous system before turning our attention to the peripheral nervous system and additional issues like neural development.

The Central Nervous System

The nervous system is divided into two subunits. The *central nervous system (CNS)* includes the brain and the spinal cord. The second part is the peripheral nervous system, which we will examine later in the chapter. Before we go any further, we need to be sure you understand a couple of terms correctly. As we talk about the nervous system, be careful not to confuse *nerve* and *neuron.* A neuron is a single neural cell; a *nerve* is a bundle of axons running together like a multiwire cable. However, the term *nerve* is used only in the peripheral nervous system; inside the CNS, bundles of axons are called *tracts.* Most of the neurons' cell bodies are also clustered in groups; a group of cell bodies is called a *nucleus* in the CNS and a *ganglion* in the peripheral nervous system. Table 3.1 should help you keep these terms straight.

TABLE 3.1 Terms for Axons and Cell Bodies in the Nervous System.

	Peripheral	Central
Bundle of axons	Nerve	Tract
Group of cell bodies	Ganglion	Nucleus

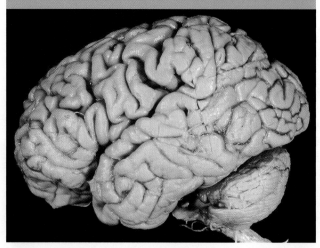

FIGURE 3.2 View of a Human Brain.

SOURCE: © Dr. Fred Hossler/Visuals Unlimited.

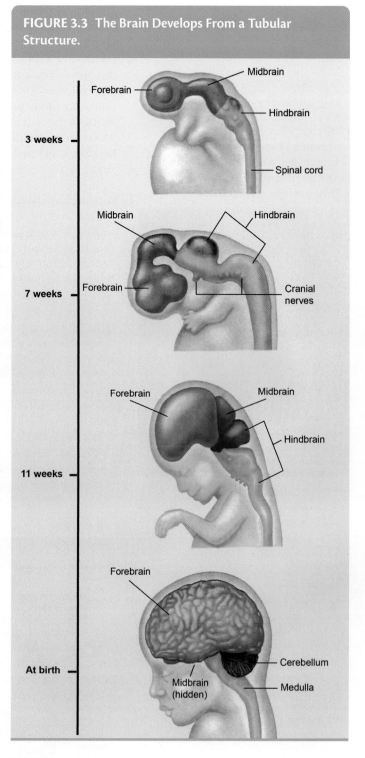

FIGURE 3.3 The Brain Develops From a Tubular Structure.

Figure 3.2 is a photograph of a human brain. It will be easier to visualize the various structures of the brain if you understand that the CNS begins as a hollow tube and preserves that shape as it develops (Figure 3.3). The upper end of the tube develops three swellings, which will become the forebrain, midbrain, and hindbrain; the lower part of the tube develops into the spinal cord. The forebrain appears to be perched on top of the lower structures as it enlarges and almost completely engulfs them. By comparing the four drawings in this series, you can see that the mature forebrain obscures much of the lower brain from view. You will get a better idea of these hidden structures later when we look at an interior view of the brain.

The Forebrain

The major structures of the forebrain are the two cerebral hemispheres, the thalamus, and the hypothalamus. The outer layer of the hemispheres, the cortex, is where the highest-level processing occurs in the brain.

The Cerebral Hemispheres

The large, wrinkled *cerebral hemispheres* dominate the brain's appearance (Figure 3.4). Not only are they large in relation to the rest of the brain, but they are also disproportionately larger than in other primates (Deacon, 1990). The *longitudinal fissure* that runs the length of the brain separates the two cerebral hemispheres, which are nearly mirror images of each other in appearance. Often the same area in each hemisphere has identical functions as well, but you will see that this is not always the case. The simplest form of *asymmetry* is that each hemisphere receives most of its sensory input from the *opposite* side of the body (or of the world, in the case of hearing and vision) and provides most of the control of the opposite side of the body.

Look again at Figures 3.2 and 3.4. The brain's surface has many ridges and grooves that give it a very wrinkled appearance; the term we use is *convoluted*. Each ridge is called a *gyrus;* the groove or space between two gyri is called a *sulcus* or, if

> One of the key strategies of the nervous system is localization of functions: specific types of information are processed in particular regions.
>
> —Eric Kandel

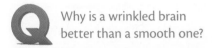

FIGURE 3.4 Human Brain Viewed From Above.

This photo shows the cerebral hemisphere and longitudinal fissure. The blood vessels have been removed from the right hemisphere.

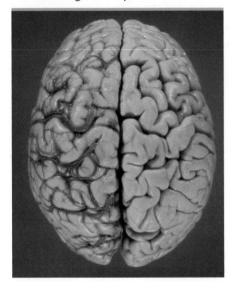

SOURCE: Photo Researchers.

1
Brain Atlas and Tutorial

it is large, a *fissure*. You can see how the gyri are structured in the cross section of a brain in Figure 3.5. The outer surface is the *cortex* (literally, "bark"), which is made up mostly of the cell bodies of neurons; because cell bodies are not myelinated, the cortex looks grayish in color, which is why it is referred to as gray matter. Remember that neural processing occurs where neurons synapse on the cell bodies of other neurons, which indicates why the cortex is so important. The cortex is only 1.5 to 4 millimeters (mm) thick, but the convolutions increase the amount of cortex by tripling the surface area. The convolutions also provide the axons with easier access to the cell bodies than if the developing cortex thickened instead of wrinkling. The axons come together in the central core of each gyrus, where their myelination gives the area a whitish appearance. Notice how the white matter of each gyrus joins with the white matter of the next gyrus, creating the large bands of axons that serve as communication routes in the brain, two of which connect between the hemispheres.

You find yourself running to your early morning test, fretting about being late while rehearsing answers to the questions you expect on the exam. You interrupt your thoughts only to greet a fellow student, doing your best to conceal your disdain because of the silly questions he asks in class. Do you ever wonder how your brain brings all this off?

It will take the rest of this book to *start* answering that question, but this is a good time to mention two ways the brain's organization helps it to be more efficient. First, the cortex in humans and most mammals is arranged in layers; the number of layers is usually six, though a particular layer may be absent in some areas. The layers stand out from each other because they are separated by fibers that serve the cell bodies, but they differ in appearance as well: They vary in type and size of cells and in the concentration of cell bodies versus axons (see Figure 3.6a). There are differences in function as well. Some researchers have concluded that layers II and III are associational, IV is sensory, and V and VI have motor functions (Buxhoeveden & Casanova, 2002).

Second, the cells of the cortex are organized into groups of 80 to 100 interconnected neurons, which are arranged in columns running perpendicular to the cortical surface (Figure 3.6b; Buxhoeveden & Casanova, 2002). They provide a

FIGURE 3.5 Section of Human Brain Showing Gyri and Sulci.

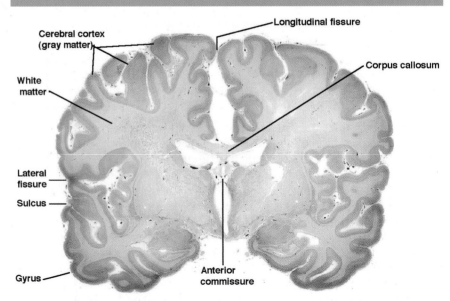

SOURCE: Courtesy of The Brain Museum. www.brainmuseum.org.

FIGURE 3.6 Layers and Columns of the Cortex.

(a)

(b)

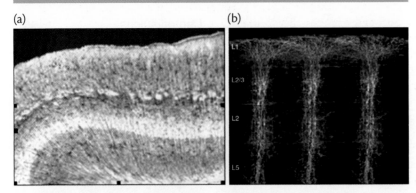

(a) Photograph of a section of cortex, revealing its layered organization.
(b) Photograph showing the columnar arrangement of cells in the cortex; the numbers on the left identify the cortical layers.

SOURCE: (a) *Cerebral Cortex*, 12(7). © 2007. Used by permission of Oxford Univeristy Press.
(b) Bloomfield Science Jerusalem, http://brain.mada.org.il/tools-e.html.

vertical unification of the cortex's horizontal layers and have been referred to as the primary information-processing unit in the cortex (Torii, Hashimoto-Torii, Levitt, & Rakic, 2009). The cells in a column have a similar function. For example, they may receive input from the same area on the skin's surface, while surrounding columns serve adjacent locations. In the visual cortex, the cells in a column may detect object edges at a particular orientation, while surrounding columns respond to edges at a slightly different orientation. Having similar functions grouped close together in well-connected columns helps the brain work quickly and efficiently.

Students often ask whether intelligent people have bigger brains. Bischoff, the leading European anatomist in the 19th century, argued that the greater average weight of men's brains was infallible proof of their intellectual superiority over women. When he died, his brain was removed and added to his extensive collection as his will had specified; ironically, it weighed only 1,245 grams (g), less than the average of about 1,250 g for women ("Proof?" 1942). There actually is a tendency for people with larger brains to be more intelligent (Willerman, Schultz, Rutledge, & Bigler, 1991), but the relationship is small and highly variable. What this means is that factors other than brain size are more important; otherwise, women would be less intelligent than men as Bischoff claimed, but we know from research that this is not the case. When we look at brain size more closely in the chapter on intelligence, you will learn that Einstein's brain was even smaller than Bischoff's.

Across species, brain size is more related to body size than to intelligence; the brains of elephants and sperm whales are five or six times larger than ours. It is a brain's complexity, not its size, that determines its intellectual power. Look at the brains in Figure 3.7, then compare them with the human brain in Figure 3.2. You can see two features that distinguish more complex, more highly evolved brains from less complex ones. One is that the higher brains are more convoluted; the greater number of gyri means more cortex. The other is that the cerebral hemispheres are larger in proportion to the lower parts of the brain. It is no accident that the cerebral hemispheres are perched atop the rest of the brain and the spinal cord. The CNS is arranged in a *hierarchy;* as you ascend from the spinal cord through the hindbrain and midbrain to the forebrain, the neural structures become more complex, and so do the behaviors they control.

FIGURE 3.7 Brains of Three Different Species.

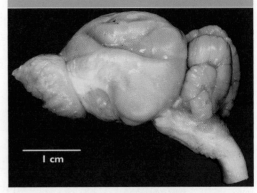

(a) Armadillo brain.

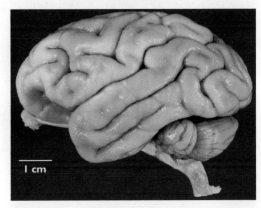

(b) Monkey brain.

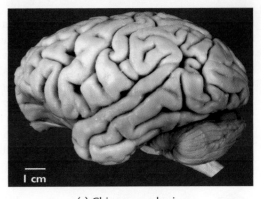

(c) Chimpanzee brain.

SOURCE: The Brain Museum. http://brainmuseum.org.

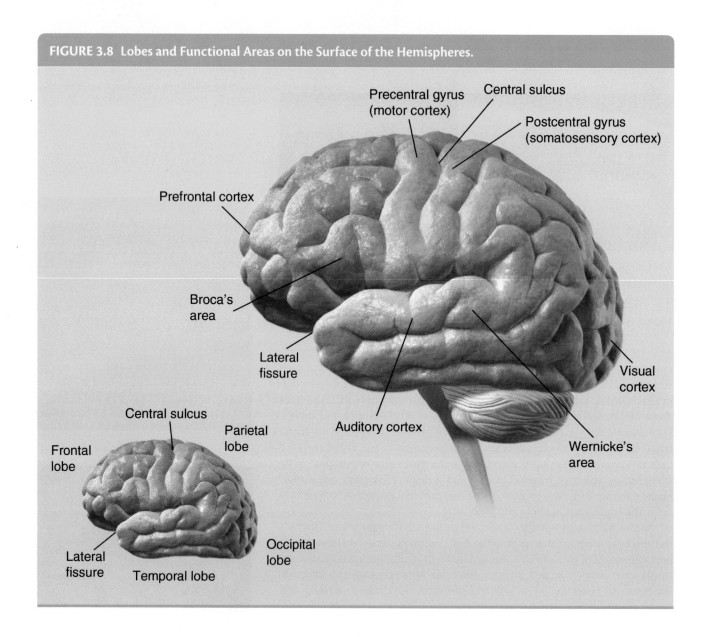

FIGURE 3.8 Lobes and Functional Areas on the Surface of the Hemispheres.

The Four Lobes

The hemispheres are divided into four lobes—frontal, parietal, occipital, and temporal—each named after the bone of the skull above it. The lobes are illustrated in Figure 3.8, along with the major functions located within them. These divisions are somewhat arbitrary, but they are very useful for locating structures and functions, so we will organize our discussion around them. Sometimes we need additional precision in locating structures, so you should get used to seeing the standard terms that are used; the most important ones are illustrated in Figure 3.9a. Also, throughout this text you will be seeing illustrations of the nervous system from a variety of perspectives; until you get more comfortable with the structure of the nervous system, it may be difficult to tell what you are seeing. The images in Figure 3.9b will serve as a guide for understanding the orientation of most of these illustrations, as well as providing a few terms of orientation that we will need.

The *frontal lobe* is the area anterior to (in front of) the *central sulcus* and superior to (above) the *lateral fissure.* The functions here are complex and include some of

 What functions are found in the frontal lobes?

FIGURE 3.9 Terms Used to Indicate Direction and Orientation in the Nervous System.

(a) *Dorsal* means toward the back, and *ventral* means toward the stomach. (This terminology was developed with other animals and becomes more meaningful when you imagine that the human is on all fours, face forward. *Anterior* means toward the front, and *posterior* means toward the rear. *Lateral* means toward the side; *medial* indicates toward the middle. *Superior* is a location above another structure, and *inferior* means below another structure.

(b) The *coronal plane* divides the brain vertically from side to side, while the *lateral plane* divides it vertically in an anterior-posterior direction, and the *horizontal plane* divides it between the top and bottom. We use these terms when we refer to an image (e.g., a lateral view) or when the brain is cut along one of these planes (e.g., a horizontal section).

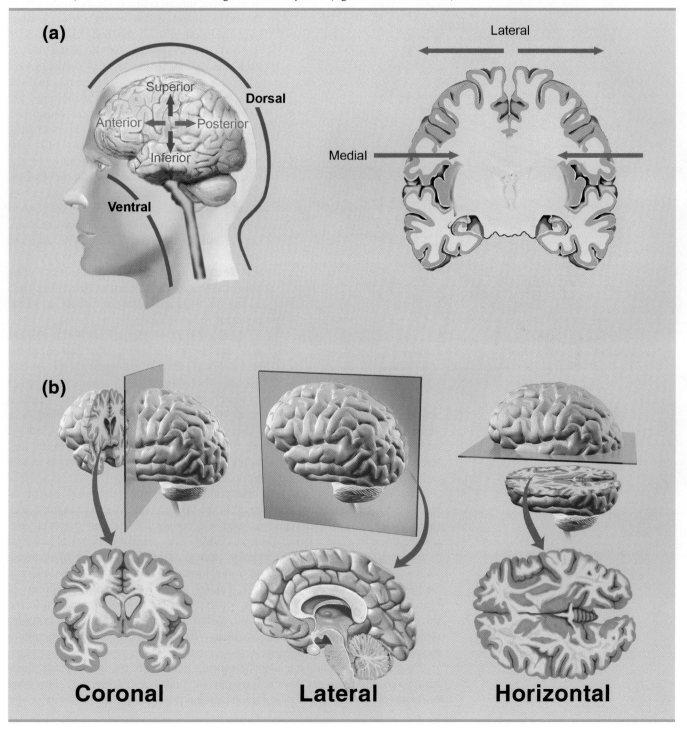

FIGURE 3.10 The Motor Cortex.

This view shows a cross section (coronal) of the precentral gyrus. The size of the homunculus indicates the relative amount of motor cortex devoted to those body areas.

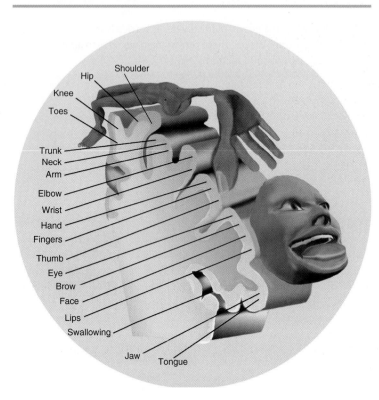

SOURCE: From Penfield/Rasmussen. *The Cerebral Cortex of Man.* © 1950 Gale, a part of Cengage Learning, Inc. Reproduced by permission. www.cengage.com/permissions.

the highest human capabilities. A considerable portion of the frontal lobes is also involved with the control of movement. And because the primary motor area is located along the posterior boundary of the frontal lobe, we will start our discussion there.

The *precentral gyrus*, which extends the length of the central sulcus, is the location of the primary *motor cortex*, which controls voluntary (nonreflexive) movement. The motor area in one hemisphere controls the opposite side of the body, though it does exert a lesser control over the same side of the body. The parts of the body are "mapped onto" the motor area of each hemisphere in the form of a *homunculus*, which means "little man." This means that the cells that control the muscles of the hand are adjacent to the cells controlling the muscles of the arm, which are next to those controlling the shoulder, and so on (see Figure 3.10). The homunculus is distorted in shape, however; the parts of the body that make precise movements, such as the hands and fingers, have more cortex devoted to their control. The *primary motor cortex,* like other functional areas of the brain, carries out its work in concert with adjacent *secondary* areas. The secondary motor areas are located just anterior to the primary area. Subcortical (below the cortex) structures, such as the basal ganglia, also contribute to motor behavior.

Looking back at Figure 3.8, locate Broca's area anterior to the motor area and along the lateral fissure. *Broca's area* controls speech production, contributing the movements involved in speech and grammatical structure. A patient with damage to this area was asked about a dental appointment; he replied, haltingly, "Yes . . . Monday . . . Dad and Dick . . . Wednesday 9 o'clock . . . 10 o'clock . . . doctors . . . and . . . teeth" (Geschwind, 1979). Similar problems occur in reading and writing. In another example of hemispheric asymmetry, language activity is mostly controlled by the left hemisphere in 9 out of 10 people.

The more anterior part of the frontal lobes—the prefrontal cortex in Figure 3.8—is functionally complex. It is the largest region in the human brain, twice as large as in chimpanzees, and it accounts for 29% of the total cortex (Andreasen et al., 1992; Deacon, 1990). The *prefrontal cortex* is involved in planning and organization, impulse control, adjusting behavior in response to rewards and punishments, and some forms of decision making (Bechara, Damasio, Tranel, & Damasio, 1997; Fuster, 1989; Kast, 2001). Symptoms of impairment are varied, depending on which part of the prefrontal area is affected (Mesulam, 1986), but malfunction often strikes at the capabilities we consider most human. Schizophrenia and depression, for example, involve dysfunction in the prefrontal cortex.

People with prefrontal damage often engage in behavior that normal individuals readily recognize will get them into trouble. In clinical interviews, they show good understanding of social and moral standards and the consequences of behavior—for example, they can describe several valid ways to develop a friendship, maintain a romantic relationship, or resolve an occupational difficulty—but they are unable to choose among the options. So in real life, they suffer loss of friends, financial disaster, and divorce (Damasio, 1994). Research indicates that prefrontal damage impairs the ability to learn from reward and punishment and to control impulses (Bechara et al., 1997).

In spite of the effects of frontal lobe damage, during the 1940s and 1950s, surgeons performed tens of thousands of *lobotomies*, a surgical procedure that disconnects the prefrontal area from the rest of the brain. Initially, the surgeries were performed on very disordered schizophrenics, but many overly enthusiastic doctors lobotomized patients with much milder problems. Walter Freeman, shown in Figure 3.11, did more than his share of the 40,000 lobotomies performed in the United States and zealously trained other psychiatrists in the technique (Valenstein, 1986). The surgery calmed agitated patients, but the benefits came at a high price; the patients often became emotionally blunted, distractible, and childlike in behavior. In a follow-up study of patient outcomes, 49% were still hospitalized, and less than a fourth of the others were living independently (A. Miller, 1967). Lack of success with lobotomy and the introduction of psychiatric drugs in the 1950s made the surgery a rare therapeutic choice. Now *psychosurgery*, the use of surgical intervention to treat cognitive and emotional disorders, is generally held in disfavor, unlike brain surgery to treat problems such as tumors. The accompanying Application describes the most famous case of accidental lobotomy.

2
History of Psychosurgery

FIGURE 3.11 Lobotomy Procedure and a Lobotomized Brain.

(a) Walter Freeman inserts his instrument between the eyelid and the eyeball, drives it through the skull with a mallet, and moves it back and forth to sever the connections between the prefrontal area and the rest of the brain. (b) A horizontal view of a brain shows the gaps (arrows) produced by a lobotomy.

(a)
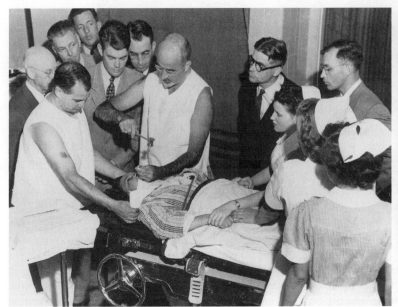

(b)
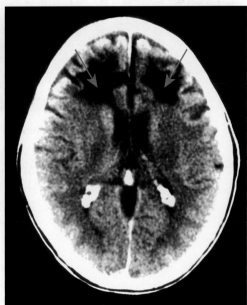

SOURCE: (a) © Bettmann/Corbis. (b) Copyright © 2007 Photo Researchers, Inc. All Rights Reserved.

Application — The Case of Phineas Gage

In 1848, Phineas Gage, a 25-year-old railroad construction fore-man in Cavendish, Vermont, was tamping explosive powder into a blasting hole when the charge ignited prematurely and drove the 3½-foot-long (1.15-m) tamping iron through his left cheek and out the top of his skull. Gage not only regained conscious-ness immediately and was able to talk and to walk with the aid of his men, but he also survived the accident with no impair-ment of speech, motor abilities, learning, memory, or intelligence. However, his personality was changed dramatically. He became irreverent and profane, and although Gage previously was the most capable man employed by the railroad, he no longer was dependable and had to be dismissed. He wandered about for a dozen years, never able to live fully independently, and died under the care of his family.

Almost a century and a half later, Hanna Damasio and her col-leagues carried out a belated postmortem examination of the skull (H. Damasio et al., 1994). Combining measurements from the skull with a three-dimensional computer rendering of a human brain, they were able to reconstruct the path of the tamping iron through Gage's brain (see the accompanying figure). They concluded that the acci-dent damaged the part of both frontal lobes involved in processing emotion and making rational decisions in personal and social matters. At the time of Gage's accident, physiologists were debating whether different parts of the brain have specific functions or are equally com-petent in carrying out functions. Gage's experience had such an impor-tant influence in tipping the balance toward localization of function that in 1998, scientists from around the world gathered in Cavendish to commemorate the 150th anniversary of the event (Vogel, 1998).

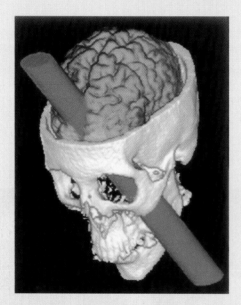
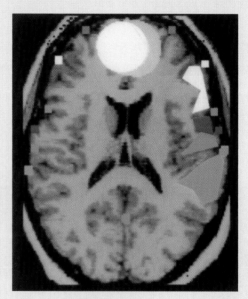

Reconstruction of the Damage to Phineas Gage's Brain.

Colors in the image on the right indicate motor, language, and body sensory areas that were unharmed.

SOURCE: Reprinted with permission from "The Return of Phineas Gage: Clues About the Brain From a Famous Patient," by H. Damasio, T. Grabowski, R. Frank, A. M. Galaburda, and A. R. Damasio, *Science*, 264, pp. 1102–1105. © 1994. Reprinted by permission from AAAS.

 What functions are found in the parietal lobes?

The *parietal lobes* are located superior to the lateral fissure and between the central sulcus and the occipital lobe. The *primary somatosensory cortex*, located on the postcentral gyrus, processes the skin senses (touch, warmth, cold, and pain) and the senses that inform us about body position and movement (see Figure 3.8 again). Like the motor cortex, the somatosensory area serves primarily the opposite side of the body. The somatosensory cortex also is organized as a homunculus, but in this case, the size of each area depends on the sensitivity in that part of the body. As you look at the senses of vision and hearing later, you will learn that this mapping is a principle of brain organization. Also, this is a good place to point out that the sensory areas of the brain are often referred to as *projection areas*, as in *somatosensory projection area.*

Each of the lobes contains *association areas,* which carry out further processing beyond what the primary area does, often combining information from other senses. Parietal lobe association areas receive input from the body senses and from vision; they help the person identify objects by touch, determine the location of the limbs, and locate objects in space. Damage to the posterior parietal cortex may produce *neglect,* a disorder in which the person ignores objects, people, and activity on the side opposite the damage. This occurs much more frequently when the damage is in the right parietal lobe. The patient may fail to shave or apply makeup on the left side of the face. In some cases, a person with a paralyzed arm or leg will deny that anything is wrong and even claim that the affected limb belongs to someone else.

The lateral fissure separates the temporal lobe from the frontal and parietal lobes. The *temporal lobes* contain the auditory projection area, visual and auditory association areas, and an additional language area (**Figure 3.8**). The *auditory cortex,* which receives sound information from the ears, lies on the superior (uppermost) gyrus of the temporal lobe, mostly hidden from view within the lateral fissure. Just posterior to the auditory cortex is *Wernicke's area,* which interprets language input arriving from the nearby auditory and visual areas; it also generates spoken language through Broca's area and written language by way of the motor cortex. When Wernicke's area is damaged, the person has trouble understanding speech or writing; the person can still speak, but the speech is mostly meaningless. Like Broca's area, this structure is found in the left hemisphere in most people.

The *inferior temporal cortex,* in the lower part of the lobe as the name implies, plays a major role in the visual identification of objects. People with damage in this area have difficulty recognizing familiar objects by sight, even though they can give detailed descriptions of the objects. They have no difficulty identifying the same items by touch. They may also fail to recognize the faces of friends and family members, though they can identify people by their voices. The neurologist Oliver Sacks (1990) described a patient who talked to parking meters, thinking they were children. Considering his strange behavior, it seems remarkable that he was unimpaired intellectually. Perhaps as you read about cases like this one and hear of patients who do things like denying ownership of their paralyzed leg, you will begin to appreciate the fact that human capabilities are somewhat independent of each other because they depend on different parts of the brain.

When the neurosurgeon Wilder Penfield (1955) stimulated patients' temporal lobes, he often elicited what appeared to be memories of visual and auditory experiences. Penfield was doing surgery to remove malfunctioning tissue that was causing epileptic seizures. Before the surgery, Penfield would stimulate the area with a weak electrical current and observe the effect; this allowed him to distinguish healthy tissue and important functional areas from the diseased tissue he wished to remove (see Figure 3.12).

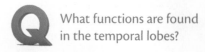

Q What functions are found in the temporal lobes?

FIGURE 3.12 Brain of One of Penfield's Patients.

The numbered tags allowed Penfield to relate areas to patients' responses.

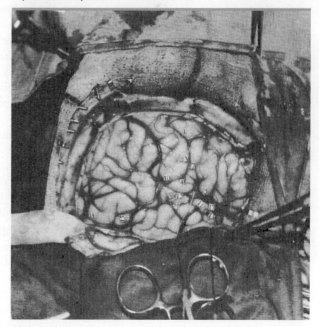

SOURCE: From *The Excitable Cortex in Conscious Man*, by Wilder Penfield, 1958. Courtesy of Dennis Coon and LIverpool University Press, Liverpool, UK © 1958. Used with permission.

The patients were awake because their verbal report was needed for carrying out this mapping; since brain tissue has no pain receptors, the patient requires only a local anesthetic for the surgery. Stimulation of primary sensory areas provoked only unorganized, meaningless sensations, such as tingling, lights, or buzzing sounds. But when Penfield stimulated the association areas of the temporal cortex, 25% of the patients reported hearing music or familiar voices or, occasionally, reliving a familiar event. One time, the patient hummed along with the music she was "hearing," and the nurse, recognizing the tune, joined in by supplying the lyrics. (Does this not sound like a scene from a Monty Python movie—a sing-along during brain surgery?) People with epileptic activity or brain damage in their temporal lobes sometimes hear familiar tunes as well. The composer Shostakovich reportedly heard music when he tilted his head, shifting the location of a tiny wartime shell fragment in his temporal lobe; he refused to have the sliver removed, saying he used the melodies when composing (Sacks, 1990). Unfortunately, Penfield made no attempt to verify whether the apparent memories were factual or a sort of electrically induced dream; however, we will see in Chapter 12 that part of the temporal lobe has an important role in memory.

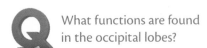

What functions are found in the occipital lobes?

Finally, the *occipital lobes* are the location of the *visual cortex,* which is where visual information is processed (see Figure 3.8). The primary projection area occupies the posterior tip of each lobe; anterior to the primary area are four secondary areas that detect individual components of a scene, such as color, movement, and form, which are then combined in association areas. Just as the somatosensory and motor areas are organized to represent the shape of the body, the visual cortex contains a map of visual space because adjacent receptors in the back of the eye send neurons to adjacent cells in the visual cortex.

Now that you are familiar with the four lobes and some of the functions located in the cortex, we will direct our tour to structures below the surface.

The Thalamus and Hypothalamus

Deep within the brain, the *thalamus* lies just below the lateral ventricles, where it receives information from all the sensory systems except olfaction (smell) and relays it to the respective cortical projection areas. (Figure 3.13 is a lateral view of a brain sliced down the middle to show the structures described in this section.) Many other neurons from the thalamus project more diffusely throughout the cortex and help arouse the cortex when appropriate. You will see additional functions for the thalamus in later chapters. Actually there are two thalami, a right and a left, lying side by side.

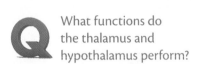

What functions do the thalamus and hypothalamus perform?

The *hypothalamus,* a smaller structure just inferior to the thalamus, plays a major role in controlling emotion and motivated behaviors such as eating, drinking, and sexual activity (Figure 3.13). The hypothalamus exerts this influence largely through its control of the autonomic nervous system, which we will consider shortly. The hypothalamus also influences the body's hormonal environment through its control over the pituitary gland. In Figure 3.13, the pituitary appears to be hanging down on its stalk just below the hypothalamus. The pituitary is known as the *master gland* because its hormones control other glands in the body. The hypothalamus, which is paired like the thalamus, contains perhaps the largest concentration of nuclei important to behavior in the entire brain.

Just posterior to the thalamus is the *pineal gland.* You can see in Figure 3.13 why it was Descartes's best candidate for the seat of the soul (see Chapter 1): a single, unpaired structure, attached by its flexible stalk just below the hemispheres. In reality, the pineal gland secretes melatonin, a hormone that induces sleep. It controls seasonal cycles in nonhuman animals and participates with other structures in controlling daily rhythms in humans.

FIGURE 3.13 Lateral View of the Interior Features of the Human Brain.

In this view, the cut has been made between the cerebral hemispheres. Everything above the midbrain is forebrain; everything below is hindbrain.

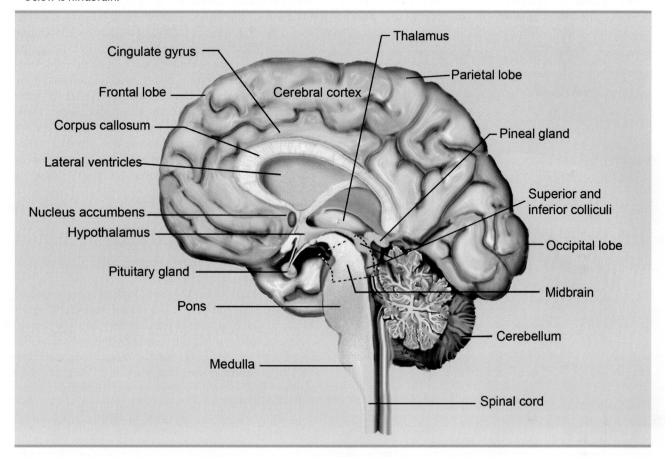

The Corpus Callosum

If you were to look inside the longitudinal fissure between the two cerebral hemispheres, you would see that the hemispheres are distinctly separate from each other. A couple of inches below the brain's surface, the longitudinal fissure ends in the *corpus callosum,* a dense band of fibers that carry information between the hemispheres. The corpus callosum is visible in Figure 3.13; you can see it from another perspective, along with a smaller band of crossing fibers, the anterior commissure, by looking back at Figure 3.6. You know that the two hemispheres carry out somewhat different functions, so you can imagine that they must communicate with each other constantly to integrate their activities. In addition, incoming information is often directed to one hemisphere—visual information appearing to one side of your field of view goes to the hemisphere on the opposite side, just as information from one side of your body does. This information is "shared" with the other hemisphere through the crossing fibers, especially the corpus callosum; the car that is too close on your left is registered in your right hemisphere, but if you are steering with your right hand, it is your left hemisphere that must react.

Occasionally, surgeons have to sever the corpus callosum in patients with incapacitating epileptic seizures that cannot be controlled by drugs. The surgery

FIGURE 3.14 A Patient With Severed Corpus Callosum Identifying Objects by Touch.

He cannot say what the object is because the right hemisphere, which receives the information from the hand, has been disconnected from the more verbal left hemisphere. Results are similar for visually presented stimuli and sound information.

FIGURE 3.15 The Ventricles of the Brain.

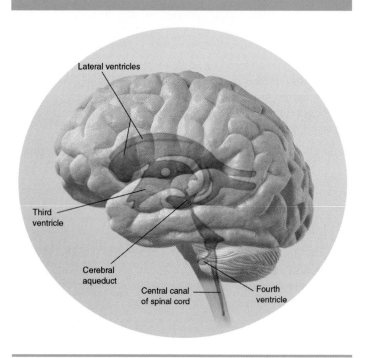

Lateral ventricles

Third ventricle

Cerebral aqueduct

Central canal of spinal cord

Fourth ventricle

prevents the out-of-control neural activity in one hemisphere from engulfing the other hemisphere as well. The patient is then able to maintain consciousness during seizures and to lead a more normal life. These patients have been very useful for studying differences in the functions of the two hemispheres, because a stimulus can be presented to one hemisphere and the information will not be shared with the other hemisphere. Studies of these individuals have helped establish, for example, that the left hemisphere is more specialized for language than the right hemisphere and the right hemisphere is better at spatial tasks and recognizing faces (Gazzaniga, 1967; Nebes, 1974). An example is shown in Figure 3.14; we will explore this topic further when we discuss consciousness in the final chapter.

The Ventricles

During development, the hollow interior of the nervous system develops into cavities called *ventricles* in the brain and the central canal in the spinal cord. The ventricles are filled with *cerebrospinal fluid,* which carries material from the blood vessels to the CNS and transports waste materials in the other direction. The *lateral ventricles* (Figures 3.13 and 3.15) extend forward deeply into the frontal lobes and in the other direction into the occipital lobes before they curve around into the temporal lobes. Below the lateral ventricles and connected to them is the *third ventricle;* it is located between the two thalami and the two halves of the hypothalamus, which form the ventricle's walls. The *fourth ventricle* is not in the forebrain, so we will locate it later.

The Midbrain and Hindbrain

The *midbrain* contains structures that have secondary roles in vision, audition, and movement (Figures 3.13 and 3.16). The *superior colliculi,* for example, help guide eye movements and fixation of gaze, and the *inferior colliculi* help locate the direction of sounds. One of the structures involved in movement is the *substantia nigra,* which projects to the basal ganglia to integrate movements; its dopamine-releasing cells degenerate in Parkinson's disease (Chapter 11). Another is the *ventral tegmental area,* which we will see in Chapter 5 plays a role in the rewarding effects of food, sex, drugs, and so on. The midbrain also contains part of the reticular formation, which is described below. Passing through the midbrain is the cerebral aqueduct, which connects the third ventricle above with the fourth ventricle below (see Figure 3.15). Notice in Figure 3.16 that the brain takes on a more obvious tubular shape here, reminding us of the CNS's origins. Considering the shape of these structures and the appearance of the cerebral hemispheres perched on top, you can see why this part of the brain is referred to as the *brain stem*.

The hindbrain is composed of the pons, the medulla, and the cerebellum (Figures 3.13 and 3.16). The *pons* contains centers related to sleep and arousal, which are part of the reticular formation. The *reticular formation* is a collection of many nuclei running through the middle of the hindbrain and the midbrain; besides its role in sleep and arousal, it contributes to attention and to aspects of motor activity, including reflexes and muscle tone. The word *pons* means "bridge" in Latin, which reflects not only its appearance but also the fact that its fibers connect the two hemispheres of the cerebellum; the pons also has pathways connecting higher areas of the brain with the brain stem. The *medulla* forms the lower part of the hindbrain; its nuclei are involved with control of essential life processes, such as cardiovascular activity and respiration (breathing).

The cerebellum is the second most distinctive-appearing brain structure (see Figures 3.2, 3.8, 3.13, and 3.20). Perched on the back of the brain stem, it is wrinkled and divided down the middle like the cerebral hemispheres—thus its name, which means "little brain." The most obvious function of the *cerebellum* is refining movements initiated by the motor cortex by controlling their speed, intensity, and direction. A person whose cerebellum is damaged has trouble making precise reaching movements and walks with difficulty because the automatic patterning of movement routines has been lost. It is not unusual for individuals with cerebellar damage to be arrested by the police because their uncoordinated gait is easily mistaken for drunkenness. The cerebellum also plays a role in motor learning, and research implicates it in other cognitive processes and in emotion (Fiez, 1996). With 70% of the brain's neurons in its fist-sized volume, it would be surprising if it did not hold a number of mysteries waiting to be solved.

We have admittedly covered a large number of structures. It may help to see them and their major functions summarized in Table 3.2. But as you review these functions, remember the caveat about localization from Chapter 1 that a behavior is seldom the province of a single brain location, but results from the interplay of a whole network of structures.

FIGURE 3.16 The Brain Stem.

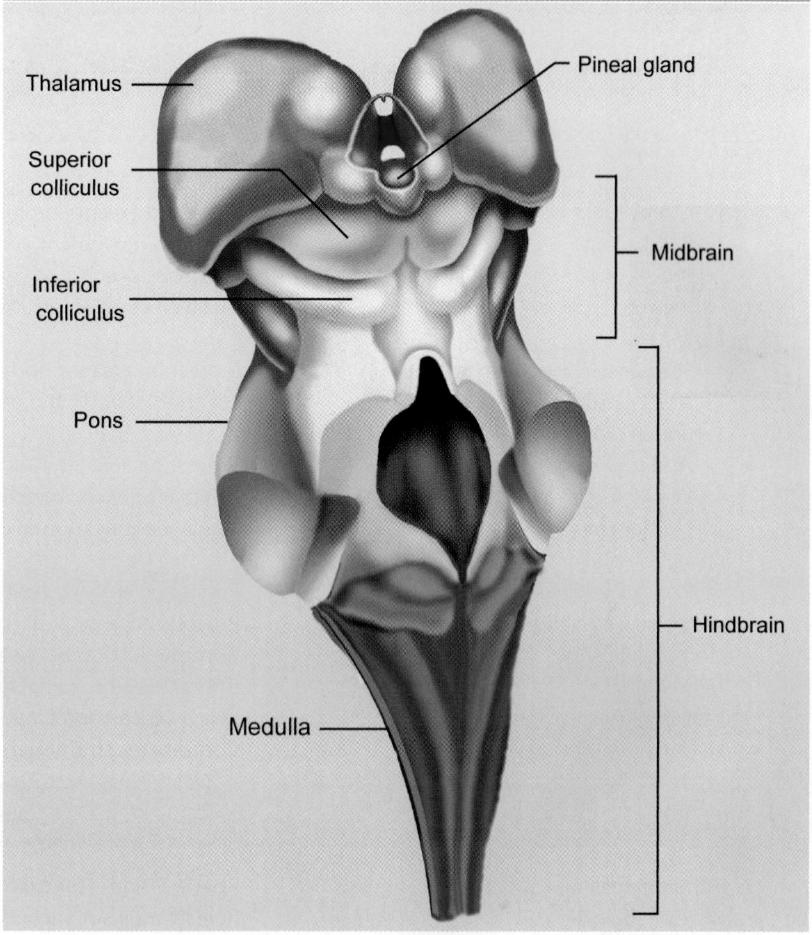

Thalamus

Superior colliculus

Inferior colliculus

Pons

Medulla

Pineal gland

Midbrain

Hindbrain

The brain stem includes posterior parts of the forebrain (thalamus, hypothalamus, etc.), the midbrain, and the hindbrain. The cerebellum has been removed to reveal the other structures. This is a dorsal view of the brain stem. Refer to Figure 3.13 for its orientation with respect to the entire brain.

3
How the Brain Works

The Spinal Cord

The *spinal cord* is a finger-sized cable of neurons that carries commands from the brain to the muscles and organs and sensory information into the brain. Its role is more complicated than that, though. It controls the rapid reflexive response

TABLE 3.2 Major Structures of the Brain and Their Functions.

Structure	Major Function
Forebrain	
Frontal lobes	
Motor cortex	Plans and executes voluntary movements
Basal ganglia	Smooths movement generated by motor cortex
Broca's area	Controls speech, adds grammar
Prefrontal cortex	Involved in planning, impulse control
Parietal lobes	
Somatosensory cortex	Projection area for body senses
Association area	Location of body and objects in space
Temporal lobes	
Auditory cortex	Projection area for auditory information
Wernicke's area	Language area involved with meaning
Inferior temporal cortex	Visual identification of objects
Occipital lobes	
Primary visual cortex	Projection area for visual information
Visual association cortex	Processes components of visual information
Corpus callosum	Communication between the hemispheres
Ventricles	Contain cerebrospinal fluid
Thalamus	Relays sensory information to cortex
Hypothalamus	Coordinates emotional and motivational functions
Midbrain	
Superior colliculi	Role in vision—for example, eye movements
Inferior colliculi	Role in audition, such as sound location
Pineal gland	Controls daily and seasonal rhythms
Substantia nigra	Integrates movement
Ventral tegmental area	Contributes to rewarding effects of food, sex, and drugs
Hindbrain	
Medulla	Reflexively controls life processes
Pons	Contains centers related to sleep and arousal
Reticular formation	Involved with sleep, arousal, attention, some motor functions
Cerebellum	Controls speed, intensity, direction of movements

when you withdraw your hand from a hot stove, and it contains pattern generators that help control routine behaviors such as walking. Notice the appearance of the interior of the spinal cord in Figure 3.17; it is arranged just the opposite of the brain, with the white matter on the outside and the gray matter in the interior. The white exterior is made up of axons—ascending sensory tracts on their way to the brain and descending motor tracts on their way to the muscles and organs.

Sensory neurons enter the spinal cord through the *dorsal root* of each spinal nerve. The sensory neurons are unipolar; clustering of their cell bodies in the dorsal root ganglion explains the dorsal root's enlargement. The sensory neuron

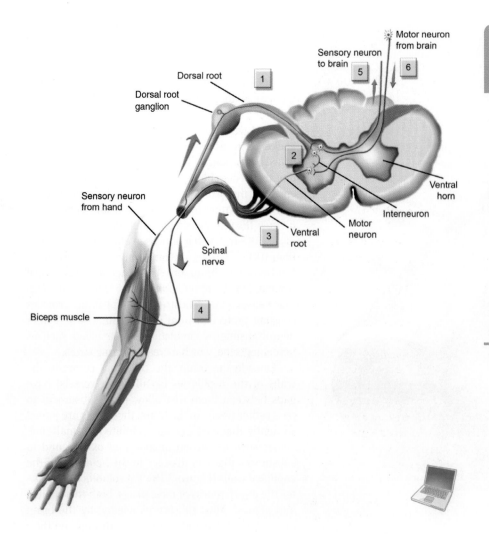

Motor neuron from brain

Sensory neuron to brain

Dorsal root

Dorsal root ganglion

1

5

6

2

Ventral horn

Interneuron

Sensory neuron from hand

Motor neuron

3

Ventral root

Spinal nerve

Biceps muscle

4

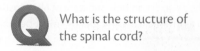

FIGURE 3.17 Horizontal Cross Section of the Spinal Cord, With Reflex Circuit.

A sensory neuron from the hand transmits signals (1) via the dorsal root of the spinal nerve into the spinal cord, where it (2) forms a reflex arc with a motor neuron that (3) exits through the ventral route and (4) activates the biceps muscle to flex the arm and withdraw the hand. The sensory input also travels (5) up to the brain to produce a sensation. (6) A motor neuron from the brain connects to the motor neuron in the ventral horn; this adds a voluntary activation of the muscle, though more slowly. (In reality, many neurons would be involved.)

Q What is the structure of the spinal cord?

in the illustration could be as much as 1 meter (m) long, with its other end out in a fingertip or a toe. The H-shaped structure in the middle of the spinal cord is made up mostly of unmyelinated cell bodies. The cell bodies of motor neurons are located in the *ventral horns*, which is why the ventral horns are enlarged. The axons of the motor neurons pass out of the spinal cord through the *ventral root*. The dorsal root and the ventral root on the same side of the cord join to form a spinal nerve that exits the spine between adjacent vertebrae (the bones that make up the spine).

Most of the motor neurons receive their input from the brain, either from the motor cortex or from nuclei that control the activity of the internal organs. Notice in the illustration, however, that in some cases sensory neurons from the dorsal side connect with motor neurons, either directly or through an interneuron. This pathway produces a simple, automatic movement in response to a sensory stimulus; this is called a *reflex*. For example, when you touch a lighted match with your hand, input travels to the spinal cord, where signals are directed out to the muscles of the arm to produce reflexive withdrawal. Many people use the term *reflex* incorrectly to refer to any action a person takes without apparent thought; however, the term is limited to behaviors that are controlled by these direct sensory-motor connections. Reflexes occur in the brain as well as in the spinal cord, and reflexes also affect the internal environment—for example, reducing blood pressure when it goes too high.

FIGURE 3.18 The Blood-Brain Barrier.

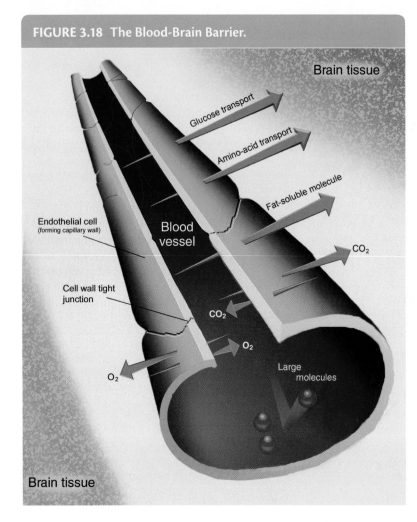

The tight junctions of the capillary walls prevent passage of large molecules into the brain. Small molecules like oxygen and carbon dioxide pass through freely, as do fat-soluble substances like most drugs. Water-soluble substances such as amino acids and glucose must be transported through.

Protecting the Central Nervous System

The brain and the spinal cord are delicate organs, vulnerable to damage from blows and jostling, to poisoning by toxins, and to disruption by mislocated or excessive neurotransmitters. Both structures are enclosed in a protective three-layered membrane called the *meninges.* The space between the meninges and the CNS is filled with cerebrospinal fluid, which cushions the neural tissue from the trauma of blows and sudden movement. The brain and spinal cord literally float in the cerebrospinal fluid, so the weight of a 1,200- to 1,400-g brain is in effect reduced to less than 100 g. The tough meninges and the cerebrospinal fluid afford the brain some protection from occasional trauma, but the *blood-brain barrier,* which limits passage between the bloodstream and the brain, provides constant protection from toxic substances and from neurotransmitters circulating in the blood such as norepinephrine, which increases during stress.

Outside the brain, the cells that compose the walls of the capillaries (small blood vessels) have gaps between them that allow most substances to pass rather freely. In the brain, these cells are joined so tightly that easy passage is limited to small molecules such as carbon dioxide and oxygen and to substances that can dissolve in the lipid (fat) of the capillary walls (Figure 3.18). Fat solubility accounts for the effectiveness of most drugs, both therapeutic and abused. Most substances needed by the brain are water soluble and cannot pass through on their own, so glucose, iron, amino acids (the building block that proteins are made of), and many vitamins must be actively carried through the walls by specialized transporters.

Not all brain areas are protected by the barrier, however. This is particularly true of brain structures surrounding the ventricles. One of them is the area postrema; when you ingest something toxic, such as an excess of alcohol, the substance passes from the bloodstream into the area postrema. Because the area postrema induces vomiting, your stomach empties quickly—hopefully before too much harm is done.

Concept Check   *Take a Minute to Check Your Knowledge and Understanding*

- ☐ What is the advantage of the convoluted structure of the cortex?
- ☐ What has been the fate of psychosurgery; what clue from past experience did doctors have that lobotomy in particular might have undesirable consequences?
- ☐ Select one of the lobes or the midbrain or the hindbrain and describe the structures and functions located there.
- ☐ Describe the pathway of a reflex, identifying the neurons and the parts of the spinal cord involved.

The Peripheral Nervous System

The *peripheral nervous system (PNS)* is made up of the *cranial nerves*, which enter and leave the underside of the brain, and the *spinal nerves*, which connect to the sides of the spinal cord at each vertebra. From a functional perspective, the PNS can be divided into the somatic nervous system and the autonomic nervous system. The *somatic nervous system* includes the motor neurons that operate the skeletal muscles—that is, the ones that move the body—and the sensory neurons that bring information into the CNS from the body and the outside world. The *autonomic nervous system (ANS)* controls smooth muscle (stomach, blood vessels, etc.), the glands, and the heart and other organs. The diagram in Figure 3.19 will help you keep track of these divisions and relate them to the CNS. We dealt with the spinal nerves when we discussed the spinal cord, and we have said all we need to for now about the somatic system, so we will give the rest of our attention to the cranial nerves and the ANS.

The Cranial Nerves

The cranial nerves enter and exit on the ventral side of the brain (Figure 3.20). While the spinal nerves are concerned exclusively with sensory and motor activities within the body, some of the cranial nerves convey sensory information to the brain from the outside world. Two of these, the olfactory nerves and the optic nerves, have the special status of often being considered part of the brain. One reason is the brainlike complexity of the olfactory bulb and of the retina at the back of the eye; another is that their receptor cells originate in the brain during development and migrate to their final locations. As a consequence, you will sometimes see the olfactory and optic nerves referred to as *tracts*.

The Autonomic Nervous System

The functions of the ANS are primarily motor; its sensory pathways provide internal information for regulating its own operations. The ANS is composed of two branches. The *sympathetic nervous system* activates the body in ways that help it cope with demands such as emotional stress and physical emergencies. Your most recent emergency may have been when you overslept on the morning of a big exam. As you raced to class, your heart and breathing sped up to provide your body the resources it needed. Your blood pressure increased as well, and your peripheral blood vessels constricted, shifting blood supply to the internal organs, including your brain. Your muscles tensed to help you fight or flee, and your sweat glands started pouring out sweat to cool your overheating body. All this activity was just the sympathetic nervous system at work. The *parasympathetic nervous system not only* slows the activity of most organs to conserve energy, but it also activates digestion to renew energy.

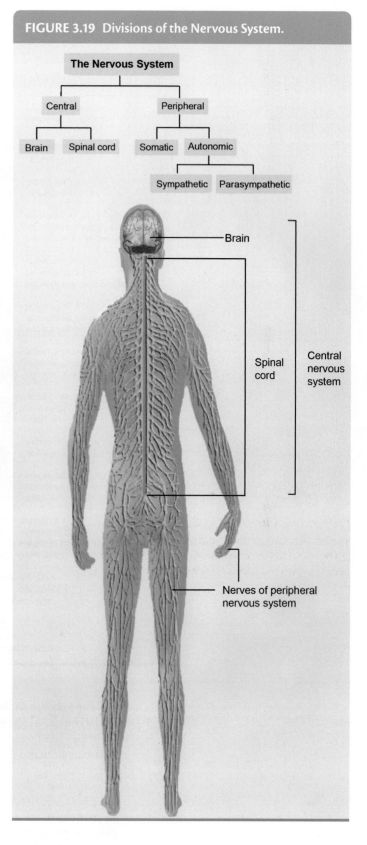

FIGURE 3.19 Divisions of the Nervous System.

 What are the functions of the autonomic nervous system?

FIGURE 3.20   Ventral View of the Brain Showing the Cranial Nerves and Their Major Functions.

Brain landmarks are labeled on the right to help you locate the nerves.

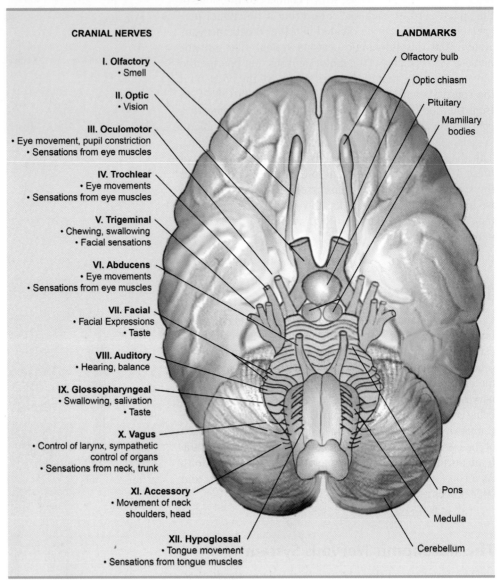

CRANIAL NERVES

I. Olfactory
• Smell

II. Optic
• Vision

III. Oculomotor
• Eye movement, pupil constriction
• Sensations from eye muscles

IV. Trochlear
• Eye movements
• Sensations from eye muscles

V. Trigeminal
• Chewing, swallowing
• Facial sensations

VI. Abducens
• Eye movements
• Sensations from eye muscles

VII. Facial
• Facial Expressions
• Taste

VIII. Auditory
• Hearing, balance

IX. Glossopharyngeal
• Swallowing, salivation
• Taste

X. Vagus
• Control of larynx, sympathetic control of organs
• Sensations from neck, trunk

XI. Accessory
• Movement of neck shoulders, head

XII. Hypoglossal
• Tongue movement
• Sensations from tongue muscles

LANDMARKS

Olfactory bulb

Optic chiasm

Pituitary

Mamillary bodies

Pons

Medulla

Cerebellum

The sympathetic branch rises from the middle (thoracic and lumbar) areas of the spinal cord (see Figure 3.21). Most sympathetic neurons pass through the *sympathetic ganglion chain*, which runs along each side of the spine; there they synapse with postsynaptic neurons that rejoin the spinal nerve and go out to the muscles or glands they serve. (The others pass directly to ganglia in the body cavity before synapsing.) Because most of the sympathetic ganglia are highly interconnected in the sympathetic ganglion chain, this system tends to respond as a unit. Thus, when you were rushing to your exam, your whole body went into hyperdrive. As you can see in the illustration, the parasympathetic branch rises from the extreme ends of the PNS—in the cranial nerves and in the spinal nerves at the lower (sacral) end of the spinal cord. The parasympathetic ganglia are not interconnected but are located on or near the muscles and glands they control; as a result, the components of the parasympathetic system operate more independently than those of the sympathetic system.

FIGURE 3.21 The Autonomic Nervous System.

A diagrammatic view of the parasympathetic and sympathetic nerves and their functions. The nerves exit both sides of the brain and spinal cord through the paired cranial and spinal nerves but are shown on one side for simplicity.

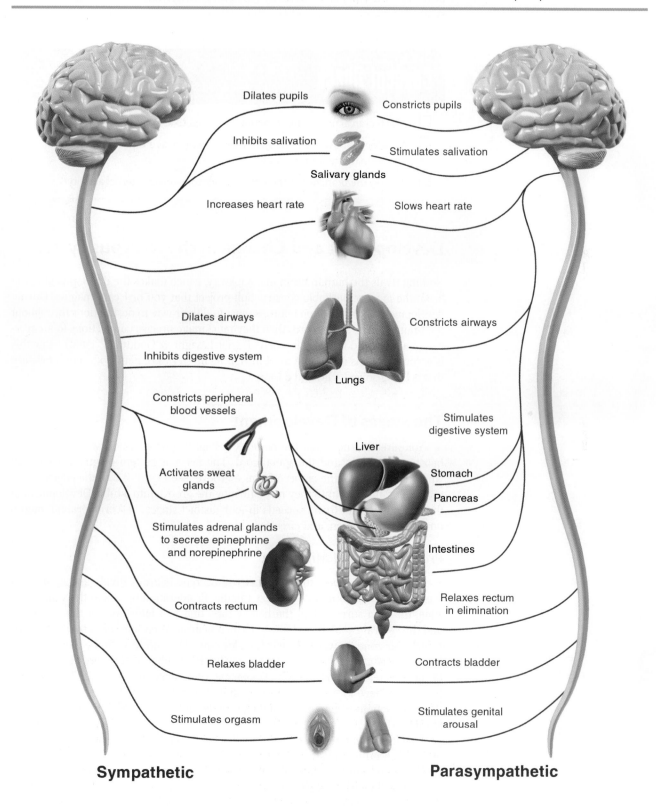

Dilates pupils

Constricts pupils

Inhibits salivation

Stimulates salivation

Salivary glands

Increases heart rate

Slows heart rate

Dilates airways

Constricts airways

Lungs

Inhibits digestive system

Constricts peripheral blood vessels

Stimulates digestive system

Liver

Activates sweat glands

Stomach

Pancreas

Stimulates adrenal glands to secrete epinephrine and norepinephrine

Intestines

Contracts rectum

Relaxes rectum in elimination

Relaxes bladder

Contracts bladder

Stimulates orgasm

Stimulates genital arousal

Sympathetic

Parasympathetic

Organs are innervated by both branches of the ANS, with the exception of the sweat glands, the adrenal glands, and the muscles that constrict blood vessels, which receive only sympathetic activation. It is not accurate to assume that one branch is active at a time and the other completely shuts down; rather, both are active to some degree all the time, and the body's general activity reflects the balance between sympathetic and parasympathetic stimulation.

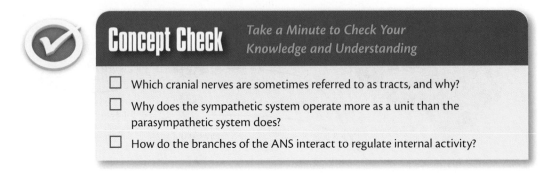

Concept Check   *Take a Minute to Check Your Knowledge and Understanding*

☐ Which cranial nerves are sometimes referred to as tracts, and why?

☐ Why does the sympathetic system operate more as a unit than the parasympathetic system does?

☐ How do the branches of the ANS interact to regulate internal activity?

Development and Change in the Nervous System

4
BrainWork News

Nothing rivals the human brain in complexity, which makes the development of the brain the most remarkable construction project that you or I can imagine. During development, its 100 billion neurons must find their way to destinations throughout the brain and the spinal cord; then they must make precise connections to an average of a thousand target cells each (Tessier-Lavigne & Goodman, 1996). How this is accomplished is one of the most intriguing mysteries of neurology, but a mystery that is being solved a little at a time.

The Stages of Development

You already know that the nervous system begins as a hollow tube that later becomes the brain and the spinal cord. The nervous system begins development when the surface of the embryo forms a groove (see Figure 3.22); the edges of this groove curl upward until they meet, turning the groove into a tube. Development of the nervous system then proceeds in four distinct stages: cell proliferation, migration, circuit formation, and circuit pruning.

Proliferation and Migration

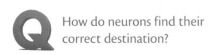

How do neurons find their correct destination?

During *proliferation*, the cells that will become neurons divide and multiply at the rate of 250,000 new cells every minute. Proliferation occurs in the ventricular zone, the area surrounding the hollow tube that will later become the ventricles and the central canal. These newly formed neurons then *migrate*, moving from the ventricular zone outward to their final location. They do so with the aid of specialized *radial glial cells* (Figure 3.23 on page 76). Remember Karen from the beginning of the chapter? This is where development went awry in her brain. The neurons that would have formed her cortex migrated only halfway; apparently, the role of the malfunctioning gene is to tell the neurons where to get off their radial glial cell scaffolds (J. W. Fox et al., 1998).

The functional role that a neuron will play depends on its location and the time of its "birth"; different structures form during different stages of fetal development. Prior to birth and for a time afterward, the neurons retain considerable functional flexibility, however. In fact, fetal brain tissue can be transplanted into a different part of an adult brain, and the transplanted neurons will form synapses and assume the function of their new location.

FIGURE 3.22 Development of the Neural Tube.

(a) Photographs of neural tube development as the embryo's surface forms a groove, which closes to form a tube. (b) Diagrammatic representation of the events. The photographs show an end view of the developing tube, while the drawings are views from above.

(a)   (b)

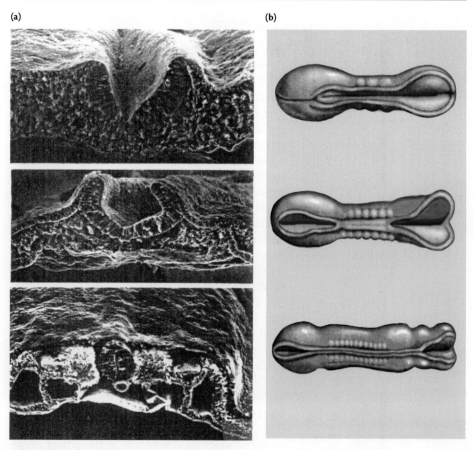

SOURCE: Photos by Kathryn Tosney.

Circuit Formation

During *circuit formation*, the axons of developing neurons grow toward their target cells and form functional connections. For example, axons of motor neurons grow toward the spinal cord, and cells in the retina of the eye send their axons to the thalamus, where they form synapses with other neurons. To find their way, axons form *growth cones* at their tip, which sample the environment for directional cues (Figure 3.24 on page 77). Chemical and molecular signposts attract or repel the advancing axon, coaxing it along the way (Tessier-Lavigne & Goodman, 1996). By pushing, pulling, and hemming neurons in from the side, the chemical and molecular forces guide the neuron to intermediate stations and past inappropriate targets until they reach their final destinations.

The path to the developing axon's destination is not necessarily direct, but thanks to changing genetic control, it is able to make direction changes along the way. This is illustrated by an axon whose destination is on the opposite side of the midline. Under the control of the gene *Robo1*, the axon is repelled by a midline chemical, and so it grows parallel to the midline. But at the appropriate location, the gene *Robo3* becomes active; the axon is attracted to the midline and turns and enters it. At that point, *Robo3* is downregulated; the axon is repelled again and, continuing in the

FIGURE 3.23 A Neuron Migrates Along Glial Scaffolding.

(a) Immature neurons migrate from the inner layer, where they were "born," to their destination between there and the outer layer. (b) A close-up of one of the neurons climbing a radial glial cell scaffold.

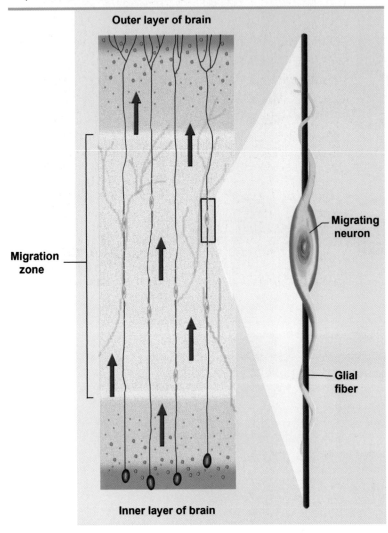

Outer layer of brain

Migration zone

Migrating neuron

Glial fiber

Inner layer of brain

SOURCE: Adapted from illustration by Lydia Kibiuk, © 1995.

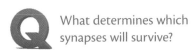

Q What determines which synapses will survive?

same direction, exits the midline and will not recross (C. G. Woods, 2004).

Circuit Pruning

The brain produces extra neurons, apparently as a means of compensating for the errors that occur in reaching targets. This overproduction is not trivial: The monkey's visual cortex contains 35% more neurons at the time of birth than in adulthood, and the number of axons crossing the corpus callosum is four times what it will be later in life (LaMantia & Rakic, 1990; R. W. Williams, Ryder, & Rakic, 1987). The next stage of neural development, *circuit pruning*, involves the elimination of excess neurons and synapses. Neurons that are unsuccessful in finding a place on a target cell, or that arrive late, die; the monkey's corpus callosum alone loses 8 million neurons a day during the first 3 weeks after birth.

In a second step of circuit pruning, the nervous system refines its organization and continues to correct errors by eliminating large numbers of excessive synapses. For example, in mature mammals, neurons from the left and right eyes project to alternating columns of cells in the visual cortex, but the connections made during development are indiscriminate. Synapses are strengthened or weakened depending on whether the presynaptic neuron and the postsynaptic neuron fire together. Because a single neuron cannot by itself cause another neuron to fire, this is likely to happen when neighboring neurons are also firing and adding summating inputs through overlapping terminals. If a neuron is *not* firing at the same time as its neighbors, it has probably made its connection in the wrong neighborhood. It is thought that the postsynaptic neuron sends feedback to the presynaptic terminals in the form of *neurotrophins*, chemicals that enhance the development and survival of neurons.

In the visual system, sensory stimulation provides neuronal activation that contributes to this refinement. However, pruning of synapses begins in some parts of the visual system even before birth. How can this stimulation occur when visual input is impossible? The answer is that waves of spontaneous neural firing sweep across the fetal retina, providing the activation that selects which synapses will survive and which will not (Hooks & Chen, 2007; Huberman, 2007). In the first few years of the rhesus monkey's life, 40% of the synapses in the primary visual cortex are eliminated, at the stunning rate of 5,000 per second (Bourgeois & Rakic, 1993). This process of producing synapses that will later be eliminated seems wasteful, but targeting neurons' destinations more precisely would require prohibitively complex chemical and molecular codes. Later, the *plasticity* (ability to be modified) of these synapses decreases; a practical example is that recovery from injury to the language areas of the brain is greatly reduced in adulthood. However, the synapses in the cortical association areas are more likely to retain their plasticity, permitting

FIGURE 3.24 Neurons With Growth Cones.

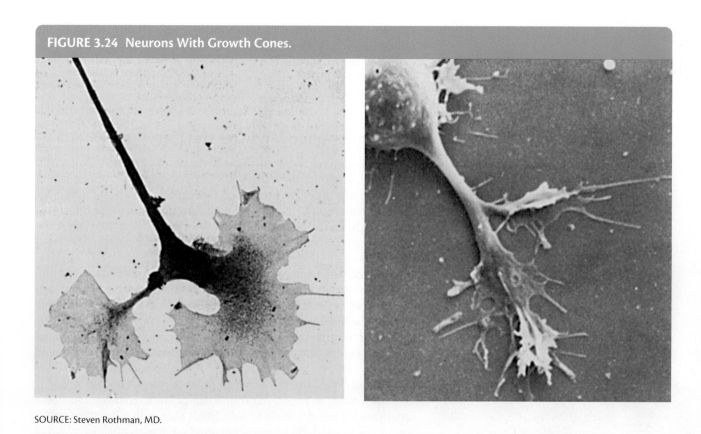

SOURCE: Steven Rothman, MD.

later modification by experience—in other words, learning (Kandel & O'Dell, 1992; W. Singer, 1995).

As impressive as the brain's ability to organize itself during development is, mistakes do occur, and for a variety of reasons. Periventricular heterotopia, the problem Karen had at the beginning of the chapter, is caused by a mutation of a gene on the X chromosome that is believed to influence the migration of neurons. *Fetal alcohol syndrome*, which often produces intellectual disability, is caused by the mother's use of alcohol during a critical period of brain development. Fetal alcohol syndrome brains are often small and malformed, and neurons are dislocated (Figure 3.25). During migration, many cortical neurons fail to line up in columns as they normally would because the radial glial cells revert to their more typical glial form prematurely; other neurons continue migrating beyond the usual boundary of the cortex (Clarren, Alvord, Sumi, Streissguth, & Smith, 1978; Gressens, Lammens, Picard, & Evrard, 1992; P. D. Lewis, 1985). Exposure to ionizing radiation, such as that produced by nuclear accidents and atomic blasts, also causes intellectual impairment by interfering with both proliferation and migration. The offspring of women who were in the 8th through 15th weeks of pregnancy during the bombing of Hiroshima and Nagasaki and during the nuclear meltdown at Chernobyl were the most vulnerable, because the rates of proliferation and migration are highest then (Schull, Norton, & Jensh, 1990).

5
Fetal Alcohol Syndrome

An additional step is required for full maturation of the nervous system—myelination. In the brain, it begins with the lower structures and then proceeds to the cerebral hemispheres, moving from occipital lobes to frontal lobes. Myelination starts around the end of the third trimester of fetal development but is not complete until late adolescence or beyond (Sowell, Thompson, Holmes, Jernigan, & Toga, 1999). This slow process has behavioral implications—for instance, contributing to

FIGURE 3.25 Fetal Alcolohol Syndrome in the Mouse Brain.

(a) In this cross section of the normal cortex, the neurons (the dark spots) tend to line up in vertical columns. (b) In the alcohol-exposed brain, the neurons are arranged randomly.

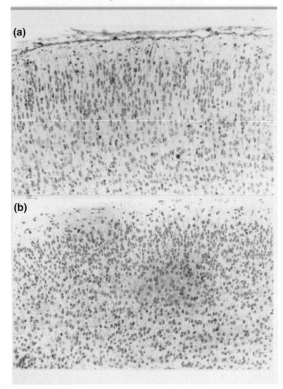

(a)

(b)

SOURCE: From "Ethanol Induced Disturbances of Gliogenesis and Neurogenesis in the Developing Murine Brain: An *in vitro* and an *in vivo* Immunohistochemical and Ultrastructural Study," by P. Gressens, M. Lammans, J. J. Picard, and P. Evrard, *Alcohol and Alcoholism, 27,* 219–226. © 1992. Used by permission of Oxford University Press.

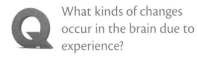

What kinds of changes occur in the brain due to experience?

the improvement through adolescence on cognitive tasks that require the frontal lobes (H. S. Levin et al., 1991). Considering the functions of the prefrontal cortex and the fact that this area is the last to mature (Sowell et al., 1999), it should come as no surprise that parents are baffled by their adolescents' behavior.

How Experience Modifies the Nervous System

Stimulation continues to shape synaptic construction and reconstruction throughout the individual's life. For example, training rats to find their way through a maze or just exposing them to a complex living environment increases the branching of synapses in the cortex (Greenough, 1975). Humans develop more synapses as they age, even while losing neurons (Buell & Coleman, 1979), presumably as the result of experience.

Much of the change resulting from experience in the mature brain involves *reorganization,* a shift in connections that changes the function of an area of the brain. For example, in blind people who read Braille, the space in the brain devoted to the index (reading) finger increases, at the expense of the area corresponding to the other fingers on the same hand (Pascual-Leone & Torres, 1993). In a brain scan study, it was discovered that blind individuals who excel at sound localization had recruited the unused visual area of their brains to aid in the task (Gougoux, Zatorre, Lassonde, Voss, & Lepore, 2005). Some of these changes can occur rapidly, as we see in a study of individuals born with a condition called *syndactyly,* in which their fingers are attached to each other by a web of skin. Use of the fingers is severely limited, and the fingers are represented by overlapping areas in the somatosensory cortex. Figure 3.26 shows that after surgery, the representations of the fingers in the cortex became separate and distinct in just 7 days (Mogilner et al., 1993).

The 19th-century philosopher and psychologist William James speculated that if a surgeon could switch your optic nerves with your auditory nerves, you would then see thunder and hear lightning (James, 1893). James was expressing Johannes Müller's *doctrine of specific nerve energies* from a half century earlier—that each sensory projection area produces its own unique experience regardless of the kind of stimulation it receives. This is why you "see stars" when your rollerblades shoot out from under you and the back of your head (where the visual cortex is located) hits the pavement.

But even this basic principle of brain operation can fall victim to reorganization during early development. In people blind from birth, the visual cortex has nothing to do; as a result, some of the somatosensory pathways take over part of the area, so the visual cortex is activated by touch. But does the visual cortex still produce a visual experience, or one of touch? To find out, researchers stimulated the visual cortex of blind individuals by applying an electromagnetic field to the scalp over the occipital area (L. G. Cohen et al., 1997). In sighted people, this disrupts visual performance, but in the blind individuals, the procedure distorted their sense of touch and interfered with their ability to identify Braille letters. Apparently, their visual area was actually processing information about touch in a meaningful way!

Reorganization does not always produce a beneficial outcome. When kittens were reared in an environment with no visual stimulation except horizontal stripes or vertical stripes, they lost their ability to respond to objects in the other orientation. A cat reared, for example, with vertical stripes would play with a rod held vertically and ignore the rod when it was horizontal. Electrical recording

FIGURE 3.26 Changes in the Somatosensory Area Following Surgery for Syndactyly.

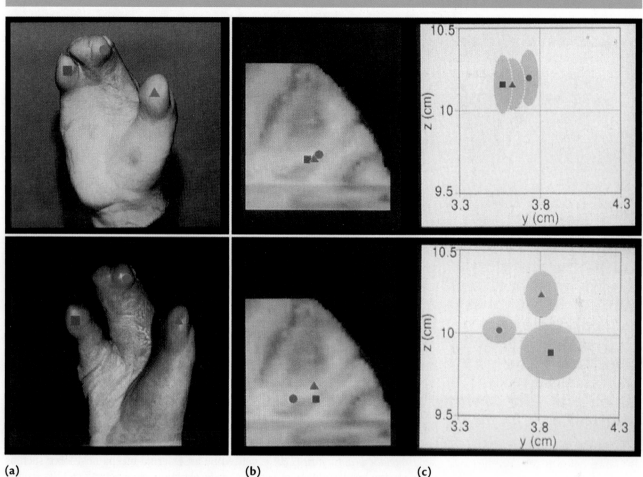

(a)   (b)   (c)

(a) The hand before (top) and after (bottom) surgery. (b) Images (coronal) showing brain areas responsive to stimulation of the fingers before and after surgery. (c) Graphic representation of the relative size and location of the responsive areas.

SOURCE: From "Somatosensory Cortical Plasticity in Adult Humans Revealed By Magnetoencephalography," by A. Mogilner et al., 1993, *Proceedings of the National Academy of Science, 90*, pp. 3593–3597.

indicated that the cells in the visual cortex that would have responded to other orientations had reorganized their connections in response to the limited stimulation. In Chapter 11, you will see that people who have a limb amputated often experience *phantom pain*, pain that seems to be located in the missing limb. It appears to be caused by sensory neurons from a nearby part of the body growing into the somatosensory area that had served the lost limb (Flor et al., 1995).

Damage and Recovery in the Central Nervous System

One reason neuroscientists are interested in the development of the nervous system is because they hope to find clues about how to repair the nervous system when it is damaged by injury, disease, or developmental error. It is difficult to convey the impairment and suffering that results from brain disorders, but the staggering financial costs in Table 3.3 will give you some idea. We will focus mostly on stroke and trauma and leave the other sources of injury for later chapters.

> In the adult centres the nerve paths are something fixed, ended and immutable.
>
> Everything may die, nothing may be regenerated. It is for the science of the future to change, if possible, this harsh decree.
>
> —Santiago Ramón y Cajal, 1928

TABLE 3.3 Annual Costs of Brain Damage and Disorders in the United States.

Damage or Disorder	Cost (in billion dollars)
Psychiatric disorders (schizophrenia, depression, anxiety disorders)	192.85
Head and spinal cord injuries	94.91
Stroke	27.03
Alzheimer's disease	170.86
Addictions	544.11
Total	1,029.26

SOURCE: Uhl and Grow (2004).

NOTE: Includes direct costs of care and treatment and indirect costs such as crime, lost wages, and so on.

Stroke, also known as cerebrovascular accident, is caused by a loss of blood flow in the brain. Most strokes are ischemic, caused by blockage of an artery by a blood clot or other obstruction; hemorrhagic strokes occur when an artery ruptures. The neurons are deprived of oxygen and glucose, of course, but most of the damage is due to excitotosis; dying neurons release excess glutamate, which overstimulates the surrounding neurons, which then die as large amounts of calcium enter the cells. Further impairment is caused by edema, an accumulation of fluid that causes increased pressure on the brain. Stroke is the third leading cause of death in the United States and a leading cause of long-term disability, including paralysis and loss of language and other functions (Heron et al., 2009).

Traumatic brain injury (TBI) is caused by an external mechanical force such as a blow to the head, sudden acceleration or deceleration, or penetration. TBIs are the cause of 52,000 deaths each year in the United States; about 35% of TBIs are caused by falls, and another 17% result from traffic accidents (Faul, Xu, Wald, & Coronado, 2010). Besides the direct damage to neurons, edema and ischemia (due to blood clots) take an additional toll. Mild traumatic injury, or concussion as it is more commonly known, is the most common TBI; it results from blows and acceleration-deceleration that occur in automobile accidents, sports activities, and battlefield explosions. Whether or not these traumas are sufficient to cause loss of consciousness, they are often followed by headache, drowsiness, and memory loss, which usually go away if the individual rests during the following 3 weeks. However, repeated concussions can cause cumulative brain damage; the expression "punch-drunk" actually refers to dementia pugilistica, the impairment suffered by boxers who didn't know when to quit. We're now finding that concussions in less violent sports can also lead to serious consequences, a concern that is very much in the news.

Limitations on Recovery

Nervous system repair is no problem for some species, particularly amphibians. For example, when Sperry (1943, 1945) severed the optic nerves of frogs, the eyes made functional reconnections to the brain even when the disconnected eye was turned upside down or transplanted into the other eye socket. *Regeneration,* the growth of severed axons, also occurs in mammals, at least in the PNS. So when you fell rollerblading, if you broke your arm so badly that a nerve was severed, the disconnected part of the cut axon would have died, but the part connected to

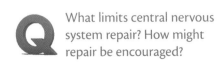

What limits central nervous system repair? How might repair be encouraged?

Is the Brain Too Fragile for Sports?

On December 17, 2009, Bengals wide receiver Chris Henry died of massive head injuries after jumping or falling from the back of a pickup truck driven by his fiancée during a domestic dispute. Over the preceding 2 years, Henry had been arrested five times for assault, driving under the influence of alcohol, and possession of marijuana, and the league had suspended him several times for violating its personal conduct policy. Examination of his brain revealed that he had chronic traumatic encephalopathy, which is caused by repeated impact; though only 26 years old, his brain had protein deposits typically seen in Alzheimer's patients and often accompanied by memory loss, depression, and erratic behavior.

In the National Football League, repeat concussions are not unusual; professional football players sometimes rack up so many they lose count. A survey commissioned by the league found that memory problems are 5 times more frequent in former players 50 and older and 19 times as likely in those aged 30 to 49. Out of 12 deceased NFL players' brains examined so far, only 1 was free of pathology. The concussions experienced in high school sports are no more benign: In a recent study, postconcussion symptoms actually worsened over the next 4 weeks in 80% of players, and one 18-year-old who died had brain changes like those seen in the professionals.

The NFL, facing potential liabilities of $5.5 million from claims made by 700 players, has formed a commission to study the issue, and the U.S. Senate is conducting hearings on concussions in high school sports. The NFL now requires that players showing signs of concussion be removed from the game or practice and be returned only after being cleared by an independent expert; several states are considering laws that would institute the same requirements in high school sports. Chris Henry never had a diagnosed concussion, however, so these changes may afford too little protection for the most vulnerable organ in the body.

SOURCES: Carrera, K. (2010, May 21). In hearing before Education and Labor Committee, high school athletes' concussions take center stage. *The Washington Post.* Retrieved June 28, 2010, from www.washingtonpost.com/wp-dyn/content/article/2010/05/20/AR2010052005129.html

Miller, G. (2009). A late hit for pro football players. *Science, 325,* 670–672.

Schwarz, A. (2010, April 5). Case will test N.F.L. teams' liability in dementia. *The New York Times.* Retrieved April 6, 2010, from www.nytimes.com/2010/04/06/sports/football/06worker.html

Schwarz, A. (2010, June 28). Former Bengal Henry found to have had brain damage. *The New York Times.* Retrieved June 30, 2010, from http://www.nytimes.com/2010/06/29/sports/football/29henry.html

---

the cell body would have survived and regrown. Myelin provides a guide tube for the sprouting end of a severed neuron to grow through (W. J. Freed, de Medinaceli, & Wyatt, 1985), and the extending axon is guided to its destination much as it would be during development (Horner & Gage, 2000).

But in the mammalian CNS, damaged neurons encounter a hostile environment. If your rollerblading accident severed neurons in your spinal cord, the axon stumps would sprout new growth, but they would make little progress toward their former target. This is partly because the CNS in adult mammals no longer produces the chemical and molecular conditions that stimulate and guide neuronal growth. In addition, scar tissue produced by glial cells blocks the original pathway, glial cells also produce axon growth inhibitors, and immune cells move into the area and possibly interfere with regrowth (D. F. Chen, Schneider, Martinou, & Tonegawa, 1997; Horner & Gage, 2000; Thallmair et al., 1998).

Another way the nervous system could repair itself is by *neurogenesis,* the birth of new neurons. While neurogenesis has been detected in several areas of the brain (C. D. Fowler, Liu, & Wang, 2007), it appears to be most extensive in two areas; one is the hippocampus, and the other is near the lateral ventricles, supplying the olfactory bulb (Gage, 2000). Interfering with neuron replacement impairs odor discrimination (Gheusi et al., 2000) and types of learning that require the hippocampus (Shors et al., 2001; J. S. Snyder, Kee, & Wojtowicz, 2001), so neuron replacement apparently supports the normal functions of these structures. There is no guarantee that this neurogenesis contributes to brain repair following injury. On the other hand, neural precursor cells migrate to damaged areas in rats' brains following experimentally induced stroke and appear to replace damaged neurons (Parent, 2003). Furthermore, increased neurogenesis has been observed at damage sites in the brains of deceased Alzheimer's patients and Huntington's disease patients (Curtis et al., 2003; Jin et al., 2004). Results such as these suggest that if we could enhance neurogenesis, it might provide a means of self-repair. In the meantime, some researchers are experimenting with artificial prostheses for the brain.

Compensation and Reorganization

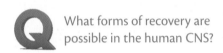 What forms of recovery are possible in the human CNS?

Although axons do not regenerate and neuron replacement is limited at best, considerable recovery of function can occur in the damaged mammalian CNS. Much of the improvement in function following injury is nonneural in nature and comes about as swelling diminishes and glia remove dead neurons (Bach-y-Rita, 1990). The simplest recovery that is neural in nature involves *compensation* as uninjured tissue takes over the functions of lost neurons. Presynaptic neurons sprout more terminals to form additional synapses with their targets (Fritschy & Grzanna, 1992; Goodman, Bogdasarian, & Horel, 1973), and postsynaptic neurons add more receptors (Bach-y-Rita, 1990). In addition, normally silent side branches from other neurons in the area become active within minutes of the injury (Das & Gilbert, 1995; Gilbert, 1993). These synaptic changes are similar to those occurring during learning; this would explain why physical therapy can be effective in promoting recovery after brain injury.

A more dramatic form of neural recovery involves reorganization of other brain areas. Reorganization is particularly seen in recovery from language impairment (*aphasia*) that results from brain damage or surgery. In most cases, the function is apparently assumed by nearby brain areas, but there are documented cases where the right hemisphere has taken over language functioning after massive damage to the left hemisphere (Guerreiro, Castro-Caldas, & Martins, 1995). Occasionally, an entire hemisphere must be removed because it is diseased. The patients typically do not reach normal levels of performance after the surgery, but they often recover their language and other cognitive skills and motor control to a remarkable degree (Glees, 1980; Ogden, 1989). In these cases, malfunction in the removed hemisphere dated back to infancy, so presumably the reorganization began then rather than at the time of surgery in late adolescence or early adulthood.

Recovery from aphasia and periventricular heterotopia challenges our understanding of how the brain works. Hydrocephalus provides another such example. *Hydrocephalus* occurs when the circulation of cerebrospinal fluid is blocked and the accumulating fluid interferes with the brain's growth, producing severe intellectual impairment. If detected in time, the condition can be treated by installing a drain that shunts the excess fluid into the bloodstream. However, the occasional individual somehow avoids impairment without this treatment. The British neurologist John Lorber described a 26-year-old hydrocephalic whose cerebral walls (between his ventricles and the outer surface of his brain) were less than 1 mm thick, compared with the usual 45 mm. Yet he had a superior IQ of 126, had earned an honors degree in mathematics, and was socially normal (Lewin, 1980). It is unclear how these individuals can function normally in the face of such enormous brain deficits. What is clear is that somewhere in this remarkable plasticity lies the key to new revelations about brain function.

Possibilities for CNS Repair

The word impossible *is not in the vocabulary of contemporary neuroscience.*

—*Pasko Rakic*

In 1995, Christopher Reeve, the movie actor best known for his role as Superman, was paralyzed from the neck down when he was thrown from his horse during a competition (Figure 3.27). Three quarters of his spinal cord was destroyed at the level of the injury (J. W. McDonald et al., 2002). He had no motor control and almost no sensation below the neck; like 90% of similarly injured patients, he experienced no functional improvement over the next several years.

In spite of Ramón y Cajal's declaration that there is no regeneration in the CNS, scientists nearly a century later are pursuing several strategies for inducing self-repair following damage like Reeve's. These efforts include using neuron growth enhancers, counteracting the forces that inhibit regrowth, and providing guide tubes or scaffolding for axons to follow (Bonner, 2005; Horner & Gage, 2000). For example, researchers induced stroke in one hemisphere in rats, then injected a neuron growth enhancer; after 3 weeks, motor and somatosensory axons had grown from the undamaged hemisphere into the midbrain and spinal cord pathways on the damaged side, and the rats had regained most of their mobility (P. Chen, Goldberg, Kolb, Lanser, & Benowitz, 2002). Similarly, rats regrew axons and regained use of their legs when glial cells from the olfactory area were provided as scaffolding in the 3- to 4-mm gap between the cut ends of their spinal cords (J. Lu, Féron, Mackay-Sim, & Waite, 2002). However, a 3-year clinical trial with human spinal cord patients resulted in almost no improvement in function (Mackay-Sim et al., 2008). Another approach involves blocking the effect of the *Nogo* gene protein Nogo-A, which inhibits axon growth following injury. In monkeys with surgically induced spinal cord damage, a Nogo-A inhibitor produced axon growth across the injured area, and the monkeys recovered 80% of the use of their paralyzed hands (Freund et al., 2006; Freund et al., 2007). The drug company Novartis is now recruiting patients for clinical trials with humans.

The most exciting research uses stem cells to replace injured neurons. *Stem cells* are undifferentiated cells that can develop into specialized cells such as neurons, muscle, or blood. Stem cells in the embryo (Figure 3.28) are *pluripotent,* which means that they can differentiate into any cell in the body; the developing cell's fate is determined by chemical signals from its environment that turn on specific genes and silence others. Later in life, stem cells lose most of their flexibility and are confined to areas with a high demand for cell replacement, such as the skin, the intestine, and bone marrow (the source of blood cells). In the brain, stem cells are responsible for the neurogenesis that provides a continuous supply of neurons in the hippocampus and olfactory bulb. Placing embryonic stem cells into an adult nervous system encourages them to differentiate into neurons appropriate to that area. The olfactory mucosa is a source not only for the olfactory glial cells mentioned above but stem cells as well. Researchers in Portugal implanted tissue from patients' own olfactory mucosa into the damaged spinal cord (Lima et al, 2006). Although the injured area ranged from 1 to 6 cm in length and some of the patients were treated as long as 6.5 years after their injury, all showed improvement. Restored functions experienced by some of the patients included ability to transfer to and from a wheelchair, ability to step with assistance, and return of bowel control.

FIGURE 3.27 Christopher Reeve (1952–2004).

FIGURE 3.28 Embryonic Stem Cells.

Because they can develop into any type of cell, stem cells offer tremendous therapeutic possibilities.

6
Curing Paralysis

Application   Mending the Brain With Computer Chips

Faced with the daunting challenge of coaxing axons to regrow and stem cells to take over the duties of brain cells, some researchers are turning to computer chips, and their results have been encouraging.

For example, Eberhard Fetz and his colleagues used a local anesthetic to block the nerves controlling monkeys' right hand and wrist, to produce temporary paralysis simulating spinal cord injury (Moritz, Perlmutter, & Fetz, 2008). Then they used a computer chip to process and amplify the signals from a dozen electrodes inserted into the hand and wrist areas of the motor cortex; these were then routed to the hand and wrist muscles, effectively bypassing the "damaged" area (see Figure a). The monkeys quickly learned to flex, extend, and rotate the wrist; the wrist movements controlled a cursor on a computer screen, which the monkey kept within a target in order to receive rewards of applesauce. The monkeys were just as capable when an electrode was recording from a cell not normally involved in wrist movements, which indicates the high level of functional plasticity in the brain.

This approach cannot produce entirely normal movement, in part because the spinal cord contains central pattern generators that produce rhythmic movements required for complex actions such as walking. To meet this need, a U.S.–Canadian team has designed a central pattern generator on a computer chip; its artificial neural network of just 10 simulated neurons and 190 synapses permitted cats whose hind legs had been surgically paralyzed to walk again (Vogelstein, Tenore, Guevremont, Etienne-Cummings, & Mushahwar, 2008).

It will be a long time before these research efforts benefit human patients. However, the ability to control external devices with thought appears simpler and may not be far down the road. After implantation of an array of 96 electrodes in the arm area of his motor cortex, a 25-year-old paraplegic (both arms and legs paralyzed) was able to move a cursor on a computer screen to read emails and to play a simple computer game; when connected to a robotic arm, he was able pick up an object and move it to another location (see Figure b) (Hochberg et al., 2006). The BrainGate device is now entering clinical trials with volunteers to determine safety and feasibility ("Brain-Computer Interface," 2009).

7
BrainGate Video

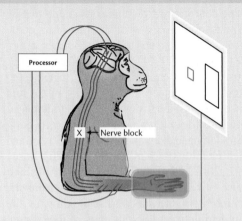

(a) A Monkey Controls Its Paralyzed Wrist With Output From Its Motor Cortex.

SOURCE: Adapted from Figure 1a of "Direct Control of Paralysed Muscles by Cortical Neurons," by C. T. Moritz, S. I. Permutter, and E. E. Fetz, *Nature*, 456, p. 639.

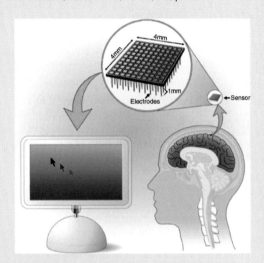

(b) The BrainGate Chip Allows the Paralyzed Indvidual to Control External Devices.

SOURCE: Brown Institute for Brain Science.

Five years after his injury, Christopher Reeve undertook an intensive rehabilitation program called activity-based recovery (J. W. McDonald et al., 2002). The therapy involved using electrical stimulation to exercise critical muscle groups. For example, electrodes placed over three muscles of each leg were activated sequentially to allow him to pedal a customized exercise bike. This effort appeared futile; no spinal cord patient classified as Grade A (the category of greatest impairment) had ever recovered more than one grade after 2 years. But after 3 years of this therapy, two thirds of his touch sensation had returned, and he was able to walk in an aquatherapy pool and make swimming movements with his arms; as a result, he was reclassified to Grade C. It even seemed possible that Reeve might achieve his goal of walking again, but he died of heart failure in October 2004. The Christopher Reeve Paralysis Foundation continues his work toward lifting Ramón y Cajal's decree. But since that day appears far off, researchers in the meantime are merging neurons with electronics to give patients back control of their bodies (see the accompanying Application).

Concept Check *Take a Minute to Check Your Knowledge and Understanding*

- ☐ Describe the four steps of nervous system development and the fifth step of maturation.
- ☐ Give three examples of changes in the brain resulting from experience.
- ☐ What are the obstacles to recovery from injury in the CNS and the strategies for overcoming them?

In Perspective

I could end this chapter by talking about how much we know about the brain and its functions. Or I could tell you about how little we know. Either point of view would be correct; it is the classic case of whether the glass is half full or half empty. As I said in Chapter 1, we made remarkable progress during the past decade. We know the functions of most areas of the brain. We have a good idea how the brain develops and how neurons find their way to their destinations and make functional connections. And we're getting closer to understanding how the neurons form complex networks that carry out the brain's work.

But we do not know just how the brain combines activity from widespread areas to bring about an action or a decision or an experience. We don't know what a thought is. And we don't know how to fix a broken brain. But, of course, there is hope, and for good reason. You will see in the following chapters that our knowledge is vast and that we have a solid foundation for making greater advances in the current decade.

Summary

The Central Nervous System

- The CNS consists of the brain and the spinal cord.
- The CNS is arranged in a hierarchy, with physically higher structures carrying out more sophisticated functions.
- The cortex is the location of the most sophisticated functions; the convoluted structure of the cerebral hemispheres provides for the maximum amount of cortex.
- See Table 3.2 for the major structures of the brain and their functions.
- Although localization is an important functional principle in the brain, most functions depend on the interaction of several brain areas.
- The spinal cord contains pathways between the brain and the body below the head and provides for sensory-motor reflexes.
- The meninges and the cerebrospinal fluid protect the brain from trauma; the blood-brain barrier blocks toxins and blood-borne neurotransmitters from entering the brain.

The Peripheral Nervous System

- See Figure 3.19 for a summary of the divisions of the nervous system.
  - The PNS consists of the cranial and spinal nerves or, alternatively, the somatic nervous system and the ANS.
  - The somatic nervous system consists of the sensory nerves and the nerves controlling the skeletal muscles.
  - The sympathetic branch of the ANS prepares the body for action; the parasympathetic branch conserves and renews energy.
- Interconnection in the sympathetic ganglion chain means that the sympathetic nervous system tends to function as a whole, unlike the parasympathetic branch.

Development and Change in the Nervous System

- Prenatal development of the nervous system involves
  - *proliferation,* the multiplication of neurons by division;
  - *migration,* in which neurons travel to their destination;
  - *circuit formation,* the growth of axons to, and their connection to, their targets; and
  - *circuit pruning,* the elimination of excess neurons and incorrect synapses.
- Myelination continues through adolescence or later, with higher brain levels myelinating last.
- Experience can produce changes in brain structure and function.
- Although some recovery of function occurs in the mammalian CNS, there is little or no true repair of damage by either neurogenesis or regeneration; enhancing repair is a major research focus. ■

Study Resources

F   For Further Thought

- Patients with damage to the right parietal lobe, the temporal lobe, or the prefrontal cortex may have little or no impairment in their intellectual capabilities, yet they show deficits in behavior that seem inconsistent for an otherwise intelligent individual. Does this modify your ideas about how we govern our behavior?
- Like the heroes in the 1966 science fiction movie *Fantastic Voyage,* you and your crew will enter a small submarine, be shrunk to microscopic size, and injected into the carotid artery of an eminent scientist who is in a coma. Your mission is to navigate through the bloodstream to deliver a lifesaving drug to a specific area in the scientist's brain. The drug can be designed to your specifications, and you can decide where in the vascular system you will release it. What are some of the strategies you could consider to ensure that the drug will enter the brain?
- What strategy do you think has the greatest potential for restoring function in brain-damaged patients? Why?

T   Testing Your Understanding

1. Describe the specific behaviors you would expect to see in a person with prefrontal cortex damage.

2. Describe compensation and reorganization in recovery from brain damage, giving examples.

3. In what ways does the brain show plasticity after birth?

Select the best answer:

1. Groups of cell bodies in the CNS are called
   a. tracts.   b. ganglia.
   c. nerves.   d. nuclei.

2. The prefrontal cortex is involved in all but which one of the following functions?

a. Responding to rewards
b. Orienting the body in space
c. Making decisions
d. Behaving in socially appropriate ways

3. Because the speech center is usually located in the left hemisphere of the brain, a person with the corpus callosum severed is unable to describe stimuli that are
a. seen in the left visual field.
b. seen in the right visual field.
c. presented directly in front of him or her.
d. felt with the right hand.

4. A person with damage to the inferior temporal cortex would most likely be unable to
a. see.
b. remember previously seen objects.
c. recognize familiar objects visually.
d. solve visual problems, such as mazes.

5. A particular behavior is typically controlled by
a. a single structure.
b. one or two structures working together.
c. a network of structures.
d. the entire brain.

6. When the police have a drunk-driving suspect walk a straight line and touch his nose with his finger, they are assessing the effect of alcohol on the
a. motor cortex.   b. corpus callosum.
c. cerebellum.   d. medulla.

7. Cardiovascular activity and respiration are controlled by the
a. pons.   b. medulla.
c. thalamus.   d. reticular formation.

8. All the following are involved in producing movement, except the
a. hippocampus.   b. cerebellum.
c. frontal lobes.   d. basal ganglia.

9. Damage would be most devastating to humans if it destroyed the
a. pineal gland.   b. inferior colliculi.
c. corpus callosum.   d. medulla.

10. If the ventral root of a spinal nerve is severed, the person will experience
a. loss of sensory input from a part of the body.
b. loss of motor control of a part of the body.

c. loss of both sensory input and motor control.
d. none of the above.

11. During a difficult exam, your heart races, your mouth is dry, and your hands are icy. In your room after the exam is over, you fall limply into a deep sleep. Activation has shifted from primarily _____ to primarily _____.
a. somatic, autonomic
b. autonomic, somatic
c. parasympathetic, sympathetic
d. sympathetic, parasympathetic

12. In the circuit formation stage of nervous system development,
a. correct connection of each neuron is necessary, since barely enough neurons are produced.
b. axons grow to their targets and form connections.
c. neurons continue dividing around a central neuron, and those neurons form a circuit.
d. neurons that fail to make functional connections die.

13. Fetal alcohol syndrome involves
a. loss of myelin.
b. overproduction of neurons.
c. errors in neuron migration.
d. excessive growth of glial cells.

14. The study in which kittens reared in an environment with only horizontal or vertical lines were later able to respond only to stimuli at the same orientation is an example of apparent
a. compensation.   b. reorientation.
c. reorganization.   d. regeneration.

15. If a peripheral nerve were transplanted into a severed spinal cord, it would
a. fail to grow across the gap.
b. grow across the gap but fail to make connections.
c. grow across the gap and make connections but fail to function.
d. bridge the gap and replace the function of the lost neurons.

1. d. 2. b. 3. a. 4. c. 5. c. 6. c. 7. b. 8. a. 9. d. 10. b. 11. d. 12. b. 13. c. 14. c. 15. a.
Answers:

Online Resources

The following resources are available at www.sagepub.com/garrett3e.

Chapter Resources

- Flash Cards
- Chapter Quiz
- Internet Resources and Exercises

On the Web

You can find links to these websites by going to the web address above and selecting this chapter's **Chapter Resources**, then choosing **Internet Resources and Exercises.**

1. **The Whole Brain Atlas** has images of normal and diseased or damaged brains.
    The **HOPES Brain Tutorial** will help you visualize how the brain is organized. The final segment, Build a Brain, is best.

2. The **History of Psychosurgery**, from trephining (drilling holes in the skull to let evil spirits out) to lobotomy to more recent experimental attempts, is the subject of this sometimes less than professional but very interesting website. At Lobotomy's Hall of Fame, you will learn, for example, that sisters of the playwright Tennessee Williams and President John F. Kennedy both had lobotomies (the story that actress Frances Farmer had a lobotomy turned out to be a fabrication).

3. **How the Brain Works** provides an interactive tour of brain functions and the areas responsible.

4. The Dana Foundation's **BrainWork** offers regular updates on the latest findings in neuroscience research.

5. The **National Organization on Fetal Alcohol Syndrome** has information and statistics on the disorder.

6. The **Miami Project to Cure Paralysis** at the University of Miami School of Medicine has summaries of basic and clinical research on central nervous system damage.
    The **Christopher and Dana Reeve Foundation** site provides information about spinal cord damage research.
    The National Institutes of Health's **Stem Cell Information** site is a good resource for information about stem cells and their potential.

7. **BrainGate Lets Your Brain Control the Computer** is a video explanation of the BrainGate system, which shows the patient controlling a computer and a prosthetic hand.

Animations

- The Spinal Cord (Figure 3.17)

Chapter Updates and Biopsychology News

For Further Reading

1. In *The New Executive Brain* (Oxford University Press, 2009), Elkhonon Goldberg draws from recent discoveries and fascinating case studies to explore how the brain engages in complex decision making, deals with ambiguity, makes moral choices, and controls emotion.

2. *The Human Brain Book* by award-winning science writer Rita Carter (DK Publishing, 2009) ranges from brain anatomy and neural transmission to explorations of behavior, sensory processing, dreaming, and genius.

3. *The Scientific American Day in the Life of Your Brain* by Judith Horstman (Scientific American, 2009) in 130 articles tackles the brain bases of creativity, hunger, sex, addictions, dreaming, biological clocks, and more.

4. *The Man Who Mistook His Wife for a Hat and Other Clinical Tales* by Oliver Sacks (Harper Perennial, 1990), a collection of case studies, is as entertaining as it is informative, as it treats the human side of brain damage and disorder.

K Key Terms

anterior 59

association area 63

auditory cortex 63

autonomic nervous system (ANS) 71

blood-brain barrier 70

Broca's area 60

central nervous system (CNS) 54

central sulcus 58

cerebellum 67

cerebral hemispheres 55

cerebrospinal fluid 66

circuit formation 75

circuit pruning 76

compensation 82

corpus callosum 65

cortex 56

cranial nerves 71

dorsal 59

dorsal root 68

fetal alcohol syndrome 77

fissure 56

frontal lobe 58

ganglion 54

growth cone 75

gyrus 55

hydrocephalus 82

hypothalamus 64

inferior 59

inferior colliculi 66

inferior temporal cortex 63

lateral 59

lateral fissure 58

lobotomy 61

longitudinal fissure 55

medial 59

medulla 67

meninges 70

midbrain 66

migrate 74

motor cortex 60

neglect 63

nerve 54

neurogenesis 81

neurotrophins 76

nucleus 54

occipital lobe 64

parasympathetic nervous system 71

parietal lobe 62

peripheral nervous system (PNS) 71

pineal gland 64

plasticity 76

pons 67

posterior 59

precentral gyrus 60

prefrontal cortex 60

primary somatosensory cortex 62

proliferation 74

psychosurgery 61

radial glial cells 74

reflex 69

regeneration 80

reorganization 78

reticular formation 67

somatic nervous system 71

spinal cord 67

spinal nerves 71

stem cell 83

stroke 80

sulcus 55

superior 59

superior colliculi 66

sympathetic ganglion chain 72

sympathetic nervous system 71

temporal lobe 63

thalamus 64

tract 54

traumatic brain injury (TBI) 80

ventral 59

ventral root 69

ventricle 66

visual cortex 64

Wernicke's area 63

The Methods and Ethics of Research

4

Science, Research, and Theory

Theory and Tentativeness in Science

Experimental Versus Correlational Studies

   CONCEPT CHECK

Research Techniques

Staining and Imaging Neurons

Light and Electron Microscopy

Measuring and Manipulating Brain Activity

   APPLICATION: BRAIN IMPLANTS THAT MOVE

Brain Imaging Techniques

   APPLICATION: SCANNING KING TUT

Investigating Heredity

   CONCEPT CHECK

Research Ethics

Plagiarism and Fabrication

Protecting the Welfare of Research Participants

Gene Therapy

   IN THE NEWS: A BRAVE NEW WORLD?

Stem Cell Therapy

   CONCEPT CHECK

In Perspective

Summary

Study Resources

In this chapter you will learn

- The value of theory in science and the relative advantages of experiments and correlational studies

- Some of the ways biopsychologists do research

- Why research in biopsychology creates ethical concerns

shanthi DeSilva developed her first infection just 2 days after her birth. There would be many more. At the age of 2, her frequent illnesses and poor growth were diagnosed as due to *severe combined immunodeficiency* (SCID)—better known as the "bubble-boy disease" after an earlier victim who had to live in a sterile environment in a plastic tent (Figure 4.1). Her immune system was so compromised that she suffered from frequent infections and gained weight slowly; because traditional enzyme treatments were inadequate, at the age of 4 her parents enrolled her in a revolutionary experimental therapy. SCID is caused by a faulty gene, so the doctors transferred healthy genes into her immune cells (Blaese et al., 1995). Ashanthi is a grown woman now and living a normal life in suburban Cleveland (Springen, 2004), her health and normal resistance to disease a silent testimonial to the power of genetic research.

Just a few years ago, cures like Ashanthi's and the gene therapies and stem cell therapies that you will read about later in this chapter seemed like miracles. These breakthroughs are products of the ingenuity of medical and neuroscience research-ers, who have built on the accumulated knowledge of their predecessors. Their accomplishments are also the result of more powerful research methods, including research design as well as technology. This is the story of the role that research methodology plays in the field of biopsychology, and of the increasing ethical implications of our advancing knowledge. But first, we need to take a few minutes to review some important points about research; you have most likely seen them before, but they are worth reemphasizing in the context of biological psychology.

FIGURE 4.1 The Original "Bubble Boy."

The most famous patient with SCID was so vulnerable to life-threatening infections that he had to live in a sterile plastic tent. Thanks to research advances, the disease is now treatable by genetic manipulation.

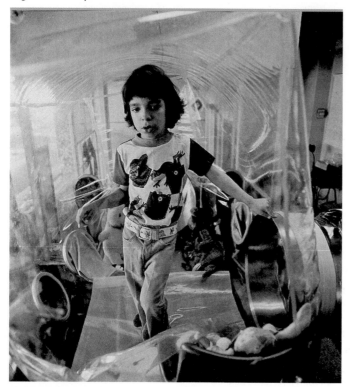

SOURCE: AP Photo/File.

Science, Research, and Theory

Science is not distinguished by the knowledge it produces but *by its method of acquiring knowledge.* We learned in Chapter 1 that scientists' primary method is empiricism; this means that they rely on observation for their information rather than on intuition, tradition, or logic (alone). Descartes started out with the traditional assumption that there was a soul, and then he located the soul in the pineal gland because it seemed the logical place for the soul to control the brain. Aristotle, using equally good logic, had located the soul in the heart because the heart is so vital to life. (He thought that the brain's function was to cool the blood!) Observation— which is a much more formal activity in science than the term suggests—is more objective than alternative ways of acquiring knowledge; this means that two observers are more likely to reach the same conclusion about what is being observed (though not necessarily about its interpretation) than two people using intuition or logic.

Biopsychologists, and scientists in general, have great confidence in observation and all the methods in their arsenal. But in scientific writings, you will often see statements beginning with "It appears that . . . ," "Perhaps . . . ," or "The results suggest. . . ." So you might well wonder, *Why do scientists always sound so tentative?*

Theory and Tentativeness in Science

One reason is that the field is very complex, so it is always possible that a study is flawed or that new data will come along that will change how we interpret previous studies. A second rea-son is that we base our conclusions on samples of subjects and samples of data from those subjects; the laws of probability tell us that even in well-designed research, we will occasionally end up with a few unusual participants in our study or there will be a slight but important shift in behavior that has nothing to do with the variable we are studying.

Q Why are scientists so tentative?

Scientists recognize that knowledge is changing rapidly and the cherished ideas of today may be discarded tomorrow. A case in point is that until recently, no one accepted that there was regrowth of severed axons or any neurogenesis in the primate central nervous system (Rakic, 1985), a belief that you now know was incorrect. You seldom hear scientists using the words *truth* and *proof,* because these terms suggest final answers. Such uncertainty may feel uncomfortable to you, but centuries of experience have shown that certainty about truth can be just as uncom-fortable, if not downright scary.

One way the researcher has of making sense out of ambiguity is through theory. A *theory* integrates and interprets diverse observations in an attempt to explain some phenomenon. For example, schizophrenia researchers noticed that people who over-dosed on the drug amphetamine were being misdiagnosed as schizophrenic when they were admitted to emergency rooms with hallucinations and paranoia. They also knew that amphetamine increases activity in neurons that release dopamine as the neurotransmitter. This led several researchers to propose that schizophrenia is due to excess dopamine activity in the brain.

A theory explains existing facts, but it also generates hypotheses that guide further research. One hypothesis that came from dopamine theory was that drugs that decrease dopamine activity would improve functioning in schizophrenics. This hypothesis was testable, which is a requirement for a good theory. The hypothesis was supported in many cases of schizophrenia, but not in others. We now realize that the dopamine theory is an incomplete explanation for schizophrenia. However, even a flawed theory inspires further research that will yield more knowledge and additional hypotheses. But remember that the best theory is still only a theory; theory and empiricism are the basis of science's ability for self-correction and its openness to change and renewal.

Now we will examine one of the knottiest issues of research, one that you will need to think about often as you evaluate the research evidence discussed through-out this text.

Experimental Versus Correlational Studies

Observation has a broad meaning in science. A biological psychologist might observe aggressive behavior in children on the playground to see if there are dif-ferences between boys and girls (*naturalistic observation*), report on the positron emission tomography scan of a patient who had violent outbursts following a car accident that caused brain injury (*case study*), use a questionnaire to find out whether some women are more aggressive during the premenstrual period (*survey*), or stimulate a part of rats' brains with electricity to see what part of the brain controls aggressive behavior (*experiment*). These different research strategies fall into the broad categories of *experimental* and *correlational studies.*

An *experiment* is a study in which the researcher manipulates a condition (the independent variable) that is expected to produce a change in the subject's behavior (the dependent variable). The experimenter also eliminates *extraneous variables* that might influence the behavior or equates them across subjects—for example, by removing environmental distractions, instructing participants not to use caffeine or other stimulants beforehand, and "running" subjects at the same time of day. In a *correlational study*, the researcher does not control an independent variable but observes whether two variables are related to each other. When we use positron

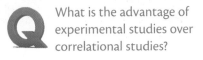

Q What is the advantage of experimental studies over correlational studies?

FIGURE 4.2 Correlational Versus Experimental Studies.

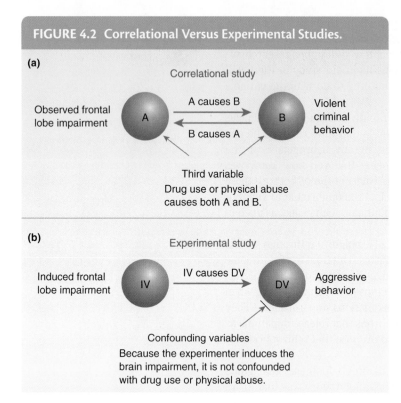

(a)

Correlational study

Observed frontal lobe impairment

A causes B

B causes A

Violent criminal behavior

Third variable
Drug use or physical abuse causes both A and B.

(b)

Experimental study

Induced frontal lobe impairment

IV causes DV

Aggressive behavior

Confounding variables
Because the experimenter induces the brain impairment, it is not confounded with drug use or physical abuse.

In a correlational study (a), we cannot tell whether *A* influences *B*, *B* influences *A*, or a third variable affects both. In an experimental study (b), the researcher manipulates the independent variable (IV), which increases assurance that it is the cause of the change in the dependent variable (DV). (The red arrows indicate possible interpretations of causation.)

emission tomography scans to determine that violent criminals more often have impaired frontal lobe activity, we are doing a correlational study; if we *induce* the impairment (independent variable) and then observe whether this increases aggression (dependent variable), we are doing an experiment.

Figure 4.2 illustrates some of the differences between a correlational and an experiment study of aggressive behavior. Based on observations that violent criminals often have impaired frontal lobe functioning, we might identify a large group of impaired individuals (using positron emission tomography scans or behavioral and cognitive tests) and see if they have a record of violent crimes. We would very likely find that they do, but Figure 4.2a reveals a problem with interpretation: For all we know, the individuals' brain damage may have been incurred in the process of committing their criminal acts rather than the other way around. Or both frontal lobe damage and violent behavior could stem from any number of third variables, such as physical abuse during childhood or long-term drug use. These variables are potentially *confounded* with each other, so we cannot separate their effects. In other words, *we cannot draw conclusions about cause and effect from a correlational study.*

What about doing this research as an experimental study? For ethical reasons, of course, we would not induce brain damage in humans, but remember that in Chapter 3 we saw that researchers used an electromagnetic field to disrupt activity in the visual cortex of blind individuals. So let's use this *transcranial magnetic stimulation* to disrupt temporarily our hypothetical volunteers' frontal lobe functioning. Granted, we won't see them become physically violent in the laboratory, but we can borrow a technique from a similar study we will see later in the chapter on emotion: We will administer several mild shocks to the subject under the pretense that the shocks are being controlled by another (fictitious) player, and we will record the intensity of shocks our participant delivers in retaliation. Because we selected our research participants and induced the brain impairment, we have eliminated the confounding variables that plagued us in the correlational study; now if we see higher levels of shock administered by subjects while their frontal activity is being disrupted, we can be fairly confident that the frontal lobe impairment is *causing* the increase in the aggressive behavior.

Of course, we could quibble about how well this study mirrors real brain damage and violent aggression; the greater control afforded by experimental studies often carries a cost of some artificiality. Experimentation is the most powerful research strategy, but correlational studies also provide unique and valuable information, such as the observation that children of schizophrenic parents have a high incidence of schizophrenia even when they are reared in normal adoptive homes. To advance our understanding in biopsychology, we need correlational studies as well as experimental research, but we equally need to be careful about interpreting their results—a point you should keep in mind as we explore the following research techniques and as we look at research in the following chapters.

Concept Check

Take a Minute to Check Your Knowledge and Understanding

☐ What is the value of empiricism? What is the value of theory?

☐ A scientist speaking to a group of students says, "I do not expect my research to find the truth." Why?

☐ What advantages do correlational studies have over experiments?

Research Techniques

The brain does not give up its secrets easily. If we remove a clock from its case and observe the gears turn and the spring expand, we can get a pretty good idea how a clock measures time; but if we open the skull, how the brain works remains just as much a mystery as before. This is where research technique comes in, extending the scientist's observation beyond what is readily accessible. Your understanding of the information that fills the rest of this book—and of the limitations of that information—will require some knowledge of how the researchers came by their conclusions. The following review of a few major research methods is abbreviated, but it will help you navigate through the rest of the book, and we will add other methods as we go along.

Staining and Imaging Neurons

It didn't take long to exhaust the possibilities for viewing the nervous system with the naked eye, but the invention of the microscope took researchers many steps beyond what the pioneers in gross anatomy could do. Unfortunately, neurons are greatly intertwined and are difficult to distinguish from each other, even when magnified. The *Golgi stain* method randomly stains about 5% of neurons, placing them in relief against the background of seeming neural chaos (Figure 4.3a). As we saw in Chapter 2, the Italian anatomist Camillo Golgi developed this technique in 1875 and, shortly after, his Spanish contemporary Santiago Ramón y Cajal used it to discover that neurons are separate cells. Golgi and Cajal jointly received the 1906 Nobel Prize in physiology and medicine for their contributions.

Other staining methods add important dimensions to the researcher's ability to study the nervous system. *Myelin stains* are taken up by the fatty myelin that wraps and insulates axons; the stain thus identifies neural pathways. In Figure 4.3b, the slice of brain tissue is heavily stained in the inner areas where many pathways converge and stained lightly or not at all in the perimeter where mostly cell bodies are located. *Nissl stains* do the opposite; they identify cell bodies of neurons (Figure 4.3c).

Later-generation techniques are used to trace pathways to determine their origin or their destination—that is, which part of the brain is communicating with another. These procedures take advantage of the fact that neurons move materials up and down the axon constantly. For example, if we inject the chemical *fluorogold* into a part of the brain, it will be taken up by the terminals of neurons and transported up the axons to the cell bodies. Under light of the appropriate wavelength, fluorogold will fluoresce—radiate light—so it will show up under a microscope and tell us which brain areas receive neural input from the area we injected. For

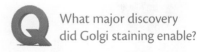

Q What major discovery did Golgi staining enable?

FIGURE 4.3 Three Staining Techniques.

(a) Golgi stains highlight individual neurons. (b) Myelin stains emphasize white matter and, therefore, neural pathways (stained blue here). (c) Nissl stains emphasize the cell bodies of neurons (stained dark).

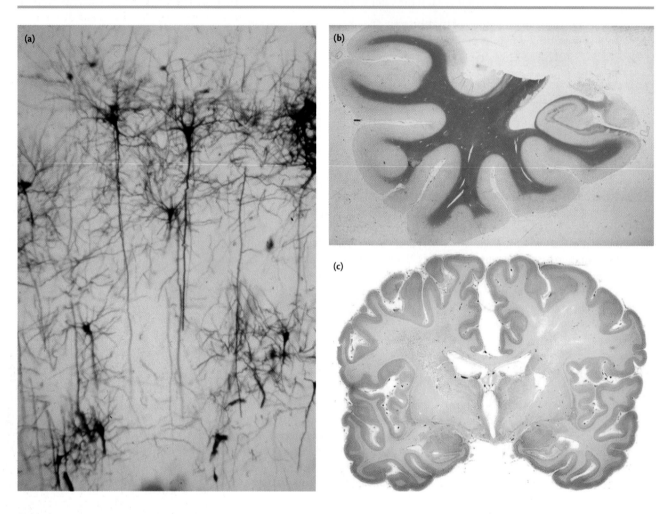

SOURCES: (a) © Dr. John D. Cunningham/Visuals Unlimited; (b) Photo Researchers; (c) © brainmuseum.org/NSF.

What advantage does autoradiography have?

example, fluorogold injected into a rat's superior colliculi will show up a few days later among the neurons at the back of the eye.

These staining and tracing procedures reveal fine anatomy, but they do not tell us anything about function. *Autoradiography* makes neurons stand out visibly just as staining does, but it also reveals which neurons are active, and this information can be correlated with the behavior the animal was engaged in. In this procedure, the animal is injected with a substance that has made radioactive, such as a type of sugar called 2-deoxyglucose (2-DG). Then, the researcher usually stimulates the animal, for instance, by presenting a visual pattern or requiring the subject to learn a task. Active neurons take up more glucose, and because 2-DG is similar to glucose, the neurons involved in the activity become radioactively "labeled." In Chapter 10 we will see an example of this technique, where vision researchers mapped the projections to the visual cortex from light-receptive cells in the eye (Tootell, Silverman, Switkes, & De Valois, 1982). After injecting monkeys with radioactive 2-DG, they presented the subjects with a geometric visual stimulus. The animals were euthanized (killed painlessly), and a section of their visual cortical tissue placed on photographic film. The radioactive areas exposed the film and produced an image of the original stimulus. This confirmed that just as the somatosensory projection area contains a map of the body, the visual cortex maps the visual-sensitive retina and, thus, the visual world (Figure 4.4a).

A variation of this method is used to determine the location and quantity of receptors for a particular drug or neurotransmitter. Candace Pert used this procedure to find out whether there are receptors in the brain for opiate drugs (a class containing opium, morphine, and heroin), which seemed like the best explanation for the drugs' potency in relieving pain (Herkenham & Pert, 1982; Pert & Snyder, 1973). First, she soaked rat brains in radioactive *naloxone,* a drug that she knew counteracts the effects of opiates, on the assumption that it does so by blocking the hypothesized receptors. She then placed the brains on photographic film. Sure enough, an image of the brain formed, highlighting the locations of opiate receptors (Figure 4.4b). This procedure identified only one type of opiate receptor, but it was an important start because it established that the receptors exist and implied that the brain makes its own opiates!

Instead of using radioactivity, *immunocytochemistry* uses antibodies attached to a dye to identify cellular components such as receptors, neurotransmitters, or enzymes. The technique takes advantage of the fact that antibodies, which attack foreign intruders in the body, can be custom designed to be specific to any cellular component. The dye, which is usually fluorescent, makes the antibodies' targets visible when the tissue is removed and examined under a microscope. Night-migrating birds use the earth's magnetic field to navigate, and earlier evidence suggested that the magnetic detectors might be *cryptochromes*, which are molecules found in some neurons in the birds' retinas. Henrik Mouritsen and his colleagues (2004) in Germany have provided strong supporting evidence. Using immunocytochemistry, they found that during the day, cryptochromes were plentiful in the retinas of both garden warblers and zebra finches; at night, however, cryptochromes diminished virtually to zero in the nonmigratory finches but *increased* in the eyes of the night-migrating warblers (Figure 4.5).

But before they could do this study, Mouritsen's team had to decide which of two kinds of cryptochrome to focus their efforts on, CRY1 or CRY2. So they used another powerful research technique that determines where particular genes are active, in this case the *cry1* and *cry2* genes. Remember from Chapter 1 that genes control the production of proteins; the instructions for protein production are carried from the nucleus into the cytoplasm of a cell by *messenger ribonucleic acid (RNA),* which is a copy of one strand of the gene's DNA. So, when we locate specific messenger RNA, we know that the gene is active in that place; this is done by in situ hybridization. *In situ hybridization* involves constructing strands of complementary DNA, which will dock with strands of messenger RNA. Because the complementary DNA is first made radioactive, autoradiography can then be used to determine the location of gene activity (see Figure 4.6). The researchers found that the protein product CRY2 was being constructed in cell nuclei, while CRY1 was being constructed outside the nucleus; because a magnetoreceptor was more likely to function outside the nucleus, they limited their study to that cryptochrome.

FIGURE 4.4 Autoradiographs.

(a) Monkeys were injected with radioactive 2-DG before they were presented with a geometric stimulus. The monkeys were euthanized, and a slice of their brain was placed on photographic film; the pattern of radioactivity produced the image you see here. (b) An autoradiograph of a horizontal slice from a rat's brain that was soaked in radioactive opiate antagonist naloxone. White areas indicated opiate receptors. The slice is at the level of the thalamus; the front of the brain is at the top of the picture.

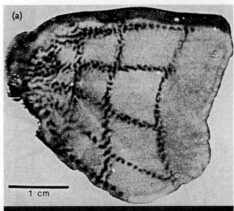

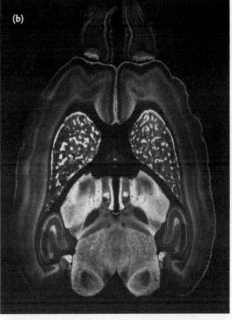

SOURCES: (a) Reprinted with permission from "Deoxyglucose Analysis of Retinotopic Organization in Primate Striate Cortex," by R. B. H. Tootell et al., *Science, 218*, pp. 902–904. Copyright 1982 American Association for the Advancement of Science (AAAS); (b) From "Light Microscopic Localization of Brain Opiate Receptors: A General Autoradiographic Method Which Preserves Tissue Quality," by M. A. Herkenham and C. B. Pert, 1982, *Journal of Neuroscience, 2*, 1129–1149.

Neurons that contain cryptochromes and are currently active are labeled in orange. The larger type of neurons (indicated by arrows) project to a brain area that responds to magnetic field stimulation.

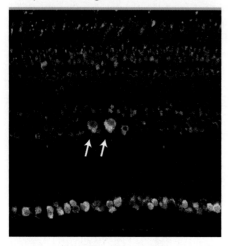

SOURCE: From "Cryptochromes and Neuronal-Activity Markers Colocalize In the Retina of Migratory Birds During Magnetic Orientation," by H. Mouritsen et al., *PNAS, 101*, pp. 14294–14299. © 2004 H. Mouritsen. Used with permission.

Light and Electron Microscopy

For over three centuries, the progress of biological research closely paralleled the development of the light microscope. The microscope evolved from a device that used a drop of water as the magnifier, through the simple microscope with a single lens, to the compound microscope with multiple lenses. At that point, investigators were able to see the gross details of neurons: cell bodies, dendrites, axons, and the largest organelles. But the capability of the light microscope is limited, not due to the skills of the lens maker but due to the nature of light; increases in magnification beyond about 1,500 times yield little additional information.

The electron microscope, on the other hand, magnifies up to about 250,000 times and can distinguish features as small as a few hundred millionths of a centimeter. The *electron microscope* works by passing a beam of electrons through a thin slice of tissue onto a photographic film; different parts of the tissue block or pass electrons to different degrees, so the electrons produce an image of the object on the film. The electron microscope's high resolution allows us to see details such as the synaptic vesicles in an axon terminal. Engineers have enhanced the technique in the *scanning electron microscope.* The beam of electrons induces the specimen to emit electrons itself, and these are captured like the conventional microscope collects reflected light. Magnification is not as great as with the electron microscope, but the images have a three-dimensional (3-D) appearance that is very helpful in visualizing details. You can see this feature in Figure 4.7, as well as in Figure 2.16.

Microscopic technology continues to evolve, for example, in the *confocal laser scanning microscope* and the *two-photon microscope.* These microscopes image specific kinds of tissue, depending on the fluorescent dye the tissue is stained with (a fluorescent dye emits light when radiated with light within a specific range of wavelengths). These microscopes have the advantage that they are not limited to very thin slices of tissue. They can be used with thicker tissue samples and can even image details in the upper layers of the exposed living brain; with optical probes, they can image neurons as deep as 1 cm below the surface. As an example, researchers using a dye specific for calcium were able to measure movement-related neural activity in the brains of mice running on a treadmill (Dombeck, Khabbaz, Collman, Adelman, & Tank, 2007).

Measuring and Manipulating Brain Activity

You learned in previous chapters that it is easy to stimulate the surface of the brain with electricity to produce movement, sensations, and even apparent memories. We can also record electrical activity from the surface of the brain or even from the scalp. Studying deeper structures will require more inventive techniques, which we will look at after we discuss electroencephalography.

FIGURE 4.6 DNA, Proteins, and In Situ Hybridization.

DNA

Messenger RNA moves out into the cytoplasm and directs construction of proteins.

Complementary radioactive DNA (red) docks with the messenger RNA.

Messenger RNA copies a strand of the DNA and then moves out into the cytoplasm, where it controls the development of proteins. Complementary radioactive DNA helps researchers locate gene activity.

Electroencephalography

In 1929, the German psychiatrist Hans Berger invented the electroencephalograph and used it to record the first electroencephalogram from his young son's brain. Since then, the technique has proved indispensable in diagnosing brain disorders such as epilepsy and brain tumors; it has also been valuable for studying brain activity during various kinds of behavior, from sleep to learning. The *electroencephalogram (EEG)* is recorded from two electrodes on the scalp over the area of interest; an electronic amplifier detects the combined electrical activity of all the neurons between the two electrodes (popularly known as "brain waves"; see Figure 4.8). Usually, the researcher applies a number of electrodes and monitors activity in multiple brain areas at the same time.

The *temporal* (time) *resolution* of the EEG is one of its best features; it can distinguish events only 1 millisecond (ms) apart in time, so it can track the brain's responses to rapidly changing events. However, its *spatial resolution*, or ability to detect precisely where in the brain the signal is coming from, is poor. This problem can be alleviated somewhat by applying electrodes directly to the brain, which removes the interference of the skull. Of course, this procedure is used only with animals or with humans undergoing surgery. So although the EEG provides relatively gross measurements, its advantages are good time resolution, ease of use, and, compared with the imaging techniques we will consider shortly, low cost.

EEG is most useful for detecting changes in arousal, as in the example in Figure 4.8. It is not good at detecting the response to a brief stimulus, such as a spoken word; its time resolution is adequate, but the "noise" of the brain's other ongoing activity drowns out the response, so the tracing looks much like the "awake" recording in the figure. However, by combining electroencephalography with the computer, the researcher can average the EEG over several presentations of the stimulus to produce an *evoked potential*, like the one in Figure 4.9. Averaging over many trials cancels out the ongoing noise, leaving only the unique response to the stimulus. In this example, Shirley Hill (1995) repeatedly presented a low-pitched tone to her research participants and occasionally interjected a high-pitched tone. Averaging showed a large dip in the electrical potential following the novel (high-pitched) stimulus. In the next chapter, you will learn that this dip is smaller in alcoholics than in nonalcoholics, as well as in the young children of alcoholics, which suggests an inherited vulnerability to alcoholism. Another example is that biopsychologists have used the technique to confirm that spoken words produce a greater response in the left hemisphere, just as you would expect, than in the right hemisphere.

Stereotaxic Techniques

When the area of interest is below the surface, the researcher must use probes that can penetrate deep into the brain. Two important aids make this task feasible. The first is a map of the brain called a *stereotaxic atlas*. A large number of brains are sliced into very thin coronal sections; drawings are prepared that show the (average) location of brain structures on each section. Figure 4.10 is one of these drawings from a stereotaxic atlas of the rat brain. Each drawing is numbered

FIGURE 4.7 Scanning Electron Micrograph of a Neuron.

Notice the depth and detail this kind of imaging provides. (The white structures on and around the cell body are glial cells.)

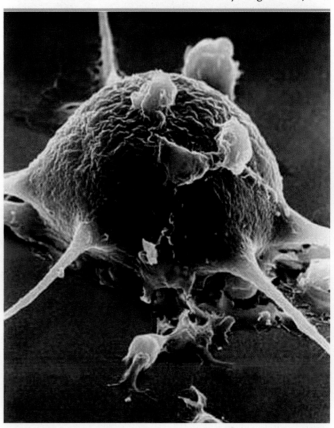

SOURCE: © Dr. Robert Berdan, 2007.

Q What can EEG and evoked potentials tell us?

FIGURE 4.8 An Electroencephalograph and a Sample EEG.

(a)

(b)

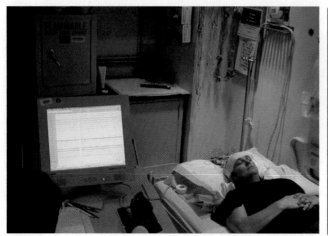

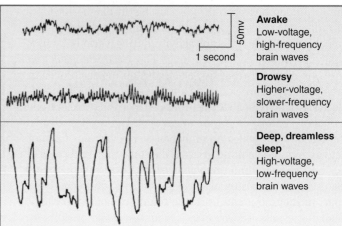

An electroencephalograph records the electrical activity of the brain through electrodes applied to the scalp (a). The up-and-down fluctuations of the tracings on the computer screen (b) indicate the EEG frequency and the height indicates the voltage. The computer does precise analyses of the signal for research or diagnostic purposes.

SOURCES: (a) Courtesy of National Institute of Neurological Disorders and Stroke; (b) From *Current Concepts: The Sleep Disorders*, by P. Hauri, 1982, Kalamazoo, MI: Upjohn.

FIGURE 4.9 Evoked Potential Produced by a Novel Tone.

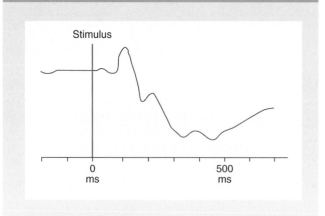

A research participant responds to a novel stimulus, such as an occasional high-pitched tone among low-pitched tones, with a large dip in the evoked potential. Without averaging over several stimulus presentations, all we would see would be an EEG like the "Awake" recording in Figure 4.8.

Q What are the different ways a stereotaxic instrument is used?

to indicate the anterior/ posterior location of the slice, and the scales on the side and the bottom of the drawing tell the researcher how far from the midline and how deep to insert the probe.

A *stereotaxic instrument* is a device used for the precise positioning in the brain of an electrode or other device. Figure 4.11 shows a stereotaxic instrument for rats; the instrument secures the anesthetized rat's head and allows the investigator to insert the probe through a small hole drilled in the skull at the precise location and depth specified by the atlas. Often the probe is a fine-wire electrode, electrically insulated except at its tip, that is used to activate the structure with very low-voltage electricity. While the still-anesthetized animal's brain is being stimulated, the researcher can monitor responses in other parts of the brain or in the body. If the animal must be awake to test the effect, the electrode can be anchored to the skull; the wound is closed, and after a couple of days of recovery, the rat's behavior can be observed during stimulation. In Chapter 5 you will see research in which animals were willing to press a lever to deliver electrical stimulation to certain parts of their own brain.

A similar electrode arrangement is used to record neural activity; the biopsychologist might subject the animal to a learning task, present visual or auditory stimuli, or introduce a member of the other sex while monitoring activity in an appropriate brain location. Electrodes ordinarily will record from all the surrounding neurons, but *microelectrodes* have tips so fine that they can record from a single neuron. The electrodes may be placed in the brain temporarily in an anesthetized animal, or they may be mounted in a socket cemented to the animal's skull to permit recording in the unanesthetized, behaving subject.

The stereotaxic instrument can also be used to insert a small-diameter tube called a *cannula* for injecting chemicals. Chemical stimulation has a special advantage over electrical stimulation in that it acts only at the dendrites and cell bodies of neurons. This means that the researcher can simulate the effects of a neurotransmitter or block a transmitter's effects at the synapses. Often the tube is not used to deliver the drug but is cemented in place and used later as a guide for inserting a smaller drug-delivery cannula; this arrangement can be used for stimulation later on multiple occasions. The same technique is used for microdialysis, in which brain fluids are extracted for analysis, but a more elaborate dual cannula is required. As Figure 4.12 shows, the brain fluids seep through a porous membrane into the lower chamber of the cannula; a biologically neutral fluid (very similar to seawater) is pumped through one of the dual tubes, and it flushes the brain fluid out the other tube for analysis. In the next chapter, you will see results from both of these techniques, when researchers deliver abused drugs to rats' brains or monitor the release of brain neurotransmitters after the animal is injected with a drug.

Ablation and Lesioning

Historically, one of the most profitable avenues of brain research has been the study of patients who have sustained brain damage. Brain damage can occur in a variety of ways: gunshot wounds, blows to the head, tumors, infection, toxins, and strokes. Although these "natural experiments" have been extremely valuable to neuroscientists, they also have major disadvantages. Most important, the damage doesn't coincide neatly with the functional area; it will affect a smaller area or overlap with other functional areas. Fortunately, the pattern of damage varies from patient to patient, so if the neuroscientist studies a large number of patients, it may be possible to identify the location of damage common to people with the same deficits.

Because of these and other difficulties in studying patients with brain damage, researchers often resort to producing the damage themselves in animals. In some

FIGURE 4.10 A Plate From a Stereotaxic Atlas.

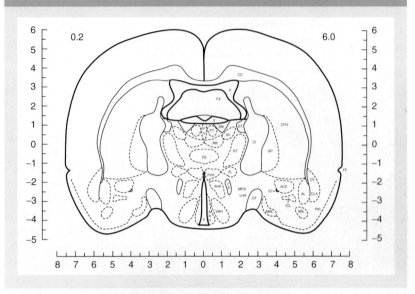

SOURCE: From Pellegrino, L. J., Pellegrino, A. S., & Cushman, A. J. (1979). *A stereotaxic atlas of the rat brain* (2nd ed.). New York: Plenum. Reprinted with kind permission from Springer Science and Business Media.

FIGURE 4.11 A Stereotaxic Instrument.

This device allows the researcher to locate an electrode precisely at the right horizontal position and depth in the animal's brain. (Although its eyes are open, the rat is deeply anesthetized.)

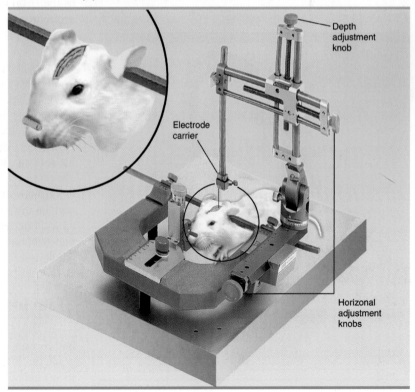

FIGURE 4.12 A Cannula for Microdialysis.

Neurochemicals in the surrounding fluid diffuse into the cannula through the porous membrane. Fluid is pumped in through the outer tube and flows out through the inside tube, carrying the neurochemicals with it.

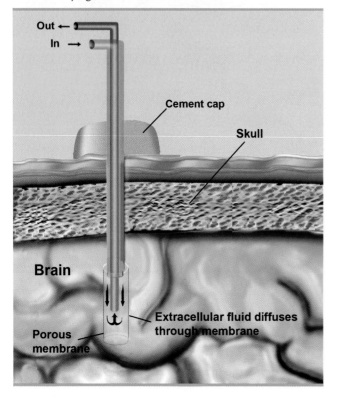

FIGURE 4.13 A Human Stereotaxic Instrument.

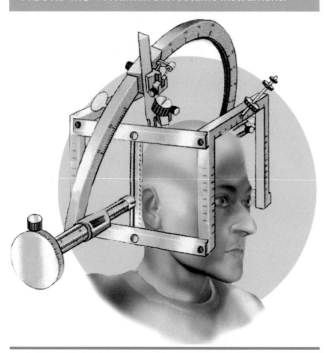

cases, a whole area of the brain may be removed; removal of brain tissue is called *ablation.* Ablation can be done with a scalpel, but *aspiration* is a more precise technique, and it allows access to deeper structures. The skull is opened, and a fine-tipped glass micropipette connected to a vacuum pump is used to suck out neural tissue. Usually, however, lesioning is preferred in place of ablation because the damage can be more precisely controlled. *Lesions,* or damage to neural tissue, can be produced by electrical current, heat, or injection of a neurotoxin (using a stereotaxic instrument) or by using a knife or a fine wire to sever connections between areas. "Reversible lesions" can be produced by chilling a brain area or by injecting certain chemicals; this means that the animal's behavior can be observed before and during treatment and again after recovery.

Occasionally there is reason to insert a cannula or an electrode into a human's brain. This is done for clinical purposes—for example, to identify functional areas by recording electrical activity prior to brain surgery, to lesion malfunctioning tissue in patients with epilepsy, or to stimulate the brain in patients with Parkinson's disease. (See the Application for additional information.) Stereotaxic atlases of the human brain are published for this purpose, and there are human stereotaxic instruments as well, usually designed to mount on the head, as shown in Figure 4.13.

Transcranial Magnetic Stimulation

Transcranial magnetic stimulation (TMS) is a relatively new noninvasive brain stimulation technique that uses a magnetic coil to induce a voltage in brain tissue. The device is held close to the scalp over the target area; coils in the shape of a circle are more powerful, while those in the shape of a figure eight focus the magnetic field more precisely (see Figure 4.14). TMS is pulsed at varying rates; frequencies of 1/s or less decrease brain excitability, and frequencies of 5/s and higher increase excitability.

TMS has demonstrated its usefulness mostly as a research instrument. By making clever combination of TMS stimulation and brain-imaging techniques (described in the next section), researchers have teased out the neural modifications accounting for recovery in stroke patients (Gerloff et al., 2006); they have also determined that making visual-spatial judgments involves not just the parietal area but a broader network including frontal regions (Sack et al., 2007). TMS is also a promising candidate as a therapeutic tool, showing potential usefulness in alleviating symptoms of Parkinson's disease, depression, and autism (Hallett, 2007).

Brain Imaging Techniques

In Broca's day, and in fact until fairly recently, a researcher had to wait for a brain-damaged patient to die in order to pinpoint the location of the damage. There was little motivation to do exhaustive observations of the patient's behavior when the patient might outlive the researcher or the body might not be available to the researcher at death. All that changed with the invention of imaging equipment that could produce a picture of the living brain showing the location of damage.

The first modern medical imaging technique came into use in the early 1970s. *Computed tomography (CT)* scanning produces a series of X-rays taken from different angles; a computer combines the series of two-dimensional horizontal cross sections, or "slices," so the researcher can scan through them as if they are a 3-D image of the entire organ (Figure 4.15). Imaging soft tissue such as the brain requires injecting a dye that will show up on an X-ray; the dye diffuses throughout the tiny blood vessels of the brain, so it is really the differing density of blood vessels that forms the image. A major drawback of earlier equipment was its extreme slowness, but newer models of CT scanners are fast, and they provide detailed images. CT scans are also popularly known as *CAT scans.*

Another imaging technique, *magnetic resonance imaging (MRI),* works by measuring the radio-frequency waves emitted by the nuclei of hydrogen atoms when they are subjected to a strong magnetic field. Most of that hydrogen is in the water that composes 78% of the brain, but the water content varies in different brain structures, so these emissions from hydrogen nuclei can be used to form a detailed image of the brain (Figure 4.16). The MRI is reasonably fast, and it can image small areas. Recent increases in power permit more versatile imaging by detecting elements other than hydrogen, including sodium, phosphorus, carbon, nitrogen, and oxygen. MRI scanners in the future should also be small enough to be portable and cost a few thousand dollars rather than a few million. A variant of MRI, *diffusion tensor imaging,* measures the movement of water molecules; because water moves more easily along the length of axons, this technique is very useful for imaging brain pathways.

FIGURE 4.14 Lindsay Oberman Uses TMS With a Subject.

Oberman and her colleagues at Beth Israel Deaconess Medical Center are using TMS to find clues to what goes wrong in the autistic brain.

SOURCE: Courtesy of National Institute of Neurological Disorders and Stroke.

Application   Brain Implants That Move

People with Lou Gehrig disease (amyotrophic lateral sclerosis, or ALS) eventually become unable to move or speak; they are mentally alert but trapped in a nonfunctioning body. Scientists are struggling to find ways to help Lou Gehrig patients and others communicate. Recently, they have had some success in implanting electrodes in the brain that allow the individual to control a mechanical arm or a computer keyboard. These electrodes are impractical in the long run, though, because they lose their signal after a few months; they move slightly from jostling or from a slight change in blood pressure, or the neurons around them die.

But Richard Andersen and Joel Burdick at Caltech have developed a tiny device that can move four electrodes individually in search of a stronger signal (Andersen et al., 2004; Graham-Rowe, 2004). Each electrode is located in a special type of crystal. A weakening neural signal activates a circuit that triggers a pulse of electricity to the crystal. This causes the crystal to expand slightly, which causes the rough edge of the crystal to ratchet the electrode downward, one micrometer at a time—less than the diameter of the brain's smallest axons. When the electrode encounters astrong signal again, the ratcheting stops; the electrode resumes delivering commands to the device

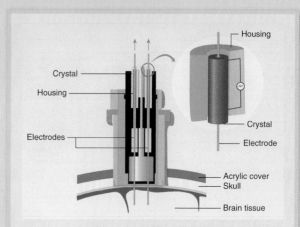

SOURCE: From "Brain Implants That Move," by D. Graham Rowe, *New Scientist,* Nov. 13, 2004, p. 25.

it is connected to. Burdick and Andersen have successfully tested the movable electrode in monkeys, and they plan to fit a paralyzed person with one of the devices soon.

FIGURE 4.15 Computed Tomography Scanning Procedure.

(a)

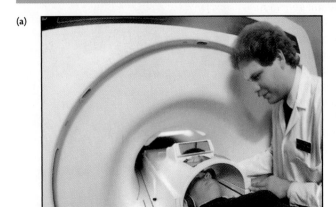

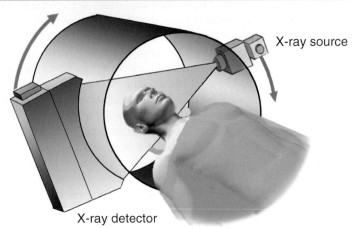

X-ray source

X-ray detector

(c)

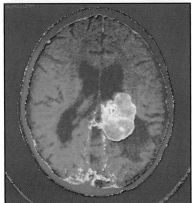

(a) The patient's head is positioned in a large cylinder, as shown here. (b) An X-ray beam and X-ray detector rotate around the patient's head, taking multiple X-rays of a horizontal slice of the patient's brain. (c) A computer combines X-rays to create an image of a horizontal slice of the brain. The scan reveals a tumor on the right side of the brain.

SOURCES: (a) Alvis Upitis/The Image Bank; (c) Dan McCoy/Rainbow.

FIGURE 4.16 Magnetic Resonance Imaging.

(a) The individual is slid into the device. (b) A sample scan, which has detected a tumor (red arrow).

(a)

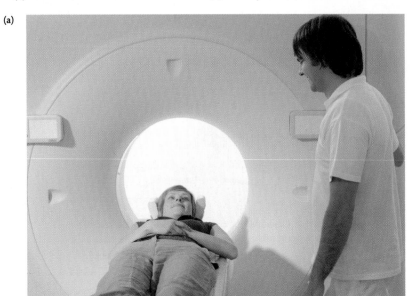

(b)

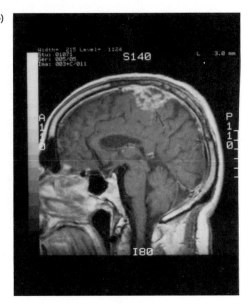

SOURCES: (a) Rayman; (b) Huntington Magnetic Resonance Center Digital Vision.

CT and MRI added tremendous capability for detecting tumors and correlating brain damage with behavioral symptoms. However, CT and MRI lack the ability to detect changing brain activity (as EEG does, for instance); the two remaining techniques add that capability.

Positron emission tomography (PET) involves injecting a radioactive substance into the bloodstream, which is taken up by parts of the brain according to how active they are. The scanner captures the positrons emitted by the radioactive substance to form an image that is color coded to show the relative amounts of activity (see Figure 4.17). Radioactive 2-DG is often the substance that is injected because increased uptake of 2-DG by active neurons provides a measure of metabolic activity. Other radioactive substances can be used to monitor blood flow or oxygen uptake, and if a neurotransmitter is made radioactive, it can be used to determine the locations and numbers of receptors for the transmitter. Usually, the researcher produces a "difference scan" by subtracting the activity occurring during a neutral control condition from the activity that occurred during the test condition; this produces an image that uses a color scale to show where activity increased or decreased in the brain.

PET equipment is expensive and requires a sophisticated staff to operate it; the facility must also be near a cyclotron, which produces the radioactive substance, and there are few of those around. The advantage that justifies this expense is that PET is able to track changing activity in the brain. The speed is not what a biopsychologist might wish, though, because PET cannot detect changes during behaviors that are briefer than 45 seconds PET also does not image the brain tissue itself, so the results are often displayed overlaid on a brain image produced by another means, such as MRI.

Q What benefits do imaging techniques add?

FIGURE 4.17 Positron Emission Tomography.

(a)

(b)

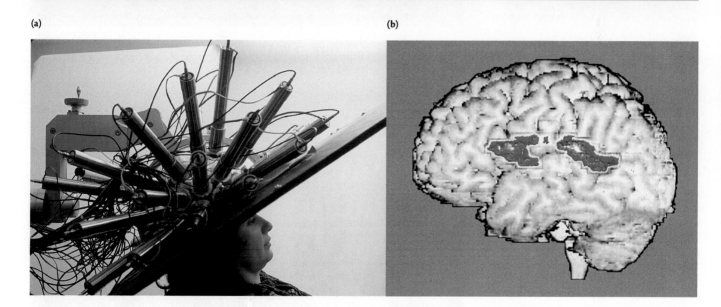

(a) The apparatus detects concentrations of radioactivity where neural activity is high; the computer produces a color-coded image like the one in (b). Traditionally, red indicates the greatest amount of activity, followed by yellow, green, and then blue. The individual was working on a verbal task, so areas involved in language processing were activated.

What advantage do PET and fMRI have over CT and MRI?

1

Brain Imaging

A modification of MRI takes advantage of the fact that oxygenated blood has magnetic properties that are different from those of blood that has given up its oxygen to cells; *functional magnetic resonance imaging (fMRI)* measures brain activation by detecting the increase in oxygen levels in active neural structures (Figure 4.18). MRI and fMRI have the advantage over PET and CT that they involve no radiation, so they are safe to use in studies that require repeated measurements. In addition, fMRI measures activity, like PET, and produces an image of the brain with good spatial resolution, like MRI. Researchers have been able to see activity in 1-millimeter-wide groups of neurons in the visual cortex (Barinaga, 1997). The fMRI machines are particularly pricey though, which limits their usefulness for research.

Brain imaging has been referred to as "the predominant technique in behavioral and cognitive neuroscience" (Friston, 2009, p. 399). That said, you will not be surprised to see many examples of its use throughout this text—though none will be as unusual as the one in the accompanying Application. However, some investigators are urging caution in interpreting results; their focus has been mostly on fMRI, but their criticisms apply more broadly. For one thing, simultaneous electrode recording of neural activity in monkeys revealed that fMRI misses a great deal of information due to its lower sensitivity as well as to chance factors such as the distance of the neurons from a blood vessel (Logothetis, 2002). A related problem is less-than-perfect test–retest reliability. For example, when researchers showed volunteers pictures of fearful faces, correlations of amygdala activity in sessions 2 weeks apart ranged from .4 to .7 (Johnstone et al., 2005).

Perhaps more importantly, some researchers have been criticized for the way they select their data. Ideally, researchers should decide in advance to obtain data from a specific area, such as the amygdala, but often researchers do not have enough information to make this selection beforehand and therefore scan the whole brain. Then they divide the scan into tiny cube-shaped areas called voxels and determine which voxels' activity is correlated with performance on a task or

---

FIGURE 4.18 An fMRI Scan.

The colored areas were more active when research participants were processing words that were later remembered than when they processed words that were not remembered.

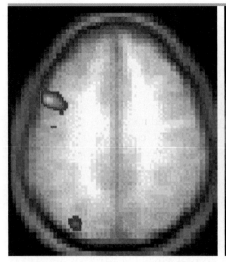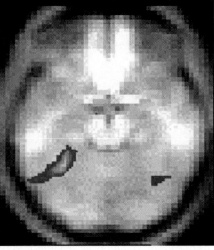

SOURCE: Reprinted with permission from "Building Memories: Remembering and Forgetting of Verbal Experiences As Predicted By Brain Activity," by A. D. Wagner et al., *Science, 281,* pp. 1188–1191. © 1998 American Association for the Advancement of Science. Reprinted by permission from AAAS.

Application   Scanning King Tut

In January 2005, King Tutankhamun, pharaoh of Egypt 3,000 years ago, was removed from his tomb in the Valley of the Kings for the first time in almost 80 years. The trip was a short one, to a nearby van where his mummified body was subjected to a CT scan to determine whether Tut's death at the age of 19 might have resulted from murderous intrigue within the royal family.

An X-ray conducted in the tomb 30 years ago found pieces of his skull inside the cranium, adding to the murder theory. But if Tut had died from a blow to the head, the bone fragments would have been caught up in the embalming material. Instead, the scan showed them lodged between the skull and the now solidified embalming fluid; more likely the archaeologist Howard Carter, who discovered Tut's tomb in 1922, damaged the skull while prying away the golden mask that was stuck to the skull by solidified resins. The scans also found no mineral deposits in the bone, which some poisons would leave behind, though other poisons could not be

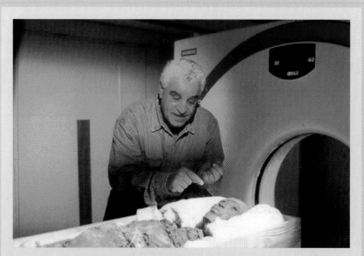

Egypt's Vice Minister of Culture Zahi Hawass Readies Tut for Scanning.

SOURCE: © AP Photo/Saedi Press.

ruled out. The most promising evidence was that one leg had been broken, within days of his death judging by the lack of healing. The break was severe enough to cause an open wound, so King Tut's death could have been caused by an infection.

SOURCE: "Press Release: Tutankhamun CT Scan," www.guardians.net/hawass/press_release_tutankhamun_ct_scan_results.htm

with a characteristic of the subjects. This involves calculating literally thousands of correlations; if you have even a basic understanding of probability, you know that this procedure guarantees finding several voxels whose activity correlates with the dependent variable even when no relationship exists. Thus, to ensure *independence* of the data, the researcher should repeat the experiment focusing only on those areas to rule out the possibility the original correlations occurred by chance.

However, researchers sometimes choose the easier approach of simply reporting the correlations they found in that one session. In an analysis of 53 studies that reported correlations of fMRI activity with social or emotional measures (such as distress from social rejection), over half used the flawed nonindependent approach, and they accounted almost entirely for the highest correlations (Vul, Harris, Winkielman, & Pashler, 2009; see also the similar results of Kriegeskorte, Simmons, Bellgowan, & Baker, 2009). Nikos Logothetis has warned that erroneous claims of brain areas specialized for everything from empathy to neuroticism is leading to a phrenology that is even more dangerous than the 19th-century variety because it is cloaked in the respectability of brain imaging (Spinney, 2002).

Still, these problems do not undermine the potential of imaging studies for understanding behavior. The fact that five people won Nobel Prizes for their work in developing scanning technology indicates the importance of imaging techniques, and you can be sure that many important future developments in neuroscience will depend on brain imaging, just as they have for the past 20 years. The techniques described here are summarized in Table 4.1.

> *The single most critical piece of equipment is still the researcher's own brain. . . . What is badly needed now, with all these scanners whirring away, is an understanding of exactly what we are observing, and seeing, and measuring, and wondering about.*
>
> —Endel Tulving

TABLE 4.1 Comparison of EEG and Imaging Techniques.

Technique	Description	Discrimination	
		Time	Spatial (mm)
EEG	Sums the electrical activity of neurons between two electrodes; detects fast-changing brain activity but is poor at localizing it	1 ms	10–15
CT	Forms 3-D image of brain by combining X-rays of cross sections of brain; images structure and damage	<1 ms	0.5
MRI	Measures variations in hydrogen concentrations in brain tissue; images structure and damage	3–5 s	1–1.5
PET	Image produced by emissions from injected substances that have been made radioactive; tracks changing activity, detects receptors, etc.	45 s	4
fMRI	Detects increases in oxygen levels during neural activity; tracks changing activity	1 s	1–2

Investigating Heredity

We looked at the interplay of heredity and environment in shaping behavior in Chapter 1; now we need to understand some of the techniques scientists use to do genetic research. The idea that behavior can be inherited is an ancient one, but most of the methods for doing genetic research were introduced or came into maturity in the past three or four decades. Until then, the work was not much more sophisticated than observing that a characteristic runs in families.

Genetic Similarities: The Correlational Approach

In a *family study,* which determines how strongly a characteristic is shared among relatives, we would find that intelligent parents usually have intelligent children. However, as one researcher put it, "Cake recipes run in families, but not because of genes" (Goodwin, 1986, p. 3). This is a good example of the problem with correlational research. People who have similar genes often share a similar environment, so the effects of heredity are confounded with the effects of environment. Still, the fact that family members are similar in a characteristic tells researchers that it would be worthwhile to pursue more complicated and costly research strategies. We will look at ways to reduce the confounding of heredity with environment, but first we need a way to quantify the results.

Quantification is a simple matter for characteristics that can be treated as present or absent, such as schizophrenia. We can say, for instance, that the rate of schizophrenia is about 1% in the general population but increases to around 13% among the offspring of a schizophrenic parent (Gottesman, 1991). For variables that are measured on a numerical scale, such as height and IQ (intelligence quotient, a measure of intelligence), we express the relationship with a statistic called the *correlation coefficient. Correlation* is the degree of relationship between two variables, measured on a scale between 0.0 and ±1.0. The strength of the relationship is indicated by the absolute value—how close the correlation is to *either* 1.0 or −1.0. A high positive correlation means that when one variable is high, the other tends to be high as well, and vice versa. A negative value indicates the opposite tendency—when one value is high, the other tends to be low not that the relationship is weaker. As examples, the correlation between the IQs of parents and their children averages about .42 across studies, and the correlation between brothers and sisters in the same family is about .47 (T. J. Bouchard & McGue, 1981). Now we can consider how to separate the effects of heredity from those of the environment.

Adoption studies eliminate much of the confounding of heredity and environment that occurs in family studies. *Adoption studies* compare adopted children's similarity to their biological parents with their similarity to their adoptive parents. This kind of study is often called a *natural experiment*, but it lacks the control of a real experiment because we do not manipulate the adoption variable. As a result, environmental confounding can still occur; for example, families that must be split up by adoption may differ from the control families in important ways. Nevertheless, the technique has yielded extremely valuable information, such as the fact that rearing children apart from their biological parents results in a drop in the correlation between their IQs from .42 to about .22 (T. J. Bouchard & McGue, 1981). The drop in correlation indicates a substantial influence of environment, while the remaining correlation indicates genetic influence.

Adoption studies do not completely control environmental influences; if adoption is delayed beyond the time of birth, the first few months of life can have an effect on later behavior, and even the prenatal environment can bring about long-term alterations in nervous system functioning. Animal researchers get around this problem by *cross-fostering,* implanting a fetus or egg into another female. Of course, this strategy is unavailable in human research, or at least it was until *in vitro* fertilization became so widespread. British researchers examined the medical records of 800 children conceived by in vitro fertilization; in one fourth of the cases, the egg or embryo was donated by another woman, so the offspring was unrelated to the mother. With this strategy, the researchers were able to conclude that the low birth weight previously observed in babies whose mothers smoked during pregnancy is environmental, while antisocial behavior (tantrums, fighting, lying, etc.) in children of smoking mothers has a genetic origin (F. Rice et al., 2009).

Twin studies assess how similar twins are in some characteristic; their similarity is then compared with that of nontwin siblings, or the similarity between identical twins is compared with the similarity between fraternal twins. Remember that fraternal twins are produced from two separately fertilized eggs (*dizygotic*), while identical twins result from a single egg that splits and develops into two individuals (*monozygotic*). Fraternal twins, like nontwin siblings, share only half their genes with each other; identical twins share 100% of their genes. With twin studies, we can compare two levels of hereditary similarity while controlling environmental similarity to a great extent (though not entirely). Because both identical twins and fraternal twins share the same environment, a greater similarity between identical twins in a characteristic is probably due to their greater genetic similarity. (Remember from Chapter 1 that a comparison of the similarity of identical twins with the similarity of fraternal twins is the basis for calculating heritability, the percentage of variation that can be attributed to heredity.) Of course, we have to select fraternal pairs that are of the same sex, because identical twins are of the same sex. A criticism of twin studies is that identical twins might be treated more similarly than fraternal twins would.

Investigations of intelligence and schizophrenia provide particularly good examples of the value of twin studies. The correlation between fraternal twins' IQs is about .60, and for identical twins it increases to around .86 (T. J. Bouchard & McGue, 1981). A useful measure for identifying genetic influence is the *concordance rate,* the frequency with which relatives are alike in a characteristic. When one fraternal twin is schizophrenic, the second twin will also be schizophrenic about 17% of the time; in identical twins the concordance almost triples, to 48% (see Figure 4.19) (Gottesman, 1991). Notice that even for identical twins, the correlation

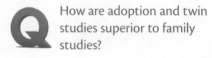

How are adoption and twin studies superior to family studies?

FIGURE 4.19 The Genain Quadruplets.

Identical quadruplets, the sisters all became schizophrenic later in life. The chances of any four unrelated individuals all being schizophrenic is 1 in 100 million. The name *Genain* is a nickname derived from the Greek word meaning "dreadful gene."

SOURCE: © AP Photo.

Q What advantage does genetic engineering have over adoption and twin studies?

falls short of a perfect 1.0 and concordance is less than 100%. Identical twins will rarely have exactly the same IQ, and the identical twin of a schizophrenic will escape schizophrenia about 52% of the time. The incomplete influence of heredity means that environmental effects are also operating. Family, adoption, and twin studies are compared in Table 4.2.

Genetic Engineering: The Experimental Approach

Although adoption and twin studies reduce confounding, they still share some of the disadvantages of correlational studies. *Genetic engineering* involves actual manipulation of the organism's genes or their functioning; studies using this technology qualify as experiments. At present, genetic engineering is employed mostly with mice, because their genetic makeup is well known and their embryos are more successfully manipulated.

An obvious way to find out what a gene does is to disable it and see what effect this has on the animal. In the *knockout* technique, a nonfunctioning mutation is introduced into the isolated gene, and the altered gene is transferred into embryos. After multiple matings, mice carrying the altered gene on both chromosomes are selected for study. Another way to disable a gene is to interfere with its messenger RNA. The *antisense RNA* procedure blocks the participation of messenger RNA in protein construction. This is accomplished by inserting strands of complementary RNA into the animal, which dock with the gene's messenger RNA (Figure 4.20). The cell recognizes this newly formed molecule as abnormal and releases an enzyme that destroys the RNA.

In *gene transfer,* a gene from another organism is inserted into the recipient's cells. An important research tool is the *transgenic animal,* created by inserting the

TABLE 4.2 Comparison of Relationship Studies.

Family Study	Adoption Study	Twin Study
• Indicates how strongly a characteristic is shared among relatives • Can show that a characteristic follows family lines • Confounds heredity and environment	• Compares adopted children with their adoptive parents and their biological parents • Confounding can occur because the adoption variable is not manipulated	• Compares similarity of identical twins with similarity of fraternal twins • Allows comparison of two levels of genetic similarity

FIGURE 4.20   Antisense RNA.

Antisense RNA is a strand of RNA that is complementary to a particular messenger RNA. The two will dock with each other, which disables the messenger RNA and halts production of its protein. The researcher observes differences in the animal to determine the gene's function.

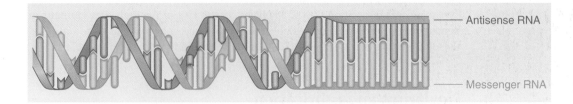

— Antisense RNA

— Messenger RNA

gene into the developing embryo. The most frequently used techniques produce an animal *mosaic,* in which the gene shows up in only some of the animal's cells, but after these animals are mated with each other, the gene is integrated into all the cells, including the sperm and the ova. Researchers use gene transfer to determine a gene's function by observing the transferred gene's effects in recipient animals. An obvious application is gene transfer to treat human disease. Transfer occurs in the developed individual, after the disease has been diagnosed, and is limited to the tissue involved in the disease. The gene is usually placed within a harmless virus, which then infects the cells.

Genetic engineering is becoming a therapeutic reality. Between 1999 and 2005 *gene therapy* was used successfully to treat at least 17 children with SCID, the disease Ashanthi had ("Gene Therapy Notches Another Victory," 2005). Researchers recently used stem cells from two 7-year-old boys to carry a corrective gene into their bloodstream and halt the progress of a demyelinating brain disorder (the disease that was the subject of the movie *Lorenzo's Oil*; Cartier et al., 2009). In Chapter 12, you will see that doctors are having some success using gene transfer to treat Alzheimer's disease. While these results are promising, the ability to manipulate our genome carries tremendous risks and raises important ethical questions.

> *There will be a gene-based treatment for essentially every disease within 50 years.*
>
> —*W. French Anderson*

Concept Check
Take a Minute to Check Your Knowledge and Understanding

- ☐ Organize your knowledge: Make a table of the staining and labeling techniques and their functions.

- ☐ What different ways could be used to determine the function of a part of the brain?

- ☐ Name three procedures that can be used to identify receptors.

- ☐ What are the advantages and disadvantages of experimental studies and correlational studies?

- ☐ Describe the different genetic manipulation strategies discussed in this chapter.

Q What are the main issues in research integrity?

Research Ethics

As important as research ethics is, the topic usually gets pushed into the background by the excitement of scientific accomplishments and therapeutic promise. To place ethics at the forefront where it belongs, the major scientific and medical organizations have adopted strict guidelines for conducting research, for the treatment of subjects, and for communicating the results of research (see, e.g., American Psychological Association, 2002; "Policies on the Use of Animals and Humans in Neuroscience Research," n.d.; "Public Health Service Policy on Humane Care and Use of Laboratory Animals," 2002; "Research Involving Human Subjects," n.d.).

Plagiarism and Fabrication

The success of research in answering questions and solving problems depends not only on the researchers' skill in designing studies and collecting data but also on their accuracy and integrity in communicating results. Unfortunately, research is sometimes intentionally misrepresented; the two cardinal sins of research are plagiarism and the fabrication of data.

Plagiarism is the theft of another's work or ideas. Plagiarism denies individuals the credit they deserve and erodes trust among the research community. The infraction may be as simple as failing to give appropriate credit through citations and references (like those you see throughout this text), but occasionally a researcher literally steals another's work. Perhaps the most blatant example is the case of the Polish medical school professor Andrzej Jendryczko, who plagiarized others' work in more than 40 of his biomedical publications (E. Marshall, 1998; Wronski, 1998). Most of the articles were retracted by the journal editors, and Jendryczko had to resign from his university as well as his institute, where he was a deputy director.

Fabrication, or faking, of results is more serious than plagiarism because it introduces erroneous information into the body of scientific knowledge. As a result, the pursuit of false leads by others consumes scarce resources and sidetracks researchers from more fruitful lines of research. More important, fabrication in clinical research can slow therapeutic progress and harm lives, so universities and agencies take research fraud seriously. In a recent case, Eric Poehlman falsified medical research data in his publications and then used the faked results to obtain $2.9 million in research funding. He agreed to repay $180,000 and to retract or correct numerous published articles, and he was sentenced to 1 year in prison ("Poehlman Sentenced to 1 Year of Prison," 2006; "Press Release: Dr. Eric T. Poehlman," 2005).

Although such cases are rare (E. Marshall, 2000b), they undermine confidence in scientific and medical research. Increasingly, concerned government agencies are taking steps to educate researchers about research ethics (R. Dalton, 2000), setting aside $1 million of grant money to support studies on research integrity (E. Marshall, 2000c) and discouraging ties between scientists and the companies whose products they are testing (Agnew, 2000).

> *Falsification is far more serious because it always corrupts the scientific record. It is a crime against science, indeed a crime against all humanity, when it legitimizes science that is false.*
>
> —David Crowe

Protecting the Welfare of Research Participants

All the scientific disciplines that use live subjects in their research have adopted strict codes for the humane treatment of both humans and animals. The specifics of the treatment of human research participants and even the legitimacy of animal research are controversial, however. These are not abstract issues. As a student, you are a consumer of the knowledge that human and animal research produce, and you benefit personally from the medical and psychological advances, so you are more than just a neutral observer.

2
Official Ethics Policies

Research With Humans

In 1953, the psychologist Albert Ax performed a study that was as significant for its ethical implications as for its scientific results. He was attempting to determine whether all emotions result in the same general bodily arousal or each emotion produces a unique pattern of activation. To do so, he measured several physiological variables sensitive to emotional arousal, such as heart rate, breathing rate, and skin temperature, while inducing anger in the individuals at one time and fear at another. If Ax had told the research participants what would happen during the study, it would have altered their behavior, so he said he was doing a study of blood pressure.

In the "anger" condition, the research participant was insulted by Ax's assistant, who complained at length that the person was not a very good experimental subject. The "fear" situation was more intense. During the recordings the individual received a mild electrical shock through the recording electrodes, while sparks jumped from nearby equipment. The experimenter acted alarmed as he explained that there was a dangerous high-voltage short circuit. Later interviews indicated that both ruses worked. Ax (1953) reported that one participant kept pleading during the fear treatment, "Please take the wires off. Oh! Please help me." Another said of the anger treatment, "I was just about to punch that character on the nose" (p. 435). You will find the results of Ax's research in Chapter 8; in the meantime, we will look at the issues of informed consent and deception in relation to his study.

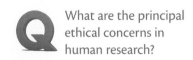 What are the principal ethical concerns in human research?

Occasionally, research involves some pain, discomfort, or even risk. Before proceeding with a study, current standards require the researcher to obtain the participant's informed consent. *Informed consent* means that the individual voluntarily agrees to participate after receiving information about any risks, discomfort, or other adverse effects that might occur. However, sometimes the nature of a study requires the researcher to use *deception,* either failing to tell the participants the exact purpose of the research or what will happen during the study or actively misinforming them. According to the American Psychological Association (APA), deception is acceptable only when the value of the study justifies it, alternative procedures are not available, and the individuals are correctly informed afterward. The APA's guidelines are also clear that psychologists should not deceive subjects about research that is reasonably expected to cause physical pain or severe emotional distress (American Psychological Association, 2002). Some researchers and subjects' rights advocates believe that deception is never justified. Ax's study would probably not be permitted today, but we will see in Chapter 8 that researchers have found interesting alternatives for doing this kind of research.

Research With Animals

Psychological and medical researchers have perhaps no more important resource than the laboratory animal. As the American Medical Association (1992) concluded, "Virtually every advance in medical science in the twentieth century, from antibiotics and vaccines to anti-depressant drugs and organ transplants, has been achieved either directly or indirectly through the use of animals in laboratory experiments" (p. 11). Psychologists have used animals to investigate behavior, aging, pain, stress, and cognitive functions such as learning and perception (Blum, 1994; F. A. King, Yarbrough, Anderson, Gordon, & Gould, 1988; N. F. Miller, 1985). It may seem that the best subjects for that purpose would be humans, but animals are useful because they live in a controlled environment and have a homogeneous history of experience, as well as a briefer development and life span. In addition, researchers feel that it is more ethical to use procedures that are painful or physically or psychologically risky on other animals rather

 What are the opposing views on animal research?

> *Every one has heard of the dog suffering under vivisection, who licked the hand of the operator: this man, unless he had a heart of stone, must have felt remorse to the last hour of his life.*
>
> —*Charles Darwin,*
> The Descent of Man *and*
> Selection in Relation to Sex*, 1871*

than humans. As a result, in the mid-1980s, U.S. scientists were using 20 million animals a year: 90% of them were rodents, mostly mice and rats, and around 3.5% were primates, mostly monkeys and chimpanzees (U.S. Congress Office of Technology Assessment, 1986).

The preference for inflicting discomfort or danger on nonhuman animals rather than on humans is based on the assumption that the suffering of animals is more acceptable than suffering in humans. Animal rights activists have called this dual ethical standard *speciesism* (H. A. Herzog, 1998), a term chosen to be intentionally similar to *racism.* Some activists work hand in hand with researchers to improve conditions for research animals; others have been more aggressive, breaking into labs, destroying equipment and records, and releasing animals (typically resulting in the animals' death). In Europe and Britain, death threats have forced some researchers to move their work behind high fences (Koenig, 1999; Schiermeier, 1998), the University of Cambridge had to abandon a primate research building, and the University of Oxford was forced to shut down construction of a research facility (Proffitt, 2004). And at the University of California at Santa Cruz, one animal researcher's home was firebombed, and another's car was burned (G. Miller, 2008).

Animal research guidelines provide for humane housing of animals, attention to their health, and minimization of discomfort and stress during research (American Psychological Association, 2002; "Guidelines for Ethical Conduct in the Care and Use of Animals," n.d.; Policies on the Use of Animals and Humans in Neuroscience Research," n.d.). But critics point out that researchers sometimes do not live up to the standards of their professional organizations. The Behavioral Biology Center in Silver Spring, Maryland, was engaged in research that involved severing the sensory nerve in one arm of monkeys to study the reorganization that occurs in the brain. The lab's contributions drew the praise of neuroscientists and led to the design of routines for extensive exercise of an afflicted limb to help people recover from brain injuries. But in 1981, a student summer employee informed the police of what he considered to be abuse of the lab's animals, and the police carried out the first raid on a research laboratory in the United States ("A Brighter Day for Edward Taub," 1997; Orlans, 1993). The director, Edward Taub, was convicted of animal abuse because of poor postoperative care, but the conviction was overturned because the state lacked jurisdiction over federally supported scientific research. The National Institutes of Health withdrew Taub's funding, and Congress enacted more stringent animal protection laws. In spite of the controversy, Taub received the William James Award from the American Psychological Society. However, he points out that the award was for work that is no longer permitted and that animal welfare rules enacted by Congress prevent him from taking measurements in the brain of the one remaining monkey for the length of time that would be needed.

The conflict between animal welfare and research needs is obviously not a simple issue and is strongly felt on both sides (Figure 4.21). Although psychologists and neuroscientists do not condone mistreatment of research animals, most of them argue that the suffering that does occur is justified by the benefits animal research has produced. The 2000 Nobel Prize in physiology or medicine was shared by three neuroscientists: Arvid Carlsson, for his discovery of the role of dopamine as a neurotransmitter in the brain; Paul Greengard, for identifying how dopamine and related neurotransmitters produce their effect on neurons; and Eric Kandel, for his work on the molecular changes that occur in the brain during learning. The work of all three prize winners relied heavily on animal research.

It is unlikely that animal research will be banned as the more extreme activists demand, but animal care and use guidelines have been tightened and outside monitoring increased, and states have passed more stringent laws. In addition, researchers have become more sensitive to the welfare of animals, adopting more

3
Benefits and Ethics of Animal Research

FIGURE 4.21 Animal Research Controversy.

The poster (a) and the demonstration (b) illustrate the contrasting views on animal research.

(a)

(b)

SOURCES: (a) Foundation for Biomedical Research; (b) Paul Conklin/PhotoEdit.

humane methods of treatment and turning to tissue cultures and computer simulations in place of live animals when possible. In a survey of articles published in major biomedical journals between 1970 and 2000, the proportion of studies using animal subjects had fallen by one third, and in the studies that did use animals, the average number had dropped by half (Carlsson, Hagelin, & Hau, 2004).

Human research has typically generated less controversy than the use of animals, largely because scientists are more restrained in their treatment of humans and humans are able to refuse to participate and to bring lawsuits. The balance of concern is shifting, though, particularly in the transition from the research lab to the clinic.

Gene Therapy

Gene therapy, the treatment of disorders by manipulating genes, has enjoyed glowing press reviews because of its potential for correcting humanity's greatest handicaps and deadly diseases. But a distinct chill fell over the research in 1999, when Jesse Gelsinger, an 18-year-old volunteer, became the first human to die as the direct result of gene research (Lehrman, 1999; E. Marshall, 2000a). The study was using a deactivated form of adenovirus, which causes the common cold, to transport a gene into the liver in an experimental attempt to correct a genetic liver

What are the problems with gene research and gene therapy?

4
Gene Therapy

FIGURE 4.22 More Protection for Human Subjects.

Mounting concerns are leading to more stringent controls on human research.

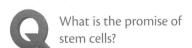

SOURCE: © Andrew Birch, 2007. Used with permission.

Q What is the promise of stem cells?

5
Stem Cell Research

enzyme deficiency. Gelsinger developed an immune reaction to the adenovirus, which resulted in his death.

The Food and Drug Administration (FDA), which was overseeing the study, reprimanded the researchers for not consulting with the FDA when most of the patients developed mild adverse reactions and for not informing the research participants that two monkeys had died in an earlier study after receiving much larger doses of adenovirus ("U.S. Government Shuts Down Pennsylvania Gene Therapy Trials," 2000). The university was assessed a $1 million fine, and the three principal investigators had restrictions placed on their human research until 2010 (Check, 2004). The case has slowed gene therapy research across the country, but a positive outcome is that it is expected to lead to stricter supervision of human studies (Figure 4.22).

There are additional concerns that gene manipulation could affect the reproductive cells and change the genome of nonconsenting future generations, a questionable outcome at least when survival is not at stake. As a result, the American Association for the Advancement of Science (2000) has called for a moratorium on research that might produce inheritable modifications. Even after the technology is deemed safe and reliable, important issues still remain. Concerned about privacy, U.K. lawmakers have made it illegal to analyze a person's DNA without that person's consent ("Sneaky DNA Analysis to Be Outlawed," 2004). And because gene therapy is very expensive, it is likely to increase inequalities further between the haves and have-nots in our society. Some worry that its application will not be limited to correcting disabilities and disease but will be used to enhance the beauty, brawn, and intelligence of the offspring of well-to-do parents. The science fiction movie *GATTACA* (whose title is a play on the four letters of the genetic code) depicts a society in which privilege and opportunity are reserved for genetically enhanced "superior" individuals. The In the News feature suggests that we may be taking a step toward a *GATTACA*-like world already. Fortunately, the U.S. Congress wisely set aside 5% of the Human Genome Project budget to fund the study of the ethical, legal, and social implications of genetic research (Jeffords & Daschle, 2001).

Stem Cell Therapy

You learned in the previous chapter that embryonic stem cells are undifferentiated cells that have the potential for developing into any other body cell. Stem cells

A Brave New World?

In the News

We may be just a bit closer than you think to a world in which a person's opportunities depend on the results of genetic testing and the ability to afford genetic enhancement. In early 2009, a U.S. fertility clinic announced that in 6 months, it would offer parents the opportunity to choose some of their child's traits through egg and sperm selection ("Designer Babies," 2009). Dr. Jeff Steinberg, who heads the clinic, was talking only about characteristics like hair and eye color, and some geneticists are not sure he can deliver even that, but ethicists are concerned that more substantive traits such as intelligence will be next. A few months later, CNN reported that for $880, Chinese parents can send their children to a summer camp for genetic testing and behavioral observation; at the end of the camp, the parents receive information about their child's projected IQ, emotional control, focus, memory, and athletic ability, along with recommendations for an appropriate career ("In China, DNA tests," 2009). One 3-year-old girl said, "I want to be the president of China. Then people will be scared of me."

have been used successfully to treat spinal cord damage (J. W. McDonald et al., 1999) and brain damage (Ren et al., 2000) in rats; they have also tracked down tumors in the brains of mice and delivered *interleukin 12*, making it easier for immune cells to kill the tumors (Ehtesham et al., 2002). In humans, heart functioning improved in patients with congestive heart failure after injection of stem cells (A. N. Patel et al., 2004). Medical researchers hope that stem cells can eventually be used to grow human organs in the laboratory to supply organ transplants and to allow genetic researchers to watch a gene produce a diseased organ rather than working backward from the diseased patient to the gene. An estimated 28 million people in the United States alone have diseases that are potentially treatable by stem cell therapy (Perry, 2000).

So if stem cells hold such wonderful potential, why are they being discussed under the topic of *ethics*? Extracting stem cells destroys the embryo, so right-to-life advocates oppose this use of human embryos, even though most are "extras" resulting from fertility treatment and would otherwise be discarded (Figure 4.23). Stem cell research was crippled for years in the United States by the Bush administration's refusal to fund research on stem cell lines derived after August 2001; earlier lines were too contaminated for human use (M. J. Martin, Muotri, Gage, & Varki, 2005). This policy was reversed in 2009 by President Obama, but the use of embryonic stem cells is still controversial in some quarters. And in spite of the European Union's efforts to achieve unity on research policy, views differ across its 27 member nations as much as they do in the United States (Drumi, 2009). Due to this controversy and the limited availability of embryonic stem cells, other sources are being sought; for example, the stem cells used to treat the heart patients were taken from their own bone marrow, and two studies have shown that stem cells can be induced from mature cells by turning on the appropriate genes (Kaji et al., 2009; Woltjen et al., 2009).

Some critics are calling for a slower pace in implementing gene and stem cell therapies, pointing to incidents like the Jesse Gelsinger case. More recently, three children being treated for SCID with gene therapy have developed leukemia, and one has died ("Therapy setback," 2005). Apparently the retrovirus used to transfer the gene triggered the leukemia by activating a gene involved in cell proliferation. Critics suggest that there are unknown dangers as well; for example, sometimes stem cells injected into animals have found their way into tissues throughout the body, and we don't know what all the consequences might be. In the case of the patients treated for heart disease, there is some question whether the stem cells repaired the patients' hearts or some other factor, such as the chemicals used in the bone marrow preparation, accounted for the improvement (Check, 2004). Many scientists are reluctant to undertake large clinical trials until there is more information about the benefits as well as the risks of stem cell therapies. When the implications of research are so far-reaching, restraint is as valuable as enthusiasm and commitment.

FIGURE 4.23 Injecting Cells Into a Damaged Brain.

(a) Although the procedure is promising, it is controversial because of where the cells come from, which is usually (b) human embryos.

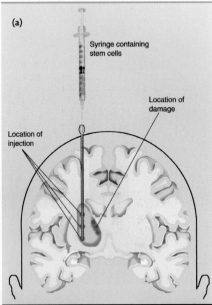

(a)

Syringe containing stem cells

Location of damage

Location of injection

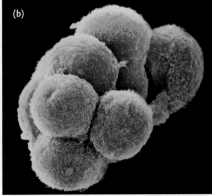

(b)

SOURCE: (a) From Coon. *Introduction to Psychology*, 9e. © 2001 Wasdworth, a part of Cengage Learning, Inc. Reproduced by permission. www.cengage.com/permissions; (b) Motta/Photo Researchers.

Concept Check   *Take a Minute to Check Your Knowledge and Understanding*

☐ What are the effects of dishonesty in research?

☐ How do researchers justify their use of animals in research?

☐ Why is there an ethical issue with human stem cell research; how might it be resolved?

In Perspective

Early progress in psychology and in biopsychology relied on the wit and perspiration of the pioneering researchers. Now they are aided by sophisticated equipment and methods that are escalating discovery at an unprecedented pace and taking research into areas that were barely conceivable a few decades ago.

Knowledge is power, and with power comes responsibility. For the scientists who study behavior, that responsibility is to the humans and animals that provide the source of our knowledge and to the people who may be healed or harmed by the new treatments resulting from research.

Summary

Science, Research, and Theory

- Researchers respect uncertainty but try to reduce it through research and the use of theory.
- Of the many research strategies at their disposal, biopsychologists favor the experimental approach because of the control it offers and the ability to determine cause and effect.
- Correlational techniques have value as well, particularly when the researcher cannot control the situation.

Research Techniques

- Staining and labeling techniques make neurons more visible, emphasize cell bodies or axons, trace pathways to or from a location, and identify active areas or specific structures such as receptors.
- Light microscopy is extremely useful, but electron microscopy reveals more detail, scanning electron microscopy adds three-dimensionality, and confocal and two-photon microscopy produce images at greater depths.
- The EEG sums the neural activity between two electrodes to assess arousal level and detect damage and some brain disorders. Evoked potentials measure averaged responses to brief stimuli.
- Brain functioning can be studied by observation of brain-damaged individuals, electrical and chemical stimulation (using stereotaxic techniques), destruction of neural tissue (ablation and lesioning), and microdialysis of brain chemicals.
- Brain imaging using CT and MRI depicts structure, for example, to assess damage, while PET and fMRI are capable of measuring activity.
- Family studies, adoption studies, and twin studies are correlational strategies for investigating heredity. Family studies determine whether a characteristic runs in families, while adoption studies assess whether adopted children are more like their adoptive parents or their biological parents in a characteristic. Twin studies compare the similarity of fraternal twins with the similarity of identical twins.
- Genetic engineering includes gene transfer and gene-disabling techniques (knockout and antisense RNA). Although experimental, it is already showing therapeutic promise.

Research Ethics

- A major concern in biopsychology is maintaining the integrity of research; plagiarism and fabrication of data are particularly serious infractions.
- Both the public and the scientific community are increasingly concerned about protecting the welfare of humans and animals in research. The various

disciplines have standards for subject welfare, but the need for more monitoring and training is evident.

- Stem cell technology is promising for treating brain and spinal cord damage and a variety of diseases, but it is controversial because obtaining stem cells usually involves destroying embryos. Gene therapy also holds much promise, but it has dangers and could be abused. ■

Study Resources

F For Further Thought

- Pay close attention as you read through this text, and you will notice that human studies are more likely than animal studies to be correlational. Why do you think this is so?
- Genetic engineering is mostly a research technique now; what practical uses can you imagine in the future?
- Is it unreasonable coercion to (a) require a student in an Introduction to Psychology course to participate in research, (b) require a student in a Research Methods course to participate in research exercises during the laboratory sessions as a part of the educational experience, or (c) offer money and a month's housing and meals to a homeless person to participate in a risky drug study?
- Do you think the rights of humans and animals are adequately protected in research? Why or why not? What do you think would be the effect of eliminating the use of animal subjects?

T Testing Your Understanding

1. Describe the four imaging techniques, including method, uses, and advantages/disadvantages.

2. Discuss the relative merits of experimental and correlational research, using family/twin/adoption studies versus genetic engineering as the example.

3. Discuss the conflicts between research needs and animal rights.

4. In spite of their promise, stem cell research and gene therapy are controversial. Why?

Select the best answer:

1. You could best identify receptors for acetylcholine by using
   a. Golgi stain.
   b. Nissl stain.
   c. immunocytochemistry.
   d. electron microscopy.

2. If you needed to measure brain activity that changes in less than 1 s, your best choice would be
   a. EEG.
   b. CT.
   c. MRI.
   d. PET.

3. Your study calls for daily measurement of activity changes in emotional areas of the brain. You would prefer to use
   a. CT.
   b. MRI.
   c. PET.
   d. fMRI.

4. Science is most distinguished from other disciplines by
   a. the topics it studies.
   b. the way it acquires knowledge.
   c. its precision of measurement.
   d. its reliance on naturalistic observation.

5. Experiments are considered superior to other research procedures because they
   a. involve control over the variable of interest.
   b. permit control of variables not of interest.
   c. permit cause-and-effect conclusions.
   d. All the above
   e. None of the above

6. A theory
   a. is the first step in research.
   b. is the final stage of research.

c. generates further research.

d. is an opinion widely accepted among researchers.

7. The best way to assess the relative contributions of heredity and environment would be to compare the similarity in behavior of
a. fraternal versus identical twins.
b. relatives versus nonrelatives.
c. siblings reared together versus those reared apart.
d. fraternal versus identical twins, half of whom have been adopted out.

8. The most sensitive way to determine whether a particular gene produces a particular behavior would be to
a. compare the behavior in identical and fraternal twins.
b. compare the behavior in people with and without the gene.
c. use genetic engineering to manipulate the gene and note the behavior change.
d. find out whether people with the behavior have the gene more often than people without the behavior.

9. Antisense RNA technology involves
a. inserting a gene into the subject's cells.
b. interfering with protein construction controlled by the gene.

c. introducing a nonfunctioning mutation into the subject's genes.
d. All the above

10. The most popular research animals among the following are
a. rats.   b. pigeons.
c. monkeys.   d. chimpanzees.

11. Speciesism refers to the belief that
a. humans are better research subjects than animals.
b. it is more ethical to do risky experiments on lower animals than on humans.
c. humans are the superior species.
d. All the above
e. None of the above

12. The biggest obstacle to using stem cells would be eliminated if researchers could
a. get adult stem cells to work as well as embryonic ones.
b. get stem cells to differentiate into neurons.
c. get stem cells to survive longer.
d. get stem cells to multiply faster.

1. c, 2. a, 3. d, 4. b, 5. d, 6. c, 7. d, 8. c, 9. b, 10. a, 11. b, 12. a.

Answers:

Online Resources

The following resources are available at www.sagepub.com/garrett3e.

Chapter Resources

- Flash Cards
- Chapter Quiz
- Internet Resources and Exercises

On the Web

Links to these websites can be found at the web address above: Select this chapter's **Chapter Resources**, then choose **Internet Resources and Exercises**.

1. **Brain Imaging** compares the advantages and disadvantages of different imaging techniques, along with sample images.

    fMRI 4 Newbies is the whimsical title of a site filled with images and information, as well

as humor (such as "Ten Things Sex and Brain Imaging Have in Common.")

2. The American Psychological Association establishes **Ethical Principles of Psychologists and Code of Conduct**, covering research and publication, therapeutic practice, and conflict of interest, as well as numerous other areas.

    The National Institutes of Health publishes its policies on the use of **human** and **animal** research.

3. You can download several *Scientific American* articles expressing contrasting opinions on the **Benefits and Ethics of Animal Research**.

    Research With Animals in Psychology is a justification of the use of animals in behavioral research by the Committee on Animal Research and Ethics of the American Psychological Association.

4. The Human Genome Project's **Gene Therapy** page is an informative source on gene therapy procedures and ethics as well as recent therapeutic developments.

5. **The International Society for Stem Cell Research** website has news, recent research, photos and movie clips, and ethics essays related to stem cell research.

Tristem Corporation is dedicated to creating stem cells from mature adult cells, and its website has informative stem cell articles and press releases along with several colorful images of stem cells.

Chapter Updates and Biopsychology News

R For Further Reading

1. Opposing views of several writers on research deception are presented in the *American Psychologist,* July 1997, 746–747, and July 1998, 803–807.

2. The *Scientific American* article "The World's First Neural Stem Cell Transplant," by K. Mossman (www.scientificamerican.com/article.cfm?id=the-worlds-first-neural-s) describes some of the promise of stem cell therapy, while "Reality Check: The Inevitable Disappointments From Stem Cells" (www.scientificamerican.com/article.cfm?id=reality-check-for-stem-cells) addresses technical, social, and political challenges that remain.

3. "Remote Control Brains," by Douglas Fox (*New Scientist,* July 27, 2007, 30–34) is a nontechnical review of the field of optogenetics, the strategy described in Chapter 2 for controlling neurons

with light; "The Optogenetic Catechism" (Gero Miesenböck, *Science, 326,* 395–399) is a more thorough treatment.

4. "Why You Should Be Skeptical of Brain Scans," by Michael Shermer (*Scientific American Mind,* Oct/Nov 2008, 66–71) is the layperson's guide to understanding why interpretations of brain scan research can be misleading.

5. The word *transgenic* is rarely seen without *mice*, because transgenic animals have almost exclusively been mice, but now the marmoset has become the first transgenic primate, enabling research with a subject considerably closer to humans ("Generation of Transgenic Non-Human Primates with Germline Transmission," by Erika Sasaki et al., *Nature,* 2009, *459,* 523–528).

K Key Terms

ablation ... 102
adoption study ... 109
antisense RNA ... 110
autoradiography ... 96
computed tomography (CT) ... 103
concordance rate ... 109
correlation ... 108
correlational study ... 93
deception ... 113
electroencephalogram (EEG) ... 99
electron microscope ... 98
evoked potential ... 99
experiment ... 93
fabrication ... 112
family study ... 108
functional magnetic resonance imaging (fMRI) ... 106
gene therapy ... 115
gene transfer ... 110
genetic engineering ... 110
Golgi stain ... 95
immunocytochemistry ... 97
informed consent ... 113
in situ hybridization ... 97
knockout ... 110
lesion ... 102
magnetic resonance imaging (MRI) ... 103
messenger ribonucleic acid (RNA) ... 97
myelin stain ... 95
Nissl stain ... 95
plagiaris ... 112
positron emission tomography (PET) ... 105
stereotaxic instrument ... 100
theory ... 93
transcranial magnetic stimulation ... 102
twin study ... 109

PART II

Motivation and Emotion: What Makes Us Go

Chapter 5. Drugs, Addiction, and Reward

Chapter 6. Motivation and the Regulation of Internal States

Chapter 7. The Biology of Sex and Gender

Chapter 8. Emotion and Health

Drugs, Addiction, and Reward

5

Psychoactive Drugs

Opiates

Depressants

Stimulants

Psychedelics

Marijuana

   IN THE NEWS: CONTROVERSY OVER MEDICAL
     MARIJUANA TAKES A NEW TURN

   CONCEPT CHECK

Addiction

The Neural Basis of Addiction and Reward

Dopamine and Reward

   APPLICATION: IS COMPULSIVE GAMBLING AN ADDICTION?

Dopamine, Learning, and Brain Plasticity

Treating Drug Addiction

   CONCEPT CHECK

The Role of Genes in Addiction

Separating Genetic and Environmental Influences

What Is Inherited?

Implications of Addiction Research

   CONCEPT CHECK

In Perspective

Summary

Study Resources

In this chapter you will learn

- The major classifications of drugs and some of their effects

- What happens in the brain during addiction

- How addiction is treated pharmacologically

- How heredity influences addiction

Honoré de Balzac (Figure 5.1) wrote a phenomenal 45 novels in 20 years. He was aided in his long writing marathons by large amounts of a stimulant drug whose effects pleased him so much that he advocated its use to others. However, he died at the age of 51 in part because of this unrelenting stimulation. What was the powerful drug that contributed both to his success and to his untimely death? According to his physician, Balzac died from a heart condition, aggravated by "the use or rather the abuse of coffee, to which he had recourse in order to counteract man's natural propensity to sleep" ("French Roast," 1996, p. 28).

There is good reason to consider caffeine an addictive drug. Coffee may have milder effects than the other drugs coming out of Colombia, but strength of effect and illegality are not the criteria for classifying a substance as addictive. As you will see, a drug's effect on the brain is the telling feature, and that is our reason for discussing drugs at this particular point: It provides the opportunity to tie together our preceding discussions of brain structures and neural (particularly synaptic) functioning.

Psychoactive Drugs

A *drug* is a substance that on entering the body changes the body or its functioning. Drugs fall into one of two general classes, according to their effect on a transmitter system. As we saw in Chapter 2, an *agonist* mimics or enhances the effect of a neurotransmitter. It can accomplish this by having the same effect on the receptor

FIGURE 5.1 Honoré de Balzac.

as the neurotransmitter, by increasing the transmitter's effect on the receptor, or by blocking the reuptake or the degradation of the transmitter. An *antagonist* may occupy the receptors without activating them, simultaneously blocking the transmitter from binding to the receptors. Or it may decrease the availability of the neurotransmitter, for example, by reducing its production or its release from the presynaptic terminals.

Psychoactive drugs are those that have psychological effects, such as anxiety relief or hallucinations. The focus of this chapter is on abused psychoactive drugs, although many of the principles discussed here are applicable more generally. We will discuss several psychotherapeutic drugs later, in the chapter on psychological disorders (see Chapter 14). The effects of abused drugs are extremely varied, but whether they arouse or relax, expand the consciousness or dull the senses, addictive drugs initially produce a sense of pleasure in one form or another. They also have several other effects in common; reviewing those effects will give us the language we need for a discussion of how the drugs work.

Most of the abused drugs produce addiction; *addiction* is identified by preoccupation with obtaining a drug, compulsive use of the drug in spite of adverse consequences, and a high tendency to relapse after quitting. Many abused drugs also produce withdrawal reactions. *Withdrawal* is a negative reaction that occurs when drug use is stopped. Withdrawal symptoms are due at least in part to the fact that the nervous system has adapted to the drug's effects, so they are typically the opposite of the effects the drug produces. For example, the relaxation, constipation, chills, and positive mood of heroin are replaced by agitation, diarrhea, fever, and depression during withdrawal.

Regular use of most abused drugs results in tolerance; *tolerance* means that the individual becomes less responsive to the drug and requires increasing amounts of the drug to produce the same results.

SOURCE: Hulton Archive.

Like withdrawal, tolerance results from compensatory adaptation in the nervous system, mostly a reduction in receptor number or sensitivity. Tolerance is one reason for overdose. Tolerance can occur to some of a drug's effects without occurring to others. Thus, if the drug abuser takes larger doses of heroin to achieve the original sense of ecstasy while the tendency to produce sleep and respiratory arrest are undiminished, overdose is nearly inevitable, and the consequences can be deadly.

Opiates

The *opiates* are drugs derived from the opium poppy (see Figure 5.2). Opiates have a variety of effects: They are *analgesic* (pain relieving) and *hypnotic* (sleep inducing), and they produce a strong *euphoria* (sense of happiness or ecstasy). Their downside is their addictive potential. *Opium* has been in use since around 4000 BCE (Berridge & Edwards, 1981); originally it was eaten, but when explorers carried the American Indians' practice of pipe smoking of tobacco back to their native countries, opium users adopted this technique. *Morphine* was extracted at the beginning of the early 1800s and has been extremely valuable as a treatment for the pain of surgery, battle wounds, and cancer. *Heroin* was synthesized from morphine in the late 1800s; at the turn of the century, it was marketed by the Bayer Drug Company of Germany as an over-the-counter analgesic until its dangers were recognized. It is now an illegal drug in the United States but is available for clinical use in Canada and Great Britain (Cherney, 1996). Codeine, another ingredient of opium, has been used as a cough suppressant, and dilute solutions of opium, in the form of paregoric and laudanum (literally, "something to be praised"), were once used to treat diarrhea and to alleviate pain, and paregoric was even used to quiet fretful children. Morphine continues to be used with cancer patients and is showing promise of safe use with milder pain in a time-release form that virtually eliminates the risk of addiction. Aside from morphine, opiates have mostly been replaced by safer synthetics, which are called *opioids* to indicate that they are not derived from opium, at least not directly. However, opioids are also subject to abuse; you probably recognize OxyContin because of its reputation for abuse rather than for its pain-relieving ability.

Heroin is the most notoriously abused opiate owing to its intense effect: a glowing, orgasm like sensation that occurs within seconds, followed by drowsy relaxation and contentment. Because heroin is highly soluble in lipids, it passes the blood-brain barrier easily; the rapid effect increases its addictive potential. The major danger of heroin use comes from overdose—either from the attempt to maintain the pleasant effects in the face of increasing tolerance or because the user unknowingly obtains the drug in a purer form than usual. In a 33-year study of 581 male heroin addicts, 49% were dead at the end of the study, with an average age at death of 46 years (Hser, Hoffman, Grella, & Anglin, 2001). Nearly a fourth of those had died of drug overdoses (mostly from heroin); 19.5% died from homicide, suicide, or accident; and 15% died from chronic liver disease. Half of the survivors who could be interviewed were still using heroin, and the high likelihood of returning to usage even after 5 years or more of abstinence suggested to the researchers that heroin addiction may be a lifelong condition. In spite of the representation of the horrors of heroin withdrawal in movies, it is best described as similar to a bad case of flu, so apparently fear of withdrawal is not the prime motivator for continued heroin addiction.

As tolerance to a drug develops, it also becomes associated with the person's drug-taking surroundings and circumstances. This learned or *conditioned tolerance* does not generalize completely to a new setting; when the person buys and takes the drug in a different neighborhood, the usual dose could lead to overdose (Macrae, Scoles, & Siegel, 1987; S. Siegel, 1984). Heroin is a particularly good example; an amount of heroin that killed 32% of rats injected in their customary drug-taking environment killed 64% of rats injected in a novel environment (S. Siegel, Hinson, Krank, & McCully, 1982).

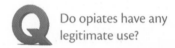

Do opiates have any legitimate use?

FIGURE 5.2 An Opium Poppy.

SOURCE: © SERDAR YAGCI/iStockphoto.

1
Effects of Heroin and Other Drugs

For years researchers puzzled at the effectiveness of opiate pain relievers. They suspected that the brain had receptors that are specific for opiate drugs, but there was no direct evidence. In one of Candace Pert's studies described in the previous chapter, she and Solomon Snyder incubated brain tissue in naloxone, an opiate antagonist that was used clinically to reverse opiate intoxication. The naloxone had been made radioactive, and it left the tissue radioactive even after thorough washing; this confirmed for the researchers that the naloxone had attached itself to receptors. So why does the brain have receptors for an abused drug? The answer is that the body produces its own *natural opioids*. Opiate drugs are effective because they mimic these *endogenous* (generated within the body) opioids, known as *endorphins*. One effect of endorphins is pain relief, as we will see in Chapter 11. Stimulation of endorphin receptors triggers some of the positive effects of opiates; others occur from indirect activation of dopamine pathways.

Depressants

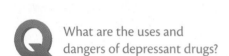

What are the uses and dangers of depressant drugs?

Depressants are drugs that reduce central nervous system activity. The group includes *sedative* (calming) drugs, *anxiolytic* (anxiety-reducing) drugs, and hypnotic drugs. Alcohol, of course, is the most common and is the most abused in this class, so we will start there.

Alcohol

"

I had booze, and when I was drinking, I felt warm and pretty and loved—at least for a while.

—*Gloria, a recovering alcoholic*

"

Ethanol, or *alcohol*, is a drug fermented from fruits, grains, and other plant products; it acts at many brain sites to produce euphoria, anxiety reduction, sedation, motor incoordination, and cognitive impairment (Koob & Bloom, 1988). It is the oldest of the abused drugs; its origin is unknown, but it was probably discovered when primitive people found that eating naturally fermented fruit had a pleasant effect. (Even elephants sometimes congregate under trees to eat fallen and fermenting fruits until they become intoxicated!) Alcohol has historically played a cultural role in celebrations and ceremonies, provided a means of achieving religious ecstasy, and, especially in primitive societies, permitted socially sanctioned temporary indulgence in hostility and sexual misbehavior. In modern societies, controlled group drinking has been replaced by uncontrolled individual abuse.

Alcohol is valued by moderate users as a social lubricant and as a disinhibitor of social constraints, owing largely to its anxiety-reducing effect. Like many drugs, its effects are complex. At low doses, say a couple of drinks, it turns off the inhibition the cortex normally exerts over behavior, so it acts as a stimulant. As intake increases, alcohol begins to have a sedative or even hypnotic effect; behavior moves from relaxation to sleep or unconsciousness. Later, after the bout of drinking has ended, the alcohol is metabolized back to a low blood level and it becomes a stimulant again. That is why a few drinks in the evening may help you get to sleep at bedtime only to awaken you later in the night.

Because it interferes with cognitive and motor functioning as well as judgment, alcohol is involved in one third of all U.S. traffic fatalities (National Highway Traffic Safety Administration, 2006). In the United States and Canada, a person is legally considered too impaired to drive when the blood alcohol concentration (BAC) reaches 0.08%. As you well know, alcohol is also closely linked with violent crime; in fact, it is involved in 38% of incarcerations for violence (K. L. Greenfeld & Henneberg, 2001). Besides affecting judgment, alcohol reduces the anxiety that normally inhibits aggression (Pihl & Peterson, 1993).

Alcohol carries with it a host of health and behavioral problems. High levels of any depressant drug have the potential to shut down the brain stem, resulting in coma or death; a blood alcohol level of 0.5% is adequate to put the drinker at risk. A common result of chronic alcoholism is cirrhosis of the liver, which in

2
Alcoholics Anonymous

its severest form is fatal. In addition, the vitamin B1 deficiency that is associated with chronic alcoholism can produce brain damage and Korsakoff's syndrome, which involves severe memory loss along with sensory and motor impairment (Figure 5.3). Binge drinkers are more likely to be impulsive and to have learning and memory impairments (Stephens & Duka, 2008). Alcohol is the fifth leading cause of premature death and disability, which has led the World Health Organization to launch a worldwide campaign to reduce alcohol's toll on health (World Health Organization, 2009). Even abstinence can be dangerous for the alcoholic. Alcohol withdrawal involves tremors, anxiety, and mood and sleep disturbances; more severe reactions are known as *delirium tremens*—hallucinations, delusions, confusion, and, in extreme cases, seizures and possible death.

Considering the health risks, violence, and disruption of homes and livelihood, alcohol is more costly to society than any of the illegal drugs. In view of all the dangers of drinking, it seems amazing now that in 1961 a speaker at a symposium of psychiatrists and physicians on drinking expressed the group's consensus that "alcohol is the safest, most available tranquilizer we have" ("Paean to Nepenthe," 1961, p. 68).

Alcohol affects a number of receptor and neurotransmitter systems. First, it inhibits the release of glutamate (Hoffman & Tabakoff, 1993; Tsai, Gastfriend, & Coyle, 1995). You may remember from Chapter 2, Table 2.2, that *glutamate* is the most prevalent excitatory neurotransmitter. Glutamate reduction produces a sedating effect; then there is a compensatory increase in the number of glutamate receptors, which probably accounts for the seizures that sometimes occur during withdrawal. Alcohol also increases the release of *gamma-aminobutyric acid (GABA)*, the most prevalent inhibitory neurotransmitter (Wan, Berton, Madamba, Francesconi, & Siggins, 1996). The combined effect at these two receptors is sedation, anxiety reduction, muscle relaxation, and inhibition of cognitive and motor skills. Alcohol also affects opiate receptors (in turn increasing dopamine release), serotonin receptors, and cannabinoid receptors, which are also excited by marijuana (Julien, 2008); these actions likely account for the pleasurable aspects of drinking.

Alcohol specifically affects the A subtype of GABA receptor; because the $GABA_A$ receptor is important in the action of other drugs, we will give it special attention. It is actually a receptor complex, composed of five different kinds of receptor sites (Figure 5.4). One receptor, of course, responds to GABA. Its activation opens the receptor's chloride channel, and the influx of chloride ions hyperpolarizes the neuron. Other receptors in the complex

FIGURE 5.3 An Alcoholic Brain and a Normal Brain.

Note that the sulci have deepened and the ventricles have enlarged in the alcoholic brain. (The scans for each individual are from different depths of the brain.)

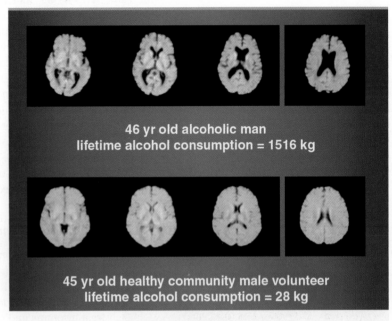

46 yr old alcoholic man
lifetime alcohol consumption = 1516 kg

45 yr old healthy community male volunteer
lifetime alcohol consumption = 28 kg

SOURCE: Durand & Barlow, *Essentials of Abnormal Psychology*, 3rd ed, credits: Dr. Adolf Pfefferbaum, Stanford U., with the support from the National Institute on Alcohol Abuse and Alcoholism and the Department of Veteran Affairs.

FIGURE 5.4 The GABA_A Receptor Complex.

The complex has receptors for GABA, barbiturates, benzodiazepines, and alcohol.

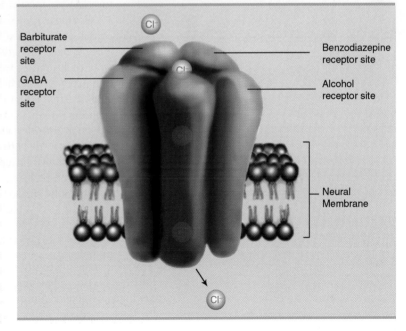

respond to alcohol, to barbiturates, and to benzodiazepines; these drugs enhance the binding of GABA to its receptor and thus its ability to open the chloride channel. Now you can understand why it is so dangerous to mix alcohol with barbiturates or benzodiazepines.

Alcohol passes easily through the placenta, raising the BAC of a fetus to about the same level as the mother's. You saw in Chapter 3 that prenatal exposure to alcohol can result in fetal alcohol syndrome (FAS; see Figure 5.5), which is the leading cause of intellectual impairment in the Western world (Abel & Sokol, 1986). Besides being intellectually impaired, FAS children are irritable and have trouble maintaining attention. Regular alcohol abuse apparently is not required to produce damage. In one study, mothers who had FAS children did not drink much more on average than the mothers of normal children, but they did report occasional binges of five or more drinks at a time (Streissguth, Barr, Bookstein, Sampson, & Olson, 1999).

Just having three or more drinks at any one time during pregnancy more than doubles the offspring's risk of drinking disorder during adulthood (Alati et al., 2006). No safe level of alcohol intake during pregnancy has been established, so most authorities recommend total abstention. (Refer to Figure 3.25 to see the developmental effects of FAS on a mouse brain.)

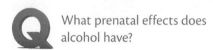

Q What prenatal effects does alcohol have?

Barbiturates and Benzodiazepines

Like alcohol, *barbiturates* in small amounts act selectively on higher cortical centers, especially those involved in inhibiting behavior; in low doses, they produce talkativeness and increased social interaction, and in higher doses, they are sedatives and hypnotics. Barbiturates have been used to treat insomnia and prevent epileptic seizures, and from 1912 to 1960, they were the drug of choice for treating anxiety and insomnia (Julien, 2008). They are not addictive in prescribed doses, but tolerance leads the person to increase the dosage, resulting in addictive symptoms similar to those of alcoholism. Like alcohol, barbiturates act at the GABA$_A$ complex, though at the barbiturate receptor, but unlike alcohol, they can open chloride channels on their own in the absence of GABA. As a result, the line between therapeutic and toxic levels is a fine one, and their use has been fraught with accidental and intentional overdose (including famous cases such as Marilyn Monroe, Judy Garland, and 1960s peace activist Abbie Hoffman). As a result, barbiturates were mostly replaced by much safer benzodiazepines.

Benzodiazepines act at the benzodiazepine receptor on the GABA$_A$ complex to produce anxiety reduction, sedation, and muscle relaxation. They reduce anxiety by suppressing activity in the *limbic system,* a network of structures we will consider in more detail in the chapter on emotion (Chapter 8). Their effect in the brain stem produces relaxation, while GABA activation in the cortex and hippocampus results in confusion and amnesia (Julien, 2008).

There are several benzodiazepine drugs, the best known of which are Valium (diazepam), Xanax (alprazolam), and Halcion (triazoplam). Because benzodiazepines are addictive and can produce mental confusion, they have been replaced in many cases by newer drugs. One of the benzodiazepines, Rohypnol (roofies or rophies), has gained notoriety as the date rape drug.

FIGURE 5.5 Child With Fetal Alcohol Syndrome.

Besides intellectual impairment and behavioral problems, FAS individuals often have facial irregularities, including a short, upturned nose that is flattened between the eyes, thin upper lip, and lack of a groove between the nose and upper lip.

SOURCE: George Steinmetz.

Stimulants

Stimulants activate the central nervous system to produce arousal, increased alertness, and elevated mood. They include a wide range of drugs, from cocaine to caffeine, which vary in the degree of risk they pose. The greatest danger lies in how they are used.

Cocaine

Cocaine, which is extracted from the South American coca plant, produces euphoria, decreases appetite, increases alertness, and relieves fatigue. It is processed with hydrochloric acid into cocaine hydrochloride, the familiar white powder that is "snorted" (inhaled) or mixed with water and injected. Pure cocaine, or *freebase*, can be extracted from cocaine hydrochloride by chemically removing the hydrochloric acid. When freebase is smoked, the cocaine enters the bloodstream and reaches the brain rapidly. A simpler chemical procedure yields *crack*, which is less pure but produces pure cocaine in the vapor when it is smoked. The low cost of crack has spread its use into poor urban communities that could not afford cocaine before.

Cocaine has not always been viewed as a dangerous drug. The coca leaf has been chewed by South American Indians for centuries as a means of enduring hardship and privation. When cocaine was isolated in the late 1800s, it was initially used as a local anesthetic. It soon found its way into over-the-counter medications (Figure 5.6), and until 1906, even Coca-Cola owed much of its refreshment to 60 milligrams of cocaine in every serving (M. S. Gold, 1997). Sigmund Freud, the father of psychoanalysis, championed the use of cocaine, giving it to his fiancée, sisters, friends, and colleagues and prescribing it to his patients. He even wrote an essay, which he called a "song of praise" to cocaine's virtues. He gave up the use of the drug, both personally and professionally, when he realized its dangers (Brecher, 1972).

Cocaine blocks the reuptake of dopamine and serotonin at synapses, potentiating their effect. Dopamine usually has an inhibitory effect, so cocaine reduces activity in much of the brain, as the positron emission tomography (PET) scans in Figure 5.7 show (London et al., 1990). Presumably, cocaine produces euphoria and excitement because dopamine removes the inhibition the cortex usually exerts on lower structures. Reduced cortical activity is typical of drugs that produce euphoria, including benzodiazepines, barbiturates, amphetamines, alcohol, and morphine, although localized activation is often reported in frontal areas (R. Z. Goldstein & Volkow, 2002; London et al., 1990). Brain metabolism rises briefly during the first week of abstinence, then falls again during prolonged withdrawal; however, during craving activity increases in several areas, as we will see later (R. Z. Goldstein & Volkow; S. Grant et al., 1996; Volkow et al., 1991, 1999).

Injection and smoking produce an immediate and intense euphoria, which increases the addictive potential of cocaine. After the end of a cocaine binge, the user crashes into a state of depression, anxiety, and cocaine craving that motivates a cycle of continued use. Withdrawal effects are typically mild, involving anxiety, lack of motivation, boredom, and lack of pleasure. Three decades ago, addiction was defined in terms of a drug's ability to produce withdrawal, and because cocaine's withdrawal symptoms are so mild, it was not believed to be addictive (Gawin, 1991). As usage increased in the population, we learned that cocaine is actually one of the most addictive of the abused drugs. The intensity of thedrug's effect makes treatment for addiction very difficult, and no treatment is generally accepted as successful. Complicating rehabilitation is the fact that cocaine addicts typically

3
Cocaine Anonymous

Q What neurotransmitter system is involved in the effects of all stimulant drugs?

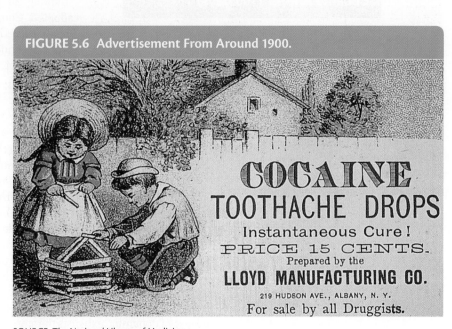

FIGURE 5.6 Advertisement From Around 1900.

COCAINE TOOTHACHE DROPS

Instantaneous Cure!

PRICE 15 CENTS.

Prepared by the

LLOYD MANUFACTURING CO.

219 HUDSON AVE., ALBANY, N. Y.

For sale by all Druggists.

SOURCE: The National Library of Medicine.

The upper two scans show activity in a cocaine-free individua. The remaining scans show reduced activity in the brain of a cocaine abuser 10 days and 100 days after last cocaine use.

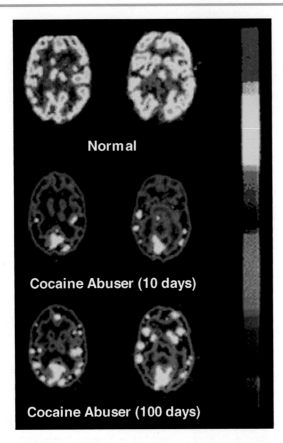

SOURCE: Photo courtesy of Nora Volkow, Ph.D. From: Nora Volkow, Ph.D. Volkow ND, Hitzemann R, Wang G-J, Fowler JS, Wolf AP, Dewey SL. Long-term frontal brain metabolic changes in cocaine abusers. *Synapse* 11:184–190, 1992.

abuse other drugs, and they have a very high rate of psychological disorders, including depression, anxiety, bipolar disorder, and posttraumatic stress disorder (Julien, 2008).

Cocaine users are markedly impaired in executive functions, including behavioral inhibition, decision making, and assessment of the emotional value of stimuli, and these deficits are accompanied by reduced activity in the prefrontal cortex (Beveridge, Gill, Hanlon, & Porrino, 2008) and gray matter reductions of 5% to 11% in the prefrontal cortex and other areas (Franklin et al., 2002). Cocaine provides a good example of selective tolerance: While increasing amounts of the drug are required to produce the desired psychological effects, the person becomes supersensitive to the effect that produces seizures. It is possible that the risks of cocaine relative to other drugs have been underestimated. In one study, rats were allowed to press a lever that caused heroin or cocaine to be injected into their bloodstream; after 1 month, 90% of rats receiving cocaine had died of self-administered overdose, compared with 36% of rats receiving heroin (Bozarth & Wise, 1985).

Like alcohol, cocaine passes through the placenta easily, where it interferes with fetal development. It is difficult to separate the effects of alcohol and cocaine on the children's development from the effects of poverty and neglect often seen in the homes. But we do have experimental evidence from animal studies that prenatal exposure to alcohol causes brain damage (Gressens et al., 1992) and that exposure to cocaine results in abnormal circuit formation among dopamine neurons (L. B. Jones et al., 2000). In addition, a Toronto group was able to control environmental factors by studying 26 cocaine-exposed children who had been adopted. Compared with control children matched for the mother's IQ and socioeconomic status, the cocaine-exposed children had lower IQs, poorer language development, and greater distractibility (Nulman et al., 2001).

Amphetamines

Amphetamines are a group of synthetic drugs that produce euphoria and increase confidence and concentration. The group includes amphetamine sulfate (marketed as Benzedrine), the three to four times more potent dextroamphetamine sulfate (marketed as Dexedrine), and the still more powerful methamphetamine (known on the street as *meth, speed, crank,* and *crystal*). Like cocaine, it can be purified to its freebase form called *ice,* which is smokable. Because they dull the appetite, reduce fatigue, and increase alertness, amphetamines have shown up in weight-loss drugs and have been used by truck drivers, pilots, and students to postpone sleep. They have been useful in treating ailments like narcolepsy, a disorder of uncontrollable daytime sleepiness.

Amphetamines increase the release of norepinephrine and dopamine. Increased release of dopamine exhausts the store of transmitter in the vesicles, which accounts for the period of depression that follows. The effects of amphetamine injection are so similar to those of cocaine that individuals cannot tell the difference between the two (Cho, 1990).

Heavy use can cause hallucinations and delusions of persecution that are so similar to the symptoms of paranoid schizophrenia that even trained professionals cannot recognize the difference (resulting in occasional emergency room mistreatment). In laboratory studies, psychotic symptoms develop after 1 to 4 days of chronic amphetamine administration. In one study, a volunteer on amphetamine was convinced that a "giant oscillator" in the ceiling was controlling his thoughts. Another believed his ex-wife had hired an assassin to kill him and was perturbed when the doctor would not guard the window while he stood watch at the door (Griffith, et al. 1972; S. H. Snyder, 1972). After an amphetamine psychosis subsides, the person may be left with a permanently increased sensitivity to the drug so that using even a small amount years later can revive symptoms (Sato, 1986).

Nicotine

Nicotine is the primary psychoactive and addictive agent in tobacco. Tobacco is ingested by smoking, chewing, and inhaling (as snuff, a finely powdered form). Nicotine has an almost unique effect (Schelling, 1992): When tobacco is smoked in short puffs, it has a stimulating effect; when inhaled deeply, it has a tranquilizing or depressant effect. In large doses, nicotine can cause nausea, vomiting, and headaches; in extremely high doses, it is powerful enough to produce convulsions and even death in laboratory animals.

4
Nicotine Effects and Addiction

The withdrawal reactions are well known because smokers "quit" so often; the most prominent symptoms are nervousness and anxiety, drowsiness, lightheadedness, and headaches. The United Kingdom annually observes a "No Smoking Day," similar to the "Great American Smokeout," in which people voluntarily abstain from smoking for a day; apparently as a result of impairment from withdrawal symptoms, workplace accidents go up by 7% (Waters, Jarvis, & Sutton, 1998). People who try to give up smoking are usually able to abstain for a while but then relapse; only about 20% of attempts to stop are successful after 2 years. Before bans on public and workplace smoking, about 80% of male smokers and two thirds of female smokers smoked at least one cigarette per waking hour (Brecher, 1972).

In part because usage is more continuous with tobacco than with other drugs, the health risks are particularly high. The health risks from smoking are not the result of nicotine but of some of the 4,000 other compounds present in tobacco smoke. For example, a metabolite of benzo-[a]pyrene damages a cancer-suppressing gene, resulting in lung cancer (Denissenko, Pao, Tang, & Pfeifer, 1996). Other cancers resulting from smoking occur in the larynx, mouth, esophagus, liver, and pancreas. Smoking can also cause Buerger's disease, constriction of the blood vessels that may lead to gangrene in the lower extremities, requiring progressively higher amputations. Although abstinence almost guarantees a halt in the disease's progress, surgeons report that it is not uncommon to find a patient smoking in the hospital bed after a second or third amputation (Brecher, 1972). Nicotine addiction is the largest cause of preventable death, accounting for a mortality of 438,000 annually in the United States and 4 million worldwide; the health and lost-productivity costs in the United States add up to $167 billion (Centers for Disease Control and Prevention, 2005; Tapper et al., 2004).

Cigarette package warnings aimed at expectant mothers are not just political propaganda. Studies show that women who smoke during pregnancy give birth to underweight infants, and a cause–effect relationship has been confirmed in experimental studies with animals (see F. Rice et al., 2009). Offspring of smoking mothers are also more likely to display conduct disorder (involving difficulty with impulse control; Fergusson, Woodward, & Horwood, 1998) and later criminal behavior (Brennan, Grekin, & Mednick, 1999). However, a study we saw in the

> *Because of the 400,000 deaths produced each year by smoking [in the U.S.], including 50,000 in non-smokers due to passive inhalation of secondhand smoke, it can reasonably be argued that nicotine is the most important drug of abuse. Heroin and cocaine combined produce no more than 6,000 deaths per year in contrast.*
> —*Charles O'Brien*

previous chapter (F. Rice et al.) indicates that these behaviors are genetic in origin. This means that the association arises because mothers with a genetic predisposition for impulse control problems are also more likely to smoke, and it is a good example of why we shouldn't be too quick to accept the "obvious" interpretation when two variables are correlated.

As you saw in Chapter 2, nicotine stimulates nicotinic acetylcholine receptors. In the periphery, it activates muscles and may cause twitching. Centrally, it produces increased alertness and faster response to stimulation. Neurons that release dopamine contain nicotinic receptors, so they are also activated, resulting in a positive mood effect (Svensson, Grenhoff, & Aston-Jones, 1986).

Caffeine

Caffeine, the active ingredient in coffee, produces arousal, increased alertness, and decreased sleepiness. It is hardly the drug that amphetamine and cocaine are, but as you saw in Balzac's case, it is subject to abuse. It blocks receptors for the neuromodulator adenosine, increasing the release of dopamine and acetylcholine (Silinsky, 1989; S. H. Snyder, 1997). Because adenosine has sedative and depressive effects, blocking its receptors contributes to arousal. Withdrawal symptoms include headaches, fatigue, anxiety, shakiness, and craving, which last about a week. Withdrawal is not a significant problem, because coffee is in plentiful supply, but heavy drinkers may wake up with a headache just from abstaining overnight. Because 80% of Americans drink coffee, researchers at the Mayo Clinic once recommended intravenous administration of caffeine to patients recovering from surgery to eliminate postoperative withdrawal headaches ("Caffeine Prevents Post-Op Headaches," 1996).

Psychedelics

Psychedelic drugs are compounds that cause perceptual distortions in the user. The term comes from the Greek words *psyche* ("mind") and *delos* ("visible"). "Visible mind" refers to the expansion of the senses and the sense of increased insight that users of these drugs report. Although the drugs are often referred to as *hallucinogenic,* they are most noted for producing perceptual distortions: Light, color, and details are intensified; objects may change shape; sounds may evoke visual experiences; and light may produce auditory sensations. Psychedelics may affect the perception of time, as well as self-perception; the body may seem to float or to change shape, size, or identity. These experiences are often accompanied by a sense of ecstasy.

5
LSD, Ecstasy, and PCP

The best-known psychedelic, *lysergic acid diethylamide (LSD),* was popularized in the student peace movement of the 1960s. LSD is structurally similar to serotonin and stimulates serotonin receptors (Jacobs, 1987); as you will see in this chapter and in later chapters, serotonin has a wide variety of psychological functions. Other serotonin-like drugs include *psilocybin* and *psilocin,* LSD-like drugs from the mushroom *Psilocybe mexicana; peyote,* the "button" on the top of the peyote cactus; and *mescaline,* the active ingredient in peyote. Peyote is used in religious ceremonies by the Native American Church, and that use is protected by the U.S. federal government and by 23 states (Julien, 2008).

Ecstasy is the street name of a drug developed as a weight-loss compound called *methylenedioxymethamphetamine* (let's just call it MDMA!); it is a popular drug among young people, especially at dance clubs and "raves." Similar to amphetamine in structure, at low doses it is a *psychomotor stimulant,* increasing energy, sociability, and sexual arousal; at higher doses it produces hallucinatory effects like those produced by LSD. MDMA stimulates the release of dopamine; one of dopamine's roles is as a psychomotor stimulant. MDMA also stimulates the release of serotonin, which probably accounts for the hallucinatory effects (Liechti & Vollenweider, 2000).

The disturbing news is that high doses of MDMA destroy serotonergic neurons in monkeys (Figure 5.8; McCann, Lowe, & Ricaurte, 1997). A study of human users found widespread reduction in serotonin functioning; this impairment decreased over a period of abstinence (McCann et al., 2005). A review of 422 studies showed persistent but small effects on cognitive performance, especially memory deficits (Rogers et al., 2009). Although health effects are usually minimal, several deaths are reported annually.

Phencyclidine (PCP) was developed as an anesthetic, but its use was abandoned because it produced disorientation and hallucinations that were almost indistinguishable from the symptoms of schizophrenia (Murray, 2002). It has found recreational popularity as *angel dust* or *crystal*. Monkeys and rats will self-administer PCP, and humans show compulsive use, indicating that PCP is addictive (Carlezon & Wise, 1996). PCP increases activity in dopamine pathways, but blocking dopamine activity does not eliminate self-administration in rats; the drug's motivating properties apparently are partly due to its inhibition of a subtype of glutamate receptors (Carlezon & Wise, 1996; E. D. French, 1994).

Scientists became interested in psychedelic drugs at the beginning of the 20th century because some of the effects resemble psychotic symptoms. This suggested that a chemical imbalance might be the cause of psychosis, so researchers tried to produce "model psychoses" that could be studied in the laboratory. Early research was unproductive, but more recent experience with PCP has led researchers to revise their theories of schizophrenia (Jentsch & Roth, 1999).

FIGURE 5.8 Brain Damage Produced by the Drug "Ecstasy."

These brain sections have been stained with a chemical that makes neurons containing serotonin turn white. Photos in the top row are from a normal monkey; those below are from a monkey given MDMA a year earlier.

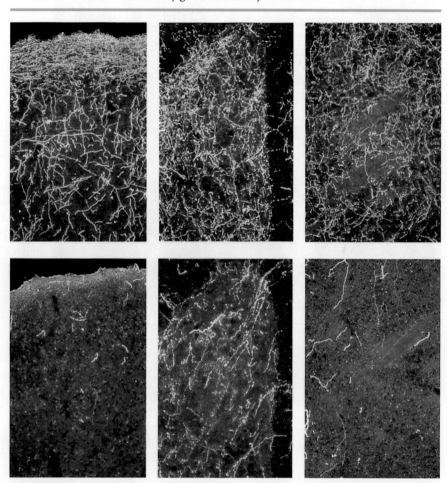

SOURCE: From "Long-Lasting Effects of Recreational Drugs of Abuse on the Central Nervous System," by U. D. McCann, K. A. Lowe, & G. A. Ricaurte, 1997, *The Neuroscientist, 3*, 399–411.

Marijuana

Marijuana is the dried and crushed leaves and flowers of the Indian hemp plant, *Cannabis sativa* (Figure 5.9). The hemp plant was heavily cultivated in the United States during World War II as a source of material for making rope, and it is still found occasionally growing wild along Midwestern roadsides. Marijuana is usually smoked but can be mixed in food and eaten. The major psychoactive ingredient is *delta-9-tetrahydrocannabinol (THC)*. THC is particularly concentrated in the dried resin from the plant, called *hashish*.

THC is a *cannabinoid*, a group of compounds that includes two known endogenous cannabinoids, anandamide and 2-arachidonyl glycerol, or 2-AG (Devane et al., 1992; di Tomaso, Beltramo, & Piomelli, 1996; Mechoulam et al., 1995).

6
All About Marijuana

FIGURE 5.9 A Marijuana Plant.

SOURCE: © Tina Lorien/iStockphoto.com.

Cannabinoid receptors are found on axon terminals; cannabinoids are released by postsynaptic neurons and act as retrograde messengers, regulating the presynaptic neuron's release of neurotransmitter (R. L. Wilson & Nicoll, 2001). The receptors are widely distributed in the brain and spinal cord, which probably accounts for the variety of effects marijuana has on behavior. The pleasurable sensation is likely due to its ability to increase dopamine levels (Tanda, Pontieri, & Di Chiara, 1997). Receptors in the frontal cortex probably account for impaired cognitive functioning and distortions of time sense and sensory perception, receptors in the hippocampus disrupt memory, and those in the basal ganglia and cerebellum impair movement and coordination (Herkenham, 1992; Howlett et al., 1990; Ong & Mackie, 1999). This is a good time to point out that although drugs may reveal a great deal about brain functioning, the pattern of effects they produce is usually unlike normal functioning; drugs affect wide areas of the brain indiscriminately, whereas normal activation tends to be more discrete and localized.

Marijuana's impact on users may be greater than previously thought. Heavier smoking is associated with memory and cognitive deficits that persist at least a month after cessation of use (Bolla, Brown, Eldreth, Tate, & Cadet, 2002). In a longitudinal study, young adults who smoked five or more joints a week had lost an average of 4 IQ points since childhood, compared with IQ gains in light users and nonusers (Fried, Watkinson, James, & Gray, 2002). The effects did not appear to be permanent, because former users did not show this effect. An Australian MRI study of long-term (>10 years), heavy (>5 joints daily) marijuana users found reduced hippocampus and amygdala volume compared to nonusers, along with reduced verbal learning and the presence of minor psychotic symptoms (Yücel et al., 2008). Because the study is correlational, we must be cautious about assuming that marijuana caused these impairments, but the authors point out that hippocampal reduction has been verified repeatedly in animal studies in which cannabis exposure was experimentally manipulated.

The effect of marijuana on prenatal development has received little attention, because babies exposed prenatally to marijuana do not show the obvious impairments caused by prenatal cocaine and alcohol. The Ottawa Prenatal Prospective Study followed prenatally exposed children for several years after birth. They had no deficits during the first 3 years of life compared with control children, but at 4 years and beyond they showed behavioral problems; decreased performance on visual perception tasks; and deficits in attention, memory, and language comprehension (Fried, 1995). These deficits are consistent with impaired functioning in prefrontal areas.

Legalization is the major controversy surrounding marijuana; it is a battle that is being waged on two very different fronts. Because of its mild effects, many contend that its use should be unrestricted. Others, citing reports that it reduces pain, the nausea of chemotherapy, and the severity of the eye disease glaucoma, believe it should be available by a doctor's prescription. The medical claims are controversial, however, because they rely largely on inadequately controlled studies. The accompanying In the News indicates just how divisive this issue has become.

Q What are the two controversies about marijuana?

Controversy Over Medical Marijuana Takes a New Turn

Cannabis was used as a medicine in China as early as 2800 BCE, and it is a part of traditional medicine in many areas of the world, but Western countries tend to be skeptical, partly because of the lack of good research. However, Western ambivalence appears to be on the wane. In the United States, where marijuana is prohibited by federal law, 17 states have legalized its medical use and 14 others have pending legislation. The result has been numerous raids on medical marijuana clinics by the Drug Enforcement Administration (DEA). But when the DEA raided a California clinic just 2 days after President Obama took office, a White House spokesperson stated that "federal resources should not be used to circumvent state laws," and Attorney General Eric Holder announced that the raids would end. And now the American Medical Association is calling for a review of marijuana's Schedule I drug classification; loosening the restrictions, the AMA says, would facilitate clinical research and the development of cannabinoid-based medicines and alternate delivery methods.

SOURCES: "13 Legal Medical Marijuana States," n.d.; Doyle, 2009; Johnson, 2009; O'Reilly, 2009.

Another controversy concerns whether marijuana is addictive. The importance of this debate is that it requires us to define just what we mean by the term. Addiction has traditionally been equated with a drug's ability to produce withdrawal symptoms; because marijuana's withdrawal symptoms are very mild, its compulsive use was attributed to "psychological dependence," a concept that is also invoked to explain the habitual use of other drugs that do not produce dramatic withdrawal symptoms, like nicotine and caffeine. Withdrawal symptoms are mild because cannabinoids dissolve in body fats and leave the body slowly. However, monkeys will press a lever to inject THC into their bloodstream in amounts similar to doses in marijuana smoke inhaled by humans (Tanda, Munzar, & Goldberg, 2000). Researchers are reluctant to attribute drug self-administration in animals to psychological dependence and usually consider it evidence of addiction. Earlier we defined addiction in terms of the drug's hold on the individual, without reference to its ability to produce withdrawal symptoms. Next, we will examine the reasons for taking this position.

Concept Check   *Take a Minute to Check Your Knowledge and Understanding*

☐ How does tolerance increase a drug's danger?

☐ Why does alcohol increase the danger of barbiturates?

☐ How are the effects of amphetamine and cocaine at the synapse alike? How are they different?

Addiction

It is an oversimplification to assume that chronic drug use is motivated primarily by the pleasurable effects of the drug; in fact, individuals who engage in compulsive drug taking often report that they no longer enjoy their drug experience. Their casual drug use has progressed into the compulsive disorder of addiction. The common belief that addiction is fueled by the drug user's desire to avoid withdrawal symptoms also has several flaws. One is that it does not explain what

 Why does the avoidance of withdrawal symptoms fail to explain addiction?

motivates the person to use the drug until addiction develops. Second, we know that many addicts go through withdrawal fairly regularly to reset their tolerance level so they can get by with lower and less costly amounts of the drug. Third, it does not explain why many addicts return to a drug after a long period of abstinence and long after withdrawal symptoms have subsided. Finally, the addictiveness of a drug is unrelated to the severity of withdrawal symptoms (Leshner, 1997). Cocaine is a good example of severe addictiveness but mild withdrawal, while a number of drugs—including some asthma inhalers, nasal decongestants, and drugs for hypertension and angina pain—produce withdrawal symptoms but are not addictive (S. E. Hyman & Malenka, 2001).

The Neural Basis of Addiction and Reward

Research indicates that addiction and withdrawal take place in different parts of the brain and that they are independent of each other. When Bozarth and Wise (1984) allowed rats to press a lever to inject morphine directly into the ventral tegmental area (Figure 5.10), the rats did so reliably, suggesting that the area is involved in addiction. Then the researchers tried to induce withdrawal by blocking the opiate receptors there with injections of naloxone, but no signs of withdrawal occurred. Rats would not press a lever for morphine injections into a nearby area called the periventricular gray, which meant that it is not involved in addiction. But when the researchers gave the rats regular morphine injections in the periventricular gray and then injected naloxone, the rats showed classic signs of withdrawal, including teeth chattering, "wet dog" shakes, and attempts to escape from the test apparatus. This independence of addiction and withdrawal does not mean that addicts never take drugs to avoid withdrawal symptoms; rather it means that withdrawal is not necessary for addiction and avoidance of withdrawal is not an explanation of addiction. Addiction depends on something else; one hypothesis is that that something is reward.

> It is as if drugs have hijacked the brain's natural motivational control circuits.
>
> —Alan Leshner

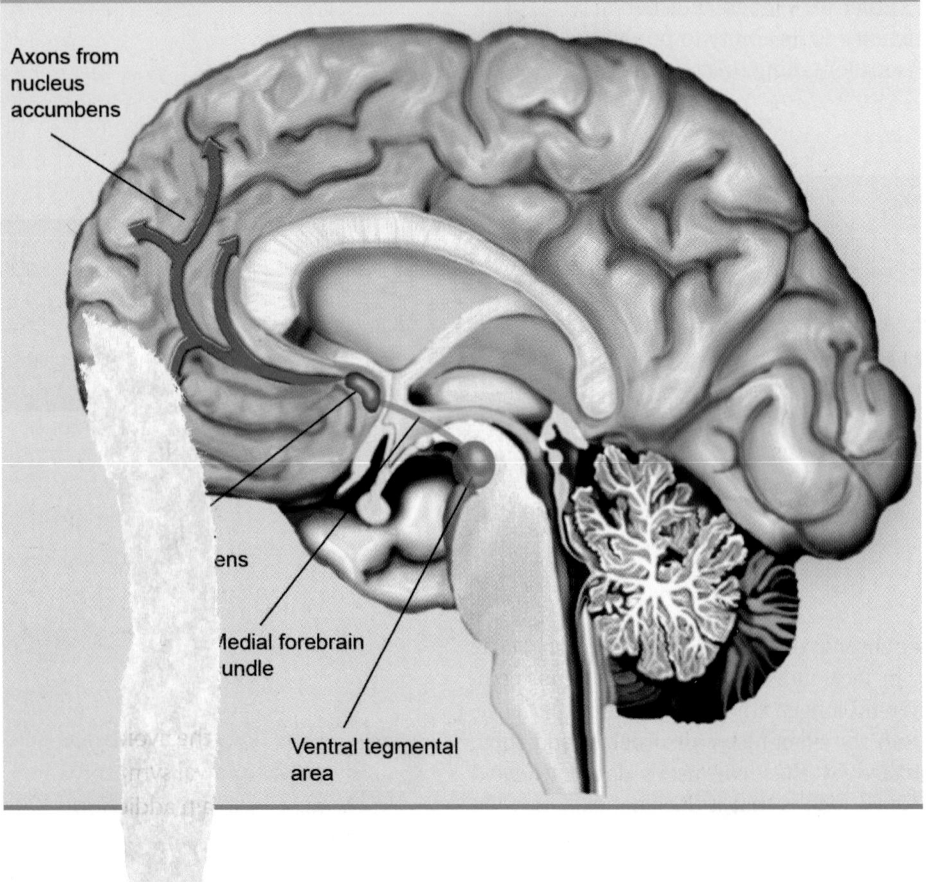

Reward refers to the positive effect an object or condition—such as a drug, food, sexual contact, or warmth—has on the user. This effect is primarily on behavior, but it is typically accompanied by feelings of pleasure. Drug researchers have traditionally identified the *mesolimbocortical dopamine system* as the location of the major drug reward system (Wise & Rompre, 1989); it takes its name from the fact that it begins in the midbrain (mesencephalon) and projects to the limbic system and prefrontal cortex. As you can see in Figure 5.10, the most important structures in the system are the *nucleus accumbens,* the *medial forebrain bundle,* and the *ventral tegmental area.* (Other structures also participate in reward, including the amygdala and the hippocampus, but they have been accorded less importance.) Rats will learn to press

FIGURE 5.10   The Mesolimbocortical Dopamine System.

Axons from nucleus accumbens

...ens

...edial forebrain ...undle

Ventral tegmental area

a lever to inject abused drugs into these areas (Bozarth & Wise, 1984; Hoebel et al., 1983), and lesioning the nucleus accumbens reduces reward effects for many drugs (Kelsey, Carlezon, & Falls, 1989).

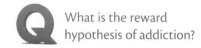

What is the reward hypothesis of addiction?

Dopamine and Reward

Virtually all the abused drugs increase dopamine levels in the nucleus accumbens, including opiates, barbiturates, alcohol, THC, PCP, MDMA, nicotine, and even caffeine (Di Chiara, 1995; Grigson, 2002). There is considerable evidence that this increase in dopamine level plays an important role in addiction. For example, rats given drugs that block dopamine activity do not learn to press a lever for amphetamine or cocaine injections; if they have learned previously, they do not continue lever pressing after receiving the dopamine-blocking drug (Wise, 2004). In PET scan studies, human volunteers who had the greatest increase in dopamine in the general area of the nucleus accumbens also experienced the most intense "highs" (Volkow, Fowler, & Wang, 2003). In one study, participants began reporting that they felt "high" when cocaine had blocked 47% of the dopamine reuptake sites in the nucleus accumbens (Volkow et al., 1997).

But the mesolimbocortical dopamine system's reward function is not limited to drugs; microdialysis studies show that food, water, and sexual stimulation also increase dopamine levels in the nucleus accumbens (Carelli, 2002; Damsma, Pfaus, Wenkstern, & Phillips, 1992). The same can be said for electrical stimulation (Fibiger, LePiane, Jakubovic, & Phillips, 1987). In *electrical stimulation of the brain (ESB),* animals and, sometimes, humans learn to press a lever to deliver mild electrical stimulation to brain areas where the stimulation is rewarding. Drugs that block dopamine receptors interfere with learning to press a lever to obtain this stimulation (Wise, 2004). ESB is thought to reflect "natural" reward processes, because effective sites are often in areas where experimenter-delivered stimulation evokes eating or sexual activity and because self-stimulation rate in the posterior hypothalamus varies with experimenter-induced sexual motivation (Caggiula, 1970). The most sensitive areas are where the density of dopaminergic neurons is greatest, especially in the medial forebrain bundle (Corbett & Wise, 1980).

Both electrical stimulation and drugs are especially powerful motivators of behavior. Animals will ignore food and water and tolerate painful shock to stimulate their brains with electricity, sometimes pressing a lever thousands of times in an hour. Humans will sacrifice their careers, relationships, and lives in the interest of acquiring and using drugs. While food and sex increase dopamine in the nucleus accumbens by 50% to 100%, drugs and electrical stimulation can have a three- to sixfold greater effect, depending on dosage (A. G. Phillips et al., 1992; Wise, 2002). Many researchers are interested in drugs and electrical stimulation because they seem to provide a more direct access to the brain's motivational systems.

But there is an intriguing paradox in the dopamine research: PET imaging reveals that chronic drug users have *diminished* dopamine release and numbers of dopamine receptors (Figure 5.11; Volkow, Fowler, Wang, & Swanson, 2004). This decreased dopamine activity may not be the *consequence* of chronic abuse. Non-drug-abusing subjects who reported the greatest "liking" for

> *Drugs of abuse create a signal in the brain that indicates, falsely, the arrival of a huge fitness benefit.*
>
> —Randolph Nesse and Kent Berridge

FIGURE 5.11 Reduced Dopamine D$_2$ Receptors in Drug Abusers.

The researchers imaged the brains using PET and an agent that binds to D$_2$ dopamine receptors. The predominance of yellow in place of red in the scans of the drug abusers' brains indicates fewer of the D$_2$ subtype of dopamine receptors than in the control subjects' brains.

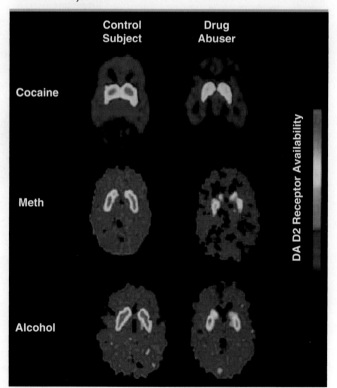

SOURCE: From "Role of Dopamine, the Frontal Cortex, and Memory Circuits in Drug Addiction: Insight From Imaging Studies," by N. D. Volkow et al., *Neurobiology of Learning and Memory, 78,* 610–624. © 2002 Nora Volkow. Used with permission.

FIGURE 5.12 Dopamine and Glutamate Neurons Converge on the Same Neurons in the Nucleus Accumbens.

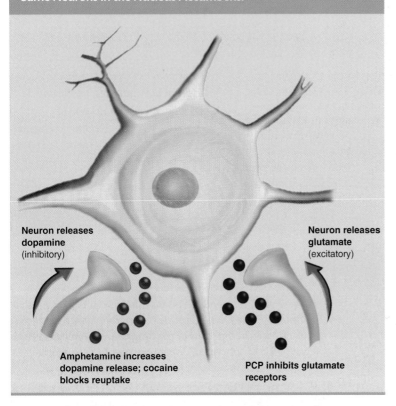

Neuron releases dopamine (inhibitory)

Neuron releases glutamate (excitatory)

Amphetamine increases dopamine release; cocaine blocks reuptake

PCP inhibits glutamate receptors

the effects of the cocaine-like stimulant methylphenidate (Ritalin) also had low numbers of the D_2 type of dopamine receptors; those with the highest numbers of receptors found the drug unpleasant (Volkow et al., 2002). Thus, lowered dopamine receptors probably precedes drug experience and creates a "reward deficiency syndrome" that accounts for addicts' lowered responsiveness to rewards in general (Volkow et al., 2003) and predisposes the individual to drug abuse. This view has received experimental support. Thanos and his colleagues trained rats to self-administer alcohol, then used a virus to insert the D_2 gene into the rats' nucleus accumbens; this increased the number of dopamine receptors, and the rats reduced their alcohol preference and alcohol intake (Thanos et al., 2001).

And now an important caveat: While most of the abused drugs trigger the release of dopamine, dopamine cannot account for all reward. The rewarding effect of alcohol, for example, depends partly on opiate receptors; in fact, opiate blockers such as naloxone and naltrexone are effective in preventing relapse in alcoholics (Garbutt, West, Carey, Lohr, & Crews, 1999). PCP produces rewarding effects by blocking glutamate receptors, likely on the same neurons that mediate dopamine-based reward (Figure 5.12; Carlezon & Wise, 1996). According to Wise (2004), dopamine is crucial for the rewarding effects of cocaine and amphetamine; important but perhaps not crucial for the effects of the opiates, nicotine, cannabis, and ethanol; and questionable in the case of benzodiazepines, barbiturates, and caffeine. (See the Application "Is Compulsive Gambling an Addiction?" for a broader view of addiction.)

Application   Is Compulsive Gambling an Addiction?

Compulsive eating, sexual activity, gambling, and computer usage are popularly referred to as "behavioral addictions." For many this is no more than a misappropriation of a specialized term, but some researchers are finding intriguing parallels between problem gambling and drug addiction.

For example, compulsive gamblers also have a high rate of alcohol and drug abuse (Petry, Stinson, & Grant, 2005). The fact that the biological relatives of gamblers have more drug problems than the rest of the population suggests a hereditary link between drug addiction and gambling, and genetic studies have found that problem gamblers share specific versions of dopamine and serotonin receptor genes (Ibáñez, Blanco, Perez de Castro, Fernandez-Piqueras, & Sáiz-Ruiz, 2003). In addition, "craving" is increased in compulsive gamblers by participation in a brief gambling episode or by a dose of amphetamine, just as drug craving is induced by exposure to drug-related stimuli or by a dose of a similar drug (Zack & Poulos, 2004).

If gambling is a form of addiction, then we would expect it to involve some of the same brain mechanisms as drug addiction. Opiate antagonists are helpful in gambling therapy, suggesting that endorphin-based reward is involved (J. E. Grant et al., 2006). Even more research implicates the dopamine system. Both gamblers and nongamblers show greater activity in dopamine pathways in response to winning than losing (Breiter, Aharon, Kahneman, Dale, & Shizgal, 2001), and like drug addicts, the rest of the time compulsive gamblers have reduced activity there as well as in prefrontal areas involved in impulse control (Potenza, 2008; Reuter et al., 2005). But the real surprise is that a rather large number of Parkinson's disease patients have reported that they became compulsive gamblers while taking dopamine agonists for their disease symptoms (Dodd et al., 2005; Mamikonyan et al., 2008). They often also experienced compulsive eating, shopping, and sexual activity, demanding sex from their partners several times a day. The symptoms cleared up when they reduced or switched their medication.

Dopamine, Learning, and Brain Plasticity

Most researchers agree that reward is an essential factor in early drug taking, but it is doubtful that reward maintains long-term drug abuse (Volkow & Fowler, 2000; Wise, 2004). Later stages of addiction, characterized by craving and withdrawal symptoms, involve potentially lifelong changes in brain functioning. Dopamine contributes to some of those changes, but not through the mechanism of reward. Our first clues came from learning research; as learning proceeds in the laboratory, a rewarding stimulus initially produces dopamine release, then ceases to do so. Instead, that capability shifts to stimuli that precede the reward, such as an auditory tone that signals the period when lever pressing will produce the reward. However, dopamine release does occur at the time the reward would be expected *if* the reward is omitted. A human fMRI study suggests that the dopamine system responds to the *unpredictability* of rewards; activity in the nucleus accumbens was greater when a drop of juice was delivered unpredictably than when it was delivered on a predictable schedule (Figure 5.13; Berns, McClure, Pagnoni, & Montague, 2001).

These observations have led to the hypothesis that dopamine not only signals rewards but also signals *errors in prediction* (Schultz, 2002). According to contemporary learning theory, learning occurs only when the reward is unexpected or better than expected, or when the reward is omitted or is worse than expected. Therefore, the ability to detect errors in prediction is critical for learning. As addiction theorists' emphasis has shifted toward dopamine's role in learning, they tend to prefer the more neutral term *reinforcer* over the term *reward*, which carries an implication of subjective pleasure. A *reinforcer* is any object or event that increases the probability of the response that precedes it. For example, the drug a rat receives after pressing a lever reinforces lever pressing.

Learning changes behavior by modifying connections in the brain, and these physical alterations can be seen in the addicted brain. For example, chronic administration of amphetamine or cocaine resulted in increased dendrite length and complexity in rats' nucleus accumbens and prefrontal cortex (T. E. Robinson, Gorny, Mitton, & Kolb, 2001; T. E. Robinson & Kolb, 1997). Prolonged drug use results in reduced dopamine activity and an accompanying deactivation of prefrontal areas involved in behavioral inhibition (Volkow, Fowler, & Wayne, 2004).

We can see the power of drug-induced learning in craving. Stimuli associated with drug use, such as drug paraphernalia, will evoke craving in addicts (Garavan et al., 2000; S. Grant et al., 1996; Maas et al., 1998). PET scans during the presentation of drug-related stimuli show a shift from reduced metabolism to hyperactivity in brain areas involved in learning and emotion (R. Z. Goldstein & Volkow, 2002; S. Grant et al., 1996; Figure 5.14). The hippocampus is important in learning and particularly in learning associations with environmental stimuli like those involved

 What alternative role besides reward has been suggested for dopamine?

FIGURE 5.13 Brain Responses to Predictable and Unpredictable Rewards.

In these fMRI scans, unexpected liquid delivery increased activity in the nucleus accumbens (NAC) (a). When delivery occurred predictably every 10 seconds, there was a smaller response in the temporal lobe (b).

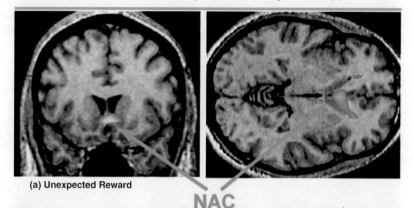

(a) Unexpected Reward

NAC

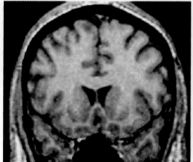

(b) Expected Reward

SOURCE: From "Predictability Modulates Human Brain Response to Reward," by Berns et al., *Journal of Neuroscience, 21*, pp. 2793–2798, fig. 3. © 2001 Society for Neuroscience. Used with permission.

FIGURE 5.14 The Brain of a Cocaine Abuser During Craving.

PET scans are shown at two depths in the brain. Notice the increased activity during presentation of cocaine-related stimuli. Frontal areas (DL, MO) and temporal areas (TL, PH) are involved in learning and emotion.

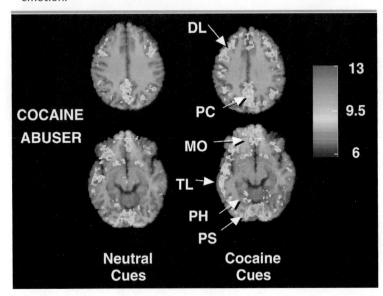

SOURCE: From "Activation of Memory Circuits During Cue-Elicited Cocaine Craving," by S. Grant et al., 1996, *Proceedings of the National Academy of Sciences, USA, 93,* pp. 12040–12045.

in drug taking. After rats have given up pressing a lever because the drug delivery mechanism has been disconnected, electrical stimulation of the hippocampus is enough to revive the lever pressing (Vorel, Liu, Hayes, Spector, & Gardner, 2001). The researchers attributed the return to lever pressing to the lengthy (30-minute) release of dopamine in the nucleus accumbens that followed hippocampal stimulation. According to many researchers, learning produces changes in the brain that explain why drugs still produce an exaggerated reaction years after the last use and why some addictions can last a lifetime.

Learning cannot explain all the changes in the addict's brain or the accompanying alterations in behavior; some drug-induced changes are better characterized as neural pathology. When T. E. Robinson et al. (2001) studied the changes in prefrontal cortex in their cocaine-abusing rats, they found malformed dendrites. This suggested a possible basis for the frontal dysfunction observed in cocaine addicts, which includes impaired judgment and decision making. Disrupted prefrontal functioning could account for the addicts' loss of control over their behavior, even while expressing a desire to abstain from drugs (Volkow et al., 2003). PET imaging has also verified dysfunction in the orbitofrontal cortex, an area that monitors the relative value of reinforcers and where pathology has also been reported in patients with obsessive-compulsive disorders (Volkow et al., 2003; Volkow & Fowler, 2000). According to Nora Volkow and her colleagues, the transition from controlled drug use to compulsive drug intake involves pathological changes in communication between prefrontal cortex and the nucleus accumbens (Kalivas, Volkow, & Seamans, 2005). The addict returns to drug taking when stress or drug-related stimuli trigger increases in dopamine release in the prefrontal cortex and glutamate release in the nucleus accumbens. The first of these increases produces a compulsive focus on drugs at the expense of other reinforcers, while the latter cranks up the drive to engage in drug seeking.

Treating Drug Addiction

Synanon, the residential community for the treatment of addictions, supplied its residents with all their food, clothing, and other necessities including, until 1970, cigarettes—which alone cost $200,000 annually (Brecher, 1972). But then Synanon's founder and head Charles Dederich had a chest X-ray that showed a cloudy area in his lungs, and he realized that residents as young as 15 were learning to smoke under his watch. He quit smoking, stopped supplying cigarettes, and banned their use on the premises. Giving up smoking was more difficult for the residents than expected. About 100 people left during the first 6 months rather than do without cigarettes. Some of the residents who quit smoking noticed that they got over withdrawal symptoms from other drugs in less than a week but the symptoms from smoking hung around for at least 6 months. As one resident said, it was easier to quit heroin than cigarettes.

FIGURE 5.15 Sigmund Freud and Relapse of Smoking Addiction.

Notice in the graph that the two legal drugs have relapse rates equal to that of heroin.

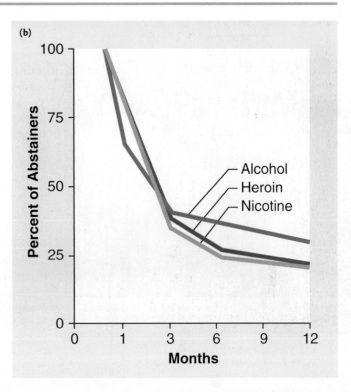

SOURCE: (a) Bettmann/CORBIS; (b) Adapted with permission from "Nicotine becomes Addictive," by R. Kanigel, 1988, *Science Illustrated*, Oct/Nov, pp. 12–14, 19–21. © 1988 Science Illustrated.

Freud had a similarly difficult experience (see Figure 5.15). He smoked as many as 20 cigars a day and commented that his passion for smoking interfered with his work. Although he quit cocaine with apparent ease, each time he gave up smoking, he relapsed. He developed cancer of the mouth and jaw, which required 33 surgeries, but he continued smoking. After replacement of his jaw with an artificial one, he was in constant pain and sometimes unable to speak, chew, or swallow, but still he smoked. He quit smoking when he died of cancer in 1939 (Brecher, 1972).

The first step in quitting drug use is detoxification. This means giving up the drug and allowing the body to cleanse itself of the drug residues. This is admittedly difficult with nicotine or opiates, but withdrawal from alcohol is potentially life threatening; medical intervention with benzodiazepines to suppress the withdrawal syndrome may be necessary (O'Brien, 1997). Still, withdrawal is often easier than the subsequent battle against relapse. Perhaps we should not be surprised about this treatment failure; addiction is potentially a chronic, lifelong disease, and its relapse rate is no higher than that of other chronic diseases such as hypertension, asthma, and type 2 diabetes (McLellan, Lewis, O'Brien, & Kleber, 2000). Fortunately, the number of treatment options is increasing; as you will see, they reflect our improving knowledge of how addiction works.

The proneness to relapse is based on changes in brain function that continue for months or years after the last use of the drug.

—*Charles O'Brien*

7

Web of Addictions

Treatment Strategies

Agonist treatments replace an addicting drug with another drug that has a similar effect; this approach is the most common defense against drug craving and

FIGURE 5.16 Effects of a GABA_A Receptor Blocker.

The two rats received the same amount of alcohol, but the one on the right received a drug that blocks the effect of alcohol at the GABA_A receptor.

SOURCE: Photo courtesy of National Institute of Mental Health/Jules Asher.

Q — What are the types of pharmacological treatment for addiction?

"

Addiction will eventually be seen as analogous to other medical illnesses— as complex constructs of genetic, environmental, and psychosocial factors that require multiple levels of intervention for their treatment and prevention.

—*Eric Nestler and George Aghajanian*

"

relapse. Nicotine gum and nicotine patches provide controlled amounts of the drug without the dangers of smoking, and their use can be systematically reduced over time. Opiate addiction is often treated with a synthetic opiate called *methadone.* This treatment is controversial because it substitutes one addiction for another, but methadone is a milder and safer drug and the person does not have to resort to crime to satisfy the habit. As a side note, methadone was developed in World War II Germany as a pain-relieving replacement for morphine, which was not available; it was called *adolphine,* after Adolph Hitler (Bellis, 1981).

Antagonist treatments, as the name implies, involve drugs that block the effects of the addicting drug. Drugs that block opiate receptors, such as naltrexone, are used to treat opiate addictions and alcoholism because they reduce the pleasurable effects of the drug. The potential for this type of treatment is dramatically illustrated in Figure 5.16 (Suzdak et al., 1986). However, antagonist treatment has a distinct disadvantage compared with agonist treatment. Because the treatment offers no replacement for the abused drug's benefit, success depends entirely on the addict's motivation to quit.

Another experimental strategy is to interfere with the dopamine reward system. Baclofen reduces dopamine activity in the ventral tegmental area by activating GABA_B receptors on dopaminergic neurons. Baclofen reduced humans' cocaine craving and rats' self-administration of cocaine, methamphetamine, heroin, alcohol, and nicotine (Paterson, Froesti, & Markou, 2005). Rimonabant, which interferes with the dopamine pathway by blocking cannabinoid receptors, is also showing promise as a multidrug treatment (Le Foll & Goldberg, 2005).

Rather than blocking the effects of a drug, *aversive treatments* cause a negative reaction when the person takes the drug. Antabuse interferes with alcohol metabolism, so drinking alcohol makes the person ill. Similarly, adding silver nitrate to chewing gum or lozenges makes tobacco taste bad. As with antagonist treatments, success depends on the addict's motivation and treatment compliance.

All of these approaches have problems: Alcoholics often fail to take Antabuse; methadone is itself addictive; naltrexone works only with a subset of addicts; and the anti-nicotine drugs Chantix and Zyban have been associated with hostility, depression, and suicidal thoughts. A very attractive alternative is antidrug vaccines. *Antidrug vaccines* consist of molecules that attach to the drug and stimulate the immune system to make antibodies that will degrade the drug. Vaccines of this sort reduce the amount of cocaine that reaches the brain by 80% (Carrera et al., 1995) and the amount of nicotine by 65% (Pentel et al., 2000). This treatment avoids the side effects that occur when receptors in the brain are manipulated. Another benefit is that the antibodies are expected to last from weeks to years, which means that therapeutic success will not depend on the addict's decision every morning to take an anti-addiction drug. The major obstacle is that only a percentage of addicts reach the necessary level of antibodies. Nevertheless, three anti-nicotine drugs and one anti-cocaine drug have completed phase II trials for safety and effectiveness, and the anti-smoking drugs could reach market during 2010 or 2011 (B. M. Kinsey, Jackson, & Orson, 2009; Martell et al., 2009).

Before we leave this topic we need to raise two additional points. One is that diminished serotonin activity has been found across several addictions, as well as a variety of other disorders. As a consequence, drugs that increase serotonin levels have shown some usefulness in treating smoking (S. M. Hall et al., 1998) and one form of alcoholism, identified in the next section as Type 1 (Dundon, Lynch, Pettinati, & Lipkin, 2004). Part of the effectiveness of the serotonin-potentiating drugs can be

TABLE 5.1 Medications Approved by the U.S. Food and Drug Administration for Treating Drug Addictions.

Drug	Medication	Action
Alcoholism	Disulfiram (Antabuse) Naltrexone Acamprosate Topiramate	Inhibits aldehyde dehydrogenase. Aversive treatment. Opioid receptor antagonist. Supposedly blocks reward; effectiveness questioned. GABA agonist, glutamate antagonist. Reduces craving, unpleasant effects of abstinence. Similar to acamprosate.
Nicotine	Nicotine gum, patch, etc. Bupropion (Zyban) Varenicline (Chantix)	Replaces nicotine. Blocks reuptake of dopamine, norepinephrine, and serotonin. Enhances general reward but reduces nicotine reward; reduces withdrawal effects.
Heroin/opiates	Naltrexone Methadone Buprenorphine Levo-alpha-acetyl-methadol (LAAM)	Opioid receptor antagonist. Competes with opiates for receptor sites. Replaces opiates at receptor. Replaces opiates at receptor. Replaces opiates at receptor.

SOURCES: Julien (2008); Volkow & Li (2004).

attributed to the fact that serotonin helps regulate activity in the mesolimbocortical dopamine system (Melichar, Daglish, & Nutt, 2001). This brings us to the second point, that the various neurotransmitter systems are highly interconnected. This provides additional windows of access to the neural mechanism we want to manipulate and may allow us to choose a more powerful drug or one with fewer side effects. Table 5.1 lists the drugs that are currently approved for the major addictions.

Effectiveness and Acceptance of Pharmacological Treatment

Pharmacological intervention increases treatment effectiveness dramatically. Methadone combined with counseling produces abstinence rates of 60% to 80% in heroin addicts, compared with 10% to 30% for programs that rely on behavioral management alone (Landry, 1997). This is not an argument for pharmacological treatment alone; drug addiction is almost always accompanied by environmental problems and emotional baggage that must be dealt with, and treating addiction as a purely biological or a purely environmental problem have not been very successful (Volkow et al., 2003).

A major difficulty for treating addiction is *comorbidity* with personality disorders, either mental or emotional. This means that drug abusers are likely to have other problems that complicate their rehabilitation. In a study of 43,000 people, 18% of drug abusers had an anxiety disorder, 20% had a mood disorder (most often depression), and 48% had a personality disorder (most often antisocial personality disorder) (B. F. Grant et al., 2004a, 2004b). These symptoms could be partly a by-product of the ravages of addiction, but drug abuse can also be the result of another disorder, for instance, when the person uses drugs as an escape or as a way of self-medicating the symptoms. However, it is more likely that the addiction and the personality disorder have a common genetic, neurological, or environmental cause.

> *Science has yet to defeat the mind/body problem—or those who view psychological problems as failures of will and values.*
>
> —Maia Szalavitz

In spite of the promise of pharmacological treatment of addiction, giving a drug to combat a drug is controversial in some segments of society. Some people believe that recovery from addiction should involve the exercise of will and that recovery should not be easy; Antabuse is okay because it causes the backslider to suffer, but methadone is not okay because it continues the pleasures of drug taking (Szalavitz, 2000). The counterargument is that the bottom line in drug treatment is effectiveness. Addiction costs an estimated $544 billion each year in the United States alone (Uhl & Grow, 2004), but every dollar invested in treatment saves $4 to $12, depending on the drug and the type of treatment (O'Brien, 1997).

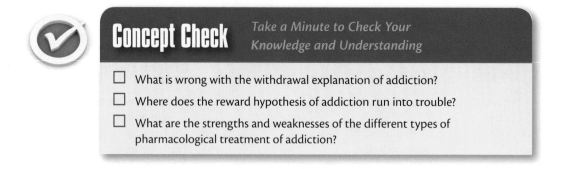

Concept Check *Take a Minute to Check Your Knowledge and Understanding*

☐ What is wrong with the withdrawal explanation of addiction?

☐ Where does the reward hypothesis of addiction run into trouble?

☐ What are the strengths and weaknesses of the different types of pharmacological treatment of addiction?

The Role of Genes in Addiction

Much of the research on what predisposes a person to addiction has focused on alcoholism to the neglect of other drugs. This is understandable, because alcoholism is such a pervasive problem in our society; also, alcoholics are readily accessible to researchers because their drug use is legal. We are beginning to accumulate the same kind of information for other drugs, but as you will see, the study of alcoholism has served as a good model for other addiction research.

Separating Genetic and Environmental Influences

The heritability of addiction was first established with alcoholism. However, for a long time heredity's role in alcoholism was controversial, because studies yielded inconsistent results. One reason is that researchers typically treated alcoholism as a unitary disorder; they would study whatever group they had access to, such as hospitalized alcoholics, and generalize to all alcoholics. An important breakthrough came when Robert Cloninger and his colleagues included all 862 men and 913 women who had been adopted by nonrelatives at an early age (average, 4 months) in Stockholm, Sweden, between 1930 and 1949 (Bohman, 1978).

They divided the alcoholics into two groups, based on drinking behaviors and personality (see Table 5.2). Type 1 alcoholics typically begin their problem drinking after the age of 25, after a long period of exposure to socially encouraged drinking, such as at lunch with coworkers; I will refer to them as *late-onset alcoholics.* They are able to abstain from drinking for long periods of time, but when they do drink, they have difficulty stopping (binge drinking), and they experience guilt about their behavior. Their associated personality traits make them cautious and emotionally dependent. Type 2 alcoholics begin drinking at a young age, so I will call them *early-onset alcoholics.* They drink frequently and feel little guilt about their drinking. They have a tendency toward antisocial behavior and often get into fights in bars and are arrested for reckless driving. They are typically impulsive and uninhibited,

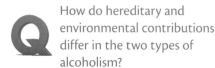
How do hereditary and environmental contributions differ in the two types of alcoholism?

TABLE 5.2 Distinguishing Characteristics of Two Types of Alcoholism.

Characteristic Features	Type of Alcoholism	
	Type 1	Type 2
Alcohol-related problems		
Usual age of onset (years)	After 25	Before 25
Spontaneous alcohol seeking (inability to abstain)	Infrequent	Frequent
Fighting and arrests when drinking	Infrequent	Frequent
Psychological dependence (loss of control)	Frequent	Infrequent
Guilt and fear about alcohol dependence	Frequent	Infrequent
Personality traits		
Novelty seeking	Low	High
Harm avoidance	High	Low
Reward dependence	High	Low

SOURCE: Cloninger (1987).

confident, and socially and emotionally detached. In other words, their behavior resembles the description of *antisocial personality disorder*. Apparently, the personality characteristics appear early; novelty seeking and low harm avoidance in 6- and 10-year-olds predicted drug and alcohol use in adolescence (Mâsse & Tremblay, 1997). Early-onset alcoholics are almost entirely male, and most of the men who are hospitalized for alcoholism fall in this category.

When all of Cloninger's adoptees were considered together, having been reared in an alcoholic home did not increase their risk for alcoholism; it appeared from these data that environmental effects were negligible. However, looking at the two groups individually revealed a different picture and explained the disagreement among earlier studies. For offspring of early-onset alcoholics, the rearing environment made no difference, but offspring of late-onset alcoholics were likely to become alcoholic only if they were reared in a home where there was alcohol abuse. Another difficulty in separating genetic and environmental influence is that environmental interactions can be highly variable, even contradictory. The Met158 version of the *COMT* gene, which is responsible for an enzyme that metabolizes dopamine, is associated with an anxious, sensitive, cautious personality. European Caucasian men tend to drink socially on a daily basis as a means of relaxing; Met158 predisposes them to late-onset alcoholism. American Plains Indians tend to drink heavily, but episodically; in this culture, the greater anxiety and cautiousness that accompanies the Met158 allele apparently confers some protection from alcoholism (Enoch, 2006).

What Is Inherited?

Twin and adoption studies indicate that the heritability for alcoholism is around 50% to 60% (Heath et al., 1997; Köhnke, 2008). Other heritabilities range from 50% for hallucinogens to 72% for cocaine (Goldman, Oroszi, & Ducci, 2005). If genetics plays such an important role in addiction, just what is it that is inherited?

Most research on the genetics of addiction implicates various neurotransmitter systems. For example, knockout mice lacking either of the two *Homer* genes, which regulate glutamate transmitter activity, are also more susceptible to cocaine's rewarding effects (Szumlinski et al., 2004). Mice lacking the *Clock* gene, which regulates sleep-wakefulness cycles, release more dopamine in the ventral tegmental area and are also more vulnerable to the effects of cocaine (McClung et al., 2005). And the gene responsible for the a4 nicotinic acetylcholine receptor apparently helps determine whether a person will become addicted to nicotine (Tapper et al., 2004). Although some genes, like *Homer*, seem to be drug specific in their effects, people often inherit a broad vulnerability to drugs; as a result, about 60% to 70% of addicts abuse three or more drugs (Cadoret, Troughton, O'Gorman, & Heywoood, 1986; S. S. Smith et al., 1992). Numerous genes have been implicated in multiple drug abuse; they are involved with the $GABA_A$ receptor, the neurotransmitter acetylcholine and its receptors, the endogenous opioid system, and the endogenous cannabinoid system (Dick & Agrawal, 2008).

Whether drug use will lead to addiction depends to a large degree on how the individual responds to the drug. For example, moderate-drinking individuals with the G allele of an opioid receptor gene report greater intoxication and feelings of euphoria than participants lacking the allele, and they are three times as likely to have a family history of alcoholism (Ray & Hutchison, 2004). On the other hand, people who are resistant to the negative effects of alcohol are also more susceptible to drinking disorders. Male college students who showed less intoxication, nausea, and motor impairment when drinking were four times more likely to be diagnosed as alcoholic 10 years later (Schuckit, 1994). This is probably because these individuals can drink more alcohol, an interpretation that is supported by the opposite effect. Many Asians react to alcohol with intense flushing, nausea, and increased heart rate; as a consequence, they drink less and they less frequently become alcoholic (T. E. Reed, 1985; Wall & Ehlers, 1995). The reason is an inheritable deficiency in *aldehyde dehydrogenase (ALDH)* enzymes, which eliminate the alcohol metabolite aldehyde in the liver. It is aldehyde that does the physical harm wreaked by alcohol, damaging the liver, muscles, heart, and brain and possibly contributing to heart attacks, Alzheimer's disease, and cancer (Melton, 2007). A person with deficient ALDH is like an alcoholic on Antabuse and experiences discomfort or illness after drinking; as a result, ALDH deficiency provides some protection against alcohol abuse. The deficiency is racially distributed; infrequent among Caucasians and Native Americans, it is found in 50% of nonalcoholic Japanese, compared with only 2% of Japanese alcoholics (Harada, Agarwal, Goedde, Tagaki, & Ishikawa, 1982), and it accounts for 20% to 30% of the differences in alcohol consumption between light-drinking and heavy-drinking Jews (Enoch, 2006). Several genes contribute to ALDH levels, and a half dozen of those are associated not only with alcohol dependence but with risk for abuse of other drugs (Dick & Agrawai, 2008). A similar genetic deficiency in metabolizing nicotine protects some people from nicotine addiction (Pianezza, Sellers, & Tyndale, 1998).

Obviously, not everyone who tries a drug becomes addicted to it; percentages run about 4% for inhalants, 9% for marijuana, 15% for alcohol, 17% for cocaine, 23% for heroin, and 32% for tobacco (Anthony, Warner, & Kessler, 1994). The genes that lead to addiction also affect the person's behavior and personality. Personality characteristics associated with drug experimentation and addiction—impulsivity, risk taking, novelty seeking, and stress responsivity—have all been linked to genes that are also implicated in drug dependence, including those involved in the dopamine, serotonin, opioid, and GABA systems (Dalley et al., 2007; Kreek, Nielsen, Butelman, & LaForge, 2005; Sinha et al., 2003).

These genes alter behavior and personality because they change the way the brain functions. One indication of this brain alteration is the increased high-frequency EEG in alcoholics; this cannot be attributed to the toxic effects of the alcohol, because

8
Additional Drug Information

it also shows up in the alcoholic's children, apparently due to one of the GABA$_A$ receptor genes (Edenberg et al., 2004). Alcoholics also show a reduced amplitude in the P300 wave, or P3, even when they are abstaining; P3 is a dip in the EEG evoked potential about 300 milliseconds after an unexpected stimulus, such as a high-pitched tone appearing in a series of lower tones. A reduced P3 is also seen in alcoholics' children and other nonalcoholic relatives, suggesting a genetic origin (Figure 5.17; Hesselbrock, Begleiter, Porjesz, O'Connor, & Bauer, 2001; S. Hill, 1995; Iacono, Carlson, Malone, & McGue, 2002). However, this characteristic is not specific to alcoholism. It is associated with a variety of disorders characterized by behavioral disinhibition, including childhood conduct disorder, adult antisocial behavior, and various forms of drug abuse. Heritabilities of P3 amplitude and disinhibition are estimated at 64% and 80%, respectively (Hicks et al., 2007). In addition, lowered P3 amplitude in boys at age 17 predicted the development of drug abuse disorders by the age of 20 (Iacono et al.).

Implications of Addiction Research

The study of drug abuse and addiction has practical societal importance, but it is worthwhile for other reasons as well, particularly in shedding light on other kinds of vulnerability and principles of behavioral inheritance. For example, alcoholism is not entirely due to genetics or to environmental influences but results from an interplay of the two. The fact that these two forces operate differently in different types of alcoholism and in different cultural settings illustrates the fact that no behavior is simple or simply explained. Even after we understand the relative roles of heredity and environment, there is further complexity, because we must also understand the mechanisms—the neurotransmitters, receptors, pathways, enzymes, and so on. In other words, the causes of addiction, and of behavior in general, are many and complex. Finally, we must look beyond simple appeals to willpower in explaining the self-defeating behavior of the addict, just as we must do when we try to understand other kinds of behavior. Our brief look at addiction is a good preparation for our inquiries into the physiological systems behind other human behaviors and misbehaviors.

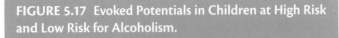

FIGURE 5.17 Evoked Potentials in Children at High Risk and Low Risk for Alcoholism.

Evoked potentials were elicited by high-pitched tones occurring among low-pitched tones. The usual dip of the P300 wave is diminished in the high-risk children.

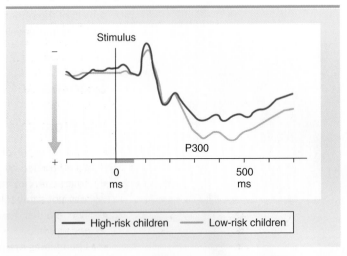

SOURCE: Reprinted by permission of Elsevier Science from S. Y. Hill, D. Muka, S. Steinhauer, and J. Locke. "P300 Amplitude Decrements in Children From Families of Alcoholic Female Probands," *Biological Psychiatry*, 38, pp. 622–632. © 1995 Society of Biological Psychiatry.

Concept Check   *Take a Minute to Check Your Knowledge and Understanding*

- ☐ How did the failure to recognize two types of alcoholism create misunderstandings about hereditary and environmental influences and gender distribution in alcoholism?
- ☐ How can lowered sensitivity to a drug increase the chances of addiction?
- ☐ What are two kinds of evidence that some people are predisposed to alcoholism from birth?

In Perspective

The costs of drug abuse include untold suffering; loss of health, productivity, and life; and billions of dollars in expenses for treatment and incarceration. The only upsides are that the study of drug abuse reveals the workings of the synapses and brain networks and helps us recognize that powerful biological forces are molding our behavior. This knowledge in turn helps us understand the behaviors that are the subject of the remaining chapters, including the disorders covered there, and guides research into developing therapeutic drugs.

Summary

Psychoactive Drugs

- Most abused drugs produce addiction, which is usually (but not always) accompanied by withdrawal symptoms when drug use is stopped.
- Tolerance can increase the dangers of drugs because life-threatening effects may not show tolerance.
- The opiates have their own receptors, which are normally stimulated by endorphins.
- The opiates are particularly addictive and dangerous.
- Depressants reduce activity in the nervous system. Some of them have important uses, but they are highly abused.
- Stimulants increase activity in the nervous system. They encompass the widest range of effects and include nicotine, most notable for its addictiveness and its association with deadly tobacco.
- Psychedelic drugs are interesting for their perceptual/hallucinatory effects, which result from their transmitter-like structures.
- Marijuana is controversial not just in terms of the legalization issue but because it raises questions about what constitutes addiction.

Addiction

- The mesolimbocortical dopamine system is implicated by several lines of research as a reward center that plays a role in drug addiction, feeding, sex, and other behaviors.
- Dopamine may also contribute to addiction through its role in learning by modifying neural functioning.
- Treatment of addiction is very difficult; effective programs combine psychological support with pharmacological strategies, including agonist, antagonist, and aversive treatments and, potentially, drug vaccines.

The Role of Genes in Addiction

- Research suggests that addiction is partially hereditary and that the inherited vulnerability may not be drug specific.
- Heredity research indicates that there are at least two kinds of alcoholism, with different genetic and environmental backgrounds.
- Alcoholics often have dopamine and serotonin irregularities that may account for the susceptibility, and some have a deficiency in evoked potentials that appears to be inherited. ■

Study Resources

For Further Thought

- Is the legality or illegality of a drug a good indication of its potential for abuse?
- Is it morally right to treat addictions with drug antagonists, aversive drugs, and antidrug vaccines? Is your opinion the same for drug agonists?

- You work for an agency that has the goal of substantially reducing the rate of drug abuse in your state through education, family support, and individualized treatment. Based on your knowledge of addiction, what should the program consist of?

Testing Your Understanding

1. Describe the two proposed roles for dopamine in addiction and give two pieces of evidence for each.

2. What are the practical and ethical considerations in using drugs to treat addiction?

3. Sally and Sam are alcoholics. Sally seldom drinks but binges when she does and feels guilty later. Sam drinks regularly and feels no remorse. What other characteristics would you expect to see in them, and what speculations can you make about their environments?

Select the best answer:

1. In the study of conditioned tolerance to heroin,
   a. human subjects failed to show the usual withdrawal symptoms.
   b. human subjects increased their drug intake.
   c. rats were unresponsive to the drug.
   d. rats tolerated the drug less in a novel environment.

2. Withdrawal from alcohol
   a. can be life threatening.
   b. is about like a bad case of flu.
   c. is slightly milder than with most drugs.
   d. is usually barely noticeable.

3. The reason alcohol, barbiturates, and benzodiazepines are deadly taken together is that they
   a. affect the thalamus to produce almost total brain shutdown.
   b. have a cumulative effect on the periaqueductal gray.

   c. affect the same receptor complex.
   d. increase dopamine release to dangerous levels.

4. Psychedelic drugs often produce hallucinations by
   a. inhibiting serotonin neurons.
   b. stimulating serotonin receptors.
   c. stimulating dopamine receptors.
   d. blocking dopamine reuptake.

5. Marijuana was the subject of disagreement among researchers because some of them
   a. believed it is more dangerous than alcohol or tobacco.
   b. believed it is highly addictive.
   c. thought it failed to meet the standard test for addictiveness.
   d. overstated its withdrawal effects.

6. Evidence that addiction does not depend on the drug's ability to produce withdrawal symptoms is that
   a. they don't usually occur together with the same drug.
   b. they are produced in different parts of the brain.
   c. either can be produced without the other in the lab.
   d. a and b.
   e. b and c.

7. When rats trained to press a lever for electrical stimulation of the brain are given a drug that blocks dopamine receptors, lever pressing
   a. increases.
   b. decreases.
   c. increases briefly then decreases.
   d. remains the same.

8. The best argument that caffeine is an addictive drug like alcohol and nicotine is that
   a. it is used regularly by most of the population.
   b. quitting produces withdrawal.
   c. it affects the same processes in the brain.
   d. it stimulates dopamine receptors directly.

9. Evidence that dopamine's contribution to addiction may be its effect on learning includes a study in which
   a. blocking dopamine receptors interfered with learning to self-inject cocaine.
   b. hippocampal stimulation released dopamine and restored learned lever pressing.
   c. rats learned to press a lever for injections of dopamine into the nucleus accumbens.
   d. rats learned a maze for food reward faster if given a dopamine uptake blocker.

10. Agonist treatments for drug addiction
    a. mimic the drug's effect.
    b. block the drug's effect.
    c. make the person sick after taking the drug.
    d. reduce anxiety so that there is less need for the drug.

11. Critics of treating drug addiction with drugs believe that
    a. getting over addiction should not be easy.
    b. it is wrong to give an addict another addictive drug.
    c. the drugs are not very effective and delay effective treatment.
    d. a and b.
    e. b and c.

12. The type of alcoholism in which the individual drinks regularly is associated with
    a. behavioral rigidity.
    b. perfectionism.
    c. feelings of guilt.
    d. antisocial personality disorder.

13. Alcoholics often
    a. have reduced serotonin and dopamine functioning.
    b. are more sensitive to the effects of alcohol.
    c. are unusually lethargic and use alcohol as a stimulant.
    d. have an inherited preference for the taste of alcohol.

 # Online Resources

The following resources are available at **www.sagepub.com/garrett3e.**

Chapter Resources

- Flash Cards
- Chapter Quiz
- Internet Resources and Exercises

On the Web

Links to these websites can be found at the web address above: Select this chapter's **Chapter Resources, then choose Internet Resources and Exercises.**

1. The National Institute on Drug Abuse's (NIDA) **Heroin** page is a good source for research and other information on heroin and its effects.
   Check **The Effects of Drugs on the Nervous System** at the Neuroscience for Kids site for information on more than a dozen drugs.

2. The **Alcoholics Anonymous** site has information about AA, testimonials from members, and a quiz for teenagers (or anybody else) to help them decide if they have a drinking problem.

3. **Cocaine Anonymous** offers news, information, a self-test for addiction, and a directory of local groups.

4. NIDA's **Tobacco/Nicotine** page provides facts, publications, and links to other sites.

5. NIDA also has information on **LSD**, **Ecstasy**, and **PCP**.

6. There's even a **Marijuana Anonymous**, and its site offers a variety of publications for the person who wants to stop using marijuana or for the student who wants to learn more. Of course, NIDA has its **Marijuana** page, too. And ProCon.org's **Medical Marijuana** page has arguments for and against the medical use of marijuana, along with discussions of legal issues and marijuana's use with each of 16 different diseases.

7. The **Web of Addictions** provides fact sheets, links to a variety of other information sites, contact information for help organizations and other

organizations concerned with drug problems, and in-depth reports on special topics.

8. Additional sites of interest:
    The **National Clearinghouse for Alcohol and Drug Information** has information on drugs and what can be done to prevent abuse; a visitor can choose material appropriate for different audiences, including by ethnic group, age, gender, and so on.
    The **NIDA** site provides news; research information; and information on prevention for parents, teachers, and students.

Chapter Updates and Biopsychology News

R — For Further Reading

1. *A Primer of Drug Action: A Comprehensive Guide to the Actions, Uses, and Side Effects of Psychoactive Drugs* by Robert M. Julien (Worth, 2008). Often used as a text in psychopharmacology and upper-level biopsychology courses, this book covers principles of drug action, properties of specific drugs, pharmacotherapy for various disorders, and societal issues. It received good reviews from students on the Amazon book site.

2. *Buzzed: The Straight Facts About the Most Used and Abused Drugs From Alcohol to Ecstasy* by Cynthia Kuhn, Scott Swartzwelder, Wilkie Wilson, Leigh Heather Wilson, and Jeremy Foster (W. W. Norton, 3rd ed., 2008). The book gives technical information about drugs written in a style appropriate for college students. It covers drug characteristics, histories of the drugs, addiction, the workings of the brain, and legal issues.

3. *The Encyclopedia of Psychoactive Substances* by Richard Rudgley (St. Martins, 2000). Formatted as a reference book, it devotes just a few pages to each of more than 100 drugs, but includes historical information as well as information about changing social attitudes. Coverage ranges from traditional drugs to exotic ones, such as hallucinogenic fish.

K — Key Terms

addiction ... 126
agonist treatment ... 143
alcohol ... 128
aldehyde dehydrogenase (ALDH) ... 148
amphetamine ... 132
analgesic ... 127
antagonist treatment ... 144
antidrug vaccine ... 144
anxiolytic ... 128
aversive treatment ... 144
barbiturate ... 130
benzodiazepine ... 130
caffeine ... 134
cannabinoids ... 135
cocaine ... 131
delirium tremens ... 129
depressant ... 128
drug ... 126
early-onset alcoholism ... 146
electrical stimulation of the brain (ESB) ... 139
endogenous ... 128

endorphins ... 128
euphoria ... 127
heroin ... 127
hypnotic ... 127
late-onset alcoholism ... 146
marijuana ... 135
medial forebrain bundle ... 138
mesolimbocortical dopamine system ... 138
methadone ... 144
nicotine ... 133
nucleus accumbens ... 138
opiate ... 127
psychedelic drug ... 134
psychoactive drug ... 126
reinforcer ... 141
reward ... 138
sedative ... 128
stimulant ... 130
tolerance ... 126
ventral tegmental area ... 138
withdrawal ... 126

Motivation and the Regulation of Internal States

6

Motivation and Homeostasis

Theoretical Approaches to Motivation

Simple Homeostatic Drives

CONCEPT CHECK

Hunger: A Complex Drive

The Role of Taste

APPLICATION: PREDATOR CONTROL THROUGH
LEARNED TASTE AVERSION

Digestion and the Two Phases of Metabolism

Signals That Start a Meal

Signals That End a Meal

Long-Term Controls

CONCEPT CHECK

Obesity

The Myths of Obesity

The Contribution of Heredity

APPLICATION: THE EPIGENETICS OF WEIGHT REGULATION

Obesity and Reduced Metabolism

Treating Obesity

CONCEPT CHECK

Anorexia and Bulimia

Environmental and Genetic Contributions

The Role of Serotonin and Dopamine

CONCEPT CHECK

In Perspective

Summary

Study Resources

In this chapter you will learn

- Some of the ways psychologists have viewed motivation

- How the concepts of drive and homeostasis explain the regulation of internal body states

- How taste helps us select a safe and nutritious diet

- How we regulate the amount of food we eat

- What some of the causes of obesity are

- What we know about anorexia and bulimia

> *I can stuff my face for a long time and I won't feel full.*
> —Christopher Theros

1
Prader-Willi Syndrome

When Christopher was born, it was obvious there was something wrong (Lyons, 2001). He was a "floppy baby," lying with his arms and legs splayed lifelessly on the bed, and he didn't cry. Doctors thought he might never walk or talk, but he seemed to progress all right until grade school, when he was diagnosed with Prader-Willi syndrome. The disorder occurs when a small section of the father's chromosome 15 fails to transfer during fertilization. The exact contribution of those genes is not known, but the symptoms are clearly defined, and Christopher had most of them. He stopped growing at 5 feet, 3 inches (1.6 meters), he had learning difficulties, and he had difficulty with impulse control.

More obviously, Christopher could never seem to recognize when he had eaten enough, so he ate constantly. He even stole his brother's paper-route money to buy snacks at the corner store. At school, he would retrieve food from the cafeteria garbage can and wolf it down; his classmates would taunt him by throwing a piece of food in the trash to watch him dive for it. The only way to protect a person like Christopher is to manage his life completely, from locking the kitchen to institutionalization. State law did not permit institutionalization for Chris, because his average-level IQ did not fit the criterion for inability to manage his affairs. He lived in a series of group homes but was thrown out of each one for rebelliousness and violence, behaviors that are characteristic of the disorder. When he died at the age of 28, he weighed 500 lb (1,100 kg; Figure 6.1).

In the previous chapter, we puzzled over why people continue to take drugs that are obviously harming them. Now we are forced to wonder why a person would be so out of control that he would literally eat himself to death. When we ask why people (and animals) do what they do, we are asking about their motivation.

Motivation and Homeostasis

Motivation, which literally means "to set in motion," refers to the set of factors that initiate, sustain, and direct behaviors. The need for the concept was prompted by psychologists' inability to explain behavior solely in terms of outside stimuli. Assuming various kinds of motivation, such as hunger or achievement need, helped make sense of differing responses to the same environmental conditions.

Keep in mind, though, that *motivation is a concept psychologists have invented and imposed on behavior;* we should not expect to find a single "motivation center" in the brain or even a network whose primary function is motivation. The fact that we sometimes cannot distinguish motivation from other aspects of behavior, like emotion, is evidence of how arbitrary the term can be. Still, it is a useful concept for organizing ideas about the sources of behavior.

After a brief overview of some of the ways psychologists have approached the problem of motivation, we will take a closer look at temperature regulation, thirst, and hunger as examples before taking up the topics of sexual behavior in the next chapter and emotion and aggression in Chapter 8.

Theoretical Approaches to Motivation

Greeks relied heavily on instinct in their attempts to explain human behavior. An *instinct* is a complex behavior that

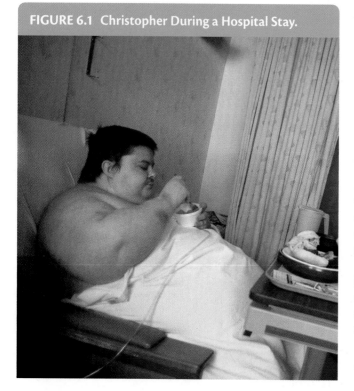

FIGURE 6.1 Christopher During a Hospital Stay.

SOURCE: Jayson Mellom /San Luis Obispo *Tribune.*

is automatic and unlearned and occurs in all the members of a species. Migration and maternal behavior are good examples of instinctive behaviors in animals. According to early instinct theorists, humans were guided by instincts, too, waging war because of an aggressive instinct, caring for their young because of a maternal instinct, and so on. At first blush, these explanations sound meaningful. But if we say that a person is combative because of an aggressive instinct, we know little more about what makes the person fight than we did before; if we cannot then analyze the supposed aggressive instinct, we have simply dodged the explanation.

The idea of instincts as an explanation of human motivation was popularized in modern times by the psychologist William McDougall (1908), who proposed that human behaviors such as reproduction, gregariousness, and parenting are instinctive. It wasn't long until one writer was able to count 10,000 names of instincts in the literature; this led him to suggest, tongue firmly in cheek, that there must also be an "instinct to produce instincts" (Bernard, 1924). Apparently the instinct explanation provided too many theorists an easy way out; the idea of human instincts fell into disrepute. Contemporary students of behavior have used stringent requirements of evidence to identify a few instincts in animals, such as homing and maternal care. But most psychologists believe that in human evolution instincts have either dropped out or weakened. At any rate, we need to be extremely careful about how we use the term *instinct* and to avoid the temptation to label any behavior that is difficult to explain as an instinct.

Drive theory has fared much better than instinct theory, at least in explaining motivation that involves physical conditions such as hunger, thirst, and body temperature. According to *drive theory,* the body maintains a condition of *homeostasis,* in which any particular system is in balance or equilibrium (C. L. Hull, 1951). Any departure from homeostasis, such as depletion of nutrients or a drop in temperature, produces an aroused condition, or *drive,* which impels the individual to engage in appropriate action such as eating, drinking, or seeking warmth. As the body's need is met, the drive and associated arousal subside. This is a temporary state, of course; soon the individual will be hungry or thirsty or cold again, and the cycle will continue.

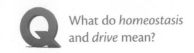

Q What do *homeostasis* and *drive* mean?

Critics of drive theory point out that it does not explain all kinds of motivation; many motivated behaviors seem to have nothing to do with satisfying tissue needs. For example, a student may be motivated by grades, some people struggle to achieve fame, and others work long hours to earn more money than they need for food and shelter. *Incentive theory* recognizes that people are motivated by external stimuli, not just internal needs (Bolles, 1975). In this respect, money and grades act as *incentives.* Incentives can even be a factor in physiological motivation; consider, for example, the effect of the smell of chocolate chip cookies baking or the sight of a sexually attractive individual.

My wife recently jumped out of an airplane for the thrill of plummeting toward the earth, hoping to be saved at the last minute by a flimsy parachute; there is no tissue need, and no obvious drive involved here. Observations like this have led to the *arousal theory,* which says that people behave in ways that keep them at their preferred level of arousal (Fiske & Maddi, 1961). Different people have different optimum levels of arousal, and some seem to have a need for varied experiences or the thrill of confronting danger (Zuckerman, 1971). This *sensation seeking* finds expression in anything from travel and unconventional dress to skydiving, drug use, armed robbery—even eating *fugu* (see Chapter 2).

SSS

In the face of challenges to drive theory, psychologists have shifted their emphasis to drives as states of the brain rather than as conditions of the tissues (Stellar & Stellar, 1985). This approach nicely accommodates sexual behavior, which troubled drive theorists because it does not involve a tissue deficit. Even eating behavior is better understood as the result of a brain state. Hunger ordinarily occurs when a lack of nutrients in the body triggers activity in the brain. However, an incentive like the smell of a steak on the grill can also cause hunger, apparently by activating the same brain mechanisms as tissue deficits do. In addition, the person feels satisfied and stops eating long before the nutrients have reached the deficient body cells. Similarly, if the brain is not "satisfied," it little matters how much the person has eaten. In other words, if the information that reserves are excessive fails to reach the brain or to have its usual effect there, the person may, like Christopher, eat to obesity and still feel hungry. In the following pages, we will look at the regulation of body temperature, fluid levels, and energy supply from the perspective of drive and homeostasis.

Simple Homeostatic Drives

To sustain life, a number of conditions, such as body temperature, fluid levels, and energy reserves, must be held within a fairly narrow range. Accomplishing that requires a *control system.* A mechanical control system that serves as a good analogy is a home heating and cooling system. Control systems have a *set point,* which is the point of equilibrium the system returns to. For the heating and cooling system, the set point is the temperature selected on the thermostat. A departure in the room temperature from the set point is analogous to a drive; the thermostat initiates an action, turning on the furnace or the air conditioner. When the room temperature returns to the preset level, the system is "satisfied" in the technical sense of the word; homeostasis has been achieved, so the system goes into a quieter state until there is another departure from the set point.

Temperature Regulation

Not only is the regulation of body temperature superficially similar to our thermostat analogy, it is almost as simple. All animals have to maintain internal temperature within certain limits to survive, and they operate more effectively within an even narrower range; this is their set point. How they respond to departures from homeostasis is much more variable than with the home heating and cooling system, however. *Ectothermic* animals, such as snakes and lizards, are unable to regulate their body temperature internally, so they adjust their temperature behaviorally by sunning themselves, finding shade, burrowing in the ground, and so on. *Endothermic* animals, which include mammals and birds, use some of the same strategies along with others that are functionally similar, such as building nests or houses, turning up the thermostat, and wearing clothing. However, endotherms are also able to use their energy reserves to maintain a nearly constant body temperature automatically. In hot weather, their temperature regulatory system reduces body heat by causing sweating, reduced metabolism, and dilation of peripheral blood vessels. In cold weather, it induces shivering, increased metabolism, and constriction of the peripheral blood vessels. To say that we make these adjustments because we *feel* hot or cold suggests that the responses are intentional behaviors, but that is not the case. So how do these behaviors occur?

 How is body temperature regulated?

In mammals, the "thermostat" is located in the *preoptic area* of the hypothalamus, which contains separate warmth-sensitive and cold-sensitive cells (Figure 6.2; Nakashima, Pierau, Simon, & Hori, 1987). Some of these neurons respond directly to the temperature of the blood flowing through the area; others receive input from temperature receptors in other parts of the body, including the skin. The preoptic area integrates information from these two sources and initiates temperature regulatory responses, such as panting, sweating, and shivering (Boulant, 1981; Kupferman, Kandel, & Iversen, 2000). We will be talking about several nuclei in the hypothalamus in this chapter, so you may want to refer to Figure 6.2 often.

Thirst

The body is about 70% water, so it seems obvious that maintaining the water balance is critical to life. Water is needed to maintain the cells of the body, to keep the blood flowing through the veins and arteries, and to digest food. You can live for weeks without eating but only for a few days without water. We constantly lose water through sweating, urination, and defecation. The design of your nose, which could have been just a pair of nostrils on your face, is testimonial to the body's efforts to conserve water. As you breathe, you exhale valuable moisture; but as your breath passes through the much cooler nose, some of the moisture condenses and is reabsorbed. The next time you get a runny nose on a cold day, you will get some idea how much water the nose recycles.

It is obvious that you drink when your mouth and throat feel dry, but at most a dry mouth and throat determine only *when* you drink, not *how much* you drink. There are two types of thirst, one generated by the water level inside the body's cells and the other reflecting the water content of the blood. Water deprivation affects both kinds of deficit, but the fluid levels in the two compartments can vary independently and so the brain manages them separately. *Osmotic thirst* occurs when the fluid content decreases inside the cells. This happens when the blood becomes more concentrated than usual, usually because the individual has not taken in enough water to compensate for food intake; eating a salty meal adds to the effect by making the blood more concentrated. As a result, water is drawn from the cells into the bloodstream by osmotic pressure. *Hypovolemic thirst* occurs when the blood volume drops due to a loss of extracellular water. This can be due to sweating, vomiting, and diarrhea. Of course, another cause is blood loss; that is why you feel thirsty after giving blood.

The reduced water content of cells that contributes to osmotic thirst is detected primarily in areas bordering the third ventricle, particularly in the *organum vasculosum lamina terminalis* (*OVLT;* see Figure 6.3). Injecting saline (salt solution) into the bloodstream draws water out of the cells and induces drinking; this effect is dramatically reduced when the OVLT is lesioned beforehand (Thrasher & Keil, 1987). The OVLT communicates the water deficit to the *median preoptic nucleus* of the hypothalamus, which initiates drinking.

Hypovolemia is detected by receptors located where the large veins enter the atrium of the heart; these receptors respond to stretching of the vascular walls by the volume of blood passing through (Fitzsimons & Moore-Gillon, 1980). The reduced blood volume in the heart that accompanies hypovolemia is signaled by the vagus nerve to the

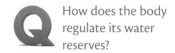

How does the body regulate its water reserves?

FIGURE 6.2 Selected Nuclei of the Hypothalamus.

The illustration shows only the right hypothalamus. The hypothalamus is a bilaterally symmetrical structure, which means that the left and right halves (separated by the third ventricle) are duplicates of each other. (The pituitary, optic chiasm, and pons have been identified for use as landmarks.)

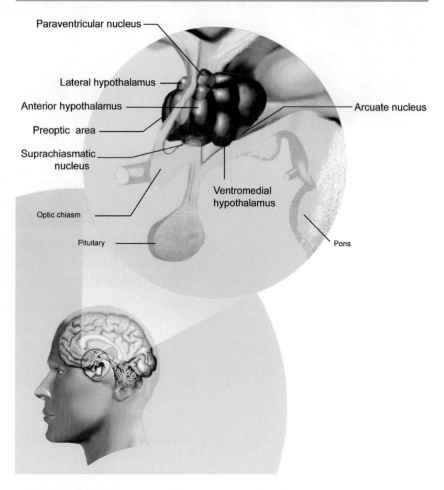

Paraventricular nucleus

Lateral hypothalamus

Anterior hypothalamus

Preoptic area

Suprachiasmatic nucleus

Arcuate nucleus

Ventromedial hypothalamus

Optic chiasm

Pituitary

Pons

SOURCE: Adapted from Nieuwenhuys, Voogd, & Van Huijzen (1988).

FIGURE 6.3 Thirst Control Signals and Brain Centers.

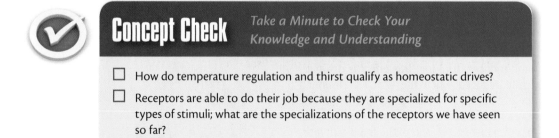

nucleus of the solitary tract (NST) in the medulla. From there, the signal goes to the median preoptic area of the hypothalamus (Figure 6.3; Stricker & Sved, 2000).

Lowered blood volume is also detected by receptors in the kidneys, which trigger the release of the hormone renin. Renin then increases production of the hormone angiotensin II. *Angiotensin II* circulating in the blood stream informs the brain of the drop in blood volume. It stimulates the *subfornical organ (SFO),* a structure bordering the third ventricle and one of the areas that is unprotected by the blood-brain barrier (Figure 6.3). Again, drinking is induced by the nearby median preoptic nucleus (Fitzsimons, 1998; Stricker & Sved, 2000). Injecting angiotensin into the SFO increases drinking; lesioning the SFO blocks this effect but has no effect on drinking in response to osmotic thirst (J. B. Simpson, Epstein, & Camardo, 1978).

Thirst is more complicated than the operation of a furnace or body temperature regulation because there is a significant time lag between drinking and the arrival of water in the tissues. The individual must stop drinking well before tissue need is satisfied. The *satiety* (satisfaction of appetite) mechanism is not well understood, but there is evidence that receptors in the stomach monitor the presence of water (Rolls, Wood, & Rolls, 1980). Also, infusion of water into the liver reduces drinking, which suggests that either water receptors or pressure receptors are there (Kozlowski & Drzewiecki, 1973). We know more about satiety when it comes to hunger and will take up the issue again later.

Concept Check *Take a Minute to Check Your Knowledge and Understanding*

☐ How do temperature regulation and thirst qualify as homeostatic drives?

☐ Receptors are able to do their job because they are specialized for specific types of stimuli; what are the specializations of the receptors we have seen so far?

Hunger: A Complex Drive

Although hunger can be described in terms of drive and homeostasis just like temperature regulation and thirst, the differences almost overshadow the similarities. Hunger is more complicated in a variety of ways. Eating provides energy for activity, fuel for maintaining body temperature, and materials needed for growth and repair of the tissues. In addition, the set point is so variable that you might think there is none. This is not surprising, because the demands on our resources change with exercise, stress, growth, and so on. A changing set point is not unique to hunger, of course. For example, our temperature set point changes daily, decreasing during our normal sleep period (even if we fly to Europe and are awake during normal sleep time). It also increases during illness to produce a fever to kill invading bacteria. What is unusual about hunger is that the set point can undergo dramatic and prolonged shifts, for instance in obesity.

Another difference is that the needs in temperature regulation and thirst are unitary, while hunger involves the need for a variety of different and specific kinds of nutrients. Making choices about what foods to eat can be more difficult than knowing when to eat and when to stop eating.

The Role of Taste

Selecting the right foods is no problem for some animals. Some *herbivores* (plant-eating animals) can get all the nutrients they need from a single source; koalas, for instance, eat only eucalyptus leaves, and giant pandas eat nothing but the shoots of the bamboo plant. *Carnivores* (meat eaters) also have it rather easy; they depend on their prey to eat a balanced diet. We are *omnivores;* we are able to get the nutrients we need from a variety of plants and animals. Being able to eat almost anything is liberating but simultaneously a burden: We must distinguish among foods that may be nutritious, nonnutritious, or toxic, and we must vary our diet among several sources to meet all our nutritional requirements. Choosing the right foods and in the right amounts can be a real challenge.

> *You are what you eat.*
>
> —*popular adage*

It is possible that you plan your diet around nutritional guidelines, but probably you rely more on what you learned at the family table about which foods and what combinations of foods make an "appropriate" meal in your culture. Have you ever wondered where these traditions came from, or why they survive when each new generation seems to delight in defying society's other customs? Long before humans understood the need for vitamins, minerals, proteins, and carbohydrates, your ancestors were using a "wisdom of the body" to choose a reasonably balanced diet that ensured their survival and your existence. That wisdom is reflected in cultural food traditions, which usually provide a balanced diet (Rozin, 1976), sometimes by dictating unattractive (to us) choices such as grub worms or cow's blood. As you will see, the internal forces that guide our selection of a balanced diet are more automatic than the term *choice* usually suggests, but they are also subtle and easily overcome by the allure of modern processed foods, which emphasize taste over nutrition.

The simplest form of dietary selection involves distinguishing between foods that are safe and nutritious and those that are either useless or dangerous. This is where the sense of taste comes in. In humans, all taste experience is the result of just five taste sensations: sour, sweet, bitter, salty, and the more recently discovered umami (Kurihara & Kashiwayanagi, 1998). The first four need no explanation; umami is often described as "meaty" or "savory." These five sensations are called *primaries;* more complex taste sensations are made up of combinations of the primaries. There may be other taste primaries; particularly, there is evidence for fat detection, but so far this has not been confirmed (Mattes, 2009).

It is easy to see why we have evolved taste receptors with these particular sensitivities, because they correspond closely to our dietary needs. We will readily eat foods that are sweet; many nutritious foods, fruits for example, have a sweet taste. We also prefer foods that are a bit salty; salt provides the sodium and chloride ions needed for cellular functioning and for neural transmission. Mountain gorillas get

FIGURE 6.4 A Microscopic Photo of a Papilla With Taste Buds.

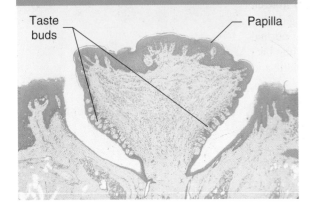

Taste buds

Papilla

SOURCE: SIU/Peter Arnold, Inc.

FIGURE 6.5 Localization of Taste Responses in the Cortex.

The color-coded areas on this fMRI scan indicate where the insular cortex was activated as the subject tasted various liquids. The locations differed among subjects but were consistent over time for each subject. The location of the insular cortex is shown in the lower image.

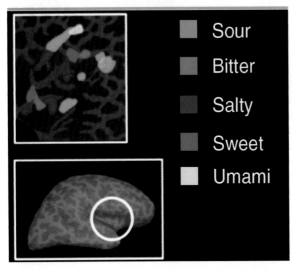

Sour
Bitter
Salty
Sweet
Umami

SOURCE: From "Functional Magnetic Resonance Tomography Correlates of Taste Perception in the Human Primary Taste Cortex," by M. A. Schoenfeld et al., 2004, *Neuroscience, 127,* 347–353.

Q In what ways does taste contribute to selection of a proper diet?

95% of their sodium by eating decaying wood—while avoiding similar wood with lower sodium content (Rothman, Van Soest, & Pell, 2006). The umami receptor responds to amino acids, including glutamate, a component in meats, cheese, soy products, and the flavor enhancer monosodium glutamate. Little is known about this fifth receptor, but it could be important in our selection of proteins. Just as we are attracted to useful foods by taste, we avoid others. Overly sour foods are likely to be spoiled, and bitter foods are likely to be toxic. You do not have to understand these relationships, much less think about them; they operate quietly, in the neural background.

Taste receptors are located on taste buds, which in turn are found on the surface of papillae; papillae are small bumps on the tongue and elsewhere in the mouth (see Figure 6.4). Taste neurons travel through the thalamus to the *insula*, the primary gustatory (taste) area in the frontal lobes. But on their way, they pass through the NST in the medulla, which we saw in Figure 6.3 in relation to drinking and will soon see plays an important role in feeding behavior. Back in Chapter 2, you saw research indicating that taste neurons encode different taste stimuli in unique time patterns of impulses, but there is another, more important, way the system informs the brain what we are tasting. Each of the primary taste stimuli is detected by receptors that are specialized for that stimulus; information from the different receptors travels to the brain via separate pathways to distinct areas in the insular cortex (Schoenfeld et al., 2004; K. Scott, 2005; see Figure 6.5). You will see in later chapters that other sensory systems exploit this *labeled line* coding of stimuli and the brain combines primary sensations into more complex sensory experiences.

Besides the nutritional and safety benefits we have already mentioned, the taste sense contributes to dietary selection in three additional ways: sensory-specific satiety, learned taste aversion, and learned taste preferences.

Sensory-Specific Satiety: Varying the Choices

One day when I was a youngster, a neighbor child joined our family for lunch. At the end of the meal, she enjoyed a bowl of my mother's homemade apple cobbler, then another, and another. Halfway through the third serving, she observed in puzzlement that the last serving wasn't nearly as good as the first. Barbara and Edmund Rolls call this experience sensory-specific satiety. *Sensory-specific satiety* means that the more of a particular food an individual eats, the less appealing the food becomes. Humans rate a food less favorably after they have consumed it, and they eat more if they are offered a variety of foods instead of a single food (Rolls, Rolls, Rowe, & Sweeney, 1981).

The effect sounds trivial, but it is not; sensory-specific satiety is the brain's way of encouraging you to vary your food choices, which is necessary for a balanced diet. Back in the 1920s, Clara Davis (1928) allowed three newly weaned infants to choose all their meals from a tray of about 20 healthy foods. They usually selected only two or three foods at one meal and continued choosing the same foods for about a week. Then they would switch to another two or three foods for a similar period. Their self-selected diet was adequate to prevent any deficiencies from developing over a period of 6 months.

Sensory-specific satiety takes place in the NST. Place a little glucose (one of the sugars) on a rat's tongue, and it produces a neural response there. But if glucose is injected into the rat's bloodstream first, sugar placed on the tongue has less effect in the NST (Giza, Scott, & Vanderweele, 1992). The brain automatically motivates the rat—or you—to switch to a new flavor and a different nutrient.

Learned Taste Aversion: Avoiding Dangerous Foods

Learned taste aversion, the avoidance of foods associated with illness or poor nutrition, was discovered when researchers were studying bait shyness in rats. Farmers know that if they put out poisoned bait in the barn, they will kill a few rats at first but the surviving rats will soon start avoiding the bait (thus the term *bait shyness*). Rats eat small amounts of a new food; that way, a poison will more likely make them ill instead of killing them, and they will avoid that food in the future. Learned aversion is studied in the laboratory by giving rats a specific food and then making them nauseous with a chemical like lithium chloride or with a dose of X-ray radiation. Later they refuse to eat that food.

Learned taste aversion helps wild animals and primitive-living humans avoid dangerous foods (see the accompanying Application). Modern-living humans experience learned taste aversions, too. In a study of people with strong aversions to particular foods, 89% could remember getting sick after eating the food, most often between the ages of 6 and 12 (Garb & Stunkard, 1974). However, in civilized settings, learned aversions have little value in identifying dangerous foods. Instead, we usually get sick following a meal because we left the food out of the refrigerator too long or because we happened to come down with stomach flu. Learned taste aversion appears to be one reason chemotherapy patients lose their appetite. Among children who were given a uniquely flavored ice cream before a chemo session, 79% later refused that flavor, compared with 33% of children receiving chemotherapy without the ice cream; the effect was just as strong 4 months later (L. L. Bernstein, 1978).

Learned taste aversion may not be very useful to modern humans for avoiding dangerous foods, but it may help us avoid nonnutritious ones. When rats are fed a diet that is deficient in a particular nutrient, such as thiamine (vitamin B), they start showing an aversion to their food; they eat less of it, and they spill the food from its container in spite of indications they are hungry, like chewing on the wire sides of their cage (Rozin, 1967). Even after recovery from the deficit, the rats prefer to go hungry rather than eat the previously deficient food. But aversion to a nutrient-deficient food is just the first step toward selecting a nutritious diet.

Learned Taste Preferences: Selecting Nutritious Foods

Although rats, and presumably humans, can detect salt, sugars, and fat directly by their taste (Beck & Galef, 1989), they must *learn* to select the foods containing other necessary nutrients. This apparently requires the development of a *learned taste preference,* which is a preference not for the nutrient itself but for the flavor of a food that contains the nutrient. In an early study, rats were fed a diet deficient in one of three vitamins (thiamine, riboflavin, or pyridoxine); later they learned to prefer a food enriched with that vitamin, which was flavored distinctively by adding anise (which tastes like licorice). When the anise was switched to the vitamin-deficient food the rats began eating that food instead (E. M. Scott & Verney, 1947). Presumably, animals learn to prefer the flavor because the nutrient makes them feel better. A diet-deficient rat enhances its chances of learning which foods are beneficial by eating a single food at a time and spacing its meals so that a nutrient has time to produce some improvement

Application   Predator Control Through Learned Taste Aversion

Learned taste aversion has been put to practical use in an unlikely context: predator control. As a novel and humane (compared to extermination) means of controlling sheep killing by wolves and coyotes, Gustavson and colleagues fed captive predators sheep carcasses laced with lithium chloride (see the photo), which made them sick. When they were placed in a pen with sheep, the wolves and coyotes avoided the sheep instead of attacking them. One coyote threw up just from smelling a lamb, and two hesitant wolves were chased away by a lamb that turned on them (Gustavson, Garcia, Hankins, & Rusiniak, 1974; Gustavson, Kelly, Sweeney, & Garcia, 1976). When the researchers placed tainted pieces of bait around a sheep ranch, sheep predation by coyotes dropped 60% compared with previous years.

One of Gustavson's Coyotes Undergoing Conditioning.

SOURCE: Janet Haas/Rainbow.

(Rozin, 1969). (Notice how similar this is to the sampling behavior of Davis's infants who were allowed to choose their own food.)

How much humans are able to make use of these abilities is unclear; certainly we often choose an unhealthy diet over a healthy one. These bad selections may not be due so much to a lack of *ability* to make good choices as it is to the distraction of tasty, high-calorie foods that are not found in nature. Even rats have trouble selecting the foods that are good for them when the competing foods are flavored with cinnamon or cocoa (Beck & Galef, 1989), and they become obese when they are offered human junk food (Rolls, Rowe, & Turner, 1980). Wisdom of the body is inadequate in the face of the temptation of french fries and ice cream.

Digestion and the Two Phases of Metabolism

Here, we confront a significant inadequacy in our thermostat analogy. To maintain consistent temperature, the thermostat calls on the furnace to cycle on and off frequently. Some species of animals do behave like the home furnace; they have to eat steadily, with only brief pauses, to provide the constant supply of nutrients the body needs. Humans do not; we eat a few discrete meals and fast in between. Eating discrete meals leaves us free to do other things with our time, but it requires a complex system for storing nutrient reserves, allocating the reserves during the fasting periods, and monitoring the reserves to determine the timing and size of the next meal.

The Digestive Process

Digestion begins in the mouth, where food is ground fine and mixed with saliva. Saliva provides lubrication and contains an enzyme that starts the breakdown of food. Digestion proceeds in the stomach as food is mixed with the gastric juices *hydrochloric acid* and *pepsin.* The partially processed food is then released gradually so that the small intestine has time to do its job. (Figure 6.6 shows the organs of the digestive system.)

The stomach provides another opportunity for screening toxic or spoiled food that gets past the taste test. If the food irritates the stomach lining sufficiently, the stomach responds by regurgitating the meal. Some toxins don't irritate the stomach, and they make their way into the bloodstream. If so, a part of the brain often takes care of this problem; the *area postrema* is one of the places in the brain that is outside the blood-brain barrier, so toxins can activate it to induce vomiting. The result can be surprisingly forceful; projectile vomiting usually means that you've got hold of something really bad. On the other hand, college students have been known to incorporate this adaptive response into a drinking game called "boot tag," the details of which I will leave to your imagination.

Digestion occurs primarily in the small intestine, particularly the initial 25 cm of the small intestine called the *duodenum.* There food is broken down into usable forms. Carbohydrates are metabolized into simple sugars, particularly *glucose.* Proteins are converted to *amino acids.* Fats are transformed into *fatty acids* and *glycerol,* either in the intestine or in the liver. The products of digestion are absorbed through the intestinal wall into the blood and transported to the liver via the *hepatic portal vein.* Digestion requires the food to be in a semiliquid mix, and the body can ill afford to give up the fluid; the large intestine's primary job is retrieving the excess water.

All this process is under the control of the autonomic nervous system, so digestion is affected by stress or excitement, as you probably well know. If too much acid is secreted into the stomach, you'll take your course exam with an upset stomach. If food moves too slowly through the system, constipation will be the result. Too fast, and there isn't time to remove the excess water, so you may be asking to leave the room in the middle of your exam to go to the bathroom. Because diarrhea causes the body to lose water, you may have to drink more liquids to avoid dehydration. You also lose electrolytes, compounds that provide the ions your neurons and other cells need, which is why your doctor may recommend a sports drink as the replacement liquid.

The Absorptive Phase

The feeding cycle is divided into two phases, the absorptive phase and the fasting phase. For a few hours following a meal, our body lives off the nutrients arriving from the digestive system; this period is called the *absorptive phase.* Following a meal, the blood level of glucose, our primary source of energy, rises. The brain detects the increased glucose and shifts the autonomic system from predominantly sympathetic activation to predominantly parasympathetic activity. As a result, the pancreas starts secreting *insulin,* a hormone that enables body cells to take up glucose for energy and certain cells to store excess nutrients. (Actually, because of conditioning, just the sight and smell of food is enough to trigger insulin secretion, increased salivation, and release of digestive fluids into the stomach. Remember the incentive theory?)

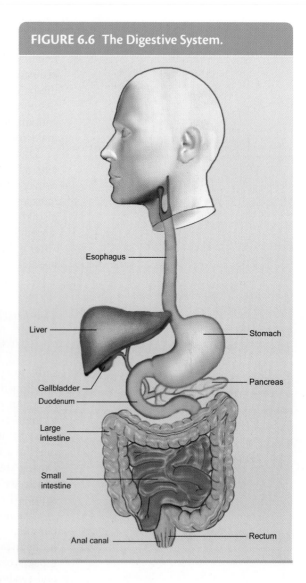

FIGURE 6.6 The Digestive System.

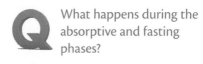
What happens during the absorptive and fasting phases?

The cells of the body outside the nervous system contain *insulin receptors,* which activate transporters that carry glucose into the cells. The cells of the nervous system have no insulin receptors; their glucose transporters can operate in the absence of insulin, thus giving the brain priority access to glucose. *Diabetes* results when the pancreas is unable to produce enough insulin (type 1 diabetes) or the body's tissues are relatively unresponsive to insulin (type 2 diabetes). The diabetic's blood contains plenty of glucose following a meal, but the cells of the body are unable to make use of it and the diabetic is chronically hungry.

During the absorptive phase, the body is also busy storing some of the nutrients as a hedge against the upcoming period of fasting. Some of the glucose is converted into *glycogen* and stored in a short-term reservoir in the liver and the muscles. Any remaining glucose is converted into fats and stored in fat cells, also known as *adipose tissue.* Fats arriving directly from the digestive system are stored there as well. Storage of both glucose and fat is under the control of insulin. After a small proportion of amino acids is used to construct proteins and peptides needed by the body, the rest is converted to fats and stored.

The Fasting Phase

Eventually the glucose level in the blood drops. Now the body must fall back on its energy stores, which is why this is called the *fasting phase.* The autonomic system shifts to sympathetic activity. The pancreas ceases secretion of insulin and starts secreting the hormone *glucagon,* which causes the liver to transform stored glycogen back into glucose. Because the insulin level is low now, this glucose is available only to the nervous system. To meet the rest of the body's needs, glucagon triggers the breakdown of stored fat into fatty acids and glycerol. The fatty acids are used by the muscles and organs, while the liver converts glycerol to more glucose for the brain. During starvation, muscle proteins can be broken down again into amino acids, which are converted into glucose by the liver. The two phases of metabolism are summarized in Figure 6.7.

The oscillations of eating and fasting and the shifts in metabolism that accompany them are orchestrated for the most part by two particularly important areas in the hypothalamus. The *lateral hypothalamus* initiates eating and controls several aspects of feeding behavior as well as metabolic responses. It controls chewing and swallowing through its brainstem connections; salivation, acid secretion, and insulin production through autonomic pathways in the medulla and spinal cord; and cortical arousal, which likely increases locomotion and the possibility of encountering food (Currie & Coscina, 1996; Saper, Chou, & Elmquist, 2002; Willie, Chemelli, Sinton, & Yanagisawa, 2001). The *paraventricular nucleus (PVN)* initiates eating, though less effectively than the lateral hypothalamus, and regulates metabolic processes such as body temperature, fat storage, and cellular metabolism (Broberger & Hökfelt, 2001; Sawchenko, 1998). You can see where these structures are located in the brain in Figure 6.8, which we will refer to throughout this discussion.

After a few hours of living off the body's stores, the falling level of nutrients signals the brain that it is time to eat again. However, by then you probably have already headed for lunch, cued by the clock rather than a brain

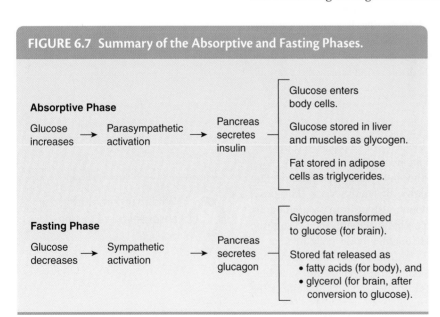

FIGURE 6.7 Summary of the Absorptive and Fasting Phases.

Absorptive Phase

Glucose increases → Parasympathetic activation → Pancreas secretes insulin

- Glucose enters body cells.
- Glucose stored in liver and muscles as glycogen.
- Fat stored in adipose cells as triglycerides.

Fasting Phase

Glucose decreases → Sympathetic activation → Pancreas secretes glucagon

- Glycogen transformed to glucose (for brain).
- Stored fat released as
  - fatty acids (for body), and
  - glycerol (for brain, after conversion to glucose).

FIGURE 6.8 Hunger Control Signals and Brain Centers.

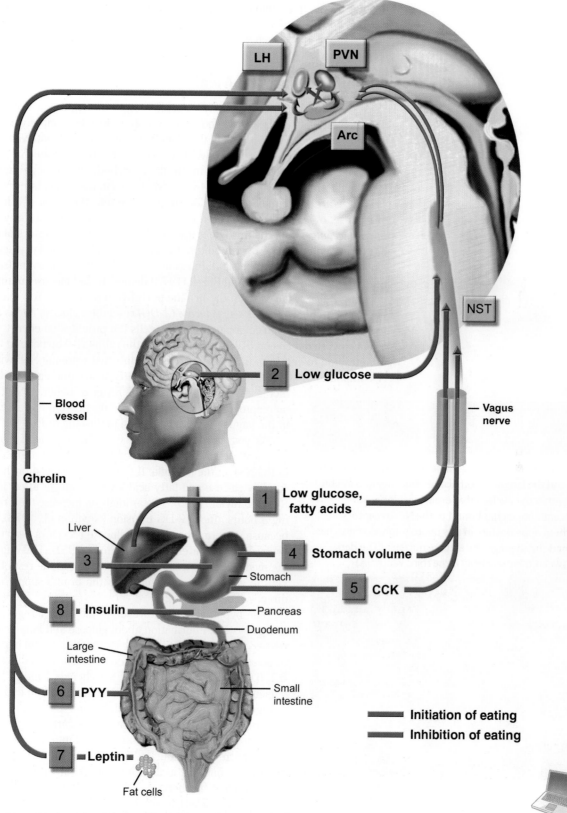

NOTE: PVN = paraventricular nucleus; LH = lateral hypothalamus; Arc = arcuate nucleus; NST = nucleus of the solitary tract.

center. In the modern, highly structured world, physiological motivations have been so incorporated into social customs that it is difficult to tell where the influence of one leaves off and the other begins. We will turn the power of research to answering the questions, "What makes a person eat?" "How does a person know when to stop eating?" and "How does a person regulate weight?" As you will learn, the answers are not simple ones; even what you see here will be an abbreviated treatment.

Signals That Start a Meal

When I ask students in class what being hungry means, the favorite answer is that their stomach feels empty. Your stomach often does feel empty when you are hungry, but we don't eat to satisfy the stomach. The stomach is not even necessary for hunger to occur; people who have their stomach removed because of cancer still report feeling hungry and still eat much like everybody else, though they have to take smaller meals (Ingelfinger, 1944). So what does make us feel hungry?

There are three major signals for hunger. One tells the brain of a low supply of glucose, or *glucoprivic hunger;* the second indicates a deficit in fatty acids, or *lipoprivic hunger;* and the third does inform us that the stomach's store of nutrients has been depleted. The liver monitors the glucose level and the fatty acids in the blood passing to it from the small intestine via the hepatic portal vein (see Figure 6.8). Novin, VanderWeele, and Rezek (1973) demonstrated glucose monitoring by injecting 2-deoxyglucose into the hepatic portal vein of rabbits. You may remember from Chapter 4 that *2-deoxyglucose (2-DG)* resembles glucose and is absorbed by cells; it takes the place of glucose in the cells but provides no energy, so it creates a glucose deficiency. The injection caused the rabbits to start eating within 10 minutes and to eat three times as much as animals that were injected with saline. A compound that blocks the metabolism of fatty acids (mercaptoacetate) also increases the amount eaten (S. Ritter & Taylor, 1990); injecting mercaptoacetate into the hepatic portal vein increases activity in the vagus nerve, sending a signal to the brain.

Q What stimuli initiate eating?

As you can see in Figure 6.8 (#1), signals of glucose and fatty acid deficits are carried by the vagus nerve from the liver to the NST in the medulla. If the NST is lesioned or the vagus is cut, low glucose and fatty acid levels no longer affect feeding (S. Ritter & Taylor, 1990). The animals do increase their rate of eating 3 hours after a 2-DG infusion (Novin et al., 1973), however, because the brain has its own glucose receptors near the fourth ventricle (R. C. Ritter, Slusser, & Stone, 1981). This suggests that the medulla keeps track of nutrient levels in the rest of the body via the vagus nerve but monitors the brain's supply of glucose directly (Figure 6.8, #2).

The hypothalamus, however, is the master regulator of the energy system. Information about glucose and fatty acid levels is relayed from the NST to the *arcuate nucleus*, a vital hypothalamic structure for monitoring the body's nutrient condition (Figures 6.2, 6.8, 6.9; Saper et al., 2002; Sawchenko, 1998). The arcuate nucleus sends neurons to the PVN and the lateral hypothalamus to regulate both feeding and metabolism.

The third major signal for hunger is *ghrelin*, a peptide that is synthesized in the stomach and released into the bloodstream as the stomach empties during fasting. Circulating ghrelin reaches the arcuate nucleus because it passes readily through the blood-brain barrier (Broberger & Hökfelt, 2001). Injecting ghrelin into rats' ventricles caused them to eat more and to gain weight four times faster than rats injected with saline (Kamegai et al., 2001). In humans, ghrelin levels in the blood rose almost 80%

FIGURE 6.9 Immunohistochemical Labeling Highlights the Arcuate Nucleus.

NPY-releasing neurons in the arcuate nucleus send output to the PVN and the lateral hypothalamus, but they also inhibit neurons within the nucleus that ordinarily block eating. A fluorescent antigen has bound to the NPY receptors, making them appear white in this photograph; doing so has also defined the shape of the arcuate nucleus. (The dark space between the two nuclei is the third ventricle.)

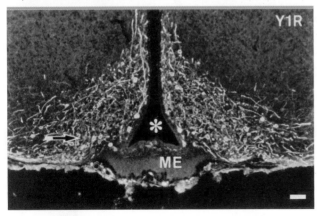

SOURCE: From Figure 1 of "Hypothalamic and Vagal Neuropeptide Circuitries Regulating Food Intake," by C. Broberger and T. Hokfelt, 2001, *Physiology and Behavior, 74*, p. 670.

before each meal and dropped sharply after eating (Figure 6.10; Cummings et al., 2001). Ghrelin may account for the uncontrollable appetite of people like Christopher; it is 2.5 times higher in individuals with Prader-Willi syndrome than in lean controls and 4.5 times higher than the depressed levels found in equally obese individuals without the syndrome (Cummings et al., 2002).

These three hunger signals target NPY neurons in the arcuate nucleus, which release *neuropeptide Y (NPY)* and *agouti-related protein (AgRP);* they excite the PVN and the lateral hypothalamus to increase eating and reduce metabolism (Figure 6.8, #3; Horvath & Diano, 2004; Kamegai et al., 2001). Rats that receive NPY injections in the paraventricular nucleus double their rate of eating and increase their rate of weight gain sixfold (B. G. Stanley, Kyrkouli, Lampert, & Leibowitz, 1986). They are so motivated for food that they will tolerate shock to the tongue to drink milk and they will drink milk adulterated with bitter quinine (Flood & Morley, 1991). The fact that their weight gain is three times greater than their increase in food intake attests to NPY's ability to reduce metabolism. During extreme deprivation, NPY conserves energy further by reducing body temperature (Billington & Levine, 1992) and suppressing sexual motivation (J. T. Clark, Kalra, & Kalra, 1985). If you think about it, sexual activity is a particularly unnecessary luxury during food shortage because it expends energy and produces offspring that compete for the limited resources.

FIGURE 6.10 Ghrelin Levels in a Human Over a 24-Hour Period.

Notice that the ghrelin level started rising just before, and peaked at, the customary meal times.

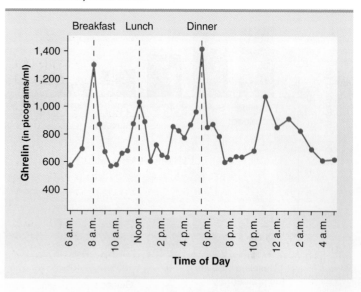

SOURCE: From "A Preprandial Rise in Plasma Ghrelin Levels Suggests a Role in Meal Initiation in Humans," by D. E. Cummings et al., *Diabetes, 50,* (2001), p. 1716, fig. 2A. © 2001 American Diabetes Association. Reprinted with permission from the American Diabetes Association.

Signals That End a Meal

Just as with drinking, there must be a satiety mechanism that ends a meal well before nutrients reach the tissues. It might seem obvious that we stop eating when we feel "full," and that answer is partly right. R. J. Phillips and Powley (1996) used a small inflatable cuff to close the connection between the stomach and intestines of rats. Infusing glucose into the stomach reduced the amount of food the rats ate later, but saline had just as much effect as glucose; this meant that in the stomach volume and not nutrient value is important. Filling the stomach activates stretch receptors that send a signal by way of the vagus nerve to the NST (Figure 6.8, #4; Broberger & Hökfelt, 2001; B. R. Olson et al., 1993).

But a full stomach cannot produce satiation by itself; otherwise, drinking water would satisfy us. Humans and other animals also adjust the amount of food they eat according to the food's nutritional value. To a small extent, this involves mouth factors and learning. A high-calorie soup produces a greater reduction in hunger if it is drunk than if it is infused into the stomach, and high-calorie drinks are more satisfying than noncaloric drinks (S. E. French & Cecil, 2001).

Optimal satiation, however, requires the interaction of mouth, stomach, and intestinal factors. When R. J. Phillips and Powley (1996) opened the cuff so the stomach's contents could flow into the intestines, nutrient value did make a difference. Glucose reduced subsequent eating more than saline did, and higher concentrations of the glucose had a greater effect (see Figure 6.11). The stomach and intestines respond to food by releasing peptides, which the brain uses to monitor nutrients. There are about a dozen different peptides that have this function; different peptides are released in response to carbohydrates, fats, proteins, or mixtures of these nutrients. They induce the pancreas, liver, and gallbladder to secrete the appropriate enzymes into the duodenum to digest the specific nutrient, and at least

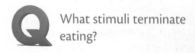

What stimuli terminate eating?

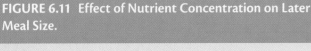

FIGURE 6.11 Effect of Nutrient Concentration on Later Meal Size.

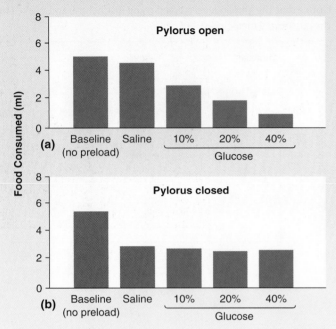

In all trials except the baseline, the stomach was preloaded with 5 milliliters of saline or glucose solution before the subject was offered a glucose solution to drink. The connection between the stomach and small intestine (the pylorus) could be closed by inflating a small cuff. (a) With the pylorus open, the amount consumed diminished as nutrient values increased. (b) With the pylorus closed, nutrient value made no difference.

SOURCE: Adapted with permission from "Gastric Volume Rather Than Nutrient Content Inhibits Food Intake," by R. J. Phillips and T. L. Powley, *American Journal of Physiology, 271*, pp. R766–R799. © 1996 American Physiological Society. Used with permission.

some of them inform the brain as to which nutrient needs are being met (S. C. Woods, 2004), either via the vagus nerve or the bloodstream.

The best known of these satiety signals is *cholecystokinin (CCK),* a peptide hormone that is released as food passes into the duodenum. CCK detects fats and causes the gall bladder to inject bile into the duodenum; the bile breaks down the fat so that it can be absorbed. When Xavier Pi-Sunyer and his colleagues injected CCK into the bloodstream of obese humans, they ate less at the next meal (Pi-Sunyer, Kissileff, Thornton, & Smith, 1982). CCK stimulates receptors on the vagus nerve; as Figure 6.8 (#5) indicates, the vagus conveys the signal to the NST, and from there it passes to the hypothalamus (S. C. Woods, 2004).

However, don't look for CCK to appear on the market as a weight loss drug. Although rats given CCK eat smaller meals, they eat more often and maintain their weight (West, Fey, & Woods, 1984). This means that there must be additional influences on food intake besides the short-term controls affecting meal size.

Long-Term Controls

Another appetite-suppressing peptide hormone that is released in the intestines in response to food is *peptide YY_{3-36} (PYY)*. PYY is carried by the bloodstream to the arcuate nucleus, where it inhibits the NPY-releasing neurons (Figure 6.8, #6; Batterham et al., 2002). Unlike CCK, PYY's nonneural route to the brain means that its action is too slow to limit the current meal; instead, it decreases caloric intake by about a third over the following 12 hours. We will see later that this hormone is receiving serious consideration as an antiobesity drug.

Over longer periods, humans and animals regulate their eating behavior by monitoring their body weight or, more precisely, their body fat. But how they sense their fat level has not always been clear. In 1952, G. R. Hervey surgically joined pairs of rats so that they shared a very small amount of blood circulation; animals joined like this are called *parabiotic.* Then Hervey operated on one member of each pair to destroy the *ventromedial hypothalamus.* This surgery increases parasympathetic activity in the vagus nerve and enhances insulin release (Weingarten, Chang, & McDonald, 1985). This creates a kind of persistent absorptive phase in which most incoming nutrients are stored rather than being available for use; as a result, the animal has to overeat to maintain normal energy level. The rat becomes obese, sometimes tripling its weight (see Figure 6.12). Hervey's lesioned rats overate and became obese as expected, but their pairmates began to undereat and lose weight. In fact, in two of the pairs the lean rat starved to death. The message was clear: The obese rat was producing a blood-borne signal that suppressed eating in the other rat, a signal to which the brain-damaged obese rat was insensitive.

What that signal was remained a mystery until recently, when researchers discovered that fat cells secrete a hormone called *leptin* that inhibits eating. The amount of leptin in the blood is proportional to body fat; it is about four times higher in obese than nonobese individuals (Considine et al., 1996). Like cholecystokinin, leptin helps regulate meal size, but it does so in response to the long-term stores of

What are the signals for controlling body weight?

fat rather than the nutrients contained in the meal. Insulin levels also are proportional to the size of fat reserves (M. W. Schwartz & Seeley, 1997).

Leptin and insulin put the brakes on feeding in the arcuate nucleus, in part by inhibiting the NPY/AgRP neurons (Berthoud & Morrison, 2008). At the same time, they activate a second population of arcuate neurons known as POMC/CART cells (because they release proopiomelanocortin and cocaine- and amphetamine-regulated transcript); these neurons reduce feeding by inhibiting the PVN and lateral hypothalamus (Figure 6.8, #7 and #8; Elmquist, 2001; Gao & Horvath, 2007; M. W. Schwartz & Morton, 2002). We now understand that when Hervey destroyed the ventromedial hypothalamus in rats, he also severed fibers passing through it and disconnected the arcuate nucleus from the PVN (see Figure 6.2 again to see how this could happen).

Table 6.1 summarizes the factors that influence hunger and feeding we have just covered.

Until now, we have been considering the ideal situation, the regulation of feeding and weight when all goes well. But in many cases, people eat too much, they eat the wrong kinds of foods, or they eat too little. As we will see, these behaviors are not just personal preferences or inconvenient quirks of behavior; too often, they are health-threatening disorders.

FIGURE 6.12 A Rat With Lesioned Ventromedial Hypothalamus.

SOURCE: Neal Miller, Yale University.

TABLE 6.1 Summary of Feeding Signals.

Stimuli	Signal Source	Pathway
Start meals		
1. Glucose, fatty acids	Detected by liver as nutrients in blood are depleted.	Signal travels via vagus nerve to NST, then to arcuate nucleus in hypothalamus.
2. Glucose (in brain)	Low level detected by glucose receptors near fourth ventricle.	Signal presumably travels from medulla to arcuate nucleus.
3. Ghrelin	Peptide released by stomach during fasting.	Circulates in blood stream to arcuate nucleus.
End meals		
4. Stomach volume	Stretch receptors in stomach detect increased volume from food.	Signal travels via vagus nerve to NST, then arcuate nucleus in hypothalamus.
5. CCK (and other nutrient indicators)	Stomach and intestines release peptides that aid digestion, signal brain of nutrient's presence.	CCK and others initiate activity in vagus to NST and hypothalamus; some may circulate in blood to brain.
Long term		
6. PYY	Released by intestines.	Travels in blood stream to arcuate nucleus; inhibits NPY neurons.
7. Leptin	Released by fat cells.	Travels in blood stream to arcuate nucleus; inhibits NPY neurons.
8. Insulin	Released by pancreas.	Travels in blood stream to arcuate nucleus; inhibits NPY neurons.

NOTE: Numbers refer to items in Figure 6.8 and in text. NST = nucleus of the solitary tract; CCK = cholecystokinin; PYY = peptide YY$_{3-36}$; NPY = neuropeptide Y.

Concept Check *Take a Minute to Check Your Knowledge and Understanding*

☐ What is the advantage of the ability to access stored nutrients between meals?

☐ How important are feeling "empty" and feeling "full" in the regulation of eating? Explain.

☐ You have lost weight during a long illness and now you are ravenous. What are your likely levels of glucose, fatty acids, ghrelin, insulin, and leptin?

Obesity

According to the National Health and Nutrition Examination Surveys, the adult obesity rate in the United States has doubled since 1980 (Ogden et al., 2006). Now two thirds of adults are overweight and a third qualify as obese. Fewer adolescents are overweight, but their percentage *tripled* during the same period. This problem is not unique to the United States; obesity is escalating at such an alarming rate in most countries that the World Health Organization has declared a global epidemic. For the first time in history, the number of people in the world who are overfed and overweight equals the number who are hungry and underweight (Figure 6.13; G. Gardner & Halweil, 2000). However, the number of people who are *mal*nourished is almost *double* the number who are *under*nourished, in part because many of those overweight are getting their calories from junk foods that are low in nutritional value.

Most researchers use the World Health Organization's BMI calculation to quantify leanness and obesity. *Body mass index (BMI)* is calculated by dividing the person's weight in kilograms by the squared height in meters. (If you're uncomfortable with metric measures, you can read your BMI from Figure 6.14). People with BMIs between 25 and 29 (shown in yellow in the table) are considered overweight, those whose BMIs are between 30 and 39 are considered obese, and those with BMIs of 40 and above qualify as morbidly obese. BMI is an inaccurate measure in some individuals; because muscle is heavier than fat, a healthy, bulked-up athlete will have a high BMI score. A more complete analysis would include a body fat measure and the waist-to-hip ratio.

Obesity is most important because of its health risks. As overweight and obesity increase, so does the incidence of a variety of diseases, including diabetes, heart disease, high blood pressure, stroke, and colon cancer (Field et al., 2001; Must et al., 1999). Obesity is also linked to cognitive decline and risk for Alzheimer's disease. Swedish researchers found that in women with lifelong obesity, for every one-point increase in BMI, there was a 13% to 16% increase in the risk of temporal lobe shrinkage due to cell loss (Gustafson, Lissner, Bengtsson, Björkelund, & Skoog, 2004). The degeneration could have been the result of impaired blood flow to the brain or excess release of the stress hormone cortisol. In the United States, obesity accounts for 9% of all medical costs, or $150 billion annually (Finkelstein, Trogdon, Cohen, & Dietz, 2009).

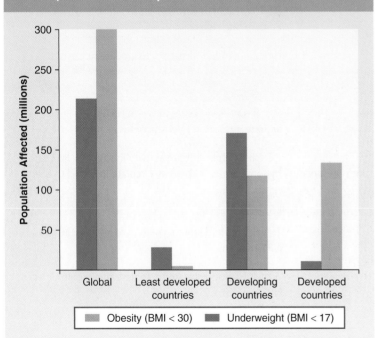

FIGURE 6.13 Underweight and Obesity According to the Country's Level of Development.

SOURCE: From "Controlling the Global Obesity Epidemic," by World Health Organization, 2003.

FIGURE 6.14 Body Mass Index Calculation Chart.

Weight in Pounds

Height	100	105	110	115	120	125	130	135	140	145	150	155	160	165	170	175	180	185	190	195	200	205
5'0"	20	21	21	22	23	24	25	26	27	28	29	30	31	32	33	34	35	36	37	38	39	40
5'1"	19	20	21	22	23	24	25	26	26	27	28	29	30	31	32	33	34	36	36	37	38	39
5'2"	18	19	20	21	22	23	24	25	26	27	27	28	29	30	31	32	33	34	35	36	37	37
5'3"	18	19	19	20	21	22	23	24	25	26	27	27	28	29	30	31	32	33	34	35	35	36
5'4"	17	18	19	20	21	21	22	23	24	26	26	27	27	28	29	30	31	32	33	33	34	35
5'5"	17	17	18	19	20	21	22	22	23	24	25	26	27	27	28	29	30	31	32	32	33	34
5'6"	16	17	18	19	19	20	21	22	23	23	24	25	26	27	27	28	29	30	31	31	32	33
5'7"	16	16	17	18	19	20	20	21	22	23	23	24	25	26	27	27	28	29	30	31	31	32
5'8"	15	16	17	17	18	19	20	21	21	22	23	24	24	25	26	27	27	28	29	30	30	31
5'9"	15	16	16	17	18	18	19	20	21	21	22	23	24	24	25	26	27	27	28	29	30	30
5'10"	14	15	16	17	17	18	19	19	20	21	22	22	23	24	24	25	26	27	27	28	29	29
5'11"	14	15	15	16	17	17	18	19	20	20	21	22	22	23	24	24	25	26	26	27	28	28
6'0"	14	14	15	16	16	17	18	18	19	20	20	21	22	22	23	24	24	25	26	26	27	28
6'1"	13	14	15	15	16	16	17	18	18	19	20	20	21	22	22	23	24	24	25	26	26	27
6'2"	13	13	14	15	15	16	17	17	18	19	19	20	21	21	22	22	23	24	24	25	26	26
6'3"	12	13	14	14	15	16	16	17	17	18	19	19	20	21	21	22	22	23	24	24	25	26
6'4"	12	13	13	14	15	15	16	16	17	18	18	19	19	20	21	21	22	23	23	24	24	25

Height in Feet and Inches

SOURCE: Adapted with permission from "Obesity: How Big a Problem?" by I. Wickelgren, *Science, 280*, pp. 1364–1367. Copyright 1998 American Association for the Advancement of Science. Reprinted with permission from AAAS.

In a study of 900,000 adults, BMIs over 25 were associated with reductions in life span up to 2 years, while individuals with BMIs between 30 and 35 lived 2 to 4 years less, and those with BMIs between 40 and 45 had their lives reduced by 8 to 10 years (Prospective Studies Collaboration, 2009). A few researchers are sounding the alarm that if something is not done to stem the rapid increase in obesity, by 2050, the dramatic increase in life expectancy over the last century will come to an end, and life expectancy might even decline (Olshansky et al., 2005).

Just as obesity is detrimental to health, dietary restriction appears to be beneficial. The restriction can take various forms, including 30% to 50% reduction in caloric intake, elimination of certain nutrients from the diet, and alternate-day feeding; these procedures have increased life span in roundworms and fruit flies by 40% to 50% and in rats and mice by 20% to 60% (Kennedy, Steffen, & Kaeberlein, 2007). The mechanism is not well understood, but it apparently involves nutrient-responsive pathways that evolved to promote growth and development in times of plenty and to shift the body into a mode of repair and preservation during famine (Kaeberlein & Kennedy, 2007). A major player appears to be *target of rapamycin kinase (TOR);* inhibition of this protein increases longevity in yeast and roundworms, and in mammals it reduces cancer risk and increases resistance to neurodegenerative disease.

Researchers now have completed studies with animals with longer life spans; after 20 years, 80% of rhesus monkeys that were limited to 30% fewer calories survived, compared to 50% of monkeys allowed to eat all they wanted (Colman et al., 2009). The diet-restricted monkeys had a lower incidence of diabetes, cancer, and cardiovascular disease, and their loss of brain gray matter was reduced as well. Of course, we do not have life-span studies with humans, but short-term treatment of obese patients produces similar physiological, metabolic, and hormonal effects (Kennedy et al., 2007). In addition, elderly adults who reduced their calorie intake by 30% for 3 months improved their verbal memory by 20% (Witte, Fobker, Gellner, Knecht, & Flöel, 2009). Most people would be unwilling to reduce their food intake that much, but biogerontologists are attempting to mimic the effects of dietary restriction with drugs as a means of combating age-related disease and deterioration (Ingram et al., 2006).

The Myths of Obesity

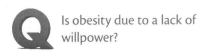

Is obesity due to a lack of willpower?

> *Most forms of obesity are likely to result not from an overwhelming lust for food or lack of willpower, but from biochemical defects at one or more points in the system responsible for the control of body weight.*
>
> —*Michael Schwartz and Randy Seeley*

Because obesity is dangerous to the person's health and the occasion for social and career discrimination, it is important to ask why people become overweight and why obesity rates are rising so dramatically. Although the causes have been difficult to document, most authorities believe that the global increase in obesity has a simple explanation: People are eating more and richer foods and exercising less (J. O. Hill & Peters, 1998; J. O. Hill, Wyatt, Reed, & Peters, 2003). The cause of obesity seems straightforward enough, then: *Energy in* exceeds *energy out,* and the person gains weight. But we would miss the point entirely if we assumed that people become obese just because they cannot resist the temptation to overeat. Research has not supported the popular opinion that obesity is completely under voluntary control (Volkow & Wise, 2005) or that it can be characterized as lack of impulse control, inability to delay gratification, or maladaptive eating style (Rodin, Schank, & Striegel-Moore, 1989). In fact, compulsive eating and drug abuse show similar responsiveness to stress, reward and craving involve the same brain areas in both groups, and obese people have deficits in D_2 receptors similar in magnitude to those of drug addicts (Wang et al., 2001).

Another popular belief is that obese children learn overindulgence from their family. Obesity does run in families, and body mass index and other measures are moderately related among family members. However, the evidence consistently points to genetic rather than environmental influences as more important (Grilo & Pogue-Geile, 1991). To the extent that environment does play a role it is, surprisingly, from outside the family.

The Contribution of Heredity

Is obesity hereditary?

Both adoption studies and twin studies demonstrate the influence of heredity on body weight. Adopted children show a moderate relationship with their biological parents' weights and BMIs, but little or no similarity to their adoptive parents (Grilo & Pogue-Geile, 1991). In a compilation of studies involving 75,000 individuals, correlations for BMI averaged .74 for identical twins and .32 for fraternal twins (Maes, Neale, & Eaves, 1997). Even when identical twins are reared apart, their correlation drops only to .62 (Grilo & Pogue-Geile, 1991), still almost double that for fraternals reared together (see Figure 6.15). The heritability of BMI is at least 50% and possibly as high as 90% (de Castro, 1993; Maes et al., 1997).

Weight regulation is complex, involving appetite, satiety, and energy management. It should not be surprising that researchers have come up with a very long list of candidate genes that might be involved in obesity. In fact, 200 different genes have been implicated, and over two dozen have been specifically linked with human obesity (Chagnon, Pérusse, Weisnagel, Rankinen, & Bouchard, 1999; Comuzzie & Allison, 1998). As illustration, I will discuss two genes in some detail.

Forty years ago, it was known that the so-called *obesity gene* on chromosome 6 and the *diabetes gene* on chromosome 4 cause obesity in mice. Mice that are homozygous for the recessive obesity gene (*ob/ob*) or the recessive diabetes gene (*db/db*) have the same symptoms: overeating, obesity, and susceptibility to diabetes (see Figure 6.16). To find out how the two genes produced these symptoms, D. L. Coleman (1973) used parabiotic pairings of the two kinds of mice and normals (Figure 6.17). When a *db/db* mouse was paired with a normal mouse, the normal mouse starved to death. The same thing happened to the *ob/ob* mouse when it was paired with the *db/db* mouse. These results suggested that the *db/db* mice were producing a fat signal but that they were not themselves sensitive to it. The *ob/ob* mouse had no effect on a normal mouse, but its own rate of weight gain slowed. The *ob/ob* mouse apparently was sensitive to a fat signal that it did not produce. It was another 20 years before researchers discovered that the fat signal in Hervey's (1952) and Coleman's studies was leptin. Following that discovery, they were able to test Coleman's hypothesis. Injecting leptin into *ob/ob* mice reduced their weight 30% in just 2 weeks, while *db/db* mice were not affected by the injections (Halaas et al., 1995).

These genes are rare in the population, however, and account for relatively few cases of obesity. On the other hand, mutations in the *MC4R* gene, which is responsible for a receptor that responds to feeding and satiety signals from the arcuate nucleus, may account for as many as 6% of cases of severe childhood obesity (Farooqi et al., 2003). The A allele of the *FTO* gene is even more common, and new research has implicated it in 13% of overweight and 20% of obese individuals. People who are homozygous for this allele are 70% more likely to be obese than people without the A allele, and those with a single copy of the allele—which is half of the population—have a 30% greater risk (Frayling et al., 2007). Unfortunately, we have no idea how this gene produces its effect. However, this is a good time to point out that not all inherited differences involve variations in DNA; the accompanying Application introduces the concept of *epigenetic* inheritance.

While we know relatively little about the functions of the more than 100 gene variations that contribute to obesity (Rankinen et al., 2005), we do know that heredity influences meal size, meal frequency, energy intake, activity level, metabolic level, and the proportion of proteins, fats, and carbohydrates consumed in the diet (C. Bouchard, 1989; de Castro, 1993). Among these, metabolic level has been investigated most, and it is our next topic.

FIGURE 6.15   Correlations of Body Mass Index Among Twins.

Notice that the correlation is higher for identical twins than for fraternal twins, even when the identicals are reared apart and the fraternals are reared together.

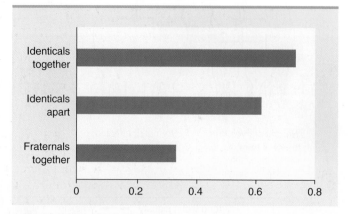

SOURCE: Based on data from Grilo and Pogue-Geile (1991).

FIGURE 6.16   The Mouse on the Right Is an *ob/ob* Mouse.

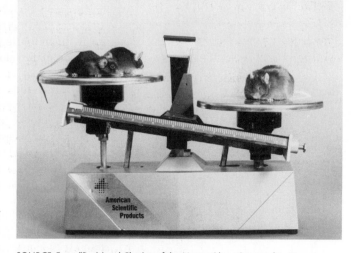

SOURCE: From "Positional Cloning of the Mouse *Obese* Gene and Its Human Homologue," by Y. Zhang et al., 1994, *Nature*, 335, pp. 11–16. Reprinted by permission of *Nature*, copyright 1994.

Obesity and Reduced Metabolism

Accounts of dieting are all too often stories of failure; overweight people report slavishly following rigorous diets without appreciable weight loss, or they lose weight and then gain it back within a year's time. One factor in the failures may be

FIGURE 6.17 Effects of Leptin on *ob/ob*, *db/db*, and Normal Mice.

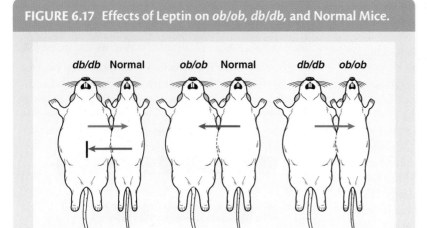

Normal starves due to signal from *db/db*; *db/db* insensitive to normal's signal.

Normal unaffected, but its signal slows *ob/ob*'s weight gain.

ob/ob starves due to signal from *db/db*.

———— Strong leptin signal   ———— Weak leptin signal

SOURCE: Based on the results of Coleman (1973).

> *What is a wisdom of the body in times of deprivation becomes a foolishness in our modern environment.*
>
> —Xavier Pi-Sunyer

dieters' misrepresentation of their efforts, whether intentional or not. One group of diet-resistant obese individuals underreported the amount of food they consumed by 47% and overreported their physical activity by 51% (Lichtman et al., 1992).

But another critical element that can make weight loss difficult is a person's rate of energy expenditure. In the average sedentary adult, about 75% of daily energy expenditure goes into resting or *basal metabolism,* the energy required to fuel the brain and other organs and to maintain body temperature; the remainder is spent about equally in physical activity and in digesting food (Bogardus et al., 1986).

Differences in basal metabolism may be a key element in explaining differences in weight. When 29 women who claimed they could not lose weight were isolated and monitored closely while they were restricted to a diet of 1,500 kilocalories (kcal; a measure of food's energy value, popularly called *calorie*), 19 did lose weight, but 10 did not (D. S. Miller & Parsonage, 1975). The 10 who failed to lose weight turned out to have a low basal metabolism rate (BMR). Heredity accounts for about 40% of people's differences in BMR (C. Bouchard, 1989). When identical twins were overfed 1,000 kcal a day for 3 months, the differences in weight gain within pairs of twins were only one third as great as the differences across pairs (C. Bouchard et al., 1990).

However, a person's metabolism can shift when the person gains or loses weight. In an unusual *experimental* manipulation, researchers had both obese and never-obese individuals either lose weight or gain weight (Leibel, Rosenbaum, & Hirsch, 1995). Those who lost weight shifted to reduced levels of energy expenditure (resting plus nonresting), and the ones who gained weight increased their energy expenditure. This was expected, because your weight affects how much energy is required to move around and even to sit or stand. However, the energy expenditure changes were greater than the weight changes would require, suggesting that the individuals' bodies were *defending* their original weight (see Keesey & Powley, 1986).

So why doesn't this defense of body weight prevent people from becoming obese? One reason is that the body defends less against weight gain than against weight loss (J. O. Hill, et al., 2003, J. O. Hill & Peters, 1998). Humans evolved in an environment in which food was sometimes scarce, so it made sense for the body to store excess nutrients during times of plenty and to protect those reserves during famine. That system is adaptive when humans are at the mercy of nature, but it is a liability when modern agriculture and global transportation provide a constant supply of more food than we need.

A second reason is that people vary tremendously in the strength of their defense response, making some people more vulnerable to becoming overweight than others. When volunteers were overfed 1,000 kcal a day, on average only 40% of the excess calories were stored as fat and the remaining 60% were burned off by increased energy expenditure (F. R. Levine, Eberhardt, & Jensen, 1999). But some individuals had smaller increases in energy expenditure, and they gained 10 times as much weight as others. Two thirds of the volunteers' increases in energy expenditure were due to nonexercise physical activity—casual walking, fidgeting,

Application   The Epigenetics of Weight Regulation

Does a woman's weight during pregnancy have an effect on her child's chance of being obese? To find out, a group of Canadian researchers located the offspring of women who had weight loss surgery that reduced their stomach volume and bypassed part of the small intestine (J. Smith et al., 2009). Surprisingly, children born after the mother's weight loss were only one third as likely to be obese as the children born before the surgery. And, compared to their siblings, they had higher levels of ghrelin, lower leptin and cholesterol levels, and decreased insulin resistance. The nature of these physiological effects, the resistance to obesity, and their duration—26 years in the oldest subjects—support the researchers' belief that these benefits were *epigenetic*.

Epigenetic characteristics are inheritable traits that are unrelated to the individual's DNA sequence. Epigenetic changes can be triggered by the environment; the most dramatic example dates back to World War II, though its implications weren't recognized until much later. During the

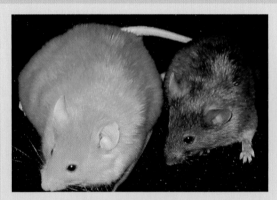

SOURCE: Courtesy of Duke University Medical Center.

winter of 1944–1945, Germany blockaded food shipments to western Holland; 20,000 people died from starvation during the *hunger winter*, and women who were pregnant during that time delivered underweight babies. A half century later, a study looked at the offspring of the hunger winter mothers; it found that the female first-generation offspring who were in the first or second trimester of gestation during the famine *also gave birth to underweight babies* (Lumey, 1992). A genetic study revealed why: In offspring exposed to famine during early gestation, a gene that controls prenatal growth was less methylated than it was in their siblings who were exposed only during late gestation (Heijmans et al., 2008). Methylation is the attachment to genes of molecules called methyl groups, and it is one of the ways epigenetic inheritance occurs. The DNA is not modified, but the gene's expression is turned off or turned on. Evidence that similar effects in mice can also be transmitted through the male offspring means that the results cannot be dismissed as due to conditions in the uterine environment (Dunn & Bale, 2009).

A graphic demonstration of methylation is an experimental study of mice that had a mutation in the *agouti* gene; the *agouti* gene is responsible for the agouti-related peptide mentioned earlier, and the mutation produces obesity along with an atypical yellow coat. Feeding pregnant females folic acid and vitamin B_{12}, which are high in methyl groups, turned off the mutant gene in some of the offspring; as you can see in the photo, those offspring had more normal coat color and were less vulnerable to obesity (Waterland & Jirtle, 2003). Giving pregnant agouti mice alcohol during the first half of pregnancy also produced more offspring with normal coat color (Kaminen-Ahola et al., 2010). In addition, their skulls were smaller and had altered facial shapes reminiscent of fetal alcohol syndrome features. Activity level was altered in 15 genes besides the agouti gene, and 3 of the downregulated genes are involved in brain development. These findings suggest that FAS is at least partially the result of epigenetic effects.

spontaneous muscle contraction, and posture maintenance. Researchers are beginning to think that spontaneous activity may be as important as basal metabolism in resisting obesity.

Prolonged weight gain may actually reset the set point at a higher level. Rolls, Rowe, et al. (1980) fattened rats on highly palatable, high-energy junk food (chocolate chip cookies, potato chips, and cheese crackers) for 90 days. Surprisingly, when the rats were returned to their usual lab chow, they did not lose weight. The rats maintained their increased weight for the 4-month duration of the study—while eating the same amount of food as the control rats. They were defending a new set point. The researchers suggested that the variety of the foods offered, the length of the fattening period, and lack of exercise all contributed to the rats' failure to defend their original weight. In view of the difficulties in shedding excess weight, one obesity researcher suggested that returning to normal weight may not be a reasonable goal, and a goal of 10% weight reduction is more practical (Pi-Sunyer, 2003).

2
Overeaters Anonymous

Q How can obesity be treated?

> *Obesity is the most dangerous epidemic facing mankind, and we are relatively unprepared for it.*
>
> —George Yancopoulos

Treating Obesity

There is no greater testimony to the difficulty of losing weight than the lengths to which patients and doctors have gone to bring about weight loss. These include wiring the jaws shut, surgically reducing the stomach's capacity, and removing fat tissue. The standard treatment for obesity, of course, is dietary restriction. However, we have seen that the body defends against weight loss, and dieters are usually frustrated. Exercise burns fat, but it takes a great deal of effort to use just a few hundred calories. On the other hand, exercise during dieting may increase resting metabolic rate or at least prevent it from dropping (Calles-Escandón & Horton, 1992). Dieters who exercise lose more weight than dieters who do not exercise (see Figure 6.18; J. O. Hill et al., 1989). In a study of formerly obese women, 90% of those who maintained their weight loss exercised, compared with 34% of those who regained their lost weight (Kayman, Bruvold, & Stern, 1990).

Another option in the treatment of obesity is medication. However, it has not been a particularly promising alternative; lack of effectiveness is one problem and because the drugs manipulate metabolic and other important body systems, they often have adverse side effects. The approval of dexfenfluramine in 1996 was the first by the Food and Drug Administration (FDA) in 20 years. But just a year later, both dexfenfluramine and the older fenfluramine (used in the now-notorious combination called fen-phen) were withdrawn from the market by the manufacturer after reports that they caused heart valve leakage (Campfield, Smith, & Burn, 1998). Orlistat, a drug that blocks fat absorption (marketed in the United States as Alli and Xenical), is under review by the FDA because of reports of liver damage among its users ("Early Communication . . . Orlistat," 2009). Likewise, sibutramine (Meridia) is receiving reconsideration because a study found an increased risk of heart attack and stroke ("Early Communication . . . Meridia," 2009).

Sibutramine merits additional attention because it suppresses appetite in part by inhibiting serotonin reuptake, and serotonin plays an interesting role in weight control. Carbohydrate regulation involves a feedback loop; eating

FIGURE 6.18 The Effect of Exercise on Weight Loss.

Dieters who also exercised lost 32% more weight than those who did not, and they lost 40% more body fat.

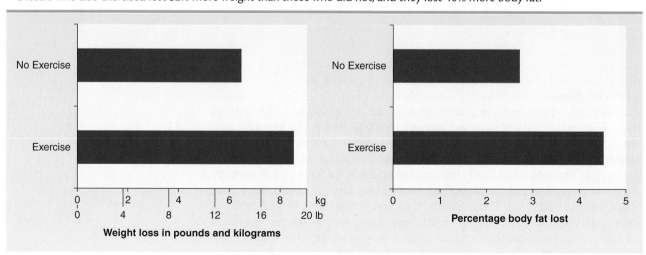

SOURCE: Based on data from J. O. Hill et al., 1989.

carbohydrates increases serotonin levels, which inhibits a person's appetite for carbohydrates (Leibowitz & Alexander, 1998), apparently by reducing NPY activity (Dryden, Wang, Frankish, Pickavance, & Williams, 1995). Drugs that block serotonin reuptake reduce carbohydrate intake, but only in the group of obese individuals who crave carbohydrates and eat a large proportion of their diet in carbohydrates (Lieberman, Wurtman, & Chew, 1986; J. J. Wurtman, Wurtman, Reynolds, Tsay, & Chew, 1987). Serotonin also enhances mood in some people, and people who have trouble maintaining weight loss often say that they use food to make themselves feel better when they are upset (Kayman et al., 1990). A high-carbohydrate meal also improves mood only in carbohydrate cravers; it actually lowers the mood of noncravers and makes them feel fatigued and sleepy (Lieberman et al., 1986). So serotonin dysregulation may be important in obesity, but only in a subset of people.

After injections of PYY reduced calorie intake in obese volunteers by 26% (Batterham et al., 2003), the pharmaceutical company Nastech developed a novel PYY nasal spray that in preliminary trials reduced appetite and calorie intake (Brandt et al., 2004). However, subsequent trials found the nasal spray ineffective, and 59% of subjects dropped out because of nausea and vomiting (Gantz et al., 2007). By now you should be asking, "Why not try the body's own hormones as weight-loss drugs?" And that is one of the directions research is taking.

Leptin is particularly attractive to obesity researchers because, unlike food restriction, it increases metabolism (N. Levin et al., 1996), and it targets fat reduction while sparing lean mass Cohen & Friedman, 2004). Leptin was administered to three severely obese children who produced no leptin at all due to a mutation in the *ob* gene (Farooqi et al., 2002). Their body weights decreased throughout treatment although they were increasing in age; more than 98% of the weight loss was in fat mass, while lean mass increased. Figure 6.19 shows how dramatic the effects were in one of the children. Unfortunately, leptin treatment benefits only the 5% to 10% of obese people who are leptin deficient; the rest are resistant to leptin's effects, apparently as a result of the high-fat diet itself (Enriori et al., 2007; Maffei et al., 1995). Many obesity researchers now believe that leptin's main role is in protecting the individual against weight loss during times of deprivation rather than against weight gain during times of plenty (Marx, 2003).

The search goes on, and one drug is showing particular promise. Exenatide (marketed as Byetta) mimics glucagon-like peptide 1 (GLP-1). Like GLP-1, it increases insulin secretion, which slows stomach emptying and leads to increased satiety and a decrease in food intake. Type 2 diabetes patients lost weight with exenatide, whereas patients treated with insulin continued to gain weight (Glass et al., 2008). However, the U.S. Food and Drug Administration is now requiring the drug to be labeled with a warning about possible kidney function problems and kidney failure ("FDA: Byetta," 2009).

A more recent therapeutic approach involves treating obesity as an addictive behavior. Compulsive overeating and drug addiction share several behavioral and neurophysiological similarities, including high relapse rate, responsiveness to stress, dopamine release in response to cues, reduced numbers of dopamine D_2 receptors (Figure 6.20) and associated decreases in metabolism in prefrontal

FIGURE 6.19 A Leptin-Deficient Boy Before and After Treatment.

(a) At age 3.5 years when treatment began and (b) at age 8 and at normal weight.

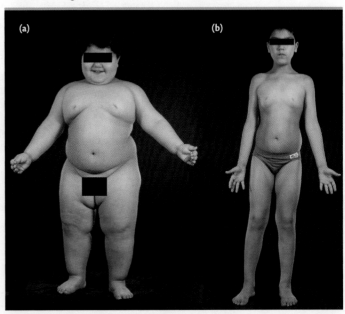

SOURCE: From "Cellular Warriors at the Battle of the Bulge," by J. Marx, 2003, *Science*, 299, pp. 846–849. Courtesy of Sadaf Farooqi and Stephen O'Rahilly, University of Cambridge.

FIGURE 6.20 Diminished Number of Dopamine D₂ Receptors in Obese Individuals.

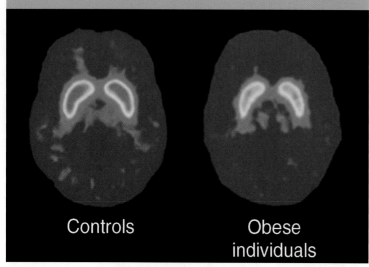

Controls   Obese individuals

SOURCE: From "Brain Dopamine and Obesity," G.-J. Wang et al., *Lancet, 357,* pp. 354–357. Copyright 2001.

FIGURE 6.21 Gastric Bypass Procedure.

A small area of the stomach is isolated from the rest. Then the small intestine is severed, and the cut end is attached to the pouch, reducing the length of the intestine and the amount of nutrient absorption.

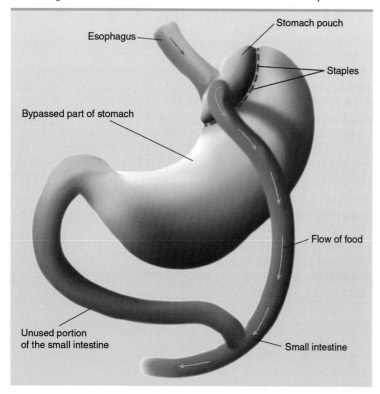

Esophagus
Stomach pouch
Staples
Bypassed part of stomach
Flow of food
Unused portion of the small intestine
Small intestine

SOURCE: Adapted from Ainsworth (2009).

areas involved in impulse control, and continued drug taking and compulsive eating when they are self-destructive and no longer pleasurable (Taylor, Curtis, & Davis, 2009; Trinko, Sears, Guarnieri, & DiLeone, 2007; Volkow, Wang, Fowler, & Telang, 2008). The various peptides that induce eating also target dopaminergic neurons in the ventral tegmental area, the nucleus accumbens, and parts of the basal ganglia; this helps account for the rewarding effects of food as well as its neurally separate motivational effects (Berthoud & Morrison, 2008; Gao & Horvath, 2007). Vigabatrin, a drug in experimental use as an addiction treatment, reduces the amount of dopamine released by cocaine and drug-related cues. When genetically obese rats were given Vigabatrin, they underwent a 19% weight loss (DeMarco et al., 2008). Contrave is a combination of Bupropion and Naltrexone, which are anti-addiction drugs (see Chapter 5). In Phase III trials, obese patients lost 50% more weight on Contrave than with placebo, and the manufacturer is applying for FDA approval ("Orexigen Therapeutics," 2009).

Unfortunately, lifestyle modification or drug therapy reduces body weight by only 5% to 10% (Mitka, 2006), and most dieters regain their weight within a year (Bray, 1992). That much weight reduction can lead to significant health gains, but it is not enough for the morbidly obese, who are increasingly turning to surgery. The most effective procedure, gastric bypass, reduces the stomach to a small pouch, which is then reconnected below the first part of the intestine (see Figure 6.21). This both limits meal size and reduces nutrient absorption in the digestive tract. The surgery costs about $27,000, which goes up to $65,000 if rehospitalization is required (Mitka), and in controlled trials the death rate is 1% (Maggard et al., 2005). However, weight loss averages 32% after 1 or 2 years and is maintained at 25% 10 years later (Sjöström et al., 2007). In addition, the weight loss engenders a cascade of additional benefits: remission of type 2 diabetes in 77% of patients, hypertension in 66%, and sleep apnea in 88% (Buchwald et al., 2004), as well as a 27% reduction in mortality at 15 years (Sjöström et al.). An additional benefit of reducing the amount of functional digestive tract and delivering nutrients to the lower part of the small intestine more rapidly is that postmeal levels of ghrelin are decreased and GLP-1 and PYY are increased (Pournaras & le Roux, 2009), so the individual doesn't feel hungry. Surgery is a last resort, though, and its downsides highlight the need for obesity prevention and new strategies for weight management.

Not everyone with an eating disorder is overweight or obese. Some try so hard to control their weight that they eat less than is needed to maintain health, or they eat normal or excess amounts and then vomit or use laxatives to avoid gaining weight. As you will see, anorexia and bulimia are as puzzling to researchers as obesity and often more deadly for the victims.

Anorexia and Bulimia

Anorexia nervosa and bulimia nervosa affect about 3% of women over their lifetime (Walsh & Devlin, 1998). Although men are also affected, women patients outnumber them 10:1; this is the most extreme gender discrepancy in medicine and psychiatry (A. E. Anderson & Holman, 1997). Because male patients are so rare, we will limit our attention to research with females.

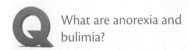

What are anorexia and bulimia?

Anorexia nervosa is known as the "starving disease" because the individual restricts food intake to maintain weight at a level so low that it is threatening to health (see Figure 6.22; Walsh & Devlin, 1998). The person may also exercise for hours a day or resort to vomiting to control weight loss. Anorexics often see themselves as fat even when they are emaciated. They are likely to deny the need for treatment and fail to comprehend the medical consequences of their disorder. There are two subgroups of anorexics. *Restrictors* rely only on reducing food intake to control their weight. *Purgers* not only restrict their calorie intake as well, but they also resort to vomiting or using laxatives.

3
Eating Disorders

If anorexia continues long enough, it leads to cessation of ovulation, loss of muscle mass, heart damage, and reduction in bone density. The death rate among anorexics is more than double that for female psychiatric patients; half of the deaths are from complications of the disease and another quarter from suicide (Sullivan, 1995). Brain scans show that the volume of gray matter in anorexics' brains is lower than normal, not only while they are underweight but up to 23 years after weight recovery (D. K. Katzman, Zipursky, Lambe, & Mikulis, 1997; Lambe, Katzman, Mikulis, Kennedy, & Zipursky, 1997). Brain scans also reveal dysfunction in areas involved in reward, emotion, and processing of bodily information, including those responsible for body image (Kaye, Fudge, & Paulus, 2009). The studies do not tell us whether these brain anomalies are due to starvation or represent conditions that precede and contribute to the anorexia. However, the poor performance of anorexics on a behavioral test of frontal cortex functioning continues after recovery and is found in healthy sisters (Attia, 2009). In addition, the presence of obsessive-compulsive traits, harm avoidance, and perfectionism during childhood as well as after recovery suggests the presence of predisposing factors before the onset of illness (Kaye et al.).

The anorexic individual's unwillingness to eat does not necessarily imply a lack of hunger. NPY and ghrelin are elevated and leptin levels are diminished (Kaye, Berrettini, Gwitsman, & George, 1990; Mantzoros, Flier, Lesem, Brewerton, & Jimerson, 1997; Shiiya et al., 2009), so apprently anorexics are "hungry" whether they realize it or not. The sight of attractive food increases their insulin levels more than it does in lean people; lean subjects eat the food when it is offered, but the anorexics do not in spite of overnight fasting, saying they aren't hungry (Broberg & Bernstein, 1989).

Bulimia nervosa also involves weight control, but the behavior is limited to bingeing and purging. If the bulimic restricts food intake, it is only for a few days at a time, and restricting takes a backseat to bingeing and purging. In fact, only 19% of bulimic women consume fewer calories than normal controls, while 44% overeat (Weltzin, Hsu, Pollice, & Kaye, 1991). Unlike anorexics, most bulimic women are of normal weight (Walsh & Devlin, 1998). However, there are indications that, like anorexics, they might also be battling hunger. Their ghrelin levels between meals are a third higher than in controls and decrease less following a meal; in addition, PYY levels do not rise as much following a meal (Kojima et al., 2005). Like anorexia, bulimia is also a dangerous disorder. Both anorexia and bulimia are difficult to treat; although three quarters of bulimics and a third of anorexics appear to be fully recovered after 8 years, a third of these relapse (D. B. Herzog et al., 1999).

Environmental and Genetic Contributions

Both anorexics and bulimics are preoccupied with weight and body shape. Because increases in anorexia and bulimia seem to have paralleled an increasing cultural emphasis on thinness and beauty, some researchers have concluded that the cause is social. The male–female difference is consistent with this argument, because women are under more pressure to be slim while men are encouraged to "bulk up." Cases are more common in Western, industrialized countries, where an impractical level of thinness is promoted by actors, models, and advertisers. Anne Becker of Harvard Medical School has been studying eating habits in the Pacific islands of Fiji since 1988 (Becker, Burwell, Gilman, Herzog, & Hamburg, 2002). Traditionally, a robust, muscled body has been valued for both sexes there. But when satellite television arrived in 1995, the tall, slim actors in shows like *Beverly Hills 90210* became teenage Fiji's new role models. By 1998, 74% of young island girls considered themselves too big or fat, even though they were not more overweight than others. Young girls who lived in homes that owned a TV were three times more likely to have an eating problem. Among 17-year-old girls, 11% admitted they had vomited to control weight, compared with just 3% in 1995.

There is little doubt that social pressure contributes to anorexia and bulimia. But the disorders are not unknown in non-Westernized societies, and anorexia has been reported for 300 years, long before the cultural emphasis on thinness. One indication that a sociocultural explanation is an oversimplification is that several studies show a genetic influence. Relatives of patients have a higher than usual incidence of the disorders, and the concordance between identical twins is much higher than between fraternals (44% vs. 12.5% for anorexia, 22.9% vs. 8.7% for bulimia; Kendler et al., 1991; Kipman, Gorwood, Mouren-Simeoni, & Ad'es, 1999). The most recent review of genetic studies of anorexia and bulimia identified a large number of candidate genes, with functions related to the hormones and neurotransmitters involved in energy regulation (Scherag, Hebebrand, & Hinney, 2009). Definitive results are hard to come by, though, in part because of the difficulty in acquiring adequate numbers of patient volunteers. Even if only a small fraction of these genes can be confirmed, it is clear that many small-effect genes will be involved.

Anorexic and bulimic patients often share a variety of other disorders with their relatives (Lilenfeld et al., 1998), and genetic studies have confirmed associations between anorexia and obsessive-compulsive disorder and between bulimia and depression (Scherag et al., 2009). Because these disorders involve neurotransmitter abnormalities and are partially hereditary, this comorbidity is another argument for a biological role in the eating disorders. Major depression is the most frequently reported comorbidity; naturally, researchers have tried to apply what they know about depression to understanding and treating anorexia and bulimia.

The Role of Serotonin and Dopamine

Because of the role serotonin has in eating and in obesity, researchers have suspected that anorexics and bulimics have lower than normal serotonin activity. This appears to be true of bulimics (Weltzin, Fernstrom, & Kaye, 1994). They also

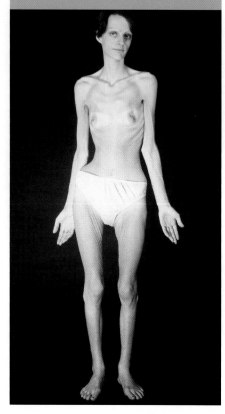

FIGURE 6.22 A Woman at a Late Stage of Anorexia.

SOURCE: Evon Agostini/Liaison Agency.

 What is the evidence for social and genetic influences in anorexia and bulimia?

have increased rates of depression, anxiety, alcoholism and other drug abuse, and impulsive behavior such as stealing and sexual activity; all of these characteristics are associated with low serotonin activity (Kendler et al., 1991; Weltzin et al.; Wiederman & Pryor, 1996; also see Chapters 5, 8, and 14). Antidepressants, which increase serotonin activity, provide at least some reduction of binge eating in bulimics, and some studies indicate improvement of other symptoms, including depression and preoccupation with food (W. H. Kaye, Klump, Frank, & Strober, 2000).

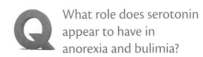

What role does serotonin appear to have in anorexia and bulimia?

Anorexics also have low serotonin levels but only while they are underweight, apparently due to malnutrition (W. H. Kaye, Ebert, Gwirtsman, & Weiss, 1984). Some researchers have suggested that restricting and purging anorexia have different neurochemical mechanisms. After they regain weight, restricters have higher serotonin activity than normal controls (W. H. Kaye et al.), which is typical of people with obsessive-compulsive disorder. Indeed, many anorexics and their relatives show symptoms of this disorder: perfectionism, rigidity, preoccupation with details, and need for order and cleanliness. Purging anorexics, on the other hand, return only to normal serotonin levels after gaining weight (W. H. Kaye et al.). They tend to be impulsive, socially outgoing, emotionally responsive, and sexually active, all characteristic of people with low serotonin activity. Interestingly, cyproheptadine, which reduces serotonin, produced improvements in restricting anorexics but impaired treatment in purgers (Halmi, Eckert, LaDu, & Cohen, 1986).

Various kinds of evidence point to a dysfunctional dopamine system in anorexics, such as lowered dopamine metabolites in the cerebrospinal fluid. This could account for the increased harm avoidance, reduced positive emotion, and diminished reward value of food seen in anorexics (W. H. Kaye et al., 2009). The drug olanzapine affects dopamine as well as serotonin receptors; on the hypothesis that it would benefit restricting anorexics through its anti-obsessional properties, and purging anorexics by reducing depression, researchers gave the drug to patient volunteers who were also receiving psychological treatment (Bissada, Tasca, Barber, & Bradwejn, 2008). The drug group gained more weight and showed more improvement in obsessive-compulsive symptoms than did a placebo group. The olanzapine group also experienced reduced anxiety and depression, but no more than the placebo group. This lack of difference could be due to the small number of subjects, which was due to more than half of the patients approached declining to participate; as mentioned before, a typical characteristic of anorexics is the inability to see any need for treatment.

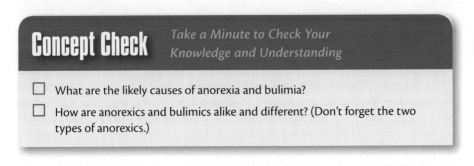

Concept Check   *Take a Minute to Check Your Knowledge and Understanding*

☐ What are the likely causes of anorexia and bulimia?

☐ How are anorexics and bulimics alike and different? (Don't forget the two types of anorexics.)

In Perspective

Temperature regulation, thirst, and hunger provide good examples of drive, homeostasis, and physiological motivation in general. Although they are explained best by drive theory, they also illustrate the point that it is ultimately the balance or imbalance in certain brain centers that determines motivated behaviors.

In addition, hunger in particular demonstrates that homeostasis alone does not explain all the facets of motivated behavior. For instance, we saw that the incentives of the sight and smell of food are enough to start the physiological processes involved in the absorptive phase. This suggests that incentives operate through physiological mechanisms and are themselves physiological in nature. We also know that there are important social influences on what and how much people eat and sensation seeking may explain why some people are gourmets or enjoy the risks of eating puffer fish.

Most of the factors that determine our eating behavior are in turn influenced by genes. If it is true that we are what we eat, it is equally true that what we eat (and how much) is the result of who we are. But we will be reminded time and again throughout this text that heredity is not destiny, that we are the products of countless interactions between our genetic propensities and the environment.

Our interest in the motivation of hunger would be mostly theoretical if it were not for eating disorders, which can have life-threatening consequences. But in spite of their importance, we are unsure about the causes of obesity, anorexia, and bulimia and of what the best treatments are. We do know that, like the motivation of hunger itself, they are complex and have a number of causes.

This chapter has given you an overview of what we mean by motivation. We will broaden that view in the next two chapters by looking at sexuality, emotion, and aggression.

Summary

Motivation and Homeostasis

- Homeostasis and drive theory are key to understanding physiological motivation, but they are not adequate alone.
- Temperature regulation involves a simple mechanism for control around a set point.
- Thirst is a bit more complex, compensating for deficits both in the cells (osmotic thirst) and outside the cells (hypovolemic thirst).

Hunger: A Complex Drive

- Hunger is a more complex motivation, involving a variety of nutrients and regulation of both short-term and long-term nutrient supplies.
- Taste helps an individual select nutritious foods, avoid dangerous ones, and vary the diet.
- The feeding cycle consists of an absorptive phase, a period of living off nutrients from the last meal, and the fasting phase, when reliance shifts to stored nutrients.
- Eating is initiated when low blood levels of glucose and fatty acids are detected in the liver. The information is sent to the medulla and to the paraventricular nucleus of the hypothalamus, where neuropeptide Y is released to initiate eating.
- Feeding stops when stretch signals from the stomach, increasing glucose levels in the liver, and cholecystokinin released in the duodenum indicate that satiety has occurred.
- How much we eat at a meal is also regulated by the amount of fat we have stored, indicated by leptin and insulin levels.

Obesity

- Obesity is associated with malnutrition and with a variety of illnesses.
- A variety of factors, many of them outside the person's control, contribute to obesity.
- Obesity is partly inheritable, and the environmental influences that exist are not from the family.

- As calorie intake decreases, metabolism decreases to defend against weight loss.
- Obesity is difficult to treat, but drugs that increase serotonin activity and leptin and the experimental drug C75 are showing promise.

Anorexia and Bulimia

- Anorexia involves restriction of food intake, and sometimes bingeing and purging, to reduce weight. Bulimia is a bingeing disease; weight increase is limited by purging or exercise.
- Social pressure and heredity both appear to be important in anorexia and bulimia.
- Serotonin appears to be low in bulimics, as it is in some obese individuals. It may be low in purging anorexics as well but high in restricting anorexics. ■

Study Resources

F   For Further Thought

- If a group of nuclei in the brain control a particular homeostatic need, what functions must those nuclei carry out?
- What do you think would happen if the brain had no way of monitoring stored fat levels?
- Several of the controls of eating seem to duplicate themselves. Is this wasteful or useful? Explain.

- What do you think a complete program of obesity treatment would look like?
- Can you propose another way to organize the three subgroups that make up anorexics and bulimics—perhaps even renaming the disorders?

T   Testing Your Understanding

1. Describe either temperature regulation or thirst in terms of homeostasis, drive, and satisfaction, including the signals and brain structures involved in the process.

2. Describe the absorptive and fasting phases of the feeding cycle; be specific about what nutrients are available, how nutrients are stored, and how they are retrieved from storage.

3. Describe obesity as a problem of metabolism.

Select the best answer:

1. A problem that makes some question drive theory is that
   a. an animal remains aroused after the need is satisfied.
   b. some people have stronger drives than others.
   c. not all motivation involves tissue needs.
   d. soon after a drive is satisfied, the system goes out of equilibrium again.

2. An animal is said to be in homeostasis when it
   a. recognizes that it is satisfied.
   b. feels a surge of pleasure from taking a drink.
   c. is in the middle of a high-calorie meal.
   d. is at its set point temperature.

3. Osmotic thirst is due to
   a. dryness of the mouth and throat.
   b. lack of fluid in the cells.
   c. reduced volume of the blood.
   d. stimulation of pressure receptors.

4. A structure in the medulla that is involved in taste as well as in hunger and eating is the
   a. NST.
   b. paraventricular nucleus.
   c. area postrema.
   d. SFO.

5. You have trouble with rabbits eating in your garden. Several sprays are available, but they are washed off each day by the sprinklers. The

solution with the best combination of kindness, effectiveness, and ease for you would be to

a. spray the plants daily with a substance that tastes too bad to eat.
b. spray the plants occasionally with a substance that makes the rabbits sick.
c. spray the plants with a poison until all the rabbits are gone.
d. forget about spraying; run outside and chase the rabbits away.

6. During the absorptive phase
a. fat is broken down into glycerol and fatty acids.
b. insulin levels are low.
c. glucagon converts glycogen to glucose.
d. glucose from the stomach is the main energy source.

7. Neurons in the arcuate nucleus release NPY, which
a. increases eating.
b. increases drinking.
c. breaks down fat.
d. causes shivering.

8. A long-term signal that influences eating is
a. glucose.
b. 2-deoxyglucose.
c. cholecystokinin.
d. leptin.

9. When we say that the body defends weight during dieting, we mean primarily that
a. the person's metabolism decreases.

b. the person eats less but selects richer foods.
c. the person eats lower-calorie foods but eats larger servings.
d. the body releases less NPY.

10. Studies comparing the weights of adopted children with their biological parents and their adoptive parents
a. show that weight is influenced most by environment.
b. show that weight is influenced most by heredity.
c. show that heredity and environment have about equal influence.
d. have not been in agreement.

11. If a *db/db* mouse is parabiotically attached to a normal mouse, the *db/db* mouse will
a. gain weight while the normal loses.
b. lose weight while the normal gains.
c. be unaffected while the normal loses.
d. be unaffected while the normal gains.

12. Serotonin appears to be
a. high in anorexics and low in bulimics.
b. low in anorexics and high in bulimics.
c. low in bulimics, high in some anorexics, and low in some anorexics.
d. high in anorexics, high in some bulimics, and low in some bulimics.

Answers: 1.c, 2.d, 3.b, 4.a, 5.b, 6.d, 7.a, 8.d, 9.a, 10.b, 11.c, 12.c.

Online Resources

The following resources are available at **www.sagepub.com/garrett3e.**

Chapter Resources

- Flash Cards
- Chapter Quiz
- Internet Resources and Exercises

On the Web

Links to these websites can be found at the web address above: select this chapter's **Chapter Resources, then choose Internet Resources and Exercises.**

1. The National Institutes of Health site **Prader-Willi Syndrome** is a valuable resource for practical as well as technical information about the disorder.
    The **Prader-Willi Association** site has information about the association, facts about the disorder, and research information.

2. You've seen Alcoholics Anonymous's 12-step program applied to all the other drug addictions; now it's being used to manage compulsive overeating. **Overeaters Anonymous** has information about its organization and links to the sites of local help organizations.

3. **NationalEatingDisorders.org** provides information about anorexia, bulimia, obesity, compulsive eating disorder, and other disorders, as well as information about treatments.
    Internet Mental Health has information on diagnosis, treatment, and research related to anorexia and bulimia (click on *Disorders* on the left side of the page).

Animations

- Hunger, Satiation, and the Regulation of Fat Reserves (Figure 6.8)

Chapter Updates and Biopsychology News

R For Further Reading

1. *Making Sense of Taste* by David V. Smith and Robert F. Margolskee (*Scientific American,* March 2001, 32–39) elaborates on taste receptors, the umami flavor, and taste processing in the brain.

2. *Common Sense About Taste: From Mammals to Insects* by David Yarmolinsky, Charles Zuker, and J. P. Ryba (*Cell,* 2009, *139*, 234–243) is a more technical review of taste emphasizing the genes and receptors that account for the different primary tastes.

3. "Extreme Obesity: A New Medical Crisis in the United States" by Donald Hensrud and Samuel Klein (*Mayo Clinic Proceedings,* 2006, 8 [10 suppl.] S5–S10) details prevalence, causes, and cost of obesity in the United States.

4. *Genetic Factors in Human Obesity* by I. S. Farooqi and S. O'Rahilly (*Obesity Reviews, 2007,* 8 [Suppl. 1] 37–40) describes the progress made over the past decade in identifying genes that contribute to obesity.

5. *Unfinished Symphony* by Jane Qiu (*Nature,* 2006, *441,* 143–145) gives a clear description of how epigenetic inheritance works and discusses possible drugs to reverse epigenetic effects and the efforts of the Human Epigenome Project to crack the epigenetic code.

6. *Full Without Food* by Claire Ainsworth (*New Scientist,* September 5, 2009, 30–33) describes surgery for obesity and emphasizes its benefits for diabetics, as well as the possibility of creating drugs that mimic the effects of GLP-1, which is increased following the surgery.

K Key Terms

absorptive phase 165
agouti-related protein (AgRP) 169
amino acids 165
angiotensin II 160
anorexia nervosa 181
arcuate nucleus 168
area postrema 165
arousal theory 157
basal metabolism 176
body mass index (BMI) 172
bulimia nervosa 181
cholecystokinin (CCK) 170
diabetes 166
diabetes gene 175
drive 157
drive theory 157
duodenum 165
epigenetic 177
fasting phase 166
fatty acids 165
ghrelin 168
glucagon 166
glucose 165
glycerol 165

glycogen 166
homeostasis 157
hypovolemic thirst 159
incentive theory 157
instinct 156
insulin 165
lateral hypothalamus 166
learned taste aversion 163
learned taste preference 163
leptin 170
median preoptic nucleus 159
motivation 156
neuropeptide Y (NPY) 169
nucleus of the solitary tract (NST) 160
obesity gene 175
organum vasculosum lamina terminalis (OVLT) ...159
osmotic thirst 159
paraventricular nucleus (PVN) 166
peptide YY$_{3-36}$ (PYY) 170
preoptic area 158
satiety 160
sensory-specific satiety 162
set point 158
subfornical organ (SFO) 160

The Biology of Sex and Gender

7

Sex as a Form of Motivation

Arousal and Satiation

The Role of Testosterone

Brain Structures and Neurotransmitters

Odors, Pheromones, and Sexual Attraction

    APPLICATION: OF LOVE AND BONDING

    CONCEPT CHECK

The Biological Determination of Sex

Chromosomes and Hormones

Prenatal Hormones and the Brain

    CONCEPT CHECK

Gender-Related Behavioral and Cognitive Differences

Some Demonstrated Male–Female Differences

Origins of Male–Female Differences

    CONCEPT CHECK

Sexual Anomalies

Male Pseudohermaphrodites

Female Pseudohermaphrodites

    IN THE NEWS: SEX, GENDER, AND SPORTS

Sex Anomalies and the Brain

Ablatio Penis: A Natural Experiment

    CONCEPT CHECK

Sexual Orientation

The Social Influence Hypothesis

Genes and Sexual Orientation

Hormonal Influence

Brain Structures

In this chapter you will learn

- How sex is similar to and different from other drives

- How hormones and brain structures control sexual development and behavior

- Some of the differences between males and females and what causes them

- How deviations in sexual development affect the body, the brain, and behavior

- How prenatal development may help explain heterosexuality and homosexuality

(Continued)

189

The Challenge of Female Homosexuality

Social Implications of the Biological Model

CONCEPT CHECK

In Perspective

Summary

Study Resources

Fourteen-year-old Jan went to her family physician complaining of a persistent hoarse voice. As is often the case, other concerns surfaced during the course of the examination. At puberty, she had failed to develop breasts or to menstruate; instead, her voice deepened, and her body became muscular. Once comfortable with her tomboyishness, she was now embarrassed by her appearance and increasingly masculine mannerisms; she withdrew from peers, and her school performance began to suffer. But there was an even more significant change at puberty: Her clitoris started growing and was 4 centimeters (1½ inches) long when she was examined by the doctor; in addition, her labia (vaginal lips) had partially closed, giving the appearance of a male scrotum. To everyone's surprise, Jan's included, the doctor discovered that she had two undescended testes in her abdomen and no ovaries. Further testing showed that her sex chromosomes were XY, which meant that genetically she was a male.

After a psychiatric evaluation, Jan's parents and doctors decided that she should be offered the opportunity to change to a male sexual identity. She immediately went home and changed into boy's clothing and got a boy's haircut. The family moved to another neighborhood where they were unknown. At the new high school, Jack became an athlete, excelled as a student, was well accepted socially, and began dating girls. Surgeons finished closing the labia and moved the testes into the newly formed scrotum. He developed into a muscular, 6-foot-tall male with a deep voice and a heavy beard. At the age of 25 he married, and he and his wife reported a mutually satisfactory sexual relationship (Imperato-McGinley, Peterson, Stoller, & Goodwin, 1979).

Humans have a great affinity for dichotomies, dividing their world into blacks and whites with few grays in between. No dichotomy is more significant for human existence than that of male and female: One's sex is often the basis for deciding how the person should behave, what the person is capable of doing, and with whom the person should fall in love. Not only are many of the differences between males and females imposed on them by society, but Jan's experience suggests that typing people as male or female may not be as simple or as appropriate as we think. We will encounter even more puzzling cases later as we take a critical look at the designation of male versus female and the expectations that go with it. In the meantime, we need to continue our discussion of motivation by considering how sex is like and unlike other drives.

Sex as a Form of Motivation

To say that sex is a motivated behavior like hunger may be stating the obvious. But theorists have had difficulty categorizing sex with other physiological drives because it does not fit the pattern of a homeostatic tissue need. If you fail to eat or if you cannot maintain body temperature within reasonable limits, you will die. But no harm will come from forgoing sex; sex ensures the survival of the species, but not of the individual.

There are, however, several similarities with other drives like hunger and thirst. They include arousal and satiation, the involvement of hormones, and control by specific areas in the brain. We will explore these similarities as well as some differences in the following pages.

How is sex like and unlike other drives?

Arousal and Satiation

The cycle of arousal and satiation is the most obvious similarity between sexual motivation and other motivated behaviors. In the 1960s, William Masters and Virginia Johnson conducted groundbreaking research on the human sexual response. Until then, research had been limited to observing sexual behavior in animals or interviewing humans about their sexual activity. Masters and Johnson (1966) observed 312 men and 382 women and recorded their physiological responses during 10,000 episodes of sexual activity in the laboratory. This kind of research was unheard of at the time; in fact, the researchers had trouble finding journals that would publish their work.

Masters and Johnson identified four phases of sexual response (see Figure 7.1). The *excitement phase* is a period of arousal and preparation for intercourse. Both sexes experience increased heart rate, respiration rate, blood pressure, and muscle tension. The male's penis becomes engorged with blood and becomes erect. The female's clitoris becomes erect as well, her vaginal lips swell and open, the vagina lubricates, her breasts enlarge, and the nipples become erect.

Hunger is a function mostly of time since the last meal. Sexual arousal, though, is more influenced by opportunity and sexual stimuli, such as explicit conversation or the presence of a sexually attractive person. (In many other animal species, sexual arousal is a regular event triggered by a surge in hormones.) Another difference between sex and other drives is that we usually are motivated to reduce hunger, thirst, and temperature deviations, but we seek sexual arousal. This difference is not unique, though; for example, we skip lunch to increase the enjoyment of a gourmet dinner.

During the *plateau phase*, the increase in sexual arousal levels off; arousal is maintained at a high level for seconds or minutes, though it is possible to prolong this period. The testes rise in the scrotum in preparation for ejaculation; vaginal lubrication increases and the vaginal entrance tightens on the penis. During *orgasm*, rhythmic contractions in the penis are accompanied by ejaculation of seminal fluid containing sperm into the vagina. Similar contractions occur in the vagina. This period lasts just a few seconds but involves an intense experience of pleasure. *Resolution* follows as arousal decreases and the body returns to its previous state.

Orgasm is similar to the pleasure one feels after eating or when warmed after a deep chill, but it is unique in its intensity; the resolution that follows is reminiscent of the period of quiet following return to homeostasis with other drives.

Males have a *refractory phase*, during which they are unable to become aroused or have another orgasm for minutes, hours, or even days, depending on the individual and the circumstances. Females do not experience a refractory period and are able to have additional orgasms anytime during the resolution phase. When comparing the sex drive with other kinds of motivation, the male refractory period has an interesting parallel with sensory-specific satiety (see Chapter 6); it is called the Coolidge effect. According to a popular but probably apocryphal

FIGURE 7.1 Phases of the Sexual Response Cycle.

This is a typical response for a male; most females are capable of multiple orgasms.

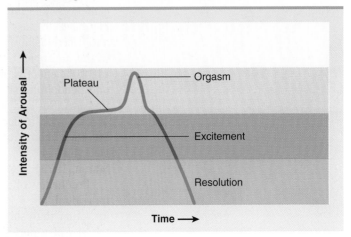

SOURCE: From *Psychology: The Adaptive Mind*, 2nd ed. by Nairne. 2000. Reprinted with permission of Wadsworth, a division of Thomson Learning.

story, President Coolidge and his wife were touring a farm when Mrs. Coolidge asked the farmer whether the flurry of sexual activity was the work of one rooster. The farmer answered yes, that the rooster copulated dozens of times each day, and Mrs. Coolidge said, "You might point that out to Mr. Coolidge." President Coolidge, so the story goes, then asked the farmer, "Is it a different hen each time?" The answer again was yes. "Tell that to Mrs. Coolidge," the president replied. Whether the story is true or not, the *Coolidge effect*—a quicker return to sexual arousal when a new partner is introduced—has been observed in a wide variety of species; we will visit the subject again shortly.

The Role of Testosterone

Q What is the role of testosterone in sexual behavior?

As important as sex is to humans, it is ironic that so much of what we know about the topic comes from the study of other species. One reason is that research into human sexual behavior was for a long time considered off-limits and funding was hard to come by. Another reason is that sexual behavior is more "accessible" in nonhuman animals; rats have sex as often as 20 times a day and are not at all embarrassed to perform in front of the experimenter. In addition, we can manipulate their sexual behavior in ways that would be considered unethical with humans. Hormonal control in particular is more often studied in animals because hormones have a clearer role in animal sexual behavior.

Castration, or removal of the gonads (testes or ovaries), is one technique used to study hormonal effects because it removes the major source of sex hormones; castration results in a loss of sexual motivation in nonhuman mammals of both sexes. Sexual behavior may not disappear completely, because the adrenal glands continue to produce both male and female hormones, though at a lesser rate than the gonads. The time course of the decline is also variable, ranging from a few weeks to 5 months in male rats (J. M. Davidson, 1966); across several species, animals who are sexually experienced are impaired the least and decline the slowest (Hart, 1968; Sachs & Meisel, 1994). Humans are less at the mercy of fluctuating hormone levels than other animals, but when they are castrated (usually for medical reasons, such as cancer), sexual interest and functioning decrease in both males and females (Bremer, 1959; Heim, 1981; Sherwin & Gelfand, 1987; Shifren et al., 2000). The decline takes longer in humans than in rats, but the rate is similarly variable.

Castration has been elected by some male criminals in the hope of controlling aggression or sexual predation, sometimes in exchange for shorter prison sentences. Castration is an extreme therapy; drugs that counter the effects of *androgens* (a class of hormones responsible for a number of male characteristics and functions) are a more attractive alternative. Those that block the production of the androgen *testosterone,* the major sex hormone in males, have been 80% to 100% effective in eliminating deviant sexual behavior such as exhibitionism and pedophilia (sexual contact with children), along with sexual fantasies and urges (A. Rösler & Witztum, 1998; Thibaut, Cordier, & Kuhn, 1996). The effects of castration indicate that testosterone is necessary for male sexual behavior, but the amount of testosterone required appears to be minimal; men with very low testosterone levels can be as sexually active as other men (Raboch & Stárka, 1973).

Frequency of sexual activity does vary with testosterone levels *within* an individual, but it looks like testosterone increases are the *result* of sexual activity rather than the cause. For example, testosterone levels are high in males at the *end* of a period in which intercourse occurred, not before (J. M. Dabbs & Mohammed, 1992; Kraemer et al., 1976). A case report (which is anecdotal and does not permit us to draw conclusions) suggests that just the anticipation of sex can increase the testosterone level. Knowing that beard growth is related to testosterone level, a researcher working in near isolation on a remote island weighed the daily clippings from his electric razor. He found that the amount of beard growth increased just before planned visits to the mainland and the opportunity for sexual activity (Anonymous, 1970)!

In most species, females are unwilling to engage in sex except during *estrus,* a period when the female is ovulating, sex hormone levels are high, and the animal is said to be in heat. Human females and females of some other primate species engage in sex throughout the reproductive cycle. Studies of sexual frequency in women have not shown a clear peak at the time of ovulation. However, initiation of sex is a better gauge of the female's sexual motivation than is her willingness to have sex; women do initiate sexual activity more often during the middle of the menstrual cycle, which is when ovulation occurs (Figure 7.2; D. B. Adams, Gold, & Burt, 1978; Harvey, 1987). The researchers attributed the effect to *estrogen,* a class of hormones responsible for a number of female characteristics and functions. Their reasons were that estrogen peaks at midcycle and the women did not increase in sexual activity if they were taking birth control pills, which level out estrogen release over the cycle.

However, testosterone peaks at the same time, and the frequency of intercourse during midcycle corresponds to the woman's testosterone level (N. M. Morris, Udry, Khan-Dawood, & Dawood, 1987). At menopause, when both estrogen and testosterone levels decline, testosterone levels show the most consistent relationship with intercourse frequency (McCoy & Davidson, 1985). How to interpret these observations is unclear, because testosterone increases in women as a *result* of sexual activity, just as it does in men (Figure 7.3; J. M. Dabbs & Mohammed, 1992). However, studies in which testosterone level was manipulated demonstrate that it also contributes to women's sexual behavior. Giving a dose of testosterone to women increases their arousal during an erotic film (Tuiten et al., 2000). More important, in women who had their ovaries removed, testosterone treatment increased sexual arousal, sexual fantasies, and intercourse frequency, but estrogen treatment did not (Sherwin & Gelfand, 1987; Shifren et al., 2000).

Brain Structures and Neurotransmitters

As neuroscientists developed a clearer understanding of the roles of various brain structures, motivation researchers began to shift their focus from drive as a tissue need to drive as a condition in particular parts of the brain. Sexual activity, like other drives and behaviors, involves a network of brain structures. This almost seems inevitable, because sexual activity involves reaction to a variety of stimuli, activation of several physiological systems, postural and movement responses, a reward experience, and so on. We do not understand yet how the sexual network operates as a whole, but we do know something about the functioning of several of its components. In this section, you will see some familiar terms, the names of hypothalamic structures you learned about

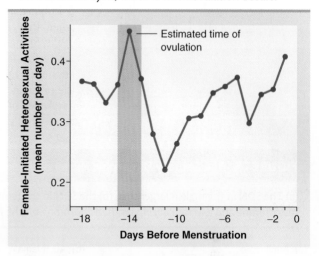

FIGURE 7.2 Female-Initiated Activity During the Menstrual Cycle.

Activity initiated by women peaks around the middle of the menstrual cycle, which is when ovulation occurs.

SOURCE: From figure 2 from "Rise in Female-Initiated Sexual Activity at Ovulation and Its Suppression by Oral Contraceptives," by D. B. Adams, A. R. Gold, & A. D. Burt, 1978, *New England Journal of Medicine, 299 (21),* pp. 1145–1150.

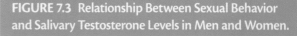

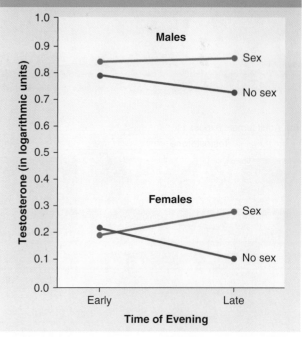

FIGURE 7.3 Relationship Between Sexual Behavior and Salivary Testosterone Levels in Men and Women.

SOURCE: From "Male and Female Salivary Testosterone Concentrations Before and After Sexual Activity," by J. M. Dabbs, Jr. and S. Mohammed, *Physiology and Behavior, 52,* pp. 195–197, Fig. 1. © 1992 Reprinted with permission from Elsevier Science.

Q What brain structures are involved in sexual behavior?

in the previous chapter. This illustrates an important principle of brain functioning, that a particular brain area, even a very small one, often has multiple functions.

The *medial preoptic area (MPOA)* of the hypothalamus is one of the more significant brain structures involved in male and female sexual behavior. (The general preoptic area can be located in Figure 6.2, in the previous chapter.) Stimulation of the MPOA increases copulation in rats of both sexes (Bloch, Butler, & Kohlert, 1996; Bloch, Butler, Kohlert, & Bloch, 1993), and the MPOA is active when they copulate spontaneously (Pfaus, Kleopoulos, Mobbs, Gibbs, & Pfaff, 1993; Shimura & Shimokochi, 1990). The MPOA appears to be more responsible for performance than for sexual motivation; when it was destroyed in male monkeys, they no longer tried to copulate, but instead they would often masturbate in the presence of a female (Slimp, Hart, & Goy, 1978).

A part of the amygdala, known as the *medial amygdala*, also contributes to sexual behavior in rats of both sexes. Located near the lateral ventricle in each temporal lobe, the *amygdala* is involved not only in sexual behavior but also in aggression and emotions. The medial amygdala is active while rats copulate (Pfaus et al., 1993), and stimulation causes the release of dopamine in the MPOA (Dominguez & Hull, 2001; Matuszewich, Lorrain, & Hull, 2000). The medial amygdala's role apparently is to respond to sexually exciting stimuli, such as the presence of a potential sex partner (de Jonge, Oldenburger, Louwerse, & Van de Poll, 1992).

In male rats, the paraventricular nucleus is important for sexual performance and particularly for penile erections (Argiolas, 1999). More significant for males is the *sexually dimorphic nucleus (SDN)*, located in the MPOA (de Jonge et al., 1989). The SDN is five times larger in male rats than in females (see Figure 7.4; Gorski, Gordon, Shryne, & Southam, 1978), and a male's level of sexual activity is related to the size of the SDN, which in turn depends on prenatal (before birth) exposure to testosterone (R. H. Anderson, Fleming, Rhees, & Kinghorn, 1986). Destruction of the SDN reduces male sexual activity (de Jonge et al., 1989). The SDN's connections to other sex-related areas of the brain suggest that it integrates sensory and hormonal information and coordinates behavioral and physiological responses to sensory cues (Roselli, Larkin, Resko, Stellflug, & Stormshak, 2004).

The *ventromedial hypothalamus* is important for sexual behavior in female rats (see Figure 6.2 for the location). Activity increases there during copulation (Pfaus et al., 1993), and its destruction reduces the female's responsiveness to a male's advances (Pfaff & Sakuma, 1979).

For obvious reasons, we know much less about the brain structures involved

---

FIGURE 7.4   The Sexually Dimorphic Nuclei of the Rat.

(a) The SDN in the male is larger than (b) the SDN in the female. (c) The effects of testosterone and a masculinizing synthetic hormone on the female SDN.

(a) Adult male

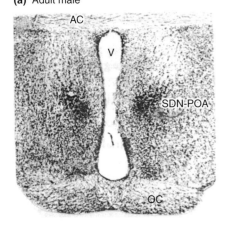

(b) Adult female

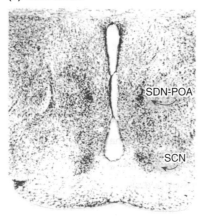

(c) Adult female exposed to:

Testosterone

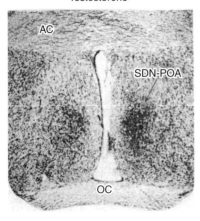

Diethylstilbestrol

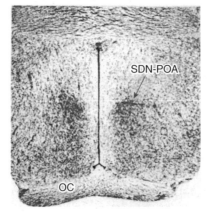

SOURCE: From "The Neuroendocrine Regulation of Sexual Behavior," by R. A. Gorski, pp 1–58, in G. Newton and A. H. Riesen (Eds.) *Advances in Psychobiology* (Vol. 2), 1974, New York: Wiley. Reprinted with permission of John Wiley & Sons, Inc.

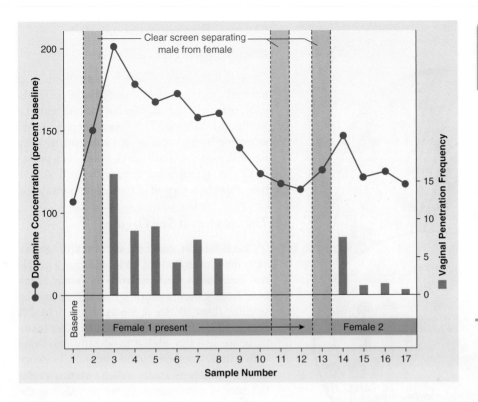

FIGURE 7.5 Dopamine Levels in the Nucleus Accumbens During the Coolidge Effect.

Activity was recorded until the male lost interest in Female 1; then, Female 2 was presented. During the periods represented by the orange bars, the female was separated from the male by a clear screen. The line graph shows dopamine levels. Bars show the number of vaginal penetrations.

SOURCE: From "Dynamic Changes in Nucleus Accumbens Dopamine Efflux During the Coolidge Effect in Male Rats," by D. F. Fiorino, A. Coury, and A. G. Phillips, 1997, *Journal of Neuroscience, 17*, p. 4852. © 1997 Society for Neuroscience. Reprinted with permission.

in human sexual behavior. We do know that a few brain structures in humans differ in size between males and females. Because their contribution to sexual behavior is not clear and the size differences may also distinguish homosexuals from heterosexuals, I will defer discussion of these structures until we take up the subject of sexual orientation.

Several neurotransmitters participate in sexual behavior. You saw in Chapter 5 that dopamine level increases in the nucleus accumbens during sexual activity and in this chapter that stimulation of the medial amygdala releases dopamine in the MPOA. Injection and microdialysis studies show that dopamine activity in the MPOA is involved in sexual motivation in males and females of several species and is critical for sexual performance in males (E. M. Hull et al., 1999). In males, initial small amounts of dopamine stimulate D_1 receptors, which activate the parasympathetic system and increase motivation and erection, while delaying ejaculation. As dopamine increases, activation of D_2 receptors shifts autonomic balance to the sympathetic system, resulting in ejaculation. D_2 activity also inhibits erection, which probably accounts in part for the sexual refractory period. Drugs that increase dopamine levels, such as those used in treating Parkinson's disease, increase sexual activity in humans (Meston & Frolich, 2000). Interestingly, dopamine release parallels sexual behavior during the Coolidge effect; as you can see in Figure 7.5, it increased in the male rat's nucleus accumbens in the presence of a female, dropped back to baseline as interest waned, then increased again with a new female (Fiorino, Coury, & Phillips, 1997). The pattern of change continued during periods when the male and female were separated by a clear panel, which indicates that the dopamine level reflects the male's interest rather than resulting from the sexual activity.

Ejaculation is also accompanied by serotonin increases in the lateral hypothalamus, which apparently contributes further to the refractory period (E. M. Hull et al., 1999). Injecting a drug that inhibits serotonin reuptake into the lateral hypothalamus increases the length of time before male rats will attempt to copulate again, and their ability to ejaculate when they do return to sexual activity. These drugs are taken by humans for a variety of problems including depression and anxiety, and both males and females often complain that their sexual ability is impaired.

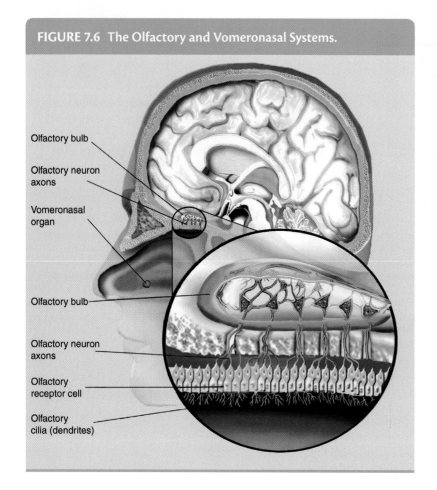

FIGURE 7.6 The Olfactory and Vomeronasal Systems.

Olfactory bulb

Olfactory neuron axons

Vomeronasal organ

Olfactory bulb

Olfactory neuron axons

Olfactory receptor cell

Olfactory cilia (dendrites)

Odors, Pheromones, and Sexual Attraction

Sexual behavior results from an interplay of internal conditions, particularly hormone levels, with external stimuli. Sexual stimuli can be anything from brightly colored plumage or an attractive body shape to particular odors. Here we will examine the role of odors and pheromones in sexual attraction, with emphasis on how important they might be for humans.

The Nose as a Sex Organ

Each human gives off a distinctive, genetically determined odor (Axel, 1995), and people can distinguish clothing worn by family members from clothing worn by strangers just by smelling them (Porter & Moore, 1981; Schaal & Porter, 1991). It is possible that, like other mammals, we use this ability to identify and bond with family members, and there is evidence that it influences mate choice; women prefer the odor of dominant men, but only during the fertile phase of their menstrual cycle (Havlicek, Roberts, & Flegr, 2005).

There is also some evidence that odor influences mate choice in a way that helps avoid genetic inbreeding while also increasing disease resistance. Women rated the odor of men's T-shirts as more pleasant when the man differed from the woman in the *major histocompatibility complex (MHC),* a group of genes that contributes to the functioning of the immune system (Wedekind, Seebeck, Bettens, & Paepke, 1995). Couples similar in MHC are less fertile and have more spontaneous abortions. The women's preference probably influenced their real-life choices; they said that the preferred odors reminded them of current or previous boyfriends. In a real-life study, women reported more sexual satisfaction and fewer outside sex partners when their romantic partners were dissimilar in MHC (Garver-Apgar, Gangestad, Thornhill, Miller, & Olp, 2006). No MHC effect was found among the men.

The effective stimuli may be odors, but often they are pheromones. *Pheromones are airborne chemicals released by an animal that have physiological or behavioral effects on another animal of the same species.* Pheromones can be very powerful, as you know if your yard has ever been besieged by all the male cats in the neighborhood when your female cat was "in heat." The female gypsy moth can attract males from as far as 2 miles away (Hopson, 1979).

To understand the role pheromones play, we need to have a basic understanding of the olfactory (smell) system. Olfaction is one of the two chemical senses, along with taste. Airborne odorous materials entering the nasal cavity must dissolve in the mucous layer overlying the receptor cells; the odorant then stimulates a receptor cell when it comes in contact with receptor sites on the cell's dendrites (see Figure 7.6). Axons from the olfactory receptors pass through openings in the base of the skull to enter the olfactory bulbs, which lie over the nasal cavity. From there, neurons follow the olfactory nerves to the nearby olfactory cortex tucked into the inner surfaces of the temporal lobes.

Humans can distinguish approximately 10,000 odors. But we do not have a different receptor for each odor, and an individual neuron cannot produce the variety of signals required to distinguish among so many different stimuli. Researchers have discovered that about 1,000 genes produce an equal number of olfactory receptor types in rats and mice; humans have between 500 and 750 odor receptor genes, although only one fourth to three fourths of these appear to be functional (Mombaerts, 1999). It is believed that each neuron has a single type of odor receptor, but whether each neuron has one or several receptor types, the brain must distinguish among the 10,000 different odors by the combination of neurons that are active.

Most pheromones are detected by the *vomeronasal organ (VNO),* a cluster of receptors also located in the nasal cavity (Figure 7.6). The two systems are separate, and the VNO's receptors are produced by a different family of genes (P. J. Hines, 1997). Not surprisingly, in animals the VNO sends its signals to the MPOA and the ventromedial nucleus of the hypothalamus, as well as to the amygdala (Keverne, 1999). The VNO is detectable in humans, but it has evolved to a diminished, often microscopic size (Garcia-Velasco & Mondragon, 1991). Although genes for the VNO have been identified in humans (Mundy & Cook, 2003), some claim the genes are no longer functional (J. Zhang & Webb, 2003). They suggest that as color vision evolved, our ancestors abandoned pheromones in favor of visual sexual signals, and this led to the VNO's functional demise (I. Rodriguez, 2004). Nevertheless, a VNO may not be required for detecting pheromones. Recent research has identified two groups of receptors that detect airborne alarm signals and other pheromones in the olfactory system of mice; humans have at least one of the receptor types and genes for the other (Brechbühl, Klaey, & Broillet, 2008; Liberles & Buck, 2006).

Interest in the possibility of human pheromones increased following reports that women living together in dorms tended to have synchronized menstrual periods and that this was caused by sweat-borne compounds that altered the frequency of luteinizing hormone release (Preti, Cutler, Garcia, Huggins, & Lawley, 1986; Preti, Wysocki, Barnhart, Sondheimer, & Leyden, 2003; Stern & McClintock, 1998). Later studies have failed to demonstrate menstrual synchrony almost as often as they have succeeded, and the results have been questioned on methodological grounds (Z. Yang & Schank, 2006), but other evidence of pheromonal influence continues to filter in.

Both men and women have reported increased sexual intercourse when using aftershave or perfume laced with underarm extracts from members of their own sex (Cutler, Friedman, & McCoy, 1998; McCoy & Pitino, 2002). Neither group increased their frequency of masturbation, so enhanced motivation for sex on their part did not appear to be the explanation. In another study, men rated the scent of women's T-shirts as more attractive when the women were in the middle of the menstrual cycle, when they would be expected to be ovulating and, therefore, fertile; this did not happen with women taking birth control pills, which suppress ovulation (Kuukasjärvi et al., 2004). PET scans suggest that these presumed pheromones activate the anterior hypothalamic area, where structures important in animal sexual behavior are located (Savic, Berglund, Gulyas, & Roland, 2001). Additional evidence for human pheromones has been found in nonsexual contexts: Subjects were able to distinguish between underarm pads worn by others during a fear-provoking film and a neutral film (Ackerl, Atzmueller, & Grammer, 2002), and fMRI showed that sniffing underarm pads worn by first-time skydivers activated the amygdala, while the sweat of people engaged in strenuous exercise did not (Mujica-Parodi et al., 2009).

In most animals attraction is fleeting, lasting only through copulation or, at best, till the end of the mating season. For a few species, though, pair bonding occurs for years or for a lifetime, as we see in the accompanying Application.

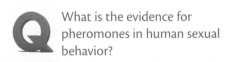

Q What is the evidence for pheromones in human sexual behavior?

Application   Of Love and Bonding

Prairie voles are a rare exception among mammals; they mate for life, and if they lose a mate they rarely take another partner. The bonding process (as reviewed in L. J. Young & Wang, 2004) begins with the release of dopamine in reward areas during mating. If dopamine activity is blocked by a receptor antagonist, partner preference fails to develop. Sexual activity also releases the neuropeptides oxytocin and vasopressin, which are likewise required for bonding to take place. Either can facilitate bonding in males or females, but oxytocin is more effective with females and vasopressin with males. Meadow voles and montane, or mountain, voles also release these neuropeptides, but they are not monogamous. At least a part of the reason lies in the *AVPR1A* gene, which is responsible for a type of vasopressin receptor. When the prairie vole's version of the gene was inserted into male meadow voles, these animals showed uncharacteristically increased preference for a female they had copulated with.

So does any of this apply to humans, who are also monogamous (more or less)? The most apparent parallel involves oxytocin. *Oxytocin not only facilitates bonding but also causes smooth muscle contractions, such as those that occur during milk ejection in lactating (breast-feeding) females.* Blood levels of oxytocin increase dramatically as males and females masturbate to orgasm (M. R. Murphy, Checkley, Seckl, & Lightman, 1990). The increase in oxytocin is probably responsible for the muscle contractions involved in orgasm and likely explains why lactating women sometimes eject a small amount of milk from their nipples during orgasm. Oxytocin also contributes to social recognition, which is necessary for developing mate preferences. Male mice without the oxytocin gene fail to recognize females from one

Prairie Voles Mate for Life and Share Parenting Duties.

SOURCE: Todd Ahern/Emory University.

encounter to the next (J. N. Ferguson et al., 2000), and human males are better at recognizing previously seen photos of women while using a nasal spray containing oxytocin (Rimmele, Hediger, Heinrichs, & Klaver, 2009); other forms of memory were unaffected in both cases. More convincing are the results of a Swedish study. Males, but not females, with two copies of a particular allele of the *AVPR1a* gene were twice as likely to report a marital crisis during the prior year, and twice as likely to be unmarried, compared to males without the allele (Walum et al., 2008).

Concept Check   *Take a Minute to Check Your Knowledge and Understanding*

☐ What change in thinking helped researchers see sex as similar to other biological drives?

☐ What roles do estrogen and testosterone play in sexual behavior in humans?

☐ In what ways do sensory stimuli influence sexual behavior?

The Biological Determination of Sex

Now we need to talk about differences between the sexes and the anomalies (exceptions) that occur. *Sex* is the term for the biological characteristics that divide humans and other animals into the categories of male and female. *Gender* refers to the behavioral characteristics associated with being male or female. For our purposes, it will be useful to make two further distinctions: *Gender role* is the set of behaviors society considers appropriate for people of a given biological sex, while *gender identity* is the person's subjective feeling of being male or female. The term *sex* cannot be used to refer to all these concepts, because the characteristics are not always consistent with each other. Thus, classifying a person as male or female can sometimes be difficult. You might think that the absolute criterion for identifying a person's sex would be a matter of chromosomes, but you will soon see that it is not that simple.

Chromosomes and Hormones

You may remember from Chapter 1 that when cells divide to produce sex cells, the pairs of chromosomes separate, and each gamete—the sperm or egg—receives only 23 chromosomes. This means that a sex cell has only one of the two sex chromosomes. An egg will always have an X chromosome, but a sperm may have either an X chromosome or a Y chromosome. The procreative function of sexual intercourse is to bring the male's sperm into contact with the female's egg, or *ovum*. When the male ejaculates into the female's vagina, the sperm use their tail-like flagella to swim through the uterus and up the fallopian tubes, where the ovum is descending. As soon as one sperm penetrates and enters the ovum, the ovum's membrane immediately becomes impenetrable so that only that sperm is allowed to fertilize the egg. The sperm makes its way to the nucleus of the ovum, where the two sets of chromosomes are combined into a full complement of 23 pairs. After fertilization, the ovum begins dividing, producing the billions of cells that make up the human fetus. If the sperm that fertilizes the ovum carries an X sex chromosome, the fetus will develop into a female; if the sperm's sex chromosome is Y, the child will be a male (see Figure 7.7).

For the first month, XX and XY fetuses are identical. Later, the primitive *gonads* (testes and ovaries, the primary reproductive organs) in the XX individual develop into *ovaries,* where the ova (eggs) develop. The *Müllerian ducts* develop into the uterus, the fallopian tubes, and the inner vagina, while the Wolffian ducts, which would become the male organs, wither and are absorbed (Figure 7.8). The undifferentiated

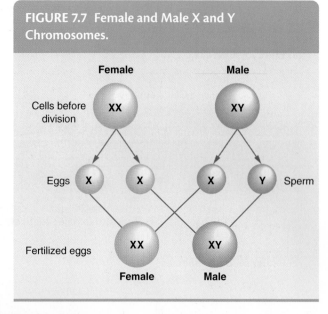

FIGURE 7.7 Female and Male X and Y Chromosomes.

Q What makes a person male or female?

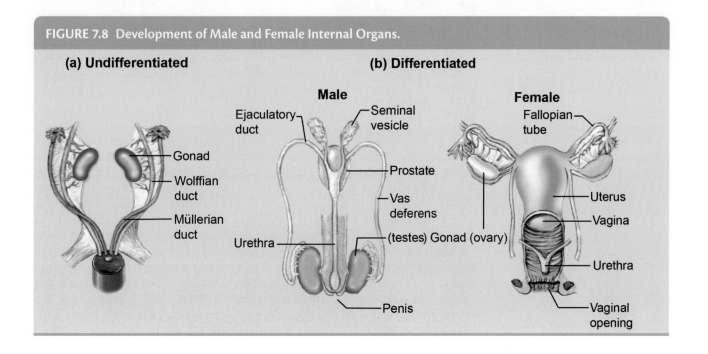

FIGURE 7.8 Development of Male and Female Internal Organs.

(a) Undifferentiated   **(b) Differentiated**

FIGURE 7.9 Differentiation of Male and Female Genitals.

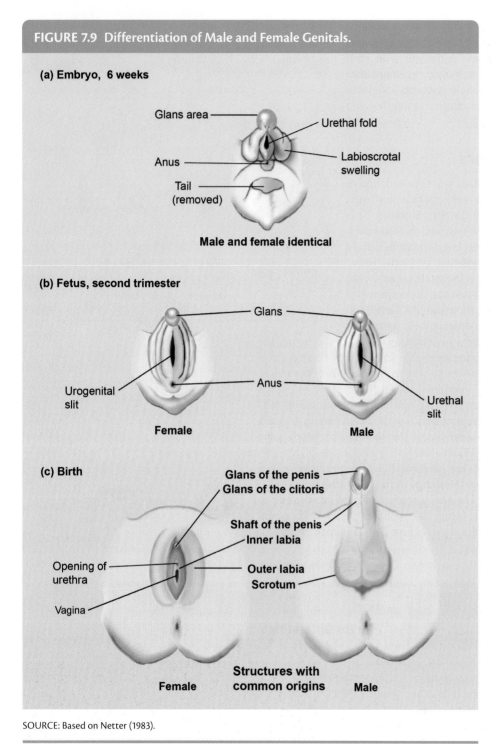

(a) Embryo, 6 weeks

Glans area

Urethal fold

Anus

Labioscrotal swelling

Tail (removed)

Male and female identical

(b) Fetus, second trimester

Glans

Urogenital slit

Anus

Urethal slit

Female

Male

(c) Birth

Glans of the penis
Glans of the clitoris

Shaft of the penis
Inner labia

Opening of urethra

Outer labia
Scrotum

Vagina

Structures with common origins

Female

Male

SOURCE: Based on Netter (1983).

external genitals become the clitoris, the outer segment of the vagina, and the labia, which partially enclose the entrance to the vagina (Figure 7.9).

If the fetus receives a Y chromosome from the father, the *SRY* gene on that chromosome produces the Sry protein, which causes the primitive gonads to develop into *testes,* the organs that will produce sperm. The testes begin secreting two types of hormones (Haqq et al., 1994). *Müllerian inhibiting hormone* defeminizes the fetus by causing the Müllerian ducts to degenerate. Testosterone, the most prominent of the androgens, masculinizes the internal organs: The *Wolffian ducts* develop into the seminal vesicles, which store semen, and the vas deferens, which carry semen from the testes to the penis. A derivative of testosterone, *dihydrotestosterone,* masculinizes the external genitals; the same structures that produce the clitoris and the labia in the female become the penis and the scrotum, into which the testes descend during childhood.

In the absence of the *SRY* gene, the primitive gonads of the XX fetus develop into ovaries. The ovaries won't begin producing estrogens until later, but the *default sex* is female, and the uterus, vagina, clitoris, and labia will all develop without benefit of hormones. You should understand that it is not entirely accurate to speak of hormones as being "male" or "female." The testes and ovaries each secrete both androgens and estrogens, although in differing amounts; the adrenal glands also secrete small amounts of both kinds of hormones.

The hormonal effects we have been discussing are called organizing effects. *Organizing effects* mostly occur prenatally and shortly after birth; they affect structure and are lifelong in nature. Organizing effects are not limited to the reproductive organs; they include sex-specific changes in the brains of males and females as well, at least in nonhuman mammals. *Activating effects* can occur at any time in the individual's life; they may come and go with hormonal fluctuations or be long lasting, but they are reversible. Some of the changes that occur during puberty are examples of activating effects.

During childhood, differences between boys and girls other than in the genitals are relatively minimal. Boys tend to be heavier and stronger, but there is considerable overlap. Boys also are usually more active and more aggressive, and interests diverge at an early age. Marked differences appear about the time the child enters puberty, usually during the preteen years. At puberty, a surge of estrogens from the ovaries and testosterone from the testes completes the process of sexual differentiation that began during prenatal development. Organizing effects include maturation of the genitals and changes in stature. Activating effects include breast development in the girl and muscle increases and beard growth in the boy. In addition, the girl's ovaries begin releasing the ova that have been there since birth (i.e., she begins to *ovulate*), and she starts to menstruate. Boys' testes start producing sperm, and ejaculation becomes possible. More important from a behavioral perspective, sexual interest increases dramatically, and in the majority of cases, preference for same-sex company shifts to an attraction to the other sex, along with an interest in sexual intimacy.

Prenatal Hormones and the Brain

Several characteristics and behaviors can be identified as male typical and others as female typical. This does not mean that the behaviors are somehow more appropriate for that sex but simply that they occur more frequently in one sex than in the other. These differences are not absolute. For example, consider the stereotypical sexual behavior of rats: The male mounts the female from behind, while the female curves her back and presents her hindquarters in a posture called *lordosis.* However, females occasionally mount other females, and males will sometimes show lordosis when approached by another male.

The same hormonal influence responsible for the development of male gonads and genitals affects behavior as well. A male rat will display lordosis and accept the sexual advances of other males if he was castrated shortly after birth or if he was given a chemical that blocks androgens just before birth and for a short time postnatally (after birth). Similarly, a female rat given testosterone during this critical period will mount other females at a higher rate than usual as an adult (Figure 7.10; Gorski, 1974). These behaviors apparently result from testosterone's influence on the size and function of several brain structures; in other words, the presence of testosterone masculinizes certain brain structures. That statement is somewhat misleading, though, because it is *estradiol,* the principal estrogen hormone, that carries out the final step of masculinization. When testosterone enters the neurons, it is converted to estradiol by a chemical process called *aromatization.* At the critical time when brain masculinization occurs, aromatase increases in the areas that are to be masculinized (Horvath & Wikler, 1999).

Until recently, it was believed that feminization of the brain, like the sex organs, required only the absence of testosterone; now we know that just as masculinization of the male brain requires estradiol, so does feminization of the female brain. Female knockout mice born unable to produce estradiol display less sexual interest and receptivity toward males or females as adults than do other mice, even when they are given replacement estrogens (Bakker, Honda, Harada, & Balthazart, 2002, 2003). Just as the male brain must be masculinized and the female brain feminized, the male brain must also be defeminized. Again, estrogens are necessary; male knockout mice lacking the estrogen receptor not only showed normal male sexual behavior but also were receptive to male advances (Kudwa, Bodo, Gustafsson, & Rissman, 2005).

This sexualization of the brain is reflected in behavioral differences, affecting not only sexual activity but also play behavior, spatial activity, and learning performance (see Collaer & Hines, 1995). Do hormones have a similar influence in humans? In the following pages we will try to answer that question.

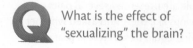

Q What is the effect of "sexualizing" the brain?

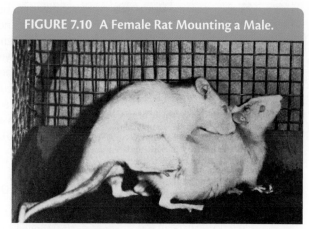

FIGURE 7.10 A Female Rat Mounting a Male.

SOURCE: From "Sex-Hormone-Dependent Brain Differentiation and Sexual Functions," in *Endocrinology of Sex* (pp. 30–37), by G. Dörner, 1974, Leipzig, Germany: J. A. Barth.

Gender-Related Behavioral and Cognitive Differences

In his popular book *Men Are From Mars, Women Are From Venus,* John Gray (1992) said that men and women communicate, think, feel, perceive, respond, love, and need differently, as if they are from different planets and speaking different languages. How different are men and women? This question is not easily answered, but it is not for lack of research on the topic. The results of studies are often ambiguous and contradictory. One reason is that different researchers measure the same characteristic in different ways. Also, the research samples are often too small to yield reliable results, and the subjects are usually not selected in a manner that ensures accurate representation of the population. Whether the differences that do exist are influenced by biology or are solely the product of experience is controversial. Contemporary parents make efforts to rear their children equally, but parents who claim to do so have been found to verbalize differently and play differently with a child dressed as a girl than when the same child is dressed as a boy (Culp, Cook, & Housley, 1983). Differential rearing, of course, could account for marked differences in behavior, temperament, and self-expectations.

FIGURE 7.11 A Spatial Rotation Task.

People are presented several pairs of drawings like these and asked whether the first could be rotated so that it looks like the second. Males are typically better at this kind of task than females. (In case you're wondering, the answer in this case is no.)

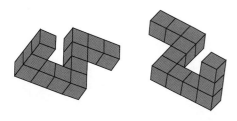

Some Demonstrated Male–Female Differences

Back in 1974, Eleanor Maccoby and Carol Jacklin reviewed over 2,000 studies and concluded that the evidence firmly supported three differences in cognitive performance and one difference in social behavior: (1) Girls have greater verbal ability than boys, (2) boys excel in visual-spatial ability, (3) boys excel in mathematical ability, and (4) boys are more aggressive than girls. Later research has supported these differences, but we should add two qualifications. First, there is considerable overlap between males and females in these characteristics. Second, the differences are rather specific. For example, females excel in verbal fluency and writing but not in reading comprehension or vocabulary (Eagly, 1995; Hedges & Nowell, 1995), and males' scores exceed females' most on tasks requiring mental rotation of a three-dimensional object (like the one in Figure 7.11) and less on other spatial tasks (Hyde, 1996).

Origins of Male–Female Differences

How do we explain the differences in verbal and spatial abilities and in aggression?

The best evidence that the three cognitive differences mentioned above are influenced by experience is that they have decreased over the years, presumably as gender roles have changed (Hedges & Nowell, 1995; Hyde, 1996; Voyer, Voyer, & Bryden, 1995). In fact, testing of 7 million students indicates that the gender difference in average mathematical performance has disappeared in the United States, although boys are still overrepresented at both the lower and higher extremes (Hyde, Lindberg, Linn, Ellis, & Williams, 2008). In addition, the dramatic variation in murder rate in different countries suggests there is also a strong cultural influence on aggression; for example, the murder rate in 2008 was 52 per 100,000 of population in Venezuela, 5.4 in the United States, 2.03 in the United Kingdom, and 0.44 in Japan ("List of Countries," n.d.).

Although environmental influences play a significant role, gender differences in cognition and behavior also owe a great deal to biology. Most often, researchers attribute the differences to the effects of estrogen and testosterone, particularly on the organizational development of the brain during gestation. This view gets support from the fact that gender differences in the volume of different brain areas correspond to the density of sex hormone receptors (J. M. Goldstein et al., 2001). Because the effects of sex hormones on brain development are most evident in people with atypical sexual development, we will hold that discussion for the next section and focus here on activational effects occurring after birth.

1
Sex Differences in the Brain

Males who produce low amounts of testosterone during the developmental years are impaired later in spatial ability (Hier & Crowley, 1982), and testosterone replacement in older men improves their spatial functioning (Janowsky, Oviatt, & Orwoll, 1994). Men who take estrogens because they identify sexually as females (*transsexuals*) increase their scores on verbal fluency tasks, but they lose spatial performance; female transsexuals taking testosterone lose verbal ability but improve in spatial performance (for references, see Hulshoff Pol et al., 2006). Men kill 30 times as often as women do (Daly & Wilson, 1988), and testosterone is usually blamed for this difference. However, whether testosterone is the cause or the result of aggression is questioned because a variety of studies show, for example, that winning a sports competition increases testosterone and losing decreases it (Archer, 1991). Aggression in males is partly inheritable; genetic effects account for about half the variance in aggression, and aggression is moderately correlated in identical twins even when they are reared apart (Rushton, Fulker, Neale, Nias, & Eysenck, 1986; Tellegen et al., 1988). The source of aggression is a complex subject, and we will deal with it more thoroughly in the next chapter.

Differences in brain anatomy and organization are also cited as bases for gender differences. Jerre Levy (1969) hypothesized that women outperform men on verbal tests because they are able to use both hemispheres of the brain in solving verbal problems rather than mostly the left hemisphere. This idea has been controversial, but in a recent fMRI study, males processed a language task primarily in the left temporal area, while females engaged both temporal areas equally (Kansaku, Yamaura, & Kitazawa, 2000). According to imaging studies, men use parietal areas to perform spatial rotations, while women rely more on frontal areas (reviewed in Andreano & Cahill, 2009); men have more cortical surface in the parietal lobes, and their scores on a spatial rotation task are correlated with the amount of cortical surface (Koscik, O'Leary, Moser, Andreasen, & Nopoulos, 2009). There are also several other indications that male and female brains work differently: Males and females have different patterns of brain activation during learning (Andreano & Cahill, 2009), pain (Naliboff et al., 2003), and stress (J. Wang et al., 2007); males are genetically more resistant to pain, and males and females respond differently to different pain medications (Bradbury, 2003); men are less affected by stress (Matud, 2004); and, as you will see in subsequent chapters, males are more susceptible to autism, Tourette's syndrome, and attention deficit hyperactivity disorder, while women are more likely to suffer from depression. There are numerous differences between the sexes in brain activation patterns and anatomy, and though some translate to behavioral differences, others do not. In those cases, the brain differences may ensure *equal* functioning in spite of hormonal differences that would disadvantage one sex (Cahill, 2006).

The value of studying these differences is not to label one sex as smarter or more aggressive but to understand what contributes to the characteristics. Keep in mind that aside from physical strength and possibly aggressiveness, the differences are small and do not justify discrimination in society or in the workplace. We are far more alike than we are different; this is a reason to use the term *other sex* instead of *opposite sex*. There are real differences, though, and an understanding of their origins could help us enhance intellectual development or reduce violence. From a scientific perspective, that knowledge also helps us understand how the brain develops, an issue that we will continue to pursue in the next two sections.

Concept Check *Take a Minute to Check Your Knowledge and Understanding*

☐ What are the origins of male–female differences in verbal and spatial abilities?

☐ What are the arguments for environmental origins and for biological origins of male–female differences in cognitive abilities and behaviors?

Sexual Anomalies

For decades, sex researchers have argued about what determines an individual's gender identity, with some believing it is formed in the first few years of life by a combination of rearing practices and genital appearance (Money & Ehrhardt, 1972) and others claiming that hormones and chromosomes are more important (M. Diamond, 1965). We cannot manipulate human development to determine what makes a person male or female, so we look to individuals on whom nature has performed "natural experiments." These lack the control of true experiments, but they can still be informative. Our earlier discussion of the effects of XY and XX chromosomes was the simple version of the sex-determination story; in reality, development sometimes takes an unexpected turn. As you will soon see, the resulting sexual anomalies not only challenge our definition of what is male and what is female, but they also tell us a great deal about the influence of biology on gender.

> *There is no one biological parameter that clearly defines sex.*
>
> —*Eric Vilain*

Male Pseudohermaphrodites

What are the characteristics of the various pseudo-hermaphrodites? What are the causes?

The Jan who became Jack at the beginning of the chapter is called a *male pseudohermaphrodite*. A hermaphrodite is a person or an animal with the sexual characteristics of both sexes. A few humans can be referred to as true hermaphrodites because they have both ovarian and testicular tissue, either as separate gonads or combined as ovotestes (J. M. Morris, 1953). They are not, however, capable of functioning sexually as both male and female, as do some simpler animals. The more common *pseudohermaphrodites* have ambiguous internal and external organs, but their gonads are consistent with their chromosomes. Pseudohermaphroditism can result from a variety of causes. The reason for Jan's unusual development was a deficiency in an enzyme (17α-hydroxysteroid) that converts testosterone into dihydrotestosterone; dihydrotestosterone masculinizes the external genitalia before birth. The large surge of testosterone at puberty enabled her body partially to carry out that process.

2

Sex Differentiation Disorders

A deficiency in another enzyme, 5α-reductase, also reduces dihy-drotestosterone levels and delays genital development; the defect is genetic and is most likely to occur when there is frequent intermarriage among relatives. Of 18 such individuals in the Dominican Republic who were reared unambiguously as girls, all but one made the transition to a male gender identity after puberty, and 15 were living or had lived with women (Imperato-McGinley, Peterson, Gautier, & Sturla, 1979). The men said they realized they were different from girls and began questioning their sex between the ages of 7 and 12. Although their transition argues for the influence of genes and hormones on gender identity, such a conclusion must be tentative because the individuals had a great deal to gain from the switch in a society that puts a high premium on maleness.

Eden Atwood (Figure 7.12) is a widely acclaimed jazz singer. She has recorded and performed all over the world and with the biggest names in jazz. Ms. Atwood is also remarkable for having been born with XY chromosomes and two testes. *Androgen insensitivity syndrome,* a form of male pseudohermaphroditism, is caused by a genetic absence of androgen receptors, which results in insensitivity to androgen. Mullerian inhibiting hormone suppresses development of most of the female internal organs, but because the individual is unaffected by androgens, the testes do not descend and the external genitals develop as more or less feminine, with a shallow vagina. If the genitals are mostly feminine, the child is reared as a girl, and at puberty her body is further feminized by estrogen from the testes and adrenal glands; estrogen supplements are often needed as well. The condition may not be recognized until menstruation fails to occur at puberty or when unsuccessful attempts to become pregnant lead to a more complete medical examination. In the absence of testosterone's influence, androgen-insensitive individuals tend to have well-developed breasts and a flawless complexion. Because these characteristics are often combined with long, slender legs, androgen-insensitive males repeatedly turn up among female fashion models (J. Diamond, 1992).

FIGURE 7.12 An XY Individual With Androgen Insensitivity.

SOURCE: Photo courtesy of Terry Cyr. Used with permission of Eden Atwood.

Female Pseudohermaphrodites

A female fetus may be partially masculinized by excess androgen and some hormone treatments during fetal development, resulting in a female pseudohermaphrodite. The internal organs are female, because no Müllerian inhibiting hormone is released, but the external genitals are virilized to some extent; that is, they have some degree of masculine appearance. In extreme cases, the clitoris is as large as a newborn male's penis, and the external labia are partially or completely fused to give the appearance of an empty scrotum.

Figure 7.13 shows an example of genital virilization in a female infant. One cause of female pseudohermaphroditism is *congenital adrenal hyperplasia (CAH),* which results from an enzyme defect that causes the individual's adrenal glands to produce large amounts of androgen during fetal development and after birth until the problem is treated. Postnatal hormone levels can be normalized by administering corticosteroids, and the parents often choose reconstructive surgery to reduce the size of the clitoris and eliminate labial fusion, giving the genitals a more feminine appearance. If masculinization is more pronounced, the parents may decide to rear the child as a boy; in that case, the surgeons usually finish closing the labia and insert artificial testes in the scrotum to enhance the masculine appearance.

3

Intersex Conditions

Parents of children with ambiguous genitalia often have difficulty knowing whether to rear them as boys or as girls. (The unusual pigmentation of the skin is due to excess excretion of sodium, or *salt wasting*, which often occurs with CAH.)

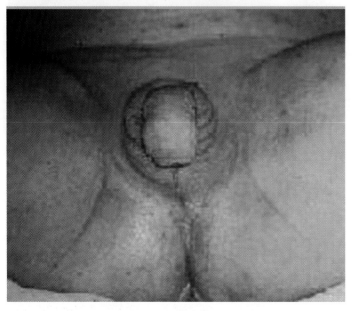

SOURCE: Used with permission of Thomas A. Wilson, MD, The School of Medicine at Stony Brook University Medical Center.

> *The evidence accumulated so far strongly suggests that man is no exception with regard to the influence of sex steroids on the developing brain and subsequent behavior.*
>
> —Anke Ehrhardt and Heino Meyer-Bahlburg

Obviously, sex cannot always be neatly divided between male and female. Some experts believe that two categories are not sufficient to describe the variations in masculinity and femininity. Anne Fausto-Sterling (1993) advocates at least five sexual categories; the ones between male and female are often referred to as *intersexes*. It would be easy to get caught up in the unusual physical characteristics of these individuals and to be distracted from our question: What makes a person male or female? This question is the topic of the accompanying In the News, as well as the next few pages.

Sex Anomalies and the Brain

As mentioned earlier, reversing the sex hormone balance during prenatal development changes the brain and later behavior in nonhuman animals. Is it possible that masculinization and feminization of the developing brain account for sex differences in behavior and cognitive abilities in humans as well? If so, then we would expect the behavior and abilities of individuals with sexual anomalies, who have experienced an excess or a deficit of androgen during prenatal development, to be at odds with their chromosomal sex.

That is indeed the case. Androgen-insensitive males are like females in that their verbal ability is higher than their spatial performance, and their spatial performance is lower than that of other males (Imperato-McGinley, Pichardo, Gautier, Voyer, & Bryden, 1991; Masica, Money, Ehrhardt, & Lewis, 1969). And though there have been contradictory results, evidence favors increased spatial ability in CAH women (Puts, McDaniel, Jordan, & Breedlove, 2008). Androgen-insensitive males also are typically feminine in behavior, have a strong childbearing urge, and are decidedly female in their sexual orientation (M. Hines, 1982; Money, Schwartz, & Lewis, 1984; J. M. Morris, 1953). Although 95% of CAH women reared as girls accept a female identity, they also show behavioral shifts in the masculine direction (Dessens, Slijper, & Drop, 2005). They have been described as tomboyish in childhood (M. Hines); to prefer boys' toys, such as trucks and building blocks (Berenbaum, Duck, & Bryk, 2000); and to draw pictures more typical of boys, using darker colors and including mechanical objects and excluding people (Iijima, Arisaka, Minamoto, & Arai, 2001). They also more often report male-dominated occupational choices (30% versus 13%), interest in rough sports (74% versus 50%), and interest in motor vehicles (14% versus 0%; Frisén et al., 2009). There is evidence that these effects are due to androgen levels before birth rather than during postnatal development (Berenbaum et al.). Homosexual or bisexual orientation has been reported to be as high as 37% (Money et al.) and at 19% in a recent larger study (Frisén et al.).

Some critics claim that humans are sexually neutral at birth and that gender identity and behavior are learned. They attribute the cognitive and behavioral effects we have just seen to feminine or ambiguous rearing in response to the child's genital appearance. (You may be beginning to appreciate the deficiencies of natural experiments.) However, some of the findings are difficult to explain from a socialization perspective. For example, the anti-miscarriage drug diethylstilbesterol (DES)

Sex, Gender, and Sports

When Caster Semenya of South Africa won the gold medal in the 800-meter race at the 2009 World Championships in Athletics, her strong performance and masculine physique aroused suspicions about her gender. Fueled by years of media reports that some female competitors might actually be men, the International Association of Athletics Federations (IAAF) and the International Olympic Committee (IOC) had introduced routine gender testing in the 1960s (J. L. Simpson et al., 1993). However, physical examination was soon rejected as unacceptable to many women, and chromosome testing turned out to be inadequate as a measure of performance advantage. For example, Barr body analysis, which identifies cells with XX chromosomes, rejects androgen-insensitive XY individuals though they receive no benefit from testosterone, but it would pass XXY males, who do benefit. The IOC and the IAAF ended routine testing in the 1990s, though both reserved the authority to request gender identification on an individual basis.

Semenya agreed to an IAAF request to undergo extensive gender testing, and a panel of experts spent months evaluating the results. In the meantime, reports were leaked that Semenya had two testes and triple the normal level of testosterone for a female. The IAAF said that if the reports turned out to be accurate, it would pay for corrective surgery; the surgery would remove the internal testes, which have a high risk for cancer, and eliminate the source of the extra testosterone. When the IAAF received the report, it did not reveal the results to the public, but nearly a year after the championships, Semenya was cleared to compete again—as a woman. Some sexual activists argue that if society would place less emphasis on gender, whether a person is male or female wouldn't matter, but this case suggests there is a need for better understanding of what it means to be male or female. In the meantime, the IAAF has realized that it does not even have a definition of gender in its rules or its gender verification policy.

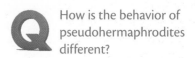

© AP Photo/Anja Niedringhaus

SOURCES: "Caster Semenya Must Wait," 2010; Hart, 2009; O'Reilly, 2010; Powers, 2010.

given to women in the 1950s and 1960s has an androgen-like effect in the brain but does not virilize the genitals, yet DES-exposed daughters reported increased homosexual fantasy and behavior (Ehrhardt et al., 1985; Meyer-Bahlburg et al., 1995). In another study, girls exposed to a similar drug and who had nonvirilized genitals scored higher in aggression than their did unexposed sisters (Reinisch, 1981). In addition, the fact that androgen-insensitive males perform even lower on spatial tests than did their unaffected sisters and female controls (Imperato-McGinley et al., 1991) can be explained by insensitivity to androgens but not by "feminine rearing."

Q How is the behavior of pseudohermaphrodites different?

Ablatio Penis: A Natural Experiment

The "neutral-at-birth" theorists claim that individuals reared in opposition to their chromosomal sex generally accept their sex of rearing and that this demonstrates that rearing has more effect on gender role behavior than chromosomes or hormones (studies reviewed in M. Diamond, 1965). Diamond, who advocates a "sexuality-at-birth" hypothesis, argues that the reason individuals with ambiguous genitals accept their assigned gender is that sex of rearing is usually decided by

> *Gender identity is sufficiently incompletely differentiated at birth as to permit successful assignment of a genetic male as a girl.*
>
> —John Money
>
> *An extensive search of the literature reveals no case where a male or female without some sort of biological abnormality . . . accepted an imposed gender role opposite to that of his or her phenotype.*
>
> —Milton Diamond

whether the genital appearance is predominantly masculine or feminine, which in turn reflects the influence of prenatal hormones. According to Diamond, there is no case in the literature where an unambiguously male or female individual was successfully reared in opposition to the biological sex. He and others (Money, Devore, & Norman, 1986) have described several instances in which individuals assigned as one sex successfully shifted to their chromosomal and gonadal sex in later years, long after Money's window for forming gender identity (the first few years of life) supposedly had closed. Failures in predicting the later gender identity of a child with ambiguous genitals has led several experts (along with the Intersex Society of North America) to advocate waiting until the child can give informed consent, or at least indicates a clear gender preference; others are reluctant to see the child subjected to the social difficulties that result from an ambiguous appearance.

In 1967, an 8-month-old boy became the most famous example of resistance to sexual reassignment when the surgeon using electrocautery to perform a circumcision turned the voltage too high and destroyed the boy's penis. At that time, it was not possible to fashion a satisfactory replacement surgically so after months of consultation and agonizing, Bruce's parents decided to let surgeons transform his genitals to feminine ones. The neutral-at-birth view was widely accepted then, and the psychologist John Money counseled the parents that they could expect their son to adopt a female gender identity (M. Diamond & Sigmundson, 1997). Bruce would be renamed Brenda, and "she" would be reared as a girl. This case study is as good an example of a "natural experiment" as we will find, for two reasons: The child was normal before the accident, and he happened to have an identical twin who served as a control.

Over the next several years, Money (1968; Money & Ehrhardt, 1972) reported that Brenda was growing up feminine, enjoying her dresses and hairdos, and choosing to help her mother in the house, while her "typical boy" brother played outside. But developmental progress was not nearly as smooth as Money claimed (M. Diamond & Sigmundson, 1997). Brenda was in fact a tomboy who played rough-and-tumble sports and fought, and preferred her brother's toys and trucks over her dolls. She looked feminine, but her movements betrayed her, and her classmates called her "caveman." When the girls barred her from the restroom because she often urinated in a standing position, she went to the boys' restroom instead. She had private doubts about her sex beginning in the second grade, and by the age of 11 had decided she was a boy. At age 14, she decided to switch to living as a male. Only then did Brenda's father tell her the story of her sexual transition in infancy. Then, said Brenda, "everything clicked. For the first time things made sense and I understood who and what I was" (p. 300).

Brenda changed her name to David and requested treatment with testosterone, removal of the breasts that had developed under estrogen treatment, and construction of a penis. The child who was isolated and teased as a girl was accepted and popular as a boy, and he attracted girlfriends. At age 25, he married Jane and adopted her three children. Although he was limited in sexual performance, he and Jane engaged in sexual play and occasional intercourse.

But life was still not ideal. He brooded about his childhood and was often angry or depressed; after 14 years, Jane told him they should separate for a while. Troubled by his past and his present, and perhaps a victim of heredity—his mother had attempted suicide, his father became an alcoholic, and his twin brother died of an overdose of antidepressants—one spring day in 2004 David Reimer took his own life (Figure 7.14; Colapinto, 2004).

Ablatio penis ("removed penis") during infancy is rare; only two other cases of female reassignment with follow-up have been reported in the literature. In one, the individual chose reassignment as a male at the age of 14 (Ochoa, 1998). In the third case, the individual reported no uncertainty about her feminine identity in adulthood (Bradley, Oliver, Chernick, & Zucker, 1998). This more positive outcome could be due to any of a number of factors, including early gender reassignment and minimal family ambivalence about the female assignment. However, in spite of her commitment to a female gender identity, she reported being a tomboy during childhood, and as an adult she chose a traditionally masculine "blue collar" occupation. At age 26, her sexual activity was evenly divided between men and women, and her sexual fantasies were predominantly about women. Meyer-Bahlburg (1999) points out that "a given gender identity can accommodate wide variations in gender role behavior" (p. 3455). We will see in the next section that this dissociation can be striking. These three cases do not permit firm conclusions, but they also do not support the view that gender behavior is primarily a result of upbringing.

FIGURE 7.14 David Reimer, 1965–2004.

SOURCE: © Reuters, Inc.

Concept Check   *Take a Minute to Check Your Knowledge and Understanding*

☐ How do the sexual anomalies require you to rethink the meaning of male and female?

☐ What reasons can you give for thinking that the brains of people with sexual anomalies have been masculinized or feminized contrary to their chromosomal sex?

Sexual Orientation

Sex researchers, along with the rest of us, spend a great deal of time arguing about why some people are attracted to members of the same sex. Whether we know it or not, we are also asking why most people are heterosexual. The answer to that question may seem obvious, but the fact that a behavior is nearly universal and widely accepted does not mean that it requires no explanation. People who are attracted to members of their own sex may be able to tell us not only about homosexuality but also about the basis for sexual orientation in general.

A word about terminology: Homosexual men are often referred to as *gay*, and homosexual women are often called *lesbians*. The term for those who are not exclusively homosexual or heterosexual is *bisexual*. The large majority of nonheterosexuals are exclusively homosexual, although bisexuality is more common among lesbians than among gays (Pillard & Bailey, 1998). The term *homosexuality* is ordinarily used to refer to regular activity or *continuing* preference. As Ellis and Ames (1987)

4
Homosexuality and Asexuality

point out, homosexual experiences are fairly common, especially in adolescence and in the absence of heterosexual opportunities, and these experiences do not make a person homosexual any more than occasional heterosexual activity makes a person heterosexual.

How many people are homosexual is uncertain. In an unusually large survey of 3,432 American men and women, 9% of men and a little over 4% of women said they had homosexual sex at least once since puberty (Michael, Gagnon, Laumann, & Kolata, 1994). About 2.8% of men and 1.4% of women thought of themselves as homosexual or bisexual. Other studies in the United States and abroad suggest lower percentages (Billy, Tanfer, Grady, & Klepinger, 1993). Interestingly, an estimated 1% of people express no interest in sex at all (Bogaert, 2004); *asexuality* is gaining acceptance as an additional category of preference.

Research does not support the belief that gay men are necessarily feminine and lesbians are masculine; only about 44% of gays and 54% of lesbians fit those descriptions (Bell, Weinberg, & Hammersmith, 1981). Even then, they usually identify with their biological sex, so gender role, gender identity, and sexual orientation are somewhat independent of each other and probably have different origins.

It is not clear what causes homosexuality, which means that we do not know how to explain heterosexuality either. There is considerable evidence for biological influences on sexual orientation, or else the topic would not appear in this chapter. But because social influences are commonly believed to be more important, we will consider this position first.

The Social Influence Hypothesis

It has been argued that homosexuality arises from parental influences or is caused by early sexual experiences. Bell and his colleagues (1981) expected to confirm these influences when they studied 979 gay and 477 heterosexual men. But they found no support for frequently hypothesized environmental influences, such as seduction by an older male or a dominant mother and a weak father.

Several developmental experiences do seem to differentiate homosexuals from heterosexuals, and these have been considered evidence for a social learning hypothesis (Van Wyk & Geist, 1984). But these experiences—such as spending more time with other-sex playmates in childhood, learning to masturbate by being masturbated by a member of the same sex, and homosexual contact by age 18—can just as easily be interpreted as reflecting an early predisposition to homosexuality. In fact, Bell and his associates (1981) concluded that "adult homosexuality "is just *a continuation of the earlier homosexual feelings and behaviors from which it can be so successfully predicted*" (p. 186; italics in the original). However, they did find more evidence for an influence of learning on bisexuality than on exclusive homosexuality. This suggests that there might be a biological influence that varies in degree, with experience making the final decision in the individuals with weaker predispositions for homosexuality.

Seventy percent of homosexuals remember feeling "different" as early as 4 or 5 years of age (Bell et al., 1981; Savin-Williams, 1996). It is difficult to evaluate these reports because yesterday's memories are easily distorted in light of today's circumstances and because we do not know how frequently heterosexuals felt the same way. However, during development, homosexuals do show a high rate of *gender nonconformity*—a tendency to engage in activities usually preferred by the other sex and an atypical preference for other-sex playmates and companions while growing up (Bell et al., 1981). If we are to entertain a biological hypothesis of sexual orientation, though, we must come up with some reasonable explanation for how it is formed and how it is altered. There are three biological approaches to the question: *genetic, hormonal,* and *neural.*

Genes and Sexual Orientation

Q What is the evidence for a biological basis for homosexuality?

Genetic studies provide the most documented and most consistent evidence for a biological basis for sexual orientation. Homosexuality is two to seven times higher among the siblings of homosexuals than it is in the population (J. M. Bailey & Bell, 1993; J. M. Bailey & Benishay, 1993; Hamer, Hu, Magnuson, Hu, & Pattatucci, 1993). Identical twins are more concordant for homosexuality than fraternal twins or nontwin siblings (Figure 7.15; J. M. Bailey & Pillard, 1991; J. M. Bailey et al., 1993; Whitman, Diamond, & Martin, 1993). Determining just how heritable sexual orientation is has turned out to be more difficult than you might expect; the main reasons are the small number of twins who are homosexual and the stigma against admitting homosexual behavior or even interest. These factors increase the importance of the volunteer effect; studies of self-volunteered subjects have produced results similar to those in Figure 7.15, and the pattern of differences shown there is generally accepted, but the actual concordances may be lower. When subjects were recruited independently of their sexual preference, identical twin concordances have ranged around 35% to 45% for men and 18% for women (Langström, Rahman, Carlström, & Lichtenstein, 2010). Interestingly, of the variability not accounted for by heredity, nonshared environmental influences were much greater than shared environmental influences. This means that family environment is less important than individual factors such as prenatal environment, which we will discuss shortly.

Hamer and his associates (1993) found that gay men had more gay relatives on the mother's side of the family than on the father's side. Because the mother contributes only X chromosomes to her sons, Hamer looked there for a gene for homosexuality. To increase the chances of locating a maternally transmitted gene, he confined his search to the 40 pairs of gay brothers who had gay relatives on the mother's side of the family. In 64% of the pairs, the brothers shared the same genetic material at one end of the X chromosome (Figure 7.16); presumably, that material would contain one or more genes for homosexuality. Certainty is difficult to come by in genetic research, though. The results were supported by a second study (Hu et al., 1995) but not by a third (G. Rice, Anderson, Risch, & Ebers, 1999); Hamer has criticized the second study for having a small sample that emphasized gays with gay paternal, rather than maternal, relatives (Hamer, 1999). More recently, a scan of the entire human genome confirmed Hamer's result only when the analysis was limited to data from the 1993 families; however, in the other families it turned up stretches of DNA on three other chromosomes that were shared between gay brothers with a higher than expected frequency (Mustanski et al., 2005).

Evidence that homosexuality is influenced by genes presents a Darwinian contradiction; how could homosexuality survive when its genes are unlikely to be passed on by the homosexual?

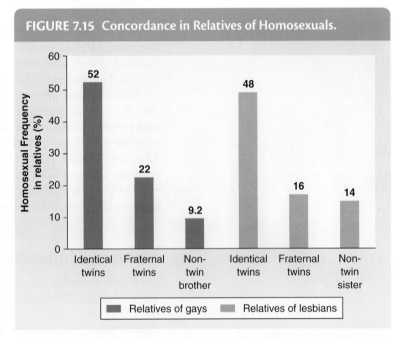

FIGURE 7.15 Concordance in Relatives of Homosexuals.

SOURCE: Based on data from J. M. Bailey and Pillard (1991) and J. M. Bailey et al. (1993).

FIGURE 7.16 Location of a Possible "Gay Gene."

The genetic material at the end of the X chromosome was the same in a high percentage of gay brothers.

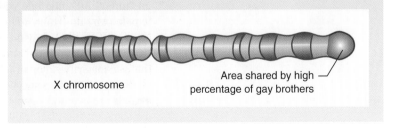

SOURCE: From *Introduction to Psychology*, 5th ed., by Plotnik. 1999. Reprinted by permission of Wadsworth, a division of Thomson Learning.

An intriguing hypothesis is that homosexuals are more likely to be "helpers" in their families, ensuring that the family members survive and pass on the genes, but research has not supported this idea. Italian researchers have offered a promising alternative; birth rate is higher in women on the mother's side of the family of male homosexuals, so they conclude that genes responsible for homosexuality also increase the women's birth rate—compensating for the homosexual's lack of productivity (Camperio Ciani, Corna, & Capiluppi, 2004; Iemmola & Camperio Ciani, 2009). A later analysis indicated that the results are best explained by two genes, at least one of which is on the X chromosome; the researchers also suggested that the genes increase attraction to men, both in men and women (Camperio Ciani, Cermelli, & Zanzotto, 2008).

Another proposal is that the genetic influence on homosexuality may sometimes occur through an epigenetic process known as *imprinting*. In females, one of each pair of X chromosomes is turned off in every cell. Which of the two X chromosomes gets turned off usually varies randomly from cell to cell, but the mothers of gay sons are more likely to favor one of the X chromosomes over the other (Bocklandt, Horvath, Vilain, & Hamer, 2006). Only 4% of women with no gay sons showed "extreme skewing"—defined as inactivation of the same chromosome in 90% of their cells—compared to 13% of women with homosexual sons and 23% of the mothers of two or more gay sons.

Hormonal Influence

If heredity influences sexual orientation, it must do so through some physiological mechanism. The most obvious possibility is the sex hormones. Early on, doctors tried to reverse male homosexuality by administering testosterone; the treatment did not affect sexual preference, but it frequently did increase the level of homosexual activity (for references, see A. C. Kinsey, Pomeroy, Martin, & Gebhard, 1953). Later studies that compared hormonal levels in homosexuals and heterosexuals did not support the hypothesis that homosexuals have a deficit or an excess of sex hormones (Gartrell, 1982; Meyer-Bahlburg, 1984).

So if there are any hormonal influences in homosexuality, they are likely to have occurred prenatally, and we turn again to animals for clues. Early hormonal manipulation results in same-sex preference later in life in rats, hamsters, ferrets, pigs, and zebra finches (for references, see LeVay, 1996). Some critics believe that this result has little meaning; they claim that homosexual behavior occurs spontaneously in animals only in the absence of members of the other sex and does not represent a shift in sexual orientation.

However, about 10% of male sheep prefer other males as sex partners, and some form pair bonds in which they take turns mounting and copulating anally with each other (Perkins & Fitzgerald, 1992). A few female gulls observed on Santa Barbara Island off the coast of California form "lesbian" pairs. They court each other, and the courting ritual occasionally ends in attempted copulation. They take turns sitting on their nest; if some of the eggs hatch because they were fertilized during an "unfaithful" interlude with a male, the two females share parenting like male–female pairs do (Hunt & Hunt, 1977; Hunt, Newman, Warner, Wingfield, & Kaiwi, 1984). The gulls' behavior could be a response to a shortage of males, but it differs from the opportunistic homosexuality usually seen when mates are unavailable, in that the majority stay paired for more than one season.

Some researchers suggest that rather than being underandrogenized, male homosexuals are overandrogenized. They base this in part on the fact that gays exaggerate some male-typical traits, such as left-handedness (reviewed in Rahman, 2005). Because male homosexuals tend to have more older brothers than heterosexual males do, some believe this exposes the later-born males to elevated testosterone in the womb (T. J. Williams et al., 2000). But the overandrogenization hypothesis runs into trouble because CAH males do not have a high level of homosexuality (Cohen-Bendahan, van de Beek, & Berenbaum, 2005) and because it is unclear why successive male births would lead to increased androgen. Ray Blanchard (2001) proposed another interpretation for the birth-order data: With each male birth, the mother develops a progressively stronger immunity to a type of protein that occurs only in males, and she produces antibodies that modify the brain's development. Blanchard believes that if homosexual males are overandrogenized, it is not the cause of their homosexuality but the brain's compensation for the antibodies' demasculinizing effects. There is no direct evidence for the immunization hypothesis, and you must be wondering whether the "older brother effect" might be social-environmental. Anthony Bogaert (2006) asked the same question and found that the presence of older nonbiological brothers did not affect orientation.

There is no direct evidence for prenatal hormone imbalances in homosexual humans, possibly because it is very difficult to determine a human's prenatal hormone environment. The finding that a deficiency in estrogen receptors resulted in female sexual behavior in male mice (Kudwa et al., 2005) suggests a potential new direction for this research. In the meantime, the case is stronger for structural and functional differences in the brain—which, in nonhuman animals at least, are influenced by prenatal and early postnatal hormones.

Brain Structures

Studies have identified three brain structures that might differ in size between gay males and heterosexuals. Simon LeVay (1991) found the *third interstitial nucleus of the anterior hypothalamus (INAH3)* to be half the size in gay men and heterosexual women as in heterosexual men (Figure 7.17; you can locate the anterior hypothalamus in Figure 6.2). The difference has also been seen in sheep (Roselli et al., 2004); another human study found a difference in the same direction, but it failed to reach statistical significance (Byne et al., 2001). In other research, the *suprachiasmatic nucleus (SCN)* was larger in gay men than in heterosexual men and contained almost twice as many cells that secrete the neuropeptide vasopressin (Swaab & Hofman, 1990). The SCN is also shown in Figure 6.2, lying, as its name implies, just above the optic chiasm. Finally, the *anterior commissure,* one of the tracts connecting the cerebral hemispheres, was larger in gay men and heterosexual women than in heterosexual men (see Figure 3.5; Allen & Gorski, 1992).

The implication of these differences, if they are supported in the future, is unclear. Although INAH3 is in an area of the brain that is involved in sexual activity in animals and various studies have shown that it is larger in males than in females, its role in humans is not known. The SCN regulates the reproductive cycle in female rats and controls daily cycles in rats and humans. When male rats were treated during the prenatal period and for the first few days after birth with a chemical that blocks the effects of testosterone, the number of vasopressin-secreting cells increased in their SCNs (Swaab, Slob, Houtsmuller, Brand, & Zhou, 1995). Given a choice between an estrous female and a sexually active male, they spent one third of their time with the male, with whom they showed lordosis and accepted mounting. The significance of a larger anterior commissure is also unclear. Assuming the brain of the homosexual male is

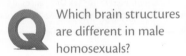

Which brain structures are different in male homosexuals?

The arrows indicate the boundaries of the structure. Note the smaller size in the homosexual brain.

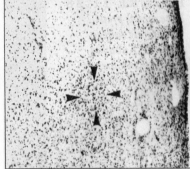

SOURCE: Reprinted with permission from S. LeVay, "A Difference in Hypothalamic Structure Between Heterosexual and Homosexual Men," *Science, 253,* pp. 1034–1047. © 1991, American Association for the Advancement of Science (AAAS).

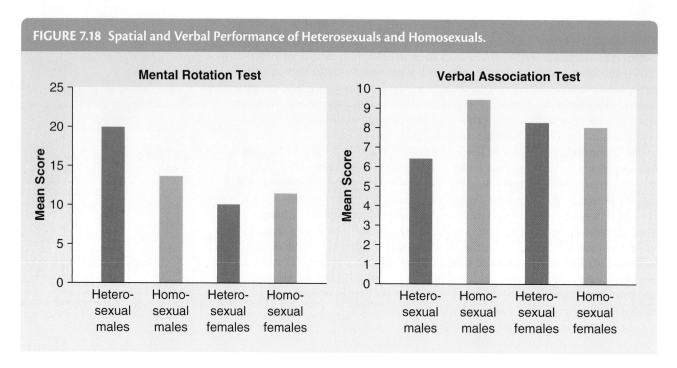

FIGURE 7.18 Spatial and Verbal Performance of Heterosexuals and Homosexuals.

SOURCE: Reprinted from "Sex Differences in Cognition: The Role of Testosterone and Sexual Orientation," by N. Neave, M. Menaged, and D. R. Weightman, *Brain and Cognition, 41*, pp. 245–262. © 1999, with permission from Elsevier Science.

> *The most powerful sex organ is between the ears, not between the legs.*
>
> —Milton Diamond

feminized, it could conceivably contribute to the greater communication between hemispheres that Levy suggested. Gay men do, in fact, perform similar to women on some cognitive tests, scoring higher on verbal tasks than heterosexual men and lower on spatial tasks (Figure 7.18; C. M. McCormick & Witelson, 1991; Neave, Menaged, & Weightman, 1999; Wegesin, 1998).

Investigations of transsexual brains also have provided some intriguing information. *Transsexual* individuals have the strong belief that they have been born into the wrong sex; they may dress and live as the other sex, and they often undergo surgery for sex reassignment. Gender identity reversal is rarer than homosexuality; estimates range between 1 and 5 per 1,000 people (Collaer & Hines, 1995). Male-to-female transsexuals—who are studied more often because they are more plentiful—have been reported to resemble females rather than males in INAH3 size (Garcia-Falgueras & Swaab, 2008), brain response to an androgen derivative found in male sweat (which we will call AND) and an estrogen-like compound found in female urine (which we will call EST; Berglund, Lindström, Dhejne-Helmy, & Savic, 2008), and brain activation patterns in response to an erotic video (Gizewski et al., 2009). However, transsexuals are not necessarily homosexual; in nine studies, the rate of homosexuality and bisexuality among male transsexuals averaged 62% (Lawrence, 2005). One brain characteristic may distinguish transsexuals: The *central bed nucleus of the stria terminalis (BSTc)* of the hypothalamus is smaller in women than in men and has been reported to be female sized in male transsexuals and male sized in the one available female transsexual (Kruijver et al., 2000; Zhou, Hofman, Gooren, & Swaab, 1995). This difference was specific to transsexualism; it occurred whether the individual was homosexual or not.

We cannot say yet why most people prefer members of the other sex or why some prefer members of their own sex. Part of the reason is that so much of our information about the human brain has been dependent on access to the brains of the deceased. As imaging techniques improve, we have the opportunity to study living brains before they are modified by a lifetime of experience or ravaged by aging, drugs, and disease. An example is a recent study assessing the effects of presumed pheromones on the brain in which the researchers used AND and EST. (Savic, Berglund, & Lindström, 2005). In heterosexuals, AND activated the preoptic area and the ventromedial nucleus in women, but not in men; EST activated the paraventricular and dorsomedial nuclei in men, but not in women. The homosexual men did not respond to EST; instead, they reacted to AND, in the same areas as the women did (Figure 7.19). In a later study, lesbian women's responses to AND and EST were more similar to those of heterosexual men than to those of heterosexual women (Berglund, Lindström, & Savic, 2006). We have plenty of evidence that sexual orientation is a brain phenomenon; now we need to learn just *how* these brain structures contribute to sexual choice.

FIGURE 7.19   Responses of Heterosexual Women, Homosexual Men, and Heterosexual Men to a Presumed Male Pheromone.

Heterosexual women and homosexual men responded to the testosterone derivative AND in the MPOA/anterior hypothalamus, while heterosexual men did not.

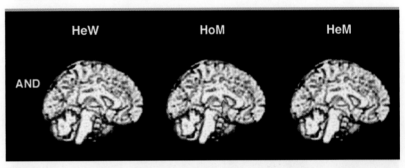

SOURCE: From "Smelling of Odorous Sex Hormone-Like Compounds Causes Sex-Differentiated Hypothalamic Activations in Humans," by I. Savic et al., *Neuron*, 31, 661–668, fig. 1. © 2002. Reprinted by permission.

The Challenge of Female Homosexuality

We have been talking mostly about gay men, because there is considerably less research on the biological and behavioral characteristics of lesbians and what there is gives us little to go on. For example, with the exception of studies of CAH women, there is little evidence that lesbians' brains have been masculinized during prenatal development. Furthermore, lesbians perform similarly to heterosexual females on verbal and spatial tests (see Figure 7.18 again; Neave et al., 1999; Wegesin, 1998). However, the high estimated heritability for homosexuality in women encourages us to continue looking for a biological basis for lesbianism. Two interesting physical characteristics seem to distinguish lesbians from heterosexuals, though they seem trivial, at least initially. First, the index-to-ring-finger ratios of lesbians are indistinguishable from those of heterosexual men (T. J. Williams et al., 2000). The second difference involves a peculiar, faint sound given off by the inner ear when it is stimulated, called *click-evoked otoacoustic emissions*. The response is weaker in lesbians and men (both heterosexual and gay) than in heterosexual women (McFadden & Pasanen, 1998). The significance of these two differences is that both of the characteristics are influenced by testosterone levels during prenatal development.

Could prenatal testosterone masculinize these two characteristics along with one or more brain structures to alter female sexual orientation while leaving spatial and verbal abilities unaffected? This does seem possible if masculinization of the cognitive abilities occurs at a different developmental time or requires a higher level of testosterone. However, this is pure speculation, and it is far too early to be drawing any conclusions; we obviously need to focus more research attention on sexual orientation in females.

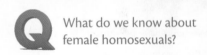

What do we know about female homosexuals?

Social Implications of the Biological Model

As is often the case, the research we have been discussing has important social implications. If homosexuality is a choice, as argued by some in the gay community,

Q Why is the search for a biological basis of homosexuality a social issue?

FIGURE 7.20 U.S. Congressman Barney Frank.

With him is Herb Moses, his companion from 1987 to 1998.

SOURCE: Photo by Burk Uzzle.

> *As for being gay, I never felt I had much choice. . . . I am who I am. I have no idea why.*
>
> —Congressman Barney Frank

then U.S. civil rights legislation does not apply to homosexuals, because protection for minorities depends on the criterion of unalterable or inborn characteristics (Ernulf, Innala, & Whitam, 1989). If homosexuality is the result of experience, then presumably a homosexual who wanted to avoid the hassle and discrimination could change through psychotherapy, behavior modification, or religious conversion.

However, some homosexuals say that they have no choice. When Congressman Barney Frank of Massachusetts (Figure 7.20) was asked if he ever considered whether switching to the straight life was a possibility, he replied, "I wished it was. But it wasn't. I can't imagine that anybody believes that a 13-year-old in 1953 thinks, 'Boy, it would be really great to be a part of this minority that everybody hates and to have a really restricted life'" (Dreifus, 1996, p. 25).

About 75% of homosexuals believe that homosexuality is inborn (Leland & Miller, 1998), but some in the gay community think that promoting this view is not in their best interest. For them, the biological model is associated too closely with the old medical "disease" explanation of homosexuality. They fear that homosexuals will be branded as defective, or even that science may find ways to identify homosexual predisposition in fetuses and that parents will have the "problem" corrected through genetic manipulation or abortion. Emotions are so strong among some homosexuals that the researcher Dick Swaab was physically attacked in Amsterdam by members of the Dutch gay movement, who felt threatened by his biological findings (Swaab, 1996).

Other gay and lesbian rights activists welcome the biological findings because they think that belief in biological causation will increase public acceptance of homosexuality. The researcher Simon LeVay, who is himself gay, supports their view. Between a third and half of the public believe that homosexuality is something a person is born with (Ellis & Ames, 1987; Ernulf et al., 1989; Leland & Miller, 1998; Schmalz, 1993), and people who believe that homosexuality is biologically based have more positive attitudes toward homosexuals than people who believe that homosexuality is learned or chosen (Ernulf et al.; Schmalz). This debate will not be settled any time soon, but most researchers believe that when we understand the origins of homosexuality and heterosexuality, they will include a combination of heredity, hormones, neural structures, and experience (LeVay, 1996).

Concept Check *Take a Minute to Check Your Knowledge and Understanding*

☐ How has the social influence hypothesis fared in explaining homosexuality?

☐ What is the evidence that homosexuality has a biological cause?

☐ Organize your knowledge: Make a table of the brain structures that may differentiate heterosexual males, heterosexual females, and homosexuals; show how the structures differ between groups and, if known, their functions.

In Perspective

The fact that sex is not motivated by any tissue deficit caused researchers to look to the brain for its basis. What they found was a model for all drives that focused on the brain rather than on tissue that lacked nutrients or water or was too cold. This view changed the approach to biological motivation, and it meant that gender identity and gender-specific behavior and abilities might all be understood from the perspective of the brain.

The fact that a person's sexual appearance, gender identity, and behavior are sometimes in contradiction with each other or with the chromosomes makes sex an elusive concept. Research is helping us understand that many differences between the sexes are cultural inventions and that many differences thought to be a matter of choice have biological origins. As a result, society is slowly coming around to the idea that distinctions should not be made on the basis of a person's sex, sexual appearance, or sexual orientation. These issues are emotional, as are the important questions behind them: Why are we attracted to a particular person? Why are we attracted to one sex and not the other? Why do we feel male or female? The emotion involved often obscures an important point: that the answers keep leading us back to the brain, which is why some have called the brain the primary sex organ.

Summary

Sex as a Form of Motivation

- Although there is no tissue deficit, sex involves arousal and satiation like other drives, as well as hormonal and neural control. Also like the other drives, sex can be thought of as a need of the brain.
- The key elements in human sexual behavior are testosterone, structures in the hypothalamus, and sensory stimuli such as certain physical characteristics and pheromones.

The Biological Determination of Sex

- Differentiation as a male or a female depends on the combination of X and Y chromosomes and the presence or absence of testosterone.
- Testosterone controls the differentiation not only of the genitals and internal sex organs but also of the brain.

Gender-Related Behavioral and Cognitive Differences

- Evidence indicates that girls exceed boys in verbal abilities and that boys are more aggressive and score higher in visual-spatial and mathematical abilities.
- With the possible exception of mathematical ability, it appears that these differences are at least partly due to differences in the brain and in hormones.

Sexual Anomalies

- People with sexual anomalies challenge our idea of male and female.
- The cognitive abilities and altered sexual preferences of people with sexual anomalies suggest that the brain is masculinized or feminized before birth.

Sexual Orientation

- The idea that sexual orientation is entirely learned has not fared well.
- Evidence indicates that homosexuality, and thus heterosexuality, is influenced by genes, prenatal hormones, and brain structures.
- The biological view is controversial among homosexuals, but most believe that it promotes greater acceptance, and research suggests that this is the case. ∎

Study Resources

F   For Further Thought

- Do you think the cognitive differences between males and females will completely disappear in time? If not, would they in an ideal society? Explain your reasons.
- Some believe that parents should have their child's ambiguous genitals corrected early, and others think it is better to see what gender identity the child develops. What do you think, and why?
- Do you think neuroscientists have made the case yet for a biological basis for homosexuality? Why or why not?

T   Testing Your Understanding

- Compare sex with other biological drives.
- Describe the processes that make a person male or female (limit your answer to typical development).
- Discuss sex as a continuum of gradations between male and female rather than a male versus female dichotomy. Give examples to illustrate.
- Identify any weak points in the evidence for a biological basis for homosexuality (ambiguous results, gaps in information, and so on) and indicate what research needs to be done to correct the weaknesses.

Select the best answer:

1. The chapter opened with the story of a boy born with female genitals. At puberty he grew a penis; developed muscles, a deep voice, and a beard; and became more masculine in behavior. The changes at puberty were _____ effects.
   a. activating
   b. organizing
   c. activating and organizing
   d. none of the above

2. Of the following, the best argument that sex is a drive like hunger and thirst is that
   a. almost everyone is interested in sex.
   b. sexual motivation is so strong.
   c. sexual behavior involves arousal and satiation.
   d. sexual interest varies from one time to another.

3. There is evidence that pheromones affect _____ in humans.
   a. menstrual cycles
   b. sexual attraction
   c. the VNO
   d. a and b
   e. a, b, and c

4. A likely result of the Coolidge effect is that an individual will
   a. be monogamous.
   b. have more sex partners.
   c. prolong a sexual encounter.
   d. prefer attractive mates.

5. The part of the sexual response cycle that most resembles homeostasis is
   a. excitement.   b. the plateau phase.
   c. orgasm.   d. resolution.

6. The increase in testosterone on nights that couples have intercourse is an example of
   a. an organizing effect.
   b. an activating effect.

c. cause.

d. effect.

7. The sex difference in the size of the sexually dimorphic nucleus is due to
   a. experience after birth.
   b. genes.
   c. sex hormone.
   d. both genes and experience.

8. The most prominent structure in the sexual behavior of female rats is the
   a. MPOA.
   b. medial amygdala.
   c. ventromedial nucleus.
   d. sexually dimorphic nucleus.

9. The chromosomal sex of a fetus is determined
   a. by the sperm.
   b. by the egg.
   c. by a combination of effects from the two.
   d. in an unpredictable manner.

10. The main point of the discussion of cognitive and behavioral differences between the sexes was to
    a. illustrate the importance of experience.
    b. make a case for masculinization and feminization of the brain.
    c. make the point that men and women are suited for different roles.
    d. explain why men usually are dominant over women.

11. The term that describes a person with XX chromosomes and masculine genitals is
    a. homosexual.
    b. male pseudohermaphrodite.

c. androgen-insensitive male.

d. female pseudohermaphrodite.

12. The best evidence that the *brains* of people with sexual anomalies have been masculinized or feminized contrary to their chromosomal sex is (are) their
    a. behavior and cognitive abilities.
    b. genital appearance.
    c. physical appearance
    d. adult hormone levels.

13. Testosterone injections in a gay man would most likely
    a. have no effect.
    b. increase his sexual activity.
    c. make him temporarily bisexual.
    d. reverse his sexual preference briefly.

14. Evidence that lesbianism is biologically influenced is
    a. that there is a high concordance among identical twins.
    b. that lesbians have a couple of distinctive physical features.
    c. that a brain structure is larger in lesbians and men than in heterosexual women.
    d. both a and b.
    e. both b and c.

Answers:
1. c, 2. c, 3. d, 4. b, 5. d, 6. d, 7. c, 8. c, 9. a, 10. b, 11. d, 12. a, 13. b, 14. d.

Online Resources

The following resources are available at www.sagepub.com/garrett3e.

Chapter Resources

- Flash Cards
- Chapter Quiz
- Internet Resources and Exercises

On the Web

Links to these websites can be found at the web address above: Select this chapter's **Chapter Resources**, then choose **Internet Resources and Exercises**.

1. **The Society for Women's Health Research** describes a number of sex differences in brain structure and cognitive function, with references.

2. The Johns Hopkins Children's Center's **Disorders of Sex Differentiation** provides information on a wide variety of anomalies of sexual development.

3. **Intersex** and **Intersex Initiatives** are two sites that provide a wealth of information about intersex conditions and treatment.

4. **Facts About Homosexuality and Mental Health** describes research and changing attitudes regarding the mental health of homosexuals. There are links to additional information.

   Asexual Visibility and Education Network (AVEN) provides information about asexuality and the opportunity to chat on a variety of related topics.

Chapter Updates and Biopsychology News

R For Further Reading

1. *Why Is Sex Fun? The Evolution of Human Sexuality* by Jared Diamond (Basic Books, 1997) takes an evolutionary approach to answer questions such as why humans have sex with no intention of procreating and why the human penis is proportionately larger than in other animals.

2. *The Sexual Brain* by Simon LeVay (MIT Press, 1993) is a well-written and technically informative coverage of topics including the evolution of sex, sexual development, and the origins of sexual orientation.

3. "His Brain, Her Brain" by Larry Cahill (*Scientific American,* May 2005, 40–47, or www.bio.uci .edu/public/press/2005/hisherbrain.pdf) is a nicely illustrated and popularly written tour of sex differences in brain and behavior.

4. *Your Sexual Mind,* a special issue of *Scientific American Mind,* contains several relevant articles. They include (with references for their original issues of publication, which might be more readily available): "The Orgasmic Mind" (brain activity during orgasm; Apr/May, 2008, 66–71); "Bisexual Species" (same-sex behavior in other animals; June/July, 2008, 68–73); "Sex and the Secret Nerve" (pheromone detection; Feb/Mar, 2007, 20–27); "Abnormal Attraction" (pedophilia; Feb/Mar, 2007, 58–63); "Misunderstood Crimes" (sex offenders; Apr/May, 2008, 78–79); and "Do Gays Have a Choice?" (gays who have adopted the straight life; Feb/Mar, 2006, 50–57).

5. In *Sexing the Body: Gender Politics and the Construction of Sexuality* by Anne Fausto-Sterling (Basic Books, 1999) and *Intersex in the Age of Ethics* by Alice Domurat Dreger (University Publishing Group, 1999), the authors argue for a more flexible view of sex and gender than our traditional either/or approach, including accepting gradations between male and female and allowing intersexed individuals to make their own gender selection.

6. *As Nature Made Him: The Boy Who Was Raised as a Girl* by John Colapinto (HarperCollins, 2000) tells the story of the boy whose penis was damaged during circumcision. Described by reviewers as "riveting," with a touching description of his suffering and of his parents' and brother's support of him.

7. "The Pheromone Myth: Sniffing Out the Truth" by Richard Doty (*New Scientist,* February 24, 2010, 28–29) is based on the author's book *The Great Pheromone Myth* (Johns Hopkins University Press, 2010), in which he dismisses pheromones as nothing more than learned odor preferences or, in some cases, the result of bad research.

K Key Terms

activating effects..200
amygdala...194
androgen insensitivity syndrome.............................205
androgens...192
anterior commissure...213
bisexual...209
castration...192
congenital adrenal hyperplasia (CAH)....................205
Coolidge effect...192
dihydrotestosterone..200
estrogen...193
estrus..193
gay..209

gender..198
gender identity..199
gender nonconformity..211
gender role...198
gonads...192
lesbian...209
major histocompatibility complex (MHC).................196
medial amygdala...194
medial preoptic area (MPOA).................................194
Müllerian ducts...199
Müllerian inhibiting hormone.................................200
organizing effects..200
ovaries...199

oxytocin 198
pheromones 196
pseudohermaphrodite 204
sex 198
sexually dimorphic nucleus (SDN) 194
suprachiasmatic nucleus (SCN) 213
testes 200

testosterone 192
third interstitial nucleus of the anterior
    hypothalamus (INAH3) 213
transsexual 214
ventromedial hypothalamus 194
vomeronasal organ (VNO) 197
Wolffian ducts 200

Emotion and Health

8

Emotion and the Nervous System

Autonomic and Muscular Involvement in Emotion

The Emotional Brain

The Prefrontal Cortex

APPLICATION: WHY I DON'T JUMP OUT OF AIRPLANES

The Amygdala

Hemispheric Specialization in Emotion

CONCEPT CHECK

Stress, Immunity, and Health

Stress as an Adaptive Response

Negative Effects of Stress

APPLICATION: ONE AFTERMATH OF
9/11 IS STRESS-RELATED BRAIN DAMAGE

Social and Personality Variables

Pain as an Adaptive Emotion

CONCEPT CHECK

Biological Origins of Aggression

Hormones and Aggression

The Brain's Role in Aggression

Serotonin and Aggression

Heredity and Environment

IN THE NEWS: AGGRESSION, GENES, AND THE LAW

CONCEPT CHECK

In Perspective

Summary

Study Resources

In this chapter you will learn

- How the brain and the rest of the body participate in emotion

- How stress affects health and immune functioning

- Why pain is an emotion as well as a sensation

- The role of hormones, brain structures, and heredity in aggression

My own brain is to me the most unaccountable of machinery—always buzzing, humming, soaring, roaring, diving, and then buried in mud. And why? What's this passion for?

—Virginia Woolf

1
Brain and Emotion Videos

When Jane was 15 months old, she was run over by a vehicle. The injuries seemed minor, and she appeared to recover fully within days of the accident. By the age of 3, however, her parents noticed that she was largely unresponsive to verbal or physical punishment. Her behavior became progressively disruptive, and by the age of 14 she had to be placed in the first of several treatment facilities. Although her intelligence was normal, she often failed to complete school assignments. She was verbally and physically abusive to others, she stole from her family and shoplifted frequently, and she engaged in early and risky sexual behavior that resulted in pregnancy at the age of 18. She showed little if any guilt or remorse; empathy was also absent, which made her dangerously insensitive to her infant's needs. Because her behavior put her at physical and financial risk, she became entirely dependent on her family and social agencies for financial support and management of her personal affairs.

MRI revealed that there was damage to Jane's prefrontal cortex, which is necessary for making judgments about behavior and its consequences. People who sustain damage to this area later in life show an understanding of moral and social rules in hypothetical situations, but they are unable to apply these rules in real-world situations, so they regularly make choices that lead to financial losses and the loss of friends and family relationships (Bechara, Damasio, Damasio, & Lee, 1999). People like Jane, whose injury occurred in infancy, cannot even verbalize these rules when confronted with a hypothetical situation, and their moral development never progresses beyond the motivation to avoid punishment; they not only make a mess of their own lives, but they also engage in behavior that harms others as well, like stealing (S. W. Anderson, Bechara, Damasio, Tranel, & Damasio, 1999).

Emotion enriches our lives with its "buzzing, humming, soaring, and roaring." It also motivates our behavior: Anger intensifies our defensive behavior, fear accelerates flight, and happiness encourages the behaviors that produce it. Emotion adds emphasis to experiences as they are processed in the brain, making them more memorable (A. K. Anderson & Phelps, 2001); as a result, we are likely to repeat the behaviors that bring joy and avoid the ones that produce danger or pain. Although Jane was intelligent, she was unable to learn from her emotional experiences because of her injury. According to Antonio Damasio (1994), reason without emotion is inadequate for making the decisions that guide our lives and, in fact, make up our lives.

Emotion and the Nervous System

If asked what *emotion* means, you would probably think first of what we call "feelings"—the sense of happiness or excitement or fear or sadness. Then you might think of the facial expressions that go along with these feelings: the curled-up corners of the mouth during a smile, the knit brow and red face of anger. Next you would probably visualize the person acting out the emotion by fleeing, striking, embracing, and so on. Emotion is all these and more; a working definition might be that *emotion* is an increase or decrease in physiological activity that is accompanied by feelings that are characteristic of the emotion and often accompanied by a characteristic behavior or facial expression. Having said that mouthful, I suspect you will understand why Joseph LeDoux (1996) wrote that we all know what emotion is until we attempt to define it. We will talk about these different facets of emotion in the following pages, along with some practical implications in the form of aggression and health.

Autonomic and Muscular Involvement in Emotion

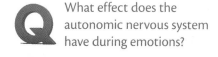

What effect does the autonomic nervous system have during emotions?

To the neuroscientist, the most obvious component of emotional response is sympathetic nervous system activation. You may remember from Chapter 3 that the sympathetic system activates the body during arousal; it increases heart rate and

FIGURE 8.1 Comparison of Sympathetic Activity During Emotional Arousal With Parasympathetic Activity During Relaxation.

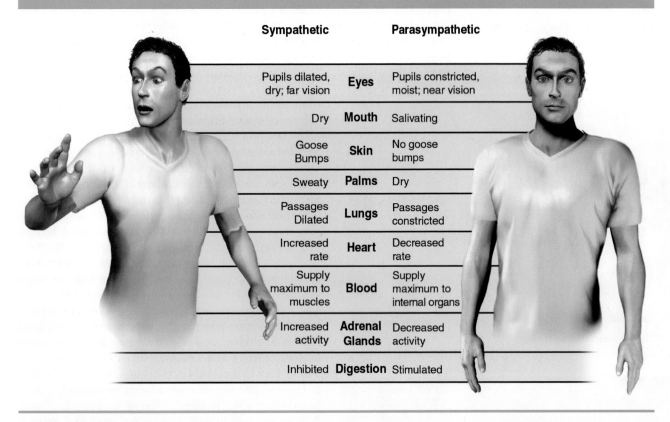

	Sympathetic		Parasympathetic
Eyes	Pupils dilated, dry; far vision		Pupils constricted, moist; near vision
Mouth	Dry		Salivating
Skin	Goose Bumps		No goose bumps
Palms	Sweaty		Dry
Lungs	Passages Dilated		Passages constricted
Heart	Increased rate		Decreased rate
Blood	Supply maximum to muscles		Supply maximum to internal organs
Adrenal Glands	Increased activity		Decreased activity
Digestion	Inhibited		Stimulated

respiration rate, increases sweat gland activity, shuts down digestion, and constricts the peripheral blood vessels, which raises the blood pressure and diverts blood to the muscles. As you will see in the section on stress, the sympathetic system also stimulates the adrenal glands to release various hormones, particularly *cortisol*. At the end of arousal, the parasympathetic system puts the brakes on most bodily activity, with the exception that it activates digestion. In other words, the sympathetic nervous system prepares the body for "fight or flight"; in contrast, the parasympathetic system generally reduces activity and conserves and restores energy (Figure 8.1).

Of course, muscular activation is involved in the external expression of emotion, such as smiling or fleeing or attacking; it is also a part of the less obvious responses of emotion, such as the bodily tension that not only prepares us to act but also produces a headache and aching muscles when we try to write a paper the night before it is due. Autonomic and muscular arousal are adaptive, because they prepare the body for an emergency and help it carry out an appropriate response. They are also a very important part of the emotion itself, though the exact nature of their contribution has been the subject of controversy. Fortunately, as you can see from the following discussion, competing theories are one of the engines driving research and scientific advancement.

The Role of Feedback From the Body

A bit over a century ago, the American psychologist William James (1893) and Danish physiologist Carl Lange independently proposed what has come to be known

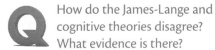

How do the James-Lange and cognitive theories disagree? What evidence is there?

> We feel sorry because we cry, angry because we strike, afraid because we tremble.
>
> —William James, 1893

as the *James-Lange theory:* Emotional experience results from the physiological arousal that precedes it, and different emotions are the result of different patterns of arousal. In our discussion of research ethics in Chapter 4, we talked about an experiment by Albert Ax (1953) in which subjects either were made angry by an insulting experimenter or were frightened by the possibility of a dangerous electric shock. Consistent with the James-Lange theory, the two emotions were accompanied by different patterns of physiological activity. Seventy years later, Stanley Schacter and Jerome Singer (1962) took a contrary position in their *cognitive theory;* they stated that the identity of the emotion is based on the cognitive assessment of the situation and physiological arousal contributes only to the emotion's intensity. Their research demonstrated how easily people could misidentify emotions depending on the environmental context. For example, young men who were interviewed by an attractive woman while crossing a swaying foot bridge 230 feet above a rocky river included more sexual content in brief stories they wrote and were more likely to call the phone number the young woman gave them than were men who were interviewed 10 minutes after crossing the bridge (Dutton & Aron, 1974).

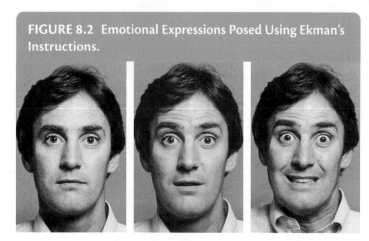

FIGURE 8.2 Emotional Expressions Posed Using Ekman's Instructions.

SOURCE: © Don Francis/Mardan Photography.

Studies like these have not determined that one theory is right and the other is wrong, but they have produced valuable insights into emotion; a good example is the contribution of facial expressions to emotional experience. Experimenters have had to be very inventive in doing facial expression studies; obviously, they can't just tell a person to smile or frown and then ask them what emotion they're feeling. Paul Ekman and his colleagues (Levenson, Ekman, & Friesen, 1990) instructed subjects to contract specific facial muscles to produce different expressions (Figure 8.2); for example, to produce an angry expression, subjects were told to pull their eyebrows down and together, raise the upper eyelid, and push the lower lip up with the lips pressed together (p. 365). The posed facial expressions for happiness, fear, anger, disgust, sadness, and surprise each resulted in the experience of the intended emotion, along with a distinct pattern of physiological arousal.

Induced facial poses also influence how the person interprets the environment. Volunteers rate a stimulus as more painful when they are making a sad face than a happy or neutral one (Salomons, Coan, Hunt, Backonja, & Davidson, 2008), and college students rate *Far Side* cartoons as more amusing when they are holding a pen between the teeth, which induces a sort of smile, than when they hold the pen between the lips, producing a frown (Strack, Martin, & Stepper, 1988). More strikingly, women who have had their corrugator muscles paralyzed by botulinum toxin (botox) treatment are unable to frown, and they report less negative mood, whether or not they perceive themselves as more attractive (M. B. Lewis & Bowler, 2009). In addition, an fMRI study showed that when these women attempt to imitate angry expressions, they produce less activation of the amygdala than women who have not had botox treatment (Figure 8.3; Hennenlotter et al., 2009).

FIGURE 8.3 Disabling Corrugator Muscle Reduces Amygdala Response to Simulated Anger.

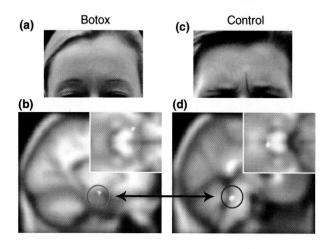

A woman treated with botox (a) is unable to produce the facial expression of anger and shows little activation of the amygdala (b). A control subject makes the expected facial expression (c) and produces much greater amygdala activation (d).

SOURCE: Adapted from "The Link Between Facial Feedback and Neural Activity Within Central Circuitries of Emotion—New Insights from Botulinum Toxin-Induced Denervation of Frown Muscles," by A. Hennenlotter, C. Dresel, F. Castrop, A. O. Ceballos Baumann, A. M. Woschläger, and B. Haslinger, 2009, *Cerebral Cortex, 19,* 537–542. Used with permission of Oxford University Press.

Some researchers suggest that feedback from emotional expressions has another role besides contributing to our emotional experience in that it also helps us understand other people's emotions; this ability to recognize others' emotions is critical to social communication and to societal success. To appreciate this point of view, you need to understand the concept of mirror neurons. *Mirror neurons* are neurons that respond both when we engage in a specific act and while observing the same act in others. They were first discovered when researchers noticed that neurons that were active while monkeys reached for food also responded when the monkeys saw the researcher picking up a piece of food (di Pellegrino, Fadiga, Fogassi, Gallese, & Rizzolatti, 1992); similar correlations have been observed in several brain areas, including those involved in emotions (see review in Bastiaansen, Thioux, & Keysers, 2009). Observation of others' emotions activate our brains' emotional areas, and the amount of activity is related to scores on a measure of empathy (Chakrabarti, Bullmore, & Baron-Cohen, 2006).

Our observation of other people's emotions is not entirely passive; we also mimic their facial expressions and their gestures, body posture, and tone of voice (Bastiaansen et al., 2009). Just as feedback during our own emotional activity adds to our own emotional experience, feedback from imitated expressions may help us empathize with the emotions of others. Indeed, interfering with the muscles required for mimicking facial expressions impairs subjects' ability to recognize happiness and disgust in photos (Oberman, Winkielman, & Ramachandran, 2007). One of the characteristics of autism is difficulty in understanding other people's emotions; autistic children imitate emotional expressions, but their mimicry is delayed by about 160 milliseconds (Oberman, Winkielman, & Ramachandran, 2009).

A system this complex requires an equally complex control system. We will turn our attention now to the brain structures responsible for emotion.

The Emotional Brain

In the late 1930s and 1940s, researchers proposed that emotions originated in the *limbic system,* a network of structures arranged around the upper brain stem (Figure 8.4). As complex as this system is with its looping interconnections, we now know that it is an oversimplification; emotion involves structures at all levels of the brain, from the prefrontal area to the brain stem (A. R. Damasio et al., 2000). Also, we know that some of the limbic structures are more involved in nonemotional functions. For example, the hippocampus and mammillary bodies have major roles in learning. The concept of a limbic system is less important as a description of

 What are some of the brain structures involved in emotions, and what are their functions?

how emotion works than for spawning a tremendous volume of research that has taken us in diverse directions, which we will explore over the next several pages.

Much of what we know about the brain's role in emotion comes from lesioning and stimulation studies with animals; this research is limited because we do not know what the animal is experiencing. Robert Heath did some of the earliest probing of the limbic system in humans in 1964 when he implanted electrodes in the brains of patients in an attempt to treat epilepsy, sleep disorders, or pain that had failed to respond to conventional treatments. Researchers knew from animal studies that the hypothalamus has primary control over the autonomic system and produces a variety of emotional expressions, such as the threatened cat's hissing and bared

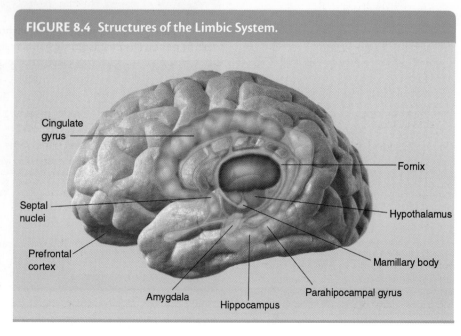

FIGURE 8.4 Structures of the Limbic System.

FIGURE 8.5 Location of the Amygdala, Insula, and Basal Ganglia.

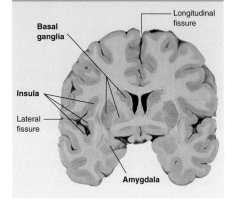

SOURCE: Photo courtesy of Dana Copeland.

teeth and claws. Stimulation of the hypothalamus in Heath's patients produced general autonomic discharge and sensations such as a pounding heart and feelings of warmth, but it also evoked feelings of fear, rage, or pleasure, depending on the location of the electrode in the hypothalamus. Septal area stimulation also produced a sense of pleasure, but in this case the feeling was accompanied by sexual fantasies and arousal. During septal stimulation one patient went from near tears while talking about his father's illness to a broad smile as he described how he planned to take his girlfriend out and seduce her. When asked why he changed the subject, he replied that the thought just came into his head.

Now researchers are more likely to use one of the scanning techniques to study the brain centers of emotion. Typically, they do magnetic resonance imaging scans to determine the location of damage in patients with emotional deficits, or they use positron emission tomography (PET) or functional magnetic resonance imaging while healthy subjects relive an emotional experience, examine facial expressions of emotion, or view an emotional video. Two of the most reliable brain–emotion associations have been the amygdala's role in fear and the location of disgust in the insular cortex and the basal ganglia (Figure 8.5; F. C. Murphy, Nimmo-Smith, & Lawrence, 2003; Phan, Wager, Taylor, & Liberzon, 2002). We will consider the amygdala in some detail later. The insula is the area we identified in Chapter 6 as the cortical projection site for taste; a number of writers have remarked on the fact that taste and disgust share the same brain area and that *dis-gust* means, roughly, *bad taste.* In Chapter 3, we identified the basal ganglia as being involved in motor functions. Interestingly, people with Huntington's disease or obsessive compulsive disorder, both of which involve abnormalities in the basal ganglia, have trouble recognizing facial expressions of disgust (Phan et al., 2002).

Another important structure in emotion is the *anterior cingulate cortex*, a part of the cingulate gyrus important in attention, cognitive processing, emotion, and possibly consciousness. You can see the cingulate gyrus in Figure 8.4 and the anterior cingulate gyrus in Figure 8.6. The anterior cingulate cortex is believed to combine emotional, attentional, and bodily information to bring about conscious emotional experience (Dalgleish, 2004). Consequently, it is involved in emotional activity regardless of which emotion is being experienced, although some studies have linked parts

FIGURE 8.6 Size Differences in the Anterior Cingulate Gyrus.

A larger anterior cingulate gyrus (highlighted in red) is associated with a higher level of the personality characteristic harm avoidance.

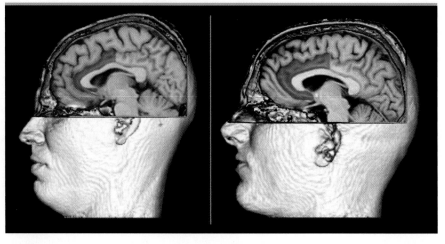

SOURCE: From "Anatomical Variability of the Anterior Cingulated Gyrus and Basic Dimensions of Human Personality," by J. Pujl et al., *NeuroImage, 15,* 847–855, fig. 1, p. 848. © 2002 with permission from Elsevier, Ltd.

FIGURE 8.7 Brain Areas Activated During Different Emotions.

Each colored square represents the location identified in a single study for a particular emotion.

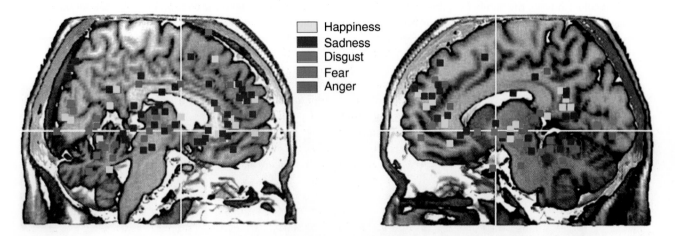

- Happiness
- Sadness
- Disgust
- Fear
- Anger

SOURCE: Reprinted from "Functional Neuroanatomy of Emotion: A Meta-Analysis of Emotion Activation Studies in PET and fMRI," *NeuroImage, 16,* by K. L. Phan, T. Wager, S. F. Taylor, and I. Liberzon, pp. 331–348. Copyright 2002, with permission from Elsevier.

of the structure to specific emotions, such as sadness and happiness (F. C. Murphy et al., 2003; Phan et al., 2002). Interestingly, an MRI investigation found that the right anterior cingulate was larger in people with high scores on *harm avoidance,* which involves worry about possible problems, fearfulness in the face of uncertainty, and shyness with strangers (Pujol et al., 2002).

Before we go too far in assigning emotions to specific brain structures, we need to understand that any emotion involves activity in many brain areas. This is well illustrated in a study that combined the results of 55 PET and fMRI investigations (Phan et al., 2002). As you can see in Figure 8.7, places activated during a specific emotion cluster somewhat in particular areas, but they are also scattered across wide areas of the brain. This is partly due to different methods of inducing the emotions in the studies, but it also reflects the complexity of emotion. With the understanding that no emotion can be relegated to a single part of the brain, we will look more closely at three areas that have particularly important roles in emotional experience and behavior: the prefrontal cortex, the amygdala, and the right hemisphere.

The Prefrontal Cortex

The prefrontal cortex (see Figure 8.4 again) is the final destination for much of the brain's information about emotion before action is taken. You saw in Chapter 3 that damage to the prefrontal area or severing its connections with the rest of the brain impairs people's ability to make rational judgments. Later in this chapter you will learn that people with deficiencies in the area are unable to restrain violent urges, and abnormalities in the prefrontal area also figure prominently in depression and schizophrenia. These deficits have a variety of causes, including injury, infection, tumors, strokes, and developmental errors. What the victims have in common is damage to the prefrontal area that includes the ventromedial cortex and the orbitofrontal cortex (see Figure 8.8).

Antoine Bechara and his colleagues took a group of patients with ventromedial damage into the laboratory to test their responses on the *gambling task* (Bechara et al., 1999). In this task, the individual chooses cards from four decks. Two "risky"

FIGURE 8.8 Location of Damage That Impairs Emotion-Based Decision Making.

In (a) the location of the ventromedial cortex and the orbitofrontal cortex is shown. In (b) you can see where damage occurred in four patients who showed judgment problems. The horizontal line shows where the scan in (c) was taken. In (b) and (c), the different colors indicate the number of patients with damage in the area, according to the code on the color bar. All shared damage in the ventromedial cortex, but some had damage in the orbitofrontal cortex as well.

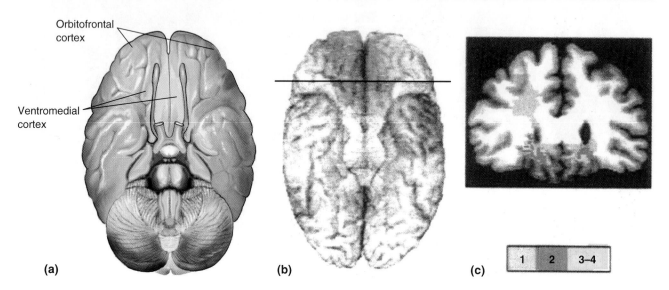

Orbitofrontal cortex

Ventromedial cortex

| 1 | 2 | 3–4 |

(a)   (b)   (c)

SOURCE: (b) and (c) from "Different Contributions of the Human Amygdala and Ventromedial Prefrontal Cortex to Decision-Making," by A. Bechara, H. Damasio, A. R. Damasio, & G. P. Lee, 1999, *Journal of Neuroscience, 19,* 5473–5481.

decks contain cards that result in high rewards of play money, along with a few cards that carry a high penalty, for an overall loss; cards in the other two "safe" decks result in lower rewards and smaller penalties for an overall gain. Initially both patients and volunteers without ventromedial damage (controls) prefer the risky decks. As they encounter more penalties, the controls gradually shift their preference to the more advantageous safe decks; the patients usually do not make the shift, even after they have figured out how the game works (Figure 8.9a).

The reason the patients failed to make good choices appears to be that they did not respond emotionally to their bad choices. As an indicator of emotion, the researchers used the *skin conductance response (SCR),* which is a measure of sweat gland activation and, thus, of sympathetic nervous system activity. The technique involves passing a very small electrical current through the individual's skin; during arousal, the skin conducts electricity more readily. Over the course of the game, the control subjects began to show anticipatory SCRs to the risky decks; that is, they had an increase in skin conductance just before drawing a card from a risky deck. In fact, they started making these emotional responses before they were able to verbalize that those stacks were risky. (This is the example of unaware emotional influence I promised earlier.) However, the patients did not make different SCRs to the four decks (Figure 8.9b). The patients were not impaired in either learning or emotional capability; like the control subjects, they produced skin conductance increases to a loud sound and to a neutral stimulus that had been paired with the loud sound. However, they were apparently unable to process the consequences of risky behavior.

In the rest of us, who like to think of ourselves as normal, the prefrontal cortex's connections to other parts of the brain determine whether we are novelty-seeking adventurers or are more restrained, as the Application explains.

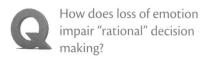

How does loss of emotion impair "rational" decision making?

FIGURE 8.9 Comparison of Gambling Task Behavior in Controls and Patients With Damage to Prefrontal Cortex.

(a) The controls shifted from preferring cards from the risky decks to preferring cards from the safe decks, but the patients did not. (b) Also, as the task progressed, only the controls showed anticipatory skin conductance responses (SCR) before choosing from the risky decks.

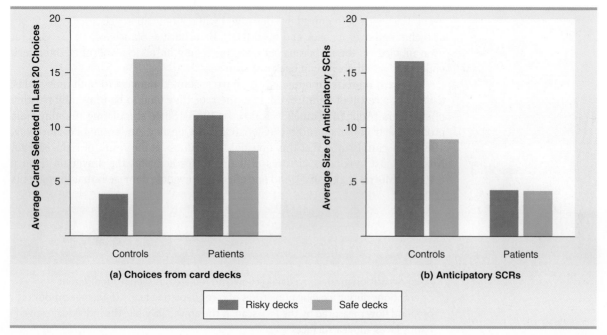

Application — Why I Don't Jump Out of Airplanes

In our discussion of sensation seeking in Chapter 6, I mentioned my wife's recent skydiving adventure. Why does she enjoy leaping out of a perfectly good airplane at 14,000 feet, while I prefer to stay on the ground and take photos of her descent? Why does she like to travel to exotic places, like Thailand and Patagonia, while I would rather stay at home and write about biopsychology? Once again, research comes to the rescue. Psychologists at the University of Arizona (M. X. Cohen, Schoene-Bake, Elger, & Weber, 2009) used a questionnaire to categorize 20 volunteers as novelty seekers (agreeing to statements like "I like to try new things just for fun.") or reward dependent (agreeing to statements like "I'd rather stay home than go out."). Novelty seekers are higher in exploratory drive and impulsivity, while those who are reward dependent are particularly sensitive to rewards and spend a lot of time pursuing activities that have been rewarding in the past.

Next, the researchers scanned the volunteers' brains using a technique called diffusion tractography, an application of MRI useful for measuring white-matter tracts. Their analysis focused on the hippocampus, amygdala, and striatum, because animal research has indicated that these structures form a looping circuit that is important in novelty seeking. There were strong white-matter connections within this loop in the human novelty seekers, but the reward-dependent volunteers' strongest connections were between the striatum and prefrontal areas. Just before my wife's sky-diving experience, the jump videographer interviewed her about why she was making the jump, then turned the camera on me and asked why I chose not to. I answered lamely, "I just don't need to." Today, armed with the results of this study, my answer would be, "My prefrontal cortex won't let me!"

The Amygdala

The prefrontal areas receive much of their emotional input from the amygdala (see Figure 8.4 again), a small limbic system structure in each temporal lobe that is involved in emotions, especially negative ones. The amygdala has other functions as well. In Chapter 12, we will see that it participates in memory formation, especially when emotion is involved. It also responds to pictures of happy faces and the recall of pleasant information and, as we saw in Chapter 7, sexually exciting stimuli. So the amygdala's role may involve responding to emotionally significant stimuli in general (Phan et al., 2002).

Although the amygdala is involved in other emotions, its role in fear and anxiety has been most thoroughly researched. *Fear* is an emotional reaction to a specific immediate threat; *anxiety* is an apprehension about a future, and often uncertain, event. Rats with both amygdalas destroyed will not only approach a sedated cat but also climb all over its back and head (D. C. Blanchard & Blanchard, 1972). One rat even nibbled on the stuporous cat's ear, provoking an attack—and after the attack ended, the rat climbed right back onto the cat.

We learn more from humans who have sustained damage to both amygdalas, usually as a result of an infection. Like the rats, they tend to be unusually trusting of strangers (Adolphs, Tranel, & Damasio, 1998). Since stimulating the amygdala produces fear in human subjects (Gloor, Olivier, Quesney, Andermann, & Horowitz, 1982), we can assume that this unusual trustfulness is the result of reduced fear. Normally, just looking at pictures of fearful faces activates the amygdala (Figure 8.10; J. S. Morris et al., 1996), and one effect of amygdala damage is that the patients

FIGURE 8.10 Activation of the Amygdala While Viewing Fearful Faces.

(a–c) Activation in the amygdala (yellow and red area) is seen from different orientations during viewing of pictures of faces depicting fear. (d) Average amounts of blood flow measured by the PET scan in the amygdala while the individual viewed happy faces and fearful faces.

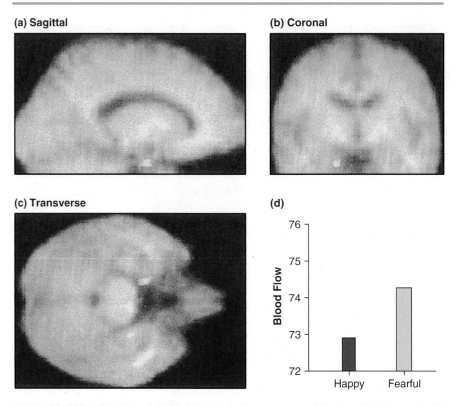

SOURCE: (a) - (c) from "A Differential Neural Response in the Human Amygdala to Fearful and Happy Facial Expressions," by J. S. Morris et al., 1996, *Nature, 383*, pp. 812–815. (d) based on data from Morris et al., 1996.

have trouble recognizing fear in other people's facial expressions (Adolphs, Tranel, Damasio, & Damasio, 1995). Not surprisingly, the amygdala is one of the sites where antianxiety drugs produce their effect. It contains receptors for benzodiazepines and opiates, and injection of either type of drug directly into the amygdala reduces signs of fear and anxiety in animals (M. Davis, 1992).

Bechara's study of prefrontal patients also included a group with damage to both amygdalas (Bechara et al., 1999). In most ways, they were like the prefrontal subjects: They were unable to shift their card selections to the safe decks, and they produced no anticipatory SCRs to their risky choices though they, too, reacted to the loud sound. The major difference was that both the normal controls and the prefrontal patients responded to monetary gains and losses with conductance increases, but the amygdale patients did not. Apparently, prefrontal patients are unable to make use of information about their emotional response to rewards and punishment, while people with damage to both amygdalas do not produce the emotional response in the first place. These abilities are needed to function successfully in a world that requires us to seek rewards and avoid punishments and to distinguish the situations in which they occur. For this reason, most people with bilateral amygdale damage have to live in supervised settings (Bechara et al., 1999).

Hemispheric Specialization in Emotion

The specialization of the cerebral hemispheres we have seen in other functions is also evident in emotion. Although both hemispheres are involved in the *experience* of emotions, the left frontal area is more active when the person is experiencing positive emotions, and the right frontal area is more active during negative emotion (R. J. Davidson, 1992). People with damage to the left hemisphere often express more anxiety and sadness about their situation, while those with right-hemisphere damage are more likely to be unperturbed or even euphoric, even when their arm or leg is paralyzed (Gainotti, 1972; Gainotti, Caltagirone, & Zoccolotti, 1993; Heller, Miller, & Nitschke, 1998). The same difference in emotions occurs when each of the cerebral hemispheres is anesthetized briefly in turn by injecting a short-acting barbiturate into the right or left carotid artery (Rossi & Rosadini, 1967). (This technique is sometimes used in evaluating patients prior to brain surgery.) In fact, when the right hemisphere is anesthetized, individuals can describe negative events in their lives but can barely recall having felt sad or angry or fearful, even with incidents as intense as their mother's death the discovery of a spouse's affair, or the wife's threatening to kill the individual (E. D. Ross, Homan, & Buck, 1994).

While both hemispheres are involved in experiencing emotion, the right is more specialized for its expression (Heller et al., 1998). Autonomic responses to emotional stimuli such as facial expressions and emotional scenes are greater when the stimuli are presented to the right hemisphere (using the strategy described on page 298; Spence, Shapiro, & Zaidel, 1996). Much of the emotional suppression in right-hemisphere-damaged patients is due to decreased autonomic response (Gainotti et al., 1993).

Perception of nonverbal aspects of emotion is impaired in right-hemisphere-damaged patients; for example, they often have difficulty recognizing emotion in others' facial expressions (Adolphs, Damasio, Tranel, & Damasio, 1996). Verbal aspects are unimpaired, however; the same patients can understand the emotion in a verbal description like "Your team's ball went through the hoop with one second left to go in the game," but they have trouble identifying the emotion in descriptions of facial or gestural expressions such as "Tears fell from her eyes," or "He shook his fist" (Blonder, Bowers, & Heilman, 1991). Right-hemisphere-damaged patients also have trouble recognizing emotion from the tone of the speaker's voice (Gorelick & Ross, 1987), and their own speech is usually emotionless as well (Heilman, Watson, & Bowers, 1983). When asked to say a neutral sentence like "The boy went to the store," in a happy, sad, or angry tone, they speak instead in a monotone and often add the designated emotion to the sentence verbally, for example, "... and he was sad."

Concept Check

Take a Minute to Check Your Knowledge and Understanding

☐ Describe the role of the autonomic nervous system in emotion (including the possible identification of emotions).

☐ Organize your knowledge: List the major parts of the brain described in this section that are involved in emotion, along with their functions.

☐ How are the effects of prefrontal and amygdala damage alike and different?

Stress, Immunity, and Health

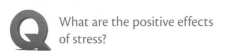

Q What are the positive effects of stress?

Stress is a rather ambiguous term that has two meanings in psychology. *Stress* is a condition in the environment that makes unusual demands on the organism, such as threat, failure, or bereavement. Stress is also an internal condition, your response to a stressful situation; you *feel* stressed, and your body reacts in several ways. Whether a situation is stressful to the person is often a matter of individual differences, either in perception of the situation or in physiological reactivity. For some, even the normal events of daily life are stressful, while others thrive on excitement and would feel stressed if they were deprived of regular challenges. In other words, stress in this sense of the term is in the eye of the beholder.

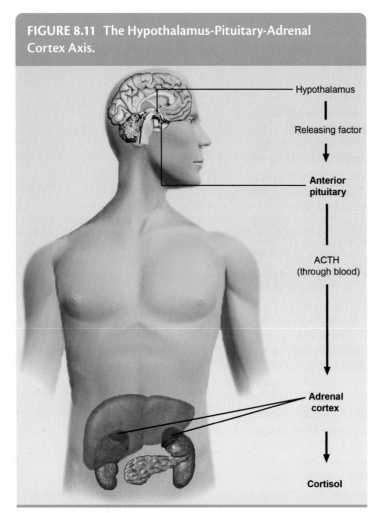

FIGURE 8.11 The Hypothalamus-Pituitary-Adrenal Cortex Axis.

Hypothalamus

Releasing factor

Anterior pituitary

ACTH
(through blood)

Adrenal cortex

Cortisol

Stress as an Adaptive Response

Ordinarily, the body's response to a stressful situation is positive and adaptive. In Chapter 3, you saw that the stress response includes activation of the sympathetic branch of the autonomic nervous system, which is largely under the control of the hypothalamus. The resulting increases in heart rate, blood flow, and respiration rate help the person deal with the stressful situation. Stress also activates the *hypothalamus-pituitary-adrenal axis,* a group of structures that help the body cope with stress. The hypothalamus activates the pituitary gland, which in turn releases hormones that stimulate the adrenal glands to release the stress hormones epinephrine, norepinephrine, and cortisol. The first two hormones increase output from the heart and liberate glucose from the muscles for additional energy. The hormone *cortisol* also increases energy levels by converting proteins to glucose, increasing fat availability, and increasing metabolism. Cortisol provides a more sustained release of energy than the sympathetic nervous system does, for coping with prolonged stress. The hypothalamuspituitary-adrenal axis is illustrated in Figure 8.11.

Brief stress increases activity in the *immune system* (Herbert et al., 1994), the cells and cell products that kill infected and malignant cells and protect the body against foreign substances, including bacteria and viruses. Of course, this is highly adaptive because it helps protect the person from any infections that might result from the threatening situation. The immune response involves two major types of cells. *Leukocytes,* or white blood cells, recognize invaders by the unique proteins that every cell has on its surface and kills them. These proteins in foreign cells are called *antigens.* A type of leukocyte called a *macrophage* ingests intruders (Figure 8.12). Then it displays the intruder's antigens on its own cell surface; this attracts *T cells,* another type of leukocyte that is specific for particular antigens, which kill the invaders. *B cells,* a third type of leukocyte, fight intruders by producing antibodies that attack a particular cell type. *Natural killer cells,* the second type of immune cells, attack and destroy certain kinds of cancer cells and cells infected with viruses; they are less specific in their targets than T or B cells. Table 8.1 summarizes the characteristics of these immune cells.

FIGURE 8.12 Macrophages Preparing to Engulf Bacteria.

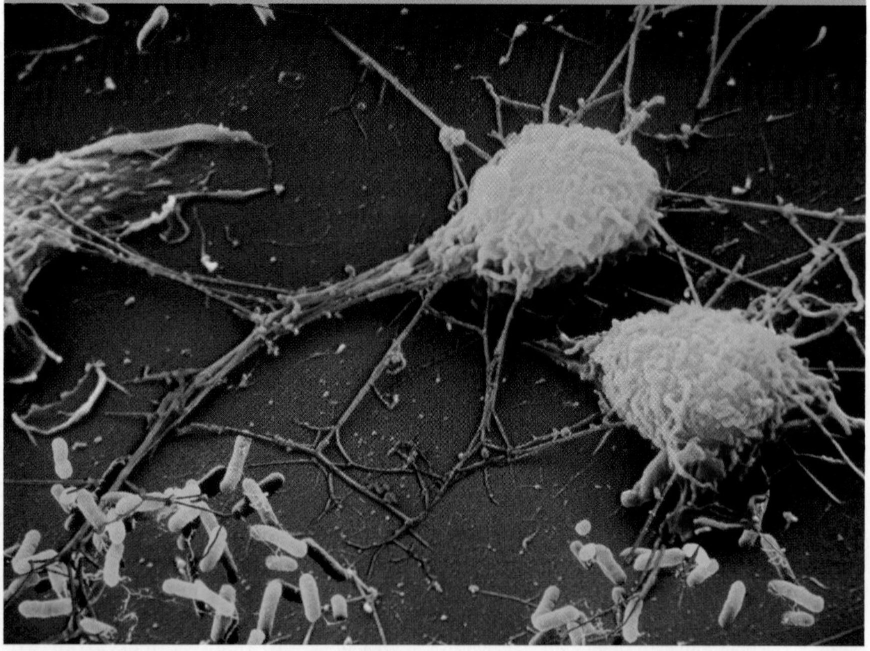

SOURCE: © Manfred Kage/Peter Arnold, Inc.

Some antibodies are transferred from mother to child during the prenatal period or postnatally through the mother's milk. Most antibodies, though, result from a direct encounter with invading cells, for example, during exposure to measles. Vaccinations work because injection of a weakened form of the disease-causing bacteria or virus triggers the B cells to make antibodies for that disease.

The preceding is a description of what happens when all goes well. In the immune deficiency disease AIDS (acquired immune deficiency syndrome), on the other hand, T cells fail to detect invaders, and the person dies of an infectious disease. In *autoimmune disorders,* the immune system goes amok and attacks the body's own cells. In the autoimmune disorder multiple sclerosis, for instance, the immune system destroys myelin in the central nervous system.

Negative Effects of Stress

We are better equipped to deal with brief stress than with prolonged stress. Chronic stress can interfere with memory, increase or decrease appetite, diminish sexual desire and performance, deplete energy, and cause mood disruptions. Although brief stress enhances immune activity, prolonged stress compromises the immune system. After the nuclear accident at the Three Mile Island electric generating plant, nearby residents had elevated stress symptoms and performed less well on tasks requiring concentration than people who lived outside the area (Baum, Gatchel,

3
Stress Less

TABLE 8.1 Major Types of Immune Cells.			
Leukocytes			**Natural Killer Cells**
Macrophages	**T Cells**	**B Cells**	
Ingest invaders, display antigens, which attract T cells.	Multiply and attack invaders.	Make antigens, which destroy intruders.	Attack cells containing viruses, certain kinds of tumor cells.

FIGURE 8.13 Relationship Between Stress and Vulnerability to Colds.

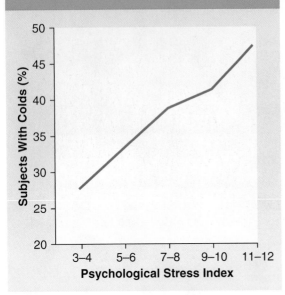

> *In [emotional] pain there is as much wisdom as in pleasure.*
>
> —*Friedrich Nietzsche*

FIGURE 8.14 Increase in Cardiac Deaths on the Day of an Earthquake.

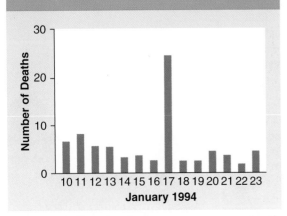

& Schaeffer, 1983). Amid concerns about continued radioactivity and the long-term effects of the initial exposure, they had reduced numbers of B cells, T cells, and natural killer cells as long as 6 years after the accident (McKinnon, Weisse, Reynolds, Bowles, & Baum, 1989).

Disease symptoms were not measured at Three Mile Island, but other studies have shown that health is compromised when stress impairs immune functioning. Recently widowed women experienced decreased immunity and marked health deterioration in the year following the spouse's death (Maddison & Viola, 1968). Also, students had reduced immune responses, more infectious illnesses, and slower wound healing at exam times than other times of the year (Glaser et al., 1987; Marucha, Kiecolt-Glaser, & Favagehi, 1998). In a rare experimental study, healthy individuals were given nasal drops containing common cold viruses and then were quarantined and observed for infections. In Figure 8.13, you can see that their chance of catching a cold depended on the level of stress they reported on a questionnaire at the beginning of the study (S. Cohen, Tyrrell, & Smith, 1991). In a follow-up study, it turned out that only stresses that had lasted longer than a month increased the risk of infection (S. Cohen et al., 1998).

The cardiovascular system is particularly vulnerable to stress. Stress increases blood pressure, and prolonged high blood pressure can damage the heart or cause a stroke. Some people are more vulnerable to health effects from stress than others. Researchers classified young children as normal reactors or excessive reactors based on their blood pressure increases while one hand was immersed in ice water. Forty-five years later, 71% of the excessive reactors had high blood pressure, compared with 19% of the normal reactors (Wood, Sheps, Elveback, & Schirger, 1984).

Stress can even produce death. This fact has not always been accepted in the scientific and medical communities, but in 1942 the physiologist Walter Cannon determined that reports of apparent stress-related deaths were legitimate. He even suggested that *voodoo death,* which has been reported to occur within hours of a person being "hexed" by a practitioner of this folk cult, is also due to stress. We now know that fear, loss of a loved one, humiliation, or even extreme joy can result in sudden cardiac death. In *sudden cardiac death,* stress causes excessive sympathetic activity that sends the heart into fibrillation, contracting so rapidly that it pumps little or no blood. When one of the largest earthquakes ever recorded in a major North American city struck the Los Angeles area in 1994, the number of deaths from heart attacks increased fivefold (Figure 8.14; Leor, Poole, & Kloner, 1996). The stress doesn't have to be as extreme as an earthquake: During the 2006 World Soccer Cup games in Germany, cardiac emergencies tripled among men in that country and almost doubled in women (Wilbert-Lampen et al., 2008); heart attacks even increase during the first few days following the autumn change to daylight saving time, as people cope with minor sleep deprivation and an earlier wake time (T. S. Janszky & Ljung, 2008).

Extreme stress can also lead to brain damage (Figure 8.15). Hippocampal volume was reduced in Vietnam combat veterans suffering from posttraumatic stress disorder (Bremner et al., 1995) and in victims of childhood abuse (Bremner et al., 1997), and cortical tissue was reduced in torture victims (T. S. Jensen et al., 1982). The abused individuals had short-term memory deficits, and some of the torture victims showed slight intellectual impairment. There is some evidence that the damage is caused by cortisol; implanting cortisol pellets in monkeys' brains damaged their hippocampi (Sapolsky, Uno, Rebert, & Finch, 1990), and elderly

FIGURE 8.15 Hippocampal Damage in a Stressed Monkey.

Compare the number of cells between the arrows in the hippocampus of a control monkey (a) and the number in the same area in a monkey that died spontaneously of apparent stress (b).

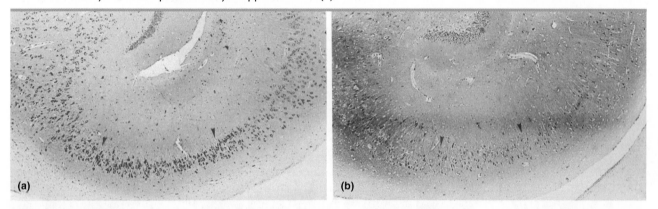

(a)   (b)

SOURCE: From "Hippocampal Damage Associated With Prolonged and Fatal Stress in Primates," by H. Uno, R. Tarara, J. G. Else, M. A. Suleman, and R. M. Sapolsky, 1999, *Journal of Neuroscience, 9,* 1705–1711.

humans who had elevated cortisol levels over a 5-year period had an average 14% decrease in hippocampal volume (Lupien et al., 1998). However, individuals with posttraumatic stress disorder have *lowered* cortisol levels. Yehuda (2001) points out that they also have an increased number and sensitivity of the glucocorticoid receptors that respond to cortisol. She suggests that posttraumatic stress disorder involves increased sensitivity to cortisol rather than an increase in cortisol level; although there is a compensatory decrease in cortisol release, it is not adequate to protect the hippocampus. Researchers are learning that stress that is inadequate to produce posttraumatic stress disorder can also damage the brain, as the Application shows.

4
Stress and Health

Several studies suggest that reducing stress can improve health. T cell counts increased in AIDS patients after 20 hours of relaxation training (Taylor, 1995); similar training was associated with reduced death rates in elderly individuals (C. N. Alexander, Langer, Newman, Chandler, & Davies, 1989) and in cancer patients (Fawzy et al., 1993; Spiegel, 1996). However, evidence that survival rate in these studies is related to immune function improvement is sketchy (Fawzy et al., 1993); it is possible that participation in these studies led the elderly subjects and cancer patients to make lifestyle changes. At any rate, it may be more practical to block stress hormones and bolster immunity chemically. Researchers at Tel Aviv University have found that the psychological stress of cancer surgery suppresses immunity, allowing the spread of cancer during the postoperative period; they have shown with animals that this effect can be reduced with drugs and are starting therapeutic trials with human patients (K. Greenfeld et al., 2007; Melamed et al., 2005).

Social and Personality Variables

Whether stress has a negative impact on health depends on a variety of factors, including social support, personality, and attitude. Social support was associated with dramatically lower death rates in several different populations (reviewed in House, Landis, & Umberson, 1988) and with lower stress and reduced stress hormone level among Three Mile Island residents (Fleming, Baum, Gisriel, & Gatchel, 1982). People who are hostile are at greater risk for heart disease (T. Q. Miller, Smith, Turner, Guijarro, & Hallet, 1996), while cancer patients who have a "fighting spirit" may live longer than patients who accept their illness or have an attitude

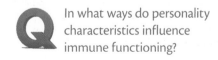

In what ways do personality characteristics influence immune functioning?

Application — One Aftermath of 9/11 Is Stress-Related Brain Damage

On September 11, 2001, terrorists flew two airliners into the World Trade Center, taking the lives of over 2,500 people. This was an obviously traumatic event for the people of the United States, and particularly for the victims' relatives and for the people living in the vicinity of the twin towers. The stress was still measurable in residents 3 years later, which led a group of researchers to ask just how serious the continuing trauma was (Ganzel, Kim, Glover, & Temple, 2008). They used fMRI scans to compare the brains of a group of volunteers living within 1.5 miles (2.4 kilometers) of the towers with those of people living 200 miles (322 kilometers) away. The near residents exhibited some signs of posttraumatic stress disorder, but the symptoms were not severe enough for a diagnosis. Yet, they had reduced gray matter volume in the hippocampus, amygdala, prefrontal cortex, anterior cingulate cortex, and insula; also, amygdala activation was greater when they viewed facial expressions of fear. The figure shows the location of these deficits. The same areas also lose gray matter with age, and the researchers suggest that much of the effect of aging on the brain is due to the lifelong accumulation of stress.

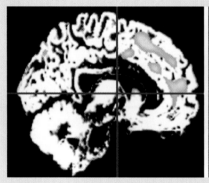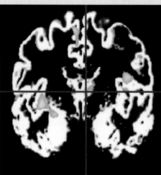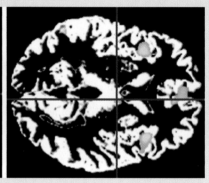

SOURCE: "Resilience After 9/11: Multimodal Neuroimaging Evidence for Stress-Related Change in the Healthy Adult Brain" by B. L. Ganzel, P. Kim, G. H. Glover, and E. Temple, 2008, *NeuroImage, 40*, 788–795.

of hopelessness (Derogatis, Abeloff, & Melisaratos, 1979; Greer, 1991; Temoshok, 1987). Because of the high association between mood disorder and cancer, some have suggested that depression is a predisposing factor; however, the opposite is more likely, because animal research indicates that immune system cytokines released by tumors can produce depressive-like behaviors (Pyter, Pineros, Galang, McClintock, & Prendergast, 2009).

Social and personality influences must work through physiological mechanisms, which, unfortunately, are seldom assessed in these studies. An exception is an investigation of individual differences in immune response. Recall that there is a greater association of positive emotion with the left prefrontal area and negative emotion with the right. Six months after volunteers were given influenza vaccinations, the ones with higher EEG activity in the left prefrontal area had a five times greater increase in antibodies than those with higher activation on the right (Figure 8.16; Rosenkranz et al., 2003). In other research, men positive for AIDS-causing HIV had HIV levels that were eight times higher if they were introverted (socially inhibited) rather than extroverted (S. W. Cole, Kemeny, Fahey, Zack & Naliboff, 2003). During treatment, their HIV levels decreased less as well, and their T cell levels did not increase at all.

Of course, because this was a correlational study, we cannot assume that introversion accounts for the differences between the two groups. If there is a negative effect of introversion, it could be that the individuals lacked social support. Another possibility is suggested by the fact that the introverted group had higher levels of autonomic activity (heart rate, blood pressure, skin conductance, etc.). Introverted individuals are high in epinephrine and norepinephrine, which cause autonomic (sympathetic) activation. Because norepinephrine increases the rate at which the HIV virus multiplies in the laboratory, it is possible that high norepinephrine levels explain both the introversion and the high HIV levels in the subjects. If this is typical, then personality characteristics and emotional states would become "markers" of certain health conditions, with both the markers and the health problems caused by physiological third variables.

Pain as an Adaptive Emotion

Eighty percent of all visits to physicians are at least partly to seek relief from pain (Gatchel, 1996), and we spend billions each year on nonprescription pain medications. These observations alone qualify pain as a major health problem.

A world without pain might sound wonderful, but in spite of the suffering it causes, pain is valuable for its adaptive benefits. It warns us that the coffee is too hot, that our shoe is rubbing a blister, that we should take our skis back to the bunny slope for more practice. People with *congenital insensitivity to pain* are born unable to sense pain; they injure themselves repeatedly because they are not motivated to avoid dangerous situations, and they die from untreated conditions like a ruptured appendix. Mild pain tells us to change our posture regularly; a woman with congenital insensitivity to pain suffered damage to her spine because she could not respond to these signals, and resulting complications led to her death (Sternbach, 1968).

Pain is one of the senses, a point we consider in more detail in Chapter 11. Here we focus on the feature that makes pain unique among the senses: It is so intimately involved with emotion that we are justified in discussing it as an emotional response. In fact, when we tell someone about a pain experience, we are usually describing an emotional reaction; it is the emotional response that makes pain adaptive.

You know it, and Harvard psychologists have confirmed it: Pain someone inflicts on you intentionally hurts more than pain you experience accidentally (K. Gray & Wegner, 2008). As Beecher (1956) observed, "The intensity of suffering is largely determined by what the pain means to the patient" (p. 1609). In our society, childbirth is considered a painful and debilitating ordeal; in other cultures, childbirth is a routine matter, and the woman returns to work in the fields almost immediately. After the landing at the Anzio beachhead in World War II, 68% of the wounded soldiers denied pain and refused morphine; only 17% of civilians with similar "wounds" from surgery accepted their pain so bravely (Beecher, 1956). The soldiers were not simply insensitive to pain, because they complained bitterly about rough treatment or inept blood draws. According to Beecher, who was the surgeon in command at Anzio, the surgery was a major annoyance for the civilians, but the soldiers' wounds meant they had escaped the battlefield alive. Spiritual context can also have a powerful influence on the meaning of pain. Each spring in some remote villages of India, a man is suspended by a rope attached to steel hooks in his back; swinging above the cheering crowd, he blesses the children and the crops. Selection for this role is an honor, and the participant seems not only to be free of pain but also in a "state of exaltation" (Kosambi, 1967; Ghosh & Sinha, 2007). Figure 8.17 shows an example of culturally sanctioned self-torture.

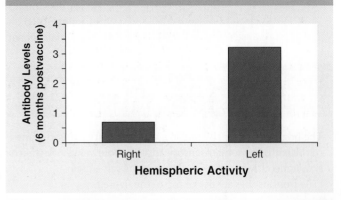

FIGURE 8.16 Differences in Postvaccine Antibody Levels in Relation to Prefrontal Hemispheric Activity.

SOURCE: From "Affective Style and in vivo Immune Response," by M. A. Rosenkranz et al., *PNAS, 100*, pp. 11148–11152. © 2003.

> *Pain is a more terrible lord of mankind than even death itself.*
>
> —Albert Schweitzer

FIGURE 8.17 Voluntary Ritualized Torture in Religious Practice.

Cultural values help determine a person's reaction to painful stimulation.

SOURCE: Alain Evrard/Photo Researchers.

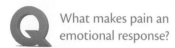
Q What makes pain an emotional response?

The pain pathway has rich interconnections with the limbic system, where pain becomes an emotional phenomenon. Besides the somatosensory area, pain particularly activates the anterior cingulate cortex, which in turn is intimately connected with other limbic structures (D. D. Price, 2000; Talbot et al., 1991). The brain scan in Figure 8.18 shows increased activity in the anterior cingulate cortex as well as the somatosensory area during painful heat stimulation. A hint that the anterior cingulate is involved in the emotional response of pain comes from microelectrode recordings in humans and monkeys; they revealed that some of the neurons respond not only to painful stimulation but to the anticipation of pain (Hutchison, Davis, Lozano, Tasker, & Dostrovsky, 1999; Koyama, Tanaka, & Mikami, 1998).

FIGURE 8.18 PET Scan of Brain During Painful Heat Stimulation.

The bright area near the midline is the cingulate gyrus; the one to the left is the somatosensory area. The four views were taken simultaneously at different depths in the same brain. (The frontal lobes are at the top of the figure.)

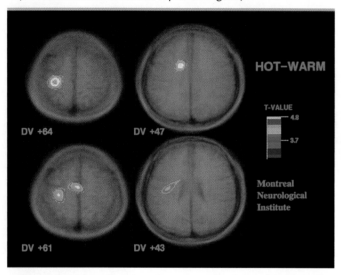

SOURCE: Reprinted with permission from "Multiple Representations of Pain in Human Cerebral Cortex," by Talbot et al., *Science, 251*, pp. 1355–1358. Copyright 1991. Reprinted by permission of AAAS.

But how can we be sure that activity in the anterior cingulate cortex represents the emotional aspect of pain? Fortunately, it is possible to separate the sensation of pain from its emotional effect. One way is to monitor changes in brain activity while the unpleasantness of pain increases with successive presentations of a painful stimulus. Another involves the use of hypnosis to increase pain unpleasantness without changing the intensity of the stimulus. With both strategies activity increases in the anterior cingulate cortex but not in the somatosensory area, suggesting that its role in pain involves emotion rather than sensation (D. D. Price, 2000; Rainville, Duncan, Price, Carrier, & Bushnell, 1997).

If pain continues, it also recruits activity in prefrontal areas where, presumably, the pain is evaluated and responses to the painful situation are planned (D. D. Price, 2000). The location of pain emotion in separate structures may explain the experience of two groups of patients. In pain insensitivity disorders, it is the emotional response that is diminished rather than the sensation of pain; the person can recognize painful stimulation, but simply is not bothered by it (Melzack, 1973; D. D. Price, 2000). The same is true for people who underwent prefrontal lobotomy back when that surgery was used to manage untreatable pain; when questioned, the patients often said they still felt the "little" pain but the "big" pain was gone.

Concept Check *Take a Minute to Check Your Knowledge and Understanding*

☐ Describe the positive and negative effects of stress, indicating why the effects become negative.

☐ Discuss the emotional aspects of pain, including the brain structures involved.

Biological Origins of Aggression

Both motivation and emotion reach a peak during aggression. Aggression can be adaptive, but it also takes many thousands of lives annually and maims countless others physically and emotionally. The systematic slaughter of millions in World War II concentration camps and the terrorist attack that destroyed the World Trade

Center are dramatic examples, but we should not allow such catastrophic events as these to blind us to the more common thread of daily aggression running throughout society.

Aggression is behavior that is intended to harm. Researchers agree that there is more than one kind of aggression, but they do not agree on what the different kinds are, partly because the forms of aggression differ among species. A distinction that has been useful in animal research is the one between predatory aggression and affective aggression. *Predatory aggression* occurs when an animal attacks and kills its prey. Predatory aggression is cold and emotionless, unlike *affective aggression,* which is characterized by its emotional arousal. Affective aggression can be further subdivided in a number of ways; here we will use the distinction between offensive and defensive aggression. An unprovoked attack on another animal is *offensive aggression,* while *defensive aggression* occurs in response to threat and is motivated by fear. Human aggression is less clearly categorized; for example, it is controversial whether humans display predatory aggression. However, most researchers do recognize a distinction between *reactive aggression,* which is impulsive, provoked, and emotional, and *proactive aggression,* which is premeditated, unprovoked, and relatively emotionless.

Hormones and Aggression

Although aggression is influenced by a person's environment, it should come as no surprise that such a powerful force has hormonal and neural roots. Hormones appear to influence offensive aggression more than the other two types, at least in rats (D. J. Albert, Walsh, & Jonik, 1993). In nonprimate animals, aggression is enhanced by testosterone in males and by both testosterone and estrogen in females.

In primates, aggressiveness increases in female monkeys during the premenstrual period (Rapkin et al., 1995), a time when estrogen and progesterone (a hormone that maintains pregnancy) are at their lowest. Studies have also reported a doubling of crimes (K. Dalton, 1961) and violent crimes (d'Orbán & Dalton, 1980) in women during this period. Anger responses to provocation increased in women during the premenstrual period, but only in those who had previously reported that they suffered from premenstrual complaints (*premenstrual syndrome,* or PMS; Van Goozen, Frijda, Wiegant, Endert, & Van de Poll, 1996). Progesterone has been reported to reduce PMS-related aggressiveness (K. Dalton, 1964, 1980), and some suggest that decreased levels of *allopregnanolone,* a metabolite of progesterone, may reduce the woman's anxiety response to stress (Monteleone et al., 2000; Rapkin et al., 1997).

Some studies have found a relationship between testosterone and violence in male prison inmates. Male prisoners convicted of violent crimes like rape and murder and prisoners rated as tougher by their peers had higher testosterone levels than other prisoners (Figure 8.19; J. M. Dabbs, Carr, Frady, & Riad, 1995; J. J. Dabbs, Frady, Carr, & Besch, 1987). In women inmates as well, there appears to be a relationship between testosterone level and aggressive

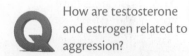

How are testosterone and estrogen related to aggression?

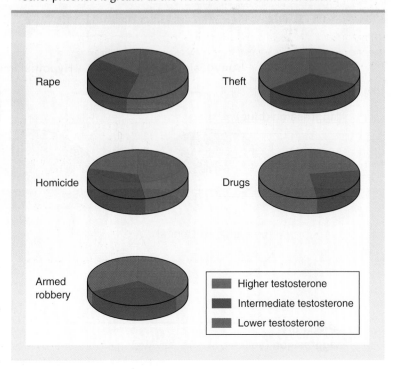

FIGURE 8.19 Testosterone Levels of Men Convicted of Various Crimes.

The proportion of men with high testosterone levels compared with other prisoners is greater as the violence of the crime increases.

SOURCE: Based on J. M. Dabbs et al. (1995).

dominance while in prison (J. J. Dabbs & Hargrove, 1997). The prison results make sense intuitively, but the function of testosterone in human aggression is open to question, just as it was in sexual behavior. The studies are correlational, so we must look elsewhere for evidence that testosterone causes aggression. There is no clear evidence that aggression in humans is affected by manipulation of testosterone levels or by disorders that increase or decrease testosterone (D. J. Albert et al., 1993). For this reason, critics argue that aggression increases testosterone level rather than the other way around, and so far the research is on their side. Not only does testosterone increase after winning a sports event (Archer, 1991; Mazur & Lamb, 1980), but it also goes up while watching one's team win a sporting event (Bernhardt, Dabbs, Fielden, & Lutter, 1998) and even after receiving the MD degree (Mazur & Lamb, 1980).

The Brain's Role in Aggression

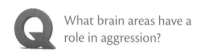

What brain areas have a role in aggression?

Research with rats and cats indicates that defensive and predatory aggression are distinct not only behaviorally but neurally (D. J. Albert et al., 1993; A. Siegel, Roeling, Gregg, & Kruk, 1999). The highly emotional nature of defensive aggression, indicated by the cat's familiar arched back, bristling fur, and hissing, contrasts sharply with the cold, emotionless stalking and killing of its prey. We know more about the brain structures involved in feline aggression and their connections than in any other animal. The two types of aggression have separate neural pathways, outlined in Figure 8.20. The defensive pathway begins in the medial nucleus of the amygdala, travels to the medial hypothalamus, and goes from there to the dorsal part of the periaqueductal gray in the brain stem. Control of predatory attack flows from the lateral nucleus and central nucleus of the amygdala to the lateral hypothalamus and the ventral periaqueductal gray (A. Siegel et al., 1999). Of course, threat does

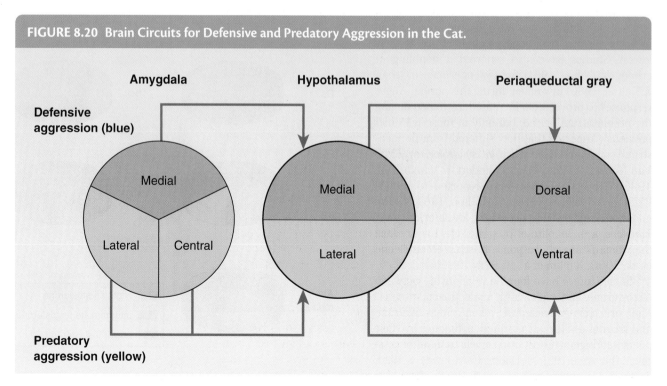

FIGURE 8.20 Brain Circuits for Defensive and Predatory Aggression in the Cat.

SOURCE: Based on A. Siegel et al. (1999).

not always result in aggression; stimulation of another area in the periaqueductal gray produces flight (S. P. Zhang, Bandler, & Carrive, 1990).

Research on human aggression is understandably constrained, but we have determined that several brain areas are ultimately involved. Tumors can cause aggression if they are in the hypothalamus or the septal region (D. J. Albert et al., 1993). Seizure activity in the area of the amygdala increases aggression, and damage to the amygdala reduces it. A PET scan study found higher activity in the right amygdala and the hypothalamus in a group of murderers of both sexes (Raine, Meloy, et al., 1998); across 13 studies, lesioning of the amygdala significantly reduced aggressive behavior in 33% to 100% of patients (Mpakopoulou, Gatos, Brotis, Paterakis, & Fountas, 2008).

There is growing evidence that the distinction between reactive and proactive aggression involves significant differences in brain functioning. PET scans have found that murderers of both sexes have reduced activity in the prefrontal cortex. However, this deficiency was limited to reactive murderers, those who had killed in a fit of rage rather than with premeditation. Reduced prefrontal gray matter volume is often associated with antisocial personality disorder (Raine, Lencz, Bihrle, LaCasse, & Colletti, 2000); people with *antisocial personality disorder* behave recklessly; violate social norms; and commit antisocial acts such as fighting, stealing, using drugs, and engaging in sexual promiscuity. Men have less gray matter in the prefrontal cortex than women, and this accounts for 77% of their greater antisocial personality/behavior (Raine, Yang, Narr, & Toga, 2009). In reactive aggressors, prefrontal deficits are accompanied by increased autonomic responsiveness to threat and hyperactivity of the amygdala (Roth & Struber, 2009).

Proactive aggression is associated with *psychopathy* (Fite, Stoppelbein, & Greening, 2009). Although not formally recognized as distinct from other antisocials, psychopaths stand out most in their lack of remorse for their behavior (Roth & Struber, 2009). Psychopaths show less autonomic responding to stress and aversive stimuli than other aggressive individuals (Patrick, 2008), and they are more likely to be fearless and to show other indications of impaired amygdala functioning (Kiehl et al., 2001; Yang, Raine, Narr, Colletti, & Toga, 2009).

Serotonin and Aggression

We have already seen some indication of the importance of serotonin in motivation. Usually its role is to suppress motivated behaviors; as a result, the motivation for food, water, sex, and drugs of abuse increases when serotonin activity is low (Pihl & Peterson, 1993). Now we will add aggression to the list.

Serotonin and the Inhibition of Aggression

Serotonin inhibits aggression, probably through its effects in the amygdala, hypothalamus, periaqueductal gray, and prefrontal area (Spoont, 1992; Vergnes, Depaulis, Boehrer, & Kempf, 1988). Researchers destroyed serotonin-producing neurons in the lateral hypothalamus of rats, which depleted their forebrain serotonin to 25% of its normal level (Vergnes et al., 1988). Afterward, the rats were dramatically more aggressive toward intruder rats.

Until recently, serotonin activity in humans had to be assessed by measuring amounts of the serotonin metabolite 5-hydroxyindoleacetic acid (5-HIAA) in the cerebrospinal fluid. Low serotonin activity is specifically associated with reactive (impulsive) aggression; 5-HIAA is lower in impulsive violent offenders than in violent offenders who planned their crimes (Linnoila et al., 1983). Now PET imaging can measure serotonin activity more directly, as well as locating it in the brain. PET studies have found that people with impulsive aggression have decreased

How does serotonin affect aggression?

The individual on the left has the short allele of the *SLC6A4* gene, which reduces serotonin activity and increases fear responses; the one on the right has two copies of the long allele.

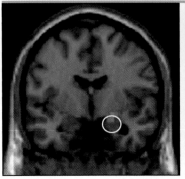 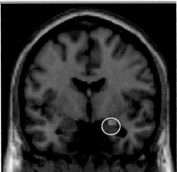

SOURCE: Reprinted with permission from "Serotonin Transporter Genetic Variation and the Response of the Human Amygdala," by Hariri et al., *Science, 297,* pp. 400–403. Copyright 2002. Reprinted by permission of AAAS.

serotonin activity in the prefrontal cortex (Meyer et al., 2008) and in the anterior cingulate gyrus, which also plays an important role in emotional regulation (Frankle et al., 2005). The amygdala is another location where a serotonin deficit likely contributes to aggression. The so-called short allele of the *SLC6A4* gene reduces serotonin and contributes to fear and anxiety; in people with this allele, the amygdala is hyperreactive to fear-inducing stimuli (Hariri et al., 2002), which we saw is a characteristic of impulsive aggression (see Figure 8.21).

To determine whether low serotonin can cause aggression in humans, Moeller and his colleagues (1996) had males drink an amino acid mixture that reduces tryptophan, the precursor for serotonin. Then the men participated in a computer game in which one response earned points exchangeable for money and a different response subtracted points from a fictitious competitor. At random times during the game, the player's screen indicated that some of his accumulated points had been deleted by the fictitious competitor. The men were more aggressive after drinking the tryptophan-depleting mixture, deleting more of the fictitious competitor's points. The social restraints on aggression are strong and the penalties are high; because low serotonin level is associated with impulsiveness, it makes sense that it would increase aggression in humans and animals.

Some research findings have reopened the possibility of a causal role for testosterone in human aggression. Higley and his colleagues (1996) suggested that high testosterone and low serotonin interact to produce aggression. Their study of freeranging male monkeys supported their position. Monkeys with high testosterone were more likely to engage in brief aggression that asserted dominance, such as threats and displacing another monkey from his position. Monkeys with low 5-HIAA levels were impulsive; they more frequently took dangerously long leaps among the treetops, and when they engaged in aggression it was more likely to accelerate into greater violence. The most aggressive monkeys of all had both low 5-HIAA and high testosterone levels. Human data suggest a similar relationship. Violent alcoholic offenders often have both low brain serotonin and high testosterone (Virkkunen, Goldman, & Linnoila, 1996; Virkkunen & Linnoila, 1993). Evidence for this relationship in humans is hard to find, but normal males with high testosterone and low serotonin levels do score higher on hostility and aggression questionnaires (Kuepper et al., 2010). If testosterone has any causal influence on human aggression—emphasizing the word *if*—it may well occur only when serotonin's inhibitory effect is reduced.

 Does testosterone increase human aggression after all?

Alcohol and Serotonin

In a study of crime in 14 countries, 62% of violent offenders were using alcohol at the time of their crime or shortly before (Murdoch, Pihl, & Ross, 1990). Evidence favors the commonsense interpretation that alcohol *facilitates* aggression rather than simply being associated with it (Bushman & Cooper, 1990). However, alcohol appears to influence aggression only in people who also have low serotonin levels, such as early-onset alcoholics (see Chapter 5), who tend to be impulsively aggressive (Buydens-Branchey, Branchey, Noumair, & Lieber, 1989; Virkkunen & Linnoila, 1990, 1997).

The dual influence of low serotonin on alcohol consumption and aggression makes for a deadly combination. After initially increasing serotonin activity, alcohol later depletes it below the original level (Pihl & Peterson, 1993); the alcohol abuser is caught in a vicious cycle as alcohol consumption increases both aggression and the craving for more alcohol. Drugs that inhibit serotonin uptake at the synapse, such as the antidepressant fluoxetine (trade name Prozac), reduce alcohol craving and intake (Naranjo, Poulos, Bremner, & Lanctot, 1994), and they also reduce hostility and aggressiveness (Coccaro & Kavoussi, 1997; Knutson et al., 1998).

Heredity and Environment

Like many other behaviors, aggression is genetically influenced. From a review of 24 studies, it was estimated that up to 50% of the variation among people in aggression is genetic in origin (D. R. Miles & Carey, 1997). A genetic influence does not necessarily mean that there is a gene for aggression; the influence may be on a broader characteristic. In fact, high scores for aggressive hostility on a questionnaire were related to which variant the subjects had of the tryptophan hydroxylase gene, which regulates the production of serotonin (Hennig, Reuter, Netter, Burk, & Landt, 2005). The genes responsible for three types of serotonin receptors have also been implicated in the genetic control of impulsive aggression (Virkkunen et al., 1996).

What roles do heredity and environment play in aggression?

With only half of the variability in aggression accounted for by heredity, there is still plenty of room for environmental influence. For example, men and women who were reared in homes with inadequate parenting have double the rate of violent criminality (Hodgins, Kratzer, & McNeil, 2001), and male and female murderers who do not have prefrontal deficits are more likely to have experienced psychological and social deprivation during childhood (Raine, Stoddard, Bihrle, & Buschbaum, 1998). But the impact of an environmental influence often depends on its interaction with the person's genetic makeup. The monoamine oxidase A (MAOA) enzyme breaks down monoamines, particularly serotonin, in the synapse; the MAOA-L allele of the MAOA gene results in deficient levels of the enzyme. A link between aggression and this allele was discovered in a Dutch family with a large number of violent males (Brunner, Nelen, Breakefield, Ropers, & van Oost, 1993). Since then it has been replicated more reliably than any other aggression–gene link, most recently when the allele was found to predict male membership in gangs and their use of weapons in fights (Beaver, DeLisi, Vaughn, & Barnes, 2010). The enzyme deficiency leads to an accumulation of serotonin, so you might expect this to deter aggression; instead, the brain compensates by reducing its sensitivity to serotonin (Manuck, Flory, Ferrell, Mann, & Muldoon, 2000). And now the interaction: Later studies found that the MAOA-L allele produced violence only in males who had been subjected to childhood physical abuse (see Figure 8.22; Caspi et al., 2002). Why are we talking only about males? Men are studied more frequently because the MAOA gene is on the X chromosome; females with the allele are less affected because it is likely to be opposed by the MAOA-H allele on the other X chromosome, which the male doesn't have.

While this research adds greatly to our understanding of the origins of violent behavior, we must be careful not to overinterpret the data, as In the News explains.

FIGURE 8.22 Genetic Influence on Violent Behavior in Victims of Childhood Maltreatment.

Low MAOA activity due to the MAOA-L allele, coupled with childhood maltreatment, results in increased violence.

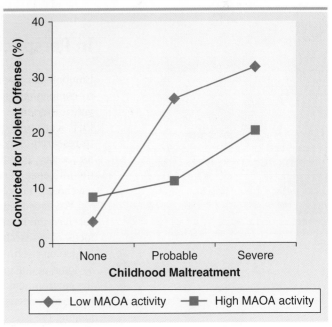

SOURCE: Based on data from "Role of Genotype in the Cycle of Violence in Maltreated Children," by A. Caspi et al., 2002, Science, 292, p. 852.

Aggression, Genes, and the Law

On March 10, 2007, an Italian court sentenced Abdelmalek Bayout to 9 years in prison for stabbing and killing a man who had insulted him. In May of 2009, the appeals court, relying on a report that Bayout has five genes that have been linked to aggression, reduced the sentence by 1 year; the judge said he found evidence of the *MAOA-L* allele particularly compelling. This was the first time that behavioral genetics had affected a sentence in a European court. The scientific community usually applauds whenever the legal system tempers its judgment with scientific evidence, but a survey by staff from the journal *Nature* found scientists less than enthusiastic about the decision. They urged caution in using genetic information when we know so little about how genes affect a particular individual, and they pointed out the important role that environment plays (childhood abuse in this case). They also noted that research has not found the link between *MAOA-L* and aggression in nonwhite populations and Bayout's racial background had not been determined. The controversy emphasizes how theoretical the field of behavioral genetics is and how far we are from making practical applications of the knowledge.

SOURCE: Feresin, E. (2009, October 30). Lighter sentence for murderer with 'bad genes.' *Nature News*. Retrieved March 17, 2010, from www.nature.com/news/2009/091030/full/news.2009.1050.html

Concept Check   *Take a Minute to Check Your Knowledge and Understanding*

- ☐ What is the evidence that the prefrontal cortex moderates aggression?
- ☐ How do serotonin, alcohol, and testosterone possibly interact to increase aggression?

In Perspective

Emotion has been difficult for neuroscientists to get a handle on because it is so complex physiologically and because so much of emotion is a subjective, private experience. With improved research strategies and new technologies like PET scans, old questions about the role of brain structures are finally yielding to research. A good example is the ability to separate the emotion of pain from its sensory aspects at the neural level. On another front, research has confirmed the influence of emotion on health, a topic that was practically relegated to fringe psychology not too long ago.

We have focused mostly on the negative aspects of emotion because they have received the most attention from researchers and we know more about them. This focus of research interest acknowledges the fact that pain, anger, and aggression helped ensure the survival of our evolutionary ancestors, but now they are viewed as some of our greatest burdens. Society not only needs to ask whether modifying the environment might reduce schoolhouse shootings or violent crime in our streets, but it also needs to appreciate the role of hormones, genes, and the brain when judging the accountability of a depressed mother who drowns all her children. In the meantime, if you find thoughts about the negative aspects of emotion a bit dismaying, you might want to take a short break while you hold your pen between your teeth; improving your corner of the world is a good place to start.

Summary

Emotion and the Nervous System

- The autonomic nervous system increases bodily arousal during an emotional event and decreases it afterward.
- The James-Lange and Schacter-Singer theories differ as to the role of bodily feedback in emotional experience. There is evidence that feedback from facial expressions contributes to emotions; also, mimicking other people's expressions may help us understand others' emotions.
- The limbic system is a network of several structures that have functions in emotion. We now know that emotion involves additional structures at all levels of the brain.
- The amygdala has a variety of functions, but its role in fear has received the most attention. Rats and humans with damage to both amygdalas lack fear and often fail to act in their own best interest.
- The prefrontal cortex combines emotional input with other information to make decisions. People with damage there have trouble following moral and social rules, and they have impaired ability to learn from the consequences of their behavior.
- Damage to the right hemisphere particularly blunts emotions and impairs the person's ability to recognize emotion in faces and in voices.

Stress, Immunity, and Health

- Stress is adaptive, mobilizing the body for action and increasing immune system activity.
- Prolonged stress interferes with mental, physical, and emotional functioning; compromises the immune system; and even damages the brain.
- Social support, personality, and attitudes are related to immune functioning and health, including cancer survival. However, these social and personality factors may not influence immune functioning; instead, they may be indicators of the individual's physiological functioning.
- Pain is also an adaptive response; it informs us of danger to the body, and the emotion that accompanies it motivates us to take action.

Biological Origins of Aggression

- Testosterone is involved in male aggression and both testosterone and estrogen in female aggression, although in humans the causal link for testosterone is questionable.
- The amygdala and hypothalamus appear to be the most important brain structures in aggression, both in lower animals and in humans. The prefrontal cortex suppresses aggression, and deficiency there is linked to violent behavior.
- Human aggression may be reactive or proactive; these two types appear to be biologically distinctive.
- Serotonin inhibits aggressive behavior, and low serotonin level is associated with aggression. Alcohol and a lowered serotonin level combine to increase aggression. Low serotonin and high testosterone levels may interact to increase aggression in humans.
- Heredity is estimated to contribute half of the variability in aggression among humans; some genetic links with aggressive behavior involve serotonin receptors and serotonin metabolism. The other half of variation is due to the environment, including inadequate parenting. Heredity and environment interact, for example in the combination of the *MAOA-L* allele with childhood abuse. ■

Study Resources

F   For Further Thought

- Do you think we rely more on bodily feedback or the stimulus situation in identifying emotions? Why?
- Stress and pain involve considerable suffering, but they are necessary. Explain. What makes the difference between good and bad stress and pain?
- You are an adviser to a government official charged with reducing aggression in your country. From what you have learned in this chapter, what would you recommend?
- The legal plea of "not guilty by reason of insanity" has historically required that the defendant did not *know* right from wrong—as evidenced, for example, by the defendant's failure to flee or try to conceal the crime. Critique this standard in terms of what you know about controlling behavior.

T   Testing Your Understanding

1. Discuss the James-Lange and cognitive theories, including evidence for the theories.

2. Explain the roles of the amygdala and the prefrontal cortex in guiding our everyday decisions and behavior.

3. Describe the role the brain plays in animal and human aggression, including structures and their functions.

Select the best answer:

1. The James-Lange theory and the cognitive theory disagree on whether
   a. specific brain centers are involved in specific emotions.
   b. there is any biological involvement in human emotions.
   c. bodily feedback determines which emotion is felt.
   d. individuals can judge their emotions accurately.

2. Some people with brain damage do not seem to learn from the consequences of their behavior and must have supervised care. Based on the location of their damage, you would expect that they would particularly be lacking in
   a. sadness.
   b. joy.
   c. fear.
   d. motivation.

3. A person with partial paralysis seems remarkably undisturbed about the impairment. The paralysis
   a. probably is on the right side of the body.
   b. probably is on the left side of the body.
   c. probably involves both sides of the body.
   d. is as likely to be on one side as the other.

4. Stress can
   a. reduce immune system function.
   b. impair health.
   c. mobilize the immune system.
   d. both a and b.
   e. a, b, and c.

5. Long-term exposure to cortisol may affect memory by
   a. reducing blood flow to the brain.
   b. destroying neurons in the hippocampus.
   c. inhibiting neurons.
   d. redirecting energy resources to the internal organs.

6. AIDS is a deficiency of the
   a. immune system.
   b. autonomic system.
   c. central nervous system.
   d. motor system.

7. Indications are that if pain did not have an emotional component, we would probably
   a. be deficient in avoiding harm.
   b. avoid harm effectively, using learning and reasoning.

c. be more aggressive.
d. generally lead happier lives.

8. A structure described in the text as being involved in both aggression and flight is the
   a. amygdala.
   b. anterior cingulate cortex.
   c. lateral hypothalamus.
   d. periaqueductal gray.

9. According to research, you would have your best chance of showing that testosterone increases aggression in humans if you injected testosterone into
   a. males rather than females.
   b. people with prefrontal damage.

c. people with low serotonin.
d. people who were being confronted by another person.

10. Based on information in the text, the chance of violent criminal behavior is increased in males who
    a. have high testosterone.
    b. were abused as children.
    c. have a gene for low MAOA.
    d. both a and b.
    e. both b and c.

Answers:
1.c, 2.c, 3.b, 4.e, 5.b, 6.a, 7.a, 8.d, 9.c, 10.e.

Online Resources

The following resources are available at **www.sagepub.com/garrett3e.**

Chapter Resources

- Flash Cards
- Chapter Quiz
- Internet Resources and Exercises

On the Web

Links to these websites can be found at the web address above: Select this chapter's **Chapter Resources,** then choose **Internet Resources and Exercises**.

1. You can link here to a variety of sites, from the University of California to YouTube, to see **Brain and Emotion Videos.**

2. You can get a feel for what an active emotion research laboratory is like by visiting lab sites at **Boston College** and **Indiana University.** You can read descriptions of their research programs and download published articles.

3. **Stress Less** features publications for sale, links to other websites on a broad variety of stress topics, and chat rooms organized by stress type.

4. The **National Center for Posttraumatic Stress Disorder** site has information on the disorder and on subtopics such as "Returning from War" and "Specific to Women."

   Various stress tests (some validated by research and some not) are available to assess the potential for stress to affect your health and well-being. You can take a **brief** test or a **longer** one for Type A personality, which research indicates contributes to heart attacks; a **test for stress** that gives you scores for quality of life, symptom distress, and level of functioning; or a modification of the historically significant **Holmes and Rahe Scale,** designed to assess health risk from recent stressful events (positive as well as negative).

Chapter Updates and Biopsychology News

For Further Reading

1. *Descartes' Error,* by Antonio Damasio (Quill, 2000), covers the various topics of emotion but develops the premise that our rational decision making is largely dependent on input from emotions.

2. *Why Zebras Don't Get Ulcers* (3rd ed.), by Robert Sapolsky (Freeman, 2004), is a lively discussion of emotion and its effects, including stress, immunity, ulcers, memory, and sex.

3. "Empathy Overkill," by Helen Thomson (*New Scientist,* March 13, 2010, 43–45), describes research with people with "mirror synesthesia," who intensely feel what they see others experiencing, and people with echopraxia, who greatly exaggerate the action mimicking that the rest of us engage in subtly.

4. "Health Psychology: Developing Biologically Plausible Models Linking the Social World and Physical Health," by Gregory Miller, Edith Chen, and Steve W. Cole (*Annual Reviews of Psychology,* 2009, 501–524), is a detailed review of the interactions between emotion and health.

K   Key Terms

affective aggression 241
aggression 241
anterior cingulate cortex 228
antisocial personality disorder 243
autoimmune disorder 235
B cell 235
cognitive theory 226
congenital insensitivity to pain 239
cortisol 234
defensive aggression 241
emotion 224
hypothalamus-pituitary-adrenal axis 234
immune system 234
James-Lange theory 226

leukocytes 235
limbic system 227
macrophage 235
mirror neurons 227
natural killer cell 235
offensive aggression 241
predatory aggression 241
proactive aggression 241
reactive aggression 241
skin conductance response (SCR) 230
stress 234
sudden cardiac death 236
T cell 235

PART III

Interacting With the World

Chapter 9. Hearing and Language

Chapter 10. Vision and Visual Perception

Chapter 11. The Body Senses and Movement

Hearing and Language

Hearing

The Stimulus for Hearing

The Auditory Mechanism

Frequency Analysis

Locating Sounds With Binaural Cues

    APPLICATION: COCHLEAR AND BRAINSTEM IMPLANTS
        TO RESTORE HEARING

    CONCEPT CHECK

Language

Broca's Area

Wernicke's Area

The Wernicke-Geschwind Model

Reading, Writing, and Their Impairment

Mechanisms of Recovery From Aphasia

A Language-Generating Mechanism?

Language in Nonhumans

Neural and Genetic Antecedents

    CONCEPT CHECK

In Perspective

Summary

Study Resources

In this chapter you will learn

- What the auditory (hearing) mechanism consists of and how it works

- How the brain processes sounds, from pure tones to speech

- Which brain structures account for language ability

- The causes of some of the major language disorders

- What we know about language abilities in nonhuman animals

The only way Heather Whitestone knew the music had started for her ballet number in the Miss America talent competition was because she could feel the vibrations through the floor. Heather was profoundly deaf, and she became the first person with deafness or any other handicap to win the title of Miss America.

> *To be deaf is a greater affliction than to be blind.*
>
> —Helen Keller

Her hearing was normal until she contracted meningitis at the age of 18 months; the problem was not the meningitis but the strong antibiotics that destroyed the sound receptor cells in her ear. A hearing aid helped some, and she read lips and received twice weekly speech therapy. By the time she reached high school, she was able to participate in mainstream classes without a sign language interpreter, and she graduated with a 3.6 grade point average. A cochlear implant was an option, but since she had been deaf so long, it was possible she would not tolerate the adjustment to the strange new sounds. Besides, she was satisfied with her level of adjustment—that is, until she married and had a son. Heather realized she could not share many of his experiences because she could not hear what he was hearing. Then one day she saw her husband running to their son when the boy hurt himself and she had not even heard him crying. It was then she decided to have the surgery.

FIGURE 9.1   Heather Whitestone Using Sign Language.

Her communication and her quality of life improved when she received a cochlear implant.

SOURCE: Getty Images.

The implant bypassed the dead receptor cells to stimulate the neurons in her inner ear directly. After the surgery, she could understand her son's speech much better, so he didn't have to repeat himself so many times. Best of all, she could hear him cry, even when he was playing outside in the backyard.

Nothing attests to the value of hearing more than the effects of losing it. Like Heather, the person is cut off from much of the discourse that our social lives depend on. There is no music, no song of birds, and no warning from thunder or car horns. When hearing is lost abruptly in later life, the effect can be so depressing that it eventually leads to suicide.

With the topics of hearing and language, we begin the discussion of how we carry on transactions with the world. This communication involves acquiring information through the senses, processing the sensory information, communicating through language, moving about in the world, and acting on the world. We have already touched on the senses of taste, smell, and pain in the context of hunger, sexual behavior, and emotion. Before we explore additional sensory capabilities, we need to establish some basic concepts.

First, a sensory system must have a specialized receptor. A *receptor* is a cell, often a specialized neuron that is suited by its structure and function to respond to a particular form of energy, such as sound. A receptor's function is to convert that energy into a neural response. You will see examples of two kinds of receptors in this and the next chapter, but receptors come in a wide array of forms to carry out their functions.

For a receptor to do its job, there must also be an adequate stimulus. An *adequate stimulus* is the energy form for which the receptor is specialized. Due to the imperfect specialization of receptors, other stimuli will often produce responses as well. For example, if you apply pressure to the side of your eyeball (through the lid), you will see a circular dark spot.

How is a stimulus translated into information the brain can use?

You will remember from Chapter 3 that, according to Müller's doctrine of specific nerve energies, the neural mechanism rather than the type of stimulus determines the kind of sensory experience you will have. A sensory system will register its peculiar type of experience even if the stimulus is inappropriate. So when the neurosurgeon stimulates the auditory cortex, the patient hears a buzzing sound or even voices or music, and when your skateboard shoots out from under you and your head hits the pavement, you really do see "stars." You will learn in this and the next two chapters that it is the *patterning* of the stimulation—that is, the amplitude and timing of neural impulses—that makes sensory information meaningful.

Most people consider audition and vision the most important senses. As a result, there has been more research on these two senses than on others and we know more about them, so we will give them the most attention. Because audition is a more mechanical sensory mechanism than vision, it is a good place to begin our formal discussion of *sensation,* the acquisition of sensory information, and *perception,* the interpretation of sensory information.

Q What is the difference between sensation and perception?

Hearing

The fact that the auditory mechanism is less complex does not mean that hearing is a simple matter. The *cochlea,* where the auditory stimulus is converted into neural impulses, contains a million moving parts. Our range of sensitivity to intensity, from the softest sound we can hear to the point where sound becomes painful, is a million to one. Our ability to hear low-intensity sounds is limited more by interference from the sound of blood coursing through our veins and arteries than by the auditory mechanism itself. In addition, we are able to hear frequencies ranging from about 20 hertz (Hz, cycles per second) up to about 20000 Hz, and we can detect a difference in frequencies of only 2 or 3 Hz. To give you some idea of what these frequencies relate to in real life, the piano—the most versatile of musical instruments—has a range of about 27 to 4000 Hz.

1
Hearing Loss

The Stimulus for Hearing

The adequate stimulus for audition is vibration in a conducting medium. Normally the conducting medium is air, but we can also hear underwater and we hear sounds conducted through our skull. The air is set to vibrating by the vibration of the sound source—a person's vocal cords, a bell that has been struck, or a stereo speaker. As the sound source vibrates, it alternately compresses and decompresses the air (Figure 9.2).

If we used a microphone to convert a sound to an electrical signal, we could display the signal on an oscilloscope, like the one we used to measure the action potential in Chapter 2; the oscilloscope would form a graph of the compressions and decompressions, and we could see what the sound "looks like." Look at Figure 9.3; the up-and-down squiggles represent the increasing and decreasing pressure of different sounds (over a brief fraction of a second). One way sounds differ is in frequency. *Frequency* refers to the number of cycles or waves of alternating compression and decompression of the vibrating medium that occur in a second (expressed in hertz). Figure 9.3a and b have the same frequency—indicated by the number of waves in the same time—so we would hear these two sounds as the same, or nearly the same, pitch. Figure 9.3c and d would also sound about the same as each other, but higher in pitch than Figure 9.3a and b. *Pitch* is our experience of the frequency of a sound. Pitch and frequency are related but do not correspond exactly, due to the characteristics of our auditory system.

Sounds also differ in amplitude. The sounds represented by Figure 9.3a and c have the same amplitude (the height of the wave), so they would sound about equally loud; Figure 9.3b and d would sound less loud but similar to each other. *Amplitude,* or *intensity,* is the term for the physical energy in a sound; *loudness* is the term for our experience of sound energy. How loud a sound is to the observer depends not only on the intensity but also on the frequency of the sound; for example, we are most sensitive to sounds between 2000 and 4000 Hz—the range within which most conversation occurs—and equally intense sounds outside this range would seem less loud to us. Similarly, the amplitude of a sound influences our experience of pitch. Because the physical stimulus and the psychological experience are not always perfectly related, we need to use the terms *intensity* versus *loudness* and *frequency* versus *pitch* carefully.

FIGURE 9.2 Alternating Compression and Decompression of Air by a Sound Source.

The surface of the speaker vibrates, alternately compressing and decompressing the surrounding air. The dark areas represent high pressure (a denser concentration of air molecules), and the light areas represent low pressure.

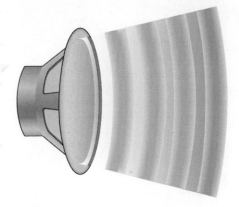

SOURCE: From *Sensation and Perception,* 5th ed., by Goldstein, 1999. Reprinted with permission of Wadsworth, a division of Thomson Learning.

Q What is the difference between frequency and pitch? Intensity and loudness?

FIGURE 9.3 Examples of Pure and Complex Sounds.

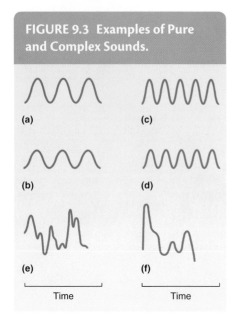

(a)

(b)

(c)

(d)

(e)

(f)

Time

Time

(a) and (b) are pure tones of the same frequency but different amplitudes, as are (c) and (d). (a) and (c) have the same amplitudes but different frequencies, as do (b) and (d). Both (e) and (f) are complex sounds—noise and a clarinet note, respectively.

2

Hearing Facts

The sounds we hear can also be classified as either pure tones or complex sounds. A pure tone, generated for example by striking a tuning fork, would produce a tracing that looked like one of the graphs in Figure 9.3a to d. Notice that these four waveforms are a very regular shape, called a sine wave. They are *pure tones:* Each has only one frequency. Figure 9.3e and f graph *complex sounds:* Each is a mixture of several frequencies. The random combination of frequencies in 9.3e would probably be described by most of us as "noise." Depending on the combination and order of frequencies, a complex sound might seem musical like the last waveform, which was produced by a clarinet (Figure 9.3f). The two waveforms may not look very different to you, but they would certainly sound different. Although what is considered *pleasantly* musical depends on experience and culture (and one's age!), we would recognize even the most foreign music as music.

The Auditory Mechanism

To hear, we must get information about the sound to the auditory cortex. This requires a series of events, including sound reception, amplification, and conversion into neural impulses that the brain can use.

The Outer and Middle Ear

The term *ear* refers generally to all the structures shown in Figure 9.4. The flap that graces the side of your head is called the outer ear or *pinna.* The outer ear captures the sound, then slightly amplifies it by funneling it from the larger area of the pinna into the smaller area of the auditory canal. It also selects for sounds in front; this makes it easier to focus on a sound, such as the conversation you're having, while excluding irrelevant sounds behind you. Dogs and cats have muscles that enable them to turn their ears toward a sound that is not directly in front of them; you may be able to wriggle your ears a bit (actually by twitching your scalp muscles), but you have to turn your head to orient toward a sound.

The first part of the middle ear, the eardrum or *tympanic membrane,* is a very thin membrane stretched across the end of the auditory canal; its vibration transmits the sound energy to the ossicles. A muscle called the *tensor tympani* can stretch the eardrum tighter or loosen it to adjust the sensitivity to changing sound levels. The tympanic membrane is very sensitive. Wilska (1935) ingeniously glued a small rod to the eardrum of a human volunteer (temporarily, of course) and used an electromagnetic coil to vibrate the rod back and forth. He determined that we can hear sounds when the eardrum moves as little as the diameter of a hydrogen atom! The experiment was remarkable for the time, and recent studies with more sophisticated equipment have shown that Wilska's measurements were surprisingly accurate (Hudspeth, 1983).

The second part of the middle ear is the *ossicles,* tiny bones that operate in lever fashion to transfer vibration from the tympanic membrane to the cochlea. The *malleus, incus,* and *stapes* are named for their shapes, as you can see from their English equivalents *hammer, anvil,* and *stirrup.* The ossicles provide additional amplification by concentrating the energy collected from the larger tympanic membrane onto the much smaller base of the stirrup, which rests on the end of the cochlea. The amplification is enough to compensate for the loss of energy as the vibration passes from air to the denser liquid inside the cochlea.

The Inner Ear

You can also see the parts of the inner ear in Figure 9.4. The semicircular canals are part of the vestibular organs and do not participate in hearing; we will talk about them in Chapter 11. The snail-shaped structure is the cochlea, where the ear's sound-analyzing structures are located. You can see from the cochlea's shape where it got

FIGURE 9.4 The Outer, Middle, and Inner Ear.

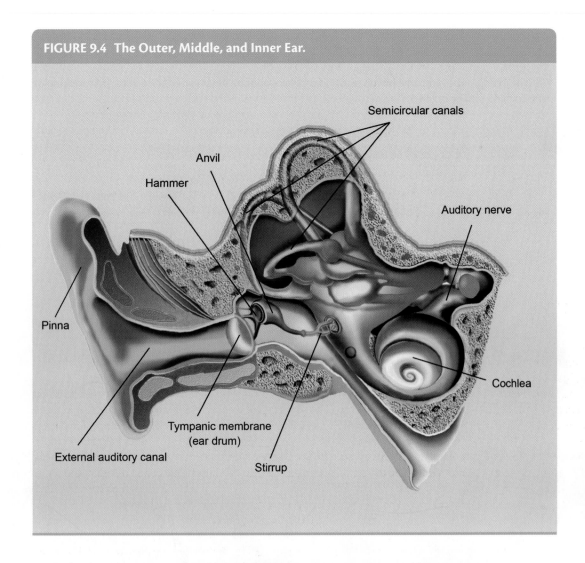

its name, which means "land snail" in Latin. Figure 9.5a shows a more highly magnified view. It is a tube that is about 35 millimeters (mm) long in humans and coiled 2½ times. It is subdivided by membranes into three fluid-filled chambers or canals (Figure 9.5b). In this illustration, the end of the cochlea has been removed, and you are looking down the three canals from the base end. The stirrup (Figure 9.5a) rests on the *oval window*, a thin, flexible membrane on the face of the vestibular canal. The *vestibular canal* (scala vestibuli) is the point of entry of sound energy into the cochlea. The vestibular canal connects with the *tympanic canal* at the far end of the cochlea through an opening called the *helicotrema*. (You might not need to remember this term, but it just *sounds* too wonderful to leave out!) The helicotrema allows the pressure waves to travel through the cochlear fluid into the tympanic canal more easily.

All this activity in the vestibular and tympanic canals literally bathes the *cochlear canal*, where the auditory receptors are located, in vibration. The vibration passes to the *organ of Corti*, the sound-analyzing structure that rests on the *basilar membrane*. The organ of Corti consists of four rows of specialized cells called hair cells, their supporting cells, and the *tectorial membrane* above the hair cells (Figure 9.5c). To visualize these structures, remember you are looking down a long tube; imagine the four rows of hair cells as picket fences or rows of telephone poles and the tectorial membrane as a shelf overlying the hair cells and extending the length of the cochlea.

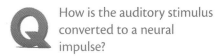

How is the auditory stimulus converted to a neural impulse?

FIGURE 9.5 Structures of the Middle and Inner Ear.

(a) The middle ear—tympanic membrane and ossicles—and the inner ear—the cochlea. (b) A section of the cochlea. (c) The organ of Corti.

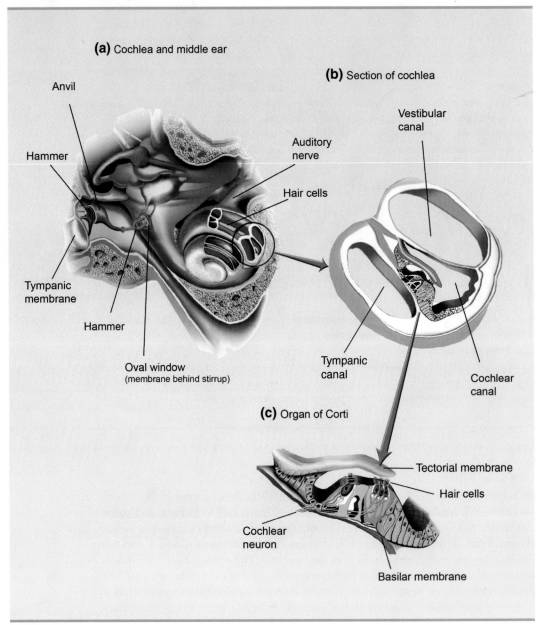

(a) Cochlea and middle ear

(b) Section of cochlea

Anvil

Hammer

Auditory nerve

Vestibular canal

Hair cells

Tympanic membrane

Hammer

Oval window
(membrane behind stirrup)

Tympanic canal

Cochlear canal

(c) Organ of Corti

Tectorial membrane

Hair cells

Cochlear neuron

Basilar membrane

The hair cells are the receptors for auditory stimulation. Vibration of the basilar membrane and the cochlear fluid bends the hair cells, opening potassium and calcium channels (not sodium channels, as in neurons) and depolarizing the hair cell membrane. This sets off impulses in the auditory neuron connected to the hair cell. When the hair cell moves back in the opposite direction, it relaxes and the potassium channels close. The hair cells are very sensitive; the amount of movement required to produce a sensation is equivalent to the Eiffel Tower leaning just the width of your thumb (Hudspeth, 1989), about the distance the tower sways in a strong wind.

The human cochlea has two sets of hair cells: a single row of about 3,500 inner hair cells and three rows of about 12,000 outer hair cells ("inner" and "outer" refer to their location relative to the center of the coiled cochlea). The less numerous *inner hair cells* receive 90% to 95% of the auditory neurons, and they provide the majority of the information about auditory stimulation (Dallos & Cheatham, 1976). A strain of mouse lacking inner hair cells due to a mutant gene is unable to hear (Deol & Gluecksohn-Waelsch, 1979). The *outer hair cells* amplify the signal produced by weak sounds and provide adjustable frequency selectivity (Ashmore, 1994; Zheng et al., 2000). How do they do this? The outer hair cells' cilia are embedded in the tectorial membrane (see Figure 9.6), and electrical stimulation causes them to lengthen or shorten very rapidly by as much as 5%. William Brownell suggested that shortening and lengthening of the outer hair cells changes the tension between the tectorial membrane and the basilar membrane and thus adjusts the rigidity of the organ of Corti (Brownell, Bader, Bertrand, & de Ribaupierre, 1985).

3
Audition and the Dancing Hair Cell

The Auditory Cortex

Neurons from the two cochleas make up part of the auditory nerves (eighth cranial nerves), one of which enters the brain on each side of the brain stem. The neurons pass through brain stem nuclei (see Figure 9.7a) to the inferior colliculi, to the medial geniculate nucleus of the thalamus, and finally to the auditory cortex in each temporal lobe. Neurons from each ear go to both temporal lobes, but there are more connections to the opposite side than to the same side. This means that a sound on your right side is registered primarily, but not exclusively, in the left hemisphere of the brain. Researchers interested in differences in function between the two hemispheres have used an interesting strategy to stimulate one side of the brain. They present an auditory stimulus through headphones to one ear and present white noise (which contains all frequencies and sounds like radio static) to the other ear to occupy the nontargeted hemisphere. This technique has helped researchers determine that the left hemisphere is dominant for language in most people and that the right hemisphere is better at other tasks, such as identifying melodies.

The auditory cortex is on the superior (upper) gyrus of the temporal lobe of each hemisphere; part of it is hidden inside the lateral fissure, as you can see in Figure 9.7b. The area is *topographically organized,* which means that neurons from adjacent receptor locations project to adjacent cells in the cortex. In this case, the projections form a sort of map of the unrolled basilar membrane (Merzenich, Knight, & Roth, 1975), just as the somatosensory cortex contains a map of the body. We will see that this organization is typical in the senses when we study vision in the next chapter.

The work of the auditory system is hardly finished when we have heard a sound. Beyond the primary auditory cortex are additional processing areas, as many as nine in some mammals; these secondary auditory areas are involved in processing complex sounds and understanding their meaning. For example, some of the cells adjacent to the monkey's primary auditory area respond selectively to calls of their own species, and some of

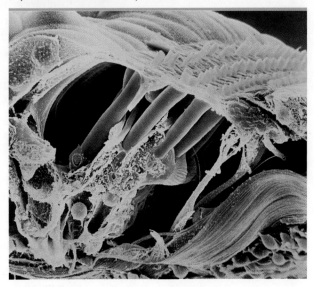

FIGURE 9.6 Electron Microscope View Showing the Hair Cells Attached to the Tectorial Membrane.

(The colors are artificial.)

SOURCE: Dr. G. Oran Bredberg/SPI/Photo Researchers.

FIGURE 9.7 The Auditory Pathway and the Auditory Cortex.

In (a) you can see that input to each ear goes to the auditory cortex in both hemispheres, but primarily to the opposite hemisphere. In (b) the temporal lobe has been pulled out to reveal the inner surface.

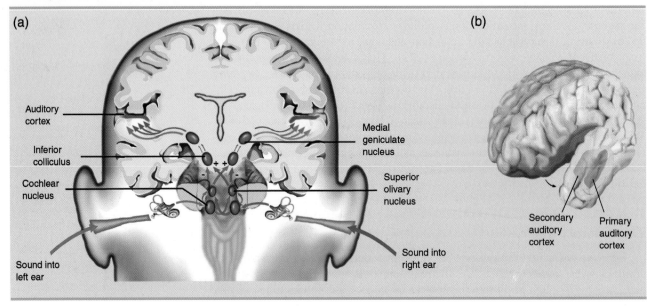

those react only to one type of call (Wollberg & Newman, 1972). The human primary auditory cortex has a secondary area surrounding it (Figure 9.7b), but auditory information also travels well beyond the auditory areas, following the *dorsal stream* or the *ventral stream* (Alain, Arnott, Hevenor, Graham, & Grady, 2001; Rauschecker & Tian, 2000).

The dorsal stream flows from the auditory cortex through the parietal area, where the brain determines the spatial location of a sound source. The information then proceeds to the frontal lobes, where it can be used for directing eye movements toward sound sources and for planning movements. The ventral stream is active when the individual is distinguishing among sounds; the call-specific cells of the monkey's auditory system are part of this system. Because of their specialties, the ventral and dorsal streams have been dubbed the *what* and *where* systems of audition. These two pathways are illustrated in Figure 9.8. We will see in the next chapter that vision has similar what and where systems.

Frequency Analysis

The sounds that are important to us, such as speech and music, vary greatly in intensity and frequency, and they change intensity and frequency rapidly. It is the task of the cochlea and the auditory cortex to analyze these complex patterns and convert the raw information into a meaningful experience.

We will concentrate on frequency analysis, which has received the lion's share of attention from researchers. More than 50 years ago, Ernest Wever (1949) described 17 versions of the two major theories of frequency analysis, which indicates the difficulty we have had in figuring out how people experience pitch. We will discuss a few versions that have been important historically. Besides introducing you to these two important theories, our discussion will describe what we know about how the auditory mechanism works and give you some idea of how theories develop in response to emerging evidence.

FIGURE 9.8   The Dorsal "Where" and Ventral "What" Streams of Auditory Processing.

The red areas were active when subjects determined the locations of sounds. Green areas were activated when they identified sounds. Localization and identification followed dorsal "where" and ventral "what" streams, respectively, with both terminating in frontal areas. (Functional MRI data was superimposed over a smoothed brain.)

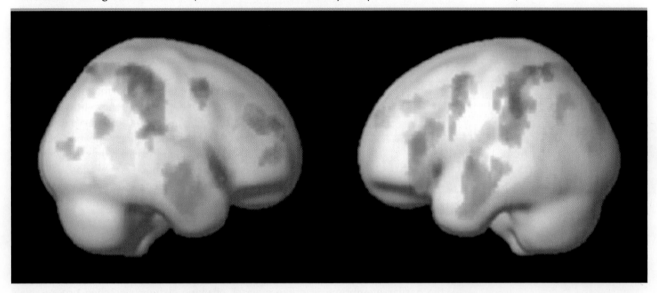

SOURCE: From P. P. Maeder, R. A. Meuli, M. Adriani, A. Bellmann, E. Fornari, J.-P. Thiran, A. Pittet, and S. Clarke, "Distinct Pathways Involved in Sound Recognition and Localization: A Human fMRI Study," 2001, *NeuroImage, 14,* 802–816.

Frequency Theories

The most obvious explanation of how the auditory system analyzes frequency is the *frequency theory,* which assumes that the auditory mechanism transmits the actual sound frequencies to the auditory cortex for analysis there. William Rutherford proposed an early version in 1886; it was called the *telephone theory* because he believed that individual neurons in the auditory nerve fired at the same frequency as the rate of vibration of the sound source. Half a century later, it was possible to test the theory with electrical recording equipment. Ernest Wever and Charles Bray (1930) performed one of the most intriguing investigations of auditory frequency analysis found in the scientific literature. They attached an electrode to the auditory nerve of anesthetized cats and recorded from the nerve while they stimulated the cat's ear with various sounds. Because the simple equipment used to record neural activity at that time was unable to respond to frequencies above 500 Hz, Wever and Bray ran the amplified neural responses into a telephone receiver in a soundproof room and listened to the output. Sounds produced by a whistle were transmitted with great fidelity. When someone spoke into the cat's ear, the speech was intelligible, and the researchers could even identify who the speaker was. The auditory nerve was "following," or firing at the same rate as the auditory stimulus. It appeared that the telephone theory was correct, but with the benefit of our more modern understanding of neural functioning, you and I know that neurons cannot fire at such high rates. (See the discussion of refractory periods in Chapter 2 if you don't remember.)

Wever and Bray were not recording from a single neuron, but from all the neurons in contact with the hook-shaped copper electrode they placed around the auditory nerve. Thus, they were monitoring the *combined* activity of hundreds of neurons. Wever explained their finding later in the *volley theory,* which states that

How do the frequency and place theories explain frequency analysis?

4
Frequency Analysis Animation

FIGURE 9.9 Illustration of Volleying in Neurons.

No single neuron can follow the frequency of the sound, but a group of neurons can.

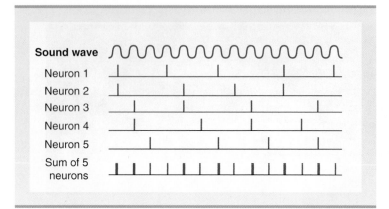

groups of neurons follow the frequency of a sound at higher frequencies where a single neuron cannot (Wever, 1949). A group of neurons is able to follow high frequencies because different neurons "take turns" firing. The term *volleying* is an analogy to the practice of soldiers with muzzle-loading rifles, who would fire in squads and then reload while the other squads were firing. Volleying is illustrated in Figure 9.9, where each of the neurons synchronizes its firing to the waves of the tone; no single neuron can fire on every wave, but some neurons will be firing on each wave. In this theory, the brain is required to combine information from many neurons to determine the frequency. In Wever and Bray's study, volleying in the auditory nerve was unable to keep up with the sound frequency beyond 5200 Hz, a figure that subsequent research has shown to be accurate (Rose, Brugge, Anderson, & Hind, 1967). So even with volleying, frequency following can account for only one fourth of the range of frequencies we hear.

Place Theory

In the 19th century, Hermann von Helmholtz (1863/1948) proposed that the basilar membrane was like a series of piano strings, stretched progressively more loosely with distance down the membrane. Then he invoked a principle from physics called *resonance* to explain how we discriminate different frequencies. Resonance is the vibration of an object in sympathy with another vibrating object. If you hold a vibrating tuning fork near the strings of a piano or a guitar, you will notice that the strings begin to vibrate slightly. A high-frequency tuning fork causes the thinner, more tautly stretched strings to vibrate more than the others, and a low-frequency tuning fork causes the thicker, looser strings to vibrate most. According to Helmholtz, resonance would cause the narrow base end of the membrane to resonate more to high-frequency sounds, the middle portion to moderate frequencies, and the wider apex (tip) to low frequencies. Helmholtz's proposal was a type of *place theory*, which states that identifying the frequency of a sound depends on the location of maximal vibration on the basilar membrane and which neurons are firing most. Place theory in its various evolving versions has been the most influential explanation of frequency analysis for a century and a half. It is an example of a theory that has become almost universally accepted but continues to be referred to as a theory.

A century later, Georg von Békésy, a communications engineer from Budapest, began a series of innovative experiments that won him the Nobel Prize for physiology in 1961. Békésy constructed mechanical models of the cochlea and also observed the responses of the basilar membrane in cochleas he removed from deceased subjects as diverse as elephants and humans. When he stimulated these cochleas with a vibrating piston, he could see under the microscope that vibrations peaked at different locations along the basilar membrane; a wavelike peak hovered near the base when the frequency was high and moved toward the apex as Békésy (1951) decreased the frequency. But Helmholtz was wrong about the basilar membrane being like a series of piano strings; Békésy (1956) determined that its frequency selectivity is due to differences in elasticity, with the membrane near the stirrup 100 times stiffer than the apical end.

Figure 9.10 shows how frequency sensitivity is distributed along the membrane's length (see Figures 9.4 and 9.5 again for the location of the basilar membrane). Recordings from single auditory neurons have confirmed that place information

FIGURE 9.10 Frequency Sensitivity on the Human Basilar Membrane.

Notice that the basilar membrane is narrow at the base end of the cochlea and widens toward the apex, the opposite of the cochlea's shape.

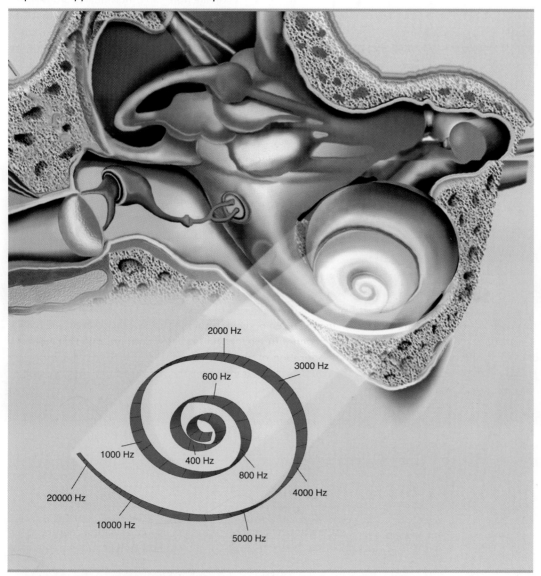

about frequency is carried from the cochlea to the cortex. You can see from the *tuning curves* in Figure 9.11 that each neuron responds most to a narrow range of frequencies (Palmer, 1987), owing to the neuron's place of origin in the cochlea. Because the auditory cortex is topographically organized, it also contains a *tonotopic map,* which means that each successive area responds to successively higher frequencies (Figure 9.12).

Each neuron responds to a range of frequencies around its "primary" frequency due to the fact that a tone produces vibrations over a wide area of the membrane. So how can neurons that make such imperfect discriminations inform the brain about the frequency of a sound with the 2- to 3-Hz sensitivity that has been observed? Most likely, the brain compares the rate of firing in neurons from adjacent places on the basilar membrane. We will see other examples of this "neural comparison" in

264   Chapter 9

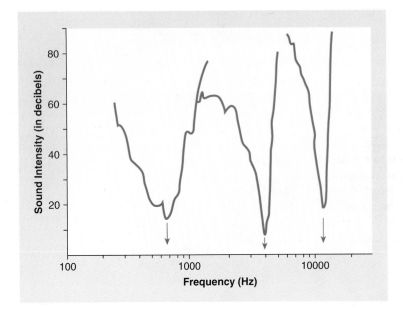

FIGURE 9.11 "Tuning Curves" of Cat Auditory Neurons.

Curves are from three individual neurons in the cat auditory cortex. Frequencies within the range of a curve will activate that neuron, but the intensity of sound required to produce a response increases as the sound moves away from the neuron's "preferred" frequency. Note that tuning is much sharper at higher frequencies.

SOURCE: Figure 11.31 from *Sensation and Perceptions* (5th ed.; p. 331) by E. Bruce Goldstein, 1999, Pacific Grove, CA: Brooks-Cole. © 1999. Reprinted by permission of Wadsworth, a division of Thomson Learning: www.thomsonrights.com. Fax 800-730-2215.

FIGURE 9.12 Tonotopic Map.

The part of the auditory cortex at the bottom of the magnified view receives neurons from the apex of the cochlea, and the other end responds to signals from the base. The auditory cortex thus forms a "map" of the basilar membrane so that each successive area responds to progressively higher frequencies.

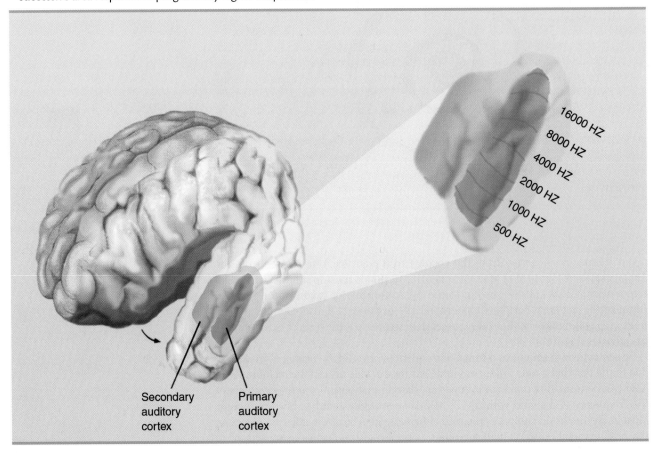

our discussion of sound localization and, in the next chapter, when we talk about color receptors.

Place analysis is the reason we can hear with some clarity through bone conduction. The vibrations enter the cochlea from all sides during bone conduction, rather than through the oval window, but Békésy (1951) demonstrated with his cochleas that this does not disrupt the tonotopic response of the basilar membrane. As he moved his vibrating piston from the base around to the side of the cochlea, or to the apex or anywhere else, the peak of vibration remained in the same location. Thomas Edison was nearly deaf, yet his second most famous invention was the phonograph. He compensated for his impaired hearing by grasping the edge of the phonograph's wooden case between his teeth and listening to the recording through bone conduction. You can still see the bite marks on one of his phonographs in the museum at his winter home and laboratory in Fort Myers, Florida. Figure 9.13 shows another bite-to-listen device, and the Application explains how doctors took advantage of place analysis to restore Heather Whitestone's ability to hear.

At low frequencies, the whole basilar membrane vibrates about equally, and researchers have been unable to find neurons that are specific for frequencies below 200 Hz (Kiang, 1965). Wever (1949) suggested a *frequency-volley-place theory:* individual neurons follow the frequency of sounds up to about 500 Hz by firing at the same rate as the sound's frequency; then between 500 and 5000 Hz, the frequency is tracked by volleying, and place analysis takes over beyond that point. However, most researchers subscribe to a simpler *frequency-place theory,* that frequency following by individual neurons accounts for frequencies up to about 200 Hz and all remaining frequencies are represented by the place of greatest activity. Whether or not volleying plays any role, it appears that place analysis alone is an inadequate explanation for auditory frequency analysis.

Fortunately, we can sum up the auditory system's handling of intensity coding much more simply. As we learned in Chapter 2, a more intense stimulus causes a neuron to fire at a higher rate. The auditory system relies on this strategy for distinguishing among different intensities of sound. However, this is not possible at lower frequencies where firing rate is the means of coding frequency. Researchers believe that at the lower frequencies, the brain relies on the number of neurons firing as increases in stimulus intensity recruit progressively higher-threshold neurons into activity.

Analyzing Complex Sounds

You may have realized that we rarely hear a pure tone. The speech, music, and noises that are so meaningful in our everyday life are complex, made up of many frequencies. Yet we have an auditory mechanism that appears to be specialized for responding to individual frequencies. But a solution to this enigma was suggested even before Helmholtz proposed his place theory. Forty years earlier, the French mathematician Fourier had demonstrated that any complex waveform—sound, electrical, or whatever—is in effect composed of two or more component sine waves. *Fourier analysis* is the analysis of a complex waveform into its sine wave components (see Figure 9.14). A few years later, Georg Ohm, better known for Ohm's law of electricity, proposed that the ear performs a Fourier analysis of a complex sound and sends information about each of the component frequencies to the cortex. Current researchers agree that the basilar membrane acts as the auditory Fourier analyzer, responding simultaneously along its length to the sound's component frequencies.

FIGURE 9.13 A Musical Toy That Works by Bone Conduction.

This is a musical lollipop holder. It has no speaker or earplug, but bite down on the lollipop and you literally hear the music in your head, thanks to bone conduction and place analysis. (Yes, it really works . . . but it won't replace your iPod!)

SOURCE: Bob Garrett.

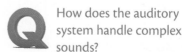 How does the auditory system handle complex sounds?

The dominant component is a relatively high-amplitude, low-frequency sine wave; the other components are progressively higher frequencies at lower intensities. If we produced sounds at each of these frequencies and amplitudes at the same time, the combined waveform when displayed on an oscilloscope would look like the waveform at the top, and the result would sound like the clarinet note.

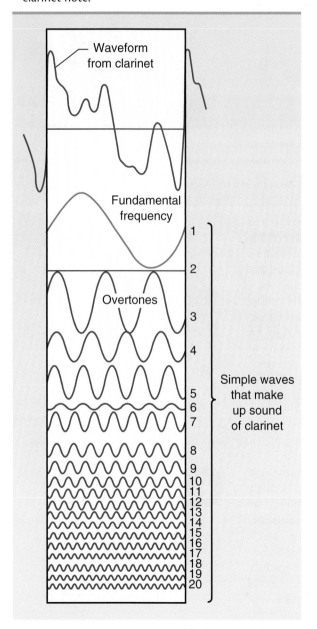

In actual experience, not only do we rarely hear a pure sound, but we also seldom hear a single complex sound alone. At a party, we hear the music playing loudly, mixed in with several conversations going on all around us. In spite of the number of complex sounds assaulting our cochleas, we are able to separate the speech of our conversation partner from the other noises in the room. We do more than that; we sample the other sounds regularly enough to enjoy the music and to hear our name brought up in a conversation across the room. The ability to sort out meaningful auditory messages from a complex background of sounds is referred to as the *cocktail party effect.*

The cocktail party effect is an example of selective attention; the brain must select the important part of the auditory environment for emphasis and suppress irrelevant background information. When we look at attention more closely in the final chapter, we will see that selective attention actually enhances activity in one part of the sensory cortex and reduces it in others. For this to be possible, we must identify sounds as distinct from each other. In this respect, it helps to think in terms of an *auditory object*—a sound that we recognize as having an identity that is distinct from other sounds. Sound localization helps, since different sounds are often in different locations. More important, though, is the ability to recognize the identity of a sound from one time to the next, for example, when you recognize your friend's voice. Recognizing the voices of individuals involves secondary auditory cortex in the superior temporal area (Kriegstein & Giraud, 2004; Petkov et al., 2008). Identifying environmental sounds, such as a dripping faucet, primarily requires posterior temporal areas, with the frontal cortex involved to a lesser extent (Figure 9.15; J. W. Lewis

Recognized sounds activated the areas in yellow; unrecognized sounds activated other areas (blue), mostly in the right hemisphere. Notice that the activity mostly occurs in the ventral "what"/frontal pathway.

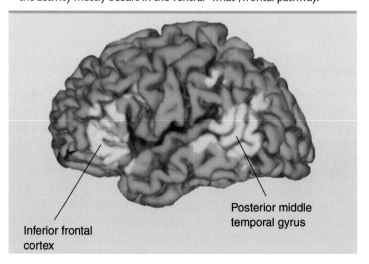

SOURCE: From "How Much Distraction Can You Hear?" by P. Milner, *Stereo Review,* June 1977, pp. 64–68. © 1977. Reprinted by permission of Stereo Review.

SOURCE: From "Human Brain Regions involved in Recognizing Environmental Sounds," by J. W. Lewis et al., 2004, *Cerebral Cortex, 14,* 1008–1021. Reprinted with permission.

Hearing and Language   267

et al., 2004). Recognizing these sound objects thus activates the ventral "what" stream. (In a few pages, you will see that these areas are also important in understanding and producing language.) Suppressing irrelevant information typically involves sending stimulation to lower levels of the sensory pathway, in this case to the cochlea itself. This descending activity probably works by changing the length of outer hair cells. Presumably, the tension changes they produce in the organ of Corti cause localized sensitivity adjustments, which suppress other sounds during attention to a particular sound (Xiao & Suga, 2002).

Locating Sounds With Binaural Cues

The most obvious way to locate a sound is to turn your head until the sound is loudest. This is not very effective because the sound may be gone before the direction is located. Three additional cues permit us to locate sounds quickly and

Application   Cochlear and Brainstem Implants to Restore Hearing

Hearing impairment can be due to *conductive deafness*, a result of interference with the conduction of vibrations through the middle ear, or *nerve deafness*, which is caused by damage to the hair cells or the auditory nerve. Ninety percent of cases of hearing impairment involve hair cell damage, and these individuals may be candidates for a cochlear implant. The implant's microphone picks up sounds and sends them to a speech processor located behind the ear; then a transmitter on the surface of the skin sends the signal to a receiver that is surgically mounted just beneath the skin. From there, the signal travels through a wire to an electrode array threaded through the cochlea (see the figure). The device sends signals representing the different frequencies of a sound to different locations along the length of the electrode. Activating different neurons with different frequencies mimics the functioning of the basilar membrane and hair cells in an unimpaired individual; in other words, it relies on the principle of place analysis. The sound representation is rather crude, but most recipients are able to hear effectively enough to use a telephone (which is more difficult than face-to-face conversation), and most children can be mainstreamed in school. In a recent survey, children with implants rated their quality of life comparably with their peers (Loy, Warner-Czyz, Tong, Tobey, & Roland, 2010). How well the implant works depends on how long the person has been deaf, partly because adjustment to an enriched auditory world can be surprisingly difficult. More important, neurons from other sensory areas can intrude into the unused auditory cortex over time; if a PET scan shows elevated activity in the auditory cortex, it means that vision or another sensory function has taken over and that a cochlear implant will not improve hearing (D. S. Lee et al., 2001). In adults, success also depends on having learned language before deafness occurred. Children, on the other hand, are able to use the implants whether they learned language previously or not (Francis, Koch, Wyatt, & Niparko, 1999).

5
Cochlear Implant

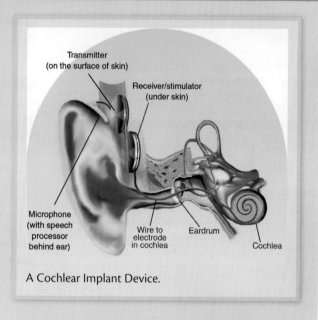

A Cochlear Implant Device.

Labels: Transmitter (on the surface of skin); Receiver/stimulator (under skin); Microphone (with speech processor behind ear); Wire to electrode in cochlea; Eardrum; Cochlea

A possible alternative is gene therapy; inoculating guinea pigs with the *ATOH1* gene, which is responsible for hair cell development, restored some hearing after their hair cells had been destroyed by a drug (Izumikawa et al., 2005). Another possibility is the use of stem cells; in the laboratory, they have been coaxed into developing into functional auditory neurons and hair cell-like cells (W. Chen et al., 2009). None of these treatments will work in some forms of cochlear damage or when the auditory nerve is nonfunctional, for example, following surgery to remove auditory nerve tumors. The auditory midbrain implant involves insertion of up to 15 electrodes into the cochlear nucleus in the brainstem. Still in its infancy, this experimental procedure has demonstrated its ability to improve lip reading by 100% and to restore speech comprehension to a useful level (Behr et al., 2007).

FIGURE 9.16 Sound Localizing Device Used by 19th-Century Sailors.

By listening through devices on a long rod, they effectively increased the distance between their ears and enhanced the binaural cues.

How does the brain determine the locations of sounds?

FIGURE 9.17 Differential Intensity and Time of Arrival as Cues for Sound Localization.

The sound is reduced in intensity and arrives later at the distant ear.

Sound is less intense at left ear because the head creates a "sound shadow."

Sound arrives at left ear later than right ear due to greater distance sound must travel.

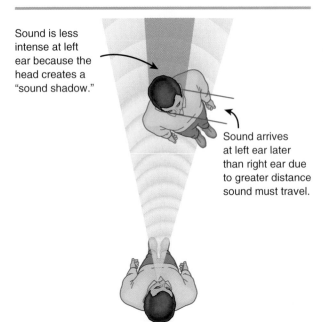

accurately, including those that are too brief to allow turning the head. All three of these cues are *binaural,* meaning that they involve the use of both ears; the brain determines the location of the sound based on differences between the sound at the two ears. These cues are useless when a sound source is in the median plane (equidistant from the person's ears), but if the sound is slightly to one side, the stimulus will differ between the ears. Animals with ears that are very close together (such as mice) are at a disadvantage in locating sounds because the differences are so small. Grasshoppers and crickets have evolved a compensation for their small head size: Their auditory organs are on their legs, as far apart as possible. Nineteenth-century sailors used a novel application of this strategy when they needed to locate a distant foghorn: They listened through tubes attached to funnels at the ends of a long rod (Figure 9.16). The following paragraphs describe the three binaural differences: *intensity, time of arrival,* and *phase.*

Binaural Cues

When a sound source is to one side or the other, the head blocks some of the sound energy. The sound shadow this creates produces a *difference in intensity,* so that the near ear receives a slightly more intense sound (Figure 9.17). Some of the neurons in the superior olivary nucleus respond to differences in intensity at the two ears. Because low-frequency sounds tend to bend around obstacles, this cue works best when the sound is above 2000 or 3000 Hz.

The second binaural cue for locating sounds is *difference in time of arrival* at the two ears. A sound that is directly to a person's left or right takes about 0.5 millisecond to travel the additional distance to the second ear (see Figure 9.17 again); humans can detect a difference in time of arrival as small as 10 microseconds (millionths of a second; Hudspeth, 2000), which means we can locate sounds even very near the midline of the head with good accuracy. We cannot distinguish such small intervals consciously, of course; this kind of precision involves automatic processing by specialized circuits, as we will see shortly.

At low frequencies, a sound arriving from one side of the body will be at a different phase of the wave at each ear, referred to as a *phase difference* (see Figure 9.18). As a result, one eardrum will be pushed in while the other is being pulled out. Some of the auditory neurons in the superior olivary nucleus respond only to phase differences. Above about 1500 Hz, a sound will have begun a new wave by the time it reaches the second ear, so phase difference is useless at higher frequencies.

Brain Circuits for Locating Sounds

Of the neural circuits for binaural sound localization, the one for time of arrival has been studied most thoroughly. The circuit has been mapped in the barn owl, which is extremely good at sound localization; in fact, it can locate a mouse in darkness just from the sounds it makes rustling through the grass. The circuit is located in the nucleus laminaris, the avian (bird) counterpart of the mammalian superior olivary nucleus. Electrical recording has revealed the function of its *coincidence detectors,* neurons that fire most when they receive input from both ears at the same time (Carr & Konishi, 1990). Figure 9.19 is a simplified diagram of a coincidence detector circuit, based on

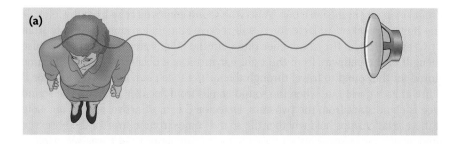

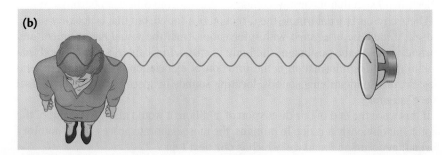

FIGURE 9.18 Phase Difference as a Cue for Sound Localization.

(a) At lower frequencies, the sound reaches each ear at a different phase of the same pressure wave; the different stimulation of the two ears can be used to locate the sound's direction. (b) At higher frequencies, the sound has begun a new wave by the time it reaches the second ear; the phase difference is useless for locating the sound.

FIGURE 9.19 A Circuit for Detecting Difference in Time of Arrival at the Two Ears.

The circuit's arrangement compensates for the greater travel time to the more distant ear. Try tracing the flow of activity through the circuit to determine which detector will fire most when sound comes from each of the speakers. (The speakers are at the same horizontal level as the ears.)

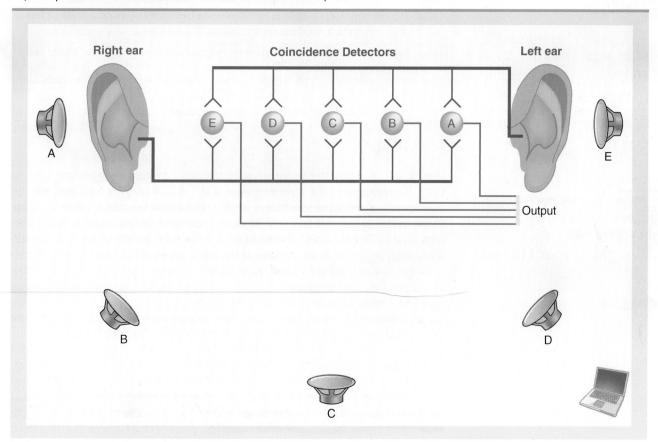

SOURCE: Based on the results of Carr and Konishi (1990).

neuronal mapping in the barn owl. When the sound comes directly from the left side of the figure (Speaker A), Detector A will receive stimulation simultaneously from the two ears and will fire at a higher rate than the other detectors. This is because the length of the pathway from the right ear imposes a delay that equals the time required for the sound to travel through the air to the left ear. Likewise, Detector B will fire at its highest rate when the sound comes from Speaker B. When the sound source is equidistant from the two ears, Detector C is most active, and so on. Note that these relationships hold whether the sound comes from in front of the observer, behind, above, or below. This circuit is a good example of what I was referring to earlier when I said that the brain "compares" inputs from different neurons.

These circuits can determine the direction of a sound, but that is not very useful by itself; it must be integrated with information from the visual environment and information about the position of the body in space. In Chapter 3 you learned that combining all this information is the function of association areas in the parietal lobes. So, unlike identifying sounds, locating sounds in space occurs in the dorsal "where" stream.

If this were the end of our discussion of audition, it would also be the end of the chapter, but obviously it is not. In humans, the most elaborate processing of auditory information occurs in language, which is our next topic.

Concept Check   *Take a Minute to Check Your Knowledge and Understanding*

- ☐ Trace an auditory stimulus from the pinna to the auditory neurons.
- ☐ Explain how, according to place theory, the frequency of sound is coded. How does the cochlea handle complex sounds?
- ☐ Explain how the circuit for detecting difference in time of arrival of sounds at the two ears works.

Language

For humans, the most important aspect of hearing is its role in processing language.

—*A. J. Hudspeth*

Few would question the importance of language in human behavior. Keep in mind the meaning of the term *language:* It is not limited to speech but includes the generation and understanding of written, spoken, and gestural communication. Communication through language has important survival value and is inestimably important to human social relationships. A person who cannot speak or write suffers a high degree of isolation; one who cannot comprehend the communications of others is worse off still. These capabilities not only require learning, but they also depend on specific structures of the brain, and damage to these structures can deprive a person of some or all of these functions.

In 1861, the French physician Paul Broca reported his observations of a patient who for 21 years had been almost unable to speak. Tan, as he was known by the hospital staff because that was one of the few sounds he could make, died shortly after he came under Broca's care. Autopsy revealed that Tan's brain damage was located in the posterior portion of the left frontal lobe. After studying eight other patients, Broca concluded that *aphasia*—language impairment caused by damage to the brain—results from damage to the frontal area anterior to the motor cortex, now known as *Broca's area.* Nine years later, a German doctor named Carl Wernicke identified a second site where damage produced a different form of aphasia. Located in the posterior portion of the left temporal lobe, this site is known

FIGURE 9.20 Language-Related Areas of the Cortex.

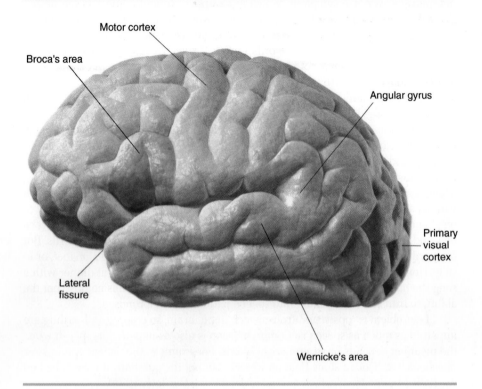

as Wernicke's area. See Figure 9.20 to locate Broca's and Wernicke's areas and the other structures to be discussed here. Most of our understanding of the brain structures involved in language comes from studies of brain-damaged individuals, so this is where we will start.

Broca's Area

Broca's aphasia is language impairment caused by damage to Broca's area and surrounding cortical and subcortical areas. The symptoms can best be understood by examining the speech of a stroke patient; as you read this interview, you will see why the disorder is also referred to as *expressive aphasia.*

Doctor: Why are you in the hospital, Mr. Ford?

Mr. Ford: Arm no good. Speech . . . can't say . . . talk, you see.

Doctor: What happened to make you lose your speech?

Mr. Ford: Head, fall, Jesus Christ, me no good, str, str . . . oh Jesus . . . stroke.

Doctor: I see. Could you tell me, Mr. Ford, what you've been doing in the hospital?

Mr. Ford: Yes, sure. Me go, er, uh, P.T. nine o'cot, speech . . . two times . . . read . . . wr . . . ripe, er rike, er, write . . . practice . . . getting better. (H. Gardner, 1975, p. 61)

Mr. Ford's speech is not nearly as impaired as Tan's; he can talk, and you can get a pretty good idea of his meaning, but he shows the classic symptoms associated with damage to Broca's area. First, his speech is *nonfluent;* although

6

Aphasia

well-practiced phrases such as "yes, sure" and "oh, Jesus" come out easily, his speech is halting, with many pauses between words. Second, he has trouble finding the right words, a symptom known as *anomia* ("without name"). Often he has *difficulty with articulation;* he mispronounces words, like "rike" for *write*. Finally, notice that his speech is *agrammatic;* it has content words (nouns and verbs) but lacks grammatical, or function, words (articles, adjectives, adverbs, prepositions, and conjunctions). The hardest phrase for a Broca's aphasic to repeat is "No ifs, ands, or buts" (Geschwind, 1972).

Thelma was similarly impaired, but I had some enjoyable conversations with her at the nursing home, mainly because I was willing to piece together her broken speech and to nod and smile when even that was impossible. She could usually manage only one or two words at a time: she showed me old photos of her parents, pointing and saying "Mother . . . Father." But like Tan, who would occasionally express his frustration with the oath *"sacre nom de Dieu!"* ("holy name of God!"), Thelma would occasionally blurt out something meaningful. After a disagreement with an aide in which she was unable to express herself effectively, she exclaimed to me, "They can say anything they want to! I know everything. I just can't say." Broca believed that Broca's aphasia impaired motor instructions for vocalizing words. But Mr. Ford was able to recite the days of the week and the letters of the alphabet, or to sing "Home on the Range," and Thelma would entertain the group at dinner with a song she had composed before she was impaired. So vocalization is not lost, but the ability to translate information into speech patterns is compromised.

The problem is "upstream" from speech in the brain, so reading and writing are impaired as much as speech is. Comprehension is also as impaired as speech when the meaning depends on grammatical words. For example, the patient can answer questions like "Does a stone float on water?" but not the question "If I say, 'The lion was killed by the tiger,' which animal is dead?" (H. Gardner, 1975).

Wernicke's Area

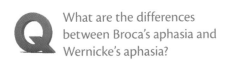

What are the differences between Broca's aphasia and Wernicke's aphasia?

In *Wernicke's aphasia,* the person has difficulty understanding and producing spoken and written language. This is often called *receptive aphasia,* but that term is misleading because the same problems with understanding language also show up in producing it. For example, the person's speech is fluent but meaningless. A patient asked to describe a picture of two boys stealing cookies behind a woman's back said, "Mother is away here working her work to get her better, but when she's looking the two boys looking in the other part. She's working another time" (Geschwind, 1979). This meaningless speech is called *word salad*, for obvious reasons.

Because the speech of the Wernicke's patient is articulate and has the proper rhythm, it sounds normal to the casual listener. The first time I met a person with Wernicke's aphasia, I was knocking on the social worker's door at the nursing home, and I thought it was because my thoughts were elsewhere that I failed to understand one of the residents when she spoke. But then my "Pardon me" elicited "She's in the frimfram," and I realized the problem was hers rather than mine. I responded with a pleasantry, and she gave a classic word-salad reply. That began a long relationship of conversations, in some ways as enjoyable as those with Thelma. The difference was that neither of us ever understood the other; another difference was that it did not matter, because she seemed strangely unaware that anything was amiss.

The Wernicke-Geschwind Model

Wernicke suggested, and Norman Geschwind later elaborated on, a model for how Broca's area and Wernicke's area interact to produce language (Geschwind,

FIGURE 9.21 The Wernicke-Geschwind Model of Language.

Verbal input arrives in the auditory cortex and then travels to Wernicke's area for interpretation. Written input arrives there via the visual cortex. If a verbal response is required, Wernicke's area sends output to Broca's area for articulation of the response, and the facial area of the motor cortex produces the speech.

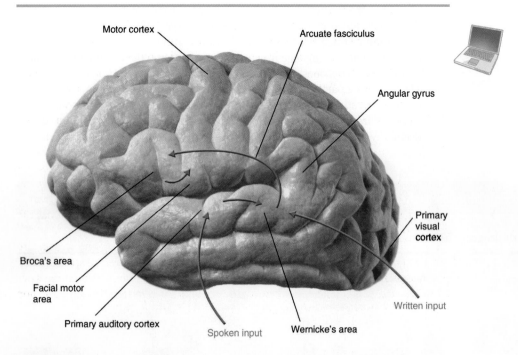

SOURCE: Adapted from "Specializations of the Human Brain," by N. Geschwind, *Scientific American*, 241 (9), pp. 180–199.

1970, 1972, 1979). The model is illustrated in Figure 9.21 and in the following examples. Answering a verbal question involves a progression of activity from the auditory cortex to Wernicke's area, and then to Broca's area. Broca's area then formulates articulation of the verbal response and sends the result to the facial area of the motor cortex, which produces the speech. If the response is to be written, Wernicke's area sends output to the angular gyrus instead, where it elicits a visual pattern. When a person reads aloud, the visual information is translated into the auditory form by the angular gyrus, then passed to Wernicke's area, where a response is generated and sent to Broca's area. The idea that visual information must be converted to an auditory form for processing arose in part from the fact that language evolved long before writing was invented, and Wernicke's area was believed to operate in an auditory fashion.

This system has long been the primary model for how language operates. It is relatively simple and seems to make sense of the various aphasias. Modern imaging techniques have confirmed the participation of Broca's and Wernicke's areas in language; one study has actually traced the progression of activity while subjects produced a verbal response to written material, from the visual cortex to Wernicke's area and then to Broca's area (Dhond, Buckner, Dale, Marinkovic, & Halgren, 2001). However, there are problems. One is that language functions are not limited to Broca's and Wernicke's areas; damage to the basal ganglia, thalamus, and subcortical white matter also produce aphasia (Hécaen & Angelergues, 1964; Mazzocchi & Vignolo, 1979; Naeser et al., 1982). Broad cortical areas also play an

Q What is the Wernicke-Geschwind model?

important role, though possibly only because they are storage sites for information. For example, using nouns (naming objects) produces activity just below the auditory cortex and Wernicke's area (H. Damasio, Grabowski, Tranel, Hichwa, & Damasio, 1996). Using verbs (describing what is happening in a picture) is impaired by damage to the left premotor cortex, which sends output to the motor cortex. This area is also activated while naming tools and by imagining hand movements (A. Martin, Wiggs, Ungerleider, & Haxby, 1996). Apparently when tool names are learned, they are stored near the brain structure that would produce the action.

Electrical stimulation studies (Mateer & Cameron, 1989; Ojemann, 1983) and studies of brain damage (Hécaen & Angelergues, 1964) have also shown that the various components of language functioning are scattered throughout all four lobes (see Figure 9.22). This does not mean that there is no specialization of the cortical areas. Damage in a particular lobe can produce a variety of symptoms, but articulation errors are still more likely to result from frontal damage and comprehension problems from damage in the temporal lobes (Hécaen & Angelergues, 1964; Mazzocchi & Vignolo, 1979).

FIGURE 9.22 Frequency of Language Deficits Resulting From Damage in Each Lobe.

The data make clear that the language functions are not restricted to specific areas.

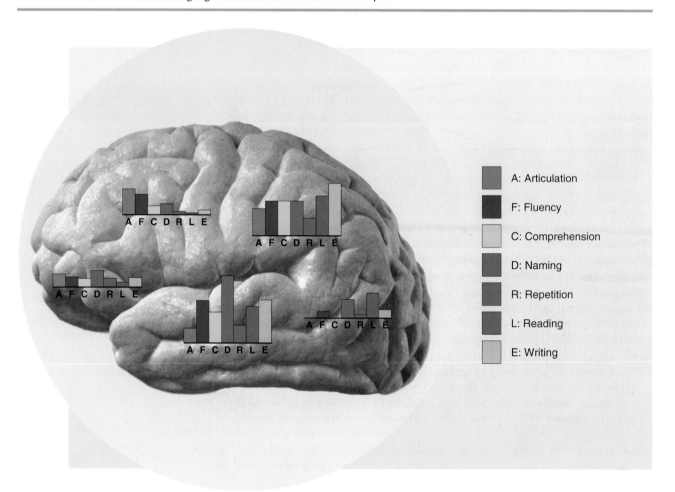

A: Articulation
F: Fluency
C: Comprehension
D: Naming
R: Repetition
L: Reading
E: Writing

SOURCE: Based on Hécaen and Angelergues (1964).

So it appears that the Wernicke-Geschwind view of a few discrete language areas is too simple. However, the theory is still a good starting point for understanding language. And if it has been misleading, it has also helped researchers organize their thinking about language and has generated volumes of research—which, after all, is how we make scientific sense of our world.

Reading, Writing, and Their Impairment

Although aphasia affects reading and writing, these functions can be impaired independently of other language abilities. *Alexia* is the inability to read, and *agraphia* is the inability to write. Presumably, they are due to disruption of pathways in the *angular gyrus* that connect the visual projection area with the auditory and visual association areas in the temporal and parietal lobes (see Figure 9.20 again). The PET scans in Figure 9.23 show that activity increases in this area during reading.

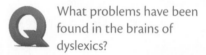

Q What problems have been found in the brains of dyslexics?

FIGURE 9.23 PET Scans During Reading.

Viewing letterlike forms (a) and strings of consonants (b) did not activate the area between the primary visual cortex and language areas, but reading pronounceable nonwords (c) and real words (d) did.

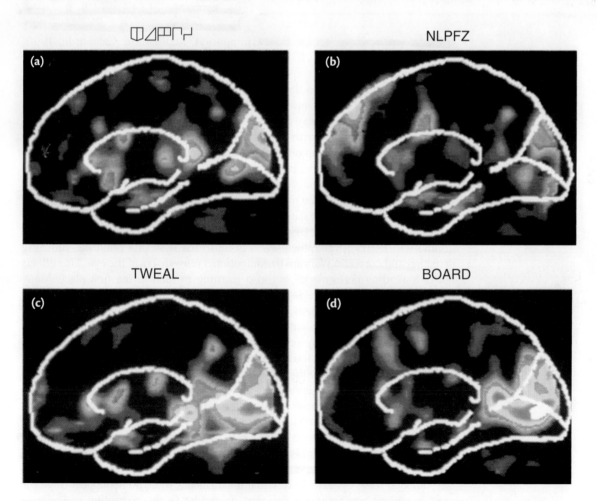

SOURCE: Reprinted with permission from S. E. Petersen, P. T. Fox, A. Z. Snyder and M. E. Raichle, "Activation of Extrastriate and Frontal Cortical Areas By Visual Words and Word-Like Stimuli," *Science*, 249, pp. 1049–1044. Copyright 1990 AAAS.

FIGURE 9.24 Developmental Anomalies in the Dyslexic Brain.

(a) Cells in the left planum temporale of a normal brain. (b) In the dyslexic brain, cells lack the normal layering and arrangement in columns, and some of the cells have migrated into the superficial layer where they would not ordinarily be found.

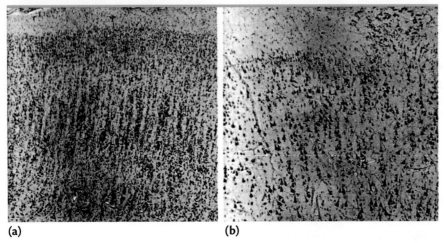

(a)   (b)

SOURCE: Courtesy A. M. Galaburda, Harvard Medical School.

7
Dyslexia

Reading and writing are also impaired in learning disorders. The most common learning disorders, are *dyslexia,* an impairment of reading; dysgraphia, difficulty in writing; and dyscalculia, a disability with arithmetic. Because of its importance and the amount of research that has been done, we will focus on dyslexia. Dyslexia can be *acquired,* through damage, but its origin is more often *developmental.* Developmental dyslexia is partially genetic, with an estimated heritability between 40% and 60% (Gayán & Olson, 2001). The genes that have been most reliably identified—*ROBO1, DCDC2, KIAA0319,* and *DYX1c1*—are all involved in neuron guidance and migration (McGrath, Smith, & Pennington, 2006; Rosen et al., 2007).

The implications of impaired brain development are far reaching. In most people, the *planum temporale,* where Wernicke's area is located, is larger in the left temporal lobe than in the right. However, in dyslexics it is more frequently equal in size or larger on the right (Humphreys, Kaufman, & Galaburda, 1990; Hynd & Semrud-Clikeman, 1989). In at least some dyslexic brains, many of the neurons in the left planum temporale lack the usual orderly arrangement, and some of them have migrated into the outermost layer (Figure 9.24; Galaburda, 1993; Humphreys et al., 1990; Kemper, 1984). The strong similarity to the defects in the brain affected by fetal alcohol syndrome that you saw in Chapter 3 suggests that this developmental error occurs prenatally.

The most familiar symptoms of dyslexia involve *visual-perceptual* difficulties. The person often reads words backwards ("now" becomes "won"), confuses mirror-image letters (*p* and *q, b* and *d*), and has trouble fixating on printed words, which seem to move around on the page. Dyslexics' brains are slow to respond (as measured by evoked potentials; see Chapter 4) to low-contrast, rapidly changing visual stimuli (Livingstone, Rosen, Drislane, & Galaburda, 1991). Presumably words jump around and letters reverse themselves because the dyslexic's brain has difficulty detecting and correcting for rapid, unintentional movements of the eye. This affects not only reading performance but also learning to read in the first place. Many dyslexics also have trouble detecting the frequency and amplitude changes that distinguish letter sounds (Stein, 2001); supposedly this impairs the dyslexic's ability to associate speech sounds with letters when learning to read and explains his or her slowness in reading nonwords.

Some researchers have focused their attention on the brain's magnocellular or large-celled systems. The visual magnocellular system, which we will learn more about in the next chapter, is responsible for visual contrast and rapid movement, and it monitors eye movements. Researchers have found that visual magnocellular cells in the thalamus are smaller in the brains of deceased dyslexics (Galaburda & Livingstone, 1993; Livingstone et al., 1991). According to the *magnocellular hypothesis* of dyslexia, deficiencies in auditory and visual magnocellular cells account for both auditory and visual impairments (Stein, 2001).

Although some dyslexics have visual impairments, these do not appear to be the most important contributor to dyslexia. According to the competing *phonological hypothesis,* the fundamental problem in dyslexia is not in visual or auditory processing but involves an impairment of phoneme processing. A phoneme is a small sound unit that distinguishes one word from another, for example, the beginning sounds that distinguish *book, took*, and *cook*. When a group of dyslexic college students was administered a battery of tests, 10 had auditory deficits and 2 had a visual magnocellular function deficit, but all 16 suffered from a phonological deficit (Ramus et al., 2003).

Functional MRI studies indicate that the problem is in the posterior language area that is believed to function as a word analysis system (as opposed to a nearby occipito-temporal area that recognizes words quickly and automatically by their form; Shaywitz et al., 1998, 2003). In unimpaired individuals, activity increases in Wernicke's area and the angular gyrus as phonologic difficulty increases (answering "Do *T* and *V* rhyme?" versus "Do *leaf* and *jeté* rhyme?"); activity does not increase in dyslexics. The incidence of dyslexia is twice as great in some cultures than in others, which seems to contradict a brain-deficit explanation of the disorder, but in fact the discrepancies support the phonological hypothesis. Italian and Spanish are phonologically simpler languages, with an almost one-to-one correspondence between phonemes and spelling. Predictably, dyslexia is much rarer in Italy and Spain than in French- and English-speaking countries, where the same spelling may have several pronunciations (*cough, tough, dough, slough*). PET imaging shows that dyslexia among Italians involves the same brain impairment as it does in French- and English-speaking dyslexics (see Figure 9.25; Paulesu et al., 2001).

FIGURE 9.25 Activation of Language Areas in Dyslexics From Three Countries.

(a) shows activation due to reading in controls; (b) shows activation due to reading in dyslexics; (c) indicates the area significantly less activated in dyslexics than in controls; (d) shows that dyslexics had the same impairment regardless of their nationality.

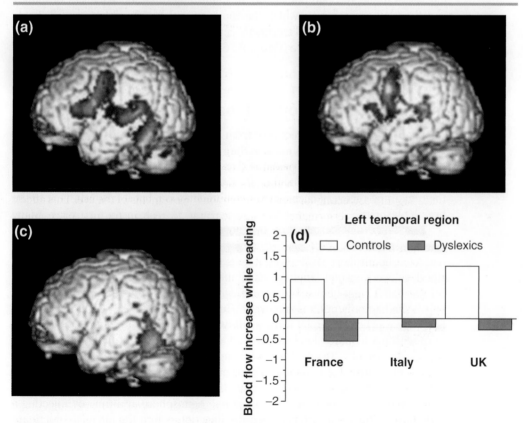

SOURCE: Reprinted with permission from E. Paulesu, J.-F. Démonet, F. Fazio, E. McCrory, V. Chanoine, N. Brunswick, S. F. Cappa, G. Cossu, M. Habib, C. D. Frith, and U. Frith, "Dyslexia: Cultural Diversity and Biological Unity," *Science, 291,* pp. 2165–2167. Copyright 2001 AAAS.

FIGURE 9.26 Effect of Training on Brain Activity in Children With Dyslexia.

Activity (a) in nondyslexic children during a rhyming task, (b) in children with dyslexia before training, and (c) in children with dyslexia after training.

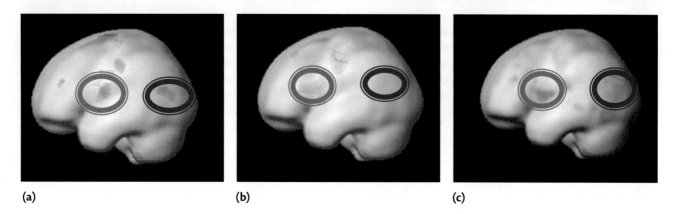

(a)   (b)   (c)

SOURCE: Reprinted with permission from J. D. E. Gabrieli, "Dyslexia: A New Synergy Between Education and Cognitive Neuroscience," *Science, 325,* 280–283. Copyright 2009 AAAS.

Dyslexia can be treated by training children in auditory and language processing, for example, through practice in distinguishing different sequences of sounds and listening to speech in which the rapid transitions have been slowed down. Although this is a behavioral treatment, activity increases in the brain areas where there had been deficits before (Figure 9.26; Temple et al., 2003).

Mechanisms of Recovery From Aphasia

There is usually some recovery from acquired aphasia during the first 1 or 2 years, more so for Broca's aphasia than for Wernicke's aphasia (I. P. Martins & Ferro, 1992). Initial improvement is due to reduction of the swelling that often accompanies brain damage rather than to any neural reorganization. Just how the remaining recovery occurs is not well understood, but it is a testament to the brain's plasticity.

The right hemisphere can take over language functions following left-hemisphere damage, as long as the injury occurs early in life. A 2-year-old girl had a left-hemisphere stroke (yes, it does occur); her language was impaired, but she developed normal language capability by the age of 7. Then at the age of 56 she had a right-hemisphere stroke, which resulted in a second aphasia from which she had only minimal recovery (Guerreiro et al., 1995). Right-hemisphere language was confirmed by fMRI in all five individuals of a group who had been born with inadequate blood supply to the language areas of the left hemisphere (Vikingstad et al., 2000). Rasmussen and Milner (1977) used the *Wada technique* and electrical stimulation to determine the location of language control in patients before removing lesioned tissue that was causing epileptic seizures. (The Wada technique involves anesthetizing one hemisphere at a time by injecting a drug into each carotid artery; when the injection is into the language-dominant hemisphere, language is impaired.) Individuals whose left-hemisphere injury occurred before the age of 5 were more likely to have language control in the right hemisphere, supporting the hypothesis of right-hemisphere compensation.

Patients whose left-hemisphere damage occurred later in life more often continued to have language control in the left hemisphere; there was, however, evidence in some cases that control had shifted into the border of the parietal lobe. Since language functions are scattered widely in the left hemisphere, perhaps the compensation involves enhancing already existing activity rather than establishing new functional areas.

The ability of the right hemisphere to assume language functions may be partly due to the fact that it normally makes several contributions to language processing. The most obvious right-hemisphere role in language is prosody; *prosody* is the use of intonation, emphasis, and rhythm to convey meaning in speech. We saw in the chapter on emotion that people with right-hemisphere damage have trouble understanding emotion when it is indicated by speaking tone and in producing emotional speech the same way. An fMRI study found that right-hemisphere activity increased while individuals detected angry, happy, sad, or neutral emotions from the intonation of words (Buchanan et al., 2000).

The right hemisphere also is important in understanding information from language that is not specifically communicated by the meaning of the words, such as when the meaning must be inferred from an entire discourse or when the meaning is figurative rather than literal. For example, interpreting the moral of a story activates the right hemisphere (Nichelli et al., 1995), as does understanding a metaphor or determining the plausibility of statements such as "Tim used feathers as paperweights" (Bottini et al., 1994). Interestingly, the right-hemisphere regions involved in all these activities correspond generally to the structures we have identified in left-hemisphere language processing.

A Language-Generating Mechanism?

When Darwin suggested that we have an instinctive tendency to speak, what he meant was that infants seem very ready to engage in language and can learn it with minimal instruction. Children learn language with such alacrity that by the age of 6 they understand about 13,000 words, and by the time they graduate from high school, their working vocabulary is at least 60,000 words (Dronkers, Pinker, & Damasio, 2000). This means that children learn a new word about every 90 waking minutes. The hearing children of deaf parents pick up language just about as fast as children with hearing parents (Lenneberg, 1969), in spite of minimal learning opportunities. Not only are preadolescent children particularly sensitive language learners, but they are also believed to be the driving force in the development of creole language (which combines elements of two languages, allowing communication between the cultures). In Nicaragua, children in the school for the deaf, where sign language is not taught, have devised their own sign language with unique gestures and grammar (Senghas, Kita, & Özyürek, 2004).

Noam Chomsky (1980) and later Steven Pinker (1994) interpreted children's readiness to learn language as evidence of a *language acquisition device,* a part of the brain hypothesized to be dedicated to learning and controlling language. Not all researchers agree with this idea, but most accept that there are biological reasons why language acquisition is so easy. This ease cuts across forms of language. For example, both hearing and deaf infants of signing parents babble in hand movements (Figure 9.27; the deaf infants' babbling proceeds into signing through the same stages and at about the same pace that children of speaking parents learn vocal language (Petitto, Holowka, Sergio, & Ostry, 2001; Petitto & Marentette, 1991). The researchers suggest that the ease of children's language acquisition is due to a brain-based sensitivity to rhythmic language patterns, a sensitivity that does not depend on the form of the language.

FIGURE 9.27 Babies of Signing Parents Babble With Their Hands.

Unlike the meaningless hand movements of other infants (which they also make at other times), their babbling is similar to their parents' signing. Babbling hand movements are slower and restricted to the space in front of the infants' bodies, and they correspond to the rhythmic patterning of adult sign-syllables.

SOURCE: Petitto, Holowka, Sergio, & Ostry, 2001. Photo courtesy of Dr. Laura-Ann Petitto, University of Toronto.

> Man has an instinctive tendency to speak, as we see in the babble of our young children.
>
> —*Charles Darwin*

Innate Brain Specializations

More than 90% of right-handed people are left-hemisphere dominant for language. This is also true for two thirds to three quarters of left-handers; the remainder are about equally divided between right-hemisphere dominant and mixed (Knecht et al., 2000; D. W. Loring et al., 1990; B. Milner, 1974). In the large majority of autopsied brains, Broca's area is larger (Falzi, Perrone, & Vignolo, 1982), and the lateral fissure (Yeni-Komshian & Benson, 1976) and planum temporale (Geschwind & Levitsky, 1968; Rubens, 1977; Wada, Clarke, & Hamm, 1975) are longer in the left than in the right hemisphere. These differences are not the result of usage. The left planum temporale is already larger by the 29th week of gestation (Wada et al., 1975; Witelson & Pallie, 1973). Newborns already show a larger increase in cerebral blood flow in response to speech rather than nonspeech sounds; as early as 3 months, speech activates the specific left-hemisphere language areas identified in adults, while prosody activates the right hemisphere (reviewed in Friederici, 2006). A striking example of newborns' sensitivity to language is that the rhythm of their crying is consistent with that of their parents' language (Mampe, Friederici, Christophe, & Wermke, 2009).

Location of Other Languages

Additional evidence for a language acquisition device comes from studies of individuals who communicate with sign language. Left-hemisphere damage impairs sign-language ability more than right-hemisphere damage (Hickok, Bellugi, & Klima, 1996), and communicating in sign language activates left-hemisphere areas similar to those involved in traditional auditory language (Figure 9.28; Neville et al., 1998; Petitto et al., 2000). This was true of both deaf and speaking signers (who all learned signed language from infancy), but the finding is especially interesting in the deaf individuals, because it cannot be the result of the brain simply using

FIGURE 9.28 Language Area in Hearing and Deaf Individuals.

(a) fMRI results while hearing subjects read written English. (b) Activation in subjects deaf from birth while processing sign language.

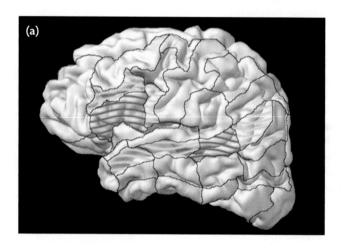
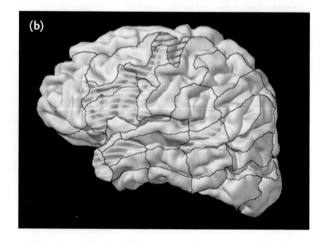

SOURCE: Reprinted with permission from H. J. Neville et al., "Cerebral Organizations for Language in Deaf and Hearing Subjects: Biological Constraints and Effects of Experience," *Proceedings of the National Academy of Sciences, USA, 95*, pp. 922–929. Copyright 1998 National Academy of Sciences, USA.

FIGURE 9.29 Brain Areas Activated by Different Languages in Bilingual Volunteers.

Green circles represent areas activated by listening to English, and yellow circles indicate activation while listening to Spanish. These images are from different subjects, selected to represent the variability among 11 subjects. Although the patterns are different, in every case the languages activate separate areas.

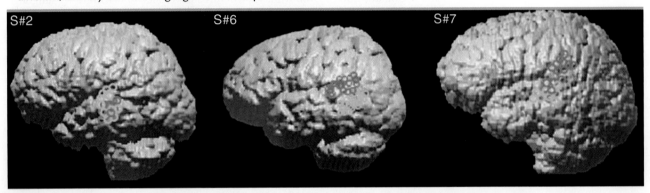

SOURCE: From "Mapping of Receptive Language Cortex in Bilingual Volunteers by Using Magnetic Source Imaging," by P. G. Simon et al., 2001, *Journal of Neurosurgery, 95*, pp. 76–81.

pathways already established by an auditory language. It is also interesting because Wernicke's area has traditionally been considered to be auditory in nature, which required the conversion of written words into an auditory form. Either the posterior language area is inherently more versatile than some theorists have thought, or the area underwent reorganization during infancy that enabled it to handle visual language. Either way, language seems to be a specialized capability of a limited subset of brain structures.

But what happens if a person learns a second language after childhood, when the brain is less plastic; will the brain then recruit other areas to handle the task? Two imaging studies indicate that this does happen, to some extent. In the first study, bilingual individuals silently "described" events from the previous day in each of their two languages; the languages activated separate areas in the frontal lobes, with centers that were 4.5 to 9 mm apart in different individuals. This was not true of subjects who learned their second language simultaneously with the first (Kim, Relkin, Lee, & Hirsch, 1997). The second study produced similar results, but in the temporal lobe, when subjects heard and read words in their two languages (Figure 9.29; Simos et al., 2001). This separation is so distinct that capability can be impaired in one of the languages while the other is unaffected (Gomez-Tortosa, Martin, Gaviria, Charbel, & Auman, 1995; M. S. Schwartz, 1994). A colleague who is originally from Lebanon told me an interesting story about his mother. She lives in the United States, and she was fluent in English until a stroke impaired her ability to speak English, but not Arabic. Her nearby family members spoke only English, so when they needed to talk with her they had to telephone a relative in another city to translate! These observations are not as inconsistent with the hypothesis of a single language acquisition device as they might seem; in both the Kim et al. (1997) and the Simos et al. (2001) studies, the second-language locations were in the same area as Broca's and Wernicke's areas, respectively.

We still cannot say that the language structures evolved specifically to serve language functions, however. You will see in the next section that some primates show similar enlargements in the left hemisphere, and their possession of language is questionable at best. Another reasonable interpretation of these data is that the structures evolved to handle rapidly changing information and fine discriminations,

8
Whistle and Click Languages

which language in its various forms requires. Another view is that the language areas are primarily specialized for different aspects of learning: the frontal area for "procedural" or how-to learning that coincides with the rules of grammar and verb tenses, and the temporal area for "declarative" or informational learning and thus, the storage of word meanings and information about irregular word forms (M. T. Ullman, 2001). Even if these structures have been "borrowed" by language and the concept of a dedicated language acquisition device isn't meaningful, it is still clear that the human brain is uniquely well fitted for creating as well as learning language. We will explore the possible evolutionary roots of this ability in the context of animal language.

Language in Nonhumans

Research has refuted most of humans' claims to uniqueness, including tool use, tool making, and self-recognition. Determining whether humans have exclusive ownership of language has been more difficult. Animal language intrigues us because we want to know whether we have any company "at the top" and because we want to trace the evolutionary roots of language. Because language leaves no fossils behind, the origin of language is "a mystery with all the fingerprints wiped off" (Terrence Deacon, quoted by Holden, 2004a). Without this evidence, an alternative is to look to the behavior and brains of our nonhuman relatives. The rationale behind animal language research is that any behavior or brain mechanism we share with genetically related animals must have originated in those common ancestors. Although dolphins, whales, and gorillas have been the subjects of research, the major contender for a copossessor of language has been the chimpanzee. The reason is that we and chimpanzees diverged from common ancestors a relatively recent 5 million years ago and we still share between 95% and 99% of our genetic material (R. J. Britten, Rowen, Williams, & Cameron, 2003; M.-C. King & Wilson, 1975).

A major obstacle has been deciding what we mean by *language*. Linguists agree that the vocalizations animals use to announce the availability of food or the presence of danger are only signals and have little to do with language. Even the human toddler's request of "milk" may initially be just a learned signal to indicate hunger and, like the monkey's alarm call, indicate no language understanding. As you will see in the following discussion, some of the results obtained in language research with animals are equally difficult to interpret.

An early study attempted to teach the home-reared chimpanzee Viki to talk, but after 6 years she had learned only "mama," "papa," and "cup" (Hayes & Hayes, 1953; Kellogg, 1968). Later, researchers concluded that chimpanzees lack the larynx for forming word sounds and, noting their tendency to communicate with a number of gestures, turned to American Sign Language. Over a 4-year period, the chimpanzee Washoe learned to use 132 signs; she was able to request food or to be tickled or to play a game, and she would sign "sorry" when she bit someone (Fouts, Fouts, & Schoenfeld, 1984). But critics argued that no chimpanzee had learned to form a sentence; they concluded that expressions such as "banana me eat banana" are just a "running-on" of words, and Washoe's signing "water bird" in the presence of a swan was not the inventive characterization of "a bird that inhabits water" but the separate identification of the bird and the water it was on (Terrace, Petitto, Sanders, & Bever, 1979).

However, animal language researchers received new encouragement when Washoe and three other trained chimps taught Washoe's adopted son Loulis 47 signs. The chimps regularly carried on sign-language conversations among themselves, most requesting hugs or tickling, asking to be chased, and signing "smile" (Fouts et al., 1984). Similarly, when Duane Rumbaugh and Sue Savage-Rumbaugh trained the pygmy chimp Mutata to communicate by pressing symbols on a panel, her son Kanzi spontaneously began to communicate with the symbols and eventually learned

What skills have chimps achieved in language studies?

9
Animal Language

FIGURE 9.30 Language Research With Chimpanzees.

(a) A researcher converses with a chimp using American Sign Language. (b) A chimp communicates through the symbol board.

SOURCES: (a) Susan Kuklin/Photo Researchers; (b) © Frans Lanting/Corbis.

150 of them without any formal instruction (Figure 9.30; S. Savage-Rumbaugh, McDonald, Sevcik, Hopkins, & Rubert, 1986; S. Savage-Rumbaugh, 1987). Kanzi uses the board to request specific food items or to be taken to specific locations on the 55-acre research preserve, asks a particular person to chase a specific other person, and responds to similar requests from trainers. His communication skills have been estimated at the level of a 2-year-old child (Savage-Rumbaugh et al., 1993). Irene Pepperberg (1993) emphasized concept learning with her African gray parrot, Alex, but his communication skills turned out to be equally interesting. Using speech, Alex could tell his trainer how many items she was holding, the color of an item, or whether two items differed in shape or color. He also could respond to complex questions, such as "What shape is the green wood?"

So do we share language ability with animals? The behavior of animals like Loulis, Kanzi, and Alex requires us to rethink our assumptions about human uniqueness, but no animal has yet turned in the critical language performance, and as far as we know, no animals in the wild have developed anything resembling a true language. But what some researchers do see in the animals' performance is evidence of evolutionary foundations of our language abilities (Gannon, Holloway, Broadfield, & Braun, 1998).

Neural and Genetic Antecedents

An approach of some researchers has been to determine whether other animals share with us any of the brain organization associated with human language. The results have been intriguing. In the chimpanzee, as with humans, there is a greater ratio of white to gray matter in the left hemisphere than the right (Cantalupo et al., 2009), and the left lateral fissure is longer and the planum temporale is larger (Gannon et al., 1998; Yeni-Komshian & Benson, 1976). Japanese macaque monkeys respond best to calls of their own species when the recorded calls are presented through headphones to the right ear (and primarily to the left hemisphere) than when they are presented to the left ear. There is no

Do other animals share our brain structures for language?

FIGURE 9.31 Chimpanzees Communicating With Face and Hand Gestures.

SOURCE: © Nigel J. Dennis / Photo Researchers, Inc.

FIGURE 9.32 Overlap Between Language Areas and Areas Involved in Imitation.

Broca's and Wernicke's areas are shown in yellow on a model of a human brain; the overlapping brown areas are also active during imitation of acts by others. Red indicates additional locations involved in imitation.

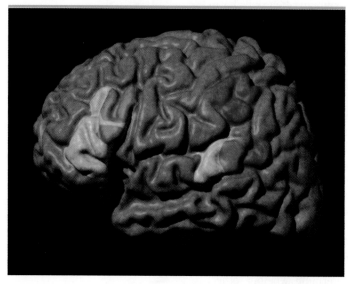

SOURCE: Image provided courtesy of Dr. Marco Iacoboni. From "The Origin of Speech," by C. Holden, 2004, *Science, 303*, p. 1318.

left-hemisphere advantage for the (nonmeaningful) calls of another monkey species (M. R. Petersen, Beecher, Zoloth, Moody, & Stebbins, 1978). Dolphins and the Rumbaughs' chimps Austin and Sherman responded more quickly when symbols or command gestures were presented to their left hemisphere (Hopkins & Morris, 1993; Morrel-Samuels & Herman, 1993). Finally, lesions on the left side of the canary brain render its attempts at song unrecognizable, while birds with right-side lesions continue to sing nearly as well as intact birds (Nottebohm, 1977).

But the presence of similar brain structures in other animals does not tell us how the animals use those structures. For example, the arcuate fasciculus, which connects Wernicke's area to Broca's area, is much smaller in chimpanzees (Rilling et al., 2008); you probably realize by now how important connectivity is to the language process. There is some thought, though, that the structures provide prelanguage communicative capabilities. One idea is that early language was gestural, consisting of hand gestures or lip smacks, tongue smacks, and teeth chatters; since chimps use all these to communicate, presumably the structures aided chimpanzees and early humans in producing and understanding these gestures (Figure 9.31; Holden, 2004a; MacNeilage, 1998). These researchers also believe that the ability to imitate gestures was critical to the development of language in humans; in fact, research indicates that children initially learn speech not by imitating sounds but by imitating the actions of the mouth (Goodell & Studdert-Kennedy, 1993), and the amount of gesturing at 14 months predicts vocabulary size at 54 months (Rowe & Goldin-Meadow, 2009). Now language theorists think that they have identified the mechanism for the imitative development of language in mirror neurons, which you learned about in Chapter 8.

Mirror neurons were first discovered in the area of the monkey brain that corresponds to Broca's area; they respond not only to monkeys' hand movements but also to communicative mouth gestures such as lip smacking (Ferrari, Gallese, Rizzolatti, & Fogassi, 2003). In humans, they are located in Broca's area and Wernicke's area and in the parietal lobe (Grèzes, Armony, Rowe, & Passingham, 2003; Holden, 2004a). Human mirror neurons are most active during imitation of another's movement (Iacoboni et al., 1999), which has encouraged the belief that they figure prominently in imitative ability and, thus, in the evolution of language (Figure 9.32). The fact that we share mirror neurons with monkeys and chimpanzees does not imply that monkeys and chimpanzees also share our language abilities. In fact, the evolutionary clues we do have suggest that language developed well after the split that led to humans and chimpanzees (Holden, 2004a). Whatever brain foundations of language we share with chimpanzees likely required extensive refinement, such as expansion of the brain, including the language areas; migration of the larynx lower in the throat, which increased vocalization range; and the development of imitative ability, which is poor in nonhuman primates (Holden, 2004a).

Suggesting that language is a product of evolution means, of course, that genes are involved (Figure 9.33). *KIAA0319*, which we identified in our discussion of dyslexia, also plays a role in the development of speech and language (M. L. Rice, Smith, & Gayán, 2009), and *CNTNAP2* has been implicated in language impairment (Vernes et al., 2008). But the most researched and best understood language gene is *FOXP2*; a mutation of this gene results in reduced gray matter in Broca's area, along with articulation difficulties, problems identifying basic speech sounds, grammatical difficulty, and trouble understanding sentences (Lai, Fisher, Hurst, Vargha-Khadem, & Monaco, 2001; Pinker, 2001; Vargha-Khadem, Gadian, Copp, & Mishkin, 2005). We also share this gene with chimpanzees, but the human version differs in two apparently very important amino acids. *FOXP2* regulates other genes, and Geneive Konopka and her colleagues (2009) have identified 116 genes that are turned off or on differently in humans compared to chimps. Researchers believe that the human version of *FOXP2* is one of the divergencies that enabled us to develop speech and that language fueled modern humans' success (Enard, Przeworski, et al., 2002). As geneticists fill in the human and animal genomes, it will be interesting to see what other clues emerge for explaining human language capability.

FIGURE 9.33 Is Language Genetic?

SOURCE: Cartoonist: Duffy, Source: *The Des Moines Register*. Reprinted with Special Permission of the North America Syndicate.

Concept Check *Take a Minute to Check Your Knowledge and Understanding*

- ☐ In what ways is the Wernicke-Geschwind model correct and incorrect?
- ☐ What are the different roles of the left and right hemispheres in language (in most people)? (See Chapter 8 for part of the answer.)
- ☐ What clues are there in animal research for the possible origins of human language?

In Perspective

My guess is that at the beginning of this chapter you would have said that vision is the most important sense. Perhaps now you can appreciate why Helen Keller thought her deafness was a greater handicap than her blindness. Hearing alerts us to danger, brings us music, and provides for the social interactions that bind humans together. Small wonder that during evolution, the body invested such resources in the intricate mechanisms of hearing.

Hearing has important adaptive functions with or without the benefit of language, but from our vantage point as language-endowed humans, it is easy to understand Hudspeth's (2000) claim that audition's most important role is in processing language. The person who is unable to talk is handicapped; the person who is unable to understand and to express language is nearly helpless. No wonder we put so much research effort into understanding how language works.

One of the most exciting directions language research has taken has involved attempts to communicate with our closest nonhuman relatives. Whether they possess language capabilities depends on how we define language. It is interesting how the capabilities we consider most characteristic of being human—such as language and consciousness—are the hardest to define. As so often happens, studying our animal relatives, however distant they may be, helps us understand ourselves.

Summary

- Sensation requires a receptor that is specialized for the particular kind of stimulus. Beyond sensation, the brain carries out further analysis, called perception.

Hearing

- The auditory mechanism responds mostly to airborne vibrations, which vary in frequency and intensity.
- Sounds are captured and amplified by the outer ear and transformed into neural impulses in the inner ear by the hair cells on the basilar membrane. The signal is then transmitted through the brain stem and the thalamus to the auditory cortex in each temporal lobe.
- Frequency discrimination depends mostly on the basilar membrane's differential vibration along its length to different frequencies, resulting in neurons from each location carrying frequency-specific information to the brain. At lower frequencies, neurons fire at the same rate as the sound's frequency; it is possible that intermediate frequencies are represented by neurons firing in volleys, though research has not indicated this is so.
- Although the cochlea specializes in responding to pure tones, the basilar membrane apparently performs a Fourier analysis on complex sounds, breaking a sound down into its component frequencies.
- When different sounds must be distinguished from each other, stimulation from the brain probably adjusts the sensitivity of the hair cells to emphasize one sound at the expense of others. Selective attention also results in differential activity in areas of the cortex.
- Locating sounds helps us approach or avoid sound sources and to attend to them in spite of competition from other sounds.
- The brain has specialized circuitry for detecting the binaural cues of differences in intensity, time of arrival, and phase at the two ears.

Language

- Researchers have identified two major language areas in the brain, with Broca's area involved with speech production and grammatical functions and Wernicke's area with comprehension.
- Damage to either area produces different symptoms of aphasia, and damage to connections with the visual cortex impairs reading and writing. Developmental dyslexia may involve planum temporale abnormalities, reduced activity in the posterior language area, or deficiencies in the auditory and visual pathways.
- Although damage to the left frontal or temporal lobes is more likely to produce the expected disruptions in language, studies have shown that control of the various components of language is distributed across the four lobes.
- Although some animals have language-like brain structures and have been taught to communicate in simple ways, it is controversial whether they possess true language. Their study suggests some possible evolutionary antecedents of language. ■

Study Resources

F   For Further Thought

- Write a modified Wernicke-Geschwind theory of language control, based on later evidence.
- Would you rather give up your hearing or your vision? Why?

- Make the argument that chimps possess language, though at a low level. Then argue the opposite, that their behavior does not rise to the level of language.

T   Testing Your Understanding

1. Describe the path that sound information takes from the outer ear to the auditory neurons, telling what happens at each point along the way.

2. State the telephone theory, the volley theory, and the place theory. Indicate a problem with each, and state the theory that is currently most widely accepted.

3. Summarize the Wernicke-Geschwind model of language function. Include structures, the effects of damage, and the steps in reading a word aloud and in repeating a word that is heard.

Select the best answer:

1. An adequate and an inadequate stimulus, such as light versus pressure on the eyeball, will produce similar experiences because
   a. they both activate visual receptors and the visual cortex.
   b. the receptors for touch and vision are similar.
   c. touch and vision receptors lie side by side in the eye.
   d. our ability to discriminate is poor.

2. Frequency is to pitch as
   a. loudness is to intensity.
   b. intensity is to loudness.
   c. stimulus is to response.
   d. response is to stimulus.

3. The sequence of sound travel in the inner ear is
   a. oval window, ossicles, basilar membrane, eardrum.
   b. ossicles, oval window, basilar membrane, eardrum.
   c. eardrum, ossicles, oval window, basilar membrane.

   d. eardrum, ossicles, basilar membrane, oval window.

4. Place analysis depends most on the physical characteristics of the
   a. hair cells.   b. basilar membrane.
   c. tectorial membrane.   d. cochlear canal.

5. The fact that neurons are limited in their rate of firing by the refractory period is most damaging to which theory?
   a. telephone   b. volley
   c. place   d. volley-place

6. The place theory's greatest problem is that
   a. neurons cannot fire as frequently as the highest frequency sounds.
   b. neurons specific for frequencies above 5,000 Hz have not been found.
   c. the whole basilar membrane vibrates about equally at low frequencies.
   d. volleying does not follow sound frequency above about 5,000 Hz.

7. An auditory neuron's tuning curve tells you
   a. which frequency it responds to.
   b. which part of the basilar membrane the neuron comes from.
   c. at what rate the neuron can fire.
   d. how much the neuron responds to different frequencies.

8. A cochlear implant works because
   a. the tympanic membrane is intact.
   b. the hair cells are intact.
   c. it stimulates the auditory cortex directly.
   d. it stimulates auditory neurons.

9. An auditory object is
   a. a vibrating object in the environment.
   b. a sound recognized as distinct from others.

c. the sound source the individual is paying attention to.

d. none of the above.

10. As a binaural sound location cue, difference in intensity works
   a. poorly at low frequencies.
   b. poorly at medium frequencies.
   c. poorly at high frequencies.
   d. about equally at all frequencies.

11. In the following diagram of coincidence detectors, which cell would respond most if the sound were directly to the person's left?

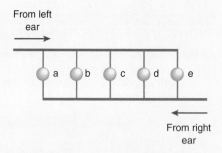

From left ear →

a b c d e

← From right ear

12. On returning home from the hospital, an elderly neighbor drags one foot when he walks and uses almost exclusively nouns and verbs in his brief sentences. You guess that he has had a mild stroke located in his
   a. left temporal lobe.   b. right temporal lobe.
   c. left frontal lobe.   d. right frontal lobe.

13. A problem in the magnocellular pathway may be associated with
   a. expressive aphasia.   b. dyscalculia.
   c. agraphia.   d. dyslexia.

14. Evidence providing some support for a language acquisition device comes from studies showing that American Sign Language activates
   a. the left hemisphere.
   b. the left and right hemispheres.
   c. both frontal lobes.
   d. the occipital lobe.

15. The most reasonable conclusion regarding language in animals is that
   a. they can use words or signs, but do not possess language.
   b. they can learn language to the level of a 6-year-old human.
   c. language is "built in" for humans, but must be learned by animals.
   d. some animals have brain structures similar to human language structures.

16. Mirror neurons' role in language development is supposedly in
   a. the repetition of word sounds.
   b. gestural aspects.
   c. development of grammar.
   d. the use of prosody.

Answers:
1.a, 2.b, 3.c, 4.b, 5.a, 6.c, 7.d, 8.d, 9.b, 10.a, 11.e, 12.c, 13.d, 14.a, 15.d, 16.b.

Online Resources

The following resources are available at **www.sagepub.com/garrett3e.**

Chapter Resources

- Flash Cards
- Chapter Quiz
- Internet Resources and Exercises

On the Web

Links to these websites can be found at the web address above: Select this chapter's **Chapter Resources**, and then choose **Internet Resources and Exercises**.

1. **National Institute on Deafness and Other Communication Disorders**, a division of the National Institutes of Health, is an excellent resource site for information on deafness, language, and speech disorders, as well as information for student and teacher activities.

   The site **Hereditary Hearing Loss** has recent research and other information on what we know about the 60-plus genes thought to cause hearing loss.

2. **Seeing, Hearing, and Smelling the World**, a site maintained by the Howard Hughes Medical Institute, has several articles on senses, including "The Quivering Bundles That Let Us Hear." Also, look for the Chapter 9 sidebar "On the Trail of Deafness Genes."

3. **Auditory Transduction (2002)** is a YouTube video that uses high-quality animation to explain how hearing works.

The **Yale Ear Lab**'s video "Dancing Outer Hair Cell" is a highly magnified view of an outer hair cell rapidly changing length in time with the rhythm of music. The video is in the upper right corner, and you may need to expand the page to see all of it.

4. **Biointeractive**, also at Howard Hughes, features an animation of the basilar membrane responding to pure tones and music. The video is very instructive, even though for simplicity's sake it suggests that basilar membrane around the area of greatest vibration does not vibrate at all.

5. **Cochlear Implant** is an animated explanation of the process of hearing and how cochlear implants restore hearing.

6. The **National Aphasia Association** has information about aphasia and about research on the disorder as well as resources.
   Stroke Family has information about recovering speech after a stroke, including free mini guides, with emphasis on how the family can help.

7. The **International Dyslexia Association** provides information on the disorder.

8. To communicate over long distances on Isla Gomera, one of the Canary Islands, people use complex whistles; you can see a video at **Free Language.** The whistles are processed in the left hemisphere language areas by whistlers, but not by others (Carreiras, Lopez, Rivero, & Corina, 2005). !Kung hunters of Africa communicate solely with clicks while stalking game. At **Nama**, part of the UCLA Phonetics Lab Archive, you can hear examples of click sounds in communication, which some believe were the earliest vocal language. Select WAV or MP3 files to play.

9. The **Chimpanzee and Human Communication Institute** is home to Loulis and three other signing chimps. You can learn about the institute and its research, watch the chimps at play on the chimp-cams, and read the chimp biographies.
   Sadly, Alex passed away in 2007, but Dr. Pepperberg continues her language research with two other parrots, Griffin and Arthur. See the **Alex Foundation**'s page About Our Research and a video of Alex performing.

Animations

- Place Analysis of Auditory Frequency (Figure 9.10)
- Sound Localization (Figure 9.19)
- The Wernicke-Geschwind Model of Language (Figure 9.21)

Chapter Updates and Biopsychology News

R For Further Reading

1. *Language Evolution* edited by Morten Christiansen and Simon Kirby (Oxford University Press, 2003) contains 17 chapters by 20 language experts and addresses issues such as the role of the *FOXP2* gene, whether animal communication sheds light on human language, the possibility that language was originally gestural, and whether there is a dedicated language mechanism in the brain.

2. *The Language Instinct* by Steven Pinker, the director of MIT's Center for Cognitive Neuroscience (Morrow, 1994), concerns the evolution of language. Pinker's expertise and lively writing style garnered one reviewer's evaluation as "an excellent book full of wit and wisdom and sound judgment."

3. "Brain and Language" by Antonio and Hanna Damasio (*Scientific American*, September 1992, 60–67) is a brief and readable introduction by two of the best-known experts.

4. *Sensation and Perception* by E. B. Goldstein (Brooks/Cole, 1999) is a textbook and makes a good reference, including nonphysiological aspects not covered here.

5. *The Design of Animal Communication,* edited by M. D. Hauser and M. Konishi (Bradford, 2003), is an excellent review of similarities and differences in human and animal communication systems.

K Key Terms

adequate stimulus 254

agraphia 275

alexia 275

amplitude 255

angular gyrus 275

aphasia 270

auditory object 266

basilar membrane 257

binaural 268

Broca's aphasia 271

cochlea 255

cochlear canal 257

cocktail party effect 266

coincidence detectors 268

complex sound 256

difference in intensity 268

difference in time of arrival 268

dyslexia 276

frequency 255

frequency theory 261

frequency-place theory 265

inner hair cells 259

intensity 255

language acquisition device 279

loudness 255

magnocellular hypothesis 276

organ of Corti 257

ossicles 256

outer hair cells 259

perception 255

phase difference 268

phonological hypothesis 276

pinna 256

pitch 255

place theory 262

planum temporale 276

prosody 279

pure tone 256

receptor 254

sensation 255

tectorial membrane 257

telephone theory 261

tonotopic map 263

topographical organization 259

tympanic membrane 256

volley theory 261

Wernicke's aphasia 272

Vision and Visual Perception

10

Light and the Visual Apparatus

The Visible Spectrum

The Eye and Its Receptors

Pathways to the Brain

    APPLICATION: RESTORING LOST VISION

    CONCEPT CHECK

Color Vision

Trichromatic Theory

Opponent Process Theory

A Combined Theory

Color Blindness

    CONCEPT CHECK

Form Vision

Contrast Enhancement and Edge Detection

Hubel and Wiesel's Theory

Spatial Frequency Theory

    CONCEPT CHECK

The Perception of Objects, Color, and Movement

The Two Pathways of Visual Analysis

Disorders of Visual Perception

The Problem of Final Integration

    APPLICATION: WHEN BINDING GOES TOO FAR

    CONCEPT CHECK

In Perspective

Summary

Study Resources

In this chapter you will learn

- The structure of the eye

- How the eye begins processing visual information even before it is sent to the brain

- The major theories of color and form vision

- How color, form, movement, and spatial location are handled in the brain

- Some of the visual disorders caused by brain damage and what they tell us about brain function

Jonathan I. was in his car when it was hit by a small truck. In the emergency room, he was told that he had a concussion. For a few days, he was unable to read, saying that the letters looked like Greek, but fortunately this *alexia* soon disappeared. Jonathan was a successful artist who had worked with the renowned Georgia O'Keeffe, and he was eager to return to his work. Driving to his studio, he noticed that everything appeared gray and misty, as if he were driving in a fog. When he arrived at his studio, he found that even his brilliantly colored paintings had become gray and lifeless.

His whole world changed. People's appearance was repulsive to him, because their skin appeared "rat-colored"; he lost interest in sex with his wife for that reason. Food was unattractive, and he came to prefer black and white foods (coffee, rice, yogurt, black olives). His enjoyment of music was diminished, too; before the accident, he used to experience *synesthesia*, in which musical tones evoked a sensation of changing colors, and this pleasure disappeared as well. Even his migraine headaches, which had been accompanied by brilliantly colored geometric hallucinations, became "dull." He retained his vivid imagery, but it too was without color.

Over the next 2 years, Jonathan seemed to forget that color once existed, and his sorrow lifted. His wife no longer appeared rat colored, and they resumed sexual activity. He turned to drawing and sculpting and to painting dancers and race horses, rendered in black and white but characterized by movement, vitality, and sensuousness. However, he preferred the colorless world of darkness and would spend half the night wandering the streets (Sacks & Wasserman, 1987).

Vision enables us to read and to absorb large amounts of complex information. It helps us navigate in the world, build structures, and avoid danger. Color helps distinguish objects from their background, and it enriches our lives with natural beauty and works of art. I suspect that in contrast to Helen Keller's belief that deafness was a greater affliction than blindness, most of you would consider vision the most important of our senses. Apparently, researchers share that opinion, because vision has received more research attention than the other senses combined. As a result, we understand a great deal about how the brain processes visual information. In addition, studies of vision are providing a valuable model for understanding complex neural processing in general.

Light and the Visual Apparatus

Vision is an impressive capability. There are approximately 126 million light receptors in the human eye and a complex network of cells connecting them to each other and to the optic nerve. The optic nerve itself boasts a million neurons, compared with 30,000 in the auditory nerve. What our brain does with the information it receives from the eye is equally remarkable. The topics of vision and visual perception form an exciting story, one of high-tech research and conflicting theories and dedicated scientists' lifelong struggles to understand our most amazing sense.

The Visible Spectrum

To understand vision, we need to start at the beginning by describing the *adequate stimulus*, as we did with audition. To say that the stimulus for vision is light seems obvious, but the point needs some elaboration. Visible light is a part of the electromagnetic spectrum. The electromagnetic spectrum includes a variety of energy forms, ranging from gamma rays at one extreme of frequency to the radiations of alternating current circuits at the other (Figure 10.1); the portion of the electromagnetic spectrum that we can see is represented by the colored area in the figure.

The visible part of the spectrum accounts for only 1/70 of the range. Most of the energies in the spectrum are not useful for producing images; for instance, AM, FM, and television waves pass right through objects. Some of the other energy forms, such as X-rays and radar, can be used for producing images, but they require powerful energy sources and special equipment for detecting the images. Heat-producing

FIGURE 10.1 The Electromagnetic Spectrum.

The visible part of the spectrum is the middle (colored) area, which has been expanded to show the color experiences usually associated with the wavelengths. Only 1/70 of the electromagnetic spectrum is visible to humans.

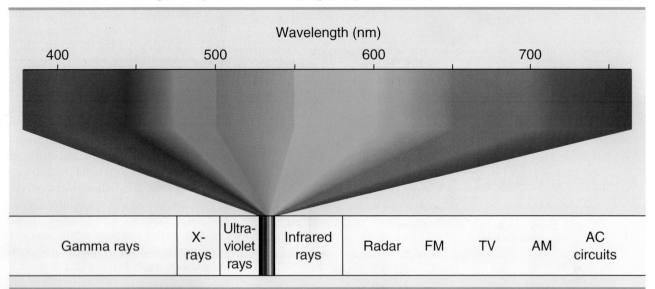

Wavelength (nm)

400   500   600   700

| Gamma rays | X-rays | Ultra-violet rays | Infrared rays | Radar | FM | TV | AM | AC circuits |

objects give off infrared energy, which some nocturnal animals, such as the sidewinder rattlesnake, can use to detect their prey in darkness. Humans can see infrared images only with the aid of specialized equipment, and this capability is very useful to the military and the police for detecting people and heat-producing vehicles and armament at night. (During the first Gulf War, the Iraqi army set up plywood silhouettes of tanks with heaters behind them to distract Allied airplanes.) But infrared images have blurred edges and fuzzy detail. The electromagnetic energy within our detectable range produces well-defined images because it is reflected from objects with minimal distortion. We are adapted to life in the daytime, and we sacrifice the ability to see in darkness in exchange for crisp, colorful images of faces and objects in daylight. In other words, our sensory equipment is specialized for detecting the energy that is most useful to us, just as the night-hunting sidewinder rattlesnake is equipped to detect the infrared radiation emitted by its prey and a bat's ears are specialized for the high-frequency sound waves it bounces off small insects.

Light is a form of oscillating energy and travels in waves just as sounds do. We could specify visible light (and the rest of the electromagnetic spectrum) in terms of frequency, just as we did with sound energy, but the numbers would be extremely large. So we describe light in terms of its wavelength—the distance the oscillating energy travels before it reverses direction. (We could do the same with sound, but those numbers would be just as inconveniently small.) The unit of measurement is the nanometer (nm), which is a billionth of a meter; visible light ranges from about 400 to 800 nanometers (nm). Notice in Figure 10.1 that different wavelengths correspond to different colors of light; for example, when light in the range of 500 to 570 nm strikes the receptors in our eye, we normally report seeing green. Later in the chapter, we will qualify this relationship when we examine why wavelength does not always correspond to the color we see.

The Eye and Its Receptors

The eye is a spherically shaped structure filled with a clear liquid (Figure 10.2). The outer covering, or sclera, is opaque except for the cornea, which is transparent.

1
Vision Info and Animations

FIGURE 10.2 The Human Eye.

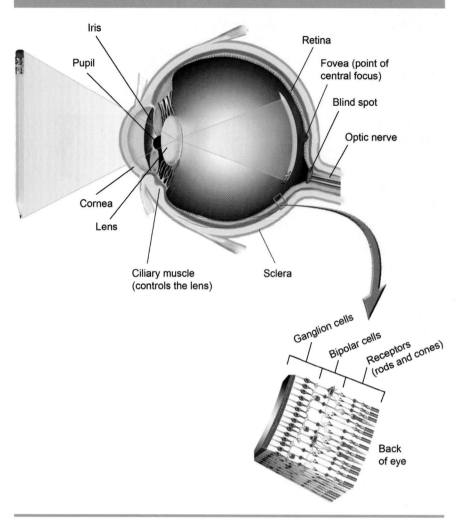

Behind the cornea is the lens. Because the lens is a flexible tissue, the muscles attached to it can stretch it out flatter to focus the image of a distant object on the retina, or they can relax to focus the image of a near object. Notice in the figure that the lens inverts the image on the retina. The lens is partly covered by the iris, which is what gives your eye its color. The iris is actually a circular muscle whose opening forms the pupil; it controls the amount of light entering the eye by contracting reflexively in bright light and relaxing in dim light. You can observe this response in yourself by watching in a mirror while you change the level of light in the room.

The *retina,* the light-sensitive structure at the rear of the eye, is made up of two main types of light-sensitive receptor cells, called rods and cones, and the neural cells that are connected to them. As you can see in Figures 10.2 and 10.3, the receptors—typically referred to as *photoreceptors*—are at the very back of the eye. Light must pass through the neural cells to reach the photoreceptors, but this presents no problem because the neural cells are transparent. The receptors connect to *bipolar cells,* which in turn connect to *ganglion cells,* whose fibers form the optic nerve. The photoreceptors are filled with light-sensitive chemicals called *photopigments;* light passing through the photopigment causes some of the molecules to break down into two components, and the ensuing chemical reaction ultimately

How does the eye detect light?

results in a neural response. The two components then recombine to maintain the supply of photopigment.

The rods and cones are named for their shapes, as you can see by looking at Figure 10.3 again. The human eye contains about 120 million rods and about 6 million cones. The two types of receptors contain different kinds of photopigments; rods and cones function similarly, but their chemical contents and their neural connections give them different specializations.

The rod photopigment is called *rhodopsin;* the name refers to its color (from the Latin *rhodon,* "rose"), not to its location in rods. Rhodopsin is more sensitive to light than cone photopigment is. For this reason, rods function better in dim light than cones do; in fact, you rely solely on your rods for vision in dim light. In very bright light, the rhodopsin in your eyes remains broken down most of the time, so the rods barely function. The delay in adjusting to a darkened movie theater is due to the time it takes the rhodopsin to resynthesize. *Iodopsin,* the cone photopigment, requires a high level of light intensity to operate, so your cones are nonfunctional in dim light but function well in daylight. Three varieties of iodopsin, located in different cones, respond differentially to different wavelengths of light; this means that cones can distinguish among different wavelengths, whereas rods differentiate only among different levels of light and dark (which is why you cannot recognize colors in dim light).

Rods and cones also differ in their location and in their amount of neural interconnection. Cones are most concentrated in the *fovea,* a 1.5-millimeter-wide area in the middle of the retina, and drop off rapidly with distance from that point. Rods are most concentrated at 20 degrees from the fovea; from that point, they decrease in number in both directions and fall to zero in the middle of the fovea. In the center of the fovea, a single cone is connected to each ganglion cell; the number of cones per ganglion cell increases with distance from the center but still remains small compared to rods. Because few cones share ganglion cells, the fovea has higher *visual acuity*, or ability to distinguish details. Many rods share each ganglion cell; this reduces their resolution but enhances their already greater sensitivity to dim light. The area of the retina from which a ganglion cell (or any other cell in the visual system) receives its input is the cell's *receptive field*. So, we can say that receptive fields are smaller in the fovea and larger in the periphery. Table 10.1 summarizes the characteristics of these two visual systems.

The receptors' response to light is different from what you have come to expect, because they are most active when they are *not* being stimulated by light. In darkness, the photoreceptor's sodium and calcium channels are open, allowing these ions to flow in freely. Thus, the membrane is partially depolarized; the receptor releases a continuous flow of glutamate, and this inhibits activity in the bipolar cells. The chemical response that occurs when light strikes the photopigments closes the sodium and calcium channels, reducing the release of glutamate in proportion to the amount of light. The bipolar cells release more neurotransmitter, which increases the firing rate in the ganglion cells. (The photoreceptors and bipolar cells do not produce action potentials.)

If you look again at Figure 10.3, you can see that the rods and cones are highly interconnected by *horizontal cells*. In addition, *amacrine cells* connect across many ganglion cells. This might suggest to you that the retina does more than transmit information about points of light to the brain. You might also suspect that a great deal of processing goes on in the retina itself. You will soon see that both of these are true. With such complexity, no wonder most vision scientists consider the retina to be part of the brain and refer to the optic nerve as a tract.

FIGURE 10.3 The Cells of the Retina.

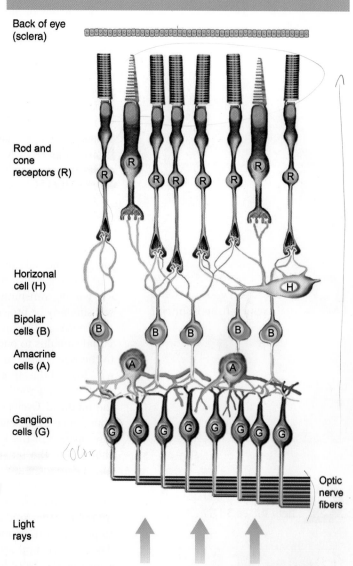

SOURCE: Adapted from "Organization of the Primate Retina," by J. E. Dwoling and B. B. Boycott, *Proceedings of the Royal Society of London, B166*, Fig. 23 on p. 104. Copyright 1966 by the Royal Society. Used with permission of the publisher and the author.

TABLE 10.1 Summary of the Characteristics of the Rod and Cone Systems.

	Rod System	Cone System
Function	Functions best in dim light, poorly or not at all in bright light. Detail vision is poor. Does not distinguish colors.	Functions best in bright light, poorly or not at all in dim light. Detail vision is good. Distinguishes among colors.
Location	Mostly in periphery of retina	Mostly in fovea and surrounding area
Receptive field	Large, due to convergence on ganglion cells; contributes to light sensitivity.	Small, with one or a few cones converging on a single ganglion cell; contributes to detail vision.

Pathways to the Brain

The axons of the ganglion cells join together and pass out of each eye to form the two optic nerves (Figure 10.4). Where the nerve exits the eye, there are no receptors, so it is referred to as the *blind spot* (see Figure 10.2). The blind spots of the two eyes fall at different points in a visual scene, so you do not notice that any of your visual world is missing; besides, your brain is good at "filling in" missing information, even when a small part of the visual system is damaged. The two optic nerves run to a point just in front of the pituitary, where they join for a short distance at the *optic chiasm* before separating again and traveling to their first synapse in the *lateral geniculate nuclei* of the thalamus. At the optic chiasm, axons from the nasal sides of the eyes cross to the other side and go to the occipital lobe in the opposite hemisphere. Neurons from the outside half of the eyes (the temporal side) do not cross over but go to the same side of the brain.

It seems like splitting the output of each eye between the two hemispheres would cause a major distortion of the image. However, if you look closely at Figure 10.4, you can see that the arrangement actually keeps related information together. Notice that the letter *A*, which appears in the person's left visual field, casts an image on the right half of *each* retina. The *visual field* is the part of the environment that is being registered on the retina. The information from the right half of each eye will be transmitted to the right hemisphere. An image in the right visual field will similarly be projected to the left hemisphere. This is how researchers who study differences in the functions of the two cerebral hemispheres are able to project a visual stimulus to one hemisphere. They present the stimulus slightly to the left or to the right of the midline, with the exposure too brief for the person to shift the eyes toward the stimulus.

There is a good reason you have two forward-facing eyes, instead of one like the mythical Cyclops or one on each side of your head like many animals. The approximately 6-centimeter separation of your eyes produces *retinal disparity,* a discrepancy in the location of an object's image on the two retinas. Figure 10.5 shows how the image of distant objects in a scene falls toward the nasal side of each retina and closer objects cast their image in the temporal half. Retinal disparity is detected in the visual cortex, where different neurons fire depending on the amount of lateral displacement. Then in the anterior parietal cortex this information is combined with information about an object's shape and location to provide three-dimensional (3-D) location of objects (J.-B. Durand et al., 2007). There are several good demonstrations of retinal disparity; the simplest is to hold your finger a foot (30 cm) in front of you and alternately close one eye and then the other while you notice how your finger moves relative to the background. The ViewMaster 3-D viewer you may have had as a child took advantage of your brain's retinal disparity detectors by presenting each eye with an image photographed at a slightly different angle. The 3-D movie Avatar used the same principle, but the two images were separated by differently polarized lenses in the special glasses you wore. A striking 3-D effect can also be obtained without a viewer from stereograms such as those at Magic Eye (see On the Web 2).

As the rest of the story of vision unfolds, you will notice three themes that will help you understand how the visual system works: inhibition, modularity, and hierarchical processing. You will learn that neural *inhibition* is just as important as excitation, because it sharpens information beyond the processing capabilities of a system that depends on excitation alone. You will also learn that, like audition and language, the visual system carries out its functions in discrete specialized structures, or *modules*, which pass information to each other in a serial, *hierarchical* fashion.

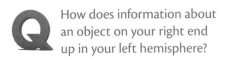

Q How does information about an object on your right end up in your left hemisphere?

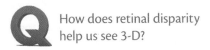

Q How does retinal disparity help us see 3-D?

2
Stereograms and Illusions

FIGURE 10.4 Projections From the Retinas to the Cerebral Hemispheres.

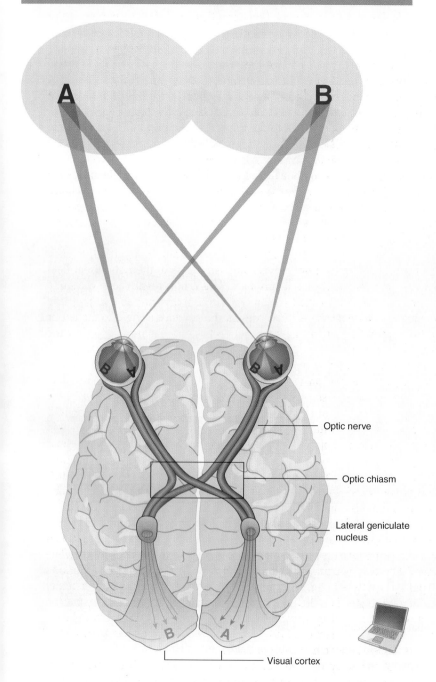

Optic nerve

Optic chiasm

Lateral geniculate nucleus

Visual cortex

FIGURE 10.5 Retinal Disparity.

The image of a focused object (A) falls on the fovea, while the image of a more distant object (B) is displaced to the inside of each retina and the image of a closer object (C) is displaced to the outside. This provides information to the brain for depth perception.

Notice how images of objects in the left and right sides of the visual field fall on the opposite sides of the two retinas; the information from the two eyes then travels to the visual cortex in the hemisphere opposite the object, where it is combined. (The distance between the two lateral geniculate nuclei is greatly exaggerated.)

The most frequent cause of blindness in developed countries is deterioration of the visual receptors due to age-related macular degeneration or the inheritable disease retinitis pigmentosa. Because the neural structures in the retina remain intact, researchers are attempting to restore sight by replacing the photoreceptors with an artificial retina. The designs vary, but the Argus Retinal Prosthesis System in the accompanying figure typifies the concept. Images from a small video camera mounted on a pair of glasses are sent to a processor and then back to the glasses, where the signal is transmitted wirelessly to a small chip implanted in the retina. The chip's 60 electrodes stimulate the retinal neurons in a pattern that roughly duplicates the image. The visual experience is crude, but it allows users to avoid obstacles and to notice when someone approaches. The artificial retina is in Federal Drug Administration-authorized human trials ("Doheny Eye Institute," 2007); at the same time, the researchers are developing a 200-electrode version, but recognizing a human face will require 1,000 electrodes (Saenz, 2010). Another device, BrainPort, uses 400 to 600 electrodes, but they are on a "lollipop" device the user places on the tongue; blind individuals begin to interpret images after just 15 minutes of practice, and they are able to find doorways and elevator buttons and pick up objects (Kendrick, 2009).

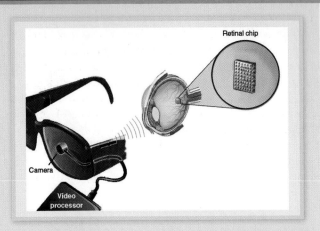

Retinal chip

Camera

Video processor

As exciting as these innovations are, repair rather than artificial replacement may provide the best long-term strategy. An obvious direction is stem cell implantation, but treatment restored sight in only one fourth of mice with retinitis pigmentosa (N.-K. Wang et al., 2010). Transplantation of fetal retinal tissue improved visual acuity in 70 percent of human patients, though vision was far from normal; one

3
Fixing Blindness

patient who had pretreatment visual acuity of 20/800 (meaning she could see at 20 feet what other people see at 800 feet), maintained improvement at 20/200 5 years later (Radtke et al., 2008). A less obvious but logical approach is application of the optogenetic technology you learned about in Chapter 2; inserting the genes for two plant rhodopsins created depolarizing and hyperpolarizing channels in the retinas of mice with degenerated retinas and converted their bipolar cells to photoreceptors; this restored ON and OFF ganglion cell responses along with some visual capability (Lagali et al., 2008; Y. Zhang, Ivanova, Bi, & Pan, 2009). Another hereditary disease of the retina, Leber's congenital amaurosis, has been treated with a healthy version of the *RPE65* gene (Simonelli et al., 2010). Nine-year-old Evan Haas, for example, can for the first time recognize faces, play baseball, read large-print books, and ride his bicycle around his neighborhood alone.

Concept Check   *Take a Minute to Check Your Knowledge and Understanding*

☐ In what ways is human vision adapted for our environment?

☐ How are the rod and cone systems specialized for different tasks? The two visual pathways?

☐ How does retinal disparity help us see 3-D?

Color Vision

In Figure 10.1, you saw that there is a correspondence between color and wavelength; this would suggest that color is a property of the light reflected from an object and, therefore, of the object itself. However, wavelength does not always predict color, as Figure 10.6 illustrates. The circle on the left and the circle on the right appear to be different colors, although they reflect exactly the same wavelengths. Just as with the auditory terms *pitch* and *loudness,* the term *color* refers to the observer's experience rather than a characteristic of the object. Thus, it is technically incorrect to say that the light is red or that a book is blue, because *red* and *blue* are experiences that are imposed by the brain. However, in the interest of simplicity, I will be rather casual about this point in future discussions, as long as we understand that "red" and "blue" are experiences rather than object characteristics. To understand the experience of color, we must now examine the neural equipment that we use to produce that experience. Our understanding of color vision has been guided over the past two centuries by two competing theories: the trichromatic theory and the opponent process theory.

FIGURE 10.6 Independence of Wavelength and Color.

Although the circles are identical, they appear to differ in color due to the color contrast with their backgrounds.

Trichromatic Theory

After observing the effect of passing light through a prism, Newton proposed in 1672 that white light is composed of seven fundamental colors that cannot themselves be resolved into other colors. If there are seven "pure" colors, this would suggest that there must be seven receptors and brain pathways for distinguishing color, just as there are five primary tastes. In 1852, Hermann von Helmholtz (whose place theory was discussed in Chapter 9) revived an idea of Thomas Young from a half century earlier. Because any color can be produced by combining different amounts of just three colors of light, Young and Helmholtz recognized that this must be due to the nature of the visual mechanism rather than the nature of light. They proposed a *trichromatic theory,* that just three color processes account for all the colors we are able to distinguish. They chose red, green, and blue as the primary colors because observers cannot resolve these colors into separate components as, for example, you can see red and blue in the color purple. When you watch television or look at a computer monitor, you see an application of trichromatic color mixing: All the colors you see on the screen are made up of tiny red, green, and blue dots of light.

3 color receptors

Opponent Process Theory

The trichromatic theory accounted for some of the observations about color perception very well, but it ran into trouble explaining why yellow also appears to observers to be a pure color. Ewald Hering (1878) "solved" this problem by adding yellow to the list of physiologically unique colors. But rather than assuming four color receptors, he asserted that there are only two—one for red *and* green and one for blue *and* yellow. *Opponent process theory* attempts to explain color vision in terms of opposing neural processes. In Hering's version, the photochemical in the red-green receptor is broken down by red light and regenerates in the presence of green light. The chemical in the second type of receptor is broken down in the presence of yellow light and regenerates in the presence of blue light.

Hering proposed this arrangement to explain the phenomenon of *complementary colors,* colors that cancel each other out to produce a neutral gray or white. (Note the spelling of this term; *complementary* means "completing.") In Figure 10.7, the visible spectrum is represented as a circle. This rearrangement of the spectrum makes sense, because violet at one end of the spectrum blends naturally into red at the

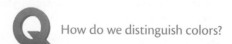 **Q** How do we distinguish colors?

FIGURE 10.7 The Color Circle.

Colors opposite each other are complementary; that is, equal amounts of light in those colors cancel each other out, producing a neutral gray.

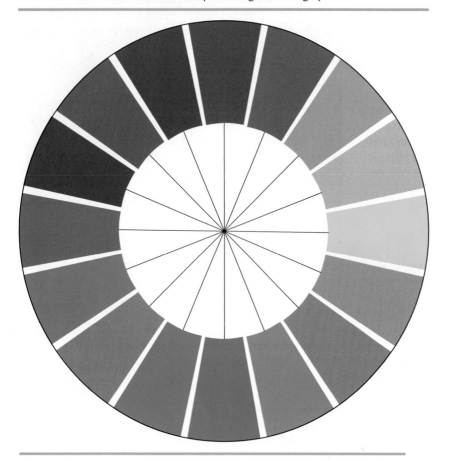

FIGURE 10.8 Complementary Colors and Negative Color Aftereffect.

Stare at the flag for about a minute, then look at a white surface (the ceiling or a sheet of paper); you should see a traditional red, white, and blue flag.

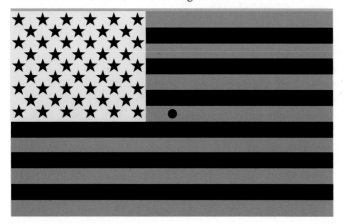

other end just as easily as the colors adjacent to each other on the spectrum blend into each other. Another reason the color circle makes sense is that any two colors opposite each other on the circle are complementary; mixing equal amounts of light from across the circle results in the sensation of a neutral gray tending toward white, depending on the brightness. An exception to this rule is the combination of red and green; they produce yellow, for reasons you will understand shortly.

Another indication of complementarity is that overstimulation of the eye with one light makes the eye more sensitive to its complement. Stare at a red stimulus for a minute, and you will begin to see a green edge around it; then look at a white wall or a sheet of paper, and you will see a green version of the original object. This experience is called a *negative color aftereffect;* the butcher decorates the inside of the meat case with parsley or other greenery to make the beef look redder. Negative color aftereffect is what one would expect if the wavelengths were affecting the same receptor in opposed directions, as Hering theorized. The flag in Figure 10.8 is a very good interactive demonstration of complementary colors and negative aftereffects.

If this discussion of color mixing seems inconsistent with what you understood in the past, it is probably because you learned the principles of color mixing in an art class. The topic of discussion there was *pigment mixing,* whereas we are talking about *mixing light.* An object appears red to us because it reflects mostly long-wavelength (red) light, while it *absorbs* other wavelengths of light. The effect of light mixing is *additive,* while pigment mixing is *subtractive;* if we mix lights, we add wavelengths to the stimulus, but as we mix paints more wavelengths are absorbed. For example, if you mix equal amounts of all wavelengths of light, the result will be white light; mixing paints in the same way produces black because each added pigment absorbs additional wavelengths of light until the result is total absorption and blackness (Figure 10.9).

Now, back to color vision theory. Although Hering's theory did a nice job of explaining complementary colors and the uniqueness of yellow, it received little acceptance. One reason was that researchers had trouble with Hering's assumption of a chemical that would break down in response to one light and regenerate in the presence of another. His theory was, in fact, in error on that point, but developments 100 years later would bring Hering's thinking back to the forefront.

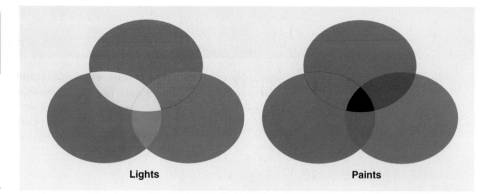

FIGURE 10.9 Mixing Lights Is Additive, Mixing Paints Is Subtractive.

The combination of all three primaries (or all colors) of light produces white; the same combination of pigments produces black, which is the lack of color.

Lights   Paints

A Combined Theory

The trichromatic and opponent process theories appear to be contradictory. Sometimes this means that one position is wrong and the other is right, but often it means that each of the competing theories is partly correct but just too simple to accommodate all the known facts. Hurvich and Jameson (1957) resolved the conflict with a compromise: They proposed that three types of color receptors—red sensitive, green sensitive, and blue sensitive—are interconnected in an opponent process fashion at the ganglion cells.

Figure 10.10 is a simplified version of how Hurvich and Jameson thought this combined color-processing strategy might work. Long-wavelength light excites "red" cones and the red-green ganglion cell to give the sensation of red. Medium-wavelength light excites the "green" cones and inhibits the red-green cell, reducing its firing rate below its spontaneous level and signaling green to the brain. Likewise, short-wavelength light excites "blue" cones and inhibits the yellow-blue ganglion cell, leading to a sensation of blue. Light midway between the sensitivities of the "red" and "green" cones would stimulate both cone types. The firing rate in the red-green ganglion cell would not change, because equal stimulation and excitation from the two cones would cancel out; however, the cones' connections to the yellow-blue ganglion cell are both excitatory, so their combined excitation would produce a sensation of yellow. According to this theory, there are three color processes at the receptors and four beyond the ganglion cells.

This scheme does explain very nicely why yellow would appear pure just like red, green, and blue do. Also, it is easy to understand why certain pairs of colors are complementary instead of producing a blended color. For example, you could have a color that is reddish blue (purple) or greenish yellow (chartreuse) but not a reddish green or a bluish yellow. Negative aftereffects can be explained as overstimulation "fatiguing" a ganglion cell's response in one direction, causing a rebound in the opposite direction and a subtle experience of the

FIGURE 10.10 Hurvich and Jameson's Proposed Interconnections of Cones Provide Four Color Responses and Complementary Colors.

"+" indicates excitation; "−"indicates inhibition. For simplicity, cells between the cones and ganglion cells are not shown.

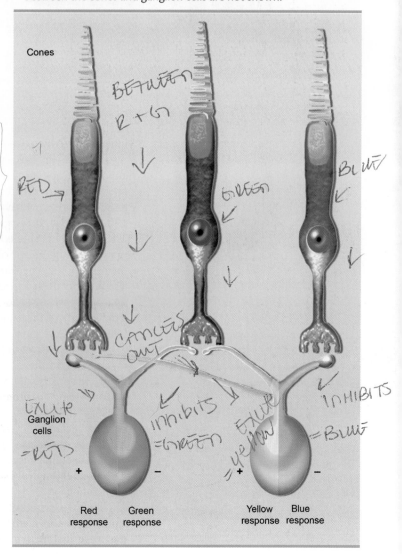

Cones

Ganglion cells

Red response   Green response   Yellow response   Blue response

FIGURE 10.11 Relative Absorption of Light of Various Wavelengths by Visual Receptors.

Note that each type of cone responds best to wavelengths corresponding to blue, green, and red light, though each responds to other wavelengths as well.

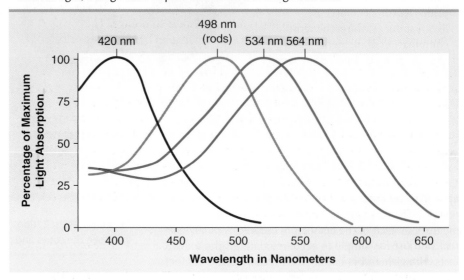

SOURCE: Adapted from "Visual Pigments of Rods and Cones in Human Retina," by Bowmaker and Dartnall, 1980, *Journal of Physiology, 298*, pp. 501–511. Copyright 1980, with permission from John Wiley & Sons, Inc.

opposing color. By the way, this is our first example of the significance of all that interconnectedness we just saw in the retina.

Evidence for this combined trichromatic/opponent process theory would be almost a decade away, however, because it depended on the development of more precise measurement capabilities. Support came in two forms. First, researchers produced direct evidence for three color receptors in the retina (K. Brown & Wald, 1964; Dartnall, Bowmaker, & Mollon, 1983; Marks, Dobelle, & MacNichol, 1964). The researchers shone light of selected wavelengths through individual receptors in eyes that had been removed from humans for medical reasons or shortly after their death; they measured the light that passed through to determine which wavelengths had been lost through absorption. The absorbed wavelengths were the ones the receptor's photochemical was sensitive to. Figure 10.11 shows the results from a study of this type. Note that there are three distinct color response curves (plus a response curve for rods), just as Hurvich and Jameson (1957) predicted.

There are three additional features of these results I want to call to your attention. Like the tuning curves for frequency we saw in the previous chapter, these curves are not very sharp; each receptor has a sensitivity peak, but its response range is broad and overlaps with that of its neighbors. This means that the medium-wavelength cones could be active because the stimulus is "green" light or because they are being stimulated with intense "blue" light. The system must *compare activity in all three types of cones* to determine which wavelengths you are seeing. This "comparison" does not occur at the level of awareness; it is an automatic neural process, like the activity of coincidence detectors in sound localization.

The two remaining features are related to the evolution of color vision. Notice that the medium- and long-wavelength curves are very close together, and very distant from the short-wavelength curve. Genetic research indicates that the genes for the photopigments in the medium- and long-wavelength cones evolved from a common precursor gene relatively recently (*only* about 35 million years ago), but the short-wavelength cones and rod receptors split off from their precursor much earlier. Indeed, the "red" and "green" genes are adjacent to each other on the X chromosome, and they are 98% identical in DNA (Gegenfurtner & Kiper, 2003). Finally, a consequence of the recent separation of the "red" and "green" genes is their functional similarity; although the "red" cones respond strongly to red, their peak sensitivity is actually closer to orange. This revelation should add even more caution to our shorthand use of color names to refer to these three types of cones.

Trichromatic vision is certainly beneficial in appreciating art, but we might well ask what evolutionary benefits compelled its development. An obvious advantage was an enhanced ability to distinguish ripe fruit and to locate young, tender leaves. Genetic research also suggests that it allowed primates to respond to colorful sexual signals in place of olfactory and pheromonal signals. Comparisons of primate genomes indicate intriguing parallels between the appearance of genes for trichromacy, the decreases in olfactory and pheromone receptor genes, and the development of visual signals such as the reddened and swollen sexual skin in female baboons and Old World monkeys (Gilad, Wiebe, Przeworski, Lancet, & Pääbo, 2004; J. Zhang & Webb, 2003).

A second major development was the confirmation in monkeys of color-opponent ganglion cells in the retina and color-opponent cells in the lateral geniculate nucleus of the thalamus (De Valois, 1960; De Valois, Abramov, & Jacobs, 1966; Gouras, 1968). Figure 10.10 illustrates two types of opponent cells, one that is excited by red and inhibited by green (R+G–) and one that is excited by yellow and inhibited by blue (Y+B–); Russell De Valois and his colleagues (1966) identified two additional types that were the inverse of the previous two: green excitatory/red inhibitory (G+R–) and blue excitatory/yellow inhibitory (B+Y–).

A surprise was that some of the color-opponent ganglion cells receive their input from cones that are arranged in two concentric circles (Gouras, 1968; Wiesel & Hubel, 1966). The cones in the center and those in the periphery have color-complementary sensitivities (see Figure 10.12). Of course, the yellow

FIGURE 10.12 Receptive Fields of Color-Opponent Ganglion Cells.

The cones in the center and the cones in the periphery respond to colors that are complementary to each other. The center cones excite the ganglion cell, and the cones in the periphery inhibit it.

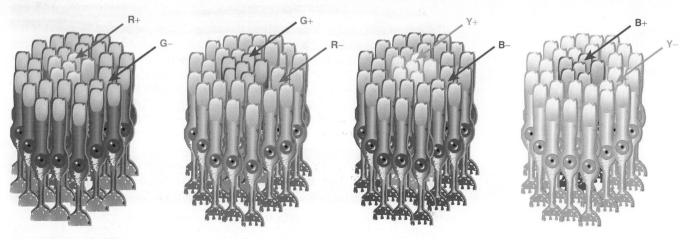

SOURCE: Based on the findings of De Valois et al. (1966).

This is one of the plates from the Ishihara test for color blindness. Most people see the number 74; the person with color deficiency sees the number 21.

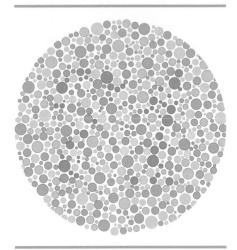

Q What is it like to be color-blind?

4
Color and Color Blindness

response is provided by the combined output of "red" and "green" cones. Why all this complexity? First, the opposition of cones at the ganglion cells provides wavelength discrimination that individual cones are incapable of producing (E. E. Goldstein, 1999). Along with this opposition, the concentric-circle receptor fields enhance information about color contrast in objects. The concept of information sharpening will become clearer when we look at how the retina distinguishes the edge of an object.

A theory is considered successful if it is consistent with the known facts, can explain those facts, and can predict new findings. The combined trichromatic and opponent process color theory meets all three criteria: (1) It is consistent with the observation that all colors can be produced by using red, green, and blue light. (2) It can explain why observers regard red, green, blue, *and* yellow as pure colors. It also explains complementary colors, negative aftereffects, and the impossibility of color experiences such as greenish red. (3) It predicted the discovery of three photopigments and of the excitatory/inhibitory neural connections at the ganglion cells.

Color Blindness

Color blindness is an intriguing curiosity; but more than that, it has played an important role in our understanding of color perception by scuttling inadequate theories and providing the inspiration for new ones. There are very few completely color-blind people—about 1 in every 100,000. They usually have an inherited lack of cones; limited to rod vision, they see in shades of gray, they are very light sensitive, and they have poor visual acuity. More typically a person is partly color-blind, due to a defect in one of the cone systems rather than a lack of cones.

There are two major types of retinal color blindness. A person who is red-green color-blind sees these two colors but is unable to distinguish between them. We know something about what color-blind people experience by noting which colors color-blind people confuse and from studying a few rare individuals who are color-blind in only one eye. When I was trying to understand a red-green color-blind colleague's experience of color, he explained that green grass was the same color as peanut butter! I don't know what peanut butter looked like to him, but he assured me that he found grass and trees "very beautiful." People in the second color-blind group do not perceive blue, so their world appears in variations of red and green. Many partly color-blind individuals are unaware that they see the world differently than the rest of us. Color vision deficiencies can be detected by having the subject match or sort colored objects or with a test like the one illustrated in Figure 10.13.

Red-green color-blind individuals show a deficiency in either the red end of the spectrum or in the green portion; this suggests that the person lacks either the appropriate cone or the photochemical. Acuity is normal in both groups, so there cannot be a lack of cones. Some are unusually sensitive to green light, and the rest are sensitive to red light; this suggests that in one case the normally red-sensitive cones are filled with green-sensitive photochemical and in the other the normally green-sensitive cones are filled with red-sensitive chemical.

Concept Check *Take a Minute to Check Your Knowledge and Understanding*

☐ Summarize the three color vision theories described here.

☐ What is the benefit of the color-opposed concentric circle receptor fields?

☐ What causes color blindness?

Form Vision

Just as the auditory cortex is organized as a map of the cochlea, the visual cortex contains a map of the retina. Russell De Valois and his colleagues demonstrated this point when they presented the image in Figure 10.14a to monkeys that had been injected with radioactive 2-deoxyglucose. The animals were sacrificed and their brains placed on photographic film. Because the more active neurons absorbed more radioactive glucose, they exposed the film more darkly in the autoradiograph in Figure 10.14b; this produced an image of the stimulus that appears to be wrapped around the monkey's occipital lobe (Tootell et al., 1982).

This result tells us that, just as there is a tonotopic map of the basilar membrane in the auditory cortex, we have a *retinotopic map* in the visual cortex, meaning that adjacent retinal receptors activate adjacent cells in the visual cortex. However, this does not tell us how we see images; transmitting an object's image to the cortex like a television picture does not amount to perception of the object. Object perception is a two-stage affair. In this section, we will discuss *form vision,* the detection of an object's boundaries and features (such as texture). We will discuss the second component, object recognition, a bit later. The story that unfolds here is about more than perception; it provides a model for understanding how the brain processes information in general. It is also a story that begins not in the cortex but in the retina itself.

Contrast Enhancement and Edge Detection

Detecting an object's boundaries is the first step in form vision. The nervous system often exaggerates especially important sensory information; in the case of boundaries, it uses lateral inhibition to enhance the contrast in brightness that defines an object's edge. To demonstrate this enhancement for yourself, look at the Hermann grid illusion in Figure 10.15a and the Mach band illusion in Figure 10.15b. The Hermann grid is the more dramatic of the two, but the Mach band illusion

5
Perception Tutorials

FIGURE 10.14 Deoxyglucose Autoradiograph Showing Retinotopic Mapping in Visual Cortex.

Monkeys were given radioactive 2-deoxyglucose, then shown the design in (a). They were sacrificed, and a section of their visual cortical tissue was placed on photographic film. The exposed film showed a pattern of activation (b) that matched the design.

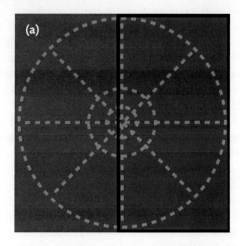
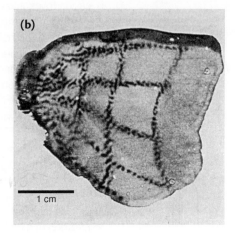

SOURCE: Reprinted with permission from R. B. H. Tootell et al., "Deoxyglucose Analysis of Retinotopic Organization in Primate Striate Cortex," *Science, 218,* pp. 902–904. Copyright 1982 American Association for the Advancement of Science (AAAS).

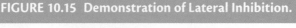

FIGURE 10.15 Demonstration of Lateral Inhibition.

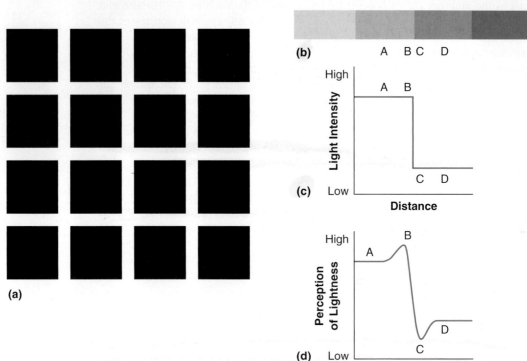

(a) In the Hermann grid illusion, lateral inhibition causes you to see small, grayish blotches at the intersections of the large squares. (b) The Mach band illusion is another example. Each band is consistent in brightness across its width, as shown in the graph of light intensity in (c). But where the bands meet, you experience a slight enhancement in brightness at the edge of the lighter band and a decrease in brightness at the edge of the darker band (e.g., at points B and C). This effect is represented graphically in (d). Exaggeration of brightness contrast at edges helps us see the boundaries of objects.

SOURCES: (a) Based on Hermann (1870). (b) From *Mach Bands: Quantitative Studies on Neural Networks in the Retina* (fig. 3.25. p. 107), by F. Ratcliff, 1965. San Francisco: Holden-Day. Copyright © Holden-Day Inc.

Deceptions of the senses are the truths of perception.
—*Johannes Purkinje*

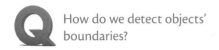

How do we detect objects' boundaries?

is easier to explain, so we will focus on it. Each bar in the Mach band image is consistent in brightness across its width, but it looks a bit darker on the left and a bit lighter on the right than it does in the middle. (If you don't see a difference at the edges, you may notice that the bars seem slightly curved. This is because the illusion suggests subtle shadowing on the left side of each bar.) An illusion is not simply an error of perception, but an exaggeration of a normal perceptual process, which makes illusions very useful in studying perception.

Figure 10.16 will help you understand how your retinas produce the illusion. In *lateral inhibition*, each neuron's activity inhibits the activity of its neighbors and in turn its activity is inhibited by them. In this case, the inhibition is delivered by horizontal cells to nearby synapses of receptors with bipolar cells. The critical point in the illustration is at ganglion cells 7 and 8. Ganglion cell 7 is inhibited less than ganglion cells 1 to 6; this is because the receptors to its right are receiving very little stimulation and producing low levels of inhibition. This lesser inhibition of ganglion cell 7 creates a sensation of a lighter band to the left of the border, as indicated at the bottom of the illustration. Similarly, ganglion cell 8 is inhibited more than its neighbors to the right, because the receptors to its left are receiving greater stimulation and producing more inhibition. As a result, the bar appears darker to the right of the border.

lateral inhibition ↓

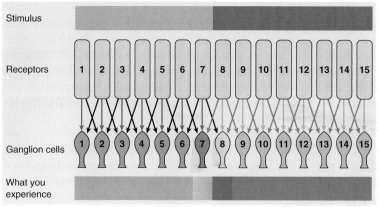

Stimulus

Receptors   1 2 3 4 5 6 7 8 9 10 11 12 13 14 15

Ganglion cells   1 2 3 4 5 6 7 8 9 10 11 12 13 14 15

What you experience

The bar at the top represents the middle two bands from Figure 10.15b. Red arrows indicate excitation, and gray arrows indicate inhibition. All ganglion cells are activated, but ganglion cell 7 is activated most; like 1 through 6, it receives more stimulation from the brighter stimulus, but it receives less inhibition from the receptors to the right. Likewise, ganglion cell 8 receives minimal stimulation, plus it receives more inhibition from the receptors to the left than do cells 9 through 15. As a result, the light bar appears lighter at its border with the dark bar, which in turn appears darker at its border. (Cells between the receptors and ganglion cells have been omitted for simplicity. Also, you would see some gradation of contrasts, because the inhibitory connections extend farther than the adjacent cell.)

Actually, this description is more appropriate for the eye of the horseshoe crab, where lateral inhibition was originally confirmed by electrical recording; in fact, the graph in Figure 10.15d was based on data from the horseshoe crab's eye (Ratliff & Hartline, 1959). The principle is the same in the mammalian eye, but each ganglion cell's receptive field is made up of several receptors arranged in circles, like the color-coded circular fields we saw earlier (Kuffler, 1953). Light in the center of the field has the opposite effect on the ganglion cell from light in the surround. In *on-center* cells, light in the center increases firing, and light in the *off surround* reduces firing below the resting levels. Other ganglion cells have an *off center* and an *on surround*. Figure 10.17 illustrates these two types of ganglion cells.

Now, for another example of the effect of interconnectedness in the retina, let's look at how the antagonistic arrangement in these ganglion cells turns them into light–dark contrast detectors (Hubel, 1982). Look at the three illustrations in Figure 10.18. Light falling across the entire field will have little or no effect on the ganglion cell's firing rate, because the excitation and inhibition cancel each other out (Figure 10.18a). Light that falls only on the *off surround* will suppress firing in the ganglion cell (Figure 10.18b). But the ganglion cell's firing will be at its maximum when the stimulus falls on all of the *on center* and only a part of the *off surround,* as in Figure 10.18c. We will see the significance of this light–dark contrast mechanism in the next section.

Hubel and Wiesel's Theory

Cells in the lateral geniculate nucleus have circular receptive fields just like the ganglion cells from which they receive their input. The receptive fields of visual neurons in the cortex, however, turn out to be surprisingly different. David Hubel and Torsten Wiesel (1959) were probing the visual cortex of anesthetized cats as they projected visual stimuli on a screen in front of the cat. Their electrode was connected to an auditory amplifier so they could listen for indications of active cells. One day, they were manipulating a glass slide with a black dot on it in the projector and getting only vague and inconsistent responses

> when suddenly over the audio monitor the cell went off like a machine gun. After some fussing and fiddling we found out what was happening. The response had nothing to do with the black dot. As the glass slide was inserted its edge was casting onto the retina a faint but sharp shadow, a straight dark line on a light background. (Hubel, 1982, p. 517)

The receptive fields and ganglion cells are shown in cross section. The connecting cells between the receptors and the ganglion cells have been omitted for simplicity.

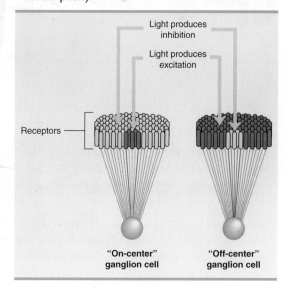

Light produces inhibition

Light produces excitation

Receptors

"On-center" ganglion cell   "Off-center" ganglion cell

FIGURE 10.18 Effect of a Luminous Border on an On-Center Ganglion Cell.

The vertical marks on the solid bar represent neural responses, and the yellow line underneath indicates when the light was on. Notice that the greatest activity occurs in the ganglion cell when light falls on all of the center but less than all of the surround.

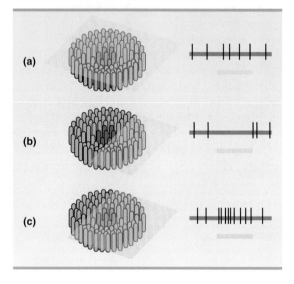

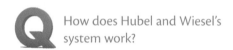

Q How does Hubel and Wiesel's system work?

Hubel and Wiesel then began exploring the receptive fields of these cortical cells by projecting bars of light on the screen. They found that an actively responding cell would decrease its responding when the stimulus was moved to another location or rotated to a slightly different angle. Figure 10.19 shows the changes in response in one cell as the orientation of the stimulus was varied. Hubel and Wiesel called these cortical cells simple cells. *Simple cells* respond to a line or an edge that is at a specific orientation and at a specific place on the retina.

How can we explain the surprising shift in specialization in these cortical cells? Imagine several contrast-detecting circular fields arranged in a straight line, like those in Figure 10.20. (Notice that the fields overlap each other; by now you shouldn't be surprised that receptors would share their output with multiple ganglion cells.). Then, connect the outputs of their ganglion cells to a single cell in the cortex—one of Hubel and Wiesel's simple cells. You now have a mechanism for detecting not just spots of light–dark contrast but a contrasting edge, such as in the border of an object that is lighter or darker than its background. Fields with on centers would detect a light edge, like the one in the figure, and a series of circular fields with off centers would detect a dark edge.

In other layers of the cortex, Hubel and Wiesel found *complex cells,* which continue to respond when a line or an edge moves to a different location, as long as it is not too far from the original site. They explained the complex cell's ability to continue responding in essentially the same way they explained the sensitivity of simple cells. They assumed that complex cells receive input from several simple cells that have the same orientation sensitivity but whose fields are next to each other on the retina. This arrangement is illustrated in Figure 10.21. Notice that as the edge moves horizontally, different simple cells will take over, but the same complex cell will continue responding. However, if the edge rotates to a different orientation, this complex cell will stop responding, and another complex cell specific for that orientation will take over. Connecting several simple cells to a single complex cell enables the complex cell not only to keep track of an edge as it moves but also to *detect* movement.

The feasibility of this kind of arrangement has received support from an interesting source—artificial neural networks. Lau et al. (2002) trained a network so that its output "neurons" gave the same responses to bar-shaped stimuli as those recorded from complex cells in cats. Then they examined the hidden layer and

FIGURE 10.19 Responses to Lines at Different Orientations in a Simple Cell Specialized for Vertical Lines.

The vertical hatchmarks represent neural responses, and the yellow line underneath indicates when the stimulus occurred. Notice that the response was greatest when the line was closest to the cell's "preferred" orientation (vertical) and least when the orientation was most discrepant. In the last example, the response was diminished because the stimulus failed to cover all of the cell's field (indicated by the stimulus being off-center of the crosshair).

SOURCE: From "Receptive Fields of Single Neurons in the Cat's Striate Cortex," by D. H. Hubel and T. N. Wiesel, 1959, *Journal of Physiology, 148,* pp. 574–591, Fig 3. © 1959 by The Physiology Society. Reprinted by permission.

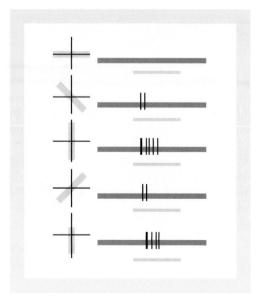

found that those "neurons" had rearranged their connections to approximate simple cells, complete with "on" and "off" regions in their receptive fields as well as directional sensitivity. In an earlier study, a neural network was trained to recognize curved visual objects (Lehky & Sejnowski, 1990). Its "neurons" spontaneously developed sensitivity to bars or edges of light even though they had never been exposed to such stimuli, suggesting that the Hubel-Wiesel model is a very versatile one.

But so far, Hubel and Wiesel had seen only the beginnings of the intricate neural organization that makes visual perception possible. They lowered electrodes perpendicularly through a monkey's striate cortex; as the electrode passed through simple and complex cells, the cells' preferred width and length changed, but they had the same orientation (Hubel & Wiesel, 1974). As the researchers moved the electrode slightly to the side, the orientation shifted slightly but systematically in a clockwise or counterclockwise direction; over a distance of 0.5 mm to 1.0 mm, the orientation would progress through the complete circle. Then the process would start over again, but the field would shift to adjacent retina. This sort of organization is typical of the cortex's efficiency in processing and transmitting information. Connections mostly run up and down in columns with much shorter lateral connections. In addition, similar functions are clustered together, increasing communication speed and reducing energy requirements.

Hubel and Wiesel shared the Nobel Prize for their work in 1981. However, their model has limitations—some would say problems. For one thing, it accounts for the detection of boundaries, but it is questionable whether edge detection cells can also handle the surface details that give depth and character to an image.

FIGURE 10.20 Hubel and Wiesel's Explanation for Responses of Simple Cells.

When the edge is in this position, the ganglion cell for each of the circular fields increases its firing. (The fields shown here have on centers.) The ganglion cells are connected to the same simple cell, which also increases firing, indicating that an edge has been detected. This particular arrangement would be specialized for a vertical light edge.

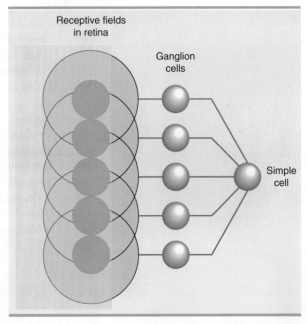

FIGURE 10.21 Hubel and Wiesel's Explanation for Responses of Complex Cells.

A complex cell receives input from several simple cells, each of which serves a group of circular fields (as in Figure 10.20). As a result, the complex cell continues to respond as the illuminated edge moves to the left or to the right. (Ganglion cells have been eliminated for simplicity.)

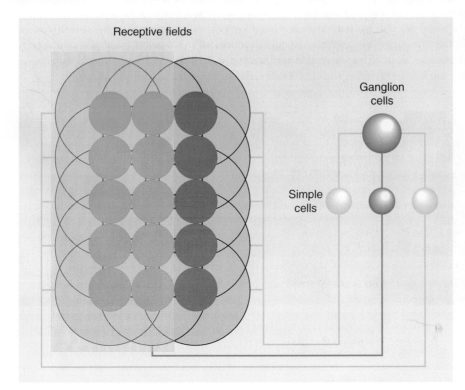

Q Is spatial frequency theory a better explanation?

Spatial Frequency Theory

Although some cortical cells respond best to edges (Albrecht, De Valois, & Thorell, 1980; De Valois, Thorell, & Albrecht, 1985; von der Heydt, Peterhans, & Dürsteler, 1992), others cells apparently are not so limited. Think of an edge as an abrupt or high-frequency change in brightness. The more gradual changes in brightness across the surface of an object are low-frequency changes. According to De Valois, some complex cells are "tuned" to respond to the high frequencies found in an object's border, while others are tuned to low frequencies, for example in the slow transition from light to shadow that gives depth to the features of a face (De Valois et al., 1985). Some cells respond better to "gratings" of alternating light and dark bars—which contain a particular combination of spatial frequencies—than they do to lines and edges. According to *spatial frequency theory,* visual cortical cells do a Fourier frequency analysis of the luminosity variations in a scene (see Chapter 9 to review Fourier analysis). According to this view, different visual cortical cells have a variety of sensitivities, not just those required to detect edges (Albrecht et al., 1980; De Valois et al., 1985).

A few photographs should help you understand what we mean by spatial frequencies, as well as the importance of low frequencies. The picture in Figure 10.22a was prepared by having a computer average the amount of light over large areas in a photograph; the result was a number of high-frequency transitions, and the image is not very meaningful. In Figure 10.22b, the computer filtered out the high frequencies, producing more gradual changes between light and dark (low frequencies). It seems paradoxical that blurring an image would make it more recognizable, but blurring eliminates the sharp boundaries. You can get the same effect from Figure 10.22a by looking at it from a distance or by squinting. In Figure 10.22c, the Spanish artist Salvadore Dali incorporated the illusion in one of his more famous

FIGURE 10.22 Illustration of High and Low Frequencies in a Visual Scene.

(a) An image limited to abrupt changes in brightness (high spatial frequencies) is not as meaningful as (b) one that has both high- and low-frequency information. (b) is the same image as (a), except that the edges have been blurred. (c) Salvadore Dali's 1976 painting *Gala Contemplating the Mediterranean Sea Which at Twenty Meters Becomes a Portrait of Abraham Lincoln.* Look closely and you will see (a) rather than Dali's wife Gala; squint your eyes and you will see (b).

(a)
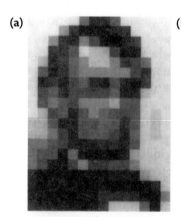

(b)
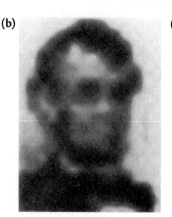

(c)

SOURCES: (a and b) Reprinted with permission from L. D. Harmon and B. Julesz, "Masking in Visual Recognition: Effects of Two-Dimensional Filtered Noise," *Science, 180,* pp. 1194–1197. Copyright 1973 American Association for the Advancement of Science (AAAS). (c) *Gala Contemplating the Mediterranean Sea Which at Twenty Meters Becomes the Portrait of Abraham Lincoln-Homage to Rothko* (Second version). 1976. Oil on canvas. 75.5 x 99.25 inches. © 2010 Salvador Dalí, Gala-Salvador Dalí Foundation/Artists Rights Society (ARS), New York.

FIGURE 10.23 The Role of High and Low Frequencies in Vision.

The original photo (a) compared with the same photo with low frequencies removed (b) and with high frequencies removed (c). These manipulations show how indispensable a range of spatial frequencies is to accurate vision.

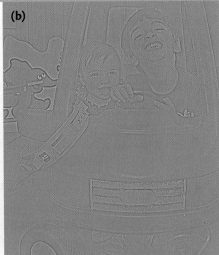
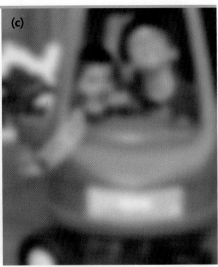

SOURCE: From "Visual Objects in Context," by M. Bar, *Nature Reviews Neuroscience*, 5, 617–629, figure in box 2, p. 621. Copyright © 2004 Moshe Bar. Used with permission.

paintings. A real-life example in Figure 10.23 suggests what our visual world might be like if we were limited to high frequencies or low frequencies.

So far, we have dealt only with the simplest aspects of visual perception. But we also are able to recognize an object as an object, assign it color under varied lighting conditions, and detect its movement. Attempting to explain these capabilities will provide challenge enough for the rest of this chapter.

Concept Check   *Take a Minute to Check Your Knowledge and Understanding*

- ☐ Explain how the opponent arrangement of a ganglion cell's field enhances brightness contrast.
- ☐ How did Hubel and Wiesel explain our ability to detect an edge, the orientation of an edge or line, and an edge or line that changed its location?
- ☐ How do Hubel and Wiesel's theory and the spatial frequency theory differ?

The Perception of Objects, Color, and Movement

One of the more interesting characteristics of the visual system is how it dissects an image into its various components and analyzes them in different parts of the brain. The separation begins in the retina and increases as visual information flows through all four lobes of the brain, with locations along the way carrying out analyses of color, movement, and other features of the visual scene. Thus, we will see how visual processing is, as I mentioned earlier, both modular and hierarchical. *Modular processing* refers to the segregation of the various components of processing into separate locations. *Hierarchical processing* means that lower levels of the nervous system analyze their information and pass the results on to the next higher level for further analysis.

Some neuroscientists reject the modular notion, arguing that any visual function is instead *distributed,* meaning that it occurs across a relatively wide area of the brain. One study found evidence that sensitivity to faces, for example, is scattered over a large area in the temporal lobe (Haxby et al., 2001). Research has not resolved the modular-distributed controversy, leaving researchers to quarrel over the interpretation of studies that seem to support one view or the other (J. D. Cohen & Tong, 2001). Vision may well involve a mix of modular and distributed functioning, rather like the arrangement we saw for language. In spite of this ambiguity, my view is that processing can still be considered modular even if components of the task are located in more than one place, as long as the pattern of activity is reasonably distinct from other processes. With this caveat in mind, we will consider what is known about the pathways and functional locations in the visual system.

The Two Pathways of Visual Analysis

Q What do the parvocellular and magnocellular systems do?

Visual information follows two routes from the retina through the brain, which make up the *parvocellular system* and the *magnocellular system* (Livingstone & Hubel, 1988; P. H. Schiller & Logothetis, 1990). Parvocellular ganglion cells are located mostly in the fovea. They have circular receptive fields that are small and color opponent, which suits them for the specialties of the *parvocellular system*, the discrimination of fine detail and color. We first mentioned the magnocellular system in the discussion of dyslexia in Chapter 9. Magnocellular ganglion cells have large circular receptive fields that are brightness opponent and respond only briefly to stimulation. As a result, the *magnocellular system* is specialized for brightness contrast and for movement.

Figure 10.24 is an interesting demonstration of the specialized nature of these two systems. The bicycle in Figure 10.24a differs from its background in color but not in brightness, so the image mostly stimulates the parvocellular system. The bicycle in Figure 10.24b differs from the background in brightness but not in color; it mostly stimulates the magnocellular system. You get a sense of depth in the second picture but not in the first, because depth perception is a capability of the magnocellular system.

We also see evidence of differences in the two systems in our everyday life. The simplest example is that at dusk our sensitivity to light increases, but we lose our ability to see color and detail. You cannot read a newspaper under such conditions or color coordinate tomorrow's outfit, because the high-resolution, color-sensitive parvocellular system is nearly nonfunctional. The magnocellular system's sensitivity to movement is most obvious in your peripheral vision.

FIGURE 10.24 Color Contrast and Brightness Contrast Stimulate Different Visual Systems.

Because (a) consists primarily of color contrast, it stimulates the parvocellular system more than the magnocellular system; (b) is made up primarily of brightness contrast and stimulates mostly the magnocellular system. As a result, we see depth in (b) but not in (a).

(a)

(b)

SOURCE: Reprinted with permission from M. Livingstone and D. Hubel, "Segregation of Form, Color, Movement, and Depth: Anatomy, Physiology, and Perception," *Science, 240,* pp. 740–749. Copyright 1988 American Association for the Advancement of Science (AAAS).

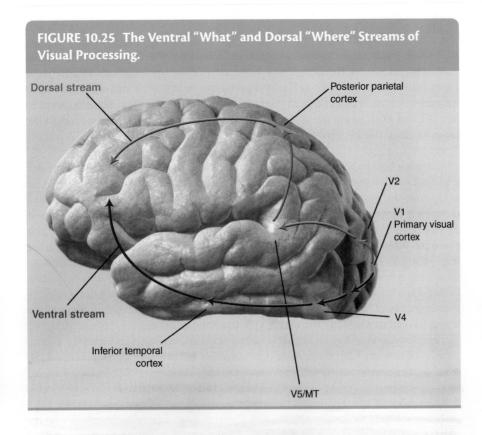

FIGURE 10.25 The Ventral "What" and Dorsal "Where" Streams of Visual Processing.

Dorsal stream

Posterior parietal cortex

V2

V1
Primary visual cortex

Ventral stream

V4

Inferior temporal cortex

V5/MT

Hold your arms outstretched to the side while you look straight ahead, and move your hands slowly forward while wriggling your fingers. When you just notice your fingers moving, stop. Notice that you can barely see your fingers but you are very sensitive to their movement.

Both pathways travel to the lateral geniculate nucleus and then to the primary visual cortex, which is also known as V1. Although the two systems are highly interconnected, the parvocellular system dominates the *ventral stream,* which flows from the visual cortex into the temporal lobes, and the magnocellular system dominates the *dorsal stream* from the visual cortex to the parietal lobes (Figure 10.25). Like the two auditory pathways, the ventral stream is often referred to as involved with the *what* of visual processing, and the dorsal stream with the *where.* Most of the research on this topic has been done with monkeys, but the two pathways have been confirmed with PET scans in humans (Ungerleider & Haxby, 1994).

Beyond V1, the ventral stream passes through V2 and into V4, which is mostly concerned with color perception. It then projects to the inferior temporal cortex, which is the lower boundary of the temporal lobe; this area shows a remarkable specialization for object recognition, which we will examine shortly.

Magnocellular neurons arrive in V1 in areas that are responsive to orientation, movement, and retinal disparity (Poggio & Poggio, 1984). The dorsal stream then proceeds through V2 to V5, also known as MT because it is on the middle temporal gyrus in the monkey; neurons there have strong directional sensitivity, which contributes to the perception of movement. The dorsal stream travels then to the posterior parietal cortex, the area just behind the somatosensory cortex; its role is primarily to locate objects in space, but the behavioral implications of its functions are far more important than that simple statement suggests.

Movement perception is a good example of how modular and distributed processing work together. V5/MT and a nearby area that receives input from MT, known as MST (for medial superior temporal area), appear to be the most important areas for perceiving movement They receive most of their input from the

Q What are the functions of the ventral and dorsal streams?

[handwritten note:] ventral - p - what
dorsal - ma - where

magnocellular pathway, including complex cells that are sensitive to movement, and are active during movement; they also respond when the motion is only implied in a photograph of an athlete in action or a picture of a cup falling off a table (reviewed in Culham, He, Dukelow, & Verstraten, 2001). At the same time, there are many other areas that are specialized for particular kinds of movement. Viewing movement of the human body or its parts activates dorsal stream areas adjacent to V5/MT and MST, in the parietal and frontal lobes, and in the ventral stream in the temoral lobes (Vaina, Solomon, Chowdhury, Sinha, & Belliveau, 2001; Wheaton, Thompson, Syngeniotis, Abbott, & Puce, 2004).

Even though images move across your retinas every time you move your eyes, you don't see the world moving around you. (Imagine trying to read, otherwise.) This is because the activity of movement-sensitive cells in MT and MST is suppressed during eye movements (Thiele, Henning, Kubischik, & Hoffmann, 2002). These cells are sensitive to movement in a particular direction, and some of them reverse their preferred direction of movement as the head moves, which allows them to continue responding to real movement of objects. The brain's movement areas are close to an area that analyzes input from the vestibular organs, which monitor body motion (Thier, Haarmeier, Chakraborty, Lindner, & Tikhonov, 2001); this is probably the reason excessive visual motion from either watching passing objects too closely or trying to read in a moving car makes you dizzy and nauseous.

The functions of the ventral and dorsal streams are best illustrated by a comparison of patients with damage in the two areas. People with damage in the temporal cortex (ventral stream) have trouble visually identifying objects, but they can walk toward or around the objects and reach for them accurately (Kosslyn, Ganis, & Thompson, 2001). People with damage to the dorsal stream have the opposite problem. They can identify objects, but they have trouble orienting their gaze to objects, reaching accurately, and shaping their hands to grasp an object using visual cues (Ungerleider & Mishkin, 1982). So the dorsal stream is also a "how" area that is important for action.

From the parietal and temporal lobes, the dorsal and ventral streams both proceed into the prefrontal cortex. One function of the prefrontal cortex is to manage this information in memory while it is being used to carry out the functions that depend on the two pathways (Courtney, Ungerleider, Keil, & Haxby, 1997; F. A. Wilson, Ó Scalaidhe, & Goldman-Rakic, 1993). As one example, we will see in Chapter 11 that the prefrontal cortex integrates information about the body and about objects around it during the planning of movements.

Disorders of Visual Perception

Because the visual system is somewhat modular, damage to a processing area can impair one aspect of visual perception while all others remain normal. This kind of deficit is often called an *agnosia*, which means "lack of knowledge." Because the disorders provide a special opportunity for understanding the neural basis of higher-order visual perception, we will orient our discussion of the perception of objects, color, movement, and spatial location around disorders of those abilities.

Object and Face Agnosia

Object agnosia is the impaired ability to recognize objects. In Chapter 3, I described Oliver Sacks's (1990) agnosic patient who patted parking meters on the head, thinking they were children; he was also surprised when carved knobs on furniture failed to return his friendly greeting. Dr. P. was intellectually intact; he continued to perform successfully as a professor of music, and he could carry on lively conversations on many topics. Patients with object agnosia are able to see an object, describe it in detail, and identify it by touch. But they are unable to identify an object by sight or even to recognize an object from a picture that they have just drawn from memory (Gurd & Marshall, 1992; Zeki, 1992).

Like Dr. P, many object agnosic patients also suffer from *prosopagnosia, an impaired ability to visually recognize familiar faces*. The problem is not memory, because they can identify individuals by their speech or mannerisms. Nor is their visual acuity impaired; they often have no difficulty recognizing facial expressions, gender, and age (Tranel, Damasio, & Damasio, 1988). However, they are unable to recognize the faces of friends and family members or even their own image in a mirror (Benton, 1980; A. R. Damasio, 1985).

Prosopagnosics do respond emotionally to photographs of familiar faces they do not recognize, as indicated by EEG evoked potentials and skin conductance response (Bauer, 1984; Renault, Signoret, Debruille, Breton, & Bolger, 1989; Tranel & Damasio, 1985). This suggests that identification and recognition are separate processes in the brain. This "hidden perception" is not without precedent. Patients blinded by damage to V1 show a surprising ability to track the movement of objects and discriminate colors, all the while claiming to be guessing (Zeki, 1992). Cortically blind individuals also can identify emotions expressed in faces they do not otherwise see (Tamietto et al., 2009), and they can avoid obstacles while walking, as the video in WWW 6 shows. This ability to respond to visual stimuli that are not consciously seen is called *blindsight*. Imaging studies have found that blindsight depends on information passing through the superior colliculus directly to extrastriate areas and possibly other pathways that bypass V1 (reviewed in de Gelder & Tamietto, 2007; Tamietto et al., 2010).

Both object agnosia and prosopagnosia are caused by damage to the inferior temporal cortex, which, of course, is part of the ventral stream (see Figure 10.25). The inferior temporal cortex has a columnar organization reminiscent of what we saw in V1; a column of object-responsive cells might respond to variations on a star-like shape, for example, and columns adjacent to one that responds to a frontal view of a face are activated by a face in profile (Tanaka, 2003). Damage usually impairs the ability to recognize both objects and faces, but the occasional case is reported of a patient with prosopagnosia alone (Benton, 1980) or of severe object agnosia with spared face identification (Behrmann, Moscovitch, & Winocur, 1994). In humans, a part of the fusiform gyrus on the underside of the temporal lobe is so important to face recognition that it is referred to as the *fusiform face area (FFA)*; damage that causes prosopagnosia is usually in the right hemisphere (see Figure 10.26; Bouvier & Engel, 2006; Gauthier, Skudlarski, Gore, & Anderson, 2000; Gauthier, Tarr, Anderson, Skudlarski, & Gore, 1999). Until 2001, researchers thought the only way prosopagnosia occurred was through brain damage, usually due to stroke. Then medical student Martina Grueter began to recognize the symptoms in her husband's behavior and made *congenital* prosopagnosia the subject of her MD thesis (Grueter, 2007). An estimated 2.5% of the population has symptoms of the disorder with no history of brain damage (Kennerknecht et al., 2006). They make errors in recognizing familiar faces, and they learn new faces slowly (Grüter, Grüter, & Carbon, 2008). Face recognition ability has a heritability of about 39% (Zhu et al., 2010), so its deficiency in the absence of brain damage probably has a genetic origin. The defect, though, does not appear to be in the fusiform face area; fMRI shows that their FFA responds just as much to faces as it does in normal individuals. Instead, connections of the FFA to more anterior temporal and frontal cortex areas are diminished (Avidan & Behrmann, 2009), indicating the importance of an extended network in face identification.

After information about edges, spatial frequencies, texture, and so on has been detected separately, it is put back together in the inferior temporal cortex. Cells have been located there in both monkeys and humans that respond selectively to geometric figures, houses, animals, hands, faces, or body parts (Figure 10.27a; Desimone, Albright, Gross, & Bruce, 1984, Downing, Jiang, Shuman, & Kanwisher, 2001; Gross, Rocha-Miranda, & Bender, 1972; Kreiman, Koch, & Fried, 2000; Sáry,

The color of the area indicates how often damage was observed there in patients with prosopagnosia.

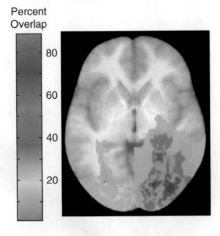

Percent Overlap

SOURCE: From "Behavioral Deficits and Cortical Damage Loci in Cerebral Achromatopsia," by S. E. Bouvier and S. A. Engel, 2006, *Cerebral Cortex, 16,* pp. 183–191.

"

I was having a wonderful conversation with a woman at a party, but then I went to get us some drinks. When I returned, I had forgotten what she looked like, and I was unable to find her the rest of the evening.

—*A young man with prosopagnosia*

"

6

Blindsight and Faceblind

FIGURE 10.27 Stimuli Used to Produce Responses in "Hand-" and "Face-" Sensitive Cells in Monkeys.

(a) The stimuli are ranked in order of increasing ability to evoke a response in a hand-sensitive cell. (b) The spikes recorded from a face-sensitive cell indicate the degree of response from the stimulus shown below.

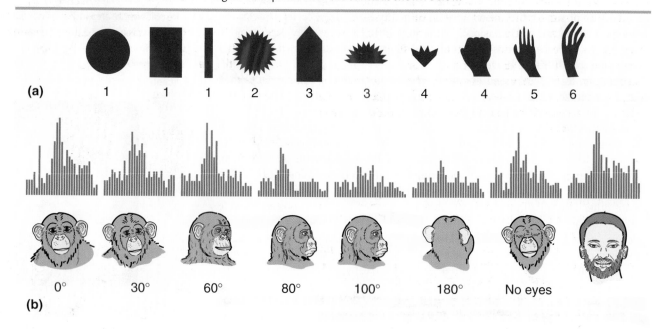

SOURCES: (a) From "Visual Properties of Neurons in Inferotemporal Cortex of the Macaque," by C. G. Gross et al., 1972, *Journal of Neurophysiology*, 35. Reprinted with permission. (b) From "Stimulus-Selective Properties of Inferior Temporal Neurons in the Macaque," by R. Desimone et al., *Journal of Neuroscience*, 4, pp. 2057 Copyright © 1984 Society for Neuroscience. Reprinted with permission.

Vogels, & Orban, 1993). Some of these cells require very specific characteristics of a stimulus, such as a face viewed in profile; others continue to respond in spite of changes in rotation, size, and color (Figure 10.27b; Miyashita, 1993; Tanaka, 1996; Vogels, 1999). Cells with such complex responsiveness likely receive their input from cells with narrower sensitivities (Tanaka, 1996), like the cells involved in detecting edges.

These capabilities may be "hardwired" at birth to some extent, but learning plays a part, at least in adding novel stimuli. When M. P. Young and Yamane (1992) showed monkeys pictures of the faces of lab workers, neurons in the inferior temporal cortex increased their firing rates according to the monkeys' familiarity with the workers. Isabel Gauthier and her colleagues (1999) trained humans to identify faces, using pictures of fictitious creatures they called "greebles" to ensure initial unfamiliarity. The fMRI scans in Figure 10.28 show that pictures of human faces activated the fusiform face area but greebles did not until the person had learned to distinguish individual creatures. A nearby area in the inferior temporal cortex, the *visual word form area (VWFA),* responds to written words as a whole. Its importance in reading was demonstrated in a patient whose VWFA was disconnected from adjacent language areas during surgery to remove defective tissue that was causing epileptic seizures (Gaillard et al., 2006). Before surgery, it took him 600 ms to recognize familiar words, regardless of length; following surgery, the time increased to 1,000 ms for three-letter words and increased by 100 ms for each additional letter, indicating that he was deciphering words letter by letter. Performance in identifying faces, tools, and houses was unaffected. Although the VWFA is typically underactivated in adult dyslexics during reading

(McCandliss & Noble, 2003), this does not mean that this inactivity is the *cause* of their dyslexia. The authors proposed that the phonological deficits we saw in the previous chapter deprive the individual of the opportunity to learn rapid word recognition. It is clear that the VWFA could not have evolved as a dedicated whole word detector, because written language is a relatively recent invention; still, it serves that function so precisely that two words evoke activity in different subareas, even when the words differ by just one letter (Glezer, Jiang, & Riesenhuber, 2009). What is intriguing about the VWFA is that, for whatever reason, the area had already evolved special capabilities that suit it for learning to identify words as if they were unique "objects."

Color Agnosia

Let's return to Jonathan I., whose plight was described in the beginning of the chapter. Jonathan's problem was *color agnosia*, which is the loss of the ability to perceive colors due to brain damage. But before we can discuss this disorder, we need to revisit the distinction between wavelength and color. Once as I walked past a colleague's slightly open office door, I was astonished to see that his face was a distinct green! Opening his door to investigate, I understood why: The light from his desk lamp was reflecting off a bright green brochure he was reading. Immediately his face appeared normal again. This ability to recognize the so-called natural color of an object in spite of the illuminating wavelength is called *color constancy*. If it were not for color constancy, objects would seem to change colors as the sun shifted its position through the day or as we went indoors into artificial light. Imagine having to survive by identifying ripe fruit if the colors kept changing.

When I reinterpreted my colleague's skin color, it was not because I *understood* that his face was bathed in green light; it occurred automatically as soon as my eyes took in the whole scene. Monkeys, who do not understand the principles of color vision, apparently have the same experience. When Zeki (1983) illuminated red, white, green, and blue patches with red light, each patch set off firing in V1 cells that preferred long-wavelength (red) light, regardless of its actual color; however, cells in V4 responded only when the patch's actual color matched the cell's color "preference." Zeki concluded that cells in V1 are *wavelength coded*, while cells in V4 are *color coded*. Schein and Desimone (1990) suggested how V4 cells provide color constancy. They have large circular receptive fields that are color opposed; so if, for example, a green light falling on the center increases the cell's firing rate, green light falling simultaneously in the surround reduces or eliminates the increase. In other words, the cells "subtract out" the color of any general illumination. It was my V4 cells that allowed me to see my colleague's face as a normal pink rather than as the green that it was reflecting.

We have no brain scan to tell us where Jonathan I.'s damage was located, but we know that cortical color blindness, or achromatopsia, occurs when people have lesions in the occipital-temporal area (Heywood & Kentridge, 2003); this, of course, is where V4 is located. Complete color blindness occurs only when the damage is bilateral, so achromatopsia is rare. Unlike Jonathan I., many patients are unaware their color vision is impaired, just as we saw in the previous chapter with Wernicke's aphasia.

FIGURE 10.28 Activity in the Fusiform Face Area While Viewing Faces and "Greebles."

Viewing faces activated a part of the fusiform gyrus (indicated by the white squares) both in "greeble novices" and in "greeble experts," who had learned to distinguish individual greebles from each other. Viewing greebles activated the area only in greeble experts.

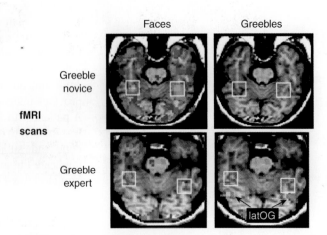

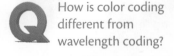

SOURCE: From "Activation of the Middle Fusiform 'Face Area' Increases With Expertise in Recognizing Novel Objects," I. Gauthier et al., *Nature Neuroscience*, 2, 568–573. Copyright © 1999 Macmillan Publishers. Used with permission.

Q How is color coding different from wavelength coding?

Movement Agnosia

Although movement is detected by neurons in V1 and beyond, area V5/MT is the place where that information is integrated; MT is also important for directing reaching movements and eye movements when tracking objects (Born & Badley, 2005; Whitney et al., 2007). A 43-year-old woman known in the literature as LM suffered a stroke that caused bilateral damage in the general area of MT; the result was *movement agnosia,* an impaired ability to detect movement (Vaina, 1998; Zihl, Cramon, & Mai, 1983). Although her vision was otherwise normal, she could distinguish between moving and stationary objects only in her peripheral view, she had difficulty making visually guided eye and finger movements, and she had trouble detecting the movement of people if there were more than two people in the room. She was often surprised to notice that an object had changed position (Zihl et al.). You might think that perceiving a change in position would be the same thing as perceiving movement, but she had no sense of the object traveling through the intermediate positions. When she poured coffee, she could not tell that the liquid was rising in the cup, so she would keep pouring until the cup overflowed! When she tried to cross a street, a car would seem far away, then suddenly very near.

Later analyses indicated that LM's most severe impairment was in her ability to detect radial movement (Vaina, 1998). We experience radial movement when the image of an approaching car expands outwardly, or radially. Radial movement also tells us that we are approaching an object when we walk or drive, because all the environmental objects around the central point appear to move outward; this effect provides information about our *heading* and is important for personal navigation. A patient with impaired perception of radial movement could not catch a ball that was thrown to him, and he frequently bumped into people in his wheel chair. Scans done while subjects perform a task involving radial movement or heading detection implicate the area MST as well as the parietal cortex (Peuskens, Sunaert, Dupont, Van Hecke, & Orban, 2001; Vaina).

Neglect and the Role of Attention in Vision

The posterior parietal cortex combines input from the visual, auditory, and somatosensory areas to help the individual locate objects in space and to orient the body in the environment. Damage impairs abilities such as reaching for objects, but it also often produces *neglect,* in which the patient ignores visual, touch, and auditory stimulation on the side opposite the injury. The term *neglect* seems particularly appropriate in patients who ignore food on the left side of the plate, shave only the right side of the face, or fail to dress the left side of the body. The manifestations are largely, but not entirely, visual, and they are more likely to occur on the left side of the body, following right-hemisphere damage. (Because the symptoms affect one side of space, the term *hemispatial neglect* is often used.)

Neglect is not due to any defect in visual processing, but rather it is due to a deficit in attention; it illustrates the fact that to the extent attention is impaired, so is visual functioning. Two patients with this condition, caused by right parietal tumors, were asked to report whether words and pictures presented simultaneously in the left and right visual fields were the same or different. They said that the task was "silly" because there was no stimulus in the left field to compare, yet they were able to answer with a high level of accuracy (Volpe, LeDoux, & Gazzaniga, 1979). Their performance is superior to that of blindsighted individuals, which suggests that neglect patients are *unaware* of the stimulus because of a deficit in *attention* to the space on one side.

Patients' drawings and paintings help us understand what they are experiencing. When asked to copy drawings, they will neglect one side while completing the other side in detail, like the example in Figure 10.29. The two portraits in

> *I knew the word "neglect" was a sort of medical term for whatever was wrong but the word bothered me because you only neglect something that is actually there, don't you? If it's not there, how can you neglect it?*
>
> —*P. P., a neglect patient*

Figure 10.30 were painted by Anton Raderscheidt 2 and 9 months after a stroke that damaged his right parietal area. Notice that the first painting has very little detail and the left half of the image is missing. In the later painting, he was using the whole canvas, and the portrait looks more normal; but notice that the left side is still much less developed than the right, with the eyeglasses and face melting into ambiguity (Jung, 1974).

The Problem of Final Integration

We have seen how the brain combines information about some aspects of an object, but many researchers wonder where *all* the information about the object is brought together; how the brain combines information from different areas into a unitary whole is known as the *binding problem*. Imagine watching a person walking across your field of view; the person is moving, shifting orientation, and changing appearance as the lighting increases and decreases under a canopy of trees. At the same time, you are walking toward the person, but your brain copes easily with the changing size of the person's image and the apparent movement of environmental objects toward you. It seems logical that a single center at the end of the visual pathway would combine all the information about shape, color, texture, and movement, constantly updating your perception of this image as the same person. In other words, the result would be a complete and dynamic awareness. Presumably, damage to that area would produce symptoms that are similar to blindsight but that affect all stimuli.

It has been suggested that our ultimate understanding of an object occurs in a part of the superior temporal gyrus that receives input from both neural streams (Baizer, Ungerleider, & Desimone, 1991) or in the part of the parietal cortex where

FIGURE 10.29 Drawings Copied by a Left-Field Neglect Patient.

Model   Patient's Copy

SOURCE: From *Brain, Mind, and Behavior* (2nd ed.; p. 300), by F. E. Bloom and A. Lazerson. © 1988 W. H. Freeman & Co.

FIGURE 10.30 Self-Portraits Demonstrating Left Visual Field Neglect.

(a) A self-portrait done 2 months after the artist's stroke, which affected the right parietal area, is incomplete, especially on the left side of the canvas. (b) One done 9 months after the stroke is more complete but still shows less attention to detail on the left side.

SOURCE: © 2010 Artists Rights Society (ARS), New York/VG Bild-Kunst, Bonn.

damage causes neglect (Driver & Mattingley, 1998). Other investigators suspect frontal areas where both streams converge. But these ideas are highly speculative, and there is no convincing evidence for a master area where all perceptual information comes together to produce awareness (Crick, 1994; Zeki, 1992). The variety of hypothesized awareness centers suggests another possibility, that visual awareness is *distributed* throughout the network of 32 areas of cortex concerned with vision and their 305 interconnecting pathways (Van Essen, Anderson, & Felleman, 1992). This thinking is exemplified on a small scale in the interaction between V5/MT and V1. After a stimulus occurs, activity continues back and forth between these areas for a few hundred milliseconds, and disrupting this interchange eliminates awareness of movement (reviewed in McKeefry, Gouws, Burton, & Morland, 2009).

Application — When Binding Goes Too Far

Before his accident, Jonathan I. saw a "tumult" of changing colors whenever he listened to music. For other people with synesthesia, each letter may have its own characteristic color, days of the week may have personalities, visual motion might produce sounds, or words might have tastes. *Synesthesia* is a condition in which stimulation in one sense triggers an experience in another sense or a concept evokes an unrelated sensory experience. Synesthesia was thought to be rare, based on the number of people who came forward to report these experiences, but when Julia Simner and her colleagues (2006) tested 1,700 individuals who were not self-referred, almost 5% showed some characteristics of synesthesia. A third of those were *projectors*, who actually experience the unrelated color or sound or taste, and the rest were *associators*, who have a persistent mental association between, for example, a word and a color but don't report that they see the color. Synesthesia is a neurological phenomenon; fMRI studies show that area V4 is active when grapheme/color synesthetes view letters and numbers and when auditory word/color synesthetes listen to spoken words (see figure; Hubbard, Arman, Ramachandran, & Boynton, 2005; Nunn et al., 2002).

7

Synesthesia and Synesthetes

There are two competing hypotheses as to why synesthetes "overbind" sensory information: Either there is excess connectivity among the involved brain areas, or there is inadequate inhibition in otherwise normal pathways in the brain. Studies favor the connectivity hypothesis; a diffusion tensor imaging MRI study (suited for imaging white matter) indicated more pronounced connections in grapheme/color synesthetes in inferior temporal, parietal, and frontal cortex—all areas involved in processing and integrating visual information (Rouw & Scholte, 2007). Synesthesia runs strongly in families, and while a whole-genome study of 43 families with multiple auditory/visual synesthetes did not find specific genes responsible, it did identify areas of linkage on four chromosomes (Asher

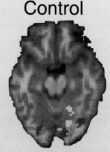

Spoken words activate left-hemisphere area V4 in a synesthete but not in a control. Why do you think this activity occurred mostly in the left hemisphere?

SOURCE: From J. A. Nunn et al., "Functional Magnetic Resonance Imaging of Synesthesia: Activation of V4/V8 by Spoken Words," *Nature Neuroscience*, 5, 371–375. Copyright © 2002 Macmillan Publishers. Used with permission.

et al., 2009). You won't be surprised to learn that these chromosomal areas contain genes for axon guidance and cortical development.

By all rights, synesthesia could be considered a brain disorder, but other than being a bit of a distraction, it is relatively benign and, like Jonathan I., many synesthetes enjoy their enriched sensory experience. Besides, they often have no idea they are any different from anybody else. Julian Asher, who led the genetic study described above, discovered his synesthesia only when, as a child attending a symphony concert with his parents, he remarked, "Oh, they turned the lights off so you could see the colors" ("Seeing Color," 2000).

Actually, the brain's task is a balancing act between combining relevant information and segregating inconsequential information, as the Application on synesthesia shows. We will spend more time examining how the brain sorts out and uses information when we talk about consciousness in the final chapter.

Concept Check   *Take a Minute to Check Your Knowledge and Understanding*

☐ Explain why higher-order processing is required to recognize the natural color of objects; how does it work?

☐ Draw a diagram of the brain, add lines showing the two major visual pathways, and label the various areas; for the higher-order processing areas, include their functions.

In Perspective

Very few subjects in the field of biological psychology can match the interest that researchers have bestowed on vision. As a result, we know more about the neuroanatomy and functioning of vision than any other neural system. Still, many challenges remain in the field of vision research.

Researchers' fascination with vision goes beyond the problems of vision itself. Our understanding of the networks of neurons and structures in the visual system provides a basis for developing theories to explain other functions as well, including the integration of complex information into a singular awareness. Whatever directions future research might take, you can be sure that vision will continue to be one of the most important topics.

Summary

Light and the Visual Apparatus

- The human eye is adapted to the part of the electromagnetic spectrum that is reflected from objects with minimal distortion. Wavelength is related to the color of light but is not synonymous with it.
- The retina contains rods, which are specialized for brightness discrimination, and cones, which are specialized for detail vision and discrimination of colors. The cells of the retina are highly interconnected to carry out some processing at that level.
- The optic nerves project to the two hemispheres so that information from the right visual field goes to the left visual cortex, and vice versa.

Color Vision

- There are three types of cones, each containing a chemical with peak sensitivity to a different segment of the electromagnetic spectrum.
- Connections of the cones to ganglion cells provide for complementary colors and for the color yellow.
- The most common cause of partial color blindness is the lack of one of the photochemicals.

Form Vision

- Form vision begins with contrast enhancement at edges by ganglion cells with light-opponent circular fields.
- These ganglion cells contribute to cortical mechanisms that detect edges (Hubel and Wiesel's theory) or that perform a Fourier analysis of a scene (spatial frequency theory).

The Perception of Objects, Color, and Movement

- The components of vision follow two somewhat separate paths through the brain.
- Structures along the way are specialized for different functions, including color, movement, object perception, and face perception.
- We do not know how or where the components of vision are combined to form the percept of a unified object. One suggestion is that this is a distributed function. ■

Study Resources

F For Further Thought

- Red and green are complementary colors and blue and yellow are complementary because their receptors have opponent connections to their ganglion cells. How would you explain the fact that bluish green and reddish yellow (orange) are also complementary?
- Considering what you know about the retina, how would you need to direct your gaze to read a book or to find a very faint star?
- Explain why the visual system analyzes an object's edges, texture, and color and then detects the object, instead of the other way around.

- Are Hubel and Wiesel's theory and the spatial frequency theory opposed or complementary theories?
- Cones in the bird retina detect blue, green, and ultraviolet. Can you imagine what they might gain from detecting ultraviolet reflections from objects in their environment? Then check www.bristol.ac.uk/biology/research/behaviour/vision/4d.html and www.vri.cz/docs/vetmed/54-8-351.pdf; you might be surprised by some of the benefits!

T Testing Your Understanding

1. Summarize the trichromatic and Hurvich-Jameson theories, indicating what facts about color vision each accounts for.

2. Compare the specialized sensitivities of simple and complex visual cortical cells; describe the interconnections among ganglion cells, simple cells, and complex cells that account for their specializations (according to Hubel and Wiesel).

3. The visual system appears to be more or less hierarchical and modular. What does this mean? (Use examples to illustrate.)

Select the best answer:

1. The receptive field of a cell in the visual system is the part of the _____ that the cell receives its input from.

a. external world b. retina
c. lateral geniculate d. cortex

2. Mixing red and green lights produces a sensation of yellow because red-sensitive and green-sensitive cones
 a. excite yellow/blue ganglion cells.
 b. have opposite effects at yellow/blue ganglion cells.
 c. excite red/green ganglion cells.
 d. have opposite effects at red/green ganglion cells.

3. If our experience of color were entirely due to the wavelength of light reflected from an object, we would not experience
 a. complementary colors. b. negative color aftereffect.
 c. primary colors. d. color constancy.

4. The parvocellular system is specialized for
 a. fine detail and movement.
 b. color and fine detail.
 c. color and movement.
 d. movement and brightness contrast.

5. Retinotopic map refers to
 a. a projection of an image on the retina by the lens.
 b. the upside-down projection of an image.
 c. the way the visual neurons connect to the cortex.
 d. the connections among the cells in the retina.

6. Cutting the optic nerve between the right eye and the chiasm would cause a loss of vision in
 a. the left visual field.
 b. the right visual field.
 c. half of each visual field.
 d. neither field, due to filling in.

7. People with red-green color blindness
 a. cannot see either red or green.
 b. see red and green as black.
 c. confuse red and green because they lack either "red" or "green" cones.
 d. confuse red and green because they lack one of the photopigments.

8. The enhanced apparent brightness of a light edge next to a dark edge is due to the fact that the neurons stimulated by the light edge are inhibited
 a. less by their "dark" neighbors.
 b. more by their "dark" neighbors.
 c. less by their "light" neighbors.
 d. more by their "light" neighbors.

9. The ability of complex visual cortical cells to track an edge as it changes position appears to be due to
 a. input from receptors with similar fields.
 b. input from ganglion cells with similar fields.
 c. input from simple cells with similar fields.
 d. input from other complex cells.

10. According to the spatial frequency theory of visual processing, edges are detected by
 a. line-detecting cells in the visual cortex.
 b. edge detectors located in the visual cortex.
 c. cells that respond to low spatial frequencies.
 d. cells that respond to high spatial frequencies.

11. The circles represent the receptive field of a ganglion cell; the rectangle represents light. Unlike the illustrations in this chapter, the receptive field has an *off center*. In which situation will the ganglion cell's rate of firing be greatest?

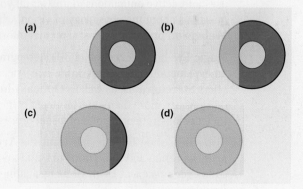

12. Studies of object, color, and movement agnosias indicate that
 a. the visual system is unstable and malfunctions with no apparent cause.
 b. components of the visual image are processed separately.
 c. color, object identification, and movement information are integrated in one place.
 d. all functions are processed in one place but the results are distributed to other parts of the brain.

13. Movement perception is the primary function in visual area
 a. V1. b. V2.
 c. V4. d. V5.

14. A person who has trouble identifying objects visually probably has damage in the
 a. temporal lobe. b. parietal lobe.
 c. occipital lobe. d. frontal lobe.

Answers:
1. b, 2. a, 3. d, 4. b, 5. c, 6. c, 7. d, 8. a, 9. c, 10. d, 11. b, 12. b, 13. d, 14. a.

Online Resources

The following resources are available at **www.sagepub.com/garrett3e**.

Chapter Resources

- Flash Cards
- Chapter Quiz
- Internet Resources and Exercises

On the Web

Links to these websites can be found at the web address above: Select this chapter's **Chapter Resources**, and then choose **Internet Resources and Exercises**.

1. **Transformations for Perception and Action** uses interactive animations to answer all your questions about the physiology of vision, while **Webvision** provides extensive textual information.

2. **Magic Eye** has a collection of 3-D stereograms and an explanation of how they work. Eye Tricks also offers **Stereograms**, as well as **Optical Illusions**; illusions are more than entertainment because they demonstrate principles of visual processing.

3. Two videos show blind people using the **Argus** artificial retina and the **BrainPort**, and another shows gene recipient **Evan Haas** navigating a visual maze.

4. **Causes of Color** is a well-designed site with numerous exhibits on color in the real world and sections on color vision; the third page under Colorblind lets you interactively simulate various forms of color blindness with your choice of photos.

   Neitzvision features color-blindness demonstrations and the research of Jay and Maureen Neitz, including a recent study in which inserting a human gene for the long-wave receptor turned dichromatic (red-green color-blind) monkeys into trichromats.

5. **Sensation and Perception** at Hanover College covers topics from receptive fields to illusions.

6. **Blindsight: Seeing Without Knowing It** is a *Scientific American* article with a fascinating video of a man using blindsight to walk down a hallway filled with obstacles.

   Faceblind is the website of prosopagnosia research centers at Harvard University and University College, London.

7. **Hearing Motion** is a video about motion synesthesia research, and **Exactly Like Breathing** is a collection of interviews of synesthetes. **Tactile-Emotion Synesthesia** is an article on that topic, while **Synesthesia** is a general article written by two researchers.

Animations

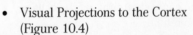

- Visual Projections to the Cortex (Figure 10.4)
- Visual Detection of Edges (Figure 10.21)

Chapter Updates and Biopsychology News

For Further Reading

1. "The Case of the Colorblind Painter" by Oliver Sacks (in Sacks's *An Anthropologist on Mars*, 1995, Vintage Books) is a compelling narrative of the case of Jonathan I.

2. "Visual Object Recognition: Do We Know More Now Than We Did 20 Years Ago?" by Jessie Peissig and Michael Tarr (*Annual Review of Psychology*, 2007, *58*, 75–96; "Mechanisms of Face Perception" by Doris Tsao and Margaret Livingstone (*Annual Review of Neuroscience*, 2008, *31*, 411–437); and "Perception of Human Motion" by Randolph Blake and Maggie Shiffrar (*Annual Review of Psychology*, 2007, *58*, 47–73) give extensive reviews of those topics.

3. *Mind Sights* by R. N. Shepard (Freeman, 1990) is a book on visual tricks and illusions. *Masters of Deception: Escher, Dali & the Artists of Optical Illusion* by Al Seckel (Sterling, 2007) is a review of illusion and perception in art and literature.

4. *The Astonishing Hypothesis* by Francis Crick (Scribner, 1994) is about the scientific search for consciousness; because the search focuses on visual awareness, it contains fascinating and readable information about vision.

K Key Terms

binding problem.. 321

blindsight 317

color agnosia.. 319

color constancy 319

complementary colors 301

complex cell 310

distributed 314

dorsal stream.. 315

form vision 307

fovea.. 296

fusiform face area (FFA).................................... 317

hierarchical processing...................................... 313

iodopsin 296

lateral inhibition 308

magnocellular system 314

modular processing 313

movement agnosia 320

negative color aftereffect.................................... 302

object agnosia.. 316

opponent process theory 301

parvocellular system 314

photopigment 296

prosopagnosia 317

receptive field.. 297

retina.. 296

retinal disparity 298

retinotopic map 307

rhodopsin 296

simple cell.. 310

spatial frequency theory.................................... 311

synesthesia 322

trichromatic theory..301

ventral stream.. 315

visual acuity.. 297

visual field.. 298

visual word form area (VWFA) 318

The Body Senses and Movement

11

The Body Senses

Proprioception

The Skin Senses

The Vestibular Sense

The Somatosensory Cortex and the Posterior Parietal Cortex

Pain and Its Disorders

    IN THE NEWS: GETTING AN ANESTHETIC SHOULDN'T HURT

    APPLICATION: TREATING PAIN IN LIMBS THAT AREN'T THERE

    CONCEPT CHECK

Movement

The Muscles

The Spinal Cord

The Brain and Movement

Disorders of Movement

    CONCEPT CHECK

In Perspective

Summary

Study Resources

In this chapter you will learn

- How the brain gets information about the body and the objects in contact with it

- What causes pain and ways it can be relieved

- How several brain structures work together to produce movement

- What some of the movement disorders are and how they impair movement

C hristina was a healthy, active woman of 27. One day she began dropping things. Then she had trouble standing or even sitting upright; soon she was bedridden, lying motionless, speaking between shallow breaths in a faint and expressionless voice, and with an equally expressionless face. A spinal tap indicated that she was suffering from neuritis, an inflammation of the nerves that is often

> *That's what they do with frogs, isn't it? They scoop out the centre, the spinal cord, they pith them. . . . That's what I am, pithed, like a frog.*
>
> —Christina

caused by a viral infection. Neurological examination showed that, although she seemed paralyzed, her motor nerves were only slightly affected. She could move, but she could not control her movements or even her posture; if she failed to watch her hands, they wandered aimlessly. She had lost all *proprioception*, the sense that collects information from our muscles and tendons and joints to tell us where our hands are and what movements our feet and legs are making.

The neuritis did not last long, but in the meantime it had damaged her nerves, and the damage was permanent. For a month, she was as floppy as a rag doll. But then she began to sit up, with an exaggeratedly erect posture, using only her vision for feedback. After a year of rehabilitation, she was able to leave the hospital, to walk and take public transportation and work at home as a computer programmer, all guided by vision. Christina never recovered from the damage to her nervous system, but she was able to make a remarkable compensation (Sacks, 1990).

In the previous two chapters, we have discussed audition and vision, sensory systems that provide information about distant objects. Now we turn our attention to the senses that inform us about the objects in direct contact with our bodies and that tell us where our body is in space, where our limbs are in relation to our body, and what is going on inside the body. Christina's case illustrates how important this information is for interacting physically with the world. The most important function of the body senses is to contribute to movement. In fact, the body senses are so intimately involved with our ability to move about in the world and to manipulate it that we sometimes hear the term *the sensorimotor system*. For that reason, we will follow our discussion of the body senses with an exploration of the topic of movement.

The Body Senses

We get information about our body from the somatosensory system and from the vestibular system. The somatosenses include proprioception; the skin senses, which tell us about conditions at the surface of our body; and the interoceptive system, concerned with sensations in our internal organs. The vestibular system informs the brain about body position and movement. The interoceptive system operates mostly in the background and participates less directly in behavior, so we will limit our attention to the other systems.

Proprioception

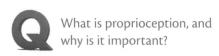

What is proprioception, and why is it important?

Proprioception (from the Latin *proprius*, "belonging to one's self") is the sense that informs us about the position and movement of our limbs and body. Its sensors report tension and length in muscles and the angle of the limbs at the joints. Proprioception is not as glamorous a sense as vision or audition, or even touch. However, without it we would have a great deal of difficulty, as Christina did, in maintaining posture, moving our limbs, and grasping objects. Ian Waterman, who is similarly afflicted, actually crumples helplessly onto the floor if someone turns the lights out (J. Cole, 1995). In other words, proprioception does more than provide information; it is critically important in the control of movement.

The Skin Senses

Why are there so many kinds of skin receptors?

The commonly accepted *skin senses* are touch, warmth, cold, and pain. This is probably an oversimplification; for example, itch was long thought to be a variant of pain, but destroying certain neurons in the spinal cord eliminates itch without affecting pain sensitivity (Sun et al., 2009). Whatever the actual number of senses, the important point here is that each is distinct from the others, with its own receptors and separate "labeled line" pathway to the brain. To demonstrate this for yourself, move the point of a lead pencil slowly across your face. You will feel the touch of the pencil pretty continuously, but the lead will feel cold only occasionally—because touch and cold are monitored by different receptors. Although their range is limited to the surface of our body, changes there are often due to external stimulation, so the skin senses inform us both about our body and the world. (We experience these sensations deeper in the body as well, but less often and with less sensitivity.) The skin sense receptors are illustrated in the diagram of a section of skin in Figure 11.1.

There are two general types of receptors. *Free nerve endings* are simply processes at the ends of neurons; they detect warmth, cold, and pain. All the other receptors are *encapsulated receptors,* which are more complex structures enclosed in a membrane; their role is to detect touch. Why are there so many receptors just for touch? Because touch is a complex sense that conveys several types of information. In the superficial layers of the skin, Meissner's corpuscles respond with a brief burst of impulses, while Merkel's disks give a more sustained response. Located near the surface of the skin as they are, they detect the texture and fine detail of objects. They also detect movement and

FIGURE 11.1 Receptors of the Skin.

The different endings of the receptors account for their varied specialties, which provide the brain with the rich information it needs to interact with the world.

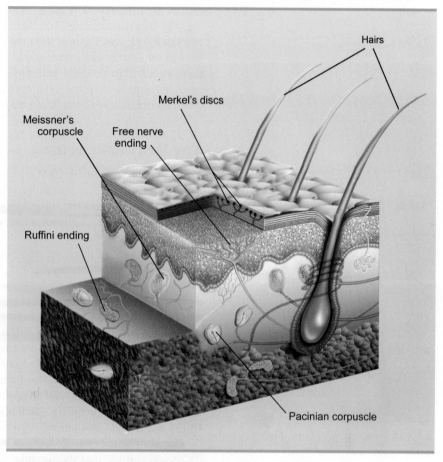

come into play when you explore an object with gentle strokes of your hand or when a blind person reads Braille. Pacinian corpuscles and Ruffini endings are located in the deeper layers, where they detect stretching of the skin and contribute to our perception of the shape of grasped objects (Gardner, Martin, & Jessell, 2000). Because the density of the skin receptors varies throughout the body, so does sensitivity—as much as 10-fold in fact. The fingertips and the lips are the most sensitive and the upper arms and calves of the legs are the least sensitive (Weinstein, 1968).

The other three skin senses are detected by free nerve endings, but this statement is a bit misleading; those nerve endings have distinctly different receptors that make the neurons stimulus specific (Basbaum, Bautista, Scherrer, & Julius, 2009). We are capable of responding to a wide range of temperatures; at least two different receptors detect different levels of warmth, and another responds to cooling of the skin. These receptors are all members of the *transient receptor potential (TRP)* family of protein ion channels. Detection of pain also requires several receptors, mostly because of the variety of pain sources; these sources are categorized as thermal, chemical, and mechanical. Two TRP receptors respond to painful heat; a receptor for painful cold has not been conclusively identified, but it is clear that the coolness receptor does not account for this sensation. Chemical receptors react to a wide range of chemical irritants. Best known is the TRPV1 heat pain receptor, which also responds to capsaicin, the ingredient in chili peppers that makes spicy foods painfully hot. Ointments containing capsaicin alleviate the joint pain of arthritis, apparently because continued stimulation of the receptors depletes the neuropeptide that the neurons use to signal pain. This receptor also responds to the pain-inducing acid released in bone cancer, and a TRPV1 antagonist relieves this pain (Ghilardi et al., 2005). The TRPM8 coolness receptor produces the cool sensation of mint in toothpaste and candies, as well as the cooling effect of menthol on the skin; menthol creams are useful for treating muscle pain and skin irritations. TRPA1 accounts for the painful irritation caused by tear gas, vehicle exhaust, hydrogen peroxide, and tobacco smoke, and it is also what makes mustard, garlic, and wasabi so pungent. As we will see shortly, the body produces its own irritants when tissues are damaged; these continue to produce pain well after the stimulus is past. The receptors for mechanical pain have not been determined, which is unfortunate because persistent hypersensitivity to touch is a major problem following tissue or nerve injury.

The Vestibular Sense

What is the function of the vestibular sense?

In Chapter 8, you saw that the cochlea in the ear is connected to a strange-looking appendage, the vestibular organs. The *vestibular sense* helps us maintain balance, and it provides information about head position and movement. The organs are the semicircular canals, the utricle, and the saccule (see Figure 11.2a). The physical arrangement of the semicircular canals makes them especially responsive to movement of the head (and body). At the base of each canal is a gelatinous (jellylike) mass called a cupula, which has a tuft of hair cells protruding into it (Figure 11.2b). During acceleration (an increase in the rate of movement), the fluid shifts in the canals and displaces the cupula; bending the hair cells in one direction depolarizes them and bending in the other direction hyperpolarizes them, increasing or decreasing the firing rate in the neurons. Deceleration has a similar effect.

The system responds only to acceleration and stops responding when speed stabilizes. Just as the coffee sloshes out of your cup when you start up from a traffic light and then levels off in the cup when you reach a stable speed, the fluid in the canals also returns to its normal position. Otherwise, you would continue to sense the movement throughout an automobile trip or, worse yet, during a 500-mile/hour (805-km/hr) flight in a jetliner!

The utricle and saccule monitor head position in relation to gravity. In Figure 11.2c, you can see that the receptors (the hair cells) are covered with a gelatinous material. When the head tilts, gravity shifts the gelatinous mass and the hair cells are depolarized or hyperpolarized, depending on the direction of tilt. The hair cell receptors in the utricle are arranged in a horizontal patch, while the saccule's receptors are on its vertical wall; thus, the two organs can detect tilt in any direction.

Consider what would happen without a vestibular system. Mr. MacGregor, one of Oliver Sacks's patients, lost his vestibular sense to the neural degeneration of Parkinson's disease (Sacks, 1990). When he walked, his body canted to the left, tilted a full 20°. Strangely, Mr. MacGregor wasn't aware of his tilt, even when his friends told him he was in danger of falling over. Once Sacks showed him a videotape, though, he was convinced. A retired carpenter, he put his expertise to work; 3 inches in front of his glasses he attached a miniature spirit level—a fluid-filled glass tube with a bubble in it that tells a carpenter when his work is level with the world. By glancing at this makeshift device occasionally, he was able to walk without any slant, and after a few weeks checking his tilt became so natural he was no longer aware he was doing it.

1
Vestibular Disorders

But the vestibular sense is not just for adjusting the body's position. When we reach for an object, we must know not just where the object is but the position of our body and the relation of our arm to our body; the brain combines information about the object's spatial location with inputs from the vestibular sense and from proprioception to tell us what arm movements are required. Proprioception also triggers reflexive eye movements that keep returning our gaze to the scene as we turn our head or as our body bobs up and down when we walk; otherwise, the world would become a meaningless blur.

FIGURE 11.2 The Vestibular Organs.

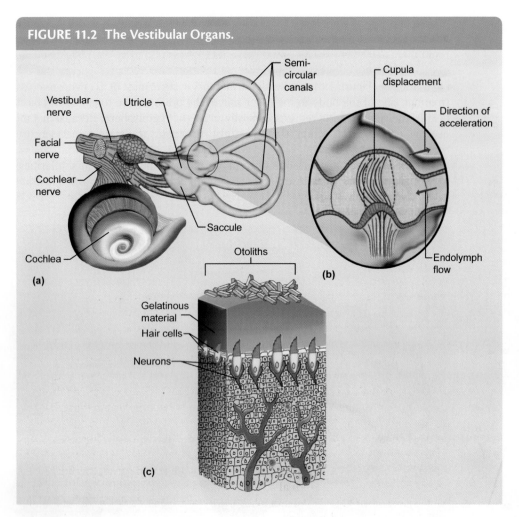

(a) The inner ear, showing the cochlea and the vestibular organs. (b) Enlarged view of a cupula in a semicircular canal. During acceleration, the flow of endolymph displaces the cupula, triggering a neural response. (c) Receptors of the utricle and saccule. Tilting the head causes the gelatinous material to shift, stimulating the hair cells. The weight of the otoliths (calcium carbonate crystals) magnifies the shift.

SOURCES: (a) Iurato (1967); (b) Based on Goldberg and Hudspeth (2000). © 2000 McGraw-Hill; (c) Based on Martini (1988).

The vestibular system sends projections to the cerebellum and the brain stem; there is also a pathway to the cortex, an area called the parieto-insular-vestibular cortex. In the previous chapter, we observed that this is the likely location where stimulation that produces excessive eye movements causes dizziness and nausea; the same thing happens with excessive body motion, for example, during a rough boat ride or from spinning around.

vestibular
↓
cerebellum + b.s.
↓
cortex.

The Somatosensory Cortex and the Posterior Parietal Cortex

The body is divided into segments called *dermatomes*, each served by a spinal nerve, as Figure 11.3 shows. The divisions are not as distinct as illustrated, because each dermatome overlaps the next by one third to one half. This way, if one nerve is

FIGURE 11.3 Dermatomes of the Human Body.

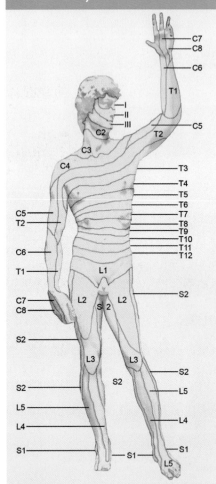

For sensory functions, the body is divided into segments called dermatomes, each served by a spinal or cranial nerve. The labels identify the nerve; letters indicate the part of the spinal cord where the nerve is located (cervical, thoracic, lumbar, sacral, or coccygeal), and the numbers indicate the nerve's position within that section. Areas I, II, and III on the face are innervated by branches of the trigeminal (fifth) cranial nerve.

injured, the area will not lose all sensation. Body sense information enters the spinal cord (via spinal nerves) or the brain (via cranial nerves) and travels to the thalamus. From there the body sense neurons go to their projection area, the *somatosensory cortex,* located in the parietal lobes just behind the primary motor cortex and the central sulcus (Figure 11.4a). As with the auditory system, most of the neurons cross from one side of the body to the other side of the brain, so the touch of an object held in the right hand is registered mostly in the left hemisphere. Because not all neurons cross over, touching the object also stimulates the right somatosensory cortex, though much less.

Sensory systems have a number of organizational and functional similarities; a comparison of the somatosensory cortex with the visual cortex will illustrate

FIGURE 11.4 The Primary and Secondary Somatosensory Areas.

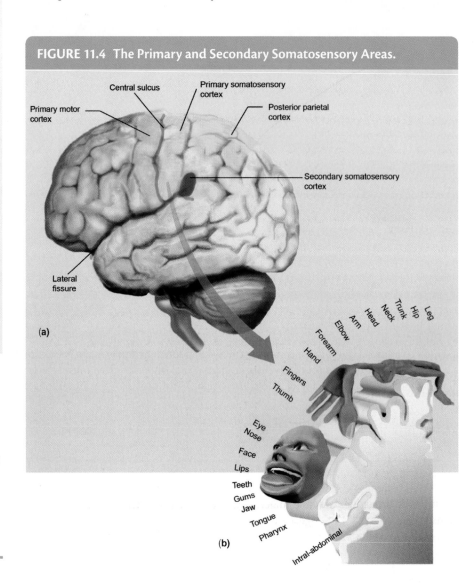

(a) The primary and secondary somatosensory cortex and the posterior parietal cortex. The primary motor cortex is shown as a landmark. (b) A slice from the somatosensory cortex, showing its somatotopic organization. The size of the body parts in the figure is proportional to the area of the cortex in which they are represented.

SOURCE: (b) Adapted from *The Cerebral Cortex of Man* by W. Penfield and T. Rasmussen, 1950, New York: Macmillan. Copyright 1950. Reprinted with permission of Gale, a division of Thomson Learning: www.thomsonrights.com. Fax 800-730-2215.

FIGURE 11.5 Receptive Fields in the Monkey Somatosensory System.

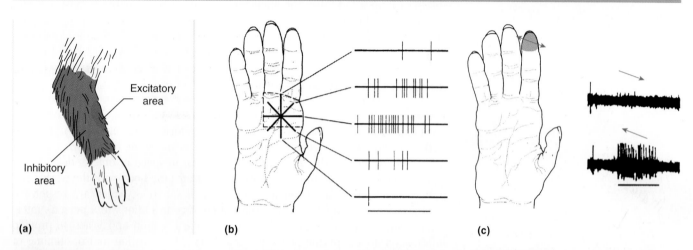

(a) Excitatory and inhibitory areas of the receptive field of a single touch neuron in the somatosensory cortex. (b) Receptive field of a somatosensory neuron that responded most to a horizontal edge. The recordings to the right indicate the strength of the neuron's response to edges of different orientations. (c) Receptive field of a neuron responsive to movement across the fingertip in one direction but not the other.

SOURCES: (a) From "Neural Mechanisms Subserving Cutaneous Sensibility, with Special Reference to the Role of Afferent Inhibition in Sensory Perception and Discriminiation," by V. B. Mountcastle and T. P. S. Powell, 1959, *Bulletin of the Johns Hopkins Hospital, 105,* p. 224, Figure 14. © The Johns Hopkins University Press. Reprinted with permission of the Johns Hopkins University Press; (b) and (c) From "Movement-Sensitive Cutaneous Receptive Fields in the Hand Area of the Post-Central Gyrus in Monkeys," by J. Hyvärinen and A. Poranen, *Journal of Physiology, 283,* pp. 523–537. Copyright 1978, with permission.

this point. First, it contains a map of the body, or homunculus, just as the visual cortex contains a map of the retina and the auditory cortex contains a map of the cochlea (Figure 11.4b). Second, some of the cortical cells have complex receptive fields on the skin. Some of them have excitatory centers and inhibitory surrounds like those we saw in the visual system (Mountcastle & Powell, 1959). Some of them are quite large, as in Figure 11.5a, while smaller excitatory-inhibitory fields sharpen the localization of excitation and help distinguish two points touching the skin. In Figure 11.5b and c, we see that other somatosensory neurons with complex fields are feature detectors; they have sensitivities for orientation, direction of movement, shape, surface curvature, or texture (Carlson, 1981; Gardner & Kandel, 2000; S. Warren, Hämäläinen, & Gardner, 1986). Apparently, these neurons combine inputs from neurons with simpler functions, just as complex visual cells integrate the inputs of multiple simple cells (Iwamura, Iriki, & Tanaka, 1994). One type of receptive field includes multiple fingers; the cells' firing rate depends on how many fingers are touched, so they give an indication of the size of a held object.

A third similarity is that somatosensory processing is hierarchical. The *primary somatosensory cortex* consists of four areas, each of which contains a map of the body and plays a role in processing sensory information from the body. The thalamus sends its output to two of these subareas, which extract some information and pass the result on to the other two areas, which in turn send their output to the secondary somatosensory cortex. At this point in processing, information from the right and left sides of the body is mostly segregated.

The *secondary somatosensory cortex* receives input from the left and the right primary somatosensory cortices, so it combines information from both sides of the body. Neurons in this area are particularly responsive to stimuli that have acquired meaning, for instance, by association with reward (Hsiao, O'Shaughnessy, & Johnson, 1993). The secondary somatosensory cortex connects to the part of the temporal lobe that

Q How are the body senses similar to other senses?

Q What processing occurs at each level of the somatosensory system?

includes the hippocampus, which is important in learning, so it may serve to determine whether a stimulus is committed to memory (Gardner & Kandel, 2000).

To pick up a forkful of the apple pie on your plate, your brain must not only receive a visual image of the pie, but it must also know where your arm and hand are in relation to your body, where your head is oriented in relation to your body, and where your eyes are oriented in relation to your head. That is where the posterior parietal cortex comes in. The primary somatosensory cortex projects to the posterior parietal cortex as well as to the secondary cortex.

As you saw in the previous chapter, the *posterior parietal cortex* is an association area that brings together the body senses, vision, and audition (K. H. Britten, 2008). See Figure 11.4 again for the location of the posterior parietal cortex in relation to the somatosensory cortex. Here, the brain determines the body's orientation in space; the location of the limbs; and the location in space of objects detected by touch, sight, and sound. In other words, it integrates the body with the world. The posterior parietal cortex is composed of several subareas, which are responsive to different sense modalities and make different contributions to a person's interaction with the world. Some cells combine proprioception and vision to provide information about specific postures, for example, the location and positioning of the arm and the hand (Bonda, Petrides, Frey, & Evans, 1995; Graziano, Cooke, & Taylor, 2000; Sakata, Takaoka, Kawarasaki, & Shibutani, 1973). Others contribute to reaching and grasping movements and eye movements toward targets of interest (Batista, Buneo, Snyder, & Andersen, 1999; Colby & Goldberg, 1999). The posterior parietal cortex's function is not solely perceptual, because many of its neurons fire before and during a movement. It does not itself produce movements but passes its information on to frontal areas that do (Colby & Goldberg, 1999).

A unified body image is critical to our ability to function and even to our self concept, so you can imagine that any disruption would have significant consequences. You saw earlier that damage to the right posterior parietal cortex can produce neglect and that some stroke patients will deny ownership of a paralyzed arm or leg. A few people with no apparent brain damage—or mental or emotional disorder, for that matter—are so convinced that a limb doesn't belong to them that they actually ask to have it amputated; this disorder is called *body integrity identity disorder,* or apotemnophilia (McGeoch et al., 2009). When the limb is touched, there is no response in the superior parietal area. Skin conductance response to stimulation is doubled in that limb, though, which indicates a high level of emotional feeling about the limb. A related phenomenon incorporates the entire body into the illusion; in an *out-of-body experience,* the individual hallucinates seeing his or her body from another location, for example, from a position above the detached body. Causes include traumatic damage and epilepsy affecting the junction between the parietal and temporal lobes; the experience can also be produced by electrical stimulation in that area (Blanke & Arzy, 2005).

Pain and Its Disorders

In Chapter 8, we learned about the emotional aspect of pain and how it motivates our behavior. Now we need to put pain in the context of the body senses and see how it works as a sensory mechanism. In spite of our attention to pain earlier, there are still a few surprises left.

Detecting Pain

Pain begins when certain free nerve endings are stimulated by intense pressure or temperature, by damage to tissue, or by various chemicals. Intense stimulation also causes the pain neurons and nonneural cells to release a wide array of signaling molecules; referred to as the "inflammatory soup," these produce the familiar

handwritten margin note:
thalamus
↓
primary motor
↓
2nd motor/PPC
↓
Avntulanted (movement)

Q What causes pain?

swelling and redness of inflammation, and they enhance excitability of the pain neurons so much that the neurons respond even to touch. This effect is adaptive because it encourages guarding of the injured area, but the resulting pain can be more troublesome than the original injury. Pain information travels to the spinal cord over large, myelinated A-delta fibers and small, lightly myelinated or unmyelinated C fibers. Because A-delta fibers transmit more rapidly than C fibers, you notice a *sharp pain* almost immediately when you are injured, followed by a longer lasting *dull pain* (Basbaum & Jessell, 2000). Sharp pain makes a good danger signal and motivates you to take quick action, while dull pain hangs around for a longer time to remind you that you have been injured.

In the spinal cord, pain neurons release glutamate and *substance P,* a neuropeptide that increases pain sensitivity (De Biasi & Rustioni, 1988; Skilling, Smullin, Beitz, & Larson, 1988). In Chapter 2, we saw that neuropeptides enhance the effect of a neurotransmitter at the synapse. Substance P is released only during more intense pain stimulation; in mice lacking receptors for substance P or the ability to produce substance P, sensitivity to moderate and intense pain is impaired, but mild pain is unaffected (Cao et al., 1998; De Felipe et al., 1998). As with the other body senses, pain information passes through the thalamus to the somatosensory cortex; however, the anterior cingulate cortex and the insula carry out additional processing of the emotional implications of pain, and the prefrontal cortex is concerned with pain of longer duration.

Treating Pain

We saw in Chapter 2 that local anesthetics—those that are applied to or injected into the painful area—block sodium channels in the pain neurons and reduce their ability to fire. General anesthetics, which may be injected or inhaled, render the patient unconscious. They work in the central nervous system, though their mechanism is not well understood. We know that they affect the functioning of several proteins, but we don't know which ones are important to the anesthesia. The most frequently used drugs are aspirin, ibuprofen, and acetaminophen (Tylenol, etc.). Aspirin and ibuprofen block enzymes required for producing prostaglandins that are released in response to injury, so they reduce inflammation as well as pain. Acetaminophen blocks the same enzymes but weakly and with no anti-inflammatory benefit, so its major effect is probably in the central nervous system. Ameliorating pain often requires more powerful drugs, and morphine is the acknowledged gold standard, but its addictiveness and patients' rapidly developing tolerance have spurred the development of numerous alternatives. The MDAN series of drugs, for example, target the mu opioid receptor while blocking the delta opioid receptor; these drugs are reportedly 50 times more potent than morphine, without producing either tolerance or addiction (Dietis et al., 2009).

Efforts are under way on a variety of other fronts, as well. Tanezumab, an antibody for nerve growth factor, has shown safety and effectiveness in clinical trials with chonic back pain and the inflammatory pain of arthritis (Cattaneo, 2010). Also, an experimental drug that blocks the TRPV1 receptor produces modest reduction of bone cancer pain in mice. However, much smaller doses significantly increase the effectiveness of morphine when the two are used in combination (Niiyama, Kawamata, Yamamoto, Furuse, & Namiki, 2009). The fact that pain ultimately occurs in the brain has inspired a novel approach that is also showing promise; chronic pain patients given continuous fMRI feedback of activity in their cingulate gyrus learned to reduce pain-related brain activity and their experience of pain (deCharms et al., 2005).

Researchers are always looking for better ways to treat pain, including delivery methods that are either more effective or patient friendly. In the News describes one effort you can really appreciate.

[handwritten margin note:] local anesthetic ↓ block Nat channels

Getting an Anesthetic Shouldn't Hurt

In the **News**

A local anesthetic takes away the pain of having your teeth filled, but that doesn't make you appreciate having your gums poked with a needle to do it. In the future, you might be able to get your lidocaine in a nasal spray. We have known for a long time that some inhaled compounds travel along the nerves to the brain, but now researchers have learned that lidocaine follows the trigeminal nerve in the other direction and accumulates in high concentrations in the jaw and around the teeth. This delivery technique should have fewer side effects and find uses outside the dentist's office in treating migraine, temporomandibular joint disorder, and chronic pain arising from the trigeminal nerve itself. Unfortunately, nasal spray anesthesia has been tried so far only on rats, so don't ask for it the next time you visit your dentist.

SOURCE: American Chemical Society. (2010, May 13). Sniff of local anesthetic in the dentist's chair could replace the needle. *ScienceDaily*. Retrieved May 30, 2010, from www.sciencedaily.com/releases/2010/05/100513104818.htm.

Internal Mechanisms of Pain Relief

People sometimes feel little or no pain in spite of serious injury; they may even fail to realize they are injured until someone calls it to their attention. In the account of his search for the mouth of the Nile river, the explorer and missionary David Livingstone (1858/1971) gave an intriguing example (see Figure 11.6):

How does the brain relieve pain?

> Starting, and looking half round, I saw the lion just in the act of springing upon me. I was upon a little height; he caught my shoulder as he sprang, and we both came to the ground below together. Growling horribly close to my ear, he shook me as a terrier dog does a rat. The shock produced a stupor similar to that which seems to be felt by a mouse after the first shake of the cat. It caused a sort of dreaminess, in which there was no sense of pain nor feeling of terror, though quite conscious of all that was happening. It was like what patients partially under the influence of chloroform describe, who see all the operation, but feel not the knife. (p. 12)

FIGURE 11.6 David Livingstone Attacked by a Lion.

Endorphins allowed him to endure the pain of a lion's attack.

SOURCE: Credit: Hulton Archive; Livingstone, 1858/1971.

You learned in earlier chapters that the reason opiate drugs are so effective at relieving pain is that they operate at receptors for the body's own pain relievers. Researchers combined the words *endogenous* ("from within") and *morphine* to come up with the name *endorphins* for this class of neurochemicals. *Endorphins function both as neurotransmitters and as hormones, and they act at opiate receptors in many parts of the nervous system.* Pain is one of the stimuli that release endorphins, but it does so only under certain conditions. Rats subjected to inescapable electric shock were highly tolerant of pain 30 minutes later; rats given an equal number of shocks that they could escape by making the correct response had only a slight increase in pain resistance (S. F. Maier, Drugan, & Grau, 1982). I am sure you can see the benefit of eliminating pain in situations of helplessness like

Livingstone's and preserving pain when it can serve as the motivation to escape. An injection of naloxone eliminates the analgesia induced by inescapable shock but not the milder analgesia that follows escapable shock; the fact that naloxone blocks opiate receptors by occupying them indicates that the analgesia of inescapable shock is endorphin based. (You may be wondering how you would determine a rat's pain resistance, since it cannot report its pain sensation. One way is to place the rat's tail under a heat lamp and record how long it takes the rat to flick its tail away.)

Several kinds of stimulation result in endorphin release and pain reduction, including physical stress (Colt, Wardlaw, & Frantz, 1981), acupuncture (L. R. Watkins & Mayer, 1982), and vaginal stimulation in rats (Komisaruk & Steinman, 1987) and women (Whipple & Komisaruk, 1988). Analgesia resulting from vaginal stimulation probably serves to reduce pain during birth or intercourse. Even the pain relief from placebo, which doctors once took as evidence that the pain was not "real," is often the result of endorphins, as revealed by both naloxone blockage of opiate receptors and by PET scans (Figure 11.7; Amanzio, Pollo, Maggi, & Benedetti, 2001; Petrovic, 2005; Petrovic, Kalso, Petersson, & Ingvar, 2002).

Not all stimulation-induced pain relief comes from endorphins. For example, naloxone does not reduce the analgesic effect of hypnosis (L. R. Watkins & Mayer, 1982), and naloxone blocks analgesia produced by acupuncture needles inserted at distant points from the pain site but not when the needles are placed near the site (L. R. Watkins & Mayer). New research indicates that the acupuncture needle causes a 24-fold increase in adenosine, which acts as a local anesthetic (N. Goldman et al., 2010). (Note, by the way, that this study's first author is a 16-year-old girl.)

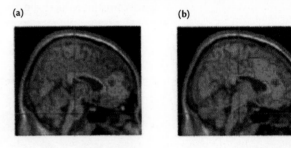

FIGURE 11.7 Activation of Opiate Receptors in the Brain by a Placebo.

(a) (b)

(a) shows activity in cortical and brain stem areas (in red) during opiate drug treatment for pain. In (b), a similar pattern of activity occurs during placebo treatment of pain. Pain alone did not produce this result. The blue dots indicate the location of the anterior cingulate cortex, which you saw in Chapter 8 is involved in emotional aspects of pain.

SOURCE: From "Placebo and Opiod Analgesia—Imaging a Shared Neuronal Network," by P. Petrovic et al., *Science*, 295, 1737–1740, fig. 4 a and b, p. 1739. © 2004. Reprinted by permission of AAAS.

The Descending Pain Inhibition Circuit

During a spring break in New Orleans, I was touring one of the Civil War–era plantation houses that the area is noted for. In a glass case was an assortment of artifacts that had been found on the plantation grounds. An odd part of the collection was a few lead rifle slugs with what were obviously deep tooth marks on them. When makeshift surgery had to be performed with only a large dose of whiskey for anesthesia, the unfortunate patient would often be given something to bite down on like a piece of leather harness or a relatively soft lead bullet. (You can probably guess what common expression this practice gave rise to.) As a toddler, when you scraped your knee you clenched your teeth and rubbed the area around the wound. And—tribute to your childhood wisdom—it really did help and got you through the pain without the benefit of either a lead bullet or whiskey. You might think that tooth clenching and rubbing simply take attention from the pain. Ronald Melzack and Patrick Wall (1965) had another idea. In their *gate control theory,* they hypothesized that pressure signals arriving in the brain trigger an inhibitory message that travels back down the spinal cord, where it closes a neural "gate" in the pain pathway.

Research has confirmed the general idea of their theory, and we now know that a descending pathway in the spinal cord is one of the ways the brain uses endorphins to control pain. Pain causes the release of endorphins in the *periaqueductal gray (PAG),* a brain stem structure with a large number of endorphin synapses (Basbaum & Fields, 1984). As you can see in Figure 11.8, endorphin release inhibits the release of substance P, closing the pain "gate" in the spinal cord. (Note that enkephalin, the type of endorphin in the spinal cord, reduces substance P release by *presynaptic* inhibition, as described in Chapter 2.) Activation of the endorphin circuit apparently has multiple neural origins, including the cingulate cortex during placebo analgesia and the amygdala in the case of fear-induced analgesia (Petrovic,

[handwritten margin note: PAG ↓ endorphin release ↓ close pathway]

FIGURE 11.8 The Descending Pain Inhibition Circuit.

Endorphin release in the periaqueductal gray inhibits the release of substance P by pain neurons in the spinal cord; this reduces the pain message reaching the brain.

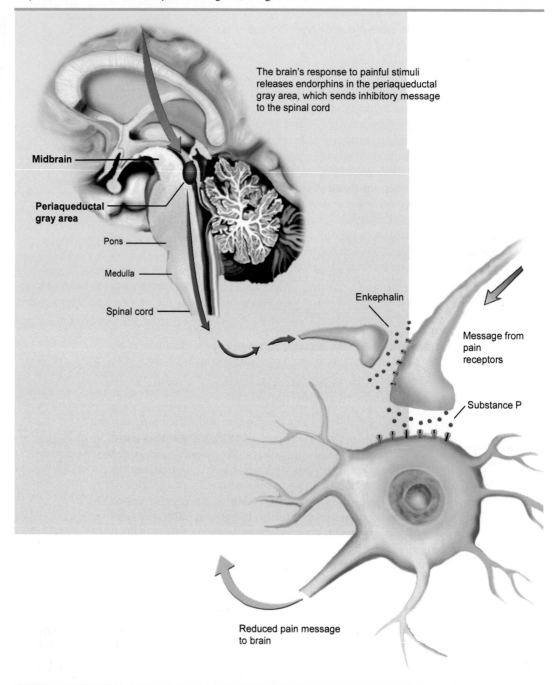

The brain's response to painful stimuli releases endorphins in the periaqueductal gray area, which sends inhibitory message to the spinal cord

Midbrain

Periaqueductal gray area

Pons

Medulla

Spinal cord

Enkephalin

Message from pain receptors

Substance P

Reduced pain message to brain

2005). Brain imaging shows that both placebo and distraction of attention reduce pain through this circuit (Petrovic, 2005; Petrovic et al., 2002; Tracey et al., 2002). Electrical stimulation of the PAG is also a very effective pain reliever, but it has the drawback of requiring brain surgery (Bittar et al., 2005). Women, unfortunately, have fewer mu opioid receptors in the PAG than men, and they receive less pain-relieving benefit from opiate drugs (Loyd, Wang, & Murphy, 2008). The PAG also contains cannabinoid receptors, which respond to endogenous cannabinoids and the active ingredient in marijuana, and they account for some forms of stimulation-induced pain relief (A. G. Hohmann et al., 2005).

Pain's Extremes: From Congenital Insensitivity to Chronic Pain

Congenital insensitivity to pain is a very rare condition. This is fortunate, because those who are afflicted not only unknowingly hurt themselves but often engage in risky and dangerous behavior that most other people avoid. One family with several afflicted members was discovered when researchers learned of a boy in northern Pakistan who entertained on the street by piercing his arms with knives; before he could be studied, he died when he celebrated his 14th birthday by jumping off a roof (Cox et al., 2006). Several genes have been identified as responsible for the different kinds of pain insensitivity. People with a mutation in the *SCN9A* gene have nonfunctioning versions of a particular type of sodium channel, so their pain neurons are disabled (Cox et al.). Those with a mutation in the gene for nerve growth factor (*NGFB*), which stimulates development of sensory nerves, have a significant loss of unmyelinated neurons and less severe loss of myelinated fibers (Einarsdottir et al., 2004). Another cause of insensitivity is illustrated in the case of a 16-year-old boy whose most significant characteristic was elevated opioid levels in his cerebrospinal fluid (Manfredi et al., 1981).

Chronic pain is much more common, and its victims often experience lifelong suffering. Studies indicate that around 15% of the population is afflicted by chronic pain (Verhaak, Kerssens, Dekker, Sorbi, & Bensing, 1998). Practitioners often attempt to distinguish chronic from acute pain in terms of duration, with standards varying from 1 month to 1 year. However, it makes more sense to define *chronic pain* as pain that persists after healing has occurred or beyond the time in which healing would be expected to occur. Chronic pain is either *nociceptive,* caused by activation of pain receptors, or *neuropathic,* caused by damage to or malfunction of either the peripheral or the central nervous system. Thresholds in the pain system are ordinarily relatively high; genes that increase sensitivity to pain make the person more susceptible to chronic pain. A different allele of the *SCN9A* sodium channel gene mentioned above lowers pain threshold and contributes to two types of neuropathic pain (Fischer & Waxman, 2010). Variations in the *COMT* gene also regulate pain sensitivity, as well as responsiveness to opioid pain relievers, and they figure in several pain syndromes (Andersen & Skorpen, 2009).

The nervous system undergoes extensive functional and structural changes during chronic pain. The pain pathways increase their sensitivity, new connections sprout where peripheral neurons make connections in the spinal cord, and normal spinal inhibitory mechanisms are depressed (Woolf & Salter, 2000). The brain also participates in these changes. Brain stem pathways become more responsive (M. Lee, Zambreanu, Menon, & Tracey, 2008); activation increases in prefrontal cortex, anterior cingulate cortex, and insula (Baliki et al., 2006); and the amount of somatosensory cortex devoted to the painful area expands (Flor, Braun, Elbert, & Birbaumer, 1997). There is considerable loss of gray matter, equivalent to 10 to 20 years of additional aging in chronic back pain patients (Figure 11.9; Apkarian, Sosa, Sonty,

FIGURE 11.9 Gray Matter Loss in Chronic Back Pain Patients.

Magnetic resonance imaging shows areas of greater loss in patients compared to controls in (a) the cortex and (b) the thalamus.

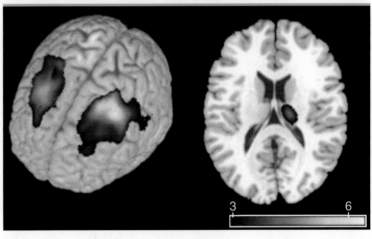

(a) (b)

SOURCE: From "Chronic Back Pain is Associated With Decreased Prefrontal and Thalamic Gray Matter Density," by Apkarian et al., *Journal of Neuroscience, 24,* 10410–10415, fig. 2. © 2004 Society for Neuroscience. Used with permission.

et al., 2004) and 9.5 years of additional aging per year of suffering with fibromyalgia (Kuchinad et al., 2007). Evidence of the functional disruption this causes is that chronic pain patients perform poorly, almost randomly, on the gambling test used to assess prefrontal impairment (described in Chapter 8; Apkarian, Sosa, Krauss, et al., 2004).

Phantom Pain

Q What causes phantom pain?

If damage to the right parietal cortex can eliminate recognition of a paralyzed left arm or leg, shouldn't removing a person's arm or leg eliminate all consciousness of the limb? Most amputees continue to experience the missing limb, not as a memory, but as vividly as if it were real (Melzack, 1992). A phantom leg seems to bend when the person sits down and then to become upright during standing; a phantom arm even feels as though it is swinging in coordination with the other arm during walking.

In 80% to 90% of amputees we see what is undeniably the strangest manifestation of chronic pain (Melzack, 1990). *Phantom pain,* pain that is experienced in a missing limb, is not simply pain at the stump but is felt in the missing arm or leg itself. The classical explanation was that the cut ends of nerves continue to send impulses to the part of the brain that once served the missing limb. Peripheral input is a factor in some cases, but anesthetizing these nerves relieves phantom pain in no more than half of patients (Birbaumer et al., 1997; Flor, 2008), so something else must be going on.

Following the clue that stimulating the face often produces sensations in a phantom arm, a team of researchers in Germany used brain imaging to map face and hand somatosensory areas in upper-limb amputees (Flor et al., 1995). In patients with phantom pain, neurons from the face area appeared to have invaded the area that normally receives input from the missing hand (see Figure 11.10). The area activated by touch on the lips was shifted an average of 2.05 centimeters (cm) in the hemisphere opposite the amputation, compared with 0.43 cm for the patients without phantom pain, a fivefold difference. It is unclear from the results whether this was due to intrusion of foreign neurons or to activation of existing neurons; facial stimulation can evoke sensations in the phantom arm within 24 hours after amputation, which indicates "unmasking" of existing but ordinarily silent inputs (Borsook et al., 1998). We usually think of neural plasticity as adaptive, but sometimes it can lead to malfunction and, in this case, truly bizarre results. As you can imagine, treating phantom pain poses special problems, so we will give it special attention in the accompanying Application.

FIGURE 11.10 Reorganization of Somatosensory Cortex in Phantom Pain Patient Following Arm Amputation.

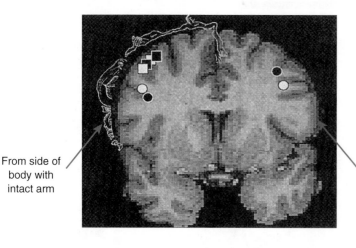

From side of body with intact arm

From side of body with amputated arm

The symbols represent the location of sensitivity to touch of the fingers (squares) and the lips (circles); black symbols are from a patient with phantom pain and white symbols from a patient without phantom pain. By looking at the homunculus superimposed on the left hemisphere (opposite the intact arm), you can see that the circles and the squares are in their normal locations. In the right hemisphere, opposite the amputated arm, lip sensitivity in the patient with the phantom pain (black circle) has migrated well into the area ordinarily serving finger sensitivity.

SOURCE: Reprinted by permission of *Nature.* Copyright 1995.

Application: Treating Pain in Limbs That Aren't There

Most treatments for phantom pain are ineffective and fail to consider the mechanisms producing the pain. Local anesthesia, surgery to sever pain pathways, and pharmacological interventions, from muscle relaxants to antidepressants, typically benefit less than 30% of patients, which is no better than results from placebo (Flor, 2008). Several studies suggest that therapies that work either prevent or reverse the cortical reorganization that occurs following amputation (Birbaumer et al., 1997). Use of a functional prosthesis (as opposed to a purely cosmetic one) prevents cortical reorganization and reduces pain occurrence (Lotze et al., 1999); newer hand prostheses that have pressure sensors in the fingers to help amputees adjust their grip to different objects should reduce phantom pain even better (Friedrich-Schiller-Universität Jena, 2010). When a prosthesis is impractical, training to discriminate frequency and location of electrical stimuli applied to the stump reverses cortical reorganization and reduces pain in 60% of patients (Flor, Denke, Schaefer, & Grüsser, 2001).

Using a mirror illusion to "replace" sensations from the missing limb has yielded some very promising results. The patient places the stump out of sight on one side of a mirror and the good limb on the other side, so the reflection of the good limb appears to replace the one that is missing (see figure). Then the patient is told to move the two limbs in unison. The first patient to use this mirror illusion immediately felt that his missing arm was moving, and the pain disappeared temporarily. After 3 weeks of brief daily practice, he reported almost complete absence of both the pain and the phantom (Ramachandran, Rogers-Ramachandran, & Cobb, 1995). In a later study, 100% of patients using the mirror box experienced decreases in pain, compared to increases in the majority of control patients (Chan et al., 2007); another study confirmed that the pain relief is accompanied by reversal of the cortical reorganization (Flor, Diers, Christmann, & Koeppe, 2006).

But how could watching the mirror image of an intact arm affect the organization and function of the *other* somatosensory area? There are various possibilities, but the answer may lie in the greater responsiveness of the mirror neuron system in phantom

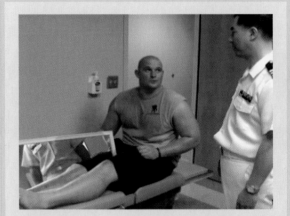

The mirror image of the patient's intact leg appears to replace the missing leg.

SOURCE: © Walter Morris/Medill News Service.

limb patients; when they watch another person's hand being stroked or pricked with a pin, they often report actually *feeling* the stimulation in their missing limb. I saw this firsthand in a colleague who had lost both legs in an accident as we watched a construction worker walking the girders of a new building. I felt a bit of anxiety and a sympathetic tension in my legs as I imagined myself up there balancing on the narrow beams, but, to my complete surprise, my friend exclaimed, "That makes my ankles hurt!" Some investigators suggest that mirror therapy works because the reflected image triggers exaggerated activity in mirror neurons in the hemisphere that once served the missing limb (Chan et al., 2007; Ramachandran & Altschuler, 2009). According to this view, because the missing limb cannot provide contradictory feedback, the mirror neuron activity is interpreted as real touch or movement, and with continued practice the brain's reorganization is reversed.

Concept Check *Take a Minute to Check Your Knowledge and Understanding*

- [ ] What is the contribution of each of the three classes of body senses?
- [ ] In what ways are the somatosensory cortex and the visual cortex organized similarly?
- [ ] In what circumstances does the brain reduce pain?

Movement

A popular view of the brain is that it is mostly preoccupied with higher cognitive processes, like thinking, learning, and language. However, a surprising proportion of the brain is devoted to planning and executing movements. We are talking about more than simply moving the body from one place to another; consider the surgeon's coordinated hand movements during a delicate operation, the control of mouth and throat muscles and diaphragm required to sing an aria, or the pass receiver's ability to follow the trajectory of the football and arrive at the right place at the right time. Studies of the control of movement provided one of the earliest windows into the brain's organization and functioning, and it is that facet of the research that will interest us most. Before we launch into that topic, we need some understanding of the equipment the brain has to work with.

The Muscles

We have three types of muscles. The ones you are most familiar with are the *skeletal muscles,* which move the body and limbs; they are also called *striated* muscles because of their striped appearance. *Smooth muscles* produce contractions in the internal organs; for example, they move food through the digestive system, constrict blood vessels, and void the bladder. *Cardiac muscles* are the muscles that make up the heart. Because our focus is on movement, we will concentrate on the skeletal muscles. Anyway, in spite of differences in appearance, the muscles function similarly.

Like other tissues of the body, a muscle is made up of many individual cells, or muscle fibers. The muscle cells are controlled by motor neurons that synapse with a muscle cell at the neuromuscular junction (Figure 11.11). The number of cells served by a single axon determines the precision of movement possible. The biceps muscles have about a hundred muscle fibers per axon, but the ratio is around 3 to 1 in the eye muscles, which must make very precise movements in tracking objects (Evarts, 1979).

A muscle fiber is made up of myosin filaments and actin filaments. When a motor neuron releases acetylcholine, the muscle fiber is depolarized, which opens calcium channels; the calcium influx initiates a series of actions by the myosin that contract the muscle. A myosin filament essentially "climbs" along the surrounding actin filaments, using small protrusions that it attaches to the actin. Movement of myosin filaments relative to the actin filaments shortens the muscle fibers and contracts the muscle.

Skeletal muscles are anchored to bones by tendons, which are fibrous bands of connective tissue. You can see in Figure 11.12 that by pulling against their attachments the muscles are able to operate the limbs like levers to produce movement. You can also see that limbs are equipped with two *antagonistic muscles,* muscles that produce opposite movements at a joint. In this case, the biceps muscle flexes the arm, and the triceps extends it. Rather than one muscle relaxing while the other does all the work, movement involves opposing contraction from both muscles. The simultaneous contraction of antagonistic muscles creates a smoother movement, allows precise stopping, and maintains a position with

FIGURE 11.11   Muscle Fibers Innervated by a Motor Neuron.

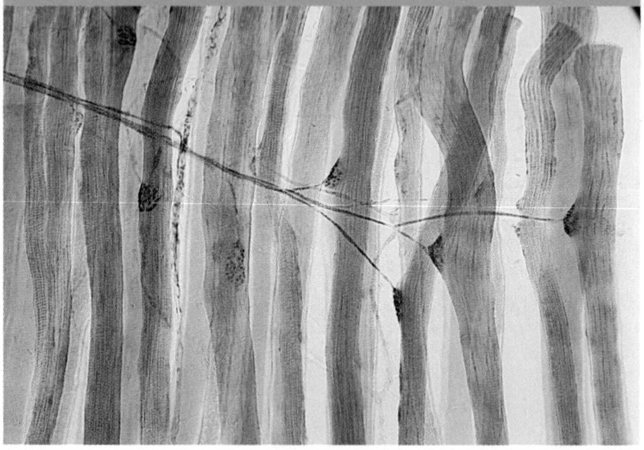

SOURCE: Ed Reschke/Peter Arnold.

FIGURE 11.12   Antagonistic Muscles of the Upper Arm.

When the biceps muscle contracts, it flexes the arm (left); contracting the triceps muscle extends the arm.

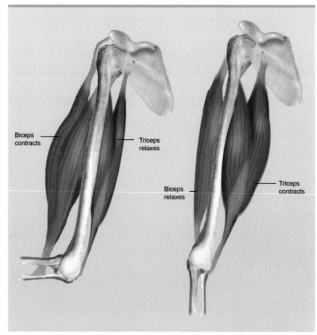

Biceps contracts

Triceps relaxes

Biceps relaxes

Triceps contracts

SOURCE: Based on Starr and Taggart (1989).

minimal tremor. Standing requires the countering effects of antagonistic muscles in the legs, as well as muscles in the torso. The amount of contraction in a muscle varies from moment to moment, so the balance between two antagonistic muscles is constantly shifting, constantly correcting. If maintaining the balance between opposed pairs of muscles required conscious, voluntary activity, we would never be able to hold a video camera still enough to get a sharp picture or even to stand without bobbling back and forth. Adjustments this fast have to be controlled by reflexes at the level of the spinal cord.

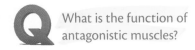

What is the function of antagonistic muscles?

The Spinal Cord

In Chapter 3, we introduced the idea of the spinal reflex. Everyone is familiar with the reflex that makes you quickly withdraw your hand from a hot stove. When you step on a sharp object, you reflexively withdraw your foot and simultaneously make a variety of reflexive postural adjustments to avoid losing your

What do spinal reflexes and central pattern generators do?

balance. The advantage of reflexes is that we can make the appropriate adjustments quickly, without the delay of having to figure out the right action.

The reflex illustrated in Figure 11.13 should also be familiar. Your doctor taps the patellar tendon, which connects the quadriceps muscle to the lower leg bone. This stretches the muscle, which is detected by muscle stretch receptors called *muscle spindles* and relayed to the spinal cord. There the sensory neurons synapse on motor neurons, which return to the quadriceps and cause it to contract and extend the lower leg. The function of the stretch reflex is not just to amuse your doctor. It enables a muscle to resist very quickly if the muscle is stretched by activity in its antagonistic partner; this helps, for example, to maintain an upright posture. It also allows a muscle to respond quickly to an increased external load, for example, when you are holding a stack of books in front of you and a friend unexpectedly drops another book on top of it. *Golgi tendon organs,* receptors that detect tension in a muscle, trigger a spinal reflex that inhibits the motor neurons and limits the contraction. This prevents muscles from contracting so much that they might be damaged.

More complex patterns of motor behavior are also controlled in the spinal cord. It has been known for some time that cats whose spinal cords have been cut, eliminating control from the brain, will make rhythmic walking movements when they are suspended with their feet on a treadmill (Grillner, 1985). This behavior depends on *central pattern generators (CPGs),* neuronal networks that produce a rhythmic pattern of motor activity, such as those involved in walking, swimming, flying, and breathing. Central pattern generators may be located in the spinal cord or in the brain. In humans, they are most obvious in infants below the age of 1 year, who also make stepping movements when held with their feet on a treadmill (Lamb & Yang, 2000). In adults, CPGs provide an important bit of automaticity to

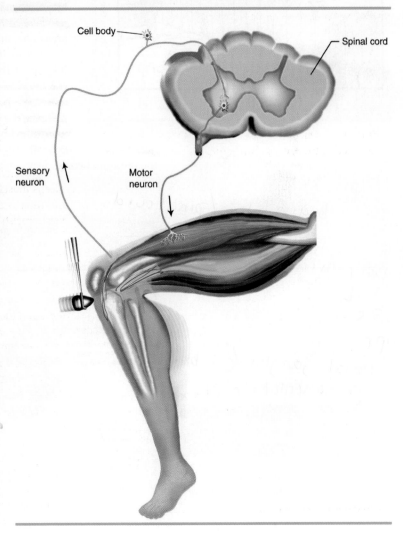

FIGURE 11.13 The Patellar Tendon Reflex, an Example of a Stretch Reflex.

The hammer stretches the tendon, causing a reflexive contraction of the extensor muscle and a kicking motion. This is a highly simplified representation; many more neurons are involved.

Cell body

Spinal cord

Sensory neuron

Motor neuron

routine movements. They can be elicited in individuals with spinal cord injury to produce rhythmic stepping movements (Dimirijevic, Gerasimenko, & Pinter, 1998), and researchers are working on ways to recruit them during therapy (Boulenguez & Vinay, 2009; Dietz, 2009; Edgerton & Roy, 2009). Spinal reflexes produce quick, reliable responses, and central pattern generators provide basic routines the brain can call up when needed, freeing the brain for more important matters (see Figure 11.14). But reflexes and central pattern generators cannot provide all our movement capabilities, so we will turn our attention to the contributions the brain makes to movement.

The Brain and Movement

In the motor system, we again see a hierarchical organization consisting of the forebrain, brain stem, and spinal cord. The motor cortex organizes complex acts and executes movements while modulating activity in the brain stem and spinal cord. The brain stem in turn modulates the activity of the spinal cord (Ghez & Krakauer, 2000). We will start with the cortex and give it most of our attention.

The motor cortex consists of the *primary motor cortex* and two major secondary motor areas, the *supplementary motor area* and the *premotor cortex* (Figure 11.15). Like the primary area, the secondary areas contain a map of the body, with greater amounts of cortex devoted to the parts of the body that produce finer movements (Figure 11.16). The sequence of processing in the motor cortex is just the opposite of what we see in the sensory areas: Planning of movement begins in the association areas, and the primary motor cortex is the final cortical motor area. Along the way, a movement is modified by inputs from the somatosensory cortex, the posterior parietal cortex, the basal ganglia, and the cerebellum. As with many other functions, the prefrontal cortex plays an executive role, so it will receive our attention first.

FIGURE 11.14 **What Life Would Be Like Without Central Pattern Generators.**

Basic lives

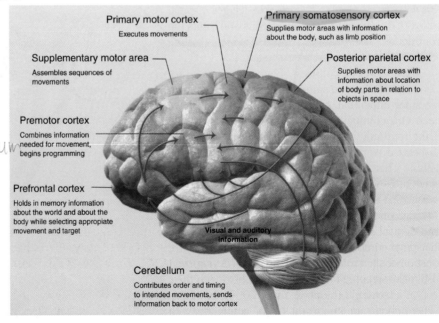

FIGURE 11.15 **The Motor Areas of the Cortex and Cerebellum.**

Connections between the primary motor cortex and the basal ganglia are not shown.

Pref

Pir iv

sui

The Prefrontal Cortex

You already know two functions of the prefrontal cortex that suit it for its role in movement control: First, it plans actions with regard to their consequences; second, it receives information from the ventral visual stream about object identity, which is useful in identifying targets of motor activity. As an initial step in motor planning, the *prefrontal cortex* integrates auditory and visual information about the world with information about the body (from the posterior parietal cortex); it then holds the information in memory while selecting the appropriate movement and its target (see Figure 11.15 again). Considering its activities, it really makes more sense to say that the role of the prefrontal cortex is not so much in planning movements as in planning *for* movements.

These functions are typically investigated in monkeys while they perform some variation of a *delayed match-to-sample task.* The monkey is presented with a visual stimulus; then, after a delay of a few seconds in which the stimulus is absent, the monkey is presented two or more stimuli and required to select the original stimulus (by reaching for it) in order to obtain a reward, such as a sip of juice. Some cells in the prefrontal cortex start firing when the first stimulus is presented and continue to fire throughout the delay, suggesting that they are "remembering" the stimulus. At response time another group of prefrontal cells starts firing before activity starts in the premotor areas; this indicates that the prefrontal cortex selects the target of behavior and the appropriate motor response (Goldman-Rakic, Bates, & Chafee, 1992; Hoshi, Shima, & Tanji, 2000; Rainer, Rao, & Miller, 1999).

FIGURE 11.16 The Primary Motor Area.

The homunculus shows the relative amount of cortex devoted to different parts of the body.

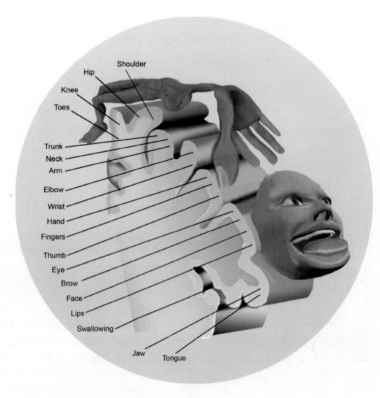

SOURCE: Adapted from *The Cerebral Cortex of Man* by W. Penfield and T. Rasmussen, 1950, New York: Macmillan. Copyright 1950. Reprinted with permission of Gale, a division of Thomson Learning: www.thomsonrights.com. Fax 800-730-2215.

The Secondary Motor Areas

The *premotor cortex* begins programming a movement by combining information from the prefrontal cortex and the posterior parietal cortex (Krakauer & Ghez, 2000). A good example comes from a study in which monkeys were cued to reach for one of two targets, A or B, in different locations and to use the left arm on some trials and the right on others. Some premotor neurons increased their firing rate only if target A was cued, and other neurons were selective for target B. Other cells fired selectively depending on which arm was to be used. Still other cells combined the information of the first two kinds of cells; they increased their firing only when a particular target was cued *and* a particular arm was to be used (Hoshi & Tanji, 2000). Two other cell types combine visual and somatosensory information to provide visual guidance of reaching and object manipulation. The first responds to visual stimuli on or near a specific part of the body, as in Figure 11.17a; another shifts the location of its visual receptive field to coincide with the location of the monkey's hand as it moves (Figure 11.17b; Graziano, Hu, & Gross, 1997; Graziano, Yap, & Gross, 1994).

A fascinating demonstration of the role these specialized cells play occurs in a bizarre phenomenon known as the "rubber hand illusion." The individual sits at a

3
Stimulating the Motor Cortex

What is the relationship between the primary motor area and the association areas?

FIGURE 11.17 Receptive Fields of Two Types of Premotor Neurons That Responded to Both Visual and Body Information.

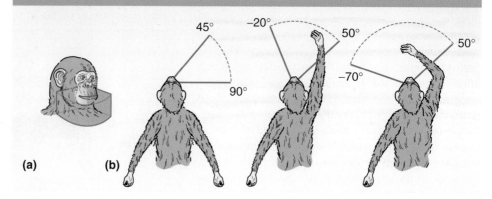

(a) The receptive field of a cell that responded when a visual stimulus was in the area outlined near the face. (b) The visual field of the second type of neuron when the arm was out of sight. Middle and right, the visual field as the monkey's arm moved forward and across.

SOURCE: Adapted from Figure 1 in "Coding of Visual Space by Premotor Neurons," by M. S. A. Graziano, G. S. Yap, and C. G. Gross, *Science, 266,* pp. 1054–1057. Copyright © 1979. Reprinted by permission of AAAS.

table with the left hand hidden from view; the experimenter strokes the hidden left hand with a brush while simultaneously stroking a rubber hand, which is in full view (Figure 11.18). After a few seconds, the sensation seems to be coming from the rubber hand, which the subject reports seems like "my hand." A recent study used functional magnetic resonance imaging (fMRI) to determine where the illusion occurs in the brain (Ehrsson, Spence, & Passingham, 2004). The posterior parietal cortex, which combines the visual and touch information, was active whether the two hands were stroked in synchrony or in asynchrony. The premotor area, on the other hand, became active only when the stroking was simultaneous, and only after the individual began to experience the illusion; moreover, the strength of activation was related to the intensity of the illusion reported by the subject.

Output from the prefrontal cortex flows to the *supplementary motor area,* which assembles sequences of movements, such as those involved in eating or playing the piano. In monkeys trained to produce several different sets of movement sequences, different neurons increase their firing during a delay period depending on which sequence has been cued for performance (Shima & Tanji, 2000; Tanji & Shima, 1994). An important form of movement sequencing is the coordination of movements between the two sides of the body. For example, when a monkey's supplementary motor cortex is damaged in one hemisphere, its hands tend to duplicate each other's actions instead of sharing the task (C. Brinkman, 1984). Humans with similar damage also have trouble carrying out tasks that require alternation of movements between the two hands (Laplane, Talairach, Meininger, Bancaud, & Orgogozo, 1977).

The Primary Motor Cortex

The *primary motor cortex* is responsible for the execution of voluntary movements; its cells fire most during the movement instead of prior to it (G. E. Alexander & Crutcher, 1990; Riehle & Requin, 1989). Individual motor cortex cells are not reserved for a specific movement but contribute their function to a range of related behaviors (Saper, Iverson, & Frackowiak, 2000); the primary motor cortex orchestrates the

FIGURE 11.18 The Premotor Cortex and the Rubber Hand Illusion.

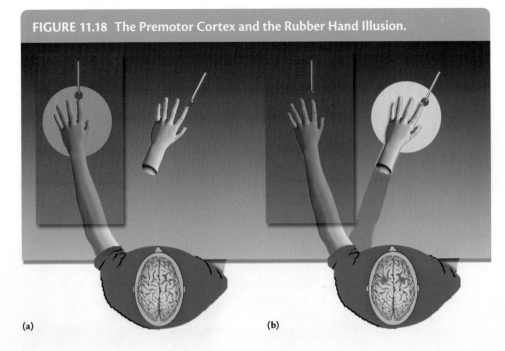

(a)   (b)

(a) The hidden left hand is stroked in synchrony with the fake hand, which is full view.
(b) After a few seconds, the individual feels that the sensation is coming from the rubber hand and reports a sense of ownership of the rubber hand. Apparently, the touch field and the visual field have become coordinated in the brain (indicated by the light blue outline and the yellow circle). fMRI recording shows that the premotor cortex is active during this illusion (indicated by the red circles).

SOURCE: From "Probing the Neural Basis of Body Ownership," by M. Botvinick, *Science*, 305, 782–783, unnumbered figure on page 782. Illustration copyright © 2004 Taina Litwak. Used with permission.

activity of these cells into a useful movement and contributes control of the movement's force and direction (Georgopoulos, Taira, & Lukashin, 1993; Maier, Bennett, Hepp-Reymond, & Lemon, 1993). This orchestration was particularly evident in a study by Graziano, Taylor, and Moore (2002). Instead of using the typical brief pulses of electricity, which produce only muscle twitches, they increased the duration to 1/2 second and saw complex, coordinated responses in the monkeys, such as grasping, moving the hand to the mouth, and opening the mouth. The primary motor cortex is able to put these complex movement sequences together with the aid of input from the secondary motor areas, the somatosensory cortex, and the posterior parietal area (see Figure 11.15 again) (Krakauer & Ghez, 2000). Presumably, information from the somatosensory and posterior parietal areas provides feedback needed for refining movements on the fly.

No one appreciates the sophistication and capabilities of the motor system like those who have lost the ability to move.

The Basal Ganglia and Cerebellum

The basal ganglia and cerebellum produce no motor acts themselves. Rather, they modulate the activity of cortical and brain stem motor systems; in that role, they are necessary for posture and smooth movement (Ghez & Krakauer, 2000). The *basal ganglia*—the caudate nucleus, putamen, and globus pallidus—use information from the primary and secondary motor areas and the somatosensory cortex to integrate and smooth movements. The basal ganglia send output directly to

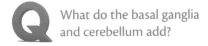
What do the basal ganglia and cerebellum add?

FIGURE 11.19 The Basal Ganglia.

The basal ganglia include the caudate nucleus, putamen, and globus pallidus.

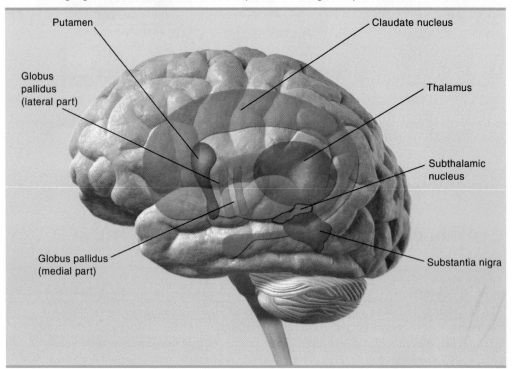

the primary motor cortex and supplementary motor area, as well as to the premotor cortex via the thalamus. As you can see in Figure 11.19, these structures border the thalamus; they apparently smooth movements through both facilitating and inhibitory outputs to the thalamus (DeLong, 2000). The basal ganglia also are especially active during complex sequences of movements (Boecker et al., 1998). It appears that they are involved in learning movement sequences so the movements can be performed as a unit (Graybiel, 1998). In fact, one of the symptoms of Parkinson's disease, which is caused by degeneration in the basal ganglia, is impaired learning, whether motor behavior is involved or not (Knowlton, Mangels, & Squire, 1996). Malfunction in the basal ganglia results in postural abnormalities and involuntary movements in Parkinson's disease and Huntington's disease.

When the *cerebellum* receives information from the motor cortex about an intended movement, it determines the order of muscular contractions and their precise timing. It also uses information from the vestibular system to maintain posture and balance, refine movements, and control eye movements that compensate for head movements (Ghez & Thach, 2000). Once an intended movement has been modified, the cerebellum sends the information back to the primary motor cortex. We can see the contribution of the cerebellum in the deficits that occur when it is damaged. For example, we begin to shape our hand for grasping while the arm is moving toward the target, but a person with cerebellar damage reaches, pauses, and then shapes the hand. A normal individual touches the nose in what appears to be a single, smooth movement; cerebellar damage results in exaggerated, wavering corrections. The effects of cerebellar damage on coordination and accuracy in limb movements is similar to the effect of alcohol; the drunk driver who is pulled over by the police has trouble walking a straight line, standing on one foot with the eyes closed, or touching the nose with the tip of the finger. People with damage to the cerebellum are often mistakenly believed to be drunk.

TABLE 11.1 The Major Brain Areas of Movement and Their Functions

Structure	Functions
Prefrontal cortex	Selects the appropriate behavior and its target, using a combination of bodily and external information.
Premotor cortex	Combines information needed for movement programming, such as the target being reaching for and its location, which arm to use, and the arm's location.
Supplementary motor area	Assembles sequences of movements, such as eating or playing the piano; coordinates movements between the two sides of the body (e.g., task sharing between the hands).
Primary motor cortex	Executes voluntary movements by organizing the activity of unspecialized cells; adds force and direction control.
Basal ganglia	Uses information from secondary areas and somatosensory cortex to integrate and smooth movements; apparently involved in learning movement sequences.
Cerebellum	Maintains balance, refines movements, controls compensatory eye movements. Involved in learning motor skills.

The cerebellum lives up to the meaning of its name, "little brain," by applying its expertise to a variety of tasks. It is necessary for learning motor skills (D. A. McCormick & Thompson, 1984), but it also participates in nonmotor learning (Canavan, Sprengelmeyer, Diener, & Hömberg, 1994) and in making time and speed judgments about auditory and visual stimuli (Keele & Ivry, 1990). Also, patients with cerebellar damage have difficulty shifting visual attention to another location in space (whether this involves eye movements or not), taking 0.8 to 1.2 s compared with 0.1 s for normal individuals (Townsend et al., 1999). We should think of the cerebellum in terms of its general functions rather than strictly as a motor organ.

Table 11.1 summarizes the structures we have discussed and their functions.

Disorders of Movement

You might think that anything as complex as the movement system would be subject to malfunction; if so, you would be correct. Predictably, movement disorders are devastating to their victims. As representatives of these diseases, we will consider Parkinson's disease, Huntington's disease, myasthenia gravis, and multiple sclerosis.

4
Movement Disorders

Parkinson's Disease

Parkinson's disease is characterized by motor tremors, rigidity, loss of balance and coordination, and difficulty in moving, especially in initiating movements (Olanow & Tatton, 1999; Youdim & Riederer, 1997). Parkinson's affects about 2% of the population (Polymeropoulos et al., 1997); muscle strength is unaffected, but the disease often becomes disabling as it progresses. The symptoms are caused by deterioration of the *substantia nigra*, whose neurons send dopamine-releasing axons to the *striatum*, which is composed of the basal ganglia's caudate nucleus and

What are the causes and effects of the movement disorders?

FIGURE 11.20 Lewy Bodies in a Brain With Parkinson's Disease.

A neuron containing two stained Lewy bodies, abnormal clumps of protein.

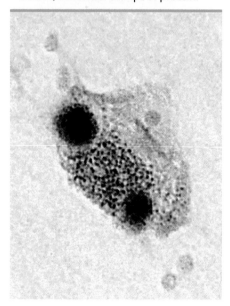

SOURCE: From "α-Synuclein in Lewy Bodies," by M. G. Spillantini et al., 1997, *Nature* 8/28/1997. Copyright © 1997. Used with permission.

putamen and the nucleus accumbens. In something less than 10% of cases, the disease is *familial,* meaning that it occurs more frequently among relatives of a person with the disease than it does in the population (Greenamyre & Hastings, 2004). If a member of a twin pair is diagnosed with Parkinson's disease before the age of 51, the chance of an identical twin also having Parkinson's is six times greater than it is for a fraternal twin (Tanner et al., 1999). The same study found no evidence of a genetic influence in individuals whose symptoms developed later in life; we will look at possible nongenetic causes shortly.

A recent whole-genome study of Parkinson's patients identified 12 genes that likely contribute to the disease (Maraganore et al., 2005). The fact that none had a large effect suggests that these genes may confer a susceptibility that interacts with environmental causes. Two of the implicated genes may play a role in the development and programmed death of dopamine-producing neurons. Two others result in deviant proteins that are components of *Lewy bodies,* abnormal clumps of protein that form within neurons. Lewy bodies are found in several brain locations in some Parkinson's patients, as well as in people with a form of Alzheimer's disease, *dementia with Lewy bodies* (see Figure 11.20; Glasson et al., 2000; Spillantini et al., 1997). Lewy bodies probably contribute to the cognitive deficits and depression that often accompany Parkinson's disease.

Several environmental influences have been implicated in Parkinson's disease. One cause is subtle brain injury; being knocked unconscious once increases the risk by 32%, and the risk rises by 174% for those knocked out several times. Other research points to a variety of toxins, including industrial chemicals, carbon monoxide, herbicides, and pesticides (Olanow & Tatton, 1999). Numerous studies show an association between pesticide use and Parkinson's, but these are correlational and a causal relationship is questioned ("Pesticide Exposure and Parkinson's Disease," 2006). However, administering the pesticide rotenone to rats for several weeks produced tremors, Lewy bodies, and loss of dopamine neurons (Betarbet et al., 2000). There are clues that some Parkinson's sufferers inherit a diminished ability to metabolize toxins (Bandmann, Vaughan, Holmans, Marsden, & Wood, 1997; C. A. D. Smith et al., 1992), so once again we have evidence for the interaction of hereditary and environmental effects.

Interestingly, the risk of Parkinson's disease is reduced as much as 80% in coffee drinkers (G. W. Ross et al., 2000). The risk also drops by 50% in smokers (Fratiglioni & Wang, 2000), but of course no benefit of smoking outweighs its dangers. Rat studies indicate that cigarette smoke may prevent the accumulation of neurotoxins (Soto-Otero, Méndez-Alvarez, Sánchez-Sellero, Cruz-Landeira, & López-Rivadulla, 2001) and that caffeine reduces the effect of neurotoxins by blocking adenosine receptors, which we saw in Chapter 5 results in increased dopamine and acetylcholine release (J.-F. Chen et al., 2001).

Parkinson's disease is typically treated by administering *levodopa (L-dopa),* which is the precursor for dopamine. Dopamine will not cross the blood-brain barrier but L-dopa will, and in the brain it is converted to dopamine. Dopamine agonists can also be helpful, and even placebos increase dopamine release (de la Fuente-Fernández et al., 2001). But these treatments increase dopamine throughout the brain, which causes side effects ranging from restlessness and involuntary movements to hallucinations. Also, as more neurons die, more drug is required, increasing the side effects. Early attempts showed that implanted embryonic neural cells could survive in the striatum and produce dopamine (Figure 11.21; C. R. Freed et al., 2001; Greene & Fahu, 2002). However, behavioral improvement was not clinically significant, and some of the patients developed involuntary movements, apparently due to excess dopamine. More recent work using adult neural stem cells resulted in more than 80% improvement in motor behavior ratings; the improvement held up for 3 years but had disappeared at the end of 5 years (Lévesque, Neuman, & Rezak, 2009). These procedures are in their infancy; however, we need to remember that the first several heart transplant operations failed but they are almost routine today.

Researchers are also using genetic interventions to increase dopamine synthesis (Jarraya et al., 2009) and to reduce the loss of dopaminergic neurons (Malagelada, Jin, Jackson-Lewis, Przedborski, & Greene, 2010). A gene transfer strategy that increases production of the inhibitory transmitter GABA is looking particularly promising (Kaplitt et al., 2007).

Frustration with therapeutic alternatives is creating something of a revival in surgical treatments, which were largely abandoned when drugs for Parkinson's disease became available (Cosgrove & Eskandar, 1998). Strategically placed lesions in the subthalamic nucleus and the globus pallidus, both in the basal ganglia (see Figure 11.19 again), have provided some improvement for patients who have difficulty using dopaminergic drugs (Cosgrove & Eskandar). These two structures produce a rhythmic bursting activity similar to the rhythm of activity in Parkinsonian tremors, which apparently explains why destroying them reduces this symptom (Perkel & Farries, 2000).

The surgeries can produce deficits of their own, such as weakness in a part of the body due to damage to adjacent areas, so researchers are experimenting with electrical stimulation through implanted electrodes. Patients receiving stimulation to the globus pallidus and the subthalamic nucleus showed improved motor functioning, along with increased metabolism in the premotor cortex and cerebellum (Deep-Brain Stimulation for Parkinson's Disease Study Group, 2001; Fukuda et al., 2001). Neuron loss in the substantia nigra decreases dopamine input to these areas; as a result, excitatory glutamate-releasing neurons are left unopposed so they overstimulate their targets. How the stimulation works is unclear, but it apparently restores inhibitory balance following the loss of dopaminergic neurons (Kaplitt et al., 2007).

Huntington's Disease

Like Parkinson's disease, *Huntington's disease* is a degenerative disorder of the motor system involving cell loss in the striatum and cortex. Years before a diagnosis, Huntington's disease begins with jerky movements that result from impaired error correction (M. A. Smith, Brandt, & Shadmehr, 2000). Later, involuntary movements appear, first as fidgeting and then as movements of the limbs and, finally, writhing of the body and facial grimacing. Because these movements sometimes resemble a dance, Huntington's disease is also called Huntington's chorea, from the Greek word *choreia*, which means "dance." Needless to say, the patient loses the ability to carry out daily activities. Death usually follows within 15 to 30 years after the onset of the disease.

Unlike Parkinson's disease, cognitive and emotional deficits are a universal characteristic of Huntington's disease. These deficits include impaired judgment, difficulty with a variety of cognitive tasks, depression, and personality changes. The motor symptoms are due to the degeneration of inhibitory GABA-releasing neurons in the striatum, while defective or degenerated neurons in the cortex probably account for the psychological symptoms (Figure 11.22; J. B. Martin, 1987; Tabrizi et al., 1999).

Huntington's disease results from a mutated form of the *huntingtin* gene (Huntington's Disease Collaborative Research Group, 1993). The loss of neurons is probably due to the accumulation of the gene's protein, also known as huntingtin, whose function is unknown (DiFiglia et al., 1997). In normal individuals, the gene has between 10 and 34 repetitions of the bases cytosine, adenine, and guanine (see Chapter 1). The more repeats the person has beyond 37, the earlier in life the person will succumb to the disease (R. R. Brinkman, Mezei, Theilmann, Almqvist, &

FIGURE 11.21 Transplanted Embryonic Cells in the Brain of a Parkinson's Patient.

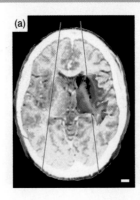

(a)   (b)

The patient died in a car accident 7 months after her surgery. (a) Her right putamen (part of the striatum) was removed and placed on a photograph of the magnetic resonance image of her brain made at the time of surgery. The red lines indicate the angle at which the needles were inserted into the brain to inject the fetal cells (right side of the brain) and as a control procedure (left side). The dark area on the putamen along the needle track is due to the staining of new dopamine cells and shows that the axons had grown 2 to 3 millimeters from the cell bodies. (b) is an enlargement of the putamen.

SOURCE: From "Transplantation of Embryonic Dopamine Neurons for Severe Parkinson's Disease," by C. R. Freed et al., 2001, *New England Journal of Medicine, 334*, 710–719, fig. 3a andb, p. 717.

> *This is a scary thing. . . . There is a test available, but I haven't had the guts to take it yet.*
>
> —Shana Martin, at risk for Huntington's disease

FIGURE 11.22 Loss of Brain Tissue in Huntington's Disease.

FIGURE 11.22 Loss of Brain Tissue in Huntington's Disease.

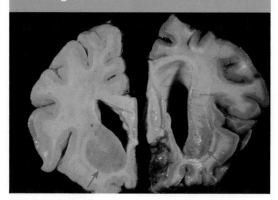

Left, a section from a normal brain; right, a section from a person with Huntington's disease. The enlarged lateral ventricle in the diseased brain is due to loss of neurons in the caudate nuclei (arrows).

SOURCE: Courtesy of Robert E. Schmidt, Washington University.

Hayden, 1997). Because the gene is dominant, a person who has a parent with Huntington's has a 50% chance of developing the disease. This is an unusual example of a human disorder resulting from a single gene.

A number of drugs are used to treat the various symptoms, including antidepressants and antipsychotics, but only one has been approved specifically for Huntington's disease by the FDA ("FDA Approves," 2008). It reduces the excess dopamine that causes the abnormal movements. Another drug that is in human trials improves cognitive functioning; it blocks glutamate receptors, which are thought to be hypersensitive in Huntington's patients, to prevent cell death from overstimulation (Kieburtz et al., 2010). Grafting of fetal striatal cells has so far produced only modest and temporary improvement (Cicchetti et al., 2009).

Autoimmune Diseases

Myasthenia gravis is a disorder of muscular weakness caused by reduced numbers or sensitivity of acetylcholine receptors. The muscle weakness can be so extreme that the patient has to be maintained on a respirator. In fact, 25 years ago the mortality rate from myesthenia gravis was about 33%; now few patients die from the disease, thanks to improved treatment (Rowland, 2000a).

The loss of receptors was demonstrated in an interesting way. The venom of the many-banded Formosan krait, a very poisonous snake from Taiwan, paralyzes prey by binding to the acetylcholine receptor. When the venom's toxin is labeled with radioactive iodine and applied to a sample of muscle tissue, it allows researchers to identify and count the acetylcholine receptors. The patients turned out to have 70% to 90% fewer receptors than normal individuals (Fambrough, Drachman, & Satyamurti, 1973). Drugs that inhibit the action of acetylcholinesterase give temporary relief from the symptoms of myesthenia gravis (Figure 11.23; Rowland, Hoefer, & Aranow, 1960). Remember that acetylcholinesterase breaks down acetylcholine at the synapse; these inhibitors increase the amount of available neurotransmitter at the neuron-muscle junction.

Although immune system therapy has sometimes been used (Shah & Lisak, 1993), removal of the thymus (thymectomy) has become a standard treatment for myasthenia gravis (Rowland, 2000a). The thymus is the major source of lymphocytes that produce antibodies. Improvement can take years, but thymectomy eliminates symptoms completely in almost 80% of patients and reduces them in

FIGURE 11.23 Effect of an Acetylcholinesterase Inhibitor on Myasthenia Gravis.

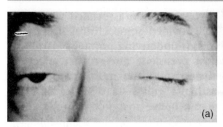

(a) Patients often have drooping eyelids, as shown here. This patient also could not move his eyes to look to the side. (b) The same patient 1 min after injection of an acetylcholinesterase inhibitor. The eyes are open and able to move freely.

SOURCE: From "Mysathenic Syndromes," by L. P. Rowland, P. F. A. Hoefer, and H. Aranow, Jr., 1960, *Research Publications-Association for Research in Nervous and Mental Disease, 38*, 547–560.

another 13% to 17% (Ashour et al., 1995; Jaretzki et al., 1988).

Multiple sclerosis is a motor disorder with many varied symptoms, caused by deterioration of myelin (demyelination) and neuron loss in the central nervous system. In Chapter 2, you saw that demyelination causes slowing or elimination of neural impulses. Demyelination thus reduces the speed and strength of movements. Even before that happens, impulses traveling in adjacent neurons, which should arrive simultaneously, become desynchronized because of differential loss of myelin. An early sign of the disorder is impairment of functions that require synchronous bursts of neural activity, like tendon reflexes and vibratory sensation (Rowland, 2000b). As the disease progresses, unmyelinated neurons die, leaving areas of *sclerosis,* or hardened scar tissue (Figure 11.24). As a result, the person experiences muscular weakness, tremor, impaired coordination, urinary incontinence, and visual problems. Recent studies indicate that neuron loss is more important than previously thought and suggest that the loss results from a degenerative process in addition to the demyelination (DeLuca, Ebers, & Esiri, 2004; De Stefano et al., 2003).

Like myasthenia gravis, multiple sclerosis is an autoimmune disease. Injecting foreign myelin protein into the brains of animals produces symptoms very similar to those of multiple sclerosis (Wekerle, 1993), and T cells that are reactive to myelin proteins (see Chapter 8) have been found in the blood of multiple sclerosis patients (Allegretta, Nicklas, Sriram, & Albertini, 1990). A genome-wide study has implicated various immune system genes in multiple sclerosis (International Multiple Sclerosis Genetics Consortium, 2007), but some environmental condition may be needed to trigger the immune attack on myelin. One possibility is that the immune system has been sensitized by an earlier viral infection; for example, studies have found antibodies for Epstein-Barr virus in multiple sclerosis patients (H. J. Wagner et al., 2000), and patients more often had mumps or measles during adolescence (Hernán, Zhang, Lipworth, Olek, & Ascherio, 2001). Several drugs are available that modify immune activity in multiple sclerosis patients; they slow the progress of the disease but do not repair the harm already done. A new direction may be indicated by a drug that has just won FDA approval; it blocks potassium channels and improves motor performance, particularly walking (Jeffrey, 2010). Use of stem cells derived from the patient's central nervous system or bone marrow halts deterioration and in some cases produces improvement. However, it is unclear how much this is due to cell replacement or to stimulation of neural protective mechanisms (Martino, Franklin, Van Evercooren, & Kerr, 2010; Rice et al., 2010).

FIGURE 11.24 The Brain of a Deceased Multiple Sclerosis Patient.

The arrows indicate areas of sclerosis, or hardened scar tissue (dark areas).

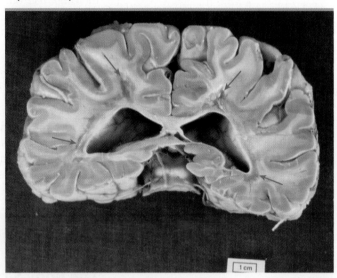

SOURCE: Photo Researchers.

Concept Check
Take a Minute to Check Your Knowledge and Understanding

- ☐ Explain how antagonistic muscles and spinal reflexes maintain posture.
- ☐ What contribution does each of the cortical motor areas make to movement?
- ☐ What are the genetic and environmental causes of the movement disorders described here?

In Perspective

Unless we have a disorder, we usually take our body senses and our capability for movement for granted. And yet just standing upright is a remarkable feat. Granted, a mechanical robot could do it easily, but only if it had a rigid body like R2D2's. If the robot had our flexibility of movement and posture, it would have to devote a fair amount of its computer brain to making split-millisecond adjustments to avoid toppling over. Then another chunk of its computer would be required just to locate a visual object in space, to reach out smoothly and quickly for the object, and to shape its hand for grasping, deciding whether to use the whole hand or the finger and thumb and how much pressure to apply, and so on. You get the idea. Better let a human do it, because all that fancy equipment comes standard on the basic model.

Now you see why so much of the brain is concerned with the sensory and motor components of movement. It is a wonder that we have enough left over for the demands of learning, intelligence, and consciousness, but as you will see in the remaining chapters, we do.

Summary

The Body Senses

- The body senses include proprioception, which tells us about the position and movement of our limbs and body; the skin senses, which inform us about the conditions in the periphery of our body; and the vestibular sense, which contributes information about head position and movement and helps us maintain balance.
- The skin senses—touch, warmth, cold, and pain—tell us about conditions at the body surface and about objects in contact with our body.
- The body senses are processed in a series of structures in the primary and secondary somatosensory cortex and in the posterior parietal cortex, with several similarities to visual processing. Pain processing also extends into additional areas.
- In their quest to find better ways of relieving pain, researchers have learned how the nervous system detects painful stimulation and found that the body has its own ways of relieving pain. Chronic pain presents particularly difficult challenges.

Movement

- There are three types of muscles: cardiac (heart); smooth (internal organs); and skeletal muscles, which move the body by tugging against their attachments to bones.
- Spinal reflexes produce quick responses and provide postural adjustments. Central pattern generators provide routines such as rhythmic walking movements.
- Cortical motor areas assess spatial and body information and construct movements by passing information through a succession of brain areas.
- The basal ganglia and cerebellum refine movements produced by the motor cortex.
- A number of diseases attack the motor system at various points of vulnerability. Major causes that have been implicated are heredity, toxins, and autoimmune disorders. ∎

Study Resources

F   For Further Thought

- Of proprioception, the vestibular sense, pain, and the other skin senses, which do you think you could most afford to give up? Why?
- If pain is beneficial, why does the body have pain relief mechanisms?
- Imagine a robot with a humanlike body. It is programmed to walk, reach, grasp, and so on.

It has visual and auditory capabilities, but no body senses. What would its movement be like?

- Judging by the examples given of movement disorders, what are the points of vulnerability in the motor system?

T   Testing Your Understanding

1. Explain how endorphins relieve pain, describing the receptors and the pathway from the periaqueductal gray; include how we determine whether pain relief is endorphin based.

2. Walking barefoot, you step on a sharp rock. You reflexively withdraw your foot, plant it firmly on the ground again, and regain your posture. Describe these behaviors in terms of the sensory/pain mechanisms and reflexes involved.

3. Trace the progress of a movement through the parietal and frontal lobes, giving the names of the structures and a general idea of the processing in each.

4. Compare the symptoms, causes, and treatment options for Parkinson's and Huntington's diseases.

Select the best answer:

1. Proprioception gives us information about
   a. conditions at the surface of our skin.
   b. conditions in the internal organs.
   c. the position and movement of our limbs and body.
   d. balance and the head's position and movement.

2. The skin senses include
   a. touch, warmth, and cold.
   b. touch, temperature, and pain.
   c. touch, temperature, movement, and pain.
   d. touch, warmth, cold, and pain.

3. Sharp pain and dull pain are due primarily to
   a. different kinds of injury.
   b. pain neurons with different characteristics.
   c. the passage of time.
   d. the person's attention to the pain.

4. According to Melzack and Wall, pressing the skin near a wound reduces pain by
   a. creating inhibition in the pain pathway.
   b. distracting attention from the injury.
   c. releasing endorphins.
   d. releasing histamine into the wound area.

5. Endorphins
   a. activate the same receptors as opiate drugs.
   b. occupy receptors for pain neurotransmitters.
   c. block reuptake of pain neurotransmitter.
   d. inhibit brain centers that process pain emotion.

6. Congenital pain insensitivity and chronic pain involve
   a. developmental alterations of brain areas responsible for the emotion of pain.
   b. alterations in the myelination of pain fibers.
   c. gene-mediated alterations of pain sensitivity.
   d. variations in the amount of substance P available.

7. Research suggests phantom pain is due to
   a. the patient's anxiety over the limb loss.
   b. memory of the pain of injury or disease that prompted the amputation.
   c. activity in severed nerve endings in the stump.
   d. neural reorganization in the somatosensory area.

8. Without a posterior parietal cortex we would be most impaired in
   a. moving.
   b. making smooth movements.
   c. orienting movements to objects in space.
   d. awareness of spontaneously occurring movements.

9. If the nerves providing sensory feedback from the legs were cut, we would
   a. have to use vision to guide our leg movements.
   b. have trouble standing upright.
   c. lose strength in our legs.
   d. a and b.
   e. b and c.

10. A monkey is presented a stimulus, then must wait a few seconds before it can reach to the correct stimulus. Activity in the secondary motor area during the delay suggests that this area
    a. prepares for the movement.
    b. initiates the movement.
    c. executes the movement.
    d. all of these.

11. Cells in the premotor cortex would be particularly involved when you
    a. remember a visual stimulus during a delay period.
    b. catch a fly ball.
    c. start to play a series of notes on the piano.
    d. execute a movement.

12. The primary motor cortex is most involved in
    a. combining sensory inputs.
    b. planning movements.
    c. preparing movements.
    d. executing movements.

13. The basal ganglia and the cerebellum produce
    a. no movements.
    b. movements requiring extra force.
    c. reflexive movements.
    d. sequences of movements.

14. Parkinson's disease is characterized most by
    a. deterioration of the myelin sheath.
    b. dancelike involuntary movements.
    c. deterioration of dopamine-releasing neurons.
    d. immune system attack on acetylcholine receptors.

15. Results of removing the thymus gland suggest that myasthenia gravis is a(n) _____ disease.
    a. genetic
    b. autoimmune
    c. virus-caused
    d. degenerative

Answers: 1.c, 2.d, 3.b, 4.a, 5.a, 6.c, 7.d, 8.c, 9.d, 10.a, 11.b, 12.d, 13.a, 14.c, 15.b

Online Resources

The following resources are available at **www.sagepub.com/garrett3e**.

Chapter Resources

- Flash Cards
- Chapter Quiz
- Internet Resources and Exercises

On the Web

Links to these websites can be found at the web address above: Select this chapter's **Chapter Resources**, and then choose **Internet Resources and Exercises**.

1. The **Vestibular Disorders Association** has information about vestibular problems and provides additional resources such as newsletters, books, and videotapes.

2. The **American Pain Foundation** offers information for pain patients, testimonials from people suffering pain from an assortment of causes, and links to numerous other pain sites.
   The **International Association for the Study of Pain** has links to more technical resources on pain.

3. **Probe the Brain** at PBS's Science Odyssey lets you "stimulate" the motor cortex and make various parts of the body move to verify that there really is a homunculus.

4. In the **Society for Neuroscience**'s Searching for Answers videos, patients and their families describe what it is like to live with Huntington's disease, Parkinson's disease, and amyotrophic lateral sclerosis (ALS, or Lou Gehrig's disease). You can view the clips or request a free DVD.

You can get information about a variety of movement disorders from the **Neuromuscular Disease Center**, **National Parkinson Foundation**, and **Huntington's Disease Association**.

In an interview with Katie Couric, actor **Michael J. Fox** talks about living with Parkinson's disease and about his views on stem cell research.

Chapter Updates and Biopsychology News

R For Further Reading

1. *Awakenings*, by Oliver Sacks (Vintage, 1999 Books), describes Dr. Sacks's early experiments in using L-dopa to treat the symptom of parkinsonism in patients with sleeping sickness. The movie with Robin Williams was based on this book.

2. *Phantoms in the Brain*, by V. S. Ramachandran and Sandra Blakeslee (Harper Perennial, 1999), called "enthralling" by the *New York Times* and "splendid" by Francis Crick, uses numerous

(often strange) cases to explain people's perception of their bodies.

3. *Wall and Melzack's Textbook of Pain*, edited by Stephen McMahon and Martin Koltzenburg (Churchill Livingstone, 5th ed., 2005), and *The Massachusetts General Hospital Handbook of Pain Management*, edited by Jane Ballantyne (Lippincott Williams and Wilkins, 3rd ed., 2005) are technical references on pain and pain management.

K Key Terms

antagonistic muscles ... 344
basal ganglia ... 349
body integrity identity disorder ... 336
cardiac muscles ... 344
central pattern generator (CPG) ... 345
chronic pain ... 341
dermatome ... 333
endorphins ... 338
familial ... 352
gate control theory ... 339
Golgi tendon organs ... 345
Huntington's disease ... 353
levodopa (L-dopa) ... 352
Lewy bodies ... 352
multiple sclerosis ... 355
muscle spindles ... 345
myasthenia gravis ... 354
out-of-body experience ... 336

Parkinson's disease ... 351
periaqueductal gray (PAG) ... 339
phantom pain ... 342
posterior parietal cortex ... 336
premotor cortex ... 347
primary motor cortex ... 348
primary somatosensory cortex ... 335
proprioception ... 330
secondary somatosensory cortex ... 335
skeletal muscles ... 344
skin senses ... 330
smooth muscles ... 344
somatosensory cortex ... 334
striatum ... 351
substance P ... 337
substantia nigra ... 351
supplementary motor area ... 348
vestibular sense ... 332

PART IV

Complex Behavior

Chapter 12. Learning and Memory

Chapter 13. Intelligence and Cognitive Functioning

Chapter 14. Psychological Disorders

Chapter 15. Sleep and Consciousness

Learning and Memory

12

Learning as the Storage of Memories

Amnesia: The Failure of Storage and Retrieval

IN THE NEWS: THE LEGACY OF HM

Mechanisms of Consolidation and Retrieval

Where Memories Are Stored

Two Kinds of Learning

Working Memory

CONCEPT CHECK

Brain Changes in Learning

Long-Term Potentiation

How LTP Happens

Neural Growth in Learning

Consolidation Revisited

Changing Our Memories

APPLICATION: TOTAL RECALL

CONCEPT CHECK

Learning Deficiencies and Disorders

Effects of Aging on Memory

Alzheimer's Disease

APPLICATION: GENETIC INTERVENTIONS FOR ALZHEIMER'S

Korsakoff's Syndrome

CONCEPT CHECK

In Perspective

Summary

Study Resources

In this chapter you will learn

- How and where memories are stored in the brain

- What changes occur in the brain during learning

- How aging and two major disorders impair learning

At the age of 7, Henry Molaison's life was forever changed by a seemingly minor incident: He was knocked down by a bicycle and was unconscious for 5 minutes. Three years later, he began to have minor seizures, and his first major seizure occurred on his 16th birthday. Still, Henry had a reasonably normal adolescence, taken up with high school, science club, hunting, and roller skating, except for a 2-year furlough from school because the other boys teased him about his seizures.

After high school, he took a job in a factory, but eventually the seizures made it impossible for him to work. He was averaging 10 small seizures a day and 1 major seizure per week. Because anticonvulsant medications were unable to control the seizures, Henry and his family decided on an experimental operation that held some promise. In 1953, when he was 27, a surgeon removed much of both of his temporal lobes, where the seizure activity was originating. The surgery worked, for the most part: With the help of medication, the petit mal seizures were mild enough not to be disturbing, and major seizures were reduced to about one a year. Henry returned to living with his parents. He helped with household chores, mowed the lawn, and spent his spare time doing difficult crossword puzzles. Later, he worked at a rehabilitation center, doing routine tasks like mounting cigarette lighters on cardboard displays.

Henry's intelligence was not impaired by the operation; his IQ test performance even went up, probably because he was freed from the interference of the abnormal brain activity. However, there was one important and unexpected effect of the surgery. Although he could recall personal and public events and remember songs from his earlier life, Henry had difficulty learning and retaining new information. He could hold new information in memory for a short while, but if he were distracted or if a few minutes passed, he could no longer recall the information. When he worked at the rehabilitation center, he could not describe the work he did. He did not remember moving into a nursing home in 1980, or even what he ate for his last meal. And although he watched television news every night, he could not remember the day's news events later or even recall the name of the president (Corkin, 1984; B. Milner, Corkin, & Teuber, 1968).

Henry's inability to form new memories was not absolute. Although he could not find his way back to the new home his family moved to after his surgery if he was more than two or three blocks away, he was able to draw a floor plan of the house, which he had navigated many times daily (Corkin, 2002). Over the years he became aware of his condition, and he was very insightful about it. In his own words,

> Every day is alone in itself, whatever enjoyment I've had, and whatever sorrow I've had. . . . Right now, I'm wondering. Have I done or said anything amiss? You see, at this moment everything looks clear to me, but what happened just before? That's what worries me. It's like waking from a dream; I just don't remember. (B. Milner, 1970, p. 37)

Over a period of 55 years, Henry would be the subject of a hundred scientific studies that he could not remember; he was known to the world as patient HM to protect his privacy. In the next several pages, you will see why many consider HM's surgery the most significant single event in the study of learning.

Learning as the Storage of Memories

Some one-celled animals "learn" surprisingly well, for example, to avoid swimming toward a light where they have received an electric shock before. I have placed the term *learn* in quotes because such simple organisms lack a nervous system; their behavior changes briefly, but if you take a lunch break during your subject's training, when you return, you will have to start all over again. Such a temporary form of learning may help an organism avoid an unsafe area long enough for the danger to pass or linger in a place where food is more abundant. But without the ability to make a more or less permanent record, you could not learn a skill, and experience would not help shape who you are. We will introduce the topic of learning by examining the problem of storage.

Amnesia: The Failure of Storage and Retrieval

HM's symptoms are referred to as *anterograde amnesia,* an impairment in forming new memories. (*Anterograde* means "moving forward.") This was not HM's only memory deficit; the surgery also caused *retrograde amnesia,* the inability to remember events prior to impairment. His retrograde amnesia extended from the time of surgery back to about the age of 16; he had a few memories from that period, but he did not remember the end of World War II or his own graduation, and when he returned for his 35th high school reunion, he recognized none of his classmates. Better memory for earlier events than for recent ones may seem implausible, but it is typical of patients who have brain damage similar to HM's. How far back the retrograde amnesia extends depends on how much damage there is and which specific structures are damaged.

HM's surgery damaged or destroyed the hippocampus, nearby structures that along with the hippocampus make up the *hippocampal formation,* and the amygdala. Figure 12.1 shows the location of these structures; because they are on or near the

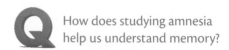

How does studying amnesia help us understand memory?

Retro
incident
antero
present.

FIGURE 12.1 Temporal Lobe Structures Involved in Amnesia.

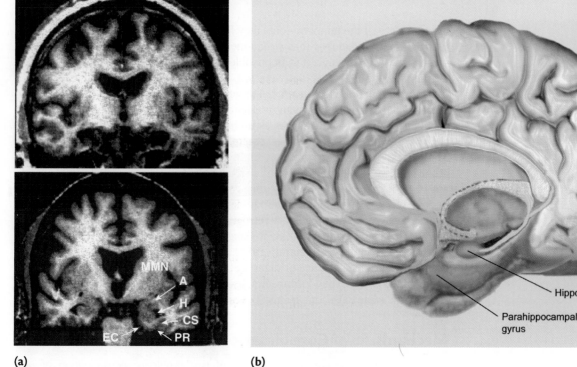

(a)   (b)

(a) HM's brain (left) and a normal brain. You can see that the amygdala (A), hippocampus (H), and other structures labeled in the normal brain are partly or completely missing in HM's brain. (b) Structures of the medial temporal lobe, which are important in learning. (The frontal lobe is to the right.)

SOURCES: (a) From "HM's Medial Temporal Lobe Lesion: Findings From Magnetic Resonance Imaging," by S. Corkin, D. G. amaral, R. G. González, K. A. Johnson, and B. T. Hyman, 1997, *Journal of Neurosicence, 17,* pp. 3964–3979. Copyright © 1997 by the Society for Neuroscience. Used with permission; (b) Adapted with permission from D. L. Schacter and A. D. Wagner, "Remembrance of Things Past," *Science, 285,* pp. 1503-1504. Illustration: K. Sutliff. © 1999 American Association for the Advancement of Science. Reprinted with permission from AAAS.

In the News

1
The Brain Observatory

The Legacy of HM

Not only did Henry Moliason devote much of his life to numerous scientific investigations, but his brain will continue to be the subject of study for many years to come. Soon after his death, Henry's preserved brain was in a plastic cooler strapped in a seat on a flight from Boston to San Diego; in the next seat was Jacopo Annese, director of the Brain Observatory at the University of California at San Diego. After several months of preparation, Annese and his colleagues dissected Henry's brain into slices as thin as the width of a hair. Each slice will be microscopically photographed with such resolution that the data from each slice would fill 200 DVDs. The data will then be combined into a three-dimensional reconstruction of the brain, which will be available online. Scientists will be able to navigate through it to the area of their interest and then zoom in to the level of individual neurons. Ironically, the man who could not remember will never be forgotten.

SOURCE: Lafee, S. (2009, November 30). H. M. recollected. *San Diego Union-Tribune.* Retrieved August 6, 2010, from http://www.signonsandiego.com/news/2009/nov/30/hm-recollected-famous-amnesic-launches-bold-new-br/

inside surface of the temporal lobe, they form part of what is known as the medial temporal lobe (remember that *medial* means "toward the middle"). Because HM's surgery was so extensive, it is impossible to tell which structures are responsible for the memory functions that were lost. Studies of patients with varying degrees of temporal lobe damage have helped determine which structures are involved in amnesia and, therefore, in memory. Henry died in 2008 at the age of 82, but he continues to make a contribution, as In the News explains.

The hippocampus consists of several substructures with different functions. The part known as *CA1* provides the primary output from the hippocampus to other brain areas; damage in that part of both hippocampi results in moderate anterograde amnesia and only minimal retrograde amnesia. If the damage includes the rest of the hippocampus, anterograde amnesia is severe. Damage of the entire hippocampal formation results in retrograde amnesia extending back 15 years or more (J. J. Reed & Squire, 1998; Rempel-Clower, Zola, Squire, & Amaral, 1996; Zola-Morgan, Squire, & Amaral, 1986). More extensive retrograde impairment occurs with broader damage or deterioration, like that seen in Alzheimer's disease, Huntington's disease, and Parkinson's disease, apparently because memory storage areas in the cortex are compromised (Squire & Alvarez, 1995).

Mechanisms of Consolidation and Retrieval

HM's memory impairment consisted of two problems: consolidation of new memories and, to a lesser extent, retrieval of older memories. *Consolidation* is the process in which the brain forms a more or less permanent physical representation of a memory. *Retrieval* is the process of accessing stored memories—in other words, the act of remembering. When a rat presses a lever to receive a food pellet or a child is bitten by a dog or you skim through the headings in this chapter, the experience is held in memory at least for a brief time. But just like the phone number that is forgotten when you get a busy signal the first time you dial, an experience does not

hippocampus
↓
CA1
↓
cortex.

> Most memories, like humans and wines, do not mature instantly. Instead they are gradually stabilized in a process referred to as consolidation.
>
> —Yadin Dudai

necessarily become a permanent memory; and if it does, the transition takes time. Until the memory is consolidated, it is particularly fragile. New memories may be disrupted just by engaging in another activity, and even older memories are vulnerable to intense experiences such as emotional trauma or electroconvulsive shock treatment (a means of inducing convulsions, usually in treating depression). Researchers divide memory into two stages, *short-term memory* and *long-term memory*. Long-term memory, at least for some kinds of learning, can be divided into two stages that have different durations and occur in different locations (see Figure 12.2), as we will see later (McGaugh, 2000).

Research techniques that allow researchers to "watch" consolidation as it takes place in the brain have identified the hippocampal formation as playing a leading role in consolidation, which would explain why patients with damage to the area have trouble learning new material. Scanning studies and an evoked potential study showed that presentation of words or pictures activates the formation, particularly the hippocampus and the *parahippocampal gyrus* (Figure 12.3; Alkire, Haier, Fallon, & Cahill, 1998; Brewer, Zhao, Desmond, Glover, & Gabrieli, 1998; Fernández et al., 1999). How well the words were remembered in later testing could be predicted from how much activation occurred in the parahippocampal area during presentation.

An animal study clearly demonstrates that the hippocampus participates in consolidation. Rats were trained in a water maze, a tank of murky water from which they could escape quickly by learning the location of a platform submerged just under the water's surface (Figure 12.4; Riedel et al., 1999). Then, for 7 days, the rats received continuous infusions into both hippocampi of a drug that blocks receptors for the neurotransmitter *glutamate*, which temporarily disabled the hippocampi. When the animals were tested 16 days after training, they performed poorly. Because the drug had adequate time to clear the rats' systems, it could not have been interfering with their performance at the time of testing.

Animals that were instead given the drug at the time of testing also had impaired recall in the water maze, indicating that the hippocampus has a role in retrieval as well as consolidation. Researchers have used PET scans to confirm that the hippocampus also retrieves memories in humans (D. L. Schacter, Alpert, Savage, Rauch, & Albert, 1996; Squire et al., 1992). Figure 12.5 shows increased activity in the hippocampi while the research participants recalled words learned during an experiment. The involvement of the hippocampus in retrieval seems inconsistent with HM's ability to recall earlier memories. But the memories that patients with hippocampal damage can recall are of events that occurred at least 2 years before their brain damage. Many researchers have concluded that the hippocampal mechanism plays a time-limited role in consolidation and retrieval, a point we will examine shortly. This diminishing role of the hippocampus would explain why older memories suffer less than recent memories after hippocampal damage.

The prefrontal area is also active during learning and retrieval, and some think that it directs the search strategy required for retrieval (Buckner & Koutstaal, 1998). Indeed, the prefrontal area is active during effortful attempts at retrieval, whereas the hippocampus is activated during successful retrieval (see Figure 12.5 again; D. L. Schacter et al., 1996). We will look at the role of the frontal area again when we consider working memory and Korsakoff's syndrome.

FIGURE 12.2 Stages of Consolidation.

Making a memory permanent involves multiple stages and different processes.

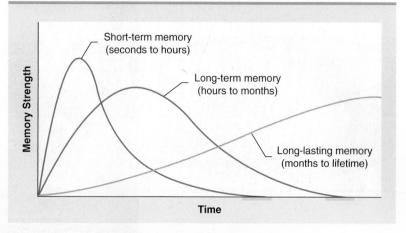

SOURCE: Reprinted with permission from J. L. McGaugh, "Memory—A Century of Consolidation," *Science, 287*, pp. 248–251. Copyright 2000 American Association for the Advancement of Science.

FIGURE 12.3 Hippocampal Activity Related to Consolidation.

The arrow is pointing to the hippocampal region. Reds and yellows indicate positive correlations of activity with recall; blues indicate negative correlations.

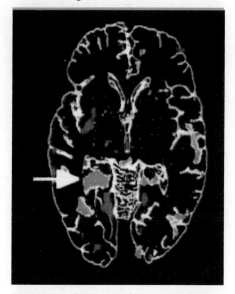

SOURCE: From "Hippocampal, But Not Amygdala Activity at Encoding Correlates With Long-Term Free Recall of Nonemotional Information," by M. T. Alkire et al., 1998, *PNAS, 95*, pp. 14506–14510, fig. 1, lower left image, p. 14507.

FIGURE 12.4 A Water Maze.

The rat learns to escape the murky water by finding the platform hidden just below the surface.

FIGURE 12.5 Hippocampal Activity in the Human Brain During Retrieval.

(a) Low Recall Minus Baseline

(b) High Recall Minus Baseline

L

+6 mm

−2 mm

$Z^3 3.00, p < .001$

$Z^3 3.00, p < .001$

(a) As participants tried to recall visually presented words that had been poorly learned (35% recall rate), the prefrontal and visual areas, but not the hippocampi, were highly activated compared with the baseline condition.

(b) However, the successful recall of well-learned words (79% recall rate) activated both hippocampal areas.

SOURCE: Reprinted with permission from D. L. Schacter et al., "Conscious Recollection and the Human Hippocampal Formation: Evidence from Positron Emission Tomography," *Proceedings of the National Academy of Sciences, USA*, 93, pp. 321–325. Copyright 1996 National Academy of Sciences, USA.

Where Memories Are Stored

The hippocampal area is not the permanent storage site for memories. If it were, patients like HM would not remember anything that happened before their damage occurred. According to most researchers, the hippocampus stores information temporarily in the hippocampal formation; then, over time, a more permanent memory is consolidated elsewhere in the brain. A study of mice that had learned a spatial discrimination task supported the hypothesis: Over 25 days of retention testing, metabolic activity progressively decreased in the hippocampus and increased in the cortical areas (Bontempi, Laurent-Demir, Destrade, & Jaffard, 1999).

Is there a place where memories are stored?

To explore further the relationship between these two areas, Remondes and Schuman (2004) severed the pathway that connects CA1 of the hippocampus with the cortex. The lesions did not impair the rats' performance in a water maze during training or 24 hours (hr) later, but after 4 weeks the rats had lost their memory for the task. The results supported the hypothesis that short-term memory depends on the hippocampus but long-term memory requires the cortex and an interaction over time between the two. To pin down the window of vulnerability of the memory, the researchers lesioned two additional groups of animals at different times *following* training. Those lesioned 24 hr after training were impaired in recall 4 weeks later, but those whose surgery was delayed until 3 weeks after training performed as well as the controls. This progression apparently occurs over a longer period of time in humans. Christine Smith and Larry Squire (2009) used fMRI to image the brain's activity while subjects recalled news events from the past

30 years. Activity was greatest in the hippocampus and related areas as subjects recalled recent events, with levels declining over a period of 12 years and stabilizing after that. At the same time, activity increased progressively with older memories in the prefrontal, temporal, and parietal cortex. So your brain works rather like your computer when it transfers volatile memory from RAM to the hard drive—it just takes a lot longer.

In Chapter 3, you learned that when Wilder Penfield (1955) stimulated association areas in the temporal lobes of surgery patients, he often evoked visual and auditory experiences that seemed like memories. We speculated that memories might be stored there, and more recent research has supported that idea, with memories for sounds activating auditory areas and memories for pictures evoking activity in the occipital region (see Figure 12.6; M. E. Wheeler, Petersen, & Buckner, 2000). You also saw in Chapter 9 that when we learn a new language, it is stored near Broca's area. Naming colors (which requires memory) activates temporal lobe areas near where we perceive color; identifying pictures of tools activates the hand motor area and an area in the left temporal lobe that is also activated by motion and by action words (A. Martin, Haxby, Lalonde, Wiggs, & Ungerleider, 1995; A. Martin et al., 1996); and spatial memories appear to be stored in the parietal area and verbal memories in the left frontal lobe (F. Rösler, Heil, & Henninghausen, 1995). Thus, all memories are not stored in a single area, nor is each memory distributed throughout the brain. Rather, different memories are located in different cortical areas, apparently according to where the information they are based on was processed.

An interesting example is the cells involved in place memory. *Place cells,* which increase their rate of firing when the individual is in a specific location in the environment, are found in the hippocampus. Each cell has a place field (overlapping somewhat with others), and together these cells form a map of the environment. This map develops during the first few minutes of exploration; the cells' fields are then remapped on entering a new environment, but they are restored on returning to the original location (Figure 12.7; Guzowski, Knierim, & Moser, 2004; Wilson & McNaughton, 1993). The fields are dependent on spatial cues in the environment; they will shift in the direction of a cue that is moved outside the rat's view, but they remain stable if the rat sees the cue being moved (Knierim, Kudrimoti, & McNaughton, 1995). Functional MRI has confirmed that humans have place cells, too; their activity is so precise that the investigators could determine the subject's "location" in a computer-generated virtual environment (Hassabis et al., 2009).

FIGURE 12.6 Functional MRI Scans of Brains During Perception and Recall.

Memories of pictures and sounds evoked responses in the same general areas (arrows) as the original stimuli.

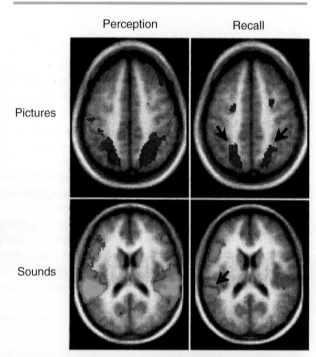

Perception   Recall

Pictures

Sounds

SOURCE: From "Memory's Echo: Vivid Remembering Reactivates Sensory-Specific Cortex," by M. E. Wheeler et al., *PNAS,* 97, 11125–11129, fig. 1c, d, e, f, p. 11127. © 2000 National Academy of Sciences, USA.

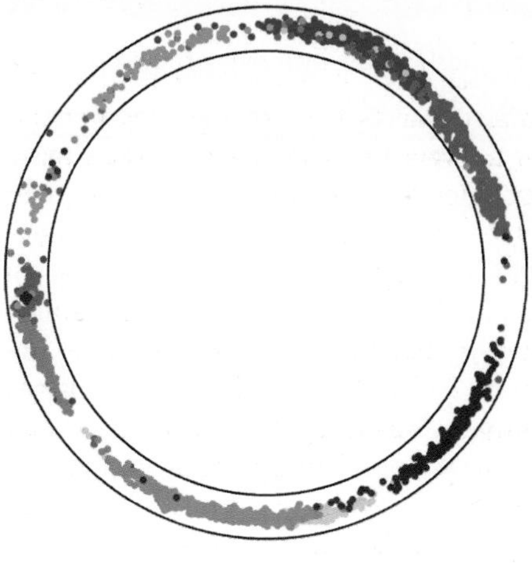

FIGURE 12.7 Recordings From Place Cells in a Rat in a Circular Runway.

The recordings are from seven different place cells, indicated by different colors. Note that each cell responds when the rat is in a particular part of the runway. (Due to cue similarities in a circular apparatus, cells occasionally respond on the opposite side of the circle.)

SOURCE: From "Neural Plasticity in the Ageing Brain," by S. N. Burke and C. A. Barnes, 2006, *Nature Reviews Neuroscience, 7,* 30–40.

Two Kinds of Learning

Learning researchers were in for a revelation when they discovered that HM could readily learn some kinds of tasks (Corkin, 1984). One was mirror drawing, in which the individual uses a pencil to trace a path around a pattern, relying solely on a view of the work surface in a mirror. HM improved in mirror-drawing ability over 3 days of training, and he learned to solve the Tower of Hanoi problem (Figure 12.8). But he could not remember learning either task, and on each day of practice he denied even having seen the Tower puzzle before (N. J. Cohen, Eichenbaum, Deacedo, & Corkin, 1985; Corkin, 1984). What this means, researchers realized, is that there are two categories of memory processing. *Declarative memory* involves learning that results in memories of facts, people, and events that a person can vebalize or declare. For example, you can remember being in class today, where you sat, who was there, and what was discussed. Declarative memory includes a variety of subtypes, such as *episodic memory* (events), *factual memory* (about facts, of course), *autobiographical memory* (information about oneself), and *spatial memory* (the location of the individual and of objects in space). *Nondeclarative memory* involves memories for behaviors; these memories result from procedural or skills learning, emotional learning, and stimulus-response conditioning. Learning mirror tracing or how to solve the Tower of Hanoi problem is an example of nondeclarative learning or, more specifically, procedural or skills learning; remembering having practiced the tasks involves declarative learning. Another way of putting it, which is admittedly a bit oversimplified, is that declarative memory is informational, while nondeclarative memory is more concerned with the control of behavior; just as we have *what* and *where* pathways in vision and audition, we have a *what* and a *how* in memory.

The main reason to distinguish between the two types of learning is that they have different origins in the brain; studying them can tell us something about how the brain carries out its tasks. For years it looked like we were limited to studying the distinction in the rare human who had brain damage in just the right place; hippocampal lesions did not seem to affect learning in rats, so researchers thought that rats did not have an equivalent of declarative memory. But it just took selecting the right tasks. R. J. McDonald and White (1993) used an apparatus called the radial arm maze, a central platform with several arms radiating from it (Figure 12.9). Rats with damage to both hippocampi could learn the simple conditioning task of going into any lighted arm for food. But if every arm was baited with food, the rats could not remember which arms they had visited and repeatedly returned to arms where the food had already been eaten.

Conversely, rats with damage to the *striatum* could remember which arms they had visited but could not learn to enter lighted arms. Because Parkinson's disease and Huntington's disease damage the basal ganglia (which include the striatum), people

What are the two kinds of learning?

FIGURE 12.8 The Tower of Hanoi Problem.

The task is to relocate the rings in order onto another post by moving them one at a time and without ever placing a larger ring over a smaller one.

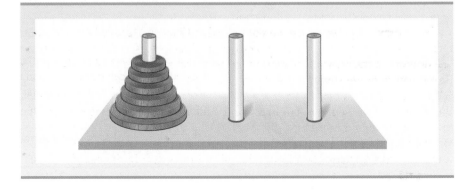

with these disorders have trouble learning procedural tasks, such as mirror tracing or the Tower of Hanoi problem (Gabrieli, 1998). Incidentally, the term *declarative* seems inappropriate with rats; researchers have often preferred the term *relational memory,* which implies that the individual must learn relationships among cues, an idea that applies equally well to humans and animals.

You already know that the amygdala is important in emotional behavior, but it also has a significant role in nondeclarative emotional learning. Bechara and his colleagues (1995) studied a patient with damage to both amygdalas and another with damage to both hippocampi. The researchers attempted to condition an emotional response in the patients by sounding a loud boat horn when a blue slide was presented but not when the slide was another color. The amygdala-damaged patient reacted emotionally to the loud noise, indicated by increased skin conductance responses (see Chapter 8). He could also tell the researchers which slide was followed by the loud noise, but the blue slide never evoked a skin conductance increase; in other words, conditioning was absent. The patient with hippocampal damage showed an emotional response and conditioning, but he could not tell the researchers which color the loud sound was paired with. This neural distinction between declarative learning and nondeclarative emotional learning may well explain how an emotional experience can have a long-lasting effect on a person's behavior even though the person does not remember the experience.

The amygdala has an additional function that cuts across learning types. Both positive and negative emotions enhance the memorability of any event; the amygdala strengthens even declarative memories about emotional events, apparently by increasing activity in the hippocampus. Electrical stimulation of the amygdala activates the hippocampus, and it enhances learning of a nonemotional task, such as a choice maze (McGaugh, Cahill, & Roozendaal, 1996). In humans, memory for both pleasant and aversive emotional material is related to the amount of activity in both amygdalas while viewing the material (Cahill et al., 1996; Hamann, Ely, Grafton, & Kilts, 1999).

Working Memory

The brain stores a tremendous amount of information, but information that is merely stored is useless. It must be available, not just when it is being recalled into awareness but when the brain needs it for carrying out a task. *Working memory* provides a temporary "register" for information while it is being used. Working memory holds a phone number you just looked up or that you recall from memory while you dial the number; it also holds information retrieved from long-term memory while it is integrated with other information for use in problem solving and decision making. Without working memory, we could not do long division, plan a chess move, or even carry on a conversation.

Think of working memory as similar to the RAM in your computer. The RAM holds information temporarily while it is being processed or used, but the information is stored elsewhere on the hard drive. But we should not take any analogy too far. Working memory has a very limited capacity (with no upgrades available), and information in working memory fades within seconds. So if you dial a new phone number and get a busy signal, you'll probably have to look the number up again. And if you have to remember the area code, too, you'd better write it down in the first place.

The *delayed match-to-sample task* described in Chapter 11 provides an excellent means of studying working memory. During the delay period, cells remain active in one or more of the association areas in the temporal and parietal lobes, depending on the nature of the stimulus (Constantinidis & Steinmetz, 1996; Fuster & Jervey, 1981; Miyashita & Chang, 1988). Cells in these areas apparently help maintain the memory of the stimulus, but they are not the location of working memory. If a distracting stimulus is introduced during the delay period, the altered firing in these locations abruptly ceases, but the animals are still able to make the correct choice (Constantinidis & Steinmetz, 1996; E. K. Miller, Erickson, & Desimone, 1996). Cells in the prefrontal cortex have several attributes that make them better candidates as

FIGURE 12.9 A Radial Arm Maze.

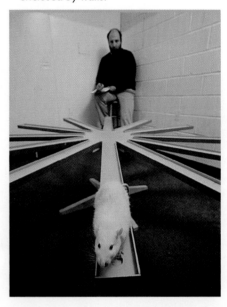

The rat learns where to find food in the maze's arms. The arms are often enclosed by walls.

SOURCE: © Hank Morgan/Photo Researchers.

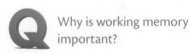 Why is working memory important?

The person recalls in almost photographic detail the total situation at the moment of shock, the expression of face, the words uttered, the position, garments, pattern of carpet, recalls them years after as though they were the experience of yesterday.

—G. M. Stratton, 1919

working memory specialists. Not only do they increase firing during a delay, but they also maintain the increase in spite of a distracting stimulus (E. K. Miller et al., 1996). Some respond selectively to the correct stimulus (di Pellegrino & Wise, 1993; E. K. Miller et al., 1996). Others respond to the correct stimulus, but only if it is presented in a particular position in the visual field; they apparently integrate information from cells that respond only to the stimulus with information from cells that respond to the location (Rao, Rainer, & Miller, 1997). Prefrontal damage impairs humans' ability to remember a stimulus during a delay (D'Esposito & Postle, 1999). All these findings suggest that the prefrontal area plays the major role in working memory.

Although the prefrontal cortex serves as a temporary memory register, its function is apparently more than that of a neural blackboard. In Chapters 3 and 8, you learned that damage to the frontal lobes impairs a person's ability to govern his or her behavior in several ways. Many researchers believe that the primary role of the prefrontal cortex in learning is as a central executive. That is, it manages certain kinds of behavioral strategies and decision making and coordinates activity in the brain areas involved in the perception and response functions of a task, all the while directing the neural traffic in working memory (Wickelgren, 1997).

prefrontal = working memory.

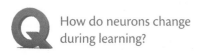

Concept Check — *Take a Minute to Check Your Knowledge and Understanding*

- ☐ What determines the symptoms and the severity of symptoms of amnesia?
- ☐ Describe the two kinds of learning and the related brain structures.
- ☐ Working memory contributes to learning and to other functions. How?

Brain Changes in Learning

Learning is a form of neural plasticity that changes behavior by remodeling neural connections. Specialized neural mechanisms have evolved to make the most of this capability. We will look at them in the context of long-term potentiation.

Long-Term Potentiation

Q How do neurons change during learning?

Over 50 years ago, Donald Hebb (1940) stated what has become known as the *Hebb rule:* If an axon of a presynaptic neuron is active while the postsynaptic neuron is firing, the synapse between them will be strengthened. We saw this principle in action during the development of the nervous system, when synaptic strengthening helped determine which neurons would survive; some of that plasticity is retained in the mature individual. Researchers have long believed that in order to understand learning as a physiological process, they would have to figure out what happens at the level of the neuron and, particularly, at the synapse. Since its discovery 4 decades ago (T. Bliss & Lømo, 1973), long-term potentiation has been the best candidate for explaining the neural changes that occur during learning.

Long-term potentiation (LTP) is an increase in synaptic strength resulting from the simultaneous activation of presynaptic neurons and postsynaptic neurons (Cooke & Bliss, 2006). LTP is usually induced in the laboratory by stimulating the presynaptic neurons with pulses of high-frequency electricity for a few seconds (W. R. Chen et al., 1996; Dudek & Bear, 1992); temporal summation of these high-frequency stimuli ensures that the postsynaptic neurons will fire along with the postsynaptic neurons. As you can see in Figure 12.10a, the postsynaptic neuron's response to a test stimulus is much stronger following induction of LTP. What is remarkable about LTP is that it can last for hours in tissue cultures and months in laboratory animals (Cooke & Bliss). LTP has been studied mostly in the hippocampus, but it also occurs in several other areas, including the visual, auditory, and motor cortex. So LTP appears to be a characteristic of much of neural tissue, at least in the areas most likely to be involved in learning.

Neural functioning requires weakening synapses as well as strengthening them. *Long-term depression (LTD)* is a decrease in the strength of synapses that occurs when stimulation of presynaptic neurons is insufficient to activate the postsynaptic neurons (S. H. Cooke & Bliss, 2006). In the laboratory, LTD is usually produced by a low frequency stimulus; you can see in Figure 12.10b that stimulation at 1 Hz for 15 minutes (min) blocked the potentiation that had been induced earlier; the result was a postsynaptic potential even smaller than the original. LTD is believed to be the mechanism the brain uses to modify memories and to clear old memories to make room for new information (Stickgold, Hobson, Fosse, & Fosse, 2001).

Activity in presynaptic neurons also influences the sensitivity of nearby synapses. If a weak synapse and a strong synapse on the same postsynaptic neuron are active simultaneously, the weak synapse will be potentiated; this effect is called *associative long-term potentiation* (Figure 12.11). Associative LTP is usually studied in isolated brain tissue with artificially created weak and strong synapses, but it has important behavioral implications, which is why it interests us. Electric shock evokes a strong response in the lateral amygdala, where fear is registered, while an auditory stimulus produces only a minimal response there. Rogan, Stäubli, and LeDoux (1997) repeatedly paired a tone with shock to the feet of rats. As a result of this procedure, the tone alone began to evoke a significantly increased response in the amygdala, as well as an emotional "freezing" response in the rats. You may recognize this scenario as an example of *classical conditioning;* we could easily change the labels in Figure 12.11 from "Strong synapse" to "Electric shock" and from "Weak synapse" to "Auditory tone." Researchers believe that associative LTP is the basis of classical conditioning, and Rogan et al.'s results support that view. LTP, LTD, and associative LTP can all be summed up in the expression "Cells that fire together wire together."

FIGURE 12.10 LTP and LTD in the Human Brain.

The graphs show excitatory postsynaptic potentials in response to a test stimulus before and after repeated stimulation. (a) 100-hertz (Hz) stimulation produced LTP. (b) 1-Hz stimulation produced LTD, which blocked the potentiation established earlier.

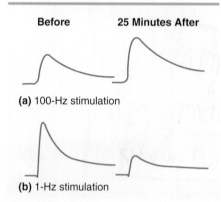

SOURCE: From "Long-Term Modifications of Synaptic Efficacy in the Human Inferior and Middle Temporal Cortex," by W. R. Chen et al., *Proceedings of the National Academy of Sciences*, USA, 93, pp. 8011–8015. Copyright 1996 National Academy of Sciences, USA. Used with permission.

FIGURE 12.11 Associative LTP.

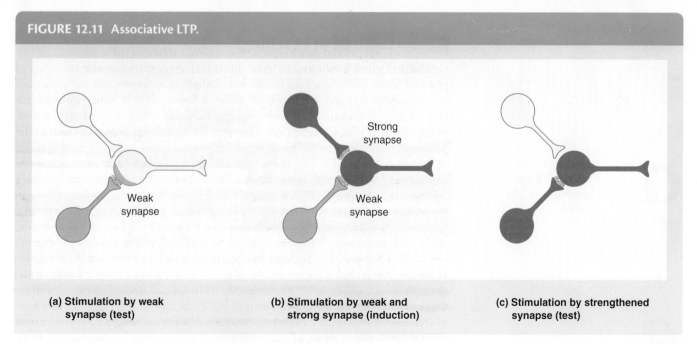

(a) Stimulation by weak synapse (test)

(b) Stimulation by weak and strong synapse (induction)

(c) Stimulation by strengthened synapse (test)

(a) Initially, the weak synapse produces only a very small excitatory postsynaptic potential. (b) Associative LTP is induced by simultaneous activation of a strong synapse along with activity in the weak synapse. (c) Later, the much larger excitatory postsynaptic potential indicates that the weak synapse has been potentiated.

How LTP Happens

The long trains of stimulation experimenters use to induce LTP and LTD seem very artificial, and they are; in the brain, these changes are more likely triggered by theta EEG. Theta rhythm is EEG activity with a frequency range of 4 to 7 Hz. This rhythm is interesting because it typically occurs in the hippocampus when an animal is experiencing a novel situation and any learning situation is somewhat novel (otherwise there would be nothing to learn). The researchers used a low-tech but effective method for producing theta in their experiment: They pinched the rats' tails. When hippocampal stimulation was timed to coincide with the peaks of theta waves, LTP could be produced by just five pulses of stimulation (Hölscher, Anwyl, & Rowan, 1997). Stimulation that coincided with the trough of theta waves reversed LTP that had been induced 30 min before. Suppressing theta EEG in the hippocampus with a sedative drug eliminated rats' ability to remember which way they turned on the previous trial in a two-choice maze (Givens & Olton, 1990). Hölscher and his colleagues believed that the theta rhythm, by responding to novel situations, might emphasize important stimuli for the brain and facilitate LTP and LTD.

LTP induction involves a cascade of events at the synapse. In most locations, the neurotransmitter involved in LTP is glutamate. Glutamate is detected by two types of receptors: the AMPA (alpha-amino-3-hydroxy-5-methyl-4-isoxazole propionic acid) receptor and the NMDA (N-methyl-d-aspartic acid) receptor. Initially, glutamate activates AMPA receptors but not NMDA receptors, because they are blocked by magnesium ions (Figure 12.12). During LTP induction, activation of the AMPA receptors by the first few pulses of stimulation partially depolarizes the membrane, which dislodges the magnesium ions. The critical NMDA receptor can then be activated, resulting in an influx of sodium and calcium ions; not only does this further depolarize the neuron, but the calcium activates CaMKII (calcium/calmodulin-dependent protein kinase Type II), an enzyme that is necessary for LTP (Lisman, Schulman, & Cline, 2002). CaMKII apparently functions as a two-way switch that changes the strength of a synapse (O'Connor, Wittenberg, & Wang, 2005).

Neural Growth in Learning

LTP induction is followed by gene activation, gene silencing, and synthesis of proteins, all of which result in functional changes in synapses and the growth of new connections (Kandel, 2001; C. A. Miller & Sweatt, 2007). When the postsynaptic neuron is activated, it releases nitric oxide gas, which is a retrograde messenger, back into the synaptic cleft. The nitric oxide diffuses across the cleft to the presynaptic neuron, where it induces the neuron to release more neurotransmitter (Schuman & Madison, 1991). The nitric oxide lasts only briefly, but the increase in neurotransmitter release is long term (O'Dell, Hawkins, Kandel, & Arancio, 1991). Within 30 min after LTP, postsynaptic neurons develop increased numbers of *dendritic spines,* outgrowths from the dendrites that partially bridge the synaptic cleft and make the synapse more sensitive (see Figure 12.13; Engert & Bonhoeffer, 1999; Maletic-Savatic, Malinow, & Svoboda, 1999). Existing spines also enlarge or split down the middle to form two spines (Matsuzaki, Honkura, Ellis-Davies, & Kasai, 2004; Toni, Buchs, Nikonenko, Bron, & Muller, 1999). Postsynaptic strength is increased further as additional AMPA receptors are transported from the dendrites into the spines (Lisman et al., 2002; Shi et al., 1999). In addition, an increase in dopamine unmasks previously silent synapses and, 12 to 18 hours later, initiates the growth of new synapses (C. H. Bailey, Kandel, & Si, 2004). One further very important change that occurs as the result of learning is the generation of new neurons in the hippocampus; though the rate of neurogenesis is relatively low, over the life span, they add up to an estimated 10% to 20% of

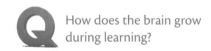

Q How does the brain grow during learning?

AMPA.
+
NDMA. (mg)
———
Glutamate
———
NO.
———
↑ NT release.

FIGURE 12.12 Participation of Glutamate Receptors in LTP.

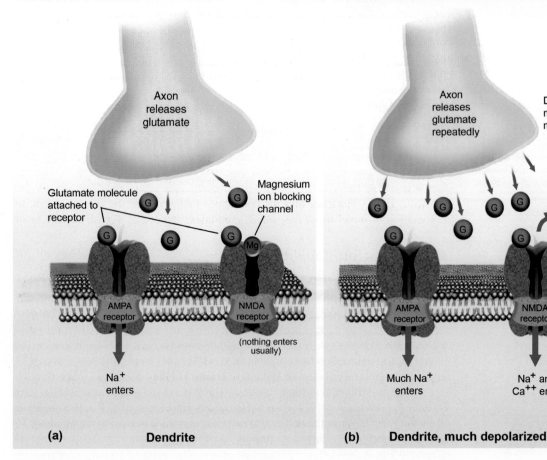

Axon releases glutamate

Glutamate molecule attached to receptor

Magnesium ion blocking channel

AMPA receptor

NMDA receptor

(nothing enters usually)

Na^+ enters

(a) **Dendrite**

Axon releases glutamate repeatedly

Displaced magnesium molecule

AMPA receptor

NMDA receptor

Much Na^+ enters

Na^+ and Ca^{++} enter

(b) **Dendrite, much depolarized**

(a) Initially, glutamate activates the AMPA receptors but not the NMDA receptors, which are blocked by magnesium ions. (b) However, if the activation is strong enough to depolarize partially the postsynaptic membrane, the magnesium ions are ejected. The NMDA receptor can then be activated, allowing sodium and calcium ions to enter.

the population (Jacobs, van Praag, & Gage, 2000). These young neurons integrate into already established neural networks, where they are more likely to participate in new learning than the older neurons (Kee, Teixeira, Wang, & Frankland, 2007).

With all that growth, you might suspect that there would be some increase in the volume of the brain areas that are involved in LTP. In fact, there is evidence that this does happen. London taxi drivers, who are noted for their ability to navigate the city's complex streets entirely from memory, spend about 2 years learning the routes before they can be licensed to operate a cab. Maguire and her colleagues (2000) used MRI to scan the brains of 16 drivers. The posterior part of their hippocampi, known to be involved in spatial navigation, was larger than in males of similar age. (Overall hippocampal volume did not change; their anterior hippocampi were smaller.) The difference was greater for cabbies who had been driving for the longest time, which we would expect if the difference were caused by experience.

FIGURE 12.13 Increase in Dendritic Spines Following LTP.

(a) A single synaptic spine on a dendrite (white) and a presynaptic terminal (red). (b) The same spine split into two following LTP.

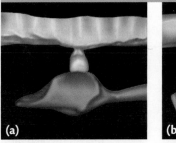

(a)

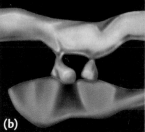

(b)

SOURCE: Based on "LTP Promotes Formation of Multiple Spine Synapses Between a Single Axon Terminal and a Dendrite," by N. Toni et al., *Nature, 402* (6706) pp. 421–425. © 1999. Reprinted by permission of Nature Publishing Group.

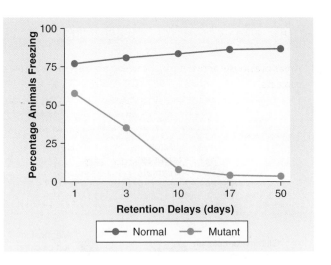

FIGURE 12.14 Retention in Normal and CaMKII Deficient Mice Over Time.

Mice were given three foot shocks in a conditioning chamber. Subgroups of mice were later tested for memory of the foot shocks (by observing emotional "freezing") at one of the retention delay times. Note that in the mice heterozygous for the mutant gene, memory had begun to decay after 3 days and they failed to form permanent memory.

SOURCE: From "αCaMKII-Dependent Plasticity in the Cortex is Required for Permanent Memory," A. Frankland et al., *Nature*, 411, 309–313, top left of fig. 1, p. 310. © 2001 Alcino Silva. Used with permission.

Q How do the roles of the hippocampus and the cortex differ?

Consolidation Revisited

For declarative memories, long-term memory consists of a stage that takes place in the medial temporal lobe, followed by a transition to a more permanent form in the cortex (refer to Figure 12.2 again for the time course of these events). A study of mice with a defective gene for CaMKII revealed some of the details of this transfer (Frankland, O'Brien, Ohno, Kirkwood, & Silva, 2001). Mice that are homozygous for the mutation produce none of the enzyme, and they show no LTP in the hippocampus and fail to learn a hippocampal-dependent task. Mice that are heterozygous—with just one copy of the defective gene—produce more CaMKII than homozygous mutants but less than normal mice. Hippocampal LTP is unaffected in these mice, but because the cortex normally has minimal CAMKII to begin with, LTP no longer occurs in the cortex. As a result, learning of a hippocampal-dependent task is normal 1 to 3 days following training but severely impaired 10 days after training and beyond (Figure 12.14). Remember that the mechanisms we are considering are concerned with declarative memory; so far, there is no clear evidence that a prolonged consolidation process occurs in nondeclarative learning (Dudai, 2004).

Although CaMKII is vital to the establishment of LTP, there is evidence that its maintenance during long-term memory depends on another enzyme known as *protein kinase M zeta*. Inhibiting CaMKII blocks the development of LTP but does not reverse LTP once it is established; on the other hand, chemical inhibition of protein kinase M zeta causes amnesia for an established conditioned response (Pastalkova et al., 2006). In fact, injection of the inhibitor into the insula, the cortical area involved in taste and in learning taste associations, eliminated a conditioned taste aversion in rats; the treatment was effective even when administered 25 days after training (Shema, Sacktor, & Dudai, 2007).

The hippocampus has the ability to acquire learning "on the fly" while the event is in progress, but a longer time is needed for long-term storage of declarative memories in the cortex. Many researchers now believe that the hippocampus transfers information to the cortex during times when the hippocampus is less occupied, even during sleep (Lisman & Morris, 2001; McClelland, McNaughton, & O'Reilly, 1995). During sleep, neurons in the rats' hippocampus and cortical areas repeat the pattern of firing that occurred during learning (Louie & Wilson, 2001; Y. Qin, McNaughton, Skaggs, & Barnes, 1997). Human EEG and PET studies showed the hippocampus repeatedly activating the cortical areas that participated in the daytime learning, and this reactivation was accompanied by significant task improvement the next morning without further practice (Figure 12.15; Maquet et al., 2000; Wierzynski, Lubenov, Gu, & Siapas, 2009). Presumably, "off-line" replay provides the cortex the opportunity to undergo LTP at the more leisurely pace it requires (Lisman & Morris,

FIGURE 12.15 PET Scans of Brain Activity During Sleep Following Learning.

Areas previously active during learning are also more active during sleep in the trained subjects, but not in the untrained subjects.

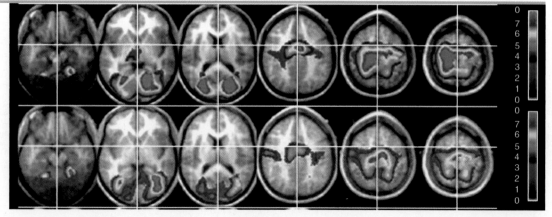

SOURCE: From "Experience-Dependent Changes in Cerebral Activation During Human REM Sleep," by P. Maquet et al., 2000, *Nature Neuroscience*, 3, 551–555, fig. 2, p. 833.

2001). During sleep more than 100 genes increase their activity; many of those have been identified as major players in protein synthesis, synaptic modification, and memory consolidation (Cirelli, Gutierrez, & Tononi, 2004).

Changing Our Memories

As hard as the brain works to make memories "permanent," it is still important that these records not be inscribed in stone. Things change; the waterhole we learned was reliable over several visits is now becoming progressively more stagnant, so we must range in other directions until we find a new source of water. And sometimes erroneous learning must be corrected; the first two redheads we knew were hot tempered, and it will take meeting additional redheads to change what we have learned. A memory needs to be stable to be useful, but at the same time it must remain malleable; there are several ways the brain accomplishes this.

Extinction

The first is *extinction*. The experimenter sounds a tone just before delivering a puff of air to your eye; after just a few trials, you blink just because you hear the tone. This doesn't happen simply because you understand that the air blast is coming; it occurs more quickly than you can make a voluntary response. Then the experimenter sounds the tone several times without administering the puff of air. Slowly the tone loses its power to make you blink. The memory is not gone; if the experimenter repeats the puff of air, you will be back to blinking every time you hear the tone. Nor is this an example of forgetting. Rather, extinction involves new learning; one indication is that, like LTP, extinction requires activation of NMDA receptors, and blocking these receptors eliminates extinction (Santini, Muller, & Quirk, 2001).

Forgetting

Most memories dissipate at least somewhat over time if they are not frequently used. We invariably regard memory loss from *forgetting* as a defect, but researchers are finding clues that the brain actively removes useless information to prevent the

Application   Total Recall

Most of us would like to remember more and forget less. But a few years ago Jill Price wrote to neuropsychologist James McGaugh at the University of California, Irvine, asking for help because she *couldn't* forget; she can remember what she did and what was happening in the world for practically every day of her life, and she is often tormented by bad memories (J. Marshall, 2008; E. S. Parker, Cahill, & McGaugh, 2006.) Two years later two men with similar memory capabilities came forward, but unlike Price, Brad Williams and Rick Baron can keep their memories at bay (Elias, 2008; D. S. Martin, 2008). Williams uses his memory in his work as a radio news reporter; Baron is unemployed but supports himself in part by winning contests that utilize his memory for facts. The researchers are eager to understand what fuels the trio's unusual abilities, because the knowledge could help the memory impaired. The interesting thing is that the three do no better than other people on memorization tests; they just don't suppress their memories once they're formed. This might relate to their somewhat enlarged prefrontal areas, but the researchers aren't ready to interpret the brain scans yet. An indication that inadequate inhibition might be involved is that all three show signs of compulsive behavior. They are devoted collectors—years of TV guides, rare record albums, hundreds of TV show tapes—and Baron arranges all the bills in his wallet according to the city of the federal reserve bank where they were issued and how the sports teams in that city did.

saturation of synapses with information that is not called up regularly or has not made connections with other stored memories. One way the brain cleans house apparently involves the enzyme protein phosphatase 1 (PP1), a product of the *PP1 gene*. To study PP1's effect, researchers created transgenic mice (see Chapter 4) with genes for a particularly active form of PP1 inhibitor (Genoux et al., 2002). The genes were inducible, which means that the researchers could activate them at any time. Mice were trained in a water maze, and then the genes were turned on in the transgenic animals; 6 weeks later, the control subjects' memory for the task was completely absent, while the transgenic mice had forgotten very little. You may remember from your introductory psychology course that for most tasks, spreading out practice sessions (*distributed practice*) leads to better learning than *massed practice*. When the inhibitor genes were turned on during training, this advantage disappeared, which suggests that the reason distributed practice is superior is that PP1's effects accumulate over massed practice trials. Another gene involved in forgetting is *Drac1(V12)*. Its protein product, Rac, causes memory to decay after learning. Interestingly, continued training suppresses Rac, which means that additional practice has a dual benefit (Shuai et al., 2010).

Efficient memory involves a balance between remembering and forgetting. Later in this chapter we will see how devastating memory impairment can be; the Application shows that there is another side to the coin as well.

Reconsolidation

Consolidation is a progressive affair extending over a relatively long period of time. During that time, the memory is vulnerable to disruption from a number of sources, including electroconvulsive shock and drugs that interfere with protein synthesis. In recent years researchers have come to realize that each time a memory is retrieved, it must be *reconsolidated,* and during that time the memory becomes even more vulnerable (Dudai, 2004). For example, Nader, Schafe, and LeDoux (2000) conditioned a fear response (freezing) to a tone in mice by pairing the tone with electric shock to the feet. Anisomycin will eliminate the fear memory if it is injected shortly after learning, but injection 24 hr after training has no effect. However, as much as 2 weeks later, anisomycin eliminated the fear learning if the researchers induced retrieval of the memory by presenting the tone again (without the shock). You might very well wonder why the brain would give up protection of a consolidated memory during retrieval. Apparently, reopening a memory provides the opportunity to refine it, correct errors, and modify your emotional response to redheaded acquaintances. (Lee, 2009). Reconsolidation may even have therapeutic usefulness. It can eliminate a conditioned fear response in humans, and (as you will see in Chapter 14) could provide an effective tool for erasing fear memories in people with posttraumatic stress disorder (D. Schiller et al., 2009).

Of course, there is no way to guarantee that the result will always be adaptive; the opportunity to correct errors also allows the introduction of new errors. We have long known that memories get "reconstructed" over time, usually by blending with other memories. Reconstruction can be a progressive affair. Evidence suggests that one reason for the "recovery" of *false* childhood memories during therapy may be therapists' repeated attempts to stimulate recall at successive sessions. Laboratory research has shown that people's agreement with memories planted by the experimenter can increase over multiple interviews (E. F. Loftus, 1997). In one study, researchers using doctored photographs found that after being questioned three times, 50% of subjects were describing a childhood ride in a hot air balloon that never happened (Wade, Garry, Read, & Lindsay, 2002). More recently, Loftus and her colleagues (D. M. Bernstein, Laney, Morris, & Loftus, 2005) were able to shift their subjects' food preferences by giving them a bogus computer analysis of their responses to a food questionnaire. For example, in a follow-up questionnaire, about 20% of the subjects agreed with the analysis that they had, in fact, been made sick by eating strawberry ice cream as children, and reported that they would avoid it in the future.

2
Learning and Memory Resources

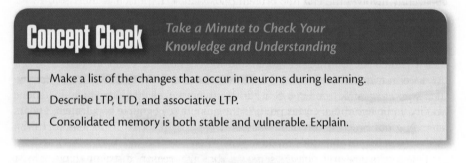

Concept Check
Take a Minute to Check Your Knowledge and Understanding

- [ ] Make a list of the changes that occur in neurons during learning.
- [ ] Describe LTP, LTD, and associative LTP.
- [ ] Consolidated memory is both stable and vulnerable. Explain.

Learning Deficiencies and Disorders

Learning may be the most complex of human functions. Not surprisingly, it is also one of the most frequently impaired. Learning can be compromised by accidents and violence that damage the structures we have been studying. But more subtle threats to learning ability come from aging and from disorders of the brain, including Alzheimer's disease and Korsakoff's syndrome, which we will discuss in this section.

Effects of Aging on Memory

Old Man: Ah, memory. It's the second thing to go.

Young Man: So what's first?

Old Man: I forget. . . .

You may or may not find humor in this old joke, but declining memory is hardly a laughing matter to the elderly. The older person might mislay car keys, forget appointments, or leave a pot on the stove for hours. Working memory and the ability to retrieve old memories and to make new memories may all be affected (Fahle & Daum, 1997; Small, Stern, Tang, & Mayeux, 1999). Memory loss is not just inconvenient and embarrassing; it is potentially dangerous, and it is disturbing because it suggests the possibility of brain degeneration.

Q Does the brain age, too?

Until fairly recently, researchers believed that declining memory and cognitive abilities were an inevitable consequence of aging. Although various kinds of cognitive deficits are typical of old age, they are not inevitable. For example, college professors in their 60s perform as well as professors in their 30s on many tests of learning and memory (Shimamura, Berry, Mangels, Rusting, & Jurica, 1995). An active lifestyle in old age has been associated with this "successful aging" (Schaie, 1994), but this fact does not necessarily tell us that staying active will stave off decline. Continued mental alertness may be the reason the person remains active, or health may be responsible for both good memory and a high activity level. However, we do know that rats reared in an enriched environment develop increased dendrites and synapses on cortical neurons (Sirevaag & Greenough, 1987). Also, we will see in the next chapter that cognitive skill training produces significant and enduring improvement in the elderly, which suggests that experience can affect the person's cognitive well-being.

For many years, researchers believed that deficits in the elderly were caused by a substantial loss of neurons, especially from the cortex and the hippocampus. However, the studies that led to this conclusion were based on flawed methods of estimating cell numbers. More recent investigations have found that the number of hippocampal neurons was not diminished in aged rats, even those with memory deficits, and that neuron loss from cortical areas was relatively minor (see M. S. Albert et al., 1999, for a review). And, as we saw in Chapter 3, the number of synapses continues to increase with age in humans (Buell & Coleman, 1979).

On the other hand, certain circuits in the hippocampus do lose synapses and NMDA receptors with aging (Gazzaley, Siegel, Kordower, Mufson, & Morrison, 1996; Geinisman, de Toledo-Morrell, Morrell, Persina, & Rossi, 1992). Probably as a result of these changes, LTP is impaired in aged rats; it develops more slowly and diminishes more rapidly (Barnes & McNaughton, 1985). The rats' memory capabilities parallel their LTP deficits: Learning is slower, and forgetting is more rapid. There is also a decrease in metabolic activity in the *entorhinal cortex,* the major input and output to the hippocampus (M. J. de Leon et al., 2001). In normal elderly individuals, metabolic activity in the entorhinal cortex predicts the amount of cognitive impairment 3 years later. Another likely cause of learning deficits is myelin loss (A. R. Jensen, 1998). Without myelin, neurons conduct more slowly and interfere with each other's activity.

One subcortical area does undergo substantial neuron loss during aging, at least in monkeys. It is the *basal forebrain region* (D. E. Smith, Roberts, Gage, & Tuszynski, 1999), whose acetylcholine-secreting neurons communicate with the hippocampus, amygdala, and cortex. Basal forebrain cell loss is much greater in Alzheimer's disease, but the less pronounced degeneration that occurs in normal aging probably contributes to memory deficits as well.

Some of the deficits in the elderly resemble those of patients with frontal lobe damage (Moscovitch & Winocur, 1995). In one study, elderly individuals participated in the "gambling task" described in Chapter 8, choosing playing cards from two "safe" decks and two "risky" decks. Like patients with prefrontal brain damage, 35% of these elderly volunteers never learned to avoid the risky decks, and another 28% were slow in doing so (Denburg, Tranel, & Bechara, 2005).

Can we improve memory in the aged? Earlier, we saw the suggestion that forgetting useless memories is adaptive; however, when useful memories are eliminated as well, forgetting becomes a deficiency. In the study described earlier, Genoux and his colleagues (2002) found that aged mice were significantly impaired on the learning task after just 1 day without practice, but performance in old mice with the enhanced PP1 inhibitor genes was still robust 4 weeks later.

Alzheimer's Disease

Substantial loss of memory and other cognitive abilities in the elderly is referred to as dementia. The most common cause of dementia is *Alzheimer's disease,* a disorder characterized by progressive brain deterioration and impaired memory and other mental abilities. Alzheimer's disease was first described by the neuro-anatomist and neurologist Alois Alzheimer in the 19th century. The earliest and most severe symptom is usually impaired declarative memory. Initially, the person is indistinguishable from a normally aging individual, though the symptoms may start earlier; the person has trouble remembering events from the day before, forgets names, and has trouble finding the right word in a conversation. Later, the person repeats questions and tells the same story again during a conversation. As time and the disease progress, the person eventually fails to recognize acquaintances and even family members. Alzheimer's disease is not just a learning disorder but a disorder of the brain, so ultimately most behaviors suffer. Language, visual-spatial functioning, and reasoning are particularly affected, and there are often behavioral problems such as aggressiveness and wandering away from home.

Alzheimer's disease is primarily a disorder of aging, although it can strike fairly early in life. It affects 10% of people over 65 years of age and nearly half of those over 85 (Evans et al., 1989). Alzheimer's researcher Zaven Khachaturian (1997) eloquently described his mother's decline: "The disease quietly loots the brain, nerve cell by nerve cell, like a burglar returning to the same house each night" (p. 21).

The Diseased Brain: Plaques and Tangles

There are two notable characteristics of the Alzheimer's brain, though they are not unique to the disease. *Plaques* are clumps of amyloid, a type of protein, that cluster among axon terminals and interfere with neural transmission (Figure 12.16a). The normal brain produces a protein made up of 40 amino acids, Aβ40; the Alzheimer's brain produces mostly Aβ42, which is 42 amino acids long and clumps easily to form the plaques. In addition, abnormal accumulations of the protein tau form *neurofibrillary tangles* inside neurons; tangles are associated with the death of brain cells (Figure 12.16b).

Figure 12.17 shows the brain of a deceased Alzheimer's patient and a normal brain. Notice the decreased size of the gyri and the increased width of the sulci

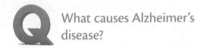

3
Alzheimer's Resources

Q What causes Alzheimer's disease?

FIGURE 12.16 Neural Abnormalities in the Brain of an Alzheimer's Patient.

(a) The round clumps in the photo are plaques, which interfere with neural transmission. (b) The dark, twisted features are neurofibrillary tangles, which are associated with death of neurons.

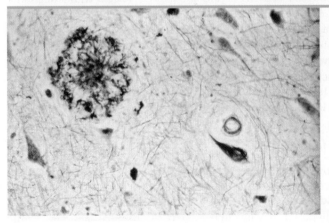

(a)

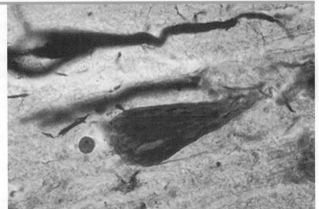

(b)

FIGURE 12.17 Alzheimer's Brain (left) and a Normal Brain.

The illustrations show the most obvious differences, the reduced size of gyri and increased size of sulci produced by cell loss in the diseased brain.

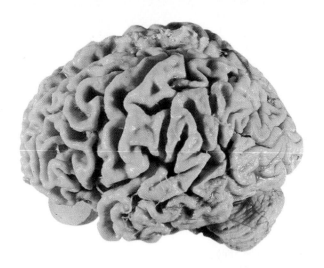 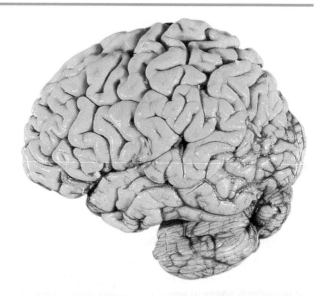

SOURCE: Photos courtesy of Dr. Robert D. Terry.

in the diseased brain. Internally, enlarged ventricles tell a similar story of severe neuron loss. Many of the lesions are located in the temporal lobes; because of their location, they effectively isolate the hippocampus from its inputs and outputs, which partly explains the early memory loss (B. T. Hyman, Van Horsen, Damasio, & Barnes, 1984). However, plaques and tangles also attack the frontal lobes, accounting for additional memory problems as well as attention and motor difficulties. The occipital lobes and parietal lobes may be involved as well; disrupted communication between the primary visual area and the visual association areas in the parietal and temporal lobes explains the visual deficits that plague some Alzheimer's sufferers.

While amyloid plaques have been considered the hallmark of Alzheimer's disease, the number of amyloid deposits is only moderately related to the degree of cognitive impairment (Selkoe, 1997), and about 25% of the elderly have plaques but suffer no dementia (Mintun et al., 2006). Attention is turning to a soluble form of amyloid known as ADDL. ADDL reaches 70-fold higher levels in Alzheimer's brains compared to the brains of controls (Gong et al., 2003); it has been linked to memory failure in mice, along with loss of synapses and failure of LTP in the hippocampus. Time will tell just how important ADDL is and what treatment options understanding it may lead to.

Alzheimer's and Heredity

Heredity is an important factor in Alzheimer's disease. The first clue to a gene location came from a comparison of Alzheimer's with Down syndrome (Lott, 1982). Down syndrome individuals also have plaques and tangles, and they invariably develop Alzheimer's disease if they live to the age of 50. Because Down syndrome is caused by an extra chromosome 21, researchers zeroed in on that chromosome; there they found mutations in the *amyloid precursor protein* (*APP*) gene (Goate et al., 1991). When aged mice were genetically engineered with an *APP* mutation

TABLE 12.1 Known Genes for Alzheimer's Disease

Gene	Chromosome	Age of Onset	Percentage of Cases
APP	21	45–66	<0.1
Presenilin 1	14	28–62	1–2
Presenilin 2	1	40–85	<0.1
ApoE4	19	>60	>50

SOURCES: Marx (1998); Selkoe (1997).

that increased plaques, both LTP and spatial learning were impaired (Chapman et al., 1999). Three additional genes that influence Alzheimer's have been confirmed, and all the genes discovered so far affect amyloid production or its deposit in the brain (Selkoe, 1997). As you can see in Table 12.1, the genes fall into two classes, those associated with early-onset Alzheimer's disease (often before the age of 60) and one found in patients with late-onset Alzheimer's.

The four genes in the table account for just a little over half the cases of Alzheimer's disease, and environmental causes seem to have little effect, so additional genes must be involved. New gene technology has made it possible for researchers to do whole-genome scans of individuals, which allows gene searches without the need for a preconceived target area. These studies have often been small, but two large studies in Germany and Australia have identified three possible genes, *CLU, CR1,* and *PICALM,* in late-onset Alzheimer's; the first two are involved in the removal of beta amyloid (Harold et al., 2009; Lambert et al., 2009). After pooling the results of 789 studies, Harvard researchers came up with a list of 13 possible genes (Bertram, McQueen, Mullin, Blacker, & Tanzi, 2007). A study of the brains of deceased Alzheimer's patients found 31 genes that were upregulated and another 87 that were downregulated in the amygdala and cingulate cortex alone (J. F. Loring, Wen, Lee, Seilhamer, & Somogyi, 2001). So not only might the number of genes be large, but the genetic causes could also be more complicated than just which genes the person has.

Treatment of Alzheimer's Disease

The annual cost of Alzheimer's disease and other dementias is estimated at $422 billion worldwide, an increase of 34% since 2005 (Wimo, Winblad, & Jönsson, 2010). Treatments that delayed nursing home placement by only 1 month would save $1 billion a year in health care costs, and a delay of 5 years would save $50 billion each year. The situation is likely to worsen significantly in the future. Between the years 2000 and 2050, the U.S. population is expected to increase by almost 50%, while the number over the age of 85 increases sixfold (Bureau of the Census, 2001). This disproportionate increase is due to better nutrition and health care and the aging of the baby boomers. As a result, the number of people over age 65 with Alzheimer's is expected to triple during that time (see Figure 12.18; L. E. Hebert, Scherr, Bienias, Bennett, & Evans, 2003).

Only four drugs are currently in use in the United States for treating Alzheimer's (Patoine & Bilanow, n.d.). Three of these are cholinesterase inhibitors; they restore acetylcholine transmission by interfering with the enzyme that breaks down

FIGURE 12.18 Projected Increases in Alzheimer's Disease in People in the United States Over Age 65.

Note that numbers begin to escalate rapidly after 2020.

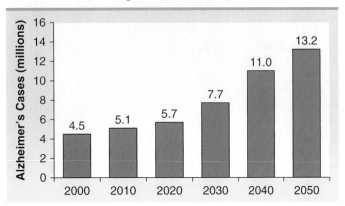

SOURCE: Based on data from L. E. Hebert, et al. (2003).

acetylcholine at the synapse. Acetylcholine-releasing neurons are significant victims of degeneration in Alzheimer's disease, and experiments show that blocking acetylcholine activity eliminates hippocampal theta and impairs learning in rats (J. A. Deutsch, 1983) and also impairs learning in humans (Newhouse, Potter, Corwin, & Lenox, 1992). The fifth drug, memantine (marketed in the United States as Namenda), is the first approved for use in patients with moderate and severe symptoms. Some of the neuron loss in Alzheimer's occurs when dying neurons trigger the release of the excitatory transmitter glutamate; the excess glutamate produces excitotoxicity, overstimulating NMDA receptors and killing neurons. Memantine limits the neuron's sensitivity to glutamate, reducing further cell death. Studies indicate moderate slowing of deterioration and improvement in symptoms ("FDA Approves Memantine," 2003; Reisberg et al., 2003). Unfortunately, these drugs provide only modest relief for the memory and behavioral symptoms of Alzheimer's, and they are little or no help when degeneration is advanced.

In their quest for more effective treatments, researchers have turned to other approaches, including immunizing the brain against Alzheimer's by injecting amyloid to produce an immune reaction. In mice engineered to develop plaques, the immune reaction cleared the plaques and protected against learning impairment (Janus et al., 2000; Morgan et al., 2000). The treatment also produced varying levels of plaque removal in people with mild to moderate Alzheimer's, but it failed to slow their progression into severe dementia (Holmes et al., 2008). This result and the low correlation between plaques and symptom severity makes sense if one accepts the hypothesis that it is the accumulation of amyloid rather than the plaques that damages neurons (Bossy-Wetzel, Schwarzenbacher, & Lipton, 2004). On the other hand, inducing the immune response with injections of immunoglobulin derived from human plasma both removes plaques and produces mental improvement (Relkin et al., 2009; Tsakanikas & Relkin, 2010). This work is now in the final stages of clinical trials aimed at securing FDA approval. Researchers are exploring other fronts as well, including the role of inflammation. Cognitive decline is four times greater in Alzheimer's sufferers with the highest levels of tumor necrosis factor alpha, which is a marker for inflammation (Holmes et al., 2009). On the other hand, anti-inflammatory drugs—even two that reached the final phase of clinical trials—have not proved to be effective treatments (Hutchison, 2010; Sano, Grossman, & Van Dyk, 2008).

Another exciting effort is on the genetic front. But rather than manipulating the genes responsible for the disease, researchers are attempting to reverse atrophy in dying neurons by implanting genes for *nerve growth factor;* nerve growth factor stimulates neuron growth and activity and helps protect neurons from dying. Aged monkeys have a 43% reduction in acetylcholine-secreting neurons in the basal forebrain (D. E. Smith et al., 1999) and a 25% reduction in acetylcholine-releasing axon terminals in the frontal and temporal cortex (Conner, Darracq, Roberts, & Tuszynski, 2001). Conner et al. (2001) grafted cells genetically engineered to produce human nerve growth factor into the basal forebrain region of aged monkeys. Three months later, the number of axons of acetylcholine-releasing neurons projecting to distant cortex equaled that of young monkeys (Figure 12.19). See the accompanying Application for examples of genetic approaches to Alzheimer's in humans.

Detecting Alzheimer's Disease

The aging individual dealing with memory problems typically wants to know, "Am I getting Alzheimer's?" No single test can diagnose Alzheimer's, but a battery

FIGURE 12.19 Effect of Nerve Growth Factor on Acetylcholine-Producing Axons in Monkeys.

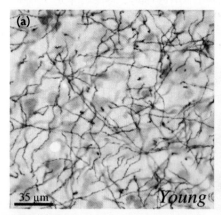

(a)

35 μm   *Young*

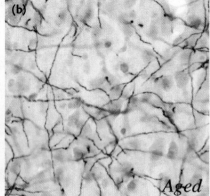

(b)

Aged

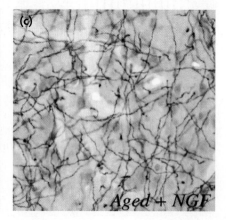

(c)

Aged + NGF

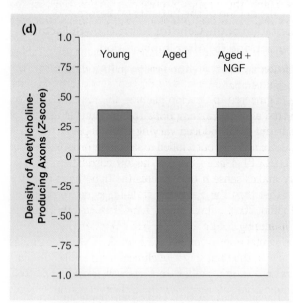

(d)

The density of axons stained with a marker for acetylcholine activity is compared in cortical tissue from (a) young and (b) aged monkeys. (c) Aged monkeys treated with nerve growth factor (NGF) genes show an increase in acetylcholine activity. (d) Comparison of the relative density of cortical acetylcholine-producing axons in the three groups.

SOURCE: From "Nontropic Actions of Neurotrophins: Subcortical Nerve Growth Factor Gene Delivery Reverses Age-Related Degeneration of Primate Cortical Cholinergic Innervation," by J. M. Conner et al., *Proceedings of the National Academy of Sciences, 98,* pp. 1941–1946. Copyright 2001 National Academy of Sciences, USA. Used with permission.

of physical, neurological, and cognitive tests can do a reasonably good job, mostly by ruling out other forms of dementia that may be more treatable, if not reversible. The physician may also order a brain scan to look for structural abnormalities, such as brain shrinkage. PET scanning techniques using two new tracers that specifically target plaques raise hopes for a truly Alzheimer's-specific test (Sabri et al., 2010; Villemagne et al., 2010). Imaging revealed neurological changes preceding the development of Alzheimer's by as much as 10 years, and extensive plaque buildup predicted the progression to Alzheimer's almost 2 years ahead of diagnosis. Early identification of Alzheimer's would allow sufferers to benefit from medication during the initial stages of cognitive impairment, and it will become more important when drugs that can treat more than the symptoms are developed. A more practical solution would be to test for biomarkers of Alzheimer's, abnormalities found in the skin, blood, and cerebrospinal fluid of patients with mild cognitive impairment. These promise 90% to 100% accuracy, as confirmed by autopsy on patient death, in predicting the progression to Alzheimer's over the next 5 to 6 years. The accuracy of clinical diagnosis is about 52%, so these tests could become routine for elderly people concerned about failing memory (De Meyer et al., 2010; Khan & Alkon, 2010; Ray et al., 2007).

There is some evidence that the disease progression might begin as much as 5 decades earlier. This finding came from a study of deceased Catholic nuns who had

Application Genetic Interventions for Alzheimer's

In 2001, surgeons at the University of California–San Diego (UCSD) began experimental gene treatments with Alzheimer's patients by injecting cells that had been augmented with a gene that controls the production of nerve growth factor. An average of 22 months after the surgeries, the patients showed an 84% reduction in loss of cognitive abilities (Tuszynski et al., 2005). And while brain metabolism typically declines in Alzheimer's patients, these had striking increases, as the accompanying PET scans show. Both the work at UCSD and a similar project sponsored by the National Institute on Aging have completed phase I human safety trials and are entering clinical trials aimed at establishing effectiveness of the treatment.

4
Alzheimer's Gene Therapy

While we can't manipulate Alzheimer's genes yet, we can detect them and some people are choosing to do so, particularly if Alzheimer's runs in their family. Although finding out that you don't have any of the known Alzheimer's genes can be very reassuring, learning that you do doesn't currently give you any options other than living a healthy lifestyle and hoping for the best. But it might make you think twice about having children. A 30-year-old woman learned she has the deadly form of the *APP* gene (see Table 12.1 again), which has plagued her family with early-onset Alzheimer's disease; her father died of Alzheimer's at the age of 42, a sister died at 38, and a brother became demented at age 35. So she had 15 of her

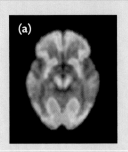
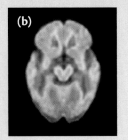

PET Scans of One of the Patients (a) Before Implantation of the Gene for Nerve Growth Factor Production and (b) After Implantation.

The greater amount of red and yellow in (b) indicates increased metabolism.

SOURCE: From "A Phase 1 Clinical Trial of Nerve Growth Factor Gene Therapy for Alzheimer Disease," by M. Tuszynski et al., 2005, *Nature Medicine, 11*, 551–555, fig. 3, p. 553.

eggs screened, and 4 that were free of *APP* were fertilized with her husband's sperm and implanted in her uterus (Verlinsky et al., 2002). Chances are she will die while her child is still young, but the child was born free of the mutation.

donated their brains to science (Iacono et al., 2009). The researchers found that the convent had autobiographical essays written by 14 of these nuns back when they joined the order in their late teens and early twenties. Some of the nuns had plaques and tangles indicative of Alzheimer's but had been unimpaired; their essays were scored 20% higher for linguistic density, defined as the number of ideas expressed for every 10 words. These sisters had larger neural cell bodies in hippocampal CA1; however, we cannot tell whether this neural growth was an adaptation to the lesions developing in their brains or was present early in life and provided protection from dementia.

Korsakoff's Syndrome

Another form of dementia is *Korsakoff's syndrome*, brain deterioration that is almost always caused by chronic alcoholism. The deterioration results from a deficiency in the vitamin *thiamine* (B₁), which has two causes: (1) the alcoholic consumes large quantities of calories in the form of alcohol in place of an adequate diet, and (2) the alcohol reduces absorption of thiamine in the stomach. The most pronounced symptom is anterograde amnesia, but retrograde amnesia is also severe; impairment is to declarative memory, while nondeclarative memory remains intact. The hippocampus and temporal lobes are unaffected; but the mammillary bodies (see Figures 3.20 and 8.4) and the medial part of the

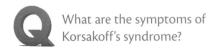

Q What are the symptoms of Korsakoff's syndrome?

thalamus are reduced in size, and structural and functional abnormalities occur in the frontal lobes (Gebhardt, Naeser, & Butters, 1984; Kopelman, 1995; Squire, Amaral, & Press, 1990). A bizarre accident demonstrated that damage limited to the thalamic and mammillary areas can cause anterograde amnesia; a 22-year-old college student received a penetrating wound to the area when his roommate accidentally thrust a toy fencing foil up his left nostril, producing an amnesia that primarily affected verbal memory (Squire, Amaral, Zola-Morgan, Kritchevsky, & Press, 1989). Thiamine therapy can relieve the symptoms of Korsakoff's syndrome somewhat if the disorder is not too advanced, but the brain damage itself is irreversible.

Some Korsakoff's patients show a particularly interesting characteristic in their behavior, called *confabulation;* they fabricate stories and facts to make up for those missing from their memories. Non-Korsakoff amnesics also confabulate, and so do normal people occasionally when their memory is vague. However, Korsakoff patients are champions at this kind of "creative remembering," especially during the volatile early period, when their symptoms have just heated up. We will talk about what causes confabulation shortly; in the meantime, try to refrain from assuming that it involves intentional deception.

For some, confabulation becomes a way of life. Mary Frances could converse fluently about her distant past as a college and high school English teacher and recite Shakespeare and poetry that she had written. But, robbed of the memory of more recent years by Korsakoff's disease, she constantly invented explanations for her nursing home surroundings. One time, she was just "visiting" at the home, and she watched patiently through the glass front doors for her brother who would pick her up shortly for an automobile trip to Florida. Another time, she complained that she was stranded in a strange place and needed to get back to her "post"; she had in fact been in the Army in World War II as a speech writer for General Clark. On another occasion, she thought that she was in prison—probably suggested by the memory that she had actually been a prisoner of war—and she was querying the nurse about what she had "done wrong."

Confabulation occurs following damage to a specific area in the frontal lobes (Turner, Cipolotti, Yousry, & Shallice, 2008). A Korsakoff's patient studied by Benson and his colleagues (1996) did poorly on cognitive tests that are sensitive to impairment in frontal lobe functioning, and a brain scan showed that activity levels were reduced in the frontal area as well as in the diencephalon, the lower part of the forebrain that includes the thalamus and hypothalamus. Four months later, he had ceased confabulating, and the scan of the frontal area had returned to normal; however, his amnesia and deficient diencephalic activity continued. Confabulating amnesic patients have more trouble than nonconfabulating amnesics in suppressing irrelevant information they have learned earlier (Schnider & Ptak, 1999). Consequently, Benson and colleagues (1996) suggest that confabulation is due to an inability to distinguish between current reality and earlier memories. We will take up this topic again in the chapter on consciousness.

5
Korsakoff's Resources

Alzheimer
Korsakoff

Concept Check *Take a Minute to Check Your Knowledge and Understanding*

☐ What changes occur in the brain during aging?

☐ What is the role of plaques and tangles in Alzheimer's disease?

☐ How are Alzheimer's disease and Korsakoff's syndrome similar and different?

In Perspective

Learning is a form of neural plasticity. However, that simple statement ignores a variety of complex features that characterize learning. For example, different kinds of learning can be impaired selectively, as we see in patients who can learn and yet have no recollection of having learned. Our exploration of learning has been an abbreviated one, in part because a number of mysteries are waiting to be solved.

In spite of all we know about the learning process, we have little ability to enhance it. Researchers tell us that blueberries can reduce learning deficits in aging rats and that wearing a nicotine patch can improve memory, but they can do disappointingly little to help the Alzheimer's patient. Curing learning disorders and improving normal learning ability are little more than aspirations today. But there is good reason to think the mysteries will be solved eventually, perhaps with your help.

Summary

Learning as the Storage of Memories

- Brain damage can cause amnesia by impairing the storage of new memories (anterograde) or the retrieval of old memories (retrograde).
- The hippocampus is involved in both consolidation and retrieval. The prefrontal area may play an executive role.
- Memories are stored near the area where the information they are based on is processed.
- There are at least two kinds of learning: declarative, mediated by the hippocampus, and nondeclarative, which involves the striatum and amygdala.
- Working memory holds new information and information retrieved from storage while it is being used.

Brain Changes in Learning

- LTP increases synaptic strength, and LTD reduces it.
- Changes at the synapse include increases in the number and sensitivity of AMPA receptors, the amount of transmitter released, and the number of dendritic spines.
- LTP is necessary for learning; diminishing it impairs learning, and increasing it enhances learning.
- The hippocampus manages new declarative memories, but they are transferred later to the cortex.

Learning Deficiencies and Disorders

- Aging usually involves some impairment of learning and memory, but in the normally aging brain, substantial loss of neurons and synapses is limited to a few areas.
- Alzheimer's disease is a hereditary disorder that impairs learning and other brain functions, largely through the destruction of acetylcholine-producing neurons. Plaques and tangles are associated with cell death but may not be the cause. Treatment usually involves increasing acetylcholine availability, but experimental treatments take other approaches.
- Korsakoff's syndrome is caused by a vitamin B deficiency resulting from alcoholism. Anterograde and retrograde amnesia are effects. ∎

Study Resources

F | For Further Thought

- If you were building an electronic learning and memory system for a robot, is there anything you would change from the human design? Why or why not?
- What are the learning and behavioral implications of impaired working memory?

- What implication does the experiment in which mice were injected with the antibiotic anisomycin at the time of retrieval have for your study conditions as you review material for an exam?
- Which direction of research for the treatment of Alzheimer's do you think holds the greatest promise? Why?

T | Testing Your Understanding

1. Discuss consolidation, including what it is, when and where it occurs, and its significance in learning and memory.

2. Make the argument that LTP provides a reasonably good explanation of learning, including some of learning's basic phenomena.

3. Compare Alzheimer's disease and Korsakoff's syndrome in terms of causes, symptoms, and brain areas affected.

Select the best answer:

1. Anterograde amnesia means that the patient has trouble remembering events that occurred
   a. more than a few minutes earlier.
   b. before the brain damage.
   c. since the brain damage.
   d. since the brain damage and for a few years before.

2. A person or animal born without the ability to consolidate would be unable to
   a. remember anything.
   b. remember for more than a few minutes.
   c. recall old memories that had been well learned.
   d. recall declarative, as opposed to nondeclarative, memories.

3. The function of the *hippocampal formation* is
   a. consolidation of new memories.
   b. retrieval of memories.
   c. as a temporary storage location.
   d. a and b.
   e. a, b, and c.

4. If HM's striatum had also been damaged, he would also not have remembered
   a. declarative memories of childhood events.
   b. skills learned before his surgery.
   c. skills learned after his surgery.
   d. emotional experiences after his surgery.
   e. all the above.

5. In the course of adding a long column of entries in your checkbook, you have to carry a 6 to the next column. If you forget the number in the process, you're having a problem with
   a. consolidation.
   b. LTP.
   c. retrieval.
   d. working memory.

6. The researcher sounds a tone, then delivers a puff of air to your eye. After several times, the tone alone causes you to blink. This behavior is probably explained by
   a. LTP.
   b. associative LTP.
   c. LTD.
   d. associative LTD.

7. LTP involves
   a. release of nitric oxide.
   b. increase in cell body size.
   c. increased number of NMDA receptors.
   d. increased sensitivity of NMDA receptors.
   e. all the above.

8. Without LTP
   a. long-term memory is impaired.
   b. working memory is impaired.

c. old memories cannot be retrieved.

d. no learning occurs.

9. The study in which the antibiotic anisomycin was injected into the brains of mice at the time of testing demonstrated that

a. protein increase improves memory.

b. memories are particularly vulnerable during recall.

c. antibiotics can improve memory.

d. once recalled, a memory takes longer to reconsolidate.

10. The aged brain is characterized by substantial ____ throughout the cortex.

a. loss of neurons   b. loss of synapses

c. decrease in metabolism   d. All the above are true.

e. None of the above is true.

11. The symptoms of Alzheimer's disease are associated with

a. plaques and tangles.

b. a single gene.

c. environmental toxins.

d. all the above.

e. none of the above.

12. The feature most common between Alzheimer's disease and Korsakoff's syndrome is the

a. symptoms.

b. age of onset.

c. degree of hereditary involvement.

d. degree of environmental contribution.

Answers: 1.c, 2.b, 3.e, 4.c, 5.d, 6.b, 7.a, 8.d, 9.b, 10.e, 11.a, 12.a.

Online Resources

The following resources are available at **www.sagepub.com/garrett3e.**

Chapter Resources

- Flash Cards
- Chapter Quiz
- Internet Resources and Exercises

On the Web

Links to these websites can be found at the web address above: Select this chapter's **Chapter Resources**; then choose **Internet Resources and Exercises**.

1. **The Brain Observatory** site describes the project to digitize HM's brain and make it available for continued scientific study.

2. The professional journal *Learning & Memory* provides free access to published articles from the preceding year and earlier.
   The **American Psychological Association** is a good source of information on learning and memory and other topics. Many of the articles are brief updates appearing in the APA *Monitor on Psychology.* Just type the name of a topic in the search window.

3. The **Alzheimer's Association** has information about the disease, help for caregivers, and descriptions of research it is funding.
   The **Fisher Center for Alzheimer's Research Foundation** provides additional information.

4. The **UCSD School of Medicine** *News* has photos and a description of the first genetic implantation for Alzheimer's disease.

5. The National Institute of Neurological Disorders and Stroke's **Wernicke-Korsakoff Syndrome Information Page** describes the disorder, treatments, and research.
   The **Family Caregiver Alliance** has a useful fact sheet on Korsakoff's syndrome, including characteristics, prevalence, diagnosis, and treatment.

Animations

Associative Long-Term Potentiation (Figure 12.11)

Glutamate's Role in Long-Term Potentiation (Figure 12.12)

Chapter Updates and Biopsychology News

R For Further Reading

1. *The Cognitive Neuroscience of Memory* by Howard Eichenbaum (Oxford University Press, 2002) is an accessible textbook that assumes little background in biology or psychology. It elaborates on major topics, including synaptic changes, consolidation, and brain mechanisms involved in memory.

2. "The Machinery of Thought" by Tim Beardsley (*Scientific American,* August, 1997, 78–83) is about the neural basis of working memory, and "The Mind and Brain of Short-Term Memory" by John Jonides et al. (*Annual Review of Psychology,* 2008, *59,* 193–224) is a more recent review of that topic.

3. "New Brain Cells Go to Work" by R. Douglas Fields (*Scientific American Mind,* August/September, 2007, 31–35) elaborates on the possible role of neurogenesis in learning.

4. "Place Cells, Grid Cells, and the Brain's Spatial Representation System" by Edvard Moser, Emilio Kropff, and May-Britt Moser (*Annual Review of Neuroscience,* 2008, *31,* 69–89) is a review of recent research on this subject.

5. "Alzheimer's: Forestalling the Darkness" by Gary Stix (*Scientific American,* June, 2010, 50–57) explores the possibility of early detection and preventive treatment, while "Recent Insights into the Molecular Genetics of Dementia" by Rosa Rademakers and Anne Rovelet-Lecrux (*Trends in Neurosciences,* 2007, *32,* 451–459) describes recent genetic findings.

K Key Terms

Alzheimer's disease 380
anterograde amnesia 365
associative long-term potentiation 373
confabulation... 387
consolidation 366
declarative memory...370
dendritic spines 374
Hebb rule.. 372
Korsakoff's syndrome 386

long-term depression (LTD) 373
long-term potentiation (LTP) 372
neurofibrillary tangles...381
nondeclarative memory 370
place cells... 369
plaques 381
retrieval 366
retrograde amnesia.. 365
working memory...371

Intelligence and Cognitive Functioning

13

The Nature of Intelligence

What Does "Intelligence" Mean?

The Structure of Intelligence

    CONCEPT CHECK

The Biological Origins of Intelligence

The Brain and Intelligence

Specific Abilities and the Brain

    APPLICATION: WE AREN'T THE ONLY TOOL USERS

Heredity and Environment

    APPLICATION: ENHANCING INTELLIGENCE AND PERFORMANCE

    CONCEPT CHECK

Deficiencies and Disorders of Intelligence

Effects of Aging on Intelligence

Intellectual Disability

Autism

    IN THE NEWS: CHILDHOOD VACCINES AND AUTISM

Attention Deficit Hyperactivity Disorder

    CONCEPT CHECK

In Perspective

Summary

Study Resources

In this chapter you will learn

- What some problems are in defining and measuring intelligence

- Some of the neural characteristics that contribute to intelligence

- The role of heredity and environment in forming intelligence

- How aging, intellectual disability, autism, and attention deficit hyperactivity disorder affect intelligence

S ome people are calling Cambridge's theoretical physicist Stephen Hawking the most brilliant person living today. Following in Einstein's footsteps, he has developed theories of the origin of the universe that are altering the way scientists think. He lectures around the world, mixing high-powered physics

His great intellect resides in a body that can communicate only by moving a cursor on a computer screen.

SOURCE: © Matt Dunham/Reuters/Corbis.

with a keen sense of humor. He has achieved all this despite having Lou Gehrig's disease (*amyotrophic lateral sclerosis*, or *ALS*), a degenerative disease that impairs voluntary movement. Confined to a wheelchair and able to make only small movements, he writes and speaks by moving a cursor on the screen of a computer equipped with a voice synthesizer (Figure 13.1).

When we consider people like Hawking, we are forced to wonder what makes one person more intelligent than another. Is it genes, upbringing, hard work, or opportunity? And in particular, is the ultra-intelligent brain in some way different? Unfortunately, we have a problem with the presumption of *who* is smarter. Is Hawking more intelligent than Einstein, or has he just had the advantage of more predecessors' accomplishments to build on? Is Marilyn vos Savant smarter than her husband because at the age of 10 she made the highest score recorded on an intelligence test (Yam, 1998) while he had trouble testing well enough to get into medical school, or is he smarter because he invented the Jarvik artificial heart? We cannot attempt to understand the biological bases of intelligence without having some appreciation of our limitations in measuring it or even defining what it is.

1
Stephen Hawking's Website

The Nature of Intelligence

There are many ideas about what intelligence is, which is the first clue that we don't have consensus about what it is. Most definitions say something to the effect that *intelligence* is the ability to reason, to understand, and to profit from experience. That is what we *think* intelligence is; the problem comes when we try to translate that abstract definition into the behaviors that would indicate the presence of intelligence. That is what we must do in order to measure intelligence, which is the first step toward determining its biological basis.

What Does "Intelligence" Mean?

Understanding how we measure intelligence is important because we are in effect defining intelligence as *what that test measures*. The measure of intelligence is typically expressed as the *intelligence quotient (IQ)*. The term originated with the scoring on early intelligence tests designed for use with children. The tests produced a score in the form of a *mental age*, which was divided by the child's chronological age and multiplied by 100. The tests were designed to produce a score of 100 for a child performing at the average for his or her chronological age. The scoring is completely different now, partly because the tests were extended to adults, who do not increase consistently in intellectual performance from year to year. The base score is still 100, a value that was arbitrarily selected and that is artificially preserved by occasional adjustments to compensate for any drift in performance in the population. Most people are near the average, as Figure 13.2 shows, with relatively few people at either of the extremes. For example, only 2% of the population score above 130 points or below 70 points.

The first intelligence test was devised by Alfred Binet in 1905, to identify French schoolchildren who needed special instruction (Binet & Simon, 1905). Predicting school performance is still what most intelligence tests do best, and intelligence tests have found their greatest use in the school setting. The correlation between IQ scores and school grades typically falls in the range of .40 to .60 (Kline, 1991). However, IQ is also related to job performance, income, socioeconomic level and, negatively, to juvenile delinquency (Neisser et al., 1996).

Critics believe that scores on traditional intelligence tests are closely related to academic performance and to higher socioeconomic levels mostly because the tests were designed to reflect that kind of success. According to these critics, the tests overemphasize verbal ability, education, and Western culture. A few tests are designed to be culture-free, like the Raven Progressive Matrices (Raven, 2003). These tests are mostly nonverbal, and the tasks require no experience with a particular culture. They have an obvious advantage for testing people from a very different culture or language background or with impaired understanding of language. Some researchers also believe that the Raven gives them a better representation of "pure" intelligence.

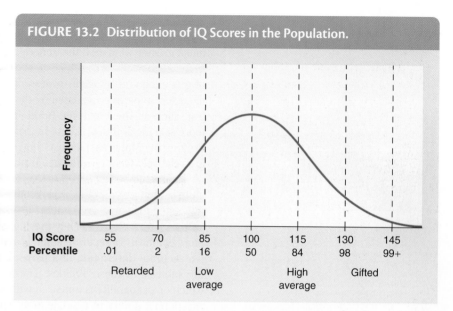

FIGURE 13.2 Distribution of IQ Scores in the Population.

SOURCE: From: NAIRNE. *Psychology*, 2E. © 2000 Wadsworth, a part of Cengage Learning, Inc. Reproduced by permission. www.cengage.com/permissions.

Claiming that true intelligence is much more than what the tests measure, these critics often point to instances where practical intelligence or "street smarts" is greater than conventional intelligence. For example, young Brazilian street vendors were adept at performing calculations as they conducted their business that they were unable to perform in a classroom setting (Carraher, Carraher, & Schliemman, 1985). In another study, expert racetrack gamblers used a highly complex algorithm involving seven variables to predict racetrack odds, but their performance was unrelated to their IQ; in fact, four of them had IQs in the low to mid-80s (Ceci & Liker, 1986). More recently, Robert Sternberg (2000) compared the scores that the presidential candidates George W. Bush, Al Gore, and Bill Bradley made on the verbal section of the Scholastic Aptitude Test (SAT) when they applied for college; the SAT has many items similar to those on conventional intelligence tests. Two of the candidates scored above average for college applicants but not markedly so, and one had a score that was below average. To Sternberg, their success raises questions about the narrowness of what intelligence tests measure.

Sternberg argues that intelligence does not exist in the sense we usually conceive of it but is "a cultural invention to account for the fact that some people are able to succeed in their environment better than others" (1988, p. 71). Perhaps intelligence is, like the mind, just a convenient abstraction we invented to describe a group of processes. If so, we should not expect to find intelligence residing in a single brain location or even in a neatly defined network of brain structures. And to the extent we find processes or structures that are directly involved in intelligence, their performance may not be highly correlated with scores on traditional intelligence tests.

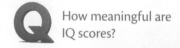

Q How meaningful are IQ scores?

(handwritten note: Lumpers v.s splitters.)

The Structure of Intelligence

Another controversy that is critical to a biological understanding of intelligence is whether intelligence is a single capability or a collection of several independent abilities. Intelligence theorists tend to fall into one of two groups, *lumpers* or *splitters*. Lumpers claim that intelligence is a single, unitary capability, which is usually

called the *general factor,* or simply *g.* General factor theorists admit that there are separate abilities that vary somewhat in strength in an individual, but they place much greater weight on the underlying g factor. They point out that a person who is high in one cognitive skill is usually high in others, so they believe that a measure of g is adequate by itself to describe a person's intellectual ability. General intelligence is sometimes assessed by the overall IQ score from a traditional intelligence test, such as the Wechsler Adult Intelligence Scale, whose 11 subtests measure more specific abilities. But many g theorists prefer to use other tests like the Raven Progressive Matrices, because they emphasize reasoning and problem solving and are relatively freer of influence from specific abilities such as verbal skills.

Splitters, on the other hand, hold that intelligence is made up of several mental abilities that are more or less independent of each other. Therefore, they are more interested in scores on the subtests of standard IQ tests or scores from tests of specific cognitive abilities. They may agree that there is a general factor, but they give more emphasis to separate abilities and to differences among them in an individual. An accurate description of a person's intelligence would require the scores on all the subtests of these abilities. These theorists point to cases of brain damage in which one capability is impaired without affecting others and to the *autistic savant's* exceptional ability in a single area. Splitters disagree with each other, though, on how many abilities there are; a review of intelligence tests identified more than 70 different abilities that can be measured by currently available tests (Carroll, 1993).

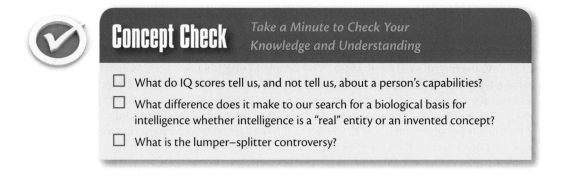

Concept Check *Take a Minute to Check Your Knowledge and Understanding*

☐ What do IQ scores tell us, and not tell us, about a person's capabilities?

☐ What difference does it make to our search for a biological basis for intelligence whether intelligence is a "real" entity or an invented concept?

☐ What is the lumper–splitter controversy?

The Biological Origins of Intelligence

With this background, we are now ready to explore the origins of intelligence. On the basis of our introduction, we will avoid two popular assumptions—that intelligence tests are the only definition of intelligence and that intelligence is a single entity. Instead, we will consider performance and achievement as additional indicators of intelligence, and we will first examine the evidence of a biological basis for a general factor and then consider the relationship between brain structures and individual abilities.

The Brain and Intelligence

Q How are intelligent brains different?

Are there identifiable ways that a more intelligent brain is different from other brains? Anyone asking this question would naturally wonder how Albert Einstein's brain was different from other people's. Fortunately, the famous scientist's brain was preserved, and it has been made available from time to time to neuroscientists (Figure 13.3). In cursory examinations, it turned out to be remarkably unremarkable. In fact, at 1,230 grams (g) it was almost 200 g lighter than the average weight of the control brains (Witelson, Kigar, & Harvey, 1999). The number of neurons did

FIGURE 13.3 Albert Einstein and His Brain.

The brain of the genius looks just like yours; it took careful study to find differences.

(a)

(b)

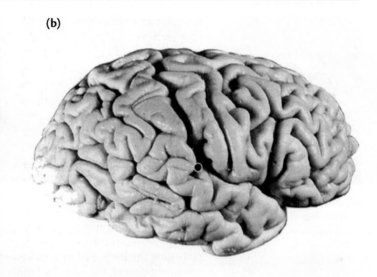

SOURCES: Historical Pictures Services, Chicago/FPG, and Sandra F. Witelson.

not differ from normal, and studies have disagreed about whether the neurons were more densely packed or the cortex was thinner, perhaps because the samples were taken from different locations (B. Anderson & Harvey, 1996; Kigar, Witelson, Glezer, & Harvey, 1997). One study found a higher ratio of glial cells to neurons in the left parietal lobe (Diamond, Scheibel, Murphy, & Harvey, 1985). The comparison brains averaged 12 years younger than Einstein's at the time of death, and we know that glial cells continue proliferating throughout life (T. Hines, 1998), but the number of glial cells was not elevated in Einstein's right parietal lobe or in either frontal sample. Each of Einstein's hemispheres was a full centimeter wider than those of control brains, due to larger parietal lobes (Witelson et al., 1999), and there were intriguing variations in some of the parietal gyri (Falk, 2009). These anomalies are interesting because the parietal lobes are involved in mathematical ability and visual-spatial processing, and Einstein reported that he performed his mathematical thinking not in words but in images. Remember, though, that in Chapter 11 we saw that London cabdrivers have enlarged hippocampi, and no one has suggested that large hippocampi explains why they became cabdrivers. Whether Einstein's large parietal lobes or intense mathematical activity came first is uncertain (assuming they are related).

It is interesting that the strongest finding in the examination of Einstein's brain seems to be related to a specific ability rather than to overall intelligence. British researcher John Duncan and his colleagues (2000) sought the location of general intelligence in the brains of more ordinary folk. They used tasks that required reasoning and that are known to correlate with general intelligence more than with any specific ability. Although verbal and spatial tasks had different patterns of activation, prefrontal activation was common to both tasks; the authors concluded that general intelligence may be located there. But general intelligence is likely more complex than what happens in one area of the brain. A team of researchers scanned the brains of 241 individuals with brain damage and correlated the location of their lesions with their scores on a set of intelligence subtests chosen to measure g (Gläscher et al., 2010). They found that g requires a distributed system, a network that spans the frontal,

> *The contrast between the popular estimate of my powers and achievements compared to the reality is simply grotesque.*
>
> —Albert Einstein

> *Words and language . . . do not seem to play any part in my thought processes.*
>
> —Albert Einstein

FIGURE 13.4 Brain Areas Involved in General Intelligence.

A view of the cortical surface (left) and subcortical areas (right) shows a g-related network extending throughout much of the brain.

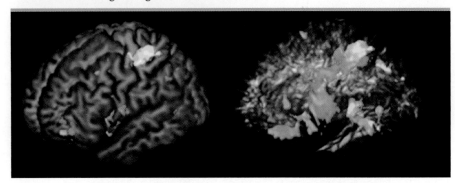

SOURCE: From "Distributed Neural System for General Intelligence Revealed by Lesion Mapping," by J. Glåscher et al., 2010, *Proceedings of the National Academy of Sciences, 90*, pp. 4705–4709.

parietal, and temporal lobes (Figure 13.4). They concluded that g involves the brain's ability to pull together different kinds of processing, such as visuospatial processing and working memory. The ability to integrate these functions depends on the quality of connections between the areas, which is highly heritable, up to 84% in some parts of the brain (Chiang et al., 2009).

Brain Size

Brain size itself does not determine intelligence. Elephants have much larger brains than we do, and not many people think elephants are smarter. What is more important is the ratio of the brain's size to body size; this ratio adjusts for the proportion of the brain needed for managing the body and tells us how much is left over for intellectual functions. As you might expect, the ratio for humans is one of the highest.

Within a species, though, the answer is a bit different. Numerous magnetic resonance imaging (MRI) studies have found correlations between brain size and measures of intelligence; a compilation of 37 studies involving 1,530 people found a modest correlation of .33 (McDaniel, 2005). Squaring the correlation coefficient tells us that about 11% of the differences among people in intelligence is related to brain size. Of course, men have larger brains than women, which we have assumed was related to men's greater body size. However, even after adjustment for body size, men's brains average about 118 g heavier than women's (Ankney, 1992). Presumably, men are no smarter than women; it is actually difficult to tell, because intelligence tests were designed to avoid gender bias. But if men's excess brain matter does not confer additional intelligence, what is its function? There are two credible hypotheses. One is that women's brains are more efficient, because of a greater density of neurons (Witelson, Glezer, & Kigar, 1995) and a higher ratio of gray matter to white matter (Gur et al., 1999). The other hypothesis is that the male's superior spatial intelligence requires greater brain capacity (D. Falk, Froese, Sade, & Dudek, 1999).

MRI studies of fraternal and identical twins found that general intelligence was correlated with both the volume of gray matter and the volume of white matter (Posthuma et al., 2002; Thompson, Cannon, et al., 2001). And, consistent with findings mentioned earlier, the volume of gray matter in the frontal area appears to be particularly important to general intelligence (Haier, Jung, Yeo, Head, & Alkire, 2004; Thompson, Cannon, et al., 2001). But before we go too far in dissecting the issue of brain size, we should keep in mind that Einstein's brain was smaller than the average female brain (Ankney, 1992). Also, how the brain matter is distributed and organized appears to be more important than size. Remember the cortical columns that we saw back in Chapter 3? In the brains of three distinguished scientists, these were smaller and packed closer together (Casanova, Switala, Trippe, & Fitzgerald, 2007). Computer modeling has shown that smaller columns provide for better discrimination between signals during information processing. Another difference showed up in two longitudinal studies of brain development. Although individuals with IQs above 120 have thinner cortices until about the age of 8, beyond that time the cortex thickens (Shaw et al., 2006) and there is a positive correlation between cortical thickness and intelligence (Figure 13.5; Karama et al., 2009). So an important feature of a smarter brain is that it has a larger processing area packed with more processing units.

FIGURE 13.5 Areas Where Cortical Thickness Is Associated With Intelligence.

Notice that critical areas are distributed throughout the brain.

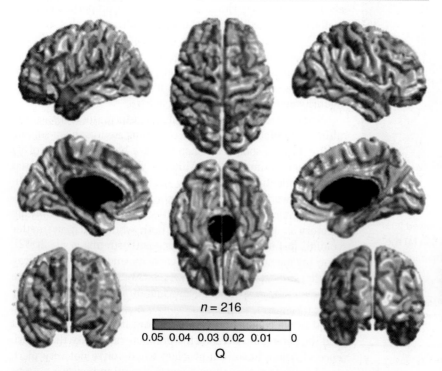

$n = 216$

0.05 0.04 0.03 0.02 0.01   0

Q

SOURCE: Montreal Neurological Institute.

Neural Conduction Speed and Processing Speed

Cognitive processes require the person to apprehend, select, and attend to meaningful items from a welter of stimuli arriving at the sensory organs. Then the person must retrieve information from memory, relate the new information to it, and then manipulate the mental representation of the combined information. All of this takes time. In 1883, Francis Galton suggested that higher intelligence depends on greater "mental speed." Because there were no intelligence tests then, he attempted to relate measures of intellectual achievement like course grades and occupational status to people's reaction times, but no relationship was found. More recently, a number of researchers have shown that IQ scores do correlate with reaction time (T. E. Reed & Jensen, 1992). The relationship is not due to the fact that most intelligence tests emphasize speed, because reaction time and IQ scores are still correlated when the IQ test is given without a time limit (A. R. Jensen, 1998).

Apparently the reason is that IQ scores are also correlated with *nerve conduction velocity*, even more than with reaction time (see Figure 13.6; McGarry-Roberts, Stelmack, & Campbell, 1992; Vernon & Mori, 1992). In addition, people who are more intelligent excel on tasks in which stimuli are presented for an extremely short interval and on tasks that require choices (A. R. Jensen, 1998). These are tasks in which processing speed is important and, presumably, higher nerve conduction velocity contributes to the more intelligent person's superior performance. We will see in the next section that faster nerve conduction velocity may make its contribution through improved processing efficiency.

FIGURE 13.6 Relationship Between IQ Scores and Nerve Conduction Velocity.

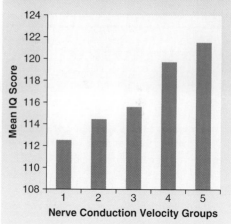

The research participants were divided into five groups according to their nerve conduction velocity. Group 1 had the lowest nerve conduction velocity and Group 5 the highest. Nerve conduction velocity was calculated by dividing the elapsed time between a visual stimulus and the occurrence of the visual evoked potential by the distance between the eyes and the back of the head.

SOURCE: Adapted from "Conduction Velocity in a Brain Nerve Pathway of Normal Adults Correlates With Intelligence Level," by T. E. Reed and A. R. Jensen, *Intelligence*, 16, pp. 259–272. Copyright 1992, with permission from Elsevier Science.

FIGURE 13.7 Greater Efficiency in the More Intelligent Brain.

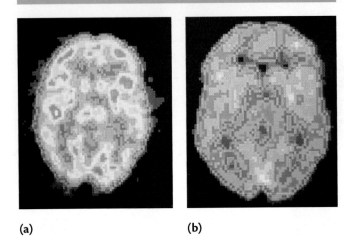

(a)   **(b)**

During an attention-demanding task, PET scans showed 20% more activity (indicated by more reds and yellows) in (a) the brain of an individual with intellectual disability than in (b) the brain of a person with above-average IQ.

SOURCE: From "Brain Size and Cerebral Glucose Metabolic Rate in Nonspecific Mental Retardation and Down Syndrome," by R. J. Haier et al., 1995, *Intelligence*, 20, pp. 191–210.

Are there separate components of intelligence?

Processing Efficiency

One way the brain could achieve greater efficiency is through enhanced myelination of its neurons (A. R. Jensen, 1998). Besides improving conduction speed, myelin insulates neurons from each other; this reduces "cross talk" that would interfere with accurate processing. Humans have a greater proportion of white matter (myelinated processes) to gray matter than other animals, and IQ is related to the degree of myelination among individuals (Willerman, Schultz, Rutledge, & Bigler, 1994). In addition, myelination, speed of information processing, and intelligence all follow a curvilinear time path, increasing from childhood to maturity and then declining in old age.

Some theorists believe that short-term memory is the ultimate limitation on human reasoning and problem-solving ability. In fact, short-term memory is a better predictor of IQ than reaction time is (L. T. Miller & Vernon, 1992). Increased nerve conduction velocity may particularly enhance the efficiency of working memory, which is a manifestation of short-term memory (A. R. Jensen, 1998; Vernon, 1987); working memory is also correlated with both white and gray matter volume, just as general intelligence is (Posthuma et al., 2002). Working memory has a limited capacity, and its contents decay rapidly. A person whose neurons conduct rapidly can complete manipulations and transfer information to long-term memory before decay occurs or short-term storage capacity is exceeded. But if nerve conduction velocity is low, the information is lost and the person must restart the process—rather like the experience of trying to solve a problem when you're not very alert, and having to review the information over and over.

Further evidence of the role of neural efficiency in intelligence is that individuals higher in IQ use less brain energy. This was indicated by a lower rate of glucose metabolism during a challenging task, playing the computer game *Tetris* (Haier, Siegel, Tang, Abel, & Buchsbaum, 1992). And, as you can see in Figure 13.7, individuals with mild intellectual disability (IQs between 50 and 70) require 20% more neural activity to perform an attention demanding task than do individuals with IQs of 115 or higher (Haier et al., 1995). You might think the more intelligent brain would be the more active one during a task, but remember we are talking about efficiency.

Specific Abilities and the Brain

Brain size, speed, and efficiency can reasonably be viewed as contributors to general intelligence; now we will consider evidence for individual components of intelligence. The statistical method called *factor analysis* has been useful in identifying possible components. The procedure involves giving a group of people several tests that measure cognitive abilities that might be related to intelligence; the tests may be intelligence tests, or they may be measures of more limited abilities such as verbal skills or reaction time. Then correlations are calculated among all combinations of the tests to locate "clusters" of abilities that are more closely related with each other than with the others. Performance on practically all tests of cognitive ability is somewhat related, which is consistent with the hypothesis of a general factor. However, factor analysis has also identified clusters of more specific abilities. Three capabilities have frequently emerged over the past 50 years as major components of intelligence: *linguistic, logical-mathematical,* and *spatial* (A. R. Jensen, 1998).

Several authors have argued that each of the cognitive abilities depends on a complex network or module in the brain that has evolved to provide that particular function. That is, they believe that the brain is hardwired for functions like mathematics and language (Dehaene, 1997; Pinker, 1994). We have already seen examples of modular functions in earlier chapters. The language (linguistic) module is made up of structures located mainly in the left frontal and temporal lobes. Spatial ability depends on the interaction of somatosensory and visual functions with parietal structures, mostly in the right hemisphere.

Mathematical ability in humans depends on two distinct areas of the brain. One is in the left frontal region, and the other is located in both parietal lobes (Dehaene, Spelke, Pinel, Stanescu, & Tsivkin, 1999; Nieder & Dehaene, 2009). When individuals performed precise calculations, activity increased in the frontal language area. This apparently is the storage site of rote arithmetic facts—times tables and information like $4 + 5 = 9$. When the research participants only estimated results, activity increased in parts of both parietal lobes (see Figure 13.8). The parietal areas probably employ a visual-spatial representation of quantity, such as finger counting and the "number line" that mathematicians often report using (illustrated in the figure), which is an almost universal stage in learning exact calculation. An EEG evoked potential study of 5-year-old children found that they use the same parietal areas as adults when estimating numbers (Temple & Posner, 1998). Studies of brain damage support these conclusions. Individuals with damage to the left frontal language area can arrange numbers in rank order and estimate results, but they cannot perform precise calculations; those with parietal damage are impaired in the opposite direction (Butterworth, 1999; Dehaene & Cohen, 1997). Presumably, the frontal and parietal areas cooperate when we perform mathematical tasks.

FIGURE 13.8 Brain Locations Involved in Mathematical Calculation.

Precise calculation involved left frontal area; estimation was accompanied by activity in both parietal lobes.

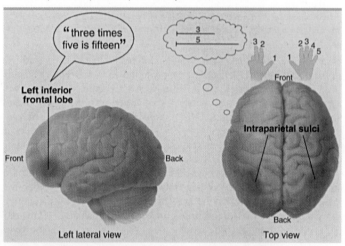

SOURCE: Reprinted with permission from B. Butterworth, "A Head For Figures," *Science*, 284, pp. 928–929. Illustration: K. Sutliff. Copyright © 1999 American Association for the Advancement of Science. Reprinted by permission of AAAS.

Neither of these areas is dedicated exclusively to numerical functions. This is consistent with the suggestion in Chapter 9 that the language areas may simply use processing strategies that make them particularly suited to the demands of the task, rather than being dedicated to language processing. However, Wynn (1998) argues that the brain has a specialized mechanism for numbers by pointing to evidence that even infants and lower primates seem to have an inborn ability for estimating quantities. The infants she studied saw two objects placed one at a time behind a screen. They acted surprised—which means that they looked for a longer time—when the screen was removed to reveal three objects or only one object instead of the expected two (Wynn, 1992).

Rhesus monkeys tested in the wild responded the same way to the task (Hauser, MacNeilage, & Ware, 1996). Monkeys can also rank order groups of objects that differ in number by touching their images on a computer monitor in the correct order (Brannon & Terrace, 1998); chimpanzees have learned to do the same with numerals (Kawai & Matsuzawa, 2000). The monkeys continued to perform at the same level of accuracy when the researchers increased the number of items presented, so the monkeys had not simply memorized the stimuli. The researchers were careful to control the spatial size of the groupings, so the monkeys could not distinguish the groups according to the space they covered. (See the accompanying Application for some thoughts on intelligence in animals.)

An impartial observer would have to say that there is evidence for characteristics that contribute to an overall intelligence and for somewhat independent capabilities as well. An either–or stance is not justified at this time; it makes sense to continue the two-pronged approach of looking for biological bases for both general intellectual ability and separate capabilities.

Application   We Aren't the Only Tool Users

Ever since people realized what surprising similarities we share with other animals, we have been fascinated with how intelligent their behavior can sometimes appear to be. Termites build great earthen mounds that are so well ventilated they stay cool under the broiling African sun. Dolphins cooperate with each other to herd schools of fish into their pod mates' waiting mouths. Kanzi communicates with symbols, Alex mastered simple concepts, and monkeys and chimps can "count."

Especially intriguing is that after decades of denying the possibility, we found other animals making and using tools. Chimpanzees have long been known to "fish" for termites by inserting a twig into the mound. To learn more about this behavior, naturalists in the Congo rain forest hid motion-activated cameras around termite nests and got a close look at the chimpanzees at work (Holden, 2004b). The chimps first use a puncturing stick to open a hole in the nest. Then they follow up with a probe, a small green twig that the termites climb onto in defense of the nest, only to be slurped up by the chimp (Figure a). Congo chimps add a twist not seen elsewhere: They fray the end of the twig by pulling it between their teeth; these tools pick up 20 times as many termites as the unmodified version (Sanz, Cali, & Morgan, 2009).

3
Animals Using Tools: Videos

When it comes to acquiring food, animals can show a great deal of what in humans would pass for ingenuity and planfulness. New Caledonia crows use twigs to fish for grubs, and in captivity they will use a short stick to retrieve a longer stick that is out of reach in order to retrieve a tasty morsel from a deep hole (Taylor, Elliffe, Hunt, & Gray, 2010). Burrowing owls spread animal dung around their burrow, but why they did it was unknown until a group of college students got curious and tackled the problem (Pain, 2005). They removed the dung from some of the burrows, and when they came back 4 days later and counted beetle shells, they found that the owls with dung around their burrows had eaten 10 times as many beetles; the owls were using the dung to lure their dinner! But the prize for planning ahead goes to Santino, a chimp in a Swedish zoo (Osvath, 2009). Santino occasionally threw stones at visitors, but zookeepers weren't concerned because there was so little ammunition on his moat-surrounded island. Then stone throwing increased dramatically, so a caretaker watched from hiding to figure out where the stones were coming from. Each morning before the zoo opened, Santino retrieved stones from the moat and placed them in piles that were handy for launching at the visitors. He was also seen dislodging chunks of concrete from a wall, after first tapping where there were cracks and listening for the hollow sound that meant a piece was loose enough to be removed.

So what do these observations tell us? First, we learn that other animals are a lot smarter than we've given them credit for. We also understand now that some of our capabilities aren't unique after all. But we also have a chance to discover where we are truly different. For example, watching someone retrieve an object with the hand or with a tool activates mirror neurons that are similarly located in humans and in monkeys (Peeters et al., 2009). But when a human sees the object being retrieved with a tool, an additional area is activated in the anterior parietal lobe (Figure b); no such area for tool use showed up in the monkeys, even the ones that were trained for 3 to 6 weeks to use the same tools. Perhaps—just perhaps—our brains are unique in possessing additional specialized structures or in the ability to develop them through experience.

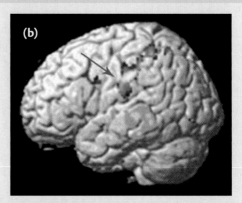

(b)

Area Activated in Humans (Arrow) but Not in Monkeys When Viewing Retrieval of an Object With a Tool.

SOURCE: From "The Representation of Tool Use in Humans and Monkeys: Common and Uniquely Human Features," by Peeters et al., 2009, *Journal of Neuroscience, 29*, pp. 11523–11539.

(a)

A Chimp Dines on Termites "Fished" From Their Mound With a Stick.

SOURCE: Corbis.

Heredity and Environment

We saw in Chapter 1 that intelligence is the most investigated of the genetically influenced behaviors. Our understanding of the genetic underpinnings of intelligence hardly lives up to the amount of effort, however; this is because intelligence is so complex and poorly understood itself and because many genes are involved. In addition, environment accounts for half of the differences among us in intelligence, yet the environmental influences have themselves not been clearly identified (Plomin, 1990).

Heritability of Intelligence

Figure 13.9 shows the IQ correlations among relatives, averaged from many studies and several thousand people. You can see that IQ is more similar in people who are more closely related (T. J. Bouchard & McGue, 1981). Separating family members early in life does not eliminate the correlation; in fact, identical twins reared apart are still more similar in IQ than fraternal twins who are reared together. Interestingly, the relative influence of heredity *increases* with age, from 41% in childhood to 55% in adolescence and 66% in adulthood (Haworth et al., 2009). Intuition tells us that environment should progressively overtake genetic effects as siblings go their separate ways and their experiences diverge, but Gray and Thompson (2004) suggest that the genes that influence intelligence also influence the individual's choices of environment and experience.

Not only do we know that intelligence has a heritability of around 50% (Plomin, 1990), but researchers have also documented genetic influence on several of the functions that contribute to it, including working memory, processing speed, and reaction time in making a choice (Ando, Ono, & Wright, 2001; Luciano et al., 2001; Posthuma, de Geus, & Boomsma, 2001). Most of the differences among individuals in the major structural contributors to intelligence are accounted for by genetic factors; estimated heritabilities in one twin study were 90% for brain volume, 82% for gray matter, and 88% for white matter (Baaré et al., 2001; see Figure 13.10).

FIGURE 13.9 Correlations of IQ Scores Among Relatives.

Percentages indicate the degree of genetic relatedness. "Together" and "apart" refer to whether related children are raised in the same household.

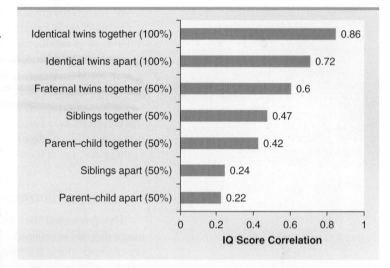

SOURCE: Data from T. J. Bouchard and McGue (1981).

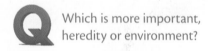 Which is more important, heredity or environment?

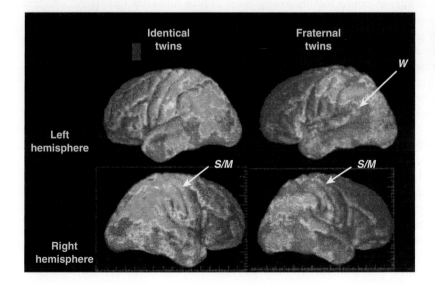

FIGURE 13.10 Genetic Control of Gray Matter Volume in Identical and Fraternal Twins.

Red indicates areas where the correlation between twins in gray matter was most significant; green indicates lesser correlation and blue no statistically detectable relationship. Notice the markedly greater similarity (greater abundance of red) between identical twins. (W indicates Wernicke's area and S/M the somatosensory and motor areas.)

SOURCE: From "Genetic Influences on Brain Structure," by Paul M. Thompson et al., 2001, *Nature Neuroscience, 4*(12), pp. 1253–1258. © 2001 Nature Publishing Group.

Also, general intelligence has higher heritability than more specific abilities, such as verbal and spatial abilities (McClearn et al., 1997), which is an additional argument for a biological basis for g (Gray & Thompson, 2004).

Locating the specific genes is another matter. A number of genes have shown a relationship with intelligence recently, but they have either failed replication or there has not yet been an attempt at replication (Gray & Thompson, 2004). A list of potential candidates has been proposed with roles, for instance, in working memory and executive attention, but they have so far not been linked with differences in intelligence. Additional leads come from two genes that underwent accelerated evolution following our separation from chimpanzees; the *ASPM* gene is a major determinant of brain size (J. Zhang, 2003), while the *PACAP precursor gene* plays a role in neurogenesis and neural signaling and may have contributed to the formation of human cognitive abilities (Y. Wang et al., 2005). In a review of the literature, two researchers compiled a list of more than 150 candidate genes that may influence some aspect of cognitive ability (Morley & Montgomery, 2001). Pinning down which ones vary with different levels of intelligence will not be easy.

The Genetic Controversy

The conclusion that intelligence is highly heritable has not been greeted with unquestioning acceptance. The heritability of intelligence is controversial on a variety of fronts; these controversies illustrate both the pervasive misunderstanding of genetic influence and the difficulty in resolving questions of heredity. Critics fear that inheritance of intelligence implies that intelligence is inborn and unchangeable (Weinberg, 1989). Nothing could be farther from the truth, however; as Weinberg points out, genes do not fix behavior but set a range within which the person may vary. By way of illustration, height is about 90% heritable (Plomin, 1990), yet the average height has increased dramatically over the past few decades, due to improved nutrition. Similarly, IQ scores are increasing at the rate of 5 to 25 points in a generation (termed "the Flynn effect"; Flynn, 1987), so test constructors must adjust the test norms occasionally to maintain an average of 100 points. Although the environmental causes have not been identified, such rapid increases cannot be due to genetic changes.

Some argue that the correlation of IQ among relatives does not mean that intelligence is inherited. They suggest, for example, that identical twins' similarity in appearance and personality lead others to treat them similarly, even when they are reared apart, and this similar treatment results in similar intellectual development. Although physical features and behavior do affect how others react to a child, there is little evidence that these responses in turn influence intelligence. To test this possibility, researchers compared the IQs of twins who had been either correctly or incorrectly perceived by their parents as fraternal or identical. If similar environmental treatment accounts for IQ similarity, then the parents' perception of their twins' classification should be more important than the twins' actual genetic classification. Instead, the studies showed that only the true genetic relationship influenced IQ similarity in the twins, not the parents' perception (Scarr & Carter-Saltzman, 1982).

In another controversy, a debate has raged since the 1930s over whether IQ differences between ethnic groups are genetically based. The question is not whether the ethnic groups differ in IQ scores, but how much the differences are due to heredity and to environment. Arthur Jensen (1969) has argued for 35 years that environment and socioeconomic differences are inadequate to account for the observed IQ differences among groups. He and Philippe Rushton cite studies indicating that IQ differences are consistent around the world, with East Asians averaging 106 points whether in the United States or in Asia, Whites about 100, and African Americans at 85 and sub-Saharan Africans around 70 (Rushton & Jensen, 2005). In addition, they say that differences in brain size correspond to

3
Flattening the Bell Curve

the IQ gaps. Their position has practical as well as theoretical implications. For example, extreme hereditarians believe that intervention efforts like Head Start not only do not work but also cannot work.

Other intelligence researchers countered that Jensen and Rushton ignored or misinterpreted the most relevant data (Nisbett, 2005). For example, African Americans with more African ancestry scored as high on cognitive tests as those with mixed ancestry (Scarr, Pakstis, Katz, & Barker, 1977). Another study suggests that socioeconomic class is more important than ethnic origin; it found that IQ runs about 20 to 30 points lower in the lowest social classes than in the highest social classes (Locurto, 1991). Higher intelligence test results are not always found for Asians (Naglieri & Ronning, 2000), and their higher academic achievement is typically attributed to cultural and motivational differences (Dandy & Nettelbeck, 2002). A task force appointed by the American Psychological Association to study the intelligence debate concluded that there is not much direct evidence regarding the genetic hypothesis of IQ differences between African Americans and Whites and what little there is does not support the hypothesis (Neisser et al., 1996).

Environmental Effects

Most intelligence researchers agree that intelligence is the result of the joint contributions of genes and environment; it has even been said that intelligence is 100% hereditary and 100% environmental, because both are necessary. However, it has been more difficult than expected to identify just which environmental conditions influence intelligence, other than those that cause brain damage. We will give environmental influences more attention here than usual because intelligence is a good arena for illustrating the difficulties in sorting out heredity from environment.

One problem is similar to the one we have encountered in identifying specific genes: The environmental influences are many and, for the most part, individually weak (A. R. Jensen, 1981). Even a twin study that looked at major environmental events such as severe infant and childhood illness failed to find an effect (Loehlin & Nichols, 1976). A second problem is that environmental influences are often hopelessly confounded with genetic effects. For example, family conditions such as socioeconomic level and parental education are moderately related to intelligence (T. J. Bouchard & Segal, 1985), but these characteristics also reflect the parents' genetic makeup, which they pass on to their children. On the other hand, illness may be an important factor in explaining worldwide differences in intelligence. Researchers at the University of New Mexico concluded that the level of infectious diseases is the best predictor of national differences in intelligence (Eppig, Fincher, & Thornhill, 2010). After controlling for confounding variables, including education, climate, and economy, the correlation between national average IQ and infectious disease was −0.76 (Figure 13.11). The investigators believe that fighting off infections during childhood robs the body of energy needed for normal brain development. Not only might disease prevalence partly account for racial differences in intelligence, but the researchers suggest that reduction in the incidence of infectious diseases explains some of the IQ increase seen in the Flynn effect.

The best way to demonstrate environmental influences is by environmental intervention. Although the Head Start program has produced long-term benefits in mathematics, educational attainment, and career accomplishments, the average increase of 7.42 IQ points compared with controls eventually disappears. The Abecedarian Project, which began at birth, produced IQ gains that were as strong

FIGURE 13.11 Worldwide Relationship Between Intelligence and Infectious Disease.

Although the relationship is not perfect (in which all the dots would fall on the line), you can see that high IQ scores are associated with low levels of infection and low IQ scores are associated with high levels of infection.

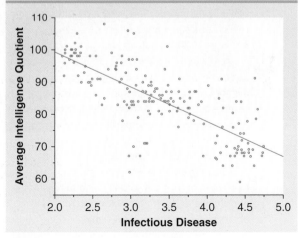

SOURCE: Adapted from C Eppig et al. (2010).

10 years later as those in the Head Start program after 2 years (Ramey et al., 2000). Apparently, intervention must occur at an earlier age; a new Early Head Start program now takes children from birth through age 5. Adoption has a better chance of demonstrating any environmental influences on intelligence, because it alters the entire environment for the child (Scarr & Weinberg, 1976). Adopted children's IQs are more highly correlated with the intelligence of their biological parents than with the intelligence of their adoptive parents (Scarr & Weinberg, 1976; Turkheimer, 1991), but this does not mean that the children's IQs do not go up or down according to the adoptive environment. When African American children were adopted from impoverished homes into middle-class homes, by the age of 6 they had increased from the 90-point average for African American children in the geographic area to 106. The beneficial effects still persisted a decade later (Scarr & Weinberg, 1976; Weinberg, Scarr, & Waldman, 1992).

Does this mean there was no genetic effect at all? No; in fact, the correlation between the children's IQs and their biological parents' educational levels (used in the absence of IQ scores) actually *increased* over the 10-year follow-up period, while correlations with their adoptive parents' educational levels *decreased* (Weinberg et al., 1992). It may puzzle you how the children's IQs could be correlated with their biological parents' intelligence if the children's IQs had moved into the adoptive parents' range. Although we usually think that correlation indicates similarity within pairs of scores (e.g., a parent's and a child's IQs), this is not necessarily the implication; rather, it means that the scores have similar rank orders in their groups. In other words, the parent with the highest IQ has the child with the highest IQ, the parent with the second highest IQ has the child with the second highest IQ, and so on. Now move the children into a new household and raise each child's IQ by

Application   Enhancing Intelligence and Performance

In 2008 the journal *Nature* asked its readers, who are mostly academics, about their attitudes on the use of cognitive-enhancing drugs. Surprisingly, 1 in 5 of the respondents said they had used drugs to stimulate their focus, concentration, or memory. But how well do these drugs work?

A survey of the literature shows that no one has found a miracle drug for boosting intelligence yet, although some drugs do enough to keep people using them (de Jongh, Bolt, Schermer, & Olivier, 2008). Stimulants such as modafinil and methylphenidate (Ritalin) improve alertness and some forms of learning; dopamine agonists such as d-amphetamine improve working memory; and the norepinephrine receptor agonist guanfacine (intended for use with attention deficit hyperactivity disorder) helps with planning and spatial memory. Of the compounds in commercial development specifically to enhance cognitive functioning, the ampakines are the most advanced; ampakines work by increasing the time AMPA receptors remain open. There are downsides, of course, mostly in the form of side effects, and half of the users in the *Nature* poll reported problems. Side effects of cognitive-enhancing drugs may include sedation, headache, nausea, sleeplessness, and reduction in blood pressure.

The enhancements are spotty, as well; one kind of functioning improves while another does not or may even be impaired. And a significant disappointment is that the drugs often provide little or no aid to people who are well rested or who are already high performers.

Eventually, it might be possible to tweak our intelligence by manipulating genes, and neuroscientists have already created 33 mutant strains of mice to help demonstrate that fact ("Small, Furry . . . and Smart," 2009). The mice are smarter at one task or another, but they have their problems, too. One strain is overly fearful, and another solves some problems well but struggles with simpler ones. We're a long way from genetically enhancing human intelligence, and a smart pill may be almost as far in the future; in fact, four companies *Science* magazine cited in 2004 as "stars" in the search for memory-enhancing drugs are now either out of business or in financial trouble (Stix, 2009). The chief scientific officer at one of the companies said, "When I began working on these learning and memory drugs I thought I'd find a drug that might be able to help my parents. Now I'm just hoping that we can find a drug in time for me" (quoted in "Small, Furry . . . and Smart," p. 864).

10 points; the parent with the highest IQ still has the child with the highest IQ and the parent with the second highest IQ . . . you get the point. The correlation tells us that the children's IQs are still tied to their parents' intelligence, as if by an elastic string that can stretch but nevertheless affects *how much* the IQ can change; that elastic string in this case is the influence of genes.

As much as we know about how the brain creates intelligence, you would think we would also know how to enhance people's cognitive functioning. The Application reveals that this is not as easy as you might think.

Concept Check   *Take a Minute to Check Your Knowledge and Understanding*

☐ What is the likely neural basis of general intelligence and of separate components of intelligence?

☐ What are the relative contributions of heredity and environment to intelligence? Why is it so difficult to identify the environmental influences?

☐ How can adopted children's IQs increase into the range of their adoptive homes and yet be more highly correlated with their parents' intelligence? (*Hint:* Draw a diagram, with made-up IQ scores of children from several families.)

Deficiencies and Disorders of Intelligence

Intelligence is as fragile as it is complex; accordingly, the list of conditions that can impair intelligence is impressively, or depressingly, long. To give you a feel for the problems that can occur in this most revered of our assets, we will add a few thoughts to what we have already covered in Chapter 12 on aging, take a brief look at intellectual disability, then spend more time on autism in recognition of its half-century-long challenge to neuroscientists' investigative skills, and finish with attention deficit hyperactivity disorder.

Effects of Aging on Intelligence

In the previous chapter, we discussed the most widely known cognitive disorder of aging, Alzheimer's disease. Here, we will limit our attention to more or less normal declines in cognitive abilities that are associated with aging. One source of normal decline is the reduced activity of numerous genes involved in long-term potentiation and memory storage due to age-related damage (T. Lu et al., 2004). Genes involved in synaptic functioning and plasticity, including those responsible for glutamate and GABA receptors and for synaptic vesicle release and recycling, are particularly affected.

How much capability is lost by the elderly?

Although intelligence and cognitive abilities do typically decline with age, the amount of loss has been overestimated. One reason is that people are often tested on rather meaningless tasks, like memorizing lists of words; older people are not necessarily motivated to perform on this kind of task. When the elderly are tested on the content of meaningful material such as television shows and conversations, the decline is moderate (Kausler, 1985). Another reason for the overestimation is that early studies were *cross-sectional:* People at one age were compared with different people at another age. You have already seen from Flynn's research that

FIGURE 13.12 Compensatory Brain Activity in High-Performing Older Adults.

A memory task activated the right prefrontal area in young and in low-performing older adults. Older adults who performed as well as the young showed activation in both prefrontal areas.

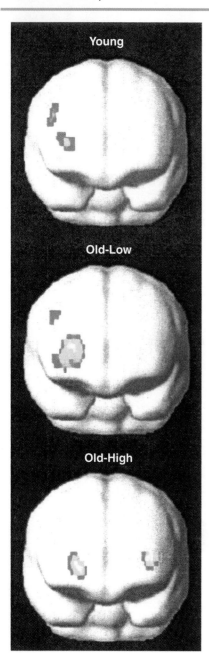

Young

Old-Low

Old-High

SOURCE: From "Aging Gracefully: Compensatory Brain Activity in High-Performing Older Adults," by R. Cabeza et al. *NeuroImage* 17: 1394–1402, fig. 2, p. 1399. © 2002 Elsevier.

more recent generations have an IQ test performance advantage over people from previous generations. When the comparison is done *longitudinally*—by following the same people through the aging process—the amount of loss diminishes (Schaie, 1994). Schaie followed 5,000 adults for 35 years. Perceptual speed dropped from age 25 on, and numeric ability dropped rather sharply after age 60. However, the other capabilities—inductive reasoning, spatial orientation, verbal ability, and verbal memory—increased until middle age before declining gradually to slightly lower than their levels at age 25.

Apparently, performance speed is particularly vulnerable during aging, and its loss turns out to be important. Schaie (1994) found that statistically removing the effects of speed from test scores significantly reduced elderly individuals' performance losses. We saw earlier that working memory is especially important to intellectual capability. A study of people ranging in age from 18 to 82 showed that speed of processing accounted for all but 1% of age-related differences in working memory (Salthouse & Babcock, 1991).

You know that a determinant of general intelligence is the ability to integrate activity between brain areas. Even in healthy aging there is a loss of coordination in the *default mode network*, portions of the frontal, parietal, and temporal lobes that are active when the brain is at rest or focused internally rather than on the outside world; activity in the default mode network is thought to represent preparedness for action (Andrews-Hanna et al., 2007; Broyd et al., 2009). To the extent this coordination diminishes, cognitive ability does as well. Imaging reveals that the loss is due to a decline in white matter connections among the areas. The brain does have ways of compensating, such as the increased metabolism that is seen in subjects with mild intellectual disability and the recruitment of frontal areas in dyslexics during reading (see Chapter 9). Studies of elderly individuals who are "aging gracefully" indicate that they are holding their own through additional neural effort (Helmuth, 2002). For example, a memory task activated only the right prefrontal cortex in young adults and in older adults who performed poorly, but older adults who performed as well as the young adults used both prefrontal areas to perform the tasks (Figure 13.12; Cabeza, Anderson, Locantore, & McIntosh, 2002). The increase in frontal activity is in proportion to decreases in other areas, which supports the idea that the shift serves a compensatory function (S. W. Davis, Dennis, Daselaar, Fleck, & Cabeza, 2008). Compensation may not be limited to frontal activity; it has also been observed in the parahippocampal area during a memory recognition task (van der Veen, Nijhuis, Tisserand, Backes, & Jolles, 2006). This less than efficient adaptation might not be the only pathway to graceful aging; elderly subjects initially coped with a task via compensation, but after just 5 hours of training on the task, they showed brain patterns like those of young subjects (Erickson et al., 2007).

Some of the loss in performance is due to nonphysical causes and is reversible; for example, older people often lack opportunity to use their skills. In one study, aged individuals were able to regain part of their lost ability through skills practice, and many of them returned to their predecline levels; they still had some advantage over controls 7 years later (Schaie, 1994). Elderly people also improved in memory test scores when their self-esteem was bolstered by presenting them with terms that depict old age in positive terms such as *wise, learned,* and *insightful* (B. Levy, 1996).

Loss in performance that has a physical basis may be reduced if not reversed. Diet appears to be one factor; for example, in a study of 6,000 people over the age of 65, cognitive decline was 13% less in those who ate two or more fish meals per week, compared with people who ate fish less than once per week (M. C. Morris, Evans, Tangney, Bienias, & Wilson, 2005). Evidence in this study and others suggested that the important factor was the overall pattern of fat intake. Other researchers have hypothesized that cognitive as well as sensory and motor decline is partly due to degradation of inhibitory activity at GABA receptors. Administration of GABA or a GABA agonist (muscimol) in the visual cortex improved the selectivity of orientation-sensitive neurons in old monkeys but not in young ones (Leventhal,

Wang, Pu, Zhou, & Ma, 2003). The authors suggest that enhancing GABA activity could improve cognitive as well as other functioning in the elderly.

Interestingly, the sex hormones provide some protection against the cognitive effects of aging. In menopausal women, estrogen replacement therapy reduces the decline in verbal and visual memory as well as lowering the risk of Alzheimer's disease (Sherwin, 2003; van Amelsvoort, Compton, & Murphy, 2001). The importance of estrogen is bolstered by the fluctuations that occur during the menstrual cycle. First, remember from Chapter 7 that women tend to be superior to men on some types of verbal tasks and that men typically outperform women on tasks requiring spatial ability. During the part of the month when estrogen is high, women perform higher on verbal tasks; then during menstruation estrogen drops and so does performance on the verbal tasks, but spatial performance improves (Kimura & Hampson, 1994; Maki, Rich, & Rosenbaum, 2002).

How does estrogen produce these effects? We are not sure, but we do know that neurons throughout the brain have estrogen receptors; estrogen levels during the menstrual cycle are correlated with cortical excitability (M. J. Smith, Keel, et al., 1999), increased glucose metabolism (Reiman, Armstrong, Matt, & Mattox, 1996), blood flow in areas involved in cognitive tasks (Dietrich et al., 2001), and responsiveness to acetylcholine, which is important in memory and cognitive functioning (O'Keane & Dinan, 1992). Finally, because untreated menopausal women are more impaired than women receiving estrogen replacement on tests of working memory, response switching, and attention, it is clear that estrogen particularly improves functioning in prefrontal areas (Keenan, Ezzat, Ginsburg, & Moore, 2001).

So what about the male of the species? Men who maintain testosterone production past the age of 50 have better preserved visual and verbal memory and visuospatial functioning (Moffat et al., 2002). The effects of replacement therapy have been variable, owing apparently to the form of the testosterone preparation used. However, a number of studies have shown improvement in spatial, verbal, and working memory (Cherrier et al., 2005; Cherrier, Craft, & Matsumoto, 2003; Gruenewald & Matsumoto, 2003). Interestingly, testosterone improves only spatial memory; additional memory improvement requires that the testosterone be delivered in the form of dihydrotestosterone, which can be converted to estrogen in the brain by the process of aromatization (Cherrier et al., 2003, 2005). When it comes to cognitive abilities, we could be tempted to consider estrogen a wonder drug.

So losses are smaller than believed, and they differ across people and across skills. In addition, practice, esteem enhancement, and an active lifestyle may slow cognitive decline during aging. Obviously, we cannot stereotype the older person as a person with diminished abilities.

Intellectual Disability

Intellectual impairment was previously referred to as retardation, but that term has taken on such negative meaning in popular usage that practitioners and authorities are shifting to the term *intellectual disability. Intellectual disability* is a limitation in intellectual functioning (reasoning, learning, problem solving) and in adaptive behavior originating before the age of 18. The criteria are arbitrary and are based on judgments about the abilities required to get along in our complex world. In 1994, the American Psychiatric Association set the criteria as a combination of an IQ below 70 points and difficulty meeting routine needs like self-care. Looking back at Figure 13.2, you can see that 2% of the population falls in this IQ range, and a lower percentage would meet both criteria. Not only is any definition arbitrary, but it is situational and cultural as well; a person considered to have an intellectual disability in our society might fare reasonably well in a simpler environment. The situational nature is illustrated by the fact that many individuals shed the label as they move from a childhood of academic failure into adulthood and demonstrate their ability to live independent lives.

4
Intellectual Disability Facts

Q What causes intellectual disabilities?

Intellectual disability has historically been divided into four categories, as shown in Table 13.1. (These are likely to change in the upcoming edition of the American Psychiatric Association's *Diagnostic and Statistical Manual.*) As you can see, most of the impaired fall in the mild category. Mildly impaired individuals are handicapped in functioning by knowing fewer facts and lacking strategies for learning and solving problems (Campione, Brown, & Ferrara, 1982). And, not surprisingly, they are slower at performing mental operations like retrieving information from memory.

Most individuals in the mild category come from families of lower socioeconomic status and have at least one relative with the diagnosis (Plomin, 1989); psychologists believe their impairment is due to a combination of environmental and hereditary causes. Recently, four gene locations have been implicated in mild intellectual disability; each location accounted for only a small amount of the variation in intelligence, but their effects were cumulative (Butcher et al., 2005). About 25% of cases can be clearly attributed to one of the 200-plus physical disorders known to cause intellectual impairment (K. G. Scott & Carran, 1987). Intellectual disability can be caused by diseases contracted during infancy, such as meningitis, and by prenatal exposure to viruses, such as rubella (German measles). As we saw in Chapter 5, maternal alcoholism is now the leading cause of intellectual disability. There are numerous other causes as well, and we will discuss a few of them in the following pages.

Down syndrome, usually caused by the presence of an extra 21st chromosome, typically results in individuals with IQs in the 40 to 55 range, although some are less impaired (Figure 13.13). Its prevalence of 1 in every 700 births makes it the most common genetic cause of intellectual disability (D. L. Nelson & Gibbs, 2004). Recall that the amyloid precursor protein gene that is involved in early-onset Alzheimer's disease is located on chromosome 21 and that it was discovered because Down syndrome individuals also develop amyloid plaques (Goate et al., 1991; Murrell, Farlow, Ghetti, & Benson, 1991). Ninety-five percent of people with Down syndrome have the entire extra chromosome, but in a few cases only an end portion is present, which is attached to another chromosome. A strain of mouse (Ts65Dn) engineered with a third copy of 55% of chromosome 21 has been a useful model for studying Down syndrome. As in people with that disorder, its glial cells secrete less of two proteins that support neuron survival. Treating pregnant Ts65Dn females with these proteins eliminates the developmental delays that would ordinarily be seen in their offspring (Toso et al., 2008). Conceivably this strategy could be used with pregnant women when amniocentesis (genetic testing of the amniotic fluid)

TABLE 13.1 Categories of Intellectual Disability.

Category	IQ	Percentage	Adaptation
Mild	50–70	85	Educable to sixth-grade level; may be self-supporting as adult, with assistance.
Moderate	35–49	10	May achieve education to second-grade level, live outside institution with family, and contribute to support.
Severe	20–34	4	Verbal communication and ability to profit from vocational training are limited.
Profound	Below 20	1	Little or no speech. Requires constant care and supervision.

FIGURE 13.13 Luke Zimmerman and Down Chromosomes.

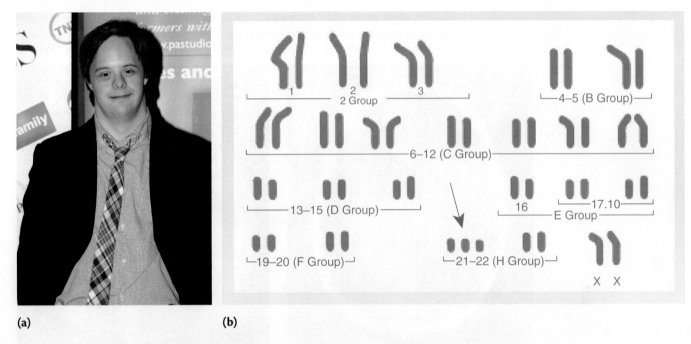

(a)   (b)

(a) Some Down syndrome sufferers are more fortunate than others. In spite of the disorder, Luke Zimmerman plays a starring role in the television series *The Secret Life of the American Teenager*. (b) Chromosomes of a person with Down syndrome (female). The arrow points to the three 21st chromosomes.

SOURCES: (a) © Chris Wolf/Getty Images; (b) © www.nads.org, 2007.

reveals that the fetus has a third 21st chromosome. Another therapeutic possibility is postnatal correction of inadequate norepinephrine release in the hippocampus, a deficiency that is believed to contribute to learning impairment. After treatment with a drug that increased norepinephrine levels, Ts65Dn mice performed as well as normal mice (Salehi et al., 2009). However, it may be an oversimplification to focus only on the effects of chromosome 21 genes. Amniocentesis in mothers carrying fetuses with three copies of the chromosome found 5 genes on chromosome 21 that were upregulated or downregulated, but 414 genes on other chromosomes that were altered (Slonim et al., 2009). The researchers are studying the protein products of these genes to improve understanding of the mechanism of the disorder and to identify prenatal biomarkers (Cho, Smith, & Diamandis, 2010).

Another frequent genetic cause of intellectual disability is *fragile X syndrome*, which is due to a mutation in the fragile X mental retardation 1 gene (*FMR1*). The *FMR1* gene normally contains between 6 and 45 repetitions of the nucleotide sequence C, G, G. If the number of CGG repeats reaches 200, the gene is turned off, and no protein is made. Fragile X syndrome is the result, with IQs typically below 75. An intermediate number of CGG repeats results in reduced protein production, accompanied by a slight increase in the chance of intellectual disability. Males are affected more often, and when fragile X does occur in females, the symptoms are usually milder. In *FMR1* knockout mice, unusual numbers of dendrites and immature spines indicate a failure to prune excess synapses (Bagni & Greenough, 2005).

Phenylketonuria is due to an inherited inability to metabolize the amino acid phenylalanine; the excess phenylalanine interferes with myelination during

FIGURE 13.14 The Hydrocephalic Brain.

(a) Normal brain; (b) Hydrocephalic brain. Notice the large lateral ventricles and the small amount of cortex around the perimeter in the hydrocephalic brain.

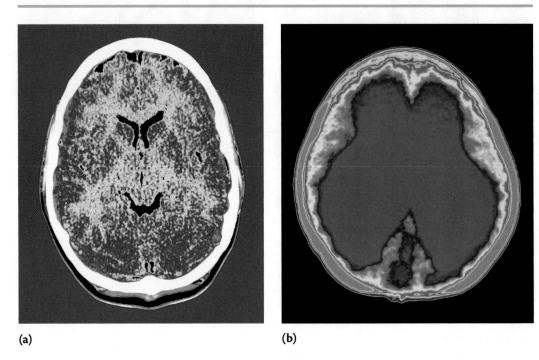

(a) **(b)**

development. Newborn infants are routinely tested for phenylalanine in the urine or blood, and intellectual disability can be prevented by avoiding foods containing phenylalanine. The artificial sweetener aspartame is a familiar example of a substance that is high in phenylalanine. Without dietary treatment, the individual is severely or profoundly disabled, with an adult IQ around 20 points.

Hydrocephalus occurs when cerebrospinal fluid builds up in the cerebral ventricles; the increased fluid volume crowds out neural tissue, usually causing intellectual disability (Figure 13.14). As we saw in Chapter 3, hydrocephalus can also be treated if caught early, by installing a shunt that prevents the accumulation of the excess cerebrospinal fluid. Also in that chapter, you learned that some individuals seem not to be harmed by the dramatic loss of cortex; in fact, half of the hydrocephalics whose ventricles fill 95% of the cranium have IQs over 100 (Lewin, 1980).

Autism

Autism is a disorder that typically includes compulsive, ritualistic behavior, impaired sociability, and intellectual disability (U. Frith, 1993; U. Frith, Morton, & Leslie, 1991). For decades, a lack of evidence for a physical cause implicated poor parenting; this along with the autistic child's social coldness and bizarre rocking, hand flapping, and head banging created an enormous burden of guilt for the parents. Autism is one of five *autism spectrum disorders. Asperger's syndrome* is the most similar to autism and the most likely to share common causes.

5
Autism Facts

People with Asperger's syndrome, like autistic individuals, are socially impaired and display repetitive movements and preoccupations with narrow interests, but their language and cognitive development and self-help skills are more normal. In the 1960s and 1970s, autism prevalence was estimated at 5 per 10,000 but has risen erratically to between 10 and 20 per 10,000; the prevalence of autistic spectrum disorders has been more consistent at around 60 per 10,000 (Newschaffer et al., 2007). At least some of the increase can be attributed to improved detection, broader diagnostic criteria, and doctors' greater willingness to use the label because of decreasing stigmatization of autism and because the diagnosis will qualify the family for increased services and financial assistance (Rutter, 2005). Whether there has been an actual increase in the disorder is unclear. When researchers in Stafford, England, repeated a study of preschoolers done in 1998 in exactly the same area and using the same diagnostic methods, they found the incidence of autism had not changed (Chakrabarti & Fombonne, 2005). However, the position of most authorities is that we simply don't know how much the rate has actually increased.

Cognitive and Social Impairment

The repetitive behaviors are also characteristics of some children with intellectual disability, and about 80% of children with autism fall into that category; autism accounts for 86% of children with IQs below 20 and 42% of those with IQs of 20 to 49 (U. Frith, 1993). Autistic individuals share a common core of impairment in *communication, imagination,* and *socialization* (U. Frith, 1993). They are mute or delayed in language development, and they have trouble understanding verbal and nonverbal communication. Their difficulty with imagination shows up in an inability to pretend or to understand make-believe situations. Their use of language is also very literal, and some autistic adults have an obsessive interest in facts.

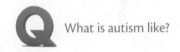 What is autism like?

These characteristics make it difficult to socialize with others, which is what sets children with autism apart the most. They usually prefer to be alone and ignore people around them (Figure 13.15). Their interaction with others often is limited to requests for things they want; they otherwise treat people as objects, sometimes even walking or climbing over them. Verbalization is usually limited, and the child often repeats what others say (echolalia).

Some researchers believe that much of the social behavior problem is that the person with autism lacks a *theory of mind,* the ability to attribute mental states to oneself and to others; in other words, the autistic person cannot infer what other people are thinking. One man with autism said that people seem to have a special sense that allows them to read other people's thoughts (Rutter, 1983), and an observant autistic youth asked, "People talk to each other with their eyes. What is it that they are saying?" (U. Frith, 1993). In a study that measured this deficiency, children watched hand puppet Anne remove a marble from a basket where puppet Sally had placed it, and put it in a box while Sally was out of the room. On Sally's return, children were asked where she would look for the marble. Normal 4-year-olds had

FIGURE 13.15 Autistic Individuals Typically Feel Threatened by Social Interaction.

Most children enjoy hiding out in a tent fashioned from sheets draped over chairs, but for this autistic child, it is a defense against social contact.

SOURCE: © Ali Jarekji/Reuters/Corbis.

no problem with this task, nor did Down syndrome children with a mental age of 5 or 6. But 80% of children with autism with an average mental age of 9 answered that Sally would look in the box (U. Frith et al., 1991).

There are two hypotheses as to how we develop a theory of mind. According to the "theory theory," we build hypotheses over time based on our experience. Simulation theory holds that we gain insight into people's thoughts and intentions by mentally mimicking the behavior of others. This view gets some support from studies of the mirror neurons talked about in Chapters 8 and 9. Individuals who score higher on a measure of empathy tend to have more activity in these mirror neurons (Gazzola, Aziz-Zadeh, & Keysers, 2006). Researchers have suggested that impaired mirror functions reduce the autistic person's ability to empathize and to learn language through imitation. Children with autism engage in less contagious yawning than other children do (Senju et al., 2007), and they show neural deficiencies during mirroring tasks. For example, they can imitate others' facial expressions, but mirror neuron activity while doing so is either delayed or nonexistent (Dapretto et al., 2005; Oberman et al., 2007). Other studies show reduced activation in the inferior frontal cortex and motor cortex, suggesting weakness in the dorsal stream connections that provide input to those areas (Nishitani, Avikainen, & Hari, 2004; Villalobos, Mizuno, Dahl, Kemmotsu, & Müller, 2005). This interpretation was supported in a study of individuals with Asperger's syndrome. When they imitated facial expressions, transmission over the dorsal stream (occipital to superior temporal to posterior parietal to frontal) was delayed 45 to 60 milliseconds compared with normal controls (Nishitani et al., 2004).

Autistic Savants and High-Functioning Autistics

A *savant* is a person with exceptional intellectual skills, beyond the level of "typical" genius, like Leonardo da Vinci or Albert Einstein. However, the term is more frequently used to describe individuals who have one or more remarkable skills but whose overall functioning is below normal; half of these individuals with islands of exceptional capabilities are *autistic savants* (Treffert & Christensen, 2006). Some can play a tune on the piano after hearing it once, another can memorize whole books, while others take cube roots of large numbers in their heads or calculate the day of the week for any date thousands of years in the past or future. A few "ordinary" individuals can perform similar feats, but the savant's performance is typically faster, more automatic, and without insight into how it is done (A. W. Snyder & Mitchell, 1999). The savant's exceptional capability may be limited in scope, however; some who are calendar calculators cannot even add or subtract with accuracy (Sacks, 1990).

The source of the autistic savant's enhanced ability is unknown. Dehaene (1997) suggests that it is due to intensely concentrated practice, but more typically the skill appears without either practice or instruction, as in the case of a 3-year-old who began drawing animated and well-proportioned horses in perfect perspective (Selfe, 1977). Ramachandran and Blakeslee (1998) suggest that a specialized area of the brain becomes enlarged at the expense of others. Allan Snyder and John Mitchell (1999) believe that these are capabilities within us all and are released when brain centers that control executive or integrative functions are compromised. This, they say, gives the savant access to speedy lower levels of processing that are unavailable to us. But, lacking the executive functions, the savants perform poorly on apparently similar tasks that require higher-order processing. The idea gains some credibility from the case of a man impaired in his left temporal and frontal areas by dementia; in spite of limited musical training, he began composing classical music, some of which was publicly performed (B. L. Miller, Boone, Cummings, Read, & Mishkin, 2000; also see Figure 13.16). The best-known savant was Kim Peek, the model used by Dustin Hoffman in his portrayal of Raymond Babbitt in the movie

6

Meet the Savants

FIGURE 13.16 Savant-Like Ability Following Brain Impairment.

(a) The scan is from a 64-year-old woman with dementia in the left frontal-temporal area, which shows less activity than the right.
(b) After the onset of her dementia, she began to do remarkable paintings like the one here.

(a)

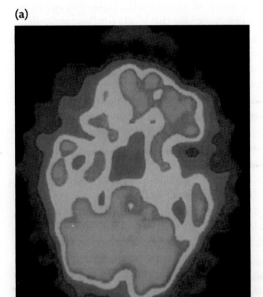

(b)

SOURCE: From "Emergence of Artistic Talent in Frontal-Temporal Dementia," by B. L. Miller et al., *Neurology, 51,* pp. 978–982. Copyright © 1998. Reprinted by permission of Wolters Kluwer.

Rain Man; Peek's brain had several anomalies, particularly in the left hemisphere (Figure 13.17; Treffert & Christensen, 2006). In a rare experimental test, Tracy Morrell (who did the study as her undergraduate honors thesis) and her colleagues used transcranial magnetic stimulation to interfere temporarily with functioning of the left frontal-temporal area in normal individuals (R. L. Young, Ridding, & Morrell, 2004). Five of the 17 participants improved in one or more specific skills, including calendar calculating, memory, drawing, and mathematics. Whatever the explanation, the phenomenon adds to the argument that intellectual ability involves multiple and somewhat independent modules.

If these savants have an island of exceptional ability, autism is an island of impairment in the *high-functioning autistic.* As an infant, Temple Grandin would stiffen and attempt to claw her way out of her affectionate mother's arms (Sacks, 1995). She was slow to develop language and social skills, and she would spend hours just dribbling sand through her fingers. A speech therapist unlocked her language capability, starting a slow emergence toward a normal life. Even so, she did not develop decent language skills until the age of 6 and did not engage in pretend play until she was 8.

As an adult, Grandin earned a doctorate in animal science; she teaches at Colorado State University and designs humane facilities for cattle, while lecturing all over the world on her area of expertise and on autism. Still, her theory of mind is poorly developed, and she must consciously review what she has learned to decide what others would do in a social situation. She says that she is baffled by relationships that are not centered around her work and that she feels like "an anthropologist on Mars."

7
Temple Grandin

FIGURE 13.17 Kim Peek, the Original Rain Man.

Kim memorized 7,600 books as well as every area code, zip code, highway, and television station in the United States, but he had an IQ of 87 and could not care for himself. He died of a heart attack in December 2009 at the age of 58.

SOURCE: From "Islands of Genius," by D. A. Treffert and G. L. Wallace, 2004, *Scientific American Mind* (January); photo by Ethan Hill.

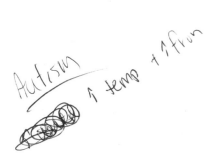

Brain Anomalies in Autism

Autism was long thought to be purely psychological in origin, because no specific brain defects had been found. The problem was blamed on a lack of maternal bonding or a disastrous experience of rejection that caused the child to retreat into a world of aloneness (U. Frith, 1993). But no evidence could be found for this kind of influence; autistic children often had exemplary homes, and children with extremely negative experiences did not become autistic. The frequent association with intellectual disability and epilepsy implied that autism was a brain disorder. Later work found subtle but widespread brain anomalies, especially in the brain stem, the cerebellum, and the temporal lobes (Happé & Frith, 1996). The location of the damage varied among individuals, which suggests that there are various pathways to autism.

Monitoring how brain development progresses revealed additional clues. The brain of individuals with autism is normal or slightly reduced in size at birth but undergoes dramatic growth during the 1st year (Redcay & Courchesne, 2005). The overgrowth occurs in frontal and temporal areas that are important for the social, emotional, and language functions that are impaired in the disorder. The excess growth ends around 3 to 5 years of age, but by then the brain has already reached normal adult size (Courchesne et al., 2007). In adulthood these areas are underactivated during tasks that would ordinarily engage them, for example, when viewing animations that require the viewer to understand the mental state of the actor (Castelli, C. Frith, Happé, & U. Frith, 2002). Also, some of the structures that are enlarged early in life are undersized in adulthood, which suggests that there is a third stage of development involving degeneration (Courchesne et al.).

It appears that some of the underactivated areas are actually functional, but the research conditions fail to enlist their activity in subjects with autism. For example, several studies have reported deficits in the fusiform face area (FFA) when subjects with autism view photographs of faces and deficits in the mirror neuron system when they observe and imitate others' movements. However, the FFA is activated if the person with autism is instructed to focus on a mark placed in the middle of the face (Hadjikhani et al., 2004; Pierce et al., 2001). A few studies have reported activation of mirror neurons, and the authors of one of these suggested that a delay in mirror activity could be one reason other imaging studies have failed to detect it (Dinstein et al., 2010). A possible explanation for inactivity and delay in otherwise functional areas is a lack of cooperation across a network. Synchronization of activity within a network—referred to as functional connectivity—is reduced in the default mode network during rest and in other networks during task performance, and the lack of connectivity is correlated with deficits in social and communication abilities and task performance (Assaf et al., 2010; Just et al., 2007). There are two possible reasons for the impaired connectivity: One is reductions in white matter connections among the frontal, parietal, and temporal lobes (Figure 13.18; Barnea-Goraly et al., 2004), and the other is a genetic weakness in synaptic connections (Garber, 2007).

Biochemical Anomalies in Autism

As researchers looked for genetic links to autism, several of the genes that turned up pointed in turn to serotonin, glutamate, GABA, and oxytocin (Freitag, Staal, Klauck,

Duketis, & Waltes, 2010; Sutcliffe, 2009). Serotonin has received considerable attention because some of the medications used to treat autistic symptoms increase serotonin activity. These include antidepressants such as fluoxetine (Prozac), which acts as a serotonin reuptake inhibitor, and the antipsychotic risperidone, which is an antagonist for dopamine as well as serotonin. Both drugs are helpful with repetitive behavior, and risperidone also reduces aggression (McDouble, Stigler, Erickson, & Posey, 2006). Drugs that antagonize glutamate activity improve social functioning, and they reduce withdrawal, hyperactivity, and inappropriate speech.

In Chapter 7 we saw that oxytocin facilitates bonding and social recognition; for that reason it is known as the "sociability molecule." Children with autism have lower levels of oxytocin than controls, and this difference is pronounced in those who are described as aloof (Modahl et al., 1998). Oxytocin has also been associated with repetitive behaviors in lower animals, so Eric Hollander and his colleagues (2003) observed the behavior of adults with autism and Asperger's syndrome while they received intravenous infusions of oxytocin. The treatment reduced the severity and the number of types of repetitive behavior. A single intravenous dose of oxytocin produced a 2-week improvement in autistic adults' ability to recognize emotions in people's tone of voice (Hollander et al., 2007), and administration in a nasal spray improved recognition of facial expressions of emotion (Guastella et al., 2010). In another study, oxytocin improved social cooperation and self-reported trust in individuals with spectrum disorders and increased the time spent looking at the eyes in photographs of faces (Andari et al., 2010). Results of an fMRI study indicated that oxytocin improves trust by reducing activity in fear areas in the amygdala and midbrain (Figure 13.19; Baumgartner, Heinrichs, Vonlanthen, Fischbacher, & Fehr, 2008).

The Environment and Autism

Parental treatment has been ruled out as the cause of autism, but a number of other environmental conditions have been proposed as contributing factors. One's suspicions immediately turn to pesticides, and there is some preliminary research to support a link. Specifically, organophosphate pesticides appear to interfere with neural development by inactivating acetylcholinesterase; in a study of children born to mothers living in the central valley of California, the likelihood of autism spectrum disorders increased with the amount of organochlorine pesticides used on nearby fields and decreased with distance from the fields (E. M. Roberts et al., 2007). Another possibility is that brain development can be disrupted by an autoimmune reaction triggered by toxins or maternal infections (R. Dietert & J. Dietert, 2008). The case for an autoimmune effect was strengthened by a study of 689,000 children in Denmark; autism risk was increased in children whose mothers had an autoimmune disease, including rheumatoid arthritis, celiac disease, and type 1 diabetes (Atladóttir et al., 2009). However, the data do not permit us to say whether the effect was due to

FIGURE 13.18 Areas of Decreased White Matter in Autistic Children.

Areas of decrease are shown in dark gray. The corpus callosum (white) and the amygdalas (checkered gray) are shown for reference.

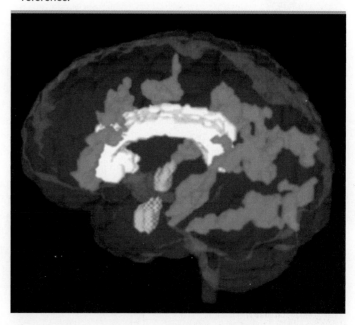

SOURCE: From "White Matter Structures in Autism: Preliminary Evidence From Diffusion Tensor Imaging," by N. Barnea-Goraly et al., 2004, *Biological Psychiatry*, 55, 323–326.

FIGURE 13.19 Reduced Response to Betrayal of Trust Following Oxytocin.

Compared to subjects receiving a placebo, those who received nasally administered oxytocin responded less to betrayal of trust in an investment simulation. Areas of comparatively reduced activity (shown in yellow) were the caudate nucleus (Cau), amygdala (Amy), and midbrain (MB).

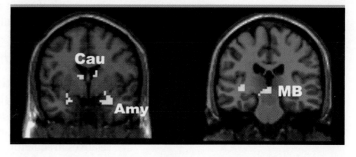

SOURCE: From "Oxytocin Shapes the Neural Circuitry of Trust and Trust Adaptation in Humans," by T. Baumgartner et al., 2008, *Neuron*, 58, 639–650, fig 4.

Childhood Vaccines and Autism

In the News

In 1998 Andrew Wakefield and several colleagues published a paper in which they linked the measles, mumps, and rubella (MMR) vaccine to autism. Years later, two reviews of all the available studies concluded that there was no credible link between autism and the MMR vaccine or the mercury-derived preservative (thimerosal) used in some vaccines (Demicheli, Jefferson, Rivetti, & Price, 2005; Immunization Safety Review Committee, 2004). In early 2010 the General Medical Council, which oversees doctors in Britain, decided that Wakefield's study was methodologically flawed and that he had acted unethically. The *Lancet*, which published the study, called the report "the most appalling catalog and litany of some of the most terrible behavior in any research" and took the rare action of withdrawing the study.

Many parents whose children's first signs of autism coincided with a round of childhood vaccinations say they find these reports unconvincing. More than 5,000 families have brought suit; a U.S. federal court has denied the claims in four test cases, but that does not signal the end of litigation or of the controversy. In the meantime there has been a disturbing decrease in the rate of childhood immunization in several countries in spite of the health risks.

SOURCES: Park, M. (2010, February 2). Medical journal retracts study linking autism to vaccine. *CNN Health*. Retrieved February 2, 2010, from http://www.cnn.com/2010/HEALTH/02/02/lancet .retraction.autism/index.html

Vaccine court finds no link to autism. (2010, March 12). *CNN Health*. Retrieved March 12, 2010, from http://www.cnn.com/2010/HEALTH/03/12/vaccine.court.ruling.autism/index.html

The Wakefield Paper as It Appears on the *Lancet* **Website.**

SOURCE: Wakefield et al. (1998). Ileal-lymphoid-nodular hyperplasia, non-specific colitis, and pervasive developmental disorder in children. *Lancet, 351*, 637–641. Retrieved from http://vaccines.procon.org/sourcefiles/ retracted-lancet-paper.pdf.

---

prenatal exposure to antibodies in the womb or to genes shared with the mother. The father can contribute to the risk of autism as well; children born to fathers over the age of 40 were almost six times more likely to be diagnosed with an autism spectrum disorder than those whose fathers were under 30 (Reichenberg et al., 2006).

Some of the environmental factors probably exert their effect by causing gene mutations. Five neurotoxins that appear to be associated with a higher incidence of autism are also known to be mutagens; these are mercury, cadmium, nickel, trichloroethylene, and vinyl chloride (Kinney, Barch, Chayka, Napoleon, & Munir, 2009). Even decreased sun exposure—due to living distant from the equator, in a city, or in an area of high precipitation—is suspect; lack of sunlight can lead to vitamin D deficiency, and vitamin D plays an important role in repairing DNA damage. We will talk about genetic influences in the next section, but first we will take a small detour to look at the environmental influence that *isn't* but nevertheless is still In the News.

Heredity and Autism

Siblings of autistics are 25 times more likely to be diagnosed with autism than other children (Abrahams & Geschwind, 2008); the number would be even higher, but parents tend to stop having children after the first autistic diagnosis. For the identical twin of an autistic, the risk of autism is at least 60%. However, nonautistic relatives frequently have autistic-like cognitive and social abnormalities. When these symptoms are also considered, the concordance for identical twins jumps to 92%, compared with 10% for fraternal pairs (A. Bailey et al., 1995). Earlier, the presence of these milder social and cognitive deficits in the parents was interpreted as evidence that they were fostering their children's symptoms psychologically. Now we understand that the reason is the number of autism-related genes the child has received from the two parents.

Autism occurs four times as frequently in males as in females (Fombonne, 2005), which suggests that some of the genes might be on the X chromosome. A few genes there were linked to autism, but the findings were not supported by later studies or their effects were negligibly small (Freitag, et al., 2010). Several other gene associations have been replicated; as you might expect, the genes contribute to neurotransmitter activity, neuron development and migration, and synapse formation (Freitag et al.; Sutcliffe, 2008). Small effect is not the only reason gene linkages are difficult to detect. Recent autism studies have found numerous copy number variations (CNVs; greater or lesser numbers of copies of a DNA sequence) in genes that are involved in neural development and synaptic transmission (Glessner et al., 2009; Pinto et al., 2010). In almost 6% of subjects these CNVs were de novo—not shared with either parent—which helps explain why some individuals with autism have no autistic relatives.

Attention Deficit Hyperactivity Disorder

Attention deficit hyperactivity disorder (ADHD) develops during childhood and is characterized by impulsiveness, inability to sustain attention, learning difficulty, and hyperactivity. Behaviorally, what we see is fidgeting and inability to sit still, difficulty organizing tasks, distractibility, forgetfulness, blurting out answers in class, and risk taking (Smalley, 1997). Although ADHD is often thought of as a learning disorder (and many ADHD children do have at least one learning disability), its effects are felt in every aspect of a person's life.

ADHD is the most common childhood-onset behavioral disorder. The National Institute of Mental Health estimates the incidence of ADHD at 3% to 5%; this means there are about 2 million affected children in the United States or, on average, 1 in every classroom of 25 to 30 students (National Institute of Mental Health, 2006). There is little consensus about the numbers, however; estimates run as high as 10% (Wender, Wolf, & Wasserstein, 2001), which fuels concerns that children are being overdiagnosed and overmedicated as an easy solution to classroom behavioral problems.

8
ADHD Facts

Between one third and two thirds of children with ADHD continue to show significant symptoms into adulthood (Wender et al., 2001). Besides having the expected difficulties with life and work, these individuals have greatly increased rates of antisocial personality disorder, criminal behavior, and drug abuse (Biederman, 2004); in fact, 35% of people seeking treatment for cocaine abuse have a history of childhood ADHD (F. R. Levin, et al., 1998). The link with drug abuse fostered concerns that treating children with stimulant drugs such as methylphenidate (Ritalin) and amphetamine was leading to addiction later in life. However, research that followed children with ADHD into adolescence and adulthood revealed that, at worst, stimulant treatment made no difference (Barkley, Fischer, Smallish, & Fletcher, 2003) and that it might even be protective against drug abuse (Figure 13.20; Wilens, Faraone, Biederman, & Gunawardene, 2003). The more positive outcome for the treated individuals could have been due to the reduction of symptoms, or it could have been the result of other factors, such as the support of parents who opted for treatment, but at least these results do not support fears that the medications act as gateway drugs.

Neurotransmitter Anomalies in ADHD

For the past 45 years, ADHD has been treated mostly with the stimulant drugs methylphenidate and amphetamines. These drugs increase dopamine and norepinephrine activity by blocking reuptake at the synapse, so both of these transmitters have been implicated. However, most research links ADHD to

FIGURE 13.20 Relative Odds of Avoiding Substance Abuse Disorder in Individuals Receiving Stimulant Treatment for ADHD as Children, Compared to Those Not Receiving Stimulant Treatment.

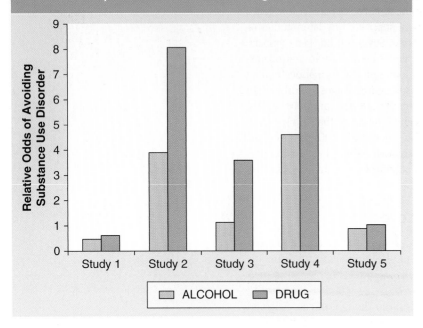

Individuals with ADHD who were treated with a stimulant drug as children were as much as 4.6 times as likely to be free of alcohol abuse disorder as individuals who did not receive stimulant treatment and up to 8.1 times as likely to be free of a drug abuse disorder. Values below 1 indicate that treated individuals in that study were at greater risk.

SOURCE: Based on data from Wilens et al. (2003).

reduced activity in dopamine pathways. One area with reduced activity is the prefrontal cortex (Ernst et al., 1998), and another is the striatum (K.-H. Krause, Dresel, Krause, King, & Tatsch, 2000). Functions of these structures include executive control, impulse inhibition, working memory, movement, learning, and reward; it is easy to see how their malfunction could contribute to ADHD symptoms. The significance of reward may be less immediately obvious, but several researchers believe that impaired reward contributes to impulsiveness because the allure of later rewards is too weak to overcome the temptation of immediate gratification (Castellanos & Tannock, 2002; Tripp & Wickens, 2009). Reduced brain activity and reward effects may both be due to the increased dopamine transporters found in the striatum of ADHD patients prior to drug treatment (Krause et al.). A second scan after 4 weeks of treatment with methylphenidate showed that transporter activity had decreased 29%. However, some 20% to 30% of patients do not respond to the traditional medications or cannot tolerate them (Biederman & Faraone, 2005). Two drugs sometimes used in their place, modafinil and atomoxetine, block norepinephrine reuptake, suggesting that transmitter's role in ADHD.

Brain Anomalies in ADHD

Studies of ADHD patients have implicated several areas of the brain, from the prefrontal cortex to the cerebellum (Castellanos et al., 2002; Raz, 2004), but the results have not always been in agreement. A review that combined the data from 21 studies concluded that children with ADHD have reduced volume in the cerebral hemispheres, especially the right; in the right caudate nucleus (a part of the striatum); and in parts of the cerebellum (Valera, Faraone, Murray, & Seidman, 2007). Elizabeth Sowell and her colleagues (2003) reported that the prefrontal and temporal areas are reduced in volume; they also interpreted a relative increase in gray matter in the temporal and inferior parietal areas as evidence of a reduction in white matter connections (Figure 13.21). The researchers suggested the disruption of an attention-inhibition network in that area, and Castellanos and his colleagues (2008) echoed that view after finding decreased connectivity in the default mode network. The idea is also supported by a study that required subjects to withhold a trained response on some trials, a task that is difficult for ADHD patients (Durston et al., 2003). On these "no-go" trials, control children activated a discrete network of structures, including prefrontal, striatal, and parietal areas, whereas ADHD children inefficiently activated a much broader area that encompassed much of the brain.

Heredity and ADHD

To say that ADHD runs in families would be an understatement: It is 5 to 6 times more frequent among patients' relatives than in the rest of the population; concordances are estimated at 79% for identical twins versus 32% for fraternals, and heritability averages 75% across studies (Biederman & Faraone, 2005; Castellanos & Tannock, 2002; Smalley, 1997). Like autism, ADHD is a complex, multisymptom disorder, and different individuals display different combinations of those symptoms. So, again, we would expect involvement of several genes, each with only a small effect and difficult to identify. That has been the case. Not surprisingly, the genes most frequently implicated in ADHD are involved in dopamine and serotonin transmission, as well as in synaptic functioning (Biederman & Faraone, 2005; Sharp, McQuillin, & Gurling, 2009; Thapar, O'Donovan, & Owen, 2005). A recently identified gene is involved in neuron survival, as well as neural transmission; *LPHN3* is expressed in several brain areas that are important in ADHD, and variations in the gene predict how well the patient will respond to stimulant medications (Arcos-Burgos et al., 2010). One study found 222 copy number variations in ADHD patients (Gai et al., 2010). The patients did not have a greater number of CNVs than controls, but theirs were located in or near genes involved in synaptic transmission, neural development, and learning and other psychological functions. Participation of a gene in more than one disorder should not be surprising, since different disorders share some symptoms. And remember that we are not talking about a gene for autism or a gene for ADHD or a gene for depression, but genes that regulate brain growth, receptor development, and so on.

The Environment and ADHD

What appears to be an environmental cause of ADHD often turns out to be an indication of a genetic predisposition in the parents. This includes correlations of ADHD with maternal smoking and stress during pregnancy (A. Rodriguez & Bohlin, 2005); parental abuse of alcohol, stimulants, and cocaine; and parental mood and anxiety disorders (Chronis et al., 2003). Confirmed environmental influences include brain injury, stroke, and pregnancy/birth complications (Biederman & Faraone, 2005; Castellanos & Tannock, 2002), along with some toxins. Lead, for example, is a known neurotoxin; although eliminating lead in gasoline and paint has reduced levels in the environment, it is still found in children's costume jewelry and imported candies as well as in the soil and water. A recent study confirmed higher blood levels of lead in children diagnosed with ADHD, and lead levels were correlated with teacher and parent ratings of symptoms (Nigg, Nikolas, Knottnerus, Cavanagh, & Friderici, 2010).

As with autism, another culprit is organophosphate pesticides; children with above average urinary levels of a metabolite of these pesticides are twice as likely to be diagnosed with ADHD compared to children with undetectable levels (M. F. Bouchard, Bellinger, Wright, & Weisskopf, 2010). Even chemicals found in cosmetics, perfumes, and shampoo are turning out to be a problem; mothers who had higher levels of phthalates in their urine during the third trimester of pregnancy reported that their children had significantly more trouble with attention, aggression, and depression (Engel et al., 2010). How phthalates cause this result is unknown, but they may disrupt thyroid hormones that are important during brain development.

FIGURE 13.21 The ADHD Brain.

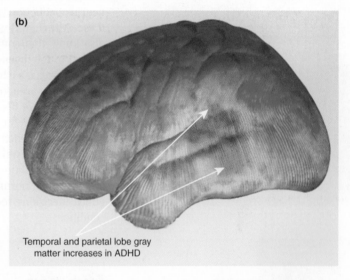

The colors indicate the relative amount of volume reduction compared with controls, with red the most, followed by yellow and green. (b) Yellow and red indicate 20% to 30% greater gray matter density compared with controls.

SOURCE: From "Cortical Abnormalities in Children and Adolescents With Attention-Deficit Hyperactivity Disorder," by E. R. Sowell et al., *Lancet, 362*, figs. 2 & 3, pp. 1702–1703. © 2003 Elsevier Ltd.

Chapter 13

Concept Check *Take a Minute to Check Your Knowledge and Understanding*

☐ Make a list of the kinds of intellectual disability described and their causes.
☐ What neural and biochemical differences have been found in autistic brains?

In Perspective

As important as the assessment of intelligence is in our society for determining our placement in school, our opportunity for continued education, and our employability and promotability, it is remarkable that there is still so much disagreement about what intelligence is. This lack of agreement makes it more difficult to study the brain functions that make up intelligence. Nevertheless, we have identified several features that appear to contribute to greater mental power; brain size, neural conduction speed, processing efficiency, and short-term memory are among these. Although it would be an error to overlocalize any function in the brain, we also know that some areas have a special role in important cognitive functions related to intelligence. It remains to be seen whether any particular characteristic of these areas, such as size, explains why some people have a particularly strong talent in one area, such as creative writing or mathematics. Some hope that we will eventually have objective brain measures that will tell us exactly how intelligent a person is or whether a child is autistic.

When we reach that point, perhaps another dream will be realized: the ability to diminish or even reverse some of the defects that rob the intellectually disabled, the autistic, and the aged of their capabilities. We may even be able to increase the intelligence of normal individuals. We can only hope that our capacity to make the ethical decisions required keeps pace with our ability to manipulate the human condition.

Summary

The Nature of Intelligence

- Intelligence is usually assessed with tests designed to measure academic ability.
- Some people show strong abilities not tapped by these tests; our understanding of intelligence should be broader than what tests measure.
- Intelligence theorists are divided over their emphasis on a general factor or multiple components of intelligence.

The Biological Origins of Intelligence

- Probable contributors to general intelligence are brain size, neural conduction speed, processing speed, quality of neural connections, and processing efficiency.
- The involvement of different brain areas suggests multiple components of intelligence.
- About half of the variation in intelligence among people is due to heredity. The closer the family relationship, the more correlated are the IQs. Apparently,

many genes are involved; there are several leads to specific genes, but little certainty.

- Research has not supported a genetic basis for ethnic differences in intelligence. Adoption has resulted in dramatic increases in IQ above the ethnic group mean.
- Although half of the variation in intelligence is due to environment, demonstrating which environmental conditions are important has been difficult. Judging by experience with Head Start and similar programs, any particular influence must be early and intense. Adoption can have dramatic effects if the difference in environments is large.

Deficiencies and Disorders of Intelligence

- Loss of intellectual functioning with age is less than previously believed and, like decreases in learning ability, it is not inevitable. Diminished speed of processing appears to be most important.
- Intellectual disability has many causes, including disease, fetal alcohol syndrome, Down syndrome, fragile X syndrome, phenylketonuria, and hydrocephalus.
- Down syndrome, caused by a third 21st chromosome, produces mild to moderate intellectual disability.
- Intellectual disability due to phenylketonuria, the inability to metabolize phenylalanine, is severe to profound.
- Hydrocephalus can usually be treated to avoid serious impairment, but there are hydrocephalic individuals with no apparent deficiencies.
- Autism is partially hereditary, with several gene locations implicated.
- Autism involves abnormalities in several brain areas and possible weak connections among them. Agents like thalidomide can damage the brain and result in autism, but damage is more often developmental in origin.
- Autism may also involve anomalies in serotonin, glutamate, GABA, and oxytocin functioning.
- ADHD is a partially genetic disorder characterized by hyperactivity, impulsiveness, and impaired attention and learning. It is associated with a variety of anomalies in functioning and in dopamine, serotonin, and possibly norepinephrine transmission. Several gene locations have been implicated.
- A variety of environmental influences have been associated with autism and ADHD. These may cause brain damage, affect gene functioning, or simply contribute along with genes. Childhood vaccines and a preservative previously used in them apparently are not at fault. ■

Study Resources

ADHD
Autism
Downs - Alz

F   For Further Thought

- Environmental influences on intelligence have been hard to identify. Does this mean that we are stuck with our genetic destiny?
- Intelligence is subject to physical disorders and genetic and environmental deviations. Speculate about why intelligence is so vulnerable.
- As you look at autism and ADHD, you see several similarities in the causes. Why, for example, do you think one child exposed to organophosphate pesticides would develop autism and another would be diagnosed with ADHD?

T   Testing Your Understanding

1. Describe the uncertainties about the measurement of intelligence and how this affects the search for biological bases of intelligence.

2. Discuss the brain characteristics that appear to contribute to general intelligence.

3. Discuss what we know about brain and biochemical differences in autistic individuals.

Select the best answer:

1. A problem with most intelligence tests is that they
   a. are not based on theory.
   b. are each based on a different theory.
   c. assess a limited group of abilities.
   d. try to cover too many abilities in one test.

2. Lumpers and splitters disagree on the significance of _____ in intelligence.
   a. heredity
   b. environment
   c. the g factor
   d. early education

3. It is likely that _____ is/are important to general intelligence.
   a. size of neurons
   b. processing speed
   c. processing efficiency
   d. a, b, and c
   e. b and c

4. Research with adults, children, chimpanzees, and monkeys suggests that we are born with
   a. a mechanism for number or quantity.
   b. the ability to do the same things as savants.
   c. many times more intellectual capacity than we use.
   d. time-limited abilities that inevitably deteriorate with age.

5. Research suggests that, normally, environmental effect on intelligence
   a. is almost nonexistent.
   b. is significant, but difficult to identify.
   c. is less important than the effect of heredity.
   d. is more important than the effect of heredity.

6. Some claim the high correlation between identical twins' IQs occurs because they evoke similar treatment from people. This was refuted by a study in which the correlation
   a. held up when the twins were reared separately.
   b. was unaffected by parents' misidentification of twins as fraternal or identical.
   c. was just as high in mixed-sex as in same-sex identical pairs.

d. increased as the twins grew older, though they lived apart.

7. The best evidence that ethnic differences in intelligence are not genetic is that
   a. the various groups perform the same on culture-free tests.
   b. no well-done research has shown an IQ difference.
   c. no genes for an ethnic difference in intelligence have been found.
   d. adoption into a more stimulating environment reduces the difference.

8. Apparently, the most critical effect on intelligence during aging is loss of
   a. speed.
   b. motivation.
   c. neurons.
   d. synapses.

9. Sam has dramatically reduced brain tissue and enlarged ventricles, but his IQ is 105; his disorder is most likely
   a. hydrocephalus.
   b. phenylketonuria.
   c. Down syndrome.
   d. autism.

10. Most mild intellectual disability is believed to be caused by
    a. an impoverished environment.
    b. brain damage sustained during birth.
    c. a combination of a large number of genes.
    d. a combination of environmental and hereditary causes.

11. Research with autism spectrum disorders suggests that autism is
    a. caused by a single gene.
    b. caused by several genes.
    c. caused by heredity alone.
    d. primarily due to environment.

12. Impaired sociability in autistics may involve low levels of
    a. risperidone.
    b. serotonin.
    c. thalidomide.
    d. oxytocin.

13. ADHD is associated with reduced or impaired
    a. gray matter.
    b. intelligence.
    c. dopamine activity.
    d. theory of mind.

Online Resources

The following resources are available at www.sagepub.com/garrett3e.

Chapter Resources

- Flash Cards
- Chapter Quiz
- Internet Resources and Exercises

On the Web

Links to these websites can be found at the web address above: Select this chapter's **Chapter Resources**, then choose **Internet Resources and Exercises**.

1. **Stephen Hawking**'s website features a brief biography, information about his professional accomplishments, and downloadable copies of public lectures.

2. At **Not Exactly Rocket Science** you can read about tool use in chimpanzees and see a video of chimps making and using fishing sticks.
    Nature's Tools: How Birds Use Them is an excerpt from a BBC wildlife film that shows a New Caledonia crow fishing for grubs.

3. *The Bell Curve* **Flattened**, an article in the online magazine Slate, refutes ideas about genetic racial differences in intelligence that were presented in the controversial book *The Bell Curve*.

4. You can get information about various kinds of intellectual disability from the **Association for Retarded Citizens**, the **American Association on Intellectual and Developmental Disabilities**, the **National Fragile X Foundation**, and the National Library of Medicine's **Williams Syndrome** page.

5. Sources of information on autism and autism spectrum disorders include the National Institute of Mental Health's website **Autism Spectrum Disorders**, Internet Mental Health's **Autistic Disorder** site, and the Autism Research Institute's page, **Autism Is Treatable!**

6. **Savant Syndrome** by the Wisconsin Medical Society is a fascinating tour of the personalities of numerous savants boasting artistic, mathematical, and memory capabilities. See What's New and Savant Profiles and, especially, the page about Kim Peek.

7. **Temple Grandin's Web Page** features her professional work along with a brief description of her.

8. The **Attention Deficit Disorder Association** has high-quality articles on ADHD, and the National Institute of Mental Health explores a number of topics at its **Attention Deficit Hyperactivity Disorder** site.

Chapter Updates and Biopsychology News

For Further Reading

1. *Possessing Genius: The Bizarre Odyssey of Einstein's Brain*, by Carolyn Abraham (St. Martin's, 2001), tells the story of the study of Einstein's brain and the controversy about how it came to be removed in the first place and about its caretaker, Thomas Harvey (coauthor of all the Einstein brain studies cited here).

2. *Frames of Mind*, by Ulric Neisser (Basic Books, 1983), is a collection of articles on the knowns and unknowns of intelligence.

3. "Representation of Number in the Brain," by Andreas Nieder and Stanislas Dehaene (*Annual Review of Neuroscience*, 2009, *32*, 185–208), reviews research on the basis of number understanding in children, adults, and nonhuman primates.

4. *Thinking in Pictures: My Life With Autism*, by Temple Grandin (Vintage, 2010), is a perspective from the point of view of an autistic and a scientist. *Born on a Blue Day: Inside the Extraordinary Mind of an Autistic Savant*, by Daniel Tammet (Free Press, 2006) chronicles his life, while his *Embracing the Wide Sky: A Tour Across the Horizons of the Mind* (Free Press, 2009) adds relevant scientific information.

Key Terms

attention deficit hyperactivity disorder (ADHD) 419
autism 412
autistic savant 414
default mode network 408
Down syndrome 410
fragile X syndrome 411
intellectual disability 409
intelligence 394
intelligence quotient (IQ) 394
phenylketonuria 411
theory of mind 413

Psychological Disorders

14

Schizophrenia
Characteristics of the Disorder
Heredity
Two Kinds of Schizophrenia
The Dopamine Hypothesis
Beyond the Dopamine Hypothesis
Brain Anomalies in Schizophrenia
   CONCEPT CHECK

Affective Disorders
Heredity
The Monoamine Hypothesis of Depression
Electroconvulsive Therapy
   APPLICATION: ELECTRICAL STIMULATION FOR DEPRESSION
Antidepressants, ECT, and Neural Plasticity
Rhythms and Affective Disorders
Bipolar Disorder
Brain Anomalies in Affective Disorder
Suicide
   CONCEPT CHECK

Anxiety Disorders
Generalized Anxiety, Panic Disorder, and Phobia
Posttraumatic Stress Disorder
   IN THE NEWS: VIRTUAL REALITY ISN'T JUST FOR VIDEO GAMES
Obsessive-Compulsive Disorder
   APPLICATION: OF HERMITS AND HOARDERS
   CONCEPT CHECK

In Perspective
Summary
Study Resources

In this chapter you will learn

- The characteristics and probable causes of schizophrenia

- How heredity and environment interact to produce disorders

- What the affective disorders are and their causes

- The symptoms and causes of the anxiety disorders

Canst thou not minister to a mind diseas'd
Pluck from the memory a rooted sorrow
Raze out the written troubles of the brain

—*Shakespeare,* Macbeth

I stood by my chair and waited for the students to take their places around the table, eagerly tying up the loose ends of conversations that the trek across campus hadn't given them time to finish. As the bell in the East College tower tolled the start of the hour and I was about to call the class to order, Ned got up from his seat and approached me.

"I forgot to give the bookstore cashier her pen after I used it to write a check. Can I take it back?"

"I think she can wait until class is over," I answered. He accepted that judgment and returned to his seat, but he seemed restless the rest of the hour. As soon as class was over, he was one of the first out of the door. I couldn't help smiling at his youthful impetuousness.

The next day I understood that Ned's behavior had a completely different origin. Around 10 o'clock the night before his dorm mates found him huddled on the stair landing, fending off an imaginary alien spaceship circling over his head and firing projectiles at him. He was taken to the hospital and sedated; then his parents took him back to his hometown, where he spent several months in a hospital psychiatric ward. He was diagnosed with paranoid schizophrenia. Fortunately, medication helped, and he was able to move to a home school with a comprehensive program of support and rehabilitation.

Ned has now spent more than half of his life at the home. A dozen years ago, he wrote to me, and we have kept up a regular correspondence since; I think his primary motivation is that he remembers his brief time in college as the happiest in his life. It is not that the home is unpleasant. He is on the baseball, basketball, and golf teams; he works part-time outside the home; and he has a girlfriend. Questions he asks in his letters reveal a healthy curiosity, usually provoked by something he has read or seen on television about the brain. Once he talked candidly about his diagnosis, and about how he prefers to believe that someone slipped him a dose of LSD on that fateful night. There is no evidence that happened, but even if it did, it only precipitated, rather than caused, the decades-long debilitation that followed. In spite of his apparent good adjustment—and I see only the face that he wants to put on his situation—the preadolescent intellectual maturity of his letters and the barely legible scrawl of his handwriting suggest the havoc that schizophrenia has wreaked in his brain. Ned is unable to function outside the home's protective environment and professional support, and he will never be able to leave.

According to recent estimates, 1 out of every 4 people in the United States suffers from a diagnosable mental disorder, and the number rises to 1 in 3 over the lifetime (Kessler, Berglund, et al., 2005; Kessler, Chiu, Demler, & Walters, 2005). The annual cost of treating these disorders is about $100 billion (Mark et al., 2007), and lost income adds another $193 billion (Kessler et al., 2008); mental disorder is the leading cause of disability among people aged 15 to 44 in the United States and Canada (Figure 14.1; World Health Organization, 2008). An obvious benefit of research is the development of improved therapeutic techniques; in addition, because the disorders involve malfunctions in neurotransmitter systems and brain structures, studying them helps researchers understand normal neural functioning as well. In this chapter we will make good use of what you have already learned about brain structure and neurotransmitter activity as we examine schizophrenia, mood disorders, and anxiety disorders, and in turn this survey will further expand your knowledge of how the brain works.

FIGURE 14.1 Psychological Disorders Impair a Person's Ability to Cope.

SOURCE: © Sheryl Griffin/iStockphoto.com.

Schizophrenia

Schizophrenia is a disabling disorder characterized by perceptual, emotional, and intellectual deficits; loss of contact with reality; and inability to function in life. It is estimated that about 3 million Americans will develop schizophrenia during their lifetimes and that around 100,000 hospitalized patients take up 20% of the psychiatric beds in the U.S. hospitals, with many more receiving outpatient care (National Institute of Mental Health, 1986; G. W. Roberts, 1990). Schizophrenia is a *psychosis,* which simply means that the individual has severe disturbances of reality, orientation, and thinking. Schizophrenia is the most severe of the mental illnesses, and it is particularly feared because of the bizarre behavior it produces in many of its victims. All social classes are equally vulnerable; though patients themselves "drift" to lower socioeconomic levels, when they are classified by their parents' socioeconomic level, the classes are proportionately represented (Huber, Gross, Schüttler, & Linz, 1980). Although schizophrenia afflicts only 1% of the population (Kessler et al., 1994), its economic burden amounts to $39 billion annually in the United States, almost half the cost of all the disorders combined (Uhl & Grow, 2004). Fortunately, schizophrenia is one of the few psychological disorders that appear to be on the decline. Smaller numbers have been thought to be due to methodological flaws in studies, but a study of all people born in Finland between 1954 and 1965 found a significant decline in each successive age group, totaling 29% for women and 33% for men (Suvisaari, Haukka, Tanskanen, & Lönnqvist, 1999).

 What is schizophrenia, and what causes it?

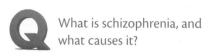 1
Schizophrenia

Characteristics of the Disorder

The term *schizophrenia* was coined in 1911 by the Swiss psychiatrist Eugen Bleuler (Figure 14.2) from the combination of two Greek words meaning "split mind." Contrary to popular belief, schizophrenia has nothing to do with multiple personality; the term refers to the distortion of thought and emotion, which are "split off" from reality. The schizophrenic has some combination of several symptoms: hallucinations (internally generated perceptual experiences, such as voices telling the person what to do); delusions (false, unfounded beliefs, such as that one is a messenger from God); paranoia, characterized by delusions of persecution; disordered thought; inappropriate emotions or lack of emotion; and social withdrawal. Note that Ned had a hallucination of a spaceship, the paranoid delusion that it was attacking him, and a possible delusion about the LSD.

Schizophrenics are usually subdivided into diagnostic categories based on which of these symptoms is predominant, such as *paranoid* or *catatonic.* However, patients often have overlapping symptoms and can receive multiple diagnoses, so there is little belief that these categories represent distinct disease processes. Also, as neuroscience progresses, we are realizing that two people can have the same symptom with different causes or the same brain defect with different symptoms. As a result, the National Institute of Mental Health is encouraging researchers to shift their focus from diagnostic categories to underlying neural and genetic mechanisms (Miller, 2010). In addition, the American Psychiatric Association and the World Health Organization are revisiting diagnostic standards that date back to the early 1970s; upcoming revisions of their diagnostic manuals will include several changes to diagnostic categories, and in particular, the schizophrenia subtypes are expected to be demoted

FIGURE 14.2 Eugen Bleuler (1857–1939).

A pioneer in the field, he introduced the term *schizophrenia.*

SOURCE: © Bettmann/Corbis.

> *I'm a paranoid schizophrenic and for us life is a living hell. . . . Society is out to kill me. . . . I tried to kill my father. I went insane and thought he ruled the world before me and caused World War Two.*
>
> —*Ross David Burke in* When the Music's Over: My Journey Into Schizophrenia

to symptom descriptions (American Psychiatric Association, 2010; World Health Organization, 2010).

Schizophrenia afflicts men and women about equally often. Men usually show the first symptoms during the teens or twenties as Ned did, while the onset for women ordinarily comes about a decade later (see Figure 14.3). *Acute* symptoms develop suddenly and are typically more responsive to treatment; the prognosis is reasonably good in spite of brief relapses. Symptoms that develop gradually and persist for a long time with poor prognosis are called *chronic*. Movies have overplayed the bizarre features of schizophrenia; many patients are able to function reasonably well, especially if they are fortunate enough to be among those who respond to the antipsychotic drugs. Among patients studied 20 years after their first admission, 22% were fully recovered, another 43% were improved, and the symptoms of the remaining 35% had remained the same or worsened; 56% were fully employed (Huber et al., 1980).

Doctors began to view mental illness as a medical problem in the late 1700s and early 1800s; at that time the mentally ill were literally released from their chains and given treatment (Figure 14.4) (Andreasen, 1984). By the turn of the century, it was widely assumed that schizophrenia had a physical basis. However, the search for biological causes produced little success. In the 1940s, the emphasis shifted to social causes of schizophrenia, especially in America, where Freud's theory of psychoanalysis was in its ascendancy and biologically oriented psychiatrists were in the minority (Andreasen, 1984; Wender, Rosenthal, Kety, Schulsinger, & Welner, 1974). Until the 1960s, research techniques were not up to the task of demonstrating the validity of the physiological position. It was then that increasing knowledge of neurotransmitters, the advent of brain scanning techniques, and improved genetic studies shifted the explanation for schizophrenia back to the realm of biology and changed the perception of mental illness in general.

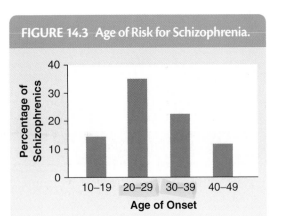

FIGURE 14.3 Age of Risk for Schizophrenia.

SOURCE: Data from Huber et al. (1980).

FIGURE 14.4 Philippe Pinel Freeing Mental Patients From Their Chains.

Patients were warehoused without treatment; sometimes care consisted of throwing in fresh straw and food once a week. Pinel was convinced that they would benefit from humane treatment and in 1794 freed the mental patients of Paris from their chains.

SOURCE: © Rapho Agence/Photo Researchers.

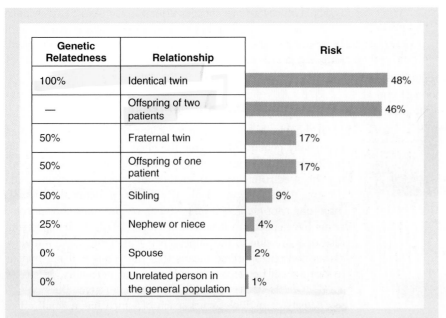

Genetic Relatedness	Relationship	Risk
100%	Identical twin	48%
—	Offspring of two patients	46%
50%	Fraternal twin	17%
50%	Offspring of one patient	17%
50%	Sibling	9%
25%	Nephew or niece	4%
0%	Spouse	2%
0%	Unrelated person in the general population	1%

FIGURE 14.5 Concordances for Schizophrenia Among Relatives.

SOURCE: From *Introduction to Psychology, Gateways to Mind and Behavior* (with InfoTrac), 9th edition by Coon, 2001. Reprinted with permission of Wadsworth, a division of Thomson Learning.

Heredity

Schizophrenia is a familial disorder, which means that the incidence of schizophrenia is higher among the relatives of schizophrenics than it is in the general population (Gottesman, McGuffin, & Farmer, 1987; Tsuang et al., 1991). Of course, this association could be due to environmental influence or to heredity; in fact, in the 1940s the genetic school and the environmental school argued for their positions from the same data (Wender et al., 1974). However, studies of twins and adoptees provided compelling evidence for a genetic influence.

Twin and Adoption Studies

In Figure 14.5, you can see that the shared incidence of schizophrenia increases with the genetic closeness of the relationship and that the concordance rate for schizophrenia is three times as high in identical twins as in fraternal twins (Lenzenweger & Gottesman, 1994). In other words, identical twins of schizophrenics are three times as likely to be schizophrenic as the fraternal twins of schizophrenics. The heritability for schizophrenia has been estimated at between .60 and .90 (Tsuang et al., 1991). This means that 10% to 40% of the variability is due to environmental factors.

Information from adoption studies gives a more impressive indication of genetic influence; these studies show that adopting out of a schizophrenic home provides little or no protection from schizophrenia. The incidence of schizophrenia *and* schizophrenia-like symptoms was 28% among individuals adopted out of Danish homes in which there was one schizophrenic parent, compared with 10% in matched adoptees from presumably normal homes (Lowing, Mirsky, & Pereira, 1983). Other studies have produced similar findings.

Discordance among identical twins has been used as an argument that schizophrenia is environmentally produced. To address this issue, Gottesman and Bertelsen (1989) compared the incidence of schizophrenia in the offspring of the schizophrenic and normal twins of schizophrenics; they found that the offspring of the unaffected identical twins were just as likely to be schizophrenic as the offspring of the affected twins (Figure 14.6). This result would not have occurred unless the

Offspring of the normal fraternal twin of a schizophrenic do not have an elevated risk. The offspring of the normal identical twin of a schizophrenic are as likely to become schizophrenic as the offspring of the schizophrenic identical twin.

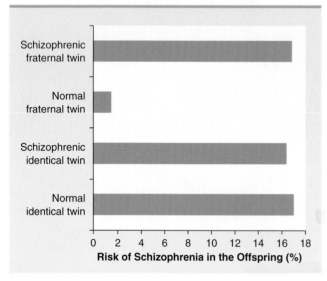

SOURCE: Based on data from Gottesman and Bertelsen (1989).

normal twins were carrying genes for schizophrenia. Discordance does raise the question, however, whether some environmental factors determine whether the person's schizophrenic genes will remain "silent."

The Search for the Schizophrenia Genes

Although we have known for a long time that schizophrenia is partially genetic, identifying the genes involved has been difficult. One reason has been researchers' inconsistency in including the spectrum disorders in their diagnosis of schizophrenia (Heston, 1970; Lowing et al., 1983). When identical twins are discordant for schizophrenia, 54% of the nonschizophrenic twins have spectrum disorders (Heston, 1970). If the spectrum disorders are due to the same genes, classifying these individuals as nonschizophrenic means that the genes will not appear to distinguish between schizophrenia and normality. A second problem is that schizophrenia apparently involves the cumulative effects of multiple genes, each of which has a small effect by itself (Fowles, 1992; Tsuang et al., 1991). A person's risk of schizophrenia presumably increases with the number of these and other genes inherited. This view is supported by the fact that risk has been found to increase with the number of relatives who are schizophrenic and with the degree of the relatives' disability (Heston, 1970; Kendler & Robinette, 1983).

Researchers have compiled a long list of candidate genes, but unfortunately, few of these have been supported in subsequent investigations. For example, a very large study of 1,870 individuals with schizophrenia and spectrum disorder did not confirm any of 14 previously identified genes (Sanders et al., 2008). An analysis of the large SzGene database identified 16 candidate genes (Table 14.1; Allen et al., 2008); they play roles in dopamine, serotonin, glutamate, and GABA transmission; neurotransmitter deactivation; neural development, axon guidance; neurodegeneration; and immune and inflammatory responses. In spite of the fact that the investigation pooled data from over 1,000 studies, most of the associations were weak, and some of the most frequently touted genes failed to make the list at all. Apparently there are so many different genetic pathways to schizophrenia that the genes may be difficult to detect except in families with several affected individuals who are closely related. Another problem is that most research has focused on identifying common genes, while a significant amount of the transmission involves rare copy number variations (International Schizophrenia Consortium, 2008; Stefansson et al., 2008).

The Vulnerability Model

Most researchers agree that genes determine only the person's vulnerability for the illness; both heredity and environment are needed to explain the *etiology* (causes) of schizophrenia (Zubin & Spring, 1977) as well as most other disorders. According to the *vulnerability model,* some threshold of causal forces must be exceeded for the illness to occur; environmental challenges combine with a person's genetic vulnerability to exceed that threshold. The environmental challenges may be external, such as bereavement, job difficulties, or divorce, or they may be internal, such as maturational changes, poor nutrition, infection, or toxic substances. There is mounting evidence that these environmental influences work in part by epigenetic means, that is by upregulating and downregulating gene functioning (Tsankova, Renthal,

TABLE 14.1 Genes Associated With Schizophrenia in a Large Database Analysis.

Genes	Related Function
Neurotransmission	
DRD1, DRD2, DRD4	Dopamine receptors
TPH1	Serotonin production
SLC6A4	Serotonin transporter
DAO	NMDA glutamate receptor function
GRIN2B	NMDA glutamate receptor subunit
DTNBP1	Glutamate release
GABRB2	$GABA_A$ receptor
COMT	Enzyme that deactivates neurotransmitters, including dopamine
Development	
MTHFR	Enzymes involved in neural development
PLXNA2	Axon guidance
Neural Damage	
APOE	Neurodegeneration
IL1B	Immune and inflammatory responses
HP	Immune and inflammatory responses
TP53	Gene mutation prevention

SOURCE: Allen et al. (2008).

Kumar, & Nestler, 2007). Vulnerability is viewed as a continuum, depending on the number of affected genes inherited. At one extreme, a small percentage of individuals will become schizophrenic under the normal physical and psychological stresses of life; at the other extreme are individuals who will not become schizophrenic under any circumstance or will do so only under the severest stress, such as the trauma of battle (Fowles, 1992).

Two Kinds of Schizophrenia

Researchers disagree not only on what the subtypes of schizophrenia are but on whether schizophrenia represents one disease or many. Whatever the answers to these questions may be, most authorities do agree that the symptoms fall into two major categories: positive and negative. *Positive symptoms* involve the presence or exaggeration of behaviors, such as delusions, hallucinations, thought disorder, and bizarre behavior. *Negative symptoms* are characterized by the absence or insufficiency of normal behaviors and include lack of affect (emotion), inability to experience pleasure, lack of motivation, poverty of speech, and impaired attention.

Crow (1985) theorized that positive and negative symptoms are due to two different syndromes of schizophrenia, with different causes and different outcomes. His Type I and Type II schizophrenias are described in Table 14.2. Research has supported this distinction in many respects. Positive symptoms are more often acute, and they are more likely to respond to antipsychotic drugs than are negative symptoms (Fowles, 1992). Negative symptoms tend to be chronic; these patients show poorer adjustment prior to the onset of the disease (Andreasen, Flaum, Swayze, Tyrrell, & Arndt, 1990); poorer prognosis after diagnosis (Dollfus et al.,

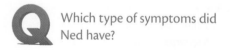

Q Which type of symptoms did Ned have?

TABLE 14.2 Positive Versus Negative Schizophrenia.

Aspect	Type I (Positive)	Type II (Negative)
Characteristic symptoms	Delusions, hallucinations, etc.	Poverty of speech, lack of affect, etc.
Response to antidopaminergic drugs	Good	Poor
Symptom outcome	Potentially reversible	Irreversible?
Intellectual impairment	Absent	Sometimes present
Suggested pathological process	Increased D_2 dopamine receptors	Cell loss in temporal lobes

SOURCE: Crow (1985).

1996); more intellectual and other cognitive deficits, suggestive of a brain disorder (Andreasen et al., 1990); and greater reduction in brain tissue (Fowles, 1992). These findings led researchers to think in terms of two more or less distinct groups of patients, a view we will modify shortly.

The Dopamine Hypothesis

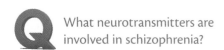 **Q** What neurotransmitters are involved in schizophrenia?

Little could be done to treat psychotic patients until the mid-1950s, when a variety of antipsychotic medications arrived on the scene. For the first time in history, the size of the hospitalized mental patient population went down. As is often the case in medicine, and more particularly in mental health, these new drugs had not been designed for this purpose—researchers had too little understanding of the disease to do so. Doctors tried chlorpromazine with a wide variety of mental illnesses because it calmed surgical patients, and it helped with schizophrenics. However, it was not clear *why* chlorpromazine worked, because tranquilizers have little or no usefulness in treating schizophrenia.

So investigators tried reverse engineering. You will remember from Chapter 5 that amphetamine overdose causes psychotic behavior indistinguishable from schizophrenia, complete with hallucinations and paranoid delusions. In time, researchers were able to determine that amphetamine produces these symptoms by increasing dopaminergic activity. This discovery eventually led to the *dopamine hypothesis,* that schizophrenia involves excessive dopamine activity in the brain. According to the theory, blockade of the D_2 type of dopamine receptors is essential for a drug to have an antipsychotic effect, and effectiveness is directly related to the drug's blocking potency. The theory has had considerable support; schizophrenic patients typically have higher dopamine activity in the striatum (Abi-Dargham et al., 2000), and drugs that block dopamine receptors are effective in treating the positive symptoms of schizophrenia (S. H. Snyder, Bannerjee, Yamamura, & Greenberg, 1974). In fact, the effective dosage for most antipsychotic drugs is directly proportional to their ability to block dopamine receptors (Figure 14.7; Seeman, Lee, Chau-Wong, & Wong, 1976).

> *What consoles me is that I am beginning to consider madness as an illness like any other, and that I accept it as such.*
>
> —*Vincent van Gogh, 1889, in a letter to his brother, Theo*

Beyond the Dopamine Hypothesis

However, 30% to 40% of schizophrenic patients were not helped by the drugs, and—troublesome for the dopamine theory—nonresponsive patients experienced just as much D_2 receptor blockade as responders. In fact, in some of them, blockade

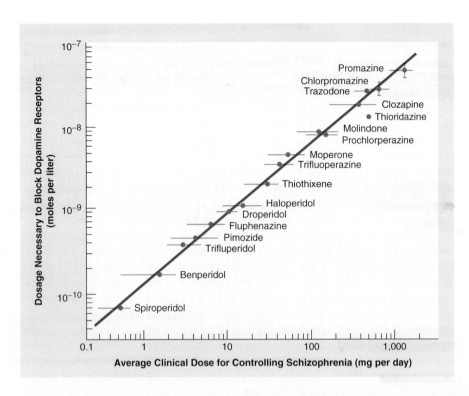

FIGURE 14.7 Relationship Between Receptor Blocking and Clinical Effectiveness of Schizophrenia Drugs.

The horizontal axis is the average daily doses prescribed by physicians; the horizontal red lines represent typical ranges of doses used. Values on the vertical axis are amounts of the drugs required to block 50% of the dopamine receptors.

SOURCE: Reprinted by permission from P. Seeman et al., "Antipsychotic Drug Doses and Neuroleptic/Dopamine Receptors," *Nature, 261,* p. 718, fig. 1. Copyright 1976 Macmillan Publishers, Ltd.

exceeded 90%, while some responders showed remarkably low levels of receptor blocking (Kane, 1987; Pilowsky et al., 1993). Furthermore, some schizophrenics appear to have a *dopamine deficiency,* especially those with chronic, treatment-resistant symptoms (Grace, 1991; Heritch, 1990; Okubo et al., 1997).

Another problem for the drugs was that the side effects could be intolerable. Prolonged use of antidopamine drugs often produces *tardive dyskinesia,* tremors and involuntary movements caused by blocking of dopamine receptors in the basal ganglia. Seventy years ago, this effect was believed to be so inevitably linked to the therapeutic benefit that the "right" dose was the one that caused some degree of motor side effects. Thus, the drugs used to treat schizophrenia became known as *neuroleptics,* because the term means "to take control of the neuron" (Julien, 2008). The effect appears to be due to a compensatory increase in the sensitivity of D_2 receptors in the basal ganglia. (This is a good illustration of the fact that drugs do not affect just the part of the brain we want to treat.)

Since the early 1990s, we have seen the introduction of several new antipsychotic drugs that are referred to as *atypical* or *second-generation.* One way atypical antipsychotics are different is that they target D_2 receptors much less, so they produce motor problems only at much higher doses. Fortunately, avoiding motor side effects does not require a therapeutic compromise. The major atypical antipsychotics are at least equivalent to the first-generation drugs, and some are 15% to 25% more effective; what is more, they often bring relief to patients who have been treatment resistant for years (Iqbal & van Praag, 1995; Leucht et al., 2009; Pickar, 1995; Siever et al., 1991). So, is the dopamine hypothesis just another example of a beautiful hypothesis slain by ugly facts? Not entirely; although atypical antipsychotics mostly target other receptors, those that lack at least a modest effect at D_2 receptors are therapeutically ineffective (H. M. Jones & Pilowsky, 2002). So, successful therapy apparently requires D_2 blockade *and* other effects.

And what are these other effects? One is glutamate increase. You may remember from Chapter 5 that the drug phencyclidine (PCP) causes some of the symptoms of schizophrenia; actually, it mimics schizophrenia far better than amphetamine does (Sawa & Snyder, 2002). The fact that one of its effects is to inhibit the NMDA

↑ D_2 = basal ganglia.

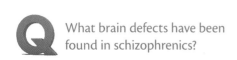

(*N*-methyl-D-aspartic acid) subtype of glutamate receptor suggested that reduced glutamate activity might be a factor in schizophrenia. Administering *glycine* activates the NMDA receptor and increases glutamate release; combining glycine with a typical antipsychotic reduces negative symptoms and improves cognition 23% more than treatment with a typical antipsychotic alone (Coyle, Tsai, & Goff, 2003; Goff et al., 1999; Heresco-Levy & Javitt, 2004). Atypical antipsychotics also increase glutamate levels (Goff et al., 2002; van der Heijden et al., 2004), apparently by downregulating the glutamate transporter gene, thus reducing reuptake (Melone et al., 2001). In one study, brain glutamate levels increased in patients whose negative symptoms improved but not in the nonresponders (Goff et al., 2002). Atypical antipsychotics affect serotonin levels (van der Heijden et al., 2004) as well as glutamate levels, and several of them block 90% of the 5-HT2 subtype of serotonin receptors (H. M. Jones & Pilowsky, 2002; Kapur, Zipursky, & Remington, 1999).

To focus on one neurotransmitter system as an explanation of schizophrenia would be misguided. Neurotransmitter systems interact in intricate ways, and changes in one would be expected to have implications for functioning in others as well. The serotonin system is one of the forces that regulate activity in the dopamine system (Iqbal & van Praag, 1995), and the glutamate system influences the number of dopamine receptors (L. Scott et al., 2002). There is also evidence that the dopamine imbalance in schizophrenia may be a result of reduced glutamate activity in the prefrontal cortex (Laruelle, Kegeles, & Abi-Dargham, 2003). At this point, it appears that the *glutamate theory,* that reduced glutamate activity is involved in schizophrenia, holds considerable theoretical and therapeutic promise, but it must account for activity in other systems if it is to survive. While we wait for that story to unfold, we have additional clues to the origins of schizophrenia from structural and functional anomalies in the brain.

Brain Anomalies in Schizophrenia

Malfunctions have been identified in virtually every part of the brain in schizophrenics. The most consistent finding has been enlargement of the ventricles; another is hypofrontality, or reduced activity in the frontal lobes. We will examine each of these defects in turn.

Brain Tissue Deficits and Ventricular Enlargement

A signature characteristic of schizophrenia is a decrease in brain tissue, both gray and white matter, with deficits reported in at least 50 different brain areas (Honea, Crow, Passingham, & Mackay, 2005). The number of sites and the variability across studies attest to the multifaceted nature of schizophrenia, but the frequency with which deficiencies are found in the frontal and temporal lobes indicates that they are particularly important. These tissue losses are accompanied by alterations in neural functioning, but not necessarily in the expected direction: Activity is decreased in dorsolateral prefrontal cortex but is increased in the orbitofrontal cortex and in a subregion of the hippocampus (Schoebel et al., 2009). In fact, the hippocampal activation is so characteristic of schizophrenia that in a group of people having brief psychotic symptoms it identified those who would later be diagnosed with full-blown schizophrenia with 70% accuracy (Schobel et al.).

An indication of the tissue deficits seen in schizophrenia is ventricular enlargement; this is because the ventricles expand to take up space normally occupied by brain cells (see Figure 14.8). Both deficiencies are usually subtle, on the order of less than a tablespoonful increase in ventricular volume (Suddath et al., 1989) and a 6-gram decrease in temporal lobe weight (Bogerts, Meertz, & Schönfeld-Bausch, 1985), but these figures belie the functional importance of the losses. In fact, an

Q What brain defects have been found in schizophrenics?

FIGURE 14.8 Ventricle Size in Normals and Schizophrenics.

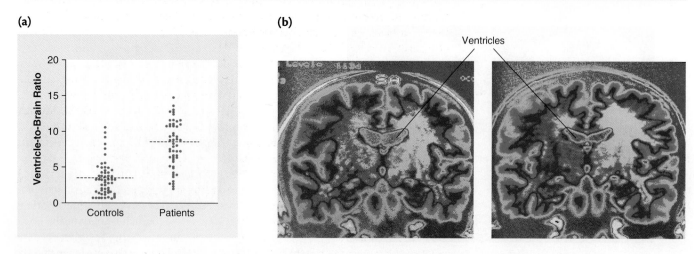

(a) Ventricle-to-brain ratios (ventricular area divided by brain area, multiplied by 100) of normal controls and chronic schizophrenics. Dotted horizontal lines indicate group means. (b) Magnetic resonance imaging scans of identical twins, one schizophrenic (left) and the other normal (right).

SOURCES: (a) From "Lateral Cerebral Ventricular Enlargement in Chronic Schizophrenia," by D. R. Weinberger et al., *Archives of General Psychiatry, 36*, pp. 735–739. Copyright 1979 American Medical Association. Reprinted with permission; (b) Copyright 1990 Massachusetts Medical Society. All rights reserved.

often distinguishing feature between identical twins discordant for schizophrenia is the size of their ventricles (Suddath, Christison, Torrey, Casanova, & Weinberger, 1990). Ventricular enlargement is not specific to schizophrenia; enlarged ventricles are also associated with several other conditions, including old age, dementia (loss of cognitive abilities), Alzheimer's disease, Huntington's chorea (D. R. Weinberger & Wyatt, 1983), and alcoholism with dementia (D. M. Smith & Atkinson, 1995). Nor are enlarged ventricles an inherent characteristic of schizophrenia, since most schizophrenics have normal ventricles. We will look more closely at the tissue deficits later when we consider their origins.

Hypofrontality

Earlier, we saw that prefrontal functioning can be assessed by using the gambling task; an alternative technique is the *Wisconsin Card Sorting Test,* which requires individuals to change strategies in midstream, first sorting cards using one criterion, then changing to another. Many schizophrenics perform poorly on the test, persisting with the previous sorting strategy. Normal individuals show increased activation in the prefrontal area during the test; schizophrenic patients typically do not, in spite of normal activation in other areas (D. R. Weinberger, Berman, & Zec, 1986). Figure 14.9 shows a normal brain practically lighting up during the test, in comparison to the schizophrenic brain, especially in the frontal area called the *dorsolateral prefrontal cortex.* This *hypofrontality* apparently involves dopamine *deficiency,* because administering amphetamine to schizophrenics increases blood flow in the prefrontal cortex and improves performance on the Wisconsin Card Sorting Test (Daniel et al., 1991). Traumatic injury to the dorsolateral prefrontal cortex causes impairments similar to the symptoms of schizophrenia: flat affect, social withdrawal, reduced intelligence and problem-solving ability, diminished motivation and work capacity, and impaired attention

FIGURE 14.9 Blood Flow in Normal and Schizophrenic Brains During Card Sorting Test.

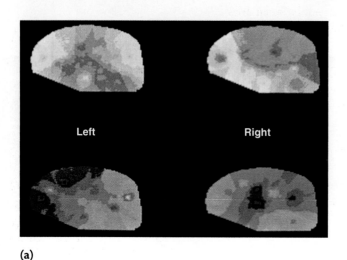

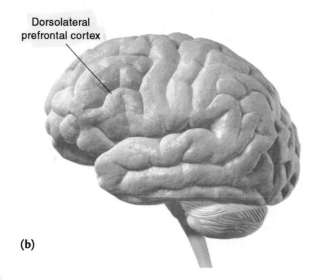

Dorsolateral prefrontal cortex

(a)   (b)

(a) The upper images are of the left and right hemispheres of a normal brain; the schizophrenic brain is below. Red and yellow represent greatest activation. Note especially the activity in the dorsolateral prefrontal cortex, whose location is identified in (b).

SOURCE: (a) From "Physiologic Dysfunction of Dorsolateral Prefrontal Cortex in Schizophrenia: I. Regional Cerebral Blood Flow Evidence," by D. R. Weinberger, K. F. Berman, and R. R. Zec, 1986, *Archives of General Psychiatry, 43*, 114–124.

and concentration (D. R. Weinberger et al., 1986). Because of the frontal lobes' involvement in planning actions, recognizing the consequences of actions, and managing working memory, it is not surprising that frontal dysfunction would cause major abnormalities in thinking and behavior.

Neural Connections and Synchrony

Recent attention has been shifting away from localized deficits and has been focusing on disrupted coordination of neural activity across brain areas. For example, in normal controls performing a working-memory task, activity in the hippocampal formation varies together with prefrontal activity, but this coordination is absent in people with schizophrenia (Meyer-Lindberg et al., 2005). The hypofrontality seen during the Wisconsin Card Sorting Test has been attributed to disrupted communication between the hippocampus and the prefrontal cortex (Weinberger, Berman, Suddath, & Torrey, 1992). Inadequate coordination between brain areas is at least partly due to white matter reduction; white matter loss has been consistently reported in the brains of people with schizophrenia, particularly in prefrontal and temporal areas (Begré & Koenig, 2008; Ellison-Wright & Bullmore, 2009).

Brain functioning is coordinated by synchronized firing that links the activity of neurons within a cortical area, across areas, and even between hemispheres. This synchronization is widely believed to be critical to perceptual binding and cognitive performance, and it is one of the functions that is disrupted in schizophrenia (Uhlhaas & Singer, 2010). Synchrony occurs in people with schizophrenia but at lower frequencies, possibly because the reduced white matter connections cannot support coordination at higher frequencies (Spencer et al., 2004). To some extent we can correlate the patterns of synchrony with the symptoms of schizophrenia; in patients with positive symptoms, for example, oscillation synchrony is enhanced within limited areas but is deficient between areas (Uhlhaas & Singer). This enhanced synchrony, which indicates hyperexcitability, is seen in the occipital area in visual hallucinators (Spencer et al.) and in the left auditory cortex in auditory hallucinators (Spencer, Niznikiewicz, Nestor, Shenton, & McCarley, 2009). At the same time, auditory hallucinators fail to show normal synchrony between frontal and temporal areas while talking (Ford, Mathalon, Whitfield, Faustman, & Roth, 2002).

It may surprise you to learn that hallucinations are associated with activity in the respective sensory areas. Scans of the brains of people with schizophrenia show that language areas are active during auditory hallucinations and visual areas are active during visual hallucinations (Figure 14.10; McGuire, Shah, & Murray, 1993; McGuire et al., 1995; Silbersweig et al., 1995). Because these areas are activated in normal individuals when they are engaged in "inner speech" (talking to oneself) and imagining visual scenes, it appears that the hallucinating schizophrenic is not simply imagining voices and images but also misperceiving self-generated thoughts.

One of the most documented symptoms of schizophrenia is the inability to suppress environmental sounds. With *auditory gating* impaired, the intrusion of traffic noise or a distant conversation is not just annoying but can be interpreted by the person with schizophrenia as threatening. This deficit is also associated with reduced synchrony across wide areas of the brain (M. H. Hall, Taylor, Salisbury, & Levy, 2010). Atypical antipsychotics improve gating, but nicotine normalizes it (Adler et al., 2004; Kumari & Postma, 2005). Seventy to eighty percent of schizophrenics smoke (J. de Leon, 1996; V. De Luca, Wong, et al., 2004), compared with 23% of the normal population, in an apparent attempt at self-medication. Besides sensory gating, nicotine improves several negative symptoms, including impaired visual tracking of moving objects, working memory, and other cognitive abilities (Sacco, Bannon, & George, 2004; Sacco et al., 2005; Tregellas, Tanabe, Martin, & Freedman, 2005). Nicotine appears to compensate for diminished functioning of nicotinic acetylcholine receptors (S. I. Deutsch et al., 2005), increase glutamate and GABA release, and increase dopamine levels in the prefrontal cortex where it is depleted in hypofrontality (Kumari & Postma, 2005; Sata et al., 2008). Three studies have linked schizophrenia with one of the genes responsible for nicotinic receptors (De Luca, Wang, et al., 2004; De Luca, Wong, et al., 2004; S. I. Deutsch et al., 2005).

Environmental Origins of the Brain Defects

An obvious potential cause of brain defects would be head injury. Several studies have reported an association between schizophrenia and brain damage that occurred within a few years prior to diagnosis (reviewed in David & Prince, 2005). However, the studies have been criticized for a number of methodological inadequacies, including reliance on patients' and relatives' memory of the injuries, casual diagnosis of schizophrenia, and failure to consider accident proneness and pre-injury symptoms as confounding factors (David & Prince; Nielsen, Mortensen, O'Callaghan, Mors, & Ewald, 2002).

Evidence is stronger for a variety of influences at the time of birth or during the prenatal period. These include both physical complications (Cannon, Jones, & Murray, 2002) and emotional stresses on the mother, such as death of the father (Huttunen, 1989) and military invasion (van Os & Selten, 1998). One indication that birth and pregnancy complications contribute to brain deficits is that they are associated with enlarged ventricles later in life (Pearlson et al., 1989). They are a possible explanation for the difference in ventricle size

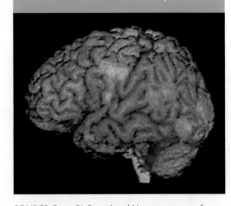

FIGURE 14.10 Brain Activation During Visual and Auditory Hallucinations in a Schizophrenic.

SOURCE: From "A Functional Neuroanatomy of Hallucinations in Schizophrenia," by D. A. Silbersweig et al., *Nature*, 378, pp. 176–179. Reprinted by permission of Nature, copyright 1995.

FIGURE 14.11 Interleukin-1β Levels in Schizophrenics and Controls.

Elevated levels of this protein at the time of a first schizophrenic episode indicated that strong immune reactions had occurred in the past.

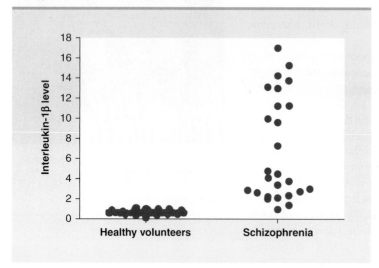

SOURCE: From "Activation of Brain Interleukin-1β in Schizophrenia," by J. Söderland et al., 2009, *Molecular Psychiatry, 14,* 1069–1071. Copyright © 2009 Nature Publishing Group. Used with permission.

between identical twins (Bracha, Torrey, Gottesman, Bigelow, & Cunniff, 1992).

It is easy to see how birth complications, such as being born with the umbilical cord around the neck, could differentiate between twins, but different experiences in the womb require some explanation. Identical twins may share the same placenta and amniotic sac or they may have their own, depending on whether the developing organism splits in two before or after the fourth day of development. Identical twins who did not share a placenta had an 11% concordance rate for schizophrenia, compared with 60% for those who shared a placenta, presumably due to the sharing of infections (J. O. Davis, Phelps, & Bracha, 1995). In spite of the importance of prenatal factors, some researchers believe that they produce schizophrenia only in individuals who are already genetically vulnerable (Schulsinger et al., 1984).

The *winter birth effect* refers to the fact that more schizophrenics are born during the winter and spring than during any other time of the year. The effect has been replicated in a large number of studies, some with more than 50,000 schizophrenic patients as subjects (Bradbury & Miller, 1985). The important factor in winter births is not cold weather, but the fact that infants born between January and May would have been in the second trimester of prenatal development in the fall or early winter, when there is a high incidence of infectious diseases (C. G. Watson, Kucala, Tilleskjor, & Jacobs, 1984). There is good evidence that the mother's exposure to *viral infections* during the fourth through sixth months of pregnancy (second trimester) increases the risk of schizophrenia. This appears to be caused not by the virus itself but by the immune reaction that it triggers. This conclusion is supported by a markedly higher level of interleukin-1β in the spinal fluid of first-episode schizophrenics, indicating that an immune response has occurred (Figure 14.11; Söderlund et al., 2009). Because the patients were infection free at the time, the infection must have occurred earlier, possibly during the prenatal period.

Several illnesses have been implicated, but the effect of influenza has been researched most frequently, and a higher incidence of schizophrenic births has been confirmed following influenza outbreaks in several countries. Figure 14.12 shows that the birth rate for schizophrenics was higher during the winter and spring in years of high influenza infection and that the peak birth rate for schizophrenics followed influenza epidemics. However, these studies could not confirm that the individual mothers had actually been exposed to the influenza virus; by analyzing the blood specimens drawn from expectant mothers, Alan Brown and his colleagues (2004) found a sevenfold increased risk for schizophrenia and spectrum disorders when influenza antibodies were present, and they estimated that influenza infection accounts for 14% of schizophrenia cases.

Prenatal starvation is another pathway to schizophrenia that until recently was the subject of controversy. The idea came about after the rate of schizophrenia doubled among the offspring of mothers who were pregnant during Hitler's 1944 to 1945 food blockade of the Netherlands (Susser et al., 1996). However, the interpretation was questionable because the sample was small and because toxins in the tulip bulbs the women ate to survive could have been to blame. But now data from a much larger sample of adults born during the 1959 to 1961 famine

FIGURE 14.12 Relationship of Schizophrenic Births to Season and Influenza Epidemics in England and Wales (1939–1960).

(a) Schizophrenic birth rates by month during years of high and low influenza incidence. (b) Schizophrenic birth rate as a function of time from beginning of epidemic.

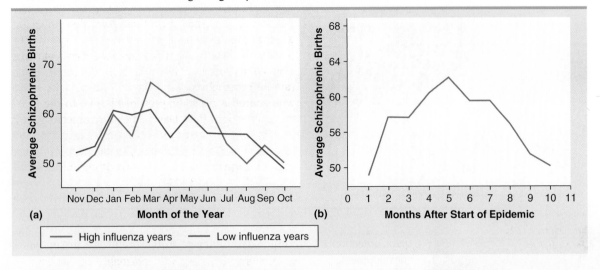

SOURCE: From "Schizophrenia Following Pre-Natal Exposure To Influenza Epidemics Between 1939 and 1960," by P. C. Sham et al., *British Journal of Psychiatry, 160,* pp. 461–466. Copyright 1992. Reprinted with permission of the publisher.

in China have confirmed the association, with an increase in schizophrenia from 0.84% to 2.15% (St. Clair et al., 2005).

Most of the environmental influences we have been discussing occur during pregnancy or birth; one, however, relates to the father. There is a greater risk of schizophrenia if the father's age at the time of conception exceeds 25, and by paternal age of 50 the risk has increased by two thirds (Miller et al., 2010). The mechanism for this effect is unknown, but chances are it is epigenetic, either due to the normal aging process or an accumulation of external environmental insults. Epigenetic effects in general can be traced to a variety of environmental influences, including toxins, diet, starvation, drugs, and stress; they likely account for most of the environmental influences we have been talking about. This is a relatively new area of investigation, so there has been little documentation of epigenetic influences in schizophrenia. However, a study is now underway to analyze the epigenome of thousands of schizophrenics for clues (National Institute of Mental Health, 2008).

Schizophrenia as a Developmental Disease

The defects in the brains of schizophrenics apparently occur early in life, some at the time of birth or before. In some schizophrenics' brains, it appears that many neurons in the temporal and frontal lobes failed to migrate to the outer areas of the cortex during the second trimester; they are disorganized and mislocated in the deeper white layers (Figure 14.13; Akbarian, Bunney, et al., 1993; Akbarian, Viñuela, et al., 1993). The hippocampus and prefrontal cortex of schizophrenics are 30% to 50% deficient in Reelin, a protein that functions as a stop factor for migrating neurons (Fatemi, Earle, & McMenomy, 2000; Guidotti et al., 2000). These observations and the association of schizophrenia with birth trauma and prenatal

FIGURE 14.13 Neural Disorganization in Schizophrenia.

The neurons in the normal hippocampus have an orderly arrangement (a), but in the brain of a schizophrenic individual you can see that they have migrated in a haphazard fashion (b).

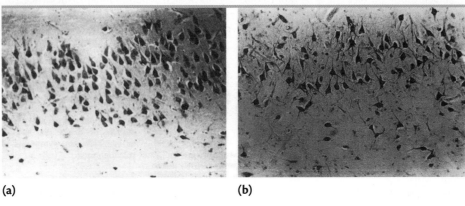

(a)   (b)

SOURCE: © Arne Scheibel, UCLA.

FIGURE 14.14 Gray Matter Loss in Schizophrenic Adolescents.

There is some loss in the brains of normal adolescents due to circuit pruning, but the rate of loss is much greater in the schizophrenic adolescents. Red and pink areas represent 3% to 5% losses annually.

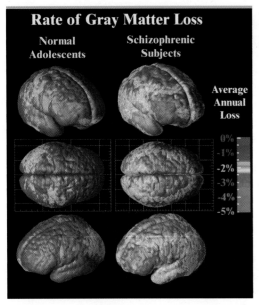

SOURCE: From "Mapping Adolescent Brain Change Reveals Dynamic Wave of Accelerated Gray Matter Loss in Very Early-Onset Schizophrenia," by P. M. Thompson, PNAS, 98, 11650–11655, fig. 1A, p. 11651, and fig. 5, p. 11653. © 2001 National Academy of Sciences, U.S.A. Used with permission.

viral infection all argue for early damage to the brain *or* a disruption of development.

This view is supported by behavioral data. Home movies of children who later became schizophrenic revealed more negative facial expressions and physical awkwardness than in their healthy siblings; the movies were rated by judges who were unaware of the children's later outcome (Walker, Lewine, & Neumann, 1996). Among New Zealanders followed from age 3 to 32, those who later developed schizophrenia had deficits in learning, attention, and problem solving during childhood, and for each year of life they fell an additional 2 to 3 months farther behind other children (Reichenberg et al., 2010).

Gray matter deficit and ventricular enlargement are ordinarily present at the time of patients' diagnosis (Degreef et al., 1992). Most of the evidence indicates that the loss of brain volume occurs rapidly and dramatically in adolescence or young adulthood and then levels off (B. T. Woods, 1998). Adolescence is a particularly significant period in the development of schizophrenia. This is a time when symptoms of schizophrenia often begin to develop and a time of brain maturation, including frontal myelination and connection of temporal limbic areas (D. R. Weinberger & Lipska, 1995). Thompson, Vidal, et al. (2001) identified a group of adolescents who had been diagnosed with early-onset schizophrenia and used MRIs to track their brain development. At the age of 13, there was little departure from the normal amount of gray matter loss that occurs with circuit pruning, but over the next 5 years, loss occurred in some areas as rapidly as 5% per year (Figure 14.14). The nature of the symptoms varied as the loss progressed from parietal to temporal to frontal areas. Studies have found no evidence of dying neurons or of the inflammation that would be expected with an ongoing degenerative disease; instead, gray matter deficits have been attributed to loss of synapses (D. A. Lewis & Levitt, 2002; D. R. Weinberger, 1987). This apparent severe pruning may reflect the elimination of circuits that have already been diminished (D. A. Lewis & Levitt, 2002) by a lack of glutamate activity (Coyle, 2006); this view is supported by the fact that the diagnosis of schizophrenia preceded significant gray matter reductions in the schizophrenic adolescents.

Concept Check
Take a Minute to Check Your Knowledge and Understanding

☐ What is the interplay between heredity and environment in schizophrenia?

☐ Describe the two kinds of schizophrenia.

☐ How are dopamine irregularities and brain deficits proposed to interact?

Affective Disorders

The affective disorders include *depression* and *mania*. Almost all of us occasionally experience *depression,* an intense feeling of sadness; we feel depressed over grades, a bad relationship, or loss of a loved one. While this *reactive* depression can be severe, major depression goes beyond the normal reaction to life's challenges. In *major depression,* a person often feels sad to the point of hopelessness for weeks at a time; loses the ability to enjoy life, relationships, and sex; and experiences loss of energy and appetite, slowness of thought, and sleep disturbance. In some cases, the person is also agitated or restless. Stress is often a contributing factor, but major depression can occur for no apparent reason. *Mania* involves excess energy and confidence that often lead to grandiose schemes; decreased need for sleep, increased sexual drive, and abuse of drugs are common.

Depression may appear alone as *unipolar depression,* or depression and mania may occur together in bipolar disorder. In *bipolar disorder,* the individual alternates between periods of depression and mania; mania can occur without periods of depression, but this is rare. Bipolar patients often show psychotic symptoms such as delusions, hallucinations, paranoia, or bizarre behavior. Two quotes provide some insight into the disorders from the patients' own perspectives (National Institute of Mental Health, 1986):

> Depression: I doubt completely my ability to do anything well. It seems as though my mind has slowed down and burned out to the point of being virtually useless . . . [I am] haunt[ed] . . . with the total, the desperate hopelessness of it all. . . . If I can't feel, move, think, or care, then what on earth is the point?

> Mania: At first when I'm high, it's tremendous . . . ideas are fast . . . like shooting stars you follow until brighter ones appear . . . all shyness disappears, the right words and gestures are suddenly there. . . . Sensuality is pervasive, the desire to seduce and be seduced is irresistible. Your marrow is infused with unbelievable feelings of ease, power, well-being, omnipotence, euphoria . . . you can do anything . . . but, somewhere this changes.

The most recent data indicate that one in five people will suffer a mood disorder in their lifetime, most likely depression (Kessler et al., 2005). Women are two to three times more likely than men to suffer from unipolar depression during their lifetime; bipolar illness occurs equally often in both sexes (Gershon, Bunney, Leckman, Van Eerdewegh, & DeBauche, 1976; P. W. Gold, Goodwin, & Chrousos, 1988) at a rate of about 4%. The risk for major depression increases with age in men, whereas women experience their peak risk between the ages of 35 and 45; the period of greatest risk for bipolar disorder is in the early 20s to around the age of 30. The financial cost of affective disorders in the United States is almost $19 billion a year (Uhl & Grow, 2004).

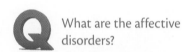
What are the affective disorders?

2
Depression Test

Heredity

As with schizophrenia, there is strong evidence that affective disorders are partially inheritable. Part of that evidence is the increased incidence of affective disorders among patients' relatives. When one identical twin has an affective disorder, the probability that the other twin will have the illness as well is about 69%, compared with 13% in fraternal twins (Gershon et al., 1976). Lack of complete concordance in identical twins indicates that there is an environmental contribution. However, the concordance rate drops surprisingly little when identical twins are reared apart (J. Price, 1968), which may mean that the most important environmental influences occur in the prenatal period or shortly after.

Heritability for depression is somewhere around 37%, with the number somewhat higher for women than for men (Kendler, Gardner, Neale, & Prescott, 2001; Sullivan, Neale, & Kendler, 2000); however, the search for the genes involved is just now beginning to yield solid results. Links to six chromosomal regions have been replicated at least once for strictly defined depression, along with eight others when the category is broadened (Camp & Cannon-Albright, 2005). There is evidence that depression has somewhat different origins in men and women. In one study, men and women shared three probable genes, but seven genes were exclusive to men and nine were exclusive to women (Zubenko, Hughes, Stiffler, Zubenko, & Kaplan, 2002). The sex disparity suggests one reason disorder genes can be difficult to locate in a clinical group, and it may explain the higher frequency of depression in women and the higher rate of suicide in men.

The genes that have surfaced in depression research not only tell us about the disorder's genetic origins but also provide a road map of its structural and functional anomalies. One gene in particular illustrates this point. People with either one or two copies of the "short" allele of the *5-HTTLPR* serotonin transporter gene have an increased vulnerability to depression, along with a 15% reduction in gray matter in the amygdala and a 25% reduction in the subgenual anterior cingulate cortex (which is located just below the anterior part of the corpus callosum; Pezawas et al., 2005). They show an exaggerated amygdala response to fearful facial expressions (Hariri et al., 2002), apparently due to a loss of feedback from the cingulate cortex that would ordinarily damp amygdala activity (Pezawas et al.). According to some studies, these deficiencies increase susceptibility to stress, which leads to depression (Figure 14.15; Canli et al., 2006; Caspi et al., 2003). However, not all studies have found this linkage (Bisch et al., 2009), which leads us to the topic of genetic *epistasis,* the suppression of one gene's effect by that of another. In this case, the *VAL66MET* allele of the gene for *brain-derived neurotrophic factor (BDNF),* a protein that encourages neuron growth and survival, protects against the *5-HTTLPR* short allele's effects on brain development (Pezawas et al., 2008).

Another serotonin-related gene, the *hTPH2* mutation, has a 10 times greater frequency in severely depressed individuals than in nondepressed controls. It produces a defective version of the enzyme that synthesizes serotonin, with the result that the brain produces less serotonin (X. Zhang et al., 2005). Depressed patients with this gene are resistant to antidepressants that block reuptake of serotonin simply because not enough serotonin is available.

A common characteristic of depression is disruption of the circadian (day-night) cycle, which is controlled by a large number of genes. One study found three of those circadian genes to be associated with depression (Gouin et al., 2010), and another identified three additional genes (Soria et al., 2010); the studies used different methods, and we will have to wait for further work to find out which associations will hold up.

In spite of similarities between depression and bipolar disorder, there is good reason to believe that they are considerably independent of each other genetically

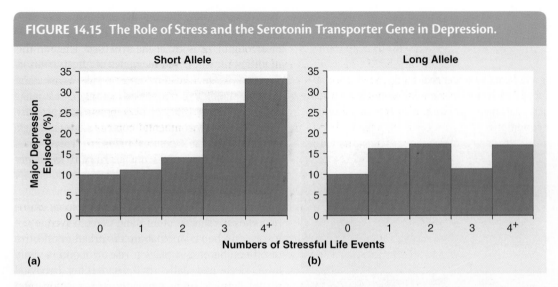

FIGURE 14.15 The Role of Stress and the Serotonin Transporter Gene in Depression.

(a) In individuals with either one or two copies of the so-called short allele, the percentage who were diagnosed at age 26 with depression increased with the number of stressful life events in the past 5 years. (b) In those with two copies of the long allele, the number of stressful events made no difference. Life events were assessed from a checklist of 14 employment, financial, housing, health, and relationship stressors.

SOURCE: From "Influence of Life Stress on Depression: Moderation By a Polymorphism in the 5-HTT Gene," by A. Caspi et al., *Science, 301*, pp. 386–389, fig. 3, p. 389. © 2003. Reprinted by permission of AAAS.

(P. W. Gold et al., 1988; Moldin, Reich, & Rice, 1991). Bipolar disorder is more heritable, with estimates of 85% and 93% (Kieseppä, Partonen, Haukka, Kaprio, & Lönnqvist, 2004; McGuffin et al., 2003). Genetic research has had no problem identifying numerous genes as candidates for explaining the origins of psychological disorders; the difficulty is in obtaining agreement on these genes across studies. A significant step has been taken by one group of researchers who surveyed four large sets of gene data; they identified 69 genes that were associated with bipolar disorder in at least 3 of the 4 sets, 23 of which also overlapped with vulnerability for substance abuse (C. Johnson, Drgon, McMahon, & Uhl, 2009). Also, it is interesting to see that mutations have been found in bipolar patients in three genes that control circadian rhythms, none of which overlapped with the six associated with depression (McGrath et al., 2009; Soria et al., 2010).

The Monoamine Hypothesis of Depression

The first effective treatment for depression was discovered accidentally, and theory again followed practice rather than the other way around. *Iproniazid* was introduced as a treatment for tuberculosis, but it was soon discovered that the drug produced elevation of mood (Crane, 1957) and was an effective antidepressant (Schildkraut, 1965). Iproniazid was later abandoned as an antidepressant because of its side effects, but its ability to increase activity at the monoamine receptors led researchers to the *monoamine hypothesis,* that depression involves reduced activity at norepinephrine and serotonin synapses. You may remember that the monoamines also include dopamine, but because dopamine agonists such as amphetamine produced inconsistent therapeutic results, researchers have limited their interest to norepinephrine and serotonin.

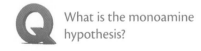

What is the monoamine hypothesis?

FIGURE 14.16 Monoamine Oxidase Levels in the Body of a Nonsmoker and a Smoker.

PET scans were done using a radioactive tracer that binds to monoamine oxidase B. Levels were reduced 33% to 46% in smokers. Monoamine oxidase reduction can have beneficial, detrimental, or neutral effects, depending on the location and other conditions.

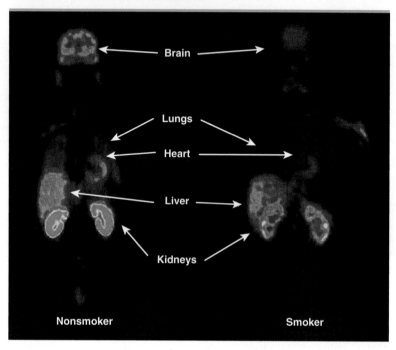

Brain

Lungs

Heart

Liver

Kidneys

Nonsmoker   Smoker

SOURCE: From "Low Monoamine Oxidase B Levels in Peripheral Organs of Smokers," by J. S. Fowler et al., *PNAS, 100*, fig. 2, p. 11602. © 2003 National Academy of Sciences, U.S.A. Used with permission.

MAO Inhibitors

tricyclic

SSRI

We now know that all the effective antidepressant drugs increase the activity of norepinephrine or serotonin, or both, at the synapses. They do this in different ways. Some block the destruction of excess monoamines in the terminals *(monoamine oxidase inhibitors)*, while others block reuptake at the synapse *(tricyclic antidepressants)*. Atypical *(second-generation)* antidepressants affect a single neurotransmitter; for example, Prozac (fluoxetine) is one of several *selective serotonin reuptake inhibitors*. These synaptic effects occur within hours, but symptom improvement takes 2 to 3 weeks.

Additional evidence to support the monoamine hypothesis is that serotonin and norepinephrine are involved in behaviors that are disturbed in affective disorders. Serotonin plays a role in mood, activity level, sleep and daily rhythms, feeding behavior, sexual activity, body temperature regulation, and cognitive function (Meltzer, 1990; Siever et al., 1991). Because the noradrenergic system is involved in responsiveness and sensitivity to the environment, reduced norepinephrine activity may contribute to the depressed individual's slowed behavior, lack of goal-directed activity, and unresponsiveness to environmental change (Siever et al., 1991).

Earlier, we saw that nicotine provides some symptom relief for schizophrenics. Nonnicotine ingredients in tobacco smoke have been found to act as monoamine oxidase inhibitors. This would explain why smoking is so frequent among depressives and why they have particular difficulty giving up smoking (J. S. Fowler et al., 1996; Khalil, Davies, & Catagnoli, 2006). I mention a therapeutic effect of smoking for the second time only to illustrate again how people may self-medicate without being aware they are doing it and why some people have so much trouble quitting; if it sounds as though the benefits of smoking outweigh the cost to the smoker's health, you should reread the section on nicotine in Chapter 5. Figure 14.16 is a dramatic demonstration of the extensive effect of smoking on monoamine oxidase inhibitor levels throughout the body.

Treatment resistance and the delay required for drugs to take effect are serious issues, especially if the patient is suicidal; experiments with ketamine, which was developed as an anesthetic but gained infamy as a club drug, suggest that these problems might be avoidable. In a study with patients who had shown resistance to at least two antidepressant drugs, ketamine produced improvement in 71% within 110 minutes of injection and remission in 29%; 35% maintained their response for a week following the single injection (Zarate et al., 2006). Relapse time is highly variable, though, and ketamine appears to be most valuable as a temporary treatment (aan het Rot et al., 2010). It also interests us because of its specific effect; it blocks the NMDA type of glutamate receptor, implicating glutamate function in depression as well as schizophrenia. About 30% to 50% of depressed patients fail to respond to drug therapy, a statistic made worse by the fact that the placebo response rate alone is 30% (Depression Guideline Panel, 1993). When the drugs do not work and the depression is severe, an alternative is electroconvulsive therapy.

Electroconvulsive Therapy

Electroconvulsive therapy (ECT) involves applying 70 to 130 volts of electricity to the head of an anesthetized patient, which produces a seizure accompanied by convulsive contractions of the neck and limbs and lasting about a half minute to a minute. (See Figure 14.17.) Without the seizure activity in the brain that produces the convulsions, the treatment does not work. Within a few minutes, the patient is conscious and coherent, though perhaps a bit confused; the patient does not remember the experience. Usually ECT is administered two to three times a week for a total of 6 to 12 treatments.

Electroconvulsive therapy is the most controversial of the psychiatric therapies. Producing convulsions by sending a jolt of electricity through the brain *sounds* inhumane, and in fact the procedures used in the early days of ECT treatment often resulted in bone fractures and long-term memory deficits. Now patients are anesthetized and given muscle relaxants that eliminate injury and reduce emotional stress. The number of treatments and the voltage have both been reduced, and stimulation is delivered in brief pulses rather than continuously, often only to the nondominant hemisphere. These changes have made ECT safer and, at the same time, more effective (Weiner & Krystal, 1994). Follow-up studies indicate that memory and cognitive impairment induced by ECT dissipates within a few months (Crowe, 1984; Weeks, Freeman, & Kendell, 1980) and that cognitive performance even improves over pretreatment levels as the depression lifts (Sackeim et al., 1993). Brain scans and autopsies of patients and actual cell counts in animal subjects show no evidence of brain damage following ECT (reviewed in Devanand, Dwork, Hutchinson, Bolwig, & Sackheim, 1994).

ECT is usually reserved for patients who do not respond to the medications or who cannot take them due to extreme side effects or because of pregnancy. In a recent analysis of 13 studies that compared ECT with antidepressant drugs, 79% of patients responded to ECT compared with 54% of patients treated with antidepressants (Pagnin, de Queiroz, Pini, & Cassano, 2004). ECT works especially well when depression or mania is accompanied by psychosis (Depression Guideline Panel, 1993; Potter & Rudorfer, 1993), and it works rapidly, which is beneficial to suicidal patients who cannot wait for weeks while a drug takes effect (Rudorfer, Henry, & Sackheim, 1997). The disadvantage of ECT is that its benefit is often short-term, but the patient can usually be maintained on drug therapy once a round of ECT has been completed.

ECT is effective with depression, mania, and schizophrenia, which suggests that its effects are complex, and research bears this out. A number of changes occur at the brain's synapses. Like the drugs, ECT increases the sensitivity of postsynaptic serotonin receptors (Mann, Arango, & Underwood, 1990); in addition, the sensitivity of autoreceptors on the terminals of norepinephrine- and dopamine-releasing neurons is reduced, so the release of those transmitters is increased. A temporary slowing of the EEG, which is correlated with therapeutic effectiveness, suggests that ECT synchronizes neuronal firing over large areas of the brain (Ishihara & Sasa, 1999; Sackeim et al., 1996). This reduced excitability is likely due to the fact that ECT increases diminished GABA concentrations (Sanacora et al., 2003). However, as you will see in the next section, both antidepressants and ECT now appear to trigger dramatic remodeling of the depressed brain. (For information about two newer forms of electrical treatment, see the accompanying Application.)

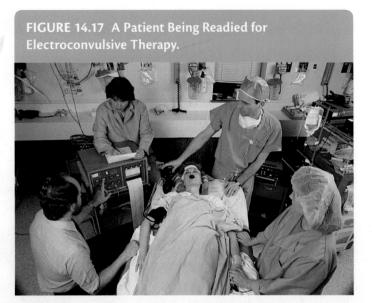

FIGURE 14.17 A Patient Being Readied for Electroconvulsive Therapy.

SOURCE: James D. Wilson/Woodfin Camp & Associates.

Application   Electrical Stimulation for Depression

Researchers often use an electromagnetic field (*transcranial magnetic stimulation*, TMS) either to disrupt or activate brain activity in order to determine the function of that area of the brain (see the figure here and descriptions on pages 94 and 102–103). TMS is finding its way into therapies for a variety of disorders, and one TMS device has been approved by the Federal Drug Administration for use with depression. Fast TMS (pulse rate above 5/second) causes neural excitation; applied over the hypoactive left dorsolateral prefrontal area daily for 4 to 6 weeks, it produces benefits comparable to those of ECT in patients with treatment-resistant severe depression (Grunhaus, Schreiber, Dolberg, Polak, & Dannon, 2003; Janicak et al., 2002; O'Reardon et al., 2007). Side effects appear to be limited to temporary scalp pain or discomfort.

A more aggressive strategy involves deep brain stimulation. Researchers in Toronto led by Helen Mayberg implanted electrodes in the hyperactive subgenual anterior cingulate cortex of six patients who were unresponsive even to ECT (Mayberg et al., 2005). The electrodes were connected to a stimulator surgically implanted under the collarbone that delivered continuous high-frequency, low-voltage pulses of electricity. All six patients experienced immediate effects from the stimulation; after 6 months, four of the six said their depression was mostly gone. Brain surgery may not be necessary; some of the same

results can be obtained using a similar implanted device to provide regular stimulation to the vagus nerve. Vagus stimulation probably works because, like ECT, it increases GABA concentrations in the cortex (George et al., 2007).

Whether these techniques will change the way therapy is done remains to be seen, but they do suggest that therapists will have alternatives available for managing the treatment-resistant patient.

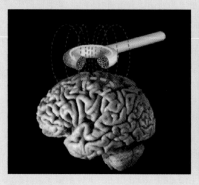 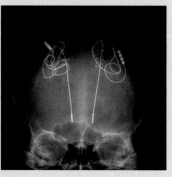

(a)   **(b)**

Transcranial Magnetic Stimulation and Deep Brain Stimulation.

(a) When the electromagnetic coil is held over the scalp, it induces an electric current in the brain tissue below. (b) An X-ray showing the location of electrodes for deep brain stimulation in a depressed patient.

SOURCES: (a) Courtesy of National Institute of Health; (b) Courtesy of Helen Mayberg.

Antidepressants, ECT, and Neural Plasticity

Although antidepressant drugs and ECT have been used to treat depression for more than half a century, we are not sure how they work. Most puzzling is the delay between neurotransmitter changes and symptom relief; hypotheses that changes in receptor sensitivity account for the delay have not been successful (Yamada, Yamada, & Higuchi, 2005). A promising lead is that antidepressant drugs, lithium, and ECT all increase neuronal birth rate in the hippocampus, at least in rodents and presumably in humans as well (Figure 14.18; Harrison, 2002; Sairanen, Lucas, Ernfors, Castrén, & Castrén, 2005). While increased neurogenesis can be detected within hours of antidepressant treatment (Sairanen et al., 2005), the time required for new hippocampal neurons to migrate to their new locations and form functional connections closely matches the delay in symptom improvement (Sairanen et al., 2005).

After new cell development was blocked by X radiation, antidepressants no longer had an effect in mice, suggesting that neurogenesis is required for antidepressive action (Santarelli et al., 2003). An increase in cell numbers is not the basis, however, because cell death also accelerates; some have suggested that the therapeutic effect is due to the greater plasticity of new cells (Gould & Gross, 2002). Although neurogenesis may provide this benefit, there is evidence that other factors contribute as well. When researchers used a drug instead of X radiation to block neurogenesis, antidepressant effect was not diminished (Bessa et al., 2009). The researchers found

FIGURE 14.18 Increased Neurogenesis in the Hippocampus During Antidepressant Treatment.

(a) Antidepressant treatment produced a 60% increase in neurogenesis, compared with administration of inert material (vehicle). (b) Brown dots are new cells (preneurons).

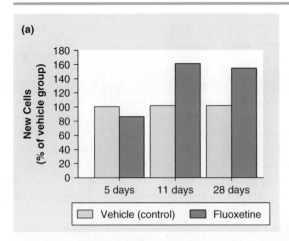

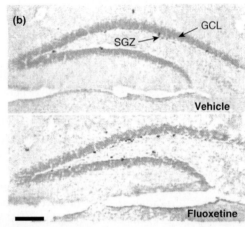

SOURCE: From "Requirement of Hippocampal Neurogenesis of the Behavioral Effects of Antidepressants," by Santarelli et al., *Science*, 301, 805–809, fig. 2a and 2b, p. 806. © 2003 AAAS. Used with permission.

evidence of increased plasticity and synaptic enhancement, and they believed this accounted for the antidepressant effect. There is additional evidence for a plasticity hypothesis: Both antidepressants and ECT modify activity in a large number of genes, especially in the hippocampus; most of those genes contribute to neurogenesis and to various aspects of neuron survival and plasticity (Altar et al., 2004; Yamada et al., 2005). Presumably, blocking neurogenesis with a drug preserved some of the effects of the antidepressant related to synaptic remodeling.

Rhythms and Affective Disorders

Depressed people often have problems with their biological rhythms. The *circadian rhythm*—the one that is a day in length—tends to be phase advanced in affective disorder patients; this means that the person feels sleepy early in the evening and then wakes up in the early morning hours, regardless of the previous evening's bedtime (Dew et al., 1996). The person also enters rapid eye movement sleep earlier in the night and spends more time in REM sleep than normal (Kupfer, 1976). As you will learn in the next chapter, *rapid eye movement (REM) sleep* is the stage of sleep during which dreaming occurs; the excess REM sleep is at the expense of the other stages of sleep. Unipolar depressed patients share this early onset of REM sleep with 70% of their relatives, and relatives with reduced REM latency are three times more likely to be depressed than relatives without reduced latency (Giles, Biggs, Rush, & Roffwarg, 1988).

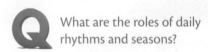

What are the roles of daily rhythms and seasons?

Circadian Rhythms and Antidepressant Therapy

Some patients who are unresponsive to medication can get relief from their depression by readjusting their circadian rhythm. They can do this simply by staying up a half hour later each night until they reach the desired bedtime. In some patients, this treatment results in a relief from depression that lasts for months (Sack, Nurnberger, Rosenthal, Ashburn, & Wehr, 1985).

Some depressed patients also benefit from a reduction in REM sleep (Wu & Bunney, 1990). This is accomplished by monitoring the person's EEG and waking the person every time the EEG indicates that sleep has moved into the REM stage. Interestingly, most drugs that act as antidepressants also suppress REM sleep (G. W. Vogel, Buffenstein, Minter, & Hennessey, 1990). Why reducing REM sleep would alleviate depression is a mystery. But before we assume that this research points to a totally different mechanism for producing depression, we should ask whether REM sleep is somehow linked to what we already know about depression. As it turns out, both serotonin and norepinephrine inhibit REM sleep.

Seasonal Affective Disorder

There is another rhythm that is important in affective disorders; some people's depression rises and falls with the seasons and is known as *seasonal affective disorder (SAD)*. Most SAD patients are more depressed during the fall and winter, then improve in the spring and summer. Others are more depressed in the summer and feel better during the winter. Members of either group may experience a mild mania-like activation called hypomania during their "good" season. While depressed, they usually sleep excessively, and they often have increased appetites, especially for carbohydrates, and gain weight. The length of day and the amount of natural light appear to be important in winter depression; symptoms improve when the patient travels farther south (or north, if the person lives in the Southern Hemisphere) even for a few days, and some report increased depression during cloudy periods in the summer or when they move to an office with fewer windows. Summer depression appears to be temperature related: traveling to a cooler climate, spending time in an air-conditioned house, and taking several cold showers a day

FIGURE 14.19 A Woman Uses a High-Intensity Light to Treat Her Seasonal Affective Disorder.

SOURCE: Dan McCoy/Rainbow.

improve the symptoms. About 10% of all cases of affective disorder are seasonal, and 71% of SAD patients are women (Faedda et al., 1993). Although seasonal influences on affective disorder have been known for 2,000 years and documented since the mid-1850s, summer depression has received relatively little attention, so we will restrict our discussion to winter depression.

A treatment for winter depression is *phototherapy*—having the patient sit in front of high-intensity lights for a couple of hours or more a day (Figure 14.19). Patients begin to respond after 2 to 4 days of treatment with light that approximates sunlight from a window on a clear spring day; they relapse in about the same amount of time following withdrawal of treatment (Rosenthal et al., 1985). The fact that midday phototherapy is effective suggests that the increased amount of light is more important than extending the length of the shortened winter day; the observation that suicide rate is related to a locale's *amount* of clear sunlight rather than the number of hours of daylight supports this conclusion (Wehr et al., 1986). Phototherapy resets the circadian rhythm (Lewy, Sack, Miller, & Hoban, 1987), so it is also helpful with circadian rhythm problems including jet lag, delayed sleep syndrome, and difficulties associated with shift work (Blehar & Rosenthal, 1989).

Lowered serotonin activity is involved in winter depression. Drugs that increase serotonin activity alleviate the depression and reduce carbohydrate craving (O'Rourke, Wurtman, Wurtman, Chebli, & Gleason, 1989). As we saw in Chapter 5, eating carbohydrates increases brain serotonin levels. So, rather than thinking that SAD patients lack willpower when they binge on junk food and gain weight, it might be more accurate to think of them as self-medicating with carbohydrates.

3
Phototherapy

Bipolar Disorder

The mystery of major depression is far from solved, but bipolar disorder is even more puzzling. Bipolar patients vary greatly in their symptoms: the depressive cycle usually lasts longer than mania, but either may predominate. Some patients cycle between depression and mania regularly, whereas others are unpredictable; cycles usually vary from weeks to months in duration, while some patients switch as frequently as every 48 hours (Bunney, Murphy, Goodwin, & Borge, 1972). Stress often precipitates the transition from depression into mania, followed by a more spontaneous change back to depression; the prospect of discharge from the hospital is particularly stressful and often will precipitate the switch into mania. However, as bipolar disorder progresses, manic episodes tend to occur independently of life's stresses (P. W. Gold et al., 1988).

Lithium, a metal administered in the form of lithium carbonate, is the medication of choice for bipolar illness; it is most effective during the manic phase, but it also prevents further depressive episodes. Examination of lithium's effects has not identified any critical neurotransmitters, partly because lithium affects several transmitter systems (Worley, Heller, Snyder, & Baraban, 1988). It may be that lithium stabilizes neurotransmitter and receptor systems to prevent the large swings seen in manic-depressive cycling; its dual role as an antidepressant argues for a normalizing effect rather than a directional one (Gitlin & Altshuler, 1997). Closer examination, however, has revealed a specific effect on mania; lithium and an alternative drug, valproate, indirectly inhibit protein kinase C (PKC), a family of enzymes that regulate neural excitability. The breast cancer drug tamoxifen is used to block estrogen receptors but it also inhibits PKC, and in a clinical trial it produced significant improvement in bipolar symptoms beginning within 5 days (Zarate et al., 2007). If the results of the just completed Phase II trial also turn out to be promising, drugmakers will be working on a compound that inhibits PKC without the side effect of altering estrogen function.

Brain Anomalies in Affective Disorder

As with schizophrenia, affective disorders are associated with structural abnormalities in several brain areas. Again, a larger ventricle size suggests loss of brain tissue, but the reductions are small and are not always found (Depue & Iacono, 1989). A review of numerous studies of depression reveals volume deficits in prefrontal areas, especially the dorsolateral cortex and the subgenual prefrontal cortex (which we will locate and discuss later), as well as in the hippocampus, but an increased volume in the amygdala (J. R. Davidson, Pizzagalli, Nitschke, & Putnam, 2002). Volume reduction apparently precedes depression rather than being a degenerative consequence of it; it is evident at the time of patients' first episode, and it can even be detected in the nondepressed offspring of patients (M. C. Chen, Hamilton, & Gotlib, 2010; Peterson et al., 2009; Zou et al., 2010).

These structural alterations are accompanied by changes in activity level. Not surprisingly, total brain activity is reduced in unipolar patients (Sackeim et al., 1990) and in bipolar patients when they are depressed (Baxter et al., 1985). Activity is particularly decreased in the caudate nucleus and the dorsolateral prefrontal cortex in both groups (Figure 14.20; Baxter et al., 1985; Baxter et al.,

4
Depression and Bipolar Support

Q What brain irregularities are involved?

FIGURE 14.20 Decreased Frontal Activity in Depression.

Blood flow was decreased (a) in the caudate nucleus and (b) in the dorsolateral prefrontal cortex (where the arrows point). The color scale is reversed in the scan in (a); yellow and red in that image indicate *decreased* activity.

SOURCES: (a) From "A Functional Anatomical Study of Unipolar Depression," by Drevets et al., *Journal of Neuroscience, 12,* 3628–3641. © 1992 Society for Neuroscience. Used with permission; (b) From "Reduction of Prefrontal Cortex Glucose Metabolism Common to Three Types of Depression," by L. R. Baxter et al., 1999, *Archives of General Psychiatry, 46*(14), 243–249.

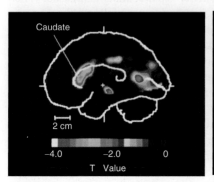

(a)

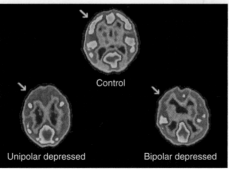

(b)

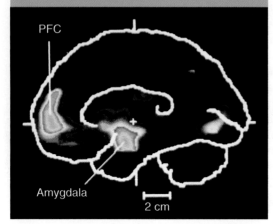

FIGURE 14.21 Increased Activity in the Ventral Prefrontal Cortex (PFC) and Amygdala in Depression.

PFC

Amygdala

2 cm

SOURCE: From "A Functional Anatomical Study of Unipolar Depression," by Drevets et al., *Journal of Neuroscience, 12*, 3628–3641. © 1992 Society for Neuroscience. Used with permission.

FIGURE 14.22 Glucose Metabolism Increase During Mania in a Rapid-Cycling Bipolar Patient.

The middle row shows the sudden increase in activity during a manic episode, just a day after the previous scan during depression. In the bottom row, the patient had retuned to the depressed state.

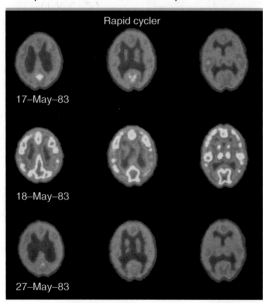

Rapid cycler

17–May–83

18–May–83

27–May–83

SOURCE: From "Cerebral Metabolic Rates for Glucose in Mood Disorders: Studies With Positron Emission Tomography and Fluorodeoxyglucose F18," by L. R. Baxter et al., 1985, *Archives of General Psychiatry, 42,* 441–447.

1989; Drevets et al., 1992). What *is* surprising is that some areas are *more* active in depressed patients. In unipolar depression, blood flow is higher in the amygdala and a frontal area connected to the amygdala called the *ventral prefrontal cortex* (Figure 14.21). The ventral prefrontal area may also be a "depression switch," because activation comes and goes with bouts of depression. The amygdala continues to be active between episodes and returns to normal only after the remission of symptoms. Activity in the amygdala corresponds to the *trait* of depression—the continuing disorder—while activation of the ventral prefrontal area indicates the *state* of depression, which subsides from time to time in some individuals (Drevets, 2001; Drevets et al., 1992; Drevets & Raichle, 1995).

It is also not surprising that when the bipolar patient begins a manic episode, brain metabolism increases from its depressed level by 4% to 36% (Figure 14.22) (Baxter et al., 1985). The *subgenual prefrontal cortex* is particularly interesting because it has been suggested as a possible "switch" controlling bipolar cycling (Figure 14.23). Its metabolic activity is reduced during both unipolar and bipolar depression, but increases during manic episodes (Drevets et al., 1997). The structure is in a good position to act as a bipolar switch, because it has extensive connections to emotion centers such as the amygdala and the lateral hypothalamus and it helps regulate neurotransmitters involved in affective disorders.

Suicide

Ninety percent of people who attempt suicide have a diagnosable psychiatric illness; mood disorder alone accounts for 60% of all completed suicides (Figure 14.24) (Mann, 2003). Bipolar patients are most at risk; about 20% of people who have been hospitalized for bipolar disorder commit suicide. According to the *stress-diathesis model,* the suicidal individual has a predisposition, known as a diathesis, and then stress such as a worsening psychiatric condition acts as an environmental "straw that breaks the camel's back" (Mann, 2003).

The predisposition is at least partly genetic; a study of depressed patients located six chromosome sites that were associated with suicidal risk but independent of susceptibility for mood disorders (Zubenko et al., 2004). Psychiatric patients who attempt suicide also are more likely to have low levels of the serotonin metabolite 5-hydroxyindoleacetic acid (5-HIAA) than nonattempters, which means that their serotonin activity is particularly decreased. When a group of patients at risk for suicide was followed for 1 year, 20% of those who were below the group median in 5-HIAA level had committed suicide; none of the patients above the median had (Träskman, Åsberg, Bertilsson, & Sjöstrand, 1981). Other studies have confirmed the association between lowered serotonin and suicidality (see Figure 14.25) (Mann, 2003; Roy, DeJong, & Linnoila, 1989; M. Stanley, Stanley, Traskman-Bendz, Mann, & Meyendorff, 1986). Lowered 5-HIAA is found in suicide attempters with a variety of disorders and probably reflects impulsiveness rather than the patient's specific psychiatric diagnosis (Mann et al., 1990; M. Stanley et al., 1986; Träskman et al., 1981); this view

FIGURE 14.23 Activity in the Subgenual Prefrontal Cortex in Depression and Mania.

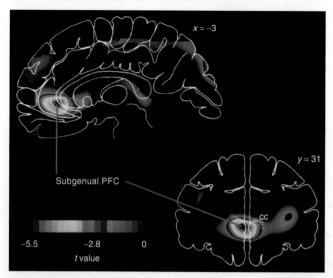

(a)

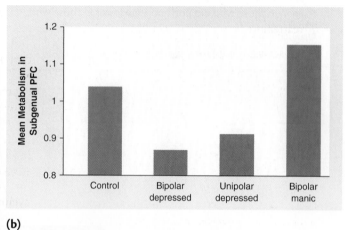

(b)

(a) The dark areas (at the end of the red lines) indicate decreased activity during depression in the subgenual prefrontal cortex.
(b) Comparison of groups shows that activity in the subgenual PFC is lower during depression and higher during mania, which suggests that it controls cycling between depression and mania.

SOURCES: (a) From "Subgenual Prefrontal Cortex Abnormalities in Mood Disorders," by W. C. Drevets et al., *Nature, 386,* 824–827. © 1997 Macmillian Publishing Inc; (b) From "Neurimaging and Neuropathological Studies of Depression: Implications for the Cognitive-Emotional Features of Mood Disorders," by W. C. Drevets, *Current Opinion in Neurobiology, 11,* 240–249, fig. 4b. © 2001 with kind permission of Elsevier.

was supported by a later study in which impulsive suicide attempters were found to have lower 5-HIAA than either nonimpulsive attempters or controls (Spreux-Varoquaux et al., 2001).

However, antidepressants *can* increase the risk of suicide. A variety of explanations have been offered, including the agitation that often accompanies selective serotonin reuptake inhibitor (SSRI) use (Fergusson et al., 2005) and disappointment over slow improvement and side effects (Mann, 2003). Particular concern about the vulnerability of children and adolescents resulted in a 22% decrease in SSRI prescriptions for youths in the United States and the Netherlands; unfortunately, this turned out to be a case of throwing out the baby with the bath water, since youthful suicides increased 14% in the United States in 1 year and 49% in the Netherlands over 2 years (Gibbons et al., 2007). Rather than reducing prescriptions wholesale, therapists need to be selective and to monitor their patients for suicidal tendencies. Researchers have just learned that two gene alleles are associated with a 15-fold increase in suicidal ideation during SSRI treatment, so prescribing selectively should become much easier (Laje et al., 2007).

FIGURE 14.24 Suicide Rates for Three Disorders in Men and Women.

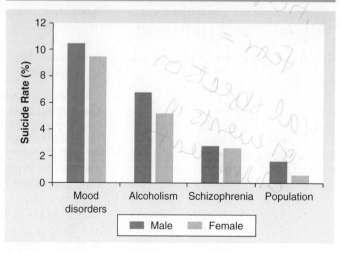

SOURCE: From "Catamnestic Long-Term Study on the Course of Life and Aging of Schizophrenics," by L. Ciompi, 1980, *Schizophrenia Bulletin, 6,* 607–618, fig 2, p. 610. Copyright © Oxford University Press. Used with permission.

FIGURE 14.25 Serotonin Levels and Suicide.

Serotonin level, as assessed by the metabolite 5-HIAA, was lower in depressed patients who attempted suicide than in those who did not, and even lower in those who reattempted.

SOURCE: Based on data from Roy, DeJong, and Linnoila (1989).

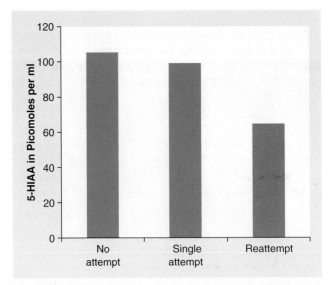

Concept Check   *Take a Minute to Check Your Knowledge and Understanding*

☐ State the monoamine hypothesis; what is the evidence for it?

☐ How is affective disorder related to circadian rhythms?

☐ What brain differences are involved in the affective disorders?

☐ What are some of the factors in suicide?

Anxiety Disorders

Anxiety disorders include several illnesses. The major ones—phobia, generalized anxiety, obsessive-compulsive disorder, and panic disorder—have lifetime risks of about 26%, 6%, 1.6%, and 4.7%, respectively (Kessler et al., 2005). But their significance lies less in their prevalence than in the disruptiveness of their symptoms. The panic disorder patient or the phobic patient may be unable to venture out of the house, much less hold down a job; the obsessive-compulsive individual is no better off, tormented by unwanted thoughts and constantly busy with checking and rituals.

Generalized Anxiety, Panic Disorder, and Phobia

Anxiety is often confused with fear; however, as we saw in Chapter 8, fear is a reaction to real objects or events present in the environment, while anxiety involves anticipation of events or an inappropriate reaction to the environment. A person with generalized anxiety has a feeling of stress and unease most of the time and overreacts to stressful conditions. In panic disorder, the person has a sudden and intense attack of anxiety, with symptoms such as rapid breathing, high heart rate, and feelings of impending disaster. A person with a phobia experiences fear or stress when confronted with a particular situation—for instance, crowds, heights, enclosed spaces, open spaces, or specific objects such as dogs or snakes.

Neurotransmitters

Benzodiazepines have been the most frequently used anxiolytic (antianxiety) drugs in the past (Costall & Naylor, 1992). You may remember from our earlier

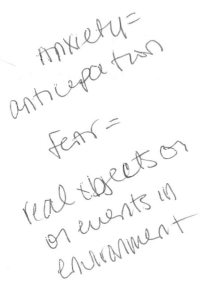

anxiety = anticipation

fear = real objects or events in environment

discussion of drugs that benzodiazepines increase receptor sensitivity to the inhibitory transmitter GABA. A deficit in benzodiazepine receptors may be one cause of anxiety disorder. Marczynski and Urbancic (1988) injected pregnant cats with a benzodiazepine tranquilizer. When the offspring were 1 year old, they were restless and appeared anxious in novel situations. When their brains were studied later, several areas of the brain had compensated for the tranquilizer by reducing the number of benzodiazepine receptors.

Anxiety also appears to involve low activity at serotonin synapses. Antianxiety drugs initially suppress serotonin activity, but then they apparently produce a compensatory increase. The idea that a serotonergic increase is involved in anxiety reduction is supported by the fact that antidepressants are becoming the drug of choice for treating anxiety (reviewed in Durand & Barlow, 2006; Stahl, 1999).

Brain Structures

A number of brain structures are activated in anxiety, including the *amygdala* and the *locus coeruleus.* Both structures participate in more specific emotions such as fear, but the locus coeruleus may be particularly important, because drugs that decrease its action are anxiolytic and drugs that increase its action increase anxiety (Charney, Woods, Krystal, & Heninger, 1990). Panic disorder patients have increased activity in the whole brain, even during symptom-free periods; activity in the *parahippocampal gyrus,* which is connected to the amygdala and the hippocampus, is lower on the left side than on the right (Reiman et al., 1986).

Posttraumatic Stress Disorder

Posttraumatic stress disorder (PTSD) is a prolonged stress reaction to a traumatic event; it is typically characterized by recurrent thoughts and images (flashbacks), nightmares, lack of concentration, and overreactivity to environmental stimuli, such as loud noises. Because of recent news coverage, we usually associate PTSD with combat experiences, but it can be triggered by all kinds of trauma, including robbery, sexual assault, hostage situations, and automobile accidents. Men are more often exposed to such traumas, but women are almost four times as likely to develop PTSD when they do experience trauma (Fullerton et al., 2001).

Whether trauma is followed by PTSD is unrelated to either the severity of the traumatic event or the individual's distress at the time (Harvey & Bryant, 2002; Shalev et al., 2000); the key apparently is the person's vulnerability. People with PTSD have smaller hippocampi than controls (Bonne et al., 2008), and this appears to be a predisposing factor rather than a result of the stress, as you might expect from your reading of Chapter 8. Mark Gilbertson and his colleagues (2002) used magnetic resonance imaging to measure hippocampal volumes in Vietnam combat veterans and their noncombat identical twins. Those who suffered from PTSD had smaller hippocampi than PTSD-free veterans, as expected, but so did the noncombat twins of the PTSD subjects (Figure 14.26). Because hippocampal reduction is often associated with childhood

Q What causes anxiety disorders?

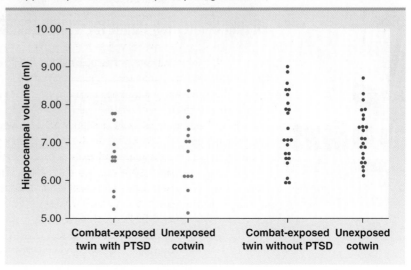

FIGURE 14.26 Hippocampal Volume Is Reduced in Combat Veterans and Their Twins.

Similar reduction in unexposed identical twins of PTSD patients suggests that hippocampal reduction is a predisposing factor.

SOURCE: From "Smaller Hippocampal Volume Predicts Pathologic Vulnerability to Psychological Trauma," by M. W. Gilbertson et al., *Nature Neuroscience, 5,* 1242–1247. Copyright © 2002 Nature Publishing Group. Used with permission.

abuse, Elizabeth Binder and her coworkers (2008) looked there for PTSD origins; they found that previously abused individuals were twice as likely to succumb to PTSD following traumatic events. PTSD has a heritability of around 30% (True et al., 1993); two of the genes are mutations of *FKBP5,* which are more common among abused individuals with PTSD and apparently add to the vulnerability (Binder et al.)

PTSD symptoms are resistant to traditional drug and psychotherapy treatments; an alternative approach is exposure therapy, which allows the individual to confront anxiety-provoking stimuli in the safety of the therapist's office. Exposure therapy is essentially an extinction process, and fear memories are notoriously resistant to extinction, especially in the 30% of people who have the *VAL66MET* allele (which we saw is also involved in depression). We know this gene plays a causal role in fear extinction, because when it was inserted into mice they showed the same increased resistance (Soliman et al., 2010). Brain imaging of human subjects during extinction trials showed why; connections between the prefrontal cortex and the amygdala that are important in fear conditioning and extinction were hypoactive in carriers of the allele.

In their search for better therapies, researchers are resorting to novel approaches; some, for example believe that therapists could take a lesson from the phenomenon of reconsolidation that you learned about in Chapter 12. A team led by Daniela Schiller (2010) used a mild electric shock to condition an emotional reaction (measured by skin conductance response) to a blue square. A day later, the response was extinguished by repeatedly presenting the blue square alone. However, two subgroups of subjects received a "reminder" of the fear memory, one 10 minutes before extinction began and the other 6 hours before; the reminder was intended to start reconsolidation, a window of opportunity that was expected to remain open during extinction for the 10-minute group but to be closed by the time the 6-hour group's extinction trials began. It worked; the skin conductance response was almost entirely absent in the 10-minute group, but it had recovered to near training levels in the two other groups. Researchers hope this technique of *fear erasure* can be used to help relieve PTSD sufferers of their lingering fear and anxiety, but you would probably be more comfortable with the strategy that is described In the News.

Virtual Reality Isn't Just for Video Games

You probably think virtual reality is the coolest gamer's device ever, but it could also be a much-needed therapeutic tool for difficult-to-treat anxiety disorders, including PTSD. A video immerses the patient in a virtual re-creation of a military patrol or a busy freeway or a darkened parking garage similar to the situation where the trauma occurred. The patient controls progress through an interactive scenario, backing off and practicing newly learned relaxation techniques when the stress gets too intense, while the therapist monitors physiological measures of stress, such as skin conductance and finger temperature.

Virtual reality is being used on an experimental basis, particularly with the estimated 13% of U.S. combat veterans who have been diagnosed with PTSD. Preliminary results indicate an 80% improvement rate, compared to 40% for drug treatment and 44% for psychotherapy (Wiederhold, 2010). Virtual reality's promise in treating PTSD led one journal to devote an entire issue to the topic.

SOURCES: Mary Ann Liebert Inc./Genetic Engineering News. (2010, February 12). Posttraumatic stress disorder: Virtual reality and other technologies offer hope. *ScienceDaily.* Retrieved August 20, 2010, from www.sciencedaily.com/releases/2010/02/100211163118.htm; *Cyberpsychology, Behavior, and Social Networking* (2010), *13*(1). Available at www.liebertonline.com/toc/cyber/13/1.

Obsessive-Compulsive Disorder

Obsessive-compulsive disorder (OCD) consists of two behaviors, obsessions and compulsions, which occur in the same person. An obsession is a recurring thought; a person may be annoyed by a tune that mentally replays over and over or by troubling thoughts such as wishing harm to another person. Normal people have similar thoughts, but for the obsessive individual, the experience is extreme and feels completely out of control. Just as the obsessive individual is a slave to thoughts, the compulsive individual is a slave to actions. He or she is compelled to engage in ritualistic behavior (such as touching a door frame three times before passing through the door), or endless bathing and hand washing, or checking to see if appliances are turned off and the door is locked (Rapoport, 1991). One psychiatrist described a patient who tired of returning home to check whether she had turned her appliances off and solved the problem by taking her coffeemaker and iron to work with her (Begley & Biddle, 1996). The playwright and humorist David Sedaris (1998) wrote that his short walk home from school during childhood took a full hour because of his compulsion to stop every few feet and press his nose to the hood of a particular car, lick a certain mailbox, or touch a specific leaf that demanded his attention. Once home, he had to make the rounds of several rooms, kissing, touching, and rearranging various objects before he could enter his own room. About a fourth of OCD patients have a family member with OCD, suggesting a genetic involvement; boys are afflicted more often than girls, but the ratio levels off in adulthood (Swedo, Rapoport, Leonard, Lenane, & Cheslow, 1989).

Brain Anomalies

Imaging studies reveal increased activity in the orbitofrontal cortex, especially the left orbital gyrus, and in the caudate nuclei (Whiteside, Port, & Abramowitz, 2004). This excess activity decreases following successful drug treatment and even behavior therapy (Figure 14.27) (Baxter et al., 1987; J. M. Schwartz et al., 1996; Swedo, Schapiro, et al., 1989). White matter reductions indicate that there are deficient connections between the cingulate gyrus and a circuit involving the basal ganglia, thalamus, and cortex, which apparently result in a loss of impulse control (Insel, 1992; Szeszko et al., 2005). An indication of dysfunction in this network is that orbitofrontal activity does not increase when OCD patients are required to reverse a previously correct choice; their unaffected relatives show the same deficit, suggesting that it is genetic (Chamberlain et al., 2008).

OCD occurs with a number of diseases that damage the basal ganglia (H. L. Leonard et al., 1992) which, you will remember, are involved in motor activity. There is growing evidence that the disorder can be triggered in children by a streptococcal infection that results in an autoimmune attack on the basal ganglia; vulnerability to the immune malfunction apparently is hereditary (P. D. Arnold & Richter, 2001; P. E. Arnold, Zai, & Richter, 2004). OCD has also been reported in cases of head injury (McKeon, McGuffin, & Robinson, 1984). The most famous obsessive-compulsive individual was the multimillionaire Howard Hughes (R. Fowler, 1986). Some signs of disorder during childhood and his mother's obsessive concern with germs suggest either genetic vulnerability or environmental influence. However, symptoms of OCD did not begin until after a series of airplane crashes and automobile accidents that left him almost unrecognizable (Figure 14.28). When a business associate died, Hughes gave explicit instructions that flowers for the funeral were to be delivered by an independent messenger who would not have any contact with the florist or with Hughes's office—even to the point of sending a bill—to prevent "backflow" of germs (Bartlett & Steele, 1979). Assistants were required to handle his papers with gloves, sometimes several pairs, and he in turn grasped them with a tissue. He instructed his assistants not to touch him, talk directly to him, or even

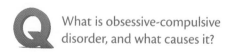

Q What is obsessive-compulsive disorder, and what causes it?

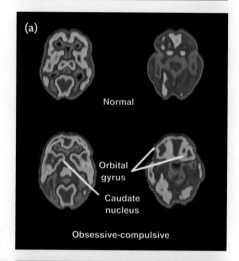

FIGURE 14.27 Brain Structures Involved in Obsessive-Compulsive Disorder.

Normal

Orbital gyrus

Caudate nucleus

Obsessive-compulsive

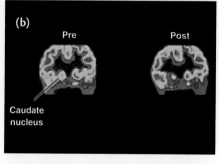

(b)

Pre   Post

Caudate nucleus

(a) The caudate nucleus (a part of the basal ganglia) and the orbital gyrus; (b) the caudate nucleus before and after behavior therapy.

SOURCES: (a) From "Local Cerebral Glucose Metabolic Rates in Obsessive Compulsive Disorder," by Baxter et al., 1987, *Archives of General Psychiatry*, 44(14), pp. 211–218; (b) From "Systematic Changes in Cerebral Glucose Metabolic Rate After Successful Behavior Modification Treatment of Obsessive-Compulsive Disorder," by J. M. Schwartz et al., 1996, *Archives of General Psychiatry*, 53(14), pp. 109–113.

look at him; his defense for this behavior was that everybody carries germs and he wanted to avoid germs (R. Fowler, 1986).

Treating Obsessive-Compulsive Disorder

Researchers believe that OCD patients are high in serotonergic activity. This was suggested by the fact that obsessive-compulsives are inhibited in action and feel guilty about aggressive impulses; sociopaths, on the other hand, feel no guilt after committing impulsive crimes, and they have lowered serotonin activity. But the only drugs that consistently improve OCD symptoms are antidepressants that inhibit serotonin reuptake (Insel, Zohar, Benkelfat, & Murphy, 1990). So, if OCD patients do have high serotonergic activity, then reuptake inhibitors must work by causing a compensatory reduction in activity; there is some evidence that treatment does decrease the sensitivity of serotonin receptors (Insel et al., 1990), but the nature of serotonin involvement remains uncertain (Graybiel & Rauch, 2000).

Patient response to serotonin reuptake inhibitors is usually only partial, and some patients do not respond at all; a third of these benefit from antipsychotics, and an antiglutamate drug produces modest relief (Abramowitz, Taylor, & McKay, 2009). In more desperate cases, surgeons have severed the connections between the orbitofrontal cortex and the anterior cingulate cortex, but the safety and effectiveness of this procedure are controversial. Deep brain stimulation has shown some promise for treating OCD, while transcranial magnetic stimulation's usefulness has not yet been demonstrated (Abramowitz et al.).

Related Disorders

The symptoms of OCD, particularly washing and "grooming" rituals and preoccupation with cleanliness, suggest to some researchers that it is a disorder of "excessive grooming" (H. L. Leonard, Lenane, Swedo, Rettew, & Rapoport, 1991; Rapoport, 1991). Dogs and cats sometimes groom their fur to the point of producing bald spots and ulcers in a disorder known as *acral lick syndrome.* Some chimpanzees and monkeys engage in excessive self-grooming and hair pulling, and 10% of birds in captivity compulsively pull out their feathers, occasionally to the point that the bird is denuded and at risk for infection and hypothermia. Clomipramine, an antidepressant that inhibits serotonin reuptake, is effective in reducing all these behaviors (Grindlinger & Ramsay, 1991; Hartman, 1995; Rapoport, 1991).

If you think that the excessive grooming idea sounds far-fetched, consider the human behaviors of nail biting and obsessive hair pulling, in which the person pulls hairs out one by one until there are visible bald spots or even complete baldness of the head, eyebrows, and eyelashes. There are several similarities between hair pulling and OCD: Both behaviors appear to be hereditary, and hair pullers have a high frequency of relatives with OCD; both symptoms also respond to serotonin reuptake inhibitors (Leonard et al., 1991; Swedo et al., 1991). However, trichotillomania and OCD sufferers appear to differ from each other in their versions of the *Sapap3* gene (Bienvenu et al., 2008). The gene's normal role is most likely a protective one, since mice that lack the gene groom to the point of self-injury; their behavior is relieved by a serotonin reuptake inhibitor (Figure 14.29) (Welch et al., 2007).

Hoarders are dedicated collectors, stashing away just about anything from string to old newspapers. Hoarders are often found among people with OCD, and the behavior is currently classified as a part of that disorder; however, researchers see enough differences between the two that hoarding will probably soon be treated as a separate disorder. The Application describes how extreme this behavior can become.

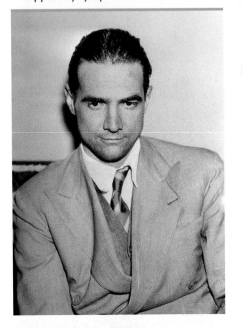

FIGURE 14.28 Howard Hughes.

Hughes was an extraordinarily successful businessman and dashing man-about-Hollywood, but spent his later years crippled by symptoms of OCD.

SOURCE: © Bettmann/Corbis.

FIGURE 14.29 Excessive Grooming and the *Sapap3* Gene.

The mouse lacking the *Sapap3* gene (above) has groomed to the point of creating lesions on the neck and face and even damaging the eye.

SOURCE: Courtesy of Jing Lu, Jeff Welch, and Guoping Feng.

Application Of Hermits and Hoarders

Hoarders is a documentary series on A&E television that profiles extreme hoarders; in one segment, the home is so cluttered that the couple is forced to take meals in bed (Weiss, 2010). In each episode, clinicians and professional organizers, with the aid of friends and family, assist the hoarder in the cleanup of his or her home. The intervention is usually precipitated by a crisis, such as a threat to remove the children for health and safety reasons. Hoarding can be so extreme that it poses broader dangers as well; for example, 14 firefighters were injured when 150 of them were called to put out a fire in a New York City apartment filled with floor-to-ceiling junk (Newman, 2006).

But the unquestionable all-time champion hoarders were the Collyer brothers. They grew up in a Fifth Avenue New York mansion with their eccentric first-cousin parents; their father, a gynecologist, canoed to work at Bellevue Hospital and the family gave up their telephone, electricity, and gas to "simplify" their lives (Weiss, 2010). Homer became an attorney, and Langley graduated from Columbia in mechanical engineering and chemistry and then tried a concert piano career. After the parents died, the two brothers became reclusive hermits. Homer was confined to the home by blindness and arthritis, but Langley busied himself with nighttime treks through Harlem, dragging a cardboard box by a rope and collecting odds and ends off the streets. On March 21, 1947, responding to a tip about an odor, police broke into the mansion through a second-story window after finding the foyer blocked by a solid wall of junk ("Collyer brothers," n.d.). They found Homer dead of starvation and no sign of Langley. The home was so filled with clutter that Langley had made tunnels through the debris in order to move from room

5
Hoarders

Searchers Climb Over Debris Piled Almost to the Ceiling in the Collyer Home.

SOURCE: Corbis.

to room; some of those tunnels were rigged with booby traps as a defense against intruders. Authorities began the task of removing the trash and junk from the home; 18 days after Homer's body was discovered, Langley was found just 10 feet away, crushed under one of his own traps. Among the 130 tons removed from the home there was so little of value that an auction brought only $1,800. For the next decade, children were forced to grow up with their mother's admonition, "Clean up your room or you'll end up like the Collyer brothers!" (Weiss, p. 251).

Another disorder associated with OCD is *Tourette's syndrome*, whose victims suffer from a variety of motor and phonic (sound) tics. They twitch and grimace, throw punches at the air, cough, grunt, bark, swear, blurt out racial slurs or sexual remarks, insult passersby, echo what others say, and mimic people's facial expressions and gestures. Both OCD and Tourette's sufferers can manage their symptoms

Q What causes the bizarre behavior of Tourette's?

for short periods; for instance, Tourette's patients are usually symptom-free while driving a car, having sex, or performing surgery (yes, some of them are surgeons!). But neither OCD nor Tourette's is a simple matter of will: Children often suppress compulsive rituals at school and "let go" at home, or they suppress tics during the day then tic during their sleep. Neurologist Oliver Sacks (1990) graphically describes a woman on the streets of New York who was imitating other people's expressions and gestures as she passed them on the sidewalk:

> Suddenly, desperately, the old woman turned aside, into an alley-way which led off the main street. And there, with all the appearances of a woman violently sick, she expelled . . . all the gestures, the postures, the expressions, the demeanours, the entire behavioural repertoires, of the past forty or fifty people she had passed. (p. 123)

Symptoms begin between the ages of 2 and 15 years and usually progress from simple to more complex tics, with increasing compulsive or ritualistic qualities. About 1 person in 2,000 is afflicted, with males outnumbering females 3:1 (R. A. Price, Kidd, Cohen, Pauls, & Leckman, 1985). Tourette's syndrome is genetically influenced, with a concordance rate of 53% for identical twins and 8% for fraternal twins (R. A. Price et al., 1985). Tourette's shares some genetic roots with OCD; a third of early-onset OCD patients also have Tourette's syndrome (do Rosario-Campos et al., 2005), and 30% of adults with Tourette's are also diagnosed with OCD (R. A. King, Leckman, Scahill, & Cohen, 1998). Recent studies have identified a mutation of the *SLITRK1* gene in Tourette's and found a mutation of the *Hdc* gene in a man and his eight offspring, all of whom have the disorder (Abelson et al., 2005; Ercan-Sencicek et al., 2010). The genes function in neural development and transmission, and both are highly active in brain areas involved in Tourette's.

Tourette's syndrome, like OCD, involves increased activity in the *basal ganglia*. But unlike OCD, the most frequently prescribed drug for Tourette's is an antidopaminergic drug, haloperidol, though newer antidopamine drugs are also being used. One effect dopamine has is motor activation, and Malison and his colleagues (1995) found that dopamine activity is elevated in Tourette's sufferers in the caudate nuclei of the basal ganglia (Figure 14.30). Three patients who were unresponsive to drugs have been treated successfully with high frequency stimulation to the thalamus (Visser-Vandewalle et al., 2003), and another with severe Tourette's showed major improvement with stimulation applied to either the thalamus or the globus pallidus of the basal ganglia (Houeto et al., 2005).

The Anxiety Disorders and Heredity

Family and twin studies indicate that the anxiety disorders are genetically influenced, with heritabilities ranging between 20% and 47%, depending on the disorder (Abramowitz et al., 2009; P. E. Arnold et al., 2004; Hettema, Neale, & Kendler, 2001). Understanding the hereditary underpinnings of anxiety is difficult because of significant genetic overlap with other disorders. More than 90% of individuals with anxiety disorders have a history of other psychiatric problems (Kaufman & Charney, 2000). The overlap with mood disorders is particularly strong; 50% to 60% of patients with major depression also have a history of one or more anxiety disorders (Kaufman & Charney), and panic disorder is found in 16% of bipolar patients (Doughty, Wells, Joyce, Olds, & Walsh, 2004). Some neural commonality between these two groups is suggested by the effectiveness of antidepressants with both mood disorders and anxiety disorders. The anxieties themselves appear to fall into four genetically related clusters, with generalized anxiety, panic, and agoraphobia (fear of crowds and open places) in one group; animal phobias and situational phobias in the second; social phobia overlapping genetically with both groups (Hettema, Prescott, Myers, Neale, & Kendler, 2005); and OCD and Tourette's syndrome composing in the fourth group.

6
Tourette's Syndrome

FIGURE 14.30 Increased Dopamine Activity in the Caudate Nuclei in Tourette's Syndrome.

These two scans have not been superimposed over images of a brain; you can refer to Figure 14.27b to see where the caudate nuclei are located in the brain.

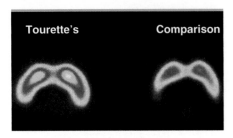

SOURCE: From "[123I] α-CIT SPECT imaging of striatal dopamine transporter binding in Tourette's disorder," by R. T. Malison et al., 1995, *American Journal of Psychiatry, 152*, 1359–1361. Copyright © 1995 American Psychiatric Publishing. Used with permission.

Genetic research has most often implicated genes responsible for serotonin production, serotonin reuptake, and various subtypes of serotonin receptors (reviewed in P. E. Arnold et al., 2004; Rothe et al., 2004; You, Hu, Chen, & Zhang, 2005). Other leads include genes for monoamine oxidase (Tadic et al., 2003), for the adenosine receptor (P. E. Arnold et al., 2004; Lam, Hong, & Tsai, 2005), and for cholecystokinin and its receptor (P. E. Arnold et al., 2004). In addition, OCD may involve genes for a glutamate receptor, brain-derived neurotrophic factor, and a protein suspected of mediating the autoimmune attack on the basal ganglia (P. E. Arnold et al., 2004; Lam, Cheng, Hong, & Tsai, 2004; Zai et al., 2004). Although OCD and Tourette's undoubtedly share a number of genes, one distinction may be a variant of the gene for the D_2 subtype of dopamine receptor, which is linked with Tourette's syndrome (C. C. Lee et al., 2005).

Concept Check   *Take a Minute to Check Your Knowledge and Understanding*

☐ What neurotransmitter deviations are involved in anxiety disorders?

☐ What brain anomalies are associated with anxiety disorders?

☐ Describe OCD and the related disorders.

In Perspective

The past three decades have seen enormous progress in research, and we now know a great deal about the physiological causes of disorders. We owe these breakthroughs to advances in imaging techniques and genetics research technology and to our greatly improved understanding of the physiology of synapses, not to mention the persistence of dedicated researchers. The result is that we can now describe at the biological level many disorders that previously were considered to be purely "psychological" in origin or that were only suspected of having an organic basis.

7
Report of the Surgeon General

In spite of these great research advances, we cannot reliably distinguish the schizophrenic brain from a normal one or diagnose depression or an anxiety disorder from a blood test. We may be able to someday, but in the meantime, we rely for diagnosis on the behavior of the individuals. We know, at least to some extent, the physiological components of mental illness, but we do not understand the unique combination that determines who will become disordered and who will not. As long as that is true, our treatments will remain a pale hope rather than a bright promise.

We have been repeatedly reminded that genetic vulnerability is not the same thing as fate. In most cases, the genes produce an illness only with the cooperation of the environment. This point is emphasized by the fact that psychotherapy plays an important role in treatment, enhancing and sometimes exceeding the benefit drugs can provide (see Durand & Barlow, 2006). While we search for genetic treatments of the disorders, we must remind ourselves once again that heredity is not destiny; improving the physical and psychological welfare of the population would go a long way toward preventing mental illness or reducing its severity.

Just before the dawning of this new age of research, one frustrated schizophrenia researcher concluded, "Almost everything remains to be done" (Heston, 1970, p. 254). Since then our knowledge of both the brain and its participation in the symptoms of mental illness has increased dramatically, but as you can see, much of our understanding remains tentative. The pace is quickening, and I am confident that in this second decade of the new millennium, we will be celebrating even more impressive advances than in the past.

Summary

Schizophrenia

- Schizophrenia is characterized by some mix of symptoms such as hallucinations, delusions, thought disorder, and withdrawal.
- Twin and adoption studies indicate that heritability is .60 to .90, and several possible gene locations have been found. Genes apparently determine the level of vulnerability.
- Schizophrenia is usually divided into positive and negative symptoms, possibly distinguished by excess dopamine activity versus brain deficits.
- Although there is evidence for the dopamine hypothesis, it is an incomplete explanation. The glutamate hypothesis is getting more attention because NMDA receptor antagonists produce symptoms of schizophrenia and drugs that activate NMDA receptors relieve them.
- The brain irregularities include ventricular enlargement (due to tissue deficits), hypofrontality, and impaired connections; these apparently arise from pre-natal insults and impaired postnatal development, in interaction with genetic vulnerability.

Affective Disorders

- The affective disorders include depression and bipolar disorder, an alternation between mania and depression.
- The affective disorders are also highly heritable.
- The most prominent explanation of affective disorders is the monoamine hypothesis.
- ECT is a controversial but very effective last-resort therapy that has value when medications fail and as a temporary suicide preventative.
- Both drugs and ECT increase neurogenesis and neural plasticity.
- People with affective disorders often have disruptions in their circadian rhythm. Others respond to seasonal changes with winter depression or summer depression.
- Bipolar disorder is less understood than unipolar depression, even though lithium is often a very effective treatment.
- A number of brain anomalies distinguish depressed people from bipolar patients and both from normal people.
- Depression, low serotonin activity, and several genes are associated with suicide risk.

Anxiety Disorders

- The anxiety disorders are partially hereditary, but we are unclear about the genes involved.
- Anxiety apparently involves low serotonin activity, while OCD patients have high serotonin activity. A deficit of benzodiazepine receptors may also be a factor in anxiety.
- Posttraumatic stress disorder is a form of anxiety that is particularly difficult to treat; it is related more to vulnerability from genes and childhood abuse than to the severity of the precipitating event.
- The amygdala and locus coeruleus are active during anxiety; OCD involves activity in the orbital frontal cortex and the caudate nuclei.
- OCD and related disorders, including Tourette's, appear related to grooming behaviors. ■

Study Resources

F For Further Thought

- Now that we are nearing the end of the text, summarize what you know about the interaction of heredity and environment. Give examples from different chapters and include the concept of vulnerability.
- Give an overall view of what produces deviant behavior (going back to earlier chapters as well as

this one). What effect does this have on your ideas about responsibility for one's behavior?
- Behavior is vulnerable to a number of disturbances, involving both genetic and environmental influences. Consider the different ways complexity of the brain contributes to this vulnerability.

T Testing Your Understanding

1. Explain the dopamine theory of schizophrenia. What are its deficiencies? What alternative or complementary explanations are available?

2. Describe the monoamine hypothesis of depression; include the evidence for it and a description of the effects of the drugs and ECT used to treat depression.

3. Describe the similarities and associations among OCD, Tourette's, and "grooming" behaviors.

Select the best answer:

1. If you were diagnosed with schizophrenia, you should prefer _____ symptoms
   a. positive
   b. negative
   c. chronic
   d. bipolar

2. The fact that schizophrenia involves multiple genes helps explain
   a. vulnerability to winter viruses.
   b. the onset late in life.
   c. positive symptoms.
   d. different degrees of vulnerability.

3. All drugs that are effective in treating schizophrenia
   a. interfere with reuptake of dopamine.
   b. have some effect at D_2 receptors.
   c. stimulate glutamate receptors.
   d. inhibit serotonin receptors.

4. Schizophrenia apparently involves
   a. tissue deficits.
   b. frontal misfunction.

   c. disrupted connections.
   d. a and b.
   e. a, b, and c.

5. One hypothesis about the timing of the onset of schizophrenia is that
   a. the individual is vulnerable to the effect of viruses then.
   b. dopamine levels decrease with age.
   c. adolescence is a period of brain maturation.
   d. there is considerable neuron death then.

6. The monoamine hypothesis states that depression results from
   a. reduced activity in norepinephrine and serotonin synapses.
   b. increased activity in norepinephrine and serotonin synapses.
   c. reduced activity in norepinephrine, serotonin, and dopamine synapses.
   d. increased activity in norepinephrine, serotonin, and dopamine synapses.

7. ECT appears to relieve depression by
   a. producing amnesia for depressing memories.
   b. the same mechanisms as antidepressant drugs.
   c. punishing depressive behavior.
   d. increasing EEG frequency.

8. A frontal area hypothesized to switch between depression and mania is the
   a. dorsolateral prefrontal cortex.
   b. caudate nucleus.
   c. ventral prefrontal cortex.
   d. subgenual prefrontal cortex.

9. Studies indicate that risk for suicide is related to
   a. low norepinephrine and serotonin.
   b. high norepinephrine and serotonin.
   c. low serotonin.
   d. low norepinephrine.

10. The anxiety disorders are associated genetically with
    a. schizophrenia and depression.
    b. schizophrenia.
    c. depression.
    d. none of these.

11. Of these, the best predictor of PTSD following trauma is:
    a. a history of childhood abuse.
    b. being male.
    c. the severity of the trauma.
    d. the intensity of the reaction to the trauma.

12. OCD can be caused by
    a. genes.
    b. diseases and head injury.
    c. example of a family member.
    d. a and b.
    e. a, b, and c.

13. OCD and Tourette's both involve compulsive rituals, probably because they involve
    a. increased dopamine.
    b. increased activity in the basal ganglia.
    c. a stressful home life
    d. all of these.

Answers: 1.a, 2.d, 3.b, 4.e, 5.c, 6.a, 7.b, 8.d, 9.c, 10.c, 11.a, 12.d, 13.b

Online Resources

The following resources are available at www.sagepub.com/garrett3e.

Chapter Resources

- Flash Cards
- Chapter Quiz
- Internet Resources and Exercises

On the Web

Links to these websites can be found at the web address above: Select this chapter's **Chapter Resources**, and then choose **Internet Resources and Exercises**.

1. Continuing Medical Education's **Schizophrenia** Health Directory has links to a large number of sites providing information on the disorder.
   The Experience of Schizophrenia is the home page of Ian Chovil, who describes his ongoing battle with the disorder.

2. The New York University School of Medicine offers an online **Depression Screening Test** to help a person assess his or her symptoms.

3. The **Society for Light Treatment and Biological Rhythms** has information about circadian rhythms, seasonal affective disorder, and the use of phototherapy.

4. The **Depression and Bipolar Support Alliance** site is a place to learn about mood disorders and ongoing research and to take a mood disorders questionnaire.

5. At *Hoarders,* you can watch entire episodes of the documentary series; if you don't have time for that, sample a few and read descriptions of the people for insights into the lives of hoarders and their families.

6. WEGO Health's **Tourette Syndrome Support** is a good resource for information on this disorder.

7. **Mental Health: A Report of the Surgeon General** concludes that mental health is an issue that must be addressed by the nation.

Chapter Updates and Biopsychology News

R For Further Reading

1. *When the Music's Over: My Journey Into Schizophrenia* by Ross Burke (Plume/Penguin, 1995) is the author's account of his battle with schizophrenia, published by his therapist after Burke ended the battle with suicide.

2. *An Unquiet Mind* by Kay Jamison (Knopf, 1995) tells the story of her continuing battle with bipolar disorder, which rendered her "ravingly psychotic" three months into her first semester as a psychology professor. With the aid of lithium, she has become an authority on mood disorders. (Kay Jamison is the writer quoted in the introduction to Chapter 1.)

3. *Essentials of Abnormal Psychology* by Mark Durand and David Barlow (fifth edition, Cengage, 2010) is a text that covers many of the topics of this chapter in greater detail.

4. "Glutamatergic Mechanisms in Schizophrenia" by G. Tsai and J. T. Cole (*Annual Review of Pharmacology and Toxicology, 42,* 165–179, 2002) reviews research that implicates the NMDA glutamate receptor in schizophrenia.

5. "All in the Mind of a Mouse" by Carina Dennis (*Nature, 438,* 151–152, 2005) is an intriguing look at the creative ways researchers are using mice to study human psychological disorders.

6. "Are bad sleeping habits driving us mad?" by Emma Young (*New Scientist,* February 18, 2009, www.newscientist.com/article/mg20126962 .100) describes evidence that lack of sleep can often be the cause rather than a symptom of disorders.

K Key Terms

acute ... 430
bipolar disorder ... 443
brain-derived neurotrophic factor (BDNF) ... 444
chronic ... 430
circadian rhythm ... 449
depression ... 443
dopamine hypothesis ... 434
electroconvulsive therapy (ECT) ... 447
glutamate theory ... 436
lithium ... 451
major depression ... 443
mania ... 443
monoamine hypothesis ... 445
negative symptoms ... 433
obsessive-compulsive disorder (OCD) ... 457
phototherapy ... 450
positive symptoms ... 433
posttraumatic stress disorder (PTSD) ... 455
psychosis ... 429
rapid eye movement (REM) sleep ... 449
schizophrenia ... 429
seasonal affective disorder (SAD) ... 450
tardive dyskinesia ... 435
Tourette's syndrome ... 459
unipolar depression ... 443
vulnerability model ... 432
winter birth effect ... 440
Wisconsin Card Sorting Test ... 437

Sleep and Consciousness

15

Sleep and Dreaming

Circadian Rhythms

Rhythms During Waking and Sleeping

The Functions of REM and Non-REM Sleep

Sleep and Memory

Brain Structures of Sleep and Waking

Sleep Disorders

   IN THE NEWS: IN THE STILL OF THE NIGHT

Sleep as a Form of Consciousness

   CONCEPT CHECK

The Neural Bases of Consciousness

Awareness

Attention

The Sense of Self

Theoretical Explanations of Consciousness

   APPLICATION: DETERMINING CONSCIOUSNESS WHEN IT COUNTS

   CONCEPT CHECK

In Perspective

Summary

Study Resources

In this chapter you will learn

- About the rhythm of sleep and waking and its neural controls

- About a shorter rhythm throughout the day, and its possible functions during sleep

- What some of the sleep disorders are and what causes them

- How researchers are tackling the problem of consciousness

- Some of the neural processes that contribute to consciousness

Kenneth Parks got up from the couch where he had been sleeping and drove 14 miles to his in-laws' home. There he struggled with his father-in-law before stabbing his mother-in-law repeatedly, killing her. He then drove to the police station, where he told the police that he thought he had "killed some people." In court, his defense was that he was sleepwalking. Based on the testimony of sleep experts and the lack of motive—Ken had an affectionate relationship with his in-laws—the jury acquitted him of murder (Broughton et al., 1994).

> *Studying the brain without studying consciousness would be like studying the stomach without studying digestion.*
>
> —*John R. Searle*

Were Ken's actions possible for someone who was sleepwalking? Was he really asleep and therefore not responsible? This case raises the question of what we mean by *consciousness*. Many psychologists, and especially neuroscientists, avoid the topic because they think that consciousness is inaccessible to research. This has not always been so; consciousness was a major concern of the fledgling discipline of psychology near the end of the 19th century. But the researchers' technique of *introspection* was subjective: The observations were open only to the individuals doing the introspecting, who often disagreed with each other. This failing encouraged the development of behaviorism, which was based on the principle that psychology should study only the relationships between external stimuli and observable responses. Behaviorism was a necessary means of cleansing psychology of its subjective methods, but its purge discarded the subject matter along with the methodology. The interests of psychologists would not shift back to include internal experience until the emergence of the field of *cognitive psychology* in the 1950s and 1960s.

Many cognitive psychologists were finding it difficult to understand psychological functions such as learning and perception without taking account of various aspects of consciousness. Still, few of them tackled the subject of consciousness itself. The problem seemed too big, there was no clear definition of consciousness, and the bias that consciousness was a problem for philosophers still lingered. Gradually, some of them began to ally themselves with philosophers, biologists, and computer experts to develop new research strategies for exploring this last frontier of psychology. The greatest inroads have been made in the study of sleep, largely because sleep is readily observable. Also, because sleep is open to study by objective techniques, it has not had the stigma among researchers that characterizes other aspects of consciousness. We will begin this last leg of our journey with the topic of sleep and dreaming.

Sleep and Dreaming

1
Sleep Info

Each night, we slip into a mysterious state that is neither entirely conscious nor unconscious. Sleep has intrigued humans throughout history: Metaphysically, dreaming suggested to our forebears that the soul took leave of the body at night to wander the world; practically, sleep is a period of enforced nonproductivity and vulnerability to predators and enemies.

In spite of thousands of research studies, we are still unclear on the most basic question—what the function of sleep is. The most obvious explanation is that sleep is *restorative.* Support for this idea comes from the observation that species with higher metabolic rates typically spend more time in sleep (Zeppelin & Rechtschaffen, 1974). A less obvious explanation is the *adaptive* hypothesis; according to this view, the amount of sleep an animal engages in depends on the availability of food and on safety considerations (Webb, 1974). Elephants, for instance, which must graze for many hours to meet their food needs, sleep briefly. Animals with low vulnerability to predators, such as the lion, sleep much of the time, as do animals that find safety by hiding, like bats and burrowing animals. Vulnerable animals that are too large to burrow or hide—for example, horses and cattle—sleep very little (see Figure 15.1). In a study of 39 species, the combined factors of body size and danger accounted for 80% of the variability in sleep time (Allison & Cicchetti, 1976).

Whatever the function of sleep may be, its importance becomes apparent when we look at the effects of sleep deprivation. These effects are nowhere more evident than in shift work. Shift workers sleep less than day workers, and as a result their work performance suffers (Tepas & Carvalhais, 1990). Also, they typically fail to adjust their sleep-wake cycle adequately, because their sleep is disturbed during the day and they conform to the rest of the world's schedule on weekends. With their work and sleep schedules at odds with their biological rhythm, shift workers find that sleep intrudes into their work and daytime arousal interferes with their sleep.

> *Early to bed and early to rise, makes a man healthy, wealthy, and wise.*
>
> —*Benjamin Franklin*

FIGURE 15.1 Time Spent in Daily Sleep for Different Animals.

Observations support the hypothesis that sleep is an adaptive response to feeding and safety needs.

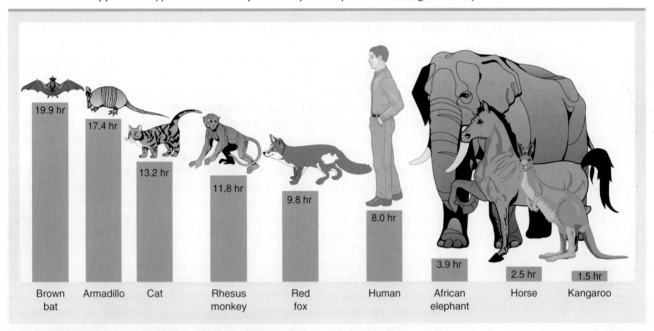

SOURCE: Based on data from "Animal Sleep: A Review of Sleep Duration Across Phylogeny, by S. S. Campbell and I. Tobler, 1984, *Neuroscience and Biobehavioral Reviews*, 8, 269–300.

In long-term sleep deprivation studies, impairment follows a rhythmic cycle—performance declines during the night, then shows some recovery during the daytime (Horne, 1988). The persistence of this rhythm represents a safety hazard of gigantic proportions when people try to function at night. The largest number of single-vehicle traffic accidents attributed to "falling asleep at the wheel" occur around 2 a.m. (Mitler et al., 1988), and the number of work errors peaks at the same time (Broughton, 1975). In addition, the Three Mile Island nuclear plant accident took place at 4 a.m.; the Chernobyl nuclear plant meltdown began at 1:23 a.m.; the Bhopal, India, chemical plant leakage, which poisoned more than 2,000 people, began shortly after midnight; and the *Exxon Valdez* ran aground at 12:04 a.m., spilling 11 million gallons of oil into fragile Alaskan waters (Alaska Oil Spill Commission, 1990; Mapes, 1990; Mitler et al., 1988).

Travel across time zones also disrupts sleep and impairs performance, particularly when you travel eastward. It is difficult to quantify the effects of *jet lag*, but three researchers have attempted to do so in a novel way by comparing the performance of baseball teams. When East Coast and West Coast teams played at home, their percentage of wins was nearly identical—50% and 49%, respectively. When they traveled across the continent but had time to adjust to the new time zone, they showed a typical visitor's disadvantage, winning 45.9% of their games. Teams traveling west without time to adjust won about the same, 43.8% of their games, while teams traveling east won only 37.1% (Recht, Lew, & Schwartz, 1995). The quality of sleep is better when you extend the day's length by traveling west, rather than shorten it as you do when you travel east. One way of looking at this effect is that it is easier to stay awake past your bedtime than it is to go to sleep when you are not sleepy. We will examine a more specific explanation when we consider circadian rhythms.

Circadian Rhythms

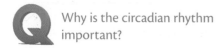

Q Why is the circadian rhythm important?

We saw in Chapter 14 that a *circadian rhythm* is a rhythm that is about a day in length; the term *circadian* comes from the Latin *circa*, meaning "approximately," and *dia*, meaning "day." We operate on a 24-hour (hr) cycle, in synchrony with the solar day. We sleep once every 24 hr, and body temperature, alertness, urine production, steroid secretion, and a variety of other activities decrease during our normal sleep period and increase during our normal waking period, even when we reverse our sleep-wake schedule temporarily.

The main biological clock that controls these rhythms in mammals is the *suprachiasmatic nucleus (SCN)* of the hypothalamus. Lesioning the SCN in rats abolishes the normal 24-hr rhythms of sleep, activity, body temperature, drinking, and steroid secretion (Abe, Kroning, Greer, & Critchlow, 1979; Stephan & Nunez, 1977). The SCN is what is known as a *pacemaker*, because it keeps time and regulates the activity of other cells. We know that the rhythm arises in the SCN, because rhythmic activity continues in isolated SCN cells (Earnest, Liang, Ratcliff, & Cassone, 1999; Inouye & Kawamura, 1979). Lesioned animals do not stop sleeping, but instead of following the usual day-night cycle, they sleep in naps scattered throughout the 24-hr period. So the SCN controls the timing of sleep, but sleep itself is controlled by other brain structures, which we will discuss later. The SCN is shown in Figure 15.2 (you can check Figure 6.2 to see its location.).

The SCN is *entrained* to the solar day by cues called *zeitgebers* ("time-givers"). If humans are kept in isolation from all time cues in an underground bunker or a cave, they usually lose their synchrony with the day-night cycle; in many studies, zeitgeber-deprived individuals "drifted" to a day that was about 25 hr long, with a progressively increasing delay in sleep onset (see Figure 15.3; Aschoff, 1984). For a long time, researchers believed that alarm clocks and the activity of others were the

FIGURE 15.2 The Suprachiasmatic Nucleus.

(a) The nuclei, indicated by the arrows, took up more radioactive 2-deoxy-glucose in the scan because the rat was injected during the "light-on" period of the day; (b) the rat was injected during the "light-off" period.

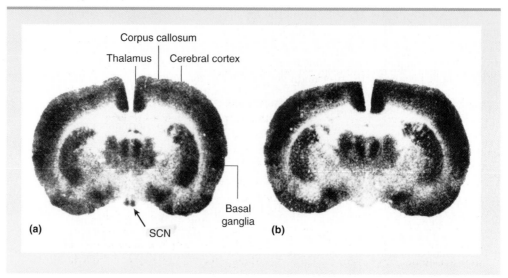

SOURCE: Reprinted with permission from W. J. Schwartz and H. Gainer, "Suprachiasmatic Nucleus: Use of ^{14}C-labeled Deoxyglucose Uptake As a Functional Marker," *Science, 197*, 1089–1091. Copyright 1977 American Association for the Advancement of Science.

most important influences that entrain our activity to the 24-hr day; but research points more convincingly to light as the primary zeitgeber.

The *difference* in light intensity between the light and dark periods is important for entraining the day-night cycle. One group of night workers worked under bright lights and slept in complete darkness during the day (*light discrepant*); a second group worked under normal light and slept in the semidarkness that is typical of the day sleeper (*similar light*). The light-discrepant workers scored higher in performance and alertness than the similar-light workers. Their physiological measures also synchronized with the new sleep-wake cycle; for example, their body temperature dropped to its low value around 3:00 p.m., when they were asleep, but the similar-light group's low continued to occur at 3:30 a.m. in spite of being awake and working (Czeisler et al., 1990).

Just how much we rely on a regular lighting schedule for entrainment was underscored in a study of four Greenpeace volunteers living in isolation during the 4-month darkness of the Antarctic winter; their sleep times and physiological measures free-ran on a roughly 25-hr interval, even though they had access to time information and social contact with each other (Kennaway & van Dorp, 1991). According to some, it is this "slow-running" clock that makes phase delays (going to sleep later) easier than phase advances (going to sleep earlier). So adjustment after westward travel is easier than after traveling east, and workers who rotate shifts sleep better and produce more if the rotation is to later shifts rather than to earlier ones (Czeisler, Moore-Ede, & Coleman, 1982). Some people seem to be relatively insensitive to the environmental cues that entrain most of us to a 24-hr day and operate on a 25-hr clock under normal circumstances; and like a clock that runs too slowly, their physical and cognitive functioning moves in and out of phase with the rest of the world, resulting in insomnia and impaired functioning.

Why the internal clock would operate on a 25-hr cycle is unclear, especially since animals kept in isolation typically run on a 24-hr cycle (Czeisler et al., 1999). Some believe that it has something to do with the 24.8-hr lunar cycle, which influences the tides and some biological systems (Bünning, 1973; J. E. M. Miles, Raynal, & Wilson, 1977). Czeisler and his colleagues (1999) suggested that the 25-hr cycle in isolation studies is no more than an artifact of allowing the individuals to control the room lighting. Bright light late in the day causes the cycle to lengthen, so Czeisler kept the light at a level that was too low to influence the circadian rhythm while people lived on a 28-hr sleep-wake schedule. Under that condition, their body temperature cycle averaged 24.18 hr, which led Czeisler to conclude that the *biological* rhythm is approximately 24 hr long.

The SCN regulates the pineal gland's secretion of *melatonin,* a hormone that induces sleepiness. Melatonin is often used to combat jet lag and to treat insomnia in shift workers and in the blind (Arendt, Skene, Middleton, Lockley, & Deacon, 1997). Light resets the biological clock by suppressing melatonin secretion (Boivin, Duffy, Kronauer, & Czeisler, 1996). Most totally blind individuals are not entrained to the 24-hr day and suffer from insomnia in spite of regular schedules of sleep, work, and social contact. These individuals do not experience a decrease in melatonin production when exposed to light; however, totally blind people *without* insomnia do show melatonin suppression by light, even though they are unaware of the light (Czeisler et al., 1995).

Recent studies explain how some blind individuals are able to entrain to the light-dark cycle and, thus, how the rest of us do so as well. Light information reaches the SCN by way of a direct connection from the retinas called the *retinohypothalamic pathway*; however, mice lacking rods and cones still show normal entrainment

FIGURE 15.3 Sleep and Wake Periods During Isolation From Time Cues.

Each dark bar indicates the timing and length of sleep during a day. During the unscheduled period (without time cues), the subject's activity assumed a 25-hr rhythm and began to advance around the clock. When light-dark periods were scheduled, he resumed a normal sleep and activity rhythm.

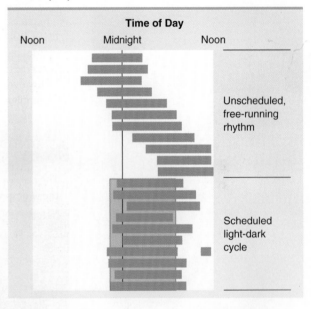

SOURCE: From *Introduction to Psychology, Gateways to Mind and Behavior* (with InfoTrac) 9th edition, by Coon, 2001. Reprinted with permission of Wadsworth, a division of Thomson Learning.

and cycling, so the signal must arise from some other retinal light receptors (M. S. Freedman et al., 1999; Lucas, Freedman, Muñoz, Garcia-Fernández, & Foster, 1999). Although ganglion cells ordinarily receive information about light from the receptors, about 1% of ganglion cells respond to light directly and send neurons into the retinohypothalamic pathway (Berson, Dunn, & Takao, 2002; Hannibal, Hindersson, Knudsen, Georg, & Fahrenkrug, 2002; Hattar, Liao, Takao, Berson, & Yau, 2002). These ganglion cells contain melanopsin, which has recently been confirmed as a light-sensitive substance, or photopigment (Dacey et al., 2005; Panda et al., 2005; Qiu et al., 2005); the melanopsin is located in their widely branching dendrites, which suits the cells for detecting the overall level of light, as opposed to contributing to image formation (Figure 15.4). Melanopsin is most sensitive to light of 480 nanometers, which is the wavelength of light that predominates at dusk and dawn (Foster, 2005), so the system apparently is most responsive to twilight in resetting the circadian clock. A recent study confirmed that human retinas have melanopsin in some of their ganglion cells (Dkhissi-Benyahya, Rieux, Hut, & Cooper, 2006).

However, synchronizing the rhythm does not account for the rhythm itself. The internal clock consists of a few genes and their protein products (Clayton,

FIGURE 15.4 Retinal Ganglion Cells Containing Melanopsin.

The cells were labeled with a fluorescent substance that reacts to melanopsin. Notice the widespread dendrites, which contain melanopsin.

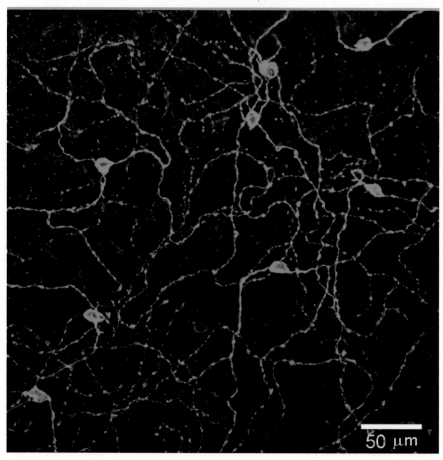

50 μm

SOURCE: From "Melanopsin-containing Retinal Ganglion Cells: Architecture, Projections, and Intrinsic Photosensitivity," by Hattar, Liao, Takao, Berson, and Yau, *Science, 295*, 1065–1070. © 2002 American Association for the Advancement of Science (AAAS). Reprinted with permission from AAAS.

Kyriacou, & Reppert, 2001; Hastings, Reddy, & Maywood, 2003; Shearman et al., 2000); the genes fall into two groups, one group that is turned on while the other is turned off. When the genes are on, their particular protein products build up. Eventually, the accumulating proteins turn their genes off, and the other set of genes is turned on. This feedback loop provides the approximately 24- or 25-hr cycle, which then must be reset each day by light. This process is not limited to neurons in the SCN; there are additional clocks, located outside the brain and controlling the activities of the body's organs (Hastings et al., 2003). These clocks operate independently of the SCN, but the SCN entrains them to the day-night cycle. Feeding is an example of an activity that is controlled independently. According to the researchers, local clocks that affect blood pressure and heart activity explain why there is a large increase in the risk of heart attack, stroke, and sudden cardiac death after waking in the morning. The clock in the SCN does not always operate properly, as we saw in the previous chapter with some depressed patients.

Rhythms During Waking and Sleeping

Riding on the day-long wave of the circadian rhythm are several *ultradian rhythms,* rhythms that are shorter than a day in length. Hormone production, urinary output, alertness, and other functions follow regular cycles throughout the day. For example, the dip in alertness and performance in the wee hours of the morning is mirrored by another in the early afternoon, which cannot be accounted for by postlunch sleepiness, because it also occurs in people who skip lunch (Broughton, 1975). Incidentally, this dip coincides with the time of siesta in many cultures and a rest period in nonhuman primates. The *basic rest and activity cycle* is a rhythm that is about 90 to 100 minutes (min) long. When people wrote down what they were thinking every 5 min for 10 hr, the contents showed that they were daydreaming on a 90-min cycle; EEG recordings verified that these were periods of decreased brain activity (Kripke & Sonnenschein, 1973).

The common view of sleep is that it is a cessation of activity that occurs when the body and brain become fatigued. Sleep, however, is an active process. This is true in two respects. First, you will soon see that sleep is a very busy time; a great deal of activity goes on in the brain. Second, sleep is not like a car running out of gas but is turned on by brain structures and later turned off by other structures.

The most important measure of sleep activity is the EEG. When a person is awake, the EEG is a mix of *alpha* and *beta* waves. Alpha is activity whose voltage fluctuates at a frequency of 8 to 12 hertz (Hz) and moderate amplitude; beta has a frequency of 13 to 30 Hz and a lower amplitude. Beta waves, which are associated with arousal and alertness, are progressively replaced by alpha waves as the person relaxes (see Figure 15.5). It may seem strange that the amplitude of the EEG is lower during arousal. Remember that the EEG is the sum of the electrical activity of all the neurons between the two recording electrodes. When a person is cognitively aroused, neurons under the electrodes are mostly desynchronized in their firing as they carry out their separate tasks; with the neurons firing at different times, the EEG has a high frequency, but the amplitude is rather low. As the person relaxes, the neurons have less processing to do and fall into a pattern of synchronized firing. The rate is low, but the cumulative amplitude of the neurons firing at the same time is high.

As the person slips into the light first stage of sleep, the EEG shifts to *theta* waves, with a frequency of 4 to 7 Hz (see Figure 15.5). About 10 min later, Stage 2 begins, indicated by the appearance of *K complexes* and *sleep spindles*. K complexes are sharp, large waves that occur about once a minute; sleep spindles are brief bursts of 12- to 14-Hz waves that appear to serve a gating function, preventing disruptive stimuli from reaching the cortex and waking the sleeper (Dang-Vu, McKinney, Buxton, Solet, & Ellenbogen, 2010). Stages 3 and 4 are known as *slow-wave sleep (SWS)* and are

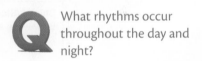

Q What rhythms occur throughout the day and night?

2
Sleep Research

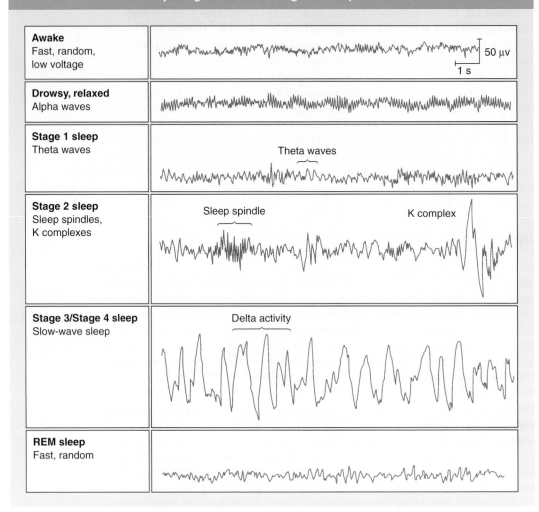

FIGURE 15.5 Electroencephalogram and the Stages of Sleep.

SOURCE: From *Current Concepts: The Sleep Disorders*, by P. Hauri, 1982, Kalamazoo, MI: Upjohn.

characterized by large, slow delta waves at a frequency of 1 to 3 Hz. The person moves around in bed during this period, turning over and changing positions. Sleepwalking, bedwetting, and night terrors, disturbances that are common in children, occur during slow-wave sleep, too. Night terrors are not nightmares but involve screaming and apparent terror, which are usually forgotten in the morning; they are not a sign of a disorder unless they continue beyond childhood. After Stage 4, the sleeper moves rather quickly back through the stages in reverse order. But rather than returning to Stage 1, the sleeper enters rapid eye movement sleep.

Rapid eye movement (REM) sleep is so called because the eyes dart back and forth horizontally during this stage. The EEG returns to a pattern similar to a relaxed waking state, but the person does not wake up; in fact, the sleeper is not easily aroused by noise but does respond to meaningful sounds, such as the sleeper's name. It is easy to see why some researchers call this stage *paradoxical sleep,* because *paradoxical* means "contradictory." During REM sleep, respiration rate and heart rate increase. Males experience genital erection, and vaginal secretion increases in females. In spite of these signs of arousal, the body is very still—in fact, in a state of muscular paralysis or *atonia*.

FIGURE 15.6 Time Spent in Various Sleep Stages During the Night.

As the night progresses, deep sleep decreases, and time in REM sleep (dark bars) increases.

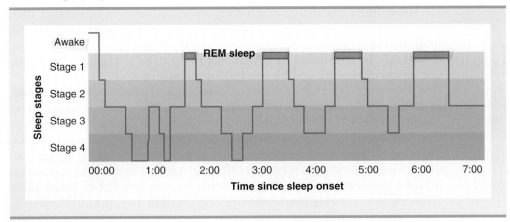

If people sleeping in the laboratory are awakened by the researcher during REM sleep, about 80% of the time they report dreaming. Dreams also occur during the other, *non-REM sleep* stages, but they are less frequent, less vivid, and less hallucinatory. Even people who say that they do not dream report dreaming when they are awakened from REM sleep; their d non-REM sleep reams are less frequent, though, and they often describe their experience as "thinking" (H. B. Lewis, Goodenough, Shapiro, & Sleser, 1966). Apparently, "nondreamers" just fail to remember their dreams in the morning; in fact, we ordinarily remember a dream only if we wake up before the short-term memory of the dream has faded (Koulack & Goodenough, 1976). A complete cycle through the stages of sleep—like the daydreaming cycle—takes about 90 min to complete. The night's sleep is a series of repetitions of this ultradian rhythm, although the length of REM sleep periods increases and the amount of slow-wave sleep decreases through the night (Figure 15.6).

The Functions of REM and Non-REM Sleep

To find out what functions REM sleep serves, researchers deprived volunteers of REM sleep; they did this by waking the research participants every time EEG and eye movement recordings indicated that they were entering a REM period. When this was done, the subjects showed a "push" for more REM sleep. They went into REM more frequently as the study progressed and had to be awakened more often; then, on uninterrupted recovery nights, they tended to make up the lost REM by increasing their REM from about 20% of total sleep time to 25% or 30% (Dement, 1960). To psychoanalytically oriented theorists, these results were evidence of a psychological need for dreaming. You are probably familiar with the theory that dreams reveal the contents of the unconscious, not through their manifest content—the story the person tells on awakening—but through symbolic representations (Freud, 1900).

What are the functions of REM and slow-wave sleep?

Most neuroscientists, on the other hand, believe that dreaming is merely the by-product of spontaneous neural activity in the brain. According to the *activation-synthesis hypothesis,* during REM sleep the forebrain integrates neural activity generated by the brain stem with information stored in memory (Hobson & McCarley, 1977); in other words, the brain engages in a sort of confabulation, using information from memory to impose meaning on nonsensical random input. This explanation does not imply that dream content is always insignificant;

there is evidence that daytime events and concerns do influence the content of a person's dreams (Webb & Cartwright, 1978). But neuroscientists consider dreams to be the least important aspect of REM sleep, and they note that after a century of intense effort, there is no agreed-on method of dream interpretation (Crick & Mitchison, 1995). Instead, neuroscientists argue that the pervasiveness of REM sleep among mammals and birds demands that any explanation for the function of REM sleep be a biological one. There are several proposals as to what biological needs might be met during REM sleep, but the number of hypotheses indicates that we are still unsure what benefit REM sleep confers.

One hypothesis is that REM sleep promotes neural development during childhood. Infant sleep starts with REM rather than non-REM, and the proportion of sleep devoted to REM is around 50% during infancy and decreases through childhood until it reaches an adult level during adolescence (Roffwarg, Muzio, & Dement, 1966). According to this hypothesis, excitation that spreads through the brain from the pons during REM sleep encourages differentiation, maturation, and myelination in higher brain centers, similar to the way spontaneous waves of excitation sweep across the retina during development to help organize its structure (Chapter 3). There is some evidence from studies of the immature visual systems of newborn cats that REM sleep, and particularly these waves from the pons, regulates the rate of neural development (Shaffery, Roffwarg, Speciale, & Marks, 1999). The fact that sleep is associated with the upregulation of a number of genes involved in neural plasticity (see Chapter 12), as well as other genes that contribute to the synthesis and maintenance of myelin and cell membranes (Cirelli et al., 2004), is certainly consistent with this neurodevelopmental hypothesis.

Early ideas about non-REM functions focused on rest and restoration, inspired by studies showing that slow-wave sleep increases following exercise; after athletes competed in a 92-kilometer race, slow-wave sleep was elevated for four consecutive nights (Shapiro, Bortz, Mitchell, Bartel, & Jooste, 1981). However, this effect appears to be due to overheating rather than fatigue. The night after people ran on treadmills, slow-wave sleep increased, at the expense of REM sleep; but if they were sprayed with water while they ran, their body temperature increased less than half as much, and there was no change in slow-wave sleep (Horne & Moore, 1985).

Horne (1988) believes that slow-wave sleep is more related to the increase in the temperature of the brain than the increase in body temperature; heating only the head and face with a hair dryer was sufficient to increase slow-wave sleep later (Horne & Harley, 1989). According to Horne (1992), slow-wave sleep promotes cerebral recovery, especially in the prefrontal cortex. Slow-wave sleep may also restore processes involved in cognitive functioning. Bonnet and Arand (1996) gave people either caffeine or a placebo before a 3.5-hr nap. The caffeine group had reduced slow-wave sleep during the nap; although they felt more vigorous and no sleepier than the placebo group, they performed less well on arithmetic and vigilance tasks during a subsequent 41-hr work period.

Sleep and Memory

In Chapter 12, you learned that a period of sleep following learning enhances later performance (see Figure 15.7). REM sleep has received the most attention; the amount of REM increases during the sleep period following learning, and REM sleep deprivation after learning reduces retention (see review in Dujardin, Guerrien, & Leconte, 1990; Karni, Tanne, Rubenstein, Askenasy, & Sagi, 1994; C. Smith, 1995). How much REM sleep increases depends on how well the subject learned (Hennevin, Hars, Maho, & Bloch, 1995). Also, if training occurs over several days, REM sleep increases daily and reaches its peak in the 24-hr period before the peak in correct performance (Dujardin et al., 1990; C. Smith, 1996).

3
NOVA Video: Sleep & Memory

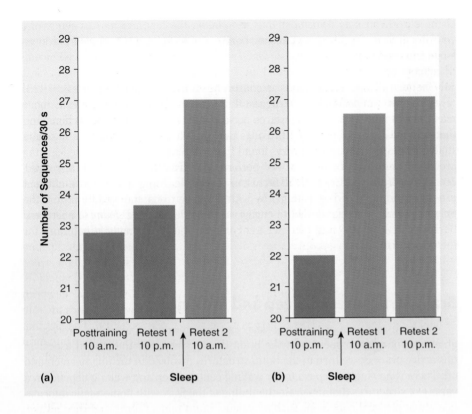

(a)   Sleep

(b)   Sleep

FIGURE 15.7 Improvement in Learning Following Sleep.

Participants learned a motor skill task and were retested twice at 10-hr intervals. There was no statistically significant improvement for individuals who remained awake during the interval (a, Retest 1), but performance improved following sleep (a, Retest 2 and b, Retest 1 and Retest 2).

FIGURE 15.8 Correlation of Slow-Wave and REM Sleep With Overnight Task Improvement.

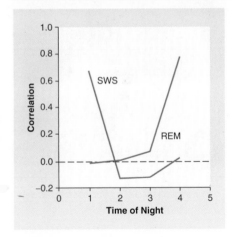

These graphs show the correlation of slow-wave sleep (SWS) and REM sleep with improvement on a visual discrimination task at the beginning of the next day's practice. They indicate that slow-wave sleep has more effect during the first quarter of the night, while REM is important during the fourth quarter.

SOURCE: Adapted with permission from Stickgold et al., "Sleep, Learning, and Dreams: Off-line Memory Reprocessing," *Science*, 294, 1052–1057. © 2001 American Association for the Advancement of Science. Reprinted with permission from AAAS.

There is increasing evidence from both animal and human studies that non-REM sleep is also important for learning (Hairston & Knight, 2004). For example, increasing slow potentials in human volunteers' brains by applying a 0.75-Hz oscillating current over the frontal and temporal area during the first period of non-REM sleep improved recall of word associations learned prior to sleep (L. Marshall, Helgadóttir, Mölle, & Born, 2006). Another study indicated that consolidation is a multistep process requiring a combination of REM and slow-wave sleep. Overnight improvement on a visual discrimination task in humans was correlated with the percentage of slow-wave sleep during the first quarter of the night *and* the percentage of REM sleep in the last quarter of the night (see Figure 15.8; Stickgold, Whidbee, Schirmer, Patel, & Hobson, 2000). These measures together accounted for 80% of the differences in learning among the research participants. Even a 60- to 90-min nap that includes both REM and slow-wave sleep produces significant improvement in performance (Mednick, Nakayama, & Stickgold, 2003). According to Ribeiro and his colleagues (2004), neuronal replay (see Chapter 12) is strongest during non-REM sleep and represents recall and amplification of the hippocampal activity that occurred during learning. Then, during REM sleep, the hippocampus upregulates genes in the cortex that are involved in synaptic plasticity, implementing the transfer of memory from the hippocampus to cortex.

Stimulating rats' reticular formation during REM sleep or just repeating the stimulus that had signaled shock in an avoidance task improved their performance the next day compared with controls; presenting either form of stimulation during slow-wave sleep did not (Hennevin et al., 1995). Close observation of hippocampal activity after learning tells us why REM sleep is important. Replay during REM sleep is synchronized with theta-frequency (3–7 Hz) activity occurring spontaneously in the hippocampus (Stickgold et al., 2001); in other words, the peaks of one wave coincide with the peaks of the other. After 4 to 7 days, the time during which memories become independent of the hippocampus, the replay shifts out of phase with the theta activity, with the peaks of one wave coinciding with the troughs of the other. You may remember that stimulating the hippocampus in synchrony with theta produces long-term potentiation and out-of-phase stimulation produces long-term depression. This suggests that a period of consolidation is followed by one of memory erasure.

The idea of memory deletion is consistent with Crick and Mitchison's (1995) *reverse-learning hypothesis*. They suggest that neural networks involved in memory must have a way to purge themselves occasionally of erroneous connections and that activity during REM sleep provides the opportunity to do this. Researchers studying computer neural networks found that when they added a reverse-learning process, it improved their networks' performance (Hopfield, Feinstein, & Palmer, 1983). According to Crick and Mitchison, reverse learning makes more efficient use of our brain, allowing us to get by with fewer neurons; they point out that the only mammals so far found not to engage in REM sleep—the *Echidna* (a nocturnal burrower in Australia) and two species of dolphin—have unusually large brains for their body size.

Brain Structures of Sleep and Waking

We have seen one of the ways sleep can be regarded as an active process: A great deal of activity goes on in the brain during sleep. For the second aspect of this active process, we turn to the brain structures involved in turning sleep on and off. There is no single sleep center or waking center; sleep and waking depend on a variety of structures that integrate the timing of the SCN with homeostatic information about physical conditions such as fatigue, brain temperature, and time awake. The network of structures governing sleep and waking is complex, so you will want to trace its connections carefully in the accompanying illustrations. We will begin with the structures that produce sleep.

Sleep Controls

Sleep is homeostatic, in that a period of deprivation is followed by a lengthened sleep period. *Adenosine* provides at least one of the mechanisms of sleep homeostasis. During wakefulness, adenosine accumulates in the *basal forebrain area;* it inhibits arousal-producing neurons there, inducing drowsiness and reducing EEG activation (Figure 15.9) (Porkka-Heiskanen et al., 1997). The accumulated adenosine dissipates during the next sleep period. We saw in Chapter 5 that caffeine counteracts drowsiness by acting as an antagonist at adenosine receptors.

Another location where adenosine increases sleep is the preoptic area of the hypothalamus (Ticho & Radulovacki, 1991). Cells in the preoptic area are involved in several functions, including regulation of body temperature. Warming this part of the hypothalamus activates sleep-related cells, inhibits waking-related cells in the basal forebrain, and enhances slow-wave EEG (Alam, Szymusiak, & McGinty, 1995; Sherin, Shiromani, McCarley, & Saper, 1996; Szymusiak, 1995). This finding has contributed to the hypothesis that one function of slow-wave sleep is to cool the brain after waking activity. Whether that is true or not, the preoptic area no doubt accounts for the sleepiness you feel in an overheated room or when you have a fever.

Neurons in a part of the preoptic area, the *ventrolateral preoptic nucleus*, double their rate of firing during sleep (J. Lu et al., 2002). They induce sleep by inhibiting neurons in arousal areas: the *tuberomammillary nucleus* in the hypothalamus and the *locus coeruleus, raphé nuclei,* and *peduculopontine and laterodorsal tegmental nuclei (PPT/LDT)* in the pons (Chou et al., 2002; Saper, Chou, & Scammell, 2001; Saper, Scammell, & Lu, 2005). Different parts of the ventrolateral preoptic nucleus induce REM and non-REM sleep (Saper et al., 2005). The nucleus also receives

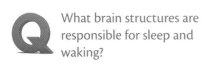

What brain structures are responsible for sleep and waking?

inhibitory inputs from the structures it innervates, which helps ensure that sleep and waking mechanisms are not turned on at the same time.

The pons also sends impulses downward to the *magnocellular nucleus* in the medulla to bring about the atonia that accompanies REM sleep. When Shouse and Siegel (1992) lesioned this nucleus in cats, the cats were no longer paralyzed during REM sleep; they seemed to be acting out their dreams (assuming that cats dream), and their movements during REM sleep often woke them up. The pons also contains adenosine receptors and is another site for the effect of your morning cup of coffee (Rainnie, Grunze, McCarley, & Greene, 1994).

Waking and Arousal

The arousal system consists of two major pathways (Figure 15.10) (Saper et al., 2005). The first arises from the PPT/LDT, whose acetylcholine-producing neurons are very active during waking. This pathway activates areas crucial for transmission of information to the cortex, including relay nuclei of the thalamus. It also shifts the EEG to asynchronized, high-frequency, low-amplitude activity by inhibiting nuclei in the thalamus that ordinarily synchronize the EEG (Hobson & Pace-Schott, 2002; Saper et al., 2001).

FIGURE 15.9 Brain Mechanisms of Sleep.

Sleep is brought about primarily by suppressing activity in arousal structures (shown in green).

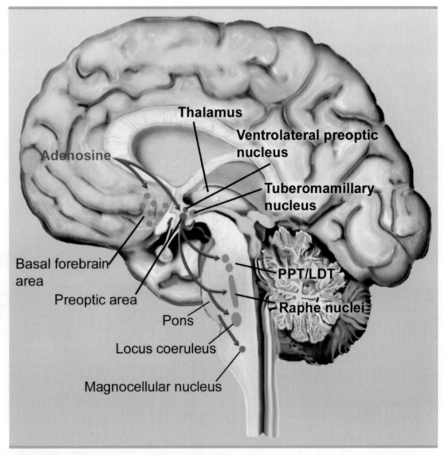

This pathway is also active when the individual shifts into each REM period. After our discussion of all the activity that goes on during sleep, it shouldn't surprise you to find arousal mechanisms active while the brain is sleeping.

The second pathway activates the cortex to facilitate the processing of inputs from the thalamus. Neurons from the locus coeruleus (which release norepinephrine) and the raphé nuclei (which release serotonin) are most active during waking, relatively quiet during non-REM sleep, and almost silent during REM sleep (Figure 15.11) (Khateb, Fort, Pegna, Jones, & Mühlethaler, 1995; Saper et al., 2005). Neurons from the tuberomammillary nucleus arouse the cortex by releasing histamine, and those from the basal forebrain area do so by releasing acetylcholine; neurons of the tuberomammillary nucleus and many in the basal forebrain area are active during both waking and REM sleep (Saper et al., 2005).

The arousing pathway is completed by neurons in the lateral hypothalamus, which send axons to the basal forebrain area, tuberomammillary nucleus, PPT/LDT, raphé nuclei, and locus coeruleus; during waking they release the peptide *orexin* to keep these arousal areas active (Figure 15.12) (Saper et al., 2001, 2005). Saper and his colleagues suggested that the presence of orexin (also known as *hypocretin*) stabilizes the sleep and waking system by preventing inappropriate switching into sleep. We will see in the section on sleep disorders what happens when this mechanism fails. As you can imagine, the orexin system has become a target for pharmaceuticals. A compound that blocks orexin receptors induces sleep without the

FIGURE 15.10 Arousal Structures of Sleep and Waking.

Several interacting structures and pathways produce waking, maintain arousal during waking, and increase arousal during REM sleep.

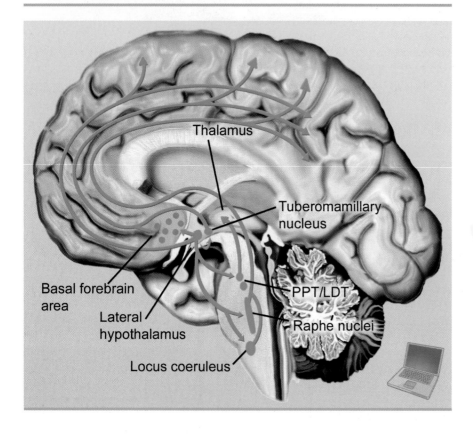

side effects of most sleep aids (Brisbare-Roch et al., 2007); Almorexant is undergoing final clinical trials and could reach market in 2012. Also, two commonly used inhalable anesthetics, isoflurane and sevoflurane, inhibit orexin receptors (Kelz et al., 2008). Eventually, a sniff of orexin delivered in a nasal spray could help students get through finals after pulling all-nighters; at least, it improved performance in sleep-deprived monkeys (Deadwyler, Porrino, Siegel, & Hampson, 2007). (I hope you're imagining the scene when the researchers administered nasal spray to the monkeys!)

The pons is the source of *PGO waves* seen during REM sleep. The name refers to the path of travel that waves of excitation take from the pons through the lateral geniculate nucleus of the thalamus to the occipital area. PGO waves are as characteristic of REM sleep as rapid eye movements are (Figure 15.13). They begin about 80 seconds before the start of a REM period and apparently are what initiate the EEG desynchrony of REM sleep (Mansari, Sakai, & Jouvet, 1989; Steriade, Paré, Bouhassira, Deschênes, & Oakson, 1989). Their arousal of the occipital area may account for the visual imagery of dreaming.

FIGURE 15.11 Firing Rates in Brain Stem Arousal Centers During Waking and Sleep.

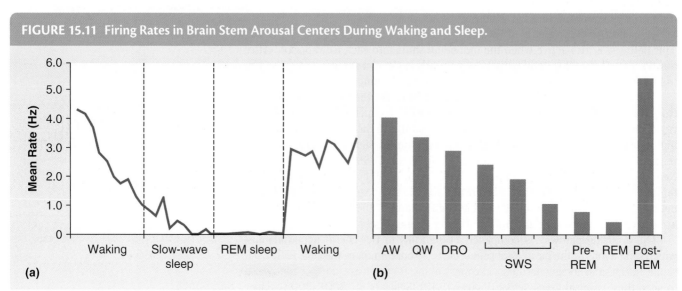

(a) Activity in the locus coeruleus; (b) activity in the raphé nuclei. AW, alert waking; QW, quiet waking; DRO, drowsy; SWS, slow-wave sleep; pre-REM, 60 seconds before REM; post-REM, first second after REM ends.

SOURCES: (a) Copyright 1981 by the Society for Neuroscience; (b) From "Activity of Serotonin-Containing Nucleus Centralis Superior (Raphe Medianus) Neurons in Freely Moving Cats," by M. E. Trulson et al., *Experimental Brain Research, 54*, 33–44, fig. 2. Copyright 1984 Springer-Verlag. Reprinted with permission.

Sleep Disorders

Insomnia

Insomnia is the inability to sleep or to obtain adequate-quality sleep, to the extent that the person feels inadequately rested. Insomnia is important not only as a nuisance but also because sleep duration has important implications for health. In a study of 1.1 million men and women, sleeping less than 6 hr a night was associated with decreased life expectancy (Kripke, Garfinkel, Wingard, Klauber, & Marler, 2002). However, the surprise in the study was that sleeping more than 8.5 hr was associated with as great an increase in risk of death as sleeping less than 4.5 hr. Lack of sleep may also be a factor in the obesity epidemic. In a long-term study of sleep behavior, people who slept less than 8 hr had a higher body mass index, along with lower leptin and higher ghrelin levels (Taheri, Lin, Austin, Young, & Mignot, 2004).

While the failure to get enough sleep is part of the lifestyle of industrialized countries, many people who try to get an adequate amount of sleep complain that they have difficulty either falling asleep or staying asleep. In a survey by the National Sleep Foundation (2002), over half the respondents reported that they had trouble sleeping or woke up unrefreshed at least a few nights a week, and a third had experienced at least one symptom of insomnia every night or almost every night in the past year. Insomnia is one of the few disorders that is essentially self-diagnosed, and several studies suggest that the reported frequencies might be misleading. But although insomniacs may overestimate the time required to get to sleep and the amount of time awake through the night (Rosa & Bonnet, 2000), there are several indications that their sleep quality suffers from hyperarousal. These include excess high-frequency EEG during non-REM sleep (Perlis, Smith, Andrews, Orff, & Giles, 2001) and disturbance of the hypothalamic-pituitary-adrenal axis (see Chapter 8), with increased secretion of cortisol and adrenocorticotropic hormone during the night (Vgontzas et al., 2001).

Insomnia can be brought on by a number of factors, such as stress, but it also occurs frequently in people with psychological problems, especially affective disorders (Benca, Obermeyer, Thisted, & Gillin, 1992). Some loss of gray matter in the orbitofrontal cortex and the parietal cortex has been

4
Sleep Disorders

FIGURE 15.12 Locations of Orexin Receptors in the Rat Brain.

The receptors appear in white. Notice how widespread they are.

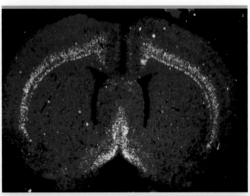

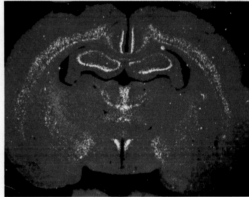

SOURCE: From "Mice Lacking the M3 Muscarinic Acetylcholine Receptor Are Hypophagic and Lean," by Yamada et al., *Nature, 410,* 207–212, © 2001. Used with permission.

FIGURE 15.13 PGO Waves, EEG Desynchrony, and Muscle Atonia.

The records are of electrical activity in the lateral geniculate nucleus (LG), eye movements (EOG, electrooculogram), electroencephalogram (EEG), and muscle tension (EMG, electromyogram). Notice that PGO waves signal the beginning of EEG desynchrony, rapid eye movements, and atonia several seconds later.

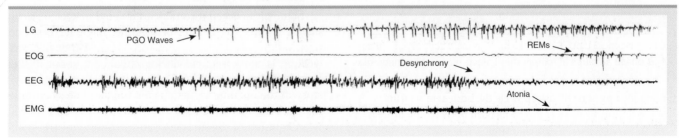

Q What are the causes of sleep disorders?

reported in insomniacs (Altena, Vrenken, Van Der Werf, van den Heuvel, & Van Someren, 2010); this could be a cause of their insomnia, or it could reflect the association with psychological disorders. Another frequent cause is the *treatment* of insomnia; most sleep medications are addictive, so attempts to do without medication or to reduce the dosage produce a rebound insomnia; this can happen after as little as three nights with some benzodiazepines (Kales, Scharf, Kales, & Soldatos, 1979). Insomnia can manifest itself as delayed sleep onset, nighttime waking, or early waking; a disruption of the circadian rhythm is often the culprit (M. Morris, Lack, & Dawson, 1990).

People with sleep difficulties often show a shift in their circadian rhythm; this can be the result of bad sleep habits, but it is more likely the cause. Normally, people fall asleep when their body temperature is decreasing in the evening and wake up when it is rising. But if your body temperature is still high at bedtime (phase delay), you will experience sleep-onset insomnia; if your body temperature rises too early (phase advance) you will wake up long before the alarm clock goes off (see Figure 15.14). Sleep is also more efficient if you go to bed when your body temperature is low; for example, one volunteer living in isolation from time cues averaged a 7.8-hr sleep length when he went to bed during his temperature minimum and a 14.4-hr sleep length when sleep began near his temperature peak (Czeisler, Weitzman, & Moore-Ede, 1980). Three mutations in circadian clock genes have been linked to advanced sleep phase syndrome (early bedtime and early waking) and delayed sleep syndrome (late bedtime and rising; Archer et al., 2003; Toh et al., 2001; Xu et al., 2005). Your *chronotype*—when your internal clock is synchronized to the 24-hr day—depends partly on your genes and partly on your environment (Roenneberg et al., 2004).

The greater ease of phase delay than phase advance led to a treatment for delayed sleep syndrome that was completely counterintuitive. The patients had a 5- to 15-year history of sleep onset insomnia so severe that they were not even going to bed until 4:15 a.m., on average. Rather than require them to retire earlier, the researchers had the patients *stay up 3 hr later* each day than the day before. After about a week of this routine—for example, going to bed at 8 a.m., 11 a.m., 2 p.m., 5 p.m., 8 p.m., and 11 p.m. on successive nights—their average sleep-onset time had shifted from 4:50 a.m. to 12:20 a.m., and their average waking time had shifted from 1:00 p.m. to 7:55 a.m. All five patients were able to give up the sleeping pills they had become dependent on, and improvement was long lasting (Czeisler et al., 1981). Phototherapy is also sometimes used to reset the circadian clock.

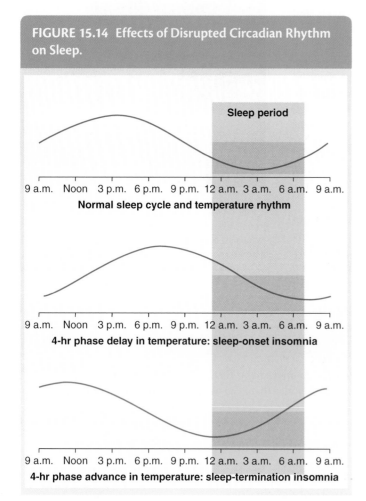

FIGURE 15.14 Effects of Disrupted Circadian Rhythm on Sleep.

9 a.m. Noon 3 p.m. 6 p.m. 9 p.m. 12 a.m. 3 a.m. 6 a.m. 9 a.m.
Normal sleep cycle and temperature rhythm

9 a.m. Noon 3 p.m. 6 p.m. 9 p.m. 12 a.m. 3 a.m. 6 a.m. 9 a.m.
4-hr phase delay in temperature: sleep-onset insomnia

9 a.m. Noon 3 p.m. 6 p.m. 9 p.m. 12 a.m. 3 a.m. 6 a.m. 9 a.m.
4-hr phase advance in temperature: sleep-termination insomnia

Sleep period

Ordinarily, a person falls asleep while the body temperature is decreasing and awakens as it is rising (a). If body temperature is phase delayed (b), the person has trouble falling asleep; if body temperature is phase advanced (c), the person wakes up early.

Sleepwalking

Some of the sleep disorders are related to specific sleep stages. As we saw before, bedwetting, night terrors, and sleepwalking occur during slow-wave sleep. Although sleepwalking is most frequent during childhood, about 3% to 8% of adults sleepwalk (A. Dalton, 2005). Kenneth Parks's story in the opening vignette is not unique. The sleepwalking defense was first used in 1846 when Albert Tirrell was acquitted of the murder of his prostitute mistress and the arson of her brothel, and the plea has been successful in a few more recent instances as well (A. Dalton, 2005).

In the Still of the Night

Shirley Koecheler raids the refrigerator at night. She would like to quit because she's gaining weight, but she isn't aware she's doing it until she wakes up in the morning to a crumb-filled bed and an uncomfortably full stomach. She even had her husband hide the Easter candy, but the next morning she found the wrappers from the chocolate bunnies in the wastebasket. Shirley's 24-year-old daughter Amy is also a sleep eater and has been since she was a toddler; the difference is that she doesn't gain weight. Anna Ryan, like Shirley, started sleep eating in adulthood; she didn't even know about the nighttime kitchen forays that added 60 pounds to her weight in a year and a half until she went to a sleep clinic to find out why she was exhausted every morning.

The victims of sleep related eating disorder are usually women; they ordinarily pass up the fruit and other healthy snacks for high-calorie junk food, but they have also been known to eat soap, Elmer's glue, frozen pizza, paper, and even egg shells. No one really knows what causes the problem, and treatment is hit-or-miss. Amy has responded well to a drug used to prevent seizures; she still sleep eats occasionally, but it's no longer a problem. It took Anna and her doctor months of trial and error to find a combination of drugs that works, but now she sleeps through the night and is losing weight.

SOURCES: Black, N., & Robertson, M. (2010). Midnight snack? Not quite." *abcNEWS/Health*, August 19. Retrieved from http://abcnews.com/Health/MedicalMysteries/story?id=5483978; Epstein, R. H. (2010). Raiding the refrigerator but still asleep. *New York Times*, April 7. Retrieved from www.nytimes.com/2010/04/07/health/07eating.html

5
Sleep Eating Video

Sleepwalking can be triggered by stress, alcohol, and sleep deprivation; Ken Parks's jury was convinced that he was not responsible because he was sleep deprived due to stress over gambling debts and the loss of his job for embezzling; there was a personal and family history of sleepwalking, sleep talking, and bedwetting; and he produced a high level of slow-wave sleep during sleep monitoring (Broughton et al., 1994).

Vulnerability for sleepwalking is at least sometimes genetic. Children of sleepwalkers are 10 times more likely to sleepwalk than children without sleepwalking relatives, and people with a version of a gene that is also implicated in narcolepsy are 3.5 times as likely to sleepwalk as others (Lecendreux et al., 2003). The gene is a member of the human leukocyte antigen (HLA) family, a group of genes that target foreign cells for attack by the immune system, and the authors suspect that cells important in sleep regulation have been attacked by the individual's immune system.

Somewhere in between, people who commit mayhem during sleepwalking and those who simply wander about the house are the individuals who suffer from *sleep-related eating disorder*, which is In the News.

Narcolepsy

Earlier, I said that we would see what happens when stabilization of the sleep switch fails; the result is *narcolepsy*, a disorder in which individuals fall asleep suddenly during the daytime and go directly into REM sleep. Another symptom of narcolepsy is *cataplexy*, in which the person has a sudden experience of one component of REM sleep, atonia, and falls to the floor paralyzed but fully awake. People with narcolepsy do not sleep more than others; rather, the boundaries are lost between sleep and waking (Nobili et al., 1996). Dogs also develop the disorder, and the study of canine narcolepsy has identified a mutated form of the gene that is responsible for the orexin receptor (see Figure 15.15) (Lin et al., 1999).

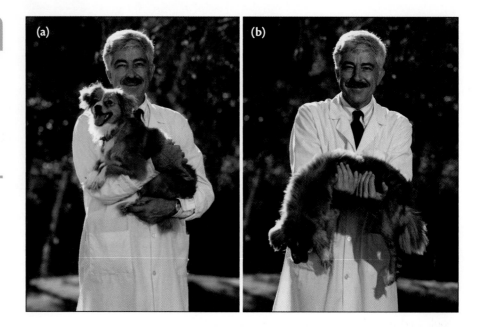

Other researchers were studying orexin's effect as a feeding stimulant in mice by disabling both copies of the gene responsible for producing orexin, but what they saw was more interesting than eating behavior (Chemelli et al., 1999). Occasionally, the mice would suddenly collapse, often while walking around or grooming; the mice were narcoleptic! Most narcoleptic humans (those with cataplexy) turned out to have the same deficiency as the mice; they had low or undetectable levels of orexin, due to a loss of orexin-secreting neurons in the hypothalamus (Higuchi et al., 2002; Kanbayashi et al., 2002). The neurons are destroyed by an autoimmune reaction, which has been traced to an allele of the *HLA* immune system gene (Hallmayer et al., 2009). Narcolepsy's concordance of 25% to 31% in identical twins (Mignot, 1998) leaves plenty of room for environmental influence, but identifying the nongenetic causes has been difficult.

REM Sleep Behavior Disorder

An apparent opposite of cataplexy is *REM sleep behavior disorder;* affected individuals are uncharacteristically physically active during REM sleep, often to the point of injuring themselves or their bed partners. A study of 93 patients, 87% of whom were male, found that 32% had injured themselves and 64% had assaulted their spouses (E. J. Olson, Boeve, & Silber, 2000). A 67-year-old man had tied himself to his bed with a rope at night for 6 years because he had a habit of leaping out of bed and landing on furniture or against the wall. One night, he was awakened by his wife's yelling because he was choking her; he was dreaming that he was wrestling a deer to the ground and was trying to break its neck (Schenck, Milner, Hurwitz, Bundlie, & Mahowald, 1989). REM sleep behavior disorder is often associated with a neurological disorder, such as Parkinson's disease or a brain stem tumor (E. J. Olson et al., 2000). Lewy bodies have been found in patients' brains, and two thirds of patients develop Parkinson's about 10 years later (Boeve et al., 2003). These findings have contributed to the hypothesis that Parkinson's disease is preceded by the development of Lewy bodies in the medulla, where inhibition of the magnocellular nucleus ordinarily produces atonia; the Lewy bodies then progress upward through the brain before reaching the substantia nigra years later, when the full-blown disease appears (Braak et al., 2003).

Sleep as a Form of Consciousness

At the beginning of this discussion, I said that sleep is neither entirely conscious nor unconscious. Francis Crick (1994), who shared a Nobel Prize for the discovery of DNA's structure in 1962 before turning to neuroscience and the study of consciousness, believes that we are in a state of diminished consciousness during REM sleep and that we are unconscious during non-REM sleep. Certainly there are some elements of consciousness in the dream state, particularly in people who are *lucid dreamers*. You have probably had the occasional experience of realizing during a bad dream that it is not actually real and will end soon. That kind of experience is common for lucid dreamers—they are often aware during a dream that they are dreaming. People can be trained to become aware of their dreaming and to signal to the researcher when they are dreaming by pressing a handheld switch (Salamy, 1970). They can even learn to *control* the content of their dreams; they may decide before sleeping what they will dream about, or they may interact with characters in their dream (Gackenbach & Bosveld, 1989). This ability tells us that the sleeping person is not necessarily as detached from reality as we have thought. This point is further illustrated by sleepwalkers, who have driven cars; wandered the streets; brandished weapons (Schenck et al., 1989); and strangled, stabbed, and beaten people to death, all presumably during non-REM sleep.

So it is not clear where or whether the transition from consciousness to nonconsciousness occurs during sleep. The idea of a dividing line is blurred even further by reports that surgical patients can sometimes remember the surgical staff's conversations while they were anesthetized, and they show some memory later for verbal material presented at the time of surgery (Andrade, 1995; Bonebakker et al., 1996). Whether you draw the line of consciousness between waking and sleeping or between REM and non-REM sleep or between sleep and coma depends more on your definition of consciousness than on any clear-cut distinctions between these conditions. Perhaps it is better to think of sleep as a different state of consciousness along a continuum of consciousness.

We can then concentrate on what the differences between waking and sleeping tell us about consciousness rather than worrying about classifications.

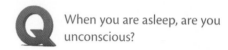 **Q** When you are asleep, are you unconscious?

> *The world shall perish not for lack of wonders, but for lack of wonder.*
>
> —*J. B. S. Haldane*

Concept Check   *Take a Minute to Check Your Knowledge and Understanding*

- ☐ Describe the circadian and ultradian rhythms discussed here.
- ☐ What, according to research, are the functions of REM and slow-wave sleep?
- ☐ Make a table showing the brain structures involved in sleep and waking, with their functions.
- ☐ Describe the sleep disorders and their causes.

The Neural Bases of Consciousness

While strict behaviorists had banned consciousness as neither observable nor necessary for explaining behavior, over the last half of the 20th century, researchers began to find that various components of consciousness *were* necessary as they studied memory, attention, mental imagery, and emotion. Still, they carefully avoided using the word *consciousness* as they talked about awareness, attention,

> *The explanation of consciousness is essential for explaining most of the features of our mental life because in one way or another they involve consciousness.*
>
> —*John R. Searle*

6
Consciousness Resources

or cognition. Then, a few respected theorists began musing about consciousness in print and even suggesting that it was an appropriate subject for neuroscientists to study. Other scientists slowly began to come out of their closets, while their more cautious colleagues warned them not to allow consciousness to become a back door for the reentry of the mind or for the proverbial homunculus, the "little man" inside the head who pulls all the levers. In the words of one team of writers, "consciousness is not some entity deep inside the brain that corresponds to the 'self,' some kernel of awareness that runs the show" (Nash, Park, & Willwerth, 1995).

So just what do we mean when by *consciousness?* Actually, the term has a variety of connotations. We use it to refer to a state—a person is conscious or unconscious, and we use it in the sense of conscious experience, or awareness of something. *Consciousness* has additional meanings for researchers, though few try to define the term; Francis Crick (1994) suggested that any attempt at definition at this point in our knowledge would be misleading and would unduly restrict thinking about the subject. While agreeing on a definition is impractical, I think most research-ers would be comfortable with the following assertions about consciousness. The person is aware, at least to some extent; as a part of awareness, the person holds some things in attention, while others recede into the background. Consciousness also involves memory, at least of the short-term variety, and fully conscious humans have a sense of self, which requires long-term memory. Consciousness varies in level, with coma and deep anesthesia on one extreme, alert wakefulness on the other, and sleep in between. There are also altered states of consciousness, includ-ing hypnosis, trances, and meditative states.

Consciousness is a phenomenon of the brain, but most researchers agree that there is no "consciousness center." As we will see later, consciousness appears to result from the interaction of widely distributed brain structures. Partly because consciousness appears to be distributed among many functions, and partly because the problem is so overwhelming, researchers have opted to begin by looking at structures responsible for the *components* of consciousness. We will consider three of those components here—*awareness, attention,* and *sense of self*—before tackling the problem of the neural bases of consciousness in general.

Awareness

As an abstract concept, awareness is difficult to define and more difficult to study. Instead, researchers have directed their attention to *awareness of something.* Taking this approach has helped identify several brain areas as potential locations of awareness. A good illustration is a study in which a tone preceded a visual stimulus and a second tone signaled a trial in which the visual stimulus would be absent; participants who became aware of the relationship of the tones to the visual stimulus showed different levels of blood flow in the left prefrontal cortex in response to the two tones (McIntosh, Rajah, & Lobaugh, 1999). The researchers suggested that the prefrontal cortex might be the key player in producing awareness. Others have proposed the hippocampus for that role because of its involvement in declarative learning, which by definition involves awareness (R. E. Clark & Squire, 1998). Others claim that the parietal lobes' ability to locate objects in space is neces-sary for combining the features of an object into a conscious whole; as evidence, they describe a patient with damage to both parietal lobes who often attributed one object's color or direction of movement to another object (L. J. Bernstein & Robertson, 1998; Treisman & Gelade, 1980).

Whether awareness occurs in one of these areas or its location depends on the nature of the task, it is clear that brain damage impaired this patient's ability to bind the spatial, color, and movement information together into an integrated percept. Increasingly, researchers are becoming convinced that binding involves synchronization of neural activity. Synchronized activity occurs mostly in the

Q How does the brain solve the "binding problem"?

gamma frequency range, between 30 and 90 Hz. Early studies found that during visual stimulation, 50% to 70% of neurons in the visual area of cats fired in synchrony at an average rate of 40 Hz (Engel, König, Kreiter, & Singer, 1991; Engel, Kreiter, König, & Singer, 1991). For an illustration of 40-Hz synchrony, see Figure 15.16. In response to a moving stimulus, activity synchronized between V1, the primary visual area, and V5, the area that detects movement (Engel, Kreiter, et al., 1991). This makes sense, because studies have indicated that visual awareness requires feedback to V1 from extrastriate areas like V5 (reviewed in Tong, 2003). (You may want to refer to Figure 10.25 for the location of V1 and V5.)

Later investigations revealed that activity is coordinated over much wider areas. For example, when researchers presented a light that had previously been paired with shock to the finger, activity became synchronized between the visual cortex and the finger area of the somatosensory cortex (Figure 15.17; Miltner, Braun, Arnold, Witte, & Taub, 1999). In the McIntosh et al. (1999) study described above, at the moment of awareness neural activity in the left prefrontal cortex became coordinated with activity in other parts of the brain, including the right prefrontal cortex, auditory association areas, visual cortex, and cerebellum. Another study better illustrates the integrative nature of this synchrony. Words were presented in various colors and at various locations on a screen; whether the subject became aware of the word's color or of its location—indicated by being able to recall it later—depended on whether a frontal or temporal area was activated during the presentation. But if the individual registered both the color and the location, additional activity occurred in a part of the parietal cortex (Uncapher, Otten, & Rugg, 2006).

Recently Gaillard and his colleagues (2009) had the rare opportunity to record EEG activity from electrodes implanted in the brains of patients being evaluated for surgery to eliminate epileptic seizures; they recorded a cascade of awareness-defining events. Whether words reached awareness or not, they evoked coordinated gamma activity in the occipital area lasting about 300 milliseconds; if the words were recognized, this localized activity was followed by synchronized activity among occipital, parietal, and temporal areas. In addition, neurons in one area appeared to be triggering firing among neurons in the other areas.

It is important to emphasize that much of our behavior is guided by processes that are outside awareness. A simple example would be our constant use of proprioceptive information to sit erect, to walk, and to reach accurately for objects in our environment. You learned in previous chapters that people with impaired facial recognition (prosopagnosics) are aroused by familiar faces that they do not otherwise recognize, that people with blindsight locate objects they deny seeing, and that patients with hippocampal damage improve over time on tasks that they deny having performed before. In one study, research participants were able to learn and use a pattern for predicting the location of a target on a computer screen, but not one of them was able to state what the pattern was—even when offered a reward of $100 for doing so. Through subtle training procedures, people have learned to associate a particular facial feature with a particular personality characteristic without being aware they had done so; in fact, when questioned, they did not believe that such a relationship existed (reviewed in Lewicki, Hill, & Czyzewska, 1992). We like to believe that our behavior is rational and guided by conscious decisions; perhaps we invent logical-sounding explanations for our behavior when we are not aware

FIGURE 15.16 Forty-Hertz Oscillations in Neurons.

Top: Recording of the combined activity of all neurons in the vicinity of the electrode. *Bottom:* Activity recorded at the same time from two neurons adjacent to the electrode. By visually lining up the peaks and valleys of the two tracings, you can see that the two neurons are firing in synchrony with all the others in the area. (The upper tracing appears smoother because it is the sum of the activity of many neurons and because random activity is equally often positive and negative and cancels itself out.)

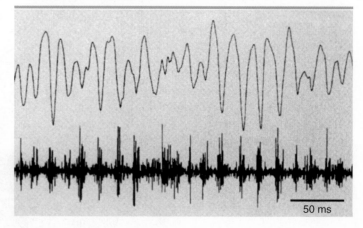

50 ms

SOURCE: Courtesy of Wolf Singer, Max-Planck-Institut für Hirnforschung.

FIGURE 15.17 Synchronized Activity Among Areas Involved in Learning.

Numbered circles indicate the location of EEG electrodes; colored areas, from anterior to posterior, are primary somatosensory cortex, secondary somatosensory cortex, and visual cortex. A light was paired several times with a shock to the middle finger. After that, presenting the light alone produced 40-Hz (average) EEG activity, which was synchronized between the visual cortex and the somatosensory cortex. The arrows indicate the pairs of electrodes between which synchrony was observed. Synchrony occurred (a) in the right hemisphere when shock had been applied to the left hand and (b) in the left hemisphere when shock had been applied to the right hand.

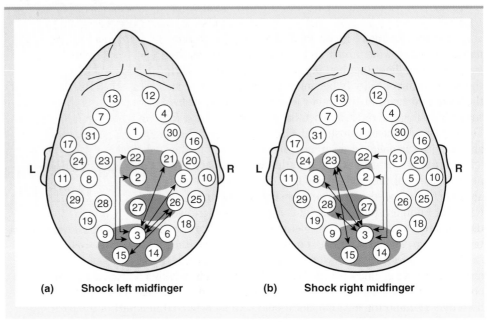

(a)   Shock left midfinger   (b)   Shock right midfinger

SOURCE: Adapted from "Coherence of Gamma-Band EEG Activity as a Basis for Associative Learning," by Miltner et al., *Nature 397*, 434–436. © 1999 Nature Publishing. Reprinted by permission.

of its true origins. So what is the benefit of conscious awareness? This is actually a matter of debate, but one apparent advantage is that it enables a consistency and a planfulness in our behavior that would not be possible otherwise.

Attention

Separating attention and awareness is difficult, and it is even controversial whether we can do so. However, it is instructive to think of awareness as referring to the *content* of consciousness and attention as a *process*—the act of attending, or the act of selecting among the contenders for our awareness. *Attention* is the brain's means of allocating its limited resources by focusing on some neural inputs to the exclusion of others. I doubt if I need to tell you how important attention is. When you are paying attention to a fascinating book, you may not notice all the hubbub around you. Some stimuli "grab" your attention, though; for example, the voice of a friend calling your name stands out above all the din. Also, what is attended to is easily remembered, and what escapes attention may be lost forever. The practical importance of attention is demonstrated in studies showing a fourfold increase in automobile accidents while drivers are using a mobile telephone (McEvoy et al.,

2005; Redelmeier & Tibshirani, 1997). This is not due to the driver having one hand off the wheel, because the risk was just as high when the driver was using a speaker phone; clearly, the problem was attention.

Although you are aware of the importance of attention, you probably do not realize just how powerful it is. An interesting demonstration is the *Cheshire cat* effect, named after the cat in *Alice in Wonderland*, who would fade from sight until only his smile remained. Have a friend stand in front of you while you hold a mirror in your left hand so that it blocks your right eye's view of your friend's face but not the left eye's (Figure 15.18). Then hold your right hand so you can see it in the mirror. (This works best if you and your friend stand in the corner of a room with blank walls on two sides.) Your hand and your friend's face will appear to be in the same position, but your friend's face, or part of it, will disappear. If you hold your hand steady, you will begin to see your friend's face again, perhaps through your "transparent" hand; move your hand slightly, and the face disappears again. By experimenting, you should be able to leave your friend with only a Cheshire cat smile. Your brain continues to receive information from both your hand and your friend's face throughout the demonstration; but because the two eyes are sending the brain conflicting information, telling it that two objects are in the same location, *binocular rivalry* occurs. The brain attends to one stimulus for a time, then switches to the other. Attention also switches when your hand or your friend's head moves and demands attention.

Attention is not just a concept; it is a physiological process, and changes in attention are accompanied by changes in neural activity. When an observer attends to an object, firing synchronizes between the brain areas involved, such as prefrontal with parietal neurons or parietal neurons with visual areas, depending on the task (Buschman & Miller, 2007; Saalmann, Pigarev, & Vidyasagar, 2007). When attention shifts—for example, during binocular rivalry, activity shifts from one group of neurons in the visual cortex to another, even though the stimulus inputs do not change (Leopold & Logothetis, 1996). When research participants focused on an object's color, PET scans showed that activity increased in visual area V4; activation shifted to the inferior temporal cortex when they attended to the object's shape, and it changed to area V5 during attention to its movement (Chawla, Rees, & Friston, 1999; Corbetta, Miezin, Dobmeyer, Shulman, & Petersen, 1990). We know that the shifts were due to attention, because activation also increased in V4 during attention to color even when the stimuli were uncolored and in V5 during attention to movement when the stimuli were stationary (Chawla et al., 1999).

So our experience of attention is a reflection of changes in brain activity. The increases in cortical activity described above are at least partly due to the modulation of activity in the thalamus, which is the gateway for sensory information to the cortex (except for olfaction). The cortex can selectively inhibit thalamic neurons and determine which information will reach it (John, 2005). When human subjects attended to a stimulus, neural responses to that stimulus increased in the lateral geniculate nucleus of the thalamus, and responses to ignored stimuli decreased (O'Connor, Fukui, Pinsk, & Kastner, 2002).

Attention involves a number of structures, and imaging suggests that they are organized into two networks: a dorsal one that allocates attention under goal-directed control and a ventral one that responds to stimulus demands (Asplund, Todd, Snyder, & Marois, 2010). However, the anterior cingulate cortex (ACC) may play an executive role (see the anterior cingulate gyrus in Figure 8.6). In individuals undergoing cingulate lesioning as a treatment for obsessive-compulsive disorder, 19% of ACC neurons either increased or decreased their firing rate during attention-demanding cognitive tests (K. D. Davis, Hutchison, Lozano, Tasker, & Dostrovsky, 2000). (You have to admire the brave souls who were willing to submit to mental arithmetic during brain surgery!) The ACC is also active during the Stroop word-color test, in which subjects must read color names as quickly as possible (Peterson et al., 1999). This is a difficult task, because some of the words

FIGURE 15.18 Setup for Demonstrating the Cheshire Cat Effect.

Your view will alternate between your hand and your friend's face.

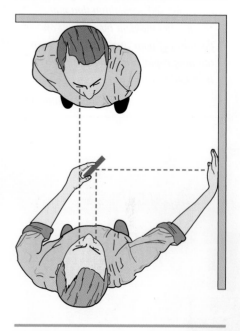

 What is the neural basis of attention?

> *Neural scientists are thus beginning to address aspects of the fundamental question of consciousness by focusing on a specific, testable problem: What neural mechanisms are responsible for focusing visual attention?*
>
> —Eric R. Kandel

are printed in a conflicting color; the researchers believe that the ACC modulates activity in attentional pathways to focus attention on the word's meaning and suppress attention to its color.

The Sense of Self

> Consciousness is a concept of your own self, something that you reconstruct moment by moment on the basis of your own body, your own autobiography and a sense of your intended future.
>
> —Antonio Damasio

Consciousness is usually studied in relation to external reality—for example, object recognition or object awareness; this is in keeping with psychology's preference during much of its history for studying phenomena that are "out there," where we can observe them objectively. But an important aspect of consciousness is what we call the self; the sense of self includes an identity—what we refer to as "I"—and the sense of *agency*, the attribution of an action or effect to ourselves rather than to another person or external force.

The sense of self is shared with few other species. We have learned this by using a cleverly simple technique developed for children. When the researcher puts a spot of rouge on a child's nose or forehead and places the child in front of a mirror, infants younger than about 15 months reach out and touch the child in the mirror or kiss it or hit it; older children will show self-recognition and use the mirror to examine the mysterious spot on their faces (Lewis & Brooks-Gunn, 1979). Chimpanzees are also able, after a time, to recognize themselves in the mirror; they examine the rouge spot, and they use the mirror to investigate parts of their body they have never seen before, like their teeth and their behinds (Figure 15.19). Elephants, orangutans, porpoises, and—get this—magpies also recognize themselves, but monkeys do not (Gallup, 1983; Plotnik, de Waal, & Reiss, 2006; Prior, Schwarz, & Güntürkün, 2008; Reiss & Marino, 2001). Although monkeys learned how a mirror works and would turn to face a person whose reflection they saw in the mirror, after 17 years of continuous exposure to a mirror in their cage, they still treated their reflections like an intruder (Gallup & Povinelli, 1998).

Investigators have had some success in identifying neural correlates of the sense of self. Frontal-temporal damage, for example, impairs episodic memory and may produce a detachment from the self (M. A. Wheeler, Stuss, & Tulving, 1997). The anterior cingulate cortex and the insula (part of the prefrontal cortex) are active when people recognize their own faces, identify memories as their own, or recognize descriptions of themselves (Devue et al., 2007; Farrer et al., 2003; Fink et al., 1998). Stroke or the dementia of old age can impair people's ability to recognize

FIGURE 15.19 A Chimp Demonstrates Concept of Self.

SOURCE: Gallup and Povinelli (1998). Photos courtesy of Cognitive Evolution Group, University of Louisiana at Lafayette.

their mirror image; the person may treat the "other" as a companion, as an intruder who must be driven from the home, or as a stalker who appears in automobile and shop windows (Feinberg, 2001).

Farrer and Frith and their colleagues suggest that the sense of agency—the ability to distinguish between the self versus another as the source of an action or event—is mediated by the anterior insula and the inferior parietal area (Farrer et al., 2003; Farrer & Frith, 2002). When their subjects perceived that they were controlling the movements of a virtual hand or a cursor on a computer screen, activity increased in the insula; when it was obvious that the experimenter was controlling the movement, activity shifted to the right inferior parietal cortex (Figure 15.20). Also, schizophrenics who believe their behavior is controlled by another person or agent show heightened parietal activity compared to other schizophrenics and normal controls (Spence et al., 1997).

We do not find in these studies, nor should we expect to find, a brain location for the self; instead, what we see is scattered bits and pieces. We should regard the self as a concept, not an entity, and the sense of self as an amalgamation of several kinds of information, mediated by networks made up of many brain areas. Body image, memory, and mirror neurons are among the contributors to the sense of self; we will examine these topics and then consider two disorders of the self.

Body Image

Body image contributes to a sense of self because we have an identification with our body and with its parts; it is *our* body, *our* hand, *our* leg. A good example of the importance of body image to the sense of self is the phantom limb phenomenon. You learned in Chapter 11 that most amputees have the illusion that their missing arm or leg is still there. The illusion occurs in 80% of amputees and may persist for the rest of the person's life, which attests to the power of the body image. Distortions in the phantom (see Figure 15.21) sometimes add credibility to this point: One man was unable to sleep on his back because his phantom arm was bent behind him, and another had to turn sideways when walking through a door because his arm was extended to the side (Melzack, 1992). Researchers once thought that phantoms occurred only after a person developed a *learned* body image, but we now know that phantoms can occur in young children and even in people born with a missing limb. Because the body image is part of the equipment we are born with, it becomes an important part of the self—even when it conflicts with reality.

The "replacement" of a limb by a phantom does not prevent a feeling of loss, which can extend to the sense of self. This point was illustrated very graphically by S. Weir Mitchell, a Civil War physician who saw numerous amputees and presented some of his observations in the fictionalized account *The Case of George Dedlow* (Mitchell, 1866). Mitchell's hero, who has lost both arms and both legs in battle, says,

> I found to my horror that at times I was less conscious of myself, of my own existence, than used to be the case. . . . I felt like asking someone constantly if I were really George Dedlow or not. (p. 8)

In Chapter 11 we saw an example of a more extensive loss in the case of Christina. Watching a home movie of herself made before the disease had destroyed her proprioceptive sense, she exclaimed,

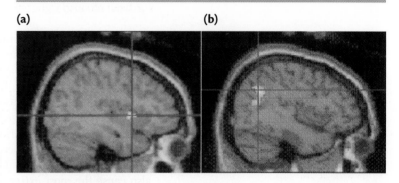

FIGURE 15.20 Brain Areas Involved in the Sense of Agency.

(a)   **(b)**

Attributing an effect (movement of a computer cursor) to oneself activated the insula (a); attributing the movement to another person activated the right inferior parietal cortex (b).

SOURCE: From "Experiencing Oneself Vs. Another Person as Being the Cause of an Action: The Neural Correlates of the Experience of Agency," by C. Farrer et al., *NeuroImage, 15,* 596–603, fig. 2 and fig. 3, p. 598. © 2002 with permission from Elsevier, Ltd.

Where does our sense of self come from?

Yes, of course, that's me! But I can't identify with that graceful girl any more! She's gone, I can't remember her, I can't even imagine her. It's like something's been scooped right out of me, right at the centre. (Sacks, 1990, p. 51)

Although information about the body is processed in the somatosensory cortex, body image is a phenomenon that develops primarily in adjacent parietal areas. An example comes from earlier attempts to eliminate phantoms by surgery; lesions in the somatosensory area had no effect, but lesioning the right posterior parietal cortex did suppress the sensations (Berlucchi & Aglioti, 1997). Damage to the right posterior insula, (part of the inferior parietal cortex located at the temporo-parietal junction) can produce a variety of symptoms, including denial that a paralyzed limb belongs to the patient or perception of an extra limb (Berlucchi & Aglioti; Karnath & Baier, 2010). In Chapter 11, we saw that damage to the inferior parietal area can also cause the most extreme of body image illusions, the out-of-body experience (Blanke & Arzy, 2005).

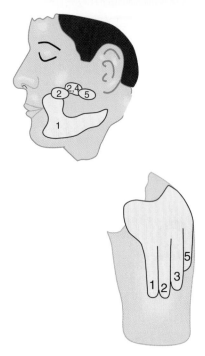

FIGURE 15.21 Maps of a Patient's Phantom Hand.

Touching the arm above the stump produced sensations of the missing hand. The same thing happened on the face, confirming what we saw in Chapter 11, that neurons from the face have invaded the hand area in the somatosensory cortex.

SOURCE: Figure 2.2 from *Phantoms in the Brain* by V. S. Ramachandran, MD, PhD, and Sandra Blakeslee. Copyright © 1998 by V. S. Ramachandran and Sandra Blakeslee. Reprinted by permission of HarperCollins Publishers, Inc.

Memory

Without long-term memory, it is doubtful there can be a self, because there is no past and no sense of who the person is. In the words of the memory researcher James McGaugh, memory "is what makes us us" (A. Wilson, 1998). Loss of short-term memory is not as disruptive; patients like HM (described in Chapter 12) have a lifetime of information about their past and about themselves as a background for interpreting current experience, even if they do not remember events that have occurred since their brain damage. However, for Korsakoff's and Alzheimer's patients, memory loss extends back several years before the onset of illness, as well as after the onset. Oliver Sacks's patient Jimmie had lost 40 years of memories to Korsakoff's disease; restless, unable to say whether he was miserable or happy, he reported that he had not felt alive for a very long time (Sacks, 1990).

Mary Frances, whom you met in Chapter 12, took another approach, explaining her situation with one false scenario after another. Another confabulator was Mr. Thompson, who took an unauthorized day's liberty from the hospital. At the end of the day, the cabdriver told the staff that he had never had so fascinating a passenger: "He seemed to have been everywhere, done everything, met everyone. I could hardly believe so much was possible in a single life" (Sacks, 1990, p. 110). According to Sacks, Mr. Thompson had to "make himself (and his world) up every moment" by turning everyone on the ward into characters in his make-believe world and weaving story after story as he attempted to create both a past and a present for himself.

The confabulated stories amnesics tell can usually be traced back to fragments of actual experiences. This is consistent with the hypothesis introduced in our earlier discussion, that confabulation is a failure to suppress irrelevant memories due to damage in the frontal area (Benson et al., 1996; Schnider, 2003; Schnider & Ptak, 1999). But, like Mr. Thompson, confabulators often prefer an embellished past to an ordinary one (Fotopoulou, Solms, & Turnbull, 2004); this together with the involvement of the anterior cingulate cortex (Turner et al., 2008) makes one suspect that the confabulation is also serving the self-image. The motivated nature of confabulation suggests the importance of real or imagined memories to the person's identity. As the movie director Luis Buñuel (1983) said as he contemplated his own failing memory,

You have to begin to lose your memory, if only in bits and pieces, to realize that memory is what makes our lives. Life without memory is no life at all. . . . Our memory is our coherence, our reason, our feeling, even our action. Without it, we are nothing.

Self, Theory of Mind, and Mirror Neurons

A sense of self requires the distinction between our self and other selves and, arguably, some understanding of other selves. We saw in the discussion of autism that an ability to attribute mental states to others is called theory of mind and that researchers who study mirror neurons believe they are critical to our development of that comprehension. They give mirror neurons considerable credit for social understanding (Gallese & Goldman, 1998), empathy (Gazzola et al., 2006), and the ability to understand the intentions of others (Iacoboni et al., 2005). When volunteers watched a video clip, their mirror neurons responded more as a model reached for a full cup beside a plate of snacks (implying the intent to eat) than when the model reached for an empty cup beside an empty plate (implying the intent to clean up). The two scenes without the model produced no differences (Figure 15.22; Iacoboni et al., 2005).

Malfunction in the mirror neuron system is one reason suggested for the autistic's failure to develop a theory of mind. Interestingly, autistics are not so much impaired in the ability to imitate as in their control of imitation, seen, for instance, in repeating what others say (echolalia). Rather than a defect in the mirror neurons, the problem may be disordered regulation of the mirror neurons' activity (J. H. Williams, Whiten, Suddendorf, & Perrett, 2001). Justin Williams and his colleagues suggest that the malfunction is in the anterior cingulate cortex. Earlier, we saw that damage to this area can diminish self-awareness; we will learn later that others give it a more critical role in consciousness.

Split Brains and Dissociative Identity Disorder: Disorders of Self

Chapter 3 describes a surgical procedure that separates the two cerebral hemispheres by cutting the corpus callosum. This surgery is used to prevent severe epileptic seizures from crossing the midline and engulfing the other side of the brain. Besides providing a unique opportunity to study the differing roles of the two hemispheres, split-brain patients also raise important questions about consciousness and the self. Gazzaniga (1970) described a patient who would sometimes find

FIGURE 15.22 Different Intentions Distinguished by Mirror Neurons.

The implied intention of the actor in the photo on the left is to drink; in the photo on the right, it is to clean up. Different neurons were active as research participants viewed these two scenes, suggesting that mirror neurons can distinguish among intentions.

SOURCE: From "Grasping the Intentions of Others With One's Own Mirror Neuron System," by M. Iacoboni, 2005, *PLoS Biology, 3*, pp. 529–535, fig. 1 upper right and lower right, p. 530.

Q   What do the disorders tell us?

his hands behaving in direct conflict with each other—for instance, one pulling up his pants while the other tried to remove them. Once the man shook his wife violently with his left hand (controlled by the more emotional right hemisphere), while his right hand tried to restrain the left. If the person with a severed corpus callosum is asked to use the right hand to form a specified design with colored blocks, performance is poor because the left hemisphere is not very good at spatial tasks; sometimes the left hand, controlled by the more spatially capable right hemisphere, joins in to set the misplaced blocks aright and has to be restrained by the experimenter.

Different researchers interpret these studies in different ways. At one extreme are those who believe that the major or language-dominant hemisphere is the arbiter of consciousness and that the minor hemisphere functions as an automaton, a nonconscious machine. At the other extreme are the researchers who believe that each hemisphere is capable of consciousness and that severing the corpus callosum divides consciousness into two selves. Sixty years of research have prompted most theorists to take positions somewhere along the continuum between those extremes.

Gazzaniga, for instance, points to the right hemisphere's differing abilities, such as the inability to form inferences, as evidence that the right hemisphere has only primitive consciousness (Gazzaniga, Ivry, & Mangun, 1998). He says the left hemisphere not only has language and inferential capability but also contains a module that he calls the "brain interpreter." The role of the *brain interpreter* is to integrate all the cognitive processes going on simultaneously in other modules of the brain. Gazzaniga was led to this notion by observing the split-brain patient PS performing one of the research tasks. PS was presented a snow scene in the left visual field and a picture of a chicken's foot in the right and asked to point to a picture that was related to what he had just seen. With the right hand he pointed to the picture of a chicken, and with the left he selected a picture of a shovel (see Figure 15.23). When asked to explain his choices he (his left hemisphere) said that the chicken went with the foot and the shovel was needed to clean out the chicken shed. Unaware that the right hemisphere had viewed a snow scene, the left hemisphere gave a reasonable but inaccurate explanation for the left hand's choice. Although the right hemisphere has less verbal capability than the left, it can respond to simple commands; if the command "Walk" is presented to the right hemisphere, the person will get up and start to walk away. When asked where he or she is going, the patient will say something like "I'm going to get a Coke." According to Gazzaniga, these confabulations are examples of the brain interpreter making sense of its inputs, even though it lacks complete information.

Perhaps researchers who view the right hemisphere's consciousness as primitive are confusing consciousness with the ability to verbalize the contents of consciousness. Assigning different levels of consciousness to the two hemispheres may be premature when our understanding of consciousness is itself so primitive. Research with split-brain patients tells us to avoid oversimplifying such a complex issue.

Another disorder of self is *dissociative identity disorder* (formerly known as multiple personality), which involves shifts in consciousness and behavior that appear to be distinct personalities or selves. Most people are familiar with this disorder from the movie *The Three Faces of Eve.* Shy and reserved, Eve White

FIGURE 15.23 Split-Brain Patient Engaged in the Task Described in the Text.

His verbal explanation of his right hand's selection was accurate, but his explanation of his left hand's choice was pure confabulation.

SOURCE: Gazzaniga (2002). Based on an illustration by John W. Karpelou, BioMedical Illustrations.

would have blackouts while her alter ego, Eve Black, would spend a night on the town dancing and drinking with strange men. The puritanical Eve White would have to deal with the hangover, explain a closetful of expensive clothes she didn't remember buying, and sometimes fend off an amorous stranger she found herself with in a bar (Lancaster, 1958; Thigpen & Cleckley, 1957). Eve, whose real name is Chris Sizemore, went on to develop 22 different personalities before she successfully integrated them into a single self (Figure 15.24) (Sizemore, 1989).

The causes of dissociative identity disorder are not understood, but 90% to 95% of patients report childhood physical and/or sexual abuse (Lowenstein & Putnam, 1990; C. A. Ross et al., 1990). In Sizemore's case, the emotional trauma came from several sources: her sense of parental rejection, fear of a scaly monster her mother invented to frighten her into being "good," and the horror of witnessing the grisly death of a sawmill worker who was cut in two by a giant saw (Lancaster, 1958). Most therapists believe that the individual creates alternate personalities ("alters") as a defense against persistent emotional stress; the alters provide escape and, often, the opportunity to engage in prohibited forms of behavior (Fike, 1990; C. A. Ross et al., 1990).

Although the disorder had been reported occasionally since the middle 1600s (E. L. Bliss, 1980), reports were rare until recent times; the number of reports jumped from 500 in 1979 to 5,000 in 1985 (Braun, 1985). While some therapists believe that dissociative identity disorder was underdiagnosed earlier (Fike, 1990; Lowenstein & Putnam, 1990; Putnam, 1991), critics say that the patients intentionally create the alternate personalities to provide an explanation for bizarre and troubling behavior, as a defense for criminal behavior, or at the urging of an overzealous therapist (Spanos, 1994). There probably are many bogus cases, but extensive documentation by therapists and inclusion of dissociative identity disorder in the *Diagnostic and Statistical Manual of Mental Disorders*, fourth edition (American Psychiatric Association, 1994), lend credibility to the diagnosis.

The earlier term, *multiple personality disorder*, inappropriately suggests that there are multiple selves or people living in one person's body, and the descriptions of patients' behavior lend themselves to that impression. Of course, if the self is just a concept, you can see that it makes no sense to say that the dissociative identity disorder patient has *multiple* selves. However, if we throw out the "multiple person" interpretation, the symptoms of multiple identity still remain, and they beg explanation of any theorist who believes that all human behavior has a physiological basis.

There is little physiological information for the neuroscientist to go on, but what does exist is intriguing. Therapists often report that alters differ from each other in handedness, reaction to medications, immune system responsiveness, allergies, and physical symptoms (N. Hall, et al., 1994; Lowenstein & Putnam, 1990; Putnam, Zahn, & Post, 1990). Laboratory studies have found differences among the alters in several physiological measures, as well as differences in heart rate, blood pressure, and cerebral blood flow responses to their own emotional memories (Reinders et al., 2006). Even if these physical differences are due to changes in arousal and muscle tension, as some researchers suggest (S. D. Miller & Triggiano, 1992; Putnam et al., 1990), they still represent an interesting physiological mechanism that requires explanation.

Bower (1994) attempted to explain the amnesia of dissociative identity disorder as an example of *state-dependent learning*. In state-dependent learning, material learned in one state is difficult to recall in the other state; the altered states can be induced in the laboratory by alcohol and other drugs and even by different

FIGURE 15.24 Chris Sizemore.

The story of her struggle with multiple personalities was the basis for the movie *The Three Faces of Eve*.

SOURCE: © AP Photo.

FIGURE 15.25 Hippocampal Activity During the Switch Between Multiple Personalities.

The scans show inhibition of the parahippocampus and hippocampus during the switch from the primary personality to the alter (a) and increased activity in the right hippocampus during the switch back (b, c). The brain levels of these three scans are shown in (d). (The brain is viewed from below, so right and left are reversed on the page.)

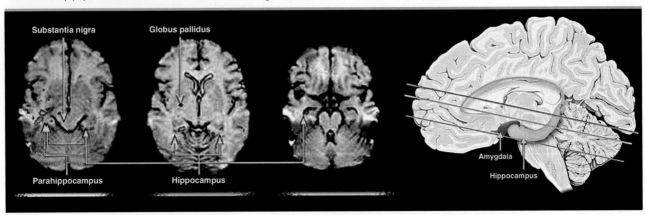

SOURCE: From "Functional Magnetic Resonance Imaging of Personality Switches in a Woman With Dissociative Identity Disorder," by Tsai et al., *Harvard Review of Psychiatry, 7*(15), pp. 119–122. © 1999. Reprinted by permission of Taylor & Francis.

moods. Bower's hypothesis is that abuse or other stresses create an altered state in which separate memories and adaptive strategies develop to the point that they form a distinct personality. Guochuan Tsai and his colleagues used functional magnetic resonance imaging to study a 33-year-old dissociative identity disorder patient as she switched between her primary and an alter personality (Tsai, Condie, Wu, & Chang, 1999). During the switch from the primary to the alter, activity was inhibited in the hippocampus and parahippocampal area, particularly on the right side; during the switch back, the right hippocampus increased in activity (see Figure 15.25). The hippocampal activity led the researchers to suggest that learning mechanisms are involved in the disorder. Imagining a new personality did not have the same effect even though it required as much effort.

Several observations are consistent with the idea that learning structures are involved in dissociation: Childhood abuse, which is frequent in patients' backgrounds, can produce hippocampal damage (Bremner et al., 1997); an association has been reported between epileptic activity in the temporal lobes (where the hippocampi are located) and identity dissociation (Mesulam, 1981); and differences in temporal lobe activity have been found between personalities in the same individual (Hughes, Kuhlman, Fichtner, & Gruenfeld, 1990; Saxe, Vasile, Hill, Bloomingdale, & Van der Kolk, 1992). But speculating about how a person can develop a whole constellation of separate memories and personality characteristics points up how inadequate our understanding of learning is. We should consider dissociative identity disorder not just a challenge but an added opportunity for studying neural functioning and cognitive processes such as learning.

> *One is always a long way from solving a problem until one actually has the answer.*
>
> —*Stephen Hawking*

Theoretical Explanations of Consciousness

Earlier, we talked about partially paralyzed patients who had symptoms of neglect, usually following right hemisphere parietal injury. Some of these patients also experience *anosognosia;* they verbally deny the paralysis and may claim that the paralyzed limb is not their own, insist that they are obeying an instruction to clap while one arm hangs motionless, and even fail to recognize paralysis in another person. By examining brain scans of neglect patients, Anna Berti and her colleagues (2005) found that the premotor area was critical for the patients' ability to recognize their paralysis. This suggests that the same part of the brain that imagines and plans movement is also responsible for awareness of what a part of the body is doing. (Remember that this is the area implicated in the rubber hand illusion.) Earlier, we saw similar results in patients who lost the ability to experience color or to identify faces. But there is good reason to believe that these "islands of consciousness" are not in themselves sufficient for producing conscious experience.

Most neurobiological theories of consciousness assume that consciousness requires a widely distributed neuronal network (Zeman, 2001). This view has arisen in part from studies that manipulate awareness of environmental stimuli by using binocular rivalry, backward masking (following a stimulus quickly with a nonmeaningful stimulus), and inattentional blindness (inserting a distracting stimulus in a visual scene). Numerous studies have shown that when stimuli become conscious, they produce widespread activity in the prefrontal and parietal cortex (Figure 15.26) (reviewed in Baars, 2005; see Dehaene et al., 2001; Sergent, Baillet, & Dehaene, 2005). Similarly, when consciousness is impaired by deep sleep or by brain injury that results in a vegetative state, auditory and pain stimuli fail to evoke activity beyond the primary areas. In addition, deep sleep, coma, vegetative states, epileptic loss of consciousness, and general anesthesia are characterized by decreased metabolism in the frontal and parietal cortex, and coordinated activity among brain areas disappears.

According to some theorists, consciousness occurs when the functioning of widespread networks becomes coordinated, enabling them to share and integrate information (Baars, 2005; Dehaene & Naccache, 2001; Tononi, 2005). Earlier we saw that gamma activity is proposed to be the mechanism that binds sensory information into awareness. Many theorists believe that gamma oscillations generated by a feedback loop between the thalamus and the cortex not only are the most likely means of achieving this coordination but are also necessary for consciousness (Ribary, 2005). Lesions that impair the functioning of this thalamo-cortical system produce a global loss of consciousness; damage to the intralaminar nuclei of the thalamus is particularly devastating, most likely because these nuclei are responsible for the ability of the thalamus and cortical areas to work as a system (Tononi, 2005).

Recently, researchers have been focusing more specifically on the default mode network (DMN) and its relationship to level of consciousness. Sleep deprivation, for example, is accompanied by some disruption in the DMN, and coordination between frontal and posterior brain areas is lost altogether in deep sleep (Gujar, Yoo, Hu, & Walker, 2010; Horovitz et al., 2009; Sämann et al., 2010). Activity in

FIGURE 15.26 Map of Evoked Potentials to Masked and Unmasked Visually Presented Words.

When briefly presented words were masked by following them with nonmeaningful visual stimuli, activation was largely confined to the primary visual area (as well as slightly delayed) and did not produce awareness. Unmasked words produced additional subsequent activity, which spread through the frontal and parietal cortex, accompanied by awareness.

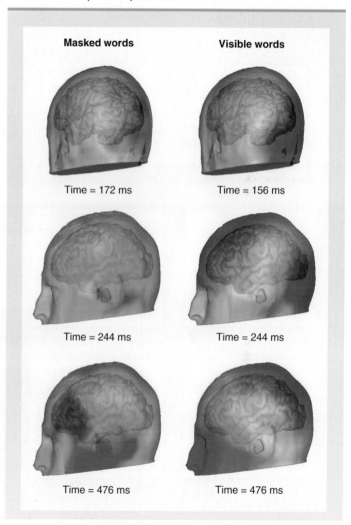

SOURCE: From "Cerebral Mechanisms of Word Masking and Unconscious Repetition Priming," by S. Dehaene et al., *Nature Neuroscience*, 4, 752–758, fig. 3, p. 755. © 2001 Macmillan Publishing. Used with permission.

7

Talking With the Experts

the DMN also corresponds to the depth of anesthesia, and during unconsciousness even transcranial magnetic stimulation cannot evoke more than localized activation (Deshpande, Kerssens, Sebel, & Hu, 2009; Ferrarelli et al., 2010). In addition, DMN activity varies with the level of consciousness in noncommunicative brain-damaged patients; locked-in syndrome patients do not differ from normal controls, but activity decreases from minimally conscious to vegetative to coma patients (Vanhaudenhuyse et al., 2010). These results are consistent with Giulio Tononi's hypothesis that consciousness depends on the brain's ability to integrate information, an ability that is impaired by diminishing connectivity (Tononi, 2008).

Distribution of consciousness means that there is no *center* of consciousness, but some researchers believe that there must be an *executive,* an area that coordinates or orchestrates the activity of all the other structures. Researchers have proposed a variety of locations for this executive, including the thalamus (Crick, 1994), the anterior cingulate cortex (Posner & Rothbart, 1998), and the claustrum (Crick & Koch, 2005). However, support for such a broad role for any structure is lacking. Just as consciousness appears to be distributed, it seems that its controls might be as well. And considering the complexity of consciousness that you see in Figure 15.27, those controls are likely to be numerous. These issues are not purely academic; as the Application shows, a better understanding of consciousness is a practical necessity.

FIGURE 15.27 Awareness and Arousal in Normal and Impaired Consciousness.

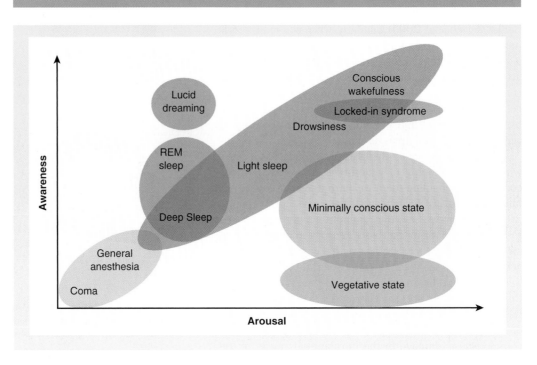

SOURCE: Adapted from "The Neural Correlate of (Un)awareness: Lessons From the Vegetative State," by S. Laureys, 2005, *Trends in Cognitive Sciences, 9,* 556–559.

Application   Determining Consciousness When It Counts

Brain damage relegates thousands of people to hospital beds and deprives them of the ability to interact with loved ones or to communicate their needs to caregivers. The worst of these are in a *coma*, unarousable and completely unresponsive to stimulation. Those who are in a *vegetative state* respond reflexively to stimulation and cycle through wakefulness and sleep, but they do not interact voluntarily with their environment. The *minimally conscious* occasionally show some voluntary responses, such as following visual movement or crying when they hear a family member's voice, but they are unable to communicate with any reliability. Patients with *locked-in syndrome*, on the other hand, are fully conscious; however, they are so paralyzed, usually due to brain stem lesions, that at best they are able to communicate only with eye movements or finger twitches.

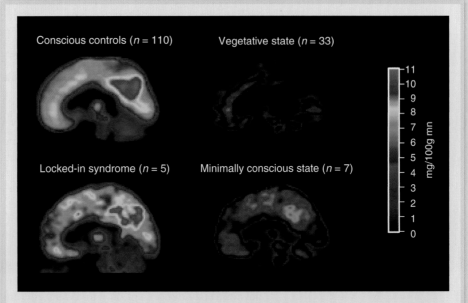

PET Scans Reveal Metabolic Activity Corresponding to Patients' Level of Consciousness.

SOURCE: From "Brain Function in Coma, Vegetative State, and Related Disorders," by S. Laureys, A. M. Owen, and N. D. Schiff, 2004, *Lancet Neurology, 3,* 537–546.

Unfortunately, physicians often rely on bedside clinical observation to diagnose their patients, rather than neurobehavioral testing. As a result, an estimated 40% of patients classified as being in a vegetative state are actually minimally conscious (Schnakers et al., 2009). One of the misdiagnosed is Rom Houben, who remained in an apparent coma for 23 years following a car crash, until Steven Laureys did a PET scan and found that Houben's brain metabolism was nearly normal (see figure) (Connolly, 2009). In spite of reports that he could type messages on a touch screen with a therapist guiding his weakened hands, Houben's paralysis has prevented him from communicating (Hensley, 2010). However, his case does make the point that accurate assessment of a patient's level of consciousness is critical not only for care decisions, such as whether pain management is appropriate, but also for guiding research to develop ways of communicating with patients and, potentially, restoring some function.

New research indicates that it may be possible to communicate with some apparently unresponsive patients. Patients were asked to imagine playing tennis or walking through their house while their brain activity was monitored with fMRI; five patients (all but one classified as being in a vegetative state) were able to produce distinctive activity in different regions of the brain (Monti et al., 2010). One of the patients, considered to have been in a vegetative state for 5 years, was able to answer questions correctly by thinking about one of the scenarios for "yes" and the other for "no." Using fMRI to communicate is not very practical, so researchers are trying to adapt the task to EEG.

There is also real hope that techniques now being used experimentally can give patients back some of their capabilities. Nicholas Schiff and his colleagues (2007) implanted electrodes in the brain of a man who had been unreliably responsive for 6 years following brain injury. With electrical stimulation to the thalamus, he can now eat without tube feeding, watch movies, and communicate with gestures and phrases up to six words. The team concluded that the thalamic stimulation compensates for lost arousal in the frontal cortex and in the anterior cingulate cortex. A less invasive approach involves the continuous injection of the dopamine receptor agonist apomorphine (Fridman et al., 2010). All patients receiving the drug improved, some within 24 hours, and seven completely recovered consciousness; all gains were sustained for at least a year, even after drug treatment was discontinued.

In Perspective

There was a time when the topic of sleep was totally mysterious and dreaming was the province of poets and shamans. Now sleep and dreaming are both yielding to the scrutiny of neuroscience. Although we are still unclear about the functions of sleep, we are learning how various structures in the brain turn it off and on, and how our body dances not only to a daily rhythm but to another that repeats itself 16 times a day and controls our fluctuations in alertness, daydreaming, and night dreaming.

Consciousness is also giving up its secrets as researchers bring modern technologies to bear on awareness, attention, and memory. In other words, what was once a taboo topic is becoming accessible to the research strategies of science and is providing a whole new arena of opportunities to observe the brain at work.

In this final chapter, we have explored a unique field of research. The study of sleep has demonstrated neuroscience's ability to unravel mysteries and dispel superstition. The investigation of consciousness has been more daring, taking scientists where none had gone before. That is the job of science, to push back darkness, whether by finding a treatment for depression or by explaining humanity's most unique characteristics and capabilities. But we have traveled a road filled with questions and uncertain facts, and the words of the schizophrenia researcher that "almost everything remains to be done" still seem appropriate. If all this ambiguity has left you with a vaguely unsatisfied feeling, that is good; you may have the makings of a neuroscientist. And we have left the most exciting discoveries for you.

> "... there is a particular problem with finding endings in science. Where do these science stories really finish? Science is truly a relay race, with each discovery handed on to the next generation."
>
> —Richard Holmes,
> *in* The Age of Wonder

Summary

Sleep and Dreaming

- Circadian rhythms are rhythms that repeat on a daily basis, affecting the timing of sleep and several bodily processes. The SCN is the most important control center but not the only one. The rhythm is primarily entrained to light.
- Several ultradian rhythms occur within the day. One involves alternating periods of arousal during waking and stages that vary in arousal during sleep.
- REM sleep is when most dreaming occurs, but it has also been implicated in neural development and learning.
- Slow-wave sleep may restore cerebral and cognitive functioning and participate in learning.

- Sleep and waking are controlled by separate networks of brain structures.
- Insomnia, sleepwalking, narcolepsy, and REM sleep behavior disorder represent the effects of psychological disturbances in some cases and malfunction of the sleep-wake mechanisms in others.
- Sleep is an active period, a state of consciousness that is neither entirely conscious nor unconscious.

The Neural Bases of Consciousness

- Any very specific definition of consciousness is premature, but normal consciousness includes awareness, attention, and a sense of self.
- How awareness comes about is unclear, but the thalamus apparently is involved, possibly as the coordinator of synchronous firing of neurons in involved areas, which is hypothesized to produce binding.
- Attention allocates the brain's resources, actually shifting neural activity among neurons or brain locations.
- Body image, memory, and mirror neurons are important contributors to a sense of self.
- Split-brain surgery provides an interesting research opportunity into consciousness, which has prompted debates about each hemisphere's contribution to consciousness and to the self.
- Dissociative identity disorder involves what appear to be distinct personalities or selves. Reports indicate that the different states include different physical and physiological characteristics.
- Consciousness appears to be distributed across numerous brain areas, perhaps coordinated by gamma activity. ■

Study Resources

F   For Further Thought

- Animals cycle on a 24-hr schedule, either sleeping at night and being active during the day or vice versa. An alternative would be to sleep when fatigue overtakes the body, regardless of the time. What advantages can you think of for a regular schedule?

- Machines and, probably, some simpler animals function just fine without awareness. Awareness probably places a significant demand on neural resources. What adaptive benefits do you see?
- Do you think we will be able to understand consciousness at the neural level? Why or why not?

T   Testing Your Understanding

1. Discuss the functions of sleep, including the REM and slow-wave stages of sleep.
2. Discuss attention as a neural phenomenon.
3. Discuss the function of confabulation in dreaming and in the behavior of split-brain patients and Korsakoff's patients.

Select the best answer:

1. The most important function of sleep is
   a. restoration of the body.
   b. restoration of the brain.
   c. safety.
   d. a, b, and c.
   e. uncertain.

2. The body's own rhythm, when the person is isolated from light, is
   a. approximately 24 hr long.
   b. approximately 25 hr long.
   c. approximately 28 hr long.
   d. unclear because of conflicting studies.

3. Jim is totally blind, but he follows a 24-hr day-night cycle like the rest of us and seems comfortably adapted to it. Animal studies suggest that he relies on
   a. a built-in rhythm in his SCN.
   b. nonvisual receptors in his eyes.
   c. clocks and social activity.
   d. a and b.
   e. b and c.

4. According to the activation-synthesis hypothesis, dreams are the result of a combination of random neural activity and
   a. external stimuli.   b. wishes.
   c. concerns from the day   d. memories.

5. Evidence that REM sleep specifically enhances consolidation is that
   a. REM increases after learning.
   b. REM deprivation interferes with learning.
   c. performance improves following REM sleep.
   d. a and b.
   e. a, b, and c.

6. An "executive" sleep and waking center is located in the
   a. rostral pons.
   b. lateral hypothalamus.
   c. preoptic area of the hypothalamus.
   d. magnocellular nucleus.
   e. none of the above.

7. The magnocellular nucleus is responsible for
   a. initiating sleep.
   b. waking the individual.
   c. switching between REM and non-REM sleep.
   d. producing atonia during REM.

8. Cataplexy is
   a. sleep without an REM component.
   b. a waking experience of atonia.

c. a more severe form of narcolepsy.
d. clinically significant insomnia.

9. The binding problem is an issue because
   a. there is no clear dividing line between consciousness and unconsciousness.
   b. we are unsure what the function of sleep is.
   c. there is no single place where all the components of an experience are integrated.
   d. we lack agreement on what consciousness is.

10. An EEG at 40 Hz is associated with
    a. binding.   b. dreaming.
    c. consolidation.   d. attention.

11. The part of the brain where attention is shifted among stimuli may be the
    a. basal   b. magnocellular
       forebrain.   nucleus.
    c. thalamus.   d. raphé nuclei.

12. An explanation offered for confabulation links it to damage to the
    a. locus coeruleus.   b. temporal lobes.
    c. pulvinar.   d. frontal areas.

13. The credibility of dissociative identity disorder is increased by
    a. the high frequency of its diagnosis.
    b. different patterns of physiological measures.
    c. patients' lack of incentive to fake the symptoms.
    d. location of the damage in a particular brain area.

14. The claustrum has been proposed as a consciousness executive largely because of its
    a. interconnections with other parts of the brain.
    b. generation of gamma activity.
    c. selection of which sensory information gets through.
    d. reduced activity during loss of consciousness.

Answers:
1.e, 2.d, 3.b, 4.d, 5.e, 6.e, 7.d, 8.b, 9.c, 10.a, 11.c, 12.d, 13.b, 14.a.

Online Resources

The following resources are available at **www.sagepub.com/garrett3e.**

Chapter Resources

- Flash Cards
- Chapter Quiz
- Internet Resources and Exercises

On the Web

Links to these websites can be found at the web address above: Select this chapter's **Chapter Resources**, and then choose **Internet Resources and Exercises**.

1. **The National Sleep Foundation** has links to sites of sleep research and support organizations,

results of the annual poll Sleep in America, and information on sleep disorders.

The National Institute of Health's **Information about Sleep** covers a broad range of topics, with references.

2. **The Sleep Well** is the website of William Dement, noted sleep researcher.

3. The NOVA video **Sleep** features research on the role of sleep in memory.

4. The **Academy of Sleep** site is a good source for information about sleep disorders and treatments.

5. **ABC News** video cameras caught Amy and Anna raiding the refrigerator during their sleep.

6. **Online Papers on Consciousness** is a directory of 2,500 articles on the topic.

7. Charlie Rose interviews neuroscientist **V. S. Ramachandran**, and the website TED presents a lecture by consciousness expert **Dan Dennett**.

Animations

Brain Mechanisms of Sleep and Arousal (Figures 15.9, 15.10)

Chapter Updates and Biopsychology News

R For Further Reading

1. *The Sleepwatchers* by William Dement (Stanford Alumni Association, 1992) is an entertaining description of sleep research by the most widely known expert.

2. *Sleep Disorders for Dummies* by Max Hirshkowitz and Patricia Smith (Wiley, 2004) is a guide for anyone who has trouble sleeping.

3. *The Quest for Consciousness: A Neurobiological Approach* by Christof Koch (Roberts & Company, 2004) is the result of Koch's collaboration with the late Francis Crick. Amazon readers gave it 4½ stars out of 5. A briefer option is Crick and Koch's "A Framework for Consciousness" (*Nature Neuroscience*, 2003, 6, 119–126).

4. *The Feeling of What Happens* by Antonio Damasio (Harcourt, 1999) discusses the contributions of the body and emotions to consciousness and the self. It is highly praised by professionals and very readable as well.

5. "Forty-Five Years of Split-Brain Research and Still Going Strong" by Michael Gazzaniga (*Nature Reviews Neuroscience*, 2005, 6, 653–659) is a useful and extensive summary of what we have learned from the research.

6. "The Patient's Journey: Living With Locked-In Syndrome" by Nick Chisholm and Grant Gillett (*British Medical Journal*, 2005, *331*, 94–97) is Nick's account of his partial recovery from locked-in syndrome, a disorder that leaves the patient completely conscious but unresponsive, unable to vocalize, open the eyes, or signal to doctors that he or she is fully aware. The article contains useful details about the disorder and the problem of distinguishing it from the persistent vegetative state.

7. "Whose Body Is It, Anyway?" by Graham Lawton (*New Scientist*, March 21, 2009, 36–37) tells you how to create the illusion you have three hands and how to get a glimpse of what an out-of-body experience is like.

K Key Terms

activation-synthesis hypothesis....475
agency....490
attention....488
basal forebrain area....478
brain interpreter....494
cataplexy....483
dissociative identity disorder....494
insomnia....481
melatonin....471
narcolepsy....483
non-REM sleep....475
orexin....479
PGO waves....480
REM sleep behavior disorder....484
slow-wave sleep (SWS)....473
ultradian rhythm....473

Glossary

ablation Removal of brain tissue.

absolute refractory period A brief period during the action potential in which the neuron cannot be fired again because the sodium channels are closed.

absorptive phase The period of a few hours following a meal during which the body relies on the nutrients arriving from the digestive system.

action potential An abrupt depolarization of the membrane that allows the neuron to communicate over long distances.

activating effects Hormonal effects on sexual development that can occur at any time in an individual's life; their duration depends on the presence of the hormone.

activation-synthesis hypothesis The hypothesis that during REM sleep the forebrain integrates neural activity generated by the brain stem with information stored in memory; an attempt to explain dreaming.

acute Referring to symptoms that develop suddenly and are usually more responsive to treatment.

addiction A preoccupation with obtaining a drug, compulsive use of the drug in spite of adverse consequences, and a high tendency to relapse after quitting.

adequate stimulus The energy form for which a receptor is specialized.

ADHD See *attention deficit hyperactivity disorder.*

adoption study In heredity research, a study that reduces environmental confounding by comparing the similarity of adopted children with their biological parents and their similarity with their adoptive parents.

affective aggression Aggression that is characterized by emotional arousal.

agency The sense that an action or effect is due to oneself rather than another person or external force.

aggression Behavior that is intended to harm.

agonist Any substance that mimics or enhances the effect of a neurotransmitter.

agonist treatment Addiction treatment that replaces the addicting drug with another drug that has a similar effect.

agouti-related protein A transmitter released by NPY/AgRP neurons in the arcuate nucleus of the hypothalamus when nutrients diminish, which stimulates feeding.

agraphia The inability to write due to brain damage.

alcohol Ethanol, a drug fermented from fruits, grains, and other plant products, which acts at many brain sites to produce euphoria, anxiety reduction, motor incoordination, and cognitive impairment.

aldehyde dehydrogenase (ALDH) A group of enzymes that metabolize the alcohol byproduct acetaldehyde into acetate. A genetic deficiency in ALDH, or inhibition of ALDH by the antialcohol drug Antabuse, makes the drinker ill.

ALDH See *aldehyde dehydrogenase.*

alexia The inability to read due to brain damage.

allele An alternate version of a gene.

all-or-none law The principle that an action potential occurs at full strength or it does not occur at all.

Alzheimer's disease A disorder characterized by progressive brain deterioration and impairment of memory and other mental abilities; the most common cause of dementia.

amino acids The building blocks of peptides, which in turn make up proteins. In digestion, the result of the breakdown of proteins.

amphetamine One of a group of synthetic drugs that produce euphoria and increase confidence and concentration.

amplitude The physical energy in a sound; the sound's intensity.

amygdala Limbic system structure located near the lateral ventricle in each temporal lobe that is involved with primarily negative emotions and with sexual behavior, aggression, and learning, especially in emotional situations.

analgesic Pain relieving.

androgen insensitivity syndrome A form of male pseudohermaphroditism, involving insensitivity to androgen as a result of a genetic absence of androgen receptors. The person has male sex chromosomes and male internal sex organs, and external sex organs that are female or ambiguous.

androgens A class of hormones responsible for a number of male characteristics and functions.

angiotensin II A hormone that signals lowered blood volume and, thus, volemic thirst to the brain.

angular gyrus A gyrus at the border of the parietal and occipital lobes containing pathways that connect the visual area with auditory, visual, and somatosensory association areas in the temporal and parietal lobes. Damage results in alexia and agraphia.

anorexia nervosa An eating disorder in which the person restricts food intake to maintain weight at a level so low that it is threatening to health.

ANS See *autonomic nervous system.*

antagonist Any substance that reduces the effect of a neurotransmitter.

antagonist treatment A form of treatment for drug addiction using drugs that block the effects of the addicting drug.

antagonistic muscles Muscles that produce opposite movements at a joint.

anterior Toward the front.

anterior cingulate cortex A part of the limbic system important in attention, cognitive processing, possibly consciousness, and emotion, including the emotion of pain.

anterior commissure One of the groups of neurons connecting the two cerebral hemispheres; according to some research, it is larger in gay men and heterosexual women than in heterosexual men.

anterograde amnesia An impairment in forming new memories.

antidrug vaccine A form of anti-addiction treatment using molecules that attach to the drug and stimulate the immune system to make antibodies that will break down the drug.

antisense RNA A technology that temporarily disables a targeted gene or reduces its effectiveness.

antisocial personality disorder A condition in which people behave recklessly; violate social norms; commit antisocial acts such as fighting, stealing, using drugs, and engaging in sexual promiscuity.

anxiolytic Anxiety reducing.

aphasia Language impairment caused by damage to the brain.

apotemnophilia See *body integrity identify disorder.*

arcuate nucleus A structure in the hypothalamus that monitors the body's nutrient condition and regulates eating behavior.

area prostrema A brain area unprotected by the blood-brain barrier; blood-borne toxins entering here induce vomiting.

arousal theory The theory that people behave in ways that keep them at their preferred level of arousal.

association area Cortical areas that carry out further processing beyond what the primary projection area does, often combining information from other senses.

associative long-term potentiation Strengthening of a weak synapse when it and a strong synapse on the same postsynaptic neuron are active simultaneously.

attention The brain's means of allocating its limited resources by focusing on some neural inputs to the exclusion of others.

attention deficit hyperactivity disorder (ADHD) Disorder that develops during childhood and is characterized by impulsiveness, inability to sustain attention, learning difficulty, and hyperactivity.

auditory cortex The area of cortex on the superior temporal gyrus, which is the primary projection area for auditory information.

auditory object A sound that we recognize as having an identity that is distinct from other sounds.

autism A disorder that typically includes compulsive, ritualistic behavior; impaired sociability; and intellectual disability.

autistic savant An individual with autism with an isolated exceptional capability.

autoimmune disorder A disorder in which the immune system attacks the body's own cells.

autonomic nervous system (ANS) One of the two branches of the peripheral nervous system; composed of the sympathetic and parasympathetic nervous systems, which control smooth muscles, glands, and the heart and other organs.

autoradiography A technique for identifying brain structures involved in an activity; it involves injecting a radioactive substance (such as 2-DG) that will be absorbed most by the more active neurons, which then will show up on X-ray film.

autoreceptor A receptor on a neuron terminal that senses the amount of transmitter in the synaptic cleft and reduces the presynaptic neuron's output when the level is excessive.

aversive treatment A form of addiction treatment that causes a negative reaction when the person takes the drug.

axon An extension from a neuron's cell body that carries information to other locations.

B cell A type of immune cell that fights intruders by producing antibodies that attack a particular intruder.

barbiturate A class of drugs that act selectively on higher cortical centers, especially those involved in inhibiting behavior, so they produce talkativeness and increased social interaction. In higher doses, they act as sedatives and hypnotics. Used to treat anxiety, aid sleep, and prevent epileptic convulsions.

basal forebrain area An area just anterior to the hypothalamus that contains both sleep-related and waking-related neurons.

basal ganglia The caudate nucleus, putamen, and globus pallidus, located subcortically in the frontal lobes; they participate in motor activity by integrating and smoothing movements using information from the primary and secondary motor areas and the somatosensory cortex.

basal metabolism The amount of energy required to fuel the brain and other organs and to maintain body temperature.

basilar membrane The membrane in the cochlea that separates the cochlear canal from the tympanic canal, and on which the organ of Corti is located.

BDNF See *brain-derived neurotrophic factor.*

benzodiazepine A class of drugs that produce anxiety reduction, sedation, and muscle relaxation by stimulating benzodiazepine receptors on the GABA$_A$ complex, facilitating GABA binding.

binaural Involving the use of both ears.

binding problem The question of how the brain combines all the information about an object into a unitary whole.

biopsychology The branch of psychology that studies the relationships between behavior and the body, particularly the brain.

bipolar disorder Depression and mania that occur together in alternation.

bisexual An individual who is not entirely heterosexual or homosexual.

blindsight The ability of cortically blind individuals to respond to visual stimuli they are unaware of seeing.

blood-brain barrier The brain's protection from toxic substances and neurotransmitters in the bloodstream; the small openings in the capillary walls prevent large molecules from passing through unless they are fat soluble or carried through by special transporters.

BMI See *body mass index.*

body integrity identity disorder The desire, in individuals with no apparent brain damage or mental illness, to have a healthy limb amputated.

body mass index (BMI) The person's weight in kilograms divided by the squared height in meters; an indication of the person's deviation from the ideal weight for the person's height.

brain-derived neurotrophic factor (BDNF) A protein that contributes to neuron growth and survival.

brain interpreter A hypothetical mechanism that integrates all the cognitive processes going on simultaneously in other modules of the brain.

Broca's aphasia Language impairment caused by damage to Broca's area and surrounding cortical and subcortical areas.

Broca's area The area anterior to the precentral gyrus (motor cortex) that sends output to the facial motor area to produce speech and also provides grammatical structure to language.

bulimia nervosa An eating disorder involving bingeing on food, followed by purging by vomiting or using laxatives.

caffeine A drug that produces arousal, increased alertness, and decreased sleepiness; the active ingredient in coffee.

CAH See *congenital adrenal hyperplasia.*

cannabinoids A group of compounds that includes the active ingredient in marijuana (tetrahydrocannabinol) and the endogenous cannabinoid receptor ligands, anandamide and 2-arachidonyl glycerol (2-Ara-Gl). Cannabinoids act as retrograde messengers.

cardiac muscles The muscles that make up the heart.

castration Removal of the gonads (testes or ovaries).

cataplexy A disorder in which a person has a sudden experience of atonia similar to that seen in REM sleep; the person may fall to the floor but remains awake.

CCK See *cholecystokinin.*

cell body The largest part of a neuron, which contains the cell's nucleus; cytoplasm; and structures that produce proteins, convert nutrients into energy, and eliminate waste materials.

central nervous system (CNS) The part of the nervous system made up of the brain and spinal cord.

central pattern generator A neuronal network that produces a rhythmic pattern of motor activity, such as that involved in walking, swimming, flying, or breathing.

central sulcus The groove between the precentral gyrus and the postcentral gyrus that separates the frontal lobe from the parietal lobe in each hemisphere.

cerebellum Structure in the hindbrain that contributes the order of muscular contractions and their precise timing to intended movements and helps maintain posture and balance. It is also necessary for learning motor skills and contributes to nonmotor learning and cognitive activities.

cerebral hemispheres The large, wrinkled structures that are the dorsal or superior part of the brain and that are covered by the cortex.

cerebrospinal fluid Fluid in the ventricles and spinal canal that carries material from the blood vessels to the central nervous system and transports waste materials in the other direction. It also helps cushion the brain and spinal cord.

cholecystokinin (CCK) A peptide hormone released as food passes into the duodenum. CCK acts as a signal to the brain that reduces meal size.

chronic Referring to symptoms that develop gradually and persist for a long time with poor response to treatment.

chronic pain Pain that persists after healing has occurred, or beyond the time in which healing would be expected to occur.

circadian rhythm A rhythm that is a day in length, such as the wake-sleep cycle.

circuit formation The third stage of nervous system development, in which the developing neurons send processes to their target cells and form functional connections.

circuit pruning The fourth stage of nervous system development, in which neurons that are unsuccessful in finding a place on the appropriate target cell or that arrive late, die and excess synapses are eliminated.

CNS See *central nervous system.*

cocaine A drug extracted from the South American coca plant, which produces euphoria, decreased appetite, increased alertness, and relief from fatigue.

cochlea The snail-shaped structure where the ear's sound-analyzing structures are located.

cochlear canal The middle canal in the cochlea; contains the organ of Corti.

cocktail party effect The ability to sort out meaningful auditory messages from a complex background of sounds.

cognitive theory Theory that states that a person relies on a cognitive assessment of the stimulus situation to identify which emotion is being experienced; physiological arousal determines the intensity of the emotional experience.

coincidence detectors Neurons that fire most when they receive input from both ears at the same time; involved in sound localization.

color agnosia Loss of the ability to perceive colors due to brain damage.

color constancy The ability to recognize the natural color of an object in spite of the illuminating wavelength.

compensation A response to nervous system injury, in which surviving presynaptic neurons sprout new terminals, postsynaptic neurons add more receptors, or surrounding tissue takes over functions.

complementary colors Colors that cancel each other out to produce a neutral gray or white.

complex cell A type of cell in the visual cortex that continues to respond (unlike simple cells) when a line or edge moves to a different location.

complex sound A sound composed of more than one pure tone.

computed tomography (CT) An imaging technique that produces a series of X-rays taken from different angles; these are combined by a computer into a three-dimensional image of the brain or other part of the body.

concordance rate The frequency that relatives are alike in a characteristic.

confabulation Fabrication of stories and facts, which are then accepted by the individual, to make up for those missing from memory.

congenital adrenal hyperplasia (CAH) A form of female pseudohermaphroditism characterized by XX chromosomes, female internal sex organs, and ambiguous external sex organs. It is caused by excess production of androgens during prenatal development.

congenital insensitivity to pain A condition present at birth in which the person is insensitive to pain.

consolidation Process in which the brain forms a permanent representation of a memory.

Coolidge effect An increase in sexual activity when the variety of sexual partners increases.

corpus callosum The largest of the groups of neurons connecting the two cerebral hemispheres.

correlation The degree to which two variables are related, such as the IQs of siblings; it is measured by the correlation coefficient, a statistic that varies between the values of 0.0 and ±1.0.

correlational study A study in which the researcher does not control an independent variable, but determines whether two variables are related to each other.

cortex The grayish 1.5- to 4-mm-thick surface of the hemispheres, composed mostly of cell bodies, where the highest-level processing occurs in the brain.

cortisol A hormone released by the adrenal glands that increases energy levels by converting proteins to glucose, increasing fat availability, and increasing metabolism. The increase is more sustained than that produced by epinephrine and norepinephrine.

cranial nerves The 12 pairs of nerves that enter and leave the underside of the brain; part of the peripheral nervous system.

CT see computed tomography.

Dale's principle The idea that a neuron is able to release only one neurotransmitter.

db See diabetes gene.

deception In research, failing to tell the participants the exact purpose of the research or what will happen during the study, or actively misinforming them.

declarative memory The memory process that records memories of facts, people, and event that the person can verbalize, or declare.

default mode network Portions of the frontal, parietal, and temporal lobes that are active when the brain is at rest or focused internally; its activity is thought to represent preparedness for action.

defensive aggression Aggression that occurs in response to threat and is motivated by fear.

delirium tremens A reaction in some cases of withdrawal from alcohol, including hallucinations, delusions, confusion, and, in extreme cases, seizures and possible death.

dendrites Extensions that branch out from the neuron cell body and receive information from other neurons.

dendritic spines Outgrowths from the dendrites that partially bridge the synaptic cleft and make the synapse more sensitive.

deoxyribonucleic acid (DNA) A double-stranded chain of chemical molecules that looks like a ladder that has been twisted around itself; genes are composed of DNA.

depressant A drug that reduces central nervous system activity.

depression An intense feeling of sadness.

dermatome A segment of the body served by a spinal nerve.

diabetes An insulin disorder in which the person produces too little insulin (Type 1), resulting in overeating with little weight gain, or the person's brain is insensitive to insulin (Type 2), resulting in overeating with weight gain.

diabetes gene (db) A gene on chromosome 4 that produces diabetes and obesity; mice with the gene are insensitive to leptin.

difference in intensity A binaural cue to the location of a sound coming from one side which results from the sound shadow created by the head; most effective above 2000 to 3000 Hz.

difference in time of arrival A binaural cue to the location of a sound coming from one side due to the time the sound requires to travel the distance between the ears.

dihydrotestosterone A derivative of testosterone that masculinizes the genitals of males.

dissociative identity disorder (DID) The disorder previously known as multiple personality, which involves shifts in consciousness and behavior that appear to be distinct personalities or selves.

distributed The term for any brain function that occurs across a relatively wide area of the brain.

DNA see deoxyribonucleic acid.

dominant The term referring to a gene that will produce its effect regardless of which gene it is paired with in the fertilized egg.

dopamine hypothesis The hypothesis that schizophrenia involves excess dopamine activity in the brain.

dorsal Toward the back side of the body.

dorsal root The branch of a spinal nerve through which neurons enter the spinal cord.

dorsal stream The visual processing pathway that extends into the parietal lobes; it is especially concerned with the location of objects in space.

Down syndrome Intellectual disability characterized by IQs in the 40 to 55 range, usually caused by the presence of an extra 21st chromosome.

drive An aroused condition resulting from a departure from homeostasis, which impels the individual to take appropriate action, such as eating.

drive theory Theory based on the assumption that the body maintains a condition of homeostasis.

drug Any substance that on entering the body changes the body or its functioning.

dualism The idea that the mind and the brain are separate.

duodenum The initial 25 cm of the small intestine, where most digestion occurs.

dyslexia An impairment of reading, which can be developmental or acquired through brain damage.

early-onset alcoholism Cloninger's Type 2 alcoholism, which involves early onset, frequent drinking without guilt, and characteristics of antisocial personality behavior.

ECT See electroconvulsive therapy.

EEG See electroencephalogram.

electrical stimulation of the brain (ESB) A procedure in which animals (or humans) learn to press a lever or perform some other action to deliver mild electrical stimulation to brain areas where the stimulation is rewarding.

electroconvulsive therapy (ECT) The application of 70 to 130 volts of electricity to the head of a lightly anesthetized patient, which produces a seizure and convulsions; a treatment for major depression.

electroencephalogram (EEG) A measure of brain activity recorded from two electrodes on the scalp over the area of interest, which are connected to an electronic amplifier; it detects the combined electrical activity of all the neurons between the two electrodes.

electron microscope A magnification system that passes a beam of electrons through a thin slice of tissue onto a photographic film, forming an image magnified up to 250,000 times. The scanning electron microscope has less magnification but produces three-dimensional images.

electrostatic pressure The force by which like-charged ions are repelled by each other and opposite-charged ions are attracted to each other.

embryo An organism in the early prenatal period; in humans, during the first 8 weeks.

emotion A state of feelings accompanied by an increase or decrease in physiological activity and, possibly, characteristic facial expression and behavior.

empiricism The procedure of obtaining information through observation.

endogenous Generated within the body; usually used to refer to natural ligands for neurotransmitter receptors.

endorphins Substances produced in the body that function both as neurotransmitters and as hormones and act on opioid receptors in many parts of the nervous system.

epigenetic Referring to inheritable characteristics that are unrelated to the DNA sequence.

EPSP See excitatory postsynaptic potential.

equipotentiality The idea that the brain functions as a whole; opposite of localization.

ESB See electrical stimulation of the brain.

estrogen A class of hormones responsible for a number of female characteristics and functions, produced by the ovaries in women and, to a lesser extent, by the adrenal glands in males and females.

estrus A period when a nonhuman female animal is ovulating and sex hormone levels are high.

euphoria A sense of happiness or ecstasy; many abused drugs produce euphoria.

evoked potential An EEG technique for measuring the brain's responses to brief stimulation; it involves presenting a stimulus repeatedly and averaging the EEG over all the presentations to cancel out random activity, leaving the electrical activity associated with the stimulus.

excitatory postsynaptic potential (EPSP) A hypopolarization of the dendrites and cell body, which makes the neuron more likely to fire.

experiment A study in which the researcher manipulates an independent variable and observes its effect on one or more dependent variables.

fabrication In research, the faking of data or results.

familial Term referring to a characteristic that occurs more frequently among relatives of a person with the characteristic than it does in the population.

family study A study of how strongly a characteristic is shared among relatives.

FAS See *fetal alcohol syndrome*.

fasting phase The period following the absorptive phase, when the glucose level in the blood drops and the body must rely on its energy stores.

fatty acids Breakdown product of fat, which supplies the muscles and organs of the body (except for the brain).

fetal alcohol syndrome (FAS) A condition caused by the mother's use of alcohol during the third trimester of pregnancy; neurons fail to migrate properly, often resulting in intellectual disability. The leading cause of intellectual disability in the Western world.

fetus An organism after the initial prenatal period; in humans, after the first 8 weeks.

FFA See *fusiform face area*.

fissure A groove between gyri of the cerebral hemispheres that is larger and deeper than a sulcus.

fMRI See *functional magnetic resonance imaging*.

force of diffusion The force that moves ions from an area of greater concentration to an area where they are less concentrated.

form vision The detection of an object's boundaries and features, such as texture.

fovea A 1.5-mm-wide area in the middle of the retina in which cones are most concentrated and visual acuity and color discrimination are greatest.

fragile X syndrome A form of intellectual disability caused by excessive CGG repeats in the *FMR1* gene; IQ is typically below 75.

frequency A characteristic of sound; the number of cycles or waves of alternating compression and decompression of the vibrating medium that occur in a second.

frequency theory Any one of a number of theories of auditory frequency analysis that state that the frequency of a sound is represented in the firing rate of each neuron or a group of neurons.

frequency-place theory The theory that frequency following accounts for the discrimination of frequencies up to about 200 Hz and higher frequencies are

represented by the place of greatest activity on the basilar membrane.

frontal lobe The area of each cerebral hemisphere anterior to the central sulcus and superior to the lateral fissure.

functional magnetic resonance imaging (fMRI) A brain-imaging procedure that measures brain activation by detecting the increase in oxygen levels in active neural structures.

fusiform face area (FFA) A part of the inferior temporal lobe important in face identification. See *prosopagnosia*.

ganglion A group of cell bodies in the peripheral nervous system.

gate control theory The idea that pressure signals arriving in the brain trigger an inhibitory message that travels back down the spinal cord, where it closes a neural "gate" in the pain pathway.

gay The term used to refer to homosexual men.

gender The behavioral characteristics associated with being male or female.

gender identity The sex a person identifies as being.

gender nonconformity A tendency to engage in activities usually preferred by the other sex and an atypical preference for other-sex playmates and companions while growing up.

gender role A set of behaviors society considers appropriate for members of the same sex.

gene The biological unit that directs cellular processes and transmits inherited characteristics.

gene therapy Treatment of a disorder by gene manipulation.

gene transfer Insertion of a gene from another organism into a recipient's cells, usually within a virus.

genetic engineering Manipulation of an organism's genes or their functioning.

genome The entire collection of genes in a species' chromosomes.

genotype The combination of genes an individual has.

ghrelin A peptide released by the stomach during fasting, which initiates eating.

glial cell Nonneural cell that provides a number of supporting functions to neurons, including myelination.

glucagon A hormone released by the pancreas that stimulates the liver to transform stored glycogen back into glucose during the fasting phase.

glucose One of the sugars; the body's main source of energy, reserved for the nervous system during the fasting phase; a major signal for hunger and satiation.

glutamate theory The theory that reduced glutamate activity is involved in schizophrenia.

glycerol A breakdown product of fats, which is converted to glucose for the brain during the fasting period.

glycogen Form in which glucose is stored in the liver and muscles during the absorptive phase; converted back to glucose for the brain during the fasting phase.

Golgi stain A staining method that randomly stains about 5% of neurons, which makes them stand out individually.

Golgi tendon organs Receptors that detect tension in a muscle.

gonads The primary reproductive organs, the testes in the male or the ovaries in the female.

graded potential A voltage change in a neuron that varies with the strength of the stimulus that initiated it.

growth cone A formation at the tip of a migrating neuron that samples the environment for directional cues.

gyrus A ridge in the cerebral cortex; the area between two sulci.

Hebb rule Principle stating that if an axon of a presynaptic neuron is active while the postsynaptic neuron is firing, the synapse between them will be strengthened.

heritability The percentage of the variation among individuals in a characteristic that can be attributed to heredity.

heroin A major drug of addiction synthesized from morphine.

heterozygous Having a pair of genes for a specific characteristic that are different from each other.

hierarchical processing A type of processing in which lower levels of the nervous system analyze their information and pass the results on to the next higher level for further analysis.

homeostasis Condition in which any particular body system is in balance or equilibrium.

homozygous Having a pair of genes for a specific characteristic that are identical with each other.

Human Genome Project An international project with the goal of mapping the location of all the genes on the human chromosomes and determining the base sequences of the genes.

Huntington's disease A degenerative disorder of the motor system involving cell loss in the striatum and cortex.

hydrocephalus A disorder in which cerebrospinal fluid fails to circulate and builds up in the cerebral ventricles, crowding out neural tissue and usually causing intellectual disability.

hyperpolarization An increase in the polarization of a neuron membrane, which is inhibitory and makes an action potential less likely to occur.

hypnotic Sleep inducing.

hypocretin See *orexin*.

hypopolarization A decrease in the polarization of a neuron membrane, which is excitatory and makes an action potential more likely to occur.

hypothalamus A subcortical structure in the forebrain just below the thalamus that plays a major role in controlling emotion and motivated behaviors such as eating, drinking, and sexual activity.

hypothalamus-pituitary-adrenal axis A group of structures that helps the body cope with stress.

hypovolemic thirst A fluid deficit that occurs when the blood volume drops due to a loss of extracellular water.

immune system The cells and cell products that kill infected and malignant cells and protect the body against foreign substances, including bacteria and viruses.

immunocytochemistry A procedure for labeling cellular components such as receptors, neurotransmitters, or enzymes, using a dye attached to an antibody designed to attach the component.

INAH 3 See *third interstitial nucleus of the anterior hypothalamus*.

incentive theory A theory that recognizes that people are motivated by external stimuli (incentives), not just internal needs.

inferior Below another structure.

inferior colliculi Part of the tectum in the brain stem that is involved in auditory functions such as locating the direction of sounds.

inferior temporal cortex Area in the lower part of the temporal lobe that plays a major role in the visual identification of objects.

informed consent Voluntary agreement to participate in a study after receiving full information about any risks, discomfort, or other adverse effects that might occur.

inhibitory postsynaptic potential (IPSP) A hyperpolarization of the dendrites and cell body, which makes a neuron less likely to fire.

inner hair cells A single row of about 3,500 hair cells located on the basilar membrane toward the inside of the cochlea's coil; they produce most, if not all, of the auditory signal.

in situ hybridization A procedure for locating gene activity, which involves constructing strands of complementary DNA that will dock with strands of messenger RNA. The complementary DNA is radioactive, so autoradiography can be used to locate the gene activity.

insomnia The inability to sleep or to obtain quality sleep, to the extent that the person feels inadequately rested.

instinct A complex behavior that is automatic and unlearned and occurs in all the members of a species.

insulin A hormone secreted by the pancreas that enables entry of glucose into cells (not including the nervous system) during the absorptive phase and facilitates storage of excess nutrients.

intellectual disability Limitation in intellectual functioning (reasoning, learning, problem solving) and in adaptive behavior originating before the age of 18.

intelligence The capacity for learning, reasoning, and understanding.

intelligence quotient (IQ) The measure typically used for intelligence.

intensity The physical energy in a sound; the sound's amplitude.

interneuron A neuron that has a short axon or no axon at all and connects one neuron to another in the same part of the central nervous system.

iodopsin The photopigment in cones.

ion An atom that is charged because it has lost or gained one or more electrons.

ionotropic receptor A receptor on a neuron membrane that opens ion channels directly and immediately to produce quick reactions.

IPSP See *inhibitory postsynaptic potential.*

IQ See *intelligence quotient.*

James-Lange theory The idea that physiological arousal precedes and is the cause of an emotional experience and the pattern of arousal identifies the emotion.

knockout Genetic engineering technique in which a nonfunctioning gene mutation is inserted during the embryonic stage.

Korsakoff's syndrome A form of dementia in which brain deterioration is almost always caused by chronic alcoholism.

language acquisition device A part of the brain hypothesized to be dedicated to learning and controlling language.

late-onset alcoholism Cloninger's Type 1 alcoholism, involving late onset, long periods of abstinence with binges, guilt over drinking, and cautious and emotionally dependent personality.

lateral Toward the side.

lateral fissure The fissure that separates the temporal lobe from the frontal and parietal lobes.

lateral hypothalamus A nucleus of the hypothalamus with roles in feeding and metabolism, aggression, and waking arousal.

lateral inhibition A method of enhancing neural information in which each neuron's activity inhibits the activity of its neighbors, and in turn its activity is inhibited by them.

L-dopa See *levodopa.*

learned taste aversion Learned avoidance of a food (based on its taste) eaten prior to becoming ill.

learned taste preference Preference for a food containing a needed nutrient (identified by the food's taste), learned, presumably, because the nutrient makes the individual feel better.

leptin Hormone secreted by fat cells, which is proportional to the percentage of body fat and which signals fat level to the brain.

lesbian The term for a homosexual woman.

lesion Damage to neural tissue. This can be brought about surgically for research or therapeutic reasons, or it can result from trauma, disease, or developmental error.

leukocytes White blood cells, which include macrophages, T cells, and B cells; part of the immune system.

levodopa (*L*-dopa) The precursor for dopamine; used to treat Parkinson's disease.

Lewy bodies Abnormal clumps of protein that form within neurons, found in some patients with Parkinson's disease and in patients with a form of Alzheimer's disease.

limbic system A group of forebrain structures arranged around the upper brain stem that have roles in emotion, motivated behavior, and learning.

lithium A metal administered in the form of lithium carbonate; the medication of choice for bipolar illness.

lobotomy A surgical procedure that disconnects the prefrontal areas from the rest of the brain; it reduces emotionality and pain, but leaves the person emotionally blunted, distractible, and childlike in behavior.

localization The idea that specific parts of the brain carry out specific functions.

longitudinal fissure The large fissure that runs the length of the brain, separating the two cerebral hemispheres.

long-term depression (LTD) Weakening of a synapse when stimulation of presynaptic neurons is insufficient to activate the postsynaptic neurons.

long-term potentiation (LTP) An increase in synaptic strength that occurs when presynaptic neurons and postsynaptic neurons are active simultaneously.

loudness The term for our *experience* of sound intensity.

LTD See *long-term depression.*

LTP See *long-term potentiation.*

macrophage A type of leukocyte that ingests intruders.

magnetic resonance imaging (MRI) An imaging technique that involves measuring the radiofrequency waves emitted by hydrogen atoms when they are subjected to a strong magnetic field. Because different structures have different concentrations of hydrogen atoms, the waves can be used to form a detailed image of the brain.

magnocellular hypothesis The hypothesis that some symptoms of dyslexia result from developmental errors in the magnocellular visual pathway, which is concerned with rapidly moving stimuli and detects eye movements.

magnocellular system A division of the visual system, extending from the retina through the visual association areas, that is specialized for brightness contrast and movement.

major depression A disorder involving feelings of sadness to the point of hopelessness for weeks at a time, along with slowness of thought, sleep disturbance, and loss of energy and appetite and the ability to enjoy life; in some cases the person is also agitated or restless.

major histocompatibility complex (MHC) A group of genes that contribute to the functioning of the immune system.

mania A disorder involving excess energy and confidence that often leads to grandiose schemes, decreased need for sleep, increased sexual drive, and, often, abuse of drugs.

marijuana The dried and crushed leaves and flowers of the Indian hemp plant *Cannabis sativa.*

materialistic monism The view that the body and the mind and everything else are physical.

medial Toward the middle.

medial amygdala Part of the amygdala that apparently responds to sexually exciting stimuli. In both male and female rats, it is active during copulation, and it causes the release of dopamine in the MPOA.

medial forebrain bundle A part of the mesolimbocortical dopamine system and a potent reward area.

medial preoptic area (MPOA) A part of the preoptic area of the hypothalamus that appears to be important for sexual performance, but not sexual motivation, in male and female rats.

median preoptic nucleus A nucleus of the hypothalamus that initiates drinking in response to osmotic and volumetric deficits.

medulla The lower part of the hindbrain; its nuclei are involved with control of essential life processes such as cardiovascular activity and respiration.

melatonin A hormone secreted by the pineal gland that induces sleepiness.

meninges A three-layered membrane that encloses and protects the brain.

mesolimbocortical dopamine system A pathway including the ventral tegmental area, medial forebrain bundle, nucleus accumbens, and projections into prefrontal areas. The pathway is important in reward effects from drugs, ESB, and activities such as eating and sex.

messenger ribonucleic acid (RNA) A copy of one strand of DNA that moves out of the nucleus to direct protein construction.

metabotropic receptor A receptor on a neuron membrane that opens ion channels slowly through a metabolic process and produces long-lasting effects.

methadone A synthetic opiate used as an agonist treatment for opiate addiction.

midbrain The middle part of the brain, consisting of the tectum (*roof*) on the dorsal side and the tegmentum on the ventral side.

migrate In brain development, movement of newly formed neurons from the ventricular zone to their final destination.

mind-brain problem The issue of what the mind is and its relationship to the brain.

mirror neurons Neurons that respond when engaging in an act and while observing the same act in others.

model A proposed mechanism for how something works.

modular processing The segregation of the various components of processing in the brain into separate locations.

monism The idea that the mind and the body consist of the same substance.

monoamine hypothesis The hypothesis that depression involves reduced activity at norepinephrine and serotonin synapses.

motivation The set of factors that initiate, sustain, and direct behavior.

motor cortex The area in the frontal lobes that controls voluntary (nonreflexive) body movements; the primary motor cortex is on the precentral gyrus.

motor neuron A neuron that carries commands to the muscles and organs.

movement agnosia Impaired ability to perceive movement.

MPOA See *medial preoptic area*.

MRI See *magnetic resonance imaging*.

Müllerian ducts Early structures that in the female develop into the uterus, fallopian tubes, and inner vagina.

Müllerian inhibiting hormone A hormone released in the male that causes the Müllerian ducts to degenerate.

multiple sclerosis A motor disorder caused by the deterioration of myelin (*demyelination*) and neuron loss in the central nervous system.

muscle spindles Receptors that detect stretching in muscles.

myasthenia gravis A disorder of muscular weakness caused by reduced numbers or sensitivity of acetylcholine receptors.

myelin A fatty tissue that wraps around an axon to insulate it from the surrounding fluid and from other neurons.

myelin stain A staining method that stains myelin, thus identifying neural pathways.

narcolepsy A disorder in which individuals fall asleep suddenly during the daytime and go directly into REM sleep.

natural killer cell A type of immune cell that attacks and destroys certain kinds of cancer cells and cells infected with viruses.

natural selection The principle that those whose genes endow them with greater speed, intelligence, or health are more likely to survive and transmit their genes to more offspring.

nature versus nurture The issue of the relative importance of heredity and environment.

negative color aftereffect The experience of a color's complement following stimulation by the color.

negative symptoms Symptoms of schizophrenia characterized by the absence or insufficiency of normal behaviors, including lack of affect (emotion), inability to experience pleasure, lack of motivation, poverty of speech, and impaired attention.

neglect A disorder in which the person ignores objects, people, and activity on the side opposite the brain damage.

nerve A bundle of axons running together in the peripheral nervous system.

neural network A group of neurons that function together to carry out a process.

neurofibrillary tangles Abnormal accumulations of the protein tau that develop inside neurons and are associated with the death of brain cells in people with Alzheimer's disease and Down syndrome.

neurogenesis The birth of new neurons.

neuron A specialized cell that conveys sensory information into the brain, carries out the operations involved in thought and feeling and action, or transmits commands out into the body to control muscles and organs; a single neural cell, in contrast to a *nerve*.

neuropeptide Y (NPY) A transmitter released by NPY/AgRP neurons in the arcuate nucleus of the hypothalamus when nutrient levels diminish; it is a powerful stimulant for eating and conserves energy.

neuroscience The multidisciplinary study of the nervous system and its role in behavior.

neurotoxin A neuron poison; substance that impairs the functioning of a neuron.

neurotransmitter A chemical substance that a neuron releases to carry a message across the synapse to the next neuron or to a muscle or organ.

neurotrophins Chemicals that enhance development and survival in neurons.

nicotine The primary psychoactive and addictive ingredient in tobacco.

Nissl stain A staining method that stains cell bodies.

node of Ranvier A gap in the myelin sheath covering an axon.

nondeclarative memory Nonstatable memories that result from procedural or skills learning, emotional learning, and simple conditioning.

non-REM sleep The periods of sleep that are not rapid eye movement sleep.

NPY See *neuropeptide Y*.

NST See *nucleus of the solitary tract*.

nucleus (1) The part of every cell that contains the chromosomes and governs activity in the cell. (2) A group of neuron cell bodies in the central nervous system.

nucleus accumbens A forebrain structure that is part of the mesolimbocortical dopamine system and a potent center for reward.

nucleus of the solitary tract (NST) A part of the medulla that monitors several signals involved in the regulation of eating.

ob See *obesity gene*.

obesity gene (*ob*) A gene on chromosome 6 that causes obesity; in mice it results in an inability to produce leptin.

object agnosia Impairment of the ability to recognize objects visually.

obsessive-compulsive disorder (OCD) A disorder consisting of obsessions (recurring thoughts) and compulsions (repetitive, ritualistic acts the person feels compelled to perform).

occipital lobe The most posterior part of each cerebral hemisphere, and the location of the visual cortex.

OCD See *obsessive-compulsive disorder*.

offensive aggression An unprovoked attack on another person or animal.

oligodendrocyte A type of glial cell that forms the myelin covering of neurons in the brain and spinal cord.

opiate Any drug derived from the opium poppy. The term is also used to refer to effects at opiate receptors, including those by endorphins.

opponent process theory A color vision theory that attempts to explain color vision in terms of opposing neural processes.

orexin A peptide released by lateral hypothalamic neurons that activates arousal centers involved in waking; also known as hypocretin.

organ of Corti The sound-analyzing structure on the basilar membrane of the cochlea; it consists of four rows of hair cells, their supporting cells, and the tectorial membrane.

organizing effects Hormonal effects of sexual development that occur during the prenatal period and shortly after birth and are permanent.

organum vasculosum lamina terminalis (OVLT) A structure bordering the third ventricle that monitors fluid content in the cells and contributes to the control of osmotic thirst.

osmotic thirst Thirst that occurs when the fluid content is low inside the body's cells.

ossicles Tiny bones in the middle ear that operate in lever fashion to transfer vibration from the tympanic membrane to the cochlea; they also produce a slight amplification of the sound.

outer hair cells Three rows of about 12,000 cells located on the basilar membrane toward the outside of the cochlea's coil; their function may be to influence inner hair cell sensitivity by adjusting the tension of the tectorial membrane.

out-of-body experience Phenomenon, usually resulting from brain damage or epilepsy, in which the person hallucinates seeing his or her detached body from another location.

ovaries The female gonads, where the ova develop.

OVLT See *organum vasculosum lamina terminalis*.

oxytocin A neuropeptide hormone and neurotransmitter involved in lactation and orgasm; dubbed the "sociability molecule" because it affects social behavior and bonding.

parasympathetic nervous system The branch of the autonomic nervous system that slows the activity of most organs to conserve energy and activates digestion to renew energy.

paraventricular nucleus (PVN) A structure in the hypothalamus that monitors several signals involved in the regulation of eating, including input from the NST.

parietal lobe The part of each cerebral hemisphere located above the lateral fissure and between the central sulcus and the occipital lobe; it contains the somatosensory cortex and visual association areas.

Parkinson's disease A movement disorder characterized by motor tremors, rigidity, loss of balance and coordination, and difficulty in moving, especially in initiating movements; it is caused by deterioration of the substantia nigra.

parvocellular system A division of the visual system, extending from the retina through the visual association areas, that is specialized for fine detail and color.

peptide YY$_{3-36}$ An appetite-suppressing peptide hormone released in the intestines in response to food.

perception The interpretation of sensory information.

periaqueductal gray (PAG) A brain-stem structure with a large number of endorphin synapses; stimulation reduces pain transmission at the spinal cord level. The PAG also produces symptoms of drug withdrawal.

peripheral nervous system (PNS) The part of the nervous system made up of the cranial nerves and spinal nerves.

PET See positron emission tomography.

PGO waves Waves of excitation that flow from the pons through the lateral geniculate nucleus of the thalamus to the occipital area and appear to initiate the EEG desynchrony of REM sleep.

phantom pain Pain that seems to be in a missing limb.

phase difference A binaural cue to the location of a sound coming from one side; at frequencies below 1500 Hz, the sound will be in a different phase of the wave at each ear.

phenotype In heredity, the characteristic of the individual.

phenylketonuria An inherited form of intellectual disability in which the body fails to metabolize the amino acid phenylalanine, which interferes with myelination during development.

pheromones Airborne chemicals released by an animal that have physiological or behavioral effects on another animal of the same species.

phonological hypothesis Hypothesis that the fundamental problem in dyslexia is impaired phoneme processing.

photopigment A light-sensitive chemical in the visual receptors that initiates the neural response.

phototherapy A treatment for winter depression involving the use of high-intensity lights for a period of time each day.

phrenology The theory in the early 1900s that "faculties" of emotion and intellect were located in precise areas of the brain and could be assessed by feeling bumps on the skull.

pineal gland A gland located just posterior to the thalamus which secretes sleep-inducing melatonin; it controls seasonal cycles in nonhuman animals and participates with other structures in controlling daily rhythms in humans.

pinna The ear flap on each side of the head; the outer ear.

pitch The experience of the frequency of a sound.

place cells Cells in the hippocampus that increase their firing rate when the individual is in a specific location in the environment.

place theory Theory that states that the frequency of a sound is identified by the location of maximal vibration on the basilar membrane and which neurons are firing most.

plagiarism The theft of another's work or ideas.

planum temporale The area in each temporal lobe that is the location in the left hemisphere of Wernicke's area and that is larger on the left in most people.

plaques Clumps of amyloid, a type of protein, that cluster among axon terminals and interfere with neural transmission in the brains of people with Alzheimer's disease and Down syndrome.

plasticity The ability to be modified; a characteristic of the nervous system.

PNS See peripheral nervous system.

polarization A difference in electrical charge between the inside and outside of a neuron.

polygenic Determined by several genes rather than a single gene.

pons A part of the brain stem that contains centers related to sleep and arousal.

positive symptoms Symptoms of schizophrenia that involve the presence or exaggeration of behaviors, such as delusions, hallucinations, thought disorder, and bizarre behavior.

positron emission tomography (PET) An imaging technique that reveals function. It involves injecting a radioactive substance into the bloodstream, which is taken up by parts of the brain according to how active they are; the scanner makes an image that is color coded to show the relative amounts of activity.

posterior Toward the rear.

posterior parietal cortex An association area that brings together the body senses, vision, and audition. It determines the body's orientation in space, the location of the limbs, and the location in space of objects detected by touch, sight, and sound.

postsynaptic Term referring to a neuron that receives transmission from another neuron.

posttraumatic stress disorder (PTSD) A prolonged stress reaction to a traumatic event; it is typically characterized by recurrent thoughts and images (flashbacks), nightmares, lack of concentration, and overreactivity to environmental stimuli, such as loud noises.

PYY See peptide YY$_{3-36}$

precentral gyrus The gyrus anterior to, and extending the length of, the central sulcus; it is the location of the primary motor cortex.

predatory aggression Aggression in which an animal attacks and kills its prey.

prefrontal cortex The most anterior cortex of the frontal lobes; it is involved in working memory, planning and organization of behavior, and regulation of behavior in response to its consequences. It also integrates information about the body with sensory information from the world to select and plan movements.

premotor cortex Area anterior to the primary motor cortex that combines information from the prefrontal cortex and the posterior parietal cortex and begins the programming of a movement.

preoptic area Structure in the hypothalamus that contains warmth-sensitive cells and cold-sensitive cells and participates in the control of body temperature. See medial preoptic area regarding regulation of sexual behavior.

presynaptic Term referring to a neuron that transmits to another neuron.

presynaptic excitation Increased release of neurotransmitter from a neuron's terminal as the result of another neuron's release of neurotransmitter onto the terminal (an axoaxonic synapse).

presynaptic inhibition Decreased release of neurotransmitter from a neuron's terminal as the result of another neuron's release of neurotransmitter onto the terminal (an axoaxonic synapse).

primary motor cortex The area on the precentral gyrus responsible for the execution of voluntary movements.

primary somatosensory cortex The first stage in the cortical-level processing of somatosensory information, which is processed through the four subareas of the primary somatosensory cortex, then passed on to the secondary somatosensory cortex.

proactive aggression Human aggression that is premeditated, unprovoked, and relatively emotionless.

proliferation The first stage of nervous system development in which cells that will become neurons multiply at the rate of 250,000 new cells every minute.

proprioception The sense that informs us about the position and movement of the parts of the body.

prosody The use of intonation, emphasis, and rhythm to convey meaning in speech.

prosopagnosia The inability to visually recognize familiar faces.

pseudohermaphrodite Individuals who have ambiguous internal and external sexual organs, but whose gonads are consistent with their chromosomes.

psychedelic drug Any compound that causes perceptual distortions in the user.

psychoactive drug Any drug that has psychological effects, such as anxiety relief or hallucinations.

psychosis A severe mental disturbance of reality, thought, and orientation.

psychosurgery The use of surgical intervention to treat cognitive and emotional disorders.

PTSD See posttraumatic stress disorder.

pure tone A sound consisting of a single frequency.

PVN See paraventricular nucleus.

radial glial cells Specialized glial cells that provide a scaffold for migrating neurons to climb to their destination.

rapid eye movement (REM) sleep The stage of sleep during which most dreaming occurs; research indicates that it is also a time of memory consolidation during which neural activity from the day is replayed.

rate law Principle that intensity of a stimulus is represented in an axon by the frequency of action potentials.

reactive aggression Human aggression that is impulsive, provoked, and emotional.

receptive field In vision, the area of the retina from which a cell in the visual system receives its input.

receptor A cell, often a specialized neuron, that is suited by its structure and function to respond to a particular form of energy, such as sound.

recessive The term referring to a gene that will have an influence only when it is paired with the same recessive gene on the other chromosome.

reflex A simple, automatic movement in response to a sensory stimulus.

regeneration The growth of severed axons; in mammals, it is limited to the peripheral nervous system.

reinforcer Any object or event that increases the probability of the response that precedes it.

relative refractory period Period during which a neuron can be fired again following an action potential, but only by an above-threshold stimulus.

REM See *rapid eye movement sleep*.

REM sleep behavior disorder A sleep disorder in which the person is physically active during REM sleep.

reorganization A shift in neural connections that changes the function of an area of the brain.

resting potential The difference in charge between the inside and outside of the membrane of a neuron at rest.

reticular formation A collection of over 90 nuclei running through the middle of the hindbrain and the midbrain with roles in sleep and arousal, attention, reflexes, and muscle tone.

retina The structure at the rear of the eye, which is made up of light-sensitive receptor cells and the neural cells that are connected to them.

retinal disparity A discrepancy in the location of an object's image on the two retinas; a cue to the distance of a focused object.

retinotopic map A map of the retina in the visual cortex, which results from adjacent receptors in the retina activating adjacent cells in the visual cortex.

retrieval The process of accessing stored memories.

retrograde amnesia The inability to remember events prior to impairment.

reuptake Process by which a neurotransmitter is taken back into the presynaptic terminals by transporters.

reward The positive effect on a user from a drug, electrical stimulation of the brain (ESB), sex, food, warmth, and so on.

rhodopsin The photopigment in rods.

SAD See *seasonal affective disorder*.

saltatory conduction Conduction in the axon in which action potentials jump from one node of Ranvier to the next.

satiety Satisfaction of appetite.

schizophrenia A disabling disorder characterized by perceptual, emotional, and intellectual deficits, loss of contact with reality, and inability to function in life.

Schwann cell A type of glial cell that forms the myelin covering on neurons outside the brain and spinal cord.

SCN See *suprachiasmatic nucleus*.

SDN See *sexually dimorphic nucleus*.

seasonal affective disorder (SAD) Depression that is seasonal, being more pronounced in the summer in some people and in the winter in others.

secondary somatosensory cortex The part of the somatosensory cortex that receives information from primary somatosensory cortex, from both sides of the body.

sedative A calming effect of a drug.

sensation The acquisition of sensory information.

sensory neuron A neuron that carries information from the body and from the outside world into the central nervous system.

sensory-specific satiety Decreased attractiveness of a food as the person or animal eats more of it.

set point A value in a control system that is the system's point of equilibrium or homeostasis; departures from this value initiate actions to restore the set-point condition.

sex The term for the biological characteristics that divide humans and other animals into the categories of male and female.

sexually dimorphic nucleus (SDN) A part of the MPOA important to male sexual behavior. It is larger in male rats, and their level of sexual activity depends on SDN size.

simple cell A cell in the visual cortex that responds to a line or an edge that is at a *specific orientation* and at a *specific place* on the retina.

skeletal muscles The muscles that move the body and limbs.

skin conductance response (SCR) A measure of sweat gland activation and thus sympathetic nervous system activity.

skin senses Touch, warmth, cold, and pain; the senses that arise from receptors in the skin.

slow-wave sleep (SWS) Stages 3 and 4 of sleep, characterized by delta EEG and increased body activity; it appears to be a period of brain recuperation and may play a role in consolidation of declarative memory.

smooth muscles Muscles that control the internal organs other than the heart.

sodium-potassium pump Large protein molecules that move sodium ions through the neuron membrane to the outside and potassium ions back inside, helping to maintain the resting potential.

somatic nervous system The division of the peripheral nervous system that carries sensory information into the central nervous system (CNS) and motor commands from the CNS to the skeletal muscles.

somatosensory cortex The area in the parietal lobes that processes the skin senses and the senses that inform us about body position and movement, or proprioception; the primary somatosensory cortex is on the postcentral gyrus.

spatial frequency theory The theory that visual cortical cells do a Fourier frequency analysis of the luminosity variations in a scene.

spatial summation The process of combining potentials that occur simultaneously at different locations on the dendrites and cell body.

spinal cord A part of the central nervous system; the spinal nerves, which communicate with the body below the head, enter and leave the spinal cord.

spinal nerves The peripheral nerves that enter and leave the spinal cord at each vertebra and communicate with the body below the head.

stem cells Undifferentiated cells that can develop into specialized cells such as neurons, muscle, or blood.

stereotaxic instrument A device used for the precise positioning in the brain of an electrode or other device, such as a cannula.

stimulant A drug that activates the nervous system to produce arousal, increased alertness, and elevated mood.

stress A condition in the environment that makes unusual demands on the organism, such as threat, failure, or bereavement; the individual's negative response to a stressful situation.

striatum The caudate nucleus and putamen, both of the basal ganglia, and the nucleus accumbens.

stroke A medical condition caused by a loss of blood flow in the brain (also known as cerebrovascular accident).

subfornical organ (SFO) One of the structures bordering the third ventricle that increases drinking when stimulated by angiotensin II.

substance P A neuropeptide involved in pain signaling.

substantia nigra The nucleus that sends dopamine-releasing neurons to the striatum and that deteriorates in Parkinson's disease.

sudden cardiac death Death occurring when stress causes excessive sympathetic activity that sends the heart into fibrillation, contracting so rapidly that little or no blood is pumped.

sulcus The groove or space between two gyri.

superior Above another structure.

superior colliculi Part of the tectum in the brain stem that is involved in visual functions such as guiding eye movements and fixation of gaze.

supplementary motor area The prefrontal area that assembles sequences of movements, such as those involved in eating or playing the piano, prior to execution by the primary motor cortex.

suprachiasmatic nucleus (SCN) A structure in the hypothalamus that (1) was found to be larger in gay men than in heterosexual men, (2) regulates the reproductive cycle in female rats, and (3) is the main biological clock, controlling several activities of the circadian rhythm.

SWS See *slow-wave sleep*.

sympathetic ganglion chain The structure running along each side of the spine through which most sympathetic neurons pass (and many synapse) on their way to and from the body's organs.

sympathetic nervous system The branch of the autonomic nervous system that activates the body in ways that help it cope with demands, such as emotional stress and physical emergencies.

synapse The connection between two neurons.

synaptic cleft The small gap between a presynaptic neuron and a postsynaptic neuron.

synesthesia A condition in which stimulation in one sense triggers an experience in another sense, or a concept evokes an unrelated sensory experience.

T cell A type of leukocyte that attacks specific invaders.

tardive dyskinesia Tremors and involuntary movements caused by blocking of dopamine receptors in the basal ganglia due to prolonged use of antidopamine drugs.

tectorial membrane A shelflike membrane overlying the hair cells and the basilar membrane in the cochlea.

telephone theory A theory of auditory frequency analysis, which stated that the auditory neurons transmit the actual sound frequencies to the cortex.

temporal lobe The part of each cerebral hemisphere ventral to the lateral fissure; it contains the auditory cortex, visual and auditory association areas, and Wernicke's area.

temporal summation The process of combining potentials that arrive a short time apart on a neuron's dendrites and cell body.

terminal A swelling on the branches at the end of a neuron that contains neurotransmitters; also called an end bulb.

testes The male gonads that produce sperm.

testosterone The major sex hormone in males; a member of the class androgens.

thalamus A forebrain structure lying just below the lateral ventricles, which receives information from all sensory systems except olfaction and relays it to the respective cortical projection areas. It has additional roles in movement, memory, and consciousness.

theory A system of statements that integrate and interpret diverse observations in an attempt to explain some phenomenon.

theory of mind The ability to attribute mental states to oneself and to others.

third interstitial nucleus of the anterior hypothalamus (INAH 3) A nucleus found to be half as large in gay men and heterosexual women as in heterosexual men.

tolerance After repeated drug use, the individual becomes less responsive and requires increasing amounts of a drug to produce the same results.

tonotopic map The form of topographic organization in the auditory cortex, such that each successive area responds to successively higher frequencies.

topographical organization Neurons from adjacent receptor locations project to adjacent cells in the cortex.

Tourette's syndrome A disorder characterized by motor and phonic (sound) tics.

tract A bundle of axons in the central nervous system.

transcranial magnetic stimulation A noninvasive stimulation technique that uses a magnetic coil to induce a voltage in brain tissue.

transsexual An individual who believes he or she has been born into the wrong sex; the person may dress and live as the other sex and may undergo surgery for sex resassignment.

traumatic brain injury (TBI) Injury caused by an external mechanical force such as a blow to the head, sudden acceleration or deceleration, or penetration.

trichromatic theory The theory that three color processes account for all the colors we are able to distinguish.

twin study In heredity research, a study that assesses how similar twins are in some characteristic; their similarity is then compared with that of nontwin siblings, or the similarity between identical twins is compared with the similarity between fraternal twins

tympanic membrane The eardrum, a very thin membrane stretched across the end of the auditory canal; its vibration transmits sound energy to the ossicles.

ultradian rhythm A rhythm with a length of less than a day, including the sleep stages and the basic rest and activity cycle during the day.

unipolar depression Depression without mania.

ventral Toward the stomach side.

ventral root The branch of each spinal nerve through which the motor neurons exit.

ventral stream The visual processing pathway that extends into the temporal lobes; it is especially concerned with the identification of objects.

ventral tegmental area A part of the mesolimbocortical dopamine system, which sends neurons to the nucleus accumbens and is a potent reward area.

ventricles Cavities in the brain filled with cerebrospinal fluid.

ventromedial hypothalamus An area in the hypothalamus important for sexual receptivity and copulation in female rats. It is also involved in eating behavior, and destruction in rats produces extreme obesity.

vesicle A membrane-enclosed container that stores neurotransmitter in the neuron terminal.

vestibular sense The sense that helps us maintain balance and that provides information about head position and movement; the receptors are located in the vestibular organs.

(VWFA) See *visual word form area*.

visual acuity Ability to distinguish visual details.

visual cortex The cortex in each occipital lobe where visual information is processed.

visual field The part of the environment that is being registered on the retina.

visual word form area (VWFA) An area in the human inferior temporal lobe involved in reading words as a whole.

VNO See *vomeronasal organ*.

volley theory Theory of auditory frequency analysis that states that groups of neurons follow the frequency of a sound when the frequency exceeds the firing rate capability of a single neuron.

voltage The difference in electrical charge between two points.

vomeronasal organ (VNO) A cluster of receptors in the nasal cavity that detect pheromones.

vulnerability The idea that genes produce susceptibility to a disorder and that environmental challenges may combine with a person's biological susceptibility to exceed the threshold required to produce the disorder.

vulnerability model The idea that environmental challenges combine with a person's genetic vulnerability for a disease to exceed the threshold for the disease.

Wernicke's aphasia Language impairment resulting from damage to Wernicke's area; the person has difficulty understanding and producing spoken and written language.

Wernicke's area The area just posterior to the auditory cortex (in the left hemisphere in most people) that interprets spoken and written language input and generates spoken and written language.

winter birth effect The tendency for more schizophrenics to be born during the winter and spring months than any other time of the year.

Wisconsin Card Sorting Test A test of prefrontal functioning that requires the individual to sort cards using one criterion and then change to another criterion.

withdrawal A negative reaction that occurs when drug use is stopped.

Wolffian ducts The early structures that in the male develop into the seminal vesicles and the vas deferens.

working memory A memory function that provides a temporary "register" for information while it is being used.

X-linked In heredity, a condition in which a gene on the X chromosome is not paired with a gene on the shorter Y chromosome, so that a single recessive gene is adequate to produce a characteristic.

zygote A fertilized egg.

References

13 legal medical marijuana states. (n.d.). ProCon.org. Retrieved November 25, 2009, from http://medicalmarijuana.procon.org/viewresource.asp?resourceID=000881

aan het Rot, M., Collins, K. A., Murrough, J. W., Perez, A. M., Reich, D. L., Charney, D. S., et al. (2010). Safety and efficacy of repeated-dose intravenous ketamine for treatment-resistant depression. *Biological Psychiatry, 15,* 139–145.

Abe, K., Kroning, J., Greer, M. A., & Critchlow, V. (1979). Effects of destruction of the suprachiasmatic nuclei on the circadian rhythms in plasma corticosterone, body temperature, feeding and plasma thyrotropin. *Neuroendocrinology, 29,* 119–131.

Abel, E. L., & Sokol, R. J. (1986). Fetal alcohol syndrome is now leading cause of mental retardation. *Lancet, 2,* 1222.

Abelson, J. F., Kwan, K. Y., O'Roak, B. J., Baek, D. Y., Stillman, A. A., Morgan, T. M., et al. (2005). Sequence variants in *SLITRK1* are associated with Tourette's syndrome. *Science, 319,* 317–320.

Abi-Dargham, A., Rodenhiser, J., Printz, D., Zea-Ponce, Y., Gil, R., Kegeles, L. S., et al. (2000). Increased baseline occupancy of D$_2$ receptors by dopamine in schizophrenia. *Proceedings of the National Academy of Sciences, USA, 97,* 8104–8109.

Abrahams, B. S., & Geschwind, D. H. (2008). Advances in autism genetics: On the threshold of a new neurobiology. *Nature Reviews Genetics, 9,* 341–355.

A brighter day for Edward Taub. (1997). *Science, 276,* 1503.

Abramowitz, J. S., Taylor, S., & McKay, D. (2009). Obsessive-compulsive disorder. *Lancet, 374,* 491–499.

Ackerl, K., Atzmueller, M., & Grammer, K. (2002). The scent of fear. *Neuroendocrinology Letters, 23,* 79–84.

Adams, D. B., Gold, A. R., & Burt, A. D. (1978). Rise in female-initiated sexual activity at ovulation and its suppression by oral contraceptives. *New England Journal of Medicine, 299,* 1145–1150.

Adler, L. E., Olincy, A., Cawthra, E. M., McRae, K. A., Harris, J. G., Nagamoto, H. T., et al. (2004). Varied effects of atypical neuroleptics on P50 auditory gating in schizophrenia patients. *American Journal of Psychiatry, 161,* 1822–1828.

Adolphs, R., Damasio, H., Tranel, D., & Damasio, A. R. (1996). Cortical systems for the recognition of emotion in facial expressions. *Journal of Neuroscience, 16,* 7678–7687.

Adolphs, R., Tranel, D., & Damasio, A. R. (1998). The human amygdala in social judgment. *Nature, 393,* 470–474.

Adolphs, R., Tranel, D., Damasio, H., & Damasio, A. (1995). Fear and the human amygdala. *Journal of Neuroscience, 15,* 5879–5891.

Agnew, B. (2000). Financial conflicts get more scrutiny in clinical trials. *Science, 289,* 1266–1267.

Ainsworth, C. (2009). Full without food: Can surgery cure obesity? *New Scientist,* September 2. Retrieved July 10, 2010, from www.newscientist.com/article/mg20327241.100-full-without-food-can-surgery-cure-obesity.html

Akbarian, S., Bunney, W. E., Potkin, S. G., Wigal, S. B., Hagman, J. O., Sandman, C. A., et al. (1993). Altered distribution of nicotinamide-adenine dinucleotide phosphate-diaphorase cells in frontal lobe of schizophrenics implies disturbances of cortical development. *Archives of General Psychiatry, 50,* 169–177.

Akbarian, S., Viñuela, A., Kim, J. J., Potkin, S. G., Bunney, W. E., & Jones, E. G. (1993). Distorted distribution of nicotinamide-adenine dinucleotide phosphate-diaphorase neurons in temporal lobe of schizophrenics implies anomalous cortical development. *Archives of General Psychiatry, 50,* 178–187.

Alain, C., Arnott, S. R., Hevenor, S., Graham, S., & Grady, C. L. (2001). "What" and "where" in the human auditory system. *Proceedings of the National Academy of Sciences, USA, 98,* 12301–12306.

Alam, N., Szymusiak, R., & McGinty, D. (1995). Local preoptic/anterior hypothalamic warming alters spontaneous and evoked neuronal activity in the magnocellular basal forebrain. *Brain Research, 696,* 221–230.

Alaska Oil Spill Commission. (1990, February). *Final report: Details about the accident.* Retrieved August 31, 2010, from www.evostc.state.ak.us/facts/details.cfm

Alati, R., Mamun, A. A., Williams, G. M., O'Callaghan, M., Najman, J. M., & Bor, W. (2006). In utero alcohol exposure and prediction of alcohol disorders in early adulthood. *Archives of General Psychiatry, 63,* 1009–1016.

Albert, D. J., Walsh, M. L., & Jonik, R. H. (1993). Aggression in humans: What is its biological foundation? *Neuroscience and Biobehavioral Reviews, 17,* 405–425.

Albert, M. S., Diamond, A. D., Fitch, R. H., Neville, H. J., Rapp, P. R., & Tallal, P. A. (1999). Cognitive development. In M. J. Zigmond, F. E. Bloom, S. C. Landis, J. L., Roberts, & L. R. Squire (Eds.), *Fundamental neuroscience* (pp. 1313–1338). New York: Academic Press.

Albrecht, D. G., De Valois, R. L., & Thorell, L. G. (1980). Visual cortical neurons: Are bars or gratings the optimal stimuli? *Science, 207,* 88–90.

Alexander, C. N., Langer, E. J., Newman, R. I., Chandler, H. M., & Davies, J. L. (1989). Transcendental meditation, mindfulness, and longevity: An experimental study with the elderly. *Journal of Personality and Social Psychology, 57,* 950–964.

Alexander, G. E., & Crutcher, M. D. (1990). Preparation for movement: Neural representations of intended direction in three motor areas of the monkey. *Journal of Neurophysiology, 64,* 133–150.

Alkire, M. T., Haier, R. J., Fallon, J. H., & Cahill, L. (1998). Hippocampal, but not amygdala, activity at encoding correlates with long-term free recall of nonemotional information. *Proceedings of the National Academy of Sciences, USA, 95,* 14506–14510.

Allegretta, M., Nicklas, J. A., Sriram, S., & Albertini, R. J. (1990). T cells responsive to myelin basic protein in patients with multiple sclerosis. *Science, 247,* 718–721.

Allen, L. S., & Gorski, R. A. (1992). Sexual orientation and the size of the anterior commissure in the human brain. *Proceedings of the National Academy of Sciences, USA, 89,* 7199–7202.

Allen, N. C., Bagade, S., McQueen, M. B., Ioannidis, J. P. A., Kavvoura, F. K., Khoury, M. J., et al. (2008). Systematic meta-analyses and field synopsis of genetic association studies in schizophrenia: The SzGene database. *Nature Genetics, 40,* 827–834.

Allison, T., & Cicchetti, D. V. (1976). Sleep in mammals: Ecological and constitutional correlates. *Science, 194,* 732–734.

Altar, C. A., Laeng, P., Jurata, L. W., Brockman, J. A., Lemire, A., Bullard, J., et al. (2004). Electroconvulsive seizures regulate gene expression of distinct neurotrophic signaling pathways. *Journal of Neuroscience, 24,* 2667–2677.

Altena, E., Vrenken, H., Van Der Werf, Y. D., van den Heuvel, O. A., & Van Someren, E. J. W. (2010). Reduced orbitofrontal and parietal gray matter in chronic insomnia: A voxel-based morphometric study. *Biological Psychiatry, 67,* 182–185.

Amanzio, M., Pollo, A., Maggi, G., & Benedetti, F. (2001). Response variability to analgesics: A role for non-specific activation of endogenous opioids. *Pain, 90,* 205–215.

American Association for the Advancement of Science. (2000). *Human inheritable genetic modifications: Findings and recommendations.* Retrieved September 12, 2007, from www .aaas.org/spp/sfrl/projects/germline/findings.htm

American Medical Association. (1992). *Use of animals in biomedical research: The challenge and response.* Chicago: Author.

American Psychiatric Association. (1994). *Diagnostic and statistical manual of mental disorders* (4th ed.). Washington, DC: Author.

American Psychiatric Association. (2010). Schizophrenia and other psychotic disorders. In *American Psychiatric Association DSM-5 Development.* Retrieved August 2, 2010, from www.dsm5.org/ProposedRevisions/Pages/SchizophreniaandOtherPsychoticDisorders.aspx

American Psychological Association. (2002). Ethical principles of psychologists and code of conduct. *American Psychologist, 57,* 1060–1073.

Andari, E., Duhamel, J.-R., Zalla, T., Herbrecht, E., Leboyer, M., & Sirigu, A. (2010). Promoting social behavior with oxytocin in high-functioning autism spectrum disorders. *Proceedings of the National Academy of Sciences, 107,* 4389–4394.

Andersen, R. A., Burdick, J. W., Musallam, S., Scherberger, H., Pesaran, B., Meeker, D., et al. (2004, September). Recording advances for neural prosthetics. In *Proceedings of the 26th Annual International Conference of the IEEE EMBS* (Vol. 2, pp. 5352–5355). San Francisco: IEEE.

Andersen, S., & Skorpen, F. (2009). Variation in the *COMT* gene: Implications for pain perception and pain treatment. *Summary Pharmacogenomics, 10,* 669–684.

Anderson, A. E., & Holman, J. E. (1997). Males with eating disorders: Challenges for treatment and research. *Psychopharmacology Bulletin, 33,* 391–397.

Anderson, A. K., & Phelps, E. A. (2001). Lesions of the human amygdala impair enhanced perception of emotionally salient events. *Nature, 411,* 305–309.

Anderson, B., & Harvey, T. (1996). Alterations in cortical thickness and neuronal density in the frontal cortex of Albert Einstein. *Neuroscience Letters, 210,* 161–164.

Anderson, R. H., Fleming, D. E., Rhees, R. W., & Kinghorn, E. (1986). Relationships between sexual activity, plasma testosterone, and the volume of the sexually dimorphic nucleus of the preoptic area in prenatally stressed and non-stressed rats. *Brain Research, 370,* 1–10.

Anderson, S. W., Bechara, A., Damasio, H., Tranel, D., & Damasio, A. R. (1999). Impairment of social and moral behavior related to early damage in human prefrontal cortex. *Nature Neuroscience, 2,* 1032–1037.

Ando, J., Ono, Y., & Wright, M. J. (2001). Genetic structure of spatial and verbal working memory. *Behavior Genetics, 31,* 615–624.

Andolfatto, P. (2005). Adaptive evolution of non-coding DNA in *Drosophila. Nature, 437,* 1149–1152.

Andrade, J. (1995). Learning during anaesthesia: A review. *British Journal of Psychology, 86,* 479–506.

Andreano, J. M., & Cahill, L. (2009). Sex influences on the neurobiology of learning and memory. *Learning and Memory, 16,* 248–266.

Andreasen, N. C. (1984). *The broken brain.* New York: Harper & Row.

Andreasen, N. C., Flaum, M., Swayze, V. W., II, Tyrrell, G., & Arndt, S. (1990). Positive and negative symptoms in schizophrenia: A critical reappraisal. *Archives of General Psychiatry, 47,* 615–621.

Andreasen, N. C., Rezai, K., Alliger, R., Swayzee, V. W., II, Flaum, M., Kirchner, et al. (1992). Hypofrontality in neuroleptic-naive patients and in patients with chronic schizophrenia: Assessment with xenon 133 single-photon emission computed tomography and the Tower of London. *Archives of General Psychiatry, 49,* 943–958.

Andrews-Hanna, J. R., Snyder, A. Z., Vincent, J. L., Lustig, C., Head, D., Raichle, M. E., et al. (2007). Disruption of large-scale brain systems in advanced aging. *Neuron, 56,* 924–935.

Ankney, C. D. (1992). Sex differences in relative brain size: The mismeasure of woman, too? *Intelligence, 16,* 329–336.

Anonymous. (1970). Effects of sexual activity on beard growth in man. *Nature, 226,* 869–870.

Anthony, J. C., Warner, L. A., & Kessler, R. C. (1994). Comparative epidemiology of dependence on tobacco, alcohol, controlled substances, and inhalants: Basic findings from the national comorbidity survey. *Experimental and Clinical Psychopharmacology, 2,* 244–268.

Apkarian, A. V., Sosa, Y., Krauss, B. R., Thomas, S. P., Fredrickson, B. E., Levy, R. E., et al. (2004). Chronic pain patients are impaired on an emotional decision-making task. *Pain, 108,* 129–136.

Apkarian, A. V., Sosa, Y., Sonty, S., Levy, R. M., Harden, R. N., Parrish, T. B., et al. (2004). Chronic back pain is associated with decreased prefrontal and thalamic gray matter density. *Journal of Neuroscience, 24,* 10410–10415.

Archer, J. (1991). The influence of testosterone on human aggression. *British Journal of Psychology, 82,* 1–28.

Archer, S. N., Robillant, D. I., Shane, D. J., Smits, M., Williams, A., Arendt, J., et al. (2003). A length polymorphism in the circadian clock gene *Per3* is linked to delayed sleep phase syndrome and extreme diurnal preference. *Sleep, 26,* 413–415.

Arcos-Burgos, M., Jain, M., Acosta, M. T., Shively, S., Stanescu, H., D., Wallis, D., et al. (2010, February 16). A common variant of the latropilin 3 gene, *LPHN3,* confers susceptibility to ADHD and predicts effectiveness of stimulant medication. *Molecular Psychiatry* [electronic version]. Retrieved July 7, 2010, from www.nature.com/mp/journal/vaop/ncurrent/pdf/mp20106a.pdf

Arendt, J., Skene, D. J., Middleton, B., Lockley, S. W., & Deacon, S. (1997). Efficacy of melatonin treatment in jet lag, shift work, and blindness. *Journal of Biological Rhythms, 12,* 604–617.

Argiolas, A. (1999). Neuropeptides and sexual behavior. *Neuroscience and Biobehavioral Reviews, 23,* 1127–1142.

Arnold, P. D., & Richter, M. A. (2001). Is obsessive-compulsive disorder an autoimmune disease? *Canadian Medical Association Journal, 165,* 1353–1358.

Arnold, P. E., Zai, G., & Richter, M. A. (2004). Genetics of anxiety disorders. *Current Psychiatry Reports, 6,* 243–254.

Aschoff, J. (1984). Circadian timing. *Annals of the New York Academy of Sciences, 423,* 442–468.

Asher, J. E., Lamb, J. A., Brocklebank, D., Cazier, J. B., Maestrini, E., Addis, L., et al. (2009). A whole-genome scan and fine-mapping linkage study of auditory-visual synesthesia reveals evidence of linkage to chromosomes 2q24, 5q33, 6p12, and 12p12. *American Journal of Human Genetics, 84,* 279–285.

Ashmore, J. F. (1994). The cellular machinery of the cochlea. *Experimental Physiology, 79,* 113–134.

Ashour, M. H., Jain, S. K., Kattan, K. M., al-Daeef, A. Q., Abdal-Jabbar, M. S., al-Tahan, A. R., et al. (1995). Maximal thymectomy for myasthenia gravis. *European Journal of Cardio-thoracic Surgery, 9,* 461–464.

Asplund, C. L., Todd, J. J., Snyder, A. P., & Marois, R. (2010). A central role for the lateral prefrontal cortex in goal-directed and stimulus-driven attention. *Nature Neuroscience, 13,* 507–512.

Assaf, M., Jagannathan, K., Calhoun, V. D., Miller, L., Stevens, M. C., Sahi, R., et al. (2010). Abnormal functional connectivity of default mode sub-networks in autism spectrum disorder patients. *NeuroImage, 53*, pp. 247–256.

Atladóttir, H. Ó., Pedersen, M. G., Thorsen, P., Mortensen, P. B., Deleuran, B., Eaton, W. W., et al. (2009). Association of family history of autoimmune diseases and autism spectrum disorders. *Pediatrics, 124*, 687–694.

Attia, E. (2009). Anorexia nervosa: Current status and future directions. *Annual Review of Medicine, 61*, 425–435.

Avidan, G., & Behrmann, M. (2009). Functional MRI reveals compromised neural integrity of the face processing network in congenital prosopagnosia. *Current biology, 19*, 1146–1150.

Ax, A. (1953). The physiological differentiation between fear and anger in humans. *Psychosomatic Medicine, 15*, 433–442.

Axel, R. (1995, October). The molecular logic of smell. *Scientific American, 273*, 154–159.

Baaré, W. F. C., Pol, H. E. H., Boomsma, D. I., Posthuma, D., de Geus, E. J. C., Schnack, H. G., et al. (2001). Quantitative genetic modeling of variation in human brain morphology. *Cerebral Cortex, 11*, 816–824.

Baars, B. J. (2005). Global workspace theory of consciousness: Toward a cognitive neuroscience of human experience. In S. Laureys (Ed.), *The boundaries of consciousness: Neurobiology and neuropathology* (pp. 45–54). New York: Elsevier.

Bach-y-Rita, P. (1990). Brain plasticity as a basis for recovery of function in humans. *Neuropsychologia, 28*, 547–554.

Bagni, C., & Greenough, W. T. (2005). From mRNP trafficking to spine dysmorphogenesis: The roots of fragile X syndrome. *Nature Reviews Neuroscience, 6*, 376–387.

Bailey, A., Le Couteur, A., Gottesman, I., Bolton, P., Simonoff, E., Yuzda, E., et al. (1995). Autism as a strongly genetic disorder: Evidence from a British twin study. *Psychological Medicine, 25*, 63–77.

Bailey, C. H., Bartsch, D., & Kandel, E. R. (1996). Toward a molecular definition of long-term memory storage. *Proceedings of the National Academy of Sciences, USA, 93*, 13445–13452.

Bailey, C. H., Kandel, E. R., & Si, K. (2004). The persistence of long-term memory: A molecular approach to self-sustaining changes in learning-induced synaptic growth. *Neuron, 44*, 49–57.

Bailey, J. M., & Bell, A. P. (1993). Familiality of female and male homosexuality. *Behavior Genetics, 23*, 313–322.

Bailey, J. M., & Benishay, D. S. (1993). Familial aggregation of female sexual orientation. *American Journal of Psychiatry, 150*, 272–277.

Bailey, J. M., & Pillard, R. C. (1991). A genetic study of male sexual orientation. *Archives of General Psychiatry, 48*, 1089–1096.

Bailey, J. M., Pillard, R. C., Neale, M. C., & Agyei, Y. (1993). Heritable factors influence sexual orientation in women. *Archives of General Psychiatry, 50*, 217–223.

Baizer, J. S., Ungerleider, L. G., & Desimone, R. (1991). Organization of visual inputs to the inferior temporal and posterior parietal cortex in macaques. *Journal of Neuroscience, 11*, 168–190.

Bakker, J., Honda, S.-I., Harada, N., & Balthazart, J. (2002). The aromatase knock-out mouse provides new evidence that estradiol is required during development in the female for the expression of sociosexual behaviors in adulthood. *Journal of Neuroscience, 22*, 9104–9112.

Bakker, J., Honda, S.-I., Harada, N., & Balthazart, J. (2003). The aromatase knock-out (ArKO) mouse provides new evidence that estrogens are required for the development of the female brain. *Annals of the New York Academy of Sciences, 1007*, 251–262.

Baliki, M. N., Chialvo, D. R., Geha, P. Y., Levy, R. M., Harden, R. N., Parrish, T. B., et al. (2006). Chronic pain and the emotional brain: Specific brain activity associated with spontaneous fluctuations of intensity of chronic back pain. *Journal of Neuroscience, 26*, 12165–12173.

Bandmann, O., Vaughan, J., Holmans, P., Marsden, C. D., & Wood, N. W. (1997). Association of slow acetylator genotype for *N*-acetyltransferase 2 with familial Parkinson's disease. *Lancet, 350*, 1136–1139.

Bar, M. (2004). Visual objects in context. *Nature Reviews Neuroscience, 5*, 617–629.

Barinaga, M. (1996). Guiding neurons to the cortex. *Science, 274*, 1100–1101.

Barinaga, M. (1997). New imaging methods provide a better view into the brain. *Science, 276*, 1974–1981.

Barkley, R. A., Fischer, M., Smallish, L., & Fletcher, K. (2003). Does the treatment of attention-deficit/hyperactivity disorder with stimulants contribute to drug use/abuse? A 13-year prospective study. *Pediatrics, 111*, 97–109.

Barnea-Goraly, N., Kwon, H., Menon, V., Eliez, S., Lotspeich, L., & Reiss, A. L. (2004). White matter structure in autism: Preliminary evidence from diffusion tensor imaging. *Biological Psychiatry, 55*, 323–326.

Barnes, C. A., & McNaughton, B. L. (1985). An age comparison of the rates of acquisition and forgetting of spatial information in relation to long-term enhancement of hippocampal synapses. *Behavioral Neuroscience, 99*, 1040–1048.

Bartlett, D. L., & Steele, J. B. (1979). *Empire: The life, legend, and madness of Howard Hughes.* New York: Norton.

Basbaum, A. I., Bautista, D. M., Scherrer, G., & Julius, D. (2009). Cellular and molecular mechanisms of pain. *Cell, 139*, 267–284.

Basbaum, A. I., & Fields, H. L. (1984). Endogenous pain control systems: Brainstem spinal pathways and endorphin circuitry. *Annual Review of Neuroscience, 7*, 309–338.

Basbaum, A. I., & Jessell, T. M. (2000). The perception of pain. In E. R. Kandel, J. H. Schwartz, & T. M. Jessell (Eds.), *Principles of neural science* (4th ed., pp. 472–491). New York: McGraw-Hill.

Bastiaansen, J. A. C. J., Thioux, M., & Keysers, C. (2009). Evidence for mirror systems in emotions. *Philosophical Transactions of the Royal Society B, 364*, 2391–2404.

Batista, A. P., Buneo, C. A., Snyder, L. H., & Andersen, R. A. (1999). Reach plans in eye-centered coordinates. *Science, 285*, 257–260.

Batterham, R. L., Cowley, M. A., Small, C. J., Herzog, H., Cohen, M. A., Dakin, C. L., et al. (2002). Gut hormone PYY_{3-36} physiologically inhibits food intake. *Nature, 418*, 650–654.

Batterham, R. L., Cohen, M. A., Ellis, S. M., Le Roux, C. W., Withers, D. J., Frost, G. S., et al. (2003). Inhibition of food intake in obese subjects by peptide YY_{3-36}. *New England Journal of Medicine, 349*, 941–948.

Bauer, R. M. (1984). Autonomic recognition of names and faces in prosopagnosia: A neuropsychological application of the guilty knowledge test. *Neuropsychologia, 22*, 457–469.

Baulac, S., Huberfeld, G., Gourinkel-An, I., Mitropoulou, G., Beranger, A., Prud'homme, J.-F., et al. (2001). First genetic evidence of GABAA receptor dysfunction in epilepsy: A mutation in the ā2-subunit gene. *Nature Genetics, 28*, 46–48.

Baum, A., Gatchel, R. J., & Schaeffer, M. A. (1983). Emotional, behavioral, and physiological effects of chronic stress at Three Mile Island. *Journal of Consulting and Clinical Psychology, 51*, 565–572.

Baumgartner, T., Heinrichs, M., Vonlanthen, A., Fischbacher, U., & Fehr, E. (2008). Oxytocin shapes the neural circuitry of trust and trust adaptation in humans. *Neuron, 58*, 639–650.

Baxter, L. R., Phelps, M. E., Mazziotta, J. C., Guze, B. H., Schwartz, J. M., & Selin, C. E. (1987). Local cerebral glucose metabolic rates in obsessivecompulsive disorder. *Archives of General Psychiatry, 44*, 211–218.

Baxter, L. R., Phelps, M. E., Mazziotta, J. C., Schwartz, J. M., Gerner, R. H., Selin, C. E., et al. (1985). Cerebral metabolic rates for glucose in mood disorders: Studies with positron emission tomography and fluorodeoxyglucose F 18. *Archives of General Psychiatry, 42*, 441–447.

Baxter, L. R., Schwartz, J. M., Phelps, M. E., Mazziotta, J. C., Guze, B. H., Selin, C. E., et al. (1989). Reduction of prefrontal cortex glucose metabolism common to three types of depression. *Archives of General Psychiatry, 46,* 243–249.

Beaver, K. M., DeLisi, M., Vaughn, M. G., & Barnes, J. C. (2010). Monoamine oxidase A genotype is associated with gang membership and weapon use. *Comprehensive Psychiatry, 51,* 130–134.

Bechara, A., Damasio, H., Damasio, A. R., & Lee, G. P. (1999). Different contributions of the human amygdala and ventromedial prefrontal cortex to decision-making. *Journal of Neuroscience, 19,* 5473–5481.

Bechara, A., Damasio, H., Tranel, D., & Damasio, A. R. (1997). Deciding advantageously before knowing the advantageous strategy. *Science, 275,* 1293–1295.

Bechara, A., Tranel, D., Damasio, H., Adolphs, R., Rockland, C., & Damasio, A. R. (1995). Double dissociation of conditioning and declarative knowledge relative to the amygdala and hippocampus in humans. *Science, 269,* 1115–1118.

Beck, A. T., & Galef, B. G. (1989). Social influences on the selection of a protein-sufficient diet by Norway rats *(Rattus norvegicus). Journal of Comparative Psychology, 103,* 132–139.

Becker, A. E., Burwell, R. A., Gilman, S. E., Herzog, D. B., & Hamburg, P. (2002). Eating behaviours and attitudes following prolonged exposure to television among ethnic Fijian adolescent girls. *British Journal of Psychiatry, 180,* 509–514.

Beecher, D. K. (1956). Relationship of significance of wound to pain experienced. *Journal of the American Medical Association, 161,* 1609–1613.

Begley, S., & Biddle, N. A. (1996, February 26). For the obsesssed, the mind can fix the brain. *Newsweek,* p. 60.

Begré, S., & Koenig, T. (2008). Cerebral disconnectivity: An early event in schizophrenia. *Neuroscientist, 14,* 19–45.

Behr, R., Müller, J., Shehata-Dieler, W., Schlake, H.-P., Helms, J., Roosen, K., et al. (2007). The high rate CIS auditory brainstem implant for restoration of hearing in NF-2 patients. *Skull Base, 17,* 91–107.

Behrmann, M., Moscovitch, M., & Winocur, G. (1994). Intact visual imagery and impaired visual perception in a patient with visual agnosia. *Journal of Experimental Psychology: Human Perception and Performance, 20,* 1068–1087.

Békésy, G. von. (1951). The mechanical properties of the ear. In S. S. Stevens (Ed.), *The handbook of experimental psychology* (pp. 1075–1115). New York: Wiley.

Békésy, G. von. (1956). Current status of theories of hearing. *Science, 123,* 779–783.

Bell, A. P., Weinberg, M. S., & Hammersmith, S. K. (1981). *Sexual preference.* Bloomington: Indiana University Press.

Bellis, D. J. (1981). *Heroin and politicians: The failure of public policy to control addiction in America.* Westport, CT: Greenwood Press.

Benca, R. M., Obermeyer, W. H., Thisted, R. A., & Gillin, J. C. (1992). Sleep and psychiatric disorders. A meta-analysis. *Archives of General Psychiatry, 49,* 651–668.

Benenson, Y., Gil, B., Ben-Dor, U., Adar, R., & Shapiro, E. (2004). An autonomous molecular computer for logical control of gene expression. *Nature, 429,* 423–429.

Bennett, M. V. L., & Zukin, S. (2004). Electrical coupling and neuronal synchronization in the mammalian brain. *Neuron, 41,* 495–511.

Benoit, E., & Dubois, J. M. (1986). Toxin I from the snake *Dendroaspis polylepsis polylepsis:* A highly specific blocker of one type of potassium channel in myelinated nerve fiber. *Brain Research, 377,* 374–377.

Benson, D. F., Djenderedjian, A., Miller, B. L., Pachana, N. A., Chang, L., Itti, L., et al. (1996). Neural basis of confabulation. *Neurology, 46,* 1239–1243.

Benton, A. L. (1980). The neuropsychology of facial recognition. *American Psychologist, 35*(10), 176–186.

Berenbaum, S. A., Duck S. C., & Bryk, K. (2000). Behavioral effects of prenatal *versus* postnatal androgen excess in children with 21-hydroxylase-deficient congenital adrenal hyperplasia. *Journal of Clinical Endocrinology & Metabolism, 85,* 727–733.

Berglund, H., Lindström, P., Dhejne-Helmy, C., & Savic, I. (2008). Male-to-female transsexuals show sex-atypical hypothalamus activation when smelling odorous steroids. *Cerebral Cortex, 18,* 1900–1908.

Berglund, H., Lindström, P., & Savic, I. (2006). Brain response to putative pheromones in lesbian women. *Proceedings of the New York Academy of Sciences, USA, 103,* 8269–8274.

Berlucchi, G., & Aglioti, S. (1997). The body in the brain: Neural bases of corporeal awareness. *Trends in Neurosciences, 20,* 560–564.

Bernard, L. L. (1924). *Instinct.* New York: Holt, Rinehart & Winston.

Bernhardt, P. C., Dabbs, J. M., Jr., Fielden, J. A., & Lutter, C. D. (1998). Testosterone changes during vicarious experiences of winning and losing among fans at sporting events. *Physiological Behavior, 65,* 59–62.

Berns, G. S., McClure, S. M., Pagnoni, G., & Montague, P. R. (2001). Predictability modulates human brain response to reward. *Journal of Neuroscience, 21,* 2793–2798.

Bernstein, D. M., Laney, C., Morris, E. K., & Loftus, E. F. (2005). False beliefs about fattening foods can have healthy consequences. *Proceedings of the National Academy of Sciences, USA, 102,* 13724–13731.

Bernstein, I. L. (1978). Learned taste aversions in children receiving chemotherapy. *Science, 200,* 1302–1303.

Bernstein, L. J., & Robertson, L. C. (1998). Illusory conjunctions of color and motion with shape following bilateral parietal lesions. *Psychological Science, 9,* 167–175.

Berridge, V., & Edwards, G. (1981). *Opium and the people: Opiate use in nineteenth-century England.* New York: St. Martin's Press.

Berson, D. M., Dunn, F. A., & Takao, M. (2002). Phototransduction by retinal ganglion cells that set the circadian clock. *Science, 295,* 1070–1073.

Berthoud, H.-R., & Morrison, C. (2008). The brain, appetite, and obesity. *Annual Review of Psychology, 59,* 55–92.

Berti, A., Bottini, M., Gandola, L, Pia, N., Smania, A., Stracciari, I., et al. (2005). Shared cortical anatomy for motor awareness and motor control. *Science, 309,* 488–491.

Bertram, L., McQueen, M. B., Mullin, K., Blacker, D., & Tanzi, R. E. (2007). Systematic meta-analyses of Alzheimer disease genetic association studies: The AlzGene database. *Nature Genetics, 39,* 17–23.

Bessa, J. M., Ferreira, D., Melo, I., Marques, F., Cerqueira, J. J., Palha, J. A., et al. (2009). The mood-improving actions of antidepressant do not depend on neurogenesis but are associated with neuronal remodeling. *Molecular Psychiatry, 14,* 764–773.

Betarbet, R., Sherer, T. B., MacKenzie, G., Garcia-Osuna, M., Panov, A. V., & Greenamyre, J. T. (2000). Chronic systemic pesticide exposure reproduces features of Parkinson's disease. *Nature Neuroscience, 3,* 1301–1306.

Beveridge, T. J. R., Gill, K. E., Hanlon, C. A., & Porrino, L. J. (2008). Parallel studies of cocaine-related neural and cognitive impairment in humans and monkeys. *Philosophical Transactions of the Royal Society B, 363,* 3257–3266.

Biederman, J. (2004). Impact of comorbidity in adults with attention-deficit/hyperactivity disorder. *Journal of Clinical Psychiatry, 65,* 3–7.

Biederman, J., & Faraone, S. V. (2005). Attention-deficit hyperactivity disorder. *Lancet, 366,* 237–248.

Bienvenu, O. J., Wang, Y., Shugart, Y. Y., Welch, J. M., Grados, M. A., Fyer, A. J., et al. (2008). *Sapap3* and pathological grooming in humans: Results from the OCD collaborative genetics study. *American Journal of Medical Genetics Part B, 150B,* 710–720.

Billington, C. J., & Levine, A. S. (1992). Hypothalamic neuropeptide Y regulation of feeding and energy metabolism. *Current Opinion in Neurobiology, 2,* 847–851.

Billy, J. O. G., Tanfer, K., Grady, W. R., & Klepinger, D. H. (1993). The sexual behavior of men in the United States. *Family Planning Perspectives, 25*(2), 52–60.

Binder, E. B., Bradley, R. G., Liu, W., Epstein, M. P., Deveau, T. C., Mercer, K. B., et al. (2008). Association of *FKBP5* polymorphisms and childhood abuse with risk of posttraumatic stress disorder symptoms in adults. *Journal of the American Medical Association, 299,* 1291–1305.

Binet, S., & Simon, T. (1905). Méthodes nouvelles pour le diagnostic du niveau intellectuel desanormaux [New methods for the diagnosis of the intellectual level of subnormals]. *L'Année Psychologique, 12,* 191–244.

Birbaumer, N., Lutzenberger, W., Montoya, P., Larbig, W., Unerti, K., Töpfner, S., et al. (1997). Effects of regional anesthesia on phantom limb pain are mirrored in changes in cortical reorganization. *Journal of Neuroscience, 17,* 5503–5508.

Bisch, N., Herrell, R., Lehner, T., Liang, K.-L., Eaves, L., Hob, J., et al. (2009). Interaction between the serotonin transporter gene (*5-HTTLPR*), stressful life events, and risk of depression: A meta-analysis. *Journal of the American Medical Association, 301,* 2462–2471.

Bissada, H., Tasca, G. A., Barber, A. M., & Bradwejn, J. (2008). Olanzapine in the treatment of low body weight and obsessive thinking in women with anorexia nervosa: A randomized, double-blind, placebo-controlled trial. *American Journal of Psychiatry, 165,* 1281–1288.

Bittar, R. G., Kar-Purkayastha, I., Owen, S. L., Bear, R. E., Green, A., Wang, S. Y., et al. (2005). Deep brain stimulation for pain relief: A meta-analysis. *Journal of Clinical Neuroscience, 12,* 515–519.

Blaese, R. M., Culver, K. W., Miller, A. D., Carter, C. S., Fleisher, T., Clerici, M., et al. (1995). T lymphocyte-directed gene therapy for ADA–SCID: Initial trial results after 4 years. *Science, 270,* 475–480.

Blanchard, D. C., & Blanchard, R. J. (1972). Innate and conditioned reactions to threat in rats with amygdaloid lesions. *Journal of Comparative and Physiological Psychology, 81,* 281–290.

Blanchard, R. (2001). Fraternal birth order and the maternal immune hypothesis of male homosexuality. *Hormones and Behavior, 40,* 105–114.

Blanke, O., & Arzy, S. (2005). The out-of-body experience: Disturbed self-processing at the temporo-parietal junction. *Neuroscientist, 11,* 16–24.

Blehar, M. C., & Rosenthal, N. E. (1989). Seasonal affective disorders and phototherapy. *Archives of General Psychiatry, 46,* 469–474.

Bleuler, E. (1911). Dementia præcox, oder die Gruppe der Schizophrenien [Premature dementia, or the group of schizophrenias]. In G. Aschaffenburg (Ed.), *Handbuch der Psychiatrie.* Leipzig, Germany: Hälfte.

Bliss, E. L. (1980). Multiple personalities: A report of 14 cases with implications for schizophrenia and hysteria. *Archives of General Psychiatry, 37,* 1388–1397.

Bliss, T. V. P., & Lømo, T. (1973). Long-lasting potentiation of synaptic transmission in the dentate area of the anaesthetized rabbit following stimulation of the perforant path. *Journal of Physiology, 232,* 331–356.

Bloch, G. J., Butler, P. C., & Kohlert, J. G. (1996). Galanin microinjected into the medial preoptic nucleus facilitates female- and male-typical sexual behaviors in the female rat. *Physiology and Behavior, 59,* 1147–1154.

Bloch, G. J., Butler, P. C., Kohlert, J. G., & Bloch, D. A. (1993). Microinjection of galanin into the medial preoptic nucleus facilitates copulatory behavior in the male rat. *Physiology and Behavior, 54,* 615–624.

Blonder, L. X., Bowers, D., & Heilman, K. M. (1991). The role of the right hemisphere in emotional communication. *Brain, 114,* 1115–1127.

Bloom, F. E., & Lazerson, A. (1988). *Brain, mind and behavior* (2nd ed.). New York: W. H. Freeman.

Blum, D. (1994). *The monkey wars.* New York: Oxford.

Bocklandt, S., Horvath, S., Vilain, E., & Hamer, D. H. (2006). Extreme skewing of X chromosome inactivation in mothers of homosexual men. *Human Genetics, 118,* 691–694.

Boecker, H., Dagher, A., Ceballos-Baumann, A. O., Passingham, R. E., Samuel, M., Friston, K. J., et al. (1998). Role of the human rostral supplementary motor area and the basal ganglia in motor sequence control: Investigations with H2 150 PET. *Journal of Neurophysiology, 79,* 1070–1080.

Boeve, B. F., Silber, M. H., Parisi, J. E., Dickson, D. W., Ferman, T. J., Benarroch, E. E., et al. (2003). Synucleinopathy pathology and REM sleep behavior disorder plus dementia or parkinsonism. *Neurology, 61,* 40–45.

Bogaert, A. F. (2004). Asexuality: Prevalence and associated factors in a national probability sample. *Journal of Sex Research, 41,* 279–287.

Bogaert, A. F. (2006). Biological versus nonbiological older brothers and men's sexual orientation. *Proceedings of the National Academy of Sciences, USA, 103,* 10771–10774.

Bogardus, C., Lillioja, S., Ravussin, E., Abbott, W., Zawadzki, J. K., Young, A., et al. (1986). Familial dependence of the resting metabolic rate. *New England Journal of Medicine, 315,* 96–100.

Bogerts, B., Meertz, E., & Schönfeldt-Bausch, R. (1985). Basal ganglia and limbic system pathology in schizophrenia: A morphometric study of brain volume and shrinkage. *Archives of General Psychiatry, 42,* 784–791.

Bohman, M. (1978). Some genetic aspects of alcoholism and criminality: A population of adoptees. *Archives of General Psychiatry, 35,* 269–276.

Boivin, D. B., Duffy, J. F., Kronauer, R. E., & Czeisler, C. A. (1996). Dose-response relationships for resetting of human circadian clock by light. *Nature, 379,* 540–542.

Bolla, K. I., Brown, K., Eldreth, D., Tate, K., & Cadet, J. L. (2002). Dose-related neurocognitive effects of marijuana use. *Neurology, 59,* 1337–1343.

Bolles, R. C. (1975). *Theory of motivation.* New York: Harper & Row.

Bonda, E., Petrides, M., Frey, S., & Evans, A. (1995). Neural correlates of mental transformations of the body-in-space. *Proceedings of the National Academy of Sciences, USA, 92,* 11180–11184.

Bonebakker, A. E., Bonke, B., Klein, J., Wolters, G., Stijnene, T., Passchier, J., et al. (1996). Information processing during general anesthesia: Evidence for unconscious memory. *Memory and Cognition, 24,* 766–776.

Bonne, O., Vythilingham, M., Inagaki, M., Wood, S., Neumeister, A., Nugent, A. C., et al. (2008). Reduced posterior hippocampal volume in posttraumatic stress disorder. *Journal of Clinical Psychiatry, 69,* 1087–1091.

Bonner, J. (2005, April 16). Paralysed dogs regain movement. *New Scientist,* p. 14.

Bonnet, M. H., & Arand, D. L. (1996). Metabolic rate and the restorative function of sleep. *Physiology and Behavior, 59,* 777–782.

Bontempi, B., Laurent-Demir, C., Destrade, C., & Jaffard, R. (1999). Time-dependent reorganization of brain circuitry underlying long-term memory storage. *Nature, 400,* 671–675.

Born, R. T., & Bradley, D. C. (2005). Structure and function of visual area MT. *Annual Review of Neuroscience, 28,* 157–189.

Borsook, D., Becerra, L., Fishman, S., Edwards, A., Jennings, C. L., Stojanovic, M., et al. (1998). Acute plasticity in the human somatosensory cortex following amputation. *Neuroreport, 9,* 1013–1017.

Bossy-Wetzel, E., Schwarzenbacher, R., & Lipton, S. A. (2004). Molecular pathways to neurodegeneration. *Nature Medicine, 10,* S2–S9.

Bottini, G., Corcoran, R., Sterzi, R., Paulesu, E., Schenone, P., Scarpa, P., et al. (1994). The role of the right hemisphere in the interpretation of figurative aspects of language: A positron emission tomography activation study. *Brain, 117,* 1241–1253.

Botvinick, M. (2004). Probing the neural basis of body ownership. *Science, 305,* 782–783.

Bouchard, C. (1989). Genetic factors in obesity. *Medical Clinics of North America, 73,* 67–81.

Bouchard, C., Tremblay, A., Després, J.-P., Nadeau, A., Lupien, P. J., Thériault, G., et al. (1990). The response to long-term overfeeding in identical twins. *New England Journal of Medicine, 322,* 1477–1482.

Bouchard, M. F., Bellinger, D. C., Wright, R. O., & Weisskopf, M. G. (2010). Attention-deficit/hyperactivity disorder and urinary metabolites of organophosphate pesticides. *Pediatrics, 125,* e1270–e1277.

Bouchard, T. J. (1994). Genes, environment, and personality. *Science, 264,* 1700–1701.

Bouchard, T. J., Jr., & McGue, M. (1981). Familial studies of intelligence: A review. *Science, 212,* 1055–1059.

Bouchard, T. J., & Segal, N. L. (1985). Environment and IQ. In B. B. Wolman (Ed.), *Handbook of intelligence: Theories, measurements, and applications* (pp. 391–464). New York: Wiley.

Boulant, J. A. (1981). Hypothalamic mechanisms in thermoregulation. *Federation Proceedings, 40,* 2843–2850.

Boulenguez, P., & Vinay, L. (2009). Strategies to restore motor functions after spinal cord injury. *Current Opinion in Neurobiology, 19,* 587–600.

Bourgeois, J.-P., & Rakic, P. (1993). Changes of synaptic density in the primary visual cortex of the macaque monkey from fetal to adult stage. *Journal of Neuroscience, 13,* 2801–2820.

Bouvier, S. E., & Engel, S. A. (2006). Behavioral deficits and cortical damage loci in cerebral achromatopsia. *Cerebral Cortex, 16,* 183–191.

Bower, K. S. (1994). Temporary emotional states act like multiple personalities. In R. M. Klein & B. K. Doane (Eds.), *Psychological concepts and dissociative disorders* (pp. 207–234). Hillsdale, NJ: Lawrence Erlbaum

Bowmaker, J. K., & Dartnall, H. J. A. (1980). Visual pigments of rods and cones in a human retina. *Journal of Physiology (London), 298,* 501–511.

Bozarth, M. A., & Wise, R. A. (1984). Anatomically distinct opiate receptor fields mediate reward and physical dependence. *Science, 224,* 516–517.

Bozarth, M. A., & Wise, R. A. (1985). Toxicity associated with long-term intravenous heroin and cocaine self-administration in the rat. *Journal of the American Medical Association, 254,* 81–83.

Braak, H., Del Tredici, K., Rüb, U., de Vos, R. A. I., Steur, E. N. H. J., & Braak, E. (2003). Staging of brain pathology related to sporadic Parkinson's disease. *Neurobiology, 24,* 197–211.

Bracha, H. S., Torrey, E. F., Gottesman, I. I., Bigelow, L. B., & Cunniff, C. (1992). Second-trimester markers of fetal size in schizophrenia: A study of monozygotic twins. *American Journal of Psychiatry, 149,* 1355–1361.

Bradbury, J. (2003). Why do men and women feel and react to pain differently? *Lancet, 361,* 2052–2053.

Bradbury, T. N., & Miller, G. A. (1985). Season of birth in schizophrenia: A review of evidence, methodology, and etiology. *Psychological Bulletin, 98,* 569–594.

Bradley, S. J., Oliver, G. D., Chernick, A. B., & Zucker, K. J. (1998). *Experiment of nature: Ablatio penis at 2 months, sex reassignment at 7 months, and a psychosexual follow-up in young adulthood.* Retrieved September 11, 2007, from www.pediatrics.org/cgi/content/full/102/1/e9

Brain-computer interface, developed at Brown, begins new clinical trial. (2009, June 10). *Brown University News Bureau.* Retrieved from http://news.brown.edu/pressreleases/2009/06/braingate2

Brandt, G., Park, A., Wynne, K., Sileno, A., Jazrawi, R., Woods, A., et al. (2004, June 16–19). *Nasal peptide YY3–36: Phase 1 dose ranging and safety studies in healthy human subjects.* Poster presented at the 86th annual meeting of the Endocrine Society, New Orleans, LA.

Brannon, E. M., & Terrace, H. S. (1998). Ordering of the numerosities 1 to 9 by monkeys. *Science, 282,* 746–749.

Braun, B. G. (1985). *Treatment of multiple personality disorder.* Washington, DC: American Psychiatric Press.

Bray, G. A. (1992). Drug treatment of obesity. *American Journal of Clinical Nutrition, 55,* 538S–544S.

Breasted, J. H. (1930). *The Edwin Smith surgical papyrus.* Chicago: University of Chicago Press.

Brechbühl, J., Klaey, M., & Broillet, M.-C. (2008). Grueneberg ganglion cells mediate alarm pheromone detection in mice. *Science, 321,* 1092–1095.

Brecher, E. M. (1972). *Licit and illicit drugs.* Boston: Little, Brown.

Breiter, H. C., Aharon, I., Kahneman, D., Dale, A., & Shizgal, P. (2001). Functional imaging of neural responses to expectancy and experience of monetary gains and losses. *Neuron, 30,* 619–639.

Breitner, J. C., Folstein, M. E., & Murphy, E. A. (1986). Familial aggregation in Alzheimer dementia-1: A model for the age-dependent expression of an autosomal dominant gene. *Journal of Psychiatric Research, 20,* 31–43.

Bremer, J. (1959). *Asexualization: A follow-up study of 244 cases.* New York: Macmillan.

Bremner, J. D., Randall, P., Scott, T. M., Bronen, R. A., Seibyl, J. P., Southwick, S. M., et al. (1995). MRI-based measurement of hippocampal volume in patients with combat related posttraumatic stress disorder. *American Journal of Psychiatry, 152,* 973–981.

Bremner, J. D., Randall, P., Vermetten, E., Staib, L., Bronen, R. A., Mazure, C., et al. (1997). Magnetic resonance imaging-based measurement of hippocampal volume in posttraumatic stress disorder related to childhood physical and sexual abuse: A preliminary report. *Biological Psychiatry, 41,* 23–32.

Brennan, P. A., Grekin, E. R., & Mednick, S. A. (1999). Maternal smoking during pregnancy and adult male criminal outcomes. *Archives of General Psychiatry, 56,* 215–219.

Brewer, J. B., Zhao, Z., Desmond, J. E., Glover, G. H., & Gabrieli, J. D. E. (1998). Making memories: Brain activity that predicts how well visual experience will be remembered. *Science, 281,* 1185–1187.

Brinkman, C. (1984). Supplementary motor area of the monkey's cerebral cortex: Short- and long-term deficits after unilateral ablation and the effects of subsequent callosal section. *Journal of Neuroscience, 4,* 918–929.

Brinkman, R. R., Mezei, M. M., Theilmann, J., Almqvist, E., & Hayden, M. R. (1997). The likelihood of being affected with Huntington disease by a particular age, for a specific CAG size. *American Journal of Human Genetics, 60,* 1202–1210.

Brisbare-Roch, C., Dingemanse, J., Koberstein, R., Hoever, P., Aissaoui, H., Flores, S., et al. (2007). Promotion of sleep by targeting the orexin system in rats, dogs and humans. *Nature Medicine, 13,* 150–155.

Britten, K. H. (2008). Mechanisms of self-motion perception. *Annual Review of Neuroscience, 31,* 389–410.

Britten, R. J. (2002). Divergence between samples of chimpanzee and human DNA sequences is 5%, counting indels. *Proceedings of the National Academy of Sciences, USA, 99,* 13633–13635.

Britten, R. J., Rowen, L., Williams, J., & Cameron, R. A. (2003). Majority of divergence between closely related DNA samples is due to indels. *Proceedings of the National Academy of Sciences, USA, 100,* 4661–4665.

Broberg, D. J., & Bernstein, I. L. (1989). Cephalic insulin release in anorexic women. *Physiology and Behavior, 45,* 871–874.

Broberger, C., & Hökfelt, T. (2001). Hypothalamic and vagal neuropeptide circuitries regulating food intake. *Physiology and Behavior, 74,* 669–682.

Broca, P. (1861). Remarques sur le siège de la faculté du langage articulé, suivies d'une observation d'aphemie (perte de la parole). *Bulletin de la Société Anatomique (Paris), 36,* 330–357.

Broughton, R. (1975). Biorhythmic variations in consciousness and psychological functions. *Canadian Psychological Review, 16,* 217–239.

Broughton, R., Billings, R., Cartwright, R., Doucette, D., Edmeads, J., Edward, H. M., et al. (1994). Homicidal somnambulism: A case report. *Sleep, 17,* 253–264.

Brown, A. S., Begg, M. D., Gravenstein, S., Schaefer, C. A., Wyatt, R. J., Bresnahan, M., et al. (2004). Serologic evidence of prenatal influenza in the etiology of schizophrenia. *Archives of General Psychiatry, 61,* 774–780.

Brown, K., & Wald, G. (1964). Visual pigments in single rods and cones of the human retina. *Science, 143,* 45–52.

Brown, M. A., & Sharp, P. E. (1995). Simulation of spatial learning in the Morris water maze by a neural network model of the hippocampal formation and nucleus accumbens. *Hippocampus, 5,* 171–188.

Brown, T. H., Chapman, P. F., Kairiss, E. W., & Keenan, C. L. (1988). Long-term synaptic potentiation. *Science, 242,* 724–727.

Brownell, W. E., Bader, C. R., Bertrand, D., & de Ribaupierre, Y. (1985). Evoked mechanical responses of isolated cochlear outer hair cells. *Science, 227,* 194–196.

Broyd, S. J., Demanuele, C., Debener, S., Helps, S. K., James, C. J., & Sonuga-Barke, E. J. S. (2009). Default-mode brain dysfunction in mental disorders: A systematic review. *Neuroscience and Biobehavioral Reviews, 33,* 279–296.

Brunner, H. G., Nelen, M., Breakefield, X. O., Ropers, H. H., & van Oost, B. A. (1993). Abnormal behavior associated with a point mutation in the structural gene for monoamine oxidase A. *Science, 262,* 578–580.

Buchanan, T. W., Lutz, K., Mirzazade, S., Specht, K., Shah, N. J., Zilles, K., et al. (2000). Recognition of emotional prosody and verbal components of spoken language: An fMRI study. *Cognitive Brain Research, 9,* 227–238.

Buchwald, H., Avidor, Y., Braunwald, E., Jensen, M. D., Pories, W., Fahrbach, K., et al. (2004). Bariatric surgery: A systematic review and meta-analysis. *Journal of the American Medical Association, 292,* 1724–1737.

Buckner, R. L., & Koutstaal, W. (1998). Functional neuroimaging studies of encoding, priming, and explicit memory retrieval. *Proceedings of the National Academy of Sciences, USA, 95,* 891–898.

Buell, S. J., & Coleman, P. D. (1979). Dendritic growth in the aged human brain and failure of growth in senile dementia. *Science, 206,* 854–856.

Bunney, W. E., Jr., Murphy, D. L., Goodwin, F. K., & Borge, G. F. (1972). The "switch process" in manic-depressive illness: I. A systematic study of sequential behavioral changes. *Archives of General Psychiatry, 27,* 295–302.

Bünning, E. (1973). *The physiological clock: Circadian rhythms and biological chronometry* (3rd ed.). New York: Springer-Verlag.

Buñuel, L. (1983). *My last sigh.* New York: Knopf.

Bureau of the Census. (2001). *Population.* Retrieved September 11, 2007, from www.census.gov/prod/2001pubs/statab/sec01.pdf

Burke, S. N., & Barnes, C. A. (2006). Neural plasticity in the ageing brain. *Nature Reviews Neuroscience, 7,* 30–40

Buschman, T. J., & Miller, E. K. (2007). Top-down versus bottom-up control of attention in the prefrontal and posterior parietal cortices. *Science, 315,* 1860–1862.

Bushman, B. J., & Cooper, H. M. (1990). Effects of alcohol on human aggression: An integrative research review. *Psychological Bulletin, 107,* 341–354.

Butcher, L. M., Meaburn, E., Knight, J., Sham, P. C., Schalkwyk, L. C., Craig, I. W., et al. (2005). SNPs, microarrays and pooled DNA: Identification of four loci associated with mild mental impairment in a sample of 6000 children. *Human Molecular Genetics, 14,* 1315–1325.

Butterworth, B. (1999). A head for figures. *Science, 284,* 928–929.

Buxhoeveden, D. P., & Casanova, M. F. (2002). The minicolumn hypothesis in neuroscience. *Brain, 125,* 935–951.

Buydens-Branchey, L., Branchey, M. H., Noumair, D., & Lieber, C. S. (1989). Age of alcoholism onset. II. Relationship to susceptibility to serotonin precursor availability. *Archives of General Psychiatry, 46,* 231–236.

Byne, W., Tobet, S., Mattiace, L. A., Lasco, M. S., Kemether, E., Edgar, M. A., et al. (2001). The interstitial nuclei of the human anterior hypothalamus: An investigation of variation with sex, sexual orientation, and HIV status. *Hormones and Behavior, 40,* 86–92.

Cabeza, R., Anderson, N. D., Locantore, J. K., & McIntosh, A. R. (2002). Aging gracefully: Compensatory brain activity in high-performing older adults. *NeuroImage, 17,* 1394–1402.

Cadoret, R. J., Troughton, E., O'Gorman, T. W., & Heywood, E. (1986). An adoption study of genetic and environmental factors in drug abuse. *Archives of General Psychiatry, 43,* 1131–1136.

Caffeine prevents post-op headaches. (1996). *United Press International, MSNBC.* Retrieved March 21, 2002, from www.msnbc.com/news/36911.asp

Caggiula, A. R. (1970). Analysis of the copulation-reward properties of posterior hypothalamic stimulation in male rats. *Journal of Comparative and Physiological Psychology, 70,* 399–412.

Cahill, L. (2006). Why sex matters for neuroscience. *Nature Reviews Neuroscience, 7,* 477–484.

Cahill, L., Haier, R. J., Fallon, J., Alkire, M. T., Tang, C., Keator, D., et al. (1996). Amygdala activity at encoding correlated with long-term, free recall of emotional information. *Proceedings of the National Academy of Sciences, USA, 93,* 8016–8021.

Calles-Escandón, J., & Horton, E. S. (1992). The thermogenic role of exercise in the treatment of morbid obesity: A critical evaluation. *American Journal of Clinical Nutrition, 55,* 533S–537S.

Camp, N. J., & Cannon-Albright, L. A. (2005). Dissecting the genetic etiology of major depressive disorder using linkage analysis. *Trends in Molecular Medicine, 11,* 138–144.

Campbell, S. S., & Tobler, I. (1984). Animal sleep: A review of sleep duration across phylogeny. *Neuroscience and Biobehavioral Reviews, 8,* 269–300.

Camperio Ciani, A., Cermelli, P., & Zanzotto, G. (2008). Sexually antagonistic selection in human male homosexuality. *PLoS ONE, 3,* e2282. Retrieved February 5, 2010, from www.plosone.org/article/info%3Adoi%2F10.1371%2Fjournal.pone.0002282

Camperio Ciani, A., Corna, F., & Capiluppi, C. (2004). Evidence for maternally inherited factors favouring male homosexuality and promoting female fecundity. *Proceedings of the Royal Society of London B, 271,* 2217–2221.

Campfield, L. A., Smith, F. J., & Burn, P. (1998). Strategies and potential molecular targets for obesity treatment. *Science, 280,* 1383–1387.

Campione, J. C., Brown, A. L., & Ferrara, R. A. (1982). Mental retardation and intelligence. In R. J. Sternberg (Ed.), *Handbook of human intelligence* (pp. 391–490). Cambridge, UK: Cambridge University Press.

Canavan, A. G. M., Sprengelmeyer, R., Diener, H.-C., & Hömberg, V. (1994). Conditional associative learning is impaired in cerebellar disease in humans. *Behavioral Neuroscience, 108,* 475–485.

Canli, T., Qiu, M., Omura, K., Congdon, E., Haas, B. W., Amin, Z., et al. (2006). Neural correlates of epigenesis. *Proceedings of the National Academy of Sciences, USA, 103,* 16033–16038.

Cannon, M., Jones, P. B., & Murray, R. M. (2002). Obstetric complications and schizophrenia: Historical and meta-analytic review. *American Journal of Psychiatry, 159,* 1080–1092.

Cannon, W. B. (1942). "Voodoo" death. *American Anthropologist, 44,* 169–181.

Cantalupo, C., Oliver, J., Smith, J., Nir, T., Taglialatela, J. P., & Hopkins, W. D. (2009). The chimpanzee brain shows human-like perisylvian asymmetries in white matter. *European Journal of Neuroscience, 30,* 431- 438.

Cao, Y. Q., Mantyh, P. W., Carlson, E. J., Gillespie, A. M., Epstein, C. J., & Basbaum, A. I. (1998). Primary afferent tachykinins are required to experience moderate-to-intense pain. *Nature, 392,* 390–394.

Carelli, R. M. (2002). Nucleus accumbens cell firing during goal-directed behaviors for cocaine vs. 'natural' reinforcement. *Physiology and Behavior, 76,* 379–387.

Cariani, P. A. (2004). Temporal codes and computations for sensory representation and scene analysis. *IEEE Transactions on Neural Networks, 15,* 1100–1111.

Carlezon, W. A., & Wise, R. A. (1996). Rewarding actions of phencyclidine and related drugs in nucleus accumbens shell and frontal cortex. *Journal of Neuroscience, 16,* 3112–3122.

Carlson, M. (1981). Characteristics of sensory deficits following lesions of Brodmann's areas 1 and 2 in the postcentral gyrus of Macaca mulatta. *Brain Research, 204,* 424–430.

Carlsson, H.-E., Hagelin, J., & Hau, J. (2004). Implementation of the Three Rs in biomedical research. *Veterinary Record, 154,* 467–470.

Carmichael, H. (2004, December 6). Medicine's next level. *Newsweek,* pp. 45–50.

Carr, C. E., & Konishi, M. (1990). A circuit for detection of interaural time differences in the brain stem of the barn owl. *Journal of Neuroscience, 10,* 3227–3246.

Carraher, T. N., Carraher, D. W., & Schliemann, A. D. (1985). Mathematics in the streets and in schools. *British Journal of Developmental Psychology, 3,* 21–29.

Carreiras, M., Lopez, J., Rivero, F., & Corina, D. (2005). Neural processing of a whistled language. *Nature, 433,* 31–32.

Carrera, M. R., Ashley, J. A., Parsons, L. H., Wirsching, P., Koob, G. F., & Janda, K. D. (1995). Suppresssion of psychoactive effects of cocaine by active immunization. *Nature, 378,* 727–730.

Carroll, J. B. (1993). *Human cognitive abilities: A survey of factor-analytic studies.* Cambridge, UK: University of Cambridge Press.

Cartier, N., Hacein-Bey-Abina, S., Bartholomae, C. C., Veres, G., Schmidt, M., Kutschera, I., et al. (2009). Hematopoietic stem cell gene therapy with a lentiviral vector in X-linked adrenoleukodystrophy. *Science, 326,* 818–823.

Casanova, M. F., Switala, A. E., Trippe, J., & Fitzgerald, M. (2007). Comparative minicolumnar morphometry of three distinguished scientists. *Autism, 11,* 557–569.

Caspi, A., McClay, J., Moffitt, T. E., Mill, J., Martin, J., Craig, I. W., et al. (2002). Role of genotype in the cycle of violence in maltreated children. *Science, 297,* 851–854.

Caspi, A., Sugden, K., Moffitt, T. E., Taylor, A., Craig, I. W., Harrington, H., et al. (2003). Influence of life stress on depression: Moderation by a polymorphism in the 5-HTT gene. *Science, 301,* 386–389.

Castellanos, F. X., Lee, P. P., Sharp, W., Jeffries, N. O., Greenstein, D. K., Clasen, L. S., et al. (2002). Developmental trajectories of brain volume abnormalities in children and adolescents with attention-deficit/hyperactivity disorder. *Journal of the American Medical Association, 288,* 1740–1748.

Castellanos, F. X., Magulies, D. S., Kelly, C., Uddin, L. Q., Ghaffari, M., Kirsch, A., et al. (2008). Cingulate-precuneus interactions: A new locus of dysfunction in adult attention-deficit/hyperactivity disorder. *Biological Psychiatry, 63,* 332–337.

Castellanos, F. X., & Tannock, R. (2002). Neuroscience of attention-deficit/hyperactivity disorder: The search for endophenotypes. *Nature Reviews Neuroscience, 3,* 617–628.

Castelli, F., Frith, C., Happé, F., & Frith, U. (2002). Autism, Asperger syndrome and brain mechanisms for the attribution of mental states to animated shapes. *Brain, 125,* 1839–1849.

Caster Semenya must wait for IAAF decision before competing. (2010, January 14). Retrieved February 23, 2010, from www.guardian.co.uk/sport/2010/jan/14/caster-semenya-iaaf-athletics-south-africa

Cattaneo, A. (2010). Tanezumab, a recombinant humanized mAb against nerve growth factor for the treatment of acute and chronic pain. *Current Opinion in Molecular Therapeutics, 12,* 94–105.

Catterall, W. A. (1984). The molecular basis of neuronal excitability. *Science, 223,* 653–661.

Ceci, S. J., & Liker, J. (1986). A day at the races: A study of IQ, expertise, and cognitive complexity. *Journal of Experimental Psychology: General, 115,* 255–266.

Centers for Disease Control and Prevention. (2005). Annual smoking-attributable mortality, years of potential life lost, and productivity losses: United States, 1997–2001. *MMWR Morbidity and Mortality Weekly Report, 54,* 625–628.

Chagnon, Y. C., Pérusse, L., Weisnagel, S. J., Rankinen, T., & Bouchard, C. (1999). The human obesity gene map: The 1999 update. *Obesity Research, 8,* 89–117.

Chakrabarti, B., Bullmore, E., & Baron-Cohen, S. (2006). Empathizing with basic emotions: Common and discrete neural substrates. *Social Neuroscience, 1,* 364–384.

Chakrabarti, S., & Fombonne, E. (2005). Pervasive developmental disorders in preschool children: Confirmation of high prevalence. *American Journal of Psychiatry, 162,* 1133–1141.

Chamberlain, S. R., Menzies, L., Hampshire, A., Suckling, J., Fineberg, N. A., de Campo, N., et al. (2008). Orbitofrontal dysfunction in patients with obsessive-compulsive disorder and their unaffected relatives. *Science, 321,* 421–422.

Chan, B. L., Witt, R., Charrow, A. P., Magee, A., Howard, R., & Pasquina, P. F. (2007). Mirror therapy for phantom limb pain. *New England Journal of Medicine, 357,* 2206–2207.

Chapman, P. F., White, G. L., Jones, M. W., Cooper-Blacketer, D., Marshall, V. J., Irizarry, M., et al. (1999). Impaired synaptic plasticity and learning in aged amyloid precursor protein transgenic mice. *Nature Neuroscience, 2,* 271–276.

Charney, D. S., Woods, S. W., Krystal, J. H., & Heninger, G. R. (1990). Serotonin function and human anxiety disorders. In P. M. Whitaker-Azmitia & S. J. Peroutka (Eds.), *Annals of the New York Academy of Sciences. Special Issue: The Neuropharmacology of Serotonin, 600,* 558–573.

Chawla, D., Rees, G., & Friston, K. J. (1999). The physiological basis of atentional modulation in extrastriate visual areas. *Nature Neuroscience, 2,* 671–675.

Check, E. (2004). Cardiologists take heart from stem-cell treatment success. *Nature, 428,* 880.

Chemelli, R. M., Willie, J. T., Sinton, C. M., Elmquist, J. K., Scammell, T., Lee, C., et al. (1999). Narcolepsy in orexin knockout mice: Molecular genetics of sleep regulation. *Cell, 98,* 437–451.

Chen, D. F., Schneider, G. E., Martinou, J.-C., & Tonegawa, S. (1997). Bcl-2 promotes regeneration of severed axons in mammalian CNS. *Nature, 385,* 434–438.

Chen, J.-F., Xu, K., Petzer, J. P., Staal, R., Xu, Y.-H., Beilstein, M., et al. (2001). Neuroprotection by caffeine and A2a adenosine receptor inactivation in a model of Parkinson's disease. *Journal of Neuroscience, 21,* 1–6.

Chen, M. C., Hamilton, J. P., & Gotlib, I. H. (2010). Decreased hippocampal volume in healthy girls at risk of depression. *Archives of General Psychiatry, 67,* 270–276.

Chen, P., Goldberg, D. E., Kolb, B., Lanser, M., & Benowitz, L. I. (2002). Inosine induces axonal rewiring and improves behavioral outcome after stroke. *Proceedings of the National Academy of Sciences, USA, 99,* 9031–9036.

Chen, W., Johnson, S. L., Marcotti, W., Andrews, P. W., Moore, H. D., & Rivolta, M. N. (2009). Human fetal auditory stem cells can be expanded in vitro and differentiate into functional auditory neurons and hair cell-like cells. *Stem Cells, 27,* 1196–1204.

Chen, W. R., Lee, S., Kato, K., Spencer, D. D., Shepherd, G. M., & Wiliamson, A. (1996). Long-term modifications of synaptic efficacy in the human inferior and middle temporal cortex. *Proceedings of the National Academy of Sciences, USA, 93,* 8011–8015.

Cherney, N. I. (1996). Opioid analgesics: Comparative features and prescribing guidelines. *Drugs, 51,* 713–737.

Cherrier, M. M., Craft, S., & Matsumoto, A. H. (2003). Cognitive changes associated with supplementation of testosterone or dihydrotestosterone in mildly hypogonadal men: A preliminary report. *Journal of Andrology, 24,* 568–576.

Cherrier, M. M., Matsumoto, A. H., Amory, J. K., Ahmed, S., Bremner, W., Peskind, E. R., et al. (2005). The role of aromatization in testosterone supplementation. *Neurology, 64,* 290–296.

Chiang, M.-C., Barysheva, M., Shattuck, D. W., Lee, A. D., Madsen, S. K., Avedissian, C., et al. (2009). Genetics of brain fiber architecture and intellectual performance. *Journal of Neuroscience, 29,* 2212–2224.

Cho, A. K. (1990). Ice: A new dosage form of an old drug. *Science, 249,* 631–634.

Cho, C.-K., Smith, C. R., & Diamandis, E. P. (2010). Amniotic fluid proteome analysis from Down syndrome pregnancies for biomarker discovery. *Journal of Proteome Research, 9,* 3574–3582.

Chomsky, N. (1980). *Rules and representations.* New York: Columbia University Press.

Chou, T. C., Bjorkum, A. A., Gaus, S. E., Lu, J., Scammell, T. E., & Saper, C. B. (2002). Afferents to the ventrolateral preoptic nucleus. *Journal of Neuroscience, 22,* 977–990.

Chronis, A. M., Lahey, B. B., Pelham, W. E., Jr., Kipp, H. L., Baumann, B. L., & Lee, S. S. (2003). Psychopathology and substance abuse in parents of young children with attention-deficit/hyperactivity disorder. *Journal of the American Academy of Child and Adolescent Psychiatry, 42,* 1424–1432.

Chuang, R. S.-I., Jaffe, H., Cribbs, L., Perez-Reyes, E., & Swartz, K. J. (1998). Inhibition of T-type voltage-gated calcium channels by a new scorpion toxin. *Nature Neuroscience, 1,* 668–674.

Cicchetti, F., Seporta, S., Hauser, R. A., Parent, M., Saint-Pierre, M., Sanberg, P. R., et al. (2009). Neural transplants in patients with Huntington's disease undergo disease-like neuronal degeneration. *Proceedings of the National Academy of Sciences, 106,* 12483–12488.

Ciompi, L. (1980). Catamnestic long-term study on the course of life and aging of schizophrenics. *Schizophrenia Bulletin, 6,* 607–618.

Cirelli, C., Gutierrez, C. M., & Tononi, G. (2004). Extensive and divergent effects of sleep and wakefulness on brain gene expression. *Neuron, 41,* 35–43.

Clark, J. T., Kalra, P. S., & Kalra, S. P. (1985). Neuropeptide Y stimulates feeding but inhibits sexual behavior in rats. *Endocrinology, 117,* 2435–2442.

Clark, R. E., & Squire, L. R. (1998). Classical conditioning and brain systems: The role of awareness. *Science, 280,* 77–81.

Clarren, S. K., Alvord, E. C., Sumi, S. M., Streissguth, A. P., & Smith, D. W. (1978). Brain malformations related to prenatal exposure to alcohol. *Journal of Pediatrics, 92,* 64–67.

Clayton, J. D., Kyriacou, C. P., & Reppert, S. M. (2001). Keeping time with the human genome. *Nature, 409,* 829–831.

Cloninger, C. R. (1987). Neurogenetic adaptive mechanisms in alcoholism. *Science, 236,* 410–416.

Coccaro, E. F., & Kavoussi, R. J. (1997). Fluoxetine and impulsive aggressive behavior in personality-disordered subjects. *Archives of General Psychiatry, 54,* 1081–1088.

Cohen-Bendahan, C. C. C., van de Beek, C., & Berenbaum, S. A. (2005). Prenatal sex hormone effects on child and adult sex-typed behavior: Methods and findings. *Neuroscience and Biobehavioral Reviews, 29,* 353–384.

Cohen, J. D., & Tong, F. (2001). The face of controversy. *Science, 293,* 2405–2407.

Cohen, L. G., Celnik, P., Pascual-Leone, A., Corwell, B., Faiz, L., Dambrosia, J., et al. (1997). Functional relevance of cross-modal plasticity in blind humans. *Nature, 389,* 180.

Cohen, M. X., Schoene-Bake, J.-C., Elger, C. E., & Weber, B. (2009). Connectivity-based segregation of the human striatum predicts personality characteristics. *Nature Neuroscience, 12,* 32–34.

Cohen, N. J., Eichenbaum, H., Deacedo, B. S., & Corkin, S. (1985). Different memory systems underlying acquisition of procedural and declarative knowledge. *Annals of the New York Academy of Sciences, 444,* 54–71.

Cohen, P., & Friedman, J. M. (2004). Leptin and the control of metabolism: Role for stearoyl-CoA desaturase-1 (SCD-1). *Journal of Nutrition, 134,* 2455S–2463S.

Cohen, S., Frank, E., Doyle, W. J., Skoner, D. P., Rabin, B. S., & Gwaltney, J. M., Jr. (1998). Types of stressors that increase susceptibility to the common cold in healthy adults. *Health Psychology, 17,* 214–223.

Cohen, S., Tyrrell, D. A., & Smith, A. P. (1991). Psychological stress and susceptibility to the common cold. *New England Journal of Medicine, 325,* 606–612.

Colapinto, J. (2004, June 3). Gender gap: What were the real reasons behind David Reimer's suicide? *Slate.* Retrieved November 26, 2007, from www.slate.com/id/2101678

Colby, C. L., & Goldberg, M. E. (1999). Space and attention in parietal cortex. *Annual Review of Neuroscience, 22,* 319–349.

Cole, J. (1995). *Pride and a daily marathon.* Cambridge: MIT Press.

Cole, S. W., Kemeny, M. E, Fahey, J. L., Zack, J. A., & Naliboff, B. D. (2003). Psychological risk factors for HIV pathogenesis: Mediation by the autonomic nervous system. *Biological Psychiatry, 54,* 1444–1456.

Coleman, D. L. (1973). Effects of parabiosis of obese with diabetes and normal mice. *Diabetologia, 9,* 294–298.

Collaer, M. L., & Hines, M. (1995). Human behavioral sex differences: A role for gonadal hormones during early development? *Psychological Bulletin, 118,* 55–107.

Collyer brothers. (n.d.). In *Wikipedia.* Retrieved August 22, 2010, from http://en.wikipedia.org/wiki/Collyer_brothers

Colman, R. J., Anderson, R. M., Johnson, S. C., Kastman, E. K., Kosmatka, K. J., Beasley, T. M., et al. (2009). Caloric restriction delays disease onset and mortality in rhesus monkeys. *Science, 325,* 201–204.

Colt, E. W. D., Wardlaw, S. L., & Frantz, A. G. (1981). The effect of running on plasma â-endorphin. *Life Sciences, 28,* 1637–1640.

Comuzzie, A. G., & Allison, D. B. (1998). The search for human obesity genes. *Science, 280,* 1374–1377.

Conner, J. M., Darracq, M. A., Roberts, J., & Tuszynski, M. H. (2001). Nontropic actions of neurotrophins: Subcortical nerve growth factor gene delivery reverses age-related degeneration of primate cortical cholinergic innervation. *Proceedings of the National Academy of Sciences, USA, 98,* 1941–1946.

Connolly, K. (2009, November 23). Trapped in his own body for 23 years: The coma victim who screamed unheard. *The Guardian.* Retrieved August 31, 2010, from www.guardian.co.uk/world/2009/nov/23/man-trapped-coma-23-years

Considine, R. V., Sinha, M. K., Heiman, M. L., Kriauciunas, A., Stephens, T. W., Nyce, M. R., et al. (1996). Serum immunoreactive-leptin concentrations in normal-weight and obese humans. *New England Journal of Medicine, 334,* 292–295.

Constantinidis, C., & Steinmetz, M. A. (1996). Neuronal activity in posterior parietal area 7a during the delay periods of a spatial memory task. *Journal of Neurophysiology, 76,* 1352–1355.

Cooke, S. F., & Bliss, T. V. P. (2006). Plasticity in the human central nervous system. *Brain, 129,* 1659–1673.

Coon, D. (2001). *Introduction to psychology: Gateways to mind and behavior* (with InfoTrac; 9th ed.) Belmont, CA: Wadsworth/ Thomson Learning.

Corbett, D., & Wise, R. A. (1980). Intracranial self-stimulation in relation to the ascending dopaminergic systems of the midbrain: A moveable electrode mapping study. *Brain Research, 185,* 1–15.

Corbetta, M., Miezin, F. M., Dobmeyer, S., Shulman, G. L., & Petersen, S. E. (1990). Attentional modulation of neural processing of shape, color, and velocity in humans. *Science, 248,* 1556–1559.

Corkin, S. (1984). Lasting consequences of bilateral medial temporal lobectomy: Clinical course and experimental findings in H. M. *Seminars in Neurology, 4,* 249–259.

Corkin, S. (2002). What's new with the amnesic patient H. M.? *Nature Reviews Neuroscience, 3,* 153–160.

Corkin, S., Amaral, D. G., González, R. G., Johnson, K. A., & Hyman, B. T. (1997). HM's medial temporal lobe lesion: Findings from magnetic resonance imaging. *Journal of Neuroscience, 17,* 3964–3979.

Cosgrove, G. R., & Eskandar, E. (1998). *Thalamotomy and pallidotomy.* Retrieved September 11, 2007, from www .neurosurgery.mgh.harvard.edu/functional/pallidt.htm

Costall, B., & Naylor, R. J. (1992). Anxiolytic potential of 5-HT3 receptor antagonists. *Pharmacology and Toxicology, 70,* 157–162.

Courchesne, E., Pierce, K., Schumann, C. M., Redcay, E., Buckwalter, J. A., Kennedy, D. P., et al. (2007). Mapping early brain development in autism. *Neuron, 56,* 399–413.

Courtney, S. M., Ungerleider, L. G., Keil, K., & Haxby, J. V. (1997). Transient and sustained activity in a distributed neural system for human working memory. *Nature, 386,* 608–611.

Cox, D. J., Merkel, R. L., Kovatchev, B., & Seward, R. (2000). Effect of stimulant medication on driving performance of young adults with attention-deficit/hyperactivity disorder: A preliminary double-blind placebo controlled trial. *Journal of Nervous and Mental Disease, 188,* 230–234.

Cox, J. J., Reimann, F., Nicholas, A. K., Thornton, G., Roberts, E., Springell, K., et al. (2006). An *SCN9A* channelopathy causes congenital inability to experience pain. *Nature, 444,* 894–898.

Coyle, J. T. (2006). Glutamate and schizophrenia: Beyond the dopamine hypothesis. *Cellular and molecular neurobiology, 26,* 365–384.

Coyle, J. T., Tsai, G., & Goff, D. (2003). Converging evidence of NMDA receptor hypofunction in the pathophysiology of schizophrenia. *Annals of the New York Academy of Sciences, 1003,* 318–327.

Crane, G. E. (1957). Iproniazid (Marsilid®) phosphate, a therapeutic agent for mental disorders and debilitating diseases. *Psychiatry Research Reports, 8,* 142–152.

Crick, F. (1994). *The astonishing hypothesis: The scientific search for the soul.* New York: Scribner.

Crick, F., & Mitchison, G. (1995). REM sleep and neural nets. *Behavioral Brain Research, 69,* 147–155.

Crick, F. C., & Koch, C. (2005). What is the function of the claustrum? *Philosophical Transactions of the Royal Society of London, B, 360,* 1270–1279.

Crow, T. J. (1985). The two-syndrome concept: Origins and current status. *Schizophrenia Bulletin, 11,* 471–486.

Crowe, R. R. (1984). Electroconvulsive therapy: A current perspective. *New England Journal of Medicine, 311,* 163–167.

Culham, J., He, S., Dukelow, S., & Verstraten, F. A. J. (2001). Visual motion and the human brain: What has neuroimaging told us? *Acta Psychologica, 107,* 69–94.

Culp, R. E., Cook, A. S., & Housley, P. C. (1983). A comparison of observed and reported adult-infant interactions: Effects of perceived sex. *Sex Roles, 9,* 475–479.

Cummings, D. E., Clement, K., Purnell, J. Q., Vaisse, C., Foster, K. E., Frayo, R. S., et al. (2002). Elevated plasma ghrelin levels in Prader-Willi syndrome. *Nature Medicine, 8,* 643–644.

Cummings, D. E., Purnell, J. Q., Frayo. R. S., Schmidova, K., Wisse, B. E., & Weigle, D. S. (2001). A preprandial rise in plasma ghrelin levels suggests a role in meal initiation in humans. *Diabetes, 50,* 1714–1719.

Currie, P. J., & Coscina, D. V. (1996). Regional hypothalamic differences in neuropeptide Y-induced feeding and energy substrate utilization. *Brain Research, 737,* 238–242.

Curtis, M. A., Penney, E. B., Pearson, A. G., van Roon-Mom, W. M. C., Butterworth, N. J., Dragunow, M., et al. (2003). Increased cell proliferation and neurogenesis in the adult human Huntington's disease brain. *Proceedings of the National Academy of Sciences, USA, 100,* 9023–9027.

Cutler, W. B., Friedmann, E., & McCoy, N. L. (1998). Pheromonal influences on sociosexual behavior in men. *Archives of Sexual Behavior, 27,* 1–13.

Czeisler, C. A., Duffy, J. F., Shanahan, T. L., Brown, E. N., Mitchell, J. F., Rimmer, D. W., et al. (1999). Stability, precision, and near-24-hour period of the human circadian pacemaker. *Science, 284,* 2177–2181.

Czeisler, C. A., Johnson, M. P., Duffy, J. F., Brown, E. N., Ronda, J. M., & Kronauer, R. E. (1990). Exposure to bright light and darkness to treat physiologic maladaptation to night work. *New England Journal of Medicine, 322,* 1253–1259.

Czeisler, C. A., Moore-Ede, M. C., & Coleman, R. M. (1982). Rotating shift work schedules that disrupt sleep are improved by applying circadian principles. *Science, 217,* 460–463.

Czeisler, C. A., Richardson, G. S., Coleman, R. M., Zimmerman, J. C., Moore-Ede, M. C., Dement, W. C., et al. (1981). Chronotherapy: Resetting the circadian clocks of patients with delayed sleep phase insomnia. *Sleep, 4,* 1–21.

Czeisler, C. A., Shanahan, T. L., Klerman, E. B., Martens, H., Brotman, D. J., Emens, J. S., et al. (1995). Suppression of melatonin secretion in some blind patients by exposure to bright light. *New England Journal of Medicine, 332,* 6–11.

Czeisler, C. A., Weitzman, E. D., & Moore-Ede, M. C. (1980). Human sleep: Its duration and organization depend on its circadian phase. *Science, 210,* 1264–1267.

Dabbs, J. J., Jr., Frady, R. L., Carr, T. S., & Besch, N. F. (1987). Saliva testosterone and criminal violence in young adult prison inmates. *Psychosomatic Medicine, 49,* 174–182.

Dabbs, J. J., Jr., & Hargrove, M. F. (1997). Age, testosterone, and behavior among female prison inmates. *Psychosomatic Medicine, 59,* 477–480.

Dabbs, J. M., Jr., Carr, T. S., Frady, R. L., & Riad, J. K. (1995). Testosterone, crime, and misbehavior among 692 male prison inmates. *Personality and Individual Differences, 18,* 627–633.

Dabbs, J. M., Jr., & Mohammed, S. (1992). Male and female salivary testosterone concentrations before and after sexual activity. *Physiology and Behavior, 52,* 195–197.

Dacey, D. M., Llao, H.-W., Peterson, B. B., Robinson, F. R., Smith, V. C., Pokorny, J., et al. (2005). Melanopsin-expressing ganglion cells in primate retina signal colour and irradiance and project to the LGN. *Nature, 433,* 749–754.

Dalgleish, T. (2004). The emotional brain. *Nature Reviews Neuroscience, 5,* 582–589.

Dalley, J. W., Fryer, T. D., Brichard, L., Robinson, E. S. J., Theobald, D. E. H., Lääne, K., et al. (2007). Nucleus accumbens D2/3 receptors predict trait impulsivity and cocaine reinforcement. *Science, 315,* 1267–1270.

Dallos, P., & Cheatham, M. A. (1976). Production of cochlear potentials by inner and outer hair cells. *Journal of the Acoustical Society of America, 60,* 510–512.

Dalton, A. (2005, May/June). Sleepstalking: A child molester pleads unconsciousness. *Legal Affairs.* Retrieved September

20, 2005, from www.legalaffairs.org/issues/May-June-2005/scene_dalton_mayjun05.msp

Dalton, K. (1961). Menstruation and crime. *British Medical Journal, 3,* 1752–1753.

Dalton, K. (1964). *The premenstrual syndrome.* Springfield, IL: Thomas.

Dalton, K. (1980). Cyclical criminal acts in premenstrual syndrome. *Lancet, 2,* 1070–1071.

Dalton, R. (2000). NIH cash tied to compulsory training in good behavior. *Nature, 408,* 629.

Daly, M., & Wilson, M. (1988). *Homicide.* New York: Aldine de Gruyter.

Damasio, A. (1994). *Descartes' error: Emotion, reason, and the human brain.* New York: Putnam.

Damasio, A. R. (1985). Prosopagnosia. *Trends in Neurosciences, 8,* 132–135.

Damasio, A. R., Grabowski, T. J., Bechara, A., Damasio, H., Ponto, L. L. B., Parvizi, J., et al. (2000). Subcortical and cortical brain activity during the feeling of self-generated emotions. *Nature Neuroscience, 3,* 1049–1056.

Damasio, H., Grabowski, T., Frank, R., Galaburda, A. M., & Damasio, A. R. (1994). The return of Phineas Gage: Clues about the brain from the skull of a famous patient. *Science, 264,* 1102–1105.

Damasio, H., Grabowski, T. J., Tranel, D., Hichwa, R. D., & Damasio, A. R. (1996). A neural basis for lexical retrieval. *Nature, 380,* 499–505.

Damsma, G., Pfaus, J. G., Wenkstern, D., & Phillips, A. G. (1992). Sexual behavior increases dopamine transmission in the nucleus accumbens and striatum of male rats: Comparison with novelty and locomotion. *Behavioral Neuroscience, 106,* 181–191.

Dandy, J., & Nettelbeck, T. (2002). The relationship between IQ, homework, aspirations and academic achievement for Chinese, Vietnamese and Anglo-Celtic Australian school children. *Educational Psychology, 22,* 267–276.

Dang-Vu, T. T., McKinney, S. M., Buxton, O. M., Solet, J. M., & Ellenbogen, J. M. (2010). Spontaneous brain rhythms predict sleep stability in the face of noise. *Current Biology, 20,* R626–R627.

Daniel, D., Weinberger, D. R., Jones, D. W., Zigun, J. R., Coppola, R., Handel, S., et al. (1991). The effect of amphetamine on regional cerebral blood flow during cognitive activation in schizophrenia. *Journal of Neuroscience, 11,* 1907–1917.

Dapretto, M., Davies, M. S., Pfeifer, J. H., Scott, A. A., Sigman, M., Bookheimer, S. Y., et al. (2006). Understanding emotions in others: Mirror neuron dysfunction in children with autism spectrum disorders. *Nature Neuroscience, 9,* 28–30.

Dartnall, H. J. A., Bowmaker, J. K., & Mollon, J. D. (1983). Human visual pigments: Microspectrophotometric results from the eyes of seven persons. *Proceedings of the Royal Society of London, B, 220,* 115–130.

Darwin, C. (1859). *On the origin of species.* London: Murray.

Das, A., & Gilbert, C. D. (1995). Long-range horizontal connections and their role in cortical reorganization revealed by optical recording of cat primary visual cortex. *Nature, 375,* 780–784.

David, A. S., & Prince, M. (2005). Psychosis following head injury: A critical review. *Journal of Neurology and Neurosurgical Psychiatry, 76,* i53–i60.

Davidson, J. M. (1966). Characteristics of sex behavior in male rats following castration. *Animal Behavior, 14,* 266–272.

Davidson, R. J. (1992). Anterior cerebral asymmetry and the nature of emotion. (1992). *Brain Cognition, 20,* 125–151.

Davidson, R. J., Pizzagalli, D., Nitschke, J. B., & Putnam, K. (2002). Depression: Perspectives from affective neuroscience. *Annual Review of Psychology, 53,* 545–574.

Davis, C. M. (1928). Self selection of diet in newly weaned infants: An experimental study. *American Journal of Diseases of Children, 36,* 651–679.

Davis, J. O., Phelps, J. A., & Bracha, H. S. (1995). Prenatal development of monozygotic twins and concordance for schizophrenia. *Schizophrenia Bulletin, 21,* 357–366.

Davis, K. D., Hutchison, W. D., Lozano, A. M., Tasker, R. R., & Dostrovsky, J. O. (2000). Human anterior cingulate cortex neurons modulated by attention-demanding tasks. *Journal of Neurophysiology, 83,* 3575–3577.

Davis, M. (1992). The role of the amygdala in fear and anxiety. *Annual Review of Neuroscience, 15,* 353–375.

Davis, S. W., Dennis, N. A., Daselaar, S. M., Fleck, M. S., & Cabeza, R. (2008). Qué PASA? The posterior-anterior shift in aging. *Cerebral Cortex, 18,* 1201–1209.

Deacon, T. W. (1990). Rethinking mammalian brain evolution. *American Zoologist, 30,* 629–705.

Deadwyler, S. A., Porrino, L., Siegel, J. M., & Hampson, R. E. (2007). Systemic and nasal delivery of orexin-A (hypocretin-1) reduces the effects of sleep deprivation on cognitive performance in nonhuman primates. *Journal of Neuroscience, 27,* 14239–14247.

de Balzac, H. (1996). The pleasures and pains of coffee (Robert Onopa, Trans.). *Michigan Quarterly Review, 35,* 273–277. (Original work published 1839)

De Biasi, S., & Rustioni, A. (1988). Glutamate and substance P coexist in primary afferent terminals in the superficial laminae of spinal cord. *Proceedings of the National Academy of Sciences, USA, 85,* 7820–7824.

de Castro, J. M. (1993). Genetic influences on daily intake and meal patterns of humans. *Physiology and Behavior, 53,* 777–782.

deCharms, R. C., Maeda, F., Glover, G. H., Ludlow, D., Pauly, J. M., Soneji, D., Gabrieli, J. D. E., & Mackey, S. C. (2005). Control over brain activation and pain learned by using real-time functional MRI. *Proceedings of the National Academy of Sciences, 102,* 18626–18631.

Deep-Brain Stimulation for Parkinson's Disease Study Group. (2001). Deep-brain stimulation of the subthalamic nucleus or the pars interna of the globus pallidus in Parkinson's disease. *New England Journal of Medicine, 345,* 956–963.

De Felipe, C., Herrero, J. F., O'Brien, J. A., Palmer, J. A., Doyle, C. A., Smith, A. J. H., et al. (1998). Altered nociception, analgesia and aggression in mice lacking the receptor for substance P. *Nature, 392,* 394–397.

de Gelder, B., & Tamietto, M. (2007). Affective blindsight. *Scholarpedia, 2,* 3555. Retrieved July 24, 2010, from www.scholarpedia.org/article/Affective_blindsight

Degreef, G., Ashtari, M., Bogerts, B., Bilder, R. M., Jody, D. N., Alvir, J., et al. (1992). Volumes of ventricular system subdivisions measured from magnetic resonance images in first-episode schizophrenic patients. *Archives of General Psychiatry, 49,* 531–537.

Dehaene, S. (1997). *The number sense: How the mind creates mathematics.* New York: Oxford University Press.

Dehaene, S., & Cohen, L. (1997). Cerebral pathways for calculation: Double dissociation between rote verbal and quantitative knowledge of arithmetic. *Cortex, 33,* 219–250.

Dehaene, S., & Naccache, L. (2001). Towards a cognitive neuroscience of consciousness: Basic evidence and a workspace framework. *Cognition, 79,* 1–37.

Dehaene, S., Naccache, L., Cohen, L., Le Bihan, D., Mangin, J.-F., Poline, J.-B., et al. (2001). Cerebral mechanisms of word masking and unconscious repetition priming. *Nature Neuroscience, 4,* 752–758.

Dehaene, S., Spelke, E., Pinel, P., Stanescu, R., & Tsivkin, S. (1999). Sources of mathematical thinking: Behavioral and brain-imaging evidence. *Science, 284,* 970–974.

de Jonge, F. H., Louwerse, A. L., Ooms, M. P., Evers, P., Endert, E., & Van de Poll, N. E. (1989). Lesions of the SDN-POA inhibit sexual behavior of male Wistar rats. *Brain Research Bulletin, 23,* 483–492.

de Jonge, F. H., Oldenburger, W. P., Louwerse, A. L., & Van de Poll, N. E. (1992). Changes in male copulatory behavior after sexual exciting stimuli: Effects of medial amygdala lesions. *Physiology and Behavior, 52,* 327–332.

de Jongh, R., Bolt, I., Schermer, M., & Olivier, B. (2008). Botox for the brain: Enhancement of cognition, mood and pro-social behavior and blunting of unwanted memories. *Neuroscience and Biobehavioral Reviews, 32,* 760–776.

de la Fuente-Fernández, R., Ruth, T. J., Sossi, V., Schulzer, M., Calne, D. B., & Stoessl, A. J. (2001). Expectation and dopamine release: Mechanism of the placebo effect in Parkinson's disease. *Science, 293,* 1164–1166.

de Leon, J. (1996). Smoking and vulnerability for schizophrenia. *Schizophrenia Bulletin, 22,* 405–409.

de Leon, M. J., Convit, A., Wolf, O. T., Tarshish, C. Y., DeSanti, S., Rusinek, H., et al. (2001). Prediction of cognitive decline in normal elderly subjects with 2-[18F]fluoro-2-deoxy-D-glucose/positronemission tomography (FDG/PET). *Proceedings of the National Academy of Sciences, USA, 98,* 10966–10971.

DeLong, M. R. (2000). The basal ganglia. In E. R. Kandel, J. H. Schwartz, & T. M. Jessell (Eds.), *Principles of neural science* (4th ed., pp. 853–867). New York: McGraw-Hill.

DeLuca, G. C., Ebers, G. C., & Esiri, M. M. (2004). Axonal loss in multiple sclerosis: A pathological survey of the corticospinal and sensory tracts. *Brain, 127,* 1009–1018.

De Luca, V., Wang, H., Squassina, A., Wong, G. W. H., Yeomans, J., & Kennedy, J. L. (2004). Linkage of M5 muscarinic and a7-nicotinic receptor genes on 15q13 to schizophrenia. *Neuropsychobiology, 50,* 124–127.

De Luca, V., Wong, A. H. C., Muller, D. J., Wong, G. W. H., Tyndale, R. F., & Kennedy, J. L. (2004). Evidence of association between smoking and a7 nicotinic receptor subunit gene in schizophrenia patients. *Neuropsychopharmacology, 29,* 1522–1526.

DeMarco, A., Dalal, R. M., Kahanda, M., Mullapudi, U., Pai, J., Hammel, C., et al. (2008). Subchronic racemic gamma vinyl-GABA produces weight loss in Sprague Dawley and Zucker fatty rats. *Synapse, 62,* 870–872.

Dement, W. (1960). The effect of dream deprivation. *Science, 131,* 1705–1707.

De Meyer, G., Shapiro, F., Vanderstichele, H., Vanmechelen, E., Engelborghs, S., et al. (2010). Diagnosis-independent Alzheimer disease biomarker signature in cognitively normal elderly people. *Archives of Neurology, 67,* 949–956.

Demicheli, V., Jefferson, T., Rivetti, A., & Price, D. (2005). *Vaccines for measles, mumps and rubella in children.* Retrieved October 10, 2005, from The Cochrane Library, www.mrw.interscience.wiley.com/cochrane/clsysrev/articles/CD004407/frame.html

Denburg, N. L., Tranel, D., & Bechara, A. (2005). The ability to decide advantageously declines prematurely in some normal older persons. *Neuropsychologia, 43,* 1099–1106.

Denissenko, M. F., Pao, A., Tang, M., & Pfeifer, G. P. (1996). Preferential formation of Benzo[a]pyrene adducts at lung cancer mutational hotspots in P53. *Science, 274,* 430–432.

Deol, M. S., & Glueecksohn-Waelsch, S. (1979). The role of inner hair cells in hearing. *Nature, 278,* 250–252.

Depression Guideline Panel. (1993). *Clinical practice guideline number 5: Depression in primary care II* (AHRQ Publication No. 93–0551). Rockville, MD: U.S. Department of Health and Human Services.

Depue, R. A., & Iacono, W. G. (1989). Neurobehavioral aspects of affective disorders. *Annual Review of Psychology, 40,* 457–492.

Derogatis, L. R., Abeloff, M. D., & Melisaratos, N. (1979). Psychological coping mechanisms and survival time in metastatic breast cancer. *Journal of the American Medical Association, 242,* 1504–1508.

Descartes, R. (1984). Treatise on man (J. Cottingham, R. Stoothoff, & D. Murdoch, Trans.). *The philosophical writings of Descartes.* New York: Cambridge University Press. (Original work published about 1662)

Deshpande, G., Kerssens, C., Sebel, P. S., & Hu, X. (2010). Altered local coherence in the default mode network due to sevoflurane anesthesia. *Brain Research, 1318,* 110–121.

Desimone, R., Albright, T. D., Gross, C. G., & Bruce, C. (1984). Stimulus-selective properties of inferior temporal neurons in the macaque. *Journal of Neuroscience, 4,* 2051–2062.

D'Esposito, M., & Postle, B. R. (1999). The dependence of span and delayed response performance on prefrontal cortex. *Neuropsychologia, 37,* 1303–1315.

Designer babies—One clinic says the future is here & now. (2009, March 3). Retrieved November 15, 2009, from http://blogs.abcnews.com/theworldnewser/2009/03/designer-babies.html

Dessens, A. B., Slijper, F. M. E., & Drop, S. L. S. (2005). Gender dysphoria and gender change in chromosomal females with congenital adrenal hyperplasia. *Archives of Sexual Behavior, 34,* 389–397.

De Stefano, N., Matthews, P. M., Filippi, M., Agosta, F., De Luca, M., Bartolozzi, M. L., et al. (2003). Evidence of early cortical atrophy in MS. *Neurology, 60,* 1157–1162.

Deutsch, J. A. (1983). The cholinergic synapse and the site of memory. In J. A. Deutsch (Ed.), *The physiological basis of memory* (pp. 367–386). New York: Academic Press.

Deutsch, S. I., Rosse, R. B., Schwartz, B. L., Weizman, A., Chilton, M., Arnold, D. S., et al. (2005). Therapeutic implications of a selective a7 nicotinic receptor abnormality in schizophrenics. *Israeli Journal of Psychiatry and Related Sciences, 42,* 33–44.

De Valois, R. L. (1960). Color vision mechanisms in the monkey. *Journal of General Physiology, 43,* 115–128.

De Valois, R. L., Abramov, I., & Jacobs, G. H. (1966). Analysis of response patterns of LGN cells. *Journal of the Optical Society of America, 56,* 966–977.

De Valois, R. L., Thorell, L. G., & Albrecht, D. G. (1985). Periodicity of striate-cortex-cell receptive fields. *Journal of the Optical Society of America, 2,* 1115–1123.

Devanand, D. P., Dwork, A. J., Hutchinson, E. R., Bolwig, T. G., & Sackeim, H. A. (1994). Does ECT alter brain structure? *American Journal of Psychiatry, 151,* 957–970.

Devane, W. A., Hanus, L., Breuer, A., Pertwee, R. G., Stevenson, L. A., Griffin, G., et al. (1992). Isolation and structure of a brain constituent that binds to the cannabinoid receptor. *Science, 258,* 1946–1949.

Devue, C., Collette, F., Balteau, E., Degueldre, C., Luxen, A., Maquet, P., et al. (2007). Here I am: The cortical correlates of visual self-recognition. *Brain Research, 1143,* 169–182.

Dew, M. A., Reynolds, C. F., III, Buysse, D. J., Houck, P. R., Hoch, C. C., Monk, T. H., et al. (1996). Electroencephalographic sleep profiles during depression. Effects of episode duration and other clinical and psychosocial factors in older adults. *Archives of General Psychiatry, 53,* 148–156.

Dhond, R. P., Buckner, R. L., Dale, A. M., Marinkovic, K., & Halgren, E. (2001). Spatiotemporal maps of brain activity underlying word generation and their modification during repetition priming. *Journal of Neuroscience, 21,* 3564–3571.

Diamond, J. (1992). Turning a man. *Discover, 13,* 71–77.

Diamond, M. (1965). A critical evaluation of the ontogeny of human sexual behavior. *Quarterly Review of Biology, 40,* 147–175.

Diamond, M., & Sigmundson, H. K. (1997). Sex reassignment at birth: Long-term review and clinical implications. *Archives of Pediatric and Adolescent Medicine, 151,* 298–304.

Diamond, M. C., Scheibel, A. B., Murphy, G. M., Jr., & Harvey, T. (1985). On the brain of a scientist: Albert Einstein. *Experimental Neurology, 88,* 198–204.

Di Chiara, G. (1995). The role of dopamine in drug abuse viewed from the perspective of its role in motivation. *Drug and Alcohol Dependence, 38,* 95–137.

Dick, D., & Agrawal, A. (2008). The genetics of alcohol and other drug dependence. *Alcohol Research and Health, 31,* 111–118.

Dietert, R. R., & Dietert, J. M. (2008). Potential for early-life immune insult including developmental immunotoxicity in autism and autism spectrum disorders: Focus on critical windows of immune vulnerability. *Journal of Toxicology and Environmental Health, Part B: Critical Reviews, 11,* 660–680.

Dietis, N., Guerrini, R., Calo, G., Salvadori, S., Rowbotham, D. J., & Lambert, D. G. (2009). Simultaneous targeting of multiple opioid receptors: A strategy to improve side-effect profile. *British Journal of Anesthesia, 103,* 38–49.

Dietrich, T., Krings, T., Neulen, J., Willmes, K., Erberich, S., Thron, A., et al. (2001). Effects of blood estrogen level on cortical activation patterns during cognitive activation as measured by functional MRI. *NeuroImage, 13,* 425–432.

Dietz, V. (2009). Body weight supported gait training: From laboratory to clinical setting. *Brain Research Bulletin, 78,* i-vi.

DiFiglia, M., Sapp, E., Chase, K. O., Davies, S. W., Bates, G. P., Vonsattel, J. P., et al. (1997). Aggregation of huntingtin in neuronal intranuclear inclusions and dystrophic neurites in brain. *Science, 277,* 1990–1993.

Di Lorenzo, P. M., & Hecht, G. S. (1993). Perceptual consequences of electrical stimulation in the gustatory system. *Behavioral Neuroscience, 107,* 130–138.

Dimitrijevic, M. R., Gerasimenko, Y., & Pinter, M. M. (1998). Evidence for a spinal central pattern generator in humans. *Annals of the New York Academy of Sciences, 860,* 360–376.

Dinstein, L., Thomas, C., Humphreys, K., Minshew, N., Behrmann, M., & Heeger, D. J. (2010). Normal movement selectivity in autism. *Neuron, 66,* 461–469.

di Pellegrino, G., Fadiga, L., Fogassi, L, Gallese, V., & Rizzolatti, G. (1992). Understanding motor events: A neurophysiological study. *Experimental Brain Research, 91,* 176–180.

di Pellegrino, G., & Wise, S. P. (1993). Effects of attention on visuomotor activity in the premotor and prefrontal cortex of a primate. *Somatosensory and Motor Research, 10,* 245–262.

di Tomaso, E. , Beltramo, M., & Piomelli, D. (1996). Brain cannabinoids in chocolate. *Nature, 382,* 677–678.

Dkhissi-Benyahya, O., Rieux, C., Hut, R. A., & Cooper, H. M. (2006). Immunohistochemical evidence of a melanopsin cone in human retina. *Investigative Ophthalmology & Visual Science, 47,* 1636–1641.

Dodd, M. L., Klos, K. J., Bower, J. H., Geda, Y. E., Josephs, K. A., & Ahlskog, M. E. (2005). Pathological gambling caused by drugs used to treat Parkinson disease. *Archives of Neurology, 62,* 1–5.

Doheny eye institute researchers to test next generation of retinal implant. (2007). Retrieved May 4, 2010, from www.doheny.org/news/retinal_implant_news.html

Dollfus, S., Everitt, B., Ribeyre, J. M., Assouly-Besse, F., Sharp, C., & Petit, M. (1996). Identifying subtypes of schizophrenia by cluster analysis. *Schizophrenia Bulletin, 22,* 545–555.

Dombeck, D. A., Khabbaz, A. N., Collman, F., Adelman,T. L., & Tank, D. W. (2007). Imaging large-scale neural activity with cellular resolution in awake, mobile, mice. *Neuron, 56,* 43–57.

Dominguez, J. M., & Hull, E. M. (2001). Stimulation of the medial amygdala enhances medial preoptic dopamine release: Implications for male rat sexual behavior. *Brain Research, 917,* 225–229.

d'Orbán, P. T., & Dalton, J. (1980). Violent crime and the menstrual cycle. *Psychological Medicine, 10,* 353–359.

Dörner, G. (1974). Sex-hormone-dependent brain differentiation and sexual functions. In G. Dörner (Ed.), *Endocrinology of sex* (pp. 30–37). Leipzig, Germany: J. A. Barth.

do Rosario-Campos, M. C., Leckman, J. F., Curi, M., Quatrano, S., Katsovitch, L., Miguel, E. C., et al. (2005). A family study of early-onset obsessive-compulsive disorder. *American Journal of Medical Genetics, Part B, 136,* 92–97.

Doughty, C. J., Wells, J. E., Joyce, P. R., & Walsh, A. E. (2004). Bipolar-panic disorder comorbidity within bipolar disorder families: A study of siblings. *Bipolar Disorders, 6,* 245–252.

Dowling, J. E., & Boycott, B. B. (1966). Organization of the primate retina. *Proceedings of the Royal Society of London, B, 166,* 80–111.

Downing, P. E., Jiang, Y., Shuman, M., & Kanwisher, N. (2001). A cortical area selective for visual processing of the human body. *Science, 293,* 2470–2473.

Doyle, M. (2009, January 29) Medical marijuana raid raises question: What's Obama policy? Retrieved November 25, 2009, from www.mcclatchydc.com/254/story/61033.html

Dreifus, C. (1996, February 4). And then there was Frank. *New York Times Magazine,* 23–25.

Drevets, W. C. (2001). Neuroimaging and neuropathological studies of depression: Implications for the cognitive-emotional features of mood disorders. *Current Opinion in Neurobiology, 11,* 240–249.

Drevets, W. C., Price, J. L., Simpson, J. R., Todd, R. D., Reich, T., Vannier, M., et al. (1997). Subgenual prefrontal cortex abnormalities in mood disorders. *Nature, 386,* 824–827.

Drevets, W. C., & Raichle, M. E. (1995). Positron emission tomographic imaging studies of human emotional disorders. In M. S. Gazzaniga (Ed.), *The cognitive neurosciences* (pp. 1153–1164). Cambridge, MA: MIT Press.

Drevets, W. C., Videen, T. O., Price, J. L., Preskorn, S. H., Carmichael, S. T., & Raichle, M. E. (1992). A functional anatomical study of unipolar depression. *Journal of Neuroscience, 12,* 3628–3641.

Driver, J., & Mattingley, J. B. (1998). Parietal neglect and visual awareness. *Nature Neuroscience, 1,* 17–22.

Dronkers, N. F., Pinker, S., & Damasio, A. (2000). Language and the aphasias. In E. R. Kandel, J. H. Schwartz, & T. M. Jessell (Eds.), *Principles of neural science* (4th ed., pp. 1169–1187). New York: McGraw-Hill.

Drumi, C. (2009). Stem cell research: Toward greater unity in Europe? *Cell, 139,* 649–651.

Dryden, S., Wang, O., Frankish, H. M., Pickavance, L., & Williams, G. (1995). The serotonin (5-HT) antagonist methysergide increases neuropeptide Y (NPY) synthesis and secretion in the hypothalamus of the rat. *Brain Research, 699,* 12–18.

Duarte, C. B., Santos, P. F., & Carvalho, A. P. (1999). Corelease of two functionally opposite neurotransmitters by retinal amacrine cells: Experimental evidence and functional significance. *Journal of Neuroscience Research, 58,* 475–479.

Dudai, Y. (2004). The neurobiology of consolidations, or, how stable is the engram? *Annual Review of Psychology, 55,* 51–86.

Dudek, S. M., & Bear, M. F. (1992). Homosynaptic long-term depression in area CA1 of hippocampus and effects of *N*-methyl-D-aspartate receptor blockade. *Proceedings of the National Academy of Sciences, USA, 89,* 4363–4367.

Dujardin, K., Guerrien, A., & Leconte, P. (1990). Sleep, brain activation and cognition. *Physiology and Behavior, 47,* 1271–1278.

Duncan, J., Seitz, R. J., Kolodny, J., Bor, D., Herzog, H., Ahmed, A., et al. (2000). A neural basis for general intelligence. *Science, 289,* 457–460.

Dundon, W., Lynch, K. G., Pettinati, H. M., & Lipkin, C. (2004). Treatment outcomes in Type A and B alcohol dependence 6 months after serotonergic pharmacotherapy. *Alcoholism: Clinical and Experimental Research, 28,* 1065–1073.

Dunn, G. A., & Bale, T. L. (2009). Maternal high-fat diet promotes body length increases and insulin insensitivity in second-generation mice. *Endocrinology, 150,* 4999–5009.

Durand, J.-B., Nelissen, K., Joly, O., Wardak, C., Todd, J. T., Norman, J. F., et al. (2007). Anterior regions of monkey parietal cortex process visual 3D shape. *Neuron, 55,* 493–505.

Durand, V. M., & Barlow, D. H. (2006). *Essentials of abnormal psychology* (4th ed.). Belmont, CA: Thomson Wadsworth.

Durston, S., Tottenham, N. T., Thomas, K. M., Davidson, M. C., Eigsti, I.-M., Yang, Y., et al. (2003). Differential patterns of striatal activation in young children with and without ADHD. *Biological Psychiatry, 53,* 871–878.

Dutton, D. G., & Aron, A. P. (1974). Some evidence for heightened sexual attraction under conditions of high anxiety. *Journal of Personality and Social Psychology, 30,* 510–517.

Eagly, A. H. (1995). The science and politics of comparing women and men. *American Psychologist, 50,* 145–158.

Early communication about an ongoing safety review of Meridia (sibutramine hydrochloride). (2009, November 20). Retrieved January 14, 2009, from www.fda.gov/Drugs/DrugSafety/PostmarketDrugSafetyInformationforPatientsandProviders/DrugSafetyInformationforHeathcareProfessionals/ucm191650.htm

Early communication about an ongoing safety review orlistat (marketed as Alli and Xenical). (2009, August 24). Retrieved January 14, 2009, from www.fda.gov/Drugs/DrugSafety/PostmarketDrugSafetyInformationforPatientsandProviders/DrugSafetyInformationforHeathcareProfessionals/ucm179166.htm

Earnest, D. J., Liang, F.-Q., Ratcliff, M., & Cassone, V. M. (1999). Immortal time: Circadian clock properties of rat suprachiasmatic cell lines. *Science, 283,* 693–695.

Edenberg, H. J., Dick, D. M., Xuei, X., Tian, H., Almasy, L., Bauer, L. O., et al. (2004). Variations in *GABRA2,* encoding the 2a-subunit of the GABAA receptor, are associated with alcohol dependence and with brain oscillations. *American Journal of Human Genetics, 74,* 705–714.

Edgerton, V. R., & Roy, R. R. (2009). Robotic training and spinal cord plasticity. *Brain Research Bulletin, 78,* 4–12.

Ehrhardt, A. A., Meyer-Bahlburg, H. F. L., Rosen, L. R., Feldman, J. F., Veridiano, N. P., Zimmerman, I., & McEwen, B. S. (1985). Sexual orientation after prenatal exposure to exogenous estrogen. *Archives of Sexual Behavior, 14,* 57–77.

Ehrsson, H. H., Spence, C., & Passingham, R. E. (2004). That's my hand! Activity in premotor cortex reflects feeling of ownership of a limb. *Science, 305,* 875–877.

Ehtesham, M., Kabos, P., Kabosova, A., Neuman, T., Black, K. L., & Yu, J. S. (2002). The use of interleukin 12-secreting neural stem cells for the treatment of intracranial glioma. *Cancer Research, 62,* 5657–5663.

Einarsdottir, E., Carlsson, A., Minde, J., Toolanen, G., Svensson, O., Solders, G., et al. (2004). A mutation in the nerve growth factor beta gene *(NGFB)* causes loss of pain perception. *Human Molecular Genetics, 13,* 799–805.

Eksioglu, Y. Z., Scheffer, I. E., Cardenas, P., Knoll, J., DiMario, F., Ramsby, G., et al. (1996). Periventricular heterotopia: An x-linked dominant epilepsy locus causing aberrant cerebral cortical development. *Neuron, 16,* 77–87.

Elias, M. (2008, May 13). Another person with super-memory skills comes forward. *USA Today.* Retrieved June 3, 2010, from www.usatoday.com/news/health/2008–05–12-super-memory_N.htm

Ellis, L., & Ames, M. A. (1987). Neurohormonal functioning and sexual orientation: A theory of homosexuality-heterosexuality. *Psychological Bulletin, 101,* 233–258.

Ellison-Wright, I., & Bullmore, E. (2009). Meta-analysis of diffusion tensor imaging studies in schizophrenia. *Schizophrenia Research, 108,* 3–10.

Elmquist, J. K. (2001). Hypothalamic pathways underlying the endocrine, autonomic, and behavioral effects of leptin. *Physiology and Behavior, 74,* 703–708.

Enard, W., Khaitovich, P., Klose, J., Zöllner, S., Heissig, F., Giavalisco, P., et al. (2002). Intra- and interspecific variation in primate gene expression patterns. *Science, 296,* 340–343.

Enard, W., Przeworski, M., Fisher, S. E., Lai, C. S. L., Wiebe, V., Kitano, T., et al. (2002). Molecular evolution of *FOXP2,* a gene involved in speech and language. *Nature, 418,* 869–872.

Engel, A. K., König, P., Kreiter, A. K., & Singer, W. (1991). Interhemispheric synchronization of oscillatory neuronal responses in cat visual cortex. *Science, 252,* 1177–1179.

Engel, A. K., Kreiter, A. K., König, P., & Singer, W. (1991). Synchronization of oscillatory neuronal responses between striate and extrastriate visual cortical areas of the cat. *Proceedings of the National Academy of Sciences, USA, 88,* 6048–6052.

Engel, S. M., Miodovnik, A., Canfield, R. L., Zhu, C., Silva, M. J., Calafat, A. M., et al. (2010). Prenatal phthalate exposure is associated with childhood behavior and executive functioning. *Environmental Health Perspectives.* Retrieved July 6, 2010, from http://ehp03.niehs.nih.gov/article/info%3Adoi%2F10.1289%2Fehp.0901470

Engert, F., & Bonhoeffer, T. (1999). Dendritic spine changes associated with hippocampal long-term synaptic plasticity. *Nature, 399,* 66–70.

Enoch, M.-A. (2006). Genetic and environmental influences on the development of alcoholism: Resilience *vs.* risk. *Annals of the New York Academy of Sciences, 1094,* 193–201.

Enriori, P. J., Evans, A. E., Sinnayah, P., Jobst, E. E., Tonelli-Lemos, L., Billes, S. K., et al. (2007). Diet-induced obesity causes severe but reversible leptin resistance in arcuate melanocortin neurons. *Cell Metabolism, 5,* 181–194.

Eppig, C., Fincher, C. L., & Thornhill, R. (2010). Parasite prevalence and the worldwide distribution of cognitive ability. *Proceedings of the Royal Society B* [electronic version]. Retrieved June 30, 2010, from http://rspb.royalsocietypublishing.org/content/early/2010/06/29/rspb.2010.0973.full.pdf+html

Ercan-Sencicek, A. G., Stillman, A. A., Ghosh, A. K., Bilguvar, K., O'Roak, B. J., Mason, C. E., et al. (2010). L-Histidine decarboxylase and Tourette's syndrome. *New England Journal of Medicine, 362,* 1901–1908.

Erickson, K. I., Colcombe, S. J., Wadhwa, R., Bherer, L., Peterson, M. S., Scalf, P. E., et al. (2007). Training-induced plasticity in older adults: Effects of training on hemispheric asymmetry. *Neurobiology of Aging, 28,* 272–283.

Ernst, M., Zametkin, A. J., Matochik, J. A., Jons, P. H., & Cohen, R. M. (1998). DOPA decarboxylase activity in attention-deficit/hyperactivity disorder adults. A [fluorine-18]fluorodopa positron emission tomographic study. *Journal of Neuroscience, 18,* 5901–5907.

Ernulf, K. E., Innala, S. M., & Whitam, F. L. (1989). Biological explanation, psychological explanation, and tolerance of homosexuals: A cross-national analysis of beliefs and attitudes. *Psychological Reports, 65,* 1003–1010.

Evans, D. A., Funkenstein, H. H., Albert, M. S., Scherr, P. A., Cook, N. R., Chown, M. J., et al. (1989). Prevalence of Alzheimer's disease in a community population of older persons. Higher than previously reported. *Journal of the American Medical Association, 262,* 2551–2556.

Evarts, E. V. (1979, March). Brain mechanisms of movement. *Scientific American, 241,* 164–179.

Faedda, G. L., Tondo, L., Teicher, M. H., Baldessarini, R. J., Gelbard, H. A., & Floris, G. F. (1993). Seasonal mood disorders: Patterns of seasonal recurrence in mania and depression. *Archives of General Psychiatry, 50,* 17–23.

Fahle, M., & Daum, I. (1997). Visual learning and memory as functions of age. *Neuropsychologia, 35,* 1583–1589.

Faigel, H. C., Szuajderman, S., Tishby, O., Turel, M., & Pinus, U. (1995). Attention-deficit disorder during adolescence: A review. *Journal of Adolescent Health, 16,* 174–184.

Falk, D. (2009). New information about Albert Einstein's brain. *Frontiers in Evolutionary Neuroscience, 1,* 1–6.

Falk, D., Froese, N., Sade, D. S., & Dudek, B. C. (1999). Sex differences in brain/body relationships of Rhesus monkeys and humans. *Journal of Human Evolution, 36,* 233–238.

Falzi, G., Perrone, P., & Vignolo, L. A. (1982). Right-left asymmetry in anterior speech region. *Archives of Neurology, 39,* 239–240.

Fambrough, D. M., Drachman, D. B., & Satyamurti, S. (1973). Neuromuscular junction in myasthenia gravis: Decreased acetylcholine receptors. *Science, 182,* 293–295.

Fancher, R. E. (1979). *Pioneers of psychology.* New York: W. W. Norton.

Farooqi, I. S., Keogh, J. M., Yeo, G. S. H., Lank, E. J., Cheetham, T., & O'Rahilly, S. (2003). Clinical spectrum of obesity and mutations in the melanocortin 4 receptor gene. *New England Journal of Medicine, 348,* 1085–1095.

Farooqi, I. S., Matarese, G., Lord, G. M., Keogh, J. M., Lawrence, E., Agwu, C., et al. (2002). Beneficial effects of leptin on obesity, T cell hyporesponsiveness, and neuroendocrine/metabolic dysfunction of human congenital leptin deficiency. *Journal of Clinical Investigation, 110,* 1093–1103.

Farrer, C., Franck, N., Georgieff, N., Frith, C. D., Decety, J., & Jeannerod, M. (2003). Modulating the experience of agency. *NeuroImage, 18,* 324–333.

Farrer, C., & Frith, C. D. (2002). Experiencing oneself vs another person as being the cause of an action: The neural correlates of the experience of agency. *NeuroImage, 15,* 596–603.

Fatemi, S. H., Earle, J. A., & McMenomy, T. (2000). Reduction in Reelin immunoreactivity in hippocampus of subjects with schizophrenia, bipolar disorder and major depression. *Molecular Psychiatry, 5,* 654–663.

Faul, M., Xu, L., Wald, M. M., & Coronado, V. G. (2010). *Traumatic brain injury in the United States.* Centers for Disease Control and Prevention. Retrieved June 27, 2010, from www.cdc.gov/traumaticbraininjury/pdf/blue_book.pdf

Fausto-Sterling, A. (1993, March/April). The five sexes: Why male and female are not enough. *The Sciences,* 20–25.

Fawzy, F. I., Fawzy, N. W., Hyun, C. S., Elashoff, R., Guthrie, D., Fahey, J. L., et al. (1993). Malignant melanoma: Effects of an early structured psychiatric intervention, coping, and affective state on recurrence and survival 6 years later. *Archives of General Psychiatry, 50,* 681–689.

FDA approves memantine (Namenda) for Alzheimer's disease. (2003, October 17). U.S. Food and Drug Administration. Retrieved August 27, 2007, from www.fda.gov/bbs/topics/NEWS/2003/NEW00961.html

FDA approves first drug for treatment of chorea in Huntington's disease [press release]. (2008, August 15). Retrieved May 29, 2010, from www.fda.gov/NewsEvents/Newsroom/PressAnnouncements/2008/ucm116936.htm

FDA: Byetta label revised to include safety information on possible kidney problems. (2009, November 2). Retrieved January 15, 2010, from www.fda.gov/NewsEvents/Newsroom/PressAnnouncements/ucm188708.htm

Feinberg, T. E. (2001). *Altered egos: How the brain creates the self.* New York: Oxford University Press.

Feresin, E. (2009, October 30). Lighter sentence for murderer with 'bad genes.' *Nature News.* Retrieved March 17, 2010, from www.nature.com/news/2009/091030/full/news.2009.1050/html

Ferguson, J. N., Young, L. J., Hearn, E. F., Matzuk, M. M., Insel, T. R., & Winslow, J. T. (2000). Social amnesia in mice lacking the oxytocin gene. *Nature Genetics, 25,* 284–288.

Fergusson, D., Doucette, S., Glass, K. C., Shapiro, S., Healy, D., Hebert, P., et al. (2005). Association between suicide attempts and selective serotonin reuptake inhibitors: Systematic review of randomised controlled trials. *British Medical Journal, 330,* 396–402.

Fergusson, D. M., Woodward, L. J., & Horwood, L. J. (1998). Maternal smoking during pregnancy and psychiatric adjustment in late adolescence. *Archives of General Psychiatry, 55,* 721–727.

Fernández, G., Effern, A., Grunwald, T., Pezer, N., Lehnertz, K., Dümpelmann, M., et al. (1999). Real-time tracking of memory formation in the human rhinal cortex and the hippocampus. *Science, 285,* 1582–1585.

Ferrari, P. F., Gallese, V., Rizzolatti, G., & Fogassi, L. (2003). Mirror neurons responding to the observation of ingestive and communicative mouth actions in the monkey ventral premotor cortex. *European Journal of Neuroscience, 17,* 1703–1714.

Ferrarelli, F., Massimini, M., Sarasso, S., Casali, A., Riedner, B. A., Angelini, G., et al. (2010). Breakdown in cortical effective connectivity during midazolam-induced loss of consciousness. *Proceedings of the National Academy of Sciences, 107,* 2681–2686.

Fertuck, H. C., & Salpeter, M. M. (1974). Localization of acetylcholine receptor by 125I-labeled alpha-bungarotoxin binding at mouse motor endplates. *Proceedings of the National Academy of Sciences, USA, 71,* 1376–1378.

Fibiger, H. C., LePiane, F. G., Jakubovic, A., & Phillips, A. G. (1987). The role of dopamine in intracranial self-stimulation of the ventral tegmental area. *Journal of Neuroscience, 7,* 3888–3896.

Field, A. E., Coakley, E. H., Must, A., Spadano, J. L., Laird, N., Dietz, W. H., et al. (2001). Impact of overweight on the risk of developing common chronic diseases during a 10-year period. *Archives of Internal Medicine, 161,* 1581–1586.

Fiez, J. (1996). Cerebellar contributions to cognition. *Neuron, 16,* 13–15.

Fike, M. L. (1990). Clinical manifestations in persons with multiple personality disorder. *American Journal of Occupational Therapy, 44,* 984–990.

Fink, G. R., Markowitsch, H. J., Reinkemeier, M., Bruckbauer, T., Kessler, J., & Heiss, W.-D. (1998). Cerebral representation of one's own past: neural networks involved in autobiographical memory. *Journal of Neuroscience, 16,* 4275–4282.

Finkelstein, E. A., Trogdon, J. G., Cohen, J. W., & Dietz, W. (2009). Annual medical spending attributable to obesity: Payer- and service-specific estimates. *Health Affairs, 28,* w822–w831.

Fiorino, D. F., Coury, A., & Phillips, A. G. (1997). Dynamic changes in nucleus accumbens dopamine efflux during the Coolidge effect in male rats. *Journal of Neuroscience, 17,* 4849–4855.

Fischer, T. Z., & Waxman, S. G. (2010). Familial pain syndromes from mutations of the Na_v 1.7 sodium channel. *Annals of the New York Academy of Sciences, 1184,* 196–207.

Fiske, D. W., & Maddi, S. R. (1961). A conceptual framework. In D. W. Fiske & S. R. Maddi (Eds.), *Functions of varied experience* (pp. 11–56). Homewood, IL: Dorsey Press.

Fite, P. J., Stoppelbein, L., & Greening, L. (2009). Proactive and reactive aggression in a child psychiatric inpatient population. *Journal of Clinical Child and Adolescent Psychology, 38,* 199–205.

Fitzsimons, J. T. (1998). Angiotensin, thirst, and sodium appetite. *Physiological Reviews, 78,* 583–686.

Fitzsimons, J. T., & Moore-Gillon, M. J. (1980). Drinking and antidiuresis in response to reductions in venous return in the dog: Neural and endocrine mechanisms. *Journal of Physiology, 308,* 403–416.

Fleischer, J. G., Gally, J. A., Edelman, G. M., & Krichmar, J. L. (2007). Retrospective and prospective responses arising in a modeled hippocampus during maze navigation by a brain-based device. *Proceedings of the National Academy of Sciences USA, 104,* 3556–3561.

Fleming, R., Baum, A., Gisriel, M. M., & Gatchel, R. J. (1982). Mediating influences of social support on stress at Three Mile Island. *Journal of Human Stress, 8,* 14–22.

Flood, J. F., & Morley, J. E. (1991). Increased food intake by neuropeptide Y is due to an increased motivation to eat. *Peptides, 12,* 1329–1332.

Flor, H. (2008). Maladaptive plasticity, memory for pain and phantom limb pain: Review and suggestions for new therapies. Expert *Review of Neurotherapeutics, 8,* 809–818.

Flor, H., Braun, C., Elbert, T., & Birbaumer, N. (1997). Extensive reorganization of primary somatosensory cortex in chronic back pain patients. *Neuroscience Letters, 224,* 5–8.

Flor, H., Denke, C., Schaefer, M., & Grüsser, S. (2001). Effect of sensory discrimination training on cortical reorganisation and phantom limb pain. *Lancet, 357,* 1763–1764.

Flor, H., Diers, M., Christmann, C., & Koeppe, C. (2006). Mirror illusions of phantom hand movements. Brain activity mapped by fMRI. *NeuroImage, 31,* S159

Flor, H., Elbert, T., Knecht, S., Wienbruch, C., Pantev, C., Birbaumer, N., et al. (1995). Phantom-limb pain as a perceptual correlate of cortical reorganization following arm amputation. *Nature, 375,* 482–484.

Flynn, J. R. (1987). Massive IQ gains in 14 nations: What IQ tests really measure. *Psychological Bulletin, 101,* 171–191.

Fombonne, E. (2005). Epidemiological studies of pervasive developmental disorders. In F. Volkmar, R. Paul, A. Klin, & D. Cohen (Eds.), *Handbook of autism and pervasive developmental disorders* (pp. 42–69). Hoboken, NJ: Wiley.

Ford, J. M., Mathalon, D. H., Whitfield, S., Faustman, W. O., & Roth, W. T. (2002). Reduced communication between frontal and temporal lobes during talking in schizophrenia. *Biological Psychiatry, 51,* 485–492.

Foster, R. G. (2005). Neurobiology: Bright blue times. *Nature, 433,* 698–699.

Fotopoulou, A., Solms, M., & Turnbull, O. (2004). Wishful reality distortions in confabulation: A case report. *Neuropsychologia, 47,* 727–744.

Fouts, R. S., Fouts, D. S., & Schoenfeld, D. (1984). Sign language conversational interactions between chimpanzees. *Sign Language Studies, 42,* 1–12.

Fowler, C. D., Liu, Y., & Wang, Z. (2007). Estrogen and adult neurogenesis in the amygdala and hypothalamus. *Brain Research Reviews, 57,* 342–351.

Fowler, J. S., Volkow, N. D., Wang, G.-J., Pappas, N., Logan, J., MacGregor, R., et al. (1996). Inhibition of monoamine oxidase B in the brains of smokers. *Nature, 379,* 733–736.

Fowler, R. (1986). Howard Hughes: A psychological autopsy. *Psychology Today, May,* 22–33.

Fowler, J. S. et al. (2003). Low monoamine oxidase B levels in peripheral organs of smokers. *Proceedings of the National Academy of Sciences, 100,* 11600–11605.

Fowles, D. C. (1992). Schizophrenia: Diathesis-stress revisited. *Annual Review of Psychology, 43,* 303–336.

Fox, J. W., Lamperti, E. D., Eksioglu, Y. Z., Hong, S. E., Feng, Y., Graham, D. A., et al. (1998). Mutations in *filamin 1* prevent migration of cerebral cortical neurons in human periventricular heterotopia. *Neuron, 21,* 1315–1325.

Francis, H. W., Koch, M. E., Wyatt, J. R., & Niparko, J. K. (1999). Trends in educational placement and cost-benefit considerations in children with cochlear implants. *Archives of Otolaryngology: Head and Neck Surgery, 125,* 499–505.

Frankland, P. W., O'Brien, C., Ohno, M., Kirkwood, A., & Silva, A. J. (2001). α-CaMKII-dependent plasticity in the cortex is required for permanent memory. *Nature, 411,* 309–313.

Frankle, W. G., Lombardo, I., New, A. S., Goodman, M., Talbot, P. S., Huang, Y., et al. (2005). Brain serotonin transporter distribution in subjects with impulsive aggressivity: A positron emission study with [*11*C]McN 5652. *American Journal of Psychiatry, 162,* 915–923.

Franklin, T. R., Acton, P. D., Maldjian, J. A., Gray, J. D., Croft, J. R., Dackis, C. A., et al. (2002). Decreased gray matter concentration in the insular, orbitofrontal, cingulate, and temporal cortices of cocaine patients. *Biological Psychiatry, 51,* 134–142.

Fratiglioni, L., & Wang, H. X. (2000). Smoking and Parkinson's and Alzheimer's disease: Review of the epidemiological studies. *Behavioral Brain Research, 113,* 117–120.

Frayling, T. M., Timson, N. J., Weedon, M. N., Zeggini, E., Freathy, R. M., Lindgren, C., et al. (2007). A common variant in the *FTO* gene is associated with body mass index and predisposes to childhood and adult obesity. *Science, 316,* 889–894.

Freed, C. R., Greene, P. E., Breeze, R. E., Tsai, W.-Y., DuMouchel, W., Kao, R., et al. (2001). Transplantation of embryonic dopamine neurons for severe Parkinson's disease. *New England Journal of Medicine, 344,* 710–719.

Freed, W. J., de Medinaceli, L., & Wyatt, R. J. (1985). Promoting functional plasticity in the damaged nervous system. *Science, 227,* 1544–1552.

Freedman, M. S., Lucas, R. J., Soni, B., von Schantz, M., Muñoz, M., David-Gray, Z., et al. (1999). Regulation of mammalian circadian behavior by non-rod, non-cone ocular photoreceptors. *Science, 284,* 502–504.

Freitag, C. M., Staal, W., Klauck, S. M., Duketis, E., & Waltes, R. (2010). Genetics of autistic disorders: Review and clinical applications. *European Child and Adolescent Psychiatry, 19,* 169–178.

French, E. D. (1994). Phencyclidine and the midbrain dopamine system: Electrophysiology and behavior. *Neurotoxicology and Teratology, 16,* 355–362.

French roast. (1996, July). *Harper's Magazine, 293,* 28–30.

French, S. J., & Cecil, J. E. (2001). Oral, gastric, and intenstinal influences on human feeding. *Physiology and Behavior, 74,* 729–734.

Freud, S. (1900). *The interpretation of dreams.* London: Hogarth.

Freund, P., Schmidlin, E., Wannier, T., Bloch, J., Mir, A., Schwab, M. E., et al. (2006). Nogo-A-specific antibody treatment enhances sprouting and functional recovery after cervical lesion in adult primates. *Nature Medicine, 12,* 790–792.

Freund, P., Wannier, T., Schmidlin, E., Bloch, J., Mir, A., Schwab, M. E., et al. (2007). Anti-Nogo-A antibody treatment enhances sprouting of corticospinal axons rostral to a unilateral cervical spinal cord lesion in adult Macaque monkey. *Journal of Comparative Neurology, 502,* 644–659.

Fridman, E. A., Krimchansky, B. Z., Bonetto, M., Galperin, T., Gamzu, E. R., Leiguarda, R. C., et al. (2010). Continuous subcutaneous apomorphine for severe disorders of consciousness after traumatic brain injury. *Brain Injury, 24,* 636–641.

Fried, P., Watkinson, B., James, D., & Gray, R. (2002). Current and former marijuana use: Preliminary findings of a longitudinal study of effects on IQ in young adults. *Canadian Medical Association Journal, 166,* 887–891.

Fried, P. A. (1995). The Ottawa Prenatal Prospective Study (OPPS): Methodological issues and findings—it's easy to throw the baby out with the bath water. *Life Sciences, 56,* 23–24.

Friederici, A. D. (2006). The neural basis of language development and its impairment. *Neuron, 52,* 941–952.

Friedrich-Schiller-Universität Jena. (2010, August 9). Prosthesis with information at its fingertips: Hand prosthesis that eases phantom pain. *Science Daily.* Retrieved August 9, 2010, from www .sciencedaily.com/releases/2010/08/100806125508.htm

Frisén, L., Nordenström, A., Falhammar, H., Filipsson, H., Holmdahl, G., Janson, P. O., et al. (2009). Gender role behavior, sexuality, and psychosocial adaptation in women with congenital adrenal hyperplasia due to *CYP21A2* deficiency. *Journal of Clinical Endocrinology and Metabolism, 94,* 3432–3439.

Friston, K. J. (2009) Modalities, modes, and models in functional neuroimaging. *Science, 326,* 399–403.

Frith, U. (1993, June). Autism. *Scientific American, 268,* 108–114.

Frith, U., Morton, J., & Leslie, A. M. (1991). The cognitive basis of a biological disorder: Autism. *Trends in Neuroscience, 14,* 433–438.

Fritsch, G., & Hitzig, E. (1870). Über die elektrische Erregbarkeit des Grosshirns [Concerning the electrical stimulability of the cerebrum]. *Archiv für Anatomie Physiologie und Wissenschaftliche Medicin, 37,* 300–332.

Fritschy, J.-M., & Grzanna, R. (1992). Degeneration of rat locus coeruleus neurons is not accompanied by an irreversible loss of ascending projections. *Annals of the New York Academy of Sciences, 648,* 275–278.

From neurons to thoughts: Exploring the new frontier. (1998). *Nature Neuroscience, 1,* 1–2.

Fukuda, M., Mentis, M. J., Ma, Y., Dhawan, V., Antonini, A., Lang, A. E., et al. (2001). Networks mediating the clinical effects of pallidal brain stimulation for Parkinson's disease. *Brain, 124,* 1601–1609.

Fullerton, C. S., Ursano, R. J., Epstein, R. S., Crowley, B., Vance, K., Kao, T.-C., et al. (2001). Gender differences in posttraumatic stress disorder after motor vehicle accidents. *American Journal of Psychiatry, 158,* 1486–1491.

Fuster, J., & Jervey, J. P. (1981). Inferotemporal neurons distinguish and retain behaviorally relevant features of visual stimuli. *Science, 212,* 952–955.

Fuster, J. M. (1989). *The prefrontal cortex: Anatomy, physiology, and neuropsychology of the frontal lobe* (2nd ed.). New York: Raven Press.

Gabrieli, J. D. E. (2009). Dyslexia: A new synergy between education and cognitive neuroscience. *Science, 325,* 280–283.

Gabrieli, J. D. E. (1998). Cognitive neuroscience of human memory. *Annual Review of Psychology, 49,* 87–115.

Gackenbach, J., & Bosveld, J. (1989). *Control your dreams.* New York: Harper & Row.

Gage, F. H. (2000). Mammalian neural stem cells. *Science, 287,* 1433–1438.

Gai, E. J., Xie, H. M., Perin, J. C., Geiger, E., Glessner, J. T., deBerardinis, R., et al. (2010). Rare structural variants found in attention-deficit hyperactivity disorder are preferentially associated with neurodevelopmental genes. *Molecular Psychiatry, 15,* 637–646.

Gaillard, R., Dehaene, S., Adam, C., Clémenceau, S., Hasboun, D., Baulac, M., et al. (2009). Converging intracranial markers of conscious access. *PLoS Biology, 7,* e1000061. Retrieved August 25, 2010, from www.plosbiology.org/article/info:doi/10.1371/journal.pbio.1000061

Gaillard, R., Naccache, L., Pinel, P., Clémenceau, S., Volle, E., Hasboun, D., et al. (2006). Direct intracranial, fMRI, and lesion evidence for the causal role of left inferotemporal cortex in reading. *Neuron, 50,* 191–204.

Gainotti, G. (1972). Emotional behavior and hemispheric side of lesion. *Cortex, 8,* 41–55.

Gainotti, G., Caltagirone, C., & Zoccolotti, P. (1993). Left/right and cortical/subcortical dichotomies in the neuropsychological study of human emotions. *Cognition and Emotion, 7,* 71–94.

Galaburda, A. M. (1993). Neurology of developmental dyslexia. *Current Opinion in Neurobiology, 3,* 237–242.

Galaburda, A., & Livingstone, M. (1993). Evidence for a magnocellular defect in developmental dyslexia. *Annals of the New York Academy of Sciences, 682,* 70–82.

Gallese, V., & Goldman, A. (1998). Mirror neurons and the simulation theory of mind-reading. *Trends in Cognitive Sciences, 2,* 493–501.

Gallup, G. G. (1983). Toward a comparative psychology of mind. In R. L. Mellgren (Ed.), *Animal cognition and behavior* (pp. 473–510). New York: North-Holland.

Gallup, G., Jr., & Povinelli, D. J. (1998). Can animals empathize? *Scientific American Presents, Winter,* 66–75.

Galton, F. (1883). *Inquiries into human faculty and its development.* London: Macmillan.

Gannon, P. J., Holloway, R. L., Broadfield, D. C., & Braun, A. R. (1998). Asymmetry of chimpanzee planum temporale: Human-like pattern of Wernicke's brain language area homolog. *Science, 279,* 220–222.

Gantz, I., Erondu, N., Mallick, M., Musser, B., Krishna, R., Tanaka, W. K., et al. (2007). Efficacy and safety of intranasal peptide YY$_{3-36}$ for weight reduction in obese adults. *Journal of Clinical Endocrinology & Metabolism, 92,* 1754–1757.

Ganzel, B. L., Kim, P., Glover, G. H., & Temple, E. (2008). Resilience after 9/11: Multimodal neuroimaging evidence for stress-related change in the healthy adult brain. *NeuroImage, 40,* 788–795.

Gao, Q., & Horvath, T. L. (2007). Neurobiology of feeding and energy expenditure. *Annual Review of Neuroscience, 30,* 367–398.

Garavan, H., Pankiewicz, J., Bloom, A., Cho, J.-K., Sperry, L., Ross, T. J., et al. (2000). Cue-induced cocaine craving: Neuroanatomical specificity for drug users and drug stimuli. *American Journal of Psychiatry, 157,* 1789–1798.

Garb, J. L., & Stunkard, A. J. (1974). Taste aversions in man. *American Journal of Psychiatry, 131,* 1204–1207.

Garber, K. (2007). Autism's cause may reside in abnormalities at the synapse. *Science, 317,* 190–191.

Garbutt, J. C., West, S. L., Carey, T. S., Lohr, K. N., & Crews, F. T. (1999). Pharmacological treatment of alcohol dependence: A review of the evidence. *Journal of the American Medical Association, 281,* 1318–1325.

Garcia-Falgueras, A., & Swaab, D. F. (2008). A sex difference in the hypothalamic uncinate nucleus: Relationship to gender identity. *Brain, 131,* 3132–3146.

Garcia-Velasco, J., & Mondragon, M. (1991). The incidence of the vomeronasal organ in 1000 human subjects and its possible clinical significance. *Journal of Steroid Biochemistry and Molecular Biology, 39,* 561–563.

Gardner, E. P., & Kandel, E. R. (2000). Touch. In E. R. Kandel, J. H. Schwartz, & T. M. Jessell (Eds.), *Principles of neural science* (4th ed., pp. 451–471). New York: McGraw-Hill.

Gardner, E. P., Martin, J. H., & Jessell, T. M. (2000). The bodily senses. In E. R. Kandel, J. H. Schwartz, & T. M. Jessell (Eds.), *Principles of neural science* (4th ed., pp. 430–450). New York: McGraw-Hill.

Gardner, G., & Halweil, B. (2000). *Overfed and underfed: The global epidemic of malnutrition.* Washington, DC: Worldwatch Institute.

Gardner, H. (1975). *The shattered mind.* New York: Alfred A. Knopf.

Gartrell, N. K. (1982). Hormones and homosexuality. In W. Paul, J. D. Weinrich, J. C. Gonsiorek, & M. E. Hotvedt (Eds.), *Homosexuality: Social, psychological, and biological issues* (pp. 169–182). Beverly Hills, CA: Sage Publications.

Garver-Apgar, C. E., Gangestad, S. W., Thornhill, R., Miller, R. D., & Olp, J. J. (2006). Major histocompatibility complex alleles, sexual responsivity, and unfaithfulness in romantic couples. *Psychological Science, 17,* 830–835.

Gatchel, R. J. (1996). Psychological disorders and chronic pain: Cause-and-effect relationships. In R. J. Gatchel & D. C. Turk (Eds.), *Psychological approaches to pain management: A practitioner's handbook* (pp. 33–52). New York: Guilford Press.

Gauthier, I., Skudlarski, P., Gore, J. C., & Anderson, A. W. (2000). Expertise for cars and birds recruits brain areas involved in face recognition. *Nature Neuroscience, 3,* 191–197.

Gauthier, I., Tarr, M. J., Anderson, A. W., Skudlarski, P., & Gore, J. C. (1999). Activation of the middle fusiform "face area" increases with expertise in recognizing novel objects. *Nature Neuroscience, 2,* 568–573.

Gawin, F. H. (1991). Cocaine addiction: Psychology and neurophysiology. *Science, 251,* 1580–1586.

Gayán, J., & Olson, R. K. (2001). Genetic and environmental influences on orthographic and phonological skills in children with reading disabilities. *Developmental Neuropsychology, 20,* 483–507.

Gazzaley, A. H., Siegel, S. J., Kordower, J. H., Mufson, E. J., & Morrison, J. H. (1996). Circuit-specific alterations of N-methyl-D-aspartate receptor subunit 1 in the dentate gyrus of aged monkeys. *Proceedings of the National Academy of Sciences, USA, 93,* 3121–3125.

Gazzaniga, M. S. (1967, August). The split brain in man. *Scientific American, 217,* 24–29.

Gazzaniga, M. S. (1970). *The bisected brain.* New York: Appleton-Century-Crofts.

Gazzaniga, M. S., Ivry, R. B., & Mangun, G. R. (1998). *Cognitive neuroscience: The biology of the mind.* New York: Norton.

Gazzaniga, M. S. (2002). The split brain revisited. *Scientific American, 12*(1), 27–31.

Gazzola, V., Aziz-Zadeh, L., & Keysers, C. (2006). Empathy and the somatotopic auditory mirror system in humans. *Current Biology, 16,* 1824–1829.

Gebhardt, C. A., Naeser, M. A., & Butters, N. (1984). Computerized measures of CT scans of alcoholics: Thalamic region related to memory. *Alcohol, 1,* 133–140.

Gefter, A. (2008, Oct 22). Creationists declare war over the brain. *New Scientist,* 46–47.

Gegenfurtner, K. R., & Kiper, D. C. (2003). Color vision. *Annual Review of Neuroscience, 26,* 181–206.

Geinisman, Y., de Toledo-Morrell, L., Morrell, F., Persina, I. S., & Rossi, M. (1992). Age-related loss of axospinous synapses formed by two afferent systems in the rat dentate gyrus as revealed by the unbiased stereological dissector technique. *Hippocampus, 2,* 437–444.

Gene therapy notches another victory. (2005, June 6). *ScienceNOW.* Retrieved July 19, 2007, from www.sciencenow.sciencemag.org/cgi/content/full/2005/606/3

Genoux, D., Haditsch, U., Knobloch, M., Michalon, A., Storm, D., & Mansuy, I. M. (2002). Protein phosphatase 1 is a molecular constraint on learning and memory. *Nature, 418,* 970–975.

George, M. S., Nahas, Z., Borckardt, J. J., Anderson, B., Burns, C., Kose, S., et al. (2007). Vagus nerve stimulation for the treatment of depression and other neuropsychiatric disorders. *Expert Review of Neurotherapeutics, 7,* 63–74.

Georgopoulos, A. P., Taira, M., & Lukashin, A. (1993). Cognitive neurophysiology of the motor cortex. *Science, 260,* 47–52.

Gerloff, C., Bushara, K., Sailer, A., Wasserman, E. M., Chen, R., Matsuoka, T., Waldvogel, D., Wittenberg, G. F., Ishii, K., Cohen, L. G., & Hallett, M. (2006). Multimodal imaging of brain reorganization in motor areas of the contralesional hemisphere of well recovered patients after capsular stroke. *Brain, 129,* 791–808.

Gershon, E. S., Bunney, W. E., Leckman, J. F., Van Eerdewegh, M., & DeBauche, B. A. (1976). The inheritance of affective disorders: A review of data and of hypotheses. *Behavior Genetics, 6,* 227–261.

Geschwind, N. (1970). The organization of language and the brain. *Science, 170,* 940–944.

Geschwind, N. (1972, April). Language and the brain. *Scientific American, 226*(4), 76–83.

Geschwind, N. (1979, September). Specializations of the human brain. *Scientific American, 241,* 180–199.

Geschwind, N., & Levitsky, W. (1968). Human brain: Left-right asymmetries in temporal speech region. *Science, 161,* 186–187.

Gheusi, G., Cremer, H., McLean, H., Chazal, G., Vincent, J.-D., & Lledo, P.-M. (2000). Importance of newly generated neurons in the adult olfactory bulb for odor discrimination. *Proceedings of the National Academy of Sciences, USA, 97,* 1823–1828.

Ghez, C., & Krakauer, J. (2000). The organization of movement. In E. R. Kandel, J. H. Schwartz, & T. M. Jessell (Eds.), *Principles of neural science* (4th ed., pp. 653–673). New York: McGraw-Hill.

Ghez, C., & Thach, W. T. (2000). The cerebellum. In E. R. Kandel, J. H. Schwartz, & T. M. Jessell (Eds.), *Principles of neural science* (4th ed., pp. 832–852). New York: McGraw-Hill.

Ghilardi, J. R., Röhrich, H., Lindsay, T. H., Sevcik, M. A., Schwei, M. J., Kubota, K., et al. (2005). Selective blockade of the capsaicin receptor TRPV1 attenuates bone cancer pain. *The Journal of Neuroscience, 25,* 3126–3131.

Ghosh, A., & Sinha, A. K. (2007). A second Charak festival from Delhi. *Anthropologist, 9,* 289–294.

Gibbons, R. D., Brown, C. H., Hur, K., Marcus, S. M., Bhaumik, D. K., Erkens, J. A., et al. (2007). Early evidence on the effects of regulators' suicidality warnings on SSRI prescriptions and suicide in children and adolescents. *American Journal of Psychiatry, 164,* 1356–1363.

Gilad, Y., Wiebe, V., Przeworski, M., Lancet, D., & Pääbo, S. (2004). Loss of olfactory receptor genes coincides with the acquisition of full trichromatic vision in primates. *Public Library of Science Biology, 2,* 120–125.

Gilbert, C. D. (1993). Rapid dynamic changes in adult cerebral cortex. *Current Opinion in Neurobiology, 3,* 100–103.

Gilbertson, M. W., Shenton, M. E., Ciszewski, A., Kasai, K., Lasko, N. B., Orr, S. P., et al. (2002). Smaller hippocampal volume predicts pathologic vulnerability to psychological trauma. *Nature Neuroscience, 5,* 1242–1247.

Giles, D. E., Biggs, M. M., Rush, A. J., & Roffwarg, H. P. (1988). Risk factors in families of unipolar depression: I. Psychiatric illness and reduced REM latency. *Journal of Affective Disorders, 14,* 51–59.

Gitlin, M. J., & Altshuler, L. L. (1997). Unanswered questions, unknown future for one of our oldest medications. *Archives of General Psychiatry, 54,* 21–23.

Givens, B. S., & Olton, D. S. (1990). Cholinergic and GABAergic modulation of medial septal area: Effect on working memory. *Behavioral Neuroscience, 104,* 849–855.

Giza, B. K., Scott, T. R., & Vanderweele, D. A. (1992). Administration of satiety factors and gustatory responsiveness in the nucleus tractus solitarius of the rat. *Brain Research Bulletin, 28,* 637–639.

Gizewski, E. R., Krause, E., Schlamann, M., Happich, F., Ladd, M. E., Forsting, M., et al. (2009). Specific cerebral activation due to visual erotic stimuli in male-to-female transsexuals compared with male and female controls: An fMRI study. *Journal of Sexual Medicine, 6,* 440–448.

Gläscher, J., Rudrauf, D., Colom, R., Paul, L. K., Tranel, D., Damasio, H., et al. (2010). Distributed neural system for general intelligence revealed by lesion mapping. *Proceedings of the National Academy of Sciences, 107,* 4705–4709.

Glaser, R., Rice, J., Sheridan, J., Fertel, R., Stout, J., Speicher, C., et al. (1987). Stress-related immune suppression: Health implications. *Brain, Behavior, and Immunity, 1,* 7–20.

Glass, L. C., Qu, Y., Lenox, S., Kim, D., Gates, J. R., Brodows, R., et al. (2008). Effects of exenatide *versus* insulin analogues on weight change in subjects with type 2 diabetes: A pooled post-hoc analysis. *Current Medical Research and Opinion, 24,* 639–644.

Glasson, B. I., Duda, J. E., Murray, I. V. J., Chen, Q., Souza, J. M., Hurtig, H. I., et al. (2000). Oxidative damage linked to neuro degeneration by selective a-synuclein nitration in synucleinopathy lesions. *Science, 290,* 985–989.

Glees, P. (1980). Functional cerebral reorganization following hemispherectomy in man and after small experimental lesions in primates. In Bach-y-Rita, P. (Ed.), *Recovery of function: Theoretical considerations for brain injury rehabilitation* (pp. 106–125). Berne, Switzerland: Hans Huber.

Glessner, J. T., Wang, K., Cai, G., Korvatska, O., Kim, C. E., Wood, S., et al. (2009). Autism genome-wide copy number variation reveals ubiquitin and neuronal genes. *Nature, 459,* 569–573.

Glezer, L. S., Jiang, X., & Riesenhuber, M. (2009). Evidence for highly selective neuronal tuning to whole words in the "visual word form area." *Neuron, 62,* 199–204.

Gloor, P., Olivier, A., Quesney, L. F., Andermann, F., & Horowitz, S. (1982). The role of the limbic system in experiential phenomena of temporal lobe epilepsy. *Annals of Neurology, 12,* 129–144.

Goate, A., Chartier-Harlin, M. C., Mullan, M., Brown, J., Crawford, F., Fidani, L., et al. (1991). Segregation of a missense mutation in the amyloid precursor protein gene with familial Alzheimer's disease. *Nature, 349,* 704–706.

Goff, D. C., Hennen, J., Lyoo, I. K., Tsai, G., Wald, L. L., Evins, A. E., et al. (2002). Modulation of brain and serum glutamatergic concentrations following a switch from conventional neuroleptics to olanzapine. *Biological Psychiatry, 51,* 493–497.

Goff, D. C., Tsai, G., Levitt, J., Amico, E., Manoach, D., Schoenfeld, D. A., et al. (1999). A placebo-controlled trial of D-cycloserine added to conventional neuroleptics in patients with schizophrenia. *Archives of General Psychiatry, 56,* 21–27.

Gold, M. S. (1997). Cocaine (and crack): Clinical aspects. In J. H. Lowinson, P. Ruiz, R. B. Millman, & J. G. Langrod (Eds.), *Substance abuse: A comprehensive textbook* (pp. 181–199). Baltimore: Williams & Wilkins.

Gold, P. W., Goodwin, F. K., & Chrousos, G. P. (1988). Clinical and biochemical manifestations of depression: Relation to the neurobiology of stress. *New England Journal of Medicine, 319,* 348–353.

Goldberg, M., & Rosenberg, H. (1987). New muscle relaxants in outpatient anesthesiology. *Dental Clinics of North America, 31,* 117–129.

Goldberg, M. E., & Hudspeth, A. J. (2000). *The vestibular system.* In E. R. Kandel, J. H. Schwartz, & T. M. Jessell (Eds.), *Principles of neural science* (4th ed., pp. 801–815). New York: McGraw-Hill.

Goldman, D., Oroszi, G., & Ducci, F. (2005). The genetics of addictions: Uncovering the genes. *Nature Reviews Genetics, 6,* 521–532.

Goldman, N., Chen, M., Fujita, T., Xu, Q., Peng, W., Liu, W., et al. (2010). Adenosine A1 receptors mediate local antinociceptive effects of acupunctures. *Nature Neuroscience, 13* [electronic version]. Retrieved May 31, 2010, from www.nature.com/neuro/journal/vaop/ncurrent/full/nn.2562.html

Goldman-Rakic, P. S., Bates, J. F., & Chafee, M. V. (1992). The prefrontal cortex and internally generated motor acts. *Current Opinion in Neurobiology, 2,* 830–835.

Goldstein, E. B. (1999). *Sensation and perception* (5th ed.). Pacific Grove, CA: Brooks-Cole.

Goldstein, J. M., Scidman, L. J., Horton, N. J., Makris, N., Kennedy, D. N., Caviness, V. S., Jr., et al. (2001). Normal sexual dimorphism of the adult human brain assessed by *in vivo* magnetic resonance imaging. *Cerebral Cortex, 11,* 490–497.

Goldstein, R. Z., & Volkow, N. D. (2002). Drug addiction and its underlying neurobiological basis: Neuroimaging evidence for the involvement of the frontal cortex. *American Journal of Psychiatry, 10,* 1642–1652.

Gomez-Tortosa, E., Martin, E. M., Gaviria, M., Charbel, F., & Auman, J. I. (1995). Selective deficit of one language in a bilingual patient following surgery in the left perisylvian area. *Brain and Language, 48,* 320–325.

Gong, Y., Chang, L., Viola, K. L., Lacor, P. N., Lambert, M. P., Finch, C. E., et al. (2003). Alzheimer's disease-affected brain: Presence of oligomeric Ab ligands (ADDLs) suggests a molecular basis for reversible memory loss. *Proceedings of the National Academy of Sciences, USA, 100,* 10417–10422.

Goodell, E. W., & Studdert-Kennedy, M. (1993). Acoustic evidence for the development of gestural coordination in the speech of 2-year-olds: A longitudinal study. *Journal of Speech and Hearing Research, 36,* 707–727.

Goodman, D. C., Bogdasarian, R. S., & Horel, J. A. (1973). Axonal sprouting of ipsilateral optic tract following opposite eye removal. *Brain, Behavior, and Evolution, 8,* 27–50.

Goodwin, D. W. (1986). Heredity and alcoholism. *Annals of Behavioral Medicine, 8,* 3–6.

Gorelick, P. B., & Ross, E. D. (1987). The aprosodias: Further functional anatomical evidence for the organisation of affective language in the right hemisphere. *Journal of Neurology, Neurosurgery, and Psychiatry, 50,* 553–560.

Gorski, R. A. (1974). The neuroendocrine regulation of sexual behavior. In G. Newton & A. H. Riesen (Eds.), *Advances in psychobiology* (Vol. 2, pp. 1–58). New York: Wiley.

Gorski, R. A., Gordon, J. H., Shryne, J. E., & Southam, A. M. (1978). Evidence for a morphological sex difference within the medial preoptic area of the rat brain. *Brain Research, 148,* 333–346.

Gottesman, I. I. (1991). *Schizophrenia genesis: The origins of madness.* New York: Freeman.

Gottesman, I. I., & Bertelsen, A. (1989). Confirming unexpressed genotypes for schizophrenia. *Archives of General Psychiatry, 46,* 867–872.

Gottesman, I. I., McGuffin, P., & Farmer, A. E. (1987). Clinical genetics as clues to the "real genetics" of schizophrenia (A decade of modest gains while playing for time). *Schizophrenia Bulletin, 13,* 12–47.

Gougoux, F., Zatorre, R. J., Lassonde, M., Voss, P., & Lepore, F. (2005). A functional neuroimaging study of sound localization: Visual cortex activity predicts performance in early-blind individuals. *Public Library of Science Biology, 3,* 1–9.

Gouin, J.-P., Connors, J., Kiecolt-Glaser, J. K., Glaser, R., Malarkey, W. B., Atkinson, C., et al. (2010). Altered expression of circadian rhythm genes among individuals with a history of depression. *Journal of Affective Disorders, 126,* 161–166.

Gould, E., & Gross, C. G. (2002). Neurogenesis in adult mammals: Some progress and problems. *Journal of Neuroscience, 22,* 619–623.

Gouras, P. (1968). Identification of cone mechanisms in monkey ganglion cells. *Journal of Physiology, 199,* 533–547.

Grace, A. A. (1991). Phasic versus tonic dopamine release and the modulation of dopamine system responsivity: A hypothesis for the etiology of schizophrenia. *Neuroscience, 41,* 1–24.

Graham-Rowe, D. (2004, November 13). Brain implants that move. *New Scientist,* 25.

Grant, B. F., Stinson, F. S., Dawson, D. A., Chou, P., Dufour, M. C., Compton, W., et al. (2004a). Prevalence and co-occurrence of substance use disorders and independent mood and anxiety disorders. *Archives of General Psychiatry, 61,* 807–816.

Grant, B. F., Stinson, F. S., Dawson, D. A., Chou, S. P., Ruan, W. J., & Pickering, R. P. (2004b). Co-occurrence of 12-month alcohol and drug use disorders and personality disorders in the United States. *Archives of General Psychiatry, 61,* 361–368.

Grant, J. E., Potenza, M. N., Hollander, E., Cunningham-Williams, R., Nurminen, T., Smits, G., et al. (2006). Multicenter investigation of the opioid antagonist Nalmefene in the treatment of pathological gambling. *American Journal of Psychiatry, 163,* 303–312.

Grant, S., London, E. D., Newlin, D. B., Villemagne, V. L., Liu, X., Contoreggi, C., et al. (1996). Activation of memory circuits during cue-elicited cocaine craving. *Proceedings of the National Academy of Sciences, USA, 93,* 12040–12045.

Gray, J. (1992). *Men are from Mars, women are from Venus: A practical guide for improving communication and getting what you want in your relationships.* New York: HarperCollins.

Gray, J. R., & Thompson, P. M. (2004). Neurobiology of intelligence: Science and ethics. *Nature Neuroscience, 5,* 471–482.

Gray, K., & Wegner, D. M. (2008). The sting of intentional pain. *Psychological Science, 19,* 1260–1262.

Graybiel, A. M. (1998). The basal ganglia and chunking of action repertoires. *Neurobiology of Learning and Memory, 70,* 119–136.

Graybiel, A. M., & Rauch, S. L. (2000). Toward a neurobiology of obsessive-compulsive disorder. *Neuron, 28,* 343–347.

Graziano, M. S. A., Cooke, D. F., & Taylor, C. S. R. (2000). Coding the location of the arm by sight. *Science, 290,* 1782–1786.

Graziano, M. S. A., Hu, X. T., & Gross, C. G. (1997). Visuospatial properties of ventral premotor cortex. *Journal of Neurophysiology, 77,* 2268–2292.

Graziano, M. S. A., Taylor, C. S. R., & Moore, T. (2002). Complex movements evoked by microstimulation of precentral cortex. *Neuron, 34,* 841–851.

Graziano, M. S. A., Yap, G. S., & Gross, C. G. (1994). Coding of visual space by premotor neurons. *Science, 266,* 1054–1057.

Greene, P. E., & Fahn, S. (2002). Status of fetal tissue transplantation for the treatment of advanced Parkinson disease. *Neurosurgery Focus, 13,* 1–4.

Greenfeld, K., Avraham, R., Benish, M., Goldfarb, Y., Rosenne, E., Shapira, Y., et al. (2007). Immune suppression while awaiting surgery and following it: Dissociations between plasma cytokine levels, their induced production, and NK cell cytotoxicity. *Brain, Behavior, and Immunity, 21,* 503–513.

Greenfeld, L. A., & Henneberg, M. A. (2001). Victim and offender self-reports of alcohol involvement in crime. *Alcohol Research and Health, 25,* 20–31.

Greenough, W. T. (1975). Experiential modification of the developing brain. *American Scientist, 63,* 37–46.

Greenamyre, J. T., & Hastings, T. G. (2004). Parkinson's—Divergent causes, convergent mechanisms. *Science, 304,* 1120–1122.

Greer, S. (1991). Psychological response to cancer and survival. *Psychological Medicine, 21,* 43–49.

Gressens, P., Lammens, M., Picard, J. J., & Evrard, P. (1992). Ethanol-induced disturbances of gliogenesis and neurogenesis in the developing murine brain: An *in vitro* and an *in vivo* immunohistochemical and ultrastructural study. *Alcohol and Alcoholism, 27,* 219–226.

Grèzes, J., Armony, J. L., Rowe, J., & Passingham, R. E. (2003). Activations related to "mirror" and "canonical" neurones in the human brain: An fMRI study. *NeuroImage, 18,* 928–937.

Griffith, J. D., Cavanaugh, J., Held, J., & Oates, J. A. (1972). Dextroamphetamine: Evaluation of psychomimetic properties in man. *Archives of General Psychiatry, 26,* 97–100.

Grigson, P. S. (2002). Like drugs for chocolate: Separate rewards modulated by common mechanisms? *Physiology and Behavior, 76,* 345–346.

Grillner, S. (1985). Neurobiological bases of rhythmic motor acts in vertebrates. *Science, 228,* 143–149.

Grilo, C. M., & Pogue-Geile, M. F. (1991). The nature of environmental influences on weight and obesity: A behavior genetic anlaysis. *Psychological Bulletin, 110,* 520–537.

Grindlinger, H. M., & Ramsay, E. (1991). Compulsive feather-picking in birds. *Archives of General Psychiatry, 48,* 857.

Gross, C. G., Rocha-Miranda, C. E., & Bender, D. B. (1972). Visual properties of neurons in inferotemporal cortex of the macaque. *Journal of Neurophysiology, 35,* 96–111.

Gruenewald, D. A., & Matsumoto, A. M. (2003). Testosterone supplementation therapy for older men: Potential benefits and risks. *Journal of the American Geriatrics Society, 51,* 101–115.

Grueter, T. (2007, August/September). Forgetting faces. *Scientific American Mind,* 69–73.

Grunhaus, L., Schreiber, S., Dolberg, O. T., Polak, D., & Dannon, P. N. (2003). A randomized controlled comparison of electroconvulsive therapy and repetitive transcranial magnetic stimulation in severe and resistant nonpsychotic major depression. *Biological Psychiatry, 53,* 324–331.

Grüter, T., Grüter, M., & Carbon, C.-C. (2008). Neural and genetic foundations of face recognition and prosopagnosia. *Journal of Neuropsychology, 2,* 79–97.

Guastella, A. J., Einfeld, S. L., Gray, K. M., Rinehart, N. J., Tonge, B. J., Lambert, T. J., et al. (2010). Intranasal oxytocin improves emotion recognition for youth with autism spectrum disorders. *Biological Psychiatry, 67,* 692–694.

Guerreiro, M., Castro-Caldas, A., & Martins, I. P. (1995). Aphasia following right hemisphere lesion in a woman with left hemisphere injury in childhood. *Brain and Language, 49,* 280–288.

Guidelines for ethical conduct in the care and use of animals. (n.d.). Retrieved March 23, 2005, from www.apa.org/science/anguide .html

Guidotti, A., Auta, J., Davis, J. M., Gerevini, V. D., Dwivedi, Y., Grayson, D. R., et al. (2000). Decrease in Reelin and glutamic acid decarboxylase67 (GAD67) expression in schizophrenia and bipolar disorder. *Archives of General Psychiatry, 57,* 1061–1069.

Gujar, N., Yoo, S.-S., Hu, P., & Walker, M. P. (2010). The unrested resting brain: Sleep deprivation alters activity within the default-mode network. *Journal of Cognitive Neuroscience, 22,* 1637–1648.

Gur, R. C., Turetsky, B. I., Matsui, M., Yan, M., Bilker, W., Hughett, P., et al. (1999). Sex differences in brain gray and white matter in healthy young adults: Correlations with cognitive performance. *Journal of Neuroscience, 19,* 4065–4072.

Gurd, J. M., & Marshall, J. C. (1992). Drawing upon the mind's eye. *Nature, 359,* 590–591.

Gusella, J. F., Wexler, N. S., Conneally, P. M., Naylor, S. L., Anderson, M. A., Tanzi, R. E., et al. (1983). A polymorphic DNA marker genetically linked to Huntington's disease. *Nature, 306,* 234–238.

Gustafson, D., Lissner, L., Bengtsson, C., Björkelund, C., & Skoog, I. (2004). A 24-year follow-up of body mass index and cerebral atrophy. *Neurology, 63,* 1876–1881.

Gustavson, C. R., Garcia, J., Hankins, W. G., & Rusiniak, K. W. (1974). Coyote predation control by aversive conditioning. *Science, 184,* 581–583.

Gustavson, C. R., Kelly, D. J., Sweeney, M., & Garcia, J. (1976). Prey lithium aversions I: Coyotes and wolves. *Behavioral Biology, 17,* 61–72.

Guzowski, J. F., Knierim, J. J., & Moser, E. I. (2004). Ensemble dynamics of hippocampal regions of CA3 and CA1. *Neuron, 44,* 581–584.

Hadjkhani, N., Joseph, R. M., Snyder, J., Chabris, C. F., Clark J., Steele, S., et al. (2004). Activation of the fusiform gyrus when individuals with autism spectrum disorder view faces. *NeuroImage, 22,* 1141–1150.

Haier, R. J., Chueh, D., Touchette, P., Lott, I., Buchsbaum, M. S., MacMillan, D., et al. (1995). Brain size and cerebral glucose metabolic rate in nonspecific mental retardation and Down syndrome. *Intelligence, 20,* 191–210.

Haier, R. J., Jung, R. E., Yeo, R. A., Head, K., & Alkire, M. T. (2004). Structural brain variation and general intelligence. *NeuroImage, 23,* 425–433.

Haier, R. J., Siegel, B., Tang, C., Abel, L., & Buchsbaum, M. S. (1992). Intelligence and changes in regional cerebral glucose metabolic rate following learning. *Intelligence, 16,* 415–426.

Hairston, I. S., & Knight, R. T. (2004). Sleep on it. *Nature, 430,* 27–28.

Halaas, J. L., Gajiwala, K. S., Maffei, M., Cohen, S. L., Chait, B. T., Rabinowitz, D., et al. (1995). Weight-reducing effects of the plasma protein encoded by the *obese* gene. *Science, 269,* 543–546.

Hall, M.-H., Taylor, G., Salisbury, D. F., & Levy, D. L. (2010). Sensory gating event-related potentials and oscillations in schizophrenia patients and their unaffected relatives. *Schizophrenia Bulletin* (published online in advance of print April 2, 2010). doi:10.1093/schbul/sbq027

Hall, N. R. S., O'Grady, M., & Calandra, D. (1994). Transformation of personality and the immune system. *Advances: The Journal of Mind-Body Health, 10,* 7–15.

Hall, S. M., Reus, V. I., Muñoz, R. F., Sees, D. O., Humfleet, G., Hartz, D. T., et al. (1998). Nortriptyline and cognitive-behavioral therapy in the treatment of cigarette smoking. *Archives of General Psychiatry, 55,* 683–690.

Hallett, M. (2007). Transcranial magnetic stimulation: A primer. *Neuron, 55,* 187–196.

Hallmayer, J., Faraco, J., Lin, L., Hesselson, S., Winkelmann, J., Kawashima, M., et al. (2009). Narcolepsy is strongly associated with the T-cell receptor alpha locus. *Nature Genetics, 41,* 708–711.

Halmi, K. A., Eckert, E., LaDu, T. J., & Cohen, J. (1986). Anorexia nervosa: Treatment efficacy of cyproheptadine and amitriptyline. *Archives of General Psychiatry, 43,* 177–181.

Hallock, R. M., & Di Lorenzo, P. M. (2006). Temporal coding in the gustatory system. *Neuroscience and Biobehavioral Reviews, 30,* 1145–1160.

Hamann, S. B., Ely, T. D., Grafton, S. T., & Kilts, C. D. (1999). Amygdala activity related to enhanced memory for pleasant and aversive stimuli. *Nature Neuroscience, 2,* 289–293.

Hamer, D. H. (1999). Genetics and male sexual orientation. *Science, 285,* 803.

Hamer, D. H., Hu, S., Magnuson, V. L., Hu, N., & Pattatucci, A. M. L. (1993). A linkage between DNA markers on the X chromosome and male sexual orientation. *Science, 261,* 321–327.

Han, X., & Boyden, E. S. (2007). Multiple-color optical activation, silencing, and desynchronization of neural activity, with single-spike temporal resolution. *PLoS ONE, 2,* e299. Retrieved January 15, 2010, from www.plosone.org/article/info%3Adoi%2F10.1371%2Fjournal.pone.0000299

Hannibal, J., Hindersson, P., Knudsen, S. M., Georg, B., & Fahrenkrug, J. (2002). The photopigment melanopsin is exclusively present in pituitary adenylate cyclase-activating polypeptide-containing retinal ganglion cells of the retinohypothalamic tract. *Journal of Neuroscience, 22:RC191,* 1–7.

Happé, F., & Frith, U. (1996). The neuropsychology of autism. *Brain, 119,* 1377–1400.

Haqq, C. M., King, C.-Y., Ukiyama, E., Falsafi, S., Haqq, T. N., Donahoe, P. K., et al. (1994). Molecular basis of mammalian sexual determination: Activation of Müllerian inhibiting substance gene expression by SRY. *Science, 266,* 1494–1500.

Harada, S., Agarwal, D. P., Goedde, H. W., Tagaki, S., & Ishikawa, B. (1982). Possible protective role against alcoholism for aldehyde dehydrogenase isozyme deficiency in Japan. *Lancet, 2,* 827.

Hariri, A. R., Mattay, V. S., Tessitore, A., Kolachana, B., Fera, F., Goldman, D., et al. (2002). Serotonin transporter genetic variation and the response of the human amygdala. *Science, 297,* 400–403.

Harmon, L. D., & Julesz, B. (1973). Masking in visual recognition: Effects of two-dimensional filtered noise. *Science, 180,* 1194–1197.

Harold, D., Abraham, R., Hollingsworth, P., Sims, R., Gerrish, A., Hamshere, M. L., et al. (2009). Genome-wide association study identifies variants at *CLU* and *PICALM* associated with Alzheimer's disease. *Nature Genetics, 41,* 1088–1093.

Harrison, P. J. (2002). The neuropathology of primary mood disorder. *Brain, 125,* 1428–1449.

Hart, B. (1968). Role of prior experience on the effects of castration on sexual behavior of male dogs. *Journal of Comparative and Physiological Psychology, 66,* 719–725.

Hart, S. (2009). IAAF offers to pay for Caster Semenya's gender surgery if she fails verification test. *Telegraph* (UK), December 11. Retrieved February 23, 2010 from www.telegraph.co.uk/sport/othersports/athletics/6791558/IAAF-offers-to-pay-for-Caster-Semenyas-gender-surgery-if-she-fails-verification-test.html

Hartman, L. (1995). Cats as possible obsessive-compulsive disorder and medication models. *American Journal of Psychiatry, 152,* 1236.

Harvey, A. G., & Bryant, R. A. (2002). Acute stress disorder: a synthesis and critique. *Psychological Bulletin, 128,* 886–902.

Harvey, S. M. (1987). Female sexual behavior: Fluctuations during the menstrual cycle. *Journal of Psychosomatic Research, 31,* 101–110.

Hassabis, D., Chu, C., Rees, G., Weiskopf, N., Molyneux, P. D., & Maguire, E. A. (2009). Decoding neuronal ensembles in the human hippocampus. *Current Biology, 19,* 546–554.

Hastings, M. H., Reddy, A. B., & Maywood, E. S. (2003). A clockwork web: Circadian timing in brain and periphery, in health and disease. *Nature Reviews Neuroscience, 4,* 649–661.

Hattar, S., Liao, H.-W., Takao, M., Berson, D. M., & Yau, K.-W. (2002). Melanopsin-containing retinal ganglion cells: Architecture, projections, and intrinsic photosensitivity. *Science, 295,* 1065–1070.

Hauri, P. (1982). *Current concepts: The sleep disorders.* Kalamazoo, MI: Upjohn.

Hauser, M. D., MacNeilage, P., & Ware, M. (1996). Numerical representations in primates. *Proceedings of the National Academy of Sciences, USA, 93,* 1514–1517.

Häusser, M., & Smith, S. L. (2007). Controlling neural circuits with light. *Nature, 446,* 617–619.

Havlicek, J., Roberts, S. C., & Flegr, J. (2005). Women's preference for dominant male odour: Effects of menstrual cycle and relationship status. *Biology Letters, 1,* 256–259.

Haworth, C. M. A., Wright, M. J., Luciano, M., Martin, N. G., de Geus, E. J. C., van Beijsterveld, C. E. M., et al. (2009). The heritability of general cognitive ability increases linearly from childhood to young adulthood. *Molecular Psychiatry* [electronic version]. Retrieved June 30, 2010, from www.nature.com/mp/journal/vaop/ncurrent/abs/mp200955a.html

Haxby, J. V., Gobbini, M. I., Furey, M. L., Ishai, A., Schouten, J. L., & Pietrini, P. (2001). Distributed and overlapping representations of faces and objects in ventral temporal cortex. *Science, 293,* 2425–2430.

Hayes, K. J., & Hayes, C. (1953). Picture perception in a home-raised chimpanzee. *Journal of Comparative and Physiological Psychology, 46,* 470–474.

Heath, A. C., Bucholz, K. K., Madden , P. A. F., Dinwiddie, S. H., Slutske, W. S., Bierut, L. J., et al. (1997). Genetic and environmental contributions to alcohol dependence risk in a national twin sample: Consistency of findings in men and women. *Psychological Medicine, 27,* 1381–1396.

Heath, R. G. (1964). *The role of pleasure in behavior.* New York: Harper & Row.

Hebb, D. O. (1940). *The organization of behavior.* New York: Wiley-Interscience.

Hebert, L. E., Scherr, P. A., Bienias, J. L., Bennett, D. A., & Evans, D. A. (2003). Alzheimer disease in the US population: Prevalence estimates using the 2000 census. *Archives of Neurology, 60,* 1119–1122.

Hécaen, H., & Angelergues, R. (1964). Localization of symptoms in aphasia. In A. V. S. de Reuck & M. O'Connor (Eds.), *Ciba Foundation symposium: Disorders of language* (pp. 223–260). Boston: Little, Brown.

Hedges, L. V., & Nowell, A. (1995). Sex differences in mental test scores, variability, and numbers of high-scoring individuals. *Science, 269,* 41–45.

Heijmans, B. T., Tobi, E.W., Stein, A. D., Putter, H., Blauw, G. J., Susser, E. S., et al. (2008). Persistent epigenetic differences associated with prenatal exposure to famine in humans. *Proceedings of the National Academy of Sciences, 105,* 17046–17049.

Heilman, K. M., Watson, R. T., & Bowers, D. (1983). Affective disorders associated with hemispheric disease. In K. M. Heilman & P. Satz (Eds.), *Neuropsychology of human emotion* (pp. 45–64). New York: Guilford Press.

Heim, N. (1981). Sexual behavior of castrated sex offenders. *Archives of Sexual Behavior, 10,* 11–19.

Heller, W., Miller, G. A., & Nitschke, J. B. (1998). Lateralization in emotion and emotional disorders. *Current Directions in Psychological Science, 1,* 26–32.

Helmholtz, H. von. (1852). On the theory of compound colors. *Philosophical Magazine, 4,* 519–534.

Helmholtz, H. von. (1948). *On the sensations of tone as a physiological basis for the theory of music* (A. J. Ellis, Trans.). New York: P. Smith. (Original work published 1863)

Helmuth, L. (2002). A generation gap in brain activity. *Science, 296,* 2131–2132.

Hennenlotter, A., Dresel, C., Castrop, F., Ceballos Baumann, A. O., Wolschläger, A. M., & Haslinger, B. (2009). The link between facial feedback and neural activity within central circuitries of emotion—new insights from botulinum toxin-induced denervation of frown muscles. *Cerebral Cortex, 19,* 537–542.

Hennevin, E., Hars, B., Maho, C., & Bloch, V. (1995). Processing of learned information in paradoxical sleep: Relevance for memory. *Behavioral Brain Research, 69,* 125–135.

Hennig, J., Reuter, M., Netter, P., Burk, C., & Landt, O. (2005). Two types of aggression are differentially related to serotonergic activity and the A779C *TPH* polymorphism. *Behavioral Neuroscience, 119,* 16–25.

Hensley, S. (2010, February 17). Book-writing man in coma fails communication test. *National Public Radio.* Retrieved August 31, 2010, from www.npr.org/blogs/health/2010/02/bookwriting_man_in_coma_flunks.html

Herbert, T. B., Cohen, S., Marsland, A. L., Bachen, E. A., Rabin, B. S., Muldoon, M. F., et al. (1994). Cardiovascular reactivity and the course of immune response to an acute psychological stressor. *Psychosomatic Medicine, 56,* 337–344.

Heresco-Levy, U., & Javitt, D. C. (2004). Comparative effects of glycine and D-cycloserine on persistent negative symptoms in schizophrenia: A retrospective analysis. *Schizophrenia Research, 66,* 89–96.

Hering, E. (1878). *Zur lehre vom lichtsinne.* Vienna, Austria: Gerold.

Heritch, A. J. (1990). Evidence for reduced and dysregulated turnover of dopamine in schizophrenia. *Schizophrenia Bulletin, 16,* 605–615.

Herkenham, M. (1992). Cannabinoid receptor localization in brain: Relationship to motor and reward systems. *Annals of the New York Academy of Sciences, 654,* 19–32.

Herkenham, M. A., & Pert, C. B. (1982). Light microscopic localization of brain opiate receptors: A general autoradiographic method which preserves tissue quality. *Journal of Neuroscience, 2,* 1129–1149.

Hermann, L. (1870). Eine Erscheinung simultanen Contrastes. *Pflügers Archiv für die gesamte Physiologie, 3,* 13–15.

Hernán, M. A., Zhang, S. M., Lipworth, L., Olek, M. J., & Ascherio, A. (2001). Multiple sclerosis and age at infection with common viruses. *Epidemiology, 12,* 301–306.

Heron, M., Hoyert, D. L., Murphy, S. L., Xu, J., Kochanek, K. D., & Tejada-Vera, B. (2009). Deaths: Final data for 2006. *National Vital Statistics Reports, 57*(14). Retrieved June 27, 2010, from www.cdc.gov/nchs/data/nvsr/nvsr57/nvsr57_14.pdf

Hervey, G. R. (1952). The effects of lesions in the hypothalamus in parabiotic rats. *Journal of Physiology, 145,* 336–352.

Herzog, D. B., Dorer, D. J., Keel, P. K., Selwyn, S. E., Ekeblad, E. R., Flores, A. T., et al. (1999). Recovery and relapse in anorexia and bulimia nervosa: A 7.5-year followup study. *Journal of the American Academy of Child and Adolescent Psychiatry, 38,* 829–837.

Herzog, H. A. (1998). Understanding animal activism. In L. A. Hart (Ed.), *Responsible conduct with animals in research* (pp. 165–184). New York: Oxford University Press.

Hesselbrock, V., Begleiter, H., Porjesz, B., O'Connor, S., & Bauer, L. (2001). P$_{300}$ event-related potential amplitude as an endophenotype of alcoholism—evidence from the collaborative study on the genetics of alcoholism. *Journal of Biomedical Science, 8,* 77–82.

Heston, L. L. (1970). The genetics of schizophrenic and schizoid disease. *Science, 167,* 249–256.

Hettema, J. M., Neale, M. C., & Kendler, K. S. (2001). A review and meta-analysis of the genetic epidemiology of anxiety disorders. *American Journal of Psychiatry, 158,* 1568–1578.

Hettema, J. M., Prescott, C. A., Myers, J. M., Neale, M. C., & Kendler, K. S. (2005). The structure of genetic and environmental risk factors for anxiety disorders in men and women. *Archives of General Psychiatry, 62,* 182–189.

Heywood, C. A., & Kentridge, R.W. (2003). Achromatopsia, color vision, and cortex. *Neurological Clinics of North America, 21,* 483–500.

Hickok, G., Bellugi, U., & Klima, E. S. (1996). The neurobiology of sign language and its implications for the neural basis of language. *Nature, 381,* 699–702.

Hicks, B. M., Bernat, E., Malone, S. M., Iacono, W. G., Patrick, C. J., Krueger, R. F., et al. (2007). Genes mediate the association between P3 amplitude and externalizing disorders. *Psychophysiology, 44,* 98–105.

Hier, D. B., & Crowley, W. F., Jr. (1982). Spatial ability in androgendeficient men. *New England Journal of Medicine, 306,* 1202–1205.

Higley, J. D., Mehlman, P. T., Poland, R. E., Taub, D. M., Vickers, J., Suomi, S. J., et al. (1996). CSF testosterone and 5-HIAA correlate with different types of aggressive behaviors. *Biological Psychiatry, 40,* 1067–1082.

Higuchi, S., Usui, A., Murasaki, M., Matsushita, S., Nishioka, N., Yoshino, A., et al. (2002). Plasma orexin-A is lower in patients with narcolepsy. *Neuroscience Letters, 318,* 61–64.

Hill, J. O., & Peters, J. C. (1998). Environmental contributions to the obesity epidemic. *Science, 280,* 1371–1374.

Hill, J. O., Schlundt, D. G., Sbrocco, T., Sharp, T., Pope-Cordle, J., Stetson, B., et al. (1989). Evaluation of an alternating-calorie diet with and without exercise in the treatment of obesity. *American Journal of Nutrition, 50,* 248–254.

Hill, J. O., Wyatt, H. R., Reed, G. W., & Peters, J. C. (2003). Obesity and the environment: Where do we go from here? *Science, 299,* 853–855.

Hill, S. (1995). Neurobiological and clinical markers for a severe form of alcoholism in women. *Alcohol Health and Research World, 19*(3), 249–259.

Hill, S. Y., Muka, D., Steinhauer, S., & Locke, J. (1995). P300 amplitude decrements in children from families of alcoholic female probands. *Biological Psychiatry, 38,* 622–632.

Hines, M. (1982). Prenatal gonadal hormones and sex differences in human behavior. *Psychological Bulletin, 92,* 56–80.

Hines, P. J. (1997). Noto bene: Unconscious odors. *Science, 278,* 79.

Hines, T. (1998). Further on Einstein's brain. *Experimental Neurology, 150,* 343–344.

Hobson, J. A., & McCarley, R. W. (1977). The brain as a dream state generator: An activation-synthesis hypothesis of the dream process. *American Journal of Psychiatry, 134,* 1335–1348.

Hobson, J. A., & Pace-Schott, E. F. (2002). The cognitive neuroscience of sleep: Neuronal systems, consciousness and learning. *Nature Reviews Neuroscience, 3,* 679–693.

Hochberg, L. R., Serruya, M. D., Friehs, G. M., Muskand, J. A., Saleh, M., Caplan, A. H., Branner, A., Chen, D., Penn, R. D., & Donoghue, J. P. (2006). Neuronal ensemble control of prosthetic devices by a human with tetraplegia. *Nature, 442,* 164–171.

Hodgins, S., Kratzer, L., & McNeil, T. F. (2001). Obstetric complications, parenting, and risk of criminal behavior. *Archives of General Psychiatry, 58,* 746–752.

Hoebel, B. G., Monaco, A., Hernandes, L., Aulisi, E., Stanley, B. G., & Lenard, L. (1983). Self-injection of amphetamine directly into the brain. *Psychopharmacology, 81,* 158–163.

Hoffman, P. L., & Tabakoff, B. (1993). Ethanol, sedative hypnotics and glutamate receptor function in brain and cultured cells. *Alcohol and Alcoholism Supplement, 2,* 345–351.

Hohmann, A. G., Suplita, R. L., Bolton, N. M., Neely, M. H., Fegley, D., Mangieri, R., et al. (2005). An endocannabinoid mechanism for stress-induced analgesia. *Nature, 435,* 1108–1112.

Hökfelt, T., Johansson, O., & Goldstein, M. (1984). Chemical anatomy of the brain. *Science, 225,* 1326–1334.

Holden, C. (2004a). The origin of speech. *Science, 303,* 1316–1319.

Holden, C. (2004b). What's in a chimp's toolbox? [Electronic version]. *Science Now.* Retrieved August 13, 2005, from sciencenow.sciencemag.org/cgi/content/full/2004/1007/2

Hollander, E., Bartz, J., Chaplin, W., Phillips, A., Sumner, J., Soorya, L., et al. (2007). Oxytocin increases retention of social cognition in autism. *Biological Psychiatry, 61,* 498–503.

Hollander, E., Novotny, S., Hanratty, M., Yaffe, R., DeCarla, C. M., Aronowitz, B. R., et al. (2003). Oxytocin infusion reduces repetitive behaviors in adults with autistic and Asperger's disorders. *Neuropsychopharmacology, 28,* 193–198.

Holmes, C., Boche, D., Wilkinson, D., Yadegarfar, G., Hopkins, V., Bayer, A., et al. (2008). Long-term effects of Aβ$_{42}$ immunisation in Alzheimer's disease: Follow-up of a randomised, placebo-controlled phase 1 trial. *Lancet, 372,* 216–223.

Holmes, C., Cunningham, C., Zotova, E., Woolford, J., Dean, C., Kerr, S., et al. (2009). Systemic inflammation and disease progression in Alzheimer disease. *Neurology, 73,* 768–774.

Hölscher, C., Anwyl, R., & Rowan, M. J. (1997). Stimulation on the positive phase of hippocampal theta rhythm induces long-term potentiation that can be depotentiated by stimulation on the negative phase in area CA1 *in vivo. Journal of Neuroscience, 17,* 6470–6477.

Honea, R., Crow, T. J., Passingham, D., & MacKay, C. E. (2005). Regional deficits in brain volume in schizophrenia: A meta-analysis of voxel-based morphometry studies. *American Journal of Psychiatry, 162,* 2233–2245.

Hooks, B. M., & Chen, C. (2007). Critical periods in the visual system: Changing views for a model of experience-dependent plasticity. *Neuron, 56,* 312–326.

Hopfield, J. J., Feinstein, D. I., & Palmer, R. G. (1983). Unlearning has a stabilizing effect in collective memories. *Nature, 304,* 158–159.

Hopkins, W. D., & Morris, R. D. (1993). Hemispheric priming as a technique in the study of lateralized cognitive processes in chimpanzees: Some recent findings. In H. L. Roitblat, L. M. Herman, & P. E. Nachtigall (Eds.), *Language and communication: Comparative perspectives* (pp. 293–309). Hillsdale, NJ: Lawrence Erlbaum.

Hopson, J. S. (1979). *Scent signals: The silent language of sex.* New York: Morrow.

Horgan, J. (1999). *The undiscovered mind.* New York: Free Press.

Horne, J. (1988). *Why we sleep: The functions of sleep in humans and other mammals.* New York: Oxford University Press.

Horne, J. (1992). Human slow wave sleep: A review and appraisal of recent findings, with implications for sleep functions, and psychiatric illness. *Experientia, 48,* 941–954.

Horne, J. A., & Harley, L. J. (1989). Human SWS following selective head heating during wakefulness. In J. Horne (Ed.), *Sleep '88* (pp. 188–190). New York: Gustav Fischer Verlag.

Horne, J. A., & Moore, V. J. (1985). Sleep EEG effects of exercise with and without additional body cooling. *Electroencephalography and Clinical Neurophysiology, 60,* 33–38.

Horner, P. J., & Gage, F. H. (2000). Regenerating the damaged central nervous system. *Nature, 407,* 963–970.

Horovitz, S. G., Braun, A. R., Carr, W. S., Picchioni, D., Balkin, T. J., Fukunaga, M., et al. (2009). Decoupling of the brain's default mode network during deep sleep. *Proceedings of the National Academy of Sciences, 106,* 11376–11381.

Horvath, T. L., & Diano, S. (2004). The floating blueprint of hypothalamic feeding circuits. *Nature Reviews Neuroscience, 5,* 662–667.

Horvath, T. L., & Wikler, K. C. (1999). Aromatase in developing sensory systems of the rat brain. *Journal of Neuroendocrinology, 11,* 77–84.

Hoshi, E., Shima, K., & Tanji, J. (2000). Neuronal activity in the primate prefrontal cortex in the process of motor selection based on two behavioral rules. *Journal of Neurophysiology, 83,* 2355–2373.

Hoshi, E., & Tanji, J. (2000). Integration of target and body-part information in the premotor cortex when planning action. *Nature, 408,* 466–470.

Houeto, J. L., Karachi, C., Mallet, L., Pillon, B., Yelnik, J., Mesnage, V., et al. (2005). Tourette's syndrome and deep brain stimulation. *Journal of Neurology, Neurosurgery, and Psychiatry, 76,* 992–995.

House, J. S., Landis, K. R., & Umberson, D. (1988). Social relationships and health. *Science, 241,* 540–545.

Howlett, A. C., Bidaut-Russell, M., Devane, W. A., Melvin, L. S., Johnson, M. R., & Herkenham, M. (1990). The cannabinoid receptor: Biochemical, anatomical and behavioral characterization. *Trends in Neurosciences, 13,* 420–423.

Hser, Y.-I., Hoffman, V., Grella, C. E., & Anglin, M. D. (2001). A 33-year follow-up of narcotics addicts. *Archives of General Psychiatry, 58,* 503–508.

Hsiao, S. S., O'Shaughnessy, D. M., & Johnson, K. O. (1993). Effects of selective attention on spatial form processing in monkey primary and secondary somatosensory cortex. *Journal of Neurophysiology, 70,* 444–447.

Hu, S., Pattatucci, A. M. L., Patterson, C., Li, L., Fulker, D. W., Cherny, S. S., et al. (1995). Linkage between sexual orientation and chromosome Xq28 in males but not in females. *Nature Genetics, 11,* 248–256.

Hubbard, E. M., Arman, A. C., Ramachandran, V. S., & Boyton, G. M. (2005). Individual differences among grapheme-color synesthetes: Brain-behavior correlations. *Neuron, 45,* 975–985.

Hubel, D. H. (1982). Exploration of the primary visual cortex, 1955–78. *Nature, 299,* 515–524.

Hubel, D. H., & Wiesel, T. N. (1959). Receptive fields of single neurons in the cat's striate cortex. *Journal of Physiology, 148,* 574–591.

Hubel, D. H., & Wiesel, T. N. (1974). Sequence regularity and geometry of orientation columns in the monkey striate cortex. *Journal of Comparative Neurology, 158,* 267–294.

Huber, G., Gross, G., Schüttler, R., & Linz, M. (1980). Longitudinal studies of schizophrenic patients. *Schizophrenia Bulletin, 6,* 593–605.

Huberman, A. D. (2007). Mechanisms of eye-specific visual circuit development. *Current opinion in neurobiology, 17,* 73–80.

Hudspeth, A. J. (1983). Mechanoelectrical transduction by hair cells in the acousticolateralis sensory system. *Annual Review of Neuroscience, 6,* 187–215.

Hudspeth, A. J. (1989). How the ear's works work. *Nature, 341,* 397–404.

Hudspeth, A. J. (2000). Hearing. In E. R. Kandel, J. H. Schwartz, & T. M. Jessell (Eds.), *Principles of neural science* (4th ed., pp. 590–613). New York: McGraw-Hill.

Hughes, J. R., Kuhlman, D. T., Fichtner, C. G., & Gruenfeld, M. J. (1990). Brain mapping in a case of multiple personality. *Clinical Electoencephalography, 21,* 200–209.

Hull, C. L. (1951). *Essentials of behavior.* New Haven, CT: Yale University Press.

Hull, E. M., Lorrain, D. S., Du, J., Matuszewich, L., Lumley, L. A., Putnam, S. K., et al. (1999). Hormone-neurotransmitter interactions in the control of sexual behavior. *Behavioural Brain Research, 105,* 105–116.

Hulshoff Pol, H. E., Cohen-Kettenis, P. T., Van Haren, N. E. M., Peper, J. S., Brans, R. G. H., Cahn, W., et al. (2006). Changing your sex changes your brain: Influences of testosterone and estrogen on adult human brain structure. *European Journal of Endocrinology, 155,* S107-S114.

Humphreys, P., Kaufman, W. E., & Galaburda, A. M. (1990). Developmental dyslexia in women: Neuropathological findings in three patients. *Annals of Neurology, 28,* 727–738.

Hunt, G. L., & Hunt, M. W. (1977). Female-female pairing in western gulls *(Larus occidentalis)* in southern California. *Science, 196,* 1466–1467.

Hunt, G. L., Jr., Newman, A. L., Warner, M. H., Wingfield, J. C., & Kaiwi, J. (1984). Comparative behavior of male-female and female-female pairs among western gulls prior to egg laying. *The Condor, 86,* 157–162.

Huntington's Disease Collaborative Research Group. (1993). A novel gene containing a trinucleotide repeat that is expanded and unstable on Huntington's disease chromosomes. *Cell, 72,* 971–983.

Hurvich, L. M, & Jameson, D. (1957). An opponent-process theory of color vision. *Psychological Review, 64,* 384–404.

Hutchison, C. (2010, March 3). Pfizer Alzheimer's drug Dimebon fails in study. *abcNEWS/Health.* Retrieved June 16, 2010, from http://abcnews.go.com/Health/Alzheimers/pfizers-promising-alzheimers-drug-fails-study/story?id=9998774

Hutchison, W. D., Davis, K. D., Lozano, A. M., Tasker, R. R., & Dostrovsky, J. O. (1999). Pain-related neurons in the human cingulate cortex. *Nature Neuroscience, 2,* 403–405.

Huttunen, M. (1989). Maternal stress during pregnancy and the behavior of the offspring. In S. Doxiadis (Ed.), *Early influences shaping the individual* (pp. 175–182). New York: Plenum Press.

Hyde, J. S. (1996). Where are the gender differences? Where are the gender similarities? In D. M. Buss & N. M. Malamuth (Eds.), *Sex, power, conflict: Evolutionary and feminist perspectives* (pp. 107–118). New York: Oxford University Press.

Hyde, J. S., Lindberg, S. M., Linn, M. C., Ellis, A. B., & Williams, C. C. (2008). Gender similarities characterize math performance. *Science, 321,* 494–495.

Hyman, B. T., Van Horsen, G. W., Damasio, A. R., & Barnes, C. L. (1984). Alzheimer's disease: Cell-specific pathology isolates the hippocampal formation. *Science, 225,* 1168–1170.

Hyman, S. E., & Malenka, R. C. (2001). Addiction and the brain: The neurobiology of compulsion and its persistence. *Nature Reviews: Neuroscience, 2,* 695–703.

Hynd, G. W., & Semrud-Clikeman, M. (1989). Dyslexia and brain morphology. *Psychological Bulletin, 106,* 447–482.

Hyvärinen, J., & Poranen, A. (1978). Movement-sensitive cutaneous receptive fields in the hand area of the post-central gyrus in monkeys. *Journal of Physiology, 283,* 523–537.

Iacoboni, M., Molnar-Szakacs, I., Gallese, V., Buccino, G., Mazziotta, J. C., & Giacomo, R. (2005). Grasping the intentions of others with one's own mirror neuron system. *Public Library of Science Biology, 3,* 529–535.

Iacoboni, M., Woods, R. P., Brass, M., Bekkering, H., Mazziotta, J. C., & Rizzolatti, G. (1999). Cortical mechanisms of human imitation. *Science, 286,* 2526–2528.

Iacono, D., Markesbery, W. R., Gross, M., Pletnikova, O., Rudow, G., Zandi, P., et al. (2009). Clinically silent AD, neuronal hypertrophy, and linguistic skills in early life. *Neurology, 73,* 665–673.

Iacono, W. G., Carlson, S. R., Malone, S. M., & McGue, M. (2002). P3 event-related potential amplitude and the risk for disinhibitory disorders in adolescent boys. *Archives of General Psychiatry, 59,* 750–757.

Ibáñez, A., Blanco, C. Perez de Castro, I., Fernandez-Piqueras, J., & Sáiz-Ruiz, J. (2003). Genetics of pathological gambling. *Journal of Gambling Studies, 19,* 11–22.

Iemmola, F., & Camperio Ciani, A. (2009). New evidence of genetic factors influencing sexual orientation in men: Gemale fecundity increase in the maternal line. *Archives of Sexual Behavior, 38,* 393–399.

Iijima, M., Arisaka, O., Minamoto, F., & Arai, Y. (2001). Sex differences in children's free drawings: A study on girls with congenital adrenal hyperplasia. *Hormones and Behavior, 40,* 99–104.

Immunization Safety Review Committee. (2004). *Immunization safety review: Vaccines and autism.* Institute of Medicine. Retrieved August 11, 2005, from www.nap.edu/catalog/10997.html

Imperato-McGinley, J., Peterson, R. E., Gautier, T., & Sturla, E. (1979). Androgen and the evolution of male-gender identity among male pseudohermaphrodites with 5a-reductase deficiency. *New England Journal of Medicine, 300,* 1233–1237.

Imperato-McGinley, J., Peterson, R. E., Stoller, R., & Goodwin, W. E. (1979). Male pseudohermaphroditism secondary to 17a-hydroxysteroid dehydrogenase deficiency: Gender role change with puberty. *Journal of Clinical Endocrinology and Metabolism, 49,* 391–395.

Imperato-McGinley, J., Pichardo, M., Gautier, T., Voyer, D., & Bryden, M. P. (1991). Cognitive abilities in androgen-insensitive subjects: Comparison with control males and females from the same kindred. *Clinical Endocrinology, 34,* 341–347.

In China, DNA tests on kids ID genetic gifts, careers. (2009, August 3). Retrieved November 15, 2009, from www.cnn.com/2009/WORLD/asiapcf/08/03/china.dna.children.ability/index.html

Ingelfinger, F. J. (1944). The late effects of total and subtotal gastrectomy. *New England Journal of Medicine, 231,* 321–327.

Ingram, D. K., Zhu, M., Mamczarz, J., Zou, S., Lane, M. A., Roth, G. S., & deCabo, R. (2006). Calorie restriction mimetics: An emerging research field. *Aging Cell, 5,* 97–108.

Inouye, S.-I. T., & Kawamura, H. (1979). Persistence of circadian rhythmicity in a mammalian hypothalamic "island" containing the suprachiasmatic nucleus. *Proceedings of the National Academy of Sciences, USA, 76,* 5962–5966.

Insel, T. R. (1992). Toward a neuroanatomy of obsessive-compulsive disorder. *Archives of General Psychiatry, 49,* 739–744.

Insel, T. R., Zohar, J., Benkelfat, C., & Murphy, D. L. (1990). Serotonin in obsessions, compulsions, and the control of aggressive impulses. *Annals of the New York Academy of Sciences. Special Issue: The Neuropharmacology of Serotonin, 600,* 574–586.

International Human Genome Sequencing Consortium. (2001). Initial sequencing and analysis of the human genome. *Nature, 409,* 860–921.

International Human Genome Sequencing Consortium. (2004). Finishing the euchromatic sequence of the human genome. *Nature, 431,* 931–945.

International Multiple Sclerosis Genetics Consortium. (2007). Risk alleles for multiple sclerosis identified by a genomewide study. *New England Journal of Medicine, 357,* 851–862.

International Schizophrenia Consortium. (2008). Rare chromosomal deletions and duplications increase risk of schizophrenia. *Nature, 455,* 237–241.

Iqbal, N., & van Praag, H. M. (1995). The role of serotonin in schizophrenia. *European Neuropsychopharmacology Supplement, 5,* 11–23.

Ishihara, K., & Sasa, M. (1999). Mechanism underlying the therapeutic effects of electroconvulsive therapy (ECT) on depression. *Japanese Journal of Pharmacology, 80,* 185–189.

Isis clinical development pipeline. (n.d.) Retrieved May 6, 2005, from www.isispharm.com/product_pipeline.html

Iurato, S. (1967). *Submicroscopic structure of the inner ear.* Oxford, UK: Pergamon Press.

Iwamura, Y., Iriki, A., & Tanaka, M. (1994). Bilateral hand representation in the postcentral somatosensory cortex. *Nature, 369,* 554–556.

Izumikawa, M., Minoda, R., Kawamoto, K., Abrashkin, K. A., Swiderski, D. L., Dolan, D. F. et al. (2005). Auditory hair cell replacement and hearing improvement by Atoh1 gene therapy in deaf mammals. *Nature Medicine, 11,* 271–276.

Jacobs, B. L. (1987). How hallucinogenic drugs work. *American Scientist, 75,* 386–392.

Jacobs, B. L., van Praag, H., & Gage, F. H. (2000). Adult brain neurogenesis and psychiatry: A novel theory of depression. *Molecular Psychiatry, 5,* 262–269.

James, W. (1893). *Psychology.* New York: Henry Holt.

Janicak, P. G., Dowd, S. M., Martis, B., Alam, D., Beedle, D., Krasuski, J., et al. (2002). Repetitive transcranial magnetic stimulation versus electroconvulsive therapy for major depression: Preliminary results of a randomized trial. *Biological Psychiatry, 51,* 659–667.

Janowsky, J. S., Oviatt, S. K., & Orwoll, E. S. (1994). Testosterone influences spatial cognition in older men. *Behavioral Neuroscience, 108,* 325–332.

Janus, C., Pearson, J., McLaurin, J., Mathews, P. M., Jiang, Y., Schmidt, S. D., et al. (2000). A b peptide immunization reduces behavioural impairment and plaques in a model of Alzheimer's disease. *Nature, 408,* 979–982.

Janszky, I., & Ljung, R. (2008). Shifts to and from daylight saving time and incidence of myocardial infarction. *New England Journal of Medicine, 359,* 1966–1968.

Jaretzki, A., III, Penn, A. S., Younger, D. S., Wolff, M., Olarte, M. R., Lovelace, R. E., et al. (1988). "Maximal" thymectomy for myasthenia gravis: Results. *Journal of Thoracic and Cardiovascular Surgery, 95,* 747–757.

Jarraya, B., Boulet, S., Ralph, G. S., Jan, C., Bonvento, G., Azzouz, M., et al. (2009). Dopamine gene therapy for Parkinson's disease in a nonhuman primate without associated dyskinesia. *Science Translational Medicine, 1,* 2ra4.

Jeffords, J. M., & Daschle, T. (2001). Political issues in the genome era. *Science, 291,* 1249–1251.

Jeffrey, S. (2010, January 22). FDA approves dalfampridine to improve walking in multiple sclerosis. Retrieved May 29, 2010 from Medscape Medical News at www.medscape.com/viewarticle/715722

Jensen, A. R. (1969). How much can we boost IQ and scholastic achievement? *Harvard Educational Review, 39,* 1–123.

Jensen, A. R. (1981). Raising the IQ: The Ramey and Haskins study. *Intelligence, 5,* 29–40.

Jensen, A. R. (1998). *The g factor.* Westport, CT: Praeger.

Jensen, T. S., Genefke, I. K., Hyldebrandt, N., Pedersen, H., Petersen, H. D., & Weile, B. (1982). Cerebral atrophy in young torture victims. *New England Journal of Medicine, 307,* 1341.

Jentsch, J. D., & Roth, R. H. (1999). The neuropsychopharmacology of phencyclidine: From NMDA receptor hypofunction to the dopamine hypothesis of schizophrenia. *Neuropsychopharmacology, 20,* 201–225.

Jin, K., Peel, A. L., Mao, X. O., Xie, L., Cottrell, B. A., Henshall, D. C., et al. A. (2004). Increased hippocampal neurogenesis in Alzheimer's disease. *Proceedings of the National Academy of Sciences, USA, 101,* 343–347.

Jo, Y.-H., & Schlichter, R. (1999). Synaptic corelease of ATP and GABA in cultured spinal neurons. *Nature Neuroscience, 2,* 241–245.

John, E. R. (2005). From synchronous neuronal discharges to subjective awareness? In S. Laureys (Ed.), *The boundaries of consciousness: Neurobiology and neuropathology* (pp. 143–171). New York: Elsevier.

Johnson, A. (2009, February 27). DEA to halt medical marijuana raids. Retrieved November 25, 2009, from www.msnbc.msn.com/id/29433708

Johnson, C., Drgon, T., McMahon, F. J., & Uhl, G. R. (2009). Convergent genome wide association results for bipolar disorder and substance dependence. *American Journal of Medical Genetics, 150B,* 182–190.

Johnstone, T., Somerville, L. H., Alexander, A. L., Oakes, T. R., Davidson, R. J., Kalin, N. H., et al. (2005). Stability of amygdala BOLD response to fearful faces over multiple scan sessions. *NeuroImage, 25,* 1112–1123.

Jones, H. M., & Pilowsky, L. S. (2002). Dopamine and antipsychotic drug action revisited. *British Journal of Psychiatry, 181,* 271–275.

Jones, L. B., Stanwood, G. D., Reinoso, B. S., Washington, R. A., Wang, H.-Y., Friedman, E., et al. (2000). *In utero* cocaine-induced dysfunction of dopamine D1 receptor signaling and abnormal differentiation of cerebral cortical neurons. *Journal of Neuroscience, 20,* 4606–4614.

Julien, R. M. (2008). *A primer of drug action* (11th ed.). New York: Worth.

Jung, R. (1974). Neuropsychologie und der Neurophysiologie des Konturund Formsehens in Zeichnung und Malerei [Neuropsychology and neurophysiology of form vision in design and painting]. In H. H. Wieck (Ed.), *Psychopathologie musicher Gestaltungen* (pp. 29–88). Stuttgart, Germany: Schattauer.

Just, M. A., Cherkassky, V. I., Keller, T. A., Kana, R. K., & Minshew, N. J. (2007). Functional and anatomical cortical underconnectivity in autism: Evidence from an fMRI study of an executive function task and corpus callosum morphometry. *Cerebral Cortex, 17,* 951–961.

Kaeberlein, M., & Kennedy, B. K. (2007). Protein translation. 2007. *Aging Cell, 6,* 731–734.

Kaji, K., Norrby, K., Paca, A., Mileikovsky, M., Mohseni, P., & Woltjen, K. (2009). Virus-free induction of pluripotency and subsequent excision of reprogramming factors. *Nature, 458,* 771–776.

Kales, A., Scharf, M. B., Kales, J. D., & Soldatos, C. R. (1979). Rebound insomnia: A potential hazard following withdrawal of certain benzodiazepines. *Journal of the American Medical Association, 241,* 1692–1695.

Kalivas, P. W., Volkow, N., & Seamans, J. (2005). Unmanageable motivation in addiction: A pathology in prefrontal-accumbens glutamate transmission. *Neuron, 45,* 647–650.

Kamegai, J., Tamura, H., Shimizu, T., Ishii, S., Sugihara, H., & Wakabayashi, I. (2001). Chronic central infusion of ghrelin increases hypothalamic neuropeptide Y and agouti-related protein mRNA levels and body weight in rats. *Diabetes, 50,* 2438–2443.

Kaminen-Ahola, N., Ahola, A., Maga, M., Mallitt, K.-A., Fahey, P., Cox, T. C., et al. (2010). Maternal ethanol consumption alters the epigenotype and the phenotype of offspring in a mouse model. *PLoS Genetics, 6,* article e1000811. Retrieved January 15, 2010, from www.plosgenetics.org/article/info%3Adoi%2F10.1371%2Fjournal.pgen.1000811

Kanbayashi, T., Inoue, Y., Chiba, S., Aizawa, R., Saito, Y., Tsukamoto, H., et al. (2002). CSF hypocretin-1 (orexin-A) concentrations in narcolepsy with and without cataplexy and idiopathic hypersomnia. *Journal of Sleep Research, 11,* 91–93.

Kandel, E. R. (2001). The molecular biology of memory storage: A dialogue between genes and synapses. *Science, 294,* 1030–1038.

Kandel, E. R., & O'Dell, T. J. (1992). Are adult learning mechanisms also used for development? *Science, 258,* 243–245.

Kandel, E. R., & Siegelbaum, S. A. (2000a). Overview of synaptic transmission. In E. R. Kandel, J. H. Schwartz, & T. M. Jessell (Eds.), *Principles of neural science* (4th ed., pp. 175–186). New York: McGraw-Hill.

Kandel, E. R., & Siegelbaum, S. A. (2000b). Synaptic integration. In E. R. Kandel, J. H. Schwartz, & T. M. Jessell (Eds.), *Principles of neural science* (4th ed., pp. 207–228). New York: McGraw-Hill.

Kane, J. M. (1987). Treatment of schizophrenia. *Schizophrenia Bulletin, 13,* 133–156.

Kanigel, R. (1988). Nicotine becomes addictive. *Science Illustrated, October/November,* 12–14, 19–21.

Kansaku, K., Yamaura, A., & Kitazawa, S. (2000). Sex differences in lateralization revealed in the posterior language areas. *Cerebral Cortex, 10,* 866–872.

Kaplitt, M. G., Feigin, A., Tong, C., Fitzsimons, H. I., Mattis, P., Lawlor, P. A., et al. (2007). Safety and tolerability of gene therapy with an adeno-associated virus (AAV) borne GAD gene for Parkinson's disease: An open label, phase I trial. *Lancet, 369,* 2097–2105.

Kapur, S., Zipursky, R. B., & Remington, G. (1999). Clinical and theoretical implications of 5-HT2 and D2 receptor occupancy of clozapine, risperidone, and olanzapine in schizophrenia. *American Journal of Psychiatry, 156,* 286–293.

Karama, S., Ad-Dab'bagh, Y., Haier, R. J., Deary, I. J., Lyttleton, O. C., Lepage, C., et al. (2009). Positive association between cognitive ability and cortical thickness in a representative US sample of healthy 6 to 18 year-olds. *Intelligence, 37,* 145–155.

Karnath, H.-O., & Baier, B. (2010). Right insula for our sense of limb ownership and self-awareness of actions. *Brain Structure and Function, 214,* 411–417.

Karni, A., Tanne, D., Rubenstein, B. S., Askenasy, J. J. M., & Sagi, D. (1994). Dependence on REM sleep of overnight improvement of a perceptual skill. *Science, 265,* 679–682.

Kast, B. (2001). Decisions, decisions . . . *Nature, 411,* 126–128.

Katz, L. C., & Shatz, L. C. (1996). Synaptic activity and the construction of cortical circuits. *Nature, 274,* 1133–1138.

Katzman, D. K., Zipursky, R. B., Lambe, E. K., & Mikulis, D. J. (1997). A longitudinal magnetic resonance imaging study of brain changes in adolescents with anorexia nervosa. *Archives of Pediatrics and Adolescent Medicine, 151,* 793–797.

Kaufman, J., & Charney, D. (2000). Comorbidity of mood and anxiety disorders. *Depression and Anxiety, 12*(Suppl. 1), 69–76.

Kausler, D. H. (1985). Episodic memory: Memorizing performance. In N. Charness (Ed.), *Aging and human performance* (pp. 101–139). New York: Wiley.

Kawai, N., & Matsuzawa, T. (2000). Numerical memory span in a chimpanzee. *Nature, 403,* 39–40.

Kaye, W. H., Berrettini, W., Gwirtsman, H., & George, D. T. (1990). Altered cerebrospinal fluid neuropeptide Y and peptide YY immunoreactivity in anorexia and bulimia nervosa. *Archives of General Psychiatry, 47,* 548–556.

Kaye, W. H., Ebert, M. H., Gwirtsman, H. E., & Weiss, S. R. (1984). Differences in brain serotonergic metabolism between nonbulimic and bulimic patients with anorexia nervosa. *American Journal of Psychiatry, 141,* 1598–1601.

Kaye, W. H., Fudge, J. L., & Paulus, M. (2009). New insights into symptoms and neurocircuit function of anorexia nervosa. *Nature Reviews of Neuroscience, 10,* 573–584

Kaye, W. H., Klump, K. L., Frank, G. K. W., & Strober, M. (2000). Anorexia and bulimia nervosa. *Annual Review of Medicine, 51,* 299–313.

Kayman, S., Bruvold, W., & Stern, J. S. (1990). Maintenance and relapse after weight loss in women: Behavioral aspects. *American Journal of Clinical Nutrition, 52,* 800–807.

Kee, N., Teixeira, C. M., Wang, A. H., & Frankland, P. W. (2007). Preferential incorporation of adult-generated granule cells into spatial memory networks in the dentate gyrus. *Nature Neuroscience, 10,* 355–362.

Keele, S. W., & Ivry, R. (1990). Does the cerebellum provide a common computation for diverse tasks? A timing hypothesis. *Annals of the New York Academy of Sciences, 608,* 179–207.

Keenan, P. A., Ezzat, W. H., Ginsburg, K., & Moore, G. J. (2001). Prefrontal cortex as the site of estrogen's effect on cognition. *Psychoneuroendocrinology, 26,* 577–590.

Keesey, R. E., & Powley, T. L. (1986). The regulation of body weight. *Annual Reviews of Psychology, 37,* 109–133.

Kellogg, W. N. (1968). Communication and language in the home-raised chimpanzee. *Science, 162,* 423–427.

Kelsey, J. E., Carlezon, W. A., Jr., & Falls, W. A. (1989). Lesions of the nucleus accumbens in rats reduce opiate reward but do not alter context-specific opiate tolerance. *Behavioral Neuroscience, 103,* 1327–1334.

Kelz, M. B., Sun, Y., Chen, J., Meng, Q. C., Moore, J. T., Veasey, S. C., et al. (2008). An essential role for orexins in emergence from general anesthesia. *Proceedings of the National Academy of Sciences, 105,* 1309–1314.

Kemper, T. L. (1984). Asymmetrical lesions in dyslexia. In N. Geschwind & A. M. Galaburda (Eds.), *Cerebral dominance: The biological foundations* (pp. 75–89). Cambridge, MA: Harvard University Press.

Kendler, K. S., Gardner, C. O., Neale, M. C., & Prescott, C. A. (2001). Genetic risk factors for major depression in men and women: Similar or different heritabilities and same or partly distinct genes? *Psychological Medicine, 31,* 605–616.

Kendler, K. S., MacLean, C., Neale, M., Kessler, R., Heath, A., & Eaves, L. (1991). The genetic epidemiology of bulimia nervosa. *American Journal of Psychiatry, 148,* 1627–1637.

Kendler, K. S., & Robinette, C. D. (1983). Schizophrenia in the National Academy of Sciences-National Research Council twin registry: A 16-year update. *American Journal of Psychiatry, 140,* 1551–1563.

Kendrick, M. (2009, October). Tasting the light. *Scientific American, 22,* 24.

Kennaway, D. J., & van Dorp, C. F. (1991). Free-running rhythms of melatonin, cortisol, electrolytes, and sleep in humans in Antarctica. *American Journal of Physiology, 260,* R1137–R1144.

Kennedy, B. K., Steffen, K. K., & Kaeberlein, M. (2007). Ruminations on dietary restriction and aging. *Cellular and Molecular Life Sciences, 64,* 1323–1328.

Kennerknecht, I., Grueter, T., Welling, B., Wentzek, S., Horst, J., Edwards, S., et al. (2006). First report of prevalence of non-syndromic hereditary prosopagnosia (HPA). *American Journal of Medical Genetics Part A, 140A,* 1617–1622.

Kessler, R. C., Berglund, P., Demler, O., Jin, R., Merikangas, K. R., & Walters, E. E. (2005). Lifetime prevalence and age-of-onset distribution of DSM-IV disorders in the national comorbidity survey replication. *Archives of General Psychiatry, 62,* 593–602.

Kessler, R. C., Chiu, W. T., Demler, O., & Walters, E. E. (2005). Prevalence, severity, and comorbidity of twelve-month DSM-IV disorders in the National Comorbidity Survey Replication (NCS-R). *Archives of General Psychiatry, 62,* 617–627.

Kessler, R. C., Heeringa, S., Lakoma, M. D., Petukhova, M., Rupp, A. E., Schoenbaum, M., et al. (2008). Individual and societal effects of mental disorders on earnings in the United States: Results from the National Comorbidity Survey Replication. *American Journal of Psychiatry, 165,* 703–711.

Kessler, R. C., McGonagle, K. A., Zhao, S., Nelson, C. B., Hughes, M., Eshleman, S., et al. (1994). Lifetime and 12-month prevalence of DSM-III-R psychiatric disorders in the United States. *Archives of General Psychiatry, 51,* 8–19.

Keverne, E. B. (1999). The vomeronasal organ. *Science, 286,* 716–720.

Khachaturian, Z. S. (1997). Plundered memories. *The Sciences, July/August,* 20–25.

Khalil, A. A., Davies, B., & Castagnoli, N., Jr. (2006). Isolation and characterization of a monoamine oxidase B selective inhibitor from tobacco smoke. *Bioorganic & Medicinal Chemistry, 14,* 3392–3398.

Khan, T. K., & Alkon, D. L. (2010). Early diagnostic accuracy and pathophysiologic relevance of an autopsy-confirmed Alzheimer's disease peripheral biomarker. *Neurobiology of Aging, 31,* 889–900.

Khateb, A., Fort, P., Pegna, A., Jones, B. E., & Mühlethaler, M. (1995). Cholinergic nucleus basalis neurons are excited by histamine in vitro. *Neuroscience, 69,* 495–506.

Kiang, N. Y.-S. (1965). *Discharge patterns of single fibers in the cat's auditory nerve.* Cambridge: MIT Press.

Kieburtz, K., McDermott, M. P., Voss, T. S., Corey-Bloom, J., Deuel, L. M., Dorsey, E. R., et al. (2010). A randomized, placebo-controlled trial of latrepirdine in Huntington disease. *Archives of Neurology, 67,* 154–160.

Kiehl, K. A., Smith, A. M., Hare, R. D., Mendrek, A., Forster, B. B., Brink, J., et al. (2001). Limbic abnormalities in affective processing by criminal psychopaths as revealed by functional magnetic resonance imaging. *Biological Psychiatry, 50,* 677–684.

Kieseppä, T., Partonen, T., Haukka, J., Kaprio, J., & Lönnqvist, J. (2004). High concordance of bipolar I disorder in a nationwide sample of twins. *American Journal of Psychiatry, 161,* 1814–1821.

Kigar, D. L., Witelson, S. F., Glezer, I. I., & Harvey, T. (1997). Estimates of cell number in temporal neocortex in the brain of Albert Einstein. *Society for Neuroscience Abstracts, 23,* 213.

Kim, K. H. S., Relkin, N. R., Lee, K.-M., & Hirsch, J. (1997). Distinct cortical areas associated with native and second languages. *Nature, 388,* 171–174.

Kimura, D., & Hampson, E. (1994). Cognitive pattern in men and women is influenced by fluctuations in sex hormones. *Current Directions in Psychological Science, 3,* 57–62.

King, F. A., Yarbrough, C. J., Anderson, D. C., Gordon, T. P., & Gould, K. G. (1988). Primates. *Science, 240,* 1475–1482.

King, M.-C., & Wilson, A. C. (1975). Evolution at two levels in humans and chimpanzees. *Science, 188,* 107–116.

King, R. A., Leckman, J. F., Scahill, L. D., & Cohen, D. J. (1998). Obsessive-compusive disorder, anxiety, and depression. In J. F. Leckman & D. J. Cohen (Eds.), *Tourette's syndrome tics, obsessions, compulsions: Developmental psychopathology and clinical care* (pp. 43–62). New York: Wiley.

Kinney, D. K., Barch, D. H., Chayka, B., Napoleon, S., & Munir, K. M. (2010). Environmental risk factors for autism: Do they help cause de novo genetic mutations that contribute to the disorder? *Medical Hypotheses, 74,* 102–106.

Kinsey, A. C., Pomeroy, W. B., Martin, C. E., & Gebhard, P. H. (1953). *Sexual behavior in the human female.* Philadelphia: Saunders.

Kinsey, B. M., Jackson, D. C., & Orson, F. M. (2009). Anti-drug vaccines to treat substance abuse. *Immunology and Cell Biology, 87,* 309–314.

Kipman, A., Gorwood, P., Mouren-Simeoni, M. C., & Ad'es, J. (1999). Genetic factors in anorexia nervosa. *European Psychiatry, 14,* 189–198.

Kirk, K. M., Bailey, J. M., Dunne, M. P., & Martin, N. G. (2000). Measurement models for sexual orientation in a community twin sample. *Behavior Genetics, 30,* 345–356.

Kline, P. (1991). *Intelligence: The psychometric view.* New York: Routledge.

Knecht, S., Dräger, B., Deppe, M., Bobe, L., Lohmann, H., Flöel, A., et al. (2000). Handedness and hemispheric language dominance in healthy humans. *Brain, 123,* 2512–2518.

Knierim, J. J., Kudrimoti, H. S., & McNaughton, B. L. (1995). Place cells, head direction cells, and the learning of landmark stability. *Journal of Neuroscience, 15,* 1648–1659.

Knowlton, B. J., Mangels, J. A., & Squire, L. R. (1996). A neostriatal habit learning system in humans. *Science, 273,* 1399–1402.

Knutson, B., Wolkowitz, O. M., Cole, S. W., Chan, T., Moore, E. A., Johnson, R. C., et al. (1998). Selective alteration of personality and social behavior by serotonergic intervention. *American Journal of Psychiatry, 155,* 373–379.

Koenig, R. (1999). European researchers grapple with animal rights. *Science, 284,* 1604–1606.

Koester, J., & Siegelbaum, S. A. (2000). Local signaling: Passive electrical properties of the neuron. In E. R. Kandel, J. H. Schwartz, & T. M. Jessell (Eds.), *Principles of neural science* (4th ed., pp. 140–149). New York: McGraw-Hill.

Köhnke, M. D. (2008). Approach to the genetics of alcoholism: A review based on pathophysiology. *Biochemical Pharmacology, 75,* 160–177.

Kojima, S., Nakahara, T., Nagai, N., Muranaga, T., Tanaka, M., Yasuhara, D., et al. (2005). Altered ghrelin and peptide YY responses to meals in bulimia nervosa. *Clinical Endocrinology, 62,* 74–78.

Komisaruk, B. R., & Steinman, J. L. (1987). Genital stimulation as a trigger for neuroendocrine and behavioral control of reproduction. *Annals of the New York Academy of Sciences, 474,* 64–75.

Konopka, G., Bomar, J. M., Winden, K., Coppola, G., Jonsson, Z. O., Gao, F., et al. (2009). Human-specific transcriptional regulation of CNS development genes by FOXP2. *Nature, 462,* 213–217.

Koob, G. F., & Bloom, F. E. (1988). Cellular and molecular mechanisms of drug dependence. *Science, 242,* 715–723.

Kopelman, M. D. (1995). The Korsakoff syndrome. *British Journal of Psychiatry, 166,* 154–173.

Kosambi, D. D. (1967, February). Living prehistory in India. *Scientific American, 216,* 105–114.

Koscik, T., O'Leary, D., Moser, D. J., Andreasen, N. C., & Nopoulos, P. (2009). Sex differences in parietal lobe morphology: Relationship to mental rotation performance. *Brain and Cognition, 69,* 451–459.

Kosslyn, S. M., Ganis, G., & Thompson, W. L. (2001). Neural foundations of imagery. *Nature Reviews Neuroscience, 2,* 635–642.

Koulack, D., & Goodenough, D. R. (1976). Dream recall and dream recall failure: An arousal-retrieval model. *Psychological Bulletin, 83,* 975–984.

Koyama, T., Tanaka, Y., & Mikami, A. (1998). Nociceptive neurons in the macaque anterior cingulate activate during anticipation of pain. *NeuroReport, 9,* 2663–2667.

Kozlowski, S., & Drzewiecki, K. (1973). The role of osmoreception in portal circulation in control of wafer intake in dogs. *Acta Physiologica Polonica, 24,* 325–330.

Kraemer, H. C., Becker, H. B., Brodie, H. K. H., Doering, C. H., Moos, R. H., & Hamburg, D. A. (1976). Orgasmic frequency and plasma testosterone levels in normal human males. *Archives of Sexual Behavior, 5,* 125–132.

Krakauer, J., & Ghez, C. (2000). Voluntary movement. In E. R. Kandel, J. H. Schwartz, & T. M. Jessell (Eds.), *Principles of neural science* (4th ed., pp. 756–781). New York: McGraw-Hill.

Krause, K.-H., Dresel, S. H., Krause, J., Kung, H. F., & Tatsch, K. (2000). Increased striatal dopamine transporter in adult patients with attention-deficit/hyperactivity disorder: Effects of methylphenidate as measured by single photon emission computed tomography. *Neuroscience Letters, 285,* 107–110.

Kreek, M. J., Nielsen, D. A., Butelman, E. R., & LaForge, K. S. (2005). Genetic influences on impulsivity, risk taking, stress responsivity and vulnerability to drug abuse and addiction. *Nature Neuroscience, 8,* 1450–1457.

Kreiman, G., Koch, C., & Fried, I. (2000). Category-specific visual responses of single neurons in the human medial temporal lobe. *Nature Neuroscience, 3,* 946–953.

Kriegeskorte, N., Simmons, W. K., Bellgowan, P. S. F., & Baker, C. I. (2009). Circular analysis in systems neuroscience: the dangers of double dipping. *Nature Neuroscience, 12,* 535–540.

Kriegstein, K. V., & Giraud, A.-L. (2004). Distinct functional substrates along the right superior temporal sulcus for the processing of voices. *NeuroImage, 22,* 948–955.

Kripke, D. F., Garfinkel, L., Wingard, D. L., Klauber, M. R., & Marler, M. R. (2002). Mortality associated with sleep duration and insomnia. *Archives of General Psychiatry, 59,* 131–136.

Kripke, D. F., & Sonnenschein, D. (1973). A 90 minute daydream cycle [Abstract]. *Sleep Research, 70,* 187.

Kruijver, F. P. M., Zhou, J.-N., Pool, C. W., Hofman, M. A., Gooren, L. J. G., & Swaab, D. F. (2000). Male-to-female transsexuals have female neuron numbers in a limbic nucleus. *Journal of Clinical Endocrinology & Metabolism, 85,* 2034–2041.

Kuchinad, A., Schweinhardt, P., Seminowicz, D. A., Wood, P. B., Chizh, B. A., & Bushnell, M. C. (2007). Accelerated brain gray matter loss in fibromyalgia patients: Premature aging of the brain? *Journal of Neuroscience, 27,* 4004–4007.

Kudwa, A. E., Bodo, C., Gustafsson, J.-Å., & Rissman, E. F. (2005). A previously uncharacterized role for estrogen receptor: Defeminization of male brain and behavior. *Proceedings of the National Academy of Sciences, USA, 102,* 4608–4612.

Kuepper, Y., Alexander, N., Osinsky, R., Mueller, E., Schmitz, A., Netter, P., et al. (2010). Aggression—interactions of serotonin and testosterone in healthy men and women. *Behavioural Brain Research, 206,* 93–100.

Kuffler, S. W. (1953). Discharge patterns and functional organization of mammalian retina. *Journal of Neurophysiology, 16,* 37–68.

Kumari, V., & Postma, P. (2005). Nicotine use in schizophrenia: The self-medication hypotheses. *Neuroscience and Biobehavioral Reviews, 29,* 1021–1034.

Kupfer, D. J. (1976). REM latency: A psychobiologic marker for primary depressive disease. *Biological Psychiatry, 11,* 159–174.

Kupferman, I., Kandel, E. R., & Iversen, S. (2000). Motivational and addictive states. In E. R. Kandel, J. H. Schwartz, & T. M. Jessell

(Eds.), *Principles of neural science* (4th ed., pp. 998–1013). New York: McGraw- Hill.

Kurihara, K., & Kashiwayanagi, M. (1998). Introductory remarks on umami taste. *Annals of the New York Academy of Sciences, 855,* 393–397.

Kuukasjärvi, S., Eriksson, C. J. P., Koskela, E., Mappes, T., Nissinen, K., & Rantala, M. J. (2004). Attractiveness of women's body odors over the menstrual cycle: The role of oral contraceptives and receiver sex. *Behavioral Ecology, 15,* 579–584.

Lagali, P. S., Balya, D., Awatramani, G. B., Münch, T. A., Kim, D. S.,Busskamp, V., et al. (2008). Light-activated channels targeted to ON bipolar cells restore visual function in retinal degeneration. *Nature Neuroscience, 11,* 667–675.

Lai, C. S. L., Fisher, S. E., Hurst, J. A., Vargha-Khadem, F., & Monaco, A. P. (2001). A forkhead-domain gene is mutated in a severe speech and language disorder. *Nature, 413,* 519–523.

Laje, G., Paddock, S., Manji, H., Rush, A. J., Wilson, A. F., Charney, D., et al. (2007). Genetic markers of suicidal ideation emerging during citalopram treatment of major depression. *American Journal of Psychiatry, 164,* 1530–1538.

Lam, P., Cheng, C. Y., Hong, C.-J., & Tsai, S.-J. (2004). Association study of a brain-derived neurotrophic factor (Va166Met) genetic polymorphism and panic disorder. *Neuropsychobiology, 49,* 178–181.

Lam, P., Hong, C.-J., & Tsai. S.-J. (2005). Association study of A2a adenosine receptor genetic polymorhism in panic disorder. *Neuroscience Letters, 378,* 98–101.

LaMantia, A. S., & Rakic, P. (1990). Axon overproduction and elimination in the corpus callosum of the developing rhesus monkey. *Journal of Neuroscience, 10,* 2156–2175.

Lamb, T., & Yang, J. E. (2000). Could different directions of infant stepping be controlled by the same locomotor central pattern generator? *Journal of Neurophysiology, 83,* 2814–2824.

Lambe, E. K., Katzman, D. K., Mikulis, D. J., Kennedy, S. H., & Zipursky, R. B. (1997). Cerebral gray matter volume deficits after weight recovery from anorexia nervosa. *Archives of General Psychiatry, 54,* 537–542.

Lambert, J.-C., Heath, S., Even, G., Campion, D., Sleegers, K., Hiltunen, M., et al. (2009). Genome-wide association study identifies variants at *CLU* and *CR1* associated with Alzheimer's disease. *Nature Genetics, 41,* 1094–1099.

Lancaster, E. (1958). *The final face of Eve.* New York: McGraw-Hill.

Landry, D. W. (1997, February). Immunotherapy for cocaine addiction. *Scientific American, 276,* 42–45.

Langström, N., Rahman, Q., Carlström, E., & Lichtenstein, P. (2010). Genetic and environmental effects on same-sex sexual behavior: A population study of twins in Sweden. *Archives of Sexual Behavior, 39,* 75–80.

Laplane, D., Talairach, J., Meininger, V., Bancaud, J., & Orgogozo, J. M. (1977). Clinical consequences of corticectomies involving the supplementary motor area in man. *Journal of the Neurological Sciences, 34,* 301–314.

Larson, G. (1995). *The Far Side gallery 5.* Kansas City, KS: Andrews & McMeel.

Laruelle, M., Kegeles, L. S., & Abi-Dargham, A. (2003). Glutamate, dopamine, and schizophrenia. *Annals of the New York Academy of Sciences, 1003,* 138–158.

Lashley, K. (1929). *Brain mechanisms and intelligence: A quantitative study of injuries to the brain.* Chicago: University of Chicago Press.

Lau, B., Stanley, G. B., & Dan, Y. (2002). Computational subunits of visual cortical neurons revealed by artificial neural networks. *Proceedings of the National Academy of Sciences, USA, 99,* 8974–8979.

Laureys, S., Owen, A. M., & Schiff, N. D. (2004). Brain function in coma, vegetative state, and related disorders. *Lancet Neurology, 3,* 537–546.

Laureys, S. (2005). The neural correlate of (un)awareness: Lessons from the vegetative state. *Trends in Cognitive Sciences, 9,* 556–559.

Lawrence, A. A. (2005). Sexuality before and after male-to-female sex reassignment surgery. *Archives of Sexual Behavior, 34,* 147–166.

Le Foll, B., & Goldberg, S. R. (2005). Cannabinoid CB₁ antagonists as promising new medications for drug dependence. *Journal of Pharmacology and Experimental Therapeutics, 312,* 875–883.

Lecendreux, M., Bassetti, C., Dauvilliers, Y., Mayer, G., Neidhart, E., & Tafti, M. (2003). HLA and genetic susceptibility to sleepwalking. *Molecular Psychiatry, 8,* 114–117.

LeDoux, J. E. (1996). *The emotional brain.* New York: Simon & Schuster.

Lee, C. C., Chou, I. C., Tsai, C. H., Wang, T. R., Li, T. C., & Tsai, F. J. (2005). Dopamine receptor D2 gene polymorphisms are associated in Taiwanese children with Tourette syndrome. *Pediatric Neurology, 33,* 272–276.

Lee, D. S., Lee, J. S., Oh, S. H., Kim, S.-K., Kim, J.-W., Chung, J.-K., et al. (2001). Cross-modal plasticity and cochlear implants. *Nature, 409,* 149–150.

Lee, J. L. C. (2009). Reconsolidation: Maintaining memory relevance. *Trends in Neurosciences, 32,* 413–420.

Lee, M., Zambreanu, L., Menon, D. K., & Tracey, I. (2008). Identifying brain activity specifically related to the maintenance and perceptual consequence of central sensitization in humans. *Journal of Neuroscience, 28,* 11642–11649.

Lehky, S. R., & Sejnowski, T. J. (1990). Neural network model of visual cortex for determining surface curvature from images of shaded surfaces. *Proceedings of the Royal Society of London, B, 240,* 251–278.

Lehrman, S. (1999). Virus treatment questioned after gene therapy death. *Nature, 401,* 517–518.

Leibel, R. L., Rosenbaum, M., & Hirsch, J. (1995). Changes in energy expenditure resulting from altered body weight. *New England Journal of Medicine, 332,* 621–628.

Leibowitz, S. F., & Alexander, J. T. (1998). Hypothalamic serotonin in control of eating behavior, meal size, and body weight. *Biological Psychiatry, 44,* 851–864.

Leland, J., & Miller, M. (1998, August 17). Can gays convert? *Newsweek,* 47–53.

Lenneberg, E. H. (1969). On explaining language. *Science, 164,* 635–643.

Lenzenweger, M. F., & Gottesman, I. I. (1994). Schizophrenia. In V. S. Ramachandran (Ed.), *Encyclopedia of human behavior.* San Diego, CA: Academic Press.

Leonard, H. L., Lenane, M. C., Swedo, S. E., Rettew, D. C., Gershon, E. S., & Rapoport, J. L. (1992). Tics and Tourette's disorder: A 2- to 7-year follow-up of 54 obsessive-compulsive children. *American Journal of Psychiatry, 149,* 1244–1251.

Leonard, H. L., Lenane, M. C., Swedo, S. E., Rettew, D. C., & Rapoport, J. L. (1991). A double-blind comparison of clomipramine and desipramine treatment of severe onychophagia (nail-biting). *Archives of General Psychiatry, 48,* 821–826.

Leopold, D. A., & Logothetis, N. K. (1996). Activity changes in early visual cortex reflect monkeys' percepts during binocular rivalry. *Nature, 379,* 549–553.

Leor, J., Poole, W. K., & Kloner, R. A. (1996). Sudden cardiac death triggered by an earthquake. *New England Journal of Medicine, 334,* 413–419.

Leshner, A. I. (1997). Addiction is a brain disease, and it matters. *Science, 278,* 45–47.

Leucht, S., Corves, C., Arbter, D., Engel, R. R., Li, C., & Davis, J. M. (2009). Second-generation versus first-generation antipsychotic drugs for schizohrenia: A meta-analysis. *Lancet, 373,* 31–41.

LeVay, S. (1991). A difference in hypothalamic structure between heterosexual and homosexual men. *Science, 253,* 1034–1037.

LeVay, S. (1996). *Queer science: The use and abuse of research into homosexuality.* Cambridge, MA: MIT Press.

Levenson, R. W., Ekman, P., & Friesen, W. V. (1990). Voluntary facial action generates emotion-specific autonomic nervous system activity. *Psychophysiology, 27,* 363–384.

Leventhal, A. G., Wang, Y., Pu, M., Zhou, Y., & Ma, Y. (2003). GABA and its agonists improved visual cortical function in senescent monkeys. *Science, 300,* 812–815.

Lévesque, M. F., Neuman, T., & Rezak, M. (2009). Therapeutic microinjection of autologous adult human neural stem cells and differentiated neurons for Parkinson's disease: Five year post-operative outcome. *Open Stem Cell Journal, 1,* 20–29.

Levin, F. R., Evans, S. M., & Kleber, H. D. (1998). Methylphenidate treatment for cocaine abusers with adult attention-deficit/hyperactivity disorder: A pilot study. *Journal of Clinical Psychiatry, 59,* 300–305.

Levin, H. S., Culhane, K. A., Hartmann, J., Evankovich, K., Mattson, A. J., Harward, H., et al. (1991). Developmental changes in performance on tests of purported frontal lobe functioning. *Developmental Neuropsychology, 7,* 377–395.

Levin, N., Nelson, C., Gurney, A., Vandlen, R., & de Sauvage, F. (1996). Decreased food intake does not completely account for adiposity reduction after ob protein infusion. *Proceedings of the National Academy of Sciences, USA, 93,* 1726–1730.

Levine, J. A., Eberhardt, N. L., & Jensen, M. D. (1999). Role of nonexercise activity thermogenesis in resistance to fat gain in humans. *Science, 283,* 212–214.

Levy, B. (1996). Improving memory in old age through implicit self-stereotyping. *Journal of Personality and Social Psychology, 71,* 1092–1107.

Levy, J. (1969). Possible basis for the evolution of lateral specialization of the human brain. *Nature, 224,* 614–615.

Levy, S., Sutton, G., Ng, P. C., Feuk, L., Halpern, A. L., et al. (2007). The diploid genome sequence of an individual human. *PLoS Biology, 5,* article e254. Retrieved January 15, 2010, from www.plosbiology.org/article/info%3Adoi%2F10.1371%2Fjournal.pbio.0050254

Lewicki, P., Hill, T., & Czyzewska, M. (1992). Nonconscious acquisition of information. *American Psychologist, 47,* 796–801.

Lewin, R. (1980). Is your brain really necessary? *Science, 210,* 1232–1234.

Lewis, D. A., & Levitt, P. (2002). Schizophrenia as a disorder of neurodevelopment. *Annual Review of Neuroscience, 25,* 409–432.

Lewis, H. B., Goodenough, D. R., Shapiro, A., & Sleser, I. (1966). Individual differences in dream recall. *Journal of Abnormal Psychology, 71,* 52–59.

Lewis, J. W., Wightman, F. L., Brefczynski, J. A., Phinney, R. E., Binder, J. R., & DeYoc, E. A. (2004). Human brain regions involved in recognizing environmental sounds. *Cerebral Cortex, 14,* 1008–1021.

Lewis, M., & Brooks-Gunn, J. (1979). *Social cognition and the acquisition of self.* New York: Plenum Press.

Lewis, M. B., & Bowler, P. J. (2009). Botulinum toxin cosmetic therapy correlates with a more positive mood. *Journal of Cosmetic Dermatology, 8,* 24–26.

Lewis, P. D. (1985). Neuropathological effects of alcohol on the developing nervous system. *Alcohol and Alcoholism, 20,* 195–200.

Lewy, A. J., Sack, R. L, Miller, L. S., & Hoban, T. M. (1987). Antidepressant and circadian phase-shifting effects of light. *Science, 235,* 352–354.

Liberles, S. D., & Buck, L. B. (2006). A second class of chemosensory receptors in the olfactory epithelium. *Nature, 442,* 645–650.

Lichtman, S. W., Pisarska, K., Berman, E. R., Pestone, M., Dowling, H., Offenbacher, E., et al. (1992). Discrepancy between self-reported and actual caloric intake and exercise in obese subjects. *New England Journal of Medicine, 327,* 1893–1898.

Lieberman, H. R., Wurtman, J. J., & Chew, B. (1986). Changes in mood after carbohydrate consumption among obese individuals. *American Journal of Clinical Nutrition, 44,* 772–778.

Liechti, M. E., & Vollenweider, F. X. (2000). Acute psychological and physiological effects of MDMA ("ecstasy") after haloperidol pretreatment in healthy humans. *European Neuropsychopharmacology, 10,* 289–295.

Lilenfeld, L. R., Kaye, W. H., Greeno, C. G., Merikangas, K. R., Plotnicov, K., Pollice, C., et al. (1998). A controlled family study of anorexia nervosa and bulimia nervosa: Psychiatric disorders in first degree relatives and effects of proband comorbidity. *Archives of General Psychiatry, 55,* 603–610.

Lima, C., Pratas-Vital, J., Escada, P., Hasse-Ferreira, A., Capucho, C., & Peduzzi, J. D. (2006). Olfactory mucosa autografts in human spinal cord injury: A pilot clinical study. *Journal of Spinal Cord Medicine, 29,* 195–203.

Lin, L., Faraco, J., Li, R., Kadotani, H., Rogers, W., Lin, X., et al. (1999). The sleep disorder canine narcolepsy is caused by a mutation in the hypocretin (orexin) receptor 2 gene. *Cell, 98,* 365–376.

Linnoila, M., Virkkunen, M., Scheinin, M., Nuutila, A., Rimon, R., & Goodwin, F. K. (1983). Low cerebrospinal fluid 5-hydroxyindoleacetic acid concentration differentiates impulsive from nonimpulsive violent behavior. *Life Sciences, 33,* 2609–2614.

Lisman, J., & Morris, R. B. M. (2001). Why is the cortex a slow learner? *Nature, 411,* 248–249.

Lisman, J., Schulman, H., & Cline, H. (2002). The molecular basis of CaMKII function in synaptic and behavioural memory. *Nature Reviews Neuroscience, 3,* 175–190.

List of Countries by International Homicide Rate. (n.d.). In *Wikipedia.* Retrieved February 13, 2010, from http://en.wikipedia.org/wiki/List_of_countries_by_intentional_homicide_rate#cite_ref-geneva_5–0

Livingstone, D. (1971). *Missionary travels.* New York: Harper & Brothers. (Original work published 1858)

Livingstone, M., & Hubel, D. (1988). Segregation of form, color, movement, and depth: Anatomy, physiology, and perception. *Science, 240,* 740–749.

Livingstone, M. S., Rosen, G. D., Drislane, F. W., & Galaburda, A. M. (1991). Physiological and anatomical evidence for a magnocellular defect in developmental dyslexia. *Proceedings of the National Academy of Sciences, USA, 88,* 7943–7947.

Locurto, C. (1991). *Sense and nonsense about IQ: The case for uniqueness.* New York: Praeger.

Loehlin, J. C., & Nichols, R. C. (1976). *Heredity, environment and personality: A study of 850 twins.* Austin: University of Texas Press.

Loewi, O. (1953). *From the workshop of discoveries.* Lawrence: University of Kansas Press.

Loftus, E. F. (1997, September). Creating false memories. *Scientific American, 277,* 70–75.

Logothetis, N. K. (2002). The neural basis of the blood-oxygen-level-dependent functional magnetic resonance imaging signal. *Philosophical Transactions of the Royal Society B, 357,* 1003–1037.

London, E. D., Cascella, N. G., Wong, D. F., Phillips, R. L., Dannals, R. F., Links, J. M., et al. (1990). Cocaine-induced reduction of glucose utilization in human brain. *Archives of General Psychiatry, 47,* 567–574.

Loring, D. W., Meador, K. J., Lee, G. P., Murro, A. M., Smith, J. R., Flanigin, H. F., et al. (1990). Cerebral language lateralization: Evidence from intracarotid amobarbital testing. *Neuropsychologia, 28,* 831–838.

Loring, J. F., Wen, X., Lee, J. M., Seilhamer, J., & Somogyi, R. (2001). A gene expression profile of Alzheimer's disease. *DNA and Cell Biology, 20,* 683–695.

Lott, I. T. (1982). Down's syndrome, aging, and Alzheimer's disease: A clinical review. *Annals of the New York Academy of Sciences, 396,* 15–27.

Lotze, M., Grodd, W., Birbaumer, N., Erb, M., Huse, E., & Flor, H. (1999). Does use of a myoelectric prosthesis prevent cortical reorganization and phantom limb pain? *Nature Neuroscience, 2,* 501–502.

Louie, K., & Wilson, M. A. (2001). Temporally structured replay of awake hippocampal ensemble activity during rapid eye movement sleep. *Neuron, 29,* 145–156.

Lowenstein, R. J., & Putnam, F. W. (1990). The clinical phenomenology of males with MPD: A report of 21 cases. *Dissociation, 3,* 135–143.

Lowing, P. A., Mirsky, A. F., & Pereira, R. (1983). The inheritance of schizophrenia spectrum disorders: A reanalysis of the Danish adoptee study data. *American Journal of Psychiatry, 140,* 1167–1171.

Loy, B., Warner-Czyz, A. D., Tong, L., Tobey, E. A., & Roland, P. S. (2010). The children speak: An examination of the quality of life of pediatric cochlear implant users. *Otolaryngology-Head and Neck Surgery, 142,* 247–253.

Loyd, D. R., Wang, X., & Murphy, A. Z. (2008). Sex differences in μ-opioid receptor expression in the rat midbrain periaqueductal gray are essential for eliciting sex differences in morphine analgesia. *Journal of Neuroscience, 28,* 14007–14017.

Lu, J., Bjorkum, A. A., Xu, M., Gaus, S. E., Shiromani, P. J., & Saper, C. B. (2002). Selective activation of the extended ventrolateral preoptic nucleus during rapid eye movement sleep. *Journal of Neuroscience, 22,* 4568–4576.

Lu, J., Féron, F., Mackay-Sim, A., & Waite, P. M. (2002). Olfactory ensheathing cells promote locomotor recovery after delayed transplantation into transected spinal cord. *Brain, 125,* 14–21.

Lu, T., Pan, Y., Kao, S.-Y., Li, C., Kohane, I., Chan, J., et al. (2004). Gene regulation and DNA damage in the ageing human brain. *Nature, 429,* 883–891.

Lucas, R. J., Freedman, M. S., Muñoz, M., Garcia-Fernández, J.-M., & Foster, R. G. (1999). Regulation of the mammalian pineal by non-rod, non-cone, ocular photoreceptors. *Science, 284,* 505–507.

Luciano, M., Wright, M., Smith, G. A., Geffen, G. M., Geffen, L. B., & Martin, N. G. (2001). Genetic covariance among measures of information processing speed, working memory, and IQ. *Behavior Genetics, 31,* 581–592.

Lumey, L. H. (1992). Decreased birthweights in infants after maternal *in utero* exposure to the Dutch famine of 1944–1945. *Paediatric and Perinatal Epidemiology, 6, 240–253.*

Lupien, S. J., de Leon, M., de Santi, S., Convit, A., Tarshish, C., Thakur, M., et al. (1998). Cortisol levels during human aging predict hippocampal atrophy and memory deficits. *Nature Neuroscience, 1,* 69–73.

Ly, D. H., Lockhart, D. J., Lerner, R. A., & Schultz, P. G. (2000). Mitotic misregulation and human aging. *Science, 287,* 2486–2492.

Lyons, S. (2001, May 20). A will to eat, a fight for life. *San Luis Obispo Tribune,* p. A1.

Maas, L. C., Lukas, S. E., Kaufman, M. J., Weiss, R. D., Daniels, S. L., Rogers, V. W., et al. (1998). Functional magnetic resonance imaging of human brain activation during cue-induced cocaine craving. *American Journal of Psychiatry, 155,* 124–126.

Maccoby, E. E., & Jacklin, C. N. (1974). *The psychology of sex differences.* Stanford, CA: Stanford University Press.

Mackay-Sim, A., Féron, F., Cochrane, J., Bassingthwaighte, L., Bayliss, C., Davies, W. et al. (2008). Autologous olfactory ensheathing cell transplantation in human paraplegia: A 3-year clinical trial. *Brain, 131,* 2376–2386.

MacNeilage, P. F. (1998). The frame/content theory of evolution of speech production. *Behavioral and Brain Sciences, 21,* 499–511.

Macrae, J. R., Scoles, M. T., & Siegel, S. (1987). The contribution of Pavlovian conditioning to drug tolerance and dependence. *British Journal of Addiction, 82,* 371–380.

Maddison, D., & Viola, A. (1968). The health of widows in the year following bereavement. *Journal of Psychosomatic Research, 12,* 297–306.

Maeder, P. P., Meuli, R. A., Adriani, M., Bellmann, A., Fornari, E., Thiran, J.-P., et al. (2001). Distinct pathways involved in sound recognition and localization: A human fMRI study. *NeuroImage, 14,* 802–816.

Maes, H. H. M., Neale, M. C., & Eaves, L. J. (1997). Genetic and environmental factors in relative body weight and human adiposity. *Behavior Genetics, 27,* 325–351.

Maffei, M., Halaas, J., Ravussin, E., Pratley, R. E., Lee, G. H., Zhang, Y., et al. (1995). Leptin levels in human and rodent: Measurement of plasma leptin and ob RNA in obese and weight-reduced subjects. *Nature Medicine, 1,* 1155–1161.

Maggard, M. A., Shugarman, L. R., Suttorp, M., Maglione, M., Sugerman, H. J., Livingston, E. H., et al. (2005). Meta-analysis: Surgical treatment of obesity. *Annals of Internal Medicine, 142,* 547–559.

Magistretti, P. J., Pellerin, L., Rothman, D. L., & Shulman, R. G. (1999). Energy on demand. *Science, 283,* 496–497.

Maguire, E. A., Gadian, D. G., Johnsrude, I. S., Good, C. D., Ashburner, J., Frackowiak, R. S. J., et al. (2000). Navigation-related structural change in the hippocampi of taxi drivers. *Proceedings of the National Academy of Sciences, USA, 97,* 4398–4403.

Maier, M. A., Bennett, K. M., Hepp-Reymond, M. C., & Lemon, R. N. (1993). Contribution of the monkey corticomotoneuronal system to the control of force in precision grip. *Journal of Neurophysiology, 69,* 772–785.

Maier, S. F., Drugan, R. C., & Grau, J. W. (1982). Controllability, coping behavior, and stress-induced analgesia in the rat. *Pain, 12,* 47–56.

Maki, P. M., Rich, J. B, & Rosenbaum, R. S. (2002). Implicit memory varies across the menstrual cycle: Estrogen effects in young women. *Neuropsychologia, 40,* 518–529.

Malagelada, C., Jin, Z. H., Jackson-Lewis, V., Przedborski, S., & Greene, L. A. (2010). Rapamycin protects against neuron death in *in vitro* and *in vivo* models of Parkinson's disease. *Journal of Neuroscience, 30,* 1166–1175.

Maletic-Savatic, M., Malinow, R., & Svoboda, K. (1999). Rapid dendritic morphogenesis in CA1 hippocampal dendrites induced by synaptic activity. *Science, 283,* 1923–1927.

Malison, R. T., McDougle, C. J., van Dyck, C. H., Scahill, L., Baldwin, R. M., Seibyl, J. P., et al. (1995). [123I] α-CIT SPECT imaging of striatal dopamine transporter binding in Tourette's disorder. *American Journal of Psychiatry, 152,* 1359–1361.

Mamikonyan, E., Siderowf, A. D., Duda, J. E., Potenza, M. N., Horn, S., Stern, M. B., et al. (2008). Long-term follow-up of impulse control disorders in Parkinson's disease. *Movement Disorders, 23,* 75–80.

Mampe, B., Friederici, A. D., Christophe, A., & Wermke, K. (2009). Newborns' cry melody is shaped by their native language. *Current Biology, 19,* 1–4.

Manfredi, M., Bini, G., Cruccu, G., Accornero,, N., Berardelli, A., & Medolago, L. (1981). Congenital absence of pain. *Archives of Neurology, 38,* 507–511.

Mann, J. J. (2003). Neurobiology of suicidal behaviour. *Nature Reviews Neuroscience, 4,* 819–828.

Mann, J. J., Arango, V., & Underwood, M. D. (1990). Serotonin and suicidal behavior. *Annals of the New York Academy of Sciences. Special Issue: The Neuropharmacology of Serotonin, 600,* 476–485.

Mansari, M., Sakai, K., & Jouvet, M. (1989). Unitary characteristics of presumptive cholinergic tegmental neurons during the sleep-waking cycle in freely moving cats. *Experimental Brain Research, 76,* 519–529.

Mantzoros, C., Flier, J. S., Lesem, M. D., Brewerton, T. D., & Jimerson, D. C. (1997). Cerebrospinal fluid leptin in anorexia nervosa: Correlation with nutritional status and potentail role in resistance to weight gain. *Journal of Clinical Endocrinology and Metabolism, 82,* 1845–1851.

Manuck, S. B., Flory, J. D., Ferrell, R. E., Mann, J. J., & Muldoon, M. F. (2000). A regulatory polymorphism of the monoamine oxidase–A gene may be associated with variability in aggression, impulsivity, and central nervous system serotonergic responsivity. *Psychiatry Research, 95,* 9–23.

Mapes, G. (1990, April 10). Beating the clock: Was it an accident Chernobyl exploded at 1:23 in the morning? *Wall Street Journal,* p. A1.

Maquet, P., Laureys, S., Peigneux, P., Fuchs, S., Petiau, C., Phillips, C., et al. (2000). Experience-dependent changes in cerebral activation during human REM sleep. *Nature Neuroscience, 3,* 831–836.

Maraganore, D. M., de Andrade, M., Lesnick, T. G., Strain, K. J., Farrer, M. J., Rocca, W. A., et al. (2005). High-resolution whole-genome association study of Parkinson disease. *American Journal of Human Genetics, 77,* 685–693.

Marczynski, T. J., & Urbancic, M. (1988). Animal models of chronic anxiety and "fearlessness." *Brain Research Bulletin, 21,* 483–490.

Mark, T. L., Levit K. R., Coffey, R. M., McKusick, D. R., Harwood, H. J., King, E. C., et al. (2007). *National expenditures for mental health services and substance abuse treatment, 1993–2003.* Rockville, MD: Substance Abuse and Mental Health Services Administration. Retrieved July 28, 2010, from www.samhsa .gov/spendingestimates/SAMHSAFINAL9303.pdf

Marks, W. B., Dobelle, W. H., & MacNichol, E. F., Jr. (1964). Visual pigments of single primate cones. *Science, 143,* 1181–1183.

Marshall, E. (1998). Medline searches turn up cases of suspected plagiarism. *Science, 279,* 473–474.

Marshall, E. (2000a). Gene therapy on trial. *Science, 288,* 951–957.

Marshall, E. (2000b). How prevalent is fraud? That's a million-dollar question. *Science, 290,* 1662–1663.

Marshall, E. (2000c). Moratorium urged on germ line gene therapy. *Science, 289,* 2023.

Marshall, J. (2008, February 16). Forgetfulness is key to a healthy mind. *New Scientist,* 29–33.

Marshall, L., Helgadóttir, H., Mölle, M., & Born, J. (2006). Boosting slow oscillations during sleep potentiates memory. *Nature, 444,* 610–613.

Martell, B. A., Orson, F. M., Poling, J., Mitchell, E., Rossen, R. D., Gardner, T., et al. (2009). Cocaine vaccine for the treatment of cocaine dependence in methadone-maintained patients. *Archives of General Psychiatry, 66,* 1116–1123.

Martin, A., Haxby, J. V., Lalonde, F. M., Wiggs, C. L., & Ungerleider, L. G. (1995). Discrete cortical regions associated with knowledge of color and knowledge of actions. *Science, 270,* 102–105.

Martin, A., Wiggs, C. L., Ungerleider, L. G., & Haxby, J. V. (1996). Neural correlates of category-specific knowledge. *Nature, 379,* 649–652.

Martin, D. S. (2008, May 16). Man's rare ability may unlock secret of memory. *CNNhealth.* Retrieved June 3, 2010, from www.cnn .com/2008/HEALTH/conditions/05/07/miraculous.memory/ index.html

Martin, J. B. (1987). Molecular genetics: Applications to the clinical neurosciences. *Science, 238,* 765–772.

Martin, M. J., Muotri, A., Gage, F., & Varki, A. (2005). Human embryonic stem cells express an immunogenic nonhuman sialic acid. *Nature medicine, 11,* 228–232.

Martini, F. (1988). *Fundamentals of anatomy and physiology* (4th ed.). Upper Saddle River, NJ: Prentice Hall.

Martino, G., Franklin, R. J. M., Van Evercooren, A. B., & Kerr, D. A. (2010). Stem cell transplantation in multiple sclerosis: Current status and future prospects. *Nature Reviews Neurology, 6,* 247–255.

Martins, I. P., & Ferro, J. M. (1992). Recovery of acquired aphasia in children. *Aphasiology, 6,* 431–438.

Marucha, P. T., Kiecolt-Glaser, J. K., & Favagehi, M. (1998). Mucosal wound healing is impaired by examination stress. *Psychosomatic Medicine, 60,* 362–365.

Marx, J. (1998). New gene tied to common form of Alzheimer's. *Science, 281,* 507–509.

Marx, J. (2003). Cellular warriors at the battle of the bulge. *Science, 299,* 846–849.

Masica, D. N., Money, J., Ehrhardt, A. A., & Lewis, V. G. (1969). IQ, fetal sex hormones and cognitive patterns: Studies in the testicular feminizing syndrome of androgen insensitivity. *Johns Hopkins Medical Journal, 124,* 34–43.

Mâsse, L. C., & Tremblay, R. E. (1997). Behavior of boys in kindergarten and the onset of substance use during adolescence. *Archives of General Psychiatry, 54,* 62–68.

Masters, W., & Johnson, V. (1966). *The human sexual response.* Boston: Little, Brown.

Mateer, C. A., & Cameron, P. A. (1989). Electrophysiological correlates of language: Stimulation mapping and evoked potential studies. In F. Boller & J. Grafman, J. (Eds.), *Handbook of neuropsychology* (Vol. 2, pp. 91–116). New York: Elsevier.

Matsui, D., Minato, T., MacDorman, K. F., & Ishiguro, H. (2005, August). Generating natural motion in an android by mapping human motion. *Proceedings of the IEEE/RSJ International Conference on Intelligent Robots and Systems.* Retrieved July 13, 2007, from www.ieeexplore.ieee.org/xpl/freeabs_all .jsp?arnumber=1545125

Matsuzaki, M., Honkura, N., Ellis-Davies, G. C. R., & Kasai, H. (2004). Structural basis of long-term potentiation in single dendritic spines. *Nature, 429,* 761–766.

Mattay, V. S., Berman, K. F., Ostrem, J. L., Esposito, G., Van Horn, J. D., Bigelow, L. B., et al. (1996). Dextroamphetamine enhances "neural network-specific" physiological signals: A positron-emission tomography rCBF study. *Journal of Neuroscience, 16,* 4816–4822.

Mattes, R. D. (2009). Is there a fatty acid taste? *Annual Review of Nutrition, 29,* 305–327.

Matud, M. P. (2004). Gender differences in stress and coping styles. *Personality and Individual Differences, 37,* 1401–1415.

Matuszewich, L., Lorrain, D. S., & Hull, E. M. (2000). Dopamine release in the medial preoptic area of female rats in response to hormonal manipulation and sexual activity. *Behavioral Neuroscience, 114,* 772–782.

Mayberg, H. S., Lozano, A. M., Voon, V., McNeely, H. E., Seminowicz, D., Hamani, C., et al. (2005). Deep brain stimulation for treatment-resistant depression. *Neuron, 45,* 651–660.

Mazur, A., & Lamb, T. A. (1980). Testosterone, status, and mood in human males. *Hormones and Behavior, 14,* 236–246.

Mazzocchi, F., & Vignolo, L. A. (1979). Localisation of lesions in aphasia: Clinical-CT scan correlations in stroke patients. *Cortex, 15,* 627–653.

McCandliss, B. D., & Noble, K. G. (2003). The development of reading impairment: A cognitive neuroscience model. *Mental Retardation and Developmental Disabilities Research Reviews, 9,* 196–205.

McCann, U. D., Lowe, K. A., & Ricaurte, G. A. (1997). Long-lasting effects of recreational drugs of abuse on the central nervous system. *Neuroscientist, 3,* 399–411.

McCann, U. D., Szabo, Z., Seckin, E., Rosenblatt, P., Mathews, W. B., Ravert, H. T., et al. (2005). Quantitative PET studies of the serotonin transporter in MDMA users and controls using [11C]McN5652 and [11C]DASB. *Neuropsychopharmacology, 30,* 1741–1750.

McClearn, G. E., Johansson, B., Berg, S., Pedersen, N. L., Ahern, F., Petrill, S. A., et al. (1997). Substantial genetic influence on cognitive abilities in twins 80 or more years old. *Science, 276,* 1560–1563.

McClelland, J. L., McNaughton, B. L., & O'Reilly, R. C. (1995). Why there are complementary learning systems in the hippocampus and neocortex: Insights from the successes and failures of connectionist models of learning and memory. *Psychological Review, 102,* 419–457.

McClung, C. A., Sidiropoulou, K., Vitaterna, M., Takahashi, J. S., White, F. J., Cooper, D. C., et al. (2005). Regulation of dopamine transmission and cocaine reward by the *Clock* gene. *Proceedings of the National Academy of Sciences, USA, 102,* 9377–9381.

McCormick, C. M., & Witelson, S. F. (1991). A cognitive profile of homosexual men compared to heterosexual men and women. *Psychoneuroendocrinology, 16,* 459–473.

McCormick, D. A., & Thompson, R. F. (1984). Cerebellum: Essential involvement in the classically conditioned eyelid response. *Science, 223,* 296–299.

McCoy, N. L., & Davidson, J. M. (1985). A longitudinal study of the effects of menopause on sexuality. *Maturitas, 7,* 203–210.

McCoy, N. L., & Pitino, L. (2002). Pheromonal influences on sociosexual behavior in young women. *Physiology and Behavior, 75,* 367–375.

McDaniel, M. A. (2005). Big-brained people are smarter: A meta-analysis of the relationship between in vivo brain volume and intelligence. *Intelligence, 33,* 337–346.

McDonald, J. W., Becker, D., Sadowsky, C. L., Jane, J. A., Conturo, T. E., & Schultz, L. M. (2002). Late recovery following spinal cord injury: Case report and review of the literature. *Journal of Neurosurgery: Spine, 97,* 252–265.

McDonald, J. W., Liu, X.-Z., Qu, Y., Liu, S., Mickey, S. K., Turetsky, D., et al. (1999). Transplanted embryonic stem cells survive, differentiate and promote recovery in injured rat spinal cord. *Nature Medicine, 5,* 1410–1412.

McDonald, R. J., & White, N. M. (1993). A triple dissociation of memory systems: Hippocampus, amygdala, and dorsal striatum. *Behavioral Neuroscience, 107,* 3–22.

McDougall, W. (1908). *An introduction to social psychology.* London: Methuen.

McDougle, C. J., Stigler, K. A., Erickson, C. A., & Posey, D. J. (2006). Pharmacology of autism. *Clinical Neuroscience Research, 6,* 179–188.

McEvoy, S. P., Stevenson, M. R., McCartt, A. T., Woodward, M., Haworth, C., & Palamara, P. (2005). Role of mobile phones in motor vehicle crashes resulting in hospital attendance: A case-crossover study. *British Medical Journal, 331,* 428–432.

McFadden, D., & Pasanen, E. G. (1998). Comparison of the auditory systems of heterosexuals and homosexuals: Click-evoked otoacoustic emissions. *Proceedings of the National Academy of Sciences, USA, 95,* 2709–2713.

McGarry-Roberts, P. A., Stelmack, R. M., & Campbell, K. B. (1992). Intelligence, reaction time, and event-related potentials. *Intelligence, 16,* 289–313.

McGaugh, J. L. (2000). Memory—A century of consolidation. *Science, 287,* 248–251.

McGaugh, J. L., Cahill, L., & Roozendaal, B. (1996). Involvement of the amygdala in memory storage: Interaction with other brain systems. *Proceedings of the National Academy of Sciences, USA, 93,* 13508–13514.

McGeoch, P. D., Brang, D., Song, T., Lee, R. R., Huang, M., & Ramachandran, V. S. (2009). Apotemnophilia—the neurological basis of a 'psychological' disorder. *Nature Precedings.* Retrieved May 30, 2010, from http://precedings.nature.com/documents/2954/version/1

McGrath, C. L., Glatt, S. J., Sklar, P., Le-Niculescu, H., Kuczenski, R., Doyle A. E., et al. (2009). Evidence for genetic association of *RORB* with bipolar disorder. *BMC Psychiatry, 9,* 70. Retrieved August 18, 2010, from www.biomedcentral.com/1471–244X/9/70

McGrath, L. M., Smith, S. D., & Pennington, B. F. (2006). Breakthroughs in the search for dyslexia candidate genes. *Trends in Molecular Medicine, 12,* 333–341.

McGue, M., & Bouchard, T. J. (1998). Genetic and environmental influences on human behavioral differences. *Annual Review of Neuroscience, 21,* 1–24.

McGuffin, P., Rijsdijk, F., Andrew, M., Sham, P., Katz, R., & Cardno, A. (2003). The heritability of bipolar affective disorder and the genetic relationship to unipolar depression. *Archives of General Psychiatry, 60,* 497–502.

McGuire, P. K., Shah, G. M. S., & Murray, R. M. (1993). Increased blood flow in Broca's area during auditory hallucinations in schizophrenia. *Lancet, 342,* 703–706.

McGuire, P. K., Silbersweig, D. A., Wright, I., Murray, R. M., David, A. S., Frackowiak, R. S. J., et al. (1995). Abnormal monitoring of inner speech: A physiological basis for auditory hallucinations. *Lancet, 346,* 596–600.

McIntosh, A. R., Rajah, M. N., & Lobaugh, N. J. (1999). Interactions of prefrontal cortex in relation to awareness in sensory learning. *Science, 284,* 1531–1533.

McKeefry, D. J., Gouws, A., Burton, M. P., & Morland, A. B. (2009). The noninvasive dissection of the human visual cortex: Using fMRI and TMS to study the organization of the visual cortex. *Neuroscientist, 15,* 489–506.

McKeon, J., McGuffin, P., & Robinson, P. (1984). Obsessive-compulsive neurosis following head injury: A report of 4 cases. *British Journal of Psychiatry, 144,* 190–192.

McKinnon, W., Weisse, C. S., Reynolds, C. P., Bowles, C. A., & Baum, A. (1989). Chronic stress, leukocyte subpopulations, and humoral response to latent viruses. *Health Psychology, 8,* 389–402.

McLellan, T. A., Lewis, D. C., O'Brien, C. P., & Kleber, H. D. (2000). Drug dependence, a chronic medical illness: Implications for treatment, insurance, and outcomes evaluation. *Journal of the American Medical Association, 284,* 1689–1695.

Mechoulam, R., Ben-Shabat, S., Hanus, L., Ligumsky, M., Kaminski, N. E., Schatz, A. R., et al. (1995). Identification of an endogenous 2-monoglyceride, present in canine gut, that binds to cannabinoid receptors. *Biochemical Pharmacology, 83,* 90.

Mednick, S., Nakayama, K., & Stickgold, R. (2003). Sleep-dependent learning: A nap is as good as a night. *Nature Neuroscience, 6,* 697–698.

Melamed, R., Rosenne, E., Shakhar, K., Schwartz, Y., Aburdarham, N., & Ben-Eliyahu, S. (2005). Marginating pulmonary-NK activity and resistance to experimental tumor metastasis: Suppression by surgery and the prophylactic use of a β-adrenergic antagonist and a prostaglandin synthesis inhibitor. *Brain, Behavior, and Immunity, 19,* 114–126.

Melichar, J. K., Daglish, M. R. C., & Nutt, D. J. (2001). Addiction and withdrawal-current views. *Current Opinion in Pharmacology, 1,* 84–90.

Melone, M., Vitellaro-Zuccarello, L., Vallejo-Illaramendi, A., Pérez-Samartin, A., Matute, C., Cozzi, A., et al. (2001). The expression of glutamate transporter GLT-1 in the rat cerebral cortex is down-regulated by the antipsychotic drug clozapine. *Molecular Psychiatry, 6,* 380–386.

Melton, L. (2007, February 12). What's your poison? *New Scientist,* 30–33.

Meltzer, H. Y. (1990). Role of serotonin in depression. *Annals of the New York Academy of Sciences. Special Issue: The Neuropharmacology of Serotonin, 600,* 486–500.

Melzack, R. (1973). *The puzzle of pain.* New York: Basic Books.

Melzack, R. (1990). Phantom limbs and the concept of a neuromatrix. *Trends in Neurosciences, 13,* 88–92.

Melzack, R. (1992, April). Phantom limbs. *Scientific American, 266,* 120–126.

Melzack, R., & Wall, P. D. (1965). Pain mechanisms: A new theory. *Science, 150,* 971–979.

Merzenich, M. M., Knight, P. L., & Roth, G. L. (1975). Representation of cochlea within primary auditory cortex in the cat. *Journal of Neurophysiology, 61,* 231–249.

Meston, C. M., & Frohlich, P. F. (2000). The neurobiology of sexual function. *Archives of General Psychiatry, 57,* 1012–1030.

Mesulam, M. M. (1981). Dissociative states with abnormal temporal lobe EEG. Multiple personality and the illusion of possession. *Archives of Neurology, 38,* 176–181.

Mesulam, M.-M. (1986). Frontal cortex and behavior. *Annals of Neurology, 19,* 320–325.

Meyer, J. H., Wilson, A. A., Rusjan, P., Clark, M., Houle, S., Woodside, S., et al. (2008). Serotonin(2A) receptor binding potential in people with aggressive and violent behaviour. *Journal of Psychiatry and Neuroscience, 33,* 499–508.

Meyer-Bahlburg, H. F. L. (1984). Psychoendocrine research on sexual orientation. Current status and future options. *Progress in Brain Research, 61,* 375–398.

Meyer-Bahlburg, H. F. L. (1999). Gender assignment and reassignment in 46,XY pseudohermaphroditism and related conditions. *Journal of Clinical Endocrinology and Metabolism, 84,* 3455–3458.

Meyer-Bahlburg, H. F. L., Ehrhardt, A. A., Rosen, L. R., Gruen, R. S., Veridiana, N. P., Vann, F. H., et al. (1995). Prenatal estrogens and the development of homosexual orientation. *Developmental Psychobiology, 31,* 12–21.

Meyer-Lindberg, A. S., Olsen, R. K., Kohn, P. D., Brown, T., Egan, M. F., Weinberger, D. R., et al. (2005). Regionally specific disturbance of dorsolateral prefrontal-hippocampal functional connectivity in schizophrenia. *Archives of General Psychiatry, 62,* 379–386.

Michael, R., Gagnon, J., Laumann, E., & Kolata, G. (1994). *Sex in America.* Boston: Little, Brown.

Mignot, E. (1998). Genetic and familial aspects of narcolepsy. *Neurology, 50*(Suppl. 1), S16–S22.

Miles, D. R., & Carey, G. (1997). Genetic and environmental architecture of human aggression. *Journal of Personal and Social Psychology, 72,* 207–217.

Miles, L. E. M., Raynal, D. M., & Wilson, M. A. (1977). Blind man living in normal society has circadian rhythms of 24.9 hours. *Science, 198,* 421–423.

Miller, A. (1967). The lobotomy patient—a decade later: A follow-up study of a research project started in 1948. *Canadian Medical Association Journal, 96,* 1095–1103.

Miller, B., Messias, E., Miettunen, J., Alaräisänen, A., Järvelin, M.-R., Koponen, H., et al. (2010). Meta-analysis of paternal age and schizophrenia risk in male versus female offspring. *Schizophrenia Bulletin* (published online in advance of print). Retrieved August 13, 2010, from http://schizophreniabulletin.oxfordjournals.org/cgi/content/abstract/sbq011v1

Miller, B. L., Boone, K., Cummings, J. L., Read, S. L., & Mishkin, F. (2000). Functional correlates of musical and visual ability in frontotemporal dementia. *British Journal of Psychiatry, 176,* 458–463.

Miller, B. L., Cummings, J., Mishkin, F., Boone, K., Prince, F., Ponton, M. et al. (1998). Emergence of artistic talent in frontotemporal dementia. *Neurology, 51,* 978–982.

Miller, C. A., & Sweatt, D. (2007). Covalent modification of DNA regulates memory formation. *Neuron, 53,* 857–869.

Miller, D. S., & Parsonage, S. (1975). Resistance to slimming: Adaptation or illusion? *Lancet, 1,* 773–775.

Miller, E. K., Erickson, C. A., & Desimone, R. (1996). Neural mechanisms of visual working memory in prefrontal cortex of the Macaque. *Journal of Neuroscience, 16,* 5151–5167.

Miller, G. (2008). Scientists targeted in California firebombings. *Science, 321,* 755.

Miller, G. (2010). Beyond *DSM*: Seeking a brain-based classification of mental illness. *Science, 327,* 1437.

Miller, L. T., & Vernon, P. A. (1992). The general factor in short-term memory, intelligence, and reaction time. *Intelligence, 16,* 5–29.

Miller, N. F. (1985). The value of behavioral research on animals. *American Psychologist, 40,* 423–440.

Miller, S. D., & Triggiano, P. J. (1992). The psychophysiological investigation of multiple personality disorder: Review and update. *American Journal of Clinical Hypnosis, 35,* 47–61.

Miller, T. Q., Smith, T. W., Turner, C. W., Guijarro, M. L., & Hallet, A. J. (1996). A meta-analytic review of research on hostility and physical health. *Psychological Bulletin, 119,* 322–348.

Milner, B. (1970). Memory and the temporal regions of the brain. In K. H. Pribram & D. E. Broadbent (Eds.), *Biology and memory* (pp. 29–50). New York: Academic Press.

Milner, B. (1974). Hemispheric specialization: Scope and limits. In F. O. Schmitt & F. G. Worden (Eds.), *The neurosciences: Third study program* (pp. 75–89). Cambridge, MA: MIT Press.

Milner, B., Corkin, S., & Teuber, H.-L. (1968). Further analysis of the hippocampal amnesic syndrome: 14-year follow-up study of HM. *Neuropsychologia, 6,* 215–234.

Milner, P. (1977). How much distraction can you hear? *Stereo Review, June,* 64–68.

Miltner, W. H. R., Braun, C., Arnold, M., Witte, H., & Taub, E. (1999). Coherence of gamma-band EEG activity as a basis for associative learning. *Nature, 397,* 434–436.

Mintun, M. A., LaRossa, G. N., Sheline, Y. I., Dence, C. S., Lee, S. Y., Mach, R. H., et al. (2006). [^{11}C]PIB in a nondemented population: Potential antecedent marker of Alzheimer disease. *Neurology, 67,* 446–452.

Mitchell, S. W. (1866, July). The case of George Dedlow. *Atlantic Monthly, 18,* 1–11.

Mitka, M. (2006). Surgery useful for morbid obesity, but safety and efficacy questions linger. *Journal of American Medical Association, 296,* 1575–1577.

Mitler, M. M., Carskadon, M. A., Czeisler, C. A., Dement, W. C., Dinges, D. F., & Graeber, R. C. (1988). Catastrophes, sleep, and public policy: Consensus report. *Sleep, 11,* 100–109.

Miyashita, Y. (1993). Inferior temporal cortex: Where visual perception meets memory. *Annual Review of Neuroscience, 16,* 245–263.

Miyashita, Y., & Chang, H. S. (1988). Neuronal correlate of pictorial short-term memory in the primate temporal cortex. *Nature, 331,* 68–70.

Modahl, C., Green, L., Fein, D., Morris, M., Waterhouse, L., Feinstein, C., et al. (1998). Plasma oxytocin levels in autistic children. *Biological Psychiatry, 43,* 270–277.

Moeller, F. G., Dougherty, D. M., Swann, A. C., Collins, D., Davis, C. M., & Cherek, D. R. (1996). Tryptophan depletion and aggressive responding in healthy males. *Psychopharmacology, 126,* 97–103.

Moffat, S. D., Zonderman, A. B., Metter, E. J., Blackman, M. R., Harman, S. M., & Resnick, S. M. (2002). Longitudinal assessment of serum-free testosterone concentration predicts memory performance and cognitive status in elderly men. *Journal of Clinical Endocrinology and Metabolism, 87,* 5001–5007.

Mogilner, A., Grossman, J. A. I., Ribary, U., Jolikot, M., Volkmann, J., Rapaport, D., et al. (1993). Somatosensory cortical plasticity in adult humans revealed by magnetoencephalography. *Proceedings of the National Academy of Sciences, USA, 90,* 3593–3597.

Moldin, S. O., Reich, T., & Rice, J. P. (1991). Current perspectives on the genetics of unipolar depression. *Behavior Genetics, 21,* 211–242.

Mombaerts, P. (1999). Seven-transmembrane proteins as odorant and chemosensory receptors. *Science, 286,* 707–711.

Money, J. (1968). *Sex errors of the body and related syndromes: A guide to counseling children, adolescents, and their families.* Baltimore: Paul H. Brookes.

Money, J., Devore, H., & Norman, B. F. (1986). Gender identity and gender transposition: Longitudinal outcome study of 32 male hermaphrodites assigned as girls. *Journal of Sex and Marital Therapy, 12,* 165–181.

Money, J., & Ehrhardt, A. A. (1972). *Man and woman, boy and girl.* Baltimore: Johns Hopkins University Press.

Money, J., Schwartz, M., & Lewis, V. G. (1984). Adult erotosexual status and fetal hormonal masculinization and demasculinization: 46, XX congenital virilizing adrenal hyperplasia and 46, XY androgen-insensitivity syndrome compared. *Psychoneuroendocrinology, 9,* 405–414.

Monteleone, P., Luisi, S., Tonetti, A., Bernardi, F., Genazzani, A. D., Luisi, M., et al. (2000). Allopregnanolone concentrations and premenstrual syndrome. *European Journal of Endocrinology, 142,* 269–273.

Monti, M. M., Vanhaudenhuyse, A., Coleman, M. R., Boly, M., Pickard, J. D., Tshibanda, L., et al. (2010). Willful modulation of brain activity in dosrders of consciousness. *New England Journal of Medicine, 362,* 579–589.

Morgan, D., Diamond, D. M., Gottschall, P. E., Ugen, K. E., Dickey, C., Hardy, J., et al. (2000). A b peptide vaccination prevents memory loss in an animal model of Alzheimer's disease. *Nature, 408,* 982–985.

Moritz, C. T., Perlmutter, S. I., & Fetz, E. E. (2008). Direct control of paralysed muscles by cortical neurons. *Nature, 456,* 639–643.

Morley, K. I., & Montgomery, G. W. (2001). The genetics of cognitive processes: Candidate genes in humans and animals. *Behavior Genetics, 31,* 511–531.

Morrel-Samuels, P., & Herman, L. M. (1993). Cognitive factors affecting comprehension of gesture language signs: A brief comparison of dolphins and humans. In H. L. Roitblat, L. M. Herman, & P. E. Nachtigall (Eds.), *Language and communicaton: Comparative perspectives* (pp. 311–327). Hillsdale, NJ: Lawrence Erlbaum.

Morris, J. M. (1953). The syndrome of testicular feminization in male pseudohermaphrodites. *American Journal of Obstetrics and Gynecology, 65,* 1192–1211.

Morris, J. S., Frith, C. D., Perrett, D. I., Rowland, D., Young, A. W., Calder, A. J., et al. (1996). A differential neural response in the human amygdala to fearful and happy facial expressions. *Nature, 383,* 812–815.

Morris, M., Lack, L., & Dawson, D. (1990). Sleep-onset insomniacs have delayed temperature rhythms. *Sleep, 13,* 1–14.

Morris, M. C., Evans, D. A., Tangney, C. C., Bienias, J. L., & Wilson, R. S. (2005). Fish consumption and cognitive decline with age in a large community study. *Archives of Neurology, 62,* 1–5.

Morris, N. M., Udry, J. R., Khan-Dawood, F., & Dawood, M. Y. (1987). Marital sex frequency and midcycle female testosterone. *Archives of Sexual Behavior, 16,* 27–37.

Moscovitch, M., & Winocur, G. (1995). Frontal lobes, memory, and aging. *Annals of the New York Academy of Sciences, 769,* 119–150.

Mountcastle, V. B., & Powell, T. P. S. (1959). Neural mechanisms subserving cutaneous sensibility, with special reference to the role of afferent inhibition in sensory perception and discrimination. *Bulletin of the Johns Hopkins Hospital, 105,* 201–232.

Mouritsen, H., Janssen-Bienhold, U., Liedvogel, M., Feenders, G., Stalleicken, J., Dirks, P., et al. (2004). Cryptocromes and neuronal-activity markers colocalize in the retina of migratory birds during magnetic orientation. *Proceedings of the National Academy of Sciences, USA, 101,* 14297.

Mpakopoulou, M., Gatos, H., Brotis, A., Paterakis, K., & Fountas, K. N. (2008). Stereotactic amygdalotomy in the management of severe aggressive behavioral disorders. *Neurosurgical Focus, 25,* E6. Published online at http://thejns.org/doi/pdf/10.3171/FOC/2008/25/7/E6

Mujica-Parodi, L. R., Strey, H. H., Frederick, B., Savoy, R., Cox, D., Botanov, Y., et al. (2009). Chemosensory cues to conspecific emotional stress activate amygdala in humans. *PLoS One, 4,* e6415. Retrieved February 14, 2010, from www.plosone.org/article/info:doi%2F10.1371%2Fjournal.pone.0006415

Mundy, N. I., & Cook, S. (2003). Positive selection during the diversification of class I vomeronasal receptor-like (V1RL) genes, putative pheromone receptor genes in human and primate evolution. *Molecular and Biological Evolution, 20,* 1805–1810.

Murdoch, D., Pihl, R. O., & Ross, D. (1990). Alcohol and crimes of violence: Present issues. *International Journal of Addiction, 25,* 1065–1081.

Murphy, F. C., Nimmo-Smith, I., & Lawrence, A. D. (2003). Functional neuroanatomy of emotions: A meta-analysis. *Cognitive, Affective, and Behavioral Neuroscience, 3,* 207–233.

Murphy, G. (1949). *Historical introduction to modern psychology.* New York: Harcourt, Brace & World.

Murphy, M. R., Checkley, S. A., Seckl, J. R., & Lightman, S. L. (1990). Naloxone inhibits oxytocin release at orgasm in man. *Journal of Clinical Endocrinology and Metabolism, 71,* 1056–1058.

Murray, J. B. (2002). Phencyclidine (PCP): A dangerous drug, but useful in schizophrenia research. *Journal of Psychology, 136,* 319–327.

Murrell, J., Farlow, M., Ghetti, B., & Benson, M. D. (1991). A mutation in the amyloid precusor protein associated with hereditary Alzheimer's disease. *Science, 254,* 97–99.

Must, A., Spadano, J., Coakley, E. H., Field, A. E., Colditz, G., & Dietz, W. H. (1999). The disease burden associated with overweight and obesity. *Journal of the American Medical Association, 282,* 1523–1529.

Mustanski, B. S., DuPree, M. G., Nievergelt, C. M., Bocklandt, S., Schork, N. J., & Hamer, D. H. (2005). A genomewide scan of male sexual orientation. *Human Genetics, 116,* 272–278.

Nader, K., Schafe, G. E., & LeDoux, J. E. (2000). Fear memories require protein synthesis in the amygdala for reconsolidation after retrieval. *Nature, 406,* 722–726.

Naeser, M. A., Alexander, M. P., Helm-Estabrooks, N., Levine, H. L., Laughlin, S. A., & Geschwind, N. (1982). Aphasia with predominantly subcortical lesion sites. *Archives of Neurology, 39,* 2–14.

Naglieri, J. A., & Ronning, M. E. (2000). Comparison of white, African American, Hispanic, and Asian children on the Naglieri nonverbal ability test. *Psychological Assessment, 12,* 328–334.

Nairne, James S. (Ed.). (2000). *Psychology: The adaptive mind* (2nd ed.). Belmont, CA: Wadsworth/Thomson Learning.

Nakashima, T., Pierau, F. K., Simon, E., & Hori, T. (1987). Comparison between hypothalamic thermoresponsive neurons from duck and rat slices. *Pflugers Archive: European Journal of Physiology, 409,* 236–243.

Naliboff, B., Berman, S., Chang, L., Derbyshire, S., Suyenobu, B., Vogt, B., et al. (2003). Sex-related differences in IBS patients: Central processing of visceral stimuli. *Gastroenterology, 124,* 1738–1747.

Naranjo, C. A., Poulos, C. X., Bremner, K. E., & Lanctot, K. L. (1994). Fluoxetine attenuates alcohol intake and desire to drink. *International Clinical Psychopharmacology, 9,* 163–172.

Nash, J. M., Park, A., & Willwerth, J. (1995, July 17). "Consciousness" may be an evanescent illusion. *Time,* p. 52.

National Highway Traffic Safety Administration. (2006). Traffic safety facts 2006 data: Alcohol-impaired driving. Retrieved November 19, 2009, from www-nrd.nhtsa.dot.gov/Pubs/810801.PDF

National Institute of Mental Health. (1986). *Schizophrenia: Questions and answers* (DHHS Publication No. ADM 86–1457). Washington, DC: Government Printing Office.

National Institute of Mental Health. (2006). *Attention deficit hyperactivity disorder.* Retrieved December 2, 2007, from www.nimh.nih.gov/health/publications/adhd/summary.shtml

National Institute of Mental Health. (2008, December 24). *Study probes environment-triggered genetic changes in schizophrenia.* Retrieved August 1, 2010, from www.nimh.nih

gov/science-news/2008/study-probes-environment-triggered-genetic-changes-in-schizophrenia.shtml

National Sleep Foundation. (2002). *2002 "Sleep in America" poll.* Retrieved November 28, 2007, from www.sleepfoundation.org/site/c.huIXKjM0IxF/b.2417355/k.143E/2002_Sleep_in_America_Poll.htm

Neave, N., Menaged, M., & Weightman, D. R. (1999). Sex differences in cognition: The role of testosterone and sexual orientation. *Brain and Cognition, 41,* 245–262.

Nebes, R. D. (1974). Hemispheric specialization in commissurotomized man. *Psychological Bulletin, 81,* 1–14.

Nedergaard, M., Ransom, B., & Goldman, S. A. (2003). New roles for astrocytes: Redefining the functional architecture of the brain. *Trends in Neurosciences, 26,* 523–530.

Neisser, U., Boodoo, G., Bouchard, T. J., Jr., Boykin, A. W., Brody, N., Ceci, S. J., et al. (1996). Intelligence: Knowns and unknowns. *American Psychologist, 51,* 77–101.

Nelson, D. L., & Gibbs, R. A. (2004). The critical region in trisomy 21. *Science, 306,* 619–621.

Nelson, L. (2004). Venomous snails: One slip, and you're dead . . . . *Nature, 429,* 798–799.

Netter, F. H. (1983). *CIBA collection of medical illustrations: Vol. 1. Nervous system.* New York: CIBA.

Neville, H. J., Bavelier, D., Corina, D., Rauschecker, J., Karni, A., Lalwani, A., et al. (1998). Cerebral organization for language in deaf and hearing subjects: Biological constraints and effects of experience. *Proceedings of the National Academy of Sciences, USA, 95,* 922–929.

Newhouse, P. A., Potter, A., Corwin, J., & Lenox, R. (1992). Acute nicotinic blockade produces cognitive impairment in normal humans. *Psychopharmacology, 108,* 480–484.

Newman, A. (2006, July 5). "Collyers' Mansion" is code for firefighters' nightmare. *New York Times.* Retrieved August 22, 2010, from www.nytimes.com/2006/07/05/nyregion/05hoard.html

Newman, E. A. (2003). New roles for astrocytes: Regulation of synaptic transmission. *Trends in Neurosciences, 26,* 536–542.

Newschaffer, C. J., Croen, L. A., Daniels, J., Giarelli, E., Grether, J. K., Levy, S. E., et al. (2007). The epidemiology of autism spectrum disorders. *Annual Review of Public Health, 28,* 235–258.

Nichelli, P., Grafman, J, Pietrini, P., Clark, K., Lee, K. Y., & Miletich, R. (1995). Where the brain appreciates the moral of a story. *Neuroreport, 6,* 2309–2313.

Nicoll, R. A., & Madison, D. V. (1982). General anesthetics hyperpolarize neurons in the vertebrate central nervous system. *Science, 217,* 1055–1057.

Nieder, A., & Dehaene, S. (2009). Representation of number in the brain. *Annual Review of Neuroscience, 32,* 185–208.

Nielsen, A. S., Mortensen, P. B., O'Callaghan, E., Mors, O., & Ewald, H. (2002). Is head injury a risk factor for schizophrenia? *Schizophrenia Research, 55,* 93–98.

Nieuwenhuys, R., Voogd, J., & vanHuijzen, C. (1988). *The human central nervous system* (3rd Rev. ed.). Berlin, Germany: Springer-Verlag.

Nigg, J. T., Nikolas, M., Knottnerus, G. M., Cavanagh, K., & Friderici, K. (2010). Confirmation and extension of association of blood lead with attention-deficit/hyperactivity disorder (ADHD) and ADHD symptom domains at population-typical exposure levels. *Journal of Child Psychology and Psychiatry, 51,* 58–65.

Niiyama, Y., Kawamata, T., Yamamoto, J., Furuse, S., & Namiki, A. (2009). SB366791, a TRPV1 antagonist, potentiates analgesic effects of systemic morphine in a murine model of bone cancer pain. *British Journal of Anesthesia, 102,* 251–258.

Nisbett, R. E. (2005). Heredity, environment, and race differences in IQ: A commentary on Rushton and Jensen. *Psychology, Public Policy, and Law, 11,* 302–310.

Nishitani, N., Avikainen, S, & Hari, R. (2004). Abnormal imitation-related cortical activation sequences in Asperger's syndrome. *Annals of Neurology, 55,* 558–562.

Nobili, L., Ferrillo, F., Besset, A., Rosadini, G., Schiavi, G., & Billiard, M. (1996). Ultradian aspects of sleep in narcolepsy. *Neurophysiologie Clinique, 26,* 51–59.

Nottebohm, F. (1977). Asymmetries in neural control of vocalization in the canary. In S. Harnad, R. W. Doty, L. Goldstein, J. Jaynes, & G. Krauthamer (Eds.), *Lateralization in the nervous system* (pp. 23–44). New York: Academic Press.

Novin, D., VanderWeele, D. A., & Rezek, M. (1973). Infusion of 2-deoxy d-glucose into the hepatic portal system causes eating: Evidence for peripheral glucoreceptors. *Science, 181,* 858–860.

Nulman, I., Rovet, J., Greenbaum, R., Loebstein, M., Wolpin, J., Pace- Asciak, P., et al. (2001). Neurodevelopment of adopted children exposed in utero to cocaine: The Toronto adoption study. *Clinical and Investigative Medicine, 24,* 129–137.

Nunn, J. A., Gregory, L. J., Brammer, M., Williams, S. C. R., Parslow, D. M., Morgan, M. J., et al. (2002). Functional magnetic resonance imaging of synesthesia: Activation of V4/V8 by spoken words. *Nature Neuroscience, 5,* 371–375.

Oberman,, L. M., Winkielman, P., & Ramachandran, V. S. (2007). Face to face: Blocking facial mimicry can selectively impair recognition of emotional expressions. *Social Neuroscience, 2,* 167–178.

Oberman,, L. M., Winkielman, P., & Ramachandran, V. S. (2009). Slow echo: Facial EMG evidence for the delay of spontaneous, but not voluntary, emotional mimicry in children with autism spectrum disorders. *Developmental Science, 12,* 510–520.

O'Brien, C. P. (1997). A range of research-based pharmacotherapies for addiction. *Science, 278,* 66–70.

Ochoa, B. (1998). Trauma of the external genitalia in children: Amputation of the penis and emasculation. *Journal of Urology, 160,* 1116–1119.

O'Connor, D. H., Fukui, M. M., Pinsk, M. A., & Kastner, S. (2002). Attention modulates responses in the human lateral geniculate nucleus. *Nature Neuroscience, 5,* 1203–1209.

O'Connor, D. H., Wittenberg, G. M., & Wang, S. S.-H. (2005). Graded bidirectional synaptic plasticity is composed of switch-like unitary events. *Proceedings of the National Academy of Sciences, USA, 102,* 9679–9684.

O'Dell, T. J., Hawkins, R. D., Kandel, E. R., & Arancio, O. (1991). Tests of the roles of two diffusible substances in long-term potentiation: Evidence for nitric oxide as a possible early retrograde messenger. *Proceedings of the National Academy of Sciences, USA, 88,* 11285–11289.

Ogden, C. L., Carroll, M. D., Curtin, L. R., McDowell, M. A., Tabak, C. J., & Flegal, K. M. (2006). Prevalence of overweight and obesity in the United States, 1999–2004. Journal of the American Medical Association, 295, 1549–1555.

Ogden, J. (1989). Visuospatial and other "right-hemispheric" functions after long recovery periods in left-hemispherectomized subjects. *Neuropsychologia, 27,* 765–776.

Ojemann, G. A. (1983). Brain organization for language from the perspective of electrical stimulation mapping. *Behavioral and Brain Sciences, 2,* 189–230.

O'Keane, V., & Dinan, T. G. (1992). Sex steroid priming effects on growth hormone response to pyridostigmine throughout the menstrual cycle. *Journal of Clinical Endocrinology and Metabolism, 75,* 11–14.

Okubo, Y., Suhara, T., Suzuki, K., Kobayashi, K., Inoue, O., Terasaki, O., et al. (1997). Decreased prefrontal dopamine D1 receptors in schizophrenia revealed by PET. *Nature, 385,* 634–636.

Olanow, C. W., & Tatton, W. G. (1999). Etiology and pathogenesis of Parkinson's disease. *Annual Review of Neuroscience, 22,* 123–144.

Oliet, S. H. R., Piet, R., & Poulain, D. A. (2001). Control of glutamate clearance and synaptic efficacy by glial coverage of neurons. *Science, 292,* 923–925.

Olshansky, S. J., Passaro, D. J., Hershow, R. C., Layden, J., Carnes, B. A., Brody, J., et al. (2005). A potential decline in life expectancy in the United States in the 21st century. *New England Journal of Medicine, 352,* 1138–1145.

Olson, B. R., Freilino, M., Hoffman, G. E., Stricker, E. M., Sved, A. F., & Verbalis, J. G. (1993). C-fos expression in rat brain and brainstem nuclei in response to treatments that alter food intake and gastric motility. *Molecular and Cellular Neuroscience, 4,* 93–106.

Olson, E. J., Boeve, B. F., & Silber, M. H. (2000). Rapid eye movement sleep behaviour disorder: Demographic, clinical and laboratory findings in 93 cases. *Brain, 123,* 331–339.

Ong, W. Y., & Mackie, K. (1999). A light and electron microscopic study of the CB1 cannabinoid receptor in primate brain. *Neuroscience, 92,* 1177–1191.

O'Reardon, J. P., Solvason H. B., Janicak, P. G., Sampson, S., Isenberg, K. E., Nahas, Z., et al. (2007). Efficacy and safety of transcranial magnetic stimulation in the acute treatment of major depression: A multisite randomized controlled trial. *Biological Psychiatry, 62,* 1208–1216.

O'Reilly, I. (2010). *Gender testing in sport: A case for treatment?* BBC News, February 15. Retrieved February 25, 2010, from http://news.bbc.co.uk/2/hi/8511176.stm

O'Reilly, K. B. (2009, November 23). AMA meeting: Delegates support review of marijuana's schedule I status. *American Medical News.* Retrieved November 25, 2009, from www.ama-assn.org/amednews/2009/11/23/prse1123.htm

Orlans, F. B. (1993). *In the name of science.* New York: Oxford University Press.

O'Rourke, D., Wurtman, J. J., Wurtman, R. J., Chebli, R., & Gleason, R. (1989). Treatment of seasonal depression with *d*-fenfluramine. *Journal of Clinical Psychiatry, 50,* 343–347.

Orexigen Therapeutics, Inc. (OREX) Contrave(R) obesity research phase 3 program meets co-primary and key secondary endpoints: Exceeds FDA efficacy benchmark for obesity treatments. (July 21, 2009). Retrieved January 21, 2010 from www.biospace.com/news_story.aspx?NewsEntityid-150551

Osvath, M. (2009). Spontaneous planning for future stone throwing by a male chimpanzee. *Current Biology, 19,* R190–R191.

Paean to Nepenthe. (1961, November 24). *Time,* p. 68.

Pagnin, D., de Queiroz, V., Pini, S., & Cassano, G. B. (2004). Efficacy of ECT in depression: A meta-analytic review. *Journal of Electroconvulsive Therapy, 20,* 13–20.

Pain, S. (2005, January 22). The curious lifestyle of the burrowing owl. *New Scientist,185,* 42–43.

Palmer, A. R. (1987). Physiology of the cochlear nerve and cochlear nucleus. In M. P. Haggard & E. F. Evans (Eds.), *Hearing* (pp. 838–855). Edinburgh, UK: Churchill Livingstone.

Panda, S., Nayak, S. K., Campo, B., Walker, J. R., Hogenesch, J. B., & Jegla, T. (2005). Illumination of the melanopsin signaling pathway. *Science, 307,* 600–604.

Pappone, P. A., & Cahalan, M. D. (1987). *Pandinus imperator* scorpion venom blocks voltage-gated potassium channels in nerve fibers. *Journal of Neuroscience, 7,* 3300–3305.

Parent, J. M. (2003). Injury-induced neurogenesis in the adult mammalian brain. *Neuroscientist, 9,* 261–272.

Parker, E. S., Cahill, L., & McGaugh, J. L. (2006). A case of unusual autobiographical remembering. *Neurocase, 12,* 35–49.

Pascual-Leone, A., & Torres, F. (1993). Plasticity of the sensorimotor cortex representation of the reading finger in Braille readers. *Brain, 116,* 39–52.

Pastalkova, E., Itskov, V., Amarasingham, A., & Buzsáki, G. (2008). Internally generated cell assembly sequences in the rat hippocampus. *Science, 321,* 1322–1327.

Pastalkova, E., Serrano, P., Pinkhasova, D., Wallace, E., Fenton, A. A., & Sacktor, C. (2006). Storage of spatial information by the maintenance mechanism of LTP. *Science, 313,* 1141–1144.

Patel, A. J., Honoré, E., Lesage, F., Fink, M., Romey, G., & Lazdunski, M. (1999). Inhalational anesthetics activate two-pore-domain background K⁺ channels. *Nature Neuroscience, 2,* 422–426.

Patel, A. N., Vina, R. F., Geffner, L., Kormos, R., Urschel, H. C., Jr., & Benetti, F. (2004, April 25). *Surgical treatment for congestive heart failure using autologous adult stem cell transplantation: A prospective randomized study.* Presented at the 84th annual meeting of the American Association for Thoracic Surgery, Toronto, Ontario, Canada.

Paterson, N. E., Froesti, W., & Markou, A. (2005). Repeated administration of the GABA$_B$ receptor agonist CGP44532 decreased nicotine self-administration, and acute administration decreased cue-induced reinstatement of nicotine-seeking in rats. *Neuropsychopharmacology, 30,* 119–128.

Patoine, B., & Bilanow, T. (n.d.) Alzheimer's approved drugs. Retrieved June 11, 2010, from the Fisher Center for Alzheimer's Research Foundation website: www.alzinfo.org/alzheimers-treatment-cognitive.asp

Patrick, C. J. (2008). Psychophysiological correlates of aggression and violence: An integrative review. *Philosophical Transactions of the Royal Society B, 363,* 2543–2555.

Paulesu, E., Démonet, J. F., Fazio, F., McCrory, E., Chanoine, V., Brunswick, N., et al. (2001). Dyslexia: Cultural diversity and biological unity. *Science, 291,* 2165–2167.

Pearlson, G. D., Kim, W. S., Kubos, K. L., Moberg, P. J., Jayaram, G., Bascom, M. J., et al. (1989). Ventricle-brain ratio, computed tomographic density, and brain area in 50 schizophrenics. *Archives of General Psychiatry, 46,* 690–697.

Peeters, R., Simone, L., Nelissen, K., Fabbri-Destro, M., Vanduffel, W., Rizzolatti, G., et al. (2009). The representation of tool use in humans and monkeys: Common and uniquely human features. *Journal of Neuroscience, 29,* 11523–11539.

Pellegrino, L. J., Pellegrino, A. S., & Cushman, A. J. (1979). *A stereotaxic atlas of the rat brain* (2nd ed.). New York: Plenum Press.

Penfield, W. (1955). The permanent record of the stream of consciousness. *Acta Psychologica, 11,* 47–69.

Penfield, W. (1958). *The excitable cortex in conscious man.* Springfield, IL: Charles C. Thomas.

Penfield, W., & Rasmussen, T. (1950). *The cerebral cortex of man.* New York: Macmillan.

Pennacchio, L. A., Ahituv, N., Moses, A. M., Prabhakar, S., Nobrega, M. A., Shoukry, M., et al. (2006). *In vivo* enhancer analysis of human conserved non-coding sequences. *Nature, 444,* 499–502.

Pentel, P. R., Malin, D. H., Ennifar, S., Hieda, Y., Keyler, D. E., Lake, J. R., et al. (2000). A nicotine conjugate vaccine reduces nicotine distribution to brain and attenuates its behavioral and cardiovascular effects in rats. *Pharmacology, Biochemistry, and Behavior, 65,* 191–198.

Pepperberg, I. M. (1993). Cognition and communication in an African Grey parrot *(Psittacus erithacus):* Studies on a nonhuman, nonprimate, nonmammalian subject. In H. L. Roitblat, L. M. Herman, & P. E. Nachtigall (Eds.), *Language and communication: Comparative perspectives* (pp. 221–248). Hillsdale, NJ: Lawrence Erlbaum.

Perkel, D. J., & Farries, M. A. (2000). Complementary "bottom-up" and "top-down" approaches to basal ganglia function. *Current Opinion in Neurobiology, 10,* 725–731.

Perkins, A., & Fitzgerald, J. A. (1992). Luteinizing hormone, testosterone, and behavioral response of male-oriented rams to estrous ewes and rams. *Journal of Animal Science, 70,* 1787–1794.

Perlis, M. L., Smith, M. T., Andrews, P. J., Orff, H., & Giles, D. E. (2001). Beta/gamma EEG activity in patients with primary and secondary insomnia and good sleeper controls. *Sleep, 24,* 110–117.

Perry, D. (2000). Patients' voices: The powerful sound in the stem cell debate. *Science, 287,* 1423.

Pert, C. B., & Snyder, S. H. (1973). Opiate receptor: Demonstration in nervous tissue. *Science, 179,* 1011–1014.

Pesticide exposure and Parkinson's disease: BfR sees association but not causal relationship. (2006, June 27). *Bundesinstitut für Risikobewertung.* Retrieved August 21, 2007, from www.bfr.bund .de/cm/289/pesticide_exposure_and_parkinsons_disease_bfr_ sees_association_but_no_causal_relationship.pdf

Petersen, M. R., Beecher, M. D., Zoloth, S. R., Moody, D. B., & Stebbins, W. C. (1978). Neural lateralization of species-specific vocalizations by Japanese Macaques *(Macaca fuscata). Science, 202,* 324–326.

Petersen, S. E., Fox, P. T., Snyder, A. Z., & Raichle, M. E. (1990). Activation of extrastriate and frontal cortical areas by visual words and word-like stimuli. *Science, 249,* 1041–1044.

Peterson, B. S., Skudlarski, P., Gatenby, J. C., Zhang, H., Anderson, A. W., & Gore, J. C. (1999). An fMRI study of Stroop word-color interference: Evidence for cingulate subregions subserving multiple distributed attentional systems. *Biological Psychiatry, 45,* 1237–1258.

Peterson, B. S., Warner, V., Bansal, R., Zhu, H., Hao, X., Liu, J., et al. (2009). Cortical thinning in persons at increased familial risk for major depression. *Proceedings of the National Academy of Sciences, 106,* 6273–6278.

Petitto, L. A., Holowka, S., Sergio, L. E., & Ostry, D. (2001). Language rhythms in baby hand movements. *Nature, 413,* 35–36.

Petitto, L. A., & Marentette, P. F. (1991). Babbling in the manual mode: Evidence for the ontogeny of language. *Science, 251,* 1493–1496.

Petitto, L. A., Zatorre, R. J., Gauna, K., Nikelski, E. J., Dostie, D., & Evans, A. C. (2000). Speech-like cerebral activity in profoundly deaf people processing signed languages: Implications for the neural basis of human language. *Proceedings of the National Academy of Sciences, USA, 97,* 13961–13966.

Petkov, C. I., Kayser, C., Steudel, T., Whittingstall, K., Augath, M., & Logothetis, N. K.(2008). A voice region in the monkey brain. *Nature Neuroscience, 11,* 367–374.

Petrovic, P. (2005). Opioid and placebo analgesia share the same network. *Seminars in Pain and Medicine, 3,* 31–36.

Petrovic, P., Kalso, E., Petersson, K. M., & Ingvar, M. (2002). Opioid and placebo analgesia—Imaging a shared neuronal network. *Science, 295,* 1737–1740.

Petry, N. M., Stinson, F. S., & Grant, B. F. (2005). Comorbidity of DSM-IV pathological gambling and other psychiatric disorders: Results from the national epidemiologic survey on alcohol and related conditions. *Journal of Clinical Psychiatry, 66,* 564–574.

Peuskens, H., Sunaert, S., Dupont, P., Van Hecke, P., & Orban, G. A. (2001). Human brain regions involved in heading estimation. *Journal of Neuroscience, 21,* 2451–2461.

Pezawas, L., Meyer-Lindberg, A., Goldman, A. L., Verchinski, B. A., Chen, G., Kolachana, B. S., et al. (2008). Evidence of biologic epistasis between BDNF and SLC6A4 and implications for depression. *Molecular Psychiatry, 13,* 709–716.

Pezawas, L., Meyer-Lindenberg, A., Drabant, E. M., Verchinski, B. A., Munoz, K. E., Kolachana, B. S., et al. (2005). 5-HTTLPR polymorphism impacts human cingulate-amygdala interactions: A genetic susceptibility mechanism for depression. *Nature Neuroscience, 8,* 828–834.

Pfaff, D. W., & Sakuma, Y. (1979). Deficit in the lordosis reflex of female rats caused by lesions in the ventromedial nucleus of the hypothalamus. *Journal of Physiology, 288,* 203–210.

Pfaus, J. G., Kleopoulos, S. P., Mobbs, C. V., Gibbs, R. B., & Pfaff, D. W. (1993). Sexual stimulation activates c-fos within estrogen-concentrating regions of the female rat forebrain. *Brain Research, 624,* 253–267.

Pfrieger, F. W., & Barres, B. A. (1997). Synaptic efficacy enhanced by glial cells in vitro. *Science, 277,* 1684–1687.

Phan, K. L., Wager, T., Taylor, S. F., & Liberzon, I. (2002). Functional neuroanatomy of emotion: A meta-analysis of emotion activation studies in PET and fMRI. *NeuroImage, 16,* 331–348.

Phillips, A. G., Coury, A., Fiorino, D., LePiane, F. G., Brown, E., & Fibiger, H. C. (1992). Self-stimulation of the ventral Tegmental area enhances dopamine release in the nucleus accumbens. *Annals of the New York Academy of Sciences, 654,* 199–206.

Phillips, R. J., & Powley, T. L. (1996). Gastric volume rather than nutrient content inhibits food intake. *American Journal of Regulatory, Integrative, and Comparative Physiology, 271,* R766–R779.

Pianezza, M. L., Sellers, E. M., & Tyndale, R. F. (1998). Nicotine metabolism defect reduces smoking. *Nature, 393,* 750.

Pickar, D. (1995). Prospects for pharmacotherapy of schizophrenia. *Lancet, 345,* 557–562.

Pierce, K., Müller, R.-A., Ambrose, J., Allen, G., & Courchesne, E. (2001). Face processing occurs outside the fusiform 'face area' in autism: Evidence from functional MRI. *Brain, 124,* 2059–2073.

Pihl, R. O., & Peterson, J. B. (1993). Alcohol, serotonin, and aggression. *Alcohol Health and Research World, 17,* 113–116.

Pillard, R. C., & Bailey, J. M. (1998). Human sexual orientation has a heritable component. *Human Biology, 70,* 347–365.

Pilowsky, L. S., Costa, D. C., Eli, P. J., Murray, R. M., Verhoeff, N. P., & Kerwin, R. W. (1993). Antipsychotic medication, D_2 dopamine receptor blockade and clinical response: A ^{123}I-IBZM SPET (single photon emission tomography) study. *Psychological Medicine, 23,* 791–799.

Pinker, S. (1994). *The language instinct.* New York: Morrow.

Pinker, S. (2001). Talk of genetics and vice versa. *Nature, 413,* 465–466.

Pinto, D., Pagnamenta, A. T., Klei, L., Anney, R., Merico, D., Regan, R., et al. (2010). Functional impact of global rare copy number variation in autism spectrum disorders. *Nature, 466,* 368–372.

Pi-Sunyer, X. (2003). A clinical view of the obesity problem. *Science, 299,* 859–860.

Pi-Sunyer, X., Kissileff, H. R., Thornton, J., & Smith, G. P. (1982). C terminal octapeptide of cholecystokinin decreases food intake in obese men. *Physiology and Behavior, 29,* 627–630.

Plomin, R. (1989). Environment and genes: Determinants of behavior. *American Psychologist, 44,* 105–111.

Plomin, R. (1990). The role of inheritance in behavior. *Science, 248,* 183–188.

Plomin, R., & McClearn, G. E. (Eds.). (1993). *Nature, nurture, and psychology.* Washington, DC: American Psychological Association.

Plomin, R., Owen, M. J., & McGuffin, P. (1994). The genetic basis of complex human behaviors. *Science, 264,* 1733–1739.

Plotnik. (1999). *Introduction to psychology* (5th ed.). Belmont, CA: Wadsworth/Thomson Learning.

Plotnik, J. M., de Waal, F. B. M., & Reiss, D. (2006). Self recognition in an Asian elephant. *Proceedings of the National Academy of Sciences, USA, 103,* 17053–17057.

Poehlman sentenced to 1 year of prison. (2006, June 28). *ScienceNOW Daily News.* Retrieved July 19, 2007 from www .sciencenow.sciencemag.org/cgi/content/full/2006/628/1

Poggio, G. F., & Poggio, T. (1984). The analysis of stereopsis. *Annual Review of Neuroscience, 7,* 379–412.

Policies on the use of animals and humans in neuroscience research. (n.d.). Retrieved March 23, 2005, from www .web.sfn.org/content/Publications/Handbookforthe UseofAnimalsinNeuroscienceResearch/Policy.htm

Polymeropoulos, M. H., Lavedan, C., Leroy, E., Ide, S. E., Dehejia, A., Dutra, A., et al. (1997). Mutation in the a-synuclein gene identified in families with Parkinson's disease. *Science, 276,* 2045–2047.

Porkka-Heiskanen, T., Strecker, R. E., Thakkar, M., Bjørkum, A. A., Greene, R. W., & McCarley, R. W. (1997). Adenosine: A mediator of the sleep-inducing effects of prolonged wakefulness. *Science, 276,* 1265–1268.

Porter, R. H., & Moore, J. D. (1981). Human kin recognition by olfactory cues. *Physiology and Behavior, 27,* 493–495.

Posner, M. I., & Rothbart, M. K. (1998). Attention, self-regulation and consciousness. *Philosophical Transactions of the Royal Society of London B, 353,* 1915–1927.

Posthuma, D., De Geus, E. J. C., Baaré, W. F. C., Pol, H. E. H., Kahn, R. S., & Boomsma, D. I. (2002). The association between brain volume and intelligence is of genetic origin. *Nature Neuroscience, 5,* 83–84.

Posthuma, D., De Geus, E. J., & Boomsma, D. I. (2001). Perceptual speed and IQ are associated through common genetic factors. *Behavior Genetics, 31,* 593–602.

Potenza, M. N. (2008). The neurobiology of pathological gambling and drug addiction: An overview and new findings. *Philosophical Transactions of the Royal Society B, 363,* 3181–3189.

Potter, W. Z., & Rudorfer, M. V. (1993). Electroconvulsive therapy: A modern medical procedure. *New England Journal of Medicine, 328,* 882–883.

Pournaras, D. J., & le Roux, C. W. (2009). Obesity, gut hormones, and bariatric surgery. *World Journal of Surgery, 33,* 1983–1988.

Powers, J. (2010, July 9). Gender issue finally resolved. *The Boston Globe.* Retrieved July 10, 2010, from www.boston.com/sports/other_sports/olympics/articles/2010/07/09/gender_issue_finally_resolved/

Prabhakar, S., Visel, Z., Akiyama, J. A., Shoukry, M., Lewis, K. D., Holt, A., et al. (2008) Human-specific gain of function in a developmental enhancer. *Science, 321,* 1346–1350.

Press release: Dr. Eric T. Poehlman. (2005, March 17). Office of Research Integrity. Retrieved October 16, 2007, from www.ori.dhhs.gov/misconduct/cases/press_release_poehlman.shtml

Preti, G., Cutler, W. B., Garcia, C. R., Huggins, G. R., & Lawley, H. J. (1986). Human axillary secretions influence women's menstrual cycles: The role of donor extract of females. *Hormones and Behavior, 20,* 474–482.

Preti, G., Wysocki, C. J., Barnhart, K. T., Sondheimer, S. J., & Leyden, J. J. (2003). Male axillary extracts contain pheromones that affect pulsatile secretion of luteinizing hormone and mood in women recipients. *Biology of Reproduction, 68,* 2107–2113.

Price, D. D. (2000). Psychological and neural mechanisms of the affective dimension of pain. *Science, 288,* 1769–1772.

Price, J. (1968). The genetics of depressive behavior. *British Journal of Psychiatry,* (Special Publication No. 2), 37–45.

Price, R. A., Kidd, K. K., Cohen, D. J., Pauls, D. L., & Leckman, J. F. (1985). A twin study of Tourette syndrome. *Archives of General Psychiatry, 42,* 815–820.

Prior, H., Schwarz, A., Güntürkün, O. (2008). Mirror-induced behavior in the magpie (*Pica pica*): Evidence of self-recognition. *PLoS Biology, 6*(8), e202. Retrieved August 30, 2010, from www.plosbiology.org/article/info:doi/10.1371/journal.pbio.0060202

Proffitt, F. (2004). Britain unveils a plan to curb aniimal-rights "extremists." *Science, 305,* 761.

Proof? The joke was on Bischoff, but too late. (1942, March). *Scientific American, 166*(3), 145.

Prospective Studies Collaboration. (2009). Body-mass index and cause-specific mortality in 900 000 adults: Collaborative analyses of 57 prospective studies. *Lancet, 373,* 1083–1096.

Public Health Service policy on humane care and use of laboratory animals. (2002). Retrieved March 23, 2005, from www.grants.nih.gov/grants/olaw/references/phspol.htm

Pujol, J., López, A., Deus, J., Cardoner, N., Vallejo, J., Capdevila, A., et al. (2002). Anatomical variability of the anterior cingulate gyrus and basic dimensions of human personality. *NeuroImage, 15,* 847–855.

Putnam, F. W. (1991). Recent research on multiple personality disorder. *Psychiatric Clinics of North America, 14,* 489–502.

Putnam, F. W., Zahn, T. P., & Post, R. M. (1990). Differential autonomic nervous system activity in multiple personality disorder. *Psychiatry Research, 31,* 251–260.

Puts, D. A., McDaniel, M. A., Jordan, C. L., Breedlove, S. M. (2008). Spatial ability and prenatal androgens: Meta-analyses of congenital adrenal hyperplasia and digit ratio (2D:4D) studies. *Archives of Sexual Behavior, 37*(1), 100–111.

Pyter, L. M., Pineros, V., Galang, J. A., McClintock, M. K., & Prendergast, B. J. (2009). Peripheral tumors induce depressive-like behaviors and cytokine production and alter hypothalamic-pituitary-adrenal axis regulation. *Proceedings of the National Academy of Sciences, 106,* 9069–9074.

Qin, Y.-L., McNaughton, B. L., Skaggs, W. E., & Barnes, C. A. (1997). Memory reprocessing in corticocortical and hippocampocortical neuronal ensembles. *Philosophical Transactions of the Royal Society of London, B, 352,* 1525–1533.

Qiu, X., Kumbalasiri, T., Carlson, S. M., Wong, K. Y., Krishna, V., Provencio, I., et al. (2005). Induction of photosensitivity by heterologous expression of melanopsin. *Nature, 433,* 745–749.

Quian Quiroga, R., Kraskov, A., Koch, C., & Fried, I. (2009). Explicit encoding of multimodal percepts by single neurons in the human brain. *Current Biology, 19*(15), 1308 -1313.

Raboch, J., & Stárka, L. (1973). Reported coital activity of men and levels of plasma testosterone. *Archives of Sexual Behavior, 2,* 309–315.

Radtke, N. D., Aramant, R. B., Petry, H. M., Green, P. T., Pidwell, D. J., & Seiler, M. J. (2008). Vision improvement in retinal degeneration patients by implantation of retina together with retinal pigment epithelium. *American Journal of Ophthalmology, 146,* 172–182.

Ragsdale, D. S., McPhee, J. C., Scheuer, T., & Catterall, W. A. (1994). Molecular determinants of state-dependent block of Na+ channels by local anesthetics. *Science, 265,* 1724–1728.

Rahman, Q. (2005). Fluctuating asymmetry, second to fourth finger length ratios and human sexual orientation. *Psychoneuroendocrinology, 30,* 382–391.

Raine, A., Lencz, T., Bihrle, S., LaCasse, L., & Colletti, P. (2000). Reduced prefrontal gray matter volume and reduced autonomic activity in antisocial personality disorder. *Archives of General Psychiatry, 57,* 119–127.

Raine, A., Meloy, J. R., Bihrle, S., Stoddard, J., LaCasse, L., & Buchsbaum, M. S. (1998). Reduced prefrontal and increased subcortical brain functioning assessed using positron emission tomography in predatory and affective murderers. *Behavioral Science and Law, 16,* 319–332.

Raine, A., Stoddard, J., Bihrle, S., & Buchsbaum, M. (1998). Prefrontal glucose deficits in murderers lacking psychosocial deprivation. *Neuropsychiatry, Neuropsychology, and Behavioral Neurology, 11,* 1–7.

Raine, A., Yang, Y., Narr, K. L., & Toga, A. W. (2009, December 22). Sex differences in orbitofrontal gray as a partial explanation for sex differences in antisocial personality. *Molecular Psychiatry.* doi:10.1038/mp.2009.136

Rainer, G. S., Rao, S. C., & Miller, E. K. (1999). Prospective coding for objects in primate prefrontal cortex. *Journal of Neuroscience, 19,* 5493–5505.

Rainnie, D. G., Grunze, H. C., McCarley, R. W., & Greene, R. W. (1994). Adenosine inhibition of mesopontine cholinergic neurons: Implications for EEG arousal. *Science, 263,* 689–692.

Rainville, P., Duncan, G. H., Price, D. D., Carrier, B., & Bushnell, M. C. (1997). Pain affect encoded in human anterior cingulate but not somatosensory cortex. *Science, 227,* 968–971.

Rakic, P. (1985). Limits of neurogenesis in primates. *Science, 227,* 1054–1056.

Ramachandran, V. S., & Altschuler, E. L. (2009). The use of visual feedback, in particular mirror visual feedback, in restoring brain function. *Brain, 132,* 1693–1710.

Ramachandran, V. S., & Blakeslee, S. (1998). *Phantoms in the brain.* New York: Morrow.

Ramachandran, V. S., Rogers-Ramachandran, D., & Cobb, S. (1995). Touching the phantom limb. *Nature, 377,* 489–490.

Ramey, C. T., Campbell, F. A., Burchinal, M., Skinner, M. L., Gardner, D. M., & Ramey, S. L. (2000). Persistent effects of early childhood education on high-risk children and their mothers. *Applied Developmental Science, 4,* 2–14.

Ramón y Cajal, S. (1928). *Degeneration and regeneration of the nervous system.* New York: Hafner.

Ramón y Cajal, S. (1989). *Recollections of my life* (E. H. Craigie & J. Cano, Trans.). Cambridge, MA: MIT Press. (Original work published 1937)

Ramus, F., Rosen, S., Dakin, S. C., Day, B. L., Castellote, J. M., White, S., et al. (2003). Theories of developmental dyslexia: Insights from a multiple case study of dyslexic adults. *Brain, 126,* 841–865.

Raninken, T., Zuberi, A., Chagnon, Y. C., Weisnagel, J., Argyropoulos, G., Walts, B., et al. (2006). The human obesity gene map: The 2005 update. *Obesity, 14,* 529–644.

Rao, S. C., Rainer, G., & Miller, E. K. (1997). Integration of what and where in the primate prefrontal cortex. *Science, 276,* 821–824.

Rapkin, A. J., Morgan, M., Goldman, L., Brann, D. W., Simone, D., & Mahesh, V. B. (1997). Progesterone metabolite allopregnanolone in women with premenstrual syndrome. *Obstetrics and Gynecology, 90,* 709–714.

Rapkin, A. J., Pollack, D. B., Raleigh, M. J., Stone, B., & McGuire, M. T. (1995). Menstrual cycle and social behavior in vervet monkeys. *Psychoneuroendocrinology, 20,* 289–297.

Rapoport, J. L. (1991). Recent advances in obsessive-compulsive disorder. *Neuropsychopharmacology, 5,* 1–10.

Rasmussen, T., & Milner, B. (1977). The role of early left-brain injury in determining lateralization of cerebral speech functions. *Annals of the New York Academy of Sciences, 299,* 355–369.

Ratliff, F., & Hartline, H. K. (1959). The responses of Limulus optic nerve fibers to patterns of illumination on the receptor mosaic. *Journal of General Physiology, 42,* 1241–1255.

Rauschecker, J. P., & Tian, B. (2000). Mechanisms and streams for processing of "what" and "where" in auditory cortex. *Proceedings of the National Academy of Sciences, USA, 97,* 11800–11806.

Raven, J. C. (2003). *Standard Progressive Matrices (Manual).* PsychCorp/Pearson: San Antonio, TX.

Ray, L. A., & Hutchison, K. E. (2004). A polymorphism of the μ-opioid receptor gene (*OPRM1*) and sensitivity to the effects of alcohol in humans. *Alcoholism Clinical and Experimental Research, 28,* 1789–1795.

Ray, S., Britschgi, M., Herbert, C., Takeda-Uchimura, Y. Boxer, A., Blennow, K., et al. (2007). Classification and prediction of clinical Alzheimer's diagnosis based on plasma signaling proteins. *Nature Medicine, 13,* 1359–1362.

Raz, A. (2004). Brain imaging data of ADHD [Electronic version]. *Psychiatric Times, 21*(9). Retrieved August 13, 2005, from www.psychiatrictimes.com/p040842.html

Recht, L. D., Lew, R. A., & Schwartz, W. J. (1995). Baseball teams beaten by jet lag. *Nature, 377,* 583.

Redcay, E., & Courchesne, E. (2005). When is the brain enlarged in autism? A meta-analysis of all brain size reports. *Biological Psychiatry, 58,* 1–9.

Redelmeier, D. A., & Tibshirani, R. J. (1997). Association between cellular telephone calls and motor vehicle collisions. *New England Journal of Medicine, 336,* 453–458.

Reed, J. J., & Squire, L. R. (1998). Retrograde amnesia for facts and events: Findings from four new cases. *Journal of Neuroscience, 18,* 3943–3954.

Reed, T. E. (1985). Ethnic differences in alcohol use, abuse, and sensitivity: A review with genetic interpretations. *Social Biology, 32,* 195–209.

Reed, T. E., & Jensen, A. R. (1992). Conduction velocity in a brain nerve pathway of normal adults correlates with intelligence level. *Intelligence, 16,* 259–272.

Reichenberg, A., Gross, R., Weiser, M., Bresnahan, M., Silverman, J., Harlap, S., et al. (2006). Advancing paternal age and autism. *Archives of General Psychiatry, 63,* 1026–1032.

Reichenberg, A., Caspi, A., Harrington, H., Houts, R., Keefe, R. S. E., Murray, R. M., et al. (2010). Static and dynamic cognitive deficits in childhood preceding adult schizophrenia: A 30-year study. *American Journal of Psychiatry, 167,* 160–169.

Reiman, E. M., Armstrong, S. M., Matt, K. S., & Mattox, J. H. (1996). The application of positron emission tomography to the study of the normal menstrual cycle. *Human Reproduction, 11,* 2799–2805.

Reiman, E. M., Raichle, M. E., Robins, E., Butler, F. K., Herscovitch, P., Fox, P., et al. (1986). The application of positron emission tomography to the study of panic disorder. *American Journal of Psychiatry, 143,* 469–477.

Reinders, A. A., Nijenhuis, E. R., Quak, J., Korf, J., Haaksma, J., Paans, A. M., et al. (2006). Psychobiological characteristics of dissociative identity disorder: A symptom provocation study. *Biological Psychiatry, 60,* 730–740.

Reinisch, J. M. (1981). Prenatal exposure to synthetic progestins increases potential for aggression in humans. *Science, 211,* 1171–1173.

Reisberg, B., Doody, R., Stöffler, A., Schmitt, F., Ferris, S., & Möbius, H. J. (2003). Memantine in moderate-to-severe Alzheimer's disease. *New England Journal of Medicine, 348,* 1333–1341.

Reiss, D., & Marino, L. (2001). Mirror self-recognition in the bottlenose dolphin: A case of cognitive convergence. *Proceedings of the National Academy of Sciences, USA, 98,* 5937–5942.

Rekling, J. C., Funk, G. D., Bayliss, D. A., Dong, X.-W., & Feldman, J. L. (2000). Synaptic control of motoneuronal excitability. *Physiological Reviews, 80,* 767–852.

Relkin, N. R., Szabo, P., Adamiak, B., Burgut, T., Monthe, C., Lent, R. W., et al. (2009). 18-month study of intravenous immunoglobulin for treatment of mild Alzheimer disease. *Neurobiology of Aging, 30,* 1728–1736.

Remondes, M., & Schuman, E. M. (2004). Role for a cortical input to the hippocampal area CA1 in the consolidation of a long-term memory. *Nature, 431,* 699–703.

Rempel-Clower, N. L., Zola, S. M., Squire, L. R., & Amaral, D. G. (1996). Three cases of enduring memory impairment after bilateral damage limited to the hippocampal formation. *Journal of Neuroscience, 16,* 5233–5255.

Ren, J., Tate, B. A., Sietsma, D., Marciniak, A., Snyder, E. Y., & Finklestein, S. P. (2000, November). *Co-administration of neural stem cells and BFGF enhances functional recovery following focal cerebral infarction in rat.* Poster session presented at the annual meeting of the Society for Neuroscience, New Orleans, LA.

Renault, B., Signoret, J.-L., Debruille, B., Breton, F., & Bolger, F. (1989). Brain potentials reveal covert facial recognition in prosopagnosia. *Neuropsychologia, 27,* 905–912.

Research involving human subjects. (n.d.). Retrieved November 3, 2009, from http://grants.nih.gov/grants/policy/hs/index.htm

Reuter, J., Raedler, T., Rose, M., Hand, I., Gläscher, J., & Büchel, C. (2005). Pathological gambling is linked to reduced activation of the mesolimbic reward system. *Nature Neuroscience, 8,* 147–148.

Ribary, U. (2005). Dynamics of thalamo-cortical network oscillations and human perception. In S. Laureys (Ed.), *The boundaries of consciousness: Neurobiology and neuropathology* (pp. 127–142). New York: Elsevier.

Ribeiro, S., Gervasoni, D., Soares, E. S., Zhou, Y., Lin, S.-C., Pantoja, J., et al. (2004). Long-lasting novelty-induced neuronal reverberation during slow-wave sleep in multiple forebrain areas. *Public Library of Science Biology, 2,* 126–137.

Rice, C. M., Mallam, E. A., Whone, A. L., Walsh, P., Brooks, D. J., Kane, N., et al. (2010). Safety and feasibility of autologous bone marrow cellular therapy in relapsing-progressive multiple sclerosis. *Clinical Pharmacology & Therapeutics, 87,* 679–685.

Rice, F., Harold, G. T., Bolvin J., Hay, D. F., van den Bree, M., & Thapar, A. (2009). Disentangling prenatal and inherited influences in humans with an experimental design. *Proceedings of the National Academy of Sciences, USA, 106,* 2464–2467.

Rice, G., Anderson, C., Risch, N., & Ebers, G. (1999). Male homosexuality: Absence of linkage to microsatellite markers at Xq28. *Science, 284,* 665–667.

Rice, M. L., Smith, S. D., & Gayán, J. (2009). Convergent genetic linkage and associations to language, speech and reading measures in families of probands with specific language impairment. *Journal of Neurodevelopmental Disorders, 1,* 264–282.

Riedel, G., Micheau, J., Lam, A. G. M., Roloff, E. V. L., Martin, S. J., Bridge, H., et al. (1999). Reversible neural inactivation reveals hippocampal participation in several memory processes. *Nature Neuroscience, 2,* 898–905.

Riehle, A., & Requin, J. (1989). Monkey primary motor and premotor cortex: Single-cell activity related to prior information about direction and extent of an intended movement. *Journal of Neurophysiology, 61,* 534–549.

Rilling, J. K., Glasser, M. F., Preuss, T. M., Ma, X., Zhao, T., Hu, X., et al. (2008). The evolution of the arcuate fasciculus revealed with comparative DTI. *Nature Neuroscience, 11,* 426–428.

Rimmele, U., Hediger, K., Heinrichs, M., & Klaver, P. (2009). Oxytocin makes a face in memory familiar. *Journal of Neuroscience, 29,* 38–42.

Ritter, R. C., Slusser, P. G., & Stone, S. (1981). Glucoreceptors controlling feeding and blood glucose: Location in the hindbrain. *Science, 213,* 451–453.

Ritter, S., & Taylor, J. S. (1990). Vagal sensory neurons are required for lipoprivic but not glucoprivic feeding in rats. *American Journal of Physiology, 258,* R1395–R1401.

Roberts, E. M., English, P. B., Grether, J. K., Windham, G. C., Somberg, L., & Wolff, C. (2007). Maternal residence near agricultural pesticide applications and autism spectrum disorders among children in the California Central Valley. *Environmental Health Perspectives, 115,* 1482–1489.

Roberts, G. W. (1990). Schizophrenia: The cellular biology of a functional psychosis. *Trends in the Neurosciences, 13,* 207–211.

Robinson, T. E., Gorny, G., Mitton, E., & Kolb, B. (2001). Cocaine self-administration alters the morphology of dendrites and dendritic spines in the nucleus accumbens and neocortex. *Synapse, 39,* 257–266.

Robinson, T. E., & Kolb, B. (1997). Persistent structural modifications in nucleus accumbens and prefrontal cortex neurons produced by previous experience with amphetamine. *Journal of Neuroscience, 17,* 8491–8497.

Rodin, J., Schank, D., & Striegel-Moore, R. (1989). Psychological features of obesity. *Medical Clinics of North America, 73,* 47–66.

Rodriguez, A., & Bohlin, G. (2005). Are maternal smoking and stress during pregnancy related to ADHD symptoms in children? *Journal of Child Psychology and Psychiatry, 46,* 246–254.

Rodriguez, I. (2004). Pheromone receptors in mammals. *Hormones and Behavior, 46,* 219–230.

Roenneberg, T., Kuehnle, T., Pramstaller, P. P., Ricken, J., Havel, M., Guth, A., et al. (2004). A marker for the end of adolescence. *Current Biology, 14,* R1038–R1039.

Roffwarg, H. P., Muzio, J. N., & Dement, W. C. (1966). Ontogenetic development of the human sleep-dream cycle. *Science, 152,* 604–619.

Rogan, M. T., Stäubli, U. V., & LeDoux, J. E. (1997). Fear conditioning induces associative long-term potentiation in the amygdala. *Nature, 390,* 604–607.

Rogers, G., Elston, J., Garside, R., Roome, C., Taylor, R., Younger, P., et al. (2009). The harmful health effects of recreational ecstasy: A systematic review of observational evidence. *Health Technology Assessment, 13*(6). Retrieved December 14, 2009, from www.hta.ac.uk/fullmono/mon1306.pdf

Rolls, B. J., Rolls, E. T., Rowe, E. A., & Sweeney, K. (1981). Sensory-specific satiety in man. *Physiology and Behavior, 27,* 137–142.

Rolls, B. J., Rowe, E. A., & Turner, R. C. (1980). Persistent obesity in rats following a period of consumption of a mixed, high-energy diet. *Journal of Physiology, 298,* 415–427.

Rolls, B. J., Wood, R. J., & Rolls, R. M. (1980). Thirst: The initiation, maintenance, and termination of drinking. In J. M. Sprague & A. N. Epstein (Eds.), *Progress in psychology and physiological psychology* (pp. 263–321). New York: Academic Press.

Rosa, R. R., & Bonnet, M. H. (2000). Reported chronic insomnia is independent of poor sleep as measured by electroencephalography. *Psychosomatic Medicine, 62,* 474–482.

Rose, J. E., Brugge, J. F., Anderson, D. J., & Hind, J. E. (1967). Phase-locked response to low-frequency tones in single auditory nerve fibers of the squirrel monkey. *Journal of Neurophysiology, 30,* 769–793.

Rose, R. J. (1995). Genes and human behavior. *Annual Review of Psychology, 46,* 625–654.

Roselli, C. E., Larkin, K., Resko, J. A., Stellflug, J. N., & Stormshak, F. (2004). The volume of a sexually dimorphic nucleus in the ovine medial preoptic area/anterior hypothalamus varies with sexual partner preference. *Endocrinology, 145,* 478–483.

Rosen, G. D., Bai, J., Wang, Y., Fiondella, C. G., Threlkeld, S. W., LoTurco, J. J., et al. (2007). Disruption of neuronal migration by RNAi of *Dyx1c1* results in neocortical and hippocampal malformations. *Cerebral Cortex, 17,* 2562–2572.

Rosenkranz, M. A., Jackson, D. C., Dalton, K. M., Dolski, I., Ryff, C. D., Singer, B. H., et al. (2003). Affective style and *in vivo* immune response: Neurobehavioral mechanisms. *Proceedings of the National Academy of Sciences, USA, 100,* 11148–11152.

Rosenthal, N. E., Sack, D. A., Carpenter, C. J., Parry, B. L., Mendelson, W. B., & Wehr, T. A. (1985). Antidepressant effects of light in seasonal affective disorder. *American Journal of Psychiatry, 142,* 163–170.

Rösler, A., & Witztum, E. (1998). Treatment of men with paraphilia with a long-acting analogue of gonadotropin-releasing hormone. *New England Journal of Medicine, 338,* 416–422.

Rösler, F., Heil, M., & Henninghausen, E. (1995). Distinct cortical activation patterns during long-term memory retrieval of verbal, spatial and color information. *Journal of Cognitive Neuroscience, 7,* 51–65.

Ross, C. A., Miller, S. C., Reagor, P., Bjornson, L., Fraser, G. A., & Anderson, G. (1990). Structured interview data on 102 cases of multiple personality disorder from four centers. *American Journal of Psychiatry, 147,* 596–601.

Ross, E. D., Homan, R. W., & Buck, R. (1994). Differential hemispheric lateralization of primary and social emotions. *Neuropsychiatry, Neuropsychology, and Behavioral Neurology, 7,* 1–19.

Ross, G. W., Abbott, R. D., Petrivotch, H., Morens, D. M., Grandinetti, A., Tung, K.-H., et al. (2000). Association of coffee and caffeine intake with the risk of Parkinson disease. *Journal of the American Medical Association, 283,* 2674–2679.

Rossi, G. S., & Rosadini, G. (1967). Experimental analysis of cerebral dominance in man. In C. Millikan & F. L. Darley (Eds.), *Brain mechanisms underlying speech and language* (pp. 167–184). New York: Grune & Stratton.

Roth, G., & Strüber. D. (2009). [Neurobiological aspects of reactive and proactive violence in antisocial personality disorder and "psychopathy"] (article in German). *Praxis der Kinderpsychologie und Kinderpsychiatrie, 58,* 587–609.

Rothe, C., Gutknecht, L., Freitag, C., Tauber, R., Mössner, R., Franke, P., et al. (2004). Association of a functional 1019C>G 5-HT1A receptor gene polymorphism with panic disorder with agoraphobia. *International Journal of Neuropsychopharmacology, 7,* 189–192.

Rothman, J. M., Van Soest, P. J., & Pell, A. N. (2006). Decaying wood is a sodium source for mountain gorillas. *Biology Letters, 2,* 321–324.

Rouw, R., & Scholte, S. (2007). Increased structural connectivity in grapheme-color synesthesia. *Nature Neuroscience, 10,* 792–797.

Rowe, M. L., & Goldin-Meadow, S. (2009). Differences in early gesture explain SES disparities in child vocabulary size at school entry. *Science, 323,* 951–953.

Rowland, L. P. (2000a). Diseases of chemical transmission at the nerve-muscle synapse: Myasthenia gravis. In E. R. Kandel, J. H. Schwartz, & T. M. Jessell (Eds.), *Principles of neural science* (4th ed., pp. 298–309). New York: McGraw-Hill.

Rowland, L. P. (2000b). Diseases of the motor unit. In E. R. Kandel, J. H. Schwartz, & T. M. Jessell (Eds.), *Principles of neural science* (4th ed., pp. 695–712). New York: McGraw-Hill.

Rowland, L. P., Hoefer, P. F. A., & Aranow, H., Jr. (1960). Myasthenic syndromes. *Research Publications—Association for Research in Nervous and Mental Disease, 38,* 547–560.

Roy, A., DeJong, J., & Linnoila, M. (1989). Cerebrospinal fluid monoamine metabolites and suicidal behavior in depressed patients: A 5-year followup study. *Archives of General Psychiatry, 46,* 609–612.

Rozin, P. (1967). Specific aversions as a component of specific hungers. *Journal of Comparative and Physiological Psychology, 64,* 237–242.

Rozin, P. (1969). Adaptive food sampling patterns in vitamin deficient rats. *Journal of Comparative and Physiological Psychology, 69,* 126–132.

Rozin, P. (1976). The selection of foods by rats, humans, and other animals. *Advances in the Study of Behavior, 6,* 21–76.

Rubens, A. B. (1977). Anatomical asymmetries of human cerebral cortex. In S. Harnad, R. W. Doty, L. Goldstein, J. Jaynes, & G. Krauthamer (Eds.), *Lateralization in the nervous system* (pp. 503–516). New York: Academic Press.

Rudorfer, M. V., Henry, M. E., & Sackheim, H. A. (1997). Electroconvulsive therapy. In A. Tasman, J. Kay, & J. A. Lieberman (Eds.), *Psychiatry* (pp. 1535–1556). Philadelphia: W. B. Saunders.

Rushton, J. P., Fulker, D. W., Neale, M. C., Nias, D. K. B., & Eysenck, H. J. (1986). Altruism and aggression: The heritability of individual differences. *Journal of Personality and Social Psychology, 50,* 1192–1198.

Rushton, J. P., & Jensen, A. R. (2005). Thirty years of research on race differences in cognitive ability. *Psychology, Public Policy, and Law, 11,* 235–294.

Rutherford, W. (1886). The sense and hearing. *Journal of Anatomy and Physiology, 21,* 166–168.

Rutter, M. (1983). Cognitive deficits in the pathogenesis of autism. *Journal of Child Psychology and Psychiatry, 24,* 513–531.

Rutter, M. (2005). Incidence of autism spectrum disorders: Changes over time and their meaning. *Acta Paediatrica, 94,* 2–15.

Saalmann, Y. B., Pigarev, I. N., & Vidyasagar, T. R. (2007). Neural mechanisms of visual attention: How top-down feedback highlights relevant locations. *Science, 316,* 1612–1615.

Sabri, O., Gertz, H., Barthel, H., Dresel, S., Heuser, I., Bartenstein, P. A., et al. (2010). Multicentre phase 2 trial on florbetaben for β-amyloid brain PET in Alzheimer disease. *Journal of Nuclear Medicine, 51*(Suppl. 2), 384.

Sacco, K. A., Bannon, K. L., & George, T. P. (2004). Nicotinic receptor mechanisms and cognition in normal states and neuropsychiatric disorders. *Journal of Psychopharmacology, 18,* 457–474.

Sacco, K. A., Termine, A., Seyal, A., Dudas, M. M., Vessicchio, J. C., Krishnan-Sarin, S., et al. (2005). Effects of cigarette smoking on spatial working memory and attentional deficits in schizophrenia. *Archives of General Psychiatry, 62,* 649–659.

Sachs, B., & Meisel, R. L. (1994). The physiology of male sexual behavior. In J. D. Neill & E. Knobil (Eds.), *The physiology of reproduction* (Vol. 2, pp. 3–106). New York: Raven Press.

Sack, A. T., Kohler, A., Bestmann, S., Linden, D. E. J., Dechent, P., Goebel, R., et al. (2007). Imaging the brain activity changes underlying impaired visuospatial judgments: Simultaneous fMRI, TMS, and behavioral studies. *Cerebral Cortex, 17,* 2841–2852.

Sack, D. A., Nurnberger, J., Rosenthal, N. E., Ashburn, E., & Wehr, T. A. (1985). Potentiation of antidepressant medications by phase advance of the sleep-wake cycle. *American Journal of Psychiatry, 142,* 606–608.

Sackeim, H. A., Luber, B., Katzman, G. P., Moeller, J. R., Prudic, J., Devanand, D. P., et al. (1996). The effects of electroconvulsive therapy on quantitative electroencephalograms: Relationship to clinical outcome. *Archives of General Psychiatry, 53,* 814–823.

Sackeim, H. A., Prohovnik, I., Moeller, J. R., Brown, R. P., Apter, S., Prudic, J., et al. (1990). Regional cerebral blood flow in mood disorders: I. Comparison of major depressives and normal controls at rest. *Archives of General Psychiatry, 47,* 60–70.

Sackeim, H. A., Prudic, J., Devanand, D. P., Kiersky, J. E., Fitzsimons, L., Moody, B. J., et al. (1993). Effects of stimulus intensity and electrode placement on the efficacy and cognitive effects of electroconvulsive therapy. *New England Journal of Medicine, 328,* 839–846.

Sacks, O. (1990). *The man who mistook his wife for a hat and other clinical tales.* New York: HarperPerennial.

Sacks, O. (1995). *An anthropologist on Mars.* New York: Vintage Books.

Sacks, O., & Wasserman, R. (1987, November 19). The case of the color-blind painter. *New York Review of Books, 34,* 25–34.

Saenz, A. (2010). *Argus III—the artificial retina is near!* Retrieved May 4, 2010, from http://singularityhub.com/2010/02/25/argus-iii-the-artificial-retina-is-near

Sairanen, M., Lucas, G., Ernfors, P., Castrén, M., & Castrén, E. (2005). Brain-derived neurotrophic factor and antidepressant drugs have different but coordinated effects on neuronal turnover, proliferation, and survival in the adult dentate gyrus. *Journal of Neuroscience, 25,* 1089–1094.

Sakata, H., Takaoka, Y., Kawarasaki, A., & Shibutani, H. (1973). Somatosensory properties of neurons in the superior parietal cortex (area 5) of the rhesus monkey. *Brain Research, 64,* 85–102.

Salamy, J. (1970). Instrumental responding to internal cues associated with REM sleep. *Psychonomic Science, 18,* 342–343.

Salehi, A., Faizi, M., Colas, D., Valletta, J., Laguna, J., Takimoto-Kimura, R., et al. (2009). Restoration of norepinephrine-modulated contextual memory in a mouse model of Down syndrome. *Science Translational Medicine, 1,* 7–17. Retrieved July 4, 2010, from http://stm.sciencemag.org/content/1/7/7ra17.full.pdf

Salomons, T. V., Coan, J. A., Hunt, S. M., Backonja, M.-M., & Davidson, R. J. (2008). Voluntary facial displays of pain increase suffering in response to nociceptive stimulation. *Journal of Pain, 9,* 443–448.

Salthouse, T. A., & Babcock, R. L. (1991). Decomposing adult age differences in working memory. *Developmental Psychology, 27,* 763–776.

Sämann, P. G., Tully, C., Spoormaker, V. I., Wetter, T. C., Holsboer, F., Wehrle, R., et al. (2010). Increased sleep pressure reduces resting state functional connectivity. *Magnetic Resonance Materials in Physics, Biology and Medicine.* Published online ahead of print. Retrieved August 30, 2010, from www.springerlink.com/content/h73225102249h637/

Sanacora, G., Mason, G. F., Rothman, D. L., Hyder, F., Ciarcia, J. J., Osroff, R. B., et al. (2003). Increased cortical GABA concentrations in depressed patients receiving ECT. *American Journal of Psychiatry, 160,* 577–579.

Sanders, A. R., Duan, J., Levinson, D. F., Shi, J., He, D., Hou, C., et al. (2008). No significant association of 14 candidate genes with schizophrenia in a large European ancestry sample:

Implications for psychiatric genetics. *American Journal of Psychiatry, 165*, 497–506.

Sano, M., Grossman, H., & Van Dyk, K. (2008). Preventing Alzheimer's disease: Separating fact from fiction. *CNS Drugs, 22*, 887–902. Presented at the annual meeting of the Society of Nuclear Medicine, June 2010, Salt Lake City, Utah.

Santarelli, L., Saxe, M., Gross, C., Surget, A., Battaglia, F., Dulawa, S., et al. (2003). Requirement of hippocampal neurogenesis for the behavioral effects of antidepressants. *Science, 301*, 805–809.

Santini, E., Muller, R. U., & Quirk, G. J. (2001). Consolidation of extinction learning involves transfer from NMDA-independent to NMDA-dependent memory. *Journal of Neuroscience, 21*, 9009–9017.

Sanz, C., Cali, J., & Morgan, D. (2009). Design complexity in termite-fishing tools of chimpanzees (*Pan troglodytes*). *Biology Letters, 5*, 293–296.

Saper, C. B., Chou, T. C., & Elmquist, J. K. (2002). The need to feed: Homeostatic and hedonic control of eating. *Neuron, 36*, 199–211.

Saper, C. B., Chou, T. C., & Scammell, T. E. (2001). The sleep switch: Hypothalamic control of sleep and wakefulness. *Trends in Neurosciences, 24,* 726–731.

Saper, C. B., Iverson, S., & Frackowiak, R. (2000). Integration of sensory and motor function. In E. R. Kandel, J. H. Schwartz, & T. M. Jessell (Eds.), *Principles of neural science* (4th ed., pp. 349–380). New York: McGraw-Hill.

Saper, C. B., Scammell, T. E., & Lu, J. (2005). Hypothalamic regulation of sleep and circadian rhythms. *Nature, 437*, 1257–1263.

Sapolsky, R. M., Uno, H., Rebert, C. S., & Finch, C. E. (1990). Hippocampal damage associated with prolonged glucocorticoid exposure in primates. *Journal of Neuroscience, 10*, 2897–2902.

Sáry, G., Vogels, R., & Orban, G. A. (1993). Cue-invariant shape selectivity of macaque inferior temporal neurons. *Science, 260*, 995–997.

Sata, R., Maloku, E., Zhubi, A., Pibiri, F., Hajos, M., Costa, E., et al. (2008). Nicotine decreases DNA methyltransferase 1 expression and glutamic acid decarboxylase 67 promoter methylation in GABAergic interneurons. *Proceedings of the National Academy of Sciences, 105*, 16356–16361.

Sato, M. (1986). Acute exacerbation of methamphetamine psychosis and lasting dopaminergic supersensitivity: A clinical survey. *Psychopharmacology Bulletin, 22*, 751–756.

Savage-Rumbaugh, E. S., Murphy, J., Sevcik, R. A., Brakke, K. E., Williams, S. L., & Rumbaugh, D. M. (1993). Language comprehension in ape and child. *Monographs of the Society for Research in Child Development, 58*, 1–222.

Savage-Rumbaugh, S. (1987). A new look at ape language: Comprehension of vocal speech and syntax. *Nebraska Symposium on Motivation, 35*, 201–255.

Savage-Rumbaugh, S., McDonald, K., Sevcik, R. A., Hopkins, W. D., & Rubert, E. (1986). Spontaneous symbol acquisition and communicative use by pygmy chimpanzees *(Pan paniscus)*. *Journal of Experimental Psychology: General, 115*, 211–235.

Savic, I., Berglund, H., Gulyas, B., & Roland, P. (2001). Smelling of odorous sex hormone-like compounds causes sex-differentiated hypothalamic activations in humans. *Neuron, 31*, 661–668.

Savic, I., Berglund, H., & Lindström, P. (2005). Brain response to putative pheromones in homosexual men. *Proceedings of the National Academy of Sciences, USA, 102*, 7356–7361.

Savin-Williams, R. C. (1996). Self-labeling and disclosure among gay, lesbian, and bisexual youths. In J. Laird & R.-J. Green (Eds.), *Lesbians and gays in couples and families: A handbook for therapists* (pp. 153–182). San Francisco: Jossey-Bass.

Sawa, A., & Snyder, S. H. (2002). Schizophrenia: Diverse approaches to a complex disease. *Science, 296*, 692–695.

Sawchenko, P. E. (1998). Toward a new neurobiology of energy balance, appetite, and obesity: The anatomists weigh in. *Journal of Comparative Neurology, 402*, 435–441.

Saxe, G. N., Vasile, R. G., Hill, T. C., Bloomingdale, K., & Van der Kolk, B. A. (1992). SPECT imaging and multiple personality disorder. *Journal of Nervous and Mental Disease, 180,* 662–663.

Scarr, S., & Carter-Saltzman, L. (1982). Genetics and intelligence. In R. J. Sternberg (Ed.), *Handbook of human intelligence* (pp. 792–896). New York: Cambridge University Press.

Scarr, S., Pakstis, A. J., Katz, S. H., & Barker, W. B. (1977). Absence of a relationship between degree of white ancestry and intellectual skills within a black population. *Human Genetics, 39,* 69–86.

Scarr, S., & Weinberg, R. A. (1976). IQ test performance of black children adopted by white families. *American Psychologist, 31,* 726–739.

Schaal, B., & Porter, R. H. (1991). "Microsmatic humans" revisited: The generation and perception of chemical signals. *Advances in the Study of Behavior, 20,* 135–199.

Schacter, D. L., Alpert, N. M., Savage, C. R., Rauch, S. L., & Albert, M. S. (1996). Conscious recollection and the human hippocampal formation: Evidence from positron emission tomography. *Proccedings of the National Academy of Sciences, USA, 93,* 321–325.

Schachter, S., & Singer, J. E. (1962). Cognitive, social, and physiological determinants of emotional state. *Psychological Review, 69,* 379–399.

Schacter, D. L., & Wagner, A. D. (1999). Remembrance of things past. *Science, 285,* 1503–1504.

Schaie, K. W. (1994). The course of adult intellectual development. *American Psychologist, 49,* 304–311.

Schein, S. J., & Desimone, R. (1990). Spectral properties of V4 neurons in the macaque. *Journal of Neuroscience, 10,* 3369–3389.

Schelling, T. C. (1992). Addictive drugs: The cigarette experience. *Science, 255,* 430–433.

Schenck, C. H., Milner, D. M., Hurwitz, T. D., Bundlie, S. R., & Mahowald, M. W. (1989). A polysomnographic and clinical report on sleep-related injury in 100 adult patients. *American Journal of Psychiatry, 146,* 1166–1173.

Scherag, S., Hebebrand, J., & Hinney, A. (2009). Eating disorders: The current status of molecular genetic research. *European Child and Adolescent Psychiatry,* doi: 10.1007/s00787–009–0085–9. Retrieved December 29, 2009, from www.springerlink.com/content/u32031x128718210/fulltext.pdf

Schiermeier, Q. (1998). Animal rights activists turn the screw. *Nature, 396,* 505.

Schiff, N. D., Giacino, J. T., Kalmar, K., Victor, J. D., Baker, K., Gerber, M., et al. (2007). Behavioral improvements with thalamic stimulation after severe traumatic brain injury. *Nature, 448,* 600–604.

Schildkraut, J. J. (1965). The catecholamine hypothesis of affective disorders: A review of supporting evidence. *American Journal of Psychiatry, 122,* 509–522.

Schiller, D., Monfils, M.-H., Raio, C. M., Johnson, D. C., LeDoux, J. E., & Phelps, E. A. (2009). Preventing the return of fear in humans using reconsolidation update mechanisms. *Nature, 463,* 49–53.

Schiller, D., Monfils, M.-H., Raio, C. M., Johnson, D. C., LeDoux, J. E., & Phelps, E. A. (2010). Preventing the return of fear in humans using reconsolidation update mechanisms. *Nature, 463,* 49–53.

Schiller, P. H., & Logothetis, N. K. (1990). The color-opponent and broadband channels of the primate visual system. *Trends in Neurosciences, 13,* 392–398.

Schmalz, J. (1993, March 5). Poll finds an even split on homosexuality's cause. *New York Times,* p. A14.

Schnakers, C., Vanhaudenhuyse, A., Giacino, J., Ventura, M., Boly, M., Majerus, S., et al. (2009). Diagnostic accuracy of the vegetative and minimally conscious state: Clinical consensus versus standardized neurobehavioral assessment. *BMC Neurology, 9,* 35. Retrieved August 30, 2010, from www.biomedcentral.com/1471–2377/9/35

Schnider, A. (2003). Spontaneous confabulation and the adaptation of thought to ongoing reality. *Nature Reviews Neuroscience, 4,* 662–671.

Schnider, A., & Ptak, R. (1999). Spontaneous confabulators fail to suppress currently irrelevant memory traces. *Nature Neuroscience, 2,* 677–681.

Schobel, S. A., Lewandowski, N. M., Corcoran, C. M., Moore, H., Brown, T., Malaspina, D., et al. (2009). Differential targeting of the CA1 subfield of the hippocampal formation by schizophrenia and related psychotic disorders. *Archives of General Psychiatry, 66,* 938–946.

Schoenfeld, M. A., Neuer, G., Tempelmann, C., Schübler, K., Noesselt, T., Hopf, J.-M., et al. (2004). Functional magnetic resonance tomography correlates of taste perception in the human primary taste cortex. *Neuroscience, 127,* 347–353.

Schuckit, M. A. (1994). Low level of response to alcohol as a predictor of future alcoholism. *American Journal of Psychiatry, 151,* 184–189.

Schull, W. J., Norton, S., & Jensh, R. P. (1990). Ionizing radiation and the developing brain. *Neurotoxicology and Teratology, 12,* 249–260.

Schulsinger, F., Parnas, J., Petersen, E. T., Schulsinger, H., Teasdale, T. W., Mednick, S. A., et al. (1984). Cerebral ventricular size in the offspring of schizophrenic mothers. *Archives of General Psychiatry, 41,* 602–606.

Schultz, W. (2002). Getting formal with dopamine and reward. *Neuron, 36,* 241–263.

Schuman, E. M., & Madison, D. V. (1991). A requirement for the intercellular messenger nitric oxide in long-term potentiation. *Science, 254,* 1503–1506.

Schwartz, J. M., Stoessel, P. W., Baxter, L. R., Martin, K. M., & Phelps, M. E. (1996). Systematic changes in cerebral glucose metabolic rate after successful behavior modification treatment of obsessive-compulsive disorder. *Archives of General Psychiatry, 53,* 109–113.

Schwartz, M. S. (1994). Ictal language shift in a polyglot. *Journal of Neurology, Neurosurgery, and Psychiatry, 57,* 121.

Schwartz, M. W., & Morton, G. J. (2002). Keeping hunger at bay. *Nature, 418,* 595–597.

Schwartz, M. W., & Seeley, R. J. (1997). The new biology of body weight regulation. *Journal of the American Dietetic Association, 97,* 54–58.

Schwartz, W. J., & Gainer, H. D. (1977). Suprachiasmatic nucleus: Use of 14C-labeled deoxyglucose uptake as a functional marker. *Science, 197,* 1089–1091.

Scott, E. M., & Verney, E. L. (1947). Self-selection of diet. VI. The nature of appetites for B vitamins. *Journal of Nutrition, 34,* 471–480.

Scott, K. (2005). Taste recognition: Food for thought. *Neuron, 48,* 455–464.

Scott, K. G., & Carran, D. T. (1987). The epidemiology and prevention of mental retardation. *American Psychologist, 42,* 801–804.

Scott, L., Kruse, M. S., Forssberg, H., Brismar, H., Greengard, P., & Aperia, A. (2002). Selective up-regulation of dopamine D1 receptors in dendritic spines by NMDA receptor activation. *Proceedings of the National Academy of Sciences, USA, 99,* 1661–1664.

Sedaris, D. (1998). *Naked.* Boston: Little, Brown.

Seeing color in sounds has genetic link. (2009, February 9). Retrieved February 11, 2009, from www.cnn.com/2009/HEALTH/02/09/synesthesia.genes/index.html

Seeman, P., Lee, T., Chau-Wong, M., & Wong, K. (1976). Antipsychotic drug doses and neuroleptic/dopamine receptors. *Nature, 261,* 717–719.

Sejnowski, T. J., & Rosenberg, C. R. (1987). Parallel networks that learn to pronounce English text. *Complex Systems, 1,* 145–168.

Selfe, L. (1977). *Nadia: A case of extraordinary drawing ability in children.* London: Academic Press.

Selkoe, D. J. (1997). Alzheimer's disease: Genotypes, phenotype, and treatments. *Science, 275,* 630–631.

Senghas, A., Kita, S., & Özyürek, A. (2004). Children creating core properties of language: Evidence from an emerging sign language in Nicaragua. *Science, 305,* 1779–1782.

Senju, A., Maeda, M., Kikuchi, Y., Hasegawa, T., Tojo, Y., & Osanai, H. (2007). Absence of contagious yawning in children with autism spectrum disorder. *Biology Letters,* DOI 10.1098/rsb1.2007.0337. Retrieved September 5, 2007, from www.journals.royalsoc.ac.uk/content/3p06538k01256183/?p=8dd8b25a566a4382ba4f6dd6201fb3ab&pi=14

Sergent, C., Baillet, S., & Dehaene, S. (2005). Timing of the brain events underlying access to consciousness during the attentional blink. *Nature Neuroscience, 8,* 1391–1400.

Shaffery, J. P., Roffwarg, H. P., Speciale, S. G., & Marks, G. A. (1999). Ponto-geniculo-occipital-wave suppression amplifies lateral geniculate nucleus cell-size changes in monocularly deprived kittens. *Brain Research. Developmental Brain Research, 114,* 109–119.

Shah, A., & Lisak, R. P. (1993). Immunopharmacologic therapy in myasthenia gravis. *Clinical Neuropharmacology, 16,* 97–103.

Shalev, A. Y., Peri, T., Brandes, D., Freedman, S., Orr, S. P., & Pittman, R. K. (2000). Auditory startle response in trauma survivors with posttraumatic stress disorder: A prospective study. *American Journal of Psychiatry, 157,* 255–261.

Sham, P. C., O'Callaghan, E., Takei, N., Murray, G. K., Hare, E. H., & Murray, R. M. (1992). Schizophrenia following pre-natal exposure to influenza epidemics between 1939 and 1960. *British Journal of Psychiatry, 160,* 461–466.

Shapiro, C. M., Bortz, R., Mitchell, D., Bartel, P., & Jooste, P. (1981). Slow-wave sleep: A recovery period after exercise. *Science, 214,* 1253–1254.

Sharp, S. I., McQuillin, A., & Gurling, H. M. D. (2009). Genetics of attention-deficit hyperactivity disorder (ADHD). *Neuropharmacology, 57,* 590–600.

Shaw, P., Greenstein, D., Lerch, J., Clasen, L., Lenroot, R., Gogtay, N., et al. (2006). Intellectual ability and cortical development in children and adolescents. *Nature, 440,* 676–679.

Shaywitz, S. E., Shaywitz, B. A., Fulbright, R. K., Skudlarski, P., Mencl, W. E., Constable, R. T., et al. (2003). Neural systems for compensation and persistence: Young adult outcome of childhood reading disability. *Biological Psychiatry, 54,* 25–33.

Shaywitz, S. E., Shaywitz, B. A., Pugh, K. R., Fulbright, R. K., Constable, R. T., Mencl, W. E., et al. (1998). Functional disruption in the organization of the brain for reading in dyslexia. *Proceedings of the National Academy of Sciences, USA, 95,* 2636–2641.

Shearman, L. P., Sriram, S., Weaver, D. R., Maywood, E. S., Chaves, I., Zheng, B., et al. (2000). Interacting molecular loops in the mammalian circadian clock. *Science, 288,* 1013–1019.

Shema, R., Sacktor, T. C., & Dudai, Y. (2007). Rapid erasure of long-term memory associations in the cortex by an inhibitor of PKM (zeta). *Science, 317,* 951–953.

Sherin, J. E., Shiromani, P. J., McCarley, R. W., & Saper, C. B. (1996). Activation of ventrolateral preoptic neurons during sleep. *Science, 271,* 216–219.

Sherwin, B. B. (2003). Estrogen and cognitive functioning in women. *Endocrine Reviews, 24,* 133–151.

Sherwin, B. B., & Gelfand, M. M. (1987). The role of androgen in the maintenance of sexual functioning in oophorectomized women. *Psychosomatic Medicine, 49,* 397–409.

Shi, S.-H., Hayashi, Y., Petralia, R. S., Zaman, S. H., Wenthold, R. J., Svoboda, K., et al. (1999). Rapid spine delivery and redistribution of AMPA receptors after synaptic NMDA receptor activation. *Science, 284,* 1811–1816.

Shifren, J. L., Braunstein, G. D., Simon, J. A., Casson, P. R., Buster, J. E., Redmond, G. P., et al. (2000). Transdermal testosterone treatment in women with impaired sexual function after oophorectomy. *New England Journal of Medicine, 343,* 682–688.

Shiiya, T., Nakazato, M., Mizuta, M., Date, Y., Mondal, M. S., Tanaka, M., et al. (2009). Plasma ghrelin levels in lean and obese humans and the effect of glucose on ghrelin secretion. *Journal of Clinical Endocrinology and Metabolism, 87,* 240–244.

Shima, K., & Tanji, J. (2000). Neuronal activity in the supplementary and presupplementary motor areas for temporal organization of multiple movements. *Journal of Neurophysiology, 84,* 2148–2160.

Shimamura, A., Berry, J. M., Mangels, J. A., Rusting, C. L., & Jurica, P. J. (1995). Memory and cognitive abilities in university professors: Evidence for successful aging. *Psychological Science, 6,* 271.

Shimura, T., & Shimokochi, M. (1990). Involvement of the lateral mesencephalic tegmentum in copulatory behavior of male rats: Neuron activity in freely moving animals. *Neuroscience Research, 9,* 173–183.

Shors, T. J., Miesegaes, G., Beylin, A., Zhao, M., Rydel, T., & Gould, E. (2001). Neurogenesis in the adult is involved in the formation of trace memories. *Nature, 410,* 372–375.

Shouse, M. N., & Siegel, J. M. (1992). Pontine regulation of REM sleep components in cats: Integrity of the pedunculopontine tegmentum (PPT) is important for phasic events but unnnecessary for atonia during REM sleep. *Brain Research, 571,* 50–63.

Shuai, Y., Lu, B., Hu, Y., Wang, L., Sun, K., & Zhong, Y. (2010). Forgetting is regulated through Rac activity in *Drosophila. Cell, 140,* 579–589.

Siegel, A., Roeling, T. A., Gregg, T. R., & Kruk, M. R. (1999). Neuropharmacology of brain-stimulation-evoked aggression. *Neuroscience and Biobehavior Review, 23,* 359–389.

Siepel, A., Bejerano, G., Pedersen, J. S., Hinrichs, A. S., Hou, M., Rosenbloom, K., et al. (2005). Evolutionarily conserved elements in vertebrate, insect, worm, and yeast genomes. *Genome Research, 15,* 1034–1050.

Siegel, S. (1984). Pavlovian conditioning and heroin overdose: Reports by overdose victims. *Bulletin of the Psychonomic Society, 22,* 428–430.

Siegel, S., Hinson, R. E., Krank, M. D., & McCully, J. (1982). Heroin "overdose" death: Contribution of drug-associated environmental cues. *Science, 216,* 436–437.

Siever, L. J., Kahn, R. S., Lawlor, B. A., Trestman, R. L., Lawrence, T. L., & Coccaro, E. F. (1991). II. Critical issues in defining the role of serotonin in psychiatric disorders. *Pharmacological Reviews, 43,* 509–525.

Silbersweig, D. A., Stern, E., Frith, C., Cahill, C., Homes, A., Grootoonk, S., et al. (1995). A functional neuroanatomy of hallucinations in schizophrenia. *Nature, 378,* 176–179.

Silinsky, E. M. (1989). Adenosine derivatives and neuronal function. *Seminars in the Neurosciences, 1,* 155–165.

Simner, J., Mulvenna, C., Sagiv, N., Tsakanikos, E., Witherby, S. A., Fraser, C., et al. (2006). Synesthesia: The prevalence of atypical cross-modal experiences. *Perception, 35,* 1024–1033.

Simonelli, F., Maguire, A. M., Testa, F., Pierce, E. A., Mingozzi, F., Bennicelli, J. L., et al. (2010). Gene therapy for Leber's congenital amaurosis is safe and effective through 1.5 years after vector administration. *Molecular Therapy, 18,* 643–650.

Simos, P. G., Castillo, E. M., Fletcher, J. M., Francis, D. J., Maestu, F., Breier, J. I., et al. (2001). Mapping of receptive language cortex in bilingual volunteers by using magnetic source imaging. *Journal of Neurosurgery, 95,* 76–81.

Simpson, J. B., Epstein, A. N., & Camardo, J. S., Jr. (1978). Localization of receptors for the dipsogenic action of angiotensin II in the subfornical organ of rat. *Journal of Comparative and Physiological Psychology, 92,* 581–601.

Simpson, J. L., Ljungqvist, A., de la Chapelle, A., Ferguson-Smith, M. A., Genel, M., Carlson, A. S., et al. (1993). Gender verification in competitive sports. *Sports Medicine, 16,* 305–315.

Singer, W. (1995). Development and plasticity of cortical processing architectures. *Science, 270,* 758–764.

Sinha, R., Talih, M., Malison, R., Cooney, N., Anderson, G. M., & Kreek, M. J. (2003). Hypothalamic-pituitary-adrenal axis and sympatho-adreno-medullary responses during stress-induced and drug cue-induced cocaine craving states. *Psychopharmacology, 170,* 62–72.

Sirevaag, A. M., & Greenough, W. T. (1987). Differential rearing effects on rat visual cortex synapses. III. Neuronal and glial nuclei, boutons, dendrites, and capillaries. *Brain Research, 424,* 320–332.

Sizemore, C. C. (1989). *A mind of my own.* New York: William Morrow.

Sjöström, L., Narbro, K., Sjöström, D., Karason, K., Larsson, B., Wedel, H., et al. (2007). Effects of bariatric surgery on mortality in Swedish obese subjects. *New England Journal of Medicine, 357,* 741–752.

Skilling, S. R., Smullin, D. H., Beitz, A. J., & Larson, A. A. (1988). Extracellular amino acid concentrations in the dorsal spinal cord of freely moving rats following veratridine and nociceptive stimulation. *Journal of Neurochemistry, 51,* 127–132.

Slimp, J. C., Hart, B. L., & Goy, R. W. (1978). Heterosexual, autosexual and social behavior of adult male rhesus monkeys with medial preopticanterior hypothalamic lesions. *Brain Research, 142,* 105–122.

Slonim, D. K., Koide, K., Johnson, K. L., Tantravahi, U., Cowan, J. M., & Jarrah, Z., et al. (2009). Functional genomic analysis of amniotic fluid cell-free mRNA suggests that oxidative stress is significant in Down syndrome fetuses. *Proceedings of the National Academy of Sciences, 106,* 9425–9429.

Small, furry . . . and smart. (2009). *Nature, 461,* 862–864.

Small, S. A., Stern, Y., Tang, M., & Mayeux, R. (1999). Selective decline in memory function among healthy elderly. *Neurology, 52,* 1392–1396.

Smalley, S. L. (1997). Genetic influences in childhood-onset psychiatric disorders: Autism and attention-deficit/hyperactivity disorder. *American Journal of Human Genetics, 60,* 1276–1282.

Smith, C. (1995). Sleep states and memory processes. *Behavioural Brain Research, 69,* 137–145.

Smith, C. (1996). Sleep states, memory processes and synaptic plasticity. *Behavioural Brain Research, 78,* 49–56.

Smith, C. A. D., Gough, A. C., Leigh, P. N., Summers, B. A., Harding, A. E., Maranganore, D. M., et al. (1992). Debrisoquine hydroxylase gene polymorphism and susceptibility to Parkinson's disease. *Lancet, 339,* 1375–1377.

Smith, C. N., & Squire, L. R. (2009). Medial temporal lobe activity during retrieval of semantic memory is related to the age of the memory. *Journal of Neuroscience, 29,* 930–938.

Smith, D. E., Roberts, J., Gage, F. H., & Tuszynski, M. H. (1999). Age-associated neuronal atrophy occurs in the primate brain and is reversible by growth factor gene therapy. *Proceedings of the National Academy of Sciences, USA, 96,* 10893–10898.

Smith, D. M., & Atkinson, R. M. (1995). Alcoholism and dementia. *International Journal of Addiction, 30,* 1843–1869.

Smith, J., Cianflone, K., Biron, S., Hould, F. S., Lebel, S., Marceau, S., et al. (2009). Effects of maternal surgical weight loss in mothers on intergenerational transmission of obesity. *Journal of Clinical Endocrinology and Metabolism, 94,* 4275–4283.

Smith, M. A., Brandt, J., & Shadmehr, R. (2000). Motor disorder in Huntington's disease begins as a dysfunction in error feedback control. *Nature, 403,* 544–549.

Smith, M. J., Keel, J. C., Greenberg, B. D., Adams, L. F., Schmidt, P. J., Rubinow, D. A., et al. (1999). Menstrual cycle effects on cortical excitability. *Neurology, 53,* 2069–2072.

Smith, S. S., O'Hara, B. F., Persico, A. M., Gorelick, D. A., Newlin, D. B., Vlahov, D., et al. (1992). Genetic vulnerability to drug abuse: The D2 dopamine receptor *Taq* I B1 restriction fragment length polymorphism appears more frequently in polysubstance abusers. *Archives of General Psychiatry, 49,* 723–727.

Sneaky DNA analysis to be outlawed. (2006, August 29). *New Scientist,* 6–7.

Snyder, A. W., & Mitchell, D. J. (1999). Is integer arithmetic fundamental to mental processing? The mind's secret arithmetic. *Proceedings of the Royal Society of London B, 266,* 587–592.

Snyder, J. S., Kee, N., & Wojtowicz, J. M. (2001). Effects of adult neurogenesis on synaptic plasticity in the rat dentate gyrus. *Journal of Neurophysiology, 85,* 2423–2431.

Snyder, S. H. (1972). Catecholamines in the brain as mediators of amphetamine psychosis. *Archives of General Psychiatry, 27,* 169–179.

Snyder, S. H. (1984). Drug and neurotransmitter receptors in the brain. *Science, 224,* 22–31.

Snyder, S. H. (1997). Knockouts anxious for new therapy. *Nature, 388,* 624.

Snyder, S. H., Banerjee, S. P., Yamamura, H. I., & Greenberg, D. (1974). Drugs, neurotransmitters, and schizophrenia. *Science, 184,* 1243–1253.

Söderland, J., Schröder, J., Nordin, C., Samuelsson, M., Walther-Jallow, L., Karlsson, H., et al. (2009). Activation of brain interleukin-1β in schizophrenia. *Molecular Psychiatry, 14,* 1069–1071.

Soliman, F., Glatt, C. E., Bath, K. G., Levita, L., Jones, R. M., Pattwell, S. S., et al. (2010). A genetic variant BDNF polymorphism alters extinction learning in both mouse and human. *Science, 327,* 863–866.

Soria, V., Martinez-Amorós, È., Crespo, J. M., Martorell, L., Vilella, E., Labad, A., et al. (2010). Differential association of circadian genes with mood disorders: CRY1 and NPAS2 are associated with unipolar major depression and CLOCK and VIP with bipolar disorder. *Neuropsychopharmacology, 35,* 1279–1289.

Soto-Otero, R., Méndez-Alvarez, E., Sánchez-Sellero, J., Cruz-Landeira, A., & López-Rivadulla, L. M. (2001). Reduction of rat brain levels of the endogenous dopaminergic proneurotoxins 1,2,3,4-tetrahydroisoquinoline and 1,2,3,4-tetrahydro-beta-carboline by cigarette smoke. *Neuroscience Letters, 298,* 187–190.

Sowell, E. R., Thompson, P. M., Holmes, C. J., Jernigan, T. L., & Toga, A. W. (1999). *In vivo* evidence for post-adolescent brain maturation in frontal and striatal regions. *Nature Neuroscience, 2,* 859–861.

Sowell, E. R., Thompson, P. M., Welcome, S. E., Henkenius, A. L., Toga, A. W., & Peterson, B. S. (2003). Cortical abnormalities in children and adolescents with attention-deficit/hyperactivity disorder. *Lancet, 362,* 1699–1707.

Spanos, N. P. (1994). Multiple identity enactments and multiple personality disorder: A sociocognitive perspective. *Psychological Bulletin, 116,* 143–165.

Spence, S., Shapiro, D., & Zaidel, E. (1996). The role of the right hemisphere in the physiological and cognitive components of emotional processing. *Psychophysiology, 33,* 112–122.

Spence, S. A., Brooks, D. J., Hirsch, S. R., Liddle, P. F., Mechan, J., & Grasby, P. (1997). A PET study of voluntary movement in schizophrenic patients experiencing passivity phenomena (delusions of alien control). *Brain, 120,* 1997–2011.

Spencer, K. M., Nestor, P. G., Perlmutter, R., Niznikiewicz, M. A., Klump, M. C., Frumin, M., et al. (2004). Neural synchrony indexes disordered perception and cognition in schizophrenia. *Proceedings of the National Academy of Sciences, USA, 101,* 17288–17293.

Spencer, K. M., Nestor, P. G., Permutter, R., Nizmikiewicz, M. A., Klump, M. C., Frumin, M., et al. (2004). Neural synchrony indexes disordered perception and cognition in schizophrenia. *Proceedings of the National Academy of Sciences, 101,* 17288–17293.

Spencer, K. M., Niznikiewicz, M. A., Nestor, P. G., Shenton, M. E., & McCarley, R. W. (2009). Left auditory cortex gamma synchronization and auditory hallucination symptoms in schizophrenia. *BMC Neuroscience.* Retrieved August 8, 2010, from www.biomedcentral.com/1471-2202/10/85

Sperry, R. W. (1943). Effect of 180 degrees rotation of the retinal field on visuomotor coordination. *Journal of Experimental Zoology, 92,* 263–279.

Sperry, R. W. (1945). Restoration of vision after crossing of optic nerves and after contralateral transplantation of eye. *Journal of Neurophysiology, 8,* 15–28.

Spiegel, D. (1996). Cancer and depression. *British Journal of Psychiatry, 168,* 109–116.

Spillantini, M. G., Schmidt, M. L., Lee, V. M.-Y., Trojanowski, J. Q., Jakes, R., & Goedert, M. (1997). a-Synuclein in Lewy bodies. *Nature, 388,* 839–840.

Spinney, L. (2002, September 21). The mind readers. *New Scientist,* 38–41.

Spoont, M. (1992). Modulatory role of serotonin in neural information processing: Implications for human psychopathology. *Psychological Bulletin, 112,* 330–350.

Spreux-Varoquaux, O., Alvarez, J.-C., Berlin, I., Batista, G., Despierre, P.-G., Gilton, A., et al. (2001). Differential abnormalities in plasma 5-HIAA and platelet serotonin concentrations in violent suicide attempters: Relationships with impulsivity and depression. *Life Sciences, 69,* 647–657.

Springen, K. (2004, December 6). Using genes as medicine. *Newsweek,* p. 55.

Spurzheim, J. G. (1908). *Phrenology* (Rev. ed.). Philadelphia: Lippincott.

Squire, L. R., & Alvarez, P. (1995). Retrograde amnesia and memory consolidation: A neurobiological perspective. *Current Opinion in Neurobiology, 5,* 169–177.

Squire, L. R., Amaral, D. G., & Press, G. A. (1990). Magnetic resonance imaging of the hippocampal formation and mammillary nuclei distinguish medial temporal lobe and diencephalic amnesia. *Journal of Neuroscience, 10,* 3106–3117.

Squire, L. R., Amaral, D. G., Zola-Morgan, S., Kritchevsky, M., & Press, G. (1989). Description of brain injury in the amnesic patient N.A. based on magnetic resonance imaging. *Experimental Neurology, 105,* 23–35.

Squire, L. R., Ojemann, J. G., Miezin, F. M., Petersen, S. E., Videen, T. O., & Raichle, M. E. (1992). Activation of the hippocampus in normal humans: A functional anatomical study of memory. *Proccedings of the National Academy of Sciences, USA, 89,* 1837–1841.

Stahl, S. M. (1999). Mergers and acquisitions among psychotropics: Antidepressant takeover of anxiety may now be complete. *Journal of Clinical Psychiatry, 60,* 282–283.

Stanley, B. G., Kyrkouli, S. E., Lampert, S., & Leibowitz, S. F. (1986). Neuropeptide Y chronically injected into the hypothalamus: A powerful neurochemical inducer of hyperphagia and obesity. *Peptides, 7,* 1189–1192.

Stanley, M., Stanley, B., Traskman-Bendz, L., Mann, J. J., & Meyendorff, E. (1986). Neurochemical findings in suicide completers and suicide attempters. *Suicide and Life-Threatening Behavior, 16,* 286–299.

Starr, C., & Taggart, R. (1989). *Biology: The unity and diversity of life.* Pacific Grove, CA: Brooks/Cole.

St. Clair, D., Xu, M., Wang, P., Yu, Y., Fang, Y., Zhang, F., et al. (2005). Rates of adult schizophrenia following prenatal exposure to the Chinese famine of 1959–1961. *Journal of the American Medical Association, 294,* 557–562.

Stefansson, H., Rujescu, D., Cichon, S., Pietiläinen, O. P. H., Ingason, A., Steinberg, S., et al. (2008). Large recurrent microdeletions associated with schizophrenia. *Nature, 455,* 232–237.

Stein, J. (2001). The magnocellular theory of developmental dyslexia. *Dyslexia, 7,* 12–36.

Stellar, J. R., & Stellar, E. (1985). *The neurobiology of motivation and reward.* New York: Springer-Verlag.

Stephan, F. K., & Nunez, A. A. (1977). Elimination of circadian rhythms in drinking, activity, sleep, and temperature by isolation of the suprachiasmatic nuclei. *Behavioral Biology, 20,* 1–16.

Stephens, D. N., & Duka, T. (2008). Cognitive and emotional consequences of binge drinking: Role of amygdala and prefrontal cortex. *Philosophical Transactions of the Royal Society B, 363,* 3169–3179.

Steriade, M., Paré, D., Bouhassira, D., Deschênes, M., & Oakson, G. (1989). Phasic activation of lateral geniculate and perigeniculate thalamic neurons during sleep with ponto-geniculo-occipital waves. *Journal of Neuroscience, 9,* 2215–2229.

Stern, K., & McClintock, M. K. (1998). Regulation of ovulation by human pheromones. *Nature, 392,* 177–179.

Sternbach, R. A. (1968). *Pain: A psychophysiological analysis.* New York: Academic Press.

Sternberg, R. J. (1988). *The triarchic mind: A new theory of human intelligence.* New York: Viking.

Sternberg, R. J. (2000). The holey grail of general intelligence. *Science, 289,* 399–401.

Stickgold, R., Hobson, J. A., Fosse, R., & Fosse, M. (2001). Sleep, learning, and dreams: Off-line memory reprocessing. *Science, 294,* 1052–1057.

Stickgold, R., Whidbee, D., Schirmer, B., Patel, V., & Hobson, J. A. (2000). Visual discrimination task improvement: A multi-step process occurring during sleep. *Journal of Cognitive Neuroscience, 12,* 246–254.

Stix, G. (2009, October). Turbocharging the brain. *Scientific American,* 46–55.

Strack, F., Martin, L. L., & Stepper, S. (1988). Inhibiting and facilitating conditions of the human smile: A nonobtrusive test of the facial feedback hypothesis. *Journal of Personality and Social Psychology, 54,* 768–777.

Streissguth, A. P., Barr, H. M., Bookstein, F. L., Sampson, P. D., & Olson, H. C. (1999). The long-term neurocognitive consequences of prenatal alcohol exposure: A 14-year study. *Psychological Science, 10,* 186–190.

Stricker, E. M., & Sved, A. F. (2000). Thirst. *Nutrition, 16,* 821–826.

Suddath, R. L., Casanova, M. F., Goldberg, T. E., Daniel, D. G., Kelsoe, J. R., & Weinberger, D. R. (1989). Temporal lobe pathology in schizophrenia: A quantitative magnetic resonance imaging study. *American Journal of Psychiatry, 146,* 464–472.

Suddath, R. L., Christison, G. W., Torrey, E. F., Casanova, M. F., & Weinberger, D. R. (1990). Anatomical abnormalities in the brains of monozygotic twins discordant for schizophrenia. *New England Journal of Medicine, 322,* 789–794.

Sullivan, P. F. (1995). Mortality in anorexia nervosa. *American Journal of Psychiatry, 152,* 1073–1074.

Sullivan, P. F., Neale, M. C., & Kendler, K. S. (2000). Genetic epidemiology of major depression: Review and meta-analysis. *American Journal of Psychiatry, 157,* 1552–1562.

Sulzer, D., & Rayport, S. (2000). Dale's principle and glutamate corelease from ventral midbrain dopamine neurons. *Amino Acids, 19,* 45–52.

Sun, Y.-G., Zhao, Z.-Q., Meng, X.-L., Yin, J., Liu, X.-Y., & Chen, Z.-F. (2009). Cellular basis of itch sensation. *Science, 325,* 1531–1534.

Susser, E., Neugebauer, R., Hoek, H. W., Brown, A. S., Lin, S., Labovitz, D., et al. (1996). Schizophrenia after prenatal famine: Further evidence. *Archives of General Psychiatry, 53,* 25–31.

Sutcliffe, J. S. (2008). Insights into the pathogenesis of autism. *Science, 321,* 208–209.

Suvisaari, J. M., Haukka, J. K., Tanskanen, A. J., & Lönnqvist, J. K. (1999). Decline in the incidence of schizophrenia in Finnish cohorts born from 1954 to 1965. *Archives of General Psychiatry, 56,* 733–740.

Suzdak, P. D., Glowa, J. R., Crawley, J. N., Schwartz, R. D., Skolnick, P., & Paul, S. M. (1986). A selective imidazobenzodiazepine antagonist of ethanol in the rat. *Science, 234,* 1243–1247.

Svensson, T. H., Grenhoff, J., & Aston-Jones, G. (1986). Midbrain dopamine neurons: Nicotinic control of firing pattern. *Society for Neuroscience Abstracts, 12,* 1154.

Swaab, D. F. (1996). Desirable biology. *Science, 382,* 682–683.

Swaab, D. F., & Hofman, M. A. (1990). An enlarged suprachiasmatic nucleus in homosexual men. *Brain Research, 537,* 141–148.

Swaab, D. F., Slob, A. K., Houtsmuller, E. J., Brand, T., & Zhou, J. N. (1995). Increased number of vasopressin neurons in the suprachiasmatic nucleus (SCN) of "bisexual" adult male rats following perinatal treatment with the aromatase blocker ATD. *Developmental Brain Research, 85,* 273–279.

Swedo, S. E., Rapoport, J. L., Leonard, H. L., Lenane, M., & Cheslow, D. (1989). Obsessive-compulsive disorder in children and adolescents: Clinical phenomenology of 70 consecutive cases. *Archives of General Psychiatry, 46,* 335–341.

Swedo, S. E., Rapoport, J. L., Leonard, H. L., Schapiro, M. B., Rapoport, S. I., & Grady, C. L. (1991). Regional cerebral glucose metabolism of women with trichotillomania. *Archives of General Psychiatry, 48,* 828–833.

Swedo, S. E., Schapiro, M. B., Grady, C. L., Cheslow, D. L., Leonard, H. L., Kumar, A., et al. (1989). Cerebral glucose metabolism in childhood-onset obsessive-compulsive disorder. *Archives of General Psychiatry, 46,* 518–523.

Szalavitz, M. (2000). Drugs to fight drugs. *HMS Beagle.* Retrieved March 8, 2002, from www.news.bmn.com/hmsbeagle/91/notes/feature1

Szeszko, P. R., Ardekani, B. A., Ashtari, M., Malhotra, A. K., Robinson, D. G., Bilder, R. M., et al. (2005). White matter abnormalities in obsessive-compulsive disorder: A diffusion tensor imaging study. *Archives of General Psychiatry, 62,* 782–790.

Szobota, S., Gorostiza, P., Del Bene, F., Wyart, C., Fortin, D. L, Kolstad, K. D., et al. (2007). Remote control of neuronal activity with a light-gated glutamate receptor. *Neuron, 54,* 535–545.

Szumlinski, K. K., Dehoff, M. H., Kang, S. H., Frys, K. A., Liminac, K. D., Klugmann, M., et al. (2004). Homer proteins regulate sensitivity to cocaine. *Neuron, 43,* 401–413.

Szymusiak, R. (1995). Magnocellular nuclei of the basal forebrain: Substrates of sleep and arousal regulation. *Sleep, 18,* 478–500.

Tabrizi, S. J., Cleeter, M. W., Xuereb, J., Taanman, J. W., Cooper, J. M., & Schapira, A. H. (1999). Biochemical abnormalities and excitotoxicity in Huntington's disease brain. *Annals of Neurology, 45,* 25–32.

Tadic, A., Rujescu, D., Szegedi, A., Giegling, I., Singer, P., Möller, H.-J., et al. (2003). Association of a *MAOA* gene variant with generalized anxiety disorder, but not with panic disorder or major depression. *American Journal of Medical Genetics Part B, 117B,* 1–6.

Taheri, S., Lin, L., Austin, D., Young, T., & Mignot, E. (2004). Short sleep duration is associated with reduced leptin, elevated ghrelin, and increased body mass index. *Public Library of Science Medicine, 1,* 210–217.

Talbot, J. D., Marrett, S., Evans, A. C., Meyer, E., Bushnell, M. C., & Duncan, G. H. (1991). Multiple representations of pain in human cerebral cortex. *Science, 251,* 1355–1358.

Tamietto, M., Castelli, L., Vighetti, S., Perozzo, P., Geminiani, G., Weiskrantz, l., et al. (2009). Unseen facial and bodily expressions trigger fast emotional reactions. *Proceedings of the National Academy of Sciences, 106,* 17661–17666.

Tamietto, M., Cauda, F., Corazzini, L. L., Savazzi, S., Marzi, C. A., Goebel, R., et al. (2010). Collicular vision guides nonconscious behavior. *Journal of Cognitive Neuroscience, 22,* 888–902.

Tanaka, K. (1996). Inferotemporal cortex and object vision. *Annual Review of Neuroscience, 19,* 109–139.

Tanaka, K. (2003). Columns for complex visual object features in the inferotemporal cortex: Clustering of cells with similar but slightly different stimulus selectivities. *Cerebral Cortex, 13,* 90–99.

Tanda, G., Munzar, P., & Goldberg, S. R. (2000). Self-administration behavior is maintained by the psychoactive ingredient of marijuana in squirrel monkeys. *Nature Neuroscience, 3,* 1073–1074.

Tanda, G., Pontieri, F. E., & Di Chiara, G. (1997). Cannabinoid and heroin activation of mesolimbic dopamine transmission by a common µ1 opioid receptor mechanism. *Science, 276,* 2048–2050.

Tanji, J., & Shima, K. (1994). Role for supplementary motor area cells in planning several movements ahead. *Nature, 371,* 413–416.

Tanner, C. M., Ottman, R., Goldman, S. M., Ellenberg, J., Chan, P., Mayeux, R., et al. (1999). Parkinson disease in twins: An etiologic study. *Journal of the American Medical Association, 281,* 341–346.

Tapper, A. R., McKinney, S. L., Nashmi, R., Schwarz, J., Deshpande, P., Labarca, C., et al. (2004). Nicotine activation of a4 receptors: Sufficient for reward, tolerance, and sensitization. *Science, 306,* 1029–1032.

Taylor, A. H., Elliffe, D., Hunt, G. R., & Gray, R. D. (2010). Complex cognition and behavioural innovation in New Caledonian crows. *Proceedings of the Royal Society B.* Retrieved July 19, 2010, from http://rspb.royalsocietypublishing.org/content/early/2010/04/16/rspb.2010.0285.full.pdf

Taylor, D. N. (1995). Effects of a behavioral stress-management program on anxiety, mood, self-esteem, and T-cell count in HIV-positive men. *Psychological Reports, 76,* 451–457.

Taylor, V. H., Curtis, C. M., & Davis, C. (2009). The obesity epidemic: the role of addiction. *Canadian Medical Association Journal.* Retrieved January 13, 2010, from www.cmaj.ca/cgi/rapidpdf/cmaj.091142v1?maxtoshow=&HITS=10&hits=10&RESULTFORMAT=1&andorexacttitle=and&andorexacttitleabs=and&andorexactfulltext=and&searchid=1&FIRSTINDEX=0&sortspec=date&resourcetype=HWCIT,HWELTR

Tellegen, A., Lykken, D. T., Bouchard, T. J., Wilcox, K. J., Segal, N. L., & Rich, S. (1988). Personality similarity in twins reared apart and together. *Journal of Personality and Social Psychology, 54,* 1031–1039.

Temoshok, L. (1987). Personality, coping style, emotion and cancer: Towards an integrative model. *Cancer Survivor, 6,* 545–567.

Temple, E., Deutsch, G. K., Poldrack, R. A., Miller, S. L., Tallal, P., Merzenich, M. M., et al. (2003). Neural deficits in children with dyslexia ameliorated by behavioral remediation: Evidence from functional MRI. *Proceedings of the National Academy of Sciences, 100,* 2860–2865,

Temple, E., & Posner, M. I. (1998). Brain mechanisms of quantity are similar in 5-year-old children and adults. *Proceedings of the National Academy of Sciences, USA, 95,* 7836–7841.

Tepas, D. I., & Carvalhais, A. B. (1990). Sleep patterns of shiftworkers. *Occupational Medicine, 5,* 199–208.

Terrace, H. S., Petitto, L. A., Sanders, R. J., & Bever, T. G. (1979). Can an ape create a sentence? *Science, 206,* 891–901.

Tessier-Lavigne, M., & Goodman, C. S. (1996). The molecular biology of axon guidance. *Science, 274,* 1123–1132.

Thallmair, M., Metz, G. A. S., Z'Graggen, W. J., Raineteau, O., Kartje, G. L., & Schwab, M. E. (1998). Neurite growth inhibitors restrict plasticity and functional recovery following corticospinal tract lesions. *Nature Neuroscience, 1*(2), 124–131.

Thanos, P. K., Volkow, N. D., Freimuth, P., Umegaki, H., Ikari, H., Roth, G., et al. (2001). Over expression of dopamine D2 receptors reduces alcohol self-administration. *Journal of Neurochemistry, 78,* 1094–1103.

Thapar, A., O'Donovan, M., & Owen, M. J. (2005). The genetics of attention deficit hyperactivity disorder. *Human Molecular Genetics, 14,* R275–R282.

Therapy setback. (2005, February 12). *New Scientist, 185,* 6.

Thibaut, F., Cordier, B., & Kuhn, J.-M. (1996). Gonadotrophin hormone releasing hormone agonist in cases of severe paraphilia: A lifetime treatment? *Psychoneuroendocrinology, 21,* 411–419.

Thiele, A., Henning, P., Kubischik, M., & Hoffmann, K.-P. (2002). Neural mechanisms of saccadic suppression. *Science, 295,* 2460–2462.

Thier, P., Haarmeier, T., Chakraborty, S., Lindner, A., & Tikhonov, A. (2001). Cortical substrates of perceptual stability during eye movements. *NeuroImage, 14,* S33–S39.

Thigpen, C. H., & Cleckley, H. M. (1957). *The three faces of Eve.* London: Secker & Warburg.

Thompson, P. M., Cannon, T. D., Narr, K. L., van Erp, T., Poutanen, V.-P., Huttunen, M., et al. (2001). Genetic influences on brain structure. *Nature Neuroscience, 4,* 1253–1258.

Thompson, P. M., Vidal, C., Giedd, J. N., Gochman, P., Blumenthal, J., Nicolson, R., et al. (2001). Mapping adolescent brain change reveals dynamic wave of accelerated gray matter loss in very early-onset schizophrenia. *Proceedings of the National Academy of Sciences, USA, 98,* 11650–11655.

Thrasher, T. N., & Keil, L. C. (1987). Regulation of drinking and vasopressin secretion: Role of organum vasculosum laminae terminalis. *American Journal of Physiology, 253,* R108–R120.

Ticho, S. R., & Radulovacki, M. (1991). Role of adenosine in sleep and temperature regulation in the preoptic area of rats. *Pharmacology and Biochemistry of Behavior, 40,* 33–40.

Toh, K. L., Jones, C. R., He, Y., Eide, E. J., Hinz, W. A., Virshup, D. M., et al. (2001). An h*Per2* phosphorylation site mutation in familial advanced sleep phase syndrome. *Science, 291,* 1040–1043.

Tong, F. (2003). Primary visual cortex and visual awareness. *Nature Reviews Neuroscience, 4,* 219–229.

Toni, N., Buchs, P.-A., Nikonenko, I., Bron, C. R., & Muller, D. (1999). LTP promotes formation of multiple spine synapses between a single axon terminal and a dendrite. *Nature, 402,* 421–425.

Tononi, G. (2005). Consciousness, information integration, and the brain. In S. Laureys (Ed.), *The boundaries of consciousness: Neurobiology and neuropathology* (pp. 109–126). New York: Elsevier.

Tononi, G. (2008). Consciousness as integrated information: A provisional manifesto. *Biological Bulletin, 215,* 216–242.

Tootell, R. B. H., Silverman, M. S., Switkes, E., & De Valois, R. L. (1982). Deoxyglucose analysis of retinotopic organization in primate striate cortex. *Science, 218,* 902–904.

Torii, M., Hashimoto-Torii, K., Levitt, P., & Rakic, P. (2009). Integration of neuronal clones in the radial cortical columns by EphA and ephrin-A signalling. *Nature, 461,* 524–530.

Toso, L., Cameroni, I., Roberson, R., Abebe, D., Bissell, S., & Spong, C. Y. (2008). Prevention of developmental delays in a Down syndrome mouse model. *Obstetrics and Gynecology, 112,* 1242–1251.

Townsend, J., Courchesne, E., Covington, J., Westerfield, M., Harris, N. S., Lyden, P., et al. (1999). Spatial attention deficits in patients with acquired or developmental cerebellar abnormality. *Journal of Neuroscience, 19,* 5632–5643.

Tracey, I., Ploghaus, A., Gati, J. S., Clare, S., Smith, S., Menon, R. S., et al. (2002). Imaging attentional modulation of pain in the periaqueductal gray in humans. *Journal of Neuroscience, 22,* 2748–2752.

Tranel, D., & Damasio, A. R. (1985). Knowledge without awareness: An autonomic index of facial recognition by prosopagnosics. *Science, 228,* 1453–1454.

Tranel, D., Damasio, A. R., & Damasio, H. (1988). Intact recognition of facial expression, gender, and age in patients with impaired recognition of face identity. *Neurology, 38,* 690–696.

Träskman, L., Asberg, M., Bertilsson, L., & Sjöstrand, L. (1981). Monoamine metabolites in CSF and suicidal behavior. *Archives of General Psychiatry, 38,* 631–635.

Trautmann, A. (1983). Tubocurarine, a partial agonist for cholinergic receptors. *Journal of Neural Transmission. Supplementum, 18,* 353–361.

Treffert, D. A., & Christensen, D. D. (2006, June/July). Inside the mind of a savant. *Scientific American Mind, 17,* 50–55.

Tregellas, J. R., Tanabe, J. L., Martin, L. F., & Freedman, R. (2005). fMRI of response to nicotine during a smooth pursuit eye movement task in schizophrenia. *American Journal of Psychiatry, 162,* 391–393.

Treisman, A. M., & Gelade, G. (1980). A feature-integration theory of attention. *Cognitive Psychology, 12,* 97–136.

Trinko, R., Sears, R. M., Guarnieri, D. J., & DiLeone, R. J. (2007). Neural mechanisms underlying obesity and drug addiction. *Physiology and Behavior, 91,* 499–505.

Tripp, G., & Wickens, J. R. (2009). Neurobiology of ADHD. *Neuropharmacology, 57,* 579–589.

True, W. R., Rice, J., Eisen, S. A., Heath, A. C., Goldberg, J., Lyons, M. J., et al. (1993). A twin study of genetic and environmental contributions to liability for posttraumatic stress symptoms. *Archives of General Psychiatry, 50,* 257–264.

Trulson, M. E., Crisp, T., & Trulson, V. M. (1984). Activity of serotonin-containing nucleus centralis superior (raphe medianus) neurons in freely moving cats. *Experimental Brain Research, 54,* 33–44.

Tsai, G., Gastfriend, D. R., & Coyle, J. T. (1995). The glutamatergic basis of human alcoholism. *American Journal of Psychiatry, 152,* 332–340.

Tsai, G. E., Condie, D., Wu, M. T., & Chang, I.-W. (1999). Functional magnetic resonance imaging of personality switches in a woman with dissociative identity disorder. *Harvard Review of Psychiatry, 7,* 119–122.

Tsakanikas, D., & Relkin, N. (2010, April). Neuropsychological outcomes following 18-months of uninterrupted intravenous immunoglobulin (IVIg) treatment in patients with Alzheimer's disease (AD). Paper presented at the annual meeting of the American Academy of Neurology, Toronto, Canada.

Tsankova, N., Renthal, W., Kumar, A., & Nestler, E. J. (2007). Epigenetic regulation in psychiatric disorders. *Nature Reviews Neuroscience, 8,* 355–367.

Tsuang, M. T., Gilbertson, M. W., & Faraone, S. V. (1991). The genetics of schizophrenia: Current knowledge and future directions. *Schizophrenia Research, 4,* 157–171.

Tuiten, A., Van Honk, J., Koppeschaar, H., Bernaards, C., Thijssen, J., & Verbaten, R. (2000). Time course of effects of testosterone administration on sexual arousal in women. *Archives of General Psychiatry, 57,* 149–153.

Turkheimer, E. (1991). Individual and group differences in adoption studies of IQ. *Psychological Bulletin, 110,* 392–405.

Turner, M. S., Cipolotti, L., Yousry, T. A., & Shallice, T. (2008). Confabulation: Damage to a specific inferior medial prefrontal system. *Cortex, 44,* 637–648.

Tuszynski, M., Thal, L., Pay, M., Salmon, D. P., U, H. S., Bakay, R., et al. (2005). A phase 1 clinical trial of nerve growth factor gene therapy for Alzheimer disease. *Nature Medicine, 11,* 551–555.

Uhl, G. R., & Grow, R. W. (2004). The burden of complex genetics in brain disorders. *Archives of General Psychiatry, 61,* 223–229.

Uhlhaas, P. J., & Singer, W. (2010). Abnormal neural oscillations and synchrony in schizophrenia. *Nature Reviews Neuroscience, 11,* 100–113.

Ullian, E. M., Sapperstein, S. K., Christopherson, K. S., & Barres, B. A. (2001). Control of synapse number by glia. *Science, 291,* 657–660.

Ullman, M. T. (2001). A neurocognitive perspective on language: The declarative/procedural model. *Nature Reviews Neuroscinece, 2,* 717–726.

Uncapher, M. R., Otten, L. J., & Rugg, M. D. (2006). Episodic encoding is more than the sum of its parts: An fMRI investigation of multifeatural contextual encoding. *Neuron, 52,* 547–556.

Ungerleider, L. G., & Haxby, J. V. (1994). "What" and "where" in the human brain. *Current Opinion in Neurobiology, 4,* 157–165.

Ungerleider, L. G., & Mishkin, M. (1982). Two cortical visual systems. In D. J. Ingle, M. A. Goodale, & R. J. W. Mansfield (Eds.), *Analysis of visual behavior* (pp. 549–586). Cambridge, MA: MIT Press.

Uno, H., Tarara, R., Else, J. G., Suleman, M. A., & Sapolsky, R. M. (1999). Hippocampal damage associated with prolonged and fatal stress in primates. *Journal of Neuroscience, 9,* 1705–1711.

U.S. Congress Office of Technology Assessment. (1986). *Alternatives to animal research in testing and education.* Washington, DC: Government Printing Office.

U.S. Government shuts down Pennsylvania gene therapy trials. (2000). *Nature, 403,* 354–355.

Vaina, L. M. (1998). Complex motion perception and its deficits. *Current Opinion in Neurobiology, 8,* 494–502.

Vaina, L. M., Solomon, J., Chowdhury, S., Sinha, P., & Belliveau, J. W. (2001). Functional neuroanatomy of biological motion perception in humans. *Proceedings of the National Academy of Sciences, USA, 98,* 11656–11661.

Valenstein, E. S. (1986). *Great and desperate cures.* New York: Basic Books.

Valera, E. M., Faraone, S. V., Murray, K. E., & Seidman, L. J. (2007). Meta-analysis of structural imaging findings in attention-deficit/hyperactivity disorder. *Biological Psychiatry, 61,* 1361–1369.

van Amelsvoort, T., Compton, J., & Murphy, D. (2001). *In vivo* assessment of the effects of estrogen on human brain. *Trends in Endocrinology & Metabolism, 12,* 273–276.

van der Heijden, F. M., Tuinier, S., Fekkes, D., Sijben, A. E., Kahn, R. S., & Verhoeven, W. M. (2004). Atypical antipsychotics and the relevance of glutamate and serotonin. *European Neuropsychopharmacology, 14,* 259–265.

van der Veen, F. M., Nijhuis, F. A. P., Tisserand, D. J., Backes, W. H., & Jolles, J. (2006). Effects of aging on recognition of intentionally and incidentally stored words: An fMRI study. *Neuropsychologia, 44,* 2477–2486.

Van Essen, D. C., Anderson, C. H., & Felleman, D. J. (1992). Information processing in the primate visual system: An integrated systems perspective. *Science, 255,* 419–423.

Van Goozen, S. H., Frijda, N. H., Wiegant, V. M., Endert, E., & Van de Poll, N. E. (1996). The premenstrual phase and reactions to aversive events: A study of hormonal influences on emotionality. *Psychoneuroendocrinology, 21,* 479–497.

Vanhaudenhuyse, A., Noirhomme, Q., Luaba, J.-F., Tshibanda, L. J.-F., Bruno, M.-A., Boveroux, P., et al. (2010). Default network connectivity reflects the level of consciousness in non-communicative brain-damaged patients. *Brain, 133,* 161–171.

van Os, J., & Selten, J. (1998). Prenatal exposure to maternal stress and subsequent schizophrenia: The May 1940 invasion of The Netherlands. *British Journal of Psychiatry, 172,* 324–326.

Van Wyk, P. H., & Geist, C. S. (1984). Psychosocial development of heterosexual, bisexual, and homosexual behavior. *Archives of Sexual Behavior, 13,* 505–544.

Vargha-Khadem, F., Gadian, D. G., Copp, A., & Mishkin, M. (2005). *FOXP2* and the neuroanatomy of speech and language. *Nature Reviews Neuroscience, 6,* 131–138.

Venter, J. C. (and 273 others). (2001). The sequence of the human genome. *Science, 291,* 1304–1351.

Vergnes, M., Depaulis, A., Boehrer, A., & Kempf, E. (1988). Selective increase of offensive behavior in the rat following intrahypothalamic 5, 7-DHT induced serotonin depletion. *Behavioral Brain Research, 29,* 85–91.

Verhaak, P. F. M., Kerssens, J. J., Dekker, J., Sorbi, M. J., & Bensing, J. M. (1998). Prevalence of chronic benign pain disorder among adults: A review of the literature. *Pain, 77,* 231–239.

Verlinsky, Y., Rechitsky, S., Verlinsky, O., Masciangelo, C., Lederer, K., & Kulieve, A. (2002). Preimplantation diagnosis for early-onset Alzheimer disease caused by V717L mutation. *Journal of the American Medical Association, 287,* 1018–1021.

Vernes, S. C., Newbury, D. F., Abrahams, B. S., Winchester, L., Nicod, J., Groszer, M., et al. (2008). A functional genetic link between distinct developmental language disorders. *New England Journal of Medicine, 359,* 2337–2345.

Vernon, P. A. (1987). New developments in reaction time research. In P. A. Vernon (Ed.), *Speed of information processing and intelligence* (pp. 1–20). Norwood, NJ: Ablex.

Vernon, P. A., & Mori, M. (1992). Intelligence, reaction times, and peripheral nerve conduction velocity. *Intelligence, 16,* 273–288.

Vgontzas, A. N., Bixler, E. O., Lin, H. M., Prolo, P., Mastorakos, G., Vela-Bueno, A., et al. (2001). Chronic insomnia is associated with nyctohemeral activation of the hypothalamic-pituitary-adrenal axis: Clinical implications. *Journal of Clinical Endocrinology and Metabolism, 86,* 3787–3794.

Vikingstad, E. M., Cao, Y., Thomas, A. J., Johnson, A., Malik, G. M., & Welch, K. M. A. (2000). Language hemispheric dominance in patients with congenital lesions of eloquent brain. *Neurosurgery, 47,* 562–570.

Villalobos, M. E., Mizuno, A., Dahl, B. C., Kemmotsu, N., & Müller, R.-A. (2005). Reduced functional connectivity between V1 and inferior frontal cortex associated with visuomotor performance in autism. *NeuroImage, 25,* 916–925.

Villemagne, V. L., Pike, K., Mulligan, R. S., Jones, G., Rowe, C. C., Ellis, K. A., et al. (2010, June 5–9). *Longitudinal assessment of Aβ burden and cognition with 11C-PiB PET in aging and Alzheimer's disease.* Presented at the annual meeting of the Society of Nuclear Medicine, Salt Lake City, Utah.

Virkkunen, M., Goldman, D., & Linnoila, M. (1996). Serotonin in alcoholic violent offenders. In *Genetics of criminal and antisocial behaviour. Ciba Foundation Symposium 194* (pp. 168–182). Chichester, UK: Wiley.

Virkkunen, M., & Linnoila, M. (1990). Serotonin in early onset, male alcoholics with violent behaviour. *Annals of Medicine, 22,* 327–331.

Virkkunen, M., & Linnoila, M. (1993). Brain serotonin, Type II alcoholism and impulsive violence. *Journal of Studies of Alcohol Supplement, 11,* 163–169.

Virkkunen, M., & Linnoila, M. (1997). Serotonin in early-onset alcoholism. *Recent Developments in Alcohol, 13,* 173–189.

Visser-Vandewalle, V., Temel, Y., Boon, P., Vreeling, F., Colle, H., Hoogland, G., et al. (2003). Chronic bilateral thalamic stimulation: A new therapeutic approach in intractable Tourette syndrome. Report of three cases. *Journal of Neurosurgery, 99,* 1094–1100.

Vogel, G. (1998). Penetrating insight into the brain. *Science, 282,* 39.

Vogel, G. W., Buffenstein, A., Minter, K., & Hennessey, A. (1990). Drug effects on REM sleep and on endogenous depression. *Neuroscience and Biobehavioral Review, 14,* 49–63.

Vogels, R. (1999). Categorization of complex visual images by rhesus monkeys. Part 2: Single-cell study. *European Journal of Neuroscience, 11,* 1239–1255.

Vogelstein, R. J., Tenore, F. V. G., Guevremont, L, Etienne-Cummings, R., & Mushahwar, V. K. (2008). A silicon central pattern generator controls locomotion in vivo. *IEEE Transactions on Biomedical Circuits and Systems, 2,* 212–222.

Volkow, N. D., & Fowler, J. S. (2000). Addiction, a disease of compulsion and drive: Involvement of the orbitofrontal cortex. *Cerebral Cortex, 10,* 318–325.

Volkow, N. D., Fowler, J. S., & Wang, G.-J. (2003). The addicted human brain: Insights from imaging studies. *Journal of Clinical Investigation, 111,* 1444–1451.

Volkow, N. D., Fowler, J. S., & Wang, G.-J. (2004). The addicted human brain viewed in the light of imaging studies: Brain circuits and treatment strategies. *Neuropharmacology, 47* (Suppl.), 3–13.

Volkow, N. D., Fowler, J. S., Wang, G.-J., & Swanson, J. M. (2004). Dopamine in drug abuse and addiction: Results from imaging studies and treatment implications. *Molecular Psychiatry, 9,* 557–569.

Volkow, N. D., Fowler, J. S., Wolf, A. P., Hitzemann, R., Dewey, S., Bendriem, B., et al. (1991). Changes in brain glucose metabolism in cocaine dependence and withdrawal. *American Journal of Psychiatry, 148,* 621–626.

Volkow, N. D., & Li, T.-K. (2004). Drug addiction: The neurobiology of behaviour gone awry. *Nature Reviews: Neuroscience, 5,* 963–970.

Volkow, N. D., Wang, G.-J., Fischman, M. W., Foltin, R. W., Fowler, J. S., Abumrad, N. N., et al. (1997). Relationship between subjective effects of cocaine and dopamine transporter occupancy. *Nature, 386,* 827–830.

Volkow, N. D., Wang, G.-J., Fowler, J. S., Hitzemann, R., Angrist, B., Gatley, S. J., et al. (1999). Association of methylphenidate-induced craving with changes in right striato-orbitofrontal metabolism in cocaine abusers: Implications in addiction. *American Journal of Psychiatry, 156,* 19–26.

Volkow, N. D., Wang, G.-J., Fowler, J., & Telang, F. (2008). Overlapping neuronal circuits in addiction and obesity: Evidence of systems pathology. *Philosophical Transactions of the Royal of London B, 363,* 3191–3200.

Volkow, N. D., Wang, G.-J., Fowler, J. S., Thanos, P., Logan, J., Gatley, S. J., et al. (2002). Brain DA D2 receptors predict reinforcing effects of stimulants in humans: Replication study. *Synapse, 46,* 79–82.

Volkow, N. D., & Wise, R. A. (2005). How can drug addiction help us understand obesity? *Nature Neuroscience, 8,* 555–560.

Volpe, B. T., LeDoux, J. E., & Gazzaniga, M. S. (1979). Information processing of visual stimuli in an "extinguished" field. *Nature, 282,* 722–724.

von der Heydt, R., Peterhans, E., & Dürsteler, M. R. (1992). Periodic pattern-selective cells in monkey visual cortex. *Journal of Neuroscience, 12,* 1416–1434.

Vorel, S. R., Liu, X., Hayes, R. J., Spector, J. A., & Gardner, E. L. (2001). Relapse to cocaine-seeking after hippocampal theta burst stimulation. *Science, 292,* 1175–1178.

Voyer, D., Voyer, S., & Bryden, M. P. (1995). Magnitude of sex differences in spatial abilities: A meta-analysis and consideration of critical variables. *Psychological Bulletin, 117,* 250–270.

Vul, E., Harris, C., Winkielman, P., & Pashler, H. (2009). Puzzlingly high correlations in fMRI studies of emotion, personality, and social cognition. *Perspectives on Psychological Science, 4,* 274–290.

Wada, J. A., Clarke, R., & Hamm, A. (1975). Cerebral hemispheric asymmetry in humans: Cortical speech zones in 100 adult and 100 infant brains. *Archives of Neurology, 32,* 239–246.

Wade, J. A., Garry, M., Read, J. D., & Lindsay, S. (2002). A picture is worth a thousand lies. *Psychonomic Bulletin Reviews, 9,* 597–603.

Wagner, A. D., Schacter, D. L., Rotte, M., Koutstaal, W., Maril, A., Dale, A. M., et al. (1998). Building memories: Remembering and forgetting of verbal experiences as predicted by brain activity. *Science, 281,* 1188–1191.

Wagner, H. J., Hennig, H., Jabs, W. J., Sickhaus, A., Wessel, K., & Wandinger, K. P. (2000). Altered prevalence and reactivity of anti-Epstein-Barr virus antibodies in patients with multiple sclerosis. *Viral Immunology, 13,* 497–502.

Wakefield, A. J., Murch, S. H., Anthony, A., Linnell, J., Casson, D. M., Malik, M., et al. (1998). Ileal-lymphoid-nodular hyperplasia, non-specific colitis, and pervasive developmental disorder in children. *Lancet, 351,* 637–641.

Walker, E. F., Lewine, R. R. J., & Neumann, C. (1996). Childhood behavioral characteristics and adult brain morphology in schizophrenia. *Schizophrenia Research, 22,* 93–101.

Wall, T. L., & Ehlers, C. L. (1995). Genetic influences affecting alcohol use among Asians. *Alcohol Health and Research World, 19,* 184–189.

Walsh, B. T., & Devlin, M. J. (1998). Eating disorders: Progress and problems. *Science, 280,* 1387–1390.

Walum, H., Westberg, L., Henningsson, S., Neiderhiser, J. M., Reiss, D., Igl, W., et al. (2008). Genetic variation in the vasopressin receptor

1a gene (*AVPR1A*) associates with pair-bonding behavior in humans. *Proceedings of the National Academy of Sciences, 105,* 14153–14156.

Wan, F.-J., Berton, F., Madamba, S. G., Francesconi, W., & Siggins, G. R. (1996). Low ethanol concentrations enhance GABAergic inhibitory postsynaptic potentials in hippocampal pyramidal neurons only after block of GABAB receptors. *Proceedings of the National Academy of Sciences, USA, 93,* 5049–5054.

Wang, G.-J., Volkow, N. D., Logan, J., Papas, N. R., Wong, C. T., Zhu, W., et al. (2001). Brain dopamine and obesity. *Lancet, 357,* 354–357.

Wang, J., Korczykowski, M., Rao, H., Fan, Y., Pluta, J., Gur, R. C., et al. (2007). Gender difference in neural response to psychological stress. *Social, Cognitive and Affective Neuroscience, 2,* 227–239.

Wang, N.-K., Tosi, J., Kasanuki, J. M., Chou, C. L., Kong, J., Parmalee, N., et al. (2010). Transplantation of reprogrammed embryonic stem cells improves visual function in a mouse model for retinitis pigmentosa. *Transplantation, 89,* 911–919.

Wang, Y., Qian, Y., Yang, S., Shi, H., Liao, C., Zheng, H.-K., et al. (2005). Accelerated evolution of the pituitary adenylate cyclase-activating polypeptide precursor gene during human origin. *Genetics, 170,* 801–806.

Warren, S., Hämäläinen, H. A., & Gardner, E. P. (1986). Objective classification of motion- and direction-sensitive neurons in primary somatosensory cortex of awake monkeys. *Journal of Neurophysiology, 56,* 598–622.

Waterland, R. A., & Jirtle, R. L. (2003). Transposable elements: Targets for early nutritional effects on epigenetic gene regulation. *Molecular and Cellular Biology, 23,* 5293–5300.

Waters, A. J., Jarvis, M. J., & Sutton, S. R. (1998). Nicotine withdrawal and accident rates. *Nature, 394,* 137.

Watkins, L. R., & Mayer, D. J. (1982). Organization of endogenous opiate and nonopiate pain control systems. *Science, 216,* 1185–1192.

Watson, C. G., Kucala, T., Tilleskjor, C., & Jacobs, L. (1984). Schizophrenic birth seasonality in relation to the incidence of infectious diseases and temperature extremes. *Archives of General Psychiatry, 41,* 85–90.

Watson, J. D., & Crick, F. H. C. (1953). Genetical implications of the structure of deoxyribonucleic acid. *Nature, 171,* 964–967.

Waxman, S. G., & Ritchie, J. M. (1985). Organization of ion channels in the myelinated nerve fiber. *Science, 228,* 1502–1507.

Webb, W. B. (1974). Sleep as an adaptive response. *Perceptual and Motor Skills, 38,* 1023–1027.

Webb, W. B., & Cartwright, R. D. (1978). Sleep and dreams. *Annual Review of Psychology, 29,* 223–252.

Wedekind, C., Seebeck, T., Bettens, F., & Paepke, A. J. (1995). MHC-dependent mate preferences in humans. *Proceedings of the Royal Society of London, B, 260,* 245–249.

Weeks, D., Freeman, C. P., & Kendell, R. E. (1980). ECT: II: Enduring cognitive deficits? *British Journal of Psychiatry, 137,* 26–37.

Wegesin, D. J. (1998). A neuropsychologic profile of homosexual and heterosexual men and women. *Archives of Sexual Behavior, 27,* 91–108.

Wehr, T. A., Jacobsen, F. M., Sack, D. A., Arendt, J., Tamarkin, L., & Rosenthal, N. E. (1986). Phototherapy of seasonal affective disorder: Time of day and suppression of melatonin are not critical for antidepressant effects. *Archives of General Psychiatry, 43,* 870–875.

Weinberg, R. A. (1989). Intelligence and IQ: Landmark issues and great debates. *American Psychologist, 44,* 98–104.

Weinberg, R. A., Scarr, S., & Waldman, I. D. (1992). The Minnesota transracial adoption study: A follow-up of IQ test performance at adolescence. *Intelligence, 16,* 117–135.

Weinberger, D. R. (1987). Implications of normal brain development for the pathogenesis of schizophrenia. *Archives of General Psychiatry, 44,* 660–669.

Weinberger, D. R., Berman, K. F., Suddath, R., & Torrey, E. F. (1992). Evidence of dysfunction of a prefrontal-limbic network in schizophrenia: A magnetic resonance imaging and regional cerebral blood flow study of discordant monozygotic twins. *American Journal of Psychiatry, 149,* 890–897.

Weinberger, D. R., Berman, K. F., & Zec, R. F. (1986). Physiologic dysfunction of dorsolateral prefrontal cortex in schizophrenia: I. Regional cerebral blood flow evidence. *Archives of General Psychiatry, 43,* 114–124.

Weinberger, D. R., & Lipska, B. K. (1995). Cortical maldevelopment, antipsychotic drugs, and schizophrenia: A search for common ground. *Schizophrenia Research, 16,* 87–110.

Weinberger, D. R., Torrey, E. F., Neophytides, A. N., & Wyatt, R. J. (1979). Lateral cerebral ventricular enlargement in chronic schizophrenia. *Archives of General Psychiatry, 36,* 735–739.

Weinberger, D. R., & Wyatt, R. J. (1983). Enlarged cerebral ventricles in schizophrenia. *Psychiatric Annals, 13,* 412–418.

Weiner, R. D., & Krystal, A. D. (1994). The present use of electroconvulsive therapy. *Annual Review of Medicine, 45,* 273–281.

Weingarten, H. P., Chang, P. K., & McDonald, T. J. (1985). Comparison of the metabolic and behavioral disturbances following paraventricular- and ventromedial-hypothalamic lesions. *Brain Research Bulletin, 14,* 551–559.

Weinstein, S. (1968). Intensive and extensive aspects of tactile sensitivity as a function of body part, sex, and laterality. In D. R. Kenshalo (Ed.), *The skin senses* (pp. 195–222). Springfield, IL: Thomas.

Weiss, K. J. (2010). Hoarding, hermitage, and the law: Why we love the Collyer brothers. *Journal of the American Academy of Psychiatry and Law, 38,* 251–257.

Weiten, W. (2001). *Psychology: Themes and variations* (with InfoTrac; 5th ed.). 2001. Belmont, CA: Wadsworth/Thomson Learning.

Wekerle, H. (1993). Experimental autoimmune encephalomyelitis as a model of immune-mediated CNS disease. *Current Opinion in Neurobiology, 3,* 779–784.

Welch, J. M., Lu, J., Rodriguiz, R. M., Trotta, N. C., Peca, J., Ding, J.-D., et al. (2007). Cortico-striatal synaptic defects and OCD-like behaviours in *Sapap3*-mutant mice. *Nature, 448,* 894–901.

Weltzin, T. E., Fernstrom, M. H., & Kaye, W. H. (1994). Serotonin and bulimia nervosa. *Nutrition Reviews, 52,* 399–408.

Weltzin, T. E., Hsu, L. K., Pollice, C., & Kaye, W. H. (1991). Feeding patterns in bulimia nervosa. *Biological Psychiatry, 30,* 1093–1110.

Wender, P. H., Rosenthal, D., Kety, S. S., Schulsinger, F., & Welner, J. (1974). Cross-fostering: A research strategy for clarifying the role of genetic and experiential factors in the etiology of schizophrenia. *Archives of General Psychiatry, 30,* 121–128.

Wender, P. H., Wolf, L. E., & Wasserstein, J. (2001). Adults with ADHD: An overview. *Annals of the New York Academy of Sciences, 931,* 1–16.

West, D. B., Fey, D., & Woods, S. C. (1984). Cholecystokinin persistently suppresses meal size but not food intake in free-feeding rats. *American Journal of Physiology, 246,* R776–R787.

Wever, E. G. (1949). *Theory of hearing.* New York: Wiley.

Wever, E. G., & Bray, C. W. (1930). The nature of acoustic response: The relation between sound frequency and frequency of impulses in the auditory nerve. *Journal of Experimental Psychology, 13,* 373–387.

Wheaton, K. J., Thompson, J. C., Syngeniotis, A., Abbott, D. F., & Puce, A. (2004). Viewing the motion of human body parts activates different regions of premotor, temporal, and parietal cortex. *NeuroImage, 22,* 277–288.

Wheeler, M. A., Stuss, D. T., & Tulving, E. (1997). Toward a theory of episodic memory: The frontal lobes and autonoetic consciousness. *Psychological Bulletin, 121,* 331–354.

Wheeler, M. E., Petersen, S. E., & Buckner, R. L. (2000). Memory's echo: Vivid remembering reactivates sensory-specific cortex. *Proceedings of the National Academy of Sciences, USA, 97,* 11125–11129.

Whipple, B., & Komisaruk, B. R. (1988). Analgesia produced in women by genital self-stimulation. *Journal of Sex Research, 24,* 130–140.

Whiteside, S. P., Port, J. D., & Abramowitz, J. S. (2004). A meta-analysis of functional neuroimaging in obsessive-compulsive disorder. *Psychiatry Research, 132,* 69–79.

Whitman, F. L., Diamond, M., & Martin, J. (1993). Homosexual orientation in twins: A report on 61 pairs and three triplet sets. *Archives of Sexual Behavior, 22,* 187–206.

Whitney, D., Elison, A., Rice, N. J., Arnold, D., Goodale, M., Walsh, V., et al. (2007). Visually guided reaching depends on motion area MT+. *Cerebral Cortex, 17,* 2644–2649.

Wickelgren, I. (1997). Getting a grasp on working memory. *Science, 275,* 1580–1582.

Wickelgren, I. (1998). Obesity: How big a problem? *Science, 280,* 1364–1367.

Wiederhold, B. K. (2010). PTSD threatens global economies. *Cyberpsychology, Behavior, and Social Networking, 13,* 1–2.

Wiederman, M. W., & Pryor, T. (1996). Substance abuse and impulsive behaviors among adolescents with eating disorders. *Addictive Behaviors, 21,* 269–272.

Wierzynski, C. M., Lubenov, E. V., Gu, M., Siapas, A. G. (2009). State-dependent spike-timing relationships between hippocampal and prefrontal circuits during sleep. *Neuron, 61,* 587–596.

Wiesel, T. N., & Hubel, D. H. (1966). Spatial and chromatic interactions in the lateral geniculate body of the rhesus monkey. *Journal of Neurophysiology, 29,* 1115–1156.

Wilbert-Lampen, U., Leistner, D., Greven,S., Tilmann, P., Sper, S., Völker, C., et al. (2008). Cardiovascular events during world cup soccer. *New England Journal of Medicine, 358,* 475–483.

Wilens, T. E., Faraone, S. V., Biederman, J., & Gunawardene, S. (2003). Does stimulant therapy of attention-deficit/hyperactivity disorder beget later substance abuse? A meta-analytic review of the literature. *Pediatrics, 111,* 179–185.

Willerman, L., Schultz, R., Rutledge, J. N., & Bigler, E. D. (1991). In vivo brain size and intelligence. *Intelligence, 15,* 223–228.

Willerman, L., Schultz, R., Rutledge, J. N., & Bigler, E. D. (1994). Brain structure and cognitive function. In C. R. Reynolds (Ed.), *Cognitive assessment: A multidisciplinary perspective* (pp. 35–55). New York: Plenum Press.

Williams, J. H. G., Whiten, A., Suddendorf, T., & Perrett, D. I. (2001). Imitation, mirror neurons and autism. *Neuroscience and Biobehavioral Reviews, 25,* 287–295.

Williams, R. W., & Herrup, K. (2001). *The control of neuron number.* Retrieved June 26, 2005, from www.nervenet.org/papers/NUMBER_REV_1988.html

Williams, R. W., Ryder, K., & Rakic, P. (1987). Emergence of cytoarchitectonic differences between areas 17 and 18 in the developing rhesus monkey. *Abstracts of the Society for Neuroscience, 13,* 1044.

Williams, T. J., Pepitone, M. E., Christensen, S. E., Cooke, B. M., Huberman, A. D., Breedlove, N. J., et al. (2000). Finger-length ratios and sexual orientation. *Nature, 404,* 455–456.

Willie, J. T., Chemelli, R. M., Sinton, C. M., & Yanagisawa, M. (2001). To eat or to sleep? Orexin in the regulation of feeding and wakefulness. *Annual Review of Neuroscience, 24,* 429–458.

Wilska, A. (1935). Methode zur Bestimmung der Horschwellenamplituden der Tromenfells bei verschededen Frequenzen. *Skandinavisches Archiv für Physiologie, 72,* 161–165.

Wilson, A. (1998, September 4). Gray matters memory: How much can we remember? And why is it necessary to forget? *Orange County Register,* p. E1.

Wilson, F. A. W., Ó Scalaidhe, S. P., & Goldman-Rakic, P. S. (1993). Dissociation of object and spatial processing domains in primate prefrontal cortex. *Science, 260,* 1955–1958.

Wilson, M. A., & McNaughton, B. L. (1993). Dynamics of the hippocampal ensemble code for space. *Science, 261,* 1055–1058.

Wilson, R. I., & Nicoll, R. A. (2001). Endogenous cannabinoids mediate retrograde signalling at hippocampal synapses. *Nature, 410,* 588–592.

Wimo, A., Winblad, B., & Jönsson, L. (2010). The worldwide societal costs of dementia: Estimates for 2009. *Alzheimer's & Dementia, 6,* 98–103.

Wise, R. A. (2002). Brain reward circuitry: Insights from unsensed incentives. *Neuron, 36,* 229–240.

Wise, R. A. (2004). Dopamine, learning, and motivation. *Nature Reviews: Neuroscience, 5,* 1–12.

Wise, R. A., & Rompre, P.-P. (1989). Brain dopamine and reward. *Annual Review of Psychology, 40,* 191–225.

Witelson, S. F., Glezer, I. I., & Kigar, D. L. (1995). Women have greater density of neurons in posterior temporal cortex. *Journal of Neuroscience, 15,* 3418–3428.

Witelson, S. F., Kigar, D. L., & Harvey, T. (1999). The exceptional brain of Albert Einstein. *Lancet, 353,* 2149–2153.

Witelson, S. F., & Pallie, W. (1973). Left hemisphere specialization for languuage in the newborn: Neuroanatomical evidence of asymmetry. *Brain, 96,* 641–646.

Witte, A. V., Fobker, M., Gellner, R., Knecht, S., & Flöel, A. (2009). Caloric restriction improves memory in elderly humans. *Proceedings of the National Academy of Sciences, 106,* 1255–1260.

Wollberg, Z., & Newman, J. D. (1972). Auditory cortex of squirrel monkey: Response patterns of single cells to species-specific vocalizations. *Science, 175,* 212–214.

Woltjen, K., Michael, I. P., Mohseni, P., Desai, R., Mileikovsky, M., Hämäläinen, R., et al. (2009). *piggyBac* transposition reprograms fibroblasts to induced pluripotent stem cells. *Nature, 458,* 766–771.

Wood, D. L., Sheps, S. G., Elveback, L. R., & Schirger, A. (1984). Cold pressor test as a predictor of hypertension. *Hypertension, 6,* 301–306.

Woods, B. T. (1998). Is schizophrenia a progressive neurodevelopmental disorder? Toward a unitary pathogenetic mechanism. *American Journal of Psychiatry, 155,* 1661–1670.

Woods, C. G. (2004). Crossing the midline. *Science, 304,* 1455–1456.

Woods, S. C. (2004). Gastrointestinal satiety signals: I. An overview of gastrointestinal signals that influence food intake. *American Journal of Physiology, 286,* G7–G13.

Woodworth, R. S. (1941). *Heredity and environment: A critical survey of recently published material on twins and foster children* (A report prepared for the Committee on Social Adjustment). New York: Social Science Research Council.

Woolf, C. J., & Salter, M. W. (2000). Neuronal plasticity: Increasing the gain in pain. *Science, 288,* 1765–1768.

World Health Organization. (2003). *Controlling the global obesity epidemic.* Retrieved November 29, 2007, from www.who.int/nutrition/topics/obesity/en/

World Health Organization. (2008). *The global burden of disease: 2004 update; Annex A.* Geneva, Switzerland: Author. Retrieved July 2, 2010, from www.who.int/healthinfo/global_burden_disease/GBD_report_2004update_AnnexA.pdf

World Health Organization. (2009). *Working document for developing a draft global strategy to reduce harmful use of alcohol.* Retrieved December 13, 2009, from www.who.int/substance_abuse/activities/msbwden.pdf

World Health Organization. (2010). Revision of the International Classification of Diseases (ICD). Retrieved August 2, 2010, from www.who.int/classifications/icd/ICDRevision/en/index.html

Worley, P. F., Heller, W. A., Snyder, S. H., & Baraban, J. M. (1988). Lithium blocks a phosphoinositide-mediated cholinergic response in hippocampal slices. *Science, 239,* 1428–1429.

Wronski, M. (1998). Plagiarism in publications by Dr. Andrzej Jendryczko. *Przegl Lek, 55,* 629–633.

Wu, J. C., & Bunney, W. E. (1990). The biological basis of an antidepressant response to sleep deprivation and relapse: Review and hypothesis. *American Journal of Psychiatry, 147,* 14–21.

Wurtman, J. J., Wurtman, R. J., Reynolds, S., Tsay, R., & Chew, B. (1987). Fenfluramine suppresses snack intake among carbohydrate cravers but not among noncarbohydrate cravers. *International Journal of Eating Disorders, 6,* 687–699.

Wynn, K. (1992). Addition and subtraction by human infants. *Nature, 358,* 749–750.

Wynn, K. (1998). Psychological foundations of number: Numerical competence in human infants. *Trends in Cognitive Sciences, 2,* 296–303.

Xiao, Z., & Suga, N. (2002). Modulation of cochlear hair cells by the auditory cortex in the mustached bat. *Nature Neuroscience, 5,* 57–63.

Xu, Y., Padiath, Q. S., Shapiro, R. E., Jones, C. R., Wu, S. C., Saigoh, N., et al. (2005). Functional consequences of a *CKI*d mutation causing familial advanced sleep phase syndrome. *Nature, 434,* 640–644.

Yam, P. (1998). Intelligence considered. *Scientific American Presents, Winter,* 6–11.

Yamada et al. (2001). Mice lacking the M3 muscarinic acetylcholine receptor are hypophagic and lean. *Nature, 410,* 207–212.

Yamada, M., Yamada, M., & Higuchi, T. (2005). Remodeling of neuronal circuits as a new hypothesis for drug efficacy. *Progress in Neuro-Psychopharmacology and Biological Psychiatry, 29,* 999–1009.

Yang, Y., Raine, A., Narr, K. L., Colletti, P., & Toga, A. W. (2009). Localization of deformations within the amygdala in individuals with psychopathy. *Archives of General Psychiatry, 66,* 986–994.

Yang, Z., & Schank, J. C. (2006). Women do not synchronize their menstrual cycles. *Human Nature, 17,* 433–447.

Yehuda, R. (2001). Are glucocorticoids responsible for putative hippocampal damage in PTSD? How and when to decide. *Hippocampus, 11,* 85–89.

Yeni-Komshian, G. H., & Benson, D. A. (1976). Anatomical study of cerebral asymmetry in the temporal lobe of humans, chimpanzees, and rhesus monkeys. *Science, 192,* 387–389.

You, J. S., Hu, S. Y., Chen, B., & Zhang, H. G. (2005). Serotonin transporter and tryptophan hydroxylase gene polymorphisms in Chinese patients with generalized anxiety disorder. *Psychiatric Genetics, 15,* 7–11.

Youdim, M. B. H., & Riederer, P. (1997, January). Understanding Parkinson's disease. *Scientific American, 276,* 52–58.

Young, L. J., & Wang, Z. (2004). The neurobiology of pair bonding. *Nature Neuroscience, 10,* 1048–1054.

Young, M. P., & Yamane, S. (1992). Sparse population coding of faces in the inferotemporal cortex. *Science, 256,* 1327–1331.

Young, R. L., Ridding, M. C., & Morrell, T. L. (2004). Switching skills on by turning off part of the brain. *Neurocase, 10,* 215–222.

Yücel, M., Solowij, N., Respondek, C., Whittle, S., Fornito, A., Pantelis, C., et al. (2008). Regional brain abnormalities associated with long-term heavy cannabis use. *Archives of General Psychiatry, 65,* 694–701.

Zack, M., & Poulos, C. X. (2004). Amphetamine primes motivation to gamble and gambling-related semantic networks in problem gamblers. *Neuropsychopharmacology, 29,* 195–207.

Zai, G., Bezchlibnyk, Y. B., Richter, M. A., Arnold, P., Burroughs, E., Barr, C. L., et al. (2004). Myelin oligodendrocyte glycoprotein (MOG) gene is associated with obsessive-compulsive disorder. *American Journal of Medical Genetics Part B, 129B,* 64–68.

Zarate, C. A., Singh, J. B., Carlson, P. J., Brutsche, N. E., Ameli, R., Luckenbaugh, D. A., et al. (2006). A randomized trial of an *N*-methyl-D-aspartate antagonist in treatment-resistant major depression. *Archives of General Psychiatry, 63,* 856–864.

Zarate, C.A., Singh, J. B., Carlson, P. J., Quiroz, J., Jolkovsky, L., Luckenbaugh, D. A., et al. (2007). Efficacy of a protein kinase C inhibitor (tamoxifen) in the treatment of acute mania: A pilot study. *Bipolar Disorder, 9,* 561–570.

Zeki, S. (1983). Colour coding in the cerebral cortex: The reaction of cells in monkey visual cortex to wavelengths and colours. *Journal of Neuroscience, 9,* 741–765.

Zeki, S. (1992, September). The visual image in mind and brain. *Scientific American, 267,* 69–76.

Zeman, A. (2001). Consciousness. *Brain, 124,* 1263–1289.

Zepelin, H., & Rechtschaffen, A. (1974). Mammalian sleep, longevity, and energy metabolism. *Brain, Behavior and Evolution, 10,* 425–470.

Zhang, J. (2003). Evolution of the human ASPM gene, a major

Zhang, J., & Webb, D. M. (2003). Evolutionary deterioration of the vomeronasal pheromone transduction pathway in catarrhine primates. *Proceedings of the National Academy of Sciences, USA, 100,* 8337–8341.

Zhang, J., & Webb, D. M. (2003). Evolutionary deterioration of the vomeronasal pheromone transduction pathway in catarrhine primates. *Proceedings of the National Academy of Sciences, 100,* 8337–8341.

Zhang, S. P., Bandler, R., & Carrive, P. (1990). Flight and immobility evoked by excitatory amino acid microinjection within distinct parts of the subtentorial midbrain periaqueductal gray of the cat. *Brain Research, 520,* 73–82.

Zhang, X., Gainetdinov, R. R., Beaulieu, J.-M., Sotnikova, T. D., Burch, L. H., Williams, R. B., et al. (2005). Loss-of-function mutation in tryptophan hydroxylase-2 identified in unipolar major depression. *Neuron, 45,* 11–16.

Zhang, Y., Ivanova, E., Bi, A., & Pan, Z.-H. (2009). Ectopic expression of multiple microbial rhodopsins restores ON and OFF light responses in retinas with photoreceptor degeneration. *Journal of Neuroscience, 29,* 9186–9196.

Zhang, Y., Proenca, R., Mafei, M., Barone, M., Leopold, L., & Friedman, J. M. (1994). Positional cloning of the mouse *obese* gene and its human homologue. *Nature, 335,* 311–317.

Zheng, J., Shen, W., He, D. Z. Z., Long, K. B., Madison, L. D., & Dallos, P. (2000). Prestin is the motor protein of cochlear outer hair cells. *Nature, 405,* 149–155.

Zhou, J. N., Hofman, M. A., Gooren, L. J., Swaab, D. F. (1995). A sex difference in the human brain and its relation to transsexuality. *Nature, 378,* 68–70.

Zhu, Q., Song, Y., Hu, S., Li, X., Tian, M., Zhen, Z., et al. (2010). Hereditability of the specific cognitive ability of face perception. *Current Biology, 20,* 137–142.

Zihl, J., von Cramon, D., & Mai, N. (1983). Selective disturbance of movement vision after bilateral brain damage. *Brain, 106,* 313–340.

Zillmer, E. A., & Spiers, M. V. (2001). *Principles of neuropsychology.* Belmont, CA: Wadsworth.

Zola-Morgan, S., Squire, L. R., & Amaral, D. G. (1986). Human amnesia and the medial temporal region: Enduring memory impairment following a bilateral lesion limited to field CA1 of the hippocampus. *Neuroscience, 6,* 2950–2967.

Zou, K., Deng, W., Li, T., Zhang, B., Jiang, L., Huang, C., et al. (2010). Changes of brain morphometry in first-episode, drug naêve, non-late-life adult patients with major depression: An optimized voxel-based morphometry study. *Biological Psychiatry, 67,* 186–188.

Zubenko, G. S., Hughes, H. B., Stiffler, J. S., Zubenko, W. N., & Kaplan, B. B. (2002). Genome survey for susceptibility loci for recurrent, early-onset major depression: Results at 10cM resolution. *American Journal of Medical Genetics, 114,* 413–422.

Zubenko, G. S., Maher, B. S., Hughes, H. B., III, Zubenko, W. N., Stiffler, J. S., & Marazita, M. L. (2004). Genome-wide linkage survey for genetic loci that affect the risk of suicide attempts in families with recurrent, early-onset, major depression. *American Journal of Medical Genetics B, 129,* 47–54.

Zubin, J., & Spring, B. (1977). Vulnerability—A new view of schizophrenia. *Journal of Abnormal Psychology, 86,* 103–126.

Zuckerman, M. (1971). Dimensions of sensation seeking. *Journal of Consulting and Clinical Psychsology, 36,* 45–52.

Chapter-Opening Photo Credits

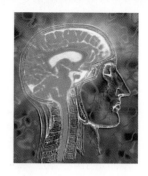

Author Index

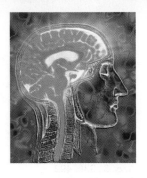

aan het Rot, M., 446
Abbott, D. F., 316
Abbott, R. D., 352
Abbott, W., 176
Abdal-Jabbar, M. S., 355
Abe, K., 470
Abede, D., 410
Abel, E. L., 130
Abel, L., 400
Abeloff, M. D., 238
Abelson, J. F., 460
Abi-Dargham, A., 434, 436
Abraham, R., 383
Abrahams, B. S., 285, 418
Abramov, I., 305
Abramowitz, J. S., 457, 458, 460
Abrashkin, K. A., 267
Abumrad, N. N., 139
Aburdarham, N., 237
Accornero, N., 341
Ackerl, K., 197
Acosta, M. T., 421
Acton, P. D., 132
Ad-Dab'bagh, Y., 398
Adam, C., 487
Adamiak, B., 384
Adams, D. B., 193
Adams, L. F., 409
Adar, R., 10
Addis, L., 322
Adelman, T. L., 98
Ad'es, J., 182
Adler, L. E., 439
Adolphs, R., 232, 233, 371
Agarwal, D. P., 148
Aghajanian, G., 144
Aglioti, S., 492
Agnew, B., 112
Agosta, F., 355
Agrawal, A., 148
Agwu, C., 179
Agyei, Y., 12, 211
Aharon, I., 140
Ahern, F., 404
Ahituv, N., 13
Ahlskog, M. E., 140
Ahmed, S., 409
Ahola, A., 177
Ainsworth, C., 180
Aissaoui, H., 480
Aizawa, R., 484
Akbarian, S., 441
Akiyama, J. A., 13
al-Daeef, A. Q., 355

al-Tahan, A. R., 355
Alain, C., 261
Alam, D., 448
Alam, N., 478
Alati, R., 130
Albert, D. J., 241, 242, 243
Albert, M. S., 367, 368, 380, 381
Albertini, R. J., 355
Albrecht, D. G., 312
Albright, T. D., 317
Alexander, A. L., 106
Alexander, C. N., 237
Alexander, G. E., 348
Alexander, J. T., 179
Alexander, M. P., 273
Alexander, N., 244
Alkire, M. T., 367, 371, 398
Alkon, D. L., 385
Allegretta, M., 355
Allen, G., 416
Allen, L. S., 213
Allen, N. C., 432, 433
Alliger, R., 60
Allison, D. B., 175
Allison, T., 468
Almasy, L., 149
Almqvist, E., 354
Alpert, N. M., 367, 368
Alschuler, E. L., 343
Altar, C. A., 449
Altena, E., 482
Altshuler, L. L., 451
Alvarez, J. -C., 453
Alvarez, P., 366
Alvir, J., 442
Alvord, E. C., 77
Amanzio, M., 339
Amaral, D. G., 365, 366, 387
Amarasingham, A., 46
Ambrose, J., 416
Ameli, R., 446
Ames, M. A., 209, 216
Amico, E., 436
Amin, Z., 444
Amory, J. K., 409
Andari, E., 417
Andermann, F., 232
Andersen, R. A., 103, 336
Andersen, A. E., 181
Anderson, A. K., 224
Anderson, A. W., 317, 318, 489
Anderson, B., 397, 448
Anderson, C., 211

Anderson, C. H., 322
Anderson, D. C., 113
Anderson, D. J., 262
Anderson, G., 495
Anderson, G. M., 148
Anderson, M. A., 13
Anderson, N. D., 408
Anderson, R. H., 194
Anderson, S. W., 224
Anderson, W. F., 111
Ando, J., 403
Andolfatto, P., 13
Andrade, J., 485
Andreano, J. M., 203
Andreasen, N. C., 60, 203, 430, 433, 434
Andrew, M., 445
Andrews, P. J., 481
Andrews, P. W., 267
Andrews-Hanna, J. R., 408
Angelergues, R., 273, 274
Angelini, G., 498
Anglin, M. D., 127
Angrist, B., 131
Ankney, C. D., 398
Anney, R., 419
Anthony, A., 418
Anthony, J. C., 148
Antonini, A., 353
Anwyl, R., 374
Aperia, A., 436
Apkarian, A. V., 341, 342
Apter, S., 451
Arai, Y., 206
Aramant, R. B., 300
Arancio, O., 374
Arand, D. L., 476
Arango, V., 447, 452
Aranow, H., Jr., 354
Arbter, D., 435
Archer, J., 203, 242
Archer, S. N., 482
Arcos-Burgos, M., 421
Ardekani, B. A., 457
Arendt, J., 450, 471, 482
Argiolas, A., 194
Argyropoulos, G., 175
Arisaka, O., 206
Aristotle, 4
Arman, A. C., 322
Armony, J. L., 284
Armstrong, S. M., 409
Arndt, S., 433, 434
Arnold, D. S., 439
Arnold, M., 487, 488

Arnold, P. D., 457
Arnold, P. E., 460, 461
Arnoled, D., 320
Arnott, S. R., 261
Aron, A. P., 226
Aronowitz, B. R., 417
Arzy, S., 336, 492
Åsberg, M., 452
Ascherio, A., 355
Aschoff, J., 470
Ashburn, E., 449
Ashburner, J., 375
Asher, J. E., 322
Ashley, J. A., 144
Ashmore, J. F., 259
Ashour, M. H., 355
Ashtari, M., 442, 457
Askenasy, J. J. M., 476
Asplund, C. L., 489
Assaf, M., 416
Assouly-Besse, F., 433
Aston-Jones, G., 134
Atkinson, C., 444
Atkinson, R. M., 437
Atladóttir, H. O., 417
Attia, E., 181
Atzmeuller, M., 197
Augath, M., 266
Auman, J. I., 281
Austin, D., 481
Auta, J., 441
Avedissian, C., 398
Avidan, G., 317
Avidor, Y., 180
Avikainen, S., 414
Avraham, R., 237
Awatramani, G. B., 300
Ax, A., 113, 226
Axel, R., 196
Aziz-Zadeh, L., 414, 493
Azzouz, M., 353

Baaré, W. F. C., 398, 400, 403
Baars, B. J., 497
Babcock, R. L., 408
Bach-y-Rita, P., 82
Bachen, E. A., 234
Backenbach, J., 485
Backes, W. H., 408
Backonja, M. -M., 226
Bader, C. R., 259
Badley, D. C., 320
Badwejn, J., 183
Baek, D. Y., 460
Bagade, S., 432, 433
Bagni, c., 411
Bai, J., 276
Baier, B., 492
Bailey, A., 418
Bailey, C. H., 14, 374
Bailey, J. M., 12, 209, 211
Baizer, J. S., 321
Bakay, R., 386
Baker, C. I., 107
Baker, K., 499
Bakker, J., 201
Baldessarini, R. J., 450

Bale, T. L., 177
Baliki, M. N., 341
Balkin, T. J., 497
Balteau, E., 490
Balthazart, J., 201
Balya, D., 300
Balzac, H. de, 126, 134
Bancaud, J., 348
Bandler, R., 243
Bandmann, O., 352
Bannerjee, S. P., 434
Bannon, K. L., 439
Bansal, R., 451
Bar, M., 313
Baraban, J. M., 451
Barber, A. M., 183
Barch, D. H., 418
Barinaga, M., 54, 106
Barker, W. B., 405
Barkley, R. A., 419
Barlow, D. H., 455, 461
Barnea-Goraly, N., 416, 417
Barnes, C. A., 369, 376, 380
Barnes, C. L., 382
Barnes, J. C., 245
Barnhart, K. T., 197
Barr, C. L., 461
Barr, H. M., 130
Barres, B. A., 35
Bartel, P., 476
Bartenstein, P. A., 385
Barthel, H., 385
Bartholomae, C. C., 111
Bartlett, D. L., 457
Bartolozzi, M. L., 355
Bartsch, D., 14
Bartz, J., 417
Barysheva, M., 398
Basbaum, A. I., 331, 337, 339
Bassetti, C., 483
Bassingthwaighte, L., 83
Bates, G. P., 353
Bates, J. F., 347
Bath, K. G., 456
Batista, A. P., 336
Batista, G., 453
Battaglia, F., 448, 449
Batterham, R. L., 170, 179
Bauer, L. O., 149
Bauer, R. M., 317
Baulac, S., 38
Baum, A., 236, 237
Baumann, B. L., 421
Baumgartner, T., 417
Bautista, D. M., 331
Bavelier, D., 280
Baxter, L. R., 451, 452, 457
Bayer, A., 384
Bayliss, C., 83
Bayliss, D. A., 43
Bear, M. F., 372
Bear, R. E., 340
Beaulineu, J. -M., 444
Beaver, K. M., 245
Becerra, L., 342
Bechara, A., 60, 61, 224, 229, 230, 233, 371, 380

Beck, A. T., 163, 164
Becker, A. E., 182
Becker, D., 82, 84
Becker, H. B., 192
Beecher, D. K., 239
Beecher, M. D., 284
Beedle, D., 448
Begg, M. D., 440
Begleiter, H., 149
Begley, S., 457
Begré, S., 438
Behr, R., 267
Behrmann, M., 317, 416
Beilstein, M., 352
Bejarano, G., 13
Békésy, G. von, 262, 265
Bekkering, H., 284
Belhar, M. C., 450
Bell, A. P., 210, 211
Bellgowan, P. S. F., 107
Bellinger, D. C., 421
Bellis, D. J., 144
Belliveau, J. W., 316
Bellugi, U., 280
Beltramo, M., 135
Ben-Dor, U., 10
Ben-Eliyahu, S., 237
Ben-Shabat, S., 135
Benarroch, E. E., 484
Benca, R. M., 481
Bender, D. B., 317
Bendriem, B., 131
Benedetti, F., 339
Benenson, Y., 10
Benetti, F., 117
Bengtsson, C., 172
Benish, M., 237
Benishay, D. S., 211
Benkelfat, C., 458
Bennett, D. A., 383
Bennett, K. M., 349
Bennett, M. V. L., 37
Bennicelli, J. L., 300
Benoit, E., 32
Benowitz, L. I., 83
Bensing, J. M., 341
Benson, D. A., 280, 283
Benson, D. F., 387
Benson, M. D., 410
Benton, A. L., 317
Beranger, A., 38
Berardelli, A., 341
Berdan, R., 99
Berenbaum, S. A., 206, 213
Berg, S., 404
Berger, H., 99
Berglund, H., 197, 214, 215
Berglund, P., 428
Berlin, I., 453
Berlucchi, G., 492
Berman, E. R., 176
Berman, K. F., 39, 437, 438
Berman, S., 203
Bernaards, C., 193
Bernard, L. L., 157
Bernardi, F., 241

Bernat, E., 149
Bernhardt, P. C., 242
Berns, G. S., 141
Bernstein, D. M., 379
Bernstein, I. L., 181
Bernstein, L. J., 486
Bernstein, L. L., 163
Berrettini, W., 181
Berridge, K., 139
Berridge, V., 127
Berry, J. M., 380
Berson, D. M., 472
Bertelsen, A., 431, 432
Berthoud, H. -R., 171, 180
Berti, A., 497
Bertilsson, L., 452
Berton, F., 129
Bertram, L., 383
Bertrand, D., 259
Besch, N. F., 241
Bessa, J. M., 448
Bestmann, S., 102
Betarbet, R., 352
Bettens, F., 196
Bever, T. G., 282
Beveridge, T. J. R., 132
Beylin, A., 81
Bezchlibnyk, Y. B., 461
Bhaumik, D. K., 453
Bherer, L., 408
Bi, A., 300
Bidaut-Russell, M., 136
Biddle, N. A., 457
Biederman, J., 419, 420, 421
Bienias, J. L., 383, 408
Bierut, L. J., 147
Bigelow, L. B., 39, 440
Biggs, M. M., 449
Bigler, E. D., 57, 400
Bihrle, S., 243, 245
Bilanow, T., 383
Bilder, R. M., 442, 457
Bilguvar, K., 460
Bilker, W., 398
Billes, S. K., 179
Billiard, M., 483
Billings, R., 483
Billington, C. J., 169
Billy, J. O. G., 210
Binder, E. B., 456
Binder, J. R., 266
Binet, S., 394
Bini, G., 341
Birbaumer, N., 79, 341, 342, 343
Biron, S., 177
Bissada, H., 183
Bissell, S., 410
Bittar, R. G., 340
Bixler, E. O., 481
Björkelund, C., 172
Bjørkum, A. A., 478
Bjornson, L., 495
Black, K. L., 117
Black, N., 483
Blacker, D., 383
Blackman, M. R., 409
Blaese, R. M., 92

Blakeslee, S., 414, 492
Blanchard, D. C., 232
Blanchard, R. J., 213, 232
Blanco, C., 140
Blanke, O., 336, 492
Blauw, G. J., 177
Blennow, K., 385
Bleuler, E., 429
Bliss, E. L., 495
Bliss, T. V. P., 372, 373
Bloch, D. A., 194
Bloch, G. J., 194
Bloch, J., 83
Bloch, V., 476, 477
Blonder, L. X., 233
Bloom, A., 141
Bloom, F. E., 128, 321
Bloomingdale, K., 496
Blum, D., 113
Blumenthal, J., 442
Bobe, L., 280
Boche, D., 384
Bocklandt, S., 211, 212
Bodo, C., 201, 213
Boecker, H., 350
Boehrer, A., 243
Boeve, B. F., 484
Bogaert, A. F., 210, 213
Bogardus, C., 176
Bogdasarian, R. S., 82
Bogerts, B., 436, 442
Bohlin, G., 421
Bohman, M., 146
Boivin, D. B., 471
Bolger, F., 317
Bolla, K. I., 136
Bolles, R. C., 157
Bolt, I., 406
Bolton, N. M., 341
Bolton, P., 418
Bolvin, J., 109, 133, 134
Bolwig, T. G., 447
Boly, M., 499
Bomar, J. M., 285
Bonda, E., 336
Bonebakker, A. E., 485
Bonetto, M., 499
Bonhoeffer, T., 374
Bonke, B., 485
Bonne, O., 455
Bonner, J., 83
Bonnet, M. H., 476, 481
Bontempi, B., 368
Bonvento, G., 353
Boodoo, G., 395, 405
Bookstein, F. L., 130
Boomsma, D. I., 398, 400, 403
Boon, P., 460
Boone, K., 414, 415, 441
Bor, D., 397
Bor, W., 130
Borckardt, J. J., 448
Borge, G. F., 451
Born, J., 477
Born, R. T., 320
Borsook, D., 342
Bortz, R., 476

Bossy-Wetzel, E., 384
Bosveld, J., 485
Botanov, Y., 197
Bottini, G., 279
Bottini, M., 497
Botvinick, M., 349
Bouchard, C., 175, 176
Bouchard, M. F., 421
Bouchard, T. J., Jr., 12, 15, 109, 203, 395,
    403, 405
Bouhassira, D., 480
Boulant, J. A., 157
Boulenguez, P., 346
Boulet, S., 353
Bourgeois, J. -P., 76
Bouvier, S. E., 317
Boveroux, P., 498
Bower, J. H., 140
Bower, K. S., 495
Bowers, D., 233
Bowler, P. J., 226
Bowles, C. A., 236
Bowmaker, J. K., 304
Boxer, A., 385
Boycott, B. B., 297
Boyden, E. S., 32
Boykin, A. W., 395, 405
Boyton, G. M., 322
Bozarth, M. A., 132, 138, 139
Braak, E., 484
Braak, H., 484
Bracha, H. S., 440
Bradbury, J., 203
Bradbury, T. N., 440
Bradley, R. G., 456
Bradley, S. J., 209
Brakke, K. E., 283
Brammer, M., 322
Branchey, M. H., 244
Brand, T., 213
Brandes, D., 455
Brandt, G., 179
Brandt, J., 353
Brang, D., 336
Brann, D. W., 241
Branner, A., 84
Brannon, E. M., 401
Brans, R. G. H., 203
Brass, M., 284
Braun, A. R., 283, 497
Braun, B. G., 495
Braun, C., 341, 487, 488
Braunstein, G. D., 192, 193
Braunwald, E., 180
Bray, C., 261
Bray, G. A., 180
Breakefield, X. O., 245
Breasted, J. G., 6
Brechbühl, J., 197
Brecher, E. M., 131, 133, 142, 143
Breedlove, N. J., 212, 215
Breedlove, S. M., 206
Breeze, R. E., 3, 352
Breier, J. I., 281
Breiter, H. C., 140
Breitner, J. C., 14
Bremer, J., 192

Bremner, J. D., 236, 496
Bremner, K. E., 245
Bremner, W., 409
Brennan, P. A., 133
Bresnahan, M., 440
Breton, F., 317
Breuer, A., 135
Brewer, J. B., 367
Brewerton, T. D., 181
Brichard, L., 148
Bridge, H., 367
Briedmann, E., 197
Brink, J., 243
Brinkman, C., 348
Brinkman, R. R., 354
Brisbare-Roch, C., 480
Brismar, H., 436
Britschgi, M., 385
Britten, K. H., 336
Britten, R. J., 14, 282
Broadfield, D. C., 283
Broberg, D. J., 181
Broberger, C., 166, 168, 169
Broca, P., 7, 270
Brocklebank, D., 322
Brockman, J. A., 449
Brodie, H. K. H., 192
Brodows, R., 179
Brody, J., 173
Brody, N., 395, 405
Broillet, M. -C., 197
Bron, C. R., 374, 375
Bronen, R. A., 236, 496
Brooks, D. J., 355, 491
Brooks-Gunn, J., 490
Brotis, A., 243
Brotman, D. J., 471
Broughton, R., 469, 483
Brown, A. L., 410
Brown, A. S., 440
Brown, C. H., 453
Brown, E., 139
Brown, E. N., 471
Brown, J., 382, 410
Brown, K., 136, 304
Brown, M. A., 47
Brown, R. P., 451
Brown, T., 436, 438
Brownell, W. E., 259
Broyd, S. J., 408
Bruce, C., 317
Bruckbauer, T., 490
Brugge, J. F., 262
Brunner, H. G., 245
Bruno, M. -A., 498
Brunswick, N., 277
Brutsche, N. E., 446
Bruvold, W., 178, 179
Bryant, R. A., 455
Bryden, M. P., 206, 207
Bryk, K., 206
Buccino, G., 493
Buchanan, T. W., 279
Büchel, C., 140
Bucholz, K. K., 147
Buchs, P. -A., 374, 375
Buchsbaum, M. S., 400

Buchwald, H., 180
Buck, L. B., 197
Buck, R., 233
Buckner, R. L., 273, 367, 369
Buckwalter, J. A., 416
Buell, S. J., 78, 380
Buffenstein, A., 450
Bullard, J., 449
Bullmore, E., 438
Bundlie, S. R., 484, 485
Buneo, C. A., 336
Bunney, W. E., Jr., 441, 443, 444, 449, 451
Bünning, E., 471
Buñuel, L., 492
Burch, L. H., 444
Burchinal, M., 406
Burdick, J. W., 103
Burgut, T., 384
Burk, C., 245
Burke, R. D., 430
Burke, S. N., 369
Burn, P., 178
Burns, C., 448
Burroughs, E., 461
Burt, A. D., 193
Burton, M. P., 322
Burwell, R. A., 182
Buschbaum, M. S., 243, 245
Buschman, T. J., 489
Bushman, B. J., 244
Bushnell, M. C., 240, 342
Busskamp, V., 300
Buster, J. E., 192, 193
Butcher, L. M., 410
Butelman, E. R., 148
Butler, F. K., 455
Butler, P. C., 194
Butters, N., 387
Butterworth, B., 401
Butterworth, N. J., 82
Buxhoeveden, D. P., 56, 57
Buxton, O. M., 473
Buydens-Branchey, L., 244
Buysse, D. J., 449
Buzsáki, G., 46
Byne, W., 213

Cabeza, R., 408
Cadet, J. L., 136
Cadoret, R. J., 148
Caggiula, A. R., 139
Cahalan, M. D., 32
Cahill, C., 439
Cahill, L., 203, 367, 371, 378
Cahn, W., 203
Cai, G., 419
Calafat, A. M., 421
Calandra, D., 495
Calder, A. J., 232
Calhoun, V. D., 416
Cali, J., 402
Calles-Escandón, J., 178
Calne, D. B., 352
Calo, G., 337
Caltagirone, C., 233
Camardo, J. S., Jr., 160
Cameron, P. A., 274

Cameron, R. A., 282
Cameroni, I., 410
Camp, N. J., 444
Campbell, F. A., 406
Campbell, K. B., 399
Campbell, S. S., 469
Camperio Ciani, A., 212
Campfield, L. A., 178
Campione, J. C., 410
Campiuon, D., 383
Campo, B., 472
Canacora, G., 447
Canavan, A. G. M., 351
Canfield, R. L., 421
Canli, T., 444
Cannon, M., 439
Cannon, T. D., 398, 403
Cannon, w., 236
Cannon-Albright, L. A., 444
Cantalupo, C., 283
Cao, Y., 278
Cao, Y. Q., 337
Capdevila, A., 229
Capiluppi, c., 212
Caplan, A. H., 84
Cappa, S. F., 277
Capucho, C., 873
Carbon, C., 317
Cardenas, P., 54
Cardno, A., 445
Cardoner, N., 229
Carelli, R. M., 139
Carey, G., 245
Carey, T. S., 140
Cariani, P. A., 45
Carlezon, W. A., Jr., 135, 139, 140
Carlson, A. S., 207
Carlson, E. J., 337
Carlson, M., 335
Carlson, P. J., 446, 451
Carlson, S. M., 472
Carlson, S. R., 149
Carlsson, A., 114, 341
Carlsson, H. -E., 115
Carlström, E., 211
Carmichael, S. T., 452
Carnes, B., 173
Carpenter, C. J., 450
Carr, C. E., 268, 269
Carr, T. S., 241
Carr, W. S., 497
Carraher, D. W., 395
Carraher, T. N., 395
Carran, D. T., 410
Carrera, K., 81
Carrera, M. R., 144
Carrive, P., 243
Carroll, J. B., 396
Carroll, M. D., 172
Carskadon, M. A., 469
Carter, C. S., 92
Carter, H., 107
Carter-Saltzman, L., 404
Cartier, N., 111
Cartwright, R., 483
Cartwright, R. D., 476
Carvalhais, A. B., 468

Carvalho, A. P., 43, 44
Casali, A., 498
Casanova, M. F., 56, 57, 398, 436, 437
Cascella, N. G., 131, 132
Caspi, A., 245, 444, 445
Cassano, G. B., 447
Casson, D. M., 418
Casson, P. R., 192, 193
Cassone, V. M., 470
Castagnoli, N., Jr., 446
Castellanos, F. X., 420, 421
Castelli, F., 416
Castelli, L., 317
Castellote, J. M., 277
Castillo, E. M., 281
Castrén, E., 448
Castrén, M., 448
Castro-Caldas, A., 82, 278
Castrop, F., 226
Cattaneo, A., 337
Catterall, W. A., 32
Cauda, F., 317
Cavanagh, K., 421
Cavanaugh, J., 133
Caviness, V. S., Jr., 203
Cawthra, E. M., 439
Cazier, J. B., 322
Ceballos-Baumann, A. O., 226, 350
Ceci, S. J., 395, 405
Cecil, J. E., 169
Celnik, P., 78
Cermelli, P., 212
Cerqueira, J. J., 448
Chabris, C. F., 416
Chafee, M. V., 347
Chagnon, Y. C., 175
Chait, B. T., 175
Chakrabarti, S., 413
Chakraborty, S., 316
Chamberlain, S. R., 457
Chan, B. L., 343
Chan, J., 407
Chan, P., 352
Chan, T., 245
Chandler, H. M., 237
Chang, H. S., 371
Chang, L., 203, 382, 387
Chang, L -W., 496
Chang, P. K., 170
Chanoine, V., 277
Chaplin, W., 417
Chapman, P. F., 383
Charbel, F., 281
Charney, D., 453, 460
Charney, D. S., 446, 455
Charrow, A. P., 343
Chartier-Harlin, M. C., 382, 410
Chase, K. O., 353
Chau-Wong, M., 434, 435
Chawla, D., 489
Chayka, B., 418
Chazal, G., 81, 102
Cheatham, M. A., 259
Chebli, R., 450
Check, E., 116, 117
Checkley, S. A., 198
Cheetham, T., 175

Chemelli, R. M., 166, 484
Chen, B., 461
Chen, C., 76
Chen, D., 84
Chen, D. F., 81
Chen, G., 444
Chen, J., 480
Chen, J. -F., 352
Chen, M., 339
Chen, M. C., 451
Chen, P., 83
Chen, Q., 352
Chen, W., 267
Chen, W. R., 372, 373
Chen, Z. -F., 330
Cheng, C. Y., 461
Cherek, D. R., 244
Cherkassky, V. I., 416
Cherney, N. I., 127
Chernick, A. B., 209
Cherny, S. S., 211
Cherrier, M. M., 409
Cheslow, D., 457
Chew, B., 179
Chialvo, D. R., 341
Chiang, M. -C., 398
Chiba, S., 484
Chilton, M., 439
Chiu, W. T., 428, 443, 454
Chizh, B. A., 342
Cho, A. K., 133
Cho, C. -K, 411
Cho, J. -K., 141
Chomsky, N., 279
Chou, C. L., 300
Chou, I. C., 461
Chou, P., 145
Chou, T. C., 166, 168, 478, 479
Chowdhury, S., 316
Chown, M. J., 381
Christensen, D. D., 414, 415
Christensen, S. E., 212, 215
Christison, G. W., 437
Christmann, C., 343
Christophe, A., 280
Chronis, A. M., 421
Chrousos, G. P., 443, 445, 451
Chuang, R. S. -I., 32
Chueh, D., 400
Chung, J. -K., 267
Cianflone, K., 177
Ciarcia, J. J., 447
Cicchetti, D. V., 468
Cicchetti, F., 354
Cichon, S., 432
Ciompi, L., 453
Cipolotti, L., 387, 492
Cirelli, C., 377, 476
Ciszewski, A., 455
Clare, S., 340
Clark, J., 416
Clark, J. T., 169
Clark, K., 279
Clark, M., 244
Clark, R. E., 486
Clarke, R., 280
Clarren, S. K., 77

Clasen, L., 398
Clayton, J. D., 472
Cleckley, H. M., 495
Cleeter, M. W., 353
Clémenceau, S., 318, 487
Clement, K., 169
Clerici, M., 92
Cline, H., 374
Cloniger, C. R., 147
Coakley, E. H., 172
Coan, J. A., 226
Cobb, S., 343
Coccaro, E. F., 245, 435, 446
Cochrane, J., 83
Coffey, R. M., 428
Cohen, D. J., 460
Cohen, J., 183
Cohen, J. D., 314
Cohen, J. W., 172
Cohen, L., 401, 497
Cohen, L. G., 78
Cohen, M. A., 170, 179
Cohen, M. X., 231
Cohen, N. J., 370
Cohen, P., 179
Cohen, R. M., 420
Cohen, S., 234, 236
Cohen, S. L., 175
Cohen-Bendahan, C. C. C., 213
Cohen-Kettenis, P. T., 203
Colapinto, J., 208
Colas, D., 411
Colby, c. L., 336
Colcombe, S. J., 408
Colditz, G., 172
Cole, S. W., 238, 245
Coleman, D. L., 175, 176
Coleman, M. R., 499
Coleman, P. D., 78, 380
Coleman, R. M., 471, 482
Collaer, M. L., 201, 214
Colle, H., 460
Collette, F., 490
Colletti, P., 243
Collins, D., 244
Collins, K. A., 446
Collman, F., 98
Colom, R., 397
Colt, E. W. D., 339
Compton, J., 409
Compton, W., 145
Comuzzie, A. G., 175
Condie, D., 496
Congdon, E., 444
Conneally, P. M., 13
Conner, J. M., 384, 385
Connolly, K., 499
Connors, J., 444
Considine, R. V., 170
Constable, R. T., 277
Constantinidis, C., 371
Contoreggi, C., 131, 141, 142
Conturo, T. E., 82, 84
Convit, A., 237, 380
Cook, A. S., 202
Cook, N. R., 381
Cook, S., 197

Cooke, B. M., 212, 215
Cooke, D. F., 336
Cooke, S. F., 372, 373
Cooke, S. H., 373
Coon, D., 431, 470
Cooney, N., 148
Cooper, D. C., 148
Cooper, H. M., 244, 472
Cooper, J., 353
Cooper-Blacketer, D., 383
Copp, A., 285
Coppola, G., 285
Coppola, R., 437
Corazzini, L. L., 317
Corbett, D., 139
Corbetta, M., 489
Corcoran, C. M., 436
Corcoran, R., 279
Cordier, B., 192
Corey-Bloom, J., 354
Corina, D., 144, 280
Corkin, S., 364, 365, 370
Corna, F., 212
Coronado, V. G., 80
Corves, C., 435
Corwell, B., 78
Corwin, J., 384
Coscina, D. V., 166
Cosgrove, G. R., 353
Cossu, G., 277
Costa, D. C., 435
Costa, E., 439
Costall, B., 454
Cottrell, B. A., 82
Courchesne, E., 351, 416
Courtney, S. M., 316
Coury, A., 139, 195
Covington, J., 351
Cowan, J. M., 411
Cowley, M. A., 170
Cox, D., 197
Cox, D. J., 39
Cox, J. J., 341
Cox, T. C., 177
Coyle, J. T., 129, 436, 442
Cozzi, A., 436
Craft, S., 409
Craig, I. W., 245, 410, 444, 445
Crane, G. E., 445
Crawford, E., 382
Crawford, F., 410
Crawley, J. N., 144
Cremer, H., 81, 102
Crespo, J. M., 444, 445
Crews, F. T., 140
Cribbs, L., 322
Crick, F., 4, 10, 322, 476, 478, 485, 486, 498
Crisp, T., 480
Critchlow, V., 470
Croen, L. A., 413
Croft, J. R., 132
Crow, T. J., 433, 434, 436
Crowe, R. R., 447
Crowley, B., 455
Crowley, W. F., Jr., 203
Cruccu, G., 341

Crutcher, M. D., 348
Cruz-Landeira, A., 352
Culham, J., 316
Culhane, K. A., 78
Culp, R. E., 202
Culver, K. W., 92
Cummings, D. E., 169
Cummings, J. L., 414, 415, 441
Cunniff, C., 440
Cunningham, J. D., 96
Cunningham-Williams, R., 140
Curi, M., 460
Currie, P. J., 166
Curtin, L. R., 172
Curtis, C. M., 180
Curtis, M. A., 82
Cushman, A. J., 101
Cutler, W. B., 197
Czeisler, C. A., 469, 471, 482
Czyzewska, M., 487

Dabbs, J. M., Jr., 192, 193, 241, 242
Dacey, D. M., 472
Dackis, C. A., 132
Dagher, A., 350
Daglish, M. R. C., 145
Dahl, B. C., 414
Dakin, C. L., 170
Dakin, S. C., 27
Dalal, R. M., 180
Dale, A., 140
Dale, A. M., 273
Dalgleish, T., 228
Dalley, J. W., 148
Dallos, P., 259
Dalton, A., 482
Dalton, J., 241
Dalton, K., 241
Dalton, K. M., 238, 239
Dalton, R., 112
Daly, M., 203
Damasio, A. R., 7, 60, 61, 62, 224, 229, 230, 232, 233, 274, 279, 317, 371, 382, 490
Damasio, H., 7, 60, 61, 62, 224, 229, 230, 233, 274, 317, 371, 397
Dambrosia, J., 78
Damsma, G., 139
Dan, Y., 310
Dandy, J., 405
Dang-Vu, T. T., 473
Daniel, D., 437
Daniel, D. G., 436
Daniels, J., 413
Daniels, S. L., 141
Dannals, R. F., 131, 132
Dannon, p. N., 448
Dapretto, M., 414
Darracq, M. A., 384, 385
Dartnall, H. J. A., 304
Darwin, C., 4, 12, 14, 114, 279
Das, A., 82
Daschle, T., 116
Daselaar, S. M., 408
Date, Y., 181
Daum, I., 379
Dauvilliers, Y., 483
David, A. S., 439

David-Gray, Z., 472
Davidson, J. M., 192, 193
Davidson, J. R., 451
Davidson, M. C., 420
Davidson, R. J., 106, 226, 233, 451
Davies, B., 446
Davies, J. L., 237
Davies, M. S., 414
Davies, S. W., 353
Davies, W., 83
Davis, C., 180
Davis, C. M., 162, 244
Davis, J. M., 435, 441
Davis, J. O., 440
Davis, K. D., 240, 489
Davis, M., 233
Davis, S. W., 408
Dawood, M. Y., 193
Dawson, D., 482
Dawson, D. A., 145
Day, B. L., 277
Ddiamond, J., 205
de Andrade, M., 352
De Biasi, S., 337
de Campo, N., 457
de Castro, J. M., 174, 175
De Felipe, C., 337
de Gelder, B., 317
De Geus, E. J. C., 398, 400, 403
de Jonge, F. H., 194
de Jongh, R., 406
de la Chapelle, A., 207
de la Fuente-Fernández, R., 352
de Leon, J., 439
de Leon, M., 237
de Leon, M. J., 380
De Luca, M., 355
De Luca, V., 439
de Medinaceli, L., 81
De Meyer, G., 385
de Queiroz, V., 447
de Riboupierre, Y., 259
de Santi, S., 237
de Sauvage, F., 179
De Stefano, N., 355
de Toledo-Morrell, L., 380
De Valois, R. L., 96, 97, 305, 307, 312
de Vos, R. A. I., 484
de Waal, F. B. M., 490
Deacedo, B. S., 370
Deacon, S., 471
Deacon, T., 282
Deacon, T. W., 55, 60
Deadwyler, S. A., 480
Deary, I. J., 398
DeBauche, B. A., 443
Debener, S., 408
deBerardinis, R., 421
Debruille, B., 317
deCabo, R., 174
DeCarla, C. M., 417
Decety, J., 490, 491
deCharms, R. C., 337
Dechent, P., 102
DeDoux, J. E., 378
Degreef, G., 442
Degueldre, C., 490

Dehaene, S., 401, 414, 497
Dehejia, A., 351
Dehoff, M. H., 148
DeJong, J., 452, 454
Dekker, J., 341
Del Bene, F., 32
Del Tredici, K., 484
Deleuran, B., 417
DeLisi, M., 245
DeLong, M. R., 350
DeLuca, G. C., 355
Demanuele, C., 408
DeMarco, A., 180
Dement, W. C., 469, 475, 476, 482
Demicheli, V., 418
Demler, O., 428, 443, 454
Démonet, J. -F., 277
Denburg, N. L., 380
Deng, W., 451
Denissenko, M. F., 133
Denke, C., 343
Dennis, N. A., 408
Depaulis, A., 243
Deppe, M., 280
Depue, R. A., 451
Derbyshire, S., 203
Derogatis, L. R., 238
DeSanti, S., 380
Descartes, R., 5, 6
Deschênes, M., 480
Deshpande, G., 498
Deshpande, P., 133, 148
Desia, R., 117
Desimone, R., 317, 319, 321, 371, 372
Desmond, J. E., 367
Despierre, P. -G., 453
D'Esposito, M., 372
Despres, J. -P., 176
Dessens, A. B., 206
Destrade, C., 368
Deuel, L., 354
Deus, J., 229
Deutsch, J. A., 384
Deutsch, S. I., 439
Devanand, D. P., 447
Devane, W. A., 135, 136
Deveau, T. C., 456
Devlin, M. J., 181, 182
Devore, H., 208
Devue, C., 490
Dew, M. A., 449
Dewey, S., 131
DeYoc, E. A., 266
Dhawan, V., 353
Dhejne-Helmy, C., 214
Dhond, R. P., 273
Dhu, C., 369
Di Chiara, G., 136, 139
Di Lorenzo, P. M., 45, 46
Di Pellegrino, G., 372
di Tomaso, E., 135
Diamandis, E. P., 411
Diamond, A. D., 380
Diamond, D. M., 384
Diamond, J., 205
Diamond, M., 207, 208, 211
Diamond, M. C., 397

Diano, S., 169
Dick, D., 148
Dick, D. M., 149
Dickey, C., 384
Dickinson, E., 54
Dickson, D. W., 484
Diener, H. -C., 351
Diers, M., 343
Dietert, J., 417
Dietert, R., 417
Dietis, N., 337
Dietrich, T., 409
Dietz, V., 346
Dietz, W. H., 172
DiFiglia, M., 353
Diketis, E., 417
DiLeone, R. J., 180
DiMario, F., 54
Dimirijevic, M. R., 346
Dimon, J. A., 192
Dinan, T. G., 409
Ding, J. -D., 458
Dingemanse, J., 480
Dinges, D. F., 469
Dinstein, L., 416
Dinwiddie, S., 147
Dirks, P., 97, 98
Djenderedjian, A., 387
Dkhissi-Benyahya, O., 472
do Rosario-Campos, M. C., 460
Dobelle, W. H., 304
Dodd, M. L., 140
Doering, C. H., 192
Dolberg, O. T., 448
Dollfus, S., 433
Dolski, I., 238, 239
Dombeck, D. A., 98
Dominguez, J. M., 194
Dong, X. -W., 43
Donoghue, J. P., 84
Doody, R., 384
Dopbmeyer, S., 367, 489
d'Orbán, P. T., 241
Dorer, D. J., 182
Dorsey, E. R., 354
Dostie, D., 280
Dostrovsky, J. O., 240, 489
Doucette, D., 483
Doucette, S., 453
Dougherty, D. M., 244
Doughty, C. J., 460
Dowd, S. M., 448
Dowling, H., 176
Downing, P. E., 317
Doyle, A. E., 445
Doyle, C. A., 337
Doyle, M., 137
Doyle, W. J., 236
Drabant, E. M., 444
Drachman, D., 354
Dräger, B., 280
Dragnunow, M., 82
Dragon, T., 445
Dreifus, C., 216
Dresel, S., 385
Dresel, S. H., 420
Dressel, C., 226

Drevets, W. C., 452, 453
Drislane, F. W., 276
Driver, J., 322
Dronkers, N. F., 279
Drop, S. L. S., 206
Drugan, R. C., 338
Drumi, C., 117
Dryden, S., 179
Drzewiecki, K., 160
Du, J., 195
Duan, J., 432
Duarte, C. B., 43, 44
Dubois, J. M., 32
Ducci, F., 147
Duck, S. C., 206
Duda, J. E., 140, 352
Dudai, Y., 366, 376, 378
Dudas, M. M., 439
Dudek, B. C., 398
Dudek, S. M., 372
Duffy, J. F., 471
Dufour, M. C., 145
Duhamel, J. -R., 417
Dujardin, K., 476
Duka, T., 129
Dukelow, S., 316
Duketis, E., 416, 419
Dulawa, S., 448, 449
DuMouchel, W., 352, 353
Dümpelmann, M., 367
Duncan, G. H., 240
Duncan, J., 397
Dundon, W., 144
Dunn, F. A., 472
Dunn, G. A., 177
Dunne, M. P., 12
Dupont, P., 320
DuPree, M. G., 211
Durand, J. -B., 298
Durand, V. M., 455, 461
Dürsteler, M. R., 312
Durston, S., 420
Dutra, A., 351
Dutton, D. G., 226
Dwivedi, Y., 441
Dwoling, J. E., 297
Dwork, A. J., 447

Eagly, A. H., 202
Earle, J. A., 441
Earnest, D. J., 470
Eaton, W. W., 417
Eaves, L., 182
Eaves, L. J., 174
Eberhardt, N. L., 176
Ebers, G., 211
Ebers, G. C., 355
Ebert, M. H., 183
Eckert, E., 183
Edelman, G. M., 47
Edenberg, H. J., 149
Edgar, M. A., 213
Edgerton, V. R., 346
Edison, T., 265
Edmeads, J., 483
Edward, H. M., 483
Edwards, A., 342

Edwards, G., 127
Edwards, S., 317
Effern, A., 367
Egan, M. F., 438
Ehlers, C. L., 148
Ehrhardt, A. A., 205, 206, 207, 208
Ehrsson, H. H., 348
Ehtesham, M., 117
Eichenbaum, H., 370
Eide, E. J., 482
Eigsti I. -M., 420
Einarsdottir, E., 341
Einfeld, S. L., 417
Einstein, A., 57, 397
Eisen, S. A., 456
Ekeblad, E., 182
Ekman, P., 226
Eksioglu, Y. Z., 54, 74
Elashoff, R., 237
Elbert, T., 79, 341, 342
Eldreth, D., 136
Elger, C. E., 231
Eli, P. J., 435
Elias, M., 378
Eliez, S., 416, 417
Elison, A., 320
Ellenberg, J., 352
Ellenbogen, J. M., 473
Elliffe, D., 402
Ellis, A. B., 203
Ellis, K. A., 385
Ellis, L., 209, 216
Ellis, S. M., 179
Ellis-Davies, G. C. R., 374
Ellison-Wright, I., 438
Elmquist, J. K., 166, 168, 171, 484
Else, J. G., 237
Elston, J., 135
Elveback, L. R., 236
Ely, T. D., 371
Emens, J. S., 471
Enard, W., 14, 285
Endert, E., 194, 241
Engel, A. K., 487
Engel, R. R., 435
Engel, S. A., 317
Engel, S. M., 421
Engelborghs, S., 385
Engert, F., 374
Ennifer, S., 144
Enoch, M. -A., 147, 148
Enriori, P. J., 179
Eppig, C., 405
Epstein, A. N., 160
Epstein, C. J., 337
Epstein, M. P., 456
Epstein, R. S., 455
Erb, M., 343
Erberich, S., 409
Ercan-Sencicel, A. G., 460
Erickson, C. A., 371, 372, 417
Erickson, K. I., 408
Eriksson, C. J. P., 197
Erkens, J. A., 453
Ernfors, P., 448
Ernst, M., 420
Ernulf, K. E., 216

Erondu, N., 179
Escada, P., 83
Esiri, M. M., 355
Eskandar, E., 353
Esposito, G., 39
Etienne-Cummings, R., 84
Evankovich, K., 78
Evans, A., 336
Evans, A. C., 240, 280
Evans, A. E., 179
Evans, D., 383
Evans, D. A., 381, 408
Evans, S. M., 419
Evarts, E. V., 344
Even, G., 383
Everitt, B., 433
Evers, P., 194
Evins, A. E., 436
Evrard, P., 77, 78, 132
Ewald, H., 439
Eysenck, H. J., 203
Ezzat, W. H., 409

Fabbri-Destro, M., 402
Faedda, G. L., 450
Fahey, J. L., 237, 238
Fahey, P., 177
Fahle, M., 379
Fahn, S., 352
Fahrback, K., 180
Fahrenkrug, J., 472
Faigel, H. C., 39
Faiz, L., 78
Faizi, M., 411
Falhammar, H., 206
Falk, D., 397, 398
Fallon, J., 371
Fallon, J. H., 367
Falls, W. A., 139
Falsafi, S., 200
Falzi, G., 280
Fambrough, D. M., 354
Fan, Y., 203
Fancher, R. E., 6
Fang, Y., 441
Faraco, J., 484
Faraone, S. V., 15, 419, 420, 421, 431, 432
Faraqco, J., 483
Farlow, M., 410
Farmer, A. E., 431
Farooqi, I. S., 175, 179
Farrer, C., 491
Farrer, M. J., 352
Farrer, V., 490
Farries, M. A., 353
Fatemi, S. H., 441
Faul, M., 80
Faustman, W. O., 439
Fausto-Sterling, A., 206
Favagehi, M., 236
Fawzy, F. I., 237
Fawzy, N. W., 237
Fazio, F., 277
Feenders, G., 97, 98
Fegley, D., 341
Fehr, E., 417
Feigin, A., 353

Fein, D., 417
Feinberg, T. E., 491
Feinstein, C., 417
Feinstein, D. L., 478
Fekkes, D., 436
Feldman, J. F., 207
Feldman, J. L., 43
Felleman, D. J., 322
Feng, G., 459
Feng, Y., 74
Fera, F., 244, 444
Feresin, E., 246
Ferguson, J. N., 198
Ferguson-Smith, M. A., 207
Fergusson, D., 453
Fergusson, D. M., 133
Ferman, T. J., 484
Fernández, G., 367
Fernandez-Piqueras, J., 140
Fernstrom, M. H., 182, 183
Féron, F., 83, 478
Ferrara, R. A., 410
Ferrarelli, F., 498
Ferrari, P. F., 284
Ferreira, D., 448
Ferrell, R. E., 245
Ferrillo, F., 483
Ferris, S., 384
Ferro, J. M., 278
Fertel, R., 236
Fertuck, H. C., 32
Fetz, E. E., 84
Feuk, L., 11
Fey, D., 170
Fibiger, H. C., 139
Fichtner, C. G., 496
Fidani, L., 382, 410
Field, A. E., 172
Fielden, J. A., 242
Fields, H. L., 339
Fiez, J., 67
Fike, M. L., 495
Filippi, M., 355
Filipsson, H., 206
Finch, C. E., 236, 382
Fincher, C. L., 405
Fineberg, N. A., 457
Fink, G. R., 490
Fink, M., 32
Finkelstein, E. A., 172
Finklestein, S. P., 117
Fiondella, C. G., 276
Fiorino, D., 139
Fiorino, D. F., 195
Fischbacher, U., 417
Fischer, M., 419
Fischman, M. W., 139
Fisher, S. E., 285
Fishman, S., 342
Fiske, D., 157
Fitch, R. H., 380
Fite, P. J., 243
Fitzgerald, J. A., 212
Fitzgerald, M., 398
Fitzsimmons, H. I., 353
Fitzsimmons, L., 447
Fitzsimons, J. T., 159, 160

Flanigin, H. F., 280
Flaum, I., 60
Flaum, M., 433, 434
Fleck, M. S., 408
Flegal, K. M., 172
Flegr, J., 196
Fleischer, J. G., 47
Fleisher, T., 92
Fleming, D. E., 194
Fleming, R., 237
Fletcher, J. M., 281
Fletcher, K., 419
Flier, J. S., 181
Flöel, A., 174, 280
Flood, J. F., 169
Flor, H., 79, 241, 342, 343
Flores, A. T., 182
Flores, S., 480
Floris, G. F., 450
Flory, J. D., 245
Flynn, J. R., 404
Fobker, M., 174
Fogassi, L., 284
Folstein, M. E., 14
Foltin, R. W., 139
Fombonne, E., 413, 419
Ford, J. M., 439
Fornito, A., 136
Forssberg, H., 436
Forster, B. B., 243
Forsting, M., 214
Fort, P., 479
Fortin, D. L., 32
Fosse, M., 373, 477
Fosse, R., 477
Foster, K. E., 169
Foster, R. G., 472
Fotopoulou, A., 492
Fountas, K. N., 243
Fouts, D. S., 282
Fouts, R. S., 282
Fowler, C. D., 81
Fowler, D., 457, 458
Fowler, J. S., 131, 139, 140, 141, 142, 145, 180, 446
Fowles, D. C., 432, 433, 434
Fox, J. W., 74
Fox, P., 455
Fox, P. T., 275
Frackowiak, H., 349
Frackowiak, R. S. J., 375
Frady, R. L., 241
Francesconi, w., 129
Francis, D. J., 281
Francis, H. W., 267
Franck, N., 490, 491
Frank, B., 215, 216
Frank, E., 236
Frank, G. K. W., 183
Frank, R., 7, 62
Franke, P., 461
Frankish, H. M., 179
Frankland, A., 376
Frankland, P. W., 375, 376
Frankle, W. G., 244
Franklin, R. J. M., 355
Franklin, T. R., 132

Frantz, A. G., 339
Fraser, C., 322
Fraser, G. A., 495
Frayling, T. M., 175
Frayo, R. S., 169
Freathy, R. M., 175
Frederick, B., 197
Fredrickson, B. E., 342
Freed, C. R., 352, 353
Freed, W. J., 81
Freedman, M. S., 472
Freedman, R., 439
Freedman, S., 455
Freeman, C. P., 447
Freeman, W., 61
Frefczynski, J. A., 266
Freid, P., 136
Freilino, M., 169
Freimuth, P., 140
Freitag, C., 461
Freitag, C. M., 416, 419
French, E. D., 135
French, S. E., 169
Freud, S., 475
Freund, P., 83
Frey, S., 336
Friderici, K., 421
Fridman, E. A., 499
Fried, I., 47, 317
Fried, P. A., 136
Friederici, A. D., 280
Friedman, E., 132
Friedman, J. M., 179
Friehs, G. M., 84
Friesén, W. V., 226
Frijda, N. H., 241
Frisen, L., 206
Friston, K. J., 106, 350, 489
Frith, C., 439
Frith, C. D., 232, 277, 416, 490, 491
Frith, U., 277, 412, 413, 414, 416
Fritsch, G., 6
Fritschy, J. -M., 82
Froese, N., 398
Froesti, W., 144
Frolich, P. F., 195
Frost, G. S., 179
Frumin, M., 439
Fryer, T. D., 148
Frys, K. A., 148
Fuchs, S., 376, 377
Fudge, J. L., 181, 183
Fujita, T., 339
Fukuda, M., 353
Fukui, M. M., 489
Fukunaga, M., 497
Fulbright, R. K., 277
Fulker, D. W., 203, 211
Fullerton, C. S., 455
Funk, G. D., 43
Funkenstein, H. H., 381
Furuse, S., 337
Fuster, J. M., 60, 371

Gabrieli, D. E., 278
Gabrieli, J. D. E., 337, 367, 371
Gadian, D. G., 285, 375

Gage, F., 117
Gage, F. H., 81, 83, 375, 380, 384
Gage, P., 7, 62
Gagnon, J., 210
Gai, E. J., 421
Gaillard, R., 318, 487
Gainer, H., 470
Gainetdinov, R. R., 444
Gainotti, G., 233
Gajiwala, K. S., 175
Galaburda, A. M., 7, 62, 276
Galang, J. A., 238
Galef, B. G., 163, 164
Gall, F., 7
Gallese, V., 284, 493
Gallup, G. G., 490
Gally, J. A., 47
Galperin, T., 499
Galton, F., 399
Galvani, L., 6
Gamzu, E. R., 499
Gandola, L., 497
Gangestad, S. W., 196
Ganis, G., 316
Gannon, P. J., 283
Gantz, I., 179
Ganzel, B. L., 238
Gao, F., 285
Gao, Q., 171, 180
Garavan, H., 141
Garb, J. L., 163
Garber, K., 416
Garbutt, J. C., 140
Garcia, C. R., 197
Garcia, J., 164
Garcia-Falgueras, A., 214
Garcia-Fernández, J. -M., 472
Garcia-Osuna, M., 352
Garcia-Velasco, J., 197
Gardmer, G., 172
Gardner, C. O., 444
Gardner, D. M., 406
Gardner, E. L., 142
Gardner, E. P., 331, 335, 336
Gardner, H., 272
Gardner, T., 144
Garfinkel, L., 481
Garry, M., 379
Garside, R., 135
Gartrell, N. K., 212
Garver-Apgar, C. E., 196
Gastfriend, D. R., 129
Gatchel, R. J., 236, 237, 239
Gatenby, J. C., 489
Gates, J. R., 179
Gati, J. S., 340
Gatley, S. J., 131, 139, 140
Gatos, H., 243
Gauna, K., 280
Gaus, S. E., 478
Gauthier, I., 317, 318
Gautier, T., 205, 206, 207
Gaviria, M., 281
Gayán, J., 276, 285
Gazzaley, A. H., 380
Gazzaniga, M. S., 66, 320, 493, 494
Gazzola, V., 414, 493

Gebbard, P. H., 212
Gebhardt, C. A., 387
Geda, Y. E., 140
Geffen, G. M., 403
Geffen, L. B., 403
Geffner, L., 117
Gegenfurtner, K. R., 305
Geha, P. Y., 341
Geiger, E., 421
Geinisman, Y., 380
Geist, C. S., 210
Gelade, G., 486
Gelbard, H. A., 450
Gelfand, M. M., 192, 193
Gellner, R., 174
Gelsinger, J., 115, 116, 117
Geminiani, G., 317
Genazzani, A. D., 241
Genefke, I. K., 236
Genel, M., 207
Genoux, D., 378, 380
Georg, B., 472
George, D. T., 181
George, M. S., 448
George, T. P., 439
Georgieff, N., 490, 491
Georgopoulos, A. P., 349
Gerasimenko, Y., 346
Gerber, M., 499
Gerevini, V. D., 441
Gerloff, C., 102
Gerner, R. H., 451, 452
Gerrish, A., 383
Gershon, E. S., 443, 444, 457
Gertz, H., 385
Gervasoni, D., 477
Geschwind, N., 60, 272, 273, 280, 418
Ghaffari, M., 420
Ghetti, B., 410
Gheusi, G., 81
Ghez, C., 346, 347, 349, 350
Ghilardi, J. R., 332
Ghosh, A., 239
Ghosh, A. K., 460
Giacino, J. T., 499
Giacomo, R., 493
Giarelli, E., 413
Giavalisco, P., 14
Gibbons, R. D., 453
Gibbs, R. A., 410
Gibbs, R. B., 194
Giedd, J. N., 442
Giegling, I., 461
Gil, B., 10
Gil, R., 434
Gilad, Y., 305
Gilbert, C. D., 82
Gilbertson, M. W., 15, 431, 432, 455
Giles, D. E., 449, 481
Gill, K. E., 132
Gillespie, A. M., 337
Gillin, J. C., 481
Gilman, S. E., 182
Gilton, A., 453
Ginsburg, K., 409
Giraud, A. -L., 266
Gisriel, M. M., 237

Gitlin, M. J., 451
Givens, B. S., 374
Giza, B. K., 163
Gizewski, E. R., 214
Gläscher, J., 140, 397
Glaser, R., 236, 444
Glass, K. C., 453
Glass, L. C., 179
Glasser, M. F., 284
Glasson, B. I., 352
Glatt, C. E., 456
Glatt, S. J., 445
Gleason, R., 450
Glees, P., 82
Glessner, J. T., 419, 421
Glezer, L. S., 319, 397, 398
Gloor, P., 232
Glover, G. H., 238, 337, 367
Glowa, J. R., 144
Glueckorn-Waelsch, S., 259
Goate, A., 382, 410
Gobbini, M. I., 314
Gochman, P., 442
Goebel, R., 102, 317
Goedde, H. W., 148
Goedert, M., 352
Goff, D. C., 436
Gogtay, N., 398
Gold, A. R., 193
Gold, M. S., 131
Gold, P. W., 443, 445, 451
Goldberg, D. E., 83
Goldberg, J., 456
Goldberg, M., 45
Goldberg, M. E., 333, 336
Goldberg, S. R., 137, 144
Goldberg, T. E., 436
Goldfarb, Y., 237
Goldin-Meadow, S., 284
Goldman, A., 493
Goldman, A. L., 444
Goldman, D., 147, 243, 244, 245, 444
Goldman, L., 241
Goldman, N., 339
Goldman, S. A., 35
Goldman, S. M., 352
Goldman-Rakic, P. S., 316, 347
Goldstein, E. B., 255, 264
Goldstein, E. E., 306
Goldstein, J. M., 203
Goldstein, M., 43
Goldstein, R. Z., 11, 131
Golgi, C., 36, 95
Gomez-Tortosa, E., 281
Gong, Y., 382
González, R. G., 365
Good, C. D., 375
Goodale, M., 320
Goodell, E. W., 284
Goodenough, D. R., 475
Goodman, C. S., 74, 75
Goodman, D. C., 82
Goodman, M., 244
Goodwin, D. W., 108
Goodwin, F. K., 243, 443, 445, 451
Goodwin, W. E., 190
Gooren, L. J. G., 214

Gordon, J. H., 194
Gordon, T. P., 113
Gore, J. C., 317, 489
Gorelick, D. A., 148
Gorelick, P. B., 233
Gorny, G., 141, 142
Gorostiza, P., 32
Gorski, R. A., 194, 200, 213
Gorwood, P., 182
Gotlib, I. H., 451
Gottesman, I. I., 108, 109, 418, 431, 432, 440
Gottschall, P. E., 384
Gough, A. C., 352
Gougoux, F., 78
Gouin, J. -P., 444
Gould, E., 81, 448
Gould, K. G., 113
Gouras, P., 305
Gourinkel-An, I., 38
Gouws, A., 322
Goy, R. W., 194
Grabowski, T. J., 7, 62, 274
Grace, A. A., 435
Grady, C. L., 261, 458
Grady, W. R., 210
Graeber, R. C., 469
Grafman, J., 279
Grafton, S. T., 371
Graham, D. A., 74
Graham, S., 261
Grammer, K., 197
Grandin, T., 415
Grandinetti, A., 352
Grant, B. F., 140, 145
Grant, J. E., 140
Grant, S., 131, 141, 142
Grasby, P., 491
Grau, J. W., 338
Gravenstein, S., 440
Gray, J., 202
Gray, J. D., 132
Gray, J. R., 403, 404
Gray, K., 239
Gray, K. M., 417
Gray, R., 136
Gray, R. D., 402
Graybiel, A. M., 350, 458
Grayson, D. R., 441
Graziano, M. S. A., 336, 347, 348, 349
Green, A., 340
Green, L., 417
Green, P. T., 300
Greenamyre, J. T., 352
Greenbaum, R., 132
Greenberg, B. D., 409
Greenberg, D., 434
Greene, L. A., 353
Greene, P. E., 35, 352
Greene, R. W., 478, 479
Greenfeld, K. L., 128
Greenfield, K., 237
Greengard, P., 114, 436
Greening, L., 243
Greeno, C. G., 182
Greenough, W. T., 78, 380, 411
Greenstein, D., 398
Greer, M. A., 470

Greer, S., 238
Gregg, T. R., 242
Gregory, L. J., 322
Grekin, E. R., 133
Grella, C. E., 127
Grenhoff, J., 134
Gressens, P., 77, 78, 132
Grether, J. K., 413
Greuter, M., 317
Greuter, T., 317
Greven, S., 236
Grèzes, J., 284
Griffin, G., 135
Griffith, J. D., 133
Grigson, P. S., 139
Grilo, C. M., 174
Grindlinger, H. M., 458
Grodd, W., 343
Grootoonk, S., 439
Gross, C., 448, 449
Gross, C. G., 317, 318, 347, 348, 448
Gross, G., 429, 430
Gross, M., 149, 386
Grossman, H., 384
Grossman, J. A. I., 78, 79
Groszer, M., 285
Grow, R. W., 2, 80, 146, 429, 443
Gruen, R. S., 207
Gruenfeld, M. J., 496
Gruenwald, D. A., 409
Grunhaus, L., 448
Grunwald, T., 367
Grunze, H. C., 479
Grüsser, S., 343
Grüter, M. C., 317
Grüter, T., 317
Grzanna, R., 82
Gu, M., 376
Guarnieri, D. J., 180
Guastella, A. J., 417
Guerreiro, M., 82, 278
Guerrien, A., 476
Guerrini, R., 337
Guevremont, L., 84
Guidotti, A., 441
Guijarro, M. L., 237
Gujar, N., 497
Gulyas, B., 197
Gunawardene, S., 419
Güntürkün, O., 490
Gur, R. C., 203, 398
Gurd, J. M., 317
Gurling, H. M. D., 421
Gurney, A., 179
Gusella, J. F., 13
Gustafson, D., 172
Gustafsson, J. -A., 201, 213
Gustavson, C. R., 164
Guth, A., 482
Guthrie, D., 237
Gutierrez, C. M., 377, 476
Gutknecht, L., 461
Guze, B. H., 451, 457
Guzowski, J. F., 369
Gwaltney, J. M., Jr., 236
Gwirtsman, H., 181
Gwirtsman, H. E., 183

Haarmeier, T., 316
Haas, B. W., 444
Haasksma, J., 495
Habib, M., 277
Hacein-Bey-Abina, S., 111
Haditsch, U., 378, 380
Hadjikhani, N., 416
Hagelin, J., 115
Hagman, J. O., 441
Haier, R. J., 367, 371, 398, 400
Hairston, I. S., 477
Hajos, M., 439
Halaas, J., 179
Halaas, J. L., 175
Haldane, J. B. S., 485
Halgren, E., 273
Hall, M. H., 439
Hall, N. R. S., 495
Hall, S. M., 144
Hallet, A. J., 237
Hallett, M., 102
Hallmayer, J., 484
Hallock, R. M., 46
Halmi, K. A., 183
Halpern, A. L., 11
Halweil, B., 172
Hämäläinen, H. A., 335
Hamlainen, R., 117
Hamann, S. B., 371
Hamburg, D. A., 192
Hamburg, P., 182
Hamer, D. H., 211, 212
Hamilton, J. P., 451
Hamm, A., 280
Hammel, C., 180
Hammersmith, S. K., 210
Hampshire, A., 457
Hampson, E., 409
Hampson, R. F., 480
Hamshere, M. L., 383
Han, X., 32
Hand, I., 140
Handel, S., 437
Hankins, W. G., 164
Hanlon, C. A., 132
Hannibal, J., 472
Hanratty, M., 417
Hanus, L., 135
Hao, X., 451
Happé, F., 416
Happich, f., 214
Haqq, C. M., 200
Haqq, T. N., 200
Harada, N., 201
Harada, S., 148
Harden, R. N., 341, 342
Harding, A. E., 352
Hardy, J., 384
Hare, E. H., 441
Hare, R. D., 24
Hargrove, M. F., 242
Hari, R., 414
Hariri, A. R., 244, 444
Harley, L. J., 476
Harman, S. M., 409
Harmon, L. O., 312
Harold, D., 383

Harold, G. T., 109, 133, 134
Harris, C., 107
Harris, J. G., 439
Harris, N. S., 351
Harrison, P. J., 448
Hars, B., 476, 477
Hart, B., 192
Hart, B. L., 194
Hartline, H. K., 309
Hartman, L., 458
Hartmann, J., 78
Hartz, D. T., 144
Harvey, A. G., 193, 455
Harvey, T., 396, 397
Harward, H., 78
Harwood, H. J., 428
Hasboun, D., 318, 487
Hasegawa, T., 414
Hashimoto-Torii, K., 57
Haslinger, B., 226
Hassabis, D., 369
Hasse-Ferreira, A., 83
Hastings, M. H., 473
Hastings, T. G., 352
Hattar, S., 472
Hau, J., 115
Haukka, J., 445
Haukka, J. K., 429
Hauri, P., 100, 474
Hauser, M. D., 401
Hauser, R. A., 354
Häusser, M., 32
Havel, M., 482
Havlicek, J., 196
Hawking, S., 496
Hawkins, R. D., 374
Haworth, C., 488
Haworth, C. M. A., 403
Haxby, J. V., 274, 314, 315, 316, 369
Hay, D. F., 109, 133
Hayashi, Y., 374
Hayden, M. R., 354
Hayes, C., 282
Hayes, K. J., 282
Hayes, R. J., 142
He, D., 432
He, D. Z. Z., 259
He, S., 316
He, Y., 482
Head, D., 408
Head, K., 398
Healy, D., 453
Hearn, E. F., 198
Heath, A., 182, 183
Heath, A. C., 147
Heath, S., 383
Heatyh, A. C., 456
Hebb, D., 372
Hebebrand, J., 182
Hebert, L. E., 383
Hebert, P., 453
Hécaen, H., 273, 274
Hecht, G. S., 45
Hedges, L. V., 202
Hediger, K., 198
Heeger, D. J., 416
Heijmans, B. T., 177

Heil, M., 369
Heilman, K. M., 233
Heim, N., 192
Heiman, M. L., 170
Heinrichs, M., 198, 417
Heiss, W. -D., 490
Heissing, F., 14
Held, J., 133
Helgadóttir, H., 477
Heller, W., 233
Heller, W. A., 451
Helm-Estabrooks, N., 273
Helmholtz, H. von, 265, 301
Helms, J., 267
Helmuth, L., 408
Helps, S. K., 408
Heninger, G. R., 455
Henkenius, A. L., 420
Henneberg, M. A., 128
Hennen, J., 436
Hennenlotter, A., 226
Hennessey, A., 450
Hennevin, E., 476, 477
Hennig, H., 355
Hennig, J., 245
Henning, P., 316
Henninghausen, E., 369
Henningsson, S., 198
Henry, M. E., 447
Henshall, D. C., 82
Hensley, S., 499
Hepp-Reymond, M. C., 349
Herbert, C., 385
Herbert, T. B., 234
Herbrecht, E., 417
Heresco-Levy, U., 436
Hering, E., 301, 302
Heritch, A. J., 435
Herkenham, M. A., 97, 136
Herman, L. M., 284
Hernán, M. A., 355
Heron, M., 80
Herrero, J. F., 337
Herrup, K., 24
Herscovitch, P., 455
Hershow, R. C., 173
Hervey, G. R., 170, 175
Herzog, D. B., 182
Herzog, H., 170, 397
Herzog, H. A., 114
Hesselbrock, V., 149
Hesselson, S., 484
Heston, L. L., 432, 461
Hettema, J. M., 460
Heuser, I., 385
Hevenor, S., 261
Heywood, C. A., 319
Heywood, E., 148
Hichwa, R. D., 274
Hickok, G., 280
Hicks, B. M., 149
Hieda, Y., 144
Hier, D. B., 203
Higley, J. D., 244
Higuchi, S., 484
Higuchi, T., 448, 449
Hill, J. O., 174, 176, 178

Hill, S., 99
Hill, S. Y., 149
Hill, T., 487
Hill, T. C., 496
Hiltunen, M., 383
Hind, J. E., 262
Hindersson, P., 472
Hines, M., 201, 206, 214
Hines, P. J., 197
Hines, T., 397
Hinney, A., 182
Hinrichs, A. S., 13
Hinson, R. E., 127
Hinz, W. A., 482
Hirsch, J., 176, 281
Hirsch, S. R., 491
Hitler, A., 144
Hitzemann, R., 131
Hitzig, E., 6
Hoban, T. M., 450
Hobson, J. A., 373, 475, 477, 479
Hoch, C. C., 449
Hochberg, L. R., 84
Hodgins, S., 245
Hoefer, P. F. A., 354
Hoek, H. W., 440
Hoever, P., 480
Hoffman, G. E., 169
Hoffman, P. L., 129
Hoffman, V., 127
Hoffmann, K. -P., 316
Hofman, M. A., 213, 214
Hogenesch, J. B., 472
Hohmann, A. G., 341
Hökfelt, T., 1, 43, 168, 169
Holden, C., 282, 284, 285, 402
Holder, E., 137
Hollander, E., 140, 417
Hollingsworth, P., 383
Holloway, R. L., 283
Holman, J. E., 181
Holmans, P., 352
Holmes, C., 384
Holmes, C. J., 77, 78
Holowka, S., 279
Holsboer, F., 497
Hölscher, C., 374
Holt, A., 13
Holton, G., 3
Homan, R. W., 233
Hömberg, V., 351
Homes, A., 439
Honda, S. I., 201
Honea, R., 436
Hong, C. -J., 461
Hong, S. E., 74
Honkura, N., 374
Honoré, E., 32
Hoogland, G., 460
Hooks, B. M., 76
Hopfield, J. J., 478
Hopkins, V., 384
Hopkins, W. D., 2, 283, 284
Hopson, J. S., 196
Hopt, J. -M., 16
Horel, J. A., 82
Horgan, J., 16

Hori, T., 158
Horn, S., 140
Horne, J., 468
Horne, J. A., 476
Horner, P. J., 81, 83
Horovitz, S. G., 497
Horowitz, S., 232
Horst, J., 317
Horton, E. S., 178
Horton, N. J., 203
Horvath, S., 212
Horvath, T. L., 169, 171, 180, 201
Horwood, L. J., 133
Hoshi, E., 347
Hou, C., 432
Hou, M., 13
Houck, P. R., 449
Houeto, J. L., 460
Hould, F. S., 177
Houle, S., 244
House, J. S., 237
Housley, P. C., 202
Houtsmuller, E. J., 213
Howard, R., 343
Howlett, A. C., 136
Hoyert, D. L., 80
Hser, Y. -I., 127
Hsiao, S. S., 335
Hsu, L. K., 181, 183
Hu, N., 211
Hu, S., 211, 317
Hu, S. Y., 461
Hu, x., 284, 498
Hu, X. T., 347
Hu, Y., 378
Huang, C., 451
Huang, M., 336
Huang, Y., 244
Hubbard, E. M., 322
Hubel, D. H., 305, 309, 310, 311, 314
Huber, G., 429
Huberfeld, G., 38
Huberman, A. D., 76, 212, 215
Hudspeth, A. J., 256, 268, 285, 333
Huggins, G. R., 197
Hughes, H. B., III, 444, 452
Hughes, J. R., 496
Hughett, P., 398
Hull, C. L., 157
Hull, E. M., 194, 195
Hulshoff Pol, H. E., 203
Humfleet, G., 144
Humphreys, K., 416
Humphreys, P., 276
Hunt, G. L., 212
Hunt, G. R., 402
Hunt, M. W., 212
Hunt, S. M., 226
Hur, K., 453
Hurst, J. A., 285
Hurtig, H. I., 352
Hurvich, L. M., 303, 304
Hurwitz, T. D., 484, 485
Huse, E., 343
Hut, R. A., 472
Hutchison, C., 384
Hutchison, K. E., 148

Hutchison, W. D., 240, 489
Huttunen, M., 398, 403, 439
Hyde, J. S., 202, 203
Hyde, M. P., 202
Hyder, F., 447
Hyldebrandt, N., 236
Hyman, B. T., 365, 382
Hyman, S. E., 138
Hynd, G. W., 276
Hyun, C. S., 237
Hyvärinen, J., 335

Iacoboni, M., 284, 493
Iacono, D., 149, 386
Iacono, W. G., 149, 451
Ibáñez, A., 140
Ide, S. E., 351
Iemmola, F., 212
Igl, W., 198
Iijima, M., 206
Ikari, H., 140
Imperato-McGinley, J., 190, 205, 206, 207
Inagaki, M., 455
Ingason, A., 432
Ingelfinger, F. J., 168
Ingram, D. K., 174
Ingvar, M., 339, 340
Innala, S. M., 216
Inoue, O., 435
Inoue, Y., 484
Inouye, S. -I. T., 470
Insel, T. R., 198, 458
Ioannidis, J. P. A., 432, 433
Iqbal, N., 435, 436
Iriki, A., 335
Irizarry, M., 383
Isenberg, K. E., 448
Ishai, A., 314
Ishiguro, H., 47
Ishihara, K., 447
Ishii, S., 168
Ishikawa, B., 148
Itskov, V., 46
Itti, L., 387
Iurato, S., 333
Ivanova, E., 300
Iverson, S., 158, 349
Ivry, R. B., 351, 494
Iwamura, Y., 335
Izumikawa, M., 267

Jabs, W. J., 355
Jacklin, C., 202
Jackson, D. C., 144, 238, 239
Jackson-Lewis, V., 353
Jacobs, B., 28
Jacobs, B. L., 134, 375
Jacobs, G. H., 305
Jacobs, L., 440
Jaffard, R., 368
Jaffe, H., 32
Jagannathan, K., 416
Jain, M., 421
Jain, S. K., 355
Jakes, R., 352
Jakubociv, A., 139
James, C. J., 408

James, D., 136
James, W., 78, 225, 226
Jameson, D., 303, 304
Jamison, K. R., 1, 2, 16
Jan, C., 353
Jane, J. A., 82, 84
Janicak, P. G., 448
Janowsky, J. S., 203
Janssen-Bienhold, U., 97, 98
Janszky, T. S., 236
Janus, C., 384
Jaretzki, A. III., 355
Jarrah, Z., 411
Jarraya, B., 353
Jarvis, M. J., 133
Javitt, D. C., 436
Jayaram, G., 439
Jazrawi, R., 179
Jeannerod, M., 490, 491
Jefferson, T., 418
Jeffords, J. M., 116
Jeffrey, S., 355
Jennings, C. L., 32
Jensen, A. R., 380, 399, 400, 404, 405
Jensen, M. D., 176, 180
Jensen, T. S., 236
Jensh, R. P., 77
Jentsch, J. D., 135
Jernigan, T. L., 77, 78
Jervey, J. P., 371
Jessell, T. M., 39, 331, 337
Jiang, L., 451
Jiang, X., 319
Jiang, Y., 317
Jimerson, D. C., 181
Jin, K., 82
Jin, R., 428
Jin, Z. H., 353
Jinag, Y., 384
Jirtle, R. L., 177
Jo, Y. -H., 43
Jobst, E. E., 179
Jocobsen, F. M., 450
Jody, D. N., 442
Johansson, B., 404
Johansson, O., 43
John, E. R., 489
Johnson, A., 137, 278
Johnson, C., 445
Johnson, D. C., 379
Johnson, K. A., 365
Johnson, K. L., 411
Johnson, K. O., 335
Johnson, M. P., 471
Johnson, M. R., 136
Johnson, R. C., 245
Johnson, S. L., 267
Johnson, V., 191
Johnsrude, I. S., 375
Johnstone, T., 106
Johsson, Z. O., 285
Jolikot, M., 78, 79
Jolkovsky, L., 451
Jolles, J., 408
Joly, O., 298
Jones, B. E., 479
Jones, C. R., 482

Jones, D. W., 437
Jones, E. G., 441
Jones, G., 385
Jones, H. M., 435, 436
Jones, L. B., 132
Jones, M. W., 383
Jones, P. B., 439
Jones, R. M., 456
Jonik, R. H., 241, 242, 243
Jons, P. H., 420
Jönsson, L., 383
Jooste, P., 476
Jordan, C. L., 206
Joseph, R. M., 416
Josephs, K. A., 140
Jouvet, M., 480
Joyce, P. R., 460
Julesz, B., 312
Julien, R. M., 129, 130, 132, 134, 145, 435
Julius, D., 331
Jung, R., 321
Jung, R. E., 398
Jurata, L. W., 449
Jurica, P. J., 380
Just, M. A., 416
Jutchinson, E. R., 447

Kabos, P., 117
Kabosova, A. L., 117
Kadotani, H., 483
Kaeberlein, M., 173, 174
Kahanda, M., 180
Kahn, R. S., 398, 400, 435, 436, 446
Kahneman, D., 140
Kaiwi, J., 212
Kaji, K., 117
Kales, A., 482
Kales, J. D., 482
Kalin, N. H., 106
Kalivas, P. W., 142
Kalmar, K., 499
Kalra, P. S., 169
Kalra, S. P., 169
Kalso, E., 339, 340
Kamegai, J., 168
Kaminen-Ahola, N., 177
Kaminski, N. E., 135
Kana, R. K., 416
Kanbayashi, T., 484
Kandel, E. R., 14, 32, 39, 55, 77, 114, 158,
    335, 336, 374, 489
Kane, J. M., 435
Kane, N., 355
Kang, S. H., 148
Kanigel, R., 143
Kansaku, K., 203
Kanwisher, N., 317
Kao, R., 352, 353
Kao, S. -Y., 407
Kao, T. -C., 455
Kaplitt, M. G., 353
Kaprio, J., 445
Kapur, S., 436
Kar-Purkayastha, I., 340
Karachi, C., 460
Karama, S., 398
Karason, K., 180

Karlsson, H., 440
Karnath, H. -O., 492
Karni, A., 280, 476
Kartje, G. L., 81
Kasai, H., 374
Kasai, K., 455
Kasanuki, J. M., 300
Kashiwayanagi, M., 161
Kast, B., 60
Kastner, S., 489
Kato, K., 372, 373
Katsovitch, L., 460
Kattan, K. M., 355
Katz, L. C., 77
Katz, R., 445
Katz, S. H., 405
Katzman, D. K., 181
Kaufman, J., 460
Kaufman, M. J., 141
Kaufman, W. E., 276
Kausler, D. H., 407
Kavoussi, R. J., 245
Kavvoura, F. K., 432, 433
Kawai, N., 401
Kawamata, T., 337
Kawamoto, K., 267
Kawamura, H., 470
Kawarasaki, A., 336
Kawashima, M., 484
Kaye, W. H., 181, 182, 183
Kayman, S., 178, 179
Kayser, C., 266
Keator, D., 371
Kee, N., 81, 375
Keel, J. C., 409
Keel, P. K., 182
Keele, S. W., 351
Keenan, P. A., 409
Keesey, R. E., 176
Kegeles, L. S., 434, 436
Keil, K., 316
Keil, L. C., 159
Keller, H., 254
Keller, T. A., 416
Kellogg, W. N., 282
Kelly, C., 420
Kelly, D. J., 164
Kelsey, J. E., 139
Kelsoe, J. R., 436
Kelz, M. B., 480
Kemeny, M. E., 238
Kemether, E., 213
Kemmotsu, N., 414
Kemper, T. L., 276
Kempf, E., 243
Kendell, R. E., 447
Kendler, K. S., 182, 183, 432, 444, 460
Kendrick, M., 300
Kennaway, D. J., 471
Kennedy, B. K., 173, 174
Kennedy, D. N., 203
Kennedy, D. P., 416
Kennedy, J. L., 439
Kennedy, S. H., 181
Kennerknecht, I., 317
Kentridge, R. W., 319
Keogh, J. M., 175

Kerr, D. A., 355
Kerssens, C., 498
Kerssens, J. J., 341
Kerwin, R. W., 435
Kessler, J., 490
Kessler, R., 182, 183
Kessler, R. C., 148, 428, 443, 454
Kety, S. S., 430, 431
Keyler, D. E., 144
Keysers, C., 414, 493
Khabbaz, A. N., 98
Khachaturian, Z., 381
Khaitovich, P., 14
Khalil, A. A., 446
Khan, T. K., 385
Khan-Dawood, F., 193
Khateb, A., 479
Khoury, M. J., 432, 433
Kiang, N. Y. -S., 265
Kidd, K. K., 460
Kieburtz, K., 354
Kiecolt-Glaser, J. K., 236, 444
Kiehl, K. A., 243
Kiersky, J. E., 447
Kieseppä, T., 445
Kigar, D. L., 396, 397, 398
Kikuchi, Y., 414
Kilts, C. D., 371
Kim, C. E., 419
Kim, D., 179, 300
Kim, J. -W., 267
Kim, J. J., 441
Kim, K. H. S., 281
Kim, P., 238
Kim, S. -K., 267
Kim, W. S., 439
Kimura, D., 409
King, C. -Y., 200
King, E. C., 428
King, F. A., 113
King, M. -C., 14, 282
King, R. A., 460
Kinghorn, E., 194
Kinney, D. K., 418
Kinsey, A. C., 212
Kinsey, B. M., 144
Kiper, D. C., 305
Kipman, A., 182
Kipp, H. L., 421
Kirchner, M., 60
Kirk, K. M., 12
Kirkwood, A., 376
Kirsch, A., 420
Kissileff, H. R., 170
Kita, S., 279
Kitano, T., 285
Kitazawa, S., 203
Klaey, M., 197
Klauber, M. R., 481
Klauck, S. M., 416, 419
Klaver, P., 198
Kleber, H. D., 143, 419
Klei, L., 419
Klein, J., 485
Kleopoulos, S. P., 194
Klepinger, D. H., 210
Klerman, E. B., 471

Klima, E. S., 280
Kline, P., 395
Kloner, R. A., 236
Klos, K. J., 140
Klose, J., 14
Klugmann, M., 148
Klump, K. L., 183
Klump, M. C., 439
Knecht, S., 79, 174, 280, 342
Knierim, J. J., 369
Knight, J., 410
Knight, P. L., 259
Knight, R. T., 477
Knobloch, M., 378, 380
Knoll, J., 54
Knottnerus, G. M., 421
Knowlton, B. J., 350
Knudsen, S. M., 472
Knutson, B., 245
Kobayashi, K., 435
Koch, C., 47, 317, 498
Koch, M. E., 267
Kochanek, K. D., 80
Koenig, R., 114
Koenig, T., 438
Koeppe, C., 343
Koester, J., 34
Kohane, I., 407
Kohler, A., 102
Kohlert, J. G., 194
Kohn, P. D., 438
Köhnke, M. D., 147
Koide, K., 411
Kojima, S., 182
Kolachana, B. S., 244, 444
Kolata, G., 210
Kolb, B., 73, 141, 142
Kolodny, J., 397
Kolstad, K. D., 32
Komisaruk, B. R., 339
Kong, J., 300
König, P., 487
Konishi, M., 268, 269
Konopka, G., 285
Koob, G. F., 128
Kopelman, M. D., 387
Koppeschaar, H., 193
Korczykowski, M., 203
Kordower, J. H., 380
Korf, J., 495
Kormos, R., 117
Korvatska, O., 419
Kosambi, D. D., 239
Koscik, T., 203
Kose, S., 448
Koskela, E., 197
Kosslyn, S. M., 316
Koulack, D., 475
Koutstaal, W., 367
Kovatchev, B., 39
Koverstein, R., 480
Koyama, T., 240
Kozlowski, S., 160
Kraemer, H. C., 192
Krakauer, J., 346, 347, 349
Krank, M. D., 127
Kraskov, A., 47

Krasuski, J., 448
Kratzer, L., 245
Krause, E., 214
Krause, J., 420
Krause, K. -H., 420
Krauss, B. R., 342
Kreek, M. J., 148
Kreiman, G., 317
Kreiter, A. K., 487
Kriauciunas, A., 170
Krichmar, J. L., 47
Kriegeskorte, N., 107
Kriegsteni, K. V., 266
Krigs, T., 409
Krimchansky, B. Z., 499
Kripke, D. F., 473, 481
Krishna, R., 179
Krishna, V., 472
Krishnan-Sarin, S., 439
Kritchevsky, M., 387
Kronauer, R. E., 470, 471
Kroning, J., 470
Krueger, R. F., 149
Kruijver, F. P. M., 214
Kruk, M. R., 242
Kruse, M. S., 436
Krystal, A. D., 447
Krystal, J. H., 455
Kubischik, M., 316
Kubos, K. L., 439
Kubota, K., 332
Kucala, T., 440
Kuchinad, A., 342
Kuczenski, R., 445
Kudrimoti, H. S., 369
Kudwa, A. E., 201, 213
Kuehnle, T., 482
Kuepper, Y., 244
Kuffler, S. W., 309
Kuhlman, D. T., 496
Kuhn, J. -M., 192
Kukkuiha, S., 176
Kumar, A., 433
Kumari, V., 439
Kumbalasiri, T., 472
Kung, H. F., 420
Kupfer, D. J., 449
Kupferman, I., 158
Kurihara, K., 161
Kutschera, I., 111
Kuukasjärvi, S., 197
Kwan, K. Y., 460
Kwon, H., 416, 417
Kyriacou, C. P., 473
Kyrkouli, S. E., 169

Lääne, K., 148
Labad, A., 444, 445
Labarca, C., 133, 148
Labovitz, D., 440
LaCasse, L., 243
Lack, L., 482
Lacor, P. N., 382
Ladd, M. E., 214
LaDu, T. J., 183
Laeng, P., 449
Lafee, S., 366

LaForge, K. S., 148
Lagali, P. S., 300
Laguna, J., 411
Lahey, B. B., 421
Lai, C. S. L., 285
Laird, N., 172
Laje, G., 453
Lake, J. R., 144
Lakin, K., 194
Lalonde, F. M., 369
Lam, A. G. M., 367
Lam, P., 461
LaMantia, A. S., 76
Lamb, J. A., 322
Lamb, T., 345
Lamb, T. A., 242
Lambe, E. K., 181
Lambert, D. G., 337
Lambert, J. -C., 383
Lambert, M. P., 382
Lambert, T. J., 417
Lammens, M., 77, 78, 132
Lampert, S., 169
Lamperti, E. D., 74
Lancaster, E., 495
Lancet, D., 305
Lanctot, K. L., 245
Landis, K. R., 237
Landry, D. W., 145
Landt, O., 245
Lane, M. A., 174
Laney, C., 379
Lang, A. E., 353
Langer, E. J., 237
Langström, N., 211
Lank, E. J., 175
Lanser, M., 83
Laplane, D., 348
Larbig, W., 342, 343
Larkin, K., 194, 213
Larson, A. A., 337
Larsson, B., 180
Laruelle, M., 436
Lasco, M. S., 213
Lashley, K., 7
Lasko, N. B., 455
Lassonde, M., 78
Lau, B., 310
Laughlin, S. A., 273
Laumann, E., 210
Laurent-Demir, C., 368
Laureys, S., 376, 377, 498, 499
Lavedan, C., 351
Lawley, H. J., 197
Lawlor, B. A., 435, 446
Lawlor, P. A., 353
Lawrence, A. A., 214
Lawrence, A. D., 228, 229
Lawrence, E., 179
Lawrence, T. L., 435, 446
Layden, J., 173
Lazdunski, M., 32
Lazerson, A., 321
Le Bihan, D., 497
Le Couteur, a., 418
Le Foll, B., 144
Le Roux, C. W., 179, 180

Le-Niculescu, H., 445
Lebel, S., 1776
Leboyer, M., 417
Lecendreux, M., 483
Leckman, J. F., 443, 444, 460
Leconte, P., 476
LeDoux, J., 224
LeDoux, J. E., 320, 373, 379
Lee, A. D., 398
Lee, C., 484
Lee, C. C., 461
Lee, D. S., 267
Lee, G. H., 179
Lee, G. P., 224, 229, 230, 233, 280
Lee, J. L. C., 379
Lee, J. M., 383
Lee, J. S., 267
Lee, K. -M., 281
Lee, K. Y., 279
Lee, M., 341
Lee, R. R., 336
Lee, S., 372, 373
Lee, S. S., 421
Lee, T., 434, 435
Lee, V. M. -Y., 352
Lehky, S. R., 311
Lehnertz, K., 367
Lehrman, S., 115
Leibel, R. L., 176
Leibowitz, S. F., 169, 179
Leigh, P. N., 352
Leiguarda, R. C., 499
Leistner, D., 236
Leland, J., 216
Lemire, A., 449
Lemon, R. N., 349
Lenane, M., 457, 458
Lencz, T., 243
Lenneberg, E. H., 279
Lenox, R., 384
Lenox, S., 179
Lenroot, R., 398
Lent, R. W., 384
Lenzenweger, M.F., 431
Leonard, H. L., 457, 458
Leopold, D. A., 489
Leor, J., 236
Lepage, C., 398
LePiane, F. G., 139
Lepore, F., 78
Lerch, J., 398
Lerner, R. A., 14
Leroy, E., 351
Lesage, F., 32
Lesem, M. D., 181
Leshner, A., 138
Leslie, A. M., 412, 414
Lesnick, T. G., 352
Leucht, S., 435
LeVay, S., 212, 213, 216
Levenson, R. W., 226
Leventhal, A. G., 408
Lévesque, M. F., 352
Levin, F. R., 419
Levin, H. S., 78
Levin, N., 179
Levine, A. S., 169

Levine, F. R., 176
Levine, H. L., 273
Levinson, D. F., 432
Levit, K. R., 428
Levita, L., 456
Levitsky, W., 280
Levitt, J., 436
Levitt, P., 57, 442
Levy, B., 408
Levy, D. L., 439
Levy, J., 203
Levy, R. E., 342
Levy, R. M., 341, 342
Levy, S., 11
Levy, S. E., 413
Lewandowski, N. M., 436
Lewicki, P., 487
Lewin, R., 82
Lewine, R. R. J., 442
Lewis, D. A., 442
Lewis, D. C., 143
Lewis, H. B., 475
Lewis, J. W., 266
Lewis, K. D., 13
Lewis, M., 490
Lewis, M. B., 226
Lewis, P. D., 77
Lewis, V. G., 206
Lewy, A. J., 450
Leyden, J. J., 197
Li, C., 407, 435
Li, L., 211
Li, R., 483
Li, T., 451
Li, T. -K., 145
Li, T. C., 461
Li, X., 317
Liang, F. -Q., 470
Liao, C., 404
Liao, H. -W., 472
Liberles, S. D., 197
Liberzon, I., 228, 229, 232
Lichtenstein, P., 211
Lichtman, S. W., 176
Liddle, P. F., 491
Lieber, C. S., 244
Lieberman, H. R., 179
Liechti, M. E., 134
Liedvogel, M., 97, 98
Lightman, S. L., 198
Ligumsky, M., 135
Liker, J., 395
Lilenfeld, L. R., 182
Lima, C., 83
Liminac, K. D., 148
Lin, H. M., 481
Lin, L., 481, 483, 484
Lin, S., 440
Lin, S. -C., 477
Lin, X., 483
Lindberg, S. M., 203
Linden, D. E. J., 102
Lindgren, C., 175
Lindner, A., 316
Lindsay, S., 379
Lindsay, T. H., 332
Lindström, P., 214, 215

Links, J. M., 131, 132
Linn, M. C., 203
Linnell, J., 418
Linnoila, M., 243, 244, 245, 452, 454
Linz, M., 429, 430
Lipkin, C., 144, 397
Lipska, B. K., 442
Lipton, S. A., 384
Lipworth, L., 355
Lisak, R. P., 354
Lisman, J., 374, 376
Lissner, L., 172
Liu, J., 451
Liu, S., 117
Liu, W., 339, 456
Liu, X., 131, 141, 142
Liu, X. -Y., 330
Liu, X. -Z., 117
Liu, Y., 81
Livingston, E. H., 180
Livingstone, D., 338
Livingstone, M. S., 276, 314
Ljung, R., 236
Ljungqvist, A., 207
Llao, H. -W., 472
Lledo, P. -M., 81, 102
Lobaugh, N. J., 486, 487
Locantore, J. K., 408
Locke, J., 149
Lockhart, D. J., 14
Lockley, S. W., 471
Locurto, C., 405
Loebstein, M., 132
Loehlin, J. C., 405
Loewi, O., 24, 36
Loftus, E. F., 379
Logan, J., 139, 140, 174, 180, 446
Logothetis, N. K., 106, 266, 314, 489
Lohmann, H., 280
Lohr, K. N., 140
Lombardo, I., 244
Lømo, T., 372
London, E. D., 131, 132, 141, 142
Long, K. B., 259
Lönnqvist, J. K., 429, 445
López, A., 229
López-Rivadulla, L. M., 352
Lorber, J., 82
Lord, G. M., 179
Loring, D. W., 280
Loring, J. F., 383
Lorrain, D. S., 194, 195
Lotspeich, L., 416, 417
Lott, I., 400
Lott, I. T., 382
Lotze, M., 343
Louie, K., 376
Louwerse, A. L., 194
Lovelace, R. E., 355
Lowe, K. A., 135
Lowenstein, R. J., 495
Lowing, P. A., 431, 432
Loy, B., 267
Loyd, D. R., 341
Lozano, A. M., 240, 448, 489
Lu, B., 378
Lu, J., 83, 458, 459, 478, 479

Lu, T., 407
Luaba, J. -F., 498
Lubenov, E. V., 376
Lucas, G., 448
Lucas, R. J., 472
Luciano, M., 403
Luckenbaugh, D. A., 446, 451
Lucretius, 4
Ludlow, D., 337
Luisi, M., 241
Luisi, S., 241
Lukas, S. E., 141
Lukashin, A., 349
Lumey, L. H., 177
Lumley, L. A., 195
Lupien, P. J., 176
Lupien, S. J., 237
Lurey, M. L., 314
Lustig, C., 408
Lutter, C. D., 242
Lutz, K., 279
Lutzenberger, W., 342, 343
Luxen, A., 490
Ly, D. H., 14
Lyden, P., 351
Lykken, D. T., 203
Lynch, K. G., 144
Lyons, M. J., 456
Lyons, S., 156
Lyoo, I. K., 436
Lyttleton, O. C., 398

Ma, X., 284
Ma, Y., 353, 409
Maas, L. C., 141
Maccoby, E., 202
MacDorman, K. F., 47
MacGregor, R., 446
MacKay, C. E., 436
Mackay-Sim, A., 83, 478
MacKenzie, G., 352
Mackey, S. C., 337
Mackie, K., 136
MacLean, C., 182, 183
MacNeilage, P. F., 284, 401
MacNichol, E. F., Jr., 304
Macrae, J. R., 127
Madamba, S. G., 129
Madden, P. A. F., 147
Maddi, S. R., 157
Maddison, D., 236
Madison, D. V., 32, 374
Madison, L. D., 259
Madsen, S. K., 398
Maeda, F., 337
Maeda, M., 414
Maeder, P. P., 260
Maes, H. H. M., 174
Maestrini, E., 322
Maestu, F., 281
Maffei, M., 175, 179
Mafulies, D. S., 420
Maga, M., 177
Magee, A., 343
Maggard, M. A., 180
Maggi, G., 339
Maglione, M., 180

Magnuson, V. L., 211
Maguire, A. M., 300
Maguire, E. A., 369, 375
Maher, B. S., 452
Mahesh, V. B., 241
Maho, C., 476, 477
Mahowald, M. W., 484, 485
Mai, N., 320
Maier, M. A., 349
Maier, S. F., 338
Majerus, S., 499
Maki, P. M., 409
Makris, N., 203
Malagelada, C., 353
Malarkey, W. B., 444
Malaspina, D., 436
Maldjian, J. A., 132
Malenka, R. C., 138
Maletic-Savatic, M., 374
Malhotra, A. K., 457
Malik, G. M., 278
Malik, M., 418
Malin, D. H., 144
Malinow, R., 374
Malison, R., 148
Mallam, E. A., 355
Mallet, L., 460
Mallick, M., 179
Mallitt, K. -A., 177
Maloku, E., 439
Malone, S. M., 149
Mamczarz, J., 174
Mamikonyan, E., 140
Mampe, B., 280
Mamun, A. A., 130
Manfredi, M., 341
Mangels, J. A., 356, 380
Mangieri, R., 341
Mangin, J. -F., 497
Mangun, G. R., 494
Manji, H., 453
Mann, J. J., 447, 452, 453
Manoach, D., 436
Mansari, M., 480
Mansuy, I. M., 379, 380
Mantyh, P. W., 337
Mantzoros, C., 181
Manuck, S. B., 245
Mao, X. O., 82
Mapes, G., 469
Mappes, T., 197
Maquet, P., 376, 377, 490
Maranganore, D. M., 352
Marazita, M. L., 452
Marceau, S., 177
Marciniak, A., 117
Marcotti, W., 267
Marcus, S. M., 453
Marczynski, T. J., 455
Marentette, P. F., 279
Marinkovic, K., 273
Marino, L., 490
Mark, T. L., 428
Markesbery, W. R., 149, 386
Markou, A., 144
Markowitsch, H. J., 490
Marks, G. A., 476

Marks, W. B., 304
Marler, M. R., 481
Marois, R., 489
Marques, F., 448
Marrett, S., 240
Marrison, C., 171
Marsden, C. D., 352
Marshall, E., 112, 115
Marshall, J., 378
Marshall, J. C., 317
Marshall, L., 477
Marshall, V. J., 383
Marsland, A. L., 234
Martell, B. A., 144
Martens, H., 471
Martin, A., 274, 369
Martin, C. E., 212
Martin, D. S., 378
Martin, E. M., 281
Martin, J., 211, 245
Martin, J. B., 353
Martin, J. H., 331
Martin, K. M., 457
Martin, L. F., 439
Martin, L. L., 226
Martin, M. J., 117
Martin, N. G., 12, 403
Martin, S. J., 367
Martinez-Amorós, È., 444, 445
Martino, G., 355
Martinou, J. -C., 81
Martins, I. P., 82, 278
Martis, B., 448
Martorell, L., 444, 445
Marucha, P. T., 236
Marx, J., 179
Marzi, C. A., 317
Masica, D. N., 206
Mason, C. E., 460
Mason, G. F., 447
Mâsse, L. C., 147
Massimini, M., 498
Masters, W., 191
Mastorakos, G., 481
Matarese, G., 179
Mateer, C. A., 274
Mathalon, D. H., 439
Mathews, P. M., 384
Mathews, W. B., 135
Matochik, J. A., 420
Matsui, D., 47
Matsui, M., 398
Matsumoto, A. H., 409
Matsumoto, A. M., 409
Matsushita, S., 484
Matsuzaki, M., 374
Matsuzawa, T., 401
Matt, K. S., 409
Mattay, V. S., 39, 244, 444
Mattes, R. D., 161
Matthews, P. M., 355
Mattiace, L. A., 213
Mattingley, J. B., 322
Mattis, P., 353
Mattox, J. H., 409
Mattson, A. J., 78
Matud, M. P., 203

Matuszewich, L., 194, 195
Matute, C., 436
Matzuk, M. M., 198
Mayberg, H. S., 448
Mayer, D. J., 339
Mayer, G., 483
Mayeux, R., 352, 379
Maywood, E. S., 473
Mazur, A., 242
Mazure, C., 236, 496
Mazziotta, J. C., 284, 451, 452, 457, 493
Mazzocchi, F., 273, 274
McCandliss, B. D., 319
McCann, U. D., 135
McCarley, R. W., 439, 475, 478, 479
McCarti, A. T., 488
McClay, J., 245
McClearn, G. E., 12, 404
McClelland, J. L., 376
McClintock, M. K., 197, 238
McClung, C. A., 148
McClure, S. M., 141
McCormick, C. M., 214
McCormick, D. A., 351
McCoy, N. L., 193, 197
McCrory, E., 277
McCully, J., 127
McDaniel, M. A., 206, 398
McDermott, M. P., 354
McDonald, J. W., 82, 84, 117
McDonald, K., 283
McDonald, R. J., 370
McDonald, T. J., 170
McDougall, W., 157
McDougle, C. J., 417
McDowell, M. A., 172
McEvoy, S. P., 488
McEwen, B. S., 207
McFadden, D., 215
McGarry-Roberts, P. A., 399
McGaugh, J. L., 367, 371, 378
McGeoch, P. D., 336
McGinty, D., 478
McGrath, C. L., 445
McGrath, L. M., 276
McGue, M., 12, 15, 109, 149, 403
McGuffin, P., 12, 15, 431, 445, 457
McGuire, M. T., 241
McGuire, P. K., 439
McIntosh, A. R., 408, 486, 487
McKay, D., 458, 460
McKeefry, D. J. ., 322
McKeon, J., 457
McKinney, S. L., 133, 148
McKinney, S. M., 473
McKinnon, W., 236
McKusick, D. R., 428
McLaurin, J., 384
McLean, H., 81, 102
McLellan, T. A., 143
McMahon, F. J., 445
McMenomy, T., 441
McNaughton, B. L., 369, 376, 380
McNeely, H. E., 448
McNeil, T. F., 245
McPhee, J. C., 32
McQueen, M. B., 383, 432, 433

McQuillin, A., 421
McRae, K. A., 439
Meaburn, E., 410
Meador, K. J., 280
Mechan, J., 491
Mechoulam, R., 135
Mednick, S., 477
Mednick, S. A., 133, 440
Medolago, L., 341
Meeker, D., 103
Meertz, F., 436
Mehlman, P. T., 244
Meininger, V., 348
Meisel, R. L., 192
Melamed, R., 237
Melichar, J. K., 145
Melisaratos, N., 238
Melo, I., 448
Melone, M., 436
Meloy, J. R., 243
Melton, L., 148
Meltzer, H. Y., 446
Melvin, L. S., 136
Melzack, R., 240, 339, 342, 491
Menaged, M., 214, 215
Menci, W. E., 277
Mendelson, W. B., 450
Méndez-Alvarez, E., 352
Mendrek, A., 243
Meng, Q. C., 480
Meng, X. -L., 330
Menon, D. K., 341
Menon, R. S., 340
Menon, V., 416, 417
Mentis, M. J., 35
Menzies, L., 457
Meogh, J. M., 179
Mercer, K. B., 456
Merico, D., 419
Merikangas, K. R., 182, 428
Merkel, R. L., 39
Merzenich, M. M., 259
Mesnage, V., 460
Meston, C. M., 195
Mesulam, M. M., 60, 496
Metter, E. J., 409
Metz, G. A. S., 81
Meuller, e., 244
Meyendorff, E., 452
Meyer, E., 240
Meyer, J. H., 244
Meyer-Bahlburg, H. F. L., 206, 207, 209, 212
Meyer-Lindberg, A. S., 438, 444
Mezei, M. M., 354
Michael, I. P., 117
Michael, R., 210
Michalon, A., 378, 380
Micheau, J., 367
Mickey, S. K., 117
Micklas, J. A., 355
Middleton, B., 471
Miesegaes, G., 81
Miezin, F. M., 367, 489
Mignot, E., 481, 484
Miguel, E. C., 460
Mikami, A., 240
Mikulis, D. J., 181

Mileikovsky, M., 117
Miler, B., 441
Miler, G. A., 440
Miler, L. T., 400
Miles, D. R., 245
Miles, J. E. M., 471
Miletich, R., 279
Mill, J., 245
Miller, A., 61
Miller, A. D., 92
Miller, B. L., 387, 414, 415
Miller, C. A., 374
Miller, D. S., 176
Miller, E. K., 347, 371, 372, 489
Miller, G., 81, 114, 429
Miller, G. A., 233
Miller, L., 416
Miller, L. S., 450
Miller, M., 216
Miller, N. F., 113
Miller, R. D., 196
Miller, S. C., 495
Miller, S. D., 495
Miller, T. Q., 237
Milner, B., 278, 280, 364
Milner, D. M., 484, 485
Milner, P., 266
Miltner, W. H. R., 487, 488
Minamoto, F., 206
Minato, T., 47
Minde, J., 341
Mingozzi, F., 300
Minoda, R., 267
Minshew, N. J., 416
Minter, K., 450
Miodovnik, A., 421
Mir, A., 83
Mirsky, A. F., 431, 432
Mirzazade, S., 279
Mishkin, F., 414, 415, 441
Mishkin, M., 285, 316
Mitchell, D., 476
Mitchell, D. J., 414
Mitchell, E., 144
Mitchell, J. F., 471
Mitchell, S. W., 491
Mitchison, G., 476, 478
Mitka, M., 180
Mitler, M. M., 469
Mitropoulou, G., 38
Mitton, E., 141, 142
Miyashita, Y., 318, 371
Mizuno, A., 414
Mizuta, M., 181
Mobbs, C. V., 19
Moberg, P. J., 439
Möbius, H. J., 384
Modahl, C., 417
Moeller, F. G., 244
Moeller, J. R., 451
Moffat, S. D., 409
Moffitt, T. E., 245, 444, 445
Mogilner, A., 78, 79
Mohammed, S., 192, 193
Mohseni, P., 117
Moldin, S. O., 445
Mölle, M., 477

Moller, H. -J., 461
Mollon, J. D., 304
Molnar-Szakacs, I., 493
Molyneux, P. D., 369
Mombaerts, P., 197
Monaco, A. P., 285
Mondal, M. S., 181
Mondragon, M., 197
Money, J., 205, 206, 208
Monfils, M. -H., 379
Monk, T. H., 449
Montague, P. R., 141
Monteleone, P., 241
Montgomery, G. W., 404
Monthe, C., 384
Monti, M. M., 499
Montoya, P., 342, 343
Moody, B. J., 447
Moody, D. B., 284
Moore, E. A., 245
Moore, G. J., 409
Moore, H., 436
Moore, H. D., 267
Moore, J. D., 196
Moore, J. T., 480
Moore, T., 349
Moore, V. J., 476
Moore-Ede, M. C., 471, 482
Moore-Gillon, M. J., 159
Moos, R. H., 192
Morens, D. M., 352
Morgan, D., 384, 402
Morgan, M., 241
Morgan, M. J., 322
Morgan, T. M., 460
Mori, M., 399
Moritz, C. T., 84
Morland, A. B., 322
Morley, J. E., 169
Morley, K. I., 404
Morrel-Samuels, P., 284
Morrell, F., 380
Morrell, T. L., 415
Morris, E. K., 379
Morris, J. M., 206
Morris, J. S., 232
Morris, M., 417, 482
Morris, M. C., 408
Morris, N. M., 193
Morris, R. B. M., 376
Morris, R. D., 284
Morrison, C., 180
Morrison, J. H., 380
Mors, O., 439
Mortensen, P. B., 417, 439
Morton, G. J., 171
Morton, J., 412, 414
Moscovitch, M., 317, 380
Moser, D. J., 203
Moser, E. I., 369
Moses, A. M., 13
Mössner, R., 461
Mountcastle, V. B., 335
Mouren-Simeoni, M. C., 182
Mouritsen, H., 97, 98
Mpakopoulou, M., 243
Mufson, E. J., 380

Mühlethaler, M., 479
Mujica-Parodi, L. R., 197
Muka, D., 149
Muldoon, M. F., 234, 245
Mullan, M., 382, 410
Mullapudi, U., 180
Muller, D., 374, 375
Muller, D. J., 439
Müller, J., 78, 267
Müller, R. -A., 414, 416
Muller, R. U., 377
Mulligan, R. S., 385
Mullin, K., 383
Mulvenna, C., 322
Münch, T. A., 300
Mundy, N. I., 197
Munir, K. M., 418
Munoz, K. E., 444
Muñoz, M., 472
Muñoz, R. F., 144
Munzar, P., 137
Muotri, A., 117
Muranaga, T., 182
Murasaki, M., 484
Murch, S. H., 418
Murdoch, D., 244
Murphy, A. Z., 341
Murphy, D., 409
Murphy, D. L., 451, 458
Murphy, E. A., 14
Murphy, F. C., 228, 229
Murphy, G., 4
Murphy, G. M., Jr., 397
Murphy, J., 283
Murphy, M. R., 198
Murphy, S. L., 80
Murray, G. K., 441
Murray, I. V. J., 352
Murray, J. B., 135
Murray, K. E., 420
Murray, R. M., 435, 439, 441
Murrell, J., 410
Murro, A. M., 280
Murrough, J. W., 446
Musallam, S., 103
Mushahwar, V. K., 84
Muskand, J. A., 84
Musser, B., 179
Must, A., 172
Mustanski, B. S., 211
Muzio, J. N., 476
Myers, J. M., 460

Nabb, J. J., 245
Naccache, L., 318, 487, 497
Nadeau, A., 176
Nader, K., 378
Naeser, M. A., 273, 387
Nagai, N., 182
Nagamoto, H. T., 439
Naglieri, J. A., 405
Nahas, Z., 448
Nairne, J. S., 395
Najman, J. M., 130
Nakahara, T., 182
Nakashima, T., 158
Nakayama, K., 477

Nakazato, M., 181
Nalibof, B. D., 238
Naliboff, B., 203
Namiki, A., 337
Napoleon, S., 418
Naranjo, C. A., 245
Narbro, K., 180
Narr, K. L., 243, 398, 403
Nash, J. M., 486
Nashmi, R., 133, 148
Nayak, S. K., 472
Naylor, R. J., 454
Naylor, S. L., 13
Neale, M. C., 12, 174, 182, 183, 203, 211, 444, 460
Neave, N., 214, 215
Nebes, R. D., 66
Nedergaard, M., 35
Neely, M. H., 341
Neiderthiser, J. M., 198
Neidhart, E., 483
Neisser, U., 395, 405
Nelen, M., 245
Nelissen, K., 298, 402
Nelson, C., 179
Nelson, D. L., 410
Nelson, L., 32
Neophytides, A. N., 437
Nesse, R., 139
Nestler, E., 144
Nestler, E. J., 433
Nestor, P. G., 439
Nettelbeck, T., 405
Netter, F. H., 200
Netter, P., 245
Neuer, G., 162
Neugebauer, R., 440
Neulen, J., 409
Neuman, T., 117, 352
Neumann, C., 442
Neumeister, A., 455
Neville, H. J., 280, 380
New, A. S., 244
Newbury, D. F., 285
Newhouse, P. A., 384
Newlin, D., 148
Newlin, D. B., 131, 141, 142
Newman, A., 459
Newman, A. L., 212
Newman, E. A., 41
Newman, J. D., 259
Newman, R. I., 237
Newschaffer, C. J., 413
Newton, G., 194
Ng, P. C., 11
Nias, D. K. B., 203
Nichelli, P., 279
Nicholas, A. K., 341
Nichols, R. C., 405
Nicod, J., 285
Nicoll, R. A., 32, 136
Nicolson, R., 442
Nieder, A., 401
Nielsen, A. S., 439
Nielsen, D. A., 148
Nietzsche, F., 236
Nieuwenhuys, R., 159

Nievergelt, C. M., 211
Nigg, J. T., 421
Niiyama, Y., 337
Nijenhuis, E. R., 495
Nijhuis, F. A. P., 408
Nikelski, E. J., 280
Nikolas, M., 421
Nikonenko, I., 374, 375
Nimmo-Smith, I., 228, 229
Niparko, J. K., 267
Nir, T., 283
Nisbett, R. E., 405
Nishioka, N., 484
Nishitani, N., 414
Nissinen, K., 197
Nitschke, J. B., 233, 451
Nizmikiewicz, M. A., 439
Nobili, L., 483
Noble, K. G., 319
Nobrega, M. A., 13
Noesselt, T., 162
Noirhomme, Q., 498
Nopoulos, P., 203
Nordenström, A., 206
Nordin, C., 440
Norman, B. F., 208
Norman, J. F., 298
Norrby, K., 117
Norton, S., 77
Nottebohm, F., 284
Noumair, D., 244
Novin, D., 168
Novotny, S., 417
Nowell, A., 202
Nugent, A. C., 455
Nulman, I., 132
Nunez, A. A., 470
Nunn, J. A., 322
Nuriminen, T., 140
Nurnberger, J., 449
Nutt, D. J., 145
Nuutila, A., 243
Nyce, M. R., 170

Ó Scalaidhe, S. P., 316
Oakes, T. R., 106
Oakson, G., 480
Oates, J. A., 133
Oberman, L. M., 414
Obermeyer, W. H., 481
O'Brien, C., 376
O'Brien, C. P., 133, 143, 146
O'Brien, J. A., 337
O'Callaghan, E., 439, 441
O'Callaghan, M., 130
Ochoa, B., 208
O'Connor, D. H., 374, 489
O'Connor, S., 149
O'Dell, T. J., 77, 374
O'Donovan, M., 421
Offenbacher, E., 176
Ogden, C. L., 172
Ogden, J., 82
O'Gorman, T. W., 148
O'Grady, M., 495
Oh, S. H., 267
O'Hara, B. F., 148

Ohm, G., 265
Ohno, M., 376
Ojemann, G. A., 274
Ojemann, J. G., 367
O'Keane, V., 409
Okubo, Y., 435
Olanow, C. W., 351, 352
Olarte, M. R., 355
Oldenburger, W. P., 194
O'Leary, D., 203
Olek, M. J., 355
Oliet, S. H. R., 41
Olincy, A., 439
Oliver, G. D., 209
Oliver, J., 283
Olivier, A., 232
Olivier, B., 406
Olp, J. J., 196
Olsen, R. K., 438
Olshansky, S. J., 173
Olson, B. R., 169
Olson, E. J., 484
Olson, H. C., 130
Olson, R. K., 276
Olton, D. S., 374
Omura, K., 444
Ong, W. Y., 136
Ono, Y., 403
Ooms, M. P., 194
O'Rahilly, S., 175
Orban, G. A., 318, 320
O'Reardon, J. P., 448
O'Reilly, K. B., 137
O'Reilly, R. C., 376
Orff, H., 481
Orgogozo, J. M., 348
Orlans, F. B., 114
O'Roak, B. J., 460
Oroszi, G., 147
O'Rourke, D., 450
Orr, S. P., 455
Orson, F. M., 144
Orwoll, E. S., 203
Osanai, H., 414
O'Shaughnessy, D. M., 335
Osinsky, R., 244
Ostrem, J. L., 39
Ostroff, R. B., 447
Ostry, D., 279
Osvath, M., 402
Otten, L. J., 487
Ottman, R., 352
Oviatt, S. K., 203
Owen, A. M., 499
Owen, M. J., 12, 15, 421
Owen, S. L., 340
Özyürek, A., 279

Pääbo, S., 305
Paans, A. M., 495
Paca, A., 117
Pace-Asciak, P., 132
Pace-Schott, E. F., 479
Pachana, N. A., 387
Paddock, S., 453
Paepke, A. J., 196
Pagnamenta, A. T., 419

Pagnin, D., 447
Pagnoni, G., 141
Pai, J., 180
Pain, S., 402
Pakstis, A. J., 405
Palamara, P., 488
Palha, J. A., 448
Pallie, W., 280
Palmer, A. R., 263
Palmer, J. A., 337
Palmer, R. G., 478
Pan, Y., 407
Pan, Z. -H., 300
Panda, S., 472
Pankiewicz, J., 141
Panov, A. V., 352
Pantelis, C., 136
Pantev, C., 79, 342
Pao, A., 133
Papas, N. R., 174, 180
Pappas, N., 446
Pappone, P. A., 32
Paré, D., 480
Parent, J. M., 82
Parent, M., 354
Parisi, J. E., 484
Park, A., 179, 486
Park, M., 418
Parker, E. S., 378
Parmalee, N., 300
Parnas, J., 440
Parrish, T. B., 341
Parry, B. L., 450
Parslow, D. M., 322
Parsonage, S., 176
Partonen, T., 445
Pasanen, E. G., 215
Pascual-Leone, A., 78
Pashler, H., 107
Pasquina, P. F., 343
Passaro, D. J., 173
Passingham, R. E., 284, 348, 350, 436
Pastalkova, E., 46
Patel, A. J., 32
Patel, A. N., 117
Patel, V., 477
Paterakis, K., 243
Paterson, N. E., 144
Patoine, B., 383
Patrick, C. J., 149, 243
Pattatucci, A. M. L., 211
Patterson, C., 211
Pattwell, S. S., 456
Paul, L. K., 397
Paul, S. M., 144
Paulesu, E., 277, 279
Pauls, D. L., 460
Paulus, M., 181, 183
Pauly, J. M., 337
Pay, M., 386
Pearlson, G. D., 439
Pearson, A. G., 82
Pearson, J., 384
Peca, J., 458
Pedersen, H., 236
Pedersen, J. S., 13
Pedersen, M. G., 417

Pedersen, N. L., 404
Peduzzi, J. D., 83
Peel, A. L., 82
Peeters, R., 402
Pegna, A., 479
Peigneux, P., 376, 377
Pelham, W. E., Jr., 421
Pell, A. N., 162
Pellegrino, A. S., 101
Pellegrino, L. J., 101
Penfield, W., 60, 63, 64, 334, 347, 369
Peng, W., 339
Penn, A. S., 355
Penn, R. D., 84
Pennacchio, L. A., 13
Penney, E. B., 82
Pennington, B. F., 276
Pentel, P. R., 144
Peper, J. S., 203
Pepitone, M. E., 212, 215
Pepperberg, I., 283
Pereira, R., 431, 432
Perez, A. M., 446
Perez de Castro, I., 140
Perez-Reyes, E., 32
Pérez-Samartin, A., 436
Peri, T., 455
Perin, J. C., 421
Perkel, D. J., 353
Perkins, A., 212
Perlis, M. L., 481
Permutter, R., 439
Permutter, S. I., 84
Perozzo, P., 317
Perrett, D. I., 232, 493
Perrone, P., 280
Perry, D., 117
Persico, A. M., 148
Persina, I. S., 380
Pert, C. B., 97, 128
Pertwee, R. G., 135
Pérusse, L., 175
Pesaran, B., 103
Peskind, E. R., 409
Pestone, M., 176
Peterhans, E., 312
Peters, J. C., 174, 176
Petersen, E. T., 440
Petersen, H. D., 236
Petersen, M. R., 284
Petersen, S. E., 275, 367, 369, 489
Peterson, B. B., 472
Peterson, B. S., 420, 451, 489
Peterson, J. B., 128, 243, 245
Peterson, M. S., 408
Peterson, R. E., 190, 205
Petersson, K. M., 339, 340
Petiau, C., 376, 377
Petitto, L. A., 279, 280, 282
Petkov, C. I., 266
Petralia, R. S., 374
Petrides, M., 336
Petrill, S. A., 404
Petrivotch, H., 352
Petrovic, P., 339, 340
Petry, H. M., 300
Petry, N. M., 140

Pettinati, H. M., 144
Petzer, J. P., 352
Peuskens, H., 320
Pezawas, L., 444
Pezer, N., 367
Pfaff, D. W., 14, 194
Pfaus, J. G., 139, 194
Pfefferbam, A., 129
Pfeifer, G. P., 133
Pfeifer, J. H., 414
Pfrieger, F. W., 35
Phan, K. L., 228, 229, 232
Phelps, E. A., 224, 379, 452
Phelps, J. A., 440
Phelps, M. E., 451, 457
Phillips, A., 417
Phillips, A. G., 139, 195
Phillips, C., 376
Phillips, R. J., 169, 170
Phillips, R. L., 131, 132
Phinney, R. E., 266
Pi-Sunyer, X., 170, 177
Pia, N., 497
Pianezza, M. L., 148
Pibiri, F., 439
Picard, J. J., 77, 78, 132
Picchioni, D., 497
Pichardo, M., 206, 207
Pickar, D., 435
Pickard, J. D., 499
Pickavance, L., 179
Pickering, R. P., 145
Pidwell, D. J., 300
Pierau, F. K., 158
Pierce, E. A., 300
Pierce, K., 416
Piet, R., 41
Pietiläinen, O. P. H., 432
Pietrini, P., 279, 314
Pigarev, I. N., 489
Pihl, R. O., 128, 243, 244, 245
Pike, K., 385
Pillard, R. C., 12, 209, 211
Pillon, B., 460
Pilowsky, L. S., 435, 436
Pinel, P., 318, 401
Pineros, V., 238
Pini, S., 447
Pinker, S., 279, 285, 401
Pinsk, M. A., 489
Pinter, M. M., 346
Pinto, D., 419
Pinus, U., 39
Piomelli, D., 135
Pisarska, K., 176
Pitino, L., 197
Pittman, R. K., 455
Pizzagalli, D., 451
Plato, 4
Pletnikova, O., 149, 386
Ploghaus, A., 340
Plomin, R., 12, 14, 15, 403, 404, 410
Plotnicov, K., 182
Plotnik, J. M., 211, 490
Pluta, J., 203
Poggio, G. F., 315
Poggio, T., 315

Pogue-Geile, M. F., 174
Pokorny, J., 472
Pol, H. E. H., 398, 400, 403
Polak, D., 448
Poland, R. E., 244
Poline, J. -B., 497
Poling, J., 144
Pollack, D. B., 241
Pollice, C., 181, 182, 183
Pollo, A., 339
Polymeropoulos, M. H., 351
Pomeroy, W. B., 212
Pontieri, F. E., 136
Pool, C. W., 214
Poole, W. K., 236
Pope-Cordle, J., 178
Poranen, A., 335
Pories, W., 180
Porjesz, B., 149
Porkka-Heiskanen, T., 478
Porrino, L., 480
Porrino, L. J., 132
Port, J. D., 457
Porter, R. H., 196
Posey, D. J., 417
Posner, M. I., 278, 401, 498
Post, R. M., 495
Posthuma, D., 398, 400, 403
Postle, B. R., 372
Postma, P., 439
Potenza, M. N., 140
Potkin, S. G., 441
Potter, A., 384
Potter, W. Z., 447
Poulain, D. A., 41
Poulos, C. X., 140, 245
Pounaras, D. J., 180
Poutanen, V. -P., 398, 403
Povinelli, D. J., 490
Powell, T. P. S., 335
Powley, T. L., 169, 170, 176
Prabhakar, S., 13
Pramstaller, P. P., 482
Pratas-Vital, J., 83
Pratley, R. E., 179
Prendergast, B. J., 238
Prescott, C. A., 444, 460
Preskorn, H., 452
Press, G. A., 387
Preti, G., 197
Preuss, T. M., 284
Price, D., 418
Price, D. D., 240
Price, J. L., 452, 453
Price, R. A., 460
Prince, M., 439
Printz, D., 434
Prior, H., 490
Proffitt, F., 114
Prohovnik I., 451
Proko, P., 481
Provencio, I., 472
Prud'homme, J. -F., 38
Prudic, J., 447, 451
Pryor, T., 183
Przedboski, S., 353
Przeworski, M., 285, 305

Ptak, R., 387, 492
Pu, M., 409
Puce, A., 316
Pujl, J., 228
Pujol, J., 229
Purkinje, J., 308
Purnell, J. Q., 169
Putnam, F. W., 495
Putnam, K., 451
Puts, D. A., 206
Putter, H., 177
Puygh, K. R., 277
Pyter, L. M., 238

Qian, Y., 404
Qin, Y. -L., 376
Qiu, M., 444
Qiu, X., 472
Qu, Y., 117, 179
Quak, J., 495
Quatrano, S., 460
Quesney, L. F., 232
Quian Quiroga, R., 47
Quirk, G. J., 377
Quiroz, J., 451

Rabin, B. S., 234, 236
Rabinowitz, D., 175
Raboch, J., 192
Radtke, N. D., 300
Radulovacki, M., 478
Raedler, T., 140
Ragsdale, D. S., 32
Rahman, Q., 211, 212
Raichle, M. E., 275, 367, 408, 452, 455
Raine, A., 243, 245
Rainer, G. S., 347, 372
Raineteau, O., 81
Rainnie, D. G., 479
Rainville, P., 240
Raio, C. M., 379
Rajah, M. N., 486, 487
Rakic, P., 57, 76, 93
Raleigh, M. J., 241
Ralph, G. S., 353
Ramachandran, V. S., 322, 336, 343, 414, 492
Ramey, C. T., 406
Ramey, S. L., 406
Ramón y Cajal, S., 36, 79, 83, 85, 95
Ramsay, E., 458
Ramsby, G., 54
Ramus, F., 277
Randall, P., 236, 496
Rankinen, T., 175
Ransom, B., 35
Rantala, M. J., 197
Rao, H., 203
Rao, S. C., 347, 372
Rapaport, D., 78, 79
Rapkin, A. J., 241
Rapoport, J. L., 457, 458
Rapoport, S. I., 458
Rapp, P. R., 380
Rasmussen, T., 60, 278, 334, 347
Ratcliff, M., 470
Ratliff, F., 308, 309

Rauch, S. L., 367, 368, 458
Rauschecker, J., 280
Ravert, H. T., 135
Ravussin, E., 176, 179
Ray, L. A., 148
Ray, S., 385
Rayport, S., 43
Raz, A., 420
Read, J. D., 379
Read, S. L., 414, 415, 441
Reagor, P., 495
Rebert, C. S., 236
Rechtschaffen, A., 468
Redcay, E., 416
Reddy, A. B., 473
Redelmeier, D. A., 489
Redmond, G. P., 192, 193
Reed, G. W., 174, 176
Reed, J. J., 366
Reed, T. E., 148, 399
Reeg, G., 489
Rees, G., 369
Reeve, C., 82, 83, 84, 85
Regan, R., 419
Reich, D. L., 446
Reich, T., 445, 453
Reiman, E. M., 409, 455
Reimann, F., 341
Reinders, A. A., 495
Reinisch, J. M., 207
Reinkemeier, M., 490
Reinoso, B. S., 132
Reisberg, B., 384
Reiss, A. L., 416
Reiss, D., 198, 490
Rekling, J. C., 43
Relkin, N. R., 281, 384
Remington, G., 436
Remondes, M., 368
Rempel-Clower, N. L., 366
Ren, J., 117
Renault, B., 317
Renthal, W., 432
Reppert, S. M., 473
Requin, J., 348
Resak, M., 352
Resko, J. A., 194
Resnick, S. M., 409
Respondek, C., 136
Rettew, D. C., 457, 458
Reus, V. I., 144
Reuter, J., 140
Reuter, M., 245
Reynolds, C. F., III, 449
Reynolds, C. P., 236
Reynolds, S., 179
Rezai, K., 60
Rezek, M., 168
Rhees, R. W., 194
Riad, J. K., 241
Ribary, U., 78, 79, 497
Ribeiro, S., 477
Ribeyre, J. M., 433
Ricaurte, G. A., 135
Rice, C. M., 355
Rice, F., 109, 133, 134
Rice, G., 211

Rice, J., 236, 456
Rice, J. P., 445
Rice, M. L., 285
Rice, N. J., 320
Rich, J. B., 409
Rich, S., 203
Richardson, G. S., 482
Richter, M. A., 457, 460, 461
Ricken, J., 482
Ridding, M. C., 415
Riedel, G., 367
Riederer, P., 351
Riedner, B. A., 498
Riehle, A., 348
Riesen, A. H., 194
Riesenhuber, M., 319
Rieux, C., 472
Rijsdijk, F., 445
Rilling, J. K., 284
Rimmele, U., 198
Rimmer, D. W., 471
Rimon, R., 243
Rinehart, N. J., 417
Risch, N., 211
Rissman, E. F., 201, 213
Ritchie, J. M., 34
Ritter, R. C., 168
Ritter, S., 168
Rivero, F., 144
Rivetti, A., 418
Rivolta, M. N., 267
Rizzolatti, G., 284, 402
Roberson, R., 410
Roberts, E., 341
Roberts, E. M., 417
Roberts, G. W., 429
Roberts, J., 380, 384, 385
Roberts, S. C., 196
Robertson, M., 483
Robgertson, L. C., 486
Robillant, D. I., 482
Robinette, C. D., 432
Robins, E., 455
Robinson, D. G., 457
Robinson, E. S. J., 148
Robinson, F. R., 472
Robinson, P., 457
Robinson, T. E., 141, 142
Rocca, W. A., 352
Rocha-Miranda, D. E., 317
Rockland, C., 371
Rodenhiser, J., 434
Rodin, J., 174
Rodríguez, A., 421
Rodriguez, I., 197
Rodriguiz, R. M., 458
Roeling, T. A., 242
Roenneberg, T., 482
Roffwarg, H. P., 449, 476
Rogan, M. T., 373
Rogers, G., 135
Rogers, V. W., 141
Rogers, W., 483
Rogers-Ramachandran, D., 343
Röhrich, H., 332
Rokoff, E. V. I., 367
Roland, P., 197

Roland, P. S., 267
Rolls, B. J., 160, 162, 164, 177
Rolls, E. T., 162
Rolls, R. M., 160
Romey, G., 32
Rompre, P. -P., 138
Ronda, J. M., 471
Ronning, M. E., 405
Roome, C., 135
Roosen, K., 267
Roozendaal, B., 371
Ropers, H. H., 245
Rosa, R. R., 481
Rosadini, G., 233, 483
Rose, J. E., 262
Rose, M., 140
Rose, R. J., 15
Roselli, C. E., 194, 213
Rosen, G. D., 276
Rosen, L. R., 207
Rosen, S., 277
Rosenbaum, M., 176
Rosenbaum, R. S., 409
Rosenberg, C. R., 47
Rosenberg, H., 45
Rosenblatt, P., 135
Rosenbloom, K., 13
Rosenkranz, M. A., 238, 239
Rosenne, E., 237
Rosenthal, D., 430, 431
Rosenthal, N. E., 449, 450
Rösler, A., 192
Rosler, F., 369
Ross, C. A., 495
Ross, D., 244
Ross, E. D., 233
Ross, G. W., 352
Ross, T. J., 141
Rosse, R. B., 439
Rossen, R. D., 144
Rossi, G. S., 233
Rossi, M., 380
Roth, G., 140, 243
Roth, G. L., 259
Roth, G. S., 174
Roth, R. H., 135
Roth, W. T., 439
Rothbart, M. K., 498
Rothe, C., 461
Rothman, D. L., 447
Rothman, J. M., 162
Rothman, S., 77
Rouw, R., 322
Rovet, J., 132
Rowan, M. J., 374
Rowbotham, D. J., 337
Rowe, C. C., 385
Rowe, D. G., 103
Rowe, E. A., 162, 164, 177
Rowe, J., 288
Rowe, M. L., 284
Rowen, L., 282
Rowland, D., 232
Rowland, L. P., 354, 355
Roy, A., 452, 454
Roy, R. R., 346
Rozin, P., 161, 163, 164

Ruan, W. J., 145
Rüb, U., 484
Rubens, A. B., 280
Rubenstein, B. S., 476
Rubert, E., 283
Rubinow, D. A., 409
Rudorfer, M. V., 447
Rudow, G., 149, 386
Rudrauf, D., 397
Rugg, M. D., 487
Rujescu, D., 432, 461
Rumbaugh, D., 282, 283
Rush, A. J., 449, 453
Rushton, J. P., 203, 404
Rusinek, H., 380
Rusiuniak, K. W., 164
Rusjan, P., 244
Rusting, C. L., 380
Rustioni, A., 337
Ruth, T. J., 352
Rutherford, W., 261
Rutledge, J. N., 57, 400
Rutter, M., 413
Rydel, T., 81
Ryder, K., 76
Ryff, C. D., 238, 239
Rynal, D. M., 471

Saalmann, Y. B., 489
Sabri, O., 385
Sacco, K. A., 439
Sachs, B., 192
Sack, A. T., 102
Sack, D. A., 449, 450
Sack, R. L., 450
Sackeim, H. A., 447, 451
Sackheim, H. A., 447
Sacks, O., 63, 64, 294, 316, 330, 332, 414,
    415, 460, 492
Sacktor, T. C., 376
Sade, D. S., 398
Sadowsky, C. L., 82, 84
Saenz, A., 300
Sagi, D., 476
Sagiv, N., 322
Sahi, R., 416
Saint-Pierre, M., 354
Sairanen, M., 448
Saito, Y., 484
Sáiz-Ruiz, J., 140
Sakai, K., 480
Sakata, H., 336
Sakuma, Y., 194
Saleh, M., 84
Salehi, A., 411
Salisbury, D. F., 439
Salmon, D. P. U. H. S., 386
Salomons, T. V., 226
Salpeter, M. M., 32
Salter, M. W., 341
Salthouse, T. A., 408
Salvadori, S., 337
Sämann, P. G., 497
Sampson, P. D., 130
Sampson, S., 448
Samuel, M., 350
Samuelsson, M., 440

Sanberg, P. R., 354
Sánchez-Sellero, J., 352
Sanders, A. R., 432
Sanders, R. J., 282
Sandman, C. A., 441
Sano, M., 384
Santarelli, L., 448, 449
Santini, E., 377
Santos, P. F., 43, 44
Sanz, C., 402
Saper, C. B., 166, 168, 349, 478, 479
Sapolsky, R. M., 236, 237
Sapp, E., 353
Sarasso, S., 498
Sáry, G., 317
Sasa, M., 447
Sata, R., 439
Sato, M., 133
Satyamurti, S., 354
Savage, C. R., 367, 368
Savage-Rumbaugh, S., 282, 283
Savazzi, S., 317
Savic, I., 197, 214, 215
Savin-Williams, R. C., 210
Savoy, R., 197
Sawa, A., 435
Sawchenko, P. E., 166, 168
Saxe, G. N., 496
Saxe, M., 448, 449
Sbrocco, T., 178
Scahill, L. D., 460
Scalf, P. E., 408
Scammell, T. E., 478, 479, 484
Scarpa, P., 279
Scarr, S., 404, 405, 406
Schaal, B., 196
Schacter, D. L., 367, 368
Schacter, S., 226
Schaefer, C. A., 440
Schaefer, M., 343
Schaeffer, M. A., 236
Schafe, G. E., 378
Schaie, K. W., 380, 408
Schalkwyk, L. C., 410
Schank, J. C., 197
Schapira, A. H., 353
Schapiro, M. B., 458
Scharf, M. B., 482
Schatz, A. R., 135
Scheffer, I. E., 54
Scheibel, A. B., 397
Schein, S. J., 319
Scheinin, M., 243
Schelling, T. C., 133
Schenck, C. H., 484, 485
Schenome, P., 279
Scherag, S., 182
Scherberger, H., 103
Schermer, M., 406
Scherr, P. A., 381, 383
Scherrer, G., 331
Scheuer, T., 32
Schiavi, G., 483
Schiermeier, Q., 114
Schiff, N. D., 499
Schiller, D., 379, 456
Schiller, P. H., 314

Schirger, A., 236
Schirmer, B., 477
Schlake, H. -P., 267
Schlamann, M., 214
Schlichter, R., 43
Schliemman, A. D., 395
Schlundt, D. G., 178
Schmalz, J., 216
Schmidlin, E., 83
Schmidova, K., 169
Schmidt, M., 111
Schmidt, M. L., 352
Schmidt, P. J., 409
Schmidt, S. D., 384
Schmitt, F., 384
Schmitz, A., 244
Schnack, H. G., 403
Schnakers, C., 499
Schneider, G. E., 81
Schnider, A., 387, 492
Schoebel, S. A., 436
Schoene-Baker, J. -C., 231
Schoenfeld, D., 282
Schoenfeld, D. A., 436
Schoenfeld, M. A., 162
Scholte, S., 322
Schönfeldt-Bausch, R., 436
Schork, N. J., 211
Schouten, J. L., 314
Schreiber, S., 448
Schroder, J., 440
Schübler, K., 162
Schuckit, M. A., 148
Schull, W. J., 77
Schulman, H., 374
Schulsinger, F., 430, 431, 440
Schulsinger, H., 440
Schultz, L. M., 82
Schultz, R., 57, 400
Schultz, R. G., 14
Schultz, W., 141
Schulzer, M., 352
Schuman, E. M., 368, 374
Schumann, C. M., 416
Schüttler, R., 429, 430
Schwab, M. E., 81, 83
Schwartz, B. L., 439
Schwartz, J. H., 39
Schwartz, J. M., 451, 452, 457
Schwartz, M., 206
Schwartz, M. S., 281
Schwartz, M. W., 171
Schwartz, R. D., 144
Schwartz, W. J., 470
Schwartz, Y., 237
Schwarz, A., 81, 490
Schwarz, J., 8, 133, 148
Schwarzenbacher, R., 384
Schwei, M. J., 332
Schweinhardt, P., 342
Schweitzer, A., 239
Scidman, L. J., 203
Scoles, M. T., 127
Scott, A. A., 414
Scott, E. M., 163
Scott, K., 162
Scott, K. G., 410

Scott, L., 436
Scott, T. R., 163
Seamans, J., 142
Searle, J. R., 468, 486
Sears, R. M., 180
Sebel, P. S., 498
Seckin, E., 135
Seckl, J. R., 198
Sedaris, D., 457
Seebeck, T., 196
Seeley, R. J., 171
Seeman, P., 434, 435
Sees, D. O., 144
Segal, N. L., 203, 405
Seidman, L. J., 420
Seiler, M. J., 300
Seilhamer, J., 383
Seitz, R. J., 397
Sejnowski, T. J., 47, 311
Selfe, L., 414
Selin, C. E., 451, 452, 457
Selkoe, D. J., 382, 383
Sellers, E. M., 148
Selten, J., 439
Selwyn, S. E., 182
Seminowicz, D. A., 342
Semrud-Clikeman, M., 276
Senghas, A., 279
Senju, A., 414
Seporta, S., 354
Sergio, L. E., 279
Serruya, M. D., 84
Sevcik, M. A., 332
Sevcik, R. A., 283
Seward, R., 39
Seyal, A., 439
Shadmehr, R., 353
Shaffery, J. P., 476
Shah, A., 354
Shah, G. M. S., 439
Shah, N. J., 279
Shakhar, K., 237
Shalev, A. Y., 455
Shallice, T., 387, 492
Sham, P. C., 410, 441, 445
Shanahan, T. L., 471
Shane, D. J., 482
Shank, D., 174
Shapira, Y., 237
Shapiro, A., 475
Shapiro, C. M., 476
Shapiro, D., 233
Shapiro, E., 10
Shapiro, F., 385
Shapiro, S., 453
Sharp, C., 433
Sharp, P. E., 47
Sharp, S. I., 421
Shattuck, D. W., 398
Shatz, L. C., 77
Shaw, P., 398
Shaywitz, B. A., 277
Shaywitz, S. E., 277
Shearman, L. P., 473
Shehata-Dieber, W., 267
Shema, R., 376
Shen, W., 259

Shenton, M. E., 439, 455
Shepherd, G. M., 372, 373
Sheps, S. G., 236
Sherer, T. B., 352
Sheridan, J., 236
Sherin, J. E., 478
Sherwin, B. B., 192, 193, 409
Shi, H., 404
Shi, J., 432
Shi, S. -H., 374
Shibutani, H., 336
Shifren, J. L., 192, 193
Shiiya, T., 181
Shildkraut, J. J., 445
Shima, K., 347, 348
Shimamura, A., 380
Shimizu, T., 168
Shimokochi, M., 194
Shimura, T., 194
Shiromani, P. J., 478
Shively, S., 421
Shizgal, P., 140
Shors, T. J., 81
Shoukry, M., 13
Shouse, M. N., 479
Shryne, J. E., 194
Shuai, Y., 378
Shugarman, L. R., 180
Shulman, G. L., 367, 489
Shuman, M., 317
Si, K., 374
Siapas, A. G., 376
Sickhaus, A., 355
Siderowf, A. D., 140
Sidiropoulou, K., 148
Siegel, A., 242
Siegel, B., 400
Siegel, J. M., 479, 480
Siegel, S., 127
Siegel, S. J., 380
Siegelbaum, S. A., 32, 34, 39
Siepel, A., 13
Sietsma, D., 117
Siever, L. J., 435, 446
Siggins, G. R., 129
Sigman, M., 414
Sigmundson, H. K., 208
Signoret, J. -L., 317
Sijben, A. E., 436
Silber, M. H., 484
Silbersweig, D. A., 439
Sileno, A., 179
Silinsky, E. M., 134
Silva, A. J., 376
Silva, M. J., 421
Silverman, M. S., 96, 97, 307
Simmons, W. K., 107
Simner, J., 322
Simon, E., 158
Simon, J. A., 192, 193
Simon, P. G., 281
Simon, T., 394
Simone, D., 241
Simone, L., 402
Simonelli, F., 300
Simonoff, E., 418
Simos, P. G., 281

Simpson, J. B., 160
Simpson, J. L., 207
Simpson, J. R., 453
Sims, R., 383
Singer, B. H., 238, 239
Singer, J., 226
Singer, P., 461
Singer, W., 77, 438, 487
Singh, J. B., 446, 451
Sinha, A. K., 239
Sinha, M. K., 170
Sinha, P., 316
Sinha, R., 148
Sinnayah, P., 179
Sinton, C. M., 166, 484
Sirevaag, A. M., 380
Sirigu, A., 417
Sizemore, C. C., 495
Sjöstrand, L., 452
Sjöström, D., 180
Sjostrom, L., 180
Skaggs, W. E., 376
Skene, D. J., 471
Skilling, S. R., 337
Skinner, M. L., 406
Sklar, P., 445
Skolnick, P., 144
Skoner, D. P., 236
Skoog, I., 172
Skorpen, F., 341
Skudlarski, P., 277, 317, 318, 489
Sleegers, K., 383
Sleser, I., 475
Slijper, F. M. E., 206
Slimp, J. C., 194
Slob, A. K., 213
Slonim, D. K., 411
Slusser, P. G., 168
Slutske, W. S., 147
Small, C. J., 170
Small, S. A., 379
Smalley, S. L., 419, 421
Smallish, L., 419
Smania, A., 497
Smart, T., 178
Smith, A. J. H., 337
Smith, A. M., 243
Smith, A. P., 236
Smith, C., 476
Smith, C. A. D., 352
Smith, C. N., 368
Smith, C. R., 411
Smith, D. E., 380, 384
Smith, D. M., 437
Smith, D. W., 77
Smith, F. J., 178
Smith, G. A., 403
Smith, G. P., 170
Smith, J., 83, 177
Smith, J. R., 280
Smith, M. A., 353
Smith, M. J., 409
Smith, M. T., 481
Smith, S., 340
Smith, S. D., 276, 285
Smith, S. L., 32
Smith, S. S., 148

Smith, T. W., 237
Smith, V. C., 472
Smits, G., 140
Smits, M., 483
Smullin, D. H., 337
Snyder, A. P., 489
Snyder, A. W., 414
Snyder, A. Z., 275, 408
Snyder, E. Y., 117
Snyder, J., 416
Snyder, J. S., 81
Snyder, L. H., 336
Snyder, S. H., 45, 97, 128, 133, 134, 434,
    435, 451
Soares, E. S., 477
Söderland, J., 440
Sokol, R. J., 130
Soldatos, C. R., 482
Solders, G., 341
Solet, J. M., 473
Soliman, F., 456
Solms, M., 492
Solomon, j., 316
Solowij, N., 136
Solvason, H. B., 448
Somerville, L. H., 106
Somogyi, R., 383
Sondheimer, S. J., 197
Soneji, D., 337
Song, T., 336
Song, Y., 317
Soni, B., 472
Sonnenschein, D., 473
Sonty, S., 341
Sonuga-Barke, E. J. S., 408
Sorbi, M. J., 341
Soria, V., 444, 445
Sorya, L., 417
Sosa, Y., 341, 342
Sossi, V., 352
Sotnikova, T. D., 444
Soto-Otero, R., 352
Southam, A. M., 194
Souza, J. M., 352
Sowell, E. R., 77, 78, 420
Spadano, J. L., 172
Spanos, N. P., 495
Specht, K., 279
Speciale, S. G., 476
Spector, J. A., 142
Speicher, C., 236
Spelke, E., 401
Spence, C., 348
Spence, S., 233
Spence, S. A., 491
Spencer, D. D., 372, 373
Spencer, K. M., 439
Sper, S., 236
Sperry, L., 141
Sperry, R. W., 80
Spiegel, D., 237
Spiers, M. V., 24
Spillantini, M. G., 352
Spinney, L., 107
Spong, C. Y., 410
Spoont, M., 243
Spoormaker, V. I., 497

Sprengelmeyer, R., 351
Spreux-Varoquaux, O., 453
Spring, B., 15, 432
Springell, K., 341
Springen, K., 92
Spurzheim, J. G., 7
Squassina, A., 439
Squire, L. R., 350, 366, 367, 368, 387, 486
Sriram, S., 355, 473
Staal, R., 352
Staal, W., 416
Stahl, S. M., 455
Staib, L., 236, 496
Stall, W., 416, 419
Stalleichen, J., 97, 98
Stanescu, H., 421
Stanescu, R., 401
Stanley, B., 452
Stanley, B. G., 169
Stanley, G. B., 310
Stanley, M., 452
Stanwood, G. D., 132
Stapp, H., 8
Stárka, L., 192
Starr, C., 344
Stäubli, U. V., 373
St. Clair, D., 441
Stebbins, W. C., 284
Steele, J. B., 457
Steele, S., 416
Stefansson, H., 432
Steffen, K. K., 173, 174
Stein, A. D., 177
Stein, J., 276
Steinberg, S., 432
Steinhauer, S., 149
Steinman, J. L., 339
Steinmetz, M. A., 371
Stellar, E., 157
Stellar, J. R., 157
Stellflug, J. N., 194, 213
Stelmack, R. M., 399
Stephan, F. K., 470
Stephens, D. N., 129
Stephens, T. W., 170
Stepper, S., 226
Steriade, M., 480
Stern, E., 439
Stern, J. S., 178, 179
Stern, K., 197
Stern, M. B., 140
Stern, Y., 379
Sternbach, R. A., 239
Sternberg, R., 395
Sterzi, R., 279
Stetson, B., 178
Steudel, T., 266
Steur, E. N. H. J., 484
Stevens, M. C., 416
Stevenson, L. A., 135
Stevenson, M. R., 488
Sticker, E. M., 160
Stickgold, R., 373, 477
Stiffler, J. S., 444, 452
Stigler, K. A., 417
Stijnene, T., 485
Stillman, A. A., 460

Stinson, F. S., 140, 145
Stix, G., 406
Stoddard, J., 243, 245
Stoessel, P. W., 457
Stoessl, A. J., 352
Stöffler, A., 384
Stojanovic, M., 342
Stoller, R., 190
Stone, B., 241
Stone, S., 168
Stoppelbein, L., 243
Storm, D., 378, 380
Stormshak, F., 194, 213
Stout, J., 236
Stracciari, I., 497
Strack, F., 226
Strain, K. J., 352
Stratton, G. M., 371
Strecker, R. E., 478
Streissguth, A. P., 77, 130
Strey, H. H., 197
Stricker, E. M., 169
Striegel-Moore, R., 174
Strober, M., 183
Strüber, D., 243
Studdert-Kennedy, M., 284
Stunkard, A. J., 163
Sturla, E., 205
Stuss, D. T., 490
Suckling, J., 457
Suddath, R. L., 436, 437, 438
Suddendorf, T., 493
Suga, N., 267
Sugden, K., 444, 445
Sugerman, H. J., 180
Sugihara, H., 168
Suhara, T., 435
Suleman, M. A., 237
Sullivan, P. F., 181, 444
Sulzer, D., 43
Sumi, S. M., 77
Summers, B. A., 352
Sumner, J., 417
Sun, K., 378
Sun, Y., 480
Sun, Y. -G., 330
Sunaert, S., 320
Suomi, S. J., 244
Suplita, R. L., 341
Surget, A., 448, 449
Susser, E., 440
Susser, E. S., 177
Sutcliffe, J. S., 417, 419
Sutton, G., 11
Sutton, S. R., 133
Suttorp, M., 180
Suvisaari, J. M., 429
Suyenobu, B., 203
Suzdak, P. D., 144
Suzuki, K., 435
Sved, A. F., 160, 169
Svensson, O., 341
Svensson, T. H., 134
Svoboda, K., 374
Swaab, D. F., 213, 214, 216
Swann, A. C., 244
Swanson, J. M., 139

Swartz, K. J., 32
Swayze, V. W., II, 60, 433, 434
Sweatt, D., 374
Swedo, S. E., 457, 458
Sweeney, K., 162
Switala, A. E., 398
Switkes, E., 96, 97, 307
Syngeniotis, A., 316
Szabo, P., 384
Szabo, Z., 135
Szalavitz, M., 145, 146
Szegedi, A., 461
Szeszko, P. R., 457
Szobota, S., 32
Szuajderman, S., 39
Szumlinski, K. K., 148
Szymusiak, R., 478

Taanman, J. W., 353
Tabak, C. J., 172
Tabakoff, B., 129
Tabrizi, S. J., 353
Tadic, A., 461
Tafti, M., 483
Tagaki, S., 148
Taggart, R., 344
Taglialatela, J. P., 283
Taheri, S., 481
Taira, M., 349
Takahashi, J. S., 148
Takao, M., 472
Takaoka, Y., 336
Takeda-Uchimura, Y., 385
Takei, N., 441
Takimoto-Kimura, R., 411
Talairach, J., 348
Talbot, J. D., 240
Talbot, P. S., 244
Talih, M., 148
Tallal, P. A., 380
Tamarkin, L., 450
Tamietto, M., 317
Tamura, H., 168
Tanabe, J. L., 439
Tanaka, K., 317, 318
Tanaka, M., 181, 182, 335
Tanaka, W. K., 179
Tanaka, Y., 240
Tanda, G., 136, 137
Tanfer, K., 210
Tang, C., 371, 400
Tang, M., 133, 379
Tangney, C. C., 408
Tanji, J., 347, 348
Tank, D. W., 98
Tanne, D., 476
Tanner, C. M., 352
Tannock, R., 420, 421
Tanskanen, A. J., 429
Tantravahi, U., 411
Tanzi, R. E., 13, 383
Tapper, A. R., 133, 148
Tarara, R., 237
Tarr, M. J., 317, 318
Tarshish, C. Y., 237, 380
Tasca, G. A., 183
Tasker, R. R., 240, 489

Tate, B. A., 117
Tate, K., 136
Tatsch, K., 420
Tatton, W. G., 351, 352
Taub, D. M., 244
Taub, E., 114, 487, 488
Tauber, R., 461
Taylor, A., 444, 445
Taylor, A. H., 402
Taylor, C. S. R., 336, 349
Taylor, D. N., 237
Taylor, G., 439
Taylor, J. S., 168
Taylor, R., 135
Taylor, S., 458, 460
Taylor, S. F., 228, 229, 232
Taylor, V. H., 180
Teasdale, T. W., 440
Teicher, M. H., 450
Teixeira, C. M., 375
Tejada-Vera, B., 80
Telang, F., 180
Tellegen, A., 203
Temel, Y., 460
Temoshok, L., 238
Tempelmann, C., 162
Temple, E., 238, 278, 401
Tenore, F. V. G., 84
Tepas, D. I., 468
Terasaki, O., 435
Termine, A., 439
Terrace, H. S., 282, 401
Tessier-Lavigne, M., 74, 75
Tessitore, A., 244, 444
Testa, F., 300
Thach, W. T., 350
Thakkar, M., 478
Thakur, M., 237
Thal, L., 386
Thallmair, M., 81
Thanos, P., 139, 140
Thapar, A., 109, 133, 134, 421
Theilmann, J., 354
Theobald, D. E. H., 148
Thériault, G., 176
Thibaut, F., 192
Thiele, A., 316
Thier, P., 316
Thigpen, C. H., 495
Thijssen, J., 193
Thisted, R. A., 481
Thomas, A. J., 278
Thomas, C., 416
Thomas, K. M., 420
Thomas, S. P., 342
Thompson, J. C., 316
Thompson, P. M., 77, 78, 398, 403, 404, 420, 442
Thompson, R. F., 351
Thompson, W. L., 316
Thorell, L. G., 312
Thornhill, R., 196, 405
Thornton, G., 341
Thornton, J., 170
Thorsen, P., 417
Thrasher, T. N., 159
Threlkeld, S. W., 276

Thron, A., 409
Tian, H., 149
Tian, M., 317
Tibshirani, R. J., 489
Ticho, S. R., 478
Tikhonov, A., 316
Tilleskjor, C., 440
Tilmann, P., 236
Timson, N. J., 175
Tishby, O., 39
Tisserand, D. J., 408
Tobet, S., 213
Tobey, E. A., 267
Tobi, E. W., 177
Tobler, I., 468
Todd, J. J., 489
Todd, J. T., 298
Todd, R. D., 453
Toga, A. W., 77, 243, 420
Toh, K. L., 482
Tojo, Y., 414
Tondo, L., 450
Tonegawa, S., 81
Tonelli-Lemos, L., 179
Tonetti, A., 241
Tong, C., 353
Tong, F., 314, 487
Tong, L., 267
Tonge, B. J., 417
Toni, N., 374, 375
Tononi, G., 377, 476, 497, 498
Toolanen, G., 341
Tootell, R. B. H., 96, 97, 307
Töpfner, S., 342
Torii, M., 57
Torres, F., 78
Torrey, E. F., 437, 438, 440
Tosi, J., 300
Toso, L., 410
Tottenham, N. T., 420
Touchette, P., 400
Townsend, J., 351
Tracey, I., 340, 341
Tranel, D., 60, 61, 224, 232, 233, 274, 317, 371, 380, 397
Träskman, L., 452
Traskman-Bendz, L., 452
Trautmann, A., 45
Treffert, D. A., 414, 415, 416
Tregellas, J. R., 439
Treisman, A. M., 486
Tremblay, A., 176
Tremblay, R. E., 147
Trestman, R. L., 435, 446
Triggiano, P. J., 495
Trinko, R., 180
Tripp, G., 420
Trippe, J., 398
Trogdon, J. G., 172
Trojanowski, J. Q., 352
Trotta, N. C., 458
Troughton, E., 148
True, W. R., 456
Trulson, M. E., 480
Trulson, V. M., 480
Tsai, C. H., 461
Tsai, F. J., 461

Tsai, G., 129, 436
Tsai, G. E., 496
Tsai, S. -J., 461
Tsai, W. -Y., 352, 353
Tsakanikas, D., 384
Tsakanikos, E., 322
Tsankova, N., 432
Tsay, R., 179
Tshibanda, L., 499
Tshibanda, L. J. -F., 498
Tsivkin, S., 401
Tsuang, M. T., 14, 15, 431, 432
Tsukamoto, H., 484
Tuinier, S., 436
Tuiten, A., 193
Tully, C., 497
Tulving, E., 107, 490
Tung, K. -H., 352
Turel, M., 39
Turetsky, B. I., 398
Turetsky, D., 117, 398
Turkheimer, E., 406
Turnbull, O., 492
Turner, C. W., 237
Turner, M. S., 387, 492
Turner, R. C., 164, 177
Tuszynski, M. H., 380, 384, 385, 386
Twain, M., 7
Tyndale, R. F., 148, 439
Tyrrell, A. D., 236
Tyrrell, G., 433, 434

Uddin, L. Q., 420
Udry, J. R., 193
Ugen, K. E., 384
Uhl, G. R., 2, 80, 146, 429, 443, 445
Uhlhaas, P. J., 438
Ukiyama, E., 200
Ullman, M. T., 282
Umberson, D., 237
Umegaki, H., 140
Uncapher, M. R., 487
Underwood, M. D., 447, 452
Unerti, K., 342
Ungerleider, L. G., 274, 315, 316, 321, 369
Uno, H., 236, 237
Urbancic, M., 455
Ursano, R. J., 455
Urschel, H. C., Jr., 117
Usui, A., 484

Vaina, L. M., 316, 320
Vaisse, C., 169
Valenstein, E. S., 61
Valera, E. M., 420
Vallejo, J., 229
Vallejo-Illaramendi, A., 436
Valletta, J., 411
van Amelsvoort, T., 409
van Beijsterveld, C. E. M., 403
van de Beek, c., 213
Van de Poll, N. E., 194, 241
van den Bree, M., 109, 133, 134
van den Heuvel, O. A., 482
van der Heijden, F. M., 436
Van der Kolk, B. A., 496
van der Veen, F. M., 408

Van Der Werf, Y. D., 482
van Dorp, C. F., 471
Van Dyk, K., 384
Van Eerdewegh, M., 443, 444
van Erp, T., 398, 403
Van Essen, D. C., 322
Van Evercooren, A. B., 355
Van Goozen, S. H., 241
Van Haren, N. E. M., 203
Van Hecke, P., 320
Van Honk, J., 193
Van Horn, J. D., 39
Van Horsen, G. W., 382
Van Huijzen, C., 159
van Oost, B. A., 245
van Os, J., 439
van Praag, H. M., 375, 435, 436
van Roon-Mom, W. M. C., 82
Van Soest, P. J., 162
Van Someren, E. J. W., 482
Van Wyk, P. H., 210
Vance, K., 455
Vanderstrichele, H., 385
Vanderweele, D. A., 163, 168
Vandlen, R., 179
Vanduffel, W., 402
Vanhaudenhuyse, A., 498, 499
Vanmechelen, E., 385
Vann, F. H., 207
Vannier, M., 453
Vargha-Khadem, F., 285
Varki, A., 117
Vasile, R. G., 496
Vaughan, J., 352
Vaughn, M. G., 245
Veasey, S. C., 480
Vela-Bueno, A., 481
Venter, J. C., 13
Ventura, M., 499
Verbalis, J. G., 169
Verbaten, R., 193
Verchinski, B. A., 444
Veres, G., 111
Vergnes, M., 243
Verhaak, P. F. M., 341
Verhoeff, N. P., 435
Verhoeven, W. M., 436
Veridiano, N. P., 207
Vermetten, E., 236, 496
Vernes, S. C., 285
Verney, E. L., 163
Vernon, P. A., 399, 400
Verstraten, F. A. J., 316
Vessicchio, J. C., 439
Vgontzas, A. N., 481
Vickers, J., 244
Victor, J. D., 499
Vidal, C., 442
Videen, T. O., 367, 452
Vidyasagar, T. R., 489
Vighetti, S., 317
Vignolo, L. A., 273, 274, 280
Vikingstad, E. M., 278
Vilain, E., 212
Vilella, E., 444, 445
Villalobos, M. E., 414
Villemagne, V. L., 131, 141, 142, 385

Vina, R. F., 117
Vinay, L., 346
Vincent, J. -D., 81, 102
Vincent, J. L., 408
Viñuela, A., 441
Viola, A., 236
Viola, K. L., 382
Virkkunen, M., 243, 244, 245
Virshup, D. M., 482
Visel, Z., 13
Visser-Vandewalle, V., 460
Vitaterna, M., 148
Vitellaro-Zuccarello, L., 436
Vlahov, D., 148
Vogel, G., 62
Vogel, G. W., 450
Vogels, R., 318
Vogelstein, R. J., 84
Vogt, B., 203
Völker, C., 236
Volkmann, J., 78, 79
Volkow, N. D., 131, 139, 140, 141, 142, 145, 174, 180, 446
Volle, E., 318, 487
Vollenweider, F. X., 134
Volpe, B. T., 320
von Cramon, D., 320
von de Heydt, R., 312
von Helmholtz, H., 6
von Schantz, M., 472
Vonlanthen, A., 417
Vonsattel, J. P., 353
Voogd, J., 159
Voon, V., 448
Vorel, S. R., 142
Voss, P., 78
Voss, T. S., 354
Voyer, D., 202, 206, 207
Voyer, S., 202
Vreeling, F., 460
Vrenken, H., 482
Vul, E., 107
Vythilingham, M., 455

Wada, J. A., 280
Wade, J. A., 379
Wadhwa, R., 408
Wager, T., 228, 229, 232
Wagner, A. D., 106
Wagner, H. J., 355
Waite, P. M., 83, 478
Wakabayashi, I., 168
Wakefield, A. J., 418
Wald, G., 304
Wald, L. L., 436
Wald, M. M., 80
Waldman, I. D., 406
Walker, E. F., 442
Walker, J. R., 472
Walker, M. P., 497
Wall, P., 339
Wall, T. L., 148
Wallace, G. L., 416
Wallis, D., 421
Walsh, A. E., 460
Walsh, B. T., 181, 182
Walsh, M. L., 241, 242, 243

Walsh, P., 355
Walsh, V., 320
Walters, E. E., 428, 443, 454
Walters, R., 416, 419
Waltes, R., 417
Walther-Jallow, L., 440
Walts, B., 175
Walum, H., 198
Wan, F. -J., 129
Wandinger, K. P., 355
Wang, A. H., 375
Wang, G. -J., 131, 139, 140, 142, 145, 174,
    180, 446
Wang, H., 439
Wang, H. -Y., 132
Wang, J., 203
Wang, K., 419
Wang, L., 378
Wang, N. -K., 300
Wang, O., 179
Wang, P., 441
Wang, S. S. -H., 374
Wang, S. Y., 340
Wang, T. R., 461
Wang, X., 341
Wang, Y., 276, 404, 409
Wang, Z., 81, 198
Wannier, T., 83
Wardak, C., 298
Wardlaw, S. L., 339
Ware, M., 401
Warner, L. A., 148
Warner, M. H., 212
Warner, V., 451
Warner-Czyz, A. D., 267
Warren, S., 335
Washington, R. A., 132
Wasserman, R., 294
Wasserstein, J., 419
Waterhouse, L., 417
Waterland, R. A., 177
Waters, A. J., 133
Watkins, L. R., 339
Watkinson, B., 136
Watson, C. G., 440
Watson, J., 10
Watson, R. T., 233
Waxman, S. G., 34
Weaver, D. R., 473
Webb, D. M., 197, 305
Webb, W. B., 468, 476
Weber, B., 231
Wedekind, C., 196
Wedel, H., 180
Weedon, M. N., 175
Weeks, D., 447
Wegesin, D. J., 215
Wegner, D. M., 239
Wehr, T. A., 449, 450
Wehrle, R., 497
Weightman, D. R., 214, 215
Weigle, D. S., 169
Weile, B., 236
Weinberg, M. S., 210
Weinberg, R. A., 404, 406
Weinberger, D. R., 436, 437, 438, 442
Weiner, R. D., 447

Weingarten, H. P., 170
Weinstein, S., 331
Weiskopf, N., 369
Weiskrantz, I., 317
Weisnagel, J., 175
Weisnagel, S. J., 175
Weiss, K. J., 459
Weiss, R. D., 141
Weiss, S. R., 183
Weisse, C. S., 236
Weisskopf, M. G., 421
Weitzman, E. D., 482
Weizman, A., 439
Wekerle, H., 355
Welch, J. M., 458, 459
Welch, K. M. A., 278
Welcome, S. E., 420
Welling, B., 317
Wells, J. E., 460
Welner, J., 430, 431
Weltzin, T. F., 181, 182, 183
Wen, X., 383
Wender, P. H., 419, 431
Wenkstern, D., 139
Wenthold, R. J., 374
Wentzek, S., 317
Wermke, K., 280
Wernicke, C., 270
Wessel, K., 355
West, D. B., 170
West, S. L., 140
Westberg, L., 198
Westerfield, M., 351
Wetter, T. C., 497
Wever, E., 261, 262, 265
Wexler, N. S., 13
Wheaton, K. J., 316
Wheeler, M. A., 490
Wheeler, M. E., 369
Whidbee, D., 477
Whipple, B., 339
Whitam, F. L., 216
White, F. J., 148
White, G. L., 383
White, N. M., 370
White, S., 277
Whiten, A., 493
Whiteside, S. P., 457
Whitfield, S., 439
Whitman, F. L., 211
Whitney, D., 320
Whittingstall, K., 266
Whittle, S., 136
Whone, A. L., 355
Wickelgren, I., 173, 372
Wickens, J. R., 420
Wiebe, V., 285, 305
Wiederhold, B. K., 456
Wiederman, M. W., 183
Wiegant, V. M., 241
Wienbruch, C., 79, 342
Wierzynski, C. M., 376
Wiesel, T., 16, 309, 310, 311
Wiesel, T. N., 305
Wieten, W., 225
Wigal, S. B., 441
Wiggs, C. L., 274, 369

Wightman, F. L., 266
Wikler, K. C., 201
Wilbert-Lampen, U., 236
Wilcox, K. J., 203
Wilens, T. E., 419
Wiliamson, A., 372, 373
Wilkinson, D., 384
Willerman, L., 57, 400
Williams, A., 482
Williams, C. C., 203
Williams, G., 179
Williams, G. M., 130
Williams, J., 282
Williams, J. H. G., 493
Williams, R. B., 444
Williams, R. W., 24, 76
Williams, S. C. R., 322
Williams, S. L., 283
Williams, T. J., 212, 215
Willie, J. T., 166, 484
Willmes, K., 409
Willwerth, J., 486
Wilska, A., 256
Wilson, A., 492
Wilson, A. A., 244
Wilson, A. C., 14, 282
Wilson, A. F., 453
Wilson, F. A. W., 316
Wilson, M., 203
Wilson, M. A., 369, 376, 471
Wilson, R. L., 136
Wilson, R. S., 408
Wimo, A., 383
Winblad, B., 383
Winchester, L., 285
Winden, K., 285
Wingard, D. L., 481
Wingfield, J. C., 212
Winkelmann, J., 484
Winkielman, P., 107, 414
Winocur, G., 317, 380
Winslow, J. T., 198
Wise, R. A., 132, 135, 138, 139, 140,
    141, 174
Wise, S. P., 372
Wisse, B. E., 169
Witelson, S. F., 214, 280, 396, 397, 398
Witherby, S. A., 322
Withers, D. J., 179
Witt, R., 343
Witte, A. V., 174
Witte, H., 487, 488
Wittenberg, G. M., 374
Witztum, E., 192
Wojtowicz, J. M., 81
Wolf, A. P., 131
Wolf, L. E., 419
Wolf, O. T., 380
Wolff, M., 355
Wolkowitz, O. M., 245
Wollberg, Z., 259
Wolpin, J., 132
Wolschläger, A. M., 226
Wolters, G., 485
Woltjen, K., 117
Wong, A. H. C., 439
Wong, C. T., 174, 180

Wong, D. F., 131, 132
Wong, G. W. H., 439
Wong, K., 434, 435
Wong, K. Y., 472
Wood, D. L., 236
Wood, N. W., 352
Wood, P. D., 342
Wood, S., 419, 455
Woods, A., 179
Woods, B. T., 442
Woods, C. G., 76
Woods, R. P., 284
Woods, S. C., 170
Woods, S. E., 455
Woodside, S., 244
Woodward, L. J., 133
Woodward, M., 488
Woodworth, R. S., 16
Woolf, C. J., 341
Woolf, V., 224
Woosa, R. J., 160
Worley, P. F., 451
Wright, M. J., 403
Wright, R. O., 421
Wronski, M., 112
Wu, J. C., 449
Wu, M. T., 496
Wundt, W., 3
Wurtman, J. J., 179, 450
Wurtman, R. J., 179, 450
Wyart, C., 32
Wyatt, H. R., 174, 176
Wyatt, J. R., 267
Wyatt, R. J., 81, 437, 440
Wynn, K., 401
Wynne, K., 179
Wysocki, C. J., 197

Xiao, Z., 267
Xie, H. M., 421
Xie, L., 82
Xu, J., 80
Xu, K., 352
Xu, L., 80
Xu, M., 441
Xu, Q., 339
Xu, Y. -H., 352
Xuei, X., 149
Xuereb, J., 353

Yadegarfar, G., 384
Yaffe, R., 417
Yam, P., 394
Yamada, M., 448, 449
Yamada, M., 448, 449

Yamamoto, J., 337
Yamamura, H. I., 434
Yamane, S., 318
Yamaura, A., 203
Yan, M., 398
Yanagisawa, M., 166, 212
Yancopoulos, G., 178
Yang, J. E., 345
Yang, S., 404
Yang, Y., 243, 420
Yang, Z., 197
Yap, G. S., 347, 348
Yarbrough, C. J., 113
Yasuhara, D., 182
Yau, K.-W., 472
Yehuda, R., 237
Yelnik, J., 460
Yeni-Komshian, G. H., 280, 283
Yeo, G. S. H., 175
Yeo, R. A., 398
Yeomans, J., 439
Yin, J., 330
Yoo, S. -S., 497
Yoshino, A., 484
You, J. S., 461
Youdim, M. B. H., 351
Young, A., 176
Young, A. W., 232
Young, L. J., 198, 1988
Young, M. P., 318
Young, R. L., 415
Young, T., 301, 481
Younger, D. S., 355
Younger, P., 135
Youstry, T. A., 387, 492
Yu, J. S., 117
Yu, Y., 441
Yücel, M., 136
Yuzda, E., 418

Zack, J. A., 238
Zack, M., 140
Zahn, T. P., 495
Zai, G., 457, 460, 461
Zaidel, E., 233
Zalla, T., 417
Zaman, S. H., 374
Zambreanu, L., 341
Zametkin, A. J., 420
Zandi, P., 149
Zanzotto, G., 212
Zarate, C. A., 446, 451
Zatorre, R. J., 78, 280
Zawadzki, J. K., 176
Zea-Ponce, Y., 434

Zec, R. F., 437, 438
Zeggini, E., 175
Zeki, S., 317, 319, 322
Zeman, A., 497
Zeppelin, H., 468
Z'Graggen, W. J., 81
Zhang, B., 451
Zhang, F., 441
Zhang, H., 489
Zhang, H. G., 461
Zhang, J., 197, 305, 404
Zhang, S. M., 355
Zhang, S. P., 243
Zhang, X., 444
Zhang, Y., 179, 300
Zhao, M., 81
Zhao, T., 284
Zhao, Z., 367
Zhao, Z. -Q., 330
Zhen, Z., 317
Zheng, H. -K., 404
Zheng, J., 259
Zhong, Y., 378
Zhou, J. -N., 214
Zhou, J. N., 213, 214
Zhou, Y., 409, 477
Zhu, C., 421
Zhu, H., 451
Zhu, M., 174
Zhu, Q., 317
Zhu, W., 174, 180
Zhubi, A., 439
Zigun, J. R., 437
Zihl, J., 320
Zilles, K., 279
Zillmer, E. A., 24
Zimmerman, I., 207
Zimmerman, J. C., 482
Zipursky, R. B., 181, 436
Zoccolotti, P., 233
Zohar, J., 458
Zola, S. M., 366
Zola-Morgan, S., 366, 387
Zöllner, S., 14
Zoloth, S. R., 284
Zonderman, A. B., 409
Zou, K., 451
Zou, S., 174
Zubenko, G. S., 444, 452
Zubenko, W. N., 444, 452
Zuberi, A., 175
Zubin, J., 15, 432
Zucker, K. J., 209
Zuckerman, M., 157
Zukin, S., 37

Subject Index

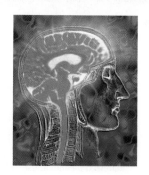

Abecedarian Project, 405–406
Ablatio penis, 207–209, 209 (figure)
Ablation, 101–102
Absolute refractory period, 31
Absorptive phase, 165–166, 166 (figure)
Acetylcholine (ACh), 41, 42, 43, 44 (table),
    45, 134, 344, 354, 384
Acquired dyslexia, 276
Acral lick syndrome, 458
Action potential, 29–31, 30 (figure), 33, 34
Activating effects, 200
Activation-synthesis hypothesis, 475–476
Activity-based recovery, 84–85
Acupuncture, 339
Acute symptoms, 430
Adaptive hypothesis (of sleep), 468
Addiction, 126, 137–138
    addiction research, implications of, 149
    agonist treatments for, 143–144
    antagonist treatments for, 144, 144 (figure)
    antidrug vaccines and, 144
    aversive treatments for, 144
    benzodiazepine drugs and, 130
    caffeine and, 126
    cocaine and, 131, 138
    comorbidities and, 145
    compensatory adaptation and, 127
    compulsive gambling and, 140
    compulsive overeating and, 179–180,
        180 (figure)
    conditioned tolerance and, 127
    craving and, 141–142
    detoxification phase and, 143
    dopamine, role of, 139–140,
        139–140 (figures), 141, 144
    genetic vs. environmental influences on,
        146–149, 147 (table), 149 (figure)
    learning/brain plasticity, dopamine and,
        141–142, 141–142 (figures)
    marijuana and, 137
    medications for treatment of,
        145–146, 145 (table)
    neural basis of, 138–139
    opiates and, 127
    overdose and, 127
    pharmacological treatment, effectiveness/
        acceptance of, 145–146
    phencyclidine/PCP and, 135
    relapse rates and, 126, 143, 143 (figure)
    reward and, 138–139
    reward deficiency syndrome and, 140
    self-medication impulse and, 145
    serotonin-potentiating drugs and, 144–145
    tolerance and, 126–127

treatments for, 142–145, 145 (table)
withdrawal reactions and, 126, 127, 138
See also Alcoholism; Psychoactive drugs;
    Substance abuse
Adenosine, 134, 339, 478–479, 479 (figure)
Adequate stimulus:
    hearing and, 254
    vision and, 294
ADHD. See attention deficit hyperactivity
    disorder
Adolescents:
    myelination process and, 77–78
    puberty, 201
Adoption, 15, 132, 406
Adoption studies, 109, 110 (table),
    147–148, 431
Adrenal glands, 74
Adulthood:
    late-life gene functioning, 14
    See also Aging
Affective aggression, 241
Affective disorders, 443
    antidepressant treatment for, 446,
        448–450, 449 (figure)
    bipolar disorder, 443, 445, 450–451,
        452, 452 (figure)
    brain anomalies in, 451–452, 451–453
        (figures)
    brain-derived neurotrophic factor and, 444
    circadian rhythm and, 444, 449–450
    deep brain stimulation and, 448
    electroconvulsive therapy and, 447,
        447 (figure), 448
    heredity and, 444–445, 445 (figure)
    lithium and, 451
    major depression, 443
    mania, 443
    monoamine hypothesis of depression,
        445–446, 446 (figure)
    norepinephrine and, 446
    phototherapy and, 450
    seasonal affective disorder, 450, 450 (figure)
    serotonin and, 446, 450
    tobacco use and, 446, 446 (figure)
    transcranial magnetic stimulation and, 448
    treatment resistance/delayed effectiveness
        issues and, 446
    unipolar depression, 443, 449
    See also Anxiety disorders; Depression;
        Psychological disorders; Suicide
Agency, 490, 491, 491 (figure)
Aggression, 202, 203, 204, 207
    affective aggression, 241
    alcohol-serotonin link in, 244–245

antisocial personality disorder and, 243
biological origins of, 240–245
brain's role in, 242–243, 242 (figure)
defensive aggression, 241
definition of, 241
genetic/environmental influences on,
    245–246, 245 (figure)
hormones, influence of, 241–242,
    241 (figure)
lesioning technique and, 243
offensive aggression, 241
predatory aggression, 241
prefrontal cortical activity and, 243, 244
proactive aggression, 241, 243
psychopathy and, 243
reactive aggression, 241, 243
serotonin effects and, 243–245, 244 (figure)
social restraints on, 244
testosterone levels and, 241–242,
    241 (figure), 244
See also Emotion; Stress
Aging:
    basal forebrain region and, 380
    gene functioning and, 14
    intelligence and, 407–409, 408 (figure)
    memory, effects on, 379–380
    pain, premature aging and, 341–342
    successful aging, 380, 408
    See also Adulthood
Agonist, 45, 126
Agonist treatments, 143–144
Agouti-related protein (AgRP), 169, 171
Agraphia, 275
AIDS (acquired immune deficiency
    syndrome), 235, 237, 238
Alcoholism, 12, 128
    aggression, alcohol-serotonin effects on,
        244–245
    antisocial behavior and, 146, 147
    binge drinking and, 129, 130
    brain stem shut-down and, 128
    cognitive/motor malfunction and, 128, 129
    cross-generational effects and,
        130, 148, 149
    delirium tremens and, 129
    early-onset alcoholism, 146–147,
        147 (table), 244
    fetal alcohol syndrome, 77, 78 (figure),
        130, 130 (figure)
    GABA$_A$ receptor complex and, 129–130,
        130 (figure)
    genetic vs. environmental influences on,
        146–149, 147 (table), 149 (figure)
    glutamate release and, 129

health/behavioral problems and,
128–129, 129 (figure)
judgment impairment and, 128
Korsakoff's syndrome and, 129, 129
(figure), 386–387
late-onset alcoholism, 146, 147, 147 (table)
maternal alcohol abuse, 421
medications for treatment of, 145 (table)
personality characteristics and, 146–147,
147 (table)
relapse rates, 143, 143 (figure)
societal costs of, 129
vitamin deficiencies and, 129
withdrawal reactions and, 129
See also Depressants; Psychoactive drugs
Aldehyde dehydrogenase (ALDH), 148
ALDH. *See* aldehyde dehydrogenase
Alexia, 275, 294
All-or-none law, 31, 38
Alleles, 11
Alzheimer's disease, 3, 14, 111, 172, 352,
352 (figure), 380, 492
amyloid precursor protein gene and,
382–383, 383 (table)
attention difficulties, 382
basal forebrain cell loss and, 380
brain size, neuron loss and, 381–382,
382 (figure)
characteristics of, 381, 382
declarative memory and, 381
detection of, 384–386
Down syndrome and, 382
early-onset Alzheimer's disease,
383, 383 (table)
estrogen replacement therapy and, 409
genetic interventions for, 386
heredity/genetic activity and, 382–383,
383 (table)
immunization and, 384
incidence of, 383, 384 (figure)
late-onset Alzheimer's disease,
383, 383 (table)
motor difficulties and, 382
nerve growth factor and, 384,
385 (figure), 386
neurofibrillary tangles in, 381,
381 (figure), 382
plaques in, 381, 381 (figure), 382
soluble amyloid and, 382
treatment of, 383–384
visual deficits and, 382
Amacrine cells, 297, 297 (figure)
American Association for the Advancement
of Science, 116
American Medical Association (AMA), 113
American Psychiatric Association, 409, 429,
430, 495
American Psychological Association (APA),
12, 113, 405
American Psychological Society (APS), 114
Amino acids, 28, 43, 44 (table), 165
Amnesia, 130, 365
anterograde amnesia, 366, 386
confabulation and, 387
dissociative identity disorder, state-
dependent learning and, 495–496,
496 (figure)

hippocampal formation and, 365–366
medial temporal lobe damage and,
365–366
retrograde amnesia, 365, 366, 386
See also Learning process; Memory
Amphetamines, 93, 131, 132–133
freebase form of, 132
hallucinations/delusions and, 133
long-term sensitivity to, 133
methamphetamine, 132
psychotic symptoms and, 133
uses for, 132, 419–420
See also Psychoactive drugs; Stimulants
Amplitude, 255, 256 (figure)
Amputations, 79, 342, 342 (figure),
343, 491, 491 (figure)
Amygdala, 136, 138, 194, 197, 232
aggression and, 242, 242 (figure), 243, 244
antianxiety medications and, 233
damage to, 232–233
emotion and, 228, 228 (figure),
231–233, 232 (figure)
memory and, 232
nondeclarative emotional learning
and, 371
trustfulness and, 232
See also Anxiety disorders; Limbic
system; Sex
Amyloid precursor protein (APP), 382–383,
383 (table)
Amyotrophic lateral sclerosis (ALS),
103, 392
Analgesics, 127, 128
Androgen insensitivity syndrome, 205,
205 (figure), 206, 207
Androgens, 192, 200, 205, 212–213
Androids, 47, 47 (figure)
Anesthetics, 131, 337, 338, 339
Angel dust, 135
Angiotensin II, 160
Angular gyrus, 275
Animal mosaic, 111
Animal research subjects, 110–111,
113–115, 115 (figure)
Animal spirits, 5
Anorexia nervosa, 181–183, 182 (figure)
ANS. *See* autonomic nervous system
Antagonist, 45, 126
Antagonist treatments, 144, 144 (figure)
Antagonistic muscles, 344–345
Anterior, 59, 59 (figure)
Anterior cingulate cortex (ACC), 227–228
(figures), 228–229, 240, 244, 339
(figure), 489, 490, 493
Anterior commissure, 213
Anterograde amnesia, 366, 386
Antidepressant medications, 245, 354, 417,
446, 448–449, 449 (figure)
Antidrug vaccines, 144
Antipsychotic medications, 354, 417,
435–436, 439
Antisense RNA procedure, 110, 111 (figure)
Antisocial personality disorder, 147, 243
Anxiety, 12, 126, 130, 131, 132
Anxiety disorders, 145, 421, 454
antianxiety drugs and, 454–455
benzodiazepine and, 454–455

brain structures and, 455
fear ensure technique and, 456
generalized anxiety, 454
heredity and, 460–461
neurotransmitters and, 454–455
panic disorder, 454, 455
phobia, 454
posttraumatic stress disorder and,
455–456, 455 (figure)
serotonin activity and, 455
virtual reality therapeutic tool and, 456
See also Affective disorders; Obsessive-
compulsive disorder; Psychological
disorders
Aphasia, 82, 270–271
Broca's aphasia, 271–272, 278
receptive aphasia, 272
recovery, mechanisms of, 278–279
Wada technique and, 278
Wernicke's aphasia, 272, 278
See also Language
Apotemnophilia, 336
Arcuate nucleus, 168, 168 (figure), 171
Area postrema, 165
Argus Retinal Prothesis System, 300
Arousal system, 479–480,
480–481 (figures)
Arousal theory, 157
Artificial neural networks, 46–47,
47 (figure)
Asexuality, 210
Asperger's syndrome, 412–413, 414
Association areas, 63, 68 (table), 346
Associative long-term potentiation, 373
Astrocytes, 35, 35 (figure)
Asymmetry, 55
Atoms, 4
Attention, 67, 488
auditory object and, 266
Cheshire cat effect and, 489, 489 (figure)
cocktail party effect, selective attention
and, 266
consciousness and, 488–490, 489 (figure)
neglect, attention in vision and, 320–321,
321 (figures)
physiological process of, 489–490
Attention deficit hyperactivity disorder
(ADHD), 203, 419
adult behaviors and, 419
brain anomalies in, 420, 421 (figure)
dopamine pathways and, 420
environmental influences in, 421
heredity and, 420–421
neurotransmitter anomalies in,
419–420
pesticide pollution and, 421
prefrontal cortex and, 420
striatum and, 420
substance abuse and, 419, 420 (figure)
treatments for, 419
Auditory cortex, 58 (figure), 63, 68 (table),
259, 260 (figure)
auditory pathways and, 259,
260 (figures), 261
dorsal stream and, 261 (figure), 261
primary auditory areas, 259,
260 (figure), 261

secondary auditory areas and, 259,
259, 260 (figure), 266
topographical organization of, 259
ventral stream and, 260 (figure), 261
See also Auditory mechanism; Hearing
Auditory mechanism, 256, 257–258
(figures)
auditory cortex and, 259–261,
260 (figures)
auditory nerve and, 257 (figure), 259
auditory pathway and, 259, 260 (figure)
basilar membrane and, 257, 258,
258 (figure), 259, 262–265,
263–265 (figures)
bone conduction and, 265, 265 (figure)
cochlea and, 255, 256–257,
257–258 (figures), 259
cochlear canal and, 257, 258 (figure)
ear and, 256, 257 (figure)
hair cells and, 257–259, 258–259
(figures), 267
helicotrema and, 257
inner ear and, 256–259, 259–260 (figure)
inner hair cells and, 259
intensity and, 255, 265
organ of Corti and, 257, 258 (figure),
259, 267
ossicles and, 256, 257–258 (figures)
outer hair cells and, 259, 267
outer/middle ear and, 256,
257–258 (figures)
oval window and, 257, 258 (figure)
potassium/calcium channels and, 258
receptors and, 254, 258
semicircular canals and, 257 (figure)
tectorial membrane and, 257,
258–259 (figures), 259
tensor tympani and, 256
tympanic canal and, 257, 258 (figure)
tympanic membrane and, 256,
257–258 (figures)
vestibular canal and, 257, 258 (figure)
See also Complex sound; Frequency
analysis; Hearing; Language
Auditory object, 266
Auditory pathways, 259, 260 (figures), 261
Autism, 203, 227, 412–413, 493
Asperger's syndrome and, 412–413, 414
autism spectrum disorders and,
412, 413
autoimmune reactions and, 417–418
biochemical anomalies in, 416–417,
417 (figure)
brain anomalies in, 416, 417 (figure)
childhood vaccinations and, 418
cognitive/social impairment and,
413–414, 413 (figure)
dorsal stream connections and, 414
environmental influences and, 417–418
functional connectivity, reduction of, 416
gene mutations and, 418
heredity and, 418–419
high-functioning autistics, 415
impaired mirror functions and, 414
incidence of, 413
pesticide pollution and, 417
theory of mind and, 413–414, 415

vitamin D deficiency and, 418
See also Intellectual disability
Autistic savants, 396, 414–415,
415–416 (figures)
Autoimmune disorders, 235
autism and, 417–418
cataplexy and, 484
movement disorders and, 354–355
multiple sclerosis and, 355, 355 (figure)
myasthenia gravis and, 354–355,
354 (figure)
Autonomic nervous system (ANS), 71,
71 (figure)
cranial nerves and, 71, 72 (figure)
digestive process and, 165
emergencies, response to, 71
emotion and, 224–227, 225 (figure)
energy conservation and, 71
hypothalamus and, 64
integrated activity of, 74
parasympathetic nervous system and,
71, 71 (figure), 72, 73 (figure), 74
sympathetic nervous system and,
71, 71 (figure), 72, 73 (figure), 74
See also Brain structure; Central nervous
system (CNS); Nervous system;
Peripheral nervous system (PNS)
Autoradiography technique, 96–97,
97 (figure)
Autoreceptors, 41, 42 (figure)
Aversive treatments, 144
Awareness, 486–488, 487–488 (figures), 497
Axon hillock, 38, 39, 39 (figure)
Axons, 25, 25–26 (figures), 29, 33, 34,
34 (figure)
axon growth inhibitors, 81, 83
bundles of, 54, 54 (figure)
circuit formation and, 75–76, 77 (figure)
growth cones and, 75, 77 (figure)
regeneration and, 80–81, 93
white matter and, 56, 68

B cells, 235, 235 (table)
Backward masking, 497, 497 (figure)
Bait shyness, 163
Barbiturates, 121, 130, 233
See also Depressants; Psychoactive drugs
Basal forebrain area, 380, 478, 479 (figure),
480 (figure)
Basal ganglia, 60, 66, 68 (table), 228,
228 (figure), 346, 349
learning process and, 370–371
movement, refinement of, 346, 349–350,
350 (figure), 351 (table)
Basal metabolism, 176
Basal metabolism rate (BMR), 176
Basilar membrane, 257, 258,
258 (figure), 259
cochlear implants and, 267
place theory and, 262–265,
263–265 (figures)
See also Auditory mechanism; Frequency
analysis
BDNF. *See* brain-derived neurotrophic factor
Behavior:
artificial neural networks and, 47
bait shyness, 163

brain localization principle and, 7–8,
7–8 (figures)
dualism and, 4
genetic influences on, 9, 10–12, 11 (figure)
mind-brain problem and, 4
motivated behaviors, 64
myelination process and, 77–78
physical model of, 4–6, 5 (figure)
polygenic characteristics and, 12
rewards/punishment and, 60
See also Biopsychology; Brain function;
Emotion; Mind-brain problem;
Motivation; Nature vs. nurture
debate
Behavioral disorders, 12, 14–16, 15 (figure)
Behavioral genetics, 13
Behaviorism, 468
Benzedrine, 132
Benzodiazepines, 130, 131, 143, 233,
454–455
See also Depressants; Psychoactive drugs
Bhopal chemical disaster, 469
Binaural cues, 268, 268–269 (figures)
Binding problem, 321–322
Binge drinking, 129, 130
Binocular rivalry, 489, 497
Biopsychology, 3, 16, 92
behavior, brain's role in, 3, 4
electrical brain, mechanistic mind and, 6
hydraulic model of brain function and,
5–6, 5 (figure)
localization issue and, 6–8, 7–8 (figures)
mind-brain problem and, 4–8
See also Brain function; Brain structure;
Nature vs. nurture debate; Nervous
system
Bipolar disorder, 132, 443, 445, 450–451,
452, 452 (figure)
Bipolar neurons, 26, 26 (figure)
Bisexuality, 209, 214
Blind spot, 298
Blindness, 78
Blindsight, 317, 487
Blood pressure, 71, 172
Blood type, 11
Blood-brain barrier, 70, 70 (figure),
127, 160, 165
Body image, 336, 491–492, 492 (figure)
Body integrity identity disorder, 336
Body mass index (BMI), 172, 173,
173 (figure)
Body senses, 329–330
body integrity identity disorder and, 336
body sense information pathway, 334
dermatomes and, 333–334, 334 (figure)
out-of-body experience and, 336
posterior parietal cortex and,
334 (figure), 336
proprioception and, 330, 332, 336
sensorimotor system and, 330
skin senses, 330–332, 331 (figure)
somatosensory cortex and, 333–336,
334–335 (figures)
somatosensory system, 330
unified body image and, 336
vestibular sense, 332–333, 333 (figure)
See also Pain

Botox treatments, 226, 226 (figure)
Boxing injury, 80
Brain damage. *See* Central nervous system damage/recovery; Traumatic brain injury (TBI)
Brain function:
    aggression and, 242–243, 242 (figure)
    behavior and, 3, 4
    concussions and, 80
    distributed functions and, 7
    efficient brain organization and, 56–57, 57 (figure)
    electrical brain, mechanistic mind and, 6
    equipotentiality concept and, 7
    hydraulic model of, 5–6, 5 (figure)
    integrated brain activity and, 7
    intelligence and, 396–405, 397–403 (figures)
    light control of neurons and, 32
    localization principle and, 6–8, 7–8 (figures), 55, 62
    mind-brain problem and, 4–8
    neural transmitters/receptors and, 12
    nonmaterial mind and, 8
    reflexes and, 69, 69 (figure)
    sound location, brain circuits for, 268–270, 269 (figure)
    vision and, 298, 299 (figures)
    See also Biopsychology; Brain structure; Learning process; Movement; Nervous system; Pain; Spinal cord; Traumatic brain injury (TBI)
Brain imaging techniques. *See* Research techniques
Brain interpreter, 494
Brain stem, 66–67, 67 (figure), 128, 130, 267, 333, 341
Brain structure:
    affective disorders and, 451–452, 451–453 (figures)
    anterior commissure, 213
    anxiety disorders and, 455
    arousal system, 479–480, 480–481 (figures)
    attention deficit hyperactivity disorder and, 420, 421 (figure)
    autism and, 416, 417 (figure)
    brain injury and, 70, 80, 80 (table)
    brain size-body size link and, 57
    cerebellum, 39, 65 (figure), 67
    cerebral hemispheres, 55–58, 56–57 (figures)
    convoluted brain surface and, 55–56, 55–56 (figures), 57
    corpus callosum, 65–66, 65–66 (figures)
    efficient brain organization and, 56–57, 57 (figure)
    emotional brain and, 227–233, 227–232 (figures)
    forebrain, 55–66, 56 (figure), 58 (figure)
    frontal lobe, 58, 58 (figure), 60–62, 60–62 (figures)
    hindbrain, 67, 67 (figure)
    intelligence and, 396–405, 397–403 (figures)
    interior features, 64–66, 65 (figure)
    learning process and, 14

lissencephalic brain, 54, 54 (figure)
lobes, 58, 58 (figure), 60–64
meninges and, 70
midbrain, 66–67, 67 (figure)
obsessive-compulsive disorder and, 457–458, 457–458 (figures)
occipital lobe, 58 (figure), 64
parietal lobes, 58 (figure), 62–63
pineal gland, 64, 65 (figure)
proteins, role of, 12
sexual activity and, 193–195, 194 (figure)
sexual orientation and, 213–215, 213–215 (figures)
sleep controls and, 478–479, 479 (figure)
spinal cord, 67–69.69 (figure)
temporal lobe, 58 (figure), 63–64
thalamus/hypothalamus, 64, 65 (figure)
third interstitial nucleus of the anterior hypothalamus, 213
ventricles, 66, 66 (figure)
vision and, 298, 299 (figures)
See also Autonomic nervous system (ANS); Brain function; Central nervous system (CNS); Consciousness; Movement; Nervous system; Peripheral nervous system (PNS); Schizophrenia; Spinal cord
Brain-derived neurotrophic factor (BDNF), 444
Broca's aphasia, 271–272, 278
Broca's area, 58 (figure), 60, 68 (table), 270, 271–272, 271 (figure), 284
Bubble-boy disease, 92, 92 (figure)
Bulimia nervosa, 181–183

Caffeine, 126, 134
    withdrawal from, 134
    See also Psychoactive drugs; Stimulants
Calcium, 32, 37, 41
Calories, 176
Cancer:
    colon cancer, 172
    mood disorders and, 238
    natural killer cells and, 235
    pain-inducing acid and, 332
    relaxation training and, 237
Cannabinoids, 129, 135–136, 341
Capillaries, 70
Carbon monoxide poisoning, 352
Cardiac muscles, 344
Cardiovascular activity, 67, 236, 236 (figure)
Case studies, 93
Castration, 192
Cataplexy, 483
Catatonic schizophrenia, 429
CCK. *See* cholecystokinin
Cell body, 24
    grouping of, 54, 54 (figure)
    myelination and, 56
    neural processing and, 56
Cell division, 9–10
Cell membrane. *See* Neuron cell membrane
Central nervous system (CNS), 54, 71 (figure)
    anesthetic drugs, 337
    blood-brain barrier and, 70, 70 (figure)
    brain size and, 57, 57 (figure)
    brain stem and, 66–67, 67 (figure)

cerebellum and, 65 (figure), 67
cerebral hemispheres and, 55–58, 56–57 (figures)
cerebrospinal fluid and, 66, 70
corpus callosum and, 65–66, 65–66 (figures)
development of, 55, 55 (figure), 66, 66 (figure)
efficient brain organization and, 56–57, 57 (figure)
forebrain structures, 55–66, 56 (figure), 58 (figure)
hierarchical arrangement of, 57–58
hindbrain structures, 65 (figure), 67, 67 (figure)
lobes in, 58–64, 58 (figure)
meninges and, 70
midbrain structures, 66–67, 67 (figure)
nucleus and, 54, 54 (figure)
pineal gland and, 64, 65 (figure)
protection of, 70
spinal cord and, 24 (figure), 34, 65 (figure), 67–69, 69 (figure)
thalamus/hypothalamus and, 64, 65 (figure)
tracts and, 54, 54 (figure)
ventricles and, 66, 66 (figure)
See also Autonomic nervous system (ANS); Brain structure; Central nervous system damage/recovery; Nervous system; Peripheral nervous system (PNS)
Central nervous system damage/recovery, 79
    activity-based recovery and, 84–85
    compensation and, 82
    computer chip implantation and, 84
    financial costs of, 79, 80 (table)
    induced self-repair strategies and, 83
    limitations on recovery and, 80–82
    myelin sheath and, 81
    neural precursor cells and, 82
    neurogenesis and, 81–82, 83
    nonneural functional improvement and, 82
    regeneration and, 80–81
    reorganization and, 82
    stem cells, pluripotent nature of, 83
    stroke and, 80, 80 (table), 82
    traumatic brain injury and, 80, 80 (table), 81
    See also Central nervous system (CNS); Nervous system development/change
Central pattern generator (CPG), 345–346, 346 (figure)
Central sulcus, 58, 58 (figure)
Cerebellum, 39, 65 (figure), 67, 68 (table)
    cognitive processes and, 67
    damage to, 67
    learning process and, 351
    movement, refinement of, 67, 346, 350–351, 351 (table)
    pons and, 65 (figure), 67, 67 (figure)
    vestibular system and, 333
    See also Central nervous system (CNS); Hindbrain

Cerebral cortex, 56, 56–57 (figures)
Cerebral hemispheres, 55, 56 (figure)
   aphasia, recovery mechanisms for,
      278–279
   asymmetry and, 55
   auditory areas in, 58 (figure), 63, 259,
      260 (figure)
   brain interpreter and, 494
   brain size and, 57, 57 (figure)
   cerebral cortex/gray matter and, 56,
      56–57 (figures)
   convoluted brain surface and, 55–56,
      55–56 (figures), 57
   corpus callosum and, 65–66, 65–66
      (figure)
   emotion, hemispheric specialization
      and, 233
   fissures and, 56
   gyrus and, 55, 56, 56 (figure), 57
   homunculus and, 60, 63
   lobes of, 58, 58 (figure), 60–64
   location of, 57
   longitudinal fissure and, 55, 56 (figure)
   motor areas in, 58 (figure), 60
   myelination process and, 56, 77–78
   size of, 55, 57
   split brain surgical procedure and,
      493–494, 494 (figure)
   sulcus, 55, 56 (figure)
   visual areas in, 57, 57–58 (figures),
      63, 64
   white matter and, 56, 56 (figure)
   See also Central nervous system (CNS);
      Forebrain; Nervous system
Cerebrospinal fluid, 66, 70
Chemical toxins, 352
Chernobyl nuclear disaster, 469
Cheshire cat effect, 489, 489 (figure)
Child abuse, 236, 245, 496
Chimpanzees, 13, 14
   brain size in, 57 (figure), 60
   communication among, 282–283,
      283 (figure), 284–285, 284 (figure)
   concept of self and, 490, 490 (figure)
   excessive self-grooming and, 458
   tool use and, 402
Cholecystokinin (CCK), 170, 171 (table)
Christopher Reeve Paralysis Foundation, 85
Chromosomes, 9–10, 9 (figure), 11, 12
   combinations of, 14
   sex, biological determination of, 199–200,
      199–200 (figures)
   See also Genes; Genetic code; Human
      Genome Project
Chronic pain, 341–342, 341 (figure)
Chronic symptoms, 430
Cingulate gyrus, 65 (figure), 227 (figure),
   228, 240
Circadian rhythm, 449, 470
   basic rest and activity cycle and, 473
   day-night cycle and, 470–472
   depression and, 444, 449–450
   insomnia and, 482, 482 (figure)
   internal clock and, 471, 472–473, 482
   light intensity and, 471
   melatonin and, 471
   pacemaker function and, 470

retinal ganglion cell light receptors and,
   472, 472(figure)
retinohypothalamic pathway, light-dark
   cycle and, 471–472
sleep phase syndrome and, 482
suprachiasmatic nucleus and, 470–471,
   470 (figure), 473
time cues, isolation from, 470–471,
   471 (figure)
ultradian rhythms and, 473
zeitgebers and, 470
   See also Sleep
Circuit formation stage, 75–76, 77 (figure)
Circuit pruning, 76–77, 78 (figure)
Classical conditioning, 373
CNS. *See* central nervous system
Cocaine, 41, 131, 421
   addictive potential and, 131, 138
   anesthetic properties of, 131
   brain imaging and, 131, 132 (figure)
   common use of, 131, 131 (figure)
   crack, 131
   euphoria and, 131
   executive functioning and, 132
   freebase form of, 131
   prenatal exposure to, 132
   psychological disorders and, 131–132
   selective tolerance and, 132
   withdrawal from, 131
   See also Psychoactive drugs; Stimulants
Cochlea, 255, 256–257, 258 (figure), 259,
   332, 333 (figure)
Cochlear canal, 257, 258 (figure)
Cochlear implants, 254, 267
Cocktail party effect, 266
Codeine, 127
Coding neural messages, 45–46, 46 (figure)
Cognitive functioning:
   aging, effects of, 379–380
   See also Intelligence
Cognitive psychology, 468
Cognitive theory of emotion, 226
Coincidence detectors, 268
Cold. *See* Skin senses
Color agnosia, 319
Color blindness, 306, 306 (figure)
Color constancy, 319
Color vision, 300–301
   absorption of light wavelengths and,
      304, 304 (figure)
   additive vs. subtractive mixing effects and,
      302, 303 (figure)
   color blindness and, 306, 306 (figure)
   color-opponent ganglion cells and,
      305–306, 305 (figure)
   combined theory of, 303–306, 303 (figure)
   comparison of cone activity and, 304
   complementary colors and, 301–302,
      302 (figure)
   concentric-circle receptor fields and, 306
   genetic basis/rationale for, 304–305, 306
   negative color aftereffect and, 302,
      302 (figure)
   observer experience of, 301, 301 (figure)
   opponent process theory and, 301–302,
      302–303 (figures)
   perception of color, 315

pigment mixing vs. light mixing and, 302
trichromatic theory and, 301
   See also Eyes; Form vision; Vision
Comatose state, 499
Communication. *See* Neuron communication;
   Neurotransmitters
Compensation, 82
Compensatory adaptation, 127
Complementary colors, 301–302, 302 (figure)
Complex cells, 310, 311 (figure)
Complex sound, 256, 256 (figure)
   analysis of, 265–267
   auditory object and, 266
   cocktail party effect and, 266
   environmental sounds, identification of,
      266–267, 266 (figure)
   Fourier analysis and, 265, 266 (figure)
   irrelevant information, suppression of, 267
   pure tone and, 265, 266
   selective attention and, 266
   sound localization and, 266, 267
   See also Auditory mechanism; Frequency
      analysis; Hearing
Compulsive gambling, 140
Compulsive overeating, 179–180, 180 (figure)
Computed tomography (CT) scanning, 103,
   104 (figure)
Computer chip implantation, 84
Computers:
   artificial neural networks and, 46–47
   virtual reality therapeutic tool and, 456
Concordance rate, 109
Concussions, 80
Conditioned tolerance, 127
Conduction speed, 33–35, 34 (figure)
Conductive deafness, 267
Cones. *See* Eyes
Confabulation, 387, 475, 492
Confounded variables, 94
Confusion, 130
Congenital adrenal hyperplasia (CAH),
   205, 206
Congenital insensitivity to pain, 239, 341
Consciousness, 4, 8, 468
   agency, sense of, 490, 491, 491 (figure)
   attention and, 488–490, 489 (figure)
   awareness and, 486–488, 487–488
      (figures), 497
   body image and, 491–492, 492 (figure)
   clinical assessment of, 499
   comatose state, 499
   confabulation and, 387, 492
   default mode network and, 497–498
   definitions of, 486
   disorders of self and, 493–496
   distributed controls of, 498, 498 (figure)
   distributed neuronal network and,
      497, 497 (figure), 498
   impairment of, 497, 498, 498 (figure)
   learning/perception and, 468
   locked-in syndrome and, 499
   long-term memory and, 492
   minimally conscious state, 499
   mirror neuron system and, 493
   neglect and, 496–497
   neural bases of, 485–486
   self, sense of, 490–491, 490 (figure), 493

sleep and, 485
split brain surgical procedure and,
      493–494, 494 (figure)
theoretical explanations of, 496–498,
      497–498 (figures)
theory of mind and, 493
vegetative state, 499
See also Self; Sleep
Consolidation process, 3–377, 366–367,
      367–368 (figures), 377 (figure), 477
Constipation, 165
Control system, 158
Control variables, 93
Conventional intelligence, 395
Coolidge effect, 191–192, 195, 195 (figure)
Coronal plane, 59, 59 (figure)
Corpus callosum, 65–66, 65–66 (figures),
      68 (table), 493–494
Correlational studies, 93–94, 94 (figure)
      correlation coefficient, 108
      genetic similarities, correlational research
            on, 108–110, 110 (table)
Cortex, 56, 56–57 (figures), 130, 131
Cortisol, 172, 234
Cranial nerves, 71, 72 (figure)
Crank, 132
Cravings, 141–142
Criminal behavior:
      alcohol-serotonin link in aggression and,
            244–245
      castration and, 192
      testosterone and, 241–242, 241 (figure)
Cryptochromes, 97, 98 (figure)
Crystal, 132, 135
Cytokines, 238

Dale's principle, 43
Date rape drug, 130
db. *See* diabetes gene.
Deafness, 254, 267
      conductive deafness, 267
      nerve deafness, 267
      sign language, 279, 279 (figure),
            280–281, 280 (figure)
      See also Auditory mechanism; Hearing;
            Hearing impairment
Decade of the Brain, 2–3, 16
Deception in research, 113
Decision process:
      emotion-based decision making, 229–230,
            230–231 (figures)
      neuronal communication and, 40
      prefrontal cortex and, 60, 62
      reason-emotion link and, 224
Declarative memory, 370, 371, 376, 381, 386
Deep brain stimulation, 448
Default mode network (DMN), 408, 420,
      497–498
Defensive aggression, 241
Delayed match-to-sample task, 347, 371–372
Delirium tremens, 129
Delusions, 133
Dementia, 380, 490
      See also Alzheimer's disease; Dementia with
            Lewy bodies; Korsakoff's syndrome
Dementia with Lewy bodies, 352
Dendrites, 25, 25–26 (figures), 26

Dendritic spines, 374, 375 (figure)
Deoxyribonucleic acid. *See* DNA
      (deoxyribonucleic acid)
Depressants, 128
      alcohol, 128–130, 129–130 (figures)
      anxiolytic drugs and, 128
      barbiturates, 130
      benzodiazepines, 130
      hypnotic drugs and, 128
      sedative drugs and, 128
      See also Psychoactive drugs
Depression, 3, 60, 131, 132, 229, 238, 352
      circadian rhythm disruption and,
            444, 449–450
      deep brain stimulation and, 448
      depression switch and, 452
      heredity and, 444–445, 445 (figure)
      major depression, 443
      monoamine hypothesis of, 445–446,
            446 (figure)
      seasonal affective disorder and, 450
      state of depression, 452
      trait of depression, 452
      transcranial magnetic stimulation
            and, 448
      unipolar depression, 443, 449, 451–452,
            451–452 (figures)
      winter depression, 450
      See also Affective disorders; Bipolar
            disorder; Suicide
Dermatomes, 333–334, 334 (figure)
Developmental changes, 14
Developmental dyslexia, 27 (figure), 276
Dexedrine, 132
Diabetes, 166, 172, 179, 180
Diabetes gene, 175, 175–176 (figures)
*Diagnostic and Statistical Manual of Mental
      Disorders,* fourth edition (DSM-IV), 495
Diarrhea, 165
Diethylstilbestrol exposure, 206–207
Difference in intensity, 268
Difference in time of arrival, 268
Diffusion. *See* Force of diffusion
Digestive process, 165, 165 (figure)
      absorptive phase of, 165–166, 166 (figure)
      amino acids and, 165
      area postrema and, 165
      autonomic nervous system and, 165
      constipation/diarrhea and, 165
      duodenum and, 165, 165 (figure)
      fasting phase of, 166, 166 (figure), 168
      fatty acids/glycerol and, 165
      glucose and, 165
      hepatic portal vein and, 165
      hydrochloric acid/pepsin and, 165
      insulin and, 165, 166, 170, 171, 171 (table)
      paraventricular nucleus and, 166,
            167 (figure)
      toxic/spoiled food and, 165
      See also Eating disorders; Hunger
Dihydrotestosterone, 200, 204
Direct observation, 6, 92
Discovery Institute, 8
Dissociative identity disorder, 494–496,
      495–496 (figures)
Distributed visual function, 314
Diversity, 13

Dizziness, 333
DNA (deoxyribonucleic acid), 10
      computer construction and, 10
      double helix structure of, 10, 10 (figure)
      gene expression, 13
      human thumb and, 13
      junk DNA, 13
      nucleotides and, 10–11
      in situ hybridization and, 97, 98 (figure)
      variation and, 11
      See also Genetic code; Human Genome Pro-
            ject; Proteins; RNA (ribonucleic acid)
Doctrine of specific nerve energies, 78
Dominant genes, 11
      hand clasping preference and, 11–12,
            11 (figure)
      See also Genes; Genetic code; Inheritance
Dopamine, 41, 44 (table), 66, 93
      addiction/reward and, 139–140,
            139–140 (figures)
      amphetamines and, 132
      attention deficit hyperactivity disorder
            and, 420
      caffeine and, 134
      chronic drug users, diminished dopamine
            release/receptors and, 139–140,
            139 (figure)
      cocaine and, 131
      Coolidge effect and, 195, 195 (figure)
      marijuana and, 136
      mesolimbocortical dopamine system,
            138, 138 (figure), 139, 145
      opiate-like effects and, 128
      schizophrenia and, 438–439
      psychedelics and, 134, 135
Dopamine hypothesis, 434, 435 (figure)
Dorsal, 59, 59 (figure)
Dorsal root, 68, 69, 69 (figure)
Dorsal stream, 260, 261 (figure),
      315, 315 (figure), 316, 414
Double helix structure, 10, 10 (figure)
Down syndrome, 382, 410–411, 411 (figure)
Dreaming, 475–476, 485
Drive, 157
Drive theory, 157
      See also Homeostatic drives; Motivation
Drugs, 126
      abused drugs, effects of, 126–127
      addiction and, 126
      agonists, 126
      antagonists, 126
      tolerance and, 126–127
      withdrawal reactions and, 126
      See also Addiction; Medications;
            Psychoactive drugs; Psychotherapeutic
            drugs; Substance abuse
Dualism, 4
Duodenum, 165, 165 (figure)
Dyscalculia, 276
Dysgraphia, 276
Dyslexia, 276
      acquired dyslexia, 276
      developmental dyslexia, 276, 276 (figure)
      incidence of, 277, 277 (figure)
      magnocellular hypothesis of, 276, 277
      phonological hypothesis of, 276–277
      planum temporale and, 276

treatment of, 278, 278 (figure)
visual-perceptual difficulties and, 276
word analysis system and, 277
See also Language
Ear, 256, 257 (figure)
See also Auditory mechanism; Hearing;
Vestibular sense
Early-onset alcoholism, 146
Eating disorders:
anorexia nervosa, 181, 182 (figure)
bulimia nervosa, 181–182
compulsive overeating, 174, 179–180,
180 (figure)
environmental/genetic contributions to, 182
serotonin/dopamine and, 182–183
sleep-related eating disorder, 483
See also Digestive process; Hunger; Obesity
ECT. *See* electroconvulsive therapy
Ecstasy, 134–135, 135 (figure)
psychomotor stimulant, 134
See also Psychedelics; Psychoactive drugs
Ectothermic animals, 158
EEG. *See* electroencephalogram
Electrical brain concept, 6
Electrical stimulation of the brain (ESB),
139, 448
Electroconvulsive therapy (ECT), 447,
447 (figure), 448
Electroencephalogram (EEG), 99,
100 (figures), 108 (table)
Electromagnetic spectrum, 294–295,
295 (figure)
Electron microscopes, 98, 99 (figure)
Electrostatic pressure, 29, 31
Embryo, 10
Embryonic stem cells, 83, 83 (figure),
116–117, 117 (figure), 352, 353 (figure)
Emergencies, 71
Emotion, 4, 224
amygdala and, 228, 228 (figure),
231–233, 232 (figure)
anterior cingulate cortex and, 227–228
(figures), 228–229
autonomic/muscular involvement in,
224–227, 225–226 (figures), 233
basal ganglia and, 228, 228 (figure)
behavior and, 224
botox treatments and, 226, 226 (figure)
brain structures and, 227–233, 227–232
(figures)
decision process and, 229–230, 230–231
(figures)
deficits in, 224, 228
definition of, 224
facial expressions and, 226, 226 (figure),
227, 228, 233
feedback systems and, 225–227
gambling task and, 229–230, 231 (table)
harm avoidance and, 229
hemispheric specialization and, 233
hypothalamus involvement in, 227–228
insular cortex and, 228, 228 (figure)
James-Lange theory and, 225–226
learning and, 224, 371
limbic system and, 227–228, 227 (figure)
mirror neurons and, 227
nervous system and, 224–233

pain, adaptive emotion of, 239–240,
239–240 (figures), 339 (figure)
prefrontal cortex and, 224, 227 (figure),
229–230, 230–231 (figures)
research on, 227–229
septal stimulation and, 228
See also Aggression; Pain; Stress
Emotional disorders, 2
Emotional trauma, 15
Empathy, 224, 227, 414
Empiricism, 6, 92, 93
Endogenous cannabinoids, 341
Endogenous opioids, 128
Endorphins, 44 (table), 128, 338–339,
340 (figure)
Endothermic animals, 158
Environmental challenges, 15–16
addiction, environmental vs. genetic
influences on, 146–149, 147(table),
149 (figure)
adoption studies and, 109, 110 (table)
intelligence, environmental effects and,
404, 405–406, 405 (figure)
Parkinson's disease and, 352
schizophrenia and, 439–441,
440–441 (figures)
twin studies and, 109–110, 110 (table)
Epigenetics, 177, 212, 441
Epilepsy, 38, 54, 63, 64, 65–66, 130, 227,
278, 336, 487, 496
Epinephrine, 44 (table), 239
EPSP. *See* excitatory postsynaptic potential
Epstin-Barr virus, 355
Equipotentiality, 7
ESB. *See* electrical stimulation of the brain
Essential life processes, 67
Estrogen, 193, 200, 201, 203, 409
Estrogen replacement therapy, 409
Estrus, 193
Ethical issues. *See* Research ethics
Etiology, 432
Euphoria, 127, 131, 132
Evoked potential, 99, 100 (figure)
Evolution, 4, 8, 14
Excitatory postsynaptic potential (EPSP),
38–39, 40
See also Neuron communication
Exercise, 178, 178 (figure)
Experience:
adoption studies and, 109, 110 (table)
synaptic construction/reorganization and,
78–79, 79 (figure)
Experimental research, 93, 94, 94 (figure)
natural experiments, 109
science, birth of, 6
See also Research methods
Exxon Valdez oil spill, 469
Eyes, 295–296, 296 (figure)
amacrine cells and, 297, 297 (figure)
bipolar cells and, 296, 297 (figure)
blind spot and, 298
eye movements, 332, 333, 336, 350, 351
fovea and, 296–297
ganglion cells and, 296, 297 (figure)
horizontal cells and, 297, 297 (figure)
lodopsin and, 296
lost vision, restoration of, 300

optic chiasm and, 298, 299 (figure)
optic nerve and, 296, 297 (figure), 298,
299 (figure)
pathways to the brain and, 298,
299 (figures)
photopigments and, 296
photoreceptors and, 296, 297 (figure), 300
receptive fields and, 297
retina, 296, 296–297 (figures), 297, 298
retinal disparity and, 298, 299 (figure)
rhodopsin and, 296, 300
rods/cones and, 296–297, 297 (figure),
297 (table)
visual acuity and, 297, 300
See also Color vision; Form vision; Vision;
Visual cortex

Fabrication of research results, 112
Face agnosia, 316–319, 317–319 (figures)
Familial disease, 352
Family studies, 108–109, 110 (table)
FAS. *See* fetal alcohol syndrome
Fasting phase, 166, 166 (figure), 168
Fat solubility, 70
Fatty acids, 165, 168, 171 (table)
Fear memory, 378–379, 456
Feedback:
artificial neural networks and, 47
biofeedback techniques, 337, 343
emotion and, 225–227
Fen-phen, 178
Fertilization, 9–10, 14
See also Sexual reproduction
Fetal alcohol syndrome (FAS), 77,
78 (figure), 130, 130 (figure), 276
Fetus, 10
FFA. *See* fusiform face area
Fissure, 56
lateral fissure, 58
longitudinal fissure, 55, 56 (figure)
Fluorogold injection technique, 95–96
fMRI. *See* functional magnetic resonance
imaging
Food and Drug Administration (FDA),
116, 178, 179
Food. *See* Digestive process; Eating
disorders; Hunger; Obesity; Taste
Force of diffusion, 29
Forebrain, 55, 68 (table)
Broca's area and, 58 (figure), 60
cerebral hemispheres and, 55–58,
56–57 (figures)
corpus callosum and, 65–66,
65–66 (figures)
lobes in, 58, 58 (figure), 60–64
medial forebrain bundle, 138
pineal gland and, 64, 65 (figure)
thalamus/hypothalamus and, 64,
65 (figure)
ventricles and, 66, 66 (figure)
Wernicke's area and, 58 (figure), 63
See also Central nervous system (CNS);
Nervous system
Form vision, 307
complex cells and, 310, 311 (figure)
contrast enhancement/edge detection and,
307–309

Hubel-Wiesel theory and, 309–311, 310–311 (figures)
lateral inhibition and, 308–309, 308–309 (figures)
light, center/surround of receptive field and, 309, 309 (figure)
light-dark contrast detectors and, 309, 310 (figure)
retinotopic map and, 307, 307 (figure)
simple cells and, 310, 310–311 (figures)
spatial frequency theory and, 312–313, 312–313 (figures)
See also Color vision; Eyes; Vision
Fourier analysis, 265, 266 (figure)
Fovea, 296–297
Fragile X syndrome, 411
Free nerve endings, 331, 331 (figure), 336
Frequency analysis, 260
bone conduction and, 265, 265 (figure)
cocktail party effect and, 266
complex sounds, analysis of, 265–267, 266 (figures)
frequency sensitivity, distribution of, 262–263, 263 (figure)
frequency theories and, 261–262
frequency-place theory and, 265
frequency-volley-place theory and, 265
intensity coding and, 265
place theory and, 262–265, 263–265 (figures)
resonance and, 262
sound localization and, 263, 265, 266
telephone theory and, 261
tonotopic map and, 263, 264 (figure)
tuning curves and, 263, 264 (figure)
volley theory and, 261–262, 262 (figure)
See also Auditory mechanism; Hearing
Frequency of sound waves, 255, 259
Frequency theories, 261–262, 262 (figure)
Frequency-place theory, 265
Frequency-volley-place theory, 265
Friendship, 61
Frontal lobe, 58, 58 (figure), 65 (figure), 68 (table)
association areas of, 63
Broca's area and, 58 (figure), 60
central sulcus and, 58
damage to, 60–61, 62, 380
functions of, 59, 60
homunculus and, 60
lateral fissure and, 58, 58 (figure)
lobotomies and, 61, 61–62 (figures)
motor cortex and, 60, 60 (figure)
myelination process and, 77–78
precentral gyrus and, 60
prefrontal cortex and, 58 (figure), 60–61
primary/secondary motor areas and, 60
See also Central nervous system (CNS); Cerebral hemispheres; Lobes
Functional magnetic resonance imaging (fMRI), 106–107, 106 (figure), 228, 229
Fusiform face area (FFA), 317, 318, 318–319 (figures), 416

Gambling, 140
Gambling task, 229–230, 231 (figure), 380

Gamma-aminobutyric acid (GABA), 44 (table), 129–130, 129 (figure)
Gang activity, 245
Ganglion, 54, 54 (figure)
basal ganglia, 60, 66
dorsal root ganglion, 69, 69 (figure)
retinal ganglion cell light receptors, 472, 472 (figure)
Gastric bypass procedure, 180, 180 (figure)
Gate control theory, 339
Gays, 209
See also Sexual orientation
Gender, 198
aggressiveness and, 202, 203, 204, 207
gender-related behavioral/cognitive differences, 202, 202 (figure)
male-female differences, origins of, 202–204
sex hormones, brain development and, 203
sex reassignment, 208, 209, 214
spatial acuity and, 202, 202 (figure), 203
sports, gender verification policies and, 207
verbal acuity and, 202, 203
See also Gender identity; Gender role; Sexual anomalies; Sexual orientation
Gender identity, 199, 206, 207, 208
Gender identity reversal, 214
Gender nonconformity, 210
Gender roles, 198, 207, 208
Gene therapy, 111, 115–116, 116 (figure)
Gene transfer technique, 110–111
General factor of intelligence, 395–396, 401
Generalized anxiety, 454, 460
Genes, 9
alleles and, 11
behavioral traits and, 12
DNA strands and, 10, 10 (figure)
dominant genes, 11
effects, varying degrees of, 14
expression of, 13, 14
hand clasping and, 11, 11 (figure)
human thumb and, 13
individuality and, 14
late-life functioning of, 14
marker genes, 13
medical disorders and, 13, 15, 15 (figure)
nucleotides and, 10
paring of, 11
protein production and, 14, 97
recessive genes, 11
See also Genetic code; Genetics; Heritability; Human Genome Project; Natural selection; Nature vs. nurture debate
Genetic code, 9
allele blending and, 11
behavior, genetic influences on, 9, 10–12, 11 (figure)
chromosomes and, 9–10, 9 (figure), 11
DNA, double helix structure of, 10, 10 (figure)
fertilization/cell division and, 9–10
gene expression and, 13
genes and, 9

genetic markers and, 13
genotypes and, 11
heterozygous individuals and, 11 (figure), 12
homozygous individuals and, 11, 11 (figure)
mitochondria and, 9
nucleotides and, 10–11
phenotypes and, 11
polygenic characteristics and, 12
variation and, 11
X chromosome and, 9, 11, 12
X-linkage and, 12
Y chromosome and, 9
See also Genes; Human Genome Project; Nature vs. nurture debate
Genetic engineering, 110–111, 111 (figure)
Genetic testing, 16
Genetics:
addiction, genetic vs. environmental influences on, 146–149, 147 (table), 149 (figure)
aggressive tendencies and, 245–246, 245 (figure)
attention deficit hyperactivity disorder and, 420–421
autism and, 418–419
concordance rate and, 109
diversity and, 13
DNA strands and, 10, 10 (figure)
family studies and, 108–109, 110 (table)
individuality and, 14
intelligence, heritability of, 403–405, 403 (figures)
obesity and, 174–175, 175–176 (figures)
sexual orientation and, 211–212, 211 (figures)
twin studies, 14–16, 15 (figure), 109–110, 110 (table)
See also Genetic code; Heritability; Nature vs. nurture debate
Genome, 12, 16
Genotypes, 11
Ghrelin, 168–169, 169 (figure), 171 (table), 180
Glandular system, 12
Glial cells, 33
axon growth inhibitors and, 81
graded potential and, 34
inter-neural connections and, 35, 35 (figure)
migrating neurons and, 74, 76 (figure)
myelination/conduction speed and, 33–35, 34 (figure)
neural activity/astrocytes and, 35, 35 (figure)
nodes of Ranvier and, 34
oligodendrocytes, 34, 34 (figure)
radial glial cells, 74, 77
reuptake process and, 41
saltatory conduction and, 34
scar tissue and, 81
synaptic activity, regulation of, 41, 43 (figure)
See also Nervous system; Neuron; Neuron cell membrane
Glucagon, 166
Glucoprivic hunger, 168
Glucose, 165, 168, 171 (table)

Glutamate, 32, 41, 43, 44 (table), 129, 337, 367, 374, 375 (figure)
  depression and, 446
  schizophrenia and, 435–436, 439, 442
Glutamate theory of schizophrenia, 436
Glycerol, 165
Glycine, 44 (table)
Glycogen, 166
Golgi stain method, 95, 96 (figure)
Golgi tendon organs, 345
Gonads, 192, 199, 199 (figure), 201
Graded potential, 31, 34
Gray matter, 56, 56 (figure), 68
Growth cones, 75, 77 (figure)
Gyrus, 55, 56, 56 (figure), 57
  angular gyrus, 275
  cingulate gyrus, 65 (figure)
  postcentral gyrus, 58 (figure), 62
  precentral gyrus, 60, 60 (figure)
  superior gyrus, 63

Hair cells, 257–259, 258–259 (figures), 267, 332
Halcion, 130
Hallucinations, 93, 126, 129, 133, 134, 439
Harm avoidance, 229
Hashish, 135
Head Start program, 405, 406
Hearing, 63, 254–255
  adequate stimulus and, 254
  amplitude and, 255, 256 (figure)
  audible frequencies and, 255
  auditory mechanism and, 256–261, 257–259 (figures)
  cochlea and, 255, 256–257, 257–258 (figures), 259
  cochlear implants and, 254, 267
  cocktail party effect and, 266
  complex sounds and, 256, 256 (figure), 265–267, 266 (figures)
  frequency analysis, 260–267, 262–266 (figures)
  frequency of sounds and, 255, 256 (figure), 259
  inferior colliculi and, 66
  intensity and, 255, 265, 268
  loudness and, 255
  pitch and, 255, 256 (figure)
  profound deafness, 254
  pure tones and, 256, 256 (figure), 265
  sound localization and, 263, 265, 266, 267–270, 268–269 (figures)
  sound vibrations and, 255, 255 (figure)
  stimulus for, 255–256
  See also Auditory cortex; Language; Vision
Hearing impairment, 267
  brainstem implants and, 267
  cochlear implants and, 254, 267
  conductive deafness, 267
  enriched auditory world, adjustment to, 267
  gene therapy and, 267
  nerve deafness and, 267
  stem cell therapy and, 267
  See also Auditory mechanism; Deafness; Hearing
Heart disease, 172
Hebb rule, 372

Hemispheres. See Cerebral hemispheres
Hepatic portal vein, 165
Herbicide pollution, 352
Heredity. See Genetics; Heritability; Inheritance
Heritability, 14–15, 15 (figure)
  genetic similarities, correlational research on, 108–110, 110 (table)
  See also Genetics; Inheritance
Hermann grid illusion, 307, 308 (figure)
Heroin, 45, 126, 127
  addiction treatment, medications for, 145 (table)
  blood-brain barrier and, 127
  lifelong addiction to, 127
  overdose and, 127
  relapse rates, 143, 143 (figure)
  withdrawal from, 127
  See also Opiates; Psychoactive drugs
Heterozygous individuals, 11 (figure), 12
Hierarchical processing, 313
High blood pressure, 172
Hindbrain, 65 (figure), 67, 67 (figure), 68 (table)
  cerebellum, 65 (figure), 67
  medulla and, 65 (figure), 67, 67 (figure)
  pons and, 65 (figure), 67, 67 (figure)
  See also Central nervous system (CNS); Nervous system
Hippocampal formation, 365–366, 368
Hippocampus, 81, 83, 130, 136, 138, 227 (figure)
  aging and, 380
  amnesia and, 365–366
  learning process and, 370, 371, 376
  memory and, 367, 367–368 (figures)
  parahippocampal gyrus and, 367
  state-dependent learning and, 495–496, 496 (figure)
  stress outcomes and, 236–237, 237 (figure)
  theta rhythm and, 374
  See also Anxiety disorders
HIV (human immunodeficiency virus), 238, 239
Hoarding, 458, 459
Homeostasis, 157, 478
Homeostatic drives, 158
  control system and, 158
  temperature regulation, 158
  thirst, 159–160, 160 (figure)
  See also Hunger; Motivation; Obesity
Homosexuality, 12, 209–210
  See also Sexual orientation
Homozygous individuals, 11, 11 (figure)
Homunculus, 60, 60 (figure), 63
Horizontal plane, 59, 59 (figure)
Hormones:
  activating effects of, 200, 201
  aggression and, 241–242, 241 (figure), 244
  aging, intelligence and, 409
  castration and, 192
  estrogen, 193, 200, 201, 203, 409
  melatonin, 64, 471
  organizing effects of, 200
  pituitary gland and, 64
  premenstrual syndrome and, 241

  sex, biological determination of, 200–201
  sexual orientation and, 212–213
  testosterone, 192–193, 193 (figure), 200, 201, 203, 241, 244, 409
Hubel-Wiesel theory, 309–311, 310–311 (figures)
Human Genome Project, 3, 12, 116
Human research subjects, 113
Hunger, 157–158, 160–161
  agouti-related protein and, 169
  carnivores and, 161
  cholecystokinin and, 170
  digestive process and, 165–168, 165–168 (figures)
  ghrelin levels and, 168–169, 169 (figure)
  glucoprivic hunger and, 168
  herbivores and, 161
  inhibition of eating and, 167 (figure), 169–170, 170 (figure), 171 (table)
  initiation of eating and, 167–169 (figures), 168–169, 171 (table)
  lipoprivic hunger and, 168
  long-term controls on, 170–171, 171 (figure, table)
  neuropeptide Y and, 169
  nutrient reserves and, 164
  omnivores and, 161
  taste, role of, 161–164, 162 (figures)
  wisdom of the body and, 161
  See also Digestive process; Eating disorders; Homeostatic drives; Motivation; Obesity
Huntington's disease, 12, 13, 14, 228, 350, 353
  brain tissue loss in, 353, 354 (figure)
  characteristics of, 353
  huntingtin gene and, 353–354
  learning process and, 370–371
  treatments for, 354
Hydraulic model of brain activity, 5–6, 5 (figure)
Hydrocephalus, 82, 412, 412 (figure)
Hyperpolarization, 38, 40
Hypocretin. See Orexin
Hypnotic drugs, 127, 130
Hypopolarization, 38, 40
Hypothalamus, 64, 65 (figure), 68 (table)
  aggression and, 242, 242 (figure), 243
  anterior hypothalamic area, 197
  arcuate nucleus and, 168, 168 (figure)
  central bed nucleus of the stria terminalis, 214
  eating/digestion and, 166, 167 (figure)
  emotional expressions and, 227–228, 227 (figure)
  lateral hypothalamus and, 166, 167 (figure)
  retinohypothalamic pathway, 471–472
  suprachiasmatic nucleus and, 213, 470–471, 470 (figure)
  temperature regulation and, 158, 159 (figure)
  thirst and, 159, 160, 160 (figure)
  ventromedial hypothalamus, 170, 194, 197
Hypothalamus-pituitary-adrenal axis, 234, 234 (figure), 481
Hypotheses, 93
Hypovolemic thirst, 159–160

Ice, 132
Idealistic monism, 4
Illegal drugs. *See* Addiction; Psychoactive drugs; Substance abuse
Imaging techniques. *See* Research techniques
Immune system, 234, 235 (table)
    antibodies and, 235, 238, 239 (figure)
    antigens and, 235
    autoimmune disorders and, 235
    B cells and, 235
    cytokines and, 238
    leukocytes and, 235
    macrophages and, 235, 235 (figure)
    natural killer cells and, 235
    relaxation training and, 237
    social/personality variables and, 237–239, 239 (figure)
    stress and, 234–235, 235–236(figures), 236, 237–239, 239 (figure)
    T cells and, 235, 237, 238
    vaccinations and, 235, 239 (figure), 418
    See also Stress
Immunizations, 235, 239 (figure), 418
Immunocytochemistry technique, 97, 98 (figure)
Impulse control, 60, 61, 174, 244, 420
In situ hybridization, 97, 98 (figure)
Incentive theory, 157
Individuality, 14
Inferior, 59, 59 (figure)
Inferior colliculi, 65 (figure), 66, 67 (figure), 68 (table)
Inferior temporal cortex, 63, 68 (table)
Information processing:
    cerebral cortex and, 57, 57 (figure)
    corpus callosum and, 65–66, 65 (figure)
    language ability, 63, 66
    pons and, 67
    visual identification, 63, 66
Informed consent, 113, 116
Infrared images, 294–295
Inheritance:
    addictions, genetic vs. environmental influences on, 146–149, 147 (table), 149 (figure)
    affective disorders and, 444–445, 445 (figure)
    aggressive tendencies and, 245–246, 245 (figure)
    anxiety disorders and, 460–461
    behavior, genetic influences on, 9, 10–12
    behavioral disorders and, 14–16, 15 (figure)
    concordance rate and, 109
    destiny vs. predisposition and, 13–16
    environmental challenges, vulnerability and, 14–16
    family studies, 108–109, 110 (table)
    genetic similarities, correlational research on, 108–110
    genotypes and, 11
    hand clasping preferences and, 11–12, 11 (figure)
    individuality and, 14
    medical disorders and, 13, 15, 15 (figure)
    phenotypes and, 11
    polygenic characteristics and, 12

schizophrenia and, 431–433, 431–432 (figures), 433 (table)
    twin studies, 14–16, 15 (figure), 109–110, 110 (table)
    X-linkage and, 12
    See also Genetic code; Genetics; Heritability; Twin studies
INAH 3. *See* third interstitial nucleus of the anterior hypothalamus
Inhibitory postsynaptic potential (IPSP), 38–39, 39 (figure), 40
    See also Neuron communication
Inner hair cells, 259
Inner speech, 439
Insomnia, 130, 481–482, 482(figure)
Instinct, 156–157
Insula, 490, 491, 492
Insular cortex, 228, 490-492, 491 (figure)
Insulin, 165, 166, 170, 171, 171 (table), 179
Intellectual disability, 409
    categories of, 410, 410 (table)
    Down syndrome and, 382, 410–411, 411 (figure)
    fragile X syndrome and, 411
    hydrocephalus and, 82, 412, 412 (figure)
    intelligence quotient and, 409, 410 (table)
    mild intellectual disability, 400, 400 (figure), 408, 410, 410 (table)
    phenylketonuria and, 411–412
    See also Attention deficit hyperactivity disorder (ADHD); Autism; Intelligence
Intelligence, 12, 14, 15, 16, 393–394
    adopted children and, 406
    aging, effects of, 407–409, 408 (figure)
    animal tool use and, 402
    autistic savants and, 396
    biological origins of, 396–407
    brain size and, 57, 57 (figure), 398, 399 (figure)
    brain structure and, 396–398, 397–398 (figures)
    compensatory brain activity and, 408, 408 (figure)
    culture-free assessment of, 395
    default mode network, loss of coordination in, 408
    definition of, 394–395
    enhanced intelligence/performance and, 406
    environmental influences and, 404, 405–407, 405 (figure)
    ethnic origin and, 404–405
    general factor/unitary capability and, 395–396, 401
    heritability of, 403–405, 403 (figures)
    independent mental abilities and, 396
    infectious disease and, 405, 405 (figure)
    intelligence quotient, 394–395, 395 (figure), 396, 403–405, 403 (figure)
    linguistic ability and, 400, 401
    long-term memory and, 400
    marijuana use and, 136
    mathematical ability and, 401
    mild intellectual disability, neural activity levels and, 400, 400 (figure), 408
    nature of, 394–396

nerve conduction/processing speeds and, 399, 399 (figure)
    practical intelligence vs. conventional intelligence, 395
    prenatal cocaine exposure and, 132
    processing efficiency and, 400, 400 (figure)
    savants and, 414
    schizophrenia and, 109–110
    sex hormones, cognitive slowing and, 409
    short-term memory, reasoning/problem-solving limitations and, 400
    socioeconomic status and, 405
    specific cognitive abilities, brain areas and, 400–402, 401 (figure)
    structure of, 395–396
    working memory and, 400, 408
    See also Intellectual disability
Intelligence quotient (IQ), 394–395, 395 (figure), 396, 403–405, 403 (figure)
Intelligent design, 8
Intensity, 255, 265, 268
International Association of Athletics Federations (IAAF), 207
International Human Genome Sequencing Consortium, 13
International Olympic Committee (IOC), 207
Interneurons, 26, 26 (figure), 27 (table)
Intersex Society of North America, 208
Intersexes, 206
Introspection, 468
Iodopsin, 296
Ion distribution mechanisms, 28–31, 29–30 (figures)
Ionizing radiation exposure, 77
Ionotropic receptors, 37
Ions, 28
IPSP. *See* inhibitory postsynaptic potential
IQ. *See* intelligence quotient

James-Lange theory of emotion, 226
    Jet lag, 469

Knee cap. *See* Patellar tendon reflex
    Knockout technique, 110, 201
    Korsakoff's syndrome, 129, 129 (figure), 386–387, 492

Language, 63, 66, 270
    agraphia and, 275
    alexia and, 275
    angular gyrus and, 275
    aphasia and, 270–272
    aphasia, recovery mechanisms for, 278–279
    bilingual language acquisition, 281, 281 (figure)
    Broca's aphasia and, 271–272
    Broca's area and, 270, 271–272, 271 (figure), 273, 284
    deafness and, 267
    dyslexia, 276–278, 276–278 (figures), 285
    gender and, 202, 203
    gestures and, 284, 284 (figure)
    innate brain specializations and, 280
    instinctive tendency for, 279
    language acquisition device and, 279–282, 280–281 (figures)

language deficits, brain damage and, 274, 274 (figure)
language impairment, reorganization and, 82
mirror neurons and, 284
neural/genetic antecedents for, 283–285, 284–285 (FIGURES)
nonhuman communication and, 282–283, 283 (figure), 284–285, 284 (figure)
phonemes and, 276–277
reading/writing, impairment of, 175–178, 275 (figure)
receptive aphasia and, 272
ryhthmic language patterns, 279
sexual orientation and, 214, 214 (figure)
sign language, 279, 279 (figure), 280–281, 280 (figure)
specialized learning areas and, 282
Wernicke-Geschwind model of, 272–275, 273–274 (figures)
Wernicke's aphasia, 272
Wernicke's area and, 270–271, 271 (figure), 272, 273, 274, 284
See also Auditory cortex; Auditory mechanism; Hearing
Language acquisition device, 279
bilingual language acquisition, 281, 281 (figure)
innate brain specializations and, 280
sign language, auditory language areas and, 280–281, 280 (figure)
See also Language
Late-onset alcoholism, 146
Lateral, 59, 59 (figure)
Lateral fissure, 58, 58 (figure)
Lateral geniculate nuclei, 298, 299 (figure), 305, 309, 315
Lateral hypothalamus, 166, 167 (figure)
Lateral inhibition, 308–309, 308–309 (figures)
Lateral plane, 59, 59 (figure)
Lateral ventricles, 65–66 (figures), 66, 81
L-dopa. See levodopa
Learned taste aversion, 163, 164
Learned taste preference, 163–164
Learning process, 14
aging, effects of, 379–380
amygdala, emotional learning and, 371
basal ganglia and, 350, 370–371
brain changes in, 372–377
cerebellum and, 67, 351
conditioned tolerance, 127
consolidation process and, 376–377, 377 (figure)
declarative learning and, 370–371, 376
deficiencies/disorders of, 379–387
dendritic spines and, 374, 375 (figure)
dissociative identity disorder, state-dependent learning and, 495–496, 496 (figure)
dopamine/brain plasticity and, 141–142, 141–142 (figures)
emotion and, 224
errors in prediction, detection of, 141
extinction of memories and, 377
forgetting and, 377–378
gender differences and, 203
Hebb rule and, 372

long-term depression and, 373, 373 (figure), 374
long-term potentiation and, 372–374, 373 (figures)
memories, changing of, 377–379
memory storage and, 364–365
neural growth in, 374–375, 375 (figure)
neural plasticity and, 76–77
nondeclarative learning, 370–371
prefrontal area and, 367
reconstructed memories and, 378–379
REM sleep and, 476–477, 477 (figure)
reverse-learning hypothesis, memory deletion and, 478
sleep and, 376–377, 377 (figure)
synaptic changes and, 82
synchronized brain activity and, 487, 488 (figure)
visual perception and, 318
See also Alzheimer's disease; Intelligence; Korsakoff's syndrome; Memory
Leber's congenital amaurosis, 300, 319 (figure)
Leptin, 170, 171, 171 (table), 175, 176 (figure), 179, 179 (figure)
Lesbians, 209, 210, 215
See also Sexual orientation
Lesioning technique, 102, 139, 227, 243
Leukocytes, 235
Levodopa (L-dopa), 352
Lewy bodies, 352, 352 (figure), 484
Light, 295, 295 (figure)
See also Circadian rhythm; Color vision; Form vision; Vision
Limbic system:
amygdala, 227–228, 228 (figure), 231–233, 232 (figure)
benzodiazepines and, 130
emotions and, 227–228, 227 (figure)
pain pathway and, 240, 240 (figure)
Lipoprivic hunger, 168
Lissencephalic brain, 54, 54 (figure)
Lithium, 448, 451
Lobes, 58–64, 58 (figure)
association areas and, 63
Broca's area and, 58 (figure), 60
frontal lobe, 58, 58 (figure), 60–62, 60–62 (figures)
myelination process and, 77–78
occipital lobes, 58 (figure), 64
parietal lobes, 58 (figure), 62–63
temporal lobes, 58 (figure), 63–64
Wernicke's area and, 58 (figure), 63
See also Central nervous system (CNS); Cerebral hemispheres; Forebrain
Lobotomies, 61, 61–62 (figures), 240
Localization principle, 6–8, 7–8 (figures), 55, 62
Locked-in syndrome, 499
Locus coeruleus, 478, 479 (figure), 480 (igure)
Long-term depression (LTD), 373, 373(figure), 374, 478
Long-term memory, 367, 367 (figure), 376, 400, 492
Long-term potentiation (LTP), 372–373
aging and, 380
associative long-term potentiation and, 373, 373 (figure)

classical conditioning and, 373
glutamate receptors and, 374, 375 (figure)
induction of, 374
theta rhythm and, 374, 478
See also Learning process
Longitudinal fissure, 55, 56 (figure)
Lou Gehrig's disease, 103, 394
Loudness, 255
LSD (lysergic acid kiethylamide), 134, 428, 429
LTD. See long-term depression
LTP. See long-term potentiation
Lucid dreamers, 485

Mach band illusion, 307–308, 308–309 (figures)
Macrophage, 235, 235 (table)
Magnetic resonance imaging (MRI), 103, 104 (figure), 105, 228, 229
Magnocellular hypothesis of dyslexia, 276, 277
Magnocellular system, 314–315, 314 (figure), 479
Major depression, 443
Major histocompatibility complex (MHC), 196
Mania, 443, 451, 452
Manic-depressive cycling, 451, 452, 452 (figure)
Marijuana, 129, 135, 136 (figure)
addiction and, 137
cannabinoids and, 135–136
cognitive deficits and, 136
effects of, 136
hashish and, 135
legalization controversy and, 136
medical marijuana, 136, 137
memory loss and, 136
pain relief and, 341
prefrontal cortex impairment and, 136
prenatal development and, 136
psychological dependence on, 137
THC and, 135, 137
withdrawal from, 137
See also Psychoactive drugs
Master gland. See Pituitary gland
Materialistic monism, 4
MDMA (methylenedioxymethamphetamine), 134–135, 135 (figure)
Medial, 59, 59 (figure)
Medial amygdala, 194
Medial forebrain bundle, 138
Medial preoptic area (MPOA), 194, 195, 197
Median preoptic nucleus, 159, 160 (figure)
Medical disorders, 13, 15, 15 (figure)
Medical marijuana, 136, 137
Medications:
acetaminophen, 337
addictions treatment drugs, 145–146, 145 (table)
agonists/antagonists and, 45
Alzheimer's disease treatments and, 383–384
anesthetics, 131, 337, 338
antianxiety drugs, 233, 454–455
antidepressant medications, 41, 245, 354, 417, 446, 448–449, 449 (figure)

antipsychotic medications, 354, 417, 435–436
aspirin, 337
autism treatments and, 417
fat solubility and, 70
HIV treatments, 238
hyperactivity medications, 39
ibuprofen, 337
lidocaine, 338
lithium, 448, 451
MDAN series of drugs, 337
morphine, 337
naloxone, 45
obesity and, 178–179, 180
opiates, 45, 338, 339 (figure), 340–341
pain medications, 337
reuptake process and, 41
ritalin, 39
stimulants, 39
Medulla, 65 (figure), 67, 67 (figure), 68 (table)
Meissner's corpuscles, 331, 331 (figure)
Melatonin, 64, 471
Memory, 3, 4, 5, 364–365
abused individuals and, 236
aging, effects of, 379–380
amnesia, storage/retrieval failure and, 365–366, 365 (figure)
amygdala and, 232, 371
autobiographical memory, 370
confabulation and, 387, 492
consolidation/retrieval processes and, 366–367, 367–368 (figures), 377 (figure), 376377
declarative memory, 370, 371, 376, 386
delayed match-to-sample task and, 371–372
deletion of memory, reverse-learning hypothesis and, 478
dreaming and, 475–476
ecstasy and, 135
emotion and, 371
episodic memory, 371
extinction of memories, 377
factual memory, 370
false childhood memories, 379
forgetting and, 377–378
glutamate and, 367
hippocampal mechanism and, 367, 367–368 (figures), 369
long-lasting memory, 367, 367 (figure)
long-term memory, 367, 367 (figure), 368, 376, 400, 492
marijuana and, 136
memories, changing of, 377–379
nondeclarative memory, 370, 371
place cells and, 369, 369 (figure)
reconsolidation of memories and, 378–379
relational memory, 371
REM sleep and, 476–478, 477 (figures)
short-term memory, 367, 367 (figure), 368, 400
storage of, 368–369, 369 (figures)
striatum, damage to, 370–371
temporal lobe and, 64
total recall and, 378

transfer of, 477
working memory, 371–372, 400, 408
See also Alzheimer's disease; Korsakoff's syndrome; Learning process
Meninges, 70
Meningitis, 254
Menopause, 193, 409
Mental disorders, 428
vulnerability model and, 15–16, 432
See also Psychological disorders
Merkel's discs, 331, 331 (figure)
Mescaline, 134
Mesolimbocortical dopamine system, 138, 138 (figure), 139, 145
Messenger RNA (mRNA), 97, 98 (figure), 110, 111 (figure)
Metabolism. *See* Digestive process; Hunger; Obesity
Metabotropic receptors, 37
Methadone, 144
Methamphetamine, 132
Midbrain, 65 (figure), 66, 67 (figure), 68 (table)
brain stem and, 66–67, 67 (figure)
inferior colliculi and, 65 (figure), 66, 67 (figure)
reticular formation and, 66, 67
substantia nigra and, 66
superior colliculi and, 65 (figure), 66, 67 (figure)
ventral tegmental area and, 66
See also Central nervous system (CNS); Nervous system
Migrating neurons, 74, 76 (figure), 441–442, 442 (figure)
Mind-brain problem, 4
atoms and, 4
dualism and, 4
electrical brain, mechanistic mind and, 6
localization issue and, 6–8, 7–8 (figures)
mind-brain debate, renewal of, 8
monism and, 4
nonmaterial mind and, 8
physical model of behavior and, 4–6, 5 (figure)
See also Biopsychology; Brain function; Nature vs. nurture debate
Minimally conscious state, 499
Mirror neurons, 227, 284, 343, 402, 414, 416, 493, 493 (figure)
Mitochondria, 9
Models, 4–5
direct observation/experimental manipulation and, 6
potential fallibility of, 6
Modular processing, 313
Monism, 4
idealistic monism, 4
localization principle and, 7, 8 (figure)
materialistic monism, 4
See also Biopsychology
Monkeys:
aggression/impulsivity in, 244
brain size in, 57 (figure)
concept of self and, 490
See also Chimpanzees

Monoamine hypothesis of depression, 445–446, 446 (figure)
Mood disorders, 12, 145, 238, 421, 452
Moon landing, 2–3, 2 (figure)
Moral development, 224
Moral standards, 60
Morphine, 45, 127, 131, 138, 337
Mortality:
stress-induced mortality, 236, 236 (figure)
See also Aging; Suicide
Motivation, 156
arousal theory and, 157
drive theory and, 157
drives, brain states and, 157
hunger and, 157–158
incentive theory and, 157
instinct and, 156–157
sensation seeking and, 157
sexual behavior and, 190–191
theoretical approaches to, 156–158
See also Homeostatic drives; Hunger
Motor cortex, 60, 60 (figure), 68 (table)
association areas and, 346
basal gaglia and, 60, 228
homunculus and, 60, 347 (figure)
premotor cortex and, 346, 346 (figure), 347–348, 348–349 (figures), 351 (table)
primary motor cortex and, 60, 346, 346–347 (figures), 348–349, 351 (table)
processing sequence in, 346
rubber hand illusion and, 348, 349 (figure)
secondary motor areas and, 60, 348–349
supplementary motor area and, 346, 346 (figure), 348, 351 (table)
voluntary movement, 60
written language and, 63
See also Frontal lobe; Movement
Motor function, 56, 57 (figure), 60
basal ganglia and, 60, 66
muscle tone and, 67
reflexes, 69, 69 (figure)
reticular formation and, 67
substantia nigra and, 66
See also Motor cortex; Movement; Sensorimotor system
Motor neurons, 24–25, 25 (figures), 27 (table), 34 (figure), 344
Movable electrode implants, 103
Movement, 344
antagonistic muscles and, 344–345
autoimmune diseases and, 354–355
basal ganglia and, 346, 349–350, 350 (figure), 351 (table)
central pattern generators, rhythmic motor activity and, 345–346
cerebellum, 346, 346 (figure), 350–351, 351 (table)
delayed match-to-sample task and, 347
disorders of movement, 351–355
Golgi tendon organs and, 345
Huntington's disease and, 353–354, 354 (figure), 3350
involuntary movements, 350
motor cortex and, 346–349, 346 (figure)
motor neurons and, 344, 344 (figure)

multiple sclerosis and, 355, 355 (figure)
muscle contraction and, 344, 345
muscle spindles and, 345
muscles and, 344–345, 344 (figure)
myasthenia gravis and, 354–355, 354 (figure)
Parkinson's disease and, 350, 351–353, 352–353 (figures)
perception of, 315–316
posterior parietal cortex and, 346, 347
posture control and, 349, 350
prefrontal cortex and, 346 (figure), 347, 351 (table)
premotor cortex and, 346, 346 (figure), 347–348, 348–349 (figures), 351 (table)
primary motor cortex and, 346, 346–347 (figures), 348–349, 351 (table)
rubber hand illusion and, 348, 349 (figure)
sequences of movement, 350
smooth movement and, 349, 350
spinal cord involvement in, 345–346, 345–346 (figures)
stretch reflex and, 345, 345 (figure)
supplementary motor area and, 346, 346 (figure), 348, 351(table)
voluntary movements, 60, 348–349
See also Motor cortex; Motor function; Motor neurons; Musculoskeletal system; Vestibular sense
Movement agnosia, 320
MPOA. See medial preoptic area.
MRI. See magnetic resonance imaging.
Müllerian ducts, 199, 199 (figure), 200
Müllerian inhibiting hormone, 200, 205
Multiple personality disorder, 495–496, 496 (figure)
Multiple sclerosis, 235, 355
autoimmune reactions and, 355
characteristics of, 355, 355 (figure)
Epstein-Barr virus and, 355
stem cell therapy and, 355
treatments for, 355
Multipolar neurons, 26
Muscle spindles, 345
Muscle tone, 67
Musculoskeletal system:
antagonistic muscles and, 344–345
emotion, response to, 224–227, 225–226 (figures)
hydraulic model and, 5
motor neurons and, 24–25, 25 (figures)
muscle cells/fibers, motor neurons and 34, 44 (figure)
muscle contraction and, 344, 345
skeletal/striated muscles and, 344–345, 344 (figure)
smooth muscles and, 344
spinal cord involvement and, 345–346, 345–346 (figures)
tendons and, 344, 344 (figure)
See also Movement
Mushrooms, 134
Myasthenia gravis, 354–355, 354 (figure)
Myelin, 33–35, 34 (figure)
brain processing efficiency and, 400
cell bodies and, 56

demyelinating disorders, 111, 355
multiple sclerosis and, 235, 355, 355 (figure)
nerve regeneration and, 81
white matter and, 56
Myelin sheath, 25, 25 (figure), 34 (figure)
Myelin stain method, 95, 96 (figure)
Myelination process and, 77–78

Narcolepsy, 132, 483–484
National Health and Nutrition Examination Surveys, 172
National Institute of Mental Health (NIMH), 419, 443
National Sleep Foundation, 481
Natural experiments, 109, 207–209
Natural killer cells, 235, 235(table)
Natural selection, 4, 14
Naturalistic observation, 93
Nature vs. nurture debate, 9
balanced roles of nature/nurture, 12
behavior, genetic influences on, 9, 10–12, 11 (figure)
environmental challenges, vulnerability and, 15–16
genetic code and, 9–12, 9–11 (figures), 13
heredity, destiny vs. predisposition and, 13–16
heritability, 14–15, 15 (figure)
Human Genome Project and, 12–13
twin studies and, 14–16, 15 (figure)
See also Biopsychology; Mind-brain problem
Nausea, 333
Negative color aftereffect, 302, 302 (figure)
Negative symptoms, 433
Neglect, 15, 63, 320–321, 321 (figures), 496–497
Nerve deafness, 267
Nerves, 54, 54 (figure)
hydraulic model and, 5
spinal nerves, 71
See also Cranial nerves; Nervous system; Neuron
Nervous system, 4, 23–24, 71 (figure)
axons and, 54, 54 (figure)
cell bodies, grouping of, 54, 54 (figure)
development of, 55, 55 (figure), 66, 66 (figure)
direction/orientation terminology, 59, 59 (figure)
glial cells and, 33–35, 34–35 (figures)
myelination/conduction speed and, 33–35, 34 (figure)
neural messages, coding of, 45–46, 46 (figure)
neural networks and, 46–47
neuron cell membrane and, 27–33, 27–30 (figures), 32 (figure)
neurons and, 24–27, 24–26 (figures), 27 (table)
neurotransmitters, 12, 25, 42–44, 44 (table)
nodes of Ranvier and, 34, 34 (figure)
potentials and, 28–33, 28 (figure)
receptors, 12
reflexes, 26

Schwann cells and, 34, 34 (figure)
spinal cord, 24 (figure), 34
See also Autonomic nervous system (ANS); Brain function; Brain structure; Central nervous system (CNS); Emotion; Nerves; Nervous system development/change; Peripheral nervous system (PNS)
Nervous system development/change, 74
circuit formation stage and, 75–76, 77 (figure)
circuit pruning and, 76–77, 78 (figure)
fetal brain development and, 74, 77
growth cones and, 75, 77 (figure)
Hebb rule and, 372
mistakes in, 77, 78 (figure)
myelination process and, 77–78
neural tube formation, 55, 55 (figure), 66, 67 (figure), 74, 75 (figure)
phantom pain and, 79
plasticity and, 76–77, 82, 141–142, 342
proliferation/migration stage and, 74, 76 (figure), 77
reorganization, stimulation/experience and, 78–79, 79(figure)
stages of development, 74–78
See also Central nervous system damage/ recovery; Nervous system
Neural networks, 46–47
Neural plasticity, 76–77, 82, 141–142, 342, 449, 476
Neural precursor cells, 82
Neural tube formation, 55, 55 (figure), 66, 67 (figure), 74, 75 (figure)
Neurofibrillary tangles, 381, 381 (figure), 382
Neurogenesis, 81–82, 83, 93, 448–449, 449 (figure)
Neurological diseases, 2, 3
Neuron, 24, 24(figure), 54
axons, 25, 25–26 (figures), 29, 33
basic structure of, 24–25, 25 (figures)
bipolar neurons, 26
cell body/organelles and, 24
dendrites and, 25, 25–26 (figures), 26
glial cells and, 33–35, 34–35 (figures)
inter-neural connections and, 35, 35 (figure)
interneurons, 26, 26 (figure), 27 (table)
light control of, 32
mirror neurons, 227
motor neurons, 24–25, 25 (figures), 27 (table)
multipolar neurons, 26
myelination/conduction speed and, 33–35, 34 (figure)
neural activity levels, 35, 35 (figure)
neural messages, coding of, 45–46, 46 (figure)
neural networks and, 46–47
neurotransmitters and, 25, 42–44, 44 (table)
nodes of Ranvier and, 34, 34 (figure)
polarization and, 28, 29
reflexes and, 26
responsibilities of, 24
saltatory conduction and, 34

sensory neurons, 25–26, 26 (figures), 27 (table)
size of, 25
terminals and, 25, 25–26 (figures), 31
unipolar neurons, 26
See also Learning process; Nervous system; Nervous system development/change; Neuron cell membrane; Neuron communication
Neuron cell membrane, 27, 27 (figure)
absolute refractory period and, 31, 33
action potential and, 29–31, 30 (figure), 33, 34
all-or-non law and, 31
axon membrane, 29
decremental/nondecremental potential and, 29, 31, 34
depolarization process and, 29, 30 (figure), 31
electrostatic pressure and, 29, 31
force of diffusion and, 29
graded/ungraded potential and, 31, 34
ion distribution mechanisms and, 28–31, 29–30 (figures)
light effects and, 32
polarization and, 28, 29
potentials of, 28–33, 28 (figure)
rate law and, 33
recovery phase and, 30 (figure), 31
refractory periods and, 31, 33
relative refractory period and, 33
resting potential and, 28–29, 30 (figure), 31
saltatory conduction and, 34
selective permeability of, 27–28
sodium-potassium pump and, 29, 29–30 (figures), 31, 32, 34
stimulus intensity and, 31, 33
terminals and, 31
voltage and, 28, 29–31, 30 (figure)
See also Nervous system; Neuron; Neuron communication; Neurotransmitters
Neuron communication, 36
autoreceptors and, 41, 42 (figure)
axon hillock and, 38, 39, 39 (figure)
bidirectional messages and, 39
chemical transmission at synapses and, 36–41, 37–38 (figures)
electrical communication and, 37
excitation/inhibition and, 38–39, 39 (figure)
glial cells and, 41, 43 (figure)
hyperpolarization and, 38
hypopolarization and, 38
information integration/decision making and, 40
ionotropic receptors and, 37
metabotropic receptors and, 37
neural messages, coding of, 45–46, 46 (figure)
neural networks and, 46–47
neural pathway, gaps in, 38
postsynaptic integration and, 39–40, 40–41 (figures)
postsynaptic neurons, 36, 36 (figure), 37, 38 (figure)
presynaptic neurons and, 36, 36 (figure), 37, 38 (figure)

reuptake process and, 41
spatial summation and, 39–40, 40–41 (figures)
synapses and, 36, 56
synaptic activity, regulation of, 41, 42–43 (figures)
synaptic activity, termination of, 41
synaptic cleft and, 36, 41
temporal summation and, 39–40, 40 (figures)
vesicles and, 37, 41
See also Neuron; Neuron cell membrane; Neurotransmitters
Neuronsnucleus and, 24
Neuropeptide Y (NPY), 44 (table), 169, 171, 179
Neuropeptides, 43, 337
Neuroscience, 1, 3, 12, 16
Decade of the Brain and, 2, 3
mind-brain problem and, 4–8
nonmaterial neuroscience, 8
See also Biopsychology; Brain function; Brain structure
Neuroscientists, 3
Neurotoxins, 32, 102, 421
Neurotransmitters, 12, 25, 42, 44 (table)
anxiety disorders and, 454–455
corelease of, two-way switches and, 43–44
Dale's principle and, 43
excitation/inhibition and, 38–39, 39 (figure)
neuropeptides and, 43, 337
postsynaptic integration and, 39–40, 40–41 (figures)
receptors for, 32, 38 (figure), 42–43
release of, 37, 38 (figure), 43–44
reuptake process and, 41
sexual behavior and, 195, 195 (figure)
synaptic activity, termination of, 41
variety of, 42
See also Neuron communication
Neurotrophins, 76
Nicotine, 133–134
addiction treatment, medications for, 145 (table)
health risks from, 133
muscle twitching and, 134
positive mood effect and, 134
prenatal use of, 133–134
relapse rates, 143, 143 (figure)
withdrawal from, 133, 142
See also Psychoactive drugs; Stimulants; Tobacco use
Night terrors, 474
Nissl stain method, 95, 96 (figure)
Nitric oxide, 44 (table), 374
Nodes of Ranvier, 34, 34 (figure)
Noncoding DNA, 12
Non-REM sleep, 475
consciousness and, 485
function of, 476
slow-wave sleep and, 476
See also REM sleep; Sleep
Nondeclarative memory, 370, 371
Nondecremental potential, 31, 34
Nonmaterial mind, 8
Norepinephrine, 44 (table), 70, 132, 239, 446
depression and, 446, 450

Nuclear radiation, 77
Nucleotides, 10–11
Nucleus accumbens, 65 (figure), 138, 139
Nucleus (cell), 24
Nucleus (central nervous system), 54, 54 (figure), 64, 67
Nucleus of the solitary tract (NST), 160, 160 (figure), 162, 163, 168

ob. See obesity gene.
Obesity, 172
basal metabolism and, 176
body mass index and, 172, 173, 173 (figure)
cholecystokinin injections and, 170
compulsive overeating and, 179–180, 180 (figure)
dietary restriction and, 173–174, 178
epigenetics of weight regulation and, 177
exercise and, 178, 178 (figure)
gastric bypass procedure and, 180, 180 (figure)
genetic influences on, 174–175, 175–176 (figures)
global statistics on, 172, 172 (figure)
health risks of, 172–173
insulin secretion and, 179
leptin and, 170, 175, 176 (figure)
leptin treatment for, 179, 179 (figure)
life expectancy and, 173
medications for, 178–179, 180
myths of, 174
obesity gene and, 175, 175–176 (figures)
reduced metabolism and, 175–177
target of rapamycin kinase and, 173
treatments for, 178–180, 178–180 (figures)
ventromedial hypothalamus and, 170
weight loss, difficulty with, 177, 178
See also Eating disorders; Hunger
Obesity gene, 175, 175–176 (figures)
Object agnosia, 316–319, 317–319 (figures)
Object recognition, 315
object agnosia and, 316–319, 317–319 (figures)
See also Form vision; Visual perception
Observation, 6
naturalistic observation, 93
objective observation, 92
See also Research methods
Obsessive-compulsive disorder (OCD), 8, 142, 228, 457
brain anomalies in, 457–458, 457–458 (figures)
excessive self-grooming and, 458
hoarding and, 458, 459
related disorders, 458–460
ritualistic behaviors and, 457, 458
Tourette's syndrome and, 459–460, 460 (figure), 461
treatments for, 458
See also Anxiety disorders
Occipital lobes, 58 (figure), 64, 65 (figure), 68 (table)
association areas of, 63
myelination process and, 77–78
visual cortex and, 58 (figure), 64
See also Central nervous system (CNS); Cerebral hemispheres; Lobes

Occupational interests, 15
OCD. *See* obsessive-compulsive disorder
Offensive aggression, 241
Ohm's law of electricity, 265
Olfactory bulb, 81, 83
Olfactory system, 196–197, 196 (figure)
Oligodendrocyte, 34, 34 (figure)
Opiates, 45, 97, 127, 233
    addiction treatment, medicines for,
        145 (table)
    addictive potential and, 127
    analgesic effect and, 127, 128
    codeine, 127
    conditioned tolerance and, 127
    euphoria and, 127
    heroin, 45, 126, 127
    hypnotic effect and, 127
    morphine, 127
    natural endogenous opioids, 128
    opium, 127
    overdose danger and, 127
    pain relief and, 338, 339 (figure), 340–341
    synthetic opioids and, 127
    See also Psychoactive drugs
Opium, 127
Opponent process theory, 301–302,
    302–303 (figures)
Optic chiasm, 298, 299 (figure)
Optic nerves, 296, 297 (figure),
    298, 299 (figure)
Orbitofrontal cortex, 142, 229, 230 (figure)
Orexin system, 479–480, 481 (figure),
    483, 484
Organ of Corti, 257, 258 (figure), 259, 267
Organelles, 24
Organizing effects, 200
Organum vasculosum lamina terminalis
    (OVLT), 159, 160 (figure)
Osmotic thirst, 159, 160
Ossicles, 256, 257–258 (figures)
Ottawa Prenatal Prospective Study, 136
Out-of-body experience, 336, 492
Outer hair cells, 259
Oval window, 258 (figure), 265, 275
Ovaries, 199
OVLT. *See* organum vasculosum lamina
    terminalis
Overdose, 127, 130
OxyContin, 127
Oxytocin, 198, 417, 417 (figure)

Pacinian corpuscles, 331, 331 (figure)
Pain, 239–240, 239–240 (figure), 336
    acetaminophen and, 337
    adenosine and, 339
    anesthetics and, 337, 338, 339
    arthritis, capsaicin and, 331–332
    aspirin/ibuprofen and, 337
    biofeedback techniques and, 337, 343
    cancer, pain-inducing acid and, 332
    chronic pain, 341, 341–342, 341 (figure)
    congenital insensitivity to pain, 239, 341
    cultural context and, 239, 239 (figure)
    detection of, 336–337
    emotional aspect of, 239–240, 239–240
        (figures), 339 (figure)
    endogenous cannabinoids and, 341

endorphins and, 338–339, 340 (figure)
gate control theory of, 339
inflammatory soup and, 336–337
intensity of suffering and, 239
internal mechanisms of pain relief,
    338–339, 338–339 (figures)
itch and, 330
limbic system and, 240, 240 (figure)
lobotomies and, 240
MDAN series of drugs and, 337
menthol applications and, 332
morphine and, 337
neuropathic pain, 341
nociceptive pain, 341
opiate drugs and, 338, 339 (figure),
    340–341
pain insensitivity disorders and, 240
pain pathways, 337, 339, 341
periaqueductal gray and, 339–341,
    340 (figure)
phantom pain, 79, 342–343, 342 (figure)
premature aging and, 341–342
sharp pain/dull pain, 337
substance P and, 337, 339, 340 (figure)
treatment for, 337
    See also Emotion; Skin senses; Stress
Panic disorder, 454, 455, 460
Paradoxical sleep, 474
Paranoia, 93
Paranoid schizophrenia, 429
Parasympathetic nervous system,
    71, 71 (figure), 72, 73 (figure), 74
    emotion, response to, 224–225, 225
        (figure)
    See also Autonomic nervous system (ANS)
Paraventricular nucleus (PVN), 166, 167
    (figure), 169, 194
Parietal lobes, 58 (figure), 62, 65 (figure),
    68 (table)
    association areas of, 63
    damage to, 63
    homunculus and, 63
    neglect disorder and, 63
    primary somatosensory cortex and,
        58 (figure), 62–63
    somatosensory projection area and, 63
    See also Central nervous system (CNS);
        Cerebral hemispheres; Lobes
Parieto-insular-vestibular cortex, 333
Parkinson's disease, 66, 102, 195, 332,
    350, 351
    adult neural stem cells and, 352–353
    characteristics of, 351–352
    cognitive deficits/depression and, 352
    dopamine agonists and, 352
    dopamine neurons and, 352, 353
    environmental influences in, 352
    familial occurrences of, 352
    genetic interventions and, 353
    learning process and, 370–371
    levodopa and, 352
    Lewy bodies and, 352, 352 (figure)
    pesticide exposure and, 352
    REM sleep behavior disorder and, 484
    risk reduction and, 352
    striatum and, 352
    substantia nigra and, 352, 353

surgical treatments for, 353
    treatments for, 352–353, 353 (figure)
    whole-genome study of, 352
Parvocellular system, 314–315, 314 (figure)
Patellar tendon reflex, 345, 345 (figure)
Peptide YY (PYY), 170, 171 (table), 179
Peptides, 43, 44 (table)
Perception, 4, 255
    See also Consciousness; Visual perception
Periaqueductal gray (PAG), 242–243,
    242 (figure), 339–341, 340 (figure)
Peripheral nervous system (PNS), 54, 71,
    71 (figure)
    cranial nerves and, 71, 72 (figure)
    ganglion and, 54, 54 (figure)
    nerves and, 54, 54 (figure)
    regeneration and, 80–81
    spinal nerves and, 71
    See also Autonomic nervous system (ANS);
        Cranial nerves; Nervous system
Peroptic area, 158, 159 (figure)
Persecution delusions, 133
Personality characteristics, 12, 15
    environmental challenges, vulnerability
        and, 15–16, 432–433
    See also Affective disorders; Psychological
        disorders
Pesticide pollution, 352, 417, 421
PET. *See* positron emission tomography
Peyote, 134
PGO waves, 480, 481 (figure)
Phantom limb phenomenon, 491, 491 (figure)
Phantom pain, 79, 342–343, 342 (figure)
Pharmaceuticals. *See* Medications
Phase difference, 268
Phencyclidine (PCP), 135, 435
Phenotypes, 11
Phenylketonuria, 411–412
Pheromones, 196–197, 215, 215 (figure)
Phobia, 454, 460
Phonological hypothesis of dyslexia, 276
Photopigment, 296
Photoreceptors, 296, 297 (figure), 300
Phototherapy, 450
Phrenology, 7, 7 (figure)
Phthalates, 421
Physical therapy, 82
Pineal gland, 5, 64, 65 (figure), 67 (figure),
    68 (table), 92
Pinna, 157 (figure), 256
Pitch, 255, 256 (figure)
Pituitary gland, 64, 65 (figure)
Place cells, 369, 369 (figure)
Place theory, 262–265, 263–265 (figures)
Plagiarism, 112
Planning/organization, 60
Planum temporale, 276
Plaques, 381, 381 (figure), 382
Plasticity, 76–77, 82, 141–142, 342, 449, 476
PNS. *See* peripheral nervous system
Polarization, 28, 29
    depolarization, 29, 30 (figure), 31, 38
    hyperpolarization, 38, 40
    hypopolarization, 38, 40
    See also Neuron cell membrane; Neuron
        communication; Sodium-potassium
        pump

Polygenic characteristics, 12
Pons, 65 (figure), 67, 67 (figure), 68 (table)
Positive symptoms, 433
Positron emission tomography (PET),
     105, 105 (figure), 131, 132 (figure),
     228, 229, 243
Postcentral gyrus, 58 (figure), 62
Posterior, 59, 59 (figure)
Posterior parietal cortex, 334 (figure), 336
     body integrity identity disorder and, 336
     body-world integrations and, 336
     eye movements and, 336
     movement, refinement of, 346, 347
     out-of-body experience and, 336
     posture, information about, 336
     reaching/grasping movements and, 336
     See also Body senses; Somatosensory
          cortex
Postsynaptic integration, 39–40,
     40–41 (figures)
Postsynaptic neurons, 36, 36 (figure),
     37, 38 (figure), 82
Posttraumatic stress disorder (PTSD),
     132, 236, 237, 379, 455–456,
     455 (figure)
Potassium. See Sodium-potassium pump
Potential:
     action potential, 29–31, 30 (figure), 33, 34
     all-of-non law and, 31
     backward-moving potentials, 33
     decremental/nondecremental potential,
          29, 31
     evoked potential, 99, 100 (figure)
     graded/ungraded potential and, 31, 34
     inhibitory postsynaptic potential, 38
     polarization and, 28
     resting potential, 28–29
     saltatory conduction and, 34
PPT/LDT, 479 (figure), 480 (figure)
Practical intelligence, 395
Prader-Willi syndrome, 156
Precentral gyrus, 60, 60 (figure)
Predatory aggression, 241
Prefrontal cortex, 58 (figure), 60–61,
     68 (table), 78, 132, 136
     aggression and, 243, 244
     delayed match-to-sample task and, 347
     emotions and, 224, 227 (figure),
          229–230, 230–231 (figures)
     learning/retrieval and, 367
     movement control and, 346 (figure),
          347, 351 (table)
     working memory and, 371–372
     See also Frontal lobe
Premenstrual syndrome (PMS), 241
Premotor cortex, 346, 346 (figure),
     347–348, 348–349 (figures),
     351 (table)
Presynaptic neurons, 36, 36 (figure)
     compensation and, 82
     neurotransmitter release and,
          37, 38 (figure)
     presynaptic excitation, 41
     presynaptic inhibition, 41
Primary motor cortex, 346, 346–347
     (figures), 348–349, 351 (table)
Primary somatosensory cortex, 58 (figure),
     62–63, 68 (table), 335, 336

Primary visual cortex, 57, 57–58 (figures),
     63, 64, 68 (table)
Proactive aggression, 241, 243
Projection areas, 63
Proliferation, 74
Proprioception, 330, 332, 336, 487
Prosody, 279
Prosopagnosia, 317, 317 (figure), 487
Proteins:
     blood-brain barrier and, 70, 70 (figure)
     brain structural development and, 12
     gene activity and, 14, 97
     junk DNA and, 13
     in situ hybridization and, 97, 98 (figure)
     See also DNA (deoxyribonucleic acid);
          Genetic code
Pruning. See Circuit pruning
Pseudohermaphrodites, 204
     female pseudohermaphrodites, 205–206,
          206 (figure)
     male pseudohermaphrodites, 204–205,
          205 (figure)
     See also Sexual anomalies
Psychedelic drugs, 134
     ecstasy, 134–135, 135 (figure)
     hallucinogenic substances, 134
     lysergic acid diethylamide/LSD, 134
     mushrooms, 134
     peyote, 134
     phencyclidine/PCP, 135
     research on psychoses and, 135
     See also Psychoactive drugs
Psychoactive drugs, 126
     depressants, 128–130
     marijuana, 135–137, 136 (figure)
     opiates, 127–128, 127 (figure)
     psychedelics, 134–135, 135(figure)
     stimulants, 130–134
     See also Addiction; Drugs; Substance
          abuse
Psychological disorders, 12, 428
     cocaine abusers and, 131–132
     dissociative identity disorder, 494–495,
          495 (figure)
     insomnia and, 481–482
     See also Affective disorders; Anxiety
          disorders; Psychotherapeutic drugs;
          Schizophrenia
Psychology, 3
     behaviorism and, 468
     cognitive psychology, 468
     consciousness and, 468
     genetics and, 12
     monism and, 4
     prescientific psychology, mind-brain
          problem and, 4
     See also Biopsychology; Consciousness;
          Psychological disorders
Psychopathy, 243
Psychosis, 429
Psychosurgery, 61
Psychotherapeutic drugs, 126
Punishment:
     prefrontal cortex and, 60, 61, 224
     See also Rewards
Pure tone, 256, 256 (figure), 265, 266
PVN. See paraventricular nucleus
PYY. See peptide YY3-36

Radial glial cells, 74, 77
Radiation exposure, 77
Raphé nucleus, 479 (figure), 480 (figure),
     481 (figure)
Rapid eye movement sleep. See REM sleep
Rate law, 33
Raven Progressive Matrices, 395
Reactive aggression, 241, 243
Reasoning, 4, 224
Receptive field, 297
Receptors, 12
     auditory stimulation and, 254, 258
     autoreceptors, 41, 42 (figure)
     ionotropic receptors, 37
     metabotropic receptors, 37
     neurotransmitter receptors, 32,
          38 (figure), 42–43
     skin senses and, 331, 331 (figure)
     subtypes of, 42
     transient receptor potential group,
          331–332
Recessive genes, 11
     hand clasping preference and, 11–12,
          11 (figure)
     See also Genes; Genetic code; Inheritance
Recognition ability, 63, 66
Reflex circuit, 69, 69 (figure)
Reflexes, 26, 67, 68, 69, 345
Refractory periods, 31
     absolute refractory period, 31, 33
     relative refractory period, 33
Regeneration, 80–81, 93
Reinforcers, 141
Relative refractory period, 33
Relaxation training, 237
Reliability:
     imaging, behavior studies and, 106–107
     See also Research methods
REM. See rapid eye movement sleep
REM sleep, 474–475, 474–475 (figures)
     consciousness and, 485
     depression and, 449–450
     function of, 475–476
     memory function and, 476–478,
          477 (figures)
     narcolepsy and, 483–484
     PGO waves and, 480, 481 (figure)
     reverse-learning process, memory deletion
          and, 478
     See also Non-REM sleep; Sleep
REM sleep behavior disorder, 484
Renaissance, 6
Reorganization, 78–79, 79 (figure)
Reproduction. See Sexual reproduction
Research ethics, 16, 112
     access inequities and, 116
     advancing scientific knowledge and, 92
     animal research subjects and, 113–115,
          115 (figure)
     deception, acceptability of, 113
     experimental studies and, 94
     gene therapy and, 111, 115–116,
          116 (figure)
     human research subjects and, 113
     informed consent and, 113, 116
     oversight issues, 116
     participants, protection of, 112–115
     plagiarism/fabrication, 112

specieism and, 114
stem cell therapy and, 116–117, 117 (figure)
Research methods, 92
  case studies and, 93
  cause/effect conclusions and, 94
  confounded variables and, 94
  control variables and, 93
  correlational studies, 93–94, 94 (figure),
    108–110, 110 (table)
  empiricism, 92, 93
  experimental studies, 93
  genetic inheritance, correlational studies
    on, 108–110, 110 (table)
  hypotheses and, 93
  naturalistic observation and, 93
  objective observation, 92
  self-correction and, 93
  surveys and, 93
  tentative conclusions and, 92–93
  theory development and, 93
  See also Research ethics; Research
    techniques
Research techniques, 95
  ablation/aspiration, 101–102
  animal mosaic and, 111
  antisense RNA procedure, 110, 111 (figure)
  autoradiography technique, function data
    and, 96–97, 97 (figure)
  brain activity manipulation/measurement,
    98–102
  brain imaging techniques, 102–107,
    108 (table)
  cannula and, 101, 102, 102 (figure)
  computed tomography scanning, 103,
    104 (figure), 105
  confocal laser scanning microscope, 98
  diffusion tensor imaging, 103
  electroencephalography, 99, 100 (figures),
    108 (table)
  electron microscopes, 98, 99 (figure)
  fluorescent dye and, 98
  fluorogold injection technique, 95–96
  functional magnetic resonance imaging,
    106–107, 106 (figure)
  gene transfer technique, 110–111
  genetic engineering, experimental
    approach of, 110–111, 111 (figure)
  genetic similarities, correlational approach
    and, 108–110, 110 (table)
  Golgi stain method, 95, 96 (figure)
  imaging techniques for neurons, 95–97,
    97 (figure)
  immunocytochemistry technique, 97
  knockout technique, 110
  lesioning, 102
  light microscope, limitations of, 98
  magnetic resonance imaging,
    103, 104 (figure), 105
  microelectrodes and, 100–101
  movable electrode implant, 103
  myelin stain method, 95, 96 (figure)
  nissl stain method, 95, 96 (figure)
  positron emission tomography,
    105, 105 (figure)
  radioactive substances and, 96–97, 105
  reliability issues and, 106–107
  scanning electron microscopes,
    98, 99 (figure)

in situ hybridization, 97, 98 (figure)
staining techniques, 95, 96 (figure)
stereotaxic atlas and, 99–100, 101 (figure)
stereotaxic instruments, 100–101,
  101–102 (figures)
transcracial magnetic stimulation (TMS),
  94, 102, 103 (figure)
depression and, 448
transgenic animals and, 110–111
two-photon microscope, 98
2-deoxyglucose injection technique, 96–97
See also Research ethics; Research
  methods
Resonance, 262
Respiration, 67
Resting potential, 28–29, 30 (figure), 31
Reticular formation, 66, 67, 68 (table)
Retina, 296, 296–297 (figures), 298, 300
  color blindness and, 306, 306 (figure)
  lateral inhibition and, 308–309,
    308–309 (figures)
  light-dark contrast detectors and, 309,
    310 (figure), 472, 472 (figure)
  visual analysis pathways and, 314–316,
    314–315 (figures)
  See also Color vision; Eyes; Form vision;
    Vision
Retinal disparity, 298, 299 (figure)
Retinohypothalamic pathway, 471–472
Retinotopic map, 307, 307 (figure)
Retrieval process, 366–367, 367–368
  (figures)
Retrograde amnesia, 365, 366, 386
Reuptake process, 41
Reverse-learning process, 478
Rewards, 138
  dopamine, role of, 139–140, 139–140
    (figures), 141
  drug-induced learning, brain plasticity
    and, 141–142, 141–142 (figures)
  impulsiveness and, 420
  neural basis of, 138–139
  predictability/unpredictability of, 141,
    141 (figure)
  prefrontal cortex and, 60, 61
  reward deficiency syndrome and, 140
  ventral tegmental area and, 66
Rhesus monkeys, 13
Rhodopsin, 296, 300
Ribonucleic acid. See RNA (ribonucleic acid)
Risk, 14, 430, 430 (figure), 452, 453 (figure)
RNA (ribonucleic acid):
  antisense RNA procedure, 110, 111 (figure)
  messenger RNA, 97, 98 (figure),
    110, 111 (figure)
Rods. See Eyes
Rohypnol (roofies/rophies), 130
Roofies/rophies, 130
Rubber hand illusion, 348, 349 (figure)
Ruffini endings, 331, 331 (figure)

SAD. See seasonal affective disorder
Saltatory conduction, 34
Satiety:
  hunger and, 169–170, 170 (figure), 179
  sensory-specific satiety, 162–163
  thirst and, 160
  See also Sex

Savants, 414
  See also Autistic savants
Scanning electron microscopes, 98, 99 (figure)
Scar tissue, 81
Schizophrenia, 12, 14, 15, 60, 61, 93, 94,
  428, 429
  acute vs. chronic symptoms and, 430
  adoption studies and, 431
  age risk for, 430, 430 (figure)
  amphetamine use and, 133
  atypical/second-generation antipsychotic
    drugs and, 435, 439
  auditory gating impairment and, 439
  brain anomalies in, 436–442
  brain defects, environmental origins of,
    439–441, 440–441 (figures)
  brain tissue deficits/ventricular
    enlargement and, 436–437,
    437 (figure), 442, 442 (figure)
  catatonic schizophrenia, 429
  categories of, 429, 433–434
  characteristics of, 429–430, 430 (figures)
  developmental model of, 441–442,
    442 (figure)
  dopamine deficiency and, 435, 437
  dopamine hypothesis and, 434,
    435 (figure)
  epigenetic influences and, 441
  gender and, 430
  glutamate theory and, 436
  hallucinations and, 439
  heredity and, 431–433, 433 (table)
  historic perspective on, 430
  hypofrontality and, 437–438, 438 (figure)
  infectious diseases/viral infections and,
    440, 440–441 (figures)
  intelligence and, 109–110
  negative symptoms/Type II schizophrenia
    and, 433–434, 434 (table)
  neural connections, synchrony and,
    438–439, 439 (figure)
  neuroleptic drugs and, 435
  paranoid schizophrenia, 429
  phencylidine research and, 135
  positive symptoms/Type I schizophrenia
    and, 433–434, 434 (table)
  prefrontal cortex and, 229
  prenatal factors and, 439–441
  prenatal starvation and, 440–441
  schizophrenia gene, search for,
    432, 433 (table)
  tardive dyskinesia and, 435
  tobacco use and, 439
  treatments for, 435–436, 447
  twin studies and, 431–432,
    431–432 (figures)
  vulnerability model and, 14, 15–16,
    432–433
  winter birth effect and, 440, 441 (figure)
  See also Psychological disorders
Scholastic Aptitude Test (SAT), 395
Schwann cells, 34, 34 (figure)
Science:
  empiricism, 6, 92, 93
  See also Biopsychology; Mind-brain
    problem; Research methods
SCN. See suprachiasmatic nucleus.
SDN. See sexually dimorphic nucleus.

Seasonal affective disorder (SAD),
    450, 450 (figure)
Seasonal cycles, 64
Secondary somatosensory cortex, 335–336
Sedatives, 128, 130, 134
Selective tolerance, 132
Self:
    consciousness and, 490–491,
        490–491 (figures), 493
    disorders of self, 493–496
    dissociative identity disorder and,
        494–496, 495–496 (figures)
    mirror neuron system and,
        493, 493 (figure)
    split brain surgical procedure and,
        493–494, 494 (figure)
    state-dependent learning and, 495–496,
        496 (figure)
    theory on mind and, 493
    See also Agency; Consciousness
Sensation, 5, 255
    See also Body senses; Sensory function;
        Sensory neurons
Sensation seeking, 157
Sensorimotor system, 330
Sensory function, 56, 57 (figure)
    parietal lobes and, 62–63
    perception and, 255
    projection areas and, 63
    reflexes and, 69, 69 (figure)
    sensation and, 255
    specific nerve energies doctrine and, 254
    See also Body senses; Hearing; Sensory
        neurons; Vision
Sensory neurons, 25–26, 26 (figures),
    27 (table)
    adequate stimulus and, 254
    nerve growth factor, 341
    patterning of stimulation and, 254
    receptors and, 254
    See also Spinal nerves
Sensory-specific satiety, 162–163
Serotonin, 41, 44 (table), 129
    affective disorders and, 445–446, 450
    aggression inhibition and, 243–245,
        244 (figure)
    alcohol-serotonin effects on aggression,
        244–245
    anxiety disorders and, 455
    cocaine and, 131
    eating disorders and, 182–183
    ejaculation and, 195
    psychedelics and, 134–135, 135 (figure)
    serotonin-potentiating drugs, addiction
        treatment and, 144–145
    suicide and, 452, 454 (figure)
    tryptophan and, 244
    waking and, 479, 480 (figure)
Set point, 158
Severe combined immunodeficiency (SCID),
    92, 92 (figure), 111, 117
Sex, 190
    activating effects of hormones and,
        200, 201
    amygdala and, 194, 197, 232
    anterior hypothalamic area and, 197
    arousal/satiation and, 191–192

biological determination of, 198–201,
    199–201 (figures)
brain, sexualization of, 201
brain structures and, 193–195, 194 (figure)
castration and, 192
chromosomal activity and, 199–200,
    199–200 (figures)
Coolidge effect and, 191–192, 195,
    195 (figure)
definition of, 198
dihydrotestosterone and, 200, 204
estrogen and, 193, 200, 201
estrus and, 193
excitement phase and, 191
female-initiated activity and, 193,
    193 (figure)
gender-related behavioral/cognitive
    differences and, 202–204,
    202 (figure)
gonads and, 199
hormonal effects on, 200–201, 201
intersexes and, 206
lifelong monogamy and, 197, 198
major histocompatibility and, 196
male/female organs, development of,
    199–200, 199–200 (figures)
medial amygdala and, 194
medial preoptic area and, 194, 195, 197
menopause and, 193
motivated behavior of, 190–191
Müllerian ducts and, 199, 200
Müllerian inhibiting hormone and, 200
neurotransmitters and, 195, 195 (figure)
olfactory system, sexual attraction
    and, 196–197
organizing effects of hormones and, 200
orgasm and, 191
ovaries and, 199, 200
ovulation and, 193
ovum and, 199
pheromones and, 196–197, 215,
    215 (figure)
plateau phase and, 191
puberty and, 201
refractory phase and, 191
resolution and, 191
sexual response cycle and, 191, 191 (figure)
sexually dimorphic nucleus and, 194
Solffian ducts and, 199, 200
sperm and, 199
testosterone and, 192–193, 193 (figure),
    200, 201
ventromedial hypothalamus and, 194
vomeronasal organ and, 196 (figure), 197
    See also Gender; Sexual anomalies; Sexual
        orientation; Sexual reproduction
Sex reassignment, 208, 209, 214
Sexual anomalies, 204
    ablatio penis and, 207–209, 209 (figure)
    androgen insensitivity syndrome,
        205, 206, 207
    brain sexualization and, 206–207
    congenital adrenal hyperplasia and,
        205, 206
    diethylstilbestrol exposure and, 206–207
    female pseudohermaphrodites, 205–206,
        206 (figure)

gender reassignment, 208, 209
intersexes and, 206
male pseudohermaphrodites, 204–205,
    205 (figure)
    See also Sex; Sexual orientation
Sexual orientation, 12, 209
    anterior commissure and, 213
    asexuality and, 210
    biological model of, 211–215, 216
    bisexuality, 209, 214
    brain structures/function and, 213–215,
        213–215 (figures)
    central bed nucleus of the stria terminalis
        and, 214
    diethylstilbestrol exposure and, 206–207
    disease approach to homosexuality
        and, 216
    female homosexuality, challenge of, 215
    gays, 209
    gender nonconformity and, 210
    genetic studies and, 211–212,
        211 (figures)
    homosexuality and, 209–210, 214
    hormonal influence and, 212–213
    inborn nature of homosexuality and,
        215–216
    lesbians, 209, 210, 215
    opportunistic homosexuality and, 212
    social influence hypothesis and, 210
    spatial/verbal performance and,
        214, 214 (figure)
    suprachiasmatic nucleus and, 213
    third interstitial nucleus of the anterior
        hypothalamus and, 213
    transsexuality and, 214
Sexual reproduction:
    cross-fostering technique and, 109
    fertilization and, 9–10, 14
    gene manipulation and, 116
    genetic testing, genetic enhancement
        practices and, 116
    instinctive behavior of, 157
    major histocompatibility complex
        and, 196
    in vitro fertilization, 109
    See also Gender; Sex
Sexually dimorphic nucleus (SDN), 194
Short-term memory, 367, 367 (figure), 400
Sign language, 279, 279 (figure)
    auditory language areas and, 280–281,
        280 (figure)
    nonhuman communication and, 282–283,
        283 (figure)
    See also Deafness; Language
Simple cells, 310, 310–311 (figures)
Skeletal muscles, 344, 344 (figure), 345
Skin conductance response (SCR), 230, 336,
    371, 456
Skin senses, 330–331
    chemical receptors, 331
    encapsulated receptors and, 331
    free nerve endings and, 331, 331 (figure)
    Meissner's corpuscles and, 331,
        331 (figure)
    Merkel's discs and, 331, 331 (figure)
    Pacinian corpuscles and, 331, 331 (figure)
    receptors of the skin, 331, 331 (figure)

Ruffini endings and, 331, 331 (figure)
temperature and, 331–332
touch and, 331
transient receptor potential group and, 331–332
See also Body senses
Sleep, 468, 473
    activation-synthesis hypothesis and, 475–476
    active process of, 473–475, 474 (figure), 478
    adaptive hypothesis and, 468
    alpha waves and, 473, 474 (figure)
    arousal system and, 479–480, 480–481 (figures)
    atonia and, 474, 479, 481 (figure), 483
    basal forebrain area and, 478, 479 (figure)
    cataplexy and, 483
    chronotype and, 482
    circadian rhythm and, 470–473
    consciousness and, 468, 485
    consolidation process and, 477, 477 (figure)
    daily sleep, comparative durations of, 468, 469 (figure)
    day-night cycle and, 470–472
    disturbances/disorders of, 474, 481–484
    dreaming and, 475–476, 485
    eating disorders and, 483
    insomnia and, 481–482, 482 (figure)
    internal clock and, 471, 472–473, 482
    jet lag and, 469
    K complexes and, 473, 474 (figure)
    learning process and, 376–377, 377 (figure), 476–477, 477 (figure)
    lucid dreamers and, 485
    melatonin and, 471
    memory function and, 476–478, 477 (figures)
    narcolepsy and, 483–484
    neural development in childhood and, 476
    non-REM sleep stages and, 475, 476, 485
    objective study of, 468
    orexin system and, 479–480, 481 (figure), 483, 484
    paradoxical sleep and, 474
    PGO waves and, 480, 481 (figure)
    REM sleep, 474–478, 474–475 (figures), 477 (figures), 485
    REM sleep behavior disorder and, 484
    restorative function of, 468, 476
    retinohypothalamic pathway and, 471–472
    reverse-learning hypothesis, memory deletion and, 478
    sleep controls, brain structures for, 478–479, 479 (figure)
    sleep deprivation, effects of, 468–469
    sleep spindles and, 473, 474 (figure)
    sleep-wake cycle, 67, 468, 471, 473
    slow-wave sleep and, 473–474, 474 (figure), 476, 477, 477 (figure), 478
    stages of sleep, duration of, 475, 475 (figure)
    suprachiasmatic nucleus and, 470–471, 470 (figure), 473
    theta waves and, 473, 474 (figure)

time cues, isolation from, 470–471, 471 (figure)
    ultradian rhythm and, 475
    See also Consciousness
Sleepwalking, 467–468, 474, 482–483, 485
Slow-wave sleep (SWS), 473–474, 474 (figure), 476, 477, 477 (figure), 478
Smooth muscles, 344
Sociopaths, 458
Sodium-potassium pump, 24, 29, 29–30 (figures), 31, 32, 34
Solffian ducts, 199, 200
Somatic nervous system, 71, 71 (figure)
Somatosensory cortex, 58 (figure), 62–63, 68 (table), 78, 79 (figure), 240, 334
    body sense information pathway and, 334, 492
    dermatomes and, 333–334, 334 (figure)
    movement, refinement of, 346
    phantom pain and, 342, 342 (figure)
    posterior parietal cortex and, 334 (figure), 336
    primary somatosensory cortex and, 335, 336
    receptive fields and, 335, 335 (figure)
    secondary somatosensory cortex and, 335–336
    somatotopic organization and, 334 (figure)
    visual cortex/auditory cortex, comparison with, 334–336, 334–335 (figures)
    See also Body senses
Somatosensory project area, 63
Somatosensory system, 330
    See also Body senses; Somatosensory cortex
Soul, 4, 5, 64, 92
Sound. *See* Auditory cortex; Auditory mechanism; Complex sound; Frequency analysis; Hearing; Hearing impairment; Sound localization; Sound vibrations
Sound localization, 267–268
    binaural cues and, 268, 268–269 (figures)
    brain circuits for, 268–270, 269 (figure)
    coincidence detectors and, 268, 269 (figure), 270
    complex sounds and, 266, 267
    distant sounds and, 268, 268 (figure)
    frequency analysis and, 263, 265, 266
    phase difference and, 268, 269 (figure)
    sound shadows, intensity differences and, 268, 268 (figure)
    time of arrival, differences in, 268, 268 (figure)
    visual/spatial information and, 270
    See also Auditory mechanism; Hearing
Sound vibrations, 255, 255 (figure)
Spatial frequency theory, 312–313, 312–313 (figures)
Spatial summation, 39–40, 40–41 (figures)
Spatial tasks, 66, 202, 202 (figure), 203, 207, 214, 214 (figure)
Specieism, 114
Speech production, 60, 63
Speed, 132
Spinal cord, 24 (figure), 34, 65 (figure), 67

brain development and, 66, 66 (figure)
    central pattern generators and, 345–346, 346 (figure)
    cerebral hemispheres and, 57
    dorsal root of spinal nerves and, 68, 69
    interior of, 68
    meninges and, 70
    movement, involvement in, 345–346, 345–346 (figures)
    reflex circuit and, 69, 69 (figure)
    reflexes and, 68, 69, 345, 345 (figure), 346
    role of, 67–68
    spinal neurons and, 68–69, 69 (figure), 71, 73 (figure)
    stretch reflex and, 345, 345 (figure)
    ventral horns and, 69, 69 (figure)
    ventral root and, 69, 69 (figure)
    white matter/gray matter and, 68
    See also Central nervous system (CNS); Nervous system
Spinal nerves, 68–69, 69 (figure), 71, 73 (figure)
Split brain surgical procedure, 493–494, 494 (figure)
Sports, 207
Sputnik, 1–2
Staining techniques, 95, 96 (figure)
Stem cell therapy, 116–117, 117 (figure), 267, 352–353, 355
Stem cells, 83, 83 (figure), 11, 116–117 (figure), 267, 300, 352, 355
Stereotaxic atlas, 99–100, 101 (figure)
Stereotaxic instruments, 100–101, 101–102 (figures)
Stimulants, 126, 130, 421
    amphetamines, 132–133, 419–420
    caffeine, 134
    cocaine, 131–132, 131–132 (figures)
    nicotine, 133–134
    See also Psychoactive drugs
Stomach. *See* Digestive process
Street smarts, 395
Stress, 234
    adaptive response of, 70, 234–235
    brain damage and, 236–238, 237 (figure)
    cardiovascular system and, 236
    child abuse and, 236, 245
    cortisol release and, 172, 236–237
    hippocampal damage and, 236–237, 237 (figure)
    hypothalamus-pituitary-adrenal axis and, 234, 234 (figure)
    immune system activity and, 234–235, 235–236 (figures), 236, 237–239, 239 (figure)
    insomnia and, 481
    maternal stress, 421
    mortality risk and, 236, 236 (figure)
    negative effects of, 235–238, 236–237 (figures)
    pain, adaptive emotion of, 239–240, 239–240 (figures)
    relaxation training and, 237
    social/personality variables and, 237–239, 239 (figure)
    torture victims and, 236

voodoo death and, 236
See also Emotion
Stress-diathesis model, 452
Stretch reflex, 345, 345 (figure)
Striatum, 352, 370, 420
Stroke, 80, 80 (table), 82, 172,
    271–272, 490
Subfornical organ (SFO), 160, 160 (figure)
Substance abuse, 2, 12
    abused drugs, effects of, 126–127
    attention deficit hyperactivity disorder
        and, 419, 420 (figure)
    compensatory adaptation and, 127
    overdose and, 127
    See also Addiction; Psychoactive drugs
Substance P, 44 (table), 337, 339,
    340 (figure)
Substantia nigra, 66, 68 (table), 350 (figure),
    352, 484
Sudden cardiac death, 236, 236 (figure)
Suicide, 452
    antidepressant drugs and, 453
    genetic factors and, 452–453
    impulsive suicide attempters, 453
    risk of, 452, 453 (figure)
    serotonin levels and, 452, 454 (figure)
    stress-diathesis model and, 452
    youthful suicides, 453
    See also Affective disorders; Depression
Sulcus, 55, 56 (figure)
    central sulcus, 58, 58 (figure)
    See also Cerebral hemispheres
Summation. *See* Spatial summation;
    Temporal summation
Superior, 59, 59 (figure)
Superior colliculi, 65 (figure), 66, 67
    (figure), 68 (table)
Supplementary motor area, 346, 346
    (figure), 348, 351 (table)
Suprachiasmatic nucleus (SCN), 213,
    470–471, 470 (figure), 473
Surveys, 93
Susceptibility, 15–16
Sweat glands, 74
Sympathetic ganglion chain, 72
Sympathetic nervous system, 71, 71 (figure),
    72, 73 (figure), 74
    emotion, response to, 224–225,
        225 (figure)
    See also Autonomic nervous system (ANS)
Synapse, 36
    chemical transmission at synapses, 36–41
    circuit pruning and, 76–77, 78 (figure)
    postsynaptic integration, 39–40,
        40–41 (figures)
    presynaptic/postsynaptic neurons and,
        36, 36 (figure), 37, 38 (figure), 41
    stimulation/experience, synaptic
        construction and, 78–79, 79 (figure)
    synaptic activity, regulation of,
        41, 42–43 (figures)
    synaptic activity, termination of, 41
    See also Neuron communication;
        Synaptic cleft
Synaptic cleft, 36, 41
Syndactyly, 78, 79 (figure)
Synesthesia, 294, 322

T cells, 235, 235 (table), 237, 238, 355
Tardive dyskinesia, 435
Target of rapamycin kinase (TOR), 173
Taste:
    bait shyness and, 163
    dietary selection and, 161, 162–164
    insular cortex and, 228
    learned taste aversion and, 163, 164
    learned taste preference and, 163–164
    localization of taste responses and, 162,
        162 (figure)
    nucleus of the solitary tract and, 162, 163
    sensory-specific satiety and, 162–163
    taste receptors, 161–162, 162 (figures)
    wisdom of the body and, 161
    See also Hunger
Tectorial membrane, 257, 258–259
    (figures), 259
Telephone theory, 261
Temperature regulation, 158
    ectothermic animals and, 158
    endothermic animals and, 158
    mammalian thermostat and, 158
    preoptic area of the hypothalamus and,
        158, 159 (figure)
    See also Homeostatic drives
Temperature sensation, 331–332
Temporal lobes, 58 (figure), 63, 68 (table)
    association areas of, 63, 64
    auditory cortex and, 58 (figure), 63
    damage to, 63, 365–366
    inferior temporal cortex and, 63
    memory and, 64
    obesity and, 172
    surgery on, 63–64
    visual identification ability and, 63, 66
    Wernicke's area and, 58 (figure), 63
    See also Central nervous system (CNS);
        Cerebral hemispheres; Lobes
Temporal summation, 39–40, 40 (figures)
Terminal, 25, 25–26 (figures), 31
Testes, 200
Testosterone, 192–193, 193 (figure), 200,
    201, 203, 241, 241 (figure), 244, 409
Thalamus, 64, 65 (figure), 67 (figure),
    68 (table)
Theories, 93
Theory of mind, 413–414, 415, 493
Third interstitial nucleus of the anterior
    hypothalamus (INAH3), 213
Thirst, 159, 160, 160 (figure)
    angiotensin II and, 160
    diarrhea and, 165
    hypovolemic thirst, 159–160
    median preoptic nucleus and, 159, 160,
        160 (figure)
    nucleus of the solitary tract and, 160,
        160 (figure)
    organum vasculosum lamina terminalis
        and, 159, 160 (figure)
    osmotic thirst, 159, 160
    renin and, 160
    satiety mechanism and, 160
    subfornical organ and, 160, 160 (figure)
    See also Homeostatic drives
Three Mile Island disaster, 235–236,
    237, 469

Tobacco use, 133–134, 421, 439, 446
    See also Nicotine
Tolerance, 126–127, 130
    conditioned tolerance, 127
    selective tolerance, 132
Tonotopic map, 263, 264 (figure)
Topographical organization, 259, 261
Torture victims, 236
Total recall, 378
Touch. *See* Skin senses; Somatosensory
    cortex
Tourette's syndrome, 203, 459–460,
    460 (figure), 461
Tower of Hanoi problem, 370, 370 (figure)
Toxicity:
    area postrema and, 70
    blood-brain barrier and, 70, 70 (figure)
Tracts, 54, 54 (figure), 71
Transcranial magnetic stimulation (TMS),
    94, 102, 103 (figure), 448, 458
Transient receptor potential (TRP) group,
    331–332
Transsexuality, 203, 214
Traumatic brain injury (TBI), 80, 80 (table),
    81, 101, 439
Trichromatic theory, 301
Tryptophan, 244
Tuberomamillary nucleus, 478, 479 (figure),
    480 (figure)
Tuning curves, 263, 264 (figure)
Twin studies:
    addiction and, 147–148
    affective disorders and, 444
    intelligence studies and, 109–110,
        110 (figure)
    nature vs. nurture debate and, 14–16,
        15 (figure)
    obesity and, 174, 175 (figure), 176
    research methods and, 109–110,
        110 (table)
    schizophrenia and, 431–432, 431–432
        (figures)
    See also Genetics; Inheritance
Two-photon microscope, 98
2-deoxyglucose injection technique,
    96–97
Tympanic membrane, 256, 257–258
    (figures)

Ultradian rhythm, 473, 475
Unipolar depression, 443, 449, 451–452,
    451–452 (figures)
Unipolar neurons, 26, 26 (figure)
Unitary capability, 395–396, 401

Vaccinations, 235, 239 (figure)
Valium, 130
Variables:
    confounded variables, 94
    control variables, 93
Variation, 11
Vegetative state, 499
Ventral, 59, 59 (figure)
Ventral horns, 69, 69 (figure)
Ventral root, 69, 69 (figure)
Ventral stream, 260, 261 (figure), 315,
    315 (figure), 316, 317

Ventral tegmental area, 66, 68 (table), 138
Ventricles, 66, 66 (figure), 68 (table)
area postrema and, 70
cerebrospinal fluid and, 66
fourth ventricle and, 66, 66 (figure)
lateral ventricles and, 65–66 (figures), 66
third ventricle and, 66, 66 (figure)
See also Brain structure; Central nervous system (CNS)
Ventrolateral preoptic nucleus, 478, 479 (figure), 480 (figure)
Ventromedial cortex, 229, 230 (figure)
Ventromedial hypothalamus, 170, 194, 197
Vesicles, 37, 41
Vestibular sense, 332, 333 (figure)
acceleration/deceleration and, 332
eye movements and, 332, 333
hair cells and, 332
head movement and, 332
parieto-insular-vestibular cortex and, 333
proprioception and, 332
saccule and, 332, 333 (figure)
semicircular canals and, 332, 333 (figure)
spatial information and, 332
utricle and, 332, 333 (figure)
See also Body senses
Virtual reality therapeutic tool, 456
Vision, 63, 255, 294
adequate stimulus and, 294
binocular rivalry and, 489
blind spot and, 298
cryptochromes and, 97, 98 (figure)
distributed visual function and, 314
eye function and, 295–297, 296–297 (figures), 297 (table)
hierarchical processing and, 298, 313–314
infrared images and, 294–295
lateral geniculate nuclei and, 298, 299 (figure), 305
light wavelengths and, 295, 295 (figure)
lost vision, restoration of, 300
modular processing and, 298, 313, 314
neural inhibition and, 298
optic chiasm and, 298, 299 (figure)
pathways to the brain, 298, 299 (figures), 314–316
retinal disparity and, 298, 299 (figure)
superior colliculi and, 66
three-dimensional location of objects and, 298
visible spectrum and, 294–295, 295 (figure)

visual analysis pathways and, 314–316, 314–315 (figures)
visual identification and, 62, 66
See also Color vision; Form vision; Hearing; Visual cortex; Visual perception
Visual acuity, 297, 300
Visual association cortex, 66, 68 (table)
Visual cortex, 57, 57–58 (figures), 63, 64, 68 (table)
circuit pruning and, 76
complex cells and, 310, 311 (figure)
dorsal stream and, 315, 315 (figure), 316
Hubel/Wiesel theory and, 309–311, 310–311 (figures)
neural reorganization and, 78–79
retinotopic map and, 307, 307 (figure)
simple cells and, 310, 310–311 (figures)
somatosensory cortex, comparison with, 334–336, 334–335 (figures)
spatial frequency theory and, 312–313, 312–313 (figures)
ventral stream and, 315, 315 (figure), 316
visual analysis pathways, 314–316, 314–315 (figures)
visual identification and, 62, 66
See also Vision; Visual perception
Visual field, 298
Visual perception, 313–314
binding problem and, 321–322
blindsight and, 317
color agnosia and, 319
color perception, 314, 315
disorders of, 316–321
distributed processing and, 314
distributed visual awareness and, 322
dorsal stream and, 315, 315 (figure), 316
final integration of information and, 321–322
fusiform face area and, 317, 318, 318–319 (figures)
genetic basis of, 317
hidden perception and, 317
hierarchical processing and, 298, 313–314
learning process and, 318, 319 (figure)
magnocellular nucleus, 479 (figure), 484
magnocellular system and, 314–315, 314 (figure)
modular processing and, 298, 313, 314
movement agnosia and, 320
movement perception, 315–316, 317
neglect, attention in vision and, 320–321, 321 (figures)
object/face agnosia and, 316–319, 317–319 (figures)

parvocellular system and, 314, 314 (figure), 315
prosopagnosia and, 317, 317 (figure)
synethesia and, 322
ventral stream and, 315, 315 (figure), 316, 317
visual analysis pathways and, 314–316, 314 (figure)
visual word form area and, 318–319
See also Vision; Visual cortex
Visual word form area (VWFA), 318–319, 319
VNO. See vomeronasal organ
Volley theory, 261–262, 262 (figure)
Voltage, 28, 29–31, 30 (figure), 39
Vomeronasal organ (VNO), 196 (figure), 197
Voodoo death, 236
Vulnerability model, 14, 15–16, 432–433

Wada technique, 278
Warmth. See Skin senses
Wechsler Adult Intelligence Scale, 396
Weight. See Obesity
Wernicke-Geschwind model of language, 272–275, 273–274 (figures)
Wernicke's aphasia, 272, 278
Wernicke's area, 58 (figure), 63, 68 (table), 270–271, 271 (figure), 284
White matter, 56, 56 (figure), 68
Winter birth effect, 440
Wisconsin Card Sorting Test, 437
Withdrawal reactions, 126
addictiveness and, 126, 127, 138
alcohol, 129
caffeine, 134
cocaine, 131
heroin, 127
marijuana, 137
nicotine, 133
Wolffian ducts, 199, 199 (figure), 200
Working memory, 371–372, 400, 408
World Health Organization (WHO), 129, 172, 429
(VWFA). See visual word form area

X chromosome, 9, 9 (figure), 11, 12
X-linkage, 12
Xanax, 130

Y chromosome, 9, 9 (figure)

Zeitgebers, 470
Zygote, 10